T0189832

Lecture Notes in Computer Science 10822

More information about this series at http://www.springer.com/series/7410

Jesper Buus Nielsen · Vincent Rijmen (Eds.)

Advances in Cryptology – EUROCRYPT 2018

37th Annual International Conference on the Theory and Applications of Cryptographic Techniques
Tel Aviv, Israel, April 29 – May 3, 2018
Proceedings, Part III

Editors
Jesper Buus Nielsen
Aarhus University
Aarhus
Denmark

Vincent Rijmen
University of Leuven
Leuven
Belgium

ISSN 0302-9743 ISSN 1611-3349 (electronic)
Lecture Notes in Computer Science
ISBN 978-3-319-78371-0 ISBN 978-3-319-78372-7 (eBook)
https://doi.org/10.1007/978-3-319-78372-7

Library of Congress Control Number: 2018937382

LNCS Sublibrary: SL4 – Security and Cryptology

Printed on acid-free paper

This Springer imprint is published by the registered company Springer International Publishing AG
part of Springer Nature
The registered company address is: Gewerbestrasse 11, 6330 Cham, Switzerland

Preface

Eurocrypt 2018, the 37th Annual International Conference on the Theory and Applications of Cryptographic Techniques, was held in Tel Aviv, Israel, from April 29 to May 3, 2018. The conference was sponsored by the International Association for Cryptologic Research (IACR). Orr Dunkelman (University of Haifa, Israel) was responsible for the local organization. He was supported by a local organizing team consisting of Technion's Hiroshi Fujiwara Cyber Security Research Center headed by Eli Biham, and most notably by Suzie Eid. We are deeply indebted to them for their support and smooth collaboration.

The conference program followed the now established parallel track system where the works of the authors were presented in two concurrently running tracks. Only the invited talks spanned over both tracks.

We received a total of 294 submissions. Each submission was anonymized for the reviewing process and was assigned to at least three of the 54 Program Committee members. Committee members were allowed to submit at most one paper, or two if both were co-authored. Submissions by committee members were held to a higher standard than normal submissions. The reviewing process included a rebuttal round for all submissions. After extensive deliberations, the Program Committee accepted 69 papers. The revised versions of these papers are included in these three-volume proceedings, organized topically within their respective track.

The committee decided to give the Best Paper Award to the papers "Simple Proofs of Sequential Work" by Bram Cohen and Krzysztof Pietrzak, "Two-Round Multiparty Secure Computation from Minimal Assumptions" by Sanjam Garg and Akshayaram Srinivasan, and "Two-Round MPC from Two-Round OT" by Fabrice Benhamouda and Huijia Lin. All three papers received invitations for the *Journal of Cryptology*.

The program also included invited talks by Anne Canteaut, titled "Desperately Seeking Sboxes", and Matthew Green, titled "Thirty Years of Digital Currency: From DigiCash to the Blockchain".

We would like to thank all the authors who submitted papers. We know that the Program Committee's decisions can be very disappointing, especially rejections of very good papers that did not find a slot in the sparse number of accepted papers. We sincerely hope that these works eventually get the attention they deserve.

We are also indebted to the members of the Program Committee and all external reviewers for their voluntary work. The Program Committee work is quite a workload. It has been an honor to work with everyone. The committee's work was tremendously simplified by Shai Halevi's submission software and his support, including running the service on IACR servers.

Finally, we thank everyone else — speakers, session chairs, and rump-session chairs — for their contribution to the program of Eurocrypt 2018. We would also like to thank the many sponsors for their generous support, including the Cryptography Research Fund that supported student speakers.

May 2018

Jesper Buus Nielsen
Vincent Rijmen

Eurocrypt 2018

The 37th Annual International Conference on the Theory and Applications of Cryptographic Techniques

Sponsored by *the International Association for Cryptologic Research*

April 29 – May 3, 2018
Tel Aviv, Israel

General Chair

Orr Dunkelman	University of Haifa, Israel

Program Co-chairs

Jesper Buus Nielsen	Aarhus University, Denmark
Vincent Rijmen	University of Leuven, Belgium

Program Committee

Martin Albrecht	Royal Holloway, UK
Joël Alwen	IST Austria, Austria, and Wickr, USA
Gilles Van Assche	STMicroelectronics, Belgium
Paulo S. L. M. Barreto	University of Washington Tacoma, USA
Nir Bitansky	Tel Aviv University, Israel
Céline Blondeau	Aalto University, Finland
Andrey Bogdanov	DTU, Denmark
Chris Brzuska	TU Hamburg, Germany, and Aalto University, Finland
Jan Camenisch	IBM Research – Zurich, Switzerland
Ignacio Cascudo	Aalborg University, Denmark
Melissa Chase	Microsoft Research, USA
Alessandro Chiesa	UC Berkeley, USA
Joan Daemen	Radboud University, The Netherlands, and STMicroelectronics, Belgium
Yevgeniy Dodis	New York University, USA
Nico Döttling	Friedrich Alexander University Erlangen-Nürnberg, Germany
Sebastian Faust	TU Darmstadt, Germany
Serge Fehr	CWI Amsterdam, The Netherlands
Georg Fuchsbauer	Inria and ENS, France
Jens Groth	University College London, UK
Jian Guo	Nanyang Technological University, Singapore

Martin Hirt	ETH Zurich, Switzerland
Dennis Hofheinz	KIT, Germany
Yuval Ishai	Technion, Israel, and UCLA, USA
Nathan Keller	Bar-Ilan University, Israel
Eike Kiltz	Ruhr-Universität Bochum, Germany
Gregor Leander	Ruhr-Universität Bochum, Germany
Yehuda Lindell	Bar-Ilan University, Israel
Mohammad Mahmoody	University of Virginia, USA
Willi Meier	FHNW, Windisch, Switzerland
Florian Mendel	Infineon Technologies, Germany
Bart Mennink	Radboud University, The Netherlands
María Naya-Plasencia	Inria, France
Svetla Nikova	KU Leuven, Belgium
Eran Omri	Ariel University, Israel
Arpita Patra	Indian Institute of Science, India
David Pointcheval	ENS/CNRS, France
Bart Preneel	KU Leuven, Belgium
Thomas Ristenpart	Cornell Tech, USA
Alon Rosen	IDC Herzliya, Israel
Mike Rosulek	Oregon State University, USA
Louis Salvail	Université de Montréal, Canada
Yu Sasaki	NTT Secure Platform Laboratories, Japan
Thomas Schneider	TU Darmstadt, Germany
Jacob C. N. Schuldt	AIST, Japan
Nigel P. Smart	KU Leuven, Belgium, and University of Bristol, UK
Adam Smith	Boston University, USA
Damien Stehlé	ENS de Lyon, France
Björn Tackmann	IBM Research – Zurich, Switzerland
Dominique Unruh	University of Tartu, Estonia
Vinod Vaikuntanathan	MIT, USA
Muthuramakrishnan Venkitasubramaniam	University of Rochester, USA
Frederik Vercauteren	KU Leuven, Belgium
Damien Vergnaud	Sorbonne Université, France
Ivan Visconti	University of Salerno, Italy
Moti Yung	Columbia University and Snap Inc., USA

Additional Reviewers

Masayuki Abe
Aysajan Abidin
Ittai Abraham
Hamza Abusalah
Divesh Aggarwal
Shashank Agrawal
Shweta Agrawal
Thomas Agrikola
Bar Alon
Abdel Aly
Prabhanjan Ananth
Elena Andreeva

Daniel Apon
Gilad Asharov
Nuttapong Attrapadung
Benedikt Auerbach
Daniel Augot
Christian Badertscher
Saikrishna Badrinarayanan
Shi Bai
Josep Balasch
Marshall Ball
Valentina Banciu
Subhadeep Banik
Zhenzhen Bao
Gilles Barthe
Lejla Batina
Balthazar Bauer
Carsten Baum
Christof Beierle
Amos Beimel
Sonia Belaid
Aner Ben-Efraim
Fabrice Benhamouda
Iddo Bentov
Itay Berman
Kavun Elif Bilge
Olivier Blazy
Jeremiah Blocki
Andrey Bogdanov
Carl Bootland
Jonathan Bootle
Raphael Bost
Leif Both
Florian Bourse
Elette Boyle
Zvika Brakerski
Christian Cachin
Ran Canetti
Anne Canteaut
Brent Carmer
Wouter Castryck
Andrea Cerulli
André Chailloux
Avik Chakraborti
Yilei Chen
Ashish Choudhury
Chitchanok Chuengsatiansup
Michele Ciampi
Thomas De Cnudde
Ran Cohen
Sandro Coretti
Jean-Sebastien Coron
Henry Corrigan-Gibbs
Ana Costache
Geoffroy Couteau
Claude Crépeau
Ben Curtis
Dana Dachman-Soled
Yuanxi Dai
Bernardo David
Alex Davidson
Jean Paul Degabriele
Akshay Degwekar
Daniel Demmler
Amit Deo
Apoorvaa Deshpande
Itai Dinur
Christoph Dobraunig
Manu Drijvers
Maria Dubovitskaya
Léo Ducas
Yfke Dulck
Pierre-Alain Dupont
François Dupressoir
Avijit Dutta
Lisa Eckey
Maria Eichlseder
Maximilian Ernst
Mohammad Etemad
Antonio Faonio
Oriol Farràs
Pooya Farshim
Manuel Fersch
Dario Fiore
Viktor Fischer
Nils Fleischhacker
Christian Forler
Tommaso Gagliardoni
Chaya Ganesh
Juan Garay
Sanjam Garg
Romain Gay
Peter Gaži
Rosario Gennaro
Satrajit Ghosh
Irene Giacomelli
Federico Giacon
Benedikt Gierlichs
Junqing Gong
Dov Gordon
Divya Gupta
Lorenzo Grassi
Hannes Gross
Vincent Grosso
Paul Grubbs
Chun Guo
Siyao Guo
Mohammad Hajiabadi
Carmit Hazay
Gottfried Herold
Felix Heuer
Thang Hoang
Viet Tung Hoang
Akinori Hosoyamada
Kristina Hostáková
Andreas Hülsing
Ilia Iliashenko
Roi Inbar
Vincenzo Iovino
Tetsu Iwata
Abhishek Jain
Martin Jepsen
Daniel Jost
Chiraag Juvekar
Seny Kamara
Chethan Kamath
Bhavana Kanukurthi
Harish Karthikeyan
Suichi Katsumata
Jonathan Katz
John Kelsey
Dakshita Khurana
Eunkyung Kim
Taechan Kim
Elena Kirshanova
Ágnes Kiss
Susumu Kiyoshima

Ilya Kizhvatov
Alexander Koch
Konrad Kohbrok
Lisa Kohl
Stefan Kölbl
Ilan Komargodski
Yashvanth Kondi
Venkata Koppula
Thorsten Kranz
Hugo Krawczyk
Marie-Sarah Lacharite
Kim Laine
Virginie Lallemand
Gaëtan Leurent
Anthony Leverrier
Xin Li
Pierre-Yvan Liardet
Benoît Libert
Huijia Lin
Guozhen Liu
Jian Liu
Chen-Da Liu-Zhang
Alex Lombardi
Julian Loss
Steve Lu
Atul Luykx
Vadim Lyubashevsky
Saeed Mahloujifar
Hemanta Maji
Mary Maller
Umberto Martínez-Peñas
Daniel Masny
Takahiro Matsuda
Christian Matt
Patrick McCorry
Pierrick Méaux
Lauren De Meyer
Peihan Miao
Brice Minaud
Esfandiar Mohammadi
Ameer Mohammed
Maria Chiara Molteni
Tal Moran
Fabrice Mouhartem
Amir Moradi
Pratyay Mukherjee
Marta Mularczyk
Mridul Nandi
Ventzislav Nikov
Tobias Nilges
Ryo Nishimaki
Anca Nitulescu
Ariel Nof
Achiya Bar On
Claudio Orlandi
Michele Orrù
Clara Paglialonga
Giorgos Panagiotakos
Omer Paneth
Louiza Papachristodoulou
Kostas Papagiannopoulos
Sunoo Park
Anat Paskin-Cherniavsky
Alain Passelègue
Kenny Paterson
Michaël Peeters
Chris Peikert
Alice Pellet–Mary
Geovandro C. C. F. Pereira
Leo Perrin
Giuseppe Persiano
Thomas Peters
Krzysztof Pietrzak
Benny Pinkas
Oxana Poburinnaya
Bertram Poettering
Antigoni Polychroniadou
Christopher Portmann
Manoj Prabhakaran
Emmanuel Prouff
Carla Ràfols
Somindu C. Ramanna
Samuel Ranellucci
Shahram Rasoolzadeh
Divya Ravi
Ling Ren
Oscar Reparaz
Silas Richelson
Peter Rindal
Michal Rolinek
Miruna Rosca
Ron Rothblum
David Roubinet
Adeline Roux-Langlois
Vladimir Rozic
Andy Rupp
Yusuke Sakai
Simona Samardjiska
Niels Samwel
Olivier Sanders
Pratik Sarkar
Alessandra Scafuro
Martin Schläffer
Dominique Schröder
Sven Schäge
Adam Sealfon
Yannick Seurin
abhi shelat
Kazumasa Shinagawa
Luisa Siniscalchi
Maciej Skórski
Fang Song
Ling Song
Katerina Sotiraki
Florian Speelman
Gabriele Spini
Kannan Srinathan
Thomas Steinke
Uri Stemmer
Igors Stepanovs
Noah Stephens-Davidowitz
Alan Szepieniec
Seth Terashima
Cihangir Tezcan
Mehdi Tibouchi
Elmar Tischhauser
Radu Titiu
Yosuke Todo
Junichi Tomida
Patrick Towa
Boaz Tsaban
Daniel Tschudi
Thomas Unterluggauer
Margarita Vald
Kerem Varici
Prashant Vasudevan

Philip Vejre
Daniele Venturi
Benoît Viguier
Fernando Virdia
Damian Vizár
Alexandre Wallet
Michael Walter
Haoyang Wang
Qingju Wang
Hoeteck Wee
Felix Wegener
Christian Weinert
Erich Wenger
Daniel Wichs
Friedrich Wiemer
David Wu
Thomas Wunderer
Sophia Yakoubov
Shota Yamada
Takashi Yamakawa
Kan Yasuda
Attila Yavuz
Scott Yilek
Eylon Yogev
Greg Zaverucha
Mark Zhandry
Ren Zhang

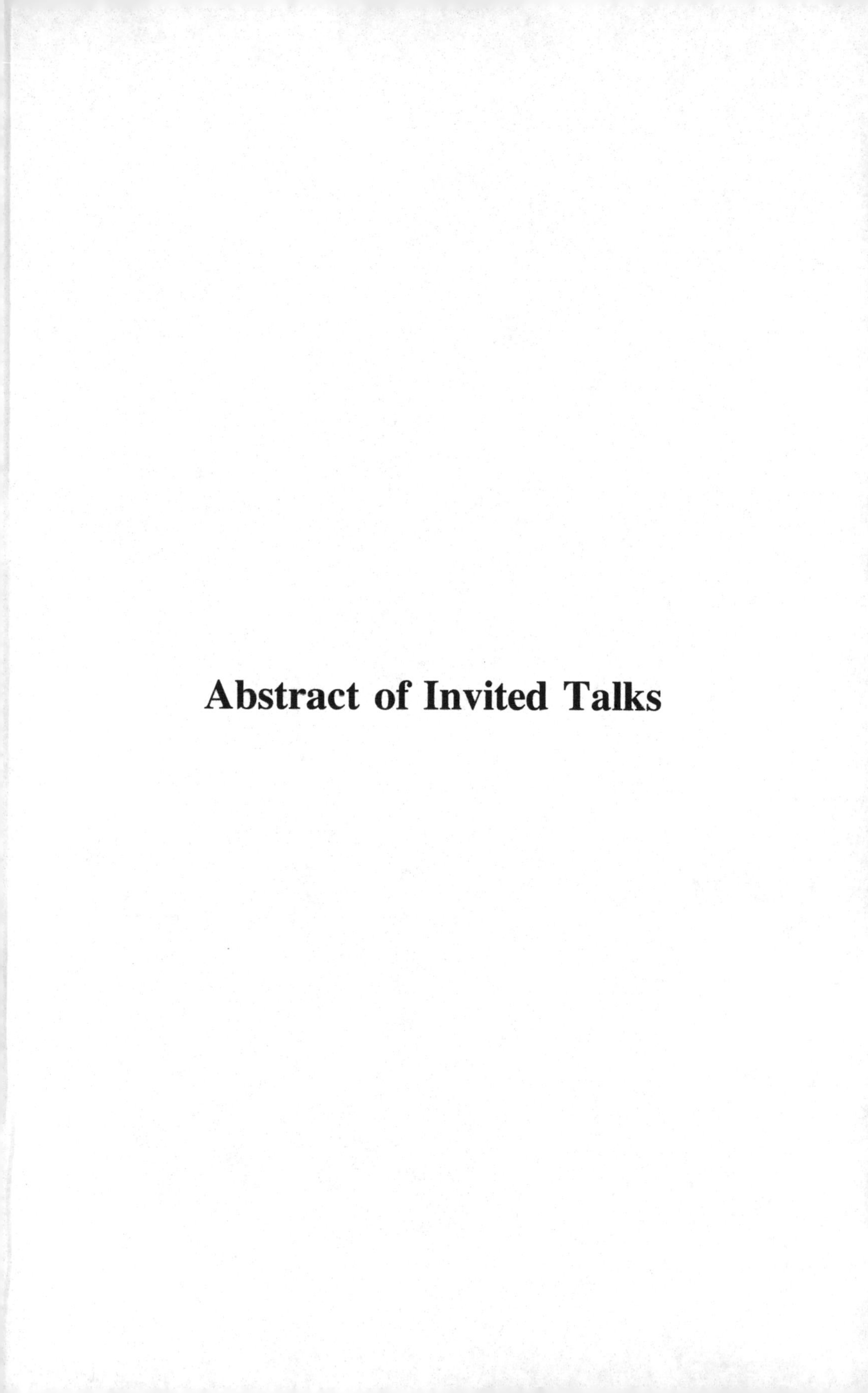

Abstract of Invited Talks

Desperately Seeking Sboxes

Anne Canteaut

Inria, Paris, France
anne.canteaut@inria.fr

Abstract. Twenty-five years ago, the definition of security criteria associated to the resistance to linear and differential cryptanalysis has initiated a long line of research in the quest for Sboxes with optimal nonlinearity and differential uniformity. Although these optimal Sboxes have been studied by many cryptographers and mathematicians, many questions remain open. The most prominent open problem is probably the determination of the optimal values of the nonlinearity and of the differential uniformity for a permutation depending on an even number of variables.

Besides those classical properties, various attacks have motivated several other criteria. Higher-order differential attacks, cube distinguishers and the more recent division property exploit some specific properties of the representation of the whole cipher as a collection of multivariate polynomials, typically the fact that some given monomials do not appear in these polynomials. This type of property is often inherited from some algebraic property of the Sbox. Similarly, the invariant subspace attack and its nonlinear counterpart also originate from specific algebraic structure in the Sbox.

Thirty Years of Digital Currency: From DigiCash to the Blockchain

Matthew Green

Johns Hopkins University
mgreen@cs.jhu.edu

Abstract. More than thirty years ago a researcher named David Chaum presented his vision for a cryptographic financial system. In the past ten years this vision has been realized. Yet despite a vast amount of popular excitement, it remains to be seen whether the development of cryptocurrencies (and their associated consensus technologies) will have a lasting positive impact—both on society and on our research community. In this talk I will examine that question. Specifically, I will review several important contributions that research cryptography has made to this field; survey the most promising deployed (or developing) technologies; and discuss the many challenges ahead.

Contents – Part III

Zero-Knowledge

Implementing Multiparty Computation

Non-interactive Zero-Knowledge

Anonymous Communication

Isogeny

Leakage

Key Exchange

Quantum

Non-maleable Codes

Provable Symmetric Cryptography

Zero-Knowledge

On the Existence of Three Round Zero-Knowledge Proofs

Nils Fleischhacker[1,2(✉)], Vipul Goyal[1], and Abhishek Jain[2]

[1] Carnegie Mellon University, Pittsburgh, USA
mail@nilsfleischhacker.de
[2] Johns Hopkins University, Baltimore, USA

Abstract. We study the round complexity of zero-knowledge (ZK) *proof* systems. While five round ZK proofs for NP are known from standard assumptions [Goldreich-Kahan, J. Cryptology'96], Katz [TCC'08] proved that four rounds are insufficient for this task w.r.t. black-box simulation. In this work, we study the feasibility of ZK proofs using *non-black-box* simulation. Our main result is that *three round* private-coin ZK proofs for NP do not exist (even w.r.t. non-black-box simulation), under certain assumptions on program obfuscation. Our approach builds upon the recent work of Kalai et al. [Crypto'17] who ruled out constant round *public-coin* ZK proofs under the same assumptions as ours.

1 Introduction

The notion of zero-knowledge (ZK) proofs [32] is fundamental in cryptography. Intuitively, ZK proofs allow one to prove a statement without revealing anything beyond the validity of the statement.

An important measure of efficiency of ZK protocols is *round complexity.* Ever since the introduction of ZK proofs nearly three decades ago, an extensive amount of research has been dedicated towards minimizing their round-complexity. Protocols with smaller round complexity are more desirable so as to minimize the effect of network latency, which in turn decreases the time complexity of the protocol.

Round-Complexity of ZK. In this work, we study the exact round complexity of ZK *proofs* that achieve soundness even against computationally unbounded adversarial provers (as opposed to *arguments* that achieve soundness only against polynomial-time adversarial provers). While initial constructions of ZK proofs required a polynomial number of rounds, the seminal work of Goldreich and Kahan [29] constructed a five round ZK proof system for NP based on collision-resistant hash functions.

N. Fleischhacker and V. Goyal—We acknowledge the generous support of Northrop Grumman.

N. Fleischhacker and A. Jain—Supported in part by a DARPA/ARL Safeware Grant W911NF-15-C-0213 and a sub-award from NSF CNS-1414023.

J. B. Nielsen and V. Rijmen (Eds.): EUROCRYPT 2018, LNCS 10822, pp. 3–33, 2018.
https://doi.org/10.1007/978-3-319-78372-7_1

In the negative direction, two-round ZK arguments for NP were ruled out by Goldreich and Oren [31]. Later, Goldreich and Krawcyzk [30] ruled out three round ZK arguments for NP where the ZK property holds w.r.t. a black-box simulator. More recently, Katz [37] proved that four round ZK proofs with black-box simulation only exist for languages whose complement is in MA.

The above state of the art motivates the following intriguing question:

Does there exist a three or four round ZK proof system for NP *using non-black-box simulation?*

In this work, we investigate precisely this question.

Private-coin vs Public-coin. In the study of ZK proofs, whether or not the verifier makes its random coins public or keeps them private has a strong bearing on the round-complexity. Indeed, constructing *public-coin* ZK proofs is viewed as a harder task. Very recently, Kalai et al. [36] ruled out constant round public-coin ZK proof systems for NP, even w.r.t. non-black-box simulation, assuming the existence of certain kinds of program obfuscation [7]. However, their approach breaks down in the *private coin* setting, where a verifier may keep its random coins used during the protocol private from the prover. This is not surprising, since five round private-coin ZK proofs are already known [29].

In this work, we investigate the feasibility of constructing private-coin ZK proofs (via non-black-box techniques) in less than five rounds. We remark that a candidate construction of three-round (private-coin) ZK proof system was given by Lepinski [40] based on a highly non-standard "knowledge-type" assumption; we discuss the bearing of our results on Lepinski's protocol (and the underlying assumption) below.

1.1 Our Results

We revisit the round complexity of zero-knowledge proof systems. As our main result, we rule out the existence of *three round* private-coin ZK proofs for languages outside BPP, under certain strong assumptions.

Theorem 1 (Informal). *Three round ZK proofs against non-uniform verifiers and distinguishers only exist for languages in* BPP*, assuming the following:*

- *Sub-exponentially secure one-way functions.*
- *Sub-exponentially secure indistinguishability obfuscation for circuits* [7,25].
- *Exponentially secure input-hiding obfuscation for multi-bit point functions* [5,11].

Our result relies on the same assumptions as those used in the recent work of Kalai et al. [36]. In their work, Kalai et al. use these assumptions to instantiate the Fiat-Shamir heuristic [24] and then rely upon its connection with public-coin ZK proofs [22] to rule out constant round public-coin ZK proofs. Naturally, this approach does not extend to the private coin setting. Nevertheless, we are able to build upon their techniques to obtain our result in Theorem 1.

Further, we note that our result contradicts the work of Lepinski [40] and thus refutes the knowledge-type assumption underlying Lepinski's protocol. We further elaborate on this in Sect. 1.3.

On our assumptions. Starting with the work of [25], several candidate constructions of indistinguishability obfuscation (iO) have been proposed over the last few years (see, e.g., [2–4,15,26,27,41–45,47]). During this time, (sub-exponentially secure) iO has also led to numerous advances in theoretical cryptography (see, e.g., [13,21,25,46]). Nevertheless, no iO scheme whose security is based on standard cryptographic assumptions is presently known.

Our second assumption on program obfuscation concerns with the notion of input-hiding obfuscation [5] for the class of multi-bit point functions $\mathcal{I}_{\alpha,\beta}$, where $\mathcal{I}_{\alpha,\beta}(\alpha) = \beta$ and 0, otherwise. Roughly speaking, an input-hiding obfuscator for this family is said to be T-secure, if any PPT adversary can succeed in guessing α with probability at most T^{-1}. For our purposes, we require T to be exponential in the security parameter. Candidate constructions of such obfuscation based on a strong variant of the DDH assumption are known from the works of [11,19] (see Sect. 2 for a more detailed discussion.)

Pessimistic Interpretation. While it is natural to be somewhat skeptical about the obfuscation assumptions we make, we note that our result implies that constructing three-round zero-knowledge proofs would require overcoming significant technical barriers. In particular, it would require disproving the existence of sub-exponentially secure iO, or the existence of exponentially secure input-hiding obfuscation for multi-bit point functions (or, less likely, disproving the existence of sub-exponentially secure one-way functions).

What about four rounds? Our result in Theorem 1 also extends to a specific relaxation of ZK, referred to as ϵ-ZK [14]. In this relaxed notion, the simulator's running time may grow polynomially with the distinguishing gap, which is allowed to be an inverse polynomial (unlike standard ZK, where the distinguishing gap must be negligible).

In a recent work, Bitansky et al. [14] construct a four round private coin ϵ-ZK proof system for NP, assuming the existence of keyless multi-collision-resistant hash functions (MCRH) [9,14,39]. Multi-collision-resistant hash functions weaken the standard notion of collision-resistant hash functions by only guaranteeing that an adversary cannot find many (rather than two) inputs that map to the same image. Presently, no constructions of keyless MCRH based on standard assumptions are known; however, unlike collision-resistant hash functions that cannot be secure against non-uniform adversaries in the keyless setting, keyless MCRH are meaningful even in the non-uniform setting if the number of required collisions are larger than the non-uniform advice to the adversary.

Their result serves as evidence that our techniques are unlikely to extend to the four round case, since otherwise it would imply the non-existence of keyless MCRH. While this is not implausible based on current evidence, in our eyes, it would be a rather surprising outcome.

It is of course possible that while four round private-coin ϵ-ZK proofs exist, four round private-coin ZK proofs do not. However, in light of the above, it seems that ruling out four round private-coin ZK proofs (w.r.t. non-black-box simulation) would require substantially new techniques.

1.2 Technical Overview

In order to rule out the existence of three-round zero knowledge proofs, we need to show that for any imaginable three round proof system, there exists a non-uniform adversarial verifier whose view cannot be efficiently simulated by any non-black-box simulator. Since a non-black-box simulator has access to the adversary's code, an immediate challenge is to "hide" the random coins of the adversarial verifier from the simulator.

Our starting approach to address this issue is to use program obfuscation. Let Π be any three-round private-coin proof system. To prove that Π is not ZK, we construct a "dummy" adversarial verifier V^* who receives as auxiliary input aux, an obfuscation of the next-message function of the honest verifier algorithm of Π. More concretely, the auxiliary input aux consists of an obfuscated program that has a key k for a pseudorandom function (PRF) hardwired in its description:

1. Upon receiving a message α from the prover, the program computes a message β of the honest verifier (as per protocol Π) using randomness $r = \mathsf{PRF}_k(\alpha)$.[1]
2. Upon receiving a protocol transcript (α, β, γ), it recomputes the randomness r used to compute β. Using the randomness r and the transcript, it honestly computes the verifier's output (i.e., whether to accept or reject the proof).

The adversarial verifier's code does not do anything intelligent on its own, and simply uses its auxiliary input aux to compute its protocol message.

Ruling out Rewinding Simulators. The above strategy for hiding the random coins of the verifier runs into the following problem: a simulator may fix the first two messages (α, β) of the protocol, and then observe the verifier's output on many different third messages to learn non-trivial information about the private randomness of the verifier. Indeed, it was recently shown in the work of Jain et al. [35] that in certain protocols, a simulator can learn the verifier's random tape by observing whether the verifier accepts or rejects in multiple trials.

A naive approach to address this problem is to simply modify the adversary and remove the protocol output from adversary's view. This can be achieved by deleting the second instruction in the obfuscated program aux. This approach, however, immediately fails because now a simulator can simply simulate a "rejecting" transcript and succeed in fooling any distinguisher.

We address this problem by using non-uniform distinguishers, in a manner similar to Goldreich and Oren [31] and the recent work of [1]. Specifically, we modify the adversarial verifier to be such that it simply outputs the protocol

[1] One may notice that this is similar to how protocols secure against "reset attacks" are constructed [6,20].

transcript at the end of the protocol. The revised auxiliary input aux only contains the first instruction described above. The PRF key k used to compute the verifier's randomness inside aux is given as non-uniform advice to the distinguisher. Note that this information is not available to the simulator. Now, given k and the protocol transcript, the distinguisher can easily decide whether or not to accept the transcript. Therefore, a simulator can no longer fool the distinguisher via a rejecting transcript.

How to rule out any Simulator? Of course the main problem remains. While the above approach constitutes a meaningful first step, we still need to formally argue that there does not exist any efficient simulator for the aforementioned adversarial verifier.

In prior works such as [31], this is achieved by showing that any efficient simulator algorithm can be used by a cheating prover to break the soundness of candidate protocol, which leads to a contradiction. It is, however, not immediately clear how to implement this strategy in our setting since a cheating prover does not have access to the code of the verifier (which is required for running the simulator algorithm).

We instead show that the existence of an efficient simulator can be used to disprove the computational soundness of a *different* protocol that is provably sound, leading to a contradiction.

Contradiction via Round Compression. We implement a compiler for compressing any three round private coin proof system into a two round *argument* system. Our round compression strategy is in fact very similar to the one developed in the recent work of Kalai et al. [36] in the context of public-coin ZK proofs. We then show that a simulator for the three round proof w.r.t. the aforementioned non-uniform verifier can be used to construct a cheating prover for the two round argument system.

We now elaborate on the round compression strategy. Consider the prover and verifier of the three-round proof to be two-stage algorithms. That is, P_1 produces the prover's first message α, V_1 is the verifier's next message function that on input α outputs the verifier's message β, P_2 on input β produces the prover's second message γ and finally V_2 is the decision procedure which uses the random tape to decide whether (α, β, γ) is an accepting transcript. The compressed two-round *argument* works as follows:

1. In the first round, the verifier obfuscates the code of a slightly modified V_1 that upon input α, computes its message β using randomness $r = \mathsf{PRF}_k(\alpha)$ generated via a hardcoded PRF key k. The verifier then sends the obfuscated program to the prover.
2. The prover now runs P_1 to get α, evaluates the obfuscated program on α to receive β and finally runs P_2 on α, β to get γ. The prover then sends α, β, γ to the verifier.
3. Finally, the verifier can use k to recompute the random tape $\mathsf{PRF}_k(\alpha)$ and run V_2 to validate the transcript.

A minor variant of the above strategy was recently used by Kalai et al. [36] in the case of public-coin ZK proofs. In their case, the obfuscated program simply corresponds to a PRF algorithm since it suffices to implement the strategy of a public-coin verifier.[2]

Now, using the above round compression strategy, we can compress any three-round proof system Π into a two-round argument system Π'. Now suppose that there exists an efficient zero-knowledge simulator Sim for Π w.r.t. the adversarial verifier V^* with auxiliary input aux, as described earlier. It is easy to see that such a simulator Sim can be used to construct an efficient cheating prover P^* for Π'. Indeed, the view of Sim in Π against V^* with aux is the same as the view of P^* against an honest verifier in Π'.

Thus, the main challenge now is to prove that our round-compression strategy indeed yields two-round arguments.

How to prove Soundness? To prove computational soundness of the two-round protocol, we proceed in two main steps:

1. First, we establish that there exists only a very small set of "bad" first messages α for which the cheating prover can even hope to be successful.
2. Second, we prove that the obfuscation sufficiently hides this small set to ensure that the cheating prover cannot find such an α.

Below, we elaborate on each of these steps.

Step 1: Upper bounding Bad α's. Imagine for a moment, that the three-round proof system is public coin. Then, for any $x \notin \mathcal{L}$ and any α, there can only exist a negligible fraction of random tapes (and therefore β) for which an accepting γ even exists. This is true because otherwise the computationally unbounded prover could simply exhaustively search for this γ once they receive β. Now, if the random tape, as in the two-round argument, is chosen pseudorandomly as a function of α, then only a very small set of α's will lead to such *bad* random tapes. This is because a distinguisher against the pseudorandom function can test for bad α's by exhaustively enumerating γ's because the PRF is assumed to be 2^n-secure. This small set would then be the set of bad α's. Clearly any successful cheating prover *must* use a bad α, since those are the only ones for which an accepting γ even exists.

In a private coin protocol, however, this notion of bad α's does not work. In fact in a private coin protocol, for any α and any random tape, an accepting γ may always *exist!* Indeed, any three-round proof system can be transformed into another proof system that has this property: the verifier in the new protocol acts exactly as the original verifier, except that it also chooses a random γ^* that it keeps private. Now, once it receives γ from the prover in the third round, the verifier accepts if either the original verifier accepts or $\gamma = \gamma^*$. Clearly in this protocol, there always exists an accepting γ but the protocol nevertheless remains sound. To break soundness, a prover must either break soundness of the original protocol or guess γ^* which is only possible with negligible probability, because the entire transcript is independent of γ^*.

[2] In particular, in the public-coin case, the obfuscated program can be interpreted as an instantiation of the random oracle in the Fiat-Shamir heuristic.

This example does not only show that the notion of bad α's from the public coin case does not work in the private-coin case, it also helps to illustrate how we can try to fix it. While an accepting γ may always exist, the prover only learns β and cannot tell which random tape was used by the verifier, beyond the obvious fact that it must have been consistent with β. Therefore, the only γ a prover can hope to use to break the soundness of the protocol are those that, for a fixed β, are accepted by *many* consistent random tapes.

We use this key observation to derive our new notion of bad α's. For any α there exists only a negligible fraction of random tapes that are consistent with a β such that there exists a γ that is accepted with high probability over all the random tapes consistent with β. This is true, because otherwise an unbounded prover could choose a random α and after receiving β, exhaustively search for all consistent random tapes and then search for the γ accepted by many of them. And then again, if the random tape, as is done in the two-round argument, is chosen pseudorandomly as a function of α, then only a very small set of α's will lead to such *bad* random tapes.

However, must a cheating prover in the two-round protocol necessarily use such a bad α to convince a verifier? *While in the public coin case this was a trivial fact, this is not at all obvious in the more general private-coin case.* Since even for "good" random tapes accepting γ's may exist, it is necessary to show that these remain hidden and cannot be used to cheat.

Here indistinguishability obfuscation comes to the rescue. Using iO and puncturable PRFs, we can show that a cheating prover must remain oblivious about which consistent random tape was used to compute β. This allows us to argue that a cheating prover cannot make use of γ's that are only accepting for a small number of consistent random tapes. Therefore, with overwhelming probability, a successful cheating prover must use a bad α.

Step 2: Hiding Bad α's. Now, it remains to argue that this set of bad α is hidden by the obfuscation. Once we have established that a cheating prover must output a bad α, the most obvious idea would be to try and lead this to a contradiction with the soundness of the three-round proof. However, to translate this into an attack, we need to use the security of the PRF. And while using iO, that means we need to puncture. Since the puncturing must be done *before* we learn α used by the cheating prover, we would incur an exponential loss in the success probability of the hypothetical three-round cheating prover. We can therefore only bring this to a contradiction if the three-round proof is exponentially sound, which would severely weaken the result. Instead, we follow the same approach as Kalai et al. [36] and "transfer" the exponential loss to another cryptographic primitive.

The idea is to use the security of another primitive to argue that bad α's are hidden. Since the goal is to argue that bad inputs to a circuit remain hidden a natural candidate for this primitive is input-hiding obfuscation. And indeed, sufficiently strong input-hiding obfuscation for multibit point functions allows to lead the existence of a cheating prover to a contradiction. Some technical issues arise in this proof due to the distribution of bad α's not being uniform.

However, using a clever trick of a "relaxed" verifier it is possible to show that the distributions are sufficiently close. In this part of the proof, we are able to adapt the elegant strategy of Kalai et al. [36] with only minor modifications.

Extension to ϵ-ZK. To extend our result to also rule out three-round ϵ-ZK proofs, we mainly need to argue that the cheating prover we described above is still successful in breaking soundness of the two-round argument, even if our starting point is an ϵ-ZK simulator instead of a regular ZK simulator.

Towards this, we note that the ϵ-ZK simulator, for every noticeable function ϵ, is required to output a distribution that is ϵ-indistinguishable from the real distribution. Thus, we can choose any small noticeable function ϵ, and then this means that, while the cheating prover against the two-round argument is no longer successful with all but negligible probability, it is still successful with probability $1 - \epsilon$. This is sufficient to break soundness and our main theorem therefore extends to ϵ-ZK proofs.

1.3 Implications to Lepinski's Protocol

Lepinski's 3-round ZK proof protocol [40] is based on a clever combination of the three round honest-verifier ZK protocol of Blum [16] for Hamiltonian Graphs and a special kind of oblivious transfer. While Lepinski chose to give a more direct description of his protocol, a more modular high-level construction is implicit in his thesis. His construction makes use of two building blocks:

1. The three round honest-verifier ZK protocol of Blum for Hamiltonian Graphs.
2. A three round string OT protocol with the following properties:
   - The protocol is "delayed input" on the sender's side. I.e., the first round of the OT can be computed independently of the sender's inputs (m_0, m_1).
   - The protocol achieves indistinguishability based security against a *computationally unbounded* malicious sender.
   - The protocol achieves simulation based security against a malicious polynomial time receiver.

Based on these assumptions a three-round ZK proof can be constructed as described below. In the description we focus on soundness 1/2. For this specific protocol, smaller soundness error can be achieved by parallel repetition without affecting the ZK property.

1. In the first round, the prover sets up the OT by sending the first message.
2. In the second round, the verifier sends the OT receiver message corresponding to their random challenge for Blum's protocol. I.e. the Blum challenge is used as the selection bit b in the OT.
3. In the third round the prover sends the first message of Blum's protocol. Additionally he sends the sender message of the OT, corresponding to the two possible prover responses to the (as of yet unknown) challenge. I.e. the prover sets m_b in the OT to be the response to challenge b.
4. Finally the verifier receives the OT message, thus learning m_b and verifies that m_b is a valid response in Blum's protocol.

It is easy to verify that this protocol is indeed sound: since the OT is secure against an unbounded sender, the prover must choose his first message without knowledge of the challenge and if the graph does not contain a Hamiltonian cycle then it can only give a valid response to one of the challenges and is thus only successful with probability 1/2. The soundness of this protocol is uncontested by our result.

To prove that the above protocol is also zero-knowledge, one can leverage the simulation-based security of the OT against malicious receivers. In particular, the ZK simulator uses the OT simulator to learn the OT selection bit, and then uses it to invoke the honest-verifier ZK simulator for Blum's protocol. This part is disputed by our result. Since the security of Blum's protocol is not in question, this means that our result disputes the existence of an OT protocol with the properties described above.

However, Lepinski implicitly gives a number-theoretic construction of such an OT protocol using a very specific "knowledge-type" assumption that is referred to as the "proof of knowledge assumption (POKA)" in this thesis. This assumption essentially states that a specific three-round public-coin proof of knowledge protocol remains a proof of knowledge even if the verifier's challenge is computed using a fixed hash function. This assumption is necessary to facilitate extraction of the receiver's selection bit in his OT protocol, which is the key to proving simulation-based security against malicious receivers.

The question, of course, remains how this protocol and the underlying assumption exactly relate to our impossibility result. For that, we should first note that Lepinski does not explicitly prove his protocol to be zero-knowledge relative to non-uniform verifiers. Since our impossibility result only rules out three-round ZK with non-uniform verifiers, our result – taken literally – does not directly apply to the protocol as stated. However, it is easy to see that Lepinski's protocol does, in fact, achieve ZK against non-uniform verifiers if the POKA assumption is suitably augmented so that it holds even against provers with arbitrary auxiliary input. This augmented assumption is therefore what is specifically refuted by our result.

In a bit more detail, what does it mean exactly to apply our result to Lepinski's protocol? As mentioned earlier, the soundness of the protocol is not in question. Therefore, the round compression part of our proof works exactly as stated, i.e., we are able to compress Lepinski's three-round proof into a two round argument. It is the second part of our result, where we show that the soundness of the two round argument and the zero-knowledge property of the three-round proof contradict each other, where we get the refutation of the POKA assumption.

Essentially, in this part of the proof, we show that in the compressed two-round argument, a malicious prover is capable of using the ZK-simulator for the three-round proof to cheat and break soundness. Since the soundness of the protocol is not in question, this means that we are refuting the existence of the ZK-simulator and thus, that the 3-round protocol can be zero-knowledge. In the generalized terms in which we described Lepinski's protocol above the

ZK-simulator only requires the HVZK-simulator of Blum's protocol and the OT-simulator to work. This means that our work specifically refutes the simulation based receiver-security of the OT protocol. If we look at our result in a bit more detail, it is also clear why this is the case. Essentially, we are constructing a malicious prover who is capable of running the ZK-simulator for the 3-round proof. For Lepinski's construction to work, this simulator must be able to extract the selection bit in the OT from the verifier's message. This means that we are constructing an algorithm capable of extracting the selection bit of the receiver while acting as a malicious sender in the OT protocol. Clearly, this immediately implies that the OT is broken.

1.4 Related Work

There is a large body of work dedicated to the study of round complexity of zero-knowledge protocols. Below, we provide a brief (and incomplete) summary of some of the prior work in this area.

ZK Proofs. Five-round ZK proofs are known based on collision-resistant hash functions [29], and four-round ϵ-ZK proofs were recently constructed based on keyless multi-collision-resistant hash functions [14]. Both of these constructions require the verifier to use private coins. There also exists a candidate for a three-round ZK proof due to Lepinski [40], which ultimately clashes with our result. Lepinski's protocol is based on a highly non-standard knowledge-type assumption which our result refutes. We explain the exact relationship and implications in Sect. 1.3.

Dwork et al. [22] (and independently, Hada and Tanaka [33]) established an intimate connection between the Fiat-Shamir paradigm [24] and constant-round public-coin ZK proofs. Using their result, [36] recently ruled out the existence of constant-round public-coin ZK proofs, under the same assumptions as in our work. Previously, such protocols were only ruled out w.r.t. black-box simulation by [30]. We refer the reader to [36] for further discussion on public-coin ZK proofs.

ZK Arguments. Four-round ZK arguments are known based on one-way functions [8,23]. Goldreich and Krawcyzk [30] ruled out the existence of three-round ZK arguments for NP w.r.t. black-box simulation. While three-round ZK arguments with non-black-box simulators were unknown for a long time, some recent works have studied them w.r.t. weaker adversaries such as uniform provers [10], or uniform verifiers [12], while finally Bitansky et al. [14] were very recently able to construct general three round ZK arguments for non-uniform provers and verifiers based on keyless multi-collision-resistant hash functions.

2 Preliminaries

We denote by $n \in \mathbb{N}$ the security parameter that is implicitly given as input to all algorithms in unary representation 1^n. We denote by $\{0,1\}^\ell$ the set of all bit-strings of length ℓ. For a finite set S, we denote the action of sampling x uniformly at random from S by $x \leftarrow_{\$} S$, and we denote the cardinality of S by $|S|$. Al algorithms are assumed to be randomized, unless explicitly stated otherwise. An algorithm is efficient or PPT if it runs in time polynomial in the security parameter. If $\mathcal{A}$ is randomized then by $y \leftarrow \mathcal{A}(x; r)$ we denote that $\mathcal{A}$ is run on input x and with random coins r and produced output y. If no randomness is specified, then it is assumed that $\mathcal{A}$ is run with freshly sampled uniform random coins, and write this as $y \leftarrow_{\$} \mathcal{A}(x)$. For a circuit C we denote by $|C|$ the size of the circuit. A function $\mathsf{negl}(n)$ is negligible if for any positive polynomial $\mathsf{poly}(n)$, there exists an $N \in \mathbb{N}$, such that for all $n > N$, $\mathsf{negl}(n) \leq \frac{1}{\mathsf{poly}(n)}$.

2.1 Interactive Proofs and Arguments

An interactive proof for an NP language $\mathcal{L}$ is an interactive protocol between two parties, a computationally unbounded prover and a polynomial-time verifier. The two parties receive a common input x and the prover tries to convince the verifier that $x \in \mathcal{L}$. Intuitively the prover should (almost) always be successful if x is indeed in $\mathcal{L}$, but should be limited in its ability to convince the verifier if $x \notin \mathcal{L}$. An interactive proof, as formally introduced by Goldwasser et al. [32] is defined as follows.

Definition 1 (Interactive Proof). *An r-round 2-Party protocol $\langle \mathsf{P}, \mathsf{V} \rangle$ between a polynomial-time verifier V and an unbounded prover P is an interactive proof with soundness error ϵ for language $\mathcal{L}$ if the following two conditions hold:*

1. Completeness: *For all $x \in \mathcal{L}$ it holds that $\Pr_{\mathsf{P},\mathsf{V}}[1 \leftarrow \langle \mathsf{P}(x), \mathsf{V}(x) \rangle] = 1 - \mathsf{negl}(n)$.*
2. Soundness: *For all $x^* \notin \mathcal{L}$ and all computationally unbounded malicious provers P^* it holds that $\Pr_{\mathsf{P}^*,\mathsf{V}}[1 \leftarrow \langle \mathsf{P}^*, \mathsf{V}(x^*) \rangle] \leq \epsilon$.*

An interactive argument is very similar to an interactive proof, except that soundness is only required to hold relative to polynomial time malicious provers. Since also the honest prover is required to run in polynomial time, it receives an NP witness for x as an additional input. Formally, this leads to the following definition.

Definition 2 (Interactive Argument). *An r-round 2-Party protocol $\langle \mathsf{P}, \mathsf{V} \rangle$ between a polynomial-time verifier V and a polynomial-time prover P is an interactive argument with soundness error ϵ for language $\mathcal{L}$ with associated relation $\mathcal{R}$ if the following two conditions hold:*

1. Completeness: *For all $(x, w) \in \mathcal{R}$ it holds that $\Pr_{\mathsf{P},\mathsf{V}}[1 \leftarrow \langle \mathsf{P}(x, w), \mathsf{V}(x) \rangle] = 1 - \mathsf{negl}(n)$.*
2. Soundness: *For all $x^* \notin \mathcal{L}$ and all polynomial-time malicious provers P^* it holds that $\Pr_{\mathsf{P}^*,\mathsf{V}}[1 \leftarrow \langle \mathsf{P}^*, \mathsf{V}(x^*) \rangle] \leq \epsilon$.*

An especially powerful class of interactive proofs and arguments are those that are zero-knowledge. Intuitively a zero-knowledge proof or argument ensures that a malicious polynomial time verifier cannot learn anything from an execution of the protocol, except that $x \in \mathcal{L}$. This was first formalized in [32] by requiring the existence of a polynomial time simulator capable of – without knowledge of an NP witness for x – simulating any interaction a malicious verifier might have with the prover. This implies that anything the verifier learns from a protocol execution it could have also learned without interacting with the prover. To obtain a contradiction in the main proof in Sect. 3 we will use the notion of non-uniform zero-knowledge, where both the malicious verifier as well as the distinguisher may be non-uniform.

Definition 3 (Non-uniform Zero-Knowledge with Auxiliary Input). *Let* $\langle \mathsf{P}, \mathsf{V} \rangle$ *be a 2-Party protocol.* $\langle \mathsf{P}, \mathsf{V} \rangle$ *is said to be non-uniformly zero-knowledge with auxiliary input, if for all (possibly malicious)* PPT *algorithms* V^* *there exists a* PPT *simulator* Sim, *such that for all* PPT *distinguishers* $\mathcal{D}$ *and all auxiliary inputs* aux *and* aux′, *it holds that for all statements* x

$$\left| \begin{array}{l} \Pr[\mathcal{D}(\langle \mathsf{P}(x,w), \mathsf{V}^*(x,\mathsf{aux})\rangle, \mathsf{aux}') = 1] \\ - \Pr[\mathcal{D}(\mathsf{Sim}(x,\mathsf{aux}), \mathsf{aux}') = 1] \end{array} \right| \leq \mathsf{negl}(n).$$

2.2 Puncturable Pseudorandom Functions

The notion of puncturable pseudorandom functions was independently introduced in [17,18,38]. A puncturable pseudorandom function allows to *puncture* a key k on some fixed input x. This punctured key should still allow to correctly evaluate the PRF on any input other than x. However, the value of the function on input x should be indistinguishable from a unform random value, *even given the punctured key.* We define a strong notion of puncturable pseudorandom functions in the following.

Definition 4 (T-Secure Puncturable Pseudorandom Functions). *A pair of probabilistic polynomial time algorithms* (PRF, Puncture) *is a T-secure puncturable pseudorandom function with key length $\kappa(n)$ input length $i(n)$ and output length $o(n)$ if the following conditions hold:*

1. Functionality Preserved Under Puncturing: *For every $n \in \mathbb{N}$, every key $k \leftarrow_\$ \{0,1\}^{\kappa(n)}$, every input $x \in \{0,1\}^{i(n)}$, every punctured key $\mathsf{k}\{x\}$, and every input $x' \in \{0,1\}^{i(n)} \setminus \{x\}$ it holds that $\mathsf{PRF}_k(x') = \mathsf{PRF}_{\mathsf{k}\{x\}}(x')$.*
2. Pseudorandomness: *For any fixed $x \in \{0,1\}^{i(n)}$ it holds that for every distinguisher $\mathcal{D}$ that runs in time at most $\mathsf{poly}(T(n))$ it holds that*

$$\left| \begin{array}{l} \Pr_{k,\mathsf{Puncture}}[\mathcal{D}(\mathsf{Puncture}(k,x), x, \mathsf{PRF}_k(x)) = 1] \\ - \Pr_{k,\mathsf{Puncture},y}[\mathcal{D}(\mathsf{Puncture}(k,x), x, y) = 1] \end{array} \right| \leq \mathsf{negl}(T(n))$$

Our impossibility result uses 2^{2n}-secure puncturable pseudorandom functions. Note that these can be constructed using the GGM construction from any subexponentially secure one-way function [28,34] by for example using keys length $\kappa(n) = n^2$.

2.3 Obfuscation

Our impossibility result uses two different kinds of obfuscation, indistinguishability obfuscation and input-hiding obfuscation for multi-input point functions. Indistinguishability obfuscation (iO) was first suggested as a notion by Barak et al. [7] as a weaker form of obfuscation. The security guarantee of iO is that the obfuscation of two functionally equivalent circuits should result in indistinguishable output distributions. That is, any polnomial-time reverse engineering cannot detect which of two equivalent implementations was the source of an obfuscated program. This security may seem rather weak at first glance. However, following the introduction of a first candidate construction by Garg et al. [25] it has been shown in several works that even this seemingly weak notion of obfuscation is a very powerful tool. We formally define indistinguishability obfuscation below.

Definition 5 (T-Secure Indistinguishability Obfuscation). *Let $\mathbb{C}$ be a family of polynomial size boolean circuits. Let iO be a probabilistic polynomial time algorithm, which takes as input a circuit $\mathbf{C} \in \mathbb{C}$ and a security parameter 1^n, and outputs a boolean circuit $\mathbf{B}$ (not necessarily in $\mathbb{C}$). iO is a T-secure indistinguishability obfuscator if the following two conditions hold:*

1. Correctness: *For every $n \in \mathbb{N}$, every circuit $\mathbf{C} \in \mathbb{C}$ with input length ℓ, every obfuscated circuit $\mathbf{B} \leftarrow \mathsf{iO}(\mathbf{C}, 1^n)$ and every $x \in \{0,1\}^\ell$ it holds that $\mathbf{B}(x) = \mathbf{C}(x)$.*
2. Indistinguishability: *For every $n \in \mathbb{N}$, every pair of circuit $\mathbf{C}_1, \mathbf{C}_2 \in \mathbb{C}$ with identical input length ℓ and $|\mathbf{C}_1| = |\mathbf{C}_2|$, and every $\mathsf{poly}(T(n))$-time distinguisher $\mathcal{D}$ it holds that*

$$\left| \Pr_{\mathsf{iO},\mathcal{D}}[\mathcal{D}(\mathsf{iO}(\mathbf{C}_1, 1^n)) = 1] - \Pr_{\mathsf{iO},\mathcal{D}}[\mathcal{D}(\mathsf{iO}(\mathbf{C}_2, 1^n)) = 1] \right| \leq \mathsf{negl}(T(n))$$

Our impossibility result uses a strong notion of 2^{2n}-secure indistinguishability obfuscation for general circuits. This notion is implied by any subexponentially secure indistinguishability obfuscator by instantiating the security parameter with $\kappa(n) = n^2$.

The second form of obfuscation used in our result is input-hiding obfuscation for multi-bit point functions. The notion of input-hiding obfuscation was first suggested by Barak et al. in [5]. An input-hiding obfuscator for a family of circuits $\mathbb{C}$ guarantees that, given an obfuscation of a circuit $\mathbf{C}$ drawn uniformly at random from $\mathbb{C}$ it is hard for an adversary to find any input on which the circuit doesn't output 0.

Definition 6 (*T*-Secure Input-Hiding Obfuscation). *Let $\mathbb{C} = \{\mathbb{C}_n\}_{n\in\mathbb{N}}$ be a family of polynomial size boolean circuits, where $\mathbb{C}_n$ is a set of circuits operating on inputs of length n. A polynomial time obfuscator* hideO *is a T-secure input hiding obfuscator for $\mathbb{C}$ if the following two conditions hold:*

1. Correctness: *For every $n \in \mathbb{N}$, every circuit $\mathbf{C} \in \mathbb{C}_n$, every obfuscated circuit $\mathbf{B} \leftarrow \mathsf{iO}(\mathbf{C}, 1^n)$ and every $x \in \{0,1\}^n$ it holds that $\mathbf{B}(x) = \mathbf{C}(x)$.*
2. Input Hiding: *For every $n \in \mathbb{N}$, and all probabilistic polynomial time adversary $\mathcal{A}$ it holds that*

$$\Pr_{\mathbf{C}\leftarrow\mathbb{C}_n,\mathsf{hideO},\mathcal{A}}[\mathbf{C}(\mathcal{A}(\mathsf{hideO}(\mathbf{C}, 1^n))) \neq 0] \leq T^{-1}(n).$$

Note that this security definition differs from previous definitions of T-security in so far as it requires the adversary to run in polynomial time (in n). Our result specifically uses input-hiding obfuscation for multi-bit point functions. A multi-bit point function is characterized by two values x and y and is defined as the function that on input x outputs y and outputs 0 on all other inputs.

Definition 7 (*T*-Secure Input-Hiding Obfuscation for Multi-bit Point Functions). *Let $I_{x,y}$ denote the multi-bit point function with $I_{x,y}(x) = y$ and $I_{x,y}(x') = 0$ for all $x' \neq x$ and let k be a function $k : \mathbb{N} \to \mathbb{N}$. A polynomial time obfuscator* hideO *is a T-secure input hiding obfuscator for (n, k)-multi-bit point functions if it is a T-secure input-hiding obfuscator for all circuit families $\mathbb{C}$ for which the following properties hold.*

1. *All circuits in $\mathbb{C} = \{\mathbb{C}_n\}_{n\in\mathbb{N}}$ describe point functions with n-bit input and $k(n)$-bit output. I.e., $\mathbb{C}_n \subseteq \{I_{x,y} \middle| x \in \{0,1\}^n \wedge y \in \{0,1\}^{k(n)}\}$.*
2. *The marginal distribution on x is uniform for a uniformly sampled circuit $I_{x,y} \leftarrow_\$ \mathbb{C}_n$.*

This notion was first studied by Bitansky and Cannetti in [11]. They also showed that an earlier candidate construction by Cannetti and Dakdouk [19] can be proven secure in the generic group model based on a strong variant of the DDH assumption. Our impossibility result requires 2^n-secure input hiding obfuscator for multi-bit point functions. This may on first glance seem problematic, since DDH (and thereby the instantiation due to Cannetti and Dakdouk [19]) can be broken in time less than 2^n even in the generic group model. However, in Definition 6 we explicitly – and in contrast to the other definitions in this section – require that the adversary runs in polynomial time. And known subexponential time attacks do not imply a polynomial time attack that is successful with probability greater than $\mathsf{poly}/2^{(n)}$.

3 Impossibility of Three-Round Zero-Knowledge Proofs

In this section we will prove our main result, i.e., that under the stated assumptions, zero-knowledge 3-round interactive proof systems for non-trivial languages cannot exist. Our result is formally stated in Theorem 2.

Theorem 2. *Let $\hat{\Pi}$ be a 3-round interactive proof system for a language $\mathcal{L} \notin$ BPP with negligible soundness error μ. Assume the existence of a 2^{2n}-secure puncturable pseudorandom function, a 2^{2n}-secure indistinguishability obfuscator, and a $\mu \cdot 2^n \mathsf{poly}(n)$-secure input-hiding obfuscator for multi-bit point functions. Then $\hat{\Pi}$ cannot be non-uniformly zero-knowledge with auxiliary input.*

Proof (Theorem 2). Let $\hat{\Pi} = \langle \hat{\mathsf{P}}, \hat{\mathsf{V}} \rangle$ be a 3-round interactive proof system as described in Theorem 2. We consider the prover and verifier as two-stage algorithms, $\hat{\mathsf{V}} = (\hat{\mathsf{V}}_1, \hat{\mathsf{V}}_2)$, $\hat{\mathsf{P}} = (\hat{\mathsf{P}}_1, \hat{\mathsf{P}}_2)$. The first stage of the prover $\alpha \leftarrow \hat{\mathsf{P}}_1(x, w; r)$ on input the statement x, witness w and random coins r outputs the prover's first message α. The first stage of the verifier $\beta \leftarrow \hat{\mathsf{V}}_1(x, \alpha; s)$ on input the statement x, the prover's first message α and random coins s outputs the verifier's message β. The second stage of the prover $\gamma \leftarrow \hat{\mathsf{P}}_2(x, w, \beta; r)$ on input the statement x, witness w, the verifier's message β and random coins r outputs the prover's second message γ. The second stage of the verifier $b \leftarrow \hat{\mathsf{V}}_2(x, \alpha, \gamma; s)$ on input the statement x, the prover's messages α, γ and random coins s outputs a bit b indicating whether the proof is accepted of not. Note that without loss of generality we assume that the second stages do not take their own messages as input and instead recompute them when necessary.

First we slightly modify the protocol $\hat{\Pi}$ into the protocol $\Pi = \langle \mathsf{P}, \mathsf{V} \rangle$. The protocol behaves exactly as $\hat{\Pi}$, except that V_1 takes as its random coins $s = \sigma \| \hat{s}$ with $|\sigma| = \lceil \log \mu^{-1} \rceil$ and after running $\hat{\beta} \leftarrow \mathsf{V}_1(x, \alpha; \hat{s})$ outputs $\beta := \sigma \| \hat{\beta}$. The prover's second stage P_2 then again behaves exactly as $\hat{\mathsf{P}}_2$, and ignores σ. The following claim is immediately apparent.

Claim 3. *If $\hat{\Pi}$ is a 3-round interactive proof system with negligible soundness error μ, then Π is also a 3-round interactive proof system for the same language with the same negligible soundness error μ.*

This modification is therefore without loss of generality and will allow us to cleanly define a relaxed version of the verifier later in the proof, leading to a much simpler proof.

Now, we use the pseudorandom function PRF the indistinguishability obfuscator iO to construct a two-round protocol $\bar{\Pi} = \langle \bar{\mathsf{P}}, \bar{\mathsf{V}} \rangle$ as depicted in Fig. 1. The circuit $\mathbf{C}_{\mathsf{V}_1}$ is defined as follows:

```
C_V1[k, x](α)
-------------
s := PRF_k(α)
β := V1(x, α; s)
return β
```

To prove Theorem 2 we will now use the following two lemmas proven in Sects. 3.1 and 3.2 respectively.

$\bar{\mathsf{P}}(x, w; r)$		$\bar{\mathsf{V}}(x)$
		$k \leftarrow_{\$} \{0,1\}^{\kappa(n)}$
$\alpha \leftarrow \mathsf{P}_1(x, w; r)$	$\xleftarrow{\quad \mathbf{B} \quad}$	$\mathbf{B} \leftarrow \mathsf{iO}(\mathbf{C}_{\mathsf{V}_1}[k, x])$
$\beta := \mathbf{B}(\alpha)$		
$\gamma \leftarrow \mathsf{P}_2(x, w, \beta; r)$	$\xrightarrow{\quad \alpha, \gamma \quad}$	$b \leftarrow \mathsf{V}_2(x, \alpha, \gamma; \mathsf{PRF}_k(x, \alpha))$
		return b

Fig. 1. The two-round argument system $\bar{\Pi} = \langle \bar{\mathsf{P}}, \bar{\mathsf{V}} \rangle$ resulting from compressing the three-round proof system $\Pi = \langle \mathsf{P}, \mathsf{V} \rangle$ into two rounds. The round compression is achieved by sending an obfuscated version of the verifier's own code to the prover as a first message. This allows the prover to compute the verifier's response to their first message without additional interaction. This construction is proven sound in Lemma 4.

Lemma 4. *Let $\hat{\Pi}$ be a 3-round interactive proof system with negligible soundness error μ as in Theorem 2. Let Π be the modified 3-round interactive proof system as described above. Assume that* PRF *is a 2^{2n}-secure puncturable pseudorandom function, and* iO *is a 2^{2n}-secure indistinguishability obfuscator. Further assume that* hideO *is a 2^n-secure input-hiding obfuscator for multi-bit point functions. Then $\bar{\Pi}$, described in Fig. 1 is a 2-round interactive argument system with negligible soundness error $\bar{\mu}$.*

Lemma 5. *Let Π be a 3-round interactive proof system for a language $\mathcal{L} \notin$* BPP. *Let $\bar{\Pi}$ be the transformed 2-round argument system described in Fig. 1 with soundness error $\bar{\mu}$. If $\bar{\mu} \leq \mathsf{negl}(n)$ then Π is not non-uniformly zero-knowledge with auxiliary input.*

Theorem 2 now follows as a simple corollary from combining Lemmas 4 and 5. By our assumption, Π has a negligibly small soundness error μ, which by Lemma 4 also implies a negligible soundness error $\bar{\mu}$ for $\bar{\Pi}$. Since a negligible soundness error of $\bar{\Pi}$ implies that Π is not non-uniformly zero-knowledge with auxiliary input, the theorem trivially follows. □

3.1 Proof of Lemma 4

Fix a modified 3-round interactive proof system $\Pi = \langle \mathsf{P}, \mathsf{V} \rangle$. Let $\mu \leq \mathsf{negl}(n)$ be the soundness error of Π. We assume without loss of generality, that all messages of the protocol have length n.

Assume towards contradiction that there exists a cheating PPT prover P^* breaking the soundness of $\bar{\Pi}$ for some $x^* \notin \mathcal{L}$ with probability $\nu = 1/\mathsf{poly}(n)$. I.e., we have that

$$\Pr_{k, \mathsf{iO}, \mathsf{P}^*}[\mathsf{V}_2(x^*, \alpha, \gamma; \mathsf{PRF}_k(\alpha)) = 1 : (\alpha, \gamma) \leftarrow \mathsf{P}^*(\mathsf{iO}(\mathbf{C}_{\mathsf{V}_1}[k, x]))] \geq \nu. \quad (1)$$

To obtain a contradiction we analyze a variant of the protocol Π that works with a relaxed verifier V'. The relaxed verifier V' works exactly as V, except that in addition to accepting whenever V does, it also accepts if $\beta = 0^{\lceil \log \nu/\mu \rceil} \| \beta'$ for some arbitrary β'. Remember, that $\beta = \sigma \| \hat{\beta}$ with $|\sigma| = \lceil \log \mu^{-1} \rceil$. I.e., V' also accepts if the first $\lceil \log \nu/\mu \rceil$ bits of σ are zero. In particular since V' always accepts if V accepts, it remains true that

$$\Pr_{k,\mathsf{iO},\mathsf{P}^*}[\mathsf{V}_2'(x^*,\alpha,\beta;\mathsf{PRF}_k(\alpha)) = 1 : \mathsf{P}^*(\mathsf{iO}(\mathbf{C}_{\mathsf{V}_1}[k,x])) = (\alpha,\gamma)] \geq \nu. \tag{2}$$

Further, using a union bound, we can bound the soundness error μ' of the relaxed 3-round protocol $\langle \mathsf{P}, \mathsf{V}' \rangle$ to be

$$\mu' \leq \mu + 2^{-\lceil \log \nu/\mu \rceil} \leq \mu + \frac{\mu}{\nu} \leq \frac{2\mu}{\nu}. \tag{3}$$

In particular, for any negligible μ, μ' remains negligible.

Let $S_{\alpha,\beta} = \{s | \mathsf{V}_1'(x^*, \alpha; s) = \beta\}$ denote the set of all random tapes that given α lead to the second message β. We define the following set of pairs (α, β), for which a malicious γ exists that will be accepted by the verifier for a large fraction of the random tapes that given α lead to β.

$$\mathsf{ACC} = \left\{ (\alpha,\beta) \middle| \exists \gamma : \Pr_{s' \leftarrow\$ S_{\alpha,\beta}}[\mathsf{V}_2'(x^*,\alpha,\gamma;s') = 1] \geq \frac{\nu}{2} \right\}.$$

Observe, that membership in ACC can be tested in time $2^{2n} \cdot \mathsf{poly}(n) = \mathcal{O}2^{2n}$ by enumerating all messages γ and all random tapes s, checking whether $\beta = \mathsf{V}_1'(x^*, \alpha; s)$ and $\mathsf{V}_2'(x^*, \alpha, \gamma; s) = 1$ and then computing the probability.[3] Given the cheating prover P^*, there exists an efficient algorithm $\mathcal{A}$ that outputs α, such that $(\alpha, \mathsf{V}_1'(\alpha; \mathsf{PRF}_k(\alpha))) \in \mathsf{ACC}$ with high probability. Formally this is stated in the following claim that is proven in Sect. 3.1.1.

Claim 6. *If there exists a malicious prover* P^* *as assumed above, then for the efficient algorithm* $\mathcal{A}$ *that on input* $\mathsf{iO}(\mathbf{C}_{\mathsf{V}_1}[k, x^*])|$ *runs* $(\alpha, \gamma) \leftarrow \mathsf{P}^*(\mathsf{iO}(\mathbf{C}_{\mathsf{V}_1}[k, x^*]))$, *discards* γ *and outputs* α *the following holds:*

$$\Pr_{k,\mathsf{iO},\mathcal{A}}[(\alpha, \mathsf{V}_1'(x^*,\alpha;\mathsf{PRF}_k(\alpha))) \in \mathsf{ACC} : \alpha \leftarrow \mathcal{A}(\mathsf{iO}(\mathbf{C}_{\mathsf{V}_1}[k,x^*]))] \geq \frac{\nu}{2} - 2^{-n}$$

Now consider the punctured version of the verifier circuit $\mathbf{C}_{\mathsf{pct}}$ defined follows:

$\mathbf{C}_{\mathsf{pct}}[k, \alpha^*, \beta^*](\alpha)$

---

if $\alpha \stackrel{?}{=} \alpha^*$
  $\beta := \beta^*$
else
  $s := \mathsf{PRF}_k(\alpha)$
  $\beta := \mathsf{V}_1'(x, \alpha; s)$
return β

---

[3] This assumes without loss of generality that $|\gamma| = |s| = n$.

We will use the following claim, which essentially states that, when given an obfuscation of the verifier's circuit punctured at α^*, the $\mathcal{A}$ from Claim 6 will output α^* with a probability slightly above random chance. The claim is proven in Sect. 3.1.2.

Claim 7. *If* PRF *is* 2^{2n}*-secure and* iO *is* 2^{2n}*-secure, then it must hold that*

$$\Pr_{k,\alpha^*,s^*,\mathsf{iO},\mathcal{A}}\left[\mathcal{A}\Big(\mathsf{iO}\big(\mathbf{C}_{\mathsf{pct}}[\mathsf{k}\{\alpha^*\},\alpha^*,\mathsf{V}_1'(x^*,\alpha;s^*)]\big)\Big)=\alpha^*\,\Big|\,\Big(\alpha^*,\mathsf{V}_1'(x^*,\alpha;s^*)\Big)\in\mathsf{ACC}\right]$$
$$\geq\frac{1}{8}\cdot 2^{-n}\cdot\frac{\nu^2}{\mu'}.$$

This property of $\mathcal{A}$ contradicts the security of the input hiding obfuscator hideO as shown in the following. We claim that

$$\Pr_{k,\alpha^*,s^*,\mathsf{hideO},\mathsf{iO},\mathcal{A}}\left[\mathcal{A}\Big(\mathsf{iO}\big(\mathbf{C}_{\mathsf{hide}}[k,\mathsf{hideO}(\alpha^*,s^*)]\big)\Big)=\alpha^*\,\Big|\,\Big(\alpha^*,\mathsf{V}_1'(x^*,\alpha;s^*)\Big)\in\mathsf{ACC}\right]$$
$$\geq\Pr_{k,\alpha^*,s^*,\mathsf{iO},\mathcal{A}}\left[\mathcal{A}\Big(\mathsf{iO}\big(\mathbf{C}_{\mathsf{pct}}[\mathsf{k}\{\alpha^*\},\alpha^*,\mathsf{V}_1'(x^*,\alpha;s^*)]\big)\Big)=\alpha^*\,\Big|\,\Big(\alpha^*,\mathsf{V}_1'(x,\alpha;s^*)\Big)\in\mathsf{ACC}\right]$$
$$-\Pr_{\alpha^*,s^*}\left[\Big(\alpha^*,\mathsf{V}_1'(x^*,\alpha;s^*)\Big)\in\mathsf{ACC}\right]\cdot\mathsf{negl}[2^{2n}]\qquad(4)$$
$$\geq\Pr_{k,\alpha^*,s^*,\mathsf{iO},\mathcal{A}}\left[\mathcal{A}\Big(\mathsf{iO}\big(\mathbf{C}_{\mathsf{pct}}[\mathsf{k}\{\alpha^*\},\alpha^*,\mathsf{V}_1'(x^*,\alpha;s^*)]\big)\Big)=\alpha^*\,\Big|\,\Big(\alpha^*,\mathsf{V}_1'(x,\alpha;s^*)\Big)\in\mathsf{ACC}\right]$$
$$-2^{-2n}\qquad(5)$$
$$\geq\frac{1}{8}\cdot 2^{-n}\frac{\nu^2}{\mu'}-2^{-2n}\geq\frac{1}{16}\cdot 2^{-n}\frac{\nu^2}{\mu'},\qquad(6)$$

where $\mathbf{C}_{\mathsf{hide}}[k,\mathbf{B}]$ is a circuit that defined as follows

```
C_hide[k, B](α*)
─────────────────
s* := B(α*)
if s* = ⊥
    s* := PRF_k(α*)
β* := V'_1(x*, α*; s*)
return β*
```

Equation 4 follows by reduction to the 2^{2n} security of the indistinguishability obfuscator as depicted in Fig. 2. Clearly, the two circuits are functionally equivalent. Further, if it holds that $(\alpha^*,\beta^*)\in\mathsf{ACC}$ then the two cases of the security definition of indistinguishability obfuscation directly correspond to the two cases of Eq. 4. The reduction $\mathcal{B}^{\mathsf{iO}}$ runs in time $\mathcal{O}2^{2n}$ and therefore, Eq. 4 follows. Equation 5 then follows simply by upper bounding the probability with 1 and the

$\mathcal{B}_1^{\mathsf{iO}}(1^n)$	$\mathcal{B}_2^{\mathsf{iO}}(\mathbf{B})$
$k \leftarrow_\$ \{0,1\}^{\kappa(n)} \alpha^* \leftarrow_\$ \{0,1\}^n$	**if** $(\alpha^*, \beta^*) \notin \mathsf{ACC}$
$s^* \leftarrow_\$ \{0,1\}^n$	$b \leftarrow_\$ \{0,1\}$
$\mathsf{k}\{\alpha^*\} := \mathsf{Puncture}(k, \alpha^*)$	**return** b
$\beta^* := \mathsf{V}_1'(x^*, \alpha^*; s^*)$	**else if** $\mathcal{A}(\mathbf{B}) = \alpha^*$
$\mathbf{C}_0 = \mathbf{C}_{\mathsf{pct}}[\mathsf{k}\{\alpha^*\}, \alpha^*, \beta^*]$	**return** 0
$\mathbf{C}_1 = \mathbf{C}_{\mathsf{hide}}[k, \mathsf{hideO}(\alpha^*, s^*)]$	**else**
return $(\mathbf{C}_0, \mathbf{C}_1)$	**return** 1

Fig. 2. The reduction from the claim of Eq. 4 to the 2^{2n} security of the indistinguishability obfuscator.

negligible function by 2^{-2n}. Finally Eq. 6 follows directly from Claim 7 and the last inequality follows by loosely upper bounding the negligible function 2^{-2n}.

Closely following [36], it remains to be shown that the distribution defined by uniformly sampling (α^*, β^*) from ACC is close to the distribution defined by uniformly sampling α^* and then sampling β^* conditioned on $(\alpha^*, \beta^*) \in \mathsf{ACC}$.

Formally, we define two distributions. Let D_0 be the distribution over pairs (α^*, β^*) defined by uniformly sampling $(\alpha^*, \beta^*) \leftarrow_\$ \mathsf{ACC}$. Let D_1 be the distribution over pairs (α^*, β^*) defined by uniformly sampling $\alpha^* \leftarrow_\$ \{0,1\}^n$ and then uniformly sampling $\beta^* \leftarrow_\$ \{\beta | (\alpha^*, \beta) \in \mathsf{ACC}\}$. We denote by $\mathsf{D}_b[\alpha^*, \beta^*]$ the probability of the pair (α^*, β^*) by distribution D_b.

Claim 8. *For any* $(\alpha^*, \beta^*) \in \{0,1\}^n \times \{0,1\}^{2n}$ *it holds that*

$$\mathsf{D}_1[\alpha^*, \beta^*] \geq \frac{\nu}{4} \mathsf{D}_0[\alpha^*, \beta^*]$$

It follows from Claim 8 that by drawing from D_1 instead of D_0, the probability of $\mathcal{A}$ outputting α^* can decrease at most by a multiplicative factor of $4/\nu$. Therefore, Claim 8 and Eq. 6 imply that there exists a PPT algorithm $\mathcal{A}$ such that

$$\Pr_{(\alpha^*, \beta^*, \mathsf{hideO}, \mathcal{A}) \leftarrow_\$ \mathsf{D}_1, \mathsf{hideO}, \mathcal{A}}[\mathcal{A}(\mathsf{hideO}(\mathbf{C}_{\mathsf{hide}}[\alpha^*, \beta^*])) = \alpha^*]$$

$$\geq \frac{\nu}{4} \cdot \left(\frac{1}{16} \cdot 2^{-n} \cdot \frac{\nu^2}{\mu'}\right) = \frac{1}{64} 2^{-n} \cdot \frac{\nu^3}{\mu'} \geq \mu^{-1} \cdot 2^{-n} \cdot \frac{\nu^3}{128}$$

Since the distribution of α^* drawn from D_1 is uniform, and ν is an inverse polynomial, this contradicts the $T = \mu \cdot 2^n \cdot \mathsf{poly}(n)$ security of the input hiding obfuscator and Lemma 4 follows. □

It remains to show that the various claims used in the above proof actually hold. The proofs for these claims are detailed in the following sections.

3.1.1 Proof of Claim 6

By definition of $\mathcal{A}$ we specifically need to show that

$$\Pr_{k,\mathsf{iO},\mathsf{P}^*}[(\alpha, \mathsf{V}_1'(x^*, \alpha; \mathsf{PRF}_k(\alpha))) \in \mathsf{ACC} : (\alpha,\gamma) \leftarrow \mathsf{P}^*(\mathsf{iO}(\mathbf{C}_{\mathsf{V}_1}[k, x^*]))] \geq \frac{\nu}{2} - 2^{-n}.$$

To do so, we will use the following claim, stating that if the cheating prover is successful in getting V_2' to accept using the random tape $\mathsf{PRF}_k(\alpha)$, then V_2' would accept with almost the same probability if the random tape were replaced with a randomly chosen $s \leftarrow_\$ S_{\alpha,\beta}$.

Claim 9. *If* PRF *is* 2^{2n}*-secure and* iO *is* 2^{2n}*-secure, then it must hold for any malicious prover* P^* *as assumed above, that*

$$\left| \Pr_{k,\mathsf{iO},\mathsf{P}^*}\left[\mathsf{V}_2'(x^*, \alpha^*, \gamma^*; \mathsf{PRF}_k(\alpha^*)) = 1 : \begin{array}{c}(\alpha^*,\gamma^*) \leftarrow \mathsf{P}^*(\mathsf{iO}(\mathbf{C}_{\mathsf{V}_1}[k, x^*]))\\ \beta^* \leftarrow \mathsf{V}_1'(x^*, \alpha^*; \mathsf{PRF}_k(\alpha^*))\end{array}\right] - \Pr_{k,s,\mathsf{iO},\mathsf{P}^*}\left[\mathsf{V}_2'(x^*, \alpha^*, \gamma^*; s') = 1 : \begin{array}{c}(\alpha^*,\gamma^*) \leftarrow \mathsf{P}^*(\mathsf{iO}(\mathbf{C}_{\mathsf{V}_1}[k, x^*]))\\ \beta^* \leftarrow \mathsf{V}_1'(x^*, \alpha^*; \mathsf{PRF}_k(\alpha^*))\\ s' \leftarrow_\$ S_{\alpha^*,\beta^*}\end{array}\right] \right| \leq 2^{-n}.$$

We observe the following

$$\nu = \Pr_{k,\mathsf{iO},\mathsf{P}^*}[\mathsf{V}_2'(x^*, \alpha, \gamma; \mathsf{PRF}_k(\alpha)) = 1 : (\alpha,\gamma) \leftarrow \mathsf{P}^*(\mathsf{iO}(\mathbf{C}_{\mathsf{V}_1}[k, x^*]))] \tag{7}$$

$$\leq \Pr_{k,s',\mathsf{iO},\mathsf{P}^*}\left[\mathsf{V}_2'(x^*, \alpha, \gamma; s') = 1 : \begin{array}{c}(\alpha,\gamma) \leftarrow \mathsf{P}^*(\mathsf{iO}(\mathbf{C}_{\mathsf{V}_1}[k, x^*]))\\ \beta \leftarrow \mathsf{V}_1'(x^*, \alpha, \mathsf{PRF}_k(\alpha))\\ s' \leftarrow_\$ S_{\alpha,\beta}\end{array}\right] + 2^{-n} \tag{8}$$

$$= \underbrace{\Pr_{k,s',\mathsf{iO},\mathsf{P}^*}\left[\mathsf{V}_2'(x^*, \alpha, \gamma; s') = 1 : \begin{array}{c}(\alpha,\gamma) \leftarrow \mathsf{P}^*(\mathsf{iO}(\mathbf{C}_{\mathsf{V}_1}[k, x^*])\\ \beta \leftarrow \mathsf{V}_1'(x^*, \alpha, \mathsf{PRF}_k(\alpha))\\ s' \leftarrow_\$ S_{\alpha,\beta}\end{array} \middle| (\alpha,\beta) \in \mathsf{ACC}\right]}_{\leq 1}$$

$$\cdot \Pr_{k,\mathsf{iO},\mathsf{P}^*}[(\alpha, \mathsf{V}_1'(x^*, \alpha; \mathsf{PRF}_k(\alpha))) \in \mathsf{ACC} : (\alpha,\gamma) \leftarrow \mathsf{P}^*(\mathsf{iO}(\mathbf{C}_{\mathsf{V}_1}[k, x^*]))]$$

$$+ \underbrace{\Pr_{k,s',\mathsf{iO},\mathsf{P}^*}\left[\mathsf{V}_2'(x^*, \alpha, \gamma; s') = 1 : \begin{array}{c}(\alpha,\gamma) \leftarrow \mathsf{P}^*(\mathsf{iO}(\mathbf{C}_{\mathsf{V}_1}[k, x^*])\\ \beta \leftarrow \mathsf{V}_1'(x^*, \alpha, \mathsf{PRF}_k(\alpha))\\ s' \leftarrow_\$ S_{\alpha,\beta}\end{array} \middle| (\alpha,\beta) \notin \mathsf{ACC}\right]}_{\leq \nu/2} \tag{9}$$

$$\cdot \underbrace{\Pr_{k,\mathsf{iO},\mathsf{P}^*}[(\alpha, \mathsf{V}_1'(x^*, \alpha; \mathsf{PRF}_k(\alpha))) \notin \mathsf{ACC} : (\alpha,\gamma) \leftarrow \mathsf{P}^*(\mathsf{iO}(\mathbf{C}_{\mathsf{V}_1}[k, x^*]))]}_{=1-\Pr_{k,\mathsf{iO},\mathsf{P}^*}[(\alpha,\mathsf{V}_1'(x^*,\alpha;\mathsf{PRF}_k(\alpha)))\in\mathsf{ACC}:(\alpha,\gamma)\leftarrow\mathsf{P}^*(\mathsf{iO}(\mathbf{C}_{\mathsf{V}_1}[k,x^*]))]}$$

$$+ 2^{-n}$$

$$\geq \left(1 - \frac{\nu}{2}\right) \Pr_{k,\mathsf{iO},\mathsf{P}^*}[(\alpha, \mathsf{V}_1'(x^*, \alpha; \mathsf{PRF}_k(\alpha))) \in \mathsf{ACC} : (\alpha, \gamma) \leftarrow \mathsf{P}^*(\mathsf{iO}(\mathbf{C}_{\mathsf{V}_1}[k, x^*]))]$$
$$+ \frac{\nu}{2} + 2^{-n} \tag{10}$$

where Eq. 7 follows from the definition of P^* and Eq. 8 follows directly from Claim 9. Equation 9 simply splits the probability into two cases and Eq. 10 upper bounds the probability of the verifier accepting in the two cases.

The above observation gives us

$$\Pr_{k,\mathsf{iO},\mathsf{P}^*}[(\alpha, \mathsf{V}_1'(x^*, \alpha; \mathsf{PRF}_k(\alpha))) \in \mathsf{ACC} : (\alpha, \gamma) \leftarrow \mathsf{P}^*(\mathsf{iO}(\mathbf{C}_{\mathsf{V}_1}[k, x^*]))]$$
$$\geq \frac{\nu - \frac{\nu}{2} - 2^{-n}}{1 - \frac{\nu}{2}} \geq \nu - \frac{\nu}{2} - 2^{-n} = \frac{\nu}{2} - 2^{-n}$$

as claimed. □

Proof of Claim 9. Let δ be any function such that

$$\left| \Pr_{k,s',\mathsf{iO},\mathsf{P}^*}\left[\mathsf{V}_2'(x^*, \alpha^*, \gamma^*; \mathsf{PRF}_k(\alpha^*)) = 1 : \begin{array}{r} (\alpha^*, \gamma^*) \leftarrow \mathsf{P}^*(\mathsf{iO}(\mathbf{C}_{\mathsf{V}_1}[k, x^*])) \\ \beta^* \leftarrow \mathsf{V}_1'(x^*, \alpha^*; \mathsf{PRF}_k(\alpha^*)) \\ s' \leftarrow_{\$} S_{\alpha^*,\beta^*} \end{array}\right] - \Pr_{k,s',\mathsf{iO},\mathsf{P}^*}\left[\mathsf{V}_2'(x^*, \alpha^*, \gamma^*; s') = 1 : \begin{array}{r} (\alpha^*, \gamma^*) \leftarrow \mathsf{P}^*(\mathsf{iO}(\mathbf{C}_{\mathsf{V}_1}[k, x^*])) \\ beta^* \leftarrow \mathsf{V}_1'(x^*, \alpha^*; \mathsf{PRF}_k(\alpha^*)) \\ s' \leftarrow_{\$} S_{\alpha^*,\beta^*} \end{array}\right] \right| > \delta(n).$$

In this case, we also have that for a uniformly chosen value α,

$$\left| \Pr_{k,s',\alpha,\mathsf{iO},\mathsf{P}^*}\left[\begin{array}{c}\mathsf{V}_2'(x^*, \alpha^*, \gamma^*; \mathsf{PRF}_k(\alpha^*)) = 1 \\ \wedge\ \alpha^* = \alpha\end{array} : \begin{array}{r} (\alpha^*, \gamma^*) \leftarrow \mathsf{P}^*(\mathsf{iO}(\mathbf{C}_{\mathsf{V}_1}[k, x^*])) \\ \beta^* \leftarrow \mathsf{V}_1'(x^*, \alpha^*; \mathsf{PRF}_k(\alpha^*)) \\ s' \leftarrow_{\$} S_{\alpha^*,\beta^*} \end{array}\right] - \Pr_{k,s',\alpha,\mathsf{iO},\mathsf{P}^*}\left[\begin{array}{c}\mathsf{V}_2'(x^*, \alpha^*, \gamma^*; s') = 1 \\ \wedge\ \alpha^* = \alpha\end{array} : \begin{array}{r} (\alpha^*, \gamma^*) \leftarrow \mathsf{P}^*(\mathsf{iO}(\mathbf{C}_{\mathsf{V}_1}[k, x^*])) \\ \beta^* \leftarrow \mathsf{V}_1'(x^*, \alpha^*; \mathsf{PRF}_k(\alpha^*)) \\ s \leftarrow_{\$} S_{\alpha^*,\beta^*} \end{array}\right] \right|$$
$$> 2^{-n} \cdot \delta(n).$$

Now consider the punctured version of the verifier circuit defined as before. By the 2^{2n} security of the obfuscator, the fact that the two circuits $\mathbf{C}_{\mathsf{V}_1}[k, x^*]$ and $\mathbf{C}_{\mathsf{pct}}[\mathsf{k}\{\alpha\}, \alpha, \mathsf{V}_1'(\alpha; \mathsf{PRF}_k(\alpha))]$ are functionally equivalent and the fact that $s \leftarrow_{\$} S_{\alpha^*,\beta^*}$ can be sampled in time $\mathcal{O}(2^n)$, it follows that

$$\left| \Pr_{k,s',\alpha,\mathsf{iO},\mathsf{P}^*}\left[\begin{matrix}\mathsf{V}_2'(x^*,\alpha^*,\gamma^*;\mathsf{PRF}_k(\alpha^*))=1\\ \wedge\ \alpha^*=\alpha\end{matrix} : \begin{matrix}\beta\leftarrow \mathsf{V}_1'(x^*,\alpha;\mathsf{PRF}_k(\alpha))\\ (\alpha^*,\gamma^*)\leftarrow \mathsf{P}^*(\mathsf{iO}(\mathbf{C}_{\mathsf{pct}}[\mathsf{k}\{\alpha\},\alpha,\beta]))\\ s'\leftarrow_\$ S_{\alpha^*,\beta}\end{matrix}\right] - \Pr_{k,s',\alpha,\mathsf{iO},\mathsf{P}^*}\left[\begin{matrix}\mathsf{V}_2'(x^*,\alpha^*,\gamma^*;s')=1\\ \wedge\ \alpha^*=\alpha\end{matrix} : \begin{matrix}\beta\leftarrow \mathsf{V}_1'(x^*,\alpha;\mathsf{PRF}_k(\alpha))\\ (\alpha^*,\gamma^*)\leftarrow \mathsf{P}^*(\mathsf{iO}(\mathbf{C}_{\mathsf{pct}}[\mathsf{k}\{\alpha\},\alpha,\beta]))\\ s'\leftarrow_\$ S_{\alpha^*,\beta}\end{matrix}\right] \right|$$
$$> 2^{-n}\cdot\delta(n) - \mathsf{negl}[2^{2n}].$$

Further, by the 2^{2n} security of the pseudorandom function and the fact that $s \leftarrow_\$ S_{\alpha^*,\beta^*}$ can be sampled in time $\mathcal{O}(2^n)$, it follows that

$$\left| \Pr_{k,s,s',\alpha,\mathsf{iO},\mathsf{P}^*}\left[\begin{matrix}\mathsf{V}_2'(x^*,\alpha^*,\gamma^*;s)=1\\ \wedge\ \alpha^*=\alpha\end{matrix} : \begin{matrix}\beta\leftarrow \mathsf{V}_1'(x^*,\alpha;s)\\ (\alpha^*,\gamma^*)\leftarrow \mathsf{P}^*(\mathsf{iO}(\mathbf{C}_{\mathsf{pct}}[\mathsf{k}\{\alpha\},\alpha,\beta]))\\ s'\leftarrow_\$ S_{\alpha^*,\beta}\end{matrix}\right] - \Pr_{k,s,s',\alpha,\mathsf{iO},\mathsf{P}^*}\left[\begin{matrix}\mathsf{V}_2'(x^*,\alpha^*,\gamma^*;s')=1\\ \wedge\ \alpha^*=\alpha\end{matrix} : \begin{matrix}\beta\leftarrow \mathsf{V}_1'(x^*,\alpha;s)\\ (\alpha^*,\gamma^*)\leftarrow \mathsf{P}^*(\mathsf{iO}(\mathbf{C}_{\mathsf{pct}}[\mathsf{k}\{\alpha\},\alpha,\beta]))\\ s'\leftarrow_\$ S_{\alpha^*,\beta}\end{matrix}\right] \right|$$
$$\geq 2^{-n}\cdot\delta(n) - \mathsf{negl}(2^{2n}) - \mathsf{negl}(2^{2n}) \geq 2^{-n}\cdot\delta(n) - 2^{-2n},$$

where the last inequality is obtained by loosely upper bounding the negligible functions. The circuit $\mathbf{C}_{\mathsf{pct}}[\mathsf{k}\{\alpha\},\alpha,\beta]$ no longer contains any information about s besides the fact that $s \in S_{\alpha^*,\beta^*}$. In the case where $\alpha^* = \alpha$, s and s' are, therefore, distributed identically and the two probabilities must in fact also be identical. Therefore, $2^{-n}\cdot\delta(n) - 2^{-2n} \leq 0$, giving us $\delta(n) \leq 2^{-n}$. The claim thus follows. □

3.1.2 Proof of Claim 7

By definition of conditional probability, we have that

$$\Pr_{k,\alpha^*,s^*,\mathsf{iO},\mathcal{A}}\left[\mathcal{A}\Big(\mathsf{iO}(\mathbf{C}_{\mathsf{pct}}[\mathsf{k}\{\alpha^*\},\alpha^*,\mathsf{V}_1'(x^*,\alpha;s^*)])\Big)=\alpha^* \,\middle|\, \Big(\alpha^*,\mathsf{V}_1'(x,\alpha;s^*)\Big)\in\mathsf{ACC}\right]$$
$$= \frac{\Pr_{k,\alpha^*,s^*,\mathsf{iO},\mathcal{A}}\left[\begin{matrix}\mathcal{A}\Big(\mathsf{iO}(\mathbf{C}_{\mathsf{pct}}[\mathsf{k}\{\alpha^*\},\alpha^*,\mathsf{V}_1'(x^*,\alpha;s^*)])\Big)=\alpha^*\\ \wedge\ \Big(\alpha^*,\mathsf{V}_1'(x,\alpha;s^*)\Big)\in\mathsf{ACC}\end{matrix}\right]}{\Pr_{\alpha^*,s^*}\left[\Big(\alpha^*,\mathsf{V}_1'(x^*,\alpha;s^*)\Big)\in\mathsf{ACC}\right]},$$

where we can easily bound $\Pr_{\alpha^*,s^*}[(\alpha^*,\mathsf{V}_1'(x^*,\alpha;s^*))\in\mathsf{ACC}] \leq 2\mu'/\nu$ using the soundness error μ' of $\langle\mathsf{P},\mathsf{V}'\rangle$. This is due to the fact that otherwise a (computationally unbounded) malicious prover could simply send a randomly sampled α^*. Upon receiving β^*, it would hold that $(\alpha^*,\beta^*)\in\mathsf{ACC}$ with probability greater than $2\mu'/\nu$. In this case, the prover could exhaustively search for a message γ^* that would lead many verifiers to accept. By definition of ACC, such a prover would win with probability greater than $(2\mu'/\nu) \cdot (\nu/2) = \mu'$, contradicting the soundness of the underlying protocol. It remains to bound the numerator, which we will do in two hops.

$$\Pr_{k,\alpha^*,s^*,\mathsf{iO},\mathcal{A}}\left[\begin{array}{c}\mathcal{A}(\mathsf{iO}(\mathbf{C}_{\mathsf{pct}}[\mathsf{k}\{\alpha^*\},\alpha^*,\mathsf{V}'_1(x^*,\alpha;s^*)])) = \alpha^* \\ \wedge\ (\alpha^*,\mathsf{V}'_1(x^*,\alpha;s^*)) \in \mathsf{ACC}\end{array}\right]$$

$$\geq \Pr_{k,\alpha^*,\mathsf{iO},\mathcal{A}}\left[\begin{array}{c}\mathcal{A}(\mathsf{iO}(\mathbf{C}_{\mathsf{pct}}[\mathsf{k}\{\alpha^*\},\alpha^*,\mathsf{V}'_1(x^*,\alpha^*;\mathsf{PRF}_k(\alpha^*))])) = \alpha^* \\ \wedge\ (\alpha^*,\mathsf{V}'_1(x^*,\alpha^*;\mathsf{PRF}_k(\alpha))) \in \mathsf{ACC}\end{array}\right] - \mathsf{negl}(2^{2n}) \quad (11)$$

$$\geq \Pr_{k,\alpha^*,\mathsf{iO},\mathcal{A}}\left[\begin{array}{c}\mathcal{A}(\mathsf{iO}(\mathbf{C}_{\mathsf{V}_1}[k,x^*])) = \alpha^* \\ \wedge\ (\alpha^*,\mathsf{V}'_1(x^*,\alpha^*;\mathsf{PRF}_k(\alpha^*))) \in \mathsf{ACC}\end{array}\right] - \mathsf{negl}(2^{2n}) - \mathsf{negl}(2^{2n}) \quad (12)$$

$$\geq \Pr_{k,\alpha^*,\mathsf{iO},\mathcal{A}}\left[\begin{array}{c}\mathcal{A}(\mathsf{iO}(\mathbf{C}_{\mathsf{V}_1}[k,x^*])) = \alpha^* \\ \wedge\ (\alpha^*,\mathsf{V}'_1(x^*,\alpha^*;\mathsf{PRF}_k(\alpha^*))) \in \mathsf{ACC}\end{array}\right] - 2^{-2n} \quad (13)$$

Equation 11 follows by reduction to the 2^{2n} security of the puncturable pseudorandom function as depicted in Fig. 3. Clearly, the two cases of the security definition for puncturable pseudorandom functions directly map to the two cases of Eq. 11. Further, the reduction $\mathcal{B}^{\mathsf{PRF}}$ runs in time $\mathcal{O}(2^{2n})$ and therefore, Eq. 11 follows.

$\mathcal{B}^{\mathsf{PRF}}(\mathsf{k}\{\alpha^*\},s^*)$

$\beta^* := \mathsf{V}'_1(x^*,\alpha;s^*)$
$\mathbf{B} \leftarrow \mathsf{iO}(\mathbf{C}_{\mathsf{pct}}[\mathsf{k}\{\alpha^*\},\alpha^*,\beta^*])$
if $\mathcal{A}(\mathbf{B}) = \alpha^* \wedge (\alpha^*,\beta^*) \in \mathsf{ACC}$
  return 1
else return 0

Fig. 3. The reduction from the claim of Eq. 11 to the 2^{2n} security of the puncturable pseudorandom function.

Equation 12 follows by reduction to the 2^{2n} security of the indistinguishability obfuscator as depicted in Fig. 4. Clearly, the two circuits are functionally equivalent and the two cases of the security definition for puncturable pseudorandom functions directly map to the two cases of Eq. 11. The reduction $\mathcal{B}^{\mathsf{iO}}$ runs in time $\mathcal{O}(2^{2n})$ and therefore, Eq. 12 follows. Finally, Eq. 13 then follows by the fact that the sum of two negligible functions is negligible and by loosely upper bounding the resulting negligible functions (note that 2^{-2n} is an inverse polynomial in 2^{2n}).

$\mathcal{B}_1^{\mathsf{iO}}(1^n)$	$\mathcal{B}_2^{\mathsf{iO}}(\mathbf{B})$
$k \leftarrow_{\$} \{0,1\}^{\kappa(n)},\ \alpha^* \leftarrow_{\$} \{0,1\}^n$	**if** $\mathcal{A}(\mathbf{B}) = \alpha^* \wedge (\alpha^*, \beta^*) \in \mathsf{ACC}$
$\mathsf{k}\{\alpha^*\} := \mathsf{Puncture}(k, \alpha^*)$	**return** 1
$\beta^* := \mathsf{V}_1'(x^*, \alpha^*; \mathsf{PRF}_k(\alpha^*))$	**else return** 0
$\mathbf{C}_0 = \mathbf{C}_{\mathsf{pct}}[\mathsf{k}\{\alpha^*\}, \alpha^*, \beta^*]$	
$\mathbf{C}_1 = \mathbf{C}_{\mathsf{V}_1}[k, x^*]$	
return $(\mathbf{C}_0, \mathbf{C}_1)$	

Fig. 4. The reduction from the claim of Eq. 12 to the 2^{2n} security of the indistinguishability obfuscator.

Using basic probability theory and Claim 6, we get

$$\begin{aligned}
&\Pr_{k,\alpha^*,\mathsf{iO},\mathcal{A}}[\mathcal{A}(\mathsf{iO}(\mathbf{C}_{\mathsf{V}_1}[k,x^*])) = \alpha^* \wedge (\alpha^*, \mathsf{V}_1'(x^*,\alpha^*;\mathsf{PRF}_k(\alpha^*))) \in \mathsf{ACC}] \\
=&\Pr_{k,\alpha^*,\mathsf{iO},\mathcal{A}}\left[\bigcup_\alpha \begin{pmatrix} \mathcal{A}(\mathsf{iO}(\mathbf{C}_{\mathsf{V}_1}[k,x])) = \alpha^* \\ \wedge (\alpha^*, \mathsf{V}_1'(x^*,\alpha^*;\mathsf{PRF}_k(\alpha^*))) \in \mathsf{ACC} \wedge \alpha^* = \alpha \end{pmatrix}\right] \\
=&\sum_\alpha \Pr_{k,\alpha^*,\mathsf{iO},\mathcal{A}}\left[\begin{matrix} \mathcal{A}(\mathsf{iO}(\mathbf{C}_{\mathsf{V}_1}[k,x])) = \alpha^* \\ \wedge (\alpha^*, \mathsf{V}_1'(x^*,\alpha^*;\mathsf{PRF}_k(\alpha^*))) \in \mathsf{ACC} \wedge \alpha^* = \alpha \end{matrix}\right] \\
=&2^{-n}\sum_\alpha \Pr_{k,\alpha^*,\mathsf{iO},\mathcal{A}}\left[\begin{matrix} \mathcal{A}(\mathsf{iO}(\mathbf{C}_{\mathsf{V}_1}[k,x])) = \alpha^* \\ \wedge\ (\alpha^*, \mathsf{V}_1'(x^*,\alpha^*;\mathsf{PRF}_k(\alpha^*))) \in \mathsf{ACC} \end{matrix}\right] \\
=&2^{-n} \Pr_{k,\mathsf{iO},\mathcal{A}}[(\alpha, \mathsf{V}_1'(x^*,\alpha;\mathsf{PRF}_k(\alpha))) \in \mathsf{ACC} : \alpha \leftarrow \mathcal{A}(\mathsf{iO}(\mathbf{C}_{\mathsf{V}_1}[k,x^*]))] \\
\geq&2^{-n}\cdot\left(\frac{\nu}{2} - 2^{-n}\right).
\end{aligned}$$

Combining this with Eq. 13, we get

$$\begin{aligned}
&\Pr_{k,\alpha^*,s^*,\mathsf{iO},\mathcal{A}}\left[\begin{matrix}\mathcal{A}(\mathsf{iO}(\mathbf{C}_{\mathsf{pct}}[\mathsf{k}\{\alpha^*\}, \alpha^*, \mathsf{V}_1'(x^*,\alpha;s^*)])) = \alpha^* \\ \wedge\ (\alpha^*, \mathsf{V}_1'(x^*,\alpha;s^*)) \in \mathsf{ACC}\end{matrix}\right] \\
\geq&2^{-n}\left(\frac{\nu}{2} - 2^{-n}\right) - 2^{-2n} = 2^{-n}\left(\frac{\nu}{2} - 2^{-n} - 2^{-n}\right) \geq 2^{-n}\cdot\frac{\nu}{4}
\end{aligned}$$

where the last inequality follows by loosely upper bounding the negligible function 2^{1-n} by the inverse polynomial $\nu/4$. Finally Claim 7 follows by

$$\Pr_{k,\alpha^*,s^*,\mathsf{iO},\mathcal{A}}\left[\mathcal{A}\Big(\mathsf{iO}\big(\mathbf{C}_{\mathsf{pct}}[\mathsf{k}\{\alpha^*\},\alpha^*,\mathsf{V}_1'(x^*,\alpha;s^*)]\big)\Big)=\alpha^*\,\Big|\,\Big(\alpha^*,\mathsf{V}_1'(x^*,\alpha;s^*)\Big)\in\mathsf{ACC}\right]$$

$$=\frac{\Pr_{k,\alpha^*,s^*,\mathsf{iO},\mathcal{A}}\left[\begin{array}{c}\mathcal{A}\Big(\mathsf{iO}\big(\mathbf{C}_{\mathsf{pct}}[\mathsf{k}\{\alpha^*\},\alpha^*,\mathsf{V}_1'(x^*,\alpha;s^*)]\big)\Big)=\alpha^*\\ \wedge\Big(\alpha^*,\mathsf{V}_1'(x^*,\alpha;s^*)\Big)\in\mathsf{ACC}\end{array}\right]}{\Pr_{\alpha^*,s^*}\left[\Big(\alpha^*,\mathsf{V}_1'(x^*,\alpha;s^*)\Big)\in\mathsf{ACC}\right]}$$

$$\geq\frac{2^{-n}\cdot\frac{\nu}{2}}{\mu'}=\frac{1}{2}\cdot 2^{-n}\frac{\nu}{\mu'}$$

□

3.1.3 Proof of Claim 8

For any α^* denote by $B_{\alpha^*} := \{\beta|(\alpha^*,\beta)\in\mathsf{ACC}\}$. By construction of the relaxed verifier we have that B_{α^*} contains at least a μ/ν fraction of all β. On the other hand, soundness of the protocol $\langle\mathsf{P},\mathsf{V}'\rangle$ guarantees, that B_α does not contain more than a $2\mu'/\nu \leq 4\mu/\nu^2$ fraction of all β. Thus, we have

$$\frac{\mu}{\nu}\leq\frac{|B_{\alpha^*}|}{2^{2n}}\leq\frac{4\mu}{\nu^2}$$

In particular, for any α and α^*, we have that

$$|B_\alpha|\geq\frac{\nu}{4}|B_{\alpha^*}|$$

which gives us

$$\mathsf{D}_0[\alpha^*,\beta^*]=\frac{1}{\mathsf{ACC}}=\frac{1}{\sum_{\alpha\in\{0,1\}^n}|B_\alpha|}\leq\frac{2}{2^n\cdot|B_{\alpha^*}|}=\frac{4}{\nu}\mathsf{D}_1[\alpha^*,\beta^*].$$

□

3.2 Proof of Lemma 5

Consider the following malicious verifier $\mathsf{V}^*=(\mathsf{V}_1^*,\mathsf{V}_2^*)$. The first stage V_1^* on input the statement x, the prover's first message α and auxiliary input aux simply interprets the auxiliary input as a circuit, evaluates it on x,α, and outputs the result $\beta\leftarrow\mathsf{aux}(x,\alpha)$. The second stage V_2^* on input the statement x, the prover's messages α,γ and auxiliary input aux recomputes $\beta\leftarrow\mathsf{aux}(x,\alpha)$ and then simply outputs α,β,γ.

Now, assume towards contradiction, that Π is zero-knowledge, i.e., in particular for V^* as described above there exists a PPT simulator Sim such that for all PPT distinguishers $\mathcal{D}$, all auxiliary inputs aux and aux', and all statements x it holds that

$$\left|\begin{array}{l}\Pr[\mathcal{D}(\langle\mathsf{P}(x,w),\mathsf{V}^*(x,\mathsf{aux})\rangle,\mathsf{aux}')=1]\\ -\Pr[\mathcal{D}(\mathsf{Sim}(x,\mathsf{aux}),\mathsf{aux}')=1]\end{array}\right|\leq\mathsf{negl}(n).$$

We will use said simulator to construct a malicious prover P^* against $\bar{\Pi}$ as follows: On input x and the verifier's message $\mathbf{B} = \mathsf{iO}(\mathbf{C}_{\mathsf{V}_1}[k])$, P^* invokes the simulator Sim on x and auxiliary input $\mathbf{B}$. The simulator will produce a transcript α, β, γ that P^* also outputs.

If $x \in \mathcal{L}$, then the zero-knowledge property and the completeness guarantee that $\bar{\mathsf{V}}_2$ will accept the proof with probability $1 - \mathsf{negl}(n)$, since otherwise we could easily construct a successful distinguisher against Sim as follows. The distinguisher $\mathcal{D}$ on input (α, β, γ) and auxiliary input aux' simply runs $\bar{\mathsf{V}}_2$ on (α, γ) and random coins aux' and outputs $b \leftarrow \bar{\mathsf{V}}_2((\alpha, \gamma); \mathsf{aux}')$. Further, even if $x \notin \mathcal{L}$, $\bar{\mathsf{V}}_2$ must still accept with all but negligible probability, since otherwise the combination of P^* and $\bar{\mathsf{V}}$ could be used to decide $\mathcal{L}$, implying that $\mathcal{L} \in \mathsf{BPP}$.

Therefore, P^* succeeds in convincing $\bar{\mathsf{V}}$ of false statements with all but negligible probability. Since this contradicts the premise that $\bar{\mu} \leq \mathsf{negl}(n)$, Sim cannot exist and therefore Π is not zero-knowledge. □

4 Extending the Lower Bound to ϵ-Zero Knowledge

In [14] Bitansky et al. introduced a weaker notion of zero-knowledge they called ϵ-zero-knowledge. In this weaker notion, the outputs of the simulator may be distinguishable with non-negligible probability, but the distinguishing advantage is upper bounded by any inverse monomial in the length of the statement. In this section we prove that our lower bound extends to this weaker notion of zero-knowledge. This is particularly interesting because Bitansky et al. [14] are able to construct a 4-round ϵ-zero-knowledge proof protocol from keyless multi-collision-resistant hash functions (MCRH). This provides evidence that our technique is unlikely to be extend to the case of 4-round proofs, since that would rule out MCRHs.

We start by defining ϵ-zero-knowledge. The definition is almost identical to regular zero-knowledge, except that the advantage of the distinguisher is not bounded by a negligible function.

Definition 8 (Non-uniform ϵ-Zero-Knowledge with Auxiliary Input). *Let $\langle \mathsf{P}, \mathsf{V} \rangle$ be a 2-Party protocol. $\langle \mathsf{P}, \mathsf{V} \rangle$ is said to be non-uniformly ϵ-zero-knowledge with auxiliary input, if for all (possibly malicious)* PPT *algorithms* V^* *there exists a* PPT *simulator* Sim, *such that for all* PPT *distinguishers $\mathcal{D}$ and all auxiliary inputs* aux *and* aux', *it holds that for all statements x with $|x| = \lambda$ and every noticeable function $\epsilon(\lambda) = \lambda^{-\mathcal{O}(1)}$*

$$\left| \begin{array}{l} \Pr[\mathcal{D}(\langle \mathsf{P}(x, w), \mathsf{V}^*(x, \mathsf{aux}) \rangle, \mathsf{aux}') = 1] \\ - \Pr\left[\mathcal{D}(\mathsf{Sim}(1^{1/\epsilon(\lambda)}, x, \mathsf{aux}), \mathsf{aux}') = 1\right] \end{array} \right| \leq \epsilon(\lambda).$$

Next, we state our generalized lemma about 3-round ϵ-zero-knowledge proofs. This lemma is a straightforward adaption of Lemma 5 to the ϵ-zero-knowledge case.

Lemma 10. *Let Π be a 3-round interactive proof system for a language $\mathcal{L} \notin$ BPP. Let $\bar{\Pi}$ be the transformed 2-round argument system described in Fig. 1 with soundness error $\bar{\mu}$. If $\bar{\mu} \leq \mathsf{negl}(n)$ then Π is not non-uniformly ϵ-zero-knowledge with auxiliary input.*

From combining Lemmas 4 and 10 a statement equivalent to Theorem 2 for ϵ-zero-knowledge follows as a simple corollary.

4.1 Proof of Lemma 10

Just like the lemma itself, the proof is a straightforward adaption of the proof for Lemma 5. We only need to make sure that the weaker requirement on the simulator does not cause the success probability of the cheating prover to deteriorate too much.

Consider the following malicious verifier $\mathsf{V}^* = (\mathsf{V}_1^*, \mathsf{V}_2^*)$. The first stage V_1^* on input the statement x, the prover's first message α and auxiliary input aux simply interprets the auxiliary input as a circuit, evaluates it on x, α, and outputs the result $\beta \leftarrow \mathsf{aux}(x, \alpha)$. The second stage V_2^* on input the statement x, the prover's messages α, γ and auxiliary input aux recomputes $\beta \leftarrow \mathsf{aux}(x, \alpha)$ and then simply outputs α, β, γ.

Now, assume towards contradiction, that Π is ϵ-zero-knowledge, i.e., in particular for V^* as described above there exists a PPT simulator Sim such that for all PPT distinguishers $\mathcal{D}$, all auxiliary inputs aux and aux', all statements x and all noticeable function $\epsilon(\lambda) = \lambda^{-\mathcal{O}(1)}$ it holds that

$$\left| \begin{array}{l} \Pr[\mathcal{D}(\langle \mathsf{P}(x,w), \mathsf{V}^*(x,\mathsf{aux})\rangle, \mathsf{aux}') = 1] \\ -\Pr\left[\mathcal{D}(\mathsf{Sim}(1^{1/\epsilon(|x|)}, x, \mathsf{aux}), \mathsf{aux}') = 1\right] \end{array} \right| \leq \epsilon(|x|).$$

We will use said simulator to construct a malicious prover P^* against $\bar{\Pi}$ as follows: On input x and the verifier's message $\mathbf{B} = \mathsf{iO}(\mathbf{C}_{\mathsf{V}_1}[k])$, P^* invokes the simulator Sim on $1/\epsilon(\lambda)$, x and auxiliary input $\mathbf{B}$. The simulator will produce a transcript α, β, γ that P^* also outputs.

If $x \in \mathcal{L}$, then the zero-knowledge property guarantees that $\bar{\mathsf{V}}_2$ will accept the proof with probability greater than $1 - |x|^{-c}$ for any constant $c \in \mathbb{N}$, since otherwise we could easily construct a successful distinguisher against Sim as follows. The distinguisher $\mathcal{D}$ on input (α, β, γ) and auxiliary input aux' simply runs $\bar{\mathsf{V}}_2$ on (α, γ) and random coins aux' and outputs $b \leftarrow \bar{\mathsf{V}}_2((\alpha, \gamma); \mathsf{aux}')$. This distinguisher would therefore be able to distinguish between a real transcript and a simulated transcript with probability greater than $|x|^{-c}$ for some constant $c \in \mathbb{N}$, thus clearly clearly contradicting the fact that Sim is a valid simulator. Further, even if $x \notin \mathcal{L}$, $\bar{\mathsf{V}}_2$ must still accept with probability at least $1 - |x|^{-c} - \mathsf{negl}(n)$, since otherwise the combination of P^* and $\bar{\mathsf{V}}$ could be used to decide $\mathcal{L}$, implying that $\mathcal{L} \in$ BPP.

Therefore, P^* succeeds in convincing $\bar{\mathsf{V}}$ of false statements with probability greater than $1 - |x|^{-c} - \mathsf{negl}(n)$ for any constant $c \in \mathbb{N}$, which is clearly non-negligible. Since this contradicts the premise that $\bar{\mu} \leq \mathsf{negl}(n)$, Sim cannot exist and therefore Π is not zero-knowledge. □

References

1. Ananth, P., Jain, A.: On secure two-party computation in three rounds. In: Kalai, Y., Reyzin, L. (eds.) TCC 2017. LNCS, vol. 10677, pp. 612–644. Springer, Cham (2017). https://doi.org/10.1007/978-3-319-70500-2_21
2. Ananth, P., Jain, A., Sahai, A.: Indistinguishability obfuscation from functional encryption for simple functions. Cryptology ePrint Archive, Report 2015/730 (2015). http://eprint.iacr.org/2015/730
3. Ananth, P., Sahai, A.: Projective arithmetic functional encryption and indistinguishability obfuscation from degree-5 multilinear maps. In: Coron, J.-S., Nielsen, J.B. (eds.) EUROCRYPT 2017. LNCS, vol. 10210, pp. 152–181. Springer, Cham (2017). https://doi.org/10.1007/978-3-319-56620-7_6
4. Applebaum, B., Brakerski, Z.: Obfuscating circuits via composite-order graded encoding. In: Dodis, Y., Nielsen, J.B. (eds.) TCC 2015. LNCS, vol. 9015, pp. 528–556. Springer, Heidelberg (2015). https://doi.org/10.1007/978-3-662-46497-7_21
5. Barak, B., Bitansky, N., Canetti, R., Kalai, Y.T., Paneth, O., Sahai, A.: Obfuscation for evasive functions. In: Lindell, Y. (ed.) TCC 2014. LNCS, vol. 8349, pp. 26–51. Springer, Heidelberg (2014). https://doi.org/10.1007/978-3-642-54242-8_2
6. Barak, B., Goldreich, O., Goldwasser, S., Lindell, Y.: Resettably-sound zero-knowledge and its applications. In: 42nd Annual Symposium on Foundations of Computer Science, pp. 116–125. IEEE Computer Society Press, Las Vegas, 14–17 October 2001
7. Barak, B., Goldreich, O., Impagliazzo, R., Rudich, S., Sahai, A., Vadhan, S., Yang, K.: On the (im)possibility of obfuscating programs. In: Kilian, J. (ed.) CRYPTO 2001. LNCS, vol. 2139, pp. 1–18. Springer, Heidelberg (2001). https://doi.org/10.1007/3-540-44647-8_1
8. Bellare, M., Jakobsson, M., Yung, M.: Round-optimal zero-knowledge arguments based on any one-way function. In: Fumy, W. (ed.) EUROCRYPT 1997. LNCS, vol. 1233, pp. 280–305. Springer, Heidelberg (1997). https://doi.org/10.1007/3-540-69053-0_20
9. Berman, I., Degwekar, A., Rothblum, R.D., Vasudevan, P.N.: Multi collision resistant hash functions and their applications. Cryptology ePrint Archive, Report 2017/489 (2017). http://eprint.iacr.org/2017/489
10. Bitansky, N., Brakerski, Z., Kalai, Y., Paneth, O., Vaikuntanathan, V.: 3-message zero knowledge against human ignorance. In: Hirt, M., Smith, A. (eds.) TCC 2016. LNCS, vol. 9985, pp. 57–83. Springer, Heidelberg (2016). https://doi.org/10.1007/978-3-662-53641-4_3
11. Bitansky, N., Canetti, R.: On strong simulation and composable point obfuscation. In: Rabin, T. (ed.) CRYPTO 2010. LNCS, vol. 6223, pp. 520–537. Springer, Heidelberg (2010). https://doi.org/10.1007/978-3-642-14623-7_28
12. Bitansky, N., Canetti, R., Paneth, O., Rosen, A.: On the existence of extractable one-way functions. In: Shmoys, D.B. (ed.) 46th Annual ACM Symposium on Theory of Computing, pp. 505–514. ACM Press, New York, 31 May–3 June 2014
13. Bitansky, N., Goldwasser, S., Jain, A., Paneth, O., Vaikuntanathan, V., Waters, B.: Time-lock puzzles from randomized encodings. In: Sudan, M. (ed.) ITCS 2016: 7th Innovations in Theoretical Computer Science, pp. 345–356. Association for Computing Machinery, Cambridge, 14–16 January 2016
14. Bitansky, N., Kalai, Y.T., Paneth, O.: Multi-collision resistance: a paradigm for keyless hash functions. Cryptology ePrint Archive, Report 2017/488 (2017). http://eprint.iacr.org/2017/488

15. Bitansky, N., Vaikuntanathan, V.: Indistinguishability obfuscation from functional encryption. In: Guruswami, V. (ed.) 56th Annual Symposium on Foundations of Computer Science, pp. 171–190. IEEE Computer Society Press, Berkeley, 17–20 October 2015
16. Blum, M.: How to prove a theorem so no one else can claim it. In: Proceedings of the International Congress of Mathematicians, vol. 1, p. 2 (1986)
17. Boneh, D., Waters, B.: Constrained pseudorandom functions and their applications. In: Sako, K., Sarkar, P. (eds.) ASIACRYPT 2013. LNCS, vol. 8270, pp. 280–300. Springer, Heidelberg (2013). https://doi.org/10.1007/978-3-642-42045-0_15
18. Boyle, E., Goldwasser, S., Ivan, I.: Functional signatures and pseudorandom functions. In: Krawczyk, H. (ed.) PKC 2014. LNCS, vol. 8383, pp. 501–519. Springer, Heidelberg (2014). https://doi.org/10.1007/978-3-642-54631-0_29
19. Canetti, R., Dakdouk, R.R.: Obfuscating point functions with multibit output. In: Smart, N. (ed.) EUROCRYPT 2008. LNCS, vol. 4965, pp. 489–508. Springer, Heidelberg (2008). https://doi.org/10.1007/978-3-540-78967-3_28
20. Canetti, R., Goldreich, O., Goldwasser, S., Micali, S.: Resettable zero-knowledge (extended abstract). In: 32nd Annual ACM Symposium on Theory of Computing, pp. 235–244. ACM Press, Portland, 21–23 May 2000
21. Cohen, A., Holmgren, J., Nishimaki, R., Vaikuntanathan, V., Wichs, D.: Watermarking cryptographic capabilities. In: Wichs, D., Mansour, Y. (eds.) 48th Annual ACM Symposium on Theory of Computing, pp. 1115–1127. ACM Press, Cambridge, 18–21 June 2016
22. Dwork, C., Naor, M., Reingold, O., Stockmeyer, L.J.: Magic functions. In: 40th Annual Symposium on Foundations of Computer Science, pp. 523–534. IEEE Computer Society Press, New York, 17–19 October 1999
23. Feige, U., Shamir, A.: Witness indistinguishable and witness hiding protocols. In: 22nd Annual ACM Symposium on Theory of Computing, pp. 416–426. ACM Press, Baltimore, 14–16 May 1990
24. Fiat, A., Shamir, A.: How to prove yourself: practical solutions to identification and signature problems. In: Odlyzko, A.M. (ed.) CRYPTO 1986. LNCS, vol. 263, pp. 186–194. Springer, Heidelberg (1987). https://doi.org/10.1007/3-540-47721-7_12
25. Garg, S., Gentry, C., Halevi, S., Raykova, M., Sahai, A., Waters, B.: Candidate indistinguishability obfuscation and functional encryption for all circuits. In: 54th Annual Symposium on Foundations of Computer Science, pp. 40–49. IEEE Computer Society Press, Berkeley, 26–29 October 2013
26. Garg, S., Miles, E., Mukherjee, P., Sahai, A., Srinivasan, A., Zhandry, M.: Secure obfuscation in a weak multilinear map model. In: Hirt, M., Smith, A. (eds.) TCC 2016. LNCS, vol. 9986, pp. 241–268. Springer, Heidelberg (2016). https://doi.org/10.1007/978-3-662-53644-5_10
27. Gentry, C., Lewko, A.B., Sahai, A., Waters, B.: Indistinguishability obfuscation from the multilinear subgroup elimination assumption. In: Guruswami, V. (ed.) 56th Annual Symposium on Foundations of Computer Science, pp. 151–170. IEEE Computer Society Press, Berkeley, 17–20 October 2015
28. Goldreich, O., Goldwasser, S., Micali, S.: How to construct random functions. J. ACM **33**(4), 792–807 (1986)
29. Goldreich, O., Kahan, A.: How to construct constant-round zero-knowledge proof systems for NP. J. Cryptol. **9**(3), 167–190 (1996)
30. Goldreich, O., Krawczyk, H.: On the composition of zero-knowledge proof systems. SIAM J. Comput. **25**(1), 169–192 (1996)

31. Goldreich, O., Oren, Y.: Definitions and properties of zero-knowledge proof systems. J. Cryptol. **7**(1), 1–32 (1994)
32. Goldwasser, S., Micali, S., Rackoff, C.: The knowledge complexity of interactive proof-systems (extended abstract). In: 17th Annual ACM Symposium on Theory of Computing, pp. 291–304. ACM Press, Providence, 6–8 May 1985
33. Hada, S., Tanaka, T.: On the existence of 3-round zero-knowledge protocols. In: Krawczyk, H. (ed.) CRYPTO 1998. LNCS, vol. 1462, pp. 408–423. Springer, Heidelberg (1998). https://doi.org/10.1007/BFb0055744
34. Håstad, J., Impagliazzo, R., Levin, L.A., Luby, M.: A pseudorandom generator from any one-way function. SIAM J. Comput. **28**(4), 1364–1396 (1999)
35. Jain, A., Kalai, Y.T., Khurana, D., Rothblum, R.: Distinguisher-dependent simulation in two rounds and its applications. In: Katz, J., Shacham, H. (eds.) CRYPTO 2017, Part II. LNCS, vol. 10402, pp. 158–189. Springer, Cham (2017). https://doi.org/10.1007/978-3-319-63715-0_6
36. Kalai, Y.T., Rothblum, G.N., Rothblum, R.D.: From obfuscation to the security of Fiat-Shamir for proofs. In: Katz, J., Shacham, H. (eds.) CRYPTO 2017, Part II. LNCS, vol. 10402, pp. 224–251. Springer, Cham (2017). https://doi.org/10.1007/978-3-319-63715-0_8
37. Katz, J.: Which languages have 4-round zero-knowledge proofs? In: Canetti, R. (ed.) TCC 2008. LNCS, vol. 4948, pp. 73–88. Springer, Heidelberg (2008). https://doi.org/10.1007/978-3-540-78524-8_5
38. Kiayias, A., Papadopoulos, S., Triandopoulos, N., Zacharias, T.: Delegatable pseudorandom functions and applications. In: Sadeghi, A.R., Gligor, V.D., Yung, M. (eds.) ACM CCS 2013: 20th Conference on Computer and Communications Security, pp. 669–684. ACM Press, Berlin, 4–8 November 2013
39. Komargodski, I., Naor, M., Yogev, E.: Collision resistant hashing for paranoids: dealing with multiple collisions. Cryptology ePrint Archive, Report 2017/486 (2017). http://eprint.iacr.org/2017/486
40. Lepinski, M.: On the existence of 3-round zero-knowledge proofs. Ph.D. thesis, Massachusetts Institute of Technology (2002)
41. Lin, H.: Indistinguishability obfuscation from constant-degree graded encoding schemes. In: Fischlin, M., Coron, J.-S. (eds.) EUROCRYPT 2016, Part I. LNCS, vol. 9665, pp. 28–57. Springer, Heidelberg (2016). https://doi.org/10.1007/978-3-662-49890-3_2
42. Lin, H.: Indistinguishability obfuscation from SXDH on 5-linear maps and locality-5 PRGs. In: Katz, J., Shacham, H. (eds.) CRYPTO 2017. LNCS, vol. 10401, pp. 599–629. Springer, Cham (2017). https://doi.org/10.1007/978-3-319-63688-7_20
43. Lin, H., Tessaro, S.: Indistinguishability obfuscation from trilinear maps and block-wise local PRGs. In: Katz, J., Shacham, H. (eds.) CRYPTO 2017, Part I. LNCS, vol. 10401, pp. 630–660. Springer, Cham (2017). https://doi.org/10.1007/978-3-319-63688-7_21
44. Lin, H., Vaikuntanathan, V.: Indistinguishability obfuscation from DDH-like assumptions on constant-degree graded encodings. In: Dinur, I. (ed.) 57th Annual Symposium on Foundations of Computer Science, pp. 11–20. IEEE Computer Society Press, New Brunswick, 9–11 October 2016
45. Pass, R., Seth, K., Telang, S.: Indistinguishability obfuscation from semantically-secure multilinear encodings. In: Garay, J.A., Gennaro, R. (eds.) CRYPTO 2014, Part I. LNCS, vol. 8616, pp. 500–517. Springer, Heidelberg (2014). https://doi.org/10.1007/978-3-662-44371-2_28

46. Sahai, A., Waters, B.: How to use indistinguishability obfuscation: deniable encryption, and more. In: Shmoys, D.B. (ed.) 46th Annual ACM Symposium on Theory of Computing, pp. 475–484. ACM Press, New York, 31 May–3 June 2014
47. Zimmerman, J.: How to obfuscate programs directly. In: Oswald, E., Fischlin, M. (eds.) EUROCRYPT 2015, Part II. LNCS, vol. 9057, pp. 439–467. Springer, Heidelberg (2015). https://doi.org/10.1007/978-3-662-46803-6_15

Statistical Witness Indistinguishability (and more) in Two Messages

Yael Tauman Kalai[1], Dakshita Khurana[2(✉)], and Amit Sahai[2]

[1] Microsoft Research, Cambridge, USA
yaelism@gmail.com

[2] Department of Computer Science, UCLA, Los Angeles, USA
{dakshita,sahai}@cs.ucla.edu

Abstract. Two-message witness indistinguishable protocols were first constructed by Dwork and Naor (FOCS 2000). They have since proven extremely useful in the design of several cryptographic primitives. However, so far no two-message arguments for NP provided *statistical privacy* against malicious verifiers. In this paper, we construct the first:

- Two-message statistical witness indistinguishable (SWI) arguments for NP.
- Two-message statistical zero-knowledge arguments for NP with super-polynomial simulation (Statistical SPS-ZK).
- Two-message statistical distributional weak zero-knowledge (SwZK) arguments for NP, where the simulator is a probabilistic polynomial time machine with oracle access to the distinguisher, and the instance is sampled by the prover in the second round.

These protocols are based on quasi-polynomial hardness of two-message oblivious transfer (OT), which in turn can be based on quasi-polynomial hardness of DDH or QR or N^{th} residuosity. We also show how such protocols can be used to build more secure forms of oblivious transfer.

Along the way, we show that the Kalai and Raz (Crypto 09) transform compressing interactive *proofs* to two-message arguments can be generalized to compress certain types of interactive *arguments*. We introduce and construct a new technical tool, which is a variant of extractable two-message statistically hiding commitments, building on the recent work of Khurana and Sahai (FOCS 17). These techniques may be of independent interest.

1 Introduction

Witness indistinguishable (WI) protocols [16] allow a prover to convince a verifier that some statement x belongs to an NP language L, with the following privacy guarantee: If there are two witnesses w_1, w_2 that both attest to the fact that $x \in L$, then a computationally bounded verifier should not be able to distinguish an honest prover using witness w_1 from an honest prover using witness w_2. WI is a relaxation of zero-knowledge that has proven to be surprisingly useful. Because WI is a relaxation, unlike zero-knowledge, there are no known lower bounds on

J. B. Nielsen and V. Rijmen (Eds.): EUROCRYPT 2018, LNCS 10822, pp. 34–65, 2018.
https://doi.org/10.1007/978-3-319-78372-7_2

the rounds of interaction needed to build WI protocols. Indeed, in an influential work, Dwork and Naor [14] introduced WI protocols that only require two messages to be exchanged between the prover and verifier, and these were further derandomized to non-interactive protocols by [6]. Due to this extremely low level of interaction, two-message WI protocols have proven to be very useful in the design of several cryptographic primitives. Later, [4,8,21,23] achieved two message or non-interactive WI protocols from other assumptions, namely assumptions on bilinear maps, indistinguishability obfuscation, and quasi-polynomial DDH, respectively.

Two-Message Statistical WI. In this work, we revisit this basic question of constructing two-message WI protocols, and ask whether it is possible to upgrade the WI privacy guarantee to hold even against *computationally unbounded* verifiers. In other words, can we construct statistical WI (SWI) protocols for NP that require only two messages to be exchanged? This is the natural analog of one of the earliest questions studied in the context of zero-knowledge protocols: Are statistical zero-knowledge arguments [10] possible for NP?

Indeed, statistical security is important because it allows for *everlasting privacy* against malicious verifiers, long after protocols have completed execution. On the other hand, soundness is usually necessary only in an online setting: In order to convince a verifier of a false statement, a cheating prover must find a way to cheat *during* the execution of the protocol.

The critical bottleneck to achieving two-message statistical WI has been proving soundness. For instance, the Dwork-Naor transformation from a non-interactive zero-knowledge (NIZK) protocol to two-message WI requires the underlying NIZK to be a proof system – that is, for the NIZK to be sound against computationally unbounded cheating provers. Of course, to achieve statistical privacy, we must necessarily sacrifice soundness against unbounded provers. Thus, remarkably, 17 years after the introduction of two-message WI protocols, until our work, there has been no construction of two-message statistical WI arguments. In fact, this question was open even for three-message protocols.

In our first result, we resolve this research question, constructing the first two-message statistical WI arguments for NP, based on standard cryptographic hardness assumptions against quasi-polynomial time adversaries (such as quasi-poly hardness of DDH, or Quadratic Residuosity, or N'th Residuosity). Because two-message WI is so widely applicable, and statistical privacy is useful in many situations where computational privacy does not suffice, we expect our two-message SWI argument to be a useful new tool in the protocol designer's toolkit.

Stronger Two-Message Statistically Private Protocols. The techniques we use to build two-message SWI also allow us to achieve other forms of statistical privacy.

One of the most popular notions of privacy in proof systems is that of zero-knowledge. This is usually formalized via simulation, by showing the existence of a *polynomial-time* simulator that simulates the view of any polynomial size (malicious) verifier. At an intuitive level, the existence of such a simulator means that any information that a polynomial size verifier learns from an honest prover, he could have generated on his own (in and indistinguishable manner), without access to such a prover. It is known [20] that zero-knowledge is impossible to achieve in just two messages. However, other weaker variants have been shown to be achievable in this setting.

Pass [29] was the first to construct a two-message argument with quasi-polynomial time simulation. In his work, the simulated proofs were indistinguishable by distinguishers running in time significantly smaller than that of the (uniform) simulator. Very recently, [27] constructed the first two-message arguments for NP achieving super-polynomial *strong* simulation, where the simulated proofs remain indistinguishable by distinguishers running in time significantly larger than that of the (uniform) simulator. These capture the intuition that for any information that a quasi-polynomial size verifier learns from an honest prover, indistinguishable information could have been generated by the verifier in a similar amount of time.

An even stronger security property would be super-polynomial *statistical* simulation, where the output of the simulator is indistinguishable from real executions of the protocol even against distinguishers that run in *an unbounded amount of time*. In this paper, we construct the first arguments satisfying this property in two messages.[1] This improves upon the work of [27] by pushing their privacy guarantees all the way to statistical.

We note that in all these arguments, the simulator works by breaking soundness of the proof, so all of the above two-message arguments are only sound against provers running in time less than that of the simulator.

Recently, [23] showed that this caveat could be overcome, by weakening the ZK requirement. Specifically, they constructed two-message arguments in the delayed-input distributional setting, with distinguisher-dependent polynomial-time simulation. These protocols only satisfy *computational privacy*, and a natural open question was to achieve *statistical privacy*. We show that our techniques can be used to get two-message arguments for NP in the delayed-input distributional setting with distinguisher-dependent simulation, where the simulator runs in polynomial time with oracle access to the distinguisher, and achieving *statistical* privacy.

Our Core Technique. Our key technique consists of compressing an interactive protocol into a two-message protocol. Specifically, we start with an interactive argument satisfying honest-verifier *statistical zero-knowledge*, and compress it into a two-message argument by proving soundness of the [25] heuristic, which builds on [7]. Actually, to obtain a two-message protocol with statistically privacy, it does not suffice to start with an honest-verifier statistical ZK protocol, but rather we need the ZK property to hold against semi-malicious verifiers.[2] We gloss over this detail in this high-level overview.

[1] Achieving such two-message arguments was believed to be impossible [12], however the work of [27] showed that the line of impossibility claims [12] for super-polynomial simulation was surmountable.

[2] A semi-malicious verifier is one who follows the prescribed algorithm but with possibly malicious randomness.

This heuristic is believed to be *insecure* when applied generally to interactive arguments (as opposed to proofs). Nevertheless, we construct a family of 4-message interactive arguments with statistical hiding guarantees, and prove that the [25]-heuristic is sound when applied to such protocols.

At the heart of our technique is the following idea: We devise protocols that are almost always statistically private (and only computationally sound), but with negligible probability, they are statistically sound. Crucially, we show that a (computationally bounded) prover cannot distinguish between the case when the protocol ends up being statistically private (which happens most of the time), and the case when the protocol ends up being statistically sound (which happens very rarely). At the heart of our construction is a new special commitment scheme, which build upon and significantly extend commitment schemes from [27]. We then show how to leverage this rare statistical soundness event, to allow the soundness of the the [25]-heuristic to kick in.

This rare event helps us achieve other extraction properties that we require in our applications. We elaborate on this below in our technical overview, providing a detailed but still informal overview of our techniques and results. Our protocols are based on standard cryptographic hardness assumptions with security against quasi-polynomial time adversaries (such as the quasi-poly hardness of DDH, or Quadratic Residuosity, or N'th Residuosity).

New Oblivious Transfer Protocols. Our techniques also have applicability to an intriguing question about oblivious transfer (OT): The works of Naor and Pinkas [28] and Aiello et al. [2] introduced influential two-message protocols for OT achieving a game-based notion of security, which offers security against computationally *unbounded* malicious receivers. A natural question is: Can we achieve a similar result offering security against computationally *unbounded* senders? Note that to achieve such a result, at least three messages must be exchanged in the OT protocol: Indeed, suppose to the contrary that there was a two-message OT protocol with security against an unbounded sender. Then the first message of the protocol sent by the receiver must statistically hide the choice bit of the receiver in order for this message to provide security against an unbounded cheating sender. However, a non-uniform cheating receiver could begin the protocol with non-uniform advice consisting of a valid first message m together with honest receiver randomness r_0 that explains m with regard to the choice bit $b = 0$, and honest receiver randomness r_1 that explains m with regard to the choice bit $b = 1$. Now this receiver would be able to recover both inputs of the honest sender by using both random values r_0 and r_1 on the sender's response message, violating OT security against a (bounded) malicious receiver.

Again remarkably, this basic question, of constructing a 3-message OT protocol with security against unbounded sender, has been open since the works of [2,28] 17 years ago. We resolve this question, by exhibiting such a 3-message OT protocol, based on standard cryptographic hardness assumptions with security against quasi-polynomial time adversaries (same assumptions as before). Such an OT protocol can also be plugged into the constructions of [23] to achieve three-message *proofs* for NP (as opposed to arguments) achieving delayed-input distributional weak ZK, witness hiding and strong witness indistinguishability.

Our techniques also apply to other well-studied questions about OT, even in the two-message setting with security against unbounded receivers. It has long been known that the two-message OT protocols of [2,28] do not rule out selective failure attacks. For example, if two OTs are run in parallel, we do not know how to rule out the possibility that the sender can cause the OTs to abort if and only if the receiver's two choice bits are equal. Intuitively, this should not be possible in a secure OT, and the "gold standard" for preventing all such attacks for OT is to prove security via simulation. For two-message OT protocols, however, only super-polynomial simulation is possible, and this was recently formally established in [3] but at the cost of sacrificing security against unbounded receivers. This sacrifice seems inherent: If an OT protocol has a super-polynomial simulator, then it seems that an unbounded malicious receiver can just "run the simulator" to extract the inputs of the sender. This presents a conundrum; perhaps simulation security and security against an unbounded malicious receiver cannot be simultaneously achieved.

In fact, we show that it *is* possible to construct a two-message OT protocol with both super-polynomial simulation security, and security against unbounded receivers.

1.1 Summary of Our Results

We construct several protocols with security properties assuming the existence of a quasi-poly secure OT, which can in turn be instantiated based on quasi-poly hardness of the DDH assumption [28], or based on the quasi-poly hardness of QR or the N'th residuosity assumption [22,24]. We first construct a two-message argument for NP with the following statistical hiding guarantees:

1. Our two-message argument is statistical witness indistinguishable. We note that prior to this work, we did not even know how to construct a 3-message *statistical WI* scheme.
2. Our two-message argument is statistical zero-knowledge with super-polynomial time simulation.[3]
3. Our two-message argument is statistical weak zero-knowledge in the delayed input setting where the simulator has oracle access to the distinguisher, and where the instance is sampled from some distribution after the verifier sent the first message.

We also obtain the following results on oblivious transfer:

1. We construct a three-message OT protocol simultaneously satisfying super-polynomial simulation security, and security against a computationally unbounded sender.

[3] We note that prior to this work, this was believed to be impossible to achieve via black-box reductions [12].

2. We construct a two-message OT protocol simultaneously satisfying super-polynomial simulation security, and security against a computationally unbounded receiver.

1.2 Other Related Work

Two message statistical witness indistinguishable arguments were constructed for specific languages admitting hash proof systems, by [18]. However, no two-message statistical WI arguments were known for all of NP.

Two main approaches for reducing rounds in interactive proof systems have appeared in the literature. The first is due to Fiat and Shamir [17], and the second is due to [25] and is based on the [7]-heuristic for converting multi-prover interactive proofs to two-message arguments. The [25]-heuristic is sound when applied to a *statistically sound* interactive proof, assuming the existence of a super-polynomial OT (or super-polynomially secure computational PIR) scheme. Very recently, [11,26] showed that the Fiat-Shamir heuristic is also sound when applied to a *statistically sound* interactive proof, assuming the existence of a symmetric encryption scheme where the key cannot be recovered even with *exponentially* small probability (even after seeing encryptions of key-dependent messages).[4]

The works of [3,23] are closely related to our work. They assume the existence of a quasi-poly secure oblivious transfer (OT) scheme, and show how to convert any 3-message public-coin protocol which is zero-knowledge against semi-malicious verifiers, into a two-message protocol, while keeping (and even improving) the secrecy guarantees. However, these works do not yield statistical privacy, which is the focus of the present work. More specifically, these works apply the [25]-heuristic to 3-message public-coin proofs that are zero-knowledge against semi-malicious verifiers, to obtain their resulting two-message protocols. We note that since they start with a statistically sound proof they obtain only *computational hiding* guarantees, and after applying the [25]-heuristic, their resulting two-message protocols are only *computationally sound* (in addition to being only computational hiding).

In contrast, in this work we construct two-message arguments with *statistical hiding* guarantees. More specifically, we do this by constructing a 4-message interactive argument with statistical hiding guarantees, and converting it into a two-message computationally sound protocol by applying the [25]-heuristic to it.

2 Overview of Techniques

Our starting point is the [25]-heuristic, which shows how to compress public coin interactive *proofs* into two-message arguments. We note that this heuristic is based on the heuristic introduced in [7] (and explored in [1]), which converts

[4] Their actual assumption is a bit more complex and we refer to [11] for details.

multi-prover interactive proofs into two-message arguments. We note that the [25]-heuristic is only known to be sound when applies to interactive proofs (and believed not to be sound when applied to general interactive arguments).

Recently, [3,23] proved that this heuristic also preserves (and even enhances) privacy. Our strategy will be to follow this blueprint, but in the statistical setting. This becomes quite tricky in the statistical setting because we do not have interactive *proofs* for NP with statistical privacy guarantees. In particular, we do not have an interactive *proof* for NP which is statistical zero-knowledge against semi-malicious verifiers (which is the privacy guarantee needed in [3,23], but in the computational setting).

However, we do have an interactive *argument* which is statistical zero-knowledge against semi-malicious verifiers. We construct such an interactive argument of a specific form, and prove that the [25]-heuristic is sound when applied to this interactive argument.

We begin by reviewing the techniques from [3,23], where we take as a running example the Blum protocol for Graph Hamiltonicity, which is known to be (computational) zero-knowledge against semi-malicious verifiers.

2.1 First Attempt: Compressing the Blum Protocol via OT

In what follows, we recall the two-message protocol from [3,23] (with computational privacy guarantees), which makes use of the following two components:

- A three-message proof for Graph Hamiltonicity, due to Blum [9]. Denote its three messages by (a, e, z), which can be parsed as $a = \{a_i\}_{i \in [\kappa]}$, $e = \{e_i\}_{i \in [\kappa]}$ and $z = \{z_i\}_{i \in [\kappa]}$. Here for each $i \in [\kappa]$, the triplet (a_i, e_i, z_i) are messages corresponding to an underlying Blum protocol with a single-bit challenge (i.e., where $e_i \in \{0, 1\}$). We also denote by f_1 and f_2 the functions that satisfy $a_i = f_1(x, w; r_i)$ and $z_i = f_2(x, w, r_i, e_i)$, for answers provided by the honest prover, and where r_i is uniformly chosen randomness.
- Any two-message oblivious transfer protocol, denoted by $(\mathsf{OT}_1, \mathsf{OT}_2)$, which is secure against malicious PPT receivers, and malicious senders running in time at most $2^{|z|}$. For receiver input b and sender input messages (M^0, M^1), we denote the two messages of the OT protocol as $\mathsf{OT}_1(b)$ and $\mathsf{OT}_2(M^0, M^1)$. We note that $\mathsf{OT}_2(M^0, M^1)$ also depends on the message $\mathsf{OT}_1(b)$ sent by the receiver. For the sake of simplicity, we omit this dependence from the notation.

Given these components, the two-message protocol $\langle P, V \rangle$ (from [3,23]) is described in Fig. 1.

Soundness. It was proven in [3,23,25] that such a transformation from any public-coin interactive proof to a two-round argument preserves soundness against adaptive PPT provers, who may choose the instance adaptively depending upon the message sent by the verifier.

Preliminary Two-Message Protocol from [24,3]

- For $i \in [\kappa]$, V picks $e_i \xleftarrow{\$} \{0,1\}$, and sends $\mathsf{OT}_{1,i}(e_i)$ in parallel. Each e_i is encrypted with a fresh OT instance.
- For $i \in [\kappa]$, P computes $a_i = f_1(x, w; r_i), z_i^{(0)} = f_2(x, w, r_i, 0), z_i^{(1)} = f_2(x, w, r_i, 1)$. The prover P then sends $a_i, \mathsf{OT}_{2,i}(z_i^{(0)}, z_i^{(1)})$ in parallel for all $i \in [\kappa]$.
- The verifier V recovers $z_i^{(e_i)}$ from the OT, and accepts if and only if for every $i \in [\kappa]$, the transcript $(a_i, e_i, z_i^{(e_i)})$ is an accepting transcript of the underlying Σ-protocol.

Fig. 1. Preliminary two-message protocol

Can We Achieve Statistical Privacy Against Malicious Verifiers? Let us now analyze the privacy of the protocol in Fig. 1. The work of [3,23] showed that the protocol in Fig. 1 satisfies computational witness indistinguishability, as well as other stronger (computational) privacy guarantees against malicious verifiers. Their proofs rely on the security of OT against malicious receivers, as well as the zero-knowledge property of the underlying Blum proof, when restricted to semi-malicious verifiers.

As we already described, the focus of this paper is achieving statistical privacy. To this end, we take a closer look at the Blum protocol.

Background. Recall that in the (parallel repetition of the) Blum protocol, for each index $i \in [\kappa]$, a_i consists of a statistically binding commitment to a random permutation π and the permuted graph $\pi(G)$, where G denotes the input instance with Hamiltonian cycle H. Then, if the verifier challenge $e_i = 0$, the prover computes z_i as a decommitment to $(\pi, \pi(G))$, and the verifier accepts if and only if the graph G was correctly permuted. On the other hand, if $e_i = 1$, the prover computes z_i as a decommitment only to the edges of the Hamiltonian Cycle $\pi(H)$ in $\pi(G)$, and the verifier accepts if and only if the revealed edges are indeed a Hamiltonian Cycle.

In an quest for statistical privacy, we notice the following properties about the protocol in Fig. 1:

1. A single parallel repetition of the underlying Blum proof only satisfies computational zero-knowledge. This is because it uses a statistically binding, computationally hiding commitment to generate the first message $\{a_i\}_{i\in[\kappa]}$. An unbounded malicious verifier that breaks the commitment in $\{a_i\}_{i\in[\kappa]}$ can in fact, extract π, and therefore obtain the witness (i.e., the Hamiltonian cycle) from any honest prover.
2. The underlying OT protocols [22,28] used in the protocol of Fig. 1 are already statistically private against malicious receivers. This implies that the messages $\{z_i^{(1-e_i)}\}_{i\in[\kappa]}$ are *statistically hidden* from any malicious verifier.

As a result of (1) above, the protocol in Fig. 1 is also only computationally private. At this point, it is clear that the main bottleneck towards achieving statistical privacy against malicious verifiers, is the computationally hiding commitment in the message $\{a_i\}_{i\in[\kappa]}$. A natural first idea is then to replace this commitment with a *statistically hiding commitment*.

To this end, we consider a modified version of the underlying Blum protocol, which is the same as the original Blum protocol, except that it uses a statistically hiding, computationally binding commitment. Such a commitment must contain two-messages in order to satisfy binding against non-uniform PPT provers. Therefore, our modified version of the Blum protocol has four messages, where in the first message, for $i \in [\kappa]$, the verifier sends the first message q_i of a statistically hiding, computationally binding commitment. Next, the prover responds with a_i consisting of the committer message in response to q_i, committing to values $(\pi_i, \pi_i(G))$. The next messages $\{e_i\}_{i\in[\kappa]}$ and $\{z_i\}_{i\in[\kappa]}$ remain the same as before. It is not hard to see that the resulting four-message modified Blum protocol satisfies *statistical zero-knowledge* against semi-malicious verifiers.

Let us again compress this four-message protocol using the same strategy as before, via two-message OT. That is, the verifier sends in parallel $\{q_i, \mathsf{OT}_{1,i}(e_i)\}_{i\in[\kappa]}$, and the prover responds with $\{a_i, \mathsf{OT}_{2,i}(z_i^{(0)}, z_i^{(1)})\}_{i\in[\kappa]}$. In this case, because of the *statistical* hiding of the commitments and the *statistical* sender security of OT, the proof in [3,23] can be easily extended to achieve *statistical* witness indistinguishability.

One may now hope that the analysis in [3,23,25] can be used to prove that the resulting protocol also remains *sound* against PPT provers. Unfortunately, as we noted above, the proof of soundness [3,23,25] crucially relies on the fact that the starting protocol is a *proof* (as opposed to an argument). More specifically, the soundness proof in previous works goes through as follows: Consider for simplicity the case of a single repetition, and suppose a cheating prover, on input the verifier message $\mathsf{OT}_1(e^*)$, outputs $x^* \notin L$, together with a message $(a^*, \mathsf{OT}_2(z^*))$, such that the verifier accepts with probability $\frac{1}{2} + \frac{1}{\mathsf{poly}(\kappa)}$. Intuitively, since for any $x^* \notin L$ and any a^*, there exists at most one unique value of receiver challenge e^*, for which there exists a z^* that causes the verifier to accept, this means that a^* consists of a commitment that *encodes* the receiver challenge e^*. By using an OT scheme that is secure against adversaries that can break the commitment within a^*, a cheating prover can be used to contradict receiver security of OT. This proves that a single parallel execution of the protocol in Fig. 1 has soundness $\frac{1}{2} + \mathsf{negl}(\kappa)$. The same argument can be generalized to prove that no adaptive PPT prover P^* can cheat with non-negligible probability when we perform κ parallel repetitions. More specifically, the reduction can use any prover that cheats with non-negligible probability to guess the κ-bit challenge e with non-negligible probability, contradicting the security of κ parallel repetitions of OT.

This proof crucially relies on the fact that the commitment is statistically binding. This is no longer true for the four-message modified version of the Blum protocol described above. In fact, the problem runs deeper: Note that what we seem to need for this approach to work is a *proof* that satisfies *statistical ZK* against semi-malicious verifiers, however, such proofs are unlikely to exist for all of NP (see, e.g. [30]). Therefore, the only remaining option, if we follow this approach, is to find a way to compress some form of statistical ZK *argument* while preserving soundness.

2.2 Compressing Interactive *Arguments* While Preserving Soundness

The problem of compressing general interactive arguments while preserving soundness has been a question of broader interest, even in the context of delegating computation. In this paper, unlike the setting of delegation, we are not concerned with the succinctness of our arguments. Yet, there are no previously known approaches to compressing any types of interactive argument systems that are not also *proofs*.

In this paper, we develop one such approach. Our high-level idea is as follows: Since we already ruled out constructing a *proof* that satisfies statistical ZK against semi-malicious verifiers, we will instead construct an *argument* that satisfies statistical ZK against semi-malicious verifiers. But this argument will have the property that with a small probability, it will in fact be a proof! Furthermore, no cheating prover will be able to differentiate the case when it is an argument from the case when it is a proof. In other words, we will ensure that any cheating prover that outputs $x^* \notin L$ together with an accepting proof with non-negligible probability in the original protocol, will continue to do so with non-negligible probability even when it is in proof mode. Upon switching to proof mode, we can apply the techniques of [25] to argue soundness and obtain a contradiction.

Our main technical tool that will help us realize the above outline will be a two-message *statistically-hiding extractable commitment* scheme, which we now describe.

Main Tool: Statistically Hiding Extractable Commitments. Our construction of statistically hiding, extractable commitments is obtained by building on the recent work of Khurana and Sahai [27].

They construct an extractable *computationally hiding* commitment scheme, which is completely insecure against unbounded malicious receivers. The underlying idea behind their work, which we will share, is the following: In their commitment scheme, with a negligible probability, 2^{-m} for $m = \Omega(\log \kappa)$, the message being committed to is transmitted to the receiver. Otherwise, with overwhelming probability $1 - 2^{-m}$, the receiver obtains an actual (statistically-binding) commitment to the message. Crucially, the committer does not know which case occurs – whether its message was transmitted to the receiver or not. In this way, their commitment can be seen as an unusually noisy *erasure channel*.

Committer Input: Message $M \in \{0,1\}^p$, where $p = \mathsf{poly}(\kappa)$.
Commit Stage:
Receiver Message.

- Pick challenge string $\mathsf{ch} \xleftarrow{\$} \{0,1\}$.
- Compute and send the first OT message $\mathsf{OT}_1(\mathsf{ch}, r_1)$ using uniform randomness r_1.

Committer Message.

- Sample a random string $r \xleftarrow{\$} \{0,1\}$. Set $M^r = M, M^{1-r} \xleftarrow{\$} \{0,1\}^p$.
- Compute $o_2 = \mathsf{OT}_2(M^0, M^1; r_2)$ with uniform randomness r_2.
- Send (r, o_2).

Reveal Stage: The committer reveals M, and both values (M^0, M^1) as well as the randomness r_2. The receiver accepts the decommitment to message M if and only if:

1. $o_2 = \mathsf{OT}_2(M^0, M^1; r_2)$,
2. $M^r = M$.

Fig. 2. Basic construction of a two-message statistically hiding commitment

Our commitment will work to achieve the same goal, but crucially we will seek to achieve a statistically hiding commitment.

The reason why the work of [27] was inherently limited to achieving only computational hiding is because of the way they implement the erasure channel described above: In their work, this was implemented using a two-message secure computation protocol, that implemented a coin-flipping procedure to provide the randomness underlying the erasure channel. Such two-message secure computation protocols only achieve computational hiding. Therefore, in our work, we must depart fundamentally from this method of implementing the erasure channel.

Basic Construction. In order to obtain a construction that essentially implements the erasure channel described above, we go back to the drawing board. Instead of implementing a sophisticated two-party computation using garbled circuits, we consider the following basic commitment scheme (Fig. 2) implemented using game-based oblivious transfer [2,22,24,28], with statistical sender security. We make the following observations about this protocol:

- Assuming statistical sender security of OT, this scheme is 1/2-hiding against malicious receivers (i.e., $r \neq \mathsf{ch}$ happens with probability $\frac{1}{2}$, and in this case the message is statistically hidden from any malicious receiver).
- Assuming computational receiver security of OT, this scheme is computationally binding. That is, no malicious PPT committer, upon generating a commitment transcript, can successfully decommit it to two different values $\widetilde{M_1} \neq \widetilde{M_2}$, except with negligible probability. This is because given such a

committer, the reduction can use this committer to *deduce* that $r \neq ch$, which should be impossible except with negligible probability[5]. A formal analysis can be found in the full version of the paper.

Our Construction. Recall that we would like a scheme where most transcripts ($1 - 2^{-m}$ fraction of them) should be statistically hiding and the message should be completely lost. Moreover, we would like a 2^{-m} fraction of transcripts to be statistically binding: in fact, it will suffice to directly reveal the message being committed in these transcripts to the receiver. Starting with the basic construction above, a natural way to achieve this is to commit to an XOR secret sharing of the message M via m parallel executions of the basic scheme described above. Formally, our construction is described in Fig. 3. This scheme satisfies the following properties:

- It remains computationally binding against malicious PPT committers, just like the basic scheme.
- Because the underlying OT is statistically hiding, our scheme is now $(1-2^{-m})$-statistically hiding against malicious receivers (i.e., it is not statistically hiding only in the case that $r \neq \mathsf{ch}$, which happens with probability 2^{-m}).
- Most importantly, because of receiver security of the OT, no malicious PPT committer can distinguish the case where $r = \mathsf{ch}$ from the case where $r \neq \mathsf{ch}$.[6]

Modifying Blum to Use Statistically Hiding Extractable Commitments. Now, instead of plugging in *any* statistically hiding commitment scheme, we plug in the extractable statistically hiding commitment scheme of Fig. 3 to generate messages $\{q_i, a_i\}_{i \in [\kappa]}$, with $m = \Omega(\log \kappa)$. This is formally described in Sect. 5.1. By statistical hiding of the commitment, the resulting protocol is a statistical ZK argument. On the other hand, by the extractability of the commitment, (more specifically in the case where $r = \mathsf{ch}$), the protocol, in fact, becomes a proof. Furthermore, no cheating PPT prover can distinguish the case when $r = \mathsf{ch}$ from when $r \neq \mathsf{ch}$. Looking ahead, like we already alluded to at the beginning of the overview, we will compress this while simultaneously ensuring that any malicious prover outputting an accepting transcript corresponding to $x \notin L$ with noticeable probability when $r \neq \mathsf{ch}$, must continue to do so even when $r \neq \mathsf{ch}$. We will now analyze the soundness of the resulting protocol.

Arguing Soundness of the Compressed Protocol. We show that the resulting protocol remains sound against cheating PPT provers. While we also achieve

[5] We note that this is different from guessing ch, which can be done with probability $\frac{1}{2}$: however, a cheating committer can not only guess ch but also *certify* via two valid decommitments to different messages that it guessed ch correctly, which is not allowed except with negligible probability.

[6] This requires a more delicate argument, as well as reliance on 2^m-security of the OT to ensure that a PPT cheating committer cannot bias r away from ch all the time.

Extraction parameter: m.
Committer Input: Message $M \in \{0,1\}^p$.
Commit Stage:
Receiver Message.

- Pick challenge string $\mathsf{ch} \xleftarrow{\$} \{0,1\}^m$.
- Sample uniform randomness $\{r_{1,i}\}_{i\in[m]}$.
- Compute and send $\{\mathsf{OT}_1(\mathsf{ch}_i, r_{1,i})\}_{i\in[m]}$ using m instances of two-message OT.

Committer Message.

- Sample a random string $r \xleftarrow{\$} \{0,1\}^m$.
  For every $i \in [m]$ and every $b \in \{0,1\}$, sample $M_i^b \xleftarrow{\$} \{0,1\}^p$ subject to $\bigoplus_{i\in[m]} M_i^{r_i} = M$.
- For every $i \in [m]$ compute $o_{2,i} = \mathsf{OT}_2(M_i^0, M_i^1; r_{2,i})$ with uniform randomness $r_{2,i}$.
- Send $(r, \{o_{2,i}\}_{i\in[m]})$.

Reveal Stage: The committer reveals M, and all values $\{M_i^0, M_i^1\}_{i\in[m]}$ as well as the randomness $r_{2,i}$. The receiver accepts the decommitment to message M if and only if:

1. For all $i \in [m]$, $o_{2,i} = \mathsf{OT}_2(M_i^0, M_i^1; r_{2,i})$,
2. $\bigoplus_{i\in[m]} M_i^{r_i} = M$.

Fig. 3. Our extractable commitments

a variant of adaptive soundness, for the purposes of this overview we restrict ourselves to proving soundness against non-adaptive provers that output the instance x before the start of the protocol.

At a high level, we will begin by noting that a cheating prover that first outputs $x \notin L$ together with an accepting proof with probability $p = \frac{1}{\mathsf{poly}(\kappa)}$, cannot distinguish the case when $r = \mathsf{ch}$ from the case when $r \neq \mathsf{ch}$ by the property of the extractable commitment. Moreover, such a prover must continue to generate accepting transcripts for $x \notin L$ with probability at least $\frac{1}{\mathsf{poly}(\kappa)}$ even in case $r = \mathsf{ch}$[7]. Although the event $r = \mathsf{ch}$ only occurs with negligible probability, we use the extractor of extcom to amplify this probability by making many queries to the prover. The extractor then outputs a transcript of the proof (corresponding to $r = \mathsf{ch}$), together with the values committed in all messages corresponding to the extractable commitment. This requires the oblivious trans-

[7] Ensuring this requires the decommit phase of the extractable commitment to be publicly verifiable, without the receiver needing to maintain any state from the commit phase. This is for technical reasons, specifically, public verifiability of the decommit phase is required to check whether a transcript is accepting or rejecting even while obtaining the receiver message for the extractable commitment, externally.

fer used for such compression to be hard against adversaries running in time large enough to enable extraction from the extcom. Additional details of our construction can be found in Sect. 5.2.

In fact, we notice that our technique is more generally applicable. In particular, we focus on applications to some natural questions about oblivious transfer.

2.3 Applications to OT

OT Secure Against Unbounded Senders. While we have long known two-message OT protocols with game-based security against unbounded malicious receivers and PPT malicious senders [2,22,24,28], the following natural, extremely related question has remained unanswered so far. Can we construct three-message oblivious transfer with game-based security against unbounded malicious senders and non-uniform PPT malicious receivers?

It is clear that a minimum of three rounds is required for this task, since in any two message protocol in the plain model secure against non-uniform receivers, the first message must unconditionally bind a malicious receiver to a single choice bit (as otherwise a cheating receiver may obtain non-uniformly, a receiver message as well as randomness that allows opening this message to two different bits). In order to achieve such oblivious transfer, we explore a very natural approach: [32] suggested the following way to information-theoretically reverse any ideal OT protocol (with receiver message denoted by OT_R and sender message denoted OT_S), by adding single round (Refer to Fig. 4).

If we did manage to somehow reverse the two-message OT protocols of [2,22,24,28] using such a reversal, then clearly we would obtain a three-message protocol with game-based security against unbounded senders and malicious PPT receivers. However, surprisingly, proving game-based security of the protocol obtained by reversing [2,22,24,28] appears highly non-trivial, and in fact it is not clear if such security can be proven at all. More specifically, the security reduction against a malicious receiver for the resulting 3 round protocol

Sender Input: Message bits x_0, x_1. **Receiver Input:** Choice bit b.

- **Sender Message.** Sample $x'_0, x'_1 \xleftarrow{\$} \{0,1\}^2$ and r_S uniformly at random. Set $c = x'_0 \oplus x'_1$, and send $m_S = \mathsf{OT}_R(c; r_S)$.
- **Receiver Message.**
  - Sample input (single-bit) messages m_0, m_1 uniformly at random such that $m_0 \oplus m_1 = b$.
  - Send $m_R = \mathsf{OT}_S(m_0, m_1; r_R)$.
- **Sender Message.**
  - Obtain output a of the two-message OT using (m_R, r_S).
  - Send $z = a \oplus x'_0, z_0 = x'_0 \oplus x_0, z_1 = x'_1 \oplus x_1$.
- **Receiver Output:** The receiver outputs $y = (z \oplus z_b \oplus m_0)$.

Fig. 4. Oblivious transfer reversal

must make use of a cheating receiver to contradict an assumption. To do this, it must obtain the sender's first message externally, but since the reduction no longer knows the randomness used for computing this message, it is unclear how such a reduction would be able to complete the third message of the protocol in Fig. 4. Indeed, this problem occurs because the original OT lacks any form of simulation security against malicious senders.

Our solution is to strengthen security of the underlying OT in order to make this transformation go through. As we already noted, this also turns out to be related to the problem of preventing selective failure attacks in 2-message OT.

We construct a two-message simulatable variant of oblivious transfer, with security against unbounded receivers, as well as (super-polynomial) simulation security against both malicious senders and malicious receivers[8].

Given such a protocol, the security reduction described above is able to use the underlying simulator to extract the inputs of the adversary, in order to complete the three-message OT reversal described in Fig. 4.

Simulation-Secure Two-Message Oblivious Transfer. The first question is, whether it is even possible to obtain two-message oblivious transfer, *with unbounded simulation security* against malicious senders as well as malicious receivers, *while preserving security against unbounded malicious receivers.* We will achieve this by bootstrapping known protocols that already satisfy super-polynomial simulation security against malicious receivers, to also add simulation security against malicious senders.

At first, such a definition may appear self-contradictory: if there exists a black-box simulator against that is able to *extract* both inputs of the malicious sender, then in a two-message protocol, an unbounded receiver may also be able to learn both inputs of the sender by running such a simulator – thereby blatantly violating sender security.

Our key differentiation between the simulator and a malicious receiver, that will block the above intuition from going through, will again be that the simulator can access the sender superpolynomially many times, while an unbounded malicious receiver will only be able to participate in (unbounded, but) polynomially many interactions with the sender.

That is, our protocol will be designed such that, with a small probability 2^{-m}, the sender will be forced to reveal both his inputs to the receiver[9]. On the other hand, with probability $1 - 2^{-m}$, the sender message that does not correspond to the receiver's choice bit, will remain statistically hidden. And again, most importantly, a malicious sender will not be able to distinguish between the case where he was forced to reveal both inputs, and the case where he was not.

[8] We note that existing two-message protocols [2,22,24,28] with security against unbounded receivers do not satisfy simulation-based security against malicious senders.

[9] This will be achieved by having the sender send a statistically private argument described in the previous section, proving that he computed the message correctly. Such an argument will also enable extraction of the witness with probability 2^{-m}.

As a result, the simulator against a malicious sender will run approximately 2^m executions with the malicious senders, waiting for an event where the sender is forced to reveal both inputs: and it will just use this execution to output the sender view. We will show, just like the case of statistically hiding extractable commitments, that a cheating sender will not be able to distinguish such views from views that did not allow extraction. Finally, when $m = \Omega(\log n)$, the resulting protocol will still satisfy statistical security against unbounded receivers, while simultaneously allowing approximately 2^m-time simulation. Please refer to Sect. 6 for formal details of our techniques.

2.4 On the Relationship with Non-malleability

Another way to interpret some of our results is via the lens of non-malleability: in any two-message protocol between Alice and Bob, where Alice sends the first message and Bob sends the second, we show how to enforce that the input used by Bob to generate his message remain independent of the input used by Alice.

One way to accomplish such a task is to set parameters so that the security of Bob's message is much weaker than that of Alice, in a way that it is possible to break security of Bob's message via brute-force, and extract Bob's input in time T, while arguing that Alice's input remained computationally hidden, even against T-time adversaries. However, this would crucially require Bob's message to only be computationally hidden, so that it would actually be recoverable via brute-force. This was used in several works, including [29] which gave the first constructions of computational zero-knowledge with superpolynomial time simulation.

In this paper, building on the recent work of [27], we essentially prove that it is possible to achieve similar guarantees while keeping Bob's message *statistically hidden*. Indeed, this is the main reason that our proofs of soundness go through.

3 Preliminaries

Notation. Throughout this paper, we will use κ to denote the security parameter, and $\mathsf{negl}(\kappa)$ to denote any function that is asymptotically smaller than $\frac{1}{\mathsf{poly}(\kappa)}$ for any polynomial $\mathsf{poly}(\cdot)$.

The statistical distance between two distributions D_1, D_2 is denoted by $\Delta(D_1, D_2)$ and defined as:

$$\Delta(D_1, D_2) = \frac{1}{2}\Sigma_{v\in V}|\Pr_{x\leftarrow D_1}[x = v] - \Pr_{x\leftarrow D_2}[x = v]|.$$

We say that two families of distributions $D_1 = \{D_{1,\kappa}\}, D_2 = \{D_{2,\kappa}\}$ are statistically indistinguishable if $\Delta(D_{1,\kappa}, D_{2,\kappa}) = \mathsf{negl}(\kappa)$. We say that two families of distributions $D_1 = \{D_{1,\kappa}\}, D_2 = \{D_{2,\kappa}\}$ are computationally indistinguishable if for all non-uniform probabilistic polynomial time distinguishers $\mathcal{D}$,

$$\left|\Pr_{r \leftarrow D_{1,\kappa}}[\mathcal{D}(r) = 1] - \Pr_{r \leftarrow D_{2,\kappa}}[\mathcal{D}(r) = 1]\right| = \mathsf{negl}(\kappa).$$

Let Π denote an execution of a protocol. We use $\mathsf{View}_A(\Pi)$ to denote the view, including the randomness and state of party A in an execution Π. We use $\mathsf{Output}_A(\Pi)$ to denote the output of party A in an execution of Π.

Remark 1. In what follows, we define several 2-party protocols. We note that in all these protocols both parties take as input the security parameter 1^κ. We omit this from the notation for the sake of brevity.

Definition 1 (Σ-protocols). *Let $L \in \mathsf{NP}$ with corresponding witness relation R_L. A protocol $\Pi = \langle P, V \rangle$ is a Σ-protocol for relation R_L if it is a three-round public-coin protocol which satisfies:*

- **Completeness:** *For all $(x, w) \in R_L$, $\Pr[\mathsf{Output}_V\langle P(x, w), V(x)\rangle = 1] = 1 - \mathsf{negl}(\kappa)$, assuming P and V follow the protocol honestly.*
- **Special Soundness:** *There exists a polynomial-time algorithm A that given any x and a pair of accepting transcripts $(a, e, z), (a, e', z')$ for x with the same first prover message, where $e \neq e'$, outputs w such that $(x, w) \in R_L$.*
- **Semi-malicious verifier zero-knowledge:** *There exists a probabilistic polynomial time simulator $\mathcal{S}_\Sigma$ such that for all $(x, w) \in R_L$, the distributions $\{\mathcal{S}_\Sigma(x, e)\}$ and $\{\mathsf{View}_V\langle P(x, w(x)), V(x, e)\rangle\}$ are statistically indistinguishable, where $\mathcal{S}_\Sigma(x, e)$ denotes the output of simulator $\mathcal{S}$ upon receiving input x and the verifier's random tape, denoted by e.*

3.1 Oblivious Transfer

Definition 2 (Oblivious Transfer). *Oblivious transfer is a protocol between two parties, a sender S with input messages (m_0, m_1) and a receiver R with input a choice bit b. The correctness requirement is that R obtains output m_b at the end of the protocol (with probability 1). We let $\langle S(m_0, m_1), R(b)\rangle$ denote an execution of the OT protocol with sender input (m_0, m_1) and receiver input bit b. We require OT that satisfies the following properties:*

- **Computational Receiver Security.** *For any non-uniform PPT sender S^* and any $(b, b') \in \{0, 1\}$, the views $\mathsf{View}_{S^*}(\langle S^*, R(b)\rangle)$ and $\mathsf{View}_{S^*}(\langle S^*, R(b')\rangle)$ are computationally indistinguishable.*
  We say that the OT scheme is T-secure if any $\mathsf{poly}(T)$-size malicious sender S^* has a distinguishing advantage less than $\frac{1}{poly(T)}$.
- **$(1 - \delta)$-Statistical Sender Security.** *For any receiver R^* that outputs receiver message m_{R^*}, there exists bit b such that for all m_0, m_1, the distribution $\mathsf{View}_{R^*}\langle S(m_0, m_1), R^*\rangle$ is $(1 - \delta)$ statistically close to $\mathsf{View}_{R^*}\langle S(m_b, m_b), R^*\rangle$.*

Such two-message protocols have been constructed based on the DDH assumption [28], and a stronger variant of smooth-projective hashing, which can be realized from DDH as well as the Nth-residuosity and Quadratic Residuosity assumptions [22,24]. Such two-message protocols can also be based on witness encryption or indistinguishability obfuscation (iO) together with one-way permutations [31].

Finally, we define bit OT as oblivious transfer where the sender inputs bits instead of strings.

Definition 3 (Bit Oblivious Transfer). *We say that an oblivious transfer protocol according to Definition 2 is a bit oblivious transfer if the senders messages* m_0, m_1 *are each in* $\{0, 1\}$.

3.2 Proof Systems

Delayed-Input Interactive Protocols. An n-message delayed-input interactive protocol for deciding a language L with associated relation R_L proceeds in the following manner:

- At the beginning of the protocol, P and V receive the size of the instance and security parameter, and execute the first $n-1$ messages.
- Before sending the last message, P receives input $(x, w) \in R_L$. P sends x to V together with the last message of the protocol. Upon receiving the last message from P, V outputs 1 or 0.

An execution of this protocol with instance x and witness w is denoted by $\langle P(x, w), V(x)\rangle$. A delayed-input interactive protocol is a protocol satisfying the completeness and soundness condition in the delayed input setting. One can consider both proofs – with soundness against unbounded (cheating) provers, and arguments – with soundness against computationally bounded (cheating) provers. In particular, a delayed-input interactive argument satisfies *adaptive soundness* against malicious PPT provers. That is, soundness is required to hold even against PPT provers who choose the statement adaptively (maliciously), depending upon the first $n-1$ messages of the protocol.

Definition 4 (Delayed-Input Interactive Arguments). *An n-message delayed-input interactive protocol (P, V) for deciding a language L is an interactive argument for L if it satisfies the following properties:*

- **Completeness:** *For every* $(x, w) \in R_L$,

$$\Pr\big[\mathsf{Output}_V\langle P(x, w), V(x)\rangle = 1\big] = 1 - \mathsf{negl}(\kappa),$$

where the probability is over the random coins of P and V, and where in the protocol V receives x together with the last message of the protocol.

- **Adaptive Soundness:** *For every (non-uniform) PPT prover P^* that given 1^κ chooses an input length 1^p, and then chooses $x \in \{0,1\}^p \setminus L$ adaptively, depending upon the transcript of the first $n-1$ messages,*

$$\Pr\big[\mathsf{Output}_V\langle P^*, V\rangle(x) = 1\big] = \mathsf{negl}(\kappa),$$

where the probability is over the random coins of V.

Witness Indistinguishability. A proof system is witness indistinguishable if for any statement with at least two witnesses, proofs computed using different witnesses are indistinguishable. In this paper, we only consider statistical witness indistinguishability, which we formally define below.

Definition 5 (Statistical Witness Indistinguishability). *A (delayed-input) interactive argument (P, V) for a language L is said to be* statistical witness-indistinguishable *if for every unbounded verifier V^*, every polynomially bounded function $n = n(\kappa) \leq \mathsf{poly}(\kappa)$, and every $(x_n, w_{1,n}, w_{2,n})$ such that $(x_n, w_{1,n}) \in R_L$ and $(x_n, w_{2,n}) \in R_L$ and $|x_n| = n$, the following two ensembles are statistically indistinguishable:*

$$\big\{\mathsf{View}_{V^*}\langle P(x_n, w_{1,n}), V^*(x_n)\rangle\big\} \text{ and } \big\{\mathsf{View}_{V^*}\langle P(x_n, w_{2,n}), V^*(x_n)\rangle\big\}$$

Delayed-Input Distributional Weak Zero Knowledge. Zero knowledge (ZK) requires that for any adversarial verifier, there exists a simulator that can produce a view that is indistinguishable from the real one to *every* distinguisher. Weak zero knowledge (WZK) relaxes the standard notion of ZK by reversing the order of quantifiers, and allowing the simulator to depend on the distinguisher.

We consider a variant of WZK, namely, distributional WZK [15,19], where the instances are chosen from some distribution over the language. Furthermore, we allow the simulator's running time to depend upon the distinguishing probability of the distinguisher. We refer to this as distributional ϵ-WZK, which says that for every $\mathcal{T}_\mathcal{D}$-time distinguisher $\mathcal{D}$ and every distinguishing advantage ϵ (think of ϵ as an inverse polynomial) there exists a simulator, that is an oracle machine running in time $\mathsf{poly}(\kappa, 1/\epsilon)$ with oracle access to the distinguisher, that generates a view that $\mathcal{D}$ cannot distinguish from the view generated by the real prover. This notion was previously considered in [13,15,23].

When considering delayed-input interactive protocols it is natural to consider a delayed input version of secrecy. In what follows, we define delayed-input distributional statistical ϵ-WZK.

Definition 6 (Delayed-Input Distributional Statistical ϵ-Weak Zero Knowledge). *A delayed-input interactive argument (P, V) for a language L is said to be delayed-input distributional statistical* ϵ-weak zero knowledge *if for every polynomially bounded function $n = n(\kappa) \leq \mathsf{poly}(\kappa)$, and for every efficiently samplable distribution $(\mathcal{X}_\kappa, \mathcal{W}_\kappa)$ on R_L, i.e., $\mathsf{Supp}(\mathcal{X}_\kappa, \mathcal{W}_\kappa) = \{(x, w) \in R_L : x \in \{0,1\}^{n(\kappa)}\}$, every unbounded verifier V^* that obtains the instance from* *the prover in the last message of the protocol, every unbounded distinguisher* $\mathcal{D}$, *and every* ϵ *(which will usually be set to* $1/\mathsf{poly}(\kappa)$ *for some polynomial* $\mathsf{poly}(\cdot)$*), there exists a simulator* $\mathcal{S}$ *that runs in time* $\mathsf{poly}(\kappa, 1/\epsilon)$ *and has oracle access to* $\mathcal{D}$ *and* V^**, such that:*

$$\Bigg| \Pr_{(x,w)\leftarrow(\mathcal{X}_\kappa,\mathcal{W}_\kappa)} \left[\mathcal{D}(x, \mathsf{View}_{V^*}[\langle P(x,w), V^*(x)\rangle] = 1\right]$$

$$- \Pr_{(x,w)\leftarrow(\mathcal{X}_\kappa,\mathcal{W}_\kappa)} \left[\mathcal{D}(x, \mathcal{S}^{V^*,\mathcal{D}}(x)) = 1\right] \Bigg| \leq \epsilon(\kappa),$$

where the probability is over the random choices of (x, w) *as well as the random coins of the parties.*

Zero-Knowledge with Super-Polynomial Simulation. We now define zero-knowledge with super-polynomial simulation in the same way as [29], except that we define *statistical* security against malicious verifiers.

Definition 7 (Statistical ZK with Super-polynomial Simulation). *We say that a delayed input two message argument* (P, V) *for an* NP *language* L *is statistical zero-knowledge with super-polynomial* T_{Sim}*-time simulation, if there exists a (uniform) simulator* $\mathcal{S}$ *that runs in time* T_{Sim}*, such that for every polynomial* $n = n(\kappa) \leq \mathsf{poly}(\kappa)$*, and for every* $(x_n, w_n) \in R_L$ *where each* $|x_n| = n$*, and every unbounded verifier* V^**, the two distributions* $\mathcal{S}^{V^*}(x_n)$ *and* $\mathsf{View}_{V^*}\langle P(x_n, w_n), V^*(x_n)\rangle$ *are statistically close.*

4 Extractable Commitments

4.1 Definitions

Our notion of extractable commitments tailors the definition in [27] to the setting of statistically hiding commitments. We begin by (re-)defining the notion of a commitment scheme. As before, we use κ to denote the security parameter, and we let $p = \mathsf{poly}(\kappa)$ be an arbitrary fixed polynomial such that the message space is $\{0,1\}^p$.

We restrict ourselves to commitments with non-interactive decommitment, and where the (honest) receiver is not required to maintain any state at the end of the commit phase in order to execute the decommit phase. Our construction will satisfy this property and this will be useful in our applications to constructing statistically private protocols.

Definition 8 *[Statistically Hiding Commitment Scheme]. A commitment* $\langle \mathcal{C}, \mathcal{R}\rangle$ *is a two-phase protocol between a committer* $\mathcal{C}$ *and receiver* $\mathcal{R}$*, consisting of a tuple of algorithms*

$$\mathsf{Commit}, \mathsf{Decommit}, \mathsf{Verify}.$$

At the beginning of the protocol, $\mathcal{C}$ obtains as input a message $M \in \{0,1\}^p$. Next, $\mathcal{C}$ and $\mathcal{R}$ execute the commit phase, and obtain a commitment transcript, denoted by τ, together with a private state for $\mathcal{C}$, denoted by $\mathsf{state}_{\mathcal{C},\tau}$. We use the notation

$$(\tau, \mathsf{state}_{\mathcal{C},\tau}) \leftarrow \mathsf{Commit}\langle \mathcal{C}(M), \mathcal{R}\rangle.$$

Later, $\mathcal{C}$ and $\mathcal{R}$ possibly engage in a decommit phase, where the committer $\mathcal{C}$ computes and sends message $y = \mathsf{Decommit}(\tau, \mathsf{state}_{\mathcal{C},\tau})$ to $\mathcal{R}$. At the end, $\mathcal{R}$ computes $\mathsf{Verify}(\tau, y)$ to output $\perp$ or a message $\widetilde{M} \in \{0,1\}^p$.[10]

A statistically hiding commitment scheme is required to satisfy three properties:

- **(Perfect) Completeness.** *If $\mathcal{C}, \mathcal{R}$ honestly follow the protocol, then for every $M \in \{0,1\}^p$:*

$$\Pr[\mathsf{Verify}(\tau, \mathsf{Decommit}(\tau, \mathsf{state}_{\mathcal{C},\tau})) = M] = 1$$

  where the probability is over $(\tau, \mathsf{state}_{\mathcal{C},\tau}) \leftarrow \mathsf{Commit}\langle \mathcal{C}(M), \mathcal{R}\rangle$.
- **Statistical Hiding.** *For every two messages $M_1, M_2 \in \{0,1\}^{2p}$, every unbounded malicious receiver $\mathcal{R}^*$ and honest committer $\mathcal{C}$, a commitment is $\delta(\kappa)$-statistically hiding if the statistical distance between the distributions $\mathsf{View}_{\mathcal{R}^*}(\mathsf{Commit}\langle \mathcal{C}(M_1), \mathcal{R}^*\rangle)$ and $\mathsf{View}_{\mathcal{R}^*}(\mathsf{Commit}\langle \mathcal{C}(M_2), \mathcal{R}^*\rangle)$ is at most $\delta(\kappa)$. The scheme is statistically hiding if $\delta(\kappa) \leq \frac{1}{\mathsf{poly}(\kappa)}$ for every polynomial $\mathsf{poly}(\cdot)$.*
- **Computational Binding.** *Consider any non-uniform PPT committer $\mathcal{C}^*$ that produces $\tau \leftarrow \mathsf{Commit}\langle \mathcal{C}^*, \mathcal{R}\rangle$, and then outputs y_1, y_2. Let $\widetilde{M}_1 = \mathsf{Verify}(\tau, y_1)$ and $\widetilde{M}_2 = \mathsf{Verify}(\tau, y_2)$. Then, we require that*

$$\Pr\big[(\widetilde{M}_1 \neq \perp) \ \wedge \ (\widetilde{M}_2 \neq \perp) \ \wedge \ (\widetilde{M}_1 \neq \widetilde{M}_2)\big] = \mathsf{negl}(\kappa),$$

  over the randomness of sampling $\tau \leftarrow \mathsf{Commit}\langle \mathcal{C}^*, \mathcal{R}\rangle$.

In the following, we define a PPT oracle-aided algorithm Samp such that for all $\mathcal{C}^*$, $\mathsf{Samp}^{\mathcal{C}^*}$ samples $\tau \leftarrow \mathsf{Commit}\langle \mathcal{C}^*, \mathcal{R}\rangle$ generated by a malicious committer $\mathcal{C}^*$ using uniform randomness for the receiver.

We also define an extractor $\mathcal{E}$ that given black-box access to $\mathcal{C}^*$, outputs *some* transcript generated by $\mathcal{C}^*$, and then without executing any decommitment phase with $\mathcal{C}^*$, outputs message $\widetilde{M}_e$: we require "correctness" of this extracted message $\widetilde{M}_e$. We also require that for any non-uniform PPT $\mathcal{C}^*$, the distribution of τ generated by $\mathsf{Samp}^{\mathcal{C}^*}$ is indistinguishable from the distribution output by $\mathcal{E}^{\mathcal{C}^*}$. This is formally defined in Definition 9.

Definition 9 *[$\mathcal{T}$-Extractable Commitment Scheme]. We say that a statistically hiding commitment scheme is $\mathcal{T}$-extractable if there exists a $\mathcal{T} \cdot \mathsf{poly}(\kappa)$-time*

---

[10] We note that in our definition, $\mathcal{R}$ does not need to keep a state from the commitment phase in order to execute the decommitment phase.

uniform oracle machine $\mathcal{E}$ *such that the following holds. Let* $\mathcal{C}^*$ *be any non-uniform PPT adversarial committer, that before starting the commitment phase, outputs auxiliary information denoted by* z*, and at the end of the commitment phase outputs auxiliary information denoted by* aux*. Then, the following holds.*

- *There exists a PPT oracle sampling algorithm* $\mathsf{Samp}^{\mathcal{C}^*}$ *that samples* $(\tau_{\mathcal{C}^*}, \mathsf{aux}) \leftarrow \mathsf{Commit}\langle \mathcal{C}^*, \mathcal{R} \rangle$*. Let* $\mathsf{Exp}_{\mathsf{Samp}^{\mathcal{C}^*}} = (\tau_{\mathcal{C}^*}, \mathsf{aux})$ *be the output of* $\mathsf{Samp}^{\mathcal{C}^*}$*.*
- $\mathcal{E}^{\mathcal{C}^*}$ *outputs* $(\tau_{\mathcal{C}^*}, \mathsf{aux}, \widetilde{M})$*, while only making oracle calls to* $\mathcal{C}^*$ *during the commit phase (without ever running the decommit phase). We denote by* $\mathsf{Exp}_{\mathcal{E}^{\mathcal{C}^*}} = (\tau_{\mathcal{C}^*}, \mathsf{aux})$*.*

We require that:

- **Indistinguishability.** *The distributions* $(\mathsf{Exp}_{\mathsf{Samp}^{\mathcal{C}^*}}, z)$ *and* $(\mathsf{Exp}_{\mathcal{E}^{\mathcal{C}^*}}, z)$ *are computationally indistinguishable.*
- **Correctness of Extraction.** *Consider any non-uniform PPT* $\mathcal{C}^*$ *and let* $(\tau, \mathsf{aux}, \widetilde{M})$ *denote the output of* $\mathcal{E}^{\mathcal{C}^*}$*. Then for any string* y_1*, denoting* $\widetilde{M}_1 = \mathsf{Verify}(\tau, y_1)$*,*

$$\Pr\left[(\widetilde{M} \neq \bot) \land (\widetilde{M}_1 \neq \bot) \land (\widetilde{M} \neq \widetilde{M}_1)\right] = \mathsf{negl}(\kappa),$$

where the probability is over $(\tau, \mathsf{aux}, \widetilde{M}) \leftarrow \mathcal{E}^{\mathcal{C}^*}$*.*

4.2 Protocol

In this section, we construct two-message statistically hiding, extractable commitments according to Definition 9. Our construction is described in Fig. 5.

Let $\mathsf{OT} = (\mathsf{OT}_1, \mathsf{OT}_2)$ denote a two-message string oblivious transfer protocol according to Definition 2. Let $\mathsf{OT}_1(b; r_1)$ denote the first message of the OT protocol with receiver input b and randomness r_1, and let $\mathsf{OT}_2(M_0, M_1; r_2)$ denote the second message of the OT protocol with sender input strings M_0, M_1 and randomness r_2.[11]

In the full version of this paper, we prove the following main theorem.

Theorem 1. *Set* $T = (2^m \cdot \kappa^{\log \kappa})$*. Assuming that the underlying OT protocol is* T*-secure against malicious senders,* $(1 - \delta_{\mathsf{OT}})$ *secure against malicious receivers according to Definition 2, the scheme in Fig. 5 is a* $(1 - 2^m - \delta_{\mathsf{OT}})$ *statistically hiding,* T*-extractable commitment scheme according to Definition 9.*

We prove this theorem by showing statistical hiding, computational binding, and extractability. The proof of statistical hiding follows by $(1 - \delta)$-statistical sender security of the OT. To prove computational binding, we build a reduction to the receiver security of OT according to Definition 2. The proof of extractability follows by building.

Extraction parameter: m.[a]
Committer Input: Message $M \in \{0,1\}^p$.
Commit Stage:
Receiver Message.

- Pick challenge string $\mathsf{ch} \xleftarrow{\$} \{0,1\}^m$.
- Sample uniform randomness $\{r_{1,i}\}_{i\in[m]}$.
- Compute and send $\{\mathsf{OT}_1(\mathsf{ch}_i, r_{1,i})\}_{i\in[m]}$ using m instances of two-message OT.

Committer Message.

- Sample a random string $r \xleftarrow{\$} \{0,1\}^m$.
  For every $i \in [m]$ and every $b \in \{0,1\}$, sample $M_i^b \xleftarrow{\$} \{0,1\}^p$ subject to $\bigoplus_{i\in[m]} M_i^{r_i} = M$.
- For every $i \in [m]$ compute $o_{2,i} = \mathsf{OT}_2(M_i^0, M_i^1; r_{2,i})$ with uniform randomness $r_{2,i}$.
- Send $(r, \{o_{2,i}\}_{i\in[m]})$.

Reveal Stage: The committer reveals M, and all values $\{M_i^0, M_i^1\}_{i\in[m]}$ as well as the randomness $r_{2,i}$. The receiver accepts the decommitment to message M if and only if:

1. For all $i \in [m]$, $o_{2,i} = \mathsf{OT}_2(M_i^0, M_i^1; r_{2,i})$,
2. $\bigoplus_{i\in[m]} M_i^{r_i} = M$.

[a] The value m will determine the running time $T = 2^m \cdot \kappa^{\log \kappa}$ of the extractor. The protocol will have statistical receiver security $1 - 2^{-m} - \delta_{\mathsf{OT}}$, when the underlying OT has statistical sender security $1 - \delta_{\mathsf{OT}}$.

Fig. 5. Extractable commitments

$\mathcal{E}^{\mathcal{C}^*}$ repeats the following $2^m \cdot \kappa^{\log \kappa}$ times. If it reaches the end of $2^m \cdot \kappa^{\log \kappa}$ iterations, it outputs $\perp$. We will call each iteration a *trial*.

1. Choose $\mathsf{ch} \xleftarrow{\$} \{0,1\}^m$. Compute $\tau_1 = \mathsf{OT}_1(\mathsf{ch}_i, R_i)$ using uniform randomness $R = \{R_i\}_{i\in[m]}$.
2. Query the oracle $\mathcal{C}^*$ in the Commit phase with τ_1, and obtain response (τ_2, aux), where τ_2 also contains r. If $\mathcal{C}^*$ aborts or sends an invalid message, do the following.
   - If this is the first iteration, output $(\tau_1, \tau_2, \mathsf{aux}, \perp)$ and stop.
   - If this is not the first iteration, go to Step 1 and start a new trial.
3. Else, $\mathcal{C}^*$ did not abort. If $r \neq \mathsf{ch}$, go to Step 1 and start a new trial.
4. Else, $\mathcal{C}^*$ did not abort and $r = \mathsf{ch}$ (this iteration is considered a success). Then use R to obtain $\{M_i^{\mathsf{ch}_i}\}_{i\in[m]}$. Next, compute $\widetilde{M} = \bigoplus_{i\in[m]}\{\widetilde{M_i^{\mathsf{ch}_i}}\}_{i\in[m]}$. Output $(R, \tau_1, \tau_2, \mathsf{aux}, \widetilde{M})$.

Fig. 6. Description of the extractor $\mathcal{E}^{\mathcal{C}^*}$

We build the following extractor $\mathcal{E}$ for Definition 9, in Fig. 6. In the figure, we denote the first message of transcript τ by τ_1 and the second message by τ_2. $\mathcal{E}$ will obtain oracle access to $\mathcal{C}^*$, and the running time of $\mathcal{E}^{\mathcal{C}^*}$ will be $T = 2^m \cdot \kappa^{\log \kappa}$.

The analysis of the extractor builds on the analysis of [27], and can be found in the full version of the paper.

5 Two-Message Arguments with Statistical Privacy

5.1 Modified Blum Protocol

We begin by describing a very simple modification to the Blum Σ-protocol for Graph Hamiltonicity. The protocol we describe will have soundness error $\frac{1}{2} - \mathsf{negl}(\kappa)$ against adaptive PPT provers, and will satisfy *statistical* zero-knowledge. Since Graph Hamiltonicity is NP-complete, this protocol can also be used to prove any statement in NP via a Karp reduction. This protocol is described in Fig. 7.

We give an overview of the protocol here. Note that the only modification to the original protocol of Blum [9] is that we use statistically hiding, extractable commitments instead of statistically binding commitments. The proofs of soundness and statistical zero-knowledge are fairly straightforward. They roughly follow the same structure as [9], replacing statistically binding commitments with statistically hiding commitments.

In the full version of the paper, we prove that the protocol in Fig. 7 satisfies soundness against PPT provers that may choose x adaptively in the second round of the protocol. We also prove that assuming that extcom is statistically hiding, the protocol in Fig. 7 satisfies statistical zero-knowledge.

5.2 Compressing Four Message Argument to a Two Message Argument

In Fig. 8, we describe the construction of a two-message argument, using extractable commitments (with two messages denoted by $\mathsf{ext\text{-}com}_1, \mathsf{ext\text{-}com}_2$) according to Definition 9. This essentially consists of compressing the modified Blum argument from Fig. 7 into a two-message argument.

Let $\mathsf{OT} = (\mathsf{OT}_1, \mathsf{OT}_2)$ denote a two-message bit oblivious transfer protocol according to Definition 2. Let $\mathsf{OT}_1(b)$ denote the first message of the OT protocol with receiver input b, and let $\mathsf{OT}_2(m_0, m_1)$ denote the second message of the OT protocol with sender input bits m_0, m_1.

Let $\Sigma = (q, a, e, z)$ denote the four messages of the modified Blum protocol from Fig. 7. Here (q, a) denote the messages of the extractable commitment. We will perform a parallel repetition of this protocol, thus for each $i \in [\kappa]$, (q_i, a_i, e_i, z_i) are messages corresponding to an underlying modified Blum protocol with a single-bit challenge (i.e., where $e_i \in \{0, 1\}$). We denote by f_1 and f_2 the functions that satisfy $a_i = f_1(x, w; r_i)$ and $z_i = f_2(x, w, r_i, e_i)$, where r_i is uniformly chosen randomness.

[11] Note that OT_2 also depends on OT_1. We omit this dependence in our notation for brevity.

Modified Blum Argument

1. **Verifier Message:** The verifier does the following:
   - Send the first message $\mathsf{extcom}_{1,i,j}$ for independent instances of the extractable commitment, where $i, j \in [p(\kappa)] \times [p(\kappa)]$.
   - Send an additional first message $\mathsf{extcom}_{1,P}$ for another independent instance of the extractable commitment.
2. **Prover Message:** The prover gets input graph $G \in \{0,1\}^{p(\kappa) \times p(\kappa)}$ represented as an adjacency matrix, with $(i,j)^{th}$ entry denoted by $G[i][j]$), Hamiltonian cycle $H \subseteq G$. Here $p(\cdot)$ is an a-priori fixed polynomial. The prover does the following:
   - Sample a random permutation π on $p(\kappa)$ nodes, and compute $c_P = \mathsf{extcom}_{2,P}(\pi)$ as a commitment to π using extcom.
   - Compute $\pi(G)$, which is the adjacency matrix corresponding to the graph G when its nodes are permuted according to π. Compute $c_{i,j} = \mathsf{extcom}_{2,i,j}(\pi(G)[i][j])$ for $(i,j) \in [p(\kappa)] \times [p(\kappa)]$.
   - Send $G, c_P, c_{i,j}$ for $(i,j) \in [p(\kappa)] \times [p(\kappa)]$.
3. **Verifier Message:** Sample and send $c \xleftarrow{\$} \{0,1\}$ to the prover.
4. **Prover Message:** The prover does the following:
   - If $c = 0$, send π and the decommitments of $\mathsf{extcom}_P, \mathsf{extcom}_{i,j}$ for $(i,j) \in [p(\kappa)] \times [p(\kappa)]$.
   - If $c = 1$, send the decommitment of $\mathsf{extcom}_{i,j}$ for all (i,j) such that $\pi(H)[i][j] = 1$.
5. **Verifier Output:** The verifier does the following:
   - If $c = 0$, accept if and only if all extcom openings were accepted and $\pi(G)$ was computed correctly by applying π on G.
   - If $c = 1$, accept if and only if all extcom openings were accepted and all the opened commitments form a Hamiltonian cycle.

Fig. 7. Modified blum SZK argument

We state our main lemma here, which we prove in the full version of the paper.

Lemma 1. *Assuming that* extcom *is a* $2^m \cdot \kappa^{\log \kappa}$*-extractable commitment scheme according to Definition 9 and that* OT *is* $2^{\kappa m} \cdot \kappa^{\log \kappa}$*-secure, the protocol in Fig. 8 satisfies soundness against PPT malicious provers.*

Furthermore, assuming that the distributions $\mathsf{Exp}_{\mathcal{E}C^*}$ *and* $\mathsf{Exp}_{\mathsf{Samp}C^*}$ *corresponding to* extcom*, Definition 9, are indistinguishable by* T'*-size distinguishers, the protocol in Fig. 8 satisfies adaptive soundness against all PPT provers, when the instance is chosen from a language that is decidable by* T'*-size circuits.*

Remark 2. Our proof also generalizes to executing only $\Omega(\log \kappa)$ parallel executions of the Blum protocol, while still yeilding negligible soundness error. Furthermore, we will see that statistical privacy guarantees will hold even when $m = \Omega(\log \kappa)$. Therefore, the protocol in Fig. 8 can be realized only relying on quasi-polynomially secure oblivious transfer according to Definition 2.

Two-Message Argument

- **Verifier Message:**
  - Pick $\{q_i\}_{i\in[\kappa]}$ and pick challenge $\{e_i\}_{i\in[\kappa]}$ for the modified Blum Protocol.
  - Compute $\{o_{1,i} = \mathsf{OT}_{1,i}(e_i)\}_{i\in[\kappa]}$.
  - Send $\{q_i, o_{1,i}\}_{i\in[\kappa]}$ in parallel.
- **Prover Message:**
  - Obtain input $x \in L$, witness w such that $R_L(x, w) = 1$.
  - Compute $\{a_i\}_{i\in[\kappa]}$ according to the strategy in Figure 7.
  - Compute $\{z_i^0\}_{i\in[\kappa]}$ according to the strategy in Figure 7, using (q_i, a_i, e_i') corresponding to verifier challenge bit $e_i' = 0$.
  - Compute $\{z_i^1\}_{i\in[\kappa]}$ according to the strategy in Figure 7, using (q_i, a_i, e_i') and corresponding to verifier challenge bit $e_i' = 1$.
  - Compute $o_{2,i} = \mathsf{OT}_{2,i}(z_i^0, z_i^1)$ and send $\{a_i, o_{2,i}\}_{i\in[\kappa]}$.
- **Verifier Output:** The verifier V recovers z_i as the output of $\mathsf{OT}_{1,i}, \mathsf{OT}_{2,i}$ for $i \in [\kappa]$, and outputs accept if for all $i \in [\kappa]$, $(q_i, a_i, e_i, z_i)_{i\in[\kappa]}$ is an accepting transcript of the underlying modified Blum protocol.

Fig. 8. Two message argument system for NP

Similar to the extractability of commitments, we also define an additional property of two-message arguments, that we call extractability. Roughly, this property requires the existence of a super-polynomial time uniform oracle machine $\mathcal{E}$ that extracts the witness used by any prover generating accepting proofs. It is somewhat more subtle to define, and we refer the reader to the full version for a formal definition. This property is useful in our applications to obtaining stronger forms of OT, and we believe will also be useful for other future applications. We show that the scheme in Fig. 8 is also extractable, where the extractor for the argument can extract a transcript with a witness, from any prover, by relying the extractor of the commitment scheme extcom.

5.3 Proofs of Privacy

Lemma 2. *The protocol in Fig. 8 satisfies statistical zero-knowledge with super-polynomial simulation, according to Definition 7.*

Proof. The simulation strategy is straightforward: the simulator obtains $\{q_i, o_{1,i}\}_{i\in[\kappa]}$ externally. It runs in super-polynomial time to break the receiver message OT_1 via brute-force to extract $\{e_i\}_{i\in[\kappa]}$. Given $\{e_i\}_{i\in[\kappa]}$, it runs the semi malicious verifier ZK simulator for modified Blum on input $\{a_i, e_i\}_{i\in[\kappa]}$. It obtains $\{a_i, z_{i,e_i}\}_{i\in[\kappa]}$ from the semi malicious verifier ZK simulator. Finally, it sends for $i \in [\kappa]$, a_i together with $\mathsf{OT}_{2,i}(z_{i,e_i}, z_{i,e_i})$.

Statistical zero-knowledge then follows because of statistical zero knowledge of the underlying four-message protocol, and from the statistical security of OT against unbounded verifiers.

This also yields the following lemma.

Lemma 3. *The protocol in Fig. 8 satisfies statistical witness indistinguishability against all malicious verifiers.*

Proof (Sketch). This claim follows by a simple hybrid argument, where in an intermediate hybrid, the challenger generates the proof via the superpolynomial simulator of Lemma 2 (without using any witness). By Lemma 2, this intermediate hybrids is statistically close to any hybrid where a specific witness is used. This proves witness indistinguishability of the protocol. Refer to [3] for a more detailed proof.

Lemma 4. *The protocol in Fig. 8 satisfies distributional statistical delayed-input ϵ-weak zero-knowledge according to Definition 6.*

Following [23], we develop an inductive analysis and a simulation strategy that learns the receiver's challenge bit-by-bit. The proof follows the strategy in [23], and can be found in the full version of the paper.

Therefore, we have the following main theorem.

Theorem 2. *Assuming quasi-polynomially secure oblivious transfer according to Definition 2, there exists a two-message argument system that satisfies statistical witness indistinguishability (Definition 5), statististical zero-knowledge with super-polynomial simulation (Definition 6), and statistical weak distributional ϵ-zero-knowledge for delayed-input statements (Definition 7).*

We also observe that all our two-message arguments can be made resettable statistical witness indistinguishable by applying [5].

6 Oblivious Transfer: Stronger Security and Reversal

In this section, we build OT protocols, in the two-message and three-message setting, that satisfy stronger security properties than previously known. Because of space restrictions, we only describe the protocols and defer proofs to the full version of the paper.

6.1 Simulation-Secure Two-Message Oblivious Transfer

We first construct an oblivious transfer protocol with unbounded simulation-based security against both malicious receivers and malicious senders. We define this variant below.

Definition 10 (Simulation-Secure Oblivious Transfer). *As in Definition 2, we let $\langle S(m_0, m_1), R(b)\rangle$ denote an execution of the OT protocol with sender input (m_0, m_1) and receiver input bit b. We consider OT that satisfies the following properties (which are both defined using simulation-based security definitions):*

- **Computational Receiver Security.** *There exists a T_{Sim}-time oracle-aided simulator Sim^{S^*} that interacts with any non-uniform malicious PPT sender S^* and outputs $\mathsf{View}(\mathsf{Sim}^{S^*})$. It also extracts and sends S^*'s inputs m_0, m_1 to an ideal functionality $\mathcal{F}_{\mathsf{ot}}$, which obtains choice bit b from the honest receiver R and outputs $\mathsf{Output}_{\mathsf{Ideal}} = m_b$ to R. Then, we require that for every non-uniform PPT S^*, the joint distributions $(\mathsf{View}(\mathsf{Sim}^{S^*}), \mathsf{Output}_{\mathsf{Ideal}})$ and $(\mathsf{View}_{S^*}\langle S^*, R(b)\rangle, \mathsf{Output}_R\langle S^*, R(b)\rangle)$ are computationally indistinguishable.*
- **Statistical Sender Security.** *There exists a (possibly unbounded) oracle-aided simulator Sim^{R^*} that interacts with any unbounded* adversarial *receiver R^*, and with an ideal functionality $\mathcal{F}_{\mathsf{ot}}$ on behalf of R^*. Here $\mathcal{F}_{\mathsf{ot}}$ is an oracle that obtains the inputs (m_0, m_1) from S and b from Sim^{R^*} (simulating the malicious receiver), and outputs m_b to Sim^{R^*}. Then we require that for all m_0, m_1, Sim^{R^*} outputs a receiver view that is statistically indistinguishable from the real view of the malicious receiver $\mathsf{View}_{R^*}\langle S(m_0, m_1, z), R^*\rangle$.*

Our construction of two-message OT satisfying Definition 10 is described in Fig. 9. It uses a two-message OT scheme according to Definition 2, whose messages are denoted by OT_1 and OT_2. It also uses a statistical SPS zero-knowledge stat-sps-zk according to Definition 7, whose first and second messages are denoted by stat-sps-zk_1 and stat-sps-zk_2.

Sender Input: Message bits x_0, x_1. **Receiver Input:** Choice bit b.

- **Receiver Message.**
  - Sample $r_R \xleftarrow{\$} \{0,1\}^*$ and send $m_R = \mathsf{OT}_1(b; r_R)$.
  - Sample and send stat-sps-zk_1.
- **Sender Message.**
  - Send $m_S = \mathsf{OT}_2(m_R, x_0, x_1; r_S)$.
  - Send stat-sps-zk_2 proving that $\exists(x_0, x_1, r_S)$ such that $m_S = \mathsf{OT}_2(m_R, x_0, x_1; r_S)$.
- **Receiver Output.**
  - If stat-sps-zk does not verify, output $\perp$ and abort.
  - Else obtain output a of the two-message OT using (m_S, r_R). Output a.

Fig. 9. Simulation secure oblivious transfer

6.2 Reversing Oblivious Transfer

We first construct an oblivious transfer protocol with unbounded simulation-based security against both malicious receivers and malicious senders. We define this variant below.

Definition 11 (Simulation-Secure Oblivious Transfer Against Unbounded Senders). *As in Definition 2, we let $\langle S(m_0, m_1), R(b)\rangle$ denote an* *execution of the OT protocol with sender input* (m_0, m_1) *and receiver input bit* b. *We consider OT that satisfies the following properties (which are both defined using real-ideal security definitions):*

- **Computational Sender Security.** *There exists an oracle-aided simulator* Sim^{R^*} *that interacts with any non-uniform malicious PPT receiver* R^* *and interacts with the ideal functionality* $\mathcal{F}_{\mathsf{ot}}$ *on behalf of* R^*. *Here* $\mathcal{F}_{\mathsf{ot}}$ *is an oracle that obtains the inputs* (m_0, m_1) *from* S *and* b *from* Sim^{R^*} *(simulating the malicious receiver), and outputs* m_b *to* Sim^{R^*}. *Then we require that for all* m_0, m_1, Sim^{R^*} *outputs a receiver view that is computationally indistinguishable from the real view of the malicious receiver* $\mathsf{View}_{R^*}(\langle S(m_0, m_1, z), R^*\rangle)$.
- **Statistical Receiver Security.** *There exists a (possibly unbounded) oracle-aided simulator* Sim^{S^*} *that interacts with any unbounded* adversarial *sender* S^*, *and with an ideal functionality* $\mathcal{F}_{\mathsf{ot}}$ *on behalf of* S^*. *Here* $\mathcal{F}_{\mathsf{ot}}$ *is an oracle that obtains the inputs* (m_0, m_1) *from* Sim^{S^*} *and* b *from* R *and outputs* $\mathsf{Output}_{\mathsf{Ideal}} = m_b$ *to* R. *Then, we require that for every unbounded* S^*, *the two joint distributions* $(\mathsf{View}(\mathsf{Sim}^{S^*}), \mathsf{Output}_{\mathsf{Ideal}})$ *and* $(\mathsf{View}_{S^*}\langle S^*, R(b)\rangle, \mathsf{Output}_{S^*}\langle S^*, R(b)\rangle)$ *are statistically indistinguishable.*

We now describe a three-message (bit) oblivious transfer protocol with simulation-based security against malicious receivers and unbounded malicious senders, according to Definition 11.

This is obtained by reversing a two-message (bit) oblivious transfer protocol with simulation security against unbounded malicious receivers and PPT malicious senders, according to Definition 10, constructed in Fig. 9. Let $\mathsf{OT}_R(b; r_R)$ denote the receiver message of such an oblivious transfer protocol computed as a function of input bit b and randomness r_R, and let $\mathsf{OT}_S(m_R, x_0, x_1; r_S)$ denote the sender message of such a protocol computed as a function of receiver message m_R, sender inputs x_0, x_1 and randomness r_S. Our protocol is described in Fig. 10.

Sender Input: Message bits x_0, x_1. **Receiver Input:** Choice bit b.

- **Sender Message.** Sample $x'_0, x'_1 \xleftarrow{\$} \{0,1\}^2$ and r_S uniformly at random. Set $c = x'_0 \oplus x'_1$, and send $m_S = \mathsf{OT}_R(c; r_S)$.
- **Receiver Message.**
  - Sample input (single-bit) messages m_0, m_1 uniformly at random such that $m_0 \oplus m_1 = b$.
  - Send $m_R = \mathsf{OT}_S(m_0, m_1; r_R)$.
- **Sender Message.**
  - Obtain output a of the two-message OT using (m_R, r_S).
  - Send $z = a \oplus x'_0$, $z_0 = x'_0 \oplus x_0$, $z_1 = x'_1 \oplus x_1$.
- **Receiver Output:** The receiver outputs $y = (z \oplus z_b \oplus m_0)$.

Fig. 10. Oblivious transfer reversal

Acknowledgements. Research of D. Khurana and A. Sahai supported in part from a UCLA Dissertation Year Fellowship, a DARPA/ARL SAFEWARE award, NSF Frontier Award 1413955, and NSF grant 1619348, a Xerox Faculty Research Award, a Google Faculty Research Award, an equipment grant from Intel, and an Okawa Foundation Research Grant. This material is based upon work supported by the Defense Advanced Research Projects Agency through the ARL under Contract W911NF-15-C-0205. The views expressed are those of the authors and do not reflect the official policy or position of the Department of Defense, the National Science Foundation, or the U.S. Government.

References

1. Aiello, W., Bhatt, S., Ostrovsky, R., Rajagopalan, S.R.: Fast verification of any remote procedure call: short witness-indistinguishable one-round proofs for NP. In: Montanari, U., Rolim, J.D.P., Welzl, E. (eds.) ICALP 2000. LNCS, vol. 1853, pp. 463–474. Springer, Heidelberg (2000). https://doi.org/10.1007/3-540-45022-X_39
2. Aiello, B., Ishai, Y., Reingold, O.: Priced oblivious transfer: how to sell digital goods. In: Pfitzmann, B. (ed.) EUROCRYPT 2001. LNCS, vol. 2045, pp. 119–135. Springer, Heidelberg (2001). https://doi.org/10.1007/3-540-44987-6_8
3. Badrinarayanan, S., Garg, S., Ishai, Y., Sahai, A., Wadia, A.: Two-message witness indistinguishability and secure computation in the plain model from new assumptions. IACR Cryptology ePrint Archive 2017, 433 (2017). http://eprint.iacr.org/2017/433
4. Badrinarayanan, S., Goyal, V., Jain, A., Khurana, D., Sahai, A.: Round optimal concurrent MPC via strong simulation. In: Kalai, Y., Reyzin, L. (eds.) TCC 2017, Part I. LNCS, vol. 10677, pp. 743–775. Springer, Cham (2017). https://doi.org/10.1007/978-3-319-70500-2_25
5. Barak, B., Goldreich, O., Goldwasser, S., Lindell, Y.: Resettably-sound zero-knowledge and its applications. In: 42nd Annual Symposium on Foundations of Computer Science, FOCS 2001, Las Vegas, Nevada, USA, 14–17 October 2001, pp. 116–125 (2001). https://doi.org/10.1109/SFCS.2001.959886
6. Barak, B., Ong, S.J., Vadhan, S.: Derandomization in cryptography. In: Boneh, D. (ed.) CRYPTO 2003. LNCS, vol. 2729, pp. 299–315. Springer, Heidelberg (2003). https://doi.org/10.1007/978-3-540-45146-4_18
7. Biehl, I., Meyer, B., Wetzel, S.: Ensuring the integrity of agent-based computations by short proofs. In: Rothermel, K., Hohl, F. (eds.) MA 1998. LNCS, vol. 1477, pp. 183–194. Springer, Heidelberg (1998). https://doi.org/10.1007/BFb0057658
8. Bitansky, N., Paneth, O.: ZAPs and non-interactive witness indistinguishability from indistinguishability obfuscation. In: Dodis, Y., Nielsen, J.B. (eds.) TCC 2015, Part II. LNCS, vol. 9015, pp. 401–427. Springer, Heidelberg (2015). https://doi.org/10.1007/978-3-662-46497-7_16
9. Blum, M.: How to prove a theorem so no one else can claim it. In: Proceedings of the International Congress of Mathematicians, Berkeley, CA, pp. 1444–1451 (1986)
10. Brassard, G., Chaum, D., Crépeau, C.: Minimum disclosure proofs of knowledge. J. Comput. Syst. Sci. **37**(2), 156–189 (1988)
11. Canetti, R., Chen, Y., Reyzin, L., Rothblum, R.D.: Fiat-Shamir and correlation intractability from strong KDM-secure encryption. Cryptology ePrint Archive, Report 2018/131 (2018). https://eprint.iacr.org/2018/131
12. Chung, K.M., Lui, E., Mahmoody, M., Pass, R.: Unprovable security of two-message zero knowledge. IACR Cryptology ePrint Archive 2012, 711 (2012)

13. Chung, K.-M., Lui, E., Pass, R.: From weak to strong zero-knowledge and applications. In: Dodis, Y., Nielsen, J.B. (eds.) TCC 2015, Part I. LNCS, vol. 9014, pp. 66–92. Springer, Heidelberg (2015). https://doi.org/10.1007/978-3-662-46494-6_4
14. Dwork, C., Naor, M.: Zaps and their applications. In: 41st Annual Symposium on Foundations of Computer Science, FOCS 2000, Redondo Beach, California, USA, 12–14 November 2000, pp. 283–293 (2000)
15. Dwork, C., Naor, M., Reingold, O., Stockmeyer, L.J.: Magic functions. In: 40th Annual Symposium on Foundations of Computer Science, FOCS 1999, New York, NY, USA, 17–18 October 1999, pp. 523–534 (1999)
16. Feige, U., Shamir, A.: Witness indistinguishable and witness hiding protocols. In: Proceedings of the 22nd Annual ACM Symposium on Theory of Computing, Baltimore, Maryland, USA, 13–17 May 1990, pp. 416–426 (1990)
17. Fiat, A., Shamir, A.: How to prove yourself: practical solutions to identification and signature problems. In: Odlyzko, A.M. (ed.) CRYPTO 1986. LNCS, vol. 263, pp. 186–194. Springer, Heidelberg (1987). https://doi.org/10.1007/3-540-47721-7_12
18. Garg, S., Ostrovsky, R., Visconti, I., Wadia, A.: Resettable statistical zero knowledge. In: Cramer, R. (ed.) TCC 2012. LNCS, vol. 7194, pp. 494–511. Springer, Heidelberg (2012). https://doi.org/10.1007/978-3-642-28914-9_28
19. Goldreich, O.: A uniform-complexity treatment of encryption and zero-knowledge. J. Cryptology **6**(1), 21–53 (1993)
20. Goldreich, O., Oren, Y.: Definitions and properties of zero-knowledge proof systems. J. Cryptology **7**(1), 1–32 (1994)
21. Groth, J., Ostrovsky, R., Sahai, A.: Non-interactive zaps and new techniques for NIZK. In: Dwork, C. (ed.) CRYPTO 2006. LNCS, vol. 4117, pp. 97–111. Springer, Heidelberg (2006). https://doi.org/10.1007/11818175_6
22. Halevi, S., Kalai, Y.T.: Smooth projective hashing and two-message oblivious transfer. J. Cryptology **25**(1), 158–193 (2012). https://doi.org/10.1007/s00145-010-9092-8
23. Jain, A., Kalai, Y.T., Khurana, D., Rothblum, R.: Distinguisher-dependent simulation in two rounds and its applications. In: Katz, J., Shacham, H. (eds.) CRYPTO 2017, Part II. LNCS, vol. 10402, pp. 158–189. Springer, Cham (2017). https://doi.org/10.1007/978-3-319-63715-0_6
24. Kalai, Y.T.: Smooth projective hashing and two-message oblivious transfer. In: Cramer, R. (ed.) EUROCRYPT 2005. LNCS, vol. 3494, pp. 78–95. Springer, Heidelberg (2005). https://doi.org/10.1007/11426639_5
25. Kalai, Y.T., Raz, R.: Probabilistically checkable arguments. In: Halevi, S. (ed.) CRYPTO 2009. LNCS, vol. 5677, pp. 143–159. Springer, Heidelberg (2009). https://doi.org/10.1007/978-3-642-03356-8_9
26. Kalai, Y.T., Rothblum, G.N., Rothblum, R.D.: From obfuscation to the security of Fiat-Shamir for proofs. In: Katz, J., Shacham, H. (eds.) CRYPTO 2017, Part II. LNCS, vol. 10402, pp. 224–251. Springer, Cham (2017). https://doi.org/10.1007/978-3-319-63715-0_8
27. Khurana, D., Sahai, A.: Two-message non-malleable commitments from standard sub-exponential assumptions. IACR Cryptology ePrint Archive 2017, 291 (2017). http://eprint.iacr.org/2017/291
28. Naor, M., Pinkas, B.: Efficient oblivious transfer protocols. In: Proceedings of the Twelfth Annual Symposium on Discrete Algorithms, Washington, DC, USA, 7–9 January 2001, pp. 448–457 (2001)
29. Pass, R.: Simulation in quasi-polynomial time, and its application to protocol composition. In: Biham, E. (ed.) EUROCRYPT 2003. LNCS, vol. 2656, pp. 160–176. Springer, Heidelberg (2003). https://doi.org/10.1007/3-540-39200-9_10

30. Sahai, A., Vadhan, S.P.: A complete problem for statistical zero knowledge. J. ACM **50**(2), 196–249 (2003). http://doi.acm.org/10.1145/636865.636868
31. Sahai, A., Waters, B.: How to use indistinguishability obfuscation: deniable encryption, and more. In: Shmoys, D.B. (ed.) Symposium on Theory of Computing, STOC 2014, New York, NY, USA, 31 May–03 June 2014, pp. 475–484. ACM (2014). http://doi.acm.org/10.1145/2591796.2591825
32. Wolf, S., Wullschleger, J.: Oblivious transfer is symmetric. In: Vaudenay, S. (ed.) EUROCRYPT 2006. LNCS, vol. 4004, pp. 222–232. Springer, Heidelberg (2006). https://doi.org/10.1007/11761679_14

An Efficiency-Preserving Transformation from Honest-Verifier Statistical Zero-Knowledge to Statistical Zero-Knowledge

Pavel Hubáček[1](✉), Alon Rosen[2], and Margarita Vald[3]

[1] Computer Science Institute, Charles University, Prague, Czech Republic
hubacek@iuuk.mff.cuni.cz
[2] IDC Herzliya, Herzliya, Israel
alon.rosen@idc.ac.il
[3] Tel Aviv University, Tel Aviv, Israel
margarita.vald@cs.tau.ac.il

Abstract. We present an unconditional transformation from any honest-verifier statistical zero-knowledge (HVSZK) protocol to standard SZK that preserves round complexity and efficiency of both the verifier and the prover. This improves over currently known transformations, which either rely on some computational assumptions or introduce significant computational overhead. Our main conceptual contribution is the introduction of instance-dependent SZK proofs for NP, which serve as a building block in our transformation. Instance-dependent SZK for NP can be constructed unconditionally based on instance-dependent commitment schemes of Ong and Vadhan (TCC'08).

As an additional contribution, we give a simple constant-round SZK protocol for Statistical-Difference resembling the textbook HVSZK proof of Sahai and Vadhan (J.ACM'03). This yields a conceptually simple constant-round protocol for all of SZK.

1 Introduction

Zero-knowledge proof systems, introduced by Goldwasser et al. [9], give any powerful prover the ability to convince a verifier about validity of a statement without revealing any additional information other than its correctness. This power has been extensively exploited in constructions of various cryptographic

P. Hubáček—This work was performed while at the Foundations and Applications of Cryptographic Theory (FACT) center, IDC Herzliya, Israel. Partially supported by the PRIMUS grant PRIMUS/17/SCI/9 and by the Center of Excellence – ITI, project P202/12/G061 of GA ČR.
A. Rosen—Work supported by ISF grant no 1399/17 and by NSF-BSF Cyber Security and Privacy grant no. 2014/632.
M. Vald—Work supported by ISF grant no 1399/17 and by Google Europe Doctoral Fellowship in Security.

J. B. Nielsen and V. Rijmen (Eds.): EUROCRYPT 2018, LNCS 10822, pp. 66–87, 2018.
https://doi.org/10.1007/978-3-319-78372-7_3

protocols. Besides the many applications, great effort was invested to improve our understanding of the limits of zero-knowledge proof systems with respect to different complexity measures such as round complexity or efficiency of prover and verifier.

Similarly to the requirement of soundness for interactive proof systems, there are many natural relaxations of zero-knowledge. In this work we study *statistical* zero-knowledge (SZK) proofs. In particular, we revisit the problem of immunizing any honest-verifier statistical zero-knowledge (HVSZK) protocol against malicious verifiers, while preserving the efficiency of the original protocol. Such transformation suggests a methodology for constructing zero-knowledge protocols: first construct an efficient proof system for the desired problem where the zero-knowledge property holds against honest verifiers, and then compile it to a full-blown zero-knowledge proof against malicious verifiers while preserving the efficiency.

Bellare et al. [3] initiated the study of general transformations from honest-verifier zero-knowledge protocols to protocols in which the zero-knowledge property holds against arbitrary verifiers. Their work presented such a transformation under the assumption of intractability of solving the discrete-logarithm problem. Later, Ostrovsky et al. [15] presented a transformation under a weaker assumption of existence of one-way permutations. Okamoto [13] further weakened the assumption to existence one-way functions. However, relying on intractability assumptions prevents the zero-knowledge property to hold against computationally unbounded verifiers which might be a desirable property in some contexts.

Until recently, unconditional transformations of honest-verifier zero-knowledge to zero-knowledge against malicious verifiers were only known via public-coin proof system. Under the restriction to constant-round public-coin protocols [4,5] gave first such unconditional transformations. The restriction to constant-round was lifted by [7] who gave a transformation achieving general statistical zero-knowledge starting from any *public-coin* honest-verifier statistical zero-knowledge protocol. Combining the transformation of [7] with the private-coin to public-coin transformation of [8,13] yields a general transformation starting from any honest-verifier protocol. However, it follows from Vadhan [17] that any transformation from honest-verifier zero-knowledge to general cheating verifier that goes through public-coin protocol must result in a significant blow-up in the prover's complexity. Moreover, the private-coin to public-coin transformation of [8,13] does not preserve the message complexity.

Ong and Vadhan [14] successfully avoided the standard private-coin to public-coin transformation by relying on their novel construction of a relaxed notion of commitments, called instance-dependent commitment. Instance-dependent commitments allow the hiding and binding properties of a commitment scheme not to hold simultaneously but rather to depend on a given instance. Specifically, they obtained a general transformation from honest-verifier statistical zero-knowledge to general statistical zero-knowledge by going via the transformation of honest-verifier statistical zero-knowledge to two-round Arthur-Merlin protocol due to Aiello and Håstad [1]. In the resulting statistical zero-knowledge protocol the verifier sends the first message of Arthur in the AM protocol and the prover then gives a statistical zero-knowledge proof for the NP statement of the form: there exists a message of Merlin that makes Arthur accept. The statistical zero-knowledge proof for this NP statement can be performed in constant number of rounds by instantiating known statistical zero-knowledge protocols for NP using the instance-dependent commitment scheme of Ong and Vadhan [14]. The transformation in [14] was the first to result in a protocol with constant number of rounds. However, the [14] transformation, as well as all of the above unconditional transformations, result in a significant blow-up in the complexity of the prover compared to the original honest-verifier protocol.

2 Our Results

We present a general efficiency-preserving compiler from any honest-verifier statistical zero-knowledge proof to a statistical zero-knowledge proof against malicious verifiers. Our compiler preserves both the round complexity and the prover's complexity of the original honest-verifier protocol. Our transformation yields a very simple constant-round statistical zero-knowledge protocol for every problem in honest-verifier statistical zero-knowledge.

Theorem 1 (honest-verifier SZK to SZK compiler). *For every promise problem $\Pi \in \mathsf{HVSZK}$, there exists a statistical zero-knowledge proof where the prover's complexity and the round complexity match the parameters of the best honest-verifier statistical zero-knowledge proof for Π.*

Applying Theorem 1 on the honest-verifier statistical zero-knowledge protocol of Sahai and Vadhan [16] for the HVSZK-complete problem STATISTICAL-DIFFERENCE yields the following:

Theorem 2 (Constant-round proof for SZK). *For every promise problem $\Pi \in \mathsf{HVSZK}$, there exists a constant-round statistical zero-knowledge proof.*

Additionally, we show how to achieve Theorem 2 via simple direct construction for STATISTICAL-DIFFERENCE. This is shown in Sect. 4.2.

Our transformation follows the classical approach of Goldreich et al. [6] to immunize protocols against malicious behavior. In the context of zero-knowledge, an honest verifier follows the protocol specification using a uniformly random tape. The standard way to preserve zero-knowledge in the presence of a malicious verifier is to enforce the honest behavior. To this end, we leverage the fact that the protocol specification is a deterministic function of the verifier's view; at each round the verifier's view consists of its random tape and the messages received up to this round. Thus, the verifier can give a zero-knowledge proof for the NP statement attesting that its messages to the prover are indeed computed according to the specifications of the protocol.

Note that the quality of the employed zero-knowledge proof for NP determines the quality of the resulting protocol. Specifically, if we use as a building block a proof for NP that is zero-knowledge against polynomial-time verifiers then the resulting protocol will be a zero-knowledge *argument*. This follows from the fact that the roles of the prover and verifier are reversed in the intermediate proof for NP and our compiler cannot guarantee soundness against unbounded provers unless the simulator for the intermediate proofs can handle unbounded verifiers. To solve this issue, we use a relaxation of statistical zero-knowledge for NP that is sufficient for our compiler to result in a statistical zero-knowledge *proof*.

Instance-dependent commitment schemes [2,10], in which the properties of the commitment protocol depend on a given instance of a language, proved to be useful in constructions of zero-knowledge protocols by Itoh et al. [10]. Recently, Ong and Vadhan [14] constructed instance-dependent (ID) commitments relative to all of SZK. The ID commitments of Ong and Vadhan are statistically binding on Yes instances of the SZK problem and statistically hiding on No instances (and vice versa due to the fact that SZK is closed under complement).

In this work, we define a relaxation of zero-knowledge proofs, called *instance-dependent zero-knowledge*, and show that it suffices for the [6] approach when constructing a compiler from honest-verifier statistical zero-knowledge to general statistical zero-knowledge. Analogously to other instance-dependent primitives, soundness and zero-knowledge do not necessary hold simultaneously in instance-dependent zero-knowledge proofs but depending on the underlying instance of the given promise problem. We believe that this primitive is of independent interest and may find further applications beyond our compiler. We instantiate the instance-dependent zero-knowledge by employing the construction of instance-dependent commitments [14] in the constant-round zero-knowledge proof of knowledge for NP of Lindell [11] (see Sect. 4.1 for details). The instantiation and our compiler do not rely on any intractability assumption.

3 Preliminaries

Throughout the rest of the paper we use the following notation and definitions. For $n \in \mathbb{N}$, let $[n]$ denote the set $\{1, \ldots, n\}$. A function $g : \mathbb{N} \rightarrow \mathbb{R}^+$ is *negligible* if it tends to 0 faster than any inverse polynomial, i.e., for all $c \in \mathbb{N}$ there exists $k_c \in \mathbb{N}$ such that for every $k > k_c$ it holds that $g(k) < k^{-c}$. We use $\mathsf{neg}(\cdot)$ to denote a negligible function if we do not need to specify its name.

A random variable X is a function from a finite set S to the nonnegative reals with the property that $\sum_{s \in S} X(s) = 1$. We write $x \leftarrow X$ to indicate that x is selected according to X. We write U_n to denote the random variable that is uniform over $\{0,1\}^n$. We use the terms random variable and probability distribution interchangeably.

A probability ensemble is a set of random variables $\{A_x\}_{x \in \{0,1\}^*}$, where A_x takes values in $\{0,1\}^{p(|x|)}$ for some polynomial p. We call such an ensemble samplable if there is a probabilistic polynomial-time algorithm such that for every x, the output of the algorithm is distributed according to A_x.

3.1 Interactive Proof Systems

Definition 1 (Interactive proof system). *A pair of interactive machines* $\langle \mathsf{P}, \mathsf{V} \rangle$ *is called an* interactive proof system *for a language* L *if* V *is a PPT machine and there exists a negligible function* $\mathsf{neg}(\cdot)$ *such that* $\forall k \in \mathbb{N}$ *the following holds:*

Completeness: *For all* $x \in L$,

$$\Pr[\langle \mathsf{P}, \mathsf{V} \rangle (x, 1^k) = 1] = 1.$$

Soundness: *For all* $x \notin L$, *and every interactive machine* P^*,

$$\Pr[\langle \mathsf{P}^*, \mathsf{V} \rangle (x, 1^k) = 1] \leq \mathsf{neg}(k).$$

Definition 2 (Proof of knowledge). *Let* $L \in \mathsf{NP}$ *and let* R_L *be its witness relation. An interactive proof system* $\langle P, V \rangle$ *for* L *is called* a proof of knowledge *(*PoK*) if it satisfies the following property:*

Knowledge Soundness: *There exists a PPT machine* E, *called the* extractor, *such that for every* P^*, *for every* $x \in L$, *auxiliary input* z, *random tape* r, *and* $k \in \mathbb{N}$

$$\Pr[\mathsf{E}^{\mathsf{P}^*}(x, z, r; 1^k) = w : (x, w) \in R_L] \geq \Pr[\langle \mathsf{P}^*(z; r), \mathsf{V} \rangle (x, 1^k) = 1] - \mathsf{neg}(k).$$

If the soundness property (resp. the knowledge soundness) in $\langle \mathsf{P}, \mathsf{V} \rangle$ holds only with respect to PPT provers, we call it an *interactive argument system* (resp. an *argument of knowledge*).

3.2 Statistical Zero-Knowledge

We use the standard definition of statistical difference of two probability distributions X, Y over universe $\mathbf{U}$, i.e.,

$$\mathrm{SD}\,(X, Y) = \max_{S \subset \mathbf{U}} |\Pr[X \in S] - \Pr[Y \in S]|.$$

Definition 3 (Promise problems). *A* promise problem *is specified by two disjoint sets of strings* $\Pi = (\Pi_Y, \Pi_N)$, *where* Π_Y *is the set of YES instances and* Π_N *is the set of NO instances. Any promise problem* Π *is associated with the following algorithmic task: given an input string that is promised to lie in* $\Pi_Y \cup \Pi_N$, *decide whether it is in* Π_Y *or in* Π_N.

Recall that the zero-knowledge property is captured via an existence of a simulator, an entity that simulates the view of the verifier in its interaction with the prover.

Definition 4 (View of an interactive protocol). *Let $\langle A, B\rangle$ be an interactive protocol. B's view of $\langle A, B\rangle$ on common input x is the random variable $(A, B)(x) = (m_1, \ldots, m_t; r)$ consisting of all the messages $m_1, \ldots, m_t$ exchanged between A and B together with the string r containing all the random bits that B has read during the interaction.*[1]

Statistical zero knowledge requires that the statistical difference between the simulator's output distribution and the verifier's view is so small that polynomially many repetitions of the protocol cannot make it noticeable. The definition allows the simulator to occasionally fail and output `fail`, and it only measures the quality of the simulation conditioned on non-failure.

Definition 5 (Honest-Verifier Statistical Zero-Knowledge). *An interactive proof system $\langle P, V\rangle$ for a promise problem Π is said to be* honest-verifier statistical zero-knowledge *if there exists a PPT* S *that fails with probability at most* $1/2$ *and a negligible function* $\mathsf{neg}(\cdot)$ *such that* $\forall x \in \Pi_Y, k \in \mathbb{N}$,

$$\mathrm{SD}\left(\widetilde{\mathsf{S}}(x, 1^k), (\mathsf{P}, \mathsf{V})(x, 1^k)\right) \leq \mathsf{neg}(k)$$

where $\widetilde{\mathsf{S}}$ *is the output distribution of* S *conditioned on not failing.* HVSZK *denotes the class of all promise problems admitting honest-verifier statistical zero-knowledge proofs.*

Zero knowledge against arbitrary verifier is captured by exhibiting a single, universal simulator S that simulates an arbitrary verifier strategy V^* by using V^* as a subroutine (denoted by $\mathsf{S}^{\mathsf{V}^*}$). That is, the simulator does not depend on or use the code of V^*, and instead only has black-box access to V^*. More formally,

Definition 6 (Statistical Zero-Knowledge). *An interactive proof system $\langle P, V\rangle$ for a promise problem Π is said to be* statistical zero-knowledge *if there exists a PPT* S *that fails with probability at most* $1/2$ *such that for every nonuniform PPT* V^* *it holds that*

$$\mathrm{SD}\left(\widetilde{\mathsf{S}}^{\mathsf{V}^*}(x, 1^k), (\mathsf{P}, \mathsf{V}^*)(x, 1^k)\right) \leq \mathsf{neg}(k) \qquad \forall x \in \Pi_Y, k \in \mathbb{N},$$

where $\widetilde{\mathsf{S}}$ *is the output distribution of* S *conditioned on not failing, and* $\mathsf{neg}(\cdot)$ *is some negligible function that may depend on* V^*. SZK *denotes the class of all promise problems admitting statistical zero-knowledge proofs.*

3.3 Instance-Dependent Commitment Schemes

Definition 7 (Instance-dependent commitment schemes). *An instance-dependent commitment scheme is a family of commitment schemes* $\{\mathsf{Com}_x\}_{x\in\{0,1\}^*}$ *with the following properties:*

[1] Note that equivalently we can define the view to be the messages from A to B and B's random bits. This is since the messages sent by B are a deterministic function of the received messages and the B's random bits.

1. *Scheme* Com_x *proceeds in two stages: a commit stage and a reveal stage. In both stages, the sender and receiver receive instance* x *as common input, and hence we denote the sender and receiver as* S_x *and* R_x*, respectively, and write* $\mathsf{Com}_x = (\mathsf{S}_x, \mathsf{R}_x, \mathsf{Open}_x)$.
2. *At the beginning of the commit stage, sender* S_x *receives a private input* $b \in \{0,1\}$*, which denotes the bit that* S_x *is supposed to commit to. At the end of the commit stage, both sender* S_x *and receiver* R_x *output a commitment* c.
3. *In the reveal stage, sender* S_x *sends a pair* (b, d)*, where* d *is the decommitment string for bit* b*. Receiver* R_x *outputs* $\mathsf{Open}_x(c, b, d) \in \{\mathtt{accept}, \mathtt{reject}\}$.
4. *The sender* S_x *and receiver* R_x *algorithms are computable in polynomial time (in* $|x|$*), given* x *as auxiliary input.*
5. *For every* $x \in \{0,1\}^*$, $\mathsf{Open}_x(c, b, d) = \mathtt{accept}$ *with probability* 1 *if both sender* S_x *and receiver* R_x *follow their prescribed strategy.*

Definition 8 (Statistical hiding). *Instance-dependent commitment scheme* $\mathsf{Com}_x = (\mathsf{S}_x, \mathsf{R}_x, \mathsf{Open}_x)$ *is* statistically hiding *on* $I \subseteq \{0,1\}^*$ *if for every* R^**, the ensembles* $\{\mathsf{view}_{\mathsf{R}^*}(\mathsf{S}_x(0), \mathsf{R}^*)\}_{x \in I}$ *and* $\{\mathsf{view}_{\mathsf{R}^*}(\mathsf{S}_x(1), \mathsf{R}^*)\}_{x \in I}$ *are statistically indistinguishable, where the random variable* $\mathsf{view}_{\mathsf{R}^*}(\mathsf{S}_x(b), \mathsf{R}^*)$ *denotes the view of* R^* *in the commit stage interacting with* $\mathsf{S}_x(b)$*. For a promise problem* $\Pi = (\Pi_Y, \Pi_N)$*, an instance-dependent commitment scheme* Com_x *for* Π *is statistically hiding on the YES instances if* Com_x *is statistically hiding on* Π_Y.

Definition 9 (Statistical binding). *Instance-dependent commitment scheme* $\mathsf{Com}_x = (\mathsf{S}_x, \mathsf{R}_x, \mathsf{Open}_x)$ *is* statistically binding *on* $I \subseteq \{0,1\}^*$ *if for every* S^**, there exists a negligible function* neg *such that for all* $x \in I$*, the malicious sender* S^* *wins in the following game with probability at most* $\mathsf{neg}(|x|)$.

- S^* *interacts with* R_x *in the commit stage obtaining commitment* c.
- *Then* S^* *outputs* d_0 *and* d_1*, and it wins if* $\mathsf{Open}_x(c, 0, d_0) = \mathsf{Open}_x(c, 1, d_1) = \mathtt{accept}$.

For a promise problem $\Pi = (\Pi_Y, \Pi_N)$*, an instance-dependent commitment scheme* Com_x *for* Π *is* statistically binding on the NO instances *if* Com_x *is statistically binding on* Π_N.

Theorem 3 ([14]). *Every problem* $\Pi = (\Pi_Y, \Pi_N) \in \mathsf{HVSZK}$ *has an instance-dependent commitment scheme that is statistically hiding on the YES instances and statistically binding on the NO instances. Moreover, the instance-dependent commitment scheme is public-coin and constant-round.*

Since HVSZK is closed under complement, for every $\Pi = (\Pi_{\mathrm{Y}}, \Pi_{\mathrm{N}}) \in \mathsf{HVSZK}$, we can also obtain instance dependent commitments in which the security properties are reversed (i.e., statistically binding on YES instances and statistically hiding a on NO instances).

4 Constant-Round Statistical Zero-Knowledge Proofs

In this section, we define a relaxation of zero-knowledge called *instance-dependent statistical zero-knowledge proofs.* We show that for the class NP it is possible to obtain constant-round instance-dependent statistical zero-knowledge proofs of knowledge without relying on computational assumptions. Next, using this relaxation of zero-knowledge for NP, we construct a constant-round statistical zero-knowledge proof for any promise problem in HVSZK.

4.1 Instance-Dependent Statistical Zero-Knowledge Proofs

Instance-dependent statistical zero-knowledge proofs are a relaxation of the standard notion of statistical zero-knowledge proofs that allows the proof to depend on a specific promise problem Π. Similarly to instance-dependent commitment schemes [2,10,12], the prover and the verifier receive an instance x of the problem Π as auxiliary input and a statement ψ to prove. The proof system is required to be sound proof of knowledge when $x \in \Pi_{\mathrm{Y}}$ and zero-knowledge when $x \in \Pi_{\mathrm{N}}$.

Looking ahead, instance-dependent zero-knowledge proofs will be used as a sub-protocol within some outer protocol. Note that there are two instances involved: (1) an instance of the promise problem Π, for which the outer protocol is constructed and (2) an instance of the language L for which the instance-dependent proof system is used.

Definition 10 (Instance-dependent statistical zero-knowledge). An instance-dependent statistical zero-knowledge proof of knowledge for language L with respect to a promise problem $\Pi = (\Pi_Y, \Pi_N)$ *is a family of protocols* $\{\langle P_x, V_x\rangle\}_{x\in\{0,1\}^*}$ *with the following properties:*

- $\langle P_x, V_x\rangle$ *is complete on all instances of* Π*, i.e., for all* $x \in \Pi_Y \cup \Pi_N$.
- $\langle P_x, V_x\rangle$ *is statistical zero-knowledge on the NO instances, i.e., for all* $x \in \Pi_N$.
- $\langle P_x, V_x\rangle$ *is a sound proof of knowledge on the YES instances, i.e., for all* $x \in \Pi_Y$.

We show that the protocol of Lindell [11] instantiated with the instance-dependent commitments of Ong and Vadhan [14] gives rise to a constant-round instance-dependent statistical zero-knowledge proof of knowledge for NP.

Theorem 4. *For every promise problem* $\Pi = (\Pi_Y, \Pi_N) \in$ HVSZK *and for every language* $L \in$ NP*, there exists a constant-round instance-dependent statistical zero-knowledge proof of knowledge for* L *with respect to* Π*. Moreover, the zero-knowledge property holds against unbounded verifiers.*

Similarly to instance-dependent commitments, for all $\Pi = (\Pi_{\mathrm{Y}}, \Pi_{\mathrm{N}}) \in$ HVSZK, we can obtain instance-dependent statistical zero-knowledge with the security properties reversed, i.e., with knowledge soundness on NO instances and statistical zero-knowledge on YES instances.

Proof (Proof of Theorem 4). Let $\Pi = (\Pi_{\mathrm{Y}}, \Pi_{\mathrm{N}}) \in$ HVSZK be some promise problem and denote by HC the Hamiltonian Cycle language. Let x be an instance of Π, let Com_x^{sb} be an instance-dependent commitment scheme that is statistically binding on Π_{Y} and statistically hiding on Π_{N}. Let Com_x^{sh} be an instance-dependent commitment scheme that is statistically binding on Π_{N} and statistically hiding on Π_{Y}. The protocol is formally presented in Fig. 1. Since HC is NP-complete, we obtain a proof system for any language in NP by a standard reduction.

Let x be an instance of Π and let Com_x^{sh} and Com_x^{sb} be instance-dependent commitment schemes.
Input: a graph $G = (V, E)$, with $n = |V|$, and security parameter 1^k.
Prover's auxiliary input: a directed Hamiltonian cycle $C \subseteq E$ in G.
The protocol $\langle \mathsf{P}_x, \mathsf{V}_x \rangle$ for proving $G \in HC$ proceeds as follows:

1. P_x sends n independent copies of the first message for the basic proof of Hamiltonicity. That is, for $1 \leq i \leq n$, P_x selects a random permutation π_i over the vertices V and interacts with V_x to commit (using Com_x^{sb}) to the entries of the adjacency matrix of the resulting permuted graph. That is, P_x commits to an n-by-n matrix so that the entry $(\pi_i(\ell), \pi_i(j))$ contains a commitment to 1 if $(\ell, j) \in E$, and it contains a commitment to 0 otherwise.
   (a) V_x samples $q_1 \leftarrow \{0,1\}^n$ and interacts with P_x in Com_x^{sh}, so that P_x learns c_1, a commitment to q_1.
   (b) P_x samples $q_2 \leftarrow \{0,1\}^n$ and interacts with V_x in Com_x^{sb}, so that V_x learns c_2, a commitment to q_2.
   (c) V_x opens the commitment c_1 by sending q_1 and a decommitment string d_1.
   (d) If $\mathsf{Open}_x^{sh}(c_1, q_1, d_1) = \mathtt{reject}$, then P_x aborts and halts. Otherwise, P_x opens the commitment c_2 by sending q_2 and a decommitment string d_2.
2. P_x computes an n bit string $q = q_1 \oplus q_2$ and sends the second message for the basic proof of Hamiltonicity for each of the n copies, where P_x uses the i-th bit of q as the verifier's query in the i-th copy. That is, for $1 \leq i \leq n$ do:
   - If $q(i) = 0$, then send π_i and open all the commitments in the adjacency matrix of the i-th instance.
   - If $q(i) = 1$, open *only* the commitments of entries $(\pi_i(\ell), \pi_i(j))$ for which $(\ell, j) \in C$.
3. V_x computes $q = q_1 \oplus q_2$. If either $\mathsf{Open}_x^{sb}(c_2, q_2, d_2) = \mathtt{reject}$ or the response of the prover is not accepting in all n copies, based on the queries according to q, then output $\mathtt{reject}$. Otherwise, output $\mathtt{accept}$.

Fig. 1. The instance-dependent statistical zero-knowledge proof of knowledge $\langle \mathsf{P}_x, \mathsf{V}_x \rangle$ for NP-complete problem Hamiltonian Cycle with respect to a promise problem $\Pi \in$ HVSZK. The protocol builds on the constant-round zero-knowledge proof of knowledge of Lindell [11] which we instantiate with instance-dependent commitments relative to an instance x of Π.

Lindell [11] showed that if the verifier commits using a statistically hiding scheme Com_x^{sh} and the prover commits using a statistically binding scheme Com_x^{sb} then the protocol in Fig. 1 is sound proof of knowledge for HC. Since Com_x^{sh} and Com_x^{sb} satisfy this requirement on Π_{Y}, we obtain that $\langle \mathsf{P}_x, \mathsf{V}_x \rangle$ is sound proof of knowledge for HC with respect to all $x \in \Pi_{\mathrm{Y}}$. Therefore, it is only left to show that $\langle \mathsf{P}_x, \mathsf{V}_x \rangle$ is statistical zero-knowledge against unbounded verifiers with respect to all $x \in \Pi_{\mathrm{N}}$.

Input: instance x of $\Pi \in$ HVSZK, a graph $G = (V, E)$, with $n = |V|$, and security parameter 1^k. Given oracle access to verifier V^*, the simulator S works as follows:

1. S chooses a random string $q \in \{0,1\}^n$. Then, for the prover's message in the i-th execution, S interacts in Com_x^{sb}, so that V^* learns a commitment to a random permutation of G if $q(i) = 0$, and to a simple n-cycle if $q(i) = 1$.
2. S honestly interacts with V^* in Com_x^{sh}, and learns the verifier's commitment c_1. S chooses a random q_2 and interacts with V^* in Com_x^{sb}, so that V^* learns c_2, a commitment to q_2.
3. S receives q_1 and the decommitment string d_1 from V^*. If $\mathsf{Open}_x^{sh}(c_1, q_1, d_1) =$ `reject`, then S simulates P_x aborting, outputs whatever V^* outputs and halts. Otherwise, S proceeds to the next step.
4. **Rewinding phase:**
   (a) S rewinds V^* back to the point before the interaction in Com_x^{sh}. S interacts honestly with V^* to produce a commitment c_2 for value $q_1 \oplus q$.
   (b) S receives q_1' and d_1' from V^* and proceeds as follows:
   - If $\mathsf{Open}_x^{sh}(c_1, q_1', d_1') =$ `reject` then abort on behalf of P_x, S outputs whatever V^* outputs and halts.
   - If $\mathsf{Open}_x^{sh}(c_1, q_1', d_1') =$ `accept` but $q_1' \neq q_1$ then S outputs `fail` and halts.
   - Otherwise, S opens the commitment c_2 and for each $i \in [n]$, opens the commitments either to the entire graph (for $q(i) = 0$) or the simple cycle (for $q(i) = 0$).
5. S outputs whatever V^* outputs.

Fig. 2. Simulator for the protocol in Fig. 1.

Note that when $x \in \Pi_{\mathrm{N}}$, the commitment Com_x^{sh} is statistically binding and Com_x^{sb} is statistically hiding. In Fig. 2, we present a simulator that produces a distribution of transcripts which is statistically close to the real distribution of transcripts.

Lemma 5. *For all $x \in \Pi_N$, every input graph $G = (V, E)$, every security parameter $k \in \mathbb{N}$, and any verifier V^*, it holds that*

$$\Pr[\mathsf{S}^{\mathsf{V}^*}(x, G, 1^k) = \mathtt{fail}] \leq \mathsf{neg}(k).$$

Proof. Given $x \in \Pi_{\mathrm{N}}$, let V^* be an arbitrary verifier. We get that

$$\begin{aligned}&\Pr[\mathsf{S}^{\mathsf{V}^*}(x, G, 1^k) = \mathtt{fail}]\\ &\quad\leq \Pr[\exists c, q_1, d_1, q_1', d_1' : \mathsf{Open}_x^{sh}(c, q_1, d_1) = \mathsf{Open}_x^{sh}(c, q_1', d_1') = \mathtt{accept}]\end{aligned}$$

which is at most negligible in the security parameter since the commitment scheme Com_x^{sh} is statistically binding for any $x \in \Pi_{\mathrm{N}}$. □

Note that the simulator S rewinds V^* such that the initially chosen string q is the coin-flipping result. In this case, S can decommit appropriately and conclude the proof. The statistical closeness of the distribution of transcripts produced by the simulator and the real distribution of transcripts follows from the statistical hiding of Com_x^{sb} combined with the statistical binding of Com_x^{sh}.

Due to statistical hiding of Com_x^{sb}, the probability over q_2 and r_2 that V^* decommits to c_1 in the main thread (before rewinding) is basically equivalent to the probability that V^* decommits to c_1 in the rewind. Thus, the only difference between the output distribution generated by S and the output distribution generated in a real proof is that in the case that $q(i) = 1$ the unopened commitments in the simulated transcript are all to 0, and not to the rest of the graph apart from the cycle. However, due to the statistical hiding property of Com_x^{sb} on $x \in \Pi_{\mathrm{N}}$, the distributions are statistically close. This completes the proof of Theorem 4. □

4.2 A Concrete Protocol for a SZK-Complete Problem

In this section, we show that $\mathsf{HVSZK} \subseteq \mathsf{SZK}[c]$, where $\mathsf{SZK}[c]$ is the class of all promise problems that admit constant-round statistical zero-knowledge proof. Concretely, in Fig. 3 we present a simple constant-round statistical zero-knowledge protocol secure against any malicious verifier for a complete problem in HVSZK, called STATISTICAL-DIFFERENCE. The constant-round protocol for any problem in HVSZK would comprise of a reduction to STATISTICAL-DIFFERENCE (which can be performed locally by both P and V) and then running our protocol.

First, we recall the STATISTICAL-DIFFERENCE problem which was shown to be HVSZK-complete by Sahai and Vadhan [16]. In this work we consider the polarized form of STATISTICAL-DIFFERENCE, that can be obtained from the basic definition in polynomial-time.

Definition 11 (Statistical-Difference). *Given $k \in \mathbb{N}$, the promise problem* STATISTICAL-DIFFERENCE *is $\mathsf{SD} = (\mathsf{SD}_Y, \mathsf{SD}_N)$, where*

$$\mathsf{SD}_Y = \{(X_0, X_1) : \mathrm{SD}(X_0, X_1) \geq 1 - 2^{-k}\},$$
$$\mathsf{SD}_N = \{(X_0, X_1) : \mathrm{SD}(X_0, X_1) \leq 2^{-k}\}.$$

Above, X_0, X_1 are circuits encoding probability distributions.

Given $\overline{X} = (X_0, X_1)$, an instance of STATISTICAL-DIFFERENCE, our protocol builds on the standard honest-verifier statistical zero-knowledge proof for STATISTICAL-DIFFERENCE of Sahai and Vadhan [16]. To force the verifier to behave as in the original honest-verifier protocol, we use (1) a constant-round instance-dependent commitment scheme $\mathsf{Com}_{\overline{X}} = (\mathsf{S}_{\overline{X}}, \mathsf{R}_{\overline{X}}, \mathsf{Open}_{\overline{X}})$ that is statistically binding on SD_Y, and (2) a constant-round instance-dependent statistical zero-knowledge proof of knowledge $\langle \mathsf{P}_{\overline{X}}, \mathsf{V}_{\overline{X}} \rangle$ for NP that is zero-knowledge on SD_N against any unbounded verifier. These building blocks are provided by Theorems 3 and 4, respectively. The protocol is formally presented in Fig. 3.

Theorem 6. *The protocol presented in Fig. 3 is constant-round statistical zero-knowledge proof for* STATISTICAL-DIFFERENCE.

By completeness of STATISTICAL-DIFFERENCE for HVSZK, we obtain a constant-round protocol secure against any verifier for every problem in the class.

Input: Given $\overline{X} = (X_0, X_1)$, a pair of circuits, and security parameter 1^k. Let $\mathsf{Com}_{\overline{X}} = (\mathsf{S}_{\overline{X}}, \mathsf{R}_{\overline{X}}, \mathsf{Open}_{\overline{X}})$ be an instance-dependent commitment scheme that is statistically binding on SD_Y and let $\langle \mathsf{P}_{\overline{X}}, \mathsf{V}_{\overline{X}} \rangle$ be an instance dependent statistical zero-knowledge proof of knowledge for **NP** with knowledge soundness on SD_Y. The protocol $\mathsf{SD} = \langle \mathsf{P_{SD}}, \mathsf{V_{SD}} \rangle$ for proving $\overline{X} \in \mathsf{SD}_Y$ proceeds as follows:

1. **Coin flipping phase:**
   (a) $\mathsf{V_{SD}}$ samples $r_0 \leftarrow \{0,1\}^n$, $b_0 \leftarrow \{0,1\}$.
   (b) $\mathsf{V_{SD}}$ and $\mathsf{P_{SD}}$ interact in $\mathsf{Com}_{\overline{X}}$, so that $\mathsf{P_{SD}}$ learns c, a commitment to the pair (r_0, b_0).
   (c) $\mathsf{P_{SD}}$ samples $r_1 \leftarrow \{0,1\}^n$, $b_1 \leftarrow \{0,1\}$ and sends them to $\mathsf{V_{SD}}$. Then, $\mathsf{V_{SD}}$ sets $b = b_0 \oplus b_1$ and $r = r_0 \oplus r_1$.
2. **Honest SD-protocol execution phase:**
   (a) $\mathsf{V_{SD}}$ sends $y = X_b(r)$ to $\mathsf{P_{SD}}$.
   (b) Let $L_{\mathsf{samp}} = \{(c, r', b', y) | \exists r, b, d : \mathsf{Open}_{\overline{X}}(c, (r, b), d) = \mathtt{accept} \wedge y = X_{b' \oplus b}(r' \oplus r)\}$. $\mathsf{V_{SD}}$ uses $\langle \mathsf{P}_{\overline{X}}, \mathsf{V}_{\overline{X}} \rangle$ to prove to $\mathsf{P_{SD}}$ that $(c, r_1, b_1, y) \in L_{\mathsf{samp}}$. Denote by θ the transcript of this proof.
   (c) If $\mathsf{V}_{\overline{X}}$ rejects θ then $\mathsf{P_{SD}}$ aborts, otherwise, it replies with $b'' \in \{0,1\}$ such that
   $$\Pr_{r \leftarrow U_n}[X_{b''}(r) = y] \geq \Pr_{r \leftarrow U_n}[X_{1-b''}(r) = y] .$$
   (d) If $b = b''$ then $\mathsf{V_{SD}}$ outputs $\mathtt{accept}$, otherwise outputs $\mathtt{reject}$.

Fig. 3. The statistical zero-knowledge proof $\langle \mathsf{P_{SD}}, \mathsf{V_{SD}} \rangle$ for STATISTICAL-DIFFERENCE. Our protocol builds on the honest-verifier statistical zero-knowledge proof of Sahai and Vadhan [16] with the following changes: (1) The verifier's randomness is picked mutually by the verifier and the prover (while maintaining the secrecy to the prover). (2) The verifier is required to provide a proof that it used the mutually chosen randomness.

Corollary 7. *There exists a constant-round statistical zero-knowledge proof for every $\Pi \in \mathsf{HVSZK}$, where the zero-knowledge holds against any malicious verifier.*

Proof of Theorem 6. Here we show that the protocol in Fig. 3 is complete, sound and achieves statistical zero-knowledge.

Completeness. Due to the perfect completeness of the $\langle \mathsf{P}_{\overline{X}}, \mathsf{V}_{\overline{X}} \rangle$ proof, it follows that the completeness error of our protocol is the same as the completeness error of the standard protocol for SD of [16], i.e., at most 2^{-k}.

Soundness. We present here a proof sketch. The full proof can be found in Sect. 5, where we present the general transformation. Given $\overline{X} = (X_0, X_1) \in \mathsf{SD}_N$, a NO instance of STATISTICAL-DIFFERENCE, let P^* be an arbitrary prover. Let $\mathsf{Com}_{\overline{X}}$ and $\langle \mathsf{P}_{\overline{X}}, \mathsf{V}_{\overline{X}} \rangle$ be as defined above. Finally, let $\mathsf{Sim}_{\overline{X}}$ be the statistical zero-knowledge simulator for $\langle \mathsf{P}_{\overline{X}}, \mathsf{V}_{\overline{X}} \rangle$.

We show that the soundness error in the above protocol is at most negligibly larger than the soundness error in the original honest-verifier protocol. This follows from the statistical zero-knowledge property against unbounded verifiers

Input: Given $\overline{X} = (X_0, X_1)$ and security parameter 1^k. Let E be the extractor of $\langle \mathsf{P}_{\overline{X}}, \mathsf{V}_{\overline{X}} \rangle$ scheme. The simulator S_{SD} with oracle access to V^* proceeds as follows:

1. Execute honestly the protocol up to the last round with $\mathsf{V}^*(x)$ in order to learn a commitment c, and a sample y. Let b_1 and r_1 be the values given to $\mathsf{V}^*(x)$ in the simulated coin-flipping phase. Participate as the honest $\mathsf{V}_{\overline{X}}$ in the proof of knowledge for the committed value in c and correctness of y. Denote this proof of knowledge θ.
2. If θ is not accepting then abort. Otherwise, use the knowledge extractor $\mathsf{E}^{\mathsf{V}^*}$ to extract the values r_0^*, b_0^*, d^*. If the extractor fails output `fail`.
3. Send $b = b_0^* \oplus b_1$ to V^*.
4. Output the simulated transcript and r_0^*, b_0^*, d^* as the randomness of V^*.

Fig. 4. Simulator $\mathsf{S}_{\mathsf{SD}}^{\mathsf{V}^*}$ for protocol $\langle \mathsf{P}_{\mathsf{SD}}, \mathsf{V}_{\mathsf{SD}} \rangle$. The simulator honestly participates in an execution with V^* but instead of sending the last message, it extracts the randomness of the verifier and uses it to generate the last message.

of $\langle \mathsf{P}_{\overline{X}}, \mathsf{V}_{\overline{X}} \rangle$, and the statistical hiding property of $\mathsf{Com}_{\overline{X}}$. Specifically, the distribution of transcripts $\langle \mathsf{P}^*, \mathsf{V}_{\mathsf{SD}} \rangle(\overline{X})$ is statistically close to the distribution of transcripts where the proof in Step 2b is performed using $\mathsf{Sim}_{\overline{X}}$ (this can be done since V is honest, and proves a true statement). Note that when Step 2b is performed using $\mathsf{Sim}_{\overline{X}}$, the acceptance probability of V is equivalent to its acceptance probability in a protocol where the proof of Step 2b is not performed at all. We can use the statistical hiding property of $\mathsf{Com}_{\overline{X}}$ to argue that the distribution of transcripts of the protocol without Step 2b is in turn statistically close to a distribution of transcripts where the verifier commits to a fixed value (r^*, b^*) and uses uniformly random r_0, b_0 to compute $y = X_{b_0 \oplus b_1}(r_0 \oplus r_1)$. However, this corresponds exactly to the original honest-verifier protocol of Sahai and Vadhan [16]. Therefore, the soundness error can be at most negligibly larger.

Statistical Zero-Knowledge. For any V^*, the simulator S_{SD} proceeds as described in Fig. 4.

Lemma 8. *For all PPT* V^*, $\overline{X} \in \mathsf{SD}_Y$, *and* $k \in \mathbb{N}$, *it holds that*

$$\Pr[\mathsf{S}_{\mathsf{SD}}^{\mathsf{V}^*}(\overline{X}, 1^k) = \mathtt{fail}] \leq 1/2.$$

Proof. Let V^* be some PPT verifier, let $\overline{X} \in \mathsf{SD}_Y$ be some input, and let k be the security parameter. Note that $\mathsf{S}_{\mathsf{SD}}^{\mathsf{V}^*}$ fails only when V^* provides an accepting proof of knowledge of the value committed in c while the extractor fails to extract this value. Therefore,

$$\Pr[\mathsf{S}_{\mathsf{SD}}^{\mathsf{V}^*}(\overline{X}, 1^k) = \mathtt{fail}]$$

$$\leq \Pr[\mathsf{V}_{\overline{X}}(c, r_1, b_1, y, \theta) = \mathtt{accept} \wedge \mathsf{E}_{\overline{X}}^{\mathsf{V}^*}(c, r_1, b_1, y, \theta) = \mathtt{fail}],$$

where (c, r_1, b_1, y, θ) is the partial transcript produced by $\mathsf{S}_{\mathsf{SD}}^{\mathsf{V}^*}(\overline{X}, 1^k)$ in Step 1 of the simulation. Since $\mathsf{S}_{\mathsf{SD}}^{\mathsf{V}^*}$ behaves in Step 1 exactly as the honest prover P_{SD}, we can switch to $(c, r_1, b_1, y, \theta) \leftarrow \langle \mathsf{P}_{\mathsf{SD}}, \mathsf{V}^* \rangle(\overline{X}, 1^k)$, and obtain the following series of inequalities.

$$\leq \Pr[\mathsf{V}_{\overline{X}}(c, r_1, b_1, y, \theta) = \mathtt{accept}] \cdot (1 - \Pr[\mathsf{E}_{\overline{X}}^{\mathsf{V}^*}(c, r_1, b_1, y, \theta) \neq \mathtt{fail}])$$

$$\leq \Pr[\mathsf{V}_{\overline{X}}(c, r_1, b_1, y, \theta) = \mathtt{accept}] \cdot (1 - \Pr[\mathsf{V}_{\overline{X}}(c, r_1, b_1, y, \theta) = \mathtt{accept}] + \mathsf{neg}(k))$$

$$< 1/2,$$

where $(c, r_1, b_1, y, \theta) \leftarrow \langle \mathsf{P}_{\mathsf{SD}}, \mathsf{V}^* \rangle(\overline{X}, 1^k)$. □

To complete the proof, we show that conditioned on not outputting `fail`, the output distribution of $\mathsf{S}_{\mathsf{SD}}^{\mathsf{V}^*}$ is statistically close to the view of V^*. Due to the statistical binding of $\mathsf{Com}_{\overline{X}}$, the extracted randomness is distributed statistically close to the randomness of V^*. Moreover, the simulated transcript in Step 1 is distributed identically to $\langle \mathsf{P}_{\mathsf{SD}}, \mathsf{V}^* \rangle$. Given this observation, it is sufficient to bound the probability that the last message of the simulated transcript differs from the last message of the real transcript (the real and the simulated transcript distributions are otherwise identical).

Lemma 9. *For all PPT* V^*, $\overline{X} \in \mathsf{SD}_Y$, *and* $k \in \mathbb{N}$, *it holds that*

$$\Pr[\widetilde{\mathsf{S}}_{\mathsf{SD}}^{\mathsf{V}^*}(\overline{X}, c, r_1, b_1, y, \theta) \neq b''] \leq \mathsf{neg}(k),$$

where $(c, r_1, b_1, y, \theta, b'') \leftarrow \langle \mathsf{P}_{\mathsf{SD}}, \mathsf{V}^* \rangle(\overline{X}, 1^k)$, *and* $\widetilde{\mathsf{S}}_{\mathsf{SD}}^{\mathsf{V}^*}(\overline{X}, c, r_1, b_1, y, \theta)$ *denotes simulator's message in Step 3 on input* $\overline{X}$ *and transcript prefix* (c, r_1, b_1, y, θ), *conditioned on not outputting* `fail`.

Proof. Let V^* be some PPT verifier, let $\overline{X} \in \mathsf{SD}_Y$ be some input, and let k be the security parameter. The claim follows from the fact that the transcripts may differ if either the statistical binding does not hold or the verifier samples a value from one of the distributions such that the probability of this value in the other distribution is higher (this event happens with 2^{-k} probability). That is,

$$\Pr[\widetilde{\mathsf{S}}_{\mathsf{SD}}^{\mathsf{V}^*}(\overline{X}, c, r_1, b_1, y, \theta) \neq b'']$$

$$\leq \Pr[c \text{ is not binding}] + \Pr[\exists r_0^*, d^* : \mathsf{Open}_{\overline{X}}(c, r_0^*, 1 - b'', d^*) = \mathtt{accept}]$$

$$\leq \mathsf{neg}(k),$$

where $(c, r_1, b_1, y, \theta, b'') \leftarrow \langle \mathsf{P}_{\mathsf{SD}}, \mathsf{V}^* \rangle(\overline{X}, 1^k)$. □

Lemma 9 completes the proof of Theorem 6. □

5 Efficient Transformation from Honest-Verifier SZK to SZK

The general transformation takes any honest-verifier statistical zero-knowledge protocol $\langle \mathsf{P}, \mathsf{V} \rangle$ for promise problem $\Pi = (\Pi_{\mathrm{Y}}, \Pi_{\mathrm{N}}) \in \mathsf{HVSZK}$, an instance

$x \in \Pi$, a constant-round instance-dependent commitment scheme Com_x that is statistically binding on Π_Y instances, and a constant-round instance-dependent statistical zero-knowledge proof of knowledge protocol $\langle \mathsf{P}_x, \mathsf{V}_x \rangle$ for NP (from Theorem 4), and constructs a statistical zero-knowledge proof for Π.

Theorem 10 (Theorem 1 restated). *For every promise problem $\Pi \in \mathsf{HVSZK}$, there exists a statistical zero-knowledge proof where the prover's complexity and the round complexity match the parameters of the best honest-verifier statistical zero-knowledge proof for Π.*

The transformation is given in Fig. 5. We establish the proof of Theorem 10 by arguing its correctness, soundness, and zero-knowledge property below.

Input: Given $x \in \Pi$ and security parameter 1^k. Let $\mathsf{Com}_x = (\mathsf{S}_x, \mathsf{R}_x, \mathsf{Open}_x)$ be a constant-round instance-dependent commitment scheme that is statistically binding on Π_Y, and let $\langle \mathsf{P}_x, \mathsf{V}_x \rangle$ be a constant-round instance-dependent statistical zero-knowledge proof of knowledge for NP with knowledge soundness on Π_Y. The protocol $\langle \mathsf{P}', \mathsf{V}' \rangle$ for proving $x \in \Pi_Y$ proceeds as follows:

1. **Coin-flipping phase:**
   (a) V' samples $r_\mathsf{V} \leftarrow \{0,1\}^{t_\mathsf{V}}$, where t_V is a bound on the running time of V.
   (b) V' and P' interact in Com_x, so that P' learns c, a commitment to r_V.
   (c) V' and P' run $\langle \mathsf{P}_x, \mathsf{V}_x \rangle$, where V' proves that it knows an opening for c. Denote the transcript this proof θ_c. P' aborts if θ_c is not accepting.
   (d) P' samples $r_\mathsf{P} \leftarrow \{0,1\}^{t_\mathsf{V}}$ and sends r_P to V' that sets $r = r_\mathsf{V} \oplus r_\mathsf{P}$.
2. **Honest-verifier protocol execution phase:** V' and P' engage in an execution of the honest-verifier protocol $\langle \mathsf{P}, \mathsf{V} \rangle$. For each round $1 \leq i \leq t$ of $\langle \mathsf{P}, \mathsf{V} \rangle$ they proceed as follows:
   (a) denote by $\tau_{i-1} = (\alpha_1, \beta_1, \ldots, \alpha_{i-1}, \beta_{i-1})$ the transcript of $\langle \mathsf{P}, \mathsf{V} \rangle$ up to round $i-1$ (included).
   (b) V' computes the i-th message $\alpha_i = \mathsf{V}_i(x, \tau_{i-1}; r)$ of V and sends α_i to P'.
   (c) Let $L_i = \{(c, r, \tau, \alpha) | \exists \tilde{r}, d : \mathsf{Open}_x(c, \tilde{r}, d) = \mathsf{accept} \wedge \alpha = \mathsf{V}_i(x, \tau; \tilde{r} \oplus r)\}$. V' proves to P' that $(c, r_\mathsf{P}, \tau_{i-1}, \alpha_i) \in L_i$ using $\langle \mathsf{P}_x, \mathsf{V}_x \rangle$. Denote the transcript of this proof θ_i. P' aborts if θ_i is not accepting.
   (d) P' computes the i-th message $\beta_i \leftarrow \mathsf{P}_i(x, \tau_{i-1}, \alpha_i)$ of P and sends β_i to V'.
3. If $\mathsf{V}_{t+1}(x, \tau_t; r) = \texttt{accept}$ then V' outputs $\texttt{accept}$, and otherwise $\texttt{reject}$.

Fig. 5. Compiled protocol $\langle \mathsf{P}', \mathsf{V}' \rangle$. A compiler from honest-verifier protocol $\langle \mathsf{P}, \mathsf{V} \rangle$ for promise problem Π to protocol $\langle \mathsf{P}', \mathsf{V}' \rangle$ that is zero-knowledge against general verifiers. For a t-round protocol $\langle \mathsf{P}, \mathsf{V} \rangle$ we denote by V_i the next-message function of V in round i computed on the input, the $(i-1)$-rounds transcript, and the random tape of V (where V_{t+1} refers to the output of V in the protocol). The next-message function is similarly defined for P.

Correctness. Correctness of the compiled protocol $\langle \mathsf{P}', \mathsf{V}' \rangle$ follows directly from correctness of the building blocks, i.e., the instance-dependent statistical zero-knowledge proof of knowledge for NP $\langle \mathsf{P}_x, \mathsf{V}_x \rangle$ and the honest-verifier statistical zero-knowledge proof $\langle \mathsf{P}, \mathsf{V} \rangle$.

Soundness. Soundness of the compiled protocol $\langle \mathsf{P}', \mathsf{V}' \rangle$ follows from the soundness of the basic honest-verifier protocol $\langle \mathsf{P}, \mathsf{V} \rangle$ combined with the instance-dependent zero-knowledge proofs for NP being statistical zero-knowledge against unbounded verifiers on Π_{N}. Moreover, the statistical hiding property of Com_x on Π_{N} allows V' to use random coins distributed almost identically as the randomness of V (the distribution of randomness might be influenced by a cheating prover only if the hiding property does not hold).

Proposition 11 (Soundness of $\langle \mathsf{P}', \mathsf{V}' \rangle$). *Let $\Pi = (\Pi_Y, \Pi_N) \in \mathsf{HVSZK}$, and let $\langle \mathsf{P}, \mathsf{V} \rangle$ be honest-verifier statistical zero-knowledge protocol for Π. For all $x \in \Pi_N$, $k \in \mathbb{N}$, and P^*, it holds that*

$$\Pr[\langle \mathsf{P}^*, V' \rangle(x, 1^k) = 1] = \eta_{\langle \mathsf{P},\mathsf{V} \rangle} + \mathsf{neg}(k),$$

where $\eta_{\langle \mathsf{P},\mathsf{V} \rangle}$ denotes the soundness error of $\langle \mathsf{P}, \mathsf{V} \rangle$.

Proof. The proof of soundness follows from a series of lemmas. First, we define protocol $\langle \mathsf{P}_r, \mathsf{V}_r \rangle$ to be the same as the compiled protocol $\langle \mathsf{P}', \mathsf{V}' \rangle$ but without the proofs of correctness provided by V'. We use $\langle \mathsf{P}_r, \mathsf{V}_r \rangle$ to argue that the-coin flipping phase alone increases the soundness error by at most a negligible amount over $\eta_{\langle \mathsf{P},\mathsf{V} \rangle}$.

Lemma 12. *For all $x \in \Pi_N$, $k \in \mathbb{N}$, and P_r^*, it holds that*

$$\Pr[\langle \mathsf{P}_r^*, \mathsf{V}_r \rangle(x, 1^k) = 1] \leq \eta_{\langle \mathsf{P},\mathsf{V} \rangle} + \mathsf{neg}(k).$$

Proof. We consider an intermediate protocol, denoted by $\langle \mathsf{P}_1, \mathsf{V}_1 \rangle$. The protocol $\langle \mathsf{P}_1, \mathsf{V}_1 \rangle$ is the same as $\langle \mathsf{P}_r, \mathsf{V}_r \rangle$ with the difference that V_1 commits to 0^{t_V} and uses a uniformly random string independent of r_P as its randomness.

First, we show that for all $x \in \Pi_{\mathrm{N}}$, $k \in \mathbb{N}$, and P_1^*, it holds that

$$\Pr[\langle \mathsf{P}_1^*, \mathsf{V}_1 \rangle(x, 1^k) = 1] \leq \eta_{\langle \mathsf{P},\mathsf{V} \rangle}.$$

This is shown by constructing a prover P^* that wins the security game for $\langle \mathsf{P}, \mathsf{V} \rangle$ with the same probability as P_1^*. The constructed P^* simulates for P_1^* the coin-flipping phase using a commitment to all-zero string, receives r_P and answers all messages from V with messages from P_1^*. It follows from construction of P_1^* that $\Pr[\langle \mathsf{P}_1^*, \mathsf{V}_1 \rangle(x, 1^k) = 1] = \Pr[\langle \mathsf{P}^*, \mathsf{V} \rangle(x, 1^k) = 1] \leq \eta_{\langle \mathsf{P},\mathsf{V} \rangle}$.

Next, we show that for all $x \in \Pi_{\mathrm{N}}$, $k \in \mathbb{N}$, and P_r^*, it holds that

$$\Pr[\langle \mathsf{P}_r^*, \mathsf{V}_r \rangle(x, 1^k) = 1] \leq \eta_{\langle \mathsf{P},\mathsf{V} \rangle} + \mathsf{neg}(k).$$

The bound follows from the statistical hiding property of Com_x on NO instances, i.e., on Π_{N}. Specifically, the transcripts of the coin-flipping phase in $\langle \mathsf{P}_r^*, \mathsf{V}_r \rangle$ and in $\langle \mathsf{P}_r^*, \mathsf{V}_1 \rangle$ are statistically indistinguishable. This completes the proof of Lemma 12. □

We now define a sequence of hybrid protocols that gradually move between the interaction in $\langle \mathsf{P}_r, \mathsf{V}_r \rangle$ (where the verifier does not provide any proof of correctness for its messages) and the interaction in $\langle \mathsf{P}', \mathsf{V}' \rangle$ (where every message of V' is followed by a proof of correctness). Let t be the number of rounds in $\langle \mathsf{P}, \mathsf{V} \rangle$, we define $t+2$ protocols as follows:

Protocol $\langle \mathsf{P}', \mathsf{V}'_0 \rangle$ is defined similarly to $\langle \mathsf{P}', \mathsf{V}' \rangle$, where V'_0 behaves as V', except that V'_0 provides simulated proofs using the simulator for $\langle \mathsf{P}_x, \mathsf{V}_x \rangle$.

Protocol $\langle \mathsf{P}', \mathsf{V}'_i \rangle$ is defined for $1 \leq i \leq t+1$. The protocol $\langle \mathsf{P}', \mathsf{V}'_i \rangle$ is the same as $\langle \mathsf{P}', \mathsf{V}'_{i-1} \rangle$, except that V'_i performs the i-th proof using the actual witness instead of the simulator.

Note that $\langle \mathsf{P}', \mathsf{V}'_{t+1} \rangle$ is equivalent to $\langle \mathsf{P}', \mathsf{V}' \rangle$. Moreover, the soundness error of $\langle \mathsf{P}', \mathsf{V}'_0 \rangle$ is equal to the soundness error of $\langle \mathsf{P}_r, \mathsf{V}_r \rangle$. This can be seen by converting any cheating prover P'^* for $\langle \mathsf{P}', \mathsf{V}'_0 \rangle$ to a cheating prover P^*_r for $\langle \mathsf{P}_r, \mathsf{V}_r \rangle$. Concretely, on input x, the constructed prover P^*_r internally runs P'^* and provides it with simulated proof after each message from V_r. It follows that $\Pr[\langle \mathsf{P}'^*, \mathsf{V}'_0 \rangle(x, 1^k) = 1] = \Pr[\langle \mathsf{P}^*_r, \mathsf{V}_r \rangle(x, 1^k) = 1] \leq \eta_{\langle \mathsf{P}, \mathsf{V} \rangle} + \mathsf{neg}(k)$.

Lemma 13. *For all $x \in \Pi_N$, for every $k \in \mathbb{N}$, any prover P'^*, and $1 \leq i \leq t+1$, it holds that*

$$\mathrm{SD}\left(\langle \mathsf{P}'^*, \mathsf{V}'_i \rangle(x, 1^k), \langle \mathsf{P}'^*, \mathsf{V}'_{i-1} \rangle(x, 1^k)\right) \leq \mathsf{neg}(k).$$

Proof. The only difference in two consecutive hybrid protocols $\langle \mathsf{P}'^*, \mathsf{V}'_{i-1} \rangle$ and $\langle \mathsf{P}'^*, \mathsf{V}'_i \rangle$ is the simulated vs. the real proof in the i-th round when executing $\langle \mathsf{P}', \mathsf{V}' \rangle$. Assume towards a contradiction that there exists $x \in \Pi_N$, a prover P'^*, and $1 \leq j \leq t+1$ such that for some polynomial p it holds that

$$\mathrm{SD}\left(\langle \mathsf{P}'^*, \mathsf{V}'_j \rangle(x, 1^k), \langle \mathsf{P}'^*, \mathsf{V}'_{j-1} \rangle(x, 1^k)\right) \geq p(k).$$

We show that there exists an unbounded verifier V^*_x, and a partial transcript (c, r, τ, α) up to round j such that $(c, r, \tau, \alpha) \in L_j$ and

$$\mathrm{SD}\left((\mathsf{P}_x, \mathsf{V}^*_x)(c, r, \tau, \alpha; 1^k), \mathsf{S}^{\mathsf{V}^*_x}(c, r, \tau, \alpha; 1^k)\right) \geq p(k).$$

We define V^*_x and the partial transcript as follows. To obtain the partial transcript, run P'^* and simulate V' honestly during the first $j-1$ rounds of $\langle \mathsf{P}', \mathsf{V}' \rangle$ and compute the j-th round message α. Let (c, r, τ, α) be the partial transcript so far. We define V^*_x to be identical to the behavior of P'^* in the proof of the j-th round. Note that we can complete the partial transcript to a full transcript of $\langle \mathsf{P}', \mathsf{V}' \rangle$ by continuing with the internal run of P'^* and providing it with simulated proofs for the remaining rounds $j+1, \ldots, t+1$, as if they were generated by the honest V'. Thus, if the proof provided at round j is simulated then the complete transcript is drawn from $\langle \mathsf{P}'^*, \mathsf{V}'_{j-1} \rangle(x, 1^k)$ and otherwise it is drawn from $\langle \mathsf{P}'^*, \mathsf{V}'_j \rangle(x, 1^k)$. Therefore, we obtain that

$$\mathrm{SD}\left((\mathsf{P}_x, \mathsf{V}^*_x)(c, r, \tau, \alpha; 1^k), \mathsf{S}^{\mathsf{V}^*_x}(c, r, \tau, \alpha; 1^k)\right)$$
$$\geq \mathrm{SD}\left(\langle \mathsf{P}'^*, \mathsf{V}'_j \rangle(x, 1^k), \langle \mathsf{P}'^*, \mathsf{V}'_{j-1} \rangle(x, 1^k)\right).$$

Hence,

$$\mathrm{SD}\left((\mathsf{P}_x, \mathsf{V}^*_x)(c, r, \tau, \alpha; 1^k), \mathsf{S}^{\mathsf{V}^*_x}(c, r, \tau, \alpha; 1^k)\right) \geq p(k),$$

contradicting the statistical zero-knowledge property (against unbounded verifiers) of $\langle \mathsf{P}_x, \mathsf{V}_x \rangle$. □

Given that we have polynomially many hybrids and they are all statistically close, Lemma 13 completes the proof of soundness. □

Statistical zero-knowledge. At a high level, the zero-knowledge property of the compiled protocol $\langle \mathsf{P}', \mathsf{V}' \rangle$ follows from the zero-knowledge property of the underlying honest-verifier protocol $\langle \mathsf{P}, \mathsf{V} \rangle$. That is, the proofs of correctness provided at each round by the verifier force the produced transcript to follow the same distribution as in the execution with an honest verifier, which ensures that the resulting protocol also achieves zero-knowledge. We formally show that the simulator given in Fig. 6 satisfies the statistical zero-knowledge requirement.

Input: Given $x \in \Pi_Y$ and security parameter 1^k. Let E be the extractor for $\langle \mathsf{P}_x, \mathsf{V}_x \rangle$ and let S^{V} be the honest-verifier simulator for $\langle \mathsf{P}, \mathsf{V} \rangle$. The simulator S with oracle access to V^*, denoted by $\mathsf{S}^{\mathsf{V}^*}$, proceeds as follows:

1. Sample $(\texttt{view}, r) \leftarrow \mathsf{S}^{\mathsf{V}}(x, 1^k)$, where $\texttt{view} = (\beta_1, \ldots \beta_t)$ and β_i is the i-th message of P in the simulated execution of $\langle \mathsf{P}, \mathsf{V} \rangle$, and r is the randomness of V.
2. Proceed with $\mathsf{V}^*(x)$ in the *coin-flipping phase* of $\langle \mathsf{P}', \mathsf{V}' \rangle$ in order to learn a commitment c. Participate as honest V_x in the proof of knowledge for the committed value in c. Denote the transcript of this proof of knowledge θ_c. If θ_c is accepting then use the knowledge extractor $\mathsf{E}^{\mathsf{V}_x}$ to extract the committed value r_{V}. If the extractor fails output `fail`.
3. Send $r_{\mathsf{P}} = r \oplus r_{\mathsf{V}}$ to V^*, and proceed to the *honest-verifier protocol execution phase.* To simulate each round $1 \leq i \leq t$ of $\langle \mathsf{P}, \mathsf{V} \rangle$ in $\langle \mathsf{P}', \mathsf{V}' \rangle$ proceed as follows:
   (a) Denote by $\tau_{i-1} = (\alpha_1, \beta_1, \ldots, \alpha_{i-1}, \beta_{i-1})$ the transcript of $\langle \mathsf{P}, \mathsf{V} \rangle$ up to round $i-1$ (included).
   (b) Upon receiving a message α_i from V^*, engage in a proof that $(c, r_{\mathsf{P}}, \tau_{i-1}, \alpha_i) \in L_i$ as the honest verifier V_x. Denote the transcript of this proof θ_i.
   (c) If V_x on θ_i rejects then abort, otherwise send β_i to V^*.
4. Output the simulated transcript and the induced randomness r.

Fig. 6. Simulator $\mathsf{S}^{\mathsf{V}^*}$ for the compiled protocol $\langle \mathsf{P}', \mathsf{V}' \rangle$. The simulator $\mathsf{S}^{\mathsf{V}^*}$ samples a simulated transcript for the honest-verifier protocol which it uses to provide answers to V^* in the *honest-verifier protocol execution phase*, as well as to force the prover's randomness in the *coin-flipping phase*.

Proposition 14. *For all PPT* V^*, $x \in \Pi_Y$, *and* $k \in \mathbb{N}$, *there exists a negligible function* $\mathsf{neg}(\cdot)$ *such that*

$$\mathrm{SD}\left(\widetilde{\mathsf{S}}^{\mathsf{V}^*}(x, 1^k), (\mathsf{P}', \mathsf{V}^*)(x, 1^k)\right) \leq \mathsf{neg}(k) ,$$

where $\widetilde{\mathsf{S}}^{\mathsf{V}^*}$ *is the output distribution of* $\mathsf{S}^{\mathsf{V}^*}$ *conditioned on not outputting* `fail`.

We prove Proposition 14 via a series of lemmas about the capability of any malicious verifier to deviate from the honest behavior, both in the real execution and in the simulated execution. We start by showing that in Step 2 of $\langle \mathsf{P}', \mathsf{V}' \rangle$ any verifier must produce a transcript distribution that is statistically close to the transcript distribution of the honest verifier.

Lemma 15. *For all PPT* V^*, $x \in \Pi_Y$, *and* $k \in \mathbb{N}$, *there exists a negligible function* $\mathsf{neg}(\cdot)$ *such that*

$$\Pr[\langle \mathsf{P}, \mathsf{V}(r) \rangle(x) \neq \tau_t \wedge \mathtt{transcript} \neq \bot] \leq \mathsf{neg}(k),$$

where $(\mathtt{transcript}, r) \leftarrow (\mathsf{P}', \mathsf{V}^*)(x, 1^k)$, *and* $\mathtt{transcript} \neq \bot$ *denotes that all the intermediate proofs of correctness in the transcript are accepting,* τ_i *is the projection of* $\mathtt{transcript}$ *on the messages in* $\langle \mathsf{P}, \mathsf{V} \rangle$ *up to round* i *(including), and* $\langle \mathsf{P}, \mathsf{V}(r) \rangle(x)$ *denotes the transcript produced in the honest execution of* $\langle \mathsf{P}, \mathsf{V} \rangle$ *on input* x *with verifier's randomness* r.

Proof. For $(\mathtt{transcript}, r) \leftarrow (\mathsf{P}', \mathsf{V}^*)(x, 1^k)$, we denote by $\langle \mathsf{P}, \mathsf{V}(r) \rangle(x)_i$ the message of V at round i. We denote by α_i the message of V^* and by θ_i the transcript of the proof at round i in $\mathtt{transcript}$.

$$\begin{aligned}
&\Pr[\langle \mathsf{P}, \mathsf{V}(r) \rangle(x) \neq \tau_t \wedge \mathtt{transcript} \neq \bot] \\
&\quad \leq \Pr[\exists i \in [t] : \alpha_i \neq \langle \mathsf{P}, \mathsf{V}(r) \rangle(x)_i \wedge \theta_i \text{ is accepting}] \\
&\quad \leq \sum_{i \in [t]} \Pr[\alpha_i \neq \langle \mathsf{P}, \mathsf{V}(r) \rangle(x)_i \wedge \theta_i \text{ is accepting}] \\
&\quad \leq \mathsf{neg}(k),
\end{aligned}$$

where $(\mathtt{transcript}, r) \leftarrow (\mathsf{P}', \mathsf{V}^*)(x, 1^k)$, and the last inequality follows from the soundness of $\langle \mathsf{P}_x, \mathsf{V}_x \rangle$ using the union bound. □

Lemma 16. *For all PPT* V^*, $x \in \Pi_Y$, *and* $k \in \mathbb{N}$, *there exists a negligible function* $\mathsf{neg}(\cdot)$ *such that*

$$\Pr\left[\mathsf{V}(x, \mathtt{view}; r) \neq \tau_t \wedge \mathtt{transcript} \neq \bot\right] \leq \mathsf{neg}(k),$$

where $(\mathtt{view}, r) \leftarrow \mathsf{S}^{\mathsf{V}}(x, 1^k)$, *and* $\mathtt{transcript}$ *is a simulated transcript produced by* $\widetilde{\mathsf{S}}^{\mathsf{V}^*}(x, 1^k)$ *using* $(\mathtt{view}, r)$ *as described in Fig. 6. We use* $\mathtt{transcript} \neq \bot$ *to denote that all the intermediate proofs of correctness in the transcript are accepting,* τ_i *is the projection of* $\mathtt{transcript}$ *on the* $\langle \mathsf{P}, \mathsf{V} \rangle$ *messages up to round* i *(included), and* $\mathsf{V}(x, \mathtt{view}; r)$ *denotes the transcript produced by* V *on input* x *with randomness* r *and receiving messages in* $\mathtt{view}$.

Proof. We denote by $\mathsf{V}(x, \mathtt{view}; r)_i$ the message of V at round i in $\langle \mathsf{P}, \mathsf{V} \rangle$, and by α_i and θ_i the message and proof of V^* in $\mathtt{transcript}$ at round i. We denote by c the commitment of V^* to r_V in $\mathtt{transcript}$.

$$
\begin{aligned}
&\Pr\left[\mathsf{V}(x, \mathtt{view}; r) \neq \tau_t \wedge \mathtt{transcript} \neq \bot\right] \\
&\quad \leq \Pr\left[\mathsf{V}(x, \mathtt{view}; r) \neq \tau_t \wedge \mathtt{transcript} \neq \bot \wedge \mathsf{E}^{\mathsf{V}^*} \neq \mathtt{fail}\right] \\
&\quad \leq \Pr\left[c \text{ is not binding }\right] + \Pr\left[\begin{smallmatrix} \exists i \in [t]: \alpha_i \neq \mathsf{V}(x, \mathtt{view}; r)_i \wedge \theta_i \text{ is accepting } \wedge \\ \mathsf{E}^{\mathsf{V}^*} \neq \mathtt{fail} \wedge \exists !\, r^*, d^*: \mathsf{Open}_x(c, r^*, d^*) = \mathsf{accept} \end{smallmatrix}\right] \\
&\quad \leq \mathsf{neg}(k) + \sum_{i \in [t]} \Pr\left[\begin{smallmatrix} \alpha_i \neq \mathsf{V}(x, \mathtt{view}; r)_i \wedge \theta_i \text{ is accepting } \wedge \\ \mathsf{E}^{\mathsf{V}^*} \neq \mathtt{fail} \wedge \exists !\, r^*, d^*: \mathsf{Open}_x(c, r^*, d^*) = \mathsf{accept} \end{smallmatrix}\right] \\
&\quad \leq \mathsf{neg}(k),
\end{aligned}
$$

where $(\mathtt{view}, r) \leftarrow \mathsf{S}^{\mathsf{V}}(x, 1^k)$, and $\mathtt{transcript}$ is a simulated transcript produced by $\widetilde{\mathsf{S}}^{\mathsf{V}^*}(x, 1^k)$ using $(\mathtt{view}, r)$ as described in Fig. 6. □

Proof (Proposition 14). For any PPT verifier V^*, conditioned on the simulator not outputting $\mathtt{fail}$, it follows from the statistical binding of Com_x together with the honest-verifier statistical zero-knowledge property provided by S^{V} that the distribution of the simulated transcript in the *coin-flipping phase* produced by $\widetilde{\mathsf{S}}^{\mathsf{V}^*}$ is statistically close to the transcript distribution of the coin-flipping phase in $\langle \mathsf{P}', \mathsf{V}^* \rangle$. In particular, the produced randomness for V^* in $\widetilde{\mathsf{S}}^{\mathsf{V}^*}$ is statistically close to uniform. From the following facts we obtain the desired:

1. From Lemma 15 it follows that only a $\mathsf{neg}(k)$ fraction of $\langle \mathsf{P}', \mathsf{V}^* \rangle$ transcripts disagree with $\langle \mathsf{P}, \mathsf{V} \rangle$ and the randomness distribution of $\langle \mathsf{P}', \mathsf{V}^* \rangle$ is uniform as in $\langle \mathsf{P}, \mathsf{V} \rangle$.
2. From Lemma 16 it follows that only a $\mathsf{neg}(k)$ fraction of transcripts produced by $\widetilde{\mathsf{S}}^{\mathsf{V}^*}$ disagree with S^{V} and the randomness distribution of $\widetilde{\mathsf{S}}^{\mathsf{V}^*}$ is statistically close to uniform, as in S^{V}.
3. The behavior of $\widetilde{\mathsf{S}}^{\mathsf{V}^*}$ in all the $\langle \mathsf{P}_x, \mathsf{V}_x \rangle$ proofs is identical to the behavior of P'.

Combining the above we obtain that for all PPT V^*, $x \in \Pi_{\mathrm{Y}}$, and $k \in \mathbb{N}$, it holds that the full transcript distribution of $\widetilde{\mathsf{S}}^{\mathsf{V}^*}(x, 1^k)$ is statistically close to the transcript distribution of $\langle \mathsf{P}', \mathsf{V}^* \rangle (x, 1^k)$. □

We complete the proof of statistical zero-knowledge by bounding the probability of $\mathsf{S}^{\mathsf{V}^*}$ outputting $\mathtt{fail}$.

Proposition 17. *For all PPT* V^**,* $x \in \Pi_Y$*, and* $k \in \mathbb{N}$*, it holds that*

$$\Pr[\mathsf{S}^{\mathsf{V}^*}(x, 1^k) = \mathtt{fail}] \leq 1/2.$$

Proof. Let V^* be any PPT verifier and let $x \in \Pi_Y$ be some input. Note that $\mathsf{S}^{\mathsf{V}^*}$ fails only when V^* provides an accepting proof of knowledge θ_c of the value committed in c while the extractor fails to extract this value. That is,

$$\Pr[\mathsf{S}^{\mathsf{V}^*}(x, 1^k) = \mathtt{fail}] \leq \Pr[\mathsf{V}_x(c, \theta_c) = \mathtt{accept} \wedge \mathsf{E}^{\mathsf{V}^*}(x, c, \theta_c) = \mathtt{fail}],$$

where $(c, \theta_c) \leftarrow \mathsf{S}^{\mathsf{V}^*}(x, 1^k)$. Since $\mathsf{S}^{\mathsf{V}^*}$ behaves exactly as P' during the commitment c and the proof θ_c in Step 2 of the simulation, we can switch to $(c, \theta_c) \leftarrow \langle \mathsf{P}', \mathsf{V}^* \rangle (x, 1^k)$ and obtain the following series of inequalities:

$$\begin{aligned} &\leq \Pr[\mathsf{V}_x(c, \theta_c) = \mathtt{accept}] \cdot (1 - \Pr[\mathsf{E}^{\mathsf{V}^*}(x, c, \theta_c) \neq \mathtt{fail}]) \\ &\leq \Pr[\mathsf{V}_x(c, \theta_c) = \mathtt{accept}] \cdot (1 - \Pr[\mathsf{V}_x(c, \theta_c) = \mathtt{accept}] + \mathsf{neg}(k)) \\ &< 1/2, \end{aligned}$$

where $(c, \theta_c) \leftarrow \langle \mathsf{P}', \mathsf{V}^* \rangle (x, 1^k)$. □

Acknowledgements. We wish to thank Salil Vadhan and the anonymous EUROCRYPT 2018 referees for their helpful advice.

References

1. Aiello, W., Håstad, J.: Statistical zero-knowledge languages can be recognized in two rounds. J. Comput. Syst. Sci. **42**(3), 327–345 (1991)
2. Bellare, M., Micali, S., Ostrovsky, R.: Perfect zero-knowledge in constant rounds. In: Proceedings of the 22nd Annual ACM Symposium on Theory of Computing, 13–17 May 1990, Baltimore, Maryland, USA, pp. 482–493 (1990)
3. Bellare, M., Micali, S., Ostrovsky, R.: The (true) complexity of statistical zero knowledge. In: Proceedings of the 22nd Annual ACM Symposium on Theory of Computing, 13–17 May 1990, Baltimore, Maryland, USA, pp. 494–502 (1990)
4. Damgård, I.B.: Interactive hashing can simplify zero-knowledge protocol design without computational assumptions. In: Stinson, D.R. (ed.) CRYPTO 1993. LNCS, vol. 773, pp. 100–109. Springer, Heidelberg (1994). https://doi.org/10.1007/3-540-48329-2_9
5. Damgård, I., Goldreich, O., Wigderson, A.: Hashing functions can simplify zero-knowledge protocol design(too). Technical report RS94-39, BRICS, November 1994
6. Goldreich, O., Micali, S., Wigderson, A.: How to play any mental game or a completeness theorem for protocols with honest majority. In: Proceedings of the 19th Annual ACM Symposium on Theory of Computing, New York, USA, pp. 218–229 (1987)
7. Goldreich, O., Sahai, A., Vadhan, S.P.: Honest-verifier statistical zero-knowledge equals general statistical zero-knowledge. In: Proceedings of the Thirtieth Annual ACM Symposium on the Theory of Computing, 23–26 May 1998, Dallas, Texas, USA, pp. 399–408 (1998)
8. Goldreich, O., Vadhan, S.P.: Comparing entropies in statistical zero knowledge with applications to the structure of SZK. In: Proceedings of the 14th Annual IEEE Conference on Computational Complexity, 4–6 May 1999, Atlanta, Georgia, USA, p. 54 (1999)

9. Goldwasser, S., Micali, S., Rackoff, C.: The knowledge complexity of interactive proof systems. SIAM J. Comput. **18**(1), 186–208 (1989)
10. Itoh, T., Ohta, Y., Shizuya, H.: A language-dependent cryptographic primitive. J. Cryptol. **10**(1), 37–50 (1997)
11. Lindell, Y.: A note on constant-round zero-knowledge proofs of knowledge. J. Cryptol. **26**(4), 638–654 (2013)
12. Micciancio, D., Vadhan, S.P.: Statistical zero-knowledge proofs with efficient provers: lattice problems and more. In: Boneh, D. (ed.) CRYPTO 2003. LNCS, vol. 2729, pp. 282–298. Springer, Heidelberg (2003). https://doi.org/10.1007/978-3-540-45146-4_17
13. Okamoto, T.: On relationships between statistical zero-knowledge proofs. In: Proceedings of the Twenty-Eighth Annual ACM Symposium on the Theory of Computing, 22–24 May 1996, Philadelphia, Pennsylvania, USA, pp. 649–658 (1996)
14. Ong, S.J., Vadhan, S.: An equivalence between zero knowledge and commitments. In: Canetti, R. (ed.) TCC 2008. LNCS, vol. 4948, pp. 482–500. Springer, Heidelberg (2008). https://doi.org/10.1007/978-3-540-78524-8_27
15. Ostrovsky, R., Venkatesan, R., Yung, M.: Interactive hashing simplifies zero-knowledge protocol design. In: Helleseth, T. (ed.) EUROCRYPT 1993. LNCS, vol. 765, pp. 267–273. Springer, Heidelberg (1994). https://doi.org/10.1007/3-540-48285-7_23
16. Sahai, A., Vadhan, S.P.: A complete problem for statistical zero knowledge. J. ACM **50**(2), 196–249 (2003)
17. Vadhan, S.P.: On transformation of interactive proofs that preserve the prover's complexity. In: Proceedings of the Thirty-Second Annual ACM Symposium on Theory of Computing, 21–23 May 2000, Portland, OR, USA, pp. 200–207 (2000)

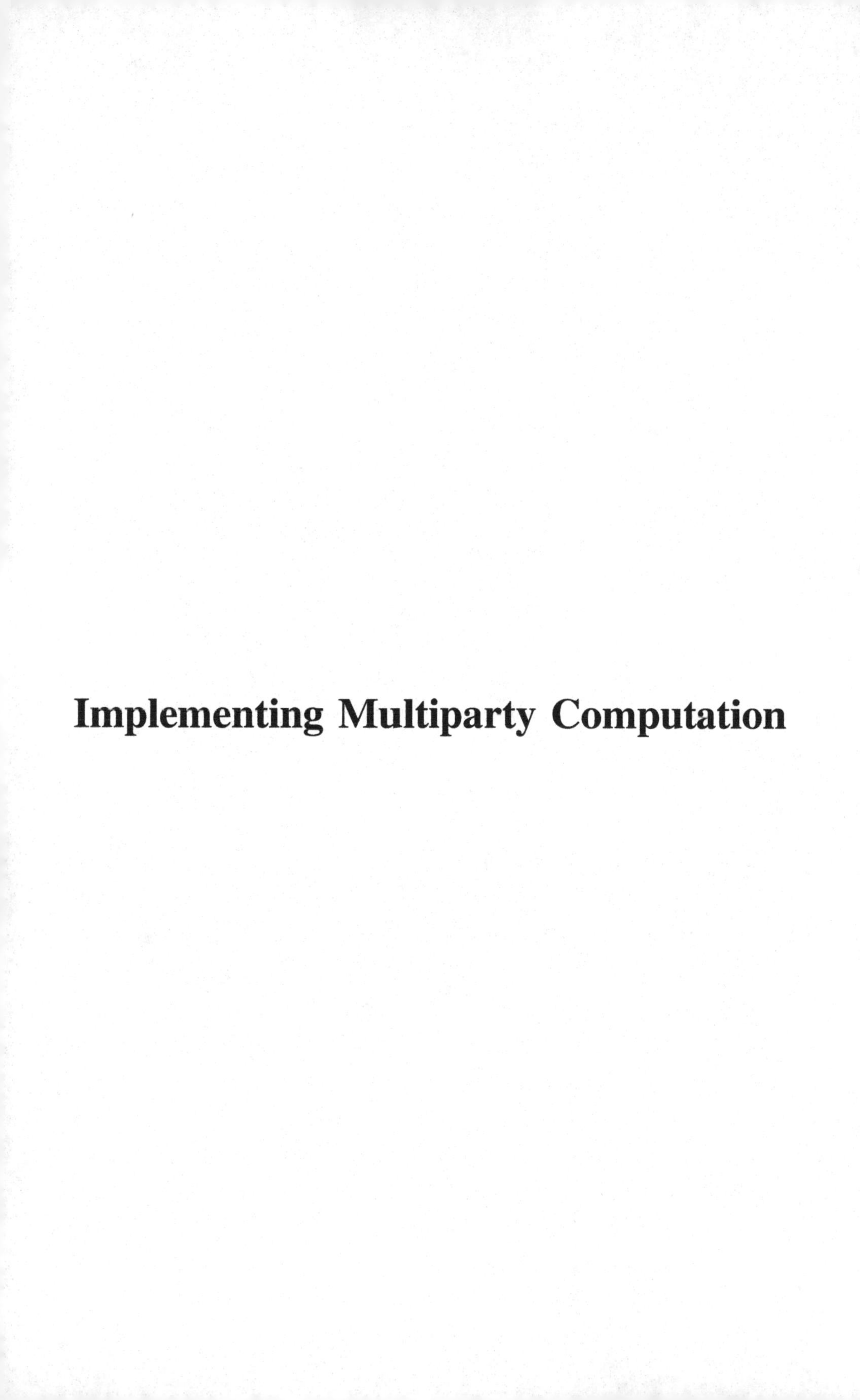

Implementing Multiparty Computation

Efficient Maliciously Secure Multiparty Computation for RAM

Marcel Keller[1](✉) and Avishay Yanai[2]

[1] University of Bristol, Bristol, UK
M.Keller@bristol.ac.uk
[2] Bar-Ilan University, Ramat Gan, Israel
Ay.Yanay@gmail.com

Abstract. A crucial issue, that mostly affects the performance of actively secure computation of RAM programs, is the task of reading/writing from/to memory in a private and authenticated manner. Previous works in the active security and multiparty settings are based purely on the SPDZ (reactive) protocol, hence, memory accesses are treated just like any input to the computation. However, a garbled-circuit-based construction (such as BMR), which benefits from a lower round complexity, must resolve the issue of converting memory data bits to their corresponding wire keys and vice versa.

In this work we propose three techniques to construct a secure memory access, each appropriates to a different level of abstraction of the underlying garbling functionality. We provide a comparison between the techniques by several metrics. To the best of our knowledge, we are the *first* to construct, prove and implement a concretely efficient garbled-circuit-based actively secure RAM computation with dishonest majority.

Our construction is based on our third (most efficient) technique, cleverly utilizing the underlying SPDZ authenticated shares (Damgård et al., Crypto 2012), yields lean circuits and a constant number of communication rounds per physical memory access. Specifically, it requires no additional circuitry on top of the ORAM's, incurs only two rounds of broadcasts between every two memory accesses and has a multiplicative overhead of 2 on top of the ORAM's storage size.

Our protocol outperforms the state of the art in this settings when deployed over WAN. Even when simulating a very conservative RTT of 100 ms our protocol is at least one order of magnitude faster than the current state of the art protocol of Keller and Scholl (Asiacrypt 2015).

This research was supported by a grant from the Ministry of Science, Technology and Space, Israel, and the UK Research Initiative in Cyber Security. This work has been supported in part by EPSRC via grants EP/M012824 and EP/N021940/1, by the European Research Council under the ERC consolidators grant agreement n. 615172 (HIPS) and by the BIU Center for Research in Applied Cryptography and Cyber Security in conjunction with the Israel National Cyber Bureau in the Prime Minister's Office.

J. B. Nielsen and V. Rijmen (Eds.): EUROCRYPT 2018, LNCS 10822, pp. 91–124, 2018.
https://doi.org/10.1007/978-3-319-78372-7_4

1 Introduction

1.1 Background

Actively secure multiparty computation (in the dishonest majority setting) allows n parties to compute an arbitrary function over their private inputs while preserving the privacy of the parties and the correctness of the computation even in the presence of a malicious adversary, who might corrupt an arbitrary strict subset of the parties.

The field of secure two-party (2PC) and multiparty (MPC) computation has a rich literature, starting with Yao [40] and Goldreich-Micali-Wigderson [16] and attracted much interest during the past decade due to advances in efficiency and fast implementations [8,23,26,37,38]. Nevertheless, almost all previous works require the parties to first "unroll" the function into an arithmetic or Boolean circuit representation and then securely evaluate the circuit gate by gate. This is in contrast to modern design of algorithms of practical interest (e.g., binary search, Dijkstra's shortest-path algorithm, Gale-Shapley stable matching, etc.) that are typically represented as Random Access Machine (RAM) programs that contain branches, recursions, loops etc., which utilize the $O(1)$ access to memory, rather than circuits. In the following we provide the necessary overview on the RAM model of computation and how it is securely realized.

RAM Model of Computation. RAM is classically modeled as a protocol that is carried out between two entities: CPU and MEMORY, which are essentially a couple of polynomial time Turing machines, such that their storage capacity is unbalanced, specifically, the CPU usually stores a small amount of data, corresponding to the state of the program, which is logarithmic in the amount of storage in MEMORY required by the program. We denote the CPU's storage by d and the MEMORY's storage by D such that $|D| = N$ and $|d| = O(\log N)$. We denote a memory block at address i by $D[i]$. During the program execution CPU typically chooses to perform one instruction I out of a final instructions set IS. A program Π and an input $\boldsymbol{x}$ are first loaded into the storage of MEMORY and then the CPU is being triggered to start working. From that point, CPU and MEMORY are engaged in a protocol with T rounds where T is the running time of Π. In the t-th round:

1. CPU computes the *CPU-step function*:

$$\mathrm{C}_{\mathrm{CPU}}(\mathsf{state}_t, b_t^{\mathsf{read}}) = (\mathsf{state}_{t+1}, i_t^{\mathsf{read}}, i_t^{\mathsf{write}}, b_t^{\mathsf{write}}) \tag{1}$$

   by executing instruction $I_t \in IS$. The input state_t is the current state of the program (registers etc.), b_t^{read} is the block that was most recently loaded from MEMORY. The outputs of the CPU-step are: The new program's state state_{t+1}, the address i_t^{read} in D to read from and the address i_t^{write} in D to write the block b_t^{write} to.
2. CPU sends $(i_t^{\mathsf{read}}, i_t^{\mathsf{write}}, b_t^{\mathsf{write}})$ to MEMORY. We define $\mathsf{access}_t \triangleq (i_t^{\mathsf{read}}, i_t^{\mathsf{write}})$.
3. MEMORY sends data block $D[i_t^{\mathsf{read}}]$ to CPU and assigns $D[i_t^{\mathsf{write}}] = b_t^{\mathsf{write}}$.

In every such a round, CPU is said to make a single request, or *logical access*, to MEMORY. The output of the protocol, denoted $\boldsymbol{y} = \Pi(D, \boldsymbol{x})$, is the result of the computation of the program Π on input $\boldsymbol{x}$ and memory D, such that CPU sets $\boldsymbol{y}$ as the last state of the program, state_{T+1}. The sequence of accesses $\{\mathsf{access}_1, \ldots, \mathsf{access}_T\}$ is called the *access pattern of Π on input $\boldsymbol{x}$ and memory D (of size N)* and denoted $\mathsf{AP}(\Pi, D, \boldsymbol{x})$. Similarly, the sequence $\{I_1, \ldots, I_T\}$ is called the *instruction pattern* and denoted $\mathsf{IP}(\Pi, D, \boldsymbol{x})$.

The general methodology of designing secure multiparty computation directly to RAM programs is by having the parties take both the role of CPU and MEMORY and sequentially evaluate sufficiently many copies of the $\mathrm{C}_{\mathrm{CPU}}$ function. Upon completing the evaluation of one function, the parties access D according to $\mathrm{C}_{\mathrm{CPU}}$'s output $(i_t^{\mathsf{read}}, i_t^{\mathsf{write}}, b_t^{\mathsf{write}})$ and obtain the input b_t^{read} to the next function.

Obviously, a secure protocol must not reveal D to the parties, otherwise it would be possible to learn information about the parties' inputs. Trivially avoiding this is by embedding two sub-procedures inside $\mathrm{C}_{\mathrm{CPU}}$, one to encrypt (and authenticate) b^{write} before it is output and one to decrypt (and verify authentication of) b^{read} before it is used by $\mathrm{C}_{\mathrm{CPU}}$. This enhanced function is denoted $\mathrm{C}_{\mathrm{CPU}^+}$. Let $\mathrm{C}^1_{\mathrm{CPU}^+}, \ldots, \mathrm{C}^T_{\mathrm{CPU}^+}$ be garbled versions of $\mathrm{C}_{\mathrm{CPU}^+}$. The parties feed their inputs $\boldsymbol{x} = x^1, \ldots, x^n$ into $\mathrm{C}^1_{\mathrm{CPU}}$, taking the place of the wires associated with state_1 and sequentially evaluate the garbled circuits to obtain $\boldsymbol{y} = \mathsf{state}_{T+1}$. This way, even an adversary who can tap (or even tamper) the memory accesses is unable to manipulate the program so it operates over forged data (since the data blocks are authenticated), yet, it might reveal information about the parties' inputs or program's state from the access pattern.

ORAM in Secure RAM Computation. Previous works on 2PC and MPC for RAM programs [1,11,14,18–20,22,24,28–31,39] use Oblivious RAM (ORAM) as an important building block. Informally speaking, an ORAM scheme is a technique to transform a program Π with runtime T and initial storage D to a new program Π' with runtime T' and initial storage D' such that the access pattern $\mathsf{AP}(\Pi', D', \boldsymbol{x})$ appears independent of both Π and $\boldsymbol{x}$, yet, both programs compute the same function, i.e. $\Pi(D, \boldsymbol{x}) = \Pi'(D', \boldsymbol{x})$ for all $\boldsymbol{x}$. All ORAM schemes that we know of work by first initializing the storage D and then *online simulate* each memory access individually (i.e. we don't know of a scheme that simulates a bunch of accesses altogether). It was shown feasible, since the work of Goldreich and Ostrovsky [17], that the simulation of a single memory access of Π (which denoted by *logical access* above) incurs $\mathsf{poly}(\log N)$ memory accesses in Π', denoted *physical accesses*, which leads to the same run time overhead, that is $T' = T \cdot \mathsf{poly}(\log N)$. In addition, we can obtain the same overhead for memory consumption of Π', that is $N' = N \cdot \mathsf{poly}(\log N)$.

The general methodology for secure computation of RAM programs using an ORAM scheme is by having the parties collaboratively compute an ORAM transformation of Π and D (via any MPC protocol) to obtain Π' and D'. This is a one-time step that incurs a computational and communication complexity that is proportional to N'. Then, they engage in a protocol of T' steps to compute $\Pi'(D', x)$ and, as before, obtain the output as $\mathsf{state}_{T'+1}$.

This way, to securely compute a program, it is no longer required to unroll it to a circuit, rather, it is enough to unroll only the ORAM scheme algorithms and the CPU-step function. Consequently, this approach may lead to concentrated research efforts to optimize a specific set of ORAM scheme algorithms instead of looking for optimizations to the circuit version of each individual program.

Oblivious vs. Non-oblivious Computation. We distinguish between oblivious and non-oblivious computation in the following sense: In *oblivious computation* the parties learn nothing about the computation (except its output and runtime). Specifically, the parties learn nothing about either the program Π, CPU's state or the input $\boldsymbol{x}$. This means that an oblivious computation is applicable for private function evaluation (PFE) in which the function itself is kept secret. On the other hand, *non-oblivious computation* allows the parties to learn which instruction is being computed in which time step, in particular, it rules out algorithms that branch on secret values (otherwise, information about the secret values might be leaked). As noted in [28], in order to hide the instruction being computed - in every time step every possible instruction must be executed. The only implementation of an oblivious computation with active security that we know of is by Keller [22]. It has a performance of 41 Hz (physical memory accesses per second) in the online phase with 1024×64 bit memory and 2 Hz for $2^{20} \times 64$ bit memory[1] for 2 parties running over a local network. On the other hand, secure non-oblivious computation (denoted "instruction-trace oblivious" in [28]) is expected to yield a much better throughput, since the parties can avoid securely evaluating the universal CPU-step circuit, but can instead simply evaluate a much smaller circuit corresponding to the current instruction.

Notwithstanding the theoretical results in this paper hold for oblivious computation, the implementation results we report hold only for the non-oblivious settings. This relaxation is justified by the fact that non-oblivious computation is applicable for plenty of useful algorithms such as graph and search algorithms.

Achieving Efficient Protocols. To achieve an efficient, actively secure RAM computation the following crucial issues are to be addressed:

1. *Round complexity.* As explained above, securely evaluating a program requires $T' = T \cdot \mathsf{poly}(\log N)$ rounds of interaction between CPU and MEMORY, corresponding to the T' physical memory accesses. Also note that the access pattern of a program is determined by the input that it is given. Now, consider the CPU-step at time t from Eq. (1), the parties need to read $D[i_t^{\mathsf{read}}]$ and map it to the input wires associated with b_t^{read}. However, they do not know from ahead (i.e. when garbling) which address i_t^{read} would be accessed in which timestep and thus cannot map the input wire labels of $\mathrm{C}_{\mathrm{CPU}}^t$ to the

[1] The decrease in throughput reflects the runtime overhead implied by the ORAM, as mentioned above, this overhead depends on the memory size N.

right memory location. Therefore, achieving a protocol with round complexity independent in T is much more challenging, in fact, there is a line of works that proposes constant-round secure RAM computation [7,11–14,19,30], however, it is highly impractical. A more reasonable path, which we follow in this paper, is to construct a scheme with a constant number of rounds *per any number of parallel physical memory accesses.* Although there exist *passively* secure implementations [28,29] that are constant-round per physical memory access, the *actively* secure implementations that we know of [22,24] have a round complexity linear in the depth of the CPU-step circuit (which depends on the ORAM implementation).

2. *Private and authenticated memory.* A natural approach suitable for securely handling memory is to choose an ORAM that encrypts its memory contents. In this approach, the parties must evaluate CPU-step circuits that include encryption/decryption and authentication/verification sub-circuits. This is undesirable since the resulting construction is non black-box in its underlying encryption/authentication primitives and, more practically, embedding encryption and authentication sub-circuits in every CPU-step circuit adds a large overhead in terms of computation, communication and space complexities for garbling and evaluating. This would be especially objectionable when the original RAM's behavior is non-cryptographic. The circuitry overhead when using the sub-circuit approach is demonstrated for several memory sizes in Table 1 where the circuit size is for a typical instruction that requires memory access[2]. *Circuit size* refers to the number of AND gates in the circuit performing a logical access, *read* and *write* are the number of bits being accessed and *encryption/authentication size* is the number of AND gates that would be necessary when incorporating the encryption and authentication procedures inside the CPU-step circuit. We measure the overhead using both our technique (described in Sect. 4.1) and a trivial solution using the AES block cipher (with circuit size of 6000 AND gates[3]), assuming blocks of $s = 40$ bits. We can see that even with our improvement (due to SPDZ representation of memory), securing memory accesses incurs an additional circuitry that is about 45 times larger than the ORAM circuit itself, therefore, we are highly motivated to find other techniques for transferring memory from storage to circuits.
3. *Memory consumption.* In actively secure BMR-based protocols memory used for storing the garbled circuit grows linearly with the number of gates and the number of parties. Let G be the number of AND gates, n the number of participants and κ the security parameter. In the online phase each party stores $4 \cdot G \cdot n \cdot \kappa$ bits that represent the garbled circuit and additional $2 \cdot G \cdot \kappa$ bits that represent its own keys (the latter are needed to verify authenticity of the keys revealed during the evaluation and deciding which garbled entry to use next). For example, the SHA1 circuit is composed of ~236K gates,

[2] The amount of memory being accessed to satisfy a CPU instruction depends on the instruction itself, for instance a SIMD instruction access more data than a SISD instruction.

[3] The state-of-the-art construction of an AES circuits incurs only 5200 AND gates.

among them ~90K are AND gates. The evaluation of SHA1 with $\kappa = 128$ by 3 parties incurs memory of size ~160 Mb and 0.5 Gb when evaluated by 10 parties. While this amount is manageable for a single execution of a circuit, it is much harder to be maintained when T garbled circuits are evaluated sequentially in an online phase, as needed in RAM computation. Thus, new techniques must be developed to address that issue.

Table 1. Proportion of additional circuitry, counted as the number of additional AND gates, for the purpose of memory encryption and authentication in a logical memory access.

Mem size	Circuit size	Read	Write	Enc/Auth. via technique, Sect. 4.1	Enc/Auth. via block cipher (AES)
2^{13}	94844	31058	29596	2426160 (2600%)	18192000 (19000%)
2^{17}	156990	92568	87634	7208080 (4500%)	54060000 (34400%)
2^{21}	269300	158508	147982	12259600 (4500%)	91944000 (34100%)
2^{25}	423014	249104	231098	19208080 (4500%)	144060000 (34000%)

1.2 Our Contribution

We construct and implement the first actively secure, garbled-circuit-based ORAM multiparty protocol. Specifically, we present the following contributions:

1. **Efficient Secure Memory Access.** We propose and compare three techniques to implement memory access in a secure computation for RAM programs. We briefly describe them in an increasing order of efficiency:
   (a) In the first technique, for each memory data item each party stores a SPDZ share of that data item. We stress that this technique has nothing to do with the SPDZ protocol, it only uses SPDZ shares representation to represent the memory content. In each access the data item is being re-shared using fresh randomness from the parties. Since SPDZ shares are also authenticated we achieve an authenticated memory as well. The re-sharing procedure is implemented as a sub-circuit, using only 2 field multiplications, which are embedded in every CPU-step circuit. For each s-bit block being accessed, the parties need to communicate $O(sn^2\kappa)$ bits (by all parties together) to reveal the appropriate keys for the input wires in the next CPU-step circuit, where κ and s are the computational and statistical security parameters respectively, s is also the size of a SPDZ share. This is because every CPU-step receives n shares, each of size s bits, and for every bit all parties need to broadcast their keys of size κ bits. The technique requires two rounds of broadcast per physical memory access, however, as explained above, embedding encryption and authentication sub-circuits has theoretical and practical disadvantages.

(b) The second technique is inspired by [1,32], in which the memory is implemented via wire soldering. That is, since every wire already carries a hidden and authentic value through the key that is revealed to the parties, the key itself could be stored in memory. This way, the parties do not need to transform wire keys to data items back and forth for every access, instead, they use wire soldering directly from the "writing circuit" to the "reading circuit". This "prunes away" the additional circuitry of the first technique, with the drawback of having each bit in the ORAM memory represented using a BMR key, i.e. $n\kappa$ bits (with n the number of parties and κ the security parameter). This technique, however, is superior in the other metrics as well, that is, it requires much less triples to be generated in the offline phase since it does not need the additional circuitry (which includes many AND gates) and has 2 communication rounds for each physical memory access, just like the first technique.
Naively generalizing the soldering of [1] to the *multiparty* settings requires each party to commit to its keys to all other parties using a xor-homomorphic commitment scheme. Instead, in this work we obviate the need of a commitment scheme and show how to use the readily available keys' shares. Moreover, we show how to do that black-box in the BMR garbling functionality, even when using the Free-XOR optimization [2], by which different garbled circuits are assigned with *different global differences*.

(c) The third technique offers a clever improvement to the soldering in that we solder only *one bit*, namely, the real value that passes through a wire, instead of the whole key that represents that value. As such, the soldering requires *no offline overhead at all*, that is, in contrast to the second technique, this technique does not invoke the multiplication command of the underlying MPC. We utilize the fact that the BMR-evaluation procedure reveals to the parties the external bit of each output wire (that is associated with a bit to be written to memory) and the fact that the permutation bits are already shared. This way the parties could obtain a share to a single bit which is the XOR (addition in the binary field) of the external and permutation bits.

Nevertheless the third technique is the most promising for it is the most efficient in all parameters (see Table 2 for a comparison), the first and second techniques are also beneficial since they work in a higher level of abstraction and assume less about the circuit-garbling functionality. In particular, the first technique can be used with any underlying circuit-based protocol for evaluating the CPU-step circuits. The second technique requires an underlying protocol that relies on the idea of two keys per wire, such as the BMR construction, however it assumes nothing about the way BMR is implemented (recall that BMR on its own uses another MPC protocol to garble the gates). On the contrary, the third technique assumes a specific implementation of BMR, which shares the wires' permutation bits among the parties. The SPDZ protocols family satisfies this last requirement and therefore we use it in our implementation.

2. **Reduced Round and Space Complexities.** As opposed to [22,24] that require communication rounds for every layer of C_{CPU}, and [29] that achieves only passive security, our protocol is constant round per physical memory access. As mentioned above, the parties can travel from one CPU-step to the next by simply performing SPDZ openings, which appears more efficient than using xor-homomorphic commitment to wire labels in a cut-and-choose based protocol such as [1] (for 2PC).
   We show that by representing memory as a "packed shares" the parties need to store *only 2 bits* per bit in the ORAM (that is, to operate an ORAM with N'-bit storage the previous parties need to store $2N'$ bits). To the best of our knowledge this is the best concrete overhead that has been achieved to date. In contrast, other BMR-based protocols, such as one instantiated using our second technique, requires each party to store $n\kappa$ bits per bit in the ORAM. We further devise a way to shrink the storage required by each party in the online phase. When using a garbling scheme that produces a garbled circuit of size independent in the number of parties (as recently proposed [3]) our optimization leads to a decrease in memory consumption of up to 2. We stress that this improvement is applicable to *all* BMR-based constructions. We present and prove security of it in Sect. 6.
3. **Implementation.** We have implemented the protocol using our most efficient memory access technique and obtained experimental access times results in both LAN and simulated WAN environments for two and three participants. In addition, we provide a comparison with the previous implementation of Keller and Scholl [6,24] that is based purely on SPDZ. Our experiments show that [24] performs better over LAN (up to a factor of two for two parties) while our work does so over WAN (by one order of magnitude for two parties), justifying our efforts to reduce communication rounds. This supports the analysis that garbled circuits are more suitable for a setting with high latency because computation on secret values (after obtaining the garbled circuit) can be entirely done locally. Note, however, that we still require communication for revealing memory addresses and transferring memory values to garbled circuit wires. This is not the case for the trivial (asymptotically more expensive) approach where the whole memory is scanned for every access. We also implemented the latter and found that our protocol breaks even at a memory size in the 1'000s for the LAN setting and in the 100'000s for the WAN setting.

We stress that even though [38] also achieves a constant-round multiparty protocol for circuit-based computation (i.e. not RAM programs) our third and most efficient technique is not directly applicable to their construction. In particular, our technique relies on the fact that all parties can identify the correctness of wire labels without communication. This is the case for BMR because every party learns both possibilities for κ bits of every wire label. This is only true for one of two parties in the above work. We therefore leave it as an open problem how to combine the two techniques and how a possible combination would compare to our work.

1.3 Related Work

Gordon et al. [18] (who followed the work of Ostrovsky and Shoup [33] that was tailored specifically for PIR) designed the first *general* two-party, semi-honest, secure computation protocol for RAM. Their work focuses on the client-server settings, where the client has a small input and the server has a large database, and require the client to maintain only a small storage (i.e. logarithmic in the size of the database). Their technique relies on the one-time-initialization of the ORAM, after which, the server stores an encrypted version of the memory, then the parties iteratively engage in a traditional, circuit-based, secure two-party computation for every ORAM instruction.

Garbled RAM, introduced by Lu and Ostrovsky [30], is an analogue object of garbled circuit with respect to RAM programs. Namely, a user can garble an arbitrary RAM program directly without converting it into a circuit first. A garbled RAM scheme can be used to garble the memory, the program and the input in a way that reveals only the evaluation outcome and nothing else. The main advantage of garbled RAM is that it leads to a constant-round two-party or multi-party protocols to both semi-honest and malicious settings. This is reflected in a series of works on variations of garbled RAM [11–14, 19], however all of these works focused on showing feasibility rather than efficiency and are impractical.

Afshar et al. [1] presented two actively secure protocols for the two-party settings: One that works in the offline-online model and one for streaming. The main idea in both of their schemes is encoding RAM memory via wire labels[4]. When the program reads from memory location ℓ, it is possible to reuse the appropriate output wire labels from the most recent circuit to write to location ℓ (which is not necessarily the previous circuit). Those protocols require the parties to coordinate before the evaluation of each CPU-step, either by soldering techniques that require XOR homomorphic commitments for aligning wire labels (based on [10, 32]) or by invocations of oblivious transfer to allow evaluation of next garbled circuits, in addition to a large amount of symmetric operations for garbling, encrypting and decrypting s copies of the circuit (since it uses the cut-and-choose technique). Overall, this would incur an additional overhead of $O(sn\kappa)$, since for each input wire, of each of the $O(s)$ garbled circuits, each party would need to commit and open its XOR homomorphic commitment, with computational security parameter κ. Moreover, the streaming version requires both the garbler and the evaluator to maintain $O(s)$ copies of the memory. That work was followed by [20, 31] to achieve a constant round protocol for ZKP of non-algebraic statements in the RAM model, but not for secure computation.

Keller and Scholl [24], showed how to implement two ORAM variants for the oblivious array and oblivious dictionary data structures, specifically, they compared their implementation for the binary Tree ORAM [34] and the Path ORAM

[4] Encoding the *state* as wire labels is simpler than encoding the *memory* since it only requires matching wire labels of output wires of one CPU-step to the input wires of the next. This can be done in the offline phase, without knowing the program or the input.

[35] using various optimizations for many parts of the ORAM algorithms. Their implementation of secure oblivious array and dictionary are purely based on the SPDZ protocol, hence, they have no use of the techniques we develop in this paper because the memory in their work is represented exactly the same as the secret state of the program is represented. Therefore, there is no requirement of conversion between those two entities (memory and state). Due to their use of a secret-sharing based MPC using the SPDZ authenticated shares representation, evaluation of multiplication gates are performed interactively such that the product results are immediately authenticated, thus, parties can use the memory as usual shared secrets and verify authenticity only once, when the evaluation is finished. The drawback in their approach is the high round complexity that is implied on top of the ORAM round complexity. In our protocol, multiplications are evaluated inside a circuit and the authentication of the result is not an integrated part of the multiplication itself (as in the SPDZ protocol).

Doerner and Shelat [9] recently published a *two-party passively* secure computation for RAM programs and reported that it outperforms previous works, even when implemented using the state-of-the-art ORAM schemes, up to large memory sizes such as 2^{32} elements of 4 bytes. Their Distributed ORAM scheme (AKA *Floram*) is derived from the Function Secret Sharing (FSS) for point functions by Boyle et al. [4,5], which resembles the trivial ORAM that read/write all memory addresses for every access in order to hide its access pattern, however, this is resolved since those $O(n)$ accesses are performed by a highly parallelizable local computation. The main advantage of Floram is that it has only $O(1)$ communication rounds for both initialization and memory access and does not require secure computation at all for the initialization. We remark that even though FSS is feasible in the multiparty setting, it does not offer the same optimizations as it does to the two party setting, thus, Floram is currently not suitable for the multiparty setting. In addition, it is not trivial to lift their scheme to have active security.

2 Preliminaries

Relying on the notation and description of the RAM model of computation presented in Sect. 1.1, we directly proceed to the definition of Oblivious RAM:

2.1 Oblivious RAM

A polynomial time algorithm C is an *Oblivious RAM (ORAM)* compiler with computational overhead $c(\cdot)$ and memory overhead $m(\cdot)$, if C, when given a security parameter κ and a deterministic RAM program Π with memory D of size N, outputs a program Π' with memory D' of size $N' = m(N) \cdot N$, such that for every input $x \in \{0,1\}^*$ the running time of $\Pi'(D', x)$ is bounded by $T' = T \cdot c(N)$ and there is a negligible function μ such that the following properties hold:

- **Correctness.** For every memory size $N \in \mathbb{N}$ and every input $x \in \{0,1\}^*$ with probability at least $1-\mu(\kappa)$, the output of the compiled program equals the output of the original program, i.e. $\Pi'(D',x) = \Pi(D,x)$.
- **Obliviousness.** For every two programs Π_1, Π_2, every D_1, D_2 of size N and every two inputs $x_1, x_2 \in \{0,1\}^*$, if the running times of $\Pi_1(D_1,x_1)$ and $\Pi_2(D_2,x_2)$ are T, then

$$\mathsf{AP}(C(\Pi_1,\kappa),D_1,x_1)) \stackrel{c}{\equiv} \mathsf{AP}(C(\Pi_2,\kappa),D_2,x_2))$$

where $\mathsf{AP}(\cdot)$ is the access pattern as defined in Sect. 1.1.

As reflected from the above definition, our ORAM scheme is required to hide only the *addresses* that CPU accesses since we handle the privacy and authenticity of the contents of the memory using other techniques. Also, note that the definition does not require to hide the runtime of the program.

2.2 Secure Computation in the RAM Model

Informally, a secure protocol for RAM programs must hide both program's access pattern and its memory contents from the parties. In addition, it must keep the memory "fresh", that is, it prevents the adversary to plug in an outdated memory block to the current CPU-step circuit.

Protocols in this model [13,14,19] typically induce two flavors of security definitions, such that their construction could be modular, i.e. first achieve a construction for the weaker security notion (usually called Unprotected Memory Access) and then enhance it with an ORAM to achieve full security. Informally, the definition of full security requires that the access pattern remains hidden, that is, the ideal adversary only obtains the runtime T of the program Π and the computation output $\boldsymbol{y}$. Given only T and $\boldsymbol{y}$, the simulator must be able to produce an indistinguishable access pattern. The weaker notion of security, as known as Unprotected Memory Access (UMA), leaks the memory contents as well as the access pattern to the adversary. In fact, UMA-secure protocols only deal with how to authentically pass a memory block written in the past to a circuit that needs to read it in a later point in time. In this work we use the same definition for full security, however, we use a different definition, called Unprotected Access Pattern (UAP) instead of the UMA. The definition of UAP is stronger than UMA since it requires the memory contents remain hidden from the adversary (and only the access pattern is leaked). Recall that since our construction is for the non-oblivious computation (see Sect. 1.1) in both security notions the adversary receives the instruction pattern as well.

Obviously, using a standard ORAM scheme we can easily transform a protocol that is UAP secure to a protocol that is fully secure [19], therefore, we may focus on the weaker notion (although our implementation achieves full security). We proceed to define both notions.

Full Security. Following the simulation paradigm [15, Chap. 7] we present the ideal and real models of executions of RAM programs.

Functionality $\mathcal{F}_{\mathrm{RAM}}$

The functionality interacts with parties $p_1, \ldots, p_n$ and the adversary $\mathcal{A}$. The program Π is known and agreed by all parties. The functionality initializes memory D of size N blocks.
Input. Party p_i has its private input x^i.
Output. Execute $\boldsymbol{y} \leftarrow \Pi(D, \boldsymbol{x})$ and send $(T, \boldsymbol{y})$ and $\mathsf{IP}(\Pi, D, \boldsymbol{x})$ to $\mathcal{A}$ (where T is the runtime of the execution). If $\mathcal{A}$ returns Abort then halt, otherwise output $(T, \boldsymbol{y})$ and $\mathsf{IP}(\Pi, D, \boldsymbol{x})$ to all honest parties.

Fig. 1. Ideal execution of $\Pi(N, \boldsymbol{x})$ with abort.

Execution in the ideal model. In an ideal execution $\mathcal{F}_{\mathrm{RAM}}$ (Fig. 1), the parties submit their inputs to a trusted party which in turn executes the program and returns the output. Let Π be a program with memory D of size N, which expects n inputs $\boldsymbol{x} = x^1, \ldots, x^n$, let $\mathcal{A}$ be a non-uniform PPT adversary and let $I \subset [n]$ be the set of indices of parties that $\mathcal{A}$ corrupts; we may refer to the set of corrupted parties by p_I. Denote the *ideal execution of* Π on $\boldsymbol{x}$, auxiliary input z to $\mathcal{A}$ and security parameter κ by the random variable $\mathbf{IDEAL}^{\mathcal{F}_{\mathrm{RAM}}}_{\mathcal{A}(z),I}(\kappa, \Pi, D, \boldsymbol{x})$, as the output set of the honest parties and the adversary $\mathcal{A}$.

Execution in the real model. In the real model there is no trusted party and the parties interact directly. The adversary $\mathcal{A}$ sends all messages in place of the corrupted parties, and may follow an arbitrary PPT strategy whereas honest parties follow the protocol. Let $\Pi, D, \mathcal{A}, I$ be as above and let $\mathcal{P}$ be a multiparty protocol for computing Π. The *real execution of* Π on input $\boldsymbol{x}$, auxiliary input z to $\mathcal{A}$ and security parameter κ, denoted by the random variable $\mathbf{REAL}^{\mathcal{P}}_{\mathcal{A}(z),I}(\kappa, \Pi, D, \boldsymbol{x})$, is defined as the outputs set of the honest parties and the adversary $\mathcal{A}$.

Definition 2.1 (Secure computation). *Protocol $\mathcal{P}$ is said to* securely compute Π with abort in the presence of malicious adversary *if for every PPT adversary $\mathcal{A}$ in the real model, there exists a PPT adversary $\mathcal{S}$ in the ideal model, such that for every $I \in [n]$, every $\boldsymbol{x}, z \in \{0,1\}^*$ and for large enough κ, the following holds*

$$\left\{\mathbf{IDEAL}^{\mathcal{F}_{\mathrm{RAM}}}_{\mathcal{S}(z),I}(\kappa, \Pi, D, \boldsymbol{x})\right\}_{\kappa,\boldsymbol{x},z} \stackrel{c}{\equiv} \left\{\mathbf{REAL}^{\mathcal{P}}_{\mathcal{A}(z),I}(\kappa, \Pi, D, \boldsymbol{x})\right\}_{\kappa,\boldsymbol{x},z}$$

Unprotected Access Pattern (UAP) Security. This notion allows the adversary to further inspect the access pattern. The ideal functionality $\mathcal{F}_{\mathsf{UAP}}$ is given in Fig. 2 and realized by protocol $\mathcal{P}_{\mathsf{UAP}}$ (Fig. 7).

Definition 2.2 (Secure computation in the UAP model). *Protocol $\mathcal{P}$ is said to* securely compute Π in the UAP model with abort in the presence of malicious adversary *if for every PPT adversary $\mathcal{A}$ for the real model, there*

Functionality $\mathcal{F}_{\mathsf{UAP}}$

The functionality interacts with parties $p_1, \ldots, p_n$ and the adversary $\mathcal{A}$. The program Π is known and agreed by all parties. The functionality initializes memory D of size N blocks.
Input. Party p_i has its private input x^i.
Output. Execute $\boldsymbol{y} \leftarrow \Pi(D, \boldsymbol{x})$ and send $\mathsf{AP}(\Pi, D, \boldsymbol{x})$, $\mathsf{IP}(\Pi, D, \boldsymbol{x})$ and $\boldsymbol{y}$ to $\mathcal{A}$. If $\mathcal{A}$ returns Abort then halt, otherwise output $\mathsf{AP}(\Pi, D, \boldsymbol{x})$, $\mathsf{IP}(\Pi, D, \boldsymbol{x})$ and $\boldsymbol{y}$ to all honest parties.

Fig. 2. Ideal execution of $\Pi(N, x)$ in the UAP model.

is a PPT adversary $\mathcal{S}$ for the ideal model, such that for every $I \in [n]$, every $\boldsymbol{x}, z \in \{0,1\}^*$ and for large enough κ

$$\left\{\mathbf{IDEAL}^{\mathcal{F}_{\mathsf{UAP}}}_{\mathcal{S}(z),I}(\kappa, \Pi, D, \boldsymbol{x})\right\}_{\kappa,\boldsymbol{x},z} \stackrel{c}{\equiv} \left\{\mathbf{REAL}^{\mathcal{P}}_{\mathcal{A}(z),I}(\kappa, \Pi, D, \boldsymbol{x})\right\}_{\kappa,\boldsymbol{x},z}$$

The transformation (or compilation) from UAP to full security is not in the scope of this paper and can be found in previous works [11–14,19]. We follow that path since it makes the security analysis simpler and modular, rather than proving full security from scratch. Therefore, functionality $\mathcal{F}_{\mathsf{UAP}}$ in Fig. 2, which is realized in protocol $\mathcal{P}_{\mathsf{UAP}}$ (Fig. 7), reveals the access pattern to the parties. By incorporating an ORAM scheme on top of our protocol that access pattern would be of no gain to the adversary for the reason that an access pattern of a program execution using an ORAM is indistinguishable from an access pattern of a randomly chosen program with the same runtime.

We note that achieving a UAP-secure protocol may be useful on its own (i.e. without lifting it up to full security) in cases where the original program Π is oblivious, that is, when the access pattern is *permitted* to be leaked to the parties.

3 Executing RAM Programs Using BMR

Our protocol follows the BMR-SPDZ approach [25,27] and adapts the free-XOR technique for the BMR garbling scheme [2]. For completeness, in the following we describe the structure of the actively secure additive secret sharing used in SPDZ-like protocols and outline the BMR-SPDZ approach.

3.1 SPDZ Secret Sharing

SPDZ-like protocols use actively secure additive secret sharing over a finite field, combined with information theoretic MACs to ensure active security. A shared secret $x \in \mathbb{F}$ is represented by

$$[\![x]\!] = ([x], [m(x)], [\alpha]) = (x^1, \ldots, x^n, m(x)^1, \ldots, m(x)^n, \alpha^1, \ldots, \alpha^n)$$

where $m(x) = x \cdot \alpha$ is a MAC on message x using a global key α. Party p_i holds: A uniformly random share x^i of x, a uniformly random share $m(x)^i$ of $m(x)$ and a uniformly random share α^i of α such that

$$x = \sum_{i=1}^{n} x^i, \qquad m(x) = \sum_{i=1}^{n} m(x)^i, \qquad \alpha = \sum_{i=1}^{n} \alpha^i$$

We denote an additive secret shared value x by $[x]$ and its authenticated shared version by $[\![x]\!]$. We also denote p_i's share by $[\![x]\!]^i = (x^i, m(x)^i)$.

When opening a shared value $[\![x]\!]$ the parties first broadcast their shares x^i and compute x. To ensure that x is correct, they then check the MAC by committing to and opening $m(x)^i - x \cdot \alpha^i$ and checking these shares sum up to zero.

3.2 The BMR-SPDZ Protocol

Unlike the two-party settings, in which we have one garbler and one evaluator, in the multiparty settings *all* parties are both garblers and evaluators such that no strict subset of parties can either influence or learn anything about the values that the wires carry. In the following we present the key points in the BMR-SPDZ approach:

Keys. Every party chooses a random key for each wire in the circuit, that is, party p_i chooses key $k_w^i \in \mathbb{F}_{2^\kappa}$ for wire w. This key is named "0-key" and denoted $k_{w,0}^i$ where $k_{w,0}^i$ is essentially the i-th coordinate of a *full* 0-key, $k_{w,0} = (k_{w,0}^1, \ldots, k_{w,0}^n) \in (\mathbb{F}_{2^\kappa})^n$.

Global difference. To enable free-XOR, each party chooses its own global-difference, that is, party p_i randomly chooses Δ_i such that the difference between its 0-key and its 1-key is Δ_i. Formally, $k_{w,1}^i = k_{w,0}^i \oplus \Delta_i$ for every w and i. Similarly Δ_i is the i-th coordinate of the full difference $\Delta = (\Delta_1, \ldots, \Delta_n)$. The value Δ_i is known only to party p_i and no strict subset of the parties (that does not include p_i) can learn it. For wire w we get that $k_{w,1} = k_{w,0} \oplus \Delta$ where $\oplus$ operates component-wise.

Permutation bits. In the course of the evaluation the parties obtain $k_{w,b}$ with either $b = 0$ or $b = 1$ for every wire w. Party p_i could easily check whether $b = 0$ or $b = 1$ by extracting the ith element from $k_{w,b}$ and compare it to $k_{w,0}^i$ and $k_{w,1}^i$. If $b = 0$ we say that the *external value* of wire w, denoted Λ_w is 0, otherwise, if $b = 1$, then $\Lambda_w = 1$. Since the *real value* that is carried by wire w, denoted by ρ_w, must be kept secret, the external value Λ_w must reveal nothing about it. To this end, a random *permutation bit*, λ_w, is assigned to each wire w in order to mask ρ_w by setting $\Lambda_w = \lambda_w \oplus \rho_w$.

Inputs. Let w be an input wire that is associated with input x^i of party p_i, then the parties open λ_w to party p_i only. Then p_i broadcasts Λ_w and k_{w,Λ_w}^i where

$\Lambda_w = \rho_w \oplus \lambda_w$ and ρ_w is its input to wire w. Then, party p_j, for all j, broadcasts its Λ_w-key k^j_{w,Λ_w} such that all parties obtain $k_{w,\Lambda_w} = (k^1_{w,\Lambda_w}, \ldots, k^n_{w,\Lambda_w})$.

Outputs. If w is an output wire then the parties open the permutation bit λ_w to everyone. This way, upon obtaining key k_{w,Λ_w} the parties learn the real value of w by $\rho_w = \Lambda_w \oplus \lambda_w$.

Encrypting a key. In the process of garbling, the parties encrypt the key of a gate's output wire using the keys of its input wires. Let $m = m^1, \ldots, m^n$ be the key to be encrypted and k_ℓ, k_r with $k_b = k^1_b, \ldots, k^n_b$ be encryption keys of the left and right input wires, where party p_i has m^i, k^i_ℓ, k^i_r. The parties produce the ciphertext $c = c^1, \ldots, c^n$ as follows: m^j is encrypted using k_ℓ, k_r to result c^j such that even a single missing coordinate of k_ℓ and k_r prevents one from decrypting c^j. To encrypt m^j, party p_i provides $F_{k^i_\ell}(j), F_{k^i_r}(j)$, where F is a pseudorandom generator and then, using a protocol for secure computation the parties evaluate and output:

$$c^j = \mathsf{Enc}_{k_\ell,k_r}(m^j) = \left(\bigoplus_{i=1}^{n} F_{k^i_\ell}(j)\right) \oplus \left(\bigoplus_{i=1}^{n} F_{k^i_r}(j)\right) \oplus m^j$$

Note that the keys k^i_ℓ, k^i_r are necessary for the decryption of c^j for every $i, j \in [n]$.

Garbled gate. A garbled version of an AND gate g with input wires u, v and output wire w, is simply a 4-entries table, each entry is an encryption of either $k_{w,0}$ or $k_{w,1}$, this depends on the permutation bits λ_u, λ_v and λ_w. We want to enable the evaluator, who holds k_{u,Λ_u} and k_{v,Λ_v} (which are translated to ρ_u and ρ_v respectively) to decrypt the ciphertext in the $(2\Lambda_u + \Lambda_v)$-th entry of the table and obtain k_{w,Λ_w} such that $\rho_w = \rho_u \cdot \rho_v$. That is, we want to have $\lambda_w \oplus \Lambda_w = (\lambda_u \oplus \Lambda_u) \cdot (\lambda_v \oplus \Lambda_v)$, thus, the $(2\Lambda_u + \Lambda_v)$-th entry conceals k_{w,Λ_w} where

$$\Lambda_w = (\lambda_u \oplus \Lambda_u) \cdot (\lambda_v \oplus \Lambda_v) \oplus \lambda_w$$

and since $k_{w,1} = k_{w,0} \oplus \Delta$ we get that the entry conceals

$$k_{w,0} + \Lambda_w \cdot \Delta = k_{w,0} + \big((\lambda_u \oplus \Lambda_u) \cdot (\lambda_v \oplus \Lambda_v) \oplus \lambda_w\big) \cdot \Delta$$

We conclude by presenting functionality $\mathcal{F}_{\mathsf{BMR}}$ (Fig. 3) for a construction of a garbled circuit. Note that the only difference between $\mathcal{F}_{\mathsf{BMR}}$ to the standard description of this functionality [2,25] is that here the functionality lets the parties learn a share to the permutation bits λ_w. This is necessary in order to obtain a neat security proof of the construction. Protocol $\mathcal{P}_{\mathsf{BMR}}$ (Fig. 5) realizes $\mathcal{F}_{\mathsf{BMR}}$ in the $\mathcal{F}_{\mathsf{MPC}}$-hybrid model (Fig. 11 in the appendix). Given a garbled circuit the parties evaluate it using the $\mathcal{E}_{\mathsf{BMR}}$ procedure described in Fig. 4.

In the presentation of the protocol (Fig. 5) and to the rest of the paper, we denote by $\langle x \rangle$ the handler (*varid*) of a variable x that is stored by $\mathcal{F}_{\mathsf{MPC}}$.

Functionality $\mathcal{F}_{\mathsf{BMR}}$

The functionality interacts with n parties $p_1, \ldots, p_n$ and exports the following instructions:

Garble On input ($\mathsf{Garble}, C, (W, P)$) from all parties where C is the circuit to garble and (W, P) is a map from a wire to a party (used to associate an input wire to the party who feeds its input to it, i.e. $W[\ell]$ is an input wire associated with party $P[\ell]$).

1. For party P_i:
   (a) Sample and output to p_i a global random difference $\Delta_i \in \mathbb{F}_{2^\kappa}$.
   (b) Sample a key $k^i_{w,0} \in \mathbb{F}_{2^\kappa}$ for all $w \in C$ (such that $k^i_{w,1} = k_{w,0} \oplus \Delta_i$).
   (c) Output $k^i_{w,0}$ to P_i for all input wires in C.
2. Samples a permutation bit $\lambda_w \in \{0, 1\}$ for all $w \in C$.
3. Create SPDZ share of λ_w for every memory access (reading and writing) wire $w \in C$ and send $[\![\lambda_w]\!]^i$ to p_i.
4. For every $1 \leq \ell \leq |W|$: Output $\lambda_{W[\ell]}$ in the clear to $P[\ell]$.
5. For every AND gate $g \in C$ with input wires u, v and output wire w, every $\Lambda_u, \Lambda_v \in \{0, 1\}$, and every $j \in [n]$, compute:

$$\tilde{g}^j_{\alpha,\beta} = \left(\bigoplus_{i=1}^{n} F_{k^i_{u,\Lambda_u}}(g, j) \bigoplus_{i=1}^{n} F_{k^i_{v,\Lambda_v}}(g, j) \right) \oplus k^j_{w,0} \\ \oplus \Delta_j \cdot \left((\lambda_u \oplus \alpha) \cdot (\lambda_v \oplus \beta) \oplus \lambda_w \right) \tag{2}$$

and store $GC = \left\{ \{(\tilde{g}^1_{\Lambda_u,\Lambda_v}, \ldots, \tilde{g}^n_{\Lambda_u,\Lambda_v})\}_{\Lambda_u,\Lambda_v \in \{0,1\}} \right\}_{g \in G}$

Open Output GC to all parties.

Fig. 3. The BMR functionality.

3.3 Towards RAM Computation

To be able to securely compute RAM programs (in the UAP model) the parties garble T circuits $GC^1, \ldots, GC^T$ and then evaluate them sequentially. To this end, we must specify how the parties obtain the keys intended for the input wires of each garbled circuit (these are the input wires associated with values state_t and b^{read}_t). This task is divided in two: First, the input wires of GC^t associated with state_t must carry the same values as the output wires associated with state_t in GC^{t-1}. Second, we need to support secure memory access, that is, the input wires of GC^t associated with b^{read}_t must carry the same values as the output wires associated with $b^{\mathsf{write}}_{t'}$ in $GC^{t'}$, where t' is the most recent timestep in which address i^{read}_{t-1} was modified. The first task could be easily achieved by changing $\mathcal{P}_{\mathsf{BMR}}$ to choose the same keys for both output and input wires that are associated to the same state in every two consecutive garbled circuits, however, this would be non-black-box in $\mathcal{F}_{\mathsf{BMR}}$ (since the functionality chooses its keys independently for every circuit). For a black-box solution we can use the techniques described in Sects. 4.2 and 4.3. We stress, though, that the two tasks are orthogonal and

Procedure $\mathcal{E}_{\mathsf{BMR}}$

Evaluate On input (Eval, K) from all parties, where $K = \{k_w\}_w$ for all input wires in C, evaluate GC as follows: Traverse C in a topological order. Let g be the current gate with input wires u, v and output wire w and $k_{u,\Lambda_u}, k_{v,\Lambda_v}$ the keys obtained for wires u and v respectively. Then,

1. If g is a XOR gate, compute the output key and its external value by $k_{w,\Lambda_w} = k_{u,\Lambda_u} \oplus k_{v,\Lambda_v}$ and $\Lambda_w = \Lambda_u \oplus \Lambda_v$ where $\oplus$ operates component-wise.
2. If g is an AND gate, compute

$$k^j_{w,\gamma} = \tilde{g}^j_{\alpha,\beta} \oplus \left(\bigoplus_{i=1}^{n} F_{k^i_{u,\alpha}}(g,j) \bigoplus_{i=1}^{n} F_{k^i_{v,\beta}}(g,j) \right)$$

for every $j \in [n]$ where $\alpha = \Lambda_u$ and $\beta = \Lambda_v$. Set $k_{w,\gamma} = (k^1_{w,\gamma}, \ldots, k^n_{w,\gamma})$. For party p_i: If $k^i_{w,\gamma} = k^i_{w,0}$ set $\Lambda_w = 0$, otherwise, if $k^i_{w,\gamma} = k^i_{w,1}$ set $\Lambda_w = 1$, otherwise abort.

Fig. 4. Evaluation of a BMR garbled circuit.

Protocol $\mathcal{P}_{\mathsf{BMR}}$

The parties initialize $\mathcal{F}_{\mathsf{MPC}}$ by calling $\mathcal{F}_{\mathsf{MPC}}.\mathsf{Init}(k)$.

Garble

1. For each $i \in [n]$ invoke $\langle \Delta_i \rangle \leftarrow \mathcal{F}_{\mathsf{MPC}}.\mathsf{Random}()$. This is party p_i's global difference. Then call $\mathcal{F}_{\mathsf{MPC}}.\mathsf{Output}(\langle \Delta_i \rangle, i)$.
2. Output $k^i_{w,0}, k^i_{w,1}$ to p_i using $\mathcal{F}_{\mathsf{MPC}}.\mathsf{Output}$ where $\langle k^i_{w,0} \rangle \leftarrow \mathcal{F}_{\mathsf{MPC}}.\mathsf{Random}()$ and $\langle k^i_{w,1} \rangle \leftarrow \mathcal{F}_{\mathsf{MPC}}.\mathsf{Add}(\langle k^0_{w,0} \rangle, \langle \Delta_i \rangle)$ for all $i \in [n]$ and $w \in W$.
3. For every wire $w \in W$ invoke $\langle \lambda_w \rangle \leftarrow \mathcal{F}_{\mathsf{MPC}}.\mathsf{RandomBit}()$.
4. Party p_i calls $\langle F_{k^i_{w,b}}(g,j) \rangle \leftarrow \mathcal{F}_{\mathsf{MPC}}.\mathsf{Input}(F_{k^i_{w,b}}(g,j))$ for every gate $g \in G, j \in [n]$ and $b \in \{0,1\}$ where w is either the first or the second input wire for gate g.
5. Compute $\langle GC \rangle = \{\{[\![\tilde{g}^j_{\alpha,\beta}]\!]\}_{j\in[n],\alpha,\beta\in\{0,1\}}\}_{g\in G}$ with $\tilde{g}^j_{\alpha,\beta}$ as in Equation 2, where additions are computed using $\mathcal{F}_{\mathsf{MPC}}.\mathsf{Add}$ and multiplications are computed using $\mathcal{F}_{\mathsf{MPC}}.\mathsf{Multiply}$.

Open Invoke $\mathcal{F}_{\mathsf{MPC}}.\mathsf{Open}$ on $\tilde{g}^j_{\alpha,\beta}$ for every $j \in [n], \alpha, \beta \in \{0,1\}$ and $g \in G$.

Fig. 5. Realizing $\mathcal{F}_{\mathsf{BMR}}$ in the $\mathcal{F}_{\mathsf{MPC}}$-hybrid model.

the techniques chosen to complete them are independent. Therefore, in the rest of the presentation we focus on realizing secure memory access (the second task) while taking for granted the traveling of the CPU's state (i.e. we may write *"the parties obtain the input wires of* state_t" without specifying how).

4 Accessing Memory

In this section we present the three techniques to achieve secure memory accesses and show how to realize $\mathcal{F}_{\mathsf{UAP}}$ in the $\mathcal{F}_{\mathsf{BMR}}$-hybrid model using the third one. We compare the performance of the techniques in Table 2. The values within the table are explained alongside the description of the techniques.

In the presentation below, we group some set of input/output wires together, according to their purpose as follows: W_{in} refers to the input wires of GC^1, which correspond to the parties' inputs where $W^i_{\mathsf{in},j}$ corresponds to the j-th bit of input x^i. $W^t_{\mathsf{read}}, W^t_{\mathsf{write}}, W^t_{\mathsf{addr,rd}}, W^t_{\mathsf{addr,wr}}$ refer to the input b^{read}_t, output b^{write}_t, addresses $i^{\mathsf{read}}_t, i^{\mathsf{write}}_t$ respectively in GC^t. In addition, W^t_{state} refers to the input state and $W^t_{\mathsf{state}'}$ refers to the output state in GC^t.

4.1 Memory via Embedded Authentication Sub-circuit

This technique assumes that the values $b^{\mathsf{read}}_t, b^{\mathsf{write}}_t$ are elements from the same field that SPDZ use as the underlying MPC protocol. This way, if SPDZ statistical parameter is s then the memory is divided into data items of s bits. We enhance C_{CPU} with two procedures: Verify and AuthShare and denote the result

Table 2. Performance of the three techniques with n parties. κ and s are the computational and statistical security parameters. The columns specify the following parameters: *Number of input wires* required in the CPU-step circuit for every input bit (of the ORAM) that is being read by that circuit. *Amount of communication* required for each memory access, this is measured in bits per input wire per party. *Number of communication rounds* required from the moment the parties obtained keys of output wires of GC^t to the moment they obtain keys to the input wires of GC^{t+1}. Communication rounds could be used for secret opening, broadcasting a value or performing multiplication over shares; among the three, only multiplication requires more work to be done in the offline phase, specifically, multiplication requires sacrificing a multiplication triple. This is reflected in the Triples column, note that we multiply a *vector* of n keys rather than a single key. *Memory overhead* specifies how many bits do we store in the memory for a single bit of ORAM memory (again, this is per party. The total memory size that the parties store should be multiplied by n). The last column specifies whether a change in the garbled circuit is needed.

	Input wires	Communication	Rounds	Triples	Memory overhead	Requires change in C_{CPU}
Embedded subcircuit (Sect. 4.1)	$4n$	4κ	2	$O(ns^2)$	2	$+2s^2$ AND gates $+4sn$ input wires per data item
Soldering (Sect. 4.2)	1	$(3n+1)\kappa$	2	$2n$	$n\kappa$	No
Shared bits (Sect. 4.3)	1	$\mathbf{2s}$	**2**	**0**	**2**	**No**

Enhanced CPU-step circuit C_{CPU^+}

Input. The parties input state and $[\![b^{\mathsf{read}}]\!]$. In addition they input 2 random values $[r_1], [r_2]$. That is, party p_i inputs $(b^{\mathsf{read}})^i, m(b^{\mathsf{read}})^i, \alpha^i, r_1^i, r_2^i$.

Output.

1. Compute $\mathsf{ver} = \mathsf{Verify}([\![b^{\mathsf{read}}]\!])$.
2. Compute $(\mathsf{state}', i^{\mathsf{read}}, i^{\mathsf{write}}, b^{\mathsf{write}}) = C_{CPU}(\mathsf{state}, b^{\mathsf{read}})$
3. Compute $(\mathsf{val}, \mathsf{mac}) = \mathsf{AuthShare}(b^{\mathsf{write}}, [\alpha], [r_1], [r_2])$.
4. Output state′, ver, val and mac.

Fig. 6. Enhanced CPU-step circuit C_{CPU^+}

by C_{CPU^+} (see Fig. 6). Note that each party has $4s$ input wires for the purpose of authentication and sharing (assuming that the global MAC key is part of the state).

Privacy. The parties maintain their memory in the form of SPDZ shares, thus, to input content from location i^{read} in memory, every party inputs its SPDZ share of this content from its own storage $D[i^{\mathsf{read}}]$. Then the secret is being constructed within the CPU-step circuit. Since this is an additive secret sharing scheme, the content is being constructed using only XOR gates, which requires no communication.

Verify Authenticity. We enhance the CPU-step with a sub-circuit that verifies the authenticity of the secret b^{read}, the sub-circuit is denoted $\mathsf{Verify}([\![v]\!])$ where v refers to b^{read}. Party p_i inputs $(v^i, m(v)^i, \alpha^i)$ and the sub-circuit computes[5]:

$$v = \sum_{i=1}^{n} v^i, \qquad m(v) = \sum_{i=1}^{n} m(v)^i, \qquad \alpha = \sum_{i=1}^{n} \alpha^i$$

and outputs $\mathsf{ver} = 1$ if $m(v) - (\alpha \cdot v) = 0$ (meaning verification succeeded) and 0 otherwise (meaning verification has not succeeded), which incurs s^2 AND gates for a single multiplication operation in addition to $2s - 1$ AND gates for deciding whether ver is 1 or 0. Note that this multiplication is over a polynomial ring over $\mathbb{F}_2$, thus the addition involves only XOR. Furthermore, we check the result directly for zero and skip the reduction modulo an irreducible polynomial (mapping from $\mathbb{F}_2[X]$ to $\mathbb{F}_{2^s}$), hence the $2s - 1$ AND gates for comparison.

Security. Obviously, since nothing is revealed except from the fact of the authenticity being correct, the adversary cannot extract any information regarding the value.

[5] Remember that x^i denotes the share of party p_i and not an exponentiation operation.

Authenticated Share. The CPU-step produces the value b^{write} to be written to the memory, which obviously could not be output in the clear, rather, it is shared between all parties. The sub-circuit $\mathsf{AuthShare}(v, [\alpha], [r_1], [r_2])$ is given the value v to share (which refers to the value b^{write}), the global MAC key α and two freshly chosen $r_1, r_2 \in \mathbb{F}_{2^s}$ from the parties such that party p_i inputs (α^i, r_1^i, r_2^i). The circuit computes

$$r_1 = \sum_{i=1}^{n} r_1^i, \qquad r_2 = \sum_{i=1}^{n} r_2^i, \qquad \alpha = \sum_{i=1}^{n} \alpha^i$$

and outputs $\mathsf{val} = (v + r_1)$ and $\mathsf{mac} = (\alpha \cdot v + r_2)$. To obtain the SPDZ sharing $[\![v]\!]$, party p_1 stores $v^1 = (\mathsf{val} - r_1^1)$ and $m(v)^1 = (\mathsf{mac} - r_2^1)$ and all other parties p_j store $v^j = (-r_1^j)$ and $m(v)^j = (-r_2^j)$.

Security. First note that $\sum_{i=1}^{n} v^i = \mathsf{val} - r_1^1 - \sum_{j=2}^{n} r_1^j = v$ and $\sum_{i=1}^{n} m(v)^i = \mathsf{mac} - r_2^1 - \sum_{j=2}^{n} r_2^j = m(v) = \alpha \cdot v$ as required. The values v and its authentication $m(v)$ are independently masked using a truly random values and thus are hidden from any strict subset of parties.

Performance. For every data item in $\mathbb{F}_{2^s}$ the parties store its MAC as well, which leads to an overhead of 2 in the memory size. To obtain the key of an input wire of the next circuit each party needs to broadcast its BMR key, which is of size κ. Since for every read of data item each party inputs 4 $\mathbb{F}_{2^s}$ elements (assuming that the global key α is part of the state), the communication complexity is $4\kappa n$ bits per input bit. The additional circuitry for authentication (and verification) is of size $2s^2 + 2s - 1$ (for two multiplications of elements from $\mathbb{F}_{2^s}$ and additional zero testing), note in Table 2 the required number of multiplications triples is multiplied by n since each AND gate manipulates keys vectors of n coordinates. To obtain the keys for the next CPU-step circuit 2 communication rounds are required, one to broadcast the external value of the input wire (which is done by the party whose input is associated with) and the other is to broadcast the appropriate keys by all parties. Note that we cannot save a communication round by broadcasting the external value when writing (rather than when reading) since the external value that the parties broadcast depends on the input wire of the circuit that is going to read it in a later, unknown point in time.

4.2 Memory via Wire Soldering

The General Technique. Wire soldering allows the parties to reuse an output wire key of one gate as an input wire key of another gate even if these two gates were not meant to be connected in garbling time. The notion of wire soldering for secure computation was introduced in [32] for the two-party settings and implemented using an additively homomorphic commitment scheme $(\mathsf{com}, \mathsf{dec})$, that is, $\mathsf{com}(a) + \mathsf{com}(b) = \mathsf{com}(a + b)$. Let u and v be wires with keys $k_{u,0}, k_{u,1}, k_{v,0}, k_{v,1}$ and permutation bits λ_u, λ_v. By soldering v to u we would like to achieve the following feature: Obtaining key k_{u,Λ_u} that carries a real value $\rho_u = \lambda_u \oplus \Lambda_u$, enables to obtain the key k_{v,Λ_v}, which carries the *same real value* $\rho_v = \lambda_v \oplus \Lambda_v = \rho_u$. It follows that if $\lambda_u = \lambda_v$ the soldering information reveals k_{v,Λ_u} (i.e. $\Lambda_u = \Lambda_v$), otherwise, if $\Lambda_u \neq \Lambda_v$, it reveals $k_{v,(1-\Lambda_u)}$.

In a circuit-based 2PC (with garbler and evaluator) this is done by having the garbler send the commitments $\mathsf{com}(k_{u,0})$, $\mathsf{com}(k_{u,1})$, $\mathsf{com}(k_{v,0})$, $\mathsf{com}(k_{v,1})$; if $\lambda_u = \lambda_v$ the garbler also sends the decommitments $s_0 = \mathsf{dec}(k_{u,0} \oplus k_{v,0})$ and $s_1 = \mathsf{dec}(k_{u,1} \oplus k_{v,1})$, otherwise (if $\lambda_u \neq \lambda_v$) the garbler sends $s_0 = \mathsf{dec}(k_{u,0} \oplus k_{v,1})$ and $s_1 = \mathsf{dec}(k_{u,1} \oplus k_{v,0})$. Given the key k_{u,Λ_u}, the evaluator computes $k_{u,\Lambda_u} \oplus s_{\Lambda_u}$ to obtain the correct key for wire k_{v,Λ_v}. To prove that the garbler hasn't inverted the truth value of the wires by choosing the wrong case above, it must also decommit to the XOR of the permutation bits $(\lambda_u \oplus \lambda_v)$. Note that the evaluator learns whether $\Lambda_u = \Lambda_v$ and thus also learns whether $\lambda_u = \lambda_v$, however, that doesn't reveal anything about the real value $\rho_u = \rho_v$ that is carried over those wires.

Soldering in Our Scheme. When $\mathcal{P}_{\mathsf{BMR}}$ (Fig. 5) uses SPDZ to garble the circuit party p_i not only obtains its own keys $k^i_{w,b}$ in the clear, but also obtains a SPDZ sharing for both a *whole* keys $k_{w,b}$ and the permutation bits λ_w for every $w \in W$ and $b \in \{0,1\}$, thus, the parties could use $\mathcal{F}_{\mathsf{MPC}}$ to perform arithmetic operations over them.

Let u and v be wires with keys $k_{u,0}, k_{u,1} = k_{u,0} \oplus \Delta^u$ and $k_{v,0}, k_{v,1} = k_{v,0} \oplus \Delta^v$ and permutation bits $\lambda_u, \lambda_v \in \{0,1\}$. The parties perform the procedure $\mathsf{Solder(u,v)}$ defined as follows: If $\Lambda_u = 0$ the parties collaboratively compute

$$s^0_{u\to v} = \Big((\lambda_u \oplus (1-\lambda_v)) \cdot (k_{u,0} \oplus k_{v,0})\Big) \oplus \Big((\lambda_u \oplus \lambda_v) \cdot (k_{u,0} \oplus k_{v,1})\Big)$$

and output $s^0_{u\to v}$ to everyone.

Otherwise, if $\Lambda_u = 1$ the parties collaboratively compute

$$s^1_{u\to v} = \Big((\lambda_u \oplus (1-\lambda_v)) \cdot (k_{u,1} \oplus k_{v,1})\Big) \oplus \Big((\lambda_u \oplus \lambda_v) \cdot (k_{u,1} \oplus k_{v,0})\Big)$$

and output $s^1_{u\to v}$ to everyone.

The information $s^{\Lambda_u}_{u\to v}$ allows the parties to solder wire v to wire u. Notice that this technique involves only one multiplication layer, since the parties simultaneously compute both multiplications and then *locally* add results in $\mathbb{F}_{2^\kappa}$.

Observe that our variation of the soldering is applicable to the multiparty settings as well, in addition, due to the already exist SPDZ shares to the full wires' keys, we don't need to rely on additional homomorphic commitment scheme and its expensive overhead. Moreover, the original soldering was thought to be a way to connect two wires *within the same circuit* (using a single global difference) while in here we show that it is applicable for wires *of different circuits* as well (that were garbled independently and with two global differences Δ^u, Δ^v).

To see why it works, without loss of generality, consider $\Lambda_u = 0$. If $\lambda_u = \lambda_v$, then given key $k_{u,0}$ that carries value $\rho_u = \lambda_u$, the parties compute

$$k_{u,0} \oplus s^0_{u \to v} = k_{u,0} \oplus \big(1 \cdot (k_{u,0} \oplus k_{v,1})\big) \oplus \big(0 \cdot (k_{u,0} \oplus k_{v,0})\big)$$
$$= k_{v,0}$$

such that $k_{u,0}$ and $k_{v,0}$ encapsulate the same real value as required. If $\lambda_u \neq \lambda_v$ we get

$$k_{u,0} \oplus s^0_{u \to v} = k_{u,0} \oplus \big(0 \cdot (k_{u,0} \oplus k_{v,0})\big) \oplus \big(1 \cdot (k_{u,0} \oplus k_{v,1})\big)$$
$$= k_{v,1}$$

such that $k_{u,0}$ and $k_{v,1}$ encapsulate the same real value as required. The same analysis holds when $\Lambda_u = 1$.

Performance. To obtain the key of the next circuit the parties simultaneously compute 2 multiplications (of n keys) over the shares in one round and then open the result in the second round, hence $2n$ multiplication triples are required in the offline phase. Multiplication requires the communication of $3n\kappa$ bits per party, opening requires κ, a total of $(3n+1)\kappa$ per party per input bit.

4.3 Memory via Free Conversion Between Keys and Shared Real Values

In this section we present a new technique, which outperforms both the embedding and soldering techniques in both the communication and memory size overheads. Essentially, it allows to freely convert between BMR wire keys and SPDZ secret shares of the real values that those keys represent. As before, it is not necessary to know which SPDZ share to convert from and to in garbling time. This allows reactive memory accesses in the sense that, during evaluation, the parties can evaluate previously garbled circuits on values read from memory at an address that was only just revealed during the evaluation phase. The latter is crucial for implementing ORAM.

Using this technique the parties need to compute only (local) additions and some SPDZ openings in order to move from the evaluation of one circuit to the next. In more detail, converting from wire keys to SPDZ shares can be done without communication at all, while the other direction requires two rounds of SPDZ opening. In any case, no multiplication is necessary, hence, no offline overhead (for triple generation) is implied. Similarly, the information required is a by-product of the BMR offline phase, hence there is no extra cost there.

When reading a bit from memory the parties need to know the external value of the wire associated with it. For a circuit-input-wire, which is associated with a particular party, that party knows the wire's permutation and external value, hence, it can broadcast the external value to the parties, who then can broadcast their appropriate share of the BMR key. In contrast, when reading from memory, the external value of the wire is shared (it is nobody's input) and reconstructed, then the parties broadcast their keys as before. The keys that the parties broadcast are stored by each party along with the garbled circuit, in the same streaming manner, and are not part of the program's memory.

Packing secret bits. Naively storing each external value as a SPDZ secret share would require s^2 bits in memory for every s-bit data block. We can reduce this overhead by packing s secret shares of bits into an s-bit secret share such that it requires only $2s$ bits in memory for every s-bit data block (s bits for the share itself and another s bits for its MAC). Packing s bits $[\![b_0]\!], \ldots, [\![b_{s-1}]\!]$ is done by computing $[\![B]\!] = \sum_{i \in [s]} b_i \cdot 2^i$ where the 2^i part is constant, so we obtain $[\![B]\!]$ by local computation only. Now, we can make operations over bits easily by inputting the entire data item and using the specific required bit. "Extracting" the j-th bit, $[\![b_j]\!]$, from $[\![B]\!]$ can be done locally as well as described below.

Writing to Memory. Recall that after issuing the **Garble** instruction in $\mathcal{F}_{\mathsf{BMR}}$ the parties hold shares to all the permutation bits of all wires. Recall that the wires groups $W^t_{\mathsf{addr,wr}}$ and W^t_{write} refer to the wires associated with the address to be written and the value to be written to that address respectively. In the protocol, the parties open the permutation bits for wires $W^t_{\mathsf{addr,wr}}$ but not for wires W^t_{write}, this means that they learn i^{write}_t in the clear, but learn nothing about b^{write}_t, rather, they only obtain the keys and their external values associated with it. That is, for $w \in W^t_{\mathsf{write}}$, the parties obtain k_{w,Λ_w} and Λ_w. To store the *real* value that is carried by wire w in memory address i the parties only need to compute $[\![\rho_w]\!] = \Lambda_w + [\![\lambda_w]\!]$. Then p_i stores $D[i] \leftarrow [\![\rho_w]\!]^i$. Furthermore, every party p_i can check whether k_{w,Λ_w} is correct because they have obtained both $k_{w,0}$ and $k_{w,1} = k_{w,0} \oplus \Delta_i$ during **Garble**. This is equivalent to checking the correctness of output wires. In order to achieve optimal memory usage, s bits $w_0, \ldots, w_{s-1}$ can be combined by (locally) computing $\sum_{i=0}^{s-1} X^i [\![\rho_{w_i}]\!]$ where X denotes a generator of the multiplicative group of a field of size 2^s.

Reading from Memory. Let i^{read}_{t-1} be the address from which the parties are instructed to read when evaluating GC^{t-1} and let $w \in W^t_{\mathsf{read}}$. We assume for a moment that secret shares packing technique above has not been applied when storing. Therefore, $D[i^{\mathsf{read}}_{t-1}]$ contains a share of the bit $[\![\rho_{w'}]\!]$ that was most recently written to i^{read}_{t-1} at a previous timestep t' with wire $w' \in W^{t'}_{\mathsf{write}}$. Party p_i holds both $k^i_{w,0}$ and $k^i_{w,1}$, but need to broadcast only one of them. Specifically, broadcast k^i_{w,Λ_w} for $\Lambda_w = \rho_w \oplus \lambda_w$. Now, since we require that $\rho_w = \rho_{w'}$ then the parties open $\Lambda_w = [\![\rho_w]\!] + [\![\lambda_w]\!] = [\![\rho_{w'}]\!] + [\![\lambda_w]\!]$ and broadcast k_{w,Λ_w}. Finally, if the parties have stored $\sum_{i=0}^{s-1} X^i [\![\rho_{w_i}]\!]$ at a particular memory address, $[\![\rho_{w_i}]\!]$ can be computed by opening $\sum_{i=0}^{s-1} X^i [\![\rho_{w_i}]\!] + \sum_{i=0}^{s-1} X^i [\![\lambda_{w_i}]\!]$. This works because any field of size 2^s has characteristic two, thus addition corresponds to bitwise XOR.

5 Realizing Functionality $\mathcal{F}_{\mathsf{UAP}}$

Protocol $\mathcal{P}_{\mathsf{UAP}}$ in Fig. 7 realizes $\mathcal{F}_{\mathsf{UAP}}$ in the $\mathcal{F}_{\mathsf{BMR}}$-hybrid model.

Protocol $\mathcal{P}_{\mathsf{UAP}}$

Party p_i allocates a memory D^i of size N.

Offline. The parties initialize $\mathcal{F}_{\mathsf{MPC}}$ by calling $\mathcal{F}_{\mathsf{MPC}}.\mathsf{Init}(k)$. The parties run $\mathcal{F}_{\mathsf{BMR}}.\mathsf{Garble}(\mathsf{C}_{\mathsf{CPU}}, (W, P))$ to garble T CPU circuits: $GC^1, \ldots, GC^T$, where (W, P) maps from input wires to their associated party in the first execution and are the empty set in all other invocations.

Input. To input $x^i = x^i_1, \ldots, x^i_\ell$ of party p_i, let w be the wire associated with input bit x^i_j. Party p_i broadcasts $\Lambda_w = x^i_j \oplus \lambda_w$ and every party p_j broadcasts k^j_{w,Λ_w}.

Output. Invoke $\mathcal{F}_{\mathsf{BMR}}.\mathsf{Open}(GC^t)$ for all $t \in [T]$.
For $t = 1$ to T:

1. The parties run $\mathcal{E}_{\mathsf{BMR}}(GC^t)$ and obtain $k_{w,\Lambda_w}, \Lambda_w$ for every $w \in W^t_{\mathsf{write}}$ and obtain $i^{\mathsf{write}}_t, i^{\mathsf{read}}_t$ in the clear.
2. For each $w \in W^t_{\mathsf{write}}$: Locally compute $[\![\rho_w]\!] = \Lambda_w + [\![\lambda_w]\!]$. Party p_i stores $[\![\rho_w]\!]^i$ in $D^i[i^{\mathsf{write}}_t]$. In addition, they check whether k_{w,Λ_w} equals the key received in $\mathcal{P}_{\mathsf{BMR}}.\mathsf{Garble}$.
3. For each $w \in W^{t+1}_{\mathsf{read}}$: Let i^{read}_t be the memory address from which w is to be fed. Then the parties open $\Lambda_w = D[i^{\mathsf{read}}_t] + [\![\lambda_w]\!]$. Given Λ_w, party p_i broadcasts k^i_{w,Λ_w} so that all parties finally hold k_{w,Λ_w}.
4. Obtain k_{w,Λ_w} for every $w \in W^{t+1}_{\mathsf{state}}$.

Finally, for every $w \in W^T_{\mathsf{state}}$, (output wires of $\mathsf{C}^T_{\mathsf{CPU}}$) the parties open λ_w and compute $\rho = \lambda_w \oplus \Lambda_w$.

Fig. 7. Realizing $\mathcal{F}_{\mathsf{UAP}}$ in the $\mathcal{F}_{\mathsf{BMR}}$-hybrid model.

5.1 Security of Protocol $\mathcal{P}_{\mathsf{UAP}}$

The security of our construction relies on the security of the underlying BMR and SPDZ protocols and the security of the transformation between the garbled wires and SPDZ shares. Informally, the latter can be seen as follows: Neither transformation reveals any secret information because one direction (writing to memory) is done locally, and the other one (reading form memory) only reveals an external value and the corresponding wire label, both of which hide the real value that is carried over the wire according to the security of the BMR protocol. For malicious security, consider that revealing the external value is done using SPDZ, which guarantees correctness by checking the MAC. Furthermore, if any party broadcasts a faulty share of the BMR key, this is guaranteed to lead to an invalid output key (and thus easy detection by honest parties) by the properties of the BMR protocol. More formally, we prove the following theorem:

Theorem 5.1. *Protocol $\mathcal{P}_{\mathsf{UAP}}$ (Fig. 7) realizes functionality $\mathcal{F}_{\mathsf{UAP}}$ (Fig. 2) in the $\mathcal{F}_{\mathsf{BMR}}$-hybrid model.*

Proof. Let $\mathcal{A}$ be an adversary controlling a subset of the parties, denoted $A = \{p_{i_1}, \ldots, p_{i_c}\}$ and denote by $\bar{A} = [n] \setminus A$ the subset of the honest parties.

We present a simulator $\mathcal{S}$ who participates in the ideal execution $\mathcal{F}_{\mathsf{UAP}}$ by taking the role of A and in an internal execution of $\mathcal{P}_{\mathsf{UAP}}$ with $\mathcal{A}$, in which $\mathcal{S}$ takes the role of $\bar{A}$ and the functionality $\mathcal{F}_{\mathsf{BMR}}$. The simulator $\mathcal{S}$ uses another simulator $\mathcal{S}_{\mathsf{BMR}}$ that when given the adversary's input/output to/from a circuit, and both keys to all wires in the garbled circuit, produces a view that is indistinguishable to the view of the adversary's evaluation of the circuit in the real execution (such a simulator was presented in [25]).

The simulator $\mathcal{S}$ does as follows:

1. **Extract $\mathcal{A}$'s inputs**
   (a) In the internal execution, garble T copies of $\mathrm{C}_{\mathrm{CPU}}$ exactly as described in $\mathcal{F}_{\mathsf{BMR}}$. In particular, for every input wire $w \in GC^1$ associated with a corrupted party $p_c \in A$ output λ_w in the clear to p_c.
   (b) Upon issuing the **Input** command in the internal execution, for an input wire w associated with a corrupted party $p_c \in A$, receive p_c's external value Λ_w and compute p_c's input to wire w by $\rho_w = \Lambda_w \oplus \lambda_w$ (the simulator $\mathcal{S}$ knows λ_w because it was garbling the circuit on behalf of $\mathcal{F}_{\mathsf{BMR}}$).
2. Engage in the ideal execution $\mathcal{F}_{\mathsf{UAP}}$ by inputting the values extracted above as the corrupted parties' input and obtain $\boldsymbol{y} = \Pi(D, \boldsymbol{x})$ along with $\mathsf{AP}(\Pi, D, \boldsymbol{x})$.
3. Open all garbled circuits $GC^1, \ldots, GC^T$ toward the adversary.
4. **Evaluation**
   (a) Invoke $\mathcal{S}_{\mathsf{BMR}}$ with $\mathcal{A}$'s inputs that were extracted earlier, the garbled circuit GC^1 and the adversary's output from GC^1: $\mathsf{access}_1 = (i_1^{\mathsf{read}}, i_1^{\mathsf{write}})$.[6] Output whatever $\mathcal{S}_{\mathsf{BMR}}$ produced.
   (b) Note that A have no inputs to circuits $GC^2, \ldots, GC^T$, thus, for every such circuit we invoke $\mathcal{S}_{\mathsf{BMR}}$ with no inputs at all. Then, for every input wire of the garbled circuit, $\mathcal{S}$ checks which of its keys $k_{w,0}$ or $k_{w,1}$ was produced as the simulated view, and then supply the correct share of the external value of that key ($\mathcal{S}$ can do this since it knows the global MAC key used for the SPDZ shares). Formally: For $t = 2$ to T:
      i. Output $\mathsf{access}_{t-1} = i_{t-1}^{\mathsf{read}}, i_{t-1}^{\mathsf{write}}$ in the clear.
      ii. Invoke $\mathcal{S}_{\mathsf{BMR}}$ with GC^t. If $t = T$ then supply $\mathcal{S}_{\mathsf{BMR}}$ with $\boldsymbol{y}$ as well.
      iii. For every input wire w of GC^t, extract from the produced view the key k_{w,Λ_w} used in the evaluation.

[6] Note that access_1 is the output to all parties, not only the adversary's, however, for the simulation purpose we use this as the adversary's output.

iv. For every input wire w of GC^t simulate the opening of $[\![\Lambda_w]\!] = [\![\lambda_w \oplus \rho_w]\!]$ (from Step 3 in $\mathcal{P}_{\mathsf{UAP}}$) as follows: Let the shares of A be Λ_w^A, then $\mathcal{S}$ chooses random shares Λ_w^h for every honest party $p_h \in \bar{A}$ such that $\Lambda_w = \Lambda_w^A + \sum_{h \in \bar{A}} \Lambda_w^h$ and use them to open Λ_w.

v. Output the view produced by $\mathcal{S}_{\mathsf{BMR}}$ in Step 4(b)ii above.

Claim. For every PPT $\mathcal{A}$ and for every $\boldsymbol{x}$ the output of $\mathcal{S}$ above is indistinguishable to the view of $\mathcal{A}$ in the real execution of $\mathcal{P}_{\mathsf{UAP}}$.

Proof. We define hybrid $\mathbf{Hyb}_t$ to be as follows: The adversary view in the real execution of $\mathcal{P}_{\mathsf{UAP}}$ for timesteps $1, \ldots, t$, followed by the simulated view for timesteps $t+1, \ldots, T$ as described in Step 4b of the simulation above. Thus, $\mathbf{Hyb}_T$ is exactly the real execution of $\mathcal{P}_{\mathsf{UAP}}$ and $\mathbf{Hyb}_0$ is exactly the output of $\mathcal{S}$ described above. Assume by contradiction that there exists a PPT $\mathcal{D}$ who can distinguish between $\mathbf{Hyb}_T$ and $\mathbf{Hyb}_0$ with non negligible probability, then we construct $\mathcal{D}'$ who distinguishes between a real execution of $\mathcal{P}_{\mathsf{BMR}}$ to the output of $\mathcal{S}_{\mathsf{BMR}}$ (for a single circuit) as follows: By the existence of $\mathcal{D}$ it is implied that there exists t' for which $\mathcal{D}$ distinguishes between $\mathbf{Hyb}_t$ and $\mathbf{Hyb}_{t+1}$ with non negligible probability. Then, given a circuit C and a view V which is either the view of the adversary in a real execution of $\mathcal{P}_{\mathsf{BMR}}$ or the output of $\mathcal{S}_{\mathsf{BMR}}$, $\mathcal{D}'$ generates a real view of the execution of $\mathcal{P}_{\mathsf{UAP}}$ for timesteps $1, \ldots, t$, then plugs V together with the opening of external values of the input wires of C, and then complete the simulation according to Step 4b above. Finally, hands the result view to $\mathcal{D}$ and outputs whatever $\mathcal{D}$ outputs. Observe that if V is a view of a real execution then the above is distributed exactly as $\mathbf{Hyb}_t$, otherwise, it is distributed exactly as $\mathbf{Hyb}_{t+1}$. It follows that $\mathcal{D}'$ distinguishes between the real execution of $\mathcal{P}_{\mathsf{BMR}}$ and the output of $\mathcal{S}_{\mathsf{BMR}}$ with non negligible probability, by contradiction of the security of $\mathcal{P}_{\mathsf{BMR}}$.

6 Optimizing BMR Evaluation

The free-XOR technique of [2] makes space and communication complexities linear in the number of AND gates (XOR gates are almost for free[7]). In this section we show how to further decrease memory consumption in the online phase by a factor of up to 2. Even though our technique could be applied to a plain BMR protocol, we present the idea over a scheme that uses the free-XOR. We stress that it is not limited to secure *RAM* computation but also applicable in BMR-based protocols, even with only a single execution.

The Evaluate instruction in $\mathcal{F}_{\mathsf{BMR}}$ (Fig. 3) that is invoked in the online phase traverses the circuit in a topological order and obtains a single output key $k_w \in (\mathbb{F}_{2^\kappa})^n$ for every wire w in the circuit, until it reaches the output wires. To check the authenticity of k_w, party p_i extracts the ith element, $k_w^i \in \mathbb{F}_{2^\kappa}$, and verifies that it is one of the keys given to him by $\mathcal{F}_{\mathsf{MPC}}$ in the offline phase, that is, $k_w^i \in \{k_{w,0}^i, k_{w_1}^i\}$. In case that $k_w^i \notin \{k_w^0, k_w^1\}$ then p_i notifies all parties with

[7] They require only a simple XOR operation.

regard to the corrupted garbled circuit and aborts. Using our technique, it is possible for p_i to discard its keys $\{k^i_{w,0}, k^i_{w_1}\}$ of all wires right after the garbled circuit construction is complete (in the offline phase), instead, it has to store only a *single* bit per wire. Since the garbled gate is of size $4nk$ and the original verification procedure requires memory of size $2k$ (i.e. party p_i stores the two keys of the output wire of the gate), this results with a decrease of memory consumption by a factor of $\frac{1}{2n}$. However, a great improvement is achieved for a more recent construction [3]. In that construction the size of a garbled gate is $4k$ (i.e. it is independent of the number of parties n), thus, memory saving is significant.

Using our technique, the evaluator is saved from loading and comparing 1.5 keys per wire in average (since in half of the wires the verification passes after the first comparison). This loading[8] and comparison time became substantial as the computation of AES has been considerably improved[9].

6.1 The Technique

Circuit garbling is done in the offline phase of the protocol using the $\mathcal{F}_{\mathsf{MPC}}$ functionality (Fig. 11). Let $\mathsf{lsb}(x)$ denote the least significant bit of x. We instruct $\mathcal{F}_{\mathsf{MPC}}$ to choose $\Delta = (\Delta_1, \ldots, \Delta_n)$ such that $\mathsf{lsb}(\Delta_i) = 1$ for every $i \in [n]$. The result is that $\mathsf{lsb}(k^i_{w,0}) \neq \mathsf{lsb}(k^i_{w,1})$ for all w and i. When garbling is completed using Δ as described, party p_i stores the bit $\delta^i_w = \mathsf{lsb}(k^i_{w,0})$ for every wire w. In addition party p_i discards all keys $k^i_{w,0}$ and $k^i_{w,1}$ for all but the output wires.

The evaluation of the circuit is done exactly as before, however, instead of verifying the key validity of the output wire of *every* gate, this is done only for *output* gates. For an inner gate with output wire w, party p_i obtains the external value Λ_w by computing $\Lambda_w = \mathsf{lsb}(k^i_w) \oplus \delta^i_w$. This way the parties learn that the key k^i_w obtained by evaluating a gate is actually the Λ-key. For output gates (i.e. gates whose output wire is also a circuit-output wire), party p_i verifies that $k^i_w \in \{k^i_{w,0}, k^i_{w,1}\}$ as before.

Forcing the last bit of a random element is featured in SPDZ-like implementation of $\mathcal{F}_{\mathsf{MPC}}$ (e.g. [23]) since they are inherently bit wise, so we can generate $k-1$ random bits and then compose the field element accordingly so its last bit is 1.

6.2 Security

Notice that we use the exact same garbling procedure as in [2] except that here the last bit of every Δ_i is known to the adversary (i.e. $\mathsf{lsb}(\Delta_i) = 1$) whereas in their scheme *all* bits of Δ are random. The security of our scheme can be easily reduced to the security of [2]. Our simulator is the same simulator as in [2]. Let the distinguisher's advantage in distinguishing between the real execution

[8] Loading time depends on the implementation, i.e. whether using dereferences or not.

[9] Using the AES-NI instruction set from Intel's Sandy Bridge microarchitecture and on, a RoundKey instruction takes a single CPU cycle and latency of 8, that is, one could reach a throughput of up to 8 RoundKey operations with the same key at the same CPU cycle [21, Chap. 5.10].

of our scheme to the ideal execution be ϵ. Then, the advantage of the *same* distinguisher in distinguishing between the real execution and the simulation of [2] is $\epsilon' = \epsilon \cdot \frac{1}{2^h}$ for h honest parties. This holds because the probability of having $\mathsf{lsb}(\Delta_i) = 1$ in the free-XOR scheme is $\frac{1}{2}$ for an honest party p_i. Recall that in the original scheme, the security depends on h keys of length k. Thus, increasing the advantage of the adversary by 2^h is negligible. Assuming that [2] is secure we conclude that our scheme is secure as well.

7 Implementation

In this section we report our results of the *first* (to the best of our knowledge) implementation of a garbled-circuit-based secure RAM computation for setting with active security and dishonest majority. We chose to implement our third technique (Sect. 4.3) as it is the most efficient technique for memory access. We have combined our new BMR implementation with the existing SPDZ system [6], and used it to implement an oblivious array[10] using Circuit ORAM [36]. The code is written in C++ using the AES-NI and AVX2 instruction sets.

Experiments. Our timing results below refer to the following experiments:

1. Circuit ORAM [36] using the BMR-SPDZ protocol with the scheme in Sect. 4.3, labeled as 'BMR, Circuit ORAM' in the figures below.
2. Circuit ORAM [36] using a pure SPDZ implementation, labeled as 'Pure SPDZ, Circuit ORAM'.
3. Path ORAM [35] using a pure SPDZ implementation [24], labeled as 'Pure SPDZ, Path ORAM'.
4. Trivial ORAM, i.e. linear scanning of the entire memory for every access, labeled as 'BMR, linear scan'.

The Path ORAM intends to optimize the *bandwidth cost* and *bandwidth blowup* where bandwidth cost refers to the average number of bits transferred for accessing a single block and bandwidth blowup is defined as bandwidth cost divided by the block size (i.e., the bit-length of a data block)[11]. The results by Keller and Scholl [24] are reported using Path ORAM, which seems preferable when round complexity is not a concern. For the sake of comparison, we have also implemented Circuit ORAM using pure SPDZ. Comparing experiments (1), (2) and (3) in Figs. 8 and 9, our approach outperforms the pure SPDZ when the parties are connected over a WAN, independently of the choice of the ORAM scheme. Furthermore, experiment (4) allows to find the *breakeven* points, that is,

[10] "Oblivious array" is the name given in [24] to the basic oblivious random memory access, which allows reading and writing with a secret index. This is in distinction to "oblivious dictionary" that allows reading according to a secret 'key' in a key-value (dictionary) data structure, where the key may be larger than the size of the memory.

[11] As defined in [36, A.2] under ORAM metrics.

to figure out up to what memory size the linear scan performs better than applying an ORAM algorithm. Given the simplicity of a linear scan, it is clear that it is faster for small enough sizes.

All experiments were performed for both LAN and WAN environment to test the influence of our approach of reducing the round complexity. We stress that our implementation is the first in this setting even when considering 2 parties only. Nevertheless, we report timing results for a protocol with 3 participants as well.

Parameters. Our security parameters are $\kappa = 128$ and $s = 40$. In all experiments, the oblivious arrays are made up of 32-bit entries, and all figures refer to the array size as the number of such entries. Therefore, our figures range from $1024 \cdot 32 \approx 32\,\text{kB}$ to $2^{25} \cdot 32 \approx 1.1\,\text{GB}$.

Our ORAM implementations (Circuit ORAM and Path ORAM) require up to three recursions such that intermediate ORAMs use 128-bit entries, and we use a linear scan for less than 256 such entries.

All reported results are measured per *logical access* to the memory (array), which, as explained before, may incorporate many physical accesses.

Environment. Our implementations were done using 4th generation Intel Core i7 with 8 cores running at 3.5 GHz, 16 GB RAM, and SSD (to store the garbled circuits) connected over a LAN (bandwidth of 1 Gbit/s and RTT of 0.1–0.2 ms).

Furthermore, we have simulated a WAN setting on the same machines by extending the round trip time to 100 ms and restricting the throughput to 50 Mbit/s. Figure 8 shows our results for the two settings with two parties while Fig. 9 shows our results for three parties. They confirm that using garbled circuits (BMR) is beneficial with high network latencies. With BMR, combining Circuit ORAM with our memory access surpasses linear scanning below a size of one million.

Offline Cost. Finally, for a more complete picture, we have estimated the offline cost in the LAN setting. Figure 10 shows the cost for one access of Circuit ORAM

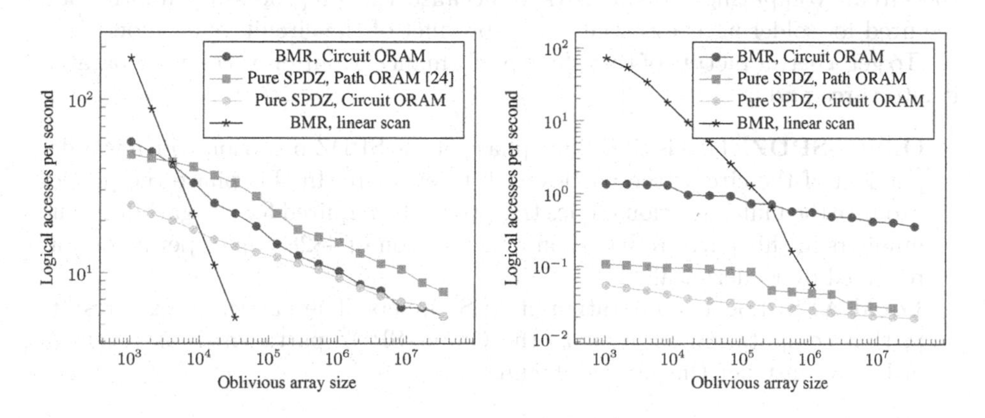

Fig. 8. Two parties over LAN (left) and over WAN (right).

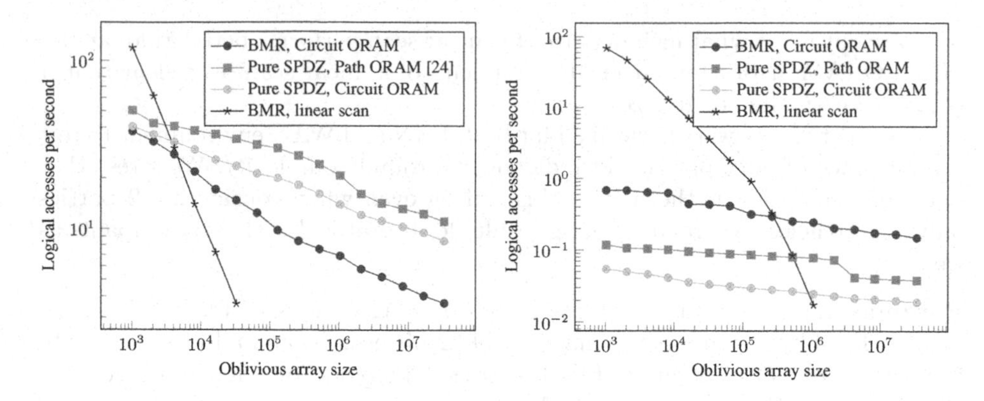

Fig. 9. Three parties over LAN (left) and over WAN (right).

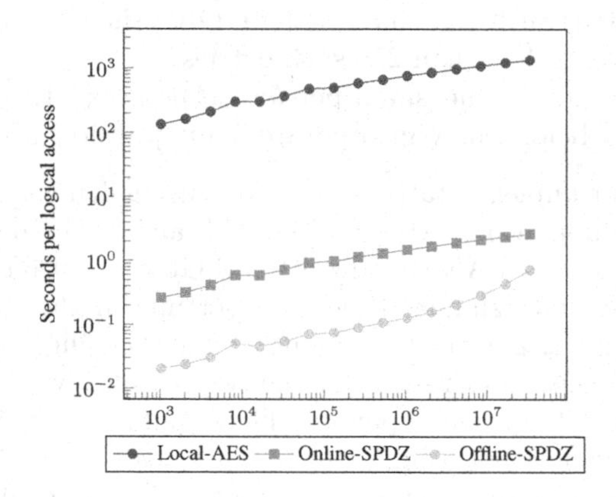

Fig. 10. Offline time per logical access with Circuit ORAM and BMR.

implemented in BMR. All figures are based on the number of AND gates in the circuit computing Circuit ORAM because the preprocessing information required for soldering is essentially a by-product of the circuit generation.

To get a better picture of the offline performance of our protocol, we separated it into three parts:

- **Offline-SPDZ.** This is the offline phase of the SPDZ protocol, which is independent of the circuit the parties wish to evaluate. In this phase the parties produce the multiplication triples that would be required for the garbling. The numbers in this part are based on a production of 4828 triples per second as reported by Keller et al. [23].
- **Local-AES.** Local computation of AES ciphers. The parties use the results of that computation as input to the Online-SPDZ part, which use them in order to construct the garbled circuit.

- **Online-SPDZ.** This is the online phase of the SPDZ protocol, in which the parties evaluate a circuit that garbles the actual circuit they want to evaluate in the BMR online phase.

In the figure we can easily observe that Offline-SPDZ dominates the cost by 3–4 orders of magnitudes because of the communication cost of MASCOT [23].

A The Generic Reactive MPC Functionality

The following functionality is used by protocols that follow the BMR-SPDZ approach.

The Generic Reactive MPC Functionality: $\mathcal{F}_{\mathsf{MPC}}$

The functionality consists of seven externally exposed commands **Initialize**, **Input-Data**, **RandomBit**, **Random**, **Add**, **Multiply**, and **Output**, and one internal subroutine **Wait**.

Initialize On input (Init, k) from all parties, the functionality activates and stores k. All additions and multiplications below will be in $\mathbb{F}_{2^\kappa}$.

Wait This waits on the adversary to return a $GO/NO\text{-}GO$ decision. If the adversary returns $NO\text{-}GO$ then the functionality aborts.

InputData On input $(\mathsf{Input}, p_i, varid, x)$ from party p_i and $(input, p_i, varid, ?)$ from all other parties, with $varid$ a fresh identifier, the functionality stores $(varid, x)$. The functionality then calls **Wait**.

RandomBit On command $(\mathsf{RandomBit}, varid)$ from all parties, with $varid$ a fresh identifier, the functionality selects a random value $r \in \{0, 1\}$ and stores $(varid, r)$. The functionality then calls **Wait**.

Random On command $(\mathsf{Random}, varid)$ from all parties, with $varid$ a fresh identifier, the functionality selects a random value $r \in \mathbb{F}_{2^\kappa}$ and stores $(varid, r)$. The functionality then calls **Wait**.

Add On command $(\mathsf{Add}, varid_1, varid_2, varid_3)$ from all parties (if $varid_1, varid_2$ are present in memory and $varid_3$ is not), the functionality retrieves $(varid_1, x)$, $(varid_2, y)$ and stores $(varid_3, x + y)$. The functionality then calls **Wait**.

Multiply On input $(\mathsf{Multiply}, varid_1, varid_2, varid_3)$ from all parties (if $varid_1, varid_2$ are present in memory and $varid_3$ is not), the functionality retrieves $(varid_1, x)$, $(varid_2, y)$ and stores $(varid_3, xy)$. The functionality then calls **Wait**.

Output On input $(\mathsf{Output}, varid, i)$ from all honest parties (if $varid$ is present in memory), the functionality retrieves $(varid, x)$ and outputs either $(varid, x)$ in the case of $i \neq 0$ or $(varid)$ if $i = 0$ to the adversary. The functionality then calls **Wait**, and only if **Wait** does not abort then it outputs x to all parties if $i = 0$, or it outputs x only to party i if $i \neq 0$.

Fig. 11. The generic reactive MPC functionality

References

1. Afshar, A., Hu, Z., Mohassel, P., Rosulek, M.: How to efficiently evaluate RAM programs with malicious security. In: Oswald, E., Fischlin, M. (eds.) EUROCRYPT 2015. LNCS, vol. 9056, pp. 702–729. Springer, Heidelberg (2015). https://doi.org/10.1007/978-3-662-46800-5_27
2. Ben-Efraim, A., Lindell, Y., Omri, E.: Optimizing semi-honest secure multiparty computation for the internet, pp. 578–590 (2016)
3. Ben-Efraim, A., Lindell, Y., Omri, E.: Efficient scalable constant-round MPC via garbled circuits. In: Takagi, T., Peyrin, T. (eds.) ASIACRYPT 2017. LNCS, vol. 10625, pp. 471–498. Springer, Cham (2017). https://doi.org/10.1007/978-3-319-70697-9_17
4. Boyle, E., Gilboa, N., Ishai, Y.: Function secret sharing. In: Oswald, E., Fischlin, M. (eds.) EUROCRYPT 2015. LNCS, vol. 9057, pp. 337–367. Springer, Heidelberg (2015). https://doi.org/10.1007/978-3-662-46803-6_12
5. Boyle, E., Gilboa, N., Ishai, Y.: Function secret sharing: improvements and extensions, pp. 1292–1303 (2016)
6. Bristol Cryptography Group: SPDZ software (2016). https://www.cs.bris.ac.uk/Research/CryptographySecurity/SPDZ/
7. Canetti, R., Holmgren, J.: Fully succinct garbled RAM, pp. 169–178 (2016)
8. Damgård, I., Pastro, V., Smart, N.P., Zakarias, S.: Multiparty computation from somewhat homomorphic encryption. In: Safavi-Naini, R., Canetti, R. (eds.) CRYPTO 2012. LNCS, vol. 7417, pp. 643–662. Springer, Heidelberg (2012). https://doi.org/10.1007/978-3-642-32009-5_38
9. Doerner, J., Shelat, A.: Scaling ORAM for secure computation. In: CCS (2017)
10. Frederiksen, T.K., Jakobsen, T.P., Nielsen, J.B., Nordholt, P.S., Orlandi, C.: MiniLEGO: efficient secure two-party computation from general assumptions. In: Johansson, T., Nguyen, P.Q. (eds.) EUROCRYPT 2013. LNCS, vol. 7881, pp. 537–556. Springer, Heidelberg (2013). https://doi.org/10.1007/978-3-642-38348-9_32
11. Garg, S., Gupta, D., Miao, P., Pandey, O.: Secure multiparty RAM computation in constant rounds. In: Hirt, M., Smith, A. (eds.) TCC 2016. LNCS, vol. 9985, pp. 491–520. Springer, Heidelberg (2016). https://doi.org/10.1007/978-3-662-53641-4_19
12. Garg, S., Lu, S., Ostrovsky, R.: Black-box garbled RAM, pp. 210–229 (2015)
13. Garg, S., Lu, S., Ostrovsky, R., Scafuro, A.: Garbled RAM from one-way functions, pp. 449–458 (2015)
14. Gentry, C., Halevi, S., Lu, S., Ostrovsky, R., Raykova, M., Wichs, D.: Garbled RAM revisited. In: Nguyen, P.Q., Oswald, E. (eds.) EUROCRYPT 2014. LNCS, vol. 8441, pp. 405–422. Springer, Heidelberg (2014). https://doi.org/10.1007/978-3-642-55220-5_23
15. Goldreich, O.: Foundations of Cryptography: Basic Applications, vol. 2. Cambridge University Press, New York (2004)
16. Goldreich, O., Micali, S., Wigderson, A.: How to play any mental game or a completeness theorem for protocols with honest majority. In: STOC, pp. 218–229 (1987)
17. Goldreich, O., Ostrovsky, R.: Software protection and simulation on oblivious RAMs. J. ACM **43**(3), 431–473 (1996)
18. Gordon, S.D., Katz, J., Kolesnikov, V., Krell, F., Malkin, T., Raykova, M., Vahlis, Y.: Secure two-party computation in sublinear (amortized) time, pp. 513–524 (2012)

19. Hazay, C., Yanai, A.: Constant-round maliciously secure two-party computation in the RAM model. In: Hirt, M., Smith, A. (eds.) TCC 2016. LNCS, vol. 9985, pp. 521–553. Springer, Heidelberg (2016). https://doi.org/10.1007/978-3-662-53641-4_20
20. Hu, Z., Mohassel, P., Rosulek, M.: Efficient zero-knowledge proofs of non-algebraic statements with sublinear amortized cost. In: Gennaro, R., Robshaw, M. (eds.) CRYPTO 2015. LNCS, vol. 9216, pp. 150–169. Springer, Heidelberg (2015). https://doi.org/10.1007/978-3-662-48000-7_8
21. Intel: Intel 64 and IA-32 Architectures Optimization Reference Manual (2016). http://www.intel.com/content/www/us/en/architecture-and-technology/64-ia-32-architectures-optimization-manual.html
22. Keller, M.: The oblivious machine - or: how to put the C into MPC. Cryptology ePrint Archive, Report 2015/467 (2015). http://eprint.iacr.org/2015/467
23. Keller, M., Orsini, E., Scholl, P.: MASCOT: faster malicious arithmetic secure computation with oblivious transfer, pp. 830–842 (2016)
24. Keller, M., Scholl, P.: Efficient, oblivious data structures for MPC. In: Sarkar, P., Iwata, T. (eds.) ASIACRYPT 2014. LNCS, vol. 8874, pp. 506–525. Springer, Heidelberg (2014). https://doi.org/10.1007/978-3-662-45608-8_27
25. Lindell, Y., Pinkas, B., Smart, N.P., Yanai, A.: Efficient constant round multi-party computation combining BMR and SPDZ. In: Gennaro, R., Robshaw, M. (eds.) CRYPTO 2015. LNCS, vol. 9216, pp. 319–338. Springer, Heidelberg (2015). https://doi.org/10.1007/978-3-662-48000-7_16
26. Lindell, Y., Riva, B.: Blazing fast 2PC in the offline/online setting with security for malicious adversaries. In: CCS, pp. 579–590 (2015)
27. Lindell, Y., Smart, N.P., Soria-Vazquez, E.: More efficient constant-round multi-party computation from BMR and SHE. In: Hirt, M., Smith, A. (eds.) TCC 2016. LNCS, vol. 9985, pp. 554–581. Springer, Heidelberg (2016). https://doi.org/10.1007/978-3-662-53641-4_21
28. Liu, C., Huang, Y., Shi, E., Katz, J., Hicks, M.W.: Automating efficient RAM-model secure computation, pp. 623–638 (2014)
29. Liu, C., Wang, X.S., Nayak, K., Huang, Y., Shi, E.: ObliVM: a programming framework for secure computation, pp. 359–376 (2015)
30. Lu, S., Ostrovsky, R.: How to garble RAM programs? In: Johansson, T., Nguyen, P.Q. (eds.) EUROCRYPT 2013. LNCS, vol. 7881, pp. 719–734. Springer, Heidelberg (2013). https://doi.org/10.1007/978-3-642-38348-9_42
31. Mohassel, P., Rosulek, M., Scafuro, A.: Sublinear zero-knowledge arguments for RAM programs. In: Coron, J.-S., Nielsen, J.B. (eds.) EUROCRYPT 2017. LNCS, vol. 10210, pp. 501–531. Springer, Cham (2017). https://doi.org/10.1007/978-3-319-56620-7_18
32. Nielsen, J.B., Orlandi, C.: LEGO for two-party secure computation. In: Reingold, O. (ed.) TCC 2009. LNCS, vol. 5444, pp. 368–386. Springer, Heidelberg (2009). https://doi.org/10.1007/978-3-642-00457-5_22
33. Ostrovsky, R., Shoup, V.: Private information storage (extended abstract), pp. 294–303 (1997)
34. Shi, E., Chan, T.-H.H., Stefanov, E., Li, M.: Oblivious RAM with $O((\log N)^3)$ worst-case cost. In: Lee, D.H., Wang, X. (eds.) ASIACRYPT 2011. LNCS, vol. 7073, pp. 197–214. Springer, Heidelberg (2011). https://doi.org/10.1007/978-3-642-25385-0_11
35. Stefanov, E., van Dijk, M., Shi, E., Fletcher, C.W., Ren, L., Yu, X., Devadas, S.: Path ORAM: an extremely simple oblivious RAM protocol, pp. 299–310 (2013)

36. Wang, X., Chan, T.-H.H., Shi, E.: Circuit ORAM: On tightness of the Goldreich-Ostrovsky lower bound, pp. 850–861 (2015)
37. Wang, X., Malozemoff, A.J., Katz, J.: Faster secure two-party computation in the single-execution setting. In: Coron, J.-S., Nielsen, J.B. (eds.) EUROCRYPT 2017. LNCS, vol. 10212, pp. 399–424. Springer, Cham (2017). https://doi.org/10.1007/978-3-319-56617-7_14
38. Wang, X., Ranellucci, S., Katz, J.: Authenticated garbling and efficient maliciously secure two-party computation. In: CCS, pp. 21–37 (2017)
39. Wang, X.S., Gordon, S.D., McIntosh, A., Katz, J.: Secure computation of MIPS machine code. In: Askoxylakis, I., Ioannidis, S., Katsikas, S., Meadows, C. (eds.) ESORICS 2016. LNCS, vol. 9879, pp. 99–117. Springer, Cham (2016). https://doi.org/10.1007/978-3-319-45741-3_6
40. Yao, A.C.-C.: How to generate and exchange secrets (extended abstract). In: FOCS, pp. 162–167 (1986)

Efficient Circuit-Based PSI via Cuckoo Hashing

Benny Pinkas[1](✉), Thomas Schneider[2], Christian Weinert[2], and Udi Wieder[3]

[1] Bar-Ilan University, Ramat Gan, Israel
benny@pinkas.net
[2] TU Darmstadt, Darmstadt, Germany
{thomas.schneider,christian.weinert}@crisp-da.de
[3] VMware Research, Palo Alto, USA
udi.wieder@gmail.com

Abstract. While there has been a lot of progress in designing efficient custom protocols for computing Private Set Intersection (PSI), there has been less research on using generic Multi-Party Computation (MPC) protocols for this task. However, there are many variants of the set intersection functionality that are not addressed by the existing custom PSI solutions and are easy to compute with generic MPC protocols (e.g., comparing the cardinality of the intersection with a threshold or measuring ad conversion rates).

Generic PSI protocols work over circuits that compute the intersection. For sets of size n, the best known circuit constructions conduct $O(n \log n)$ or $O(n \log n / \log \log n)$ comparisons (Huang et al., NDSS'12 and Pinkas et al., USENIX Security'15). In this work, we propose new circuit-based protocols for computing *variants of the intersection* with an almost linear number of comparisons. Our constructions are based on new variants of Cuckoo hashing in two dimensions.

We present an asymptotically efficient protocol as well as a protocol with better concrete efficiency. For the latter protocol, we determine the required sizes of tables and circuits experimentally, and show that the run-time is concretely better than that of existing constructions.

The protocol can be extended to a larger number of parties. The proof technique presented in the full version for analyzing Cuckoo hashing in two dimensions is new and can be generalized to analyzing standard Cuckoo hashing as well as other new variants of it.

Keywords: Private set intersection · Secure computation

1 Introduction

Private Set Intersection (PSI) refers to a protocol which enables two parties, holding respective input sets X and Y, to compute the intersection $X \cap Y$ without revealing any information about the items which are not in the intersection. The PSI functionality is useful for applications where parties need to apply a JOIN operation to private datasets. There are multiple constructions of secure

J. B. Nielsen and V. Rijmen (Eds.): EUROCRYPT 2018, LNCS 10822, pp. 125–157, 2018.
https://doi.org/10.1007/978-3-319-78372-7_5

protocols for computing PSI, but there is an advantage for computing PSI by applying a generic Multi-Party Computation (MPC) protocol to a circuit computing the intersection (see Sect. 1.1). The problem is that a naive circuit computes $O(n^2)$ comparisons, and even the most recent circuit-based constructions require $O(n \log n)$ or $O(n \log n / \log \log n)$ comparisons (see Sect. 1.4).

In this work, we present a new circuit-based protocol for computing PSI variants. In our protocol, each party first inserts its input elements into bins according to a new hashing algorithm, and then the intersection is computed by securely computing a Boolean comparison circuit over the bins. The insertion of the items is based on new Cuckoo hashing variants which guarantee that if the two parties have the same input value, then there is exactly one bin to which both parties map this value. Furthermore, the total number of bins is $O(n)$ and there are $O(1)$ items mapped to each bin, plus $\omega(1)$ items which are mapped to a special stash. Hence, the circuit that compares (1) for each bin, the items that the two parties mapped to it, and (2) all stash items to all items of the other party, computes only $\omega(n)$ comparisons.

1.1 Motivation for Circuit-Based PSI

PSI has many applications, as is detailed for example in [42]. Consequently, there has been a lot of research on efficient secure computation of PSI, as we describe in Sect. 1.4. However, most research was focused on computing the intersection itself, while there are interesting applications for the ability to securely compute arbitrary functions of the intersection. We demonstrate the need for efficient computation of PSI using generic protocols through the following arguments:

Adaptability. Assume that you are a cryptographer and were asked to propose and implement a protocol for computing PSI. One approach is to use a specialized protocol for computing PSI. Another possible approach is to use a protocol for generic secure computation, and apply it to a circuit that computes PSI. A trivial circuit performs $O(n^2)$ comparisons, while more efficient circuits, described in [26, 39], perform only $O(n \log n)$ or $O(n \log n / \log \log n)$ comparisons, respectively. The most efficient specialized PSI protocols are faster by about two orders of magnitude than circuit-based constructions (see [39]), and therefore you will probably choose to use a specialized PSI protocol. However, what happens if you are later asked to change the protocol to compute another function of the intersection? For example, output only the size of the intersection, or output 1 iff the size is greater than some threshold, or output the most "representative" item that occurs in the intersection (according to some metric). Any change to a specialized protocol will require considerable cryptographic know-how, and might not even be possible. On the other hand, the task of writing a new circuit component that computes a different function of the intersection is rather trivial, and can even be performed by undergrad students.

Consider the following function as an example of a variant of the PSI functionality for which we do not know a specialized protocol: Suppose that you want to compute the size of the intersection, but you also wish to preserve the privacy of users by ensuring differential privacy. This is done by adding some noise to the exact count before releasing it. This functionality can easily be computed by a circuit, but it is unclear how to compute it using other PSI protocols. (See [38] for constructions that add noise to the results of MPC computation in order to ensure differential privacy.)

Existing code base. Circuit-based protocols benefit from all the work that was invested in recent years in designing, implementing, and optimizing very efficient systems for generic secure computation. Users can download existing secure computation software, e.g., [13,27], and only need to design the circuit to be computed and implement the appropriate hashing technique.

Existing applications. There are existing applications that need to compute functions over the results of the set intersection. For example, Google reported [34,49] a PSI-based application for measuring ad conversion rates, namely the revenues from ad viewers who later perform a related transaction. This computation can be done by comparing the list of people who have seen an ad with those who have completed a transaction. These lists are held by the advertiser (say, Google or Facebook), and by merchants, respectively. A simple (non-private) solution is for one side to disclose its list of customers to the other side, which computes the necessary statistics. Another option is to run a secure computation over the results of the set intersection. For example, the merchant inputs pairs of the customer-identity and the value of the transactions made by this customer, and the computation calculates the total revenue from customers who have seen an ad, namely customers in the intersection of the sets known to the advertiser and the merchant. Google reported implementing this computation using a Diffie-Hellman-based PSI cardinality protocol (for computing the cardinality of the intersection) and Paillier encryption (for computing the total revenues) [28]. This protocol reveals the identities of the items in the intersection, and seems less efficient than our protocol as it uses public key operations, rather than efficient symmetric cryptographic operations.[1]

1.2 Our Contributions

This work provides the following contributions:

Circuit-based PSI protocols with almost linear overhead. We show a new circuit-based construction for computing any symmetric function on top of PSI, with an asymptotic overhead of only $\omega(n)$ comparisons. (More accurately, for any function $f \in \omega(n)$, the overhead of the construction is $o(f(n))$.) This construction is based on standard Cuckoo hashing.

[1] Facebook is running a computation of this type with companies that have transaction records for a large part of loyalty card holders in the US. According to the report in https://www.eff.org/deeplinks/2012/09/deep-dive-facebook-and-datalogix-whats-actually-getting-shared-and-how-you-can-opt, the computation is done using an insecure PSI variant based on creating pseudonyms using naive hashing of the items.

Small constants. Standard measures of asymptotic security are not always a good reflection of the actual performance on reasonable parameters. Therefore, in addition to the asymptotic improvement, we also show a concrete circuit-based PSI construction. This construction is based on a new variant of Cuckoo hashing, *two-dimensional Cuckoo hashing*, that we introduce in this work. We carefully handle implementation issues to improve the actual overhead of our protocols, and make sure that all constants are small. In particular, we ran extensive experiments to analyze the failure probabilities of the hashing scheme, and find the exact parameters that reduce this statistical failure probability to an acceptable level (e.g., 2^{-40}). Our analysis of the concrete complexities is backed by extensive experiments, which consumed about 5.5 million core hours on the Lichtenberg high performance computer of the TU Darmstadt and were used to set the parameters of the hashing scheme. Given these parameters we implemented the circuit-based PSI protocol and tested it.

Implementation and experiments. We implemented our protocols using the ABY framework for secure two-party computation [13]. Our experiments show that our protocols are considerably faster than the previously best circuit-based constructions. For example, for input sets of $n = 2^{20}$ elements of arbitrary bitlength, we improve the circuit size over the best previous construction by up to a factor of 3.8x.

New Cuckoo hashing analysis. Our two-dimensional Cuckoo hashing is based on a new Cuckoo hashing scheme that employs two tables and each item is mapped to either *two* locations in the first table, or *two* locations in the second table. This is a new Cuckoo hashing variant that has not been analyzed before. In addition to measuring its performance using simulations, we provide a probabilistic analysis of its performance. Interestingly, this analysis can also be used as a new proof technique for the success probability of standard Cuckoo hashing.

1.3 Computing Symmetric Functions

A trivial circuit for PSI that performs $O(n^2)$ comparisons between all pairs of the input items of the two parties allows the parties to set their inputs in any arbitrary order. On the other hand, there exist more efficient circuit-based PSI constructions where each party first independently orders its inputs according to some predefined algorithm: the sorting network-based construction of [26] requires each party to sort its input to the circuit, while the hashing-based construction of [39] requires the parties to map their inputs to bins using some public hash functions. (These constructions are described in Sect. 1.4.) The location of each input item thus depends on the identity of the other inputs of the input owner, and must therefore be kept hidden from the other party.

In this work, we focus on constructing a circuit that computes the intersection. The outputs of this circuit can be the items in the intersection, or some functions of the items in the intersection: say, a "1" for each intersecting item, or an arbitrary function of some data associated with the item (for example, if the items are transactions, we might want to output a financial value associated with each transaction that appears in the intersection). On top of that circuit it is possible to add circuits for computing any function that is based on the intersection. In order to preserve privacy, the output of that function must be a *symmetric* function of the items in the intersection. Namely, the output of the function must not depend on the *order* of its inputs. There are many examples of interesting symmetric functions of the intersection. (In fact, it is hard to come up with examples for interesting non-symmetric functions of the intersection, except for the intersection itself.) Examples of symmetric functions include:

- Computing the size of the intersection, i.e., PSI cardinality (PSI-CA).
- Computing a threshold function that is based on the size of the intersection. For example, outputting "1" if the size of the intersection is greater than some threshold (PSI-CAT), or outputting a rounded value of the percentage of items that are in the intersection. An extension of PSI-CAT, where the intersection is revealed only if the size of the intersection is greater than a threshold, can be used for privacy-preserving ridesharing [23].
- Computing the size of the intersection while preserving the privacy of users by ensuring differential privacy [17]. This can be done by adding some noise to the exact count.
- Computing the sum of values associated with the items in the intersection. This is used for measuring ad-generated revenue (cf. Sect. 1.1). Similarly, there could be settings where each party associates a value with each transaction, and the output is the sum of the differences between these assigned values in the intersection, or the sum of the squares of the differences, etc.

The circuits for computing all these functions are of size $O(n)$. Therefore, with our new construction, the total size of the circuits for computing these functions is $\omega(n)$, whereas circuit-based PSI protocols [26,39] had size $O(n \log n)$.

If one wishes to compute a function that is not symmetric, or wishes to output the intersection itself, then the circuit must first shuffle the values in the intersection (in order to assign a random location to each item in the intersection) and then compute the function over the shuffled values, or output the shuffled intersection. A circuit for this "shuffle" step has size $O(n \log n)$, as described in [26]. (It is unclear, though, why a circuit-based protocol should be used for computing the intersection, since this job can be done much more efficiently by specialized protocols, e.g., [31,42].)

1.4 Related Work

PSI. Work on protocols for private set intersection was presented as early as [35,46], which introduced public key-based protocols using commutative cryptography, namely the Diffie-Hellman function. A survey of PSI protocols appears in [41]. The goal of these protocols is to let one party learn the intersection itself, rather than to enable the secure computation of arbitrary functions of the intersection. Other PSI protocols are based on oblivious polynomial evaluation [20], blind RSA [11], and Bloom filters [16]. Today's most efficient PSI protocols are based on hashing the items to bins and then evaluating an oblivious pseudo-random function per bin, which is implemented using oblivious transfer (OT) extension. These protocols have linear complexity and were all implemented and evaluated, see, e.g., [31,39,41,42]. In cases where communication cost is a crucial and computation cost is a minor factor, recent solutions based on fully homomorphic encryption represent an interesting alternative [6]. PSI protocols have also been adapted to the special requirements of mobile devices [4,25,30].

Circuit-based PSI. Circuit-based PSI protocols compute the set intersection functionality by running a secure evaluation of a Boolean circuit. These protocols can easily be adapted to compute different variants of the PSI functionality. The straightforward solution to the PSI problem requires $O(n^2)$ comparisons – one comparison for each pair of items belonging to the two parties. Huang et al. [26] designed a circuit for computing PSI based on sorting networks, which computes $O(n \log n)$ comparisons and is of size $O(\sigma n \log n)$, where σ is the bitlength of the inputs. A different circuit, based on the usage of Cuckoo hashing by one party and simple hashing by the other party, was proposed in [39]. The size of that circuit is $O(\sigma n \log n / \log\log n)$. In our work we propose efficient circuits for PSI variants with an asymptotic size of $\omega(\sigma n)$ and better concrete efficiency. We give more details and a comparison of the concrete complexities of circuit-based PSI protocols in Sect. 6.2.

PSI Cardinality (PSI-CA). A specific interesting function of the intersection is its cardinality, namely $|X \cap Y|$, and is referred to as PSI-CA. There are several protocols for computing PSI-CA with linear complexity based on public key cryptography, e.g., [9] which is based on Diffie-Hellman and is essentially a variant of the DH-based PSI protocol of [35,46] (see also references given therein for other less efficient public key-based protocols); or [12] which is based on Bloom filters and the public key cryptosystem of Goldwasser-Micali. In these protocols, one of the parties learns the cardinality. As we show in our experiments in Sect. 6.3, these protocols are slower than our constructions already for relatively small set sizes ($n = 2^{12}$) in the LAN setting and for large set sizes ($n = 2^{20}$) in the WAN setting, since they are based on public key cryptography. An advantage of these protocols is that they achieve the lowest amount of communication, but it seems hard to extend them to compute arbitrary functions of the intersection. Protocols for private set intersection and union and their cardinalities with linear complexity are given in [8]. They use Bloom filters and computationally expensive additively homomorphic encryption, whereas our protocols can flexibly be adapted to different variants and are based on efficient symmetric key cryptography.

2 Preliminaries

Setting. We consider two parties, which we denote as Alice and Bob. They have input sets, X and Y, respectively, which are each of size n and each item has bitlength σ. We assume that both parties agree on a symmetric function f and would like to securely compute $f(X \cap Y)$. They also agree on a circuit that receives the items in the intersection as input and computes f.

Security Model. The secure computation literature considers *semi-honest* adversaries, which try to learn as much information as possible from a given protocol execution, but are not able to deviate from the protocol steps, and *malicious* adversaries, which are able to deviate arbitrarily from the protocol. The semi-honest adversary model is appropriate for scenarios where execution of the intended software is guaranteed via software attestation or business restrictions, and yet an untrusted third party is able to obtain the transcript of the protocol after its execution, either by stealing it or by legally enforcing its disclosure. Most protocols for private set intersection, as well as this work, focus on solutions that are secure against semi-honest adversaries. PSI protocols for the malicious setting exist, but they are less efficient than protocols for the semi-honest setting (see, e.g., [7,10,19,20,43,44]).

Secure Computation. There are two main approaches for generic secure two-party computation with security against semi-honest adversaries that allow to securely evaluate a function that is represented as a Boolean circuit: (1) Yao's garbled circuit protocol [48] has a constant round complexity and with today's most efficient optimizations provides free XOR gates [33], whereas securely evaluating an AND gate requires sending two ciphertexts [50]. (2) The GMW protocol [21] also provides free XOR gates and requires two ciphertexts of communication per AND gate using OT extension [3]. The main advantage of the GMW protocol is that *all* symmetric cryptographic operations can be pre-computed in a constant number of rounds in a setup phase, whereas the online phase is very efficient, but requires interaction for each layer of AND gates. In more detail, the setup phase is independent of the actual inputs and pre-computes multiplication triples for each AND gate using OT extension in a constant number of rounds (cf. [3]). The online phase runs from the time the inputs are provided until the result is obtained and involves sending one message for each layer of AND gates. A detailed description and a comparison between Yao and GMW is given in [45].

Cuckoo Hashing. In its simplest form, Cuckoo hashing [36] uses two hash functions h_0, h_1 to map n elements to two tables T_0, T_1, each containing $(1+\varepsilon)n$ bins. Each bin accommodates at most a single element. The scheme avoids collisions by relocating elements when a collision is found using the following procedure: Let $b \in \{0,1\}$. An element x is inserted into a bin $h_b(x)$ in table T_b. If a prior item y exists in that bin, it is evicted to bin $h_{1-b}(y)$ in T_{1-b}. The pointer b is then assigned the value $1-b$. The procedure is repeated until no more evictions are necessary, or until a threshold number of relocations has been performed. In the latter case, the last element is mapped to a special stash. It was shown in [29] that, for any constant s, the probability that the size of the stash is greater than s is at most $O(n^{-(s+1)})$. After inserting all items, each item can be found in one of two locations or in the stash. A lookup therefore requires checking only $O(1)$ locations.

Many variants of Cuckoo hashing were suggested and analyzed. See [47] for a thorough discussion and analysis of different Cuckoo hashing schemes. A variant of Cuckoo hashing that is similar to our constructions was given in [1], although in a different application domain. It considers a setting with three tables, where an item must be placed in two out of three tables. The analysis of this construction uses a different proof technique than the one we present in the full version [40], and we have not attempted to generalize their proof to a general number of item insertions (as we do for our construction). Furthermore, there is no tight analysis of the stash size in [1]. The work in [18] builds on the construction of [1] and proves that the failure probability when using a stash of size s behaves as $\tilde{O}(n^{-s})$. However, the experiments of [18, Fig. 6] reveal that the size of the stash is rather large and actually *increasing* in n within the range of 1 000 to 100 000 elements. For example, for table size $7.1n$, a stash of at least size 4 is required for inserting 10 000 elements, whereas a stash of at least size 11 is required for inserting 100 000 elements. Since each item in the stash must be compared to all items of the other party, and since these comparisons cannot use a shorter representation based on permutation-based hashing, the effect of the stash is substantial, and in the context of circuit-based PSI it is therefore preferable to use constructions that place very few or no items in the stash.

PSI based on Hashing. Some existing constructions of circuits for PSI require the parties to reorder their inputs before inputting them to the circuit: The sorting-network based construction of [26] requires the parties to sort their inputs. The hashing based construction of [39] requires that each party maps its items to bins using a hash function. It was observed as early as [20] that if the two parties agree on the same hash function and use it to map their respective input to bins, then the items that one party maps to a specific bin need to be compared only to the items that the other party maps to the same bin. However, the parties must be careful not to reveal to each other the number of items they mapped to each bin, since this data leaks information about their other items. Therefore, they agree beforehand on an upper bound m for the maximum number of items that can be mapped to a bin (such upper bounds are well known for common hashing algorithms, and can also be substantiated using simulation), and pad each bin with random dummy values until it has exactly m items in it. If both parties use the same hash algorithm, then this approach considerably reduces the overhead of the computation from $O(n^2)$ to $O(\beta \cdot m^2)$, where m is the maximum number of items mapped to any of the β bins.

When a random hash function h is used to map n items to n bins, where x is mapped to bin $h(x)$, the most occupied bin has w.h.p. $m = \frac{\ln n}{\ln \ln n}(1 + o(1))$ items [22] (a careful analysis shows, e.g., that, for $n = 2^{20}$ and an error probability of 2^{-40}, one needs to set $m = 20$). Cuckoo hashing is much more promising, since it maps n items to $2(1+\varepsilon)n$ bins, where each bin stores at most

$m = 1$ items. Cuckoo hashing typically uses two hash functions h_0, h_1, where an item x is mapped to one of the two locations $h_0(x), h_1(x)$, or to a stash of a small size. It is tempting to let both parties, Alice and Bob, map their items to bins using Cuckoo hashing, and then only compare the item that one party maps to a bin with the item that the other party maps to the same bin. The problem is that Alice might map x to $h_0(x)$ whereas Bob might map it to $h_1(x)$. They cannot use a protocol where Alice's value in bin $h_0(x)$ is compared to the two bins $h_0(x), h_1(x)$ in Bob's input, since this reveals that Alice has an item that is mapped to these two locations. The solution used in [19,39,41] is to let Alice map her items to bins using Cuckoo hashing, and Bob map his items using simple hashing. Namely, each item of Bob is mapped to both bins $h_0(x), h_1(x)$. Therefore, Bob needs to pad his bins to have $m = O(\log n/\log\log n)$ items in each bin, and the total number of comparisons is $O(n\log n/\log\log n)$.

3 Analyzing the Failure Probability

Efficient cryptographic protocols that are based on probabilistic constructions are typically secure as long as the underlying probabilistic constructions do not fail. Our work is based on variants of Cuckoo hashing, and the protocols are secure as long as the relevant tables and stashes do not overflow. (Specifically, hashing is computed using random hash functions which are chosen independently of the data. If a party observes that these functions cannot successfully hash its data, it can indeed ask to replace the hash functions, or remove some items from its input. However, the hash functions are then no longer independent of this party's input and might therefore leak some information about the input.)

There are two approaches for arguing about the failure probability of cryptographic protocols:

1. For an **asymptotic analysis**, the failure probability must be negligible in n.
2. For a **concrete analysis**, the failure probability is set to be smaller than some threshold, say $2^{-\lambda}$, where λ is a statistical security parameter.
   In typical experiments, the statistical security parameter is set to $\lambda = 40$. This means that "unfortunate" events that leak information happen with a probability of at most 2^{-40}. In particular, $\lambda = 40$ was used in all PSI constructions which are based on hashing (e.g., [16,19,31,39,41]).

With regards to the probabilistic constructions, there are different levels of analysis of the failure probability:

1. For simple constructions, it is sometimes possible to compute the **exact failure probability**. (For example, suppose that items are hashed to a table using a random hash function, and a failure happens when two items are mapped to the same location. In this case it is trivial to compute the exact failure probability.)

2. For some constructions there are known **asymptotic bounds** for the failure probability, but no concrete expressions. (For example, for Cuckoo hashing with a stash of size s, it was shown in [29] that the overflow probability is $O(n^{-(s+1)})$, but the exact constants are unknown.)[2]
3. For other constructions there is no analysis for the failure probability, even though they **perform very well in practice**. For example, Cuckoo hashing variants where items can be mapped to $d > 2$ locations, or where each bin can hold $k > 1$ items, were known to have better space utilization than standard Cuckoo hashing, but it took several years to theoretically analyze their performance [47]. There are also insertion algorithms for these Cuckoo hashing variants which are known to perform well but which have not yet been fully analyzed.

3.1 Using Probabilistic Constructions for Cryptography

Suppose that one is using a probabilistic construction (e.g., a hash table) in the design of a cryptographic protocol. An asymptotic analysis of the cryptographic protocol can be done if the hash table has either an exact analysis or an asymptotic analysis of its failure probability (items 1 and 2 in the previous list).

If the aim is a concrete analysis of the cryptographic protocol, then exact values for the parameters of the hash construction must be identified. If an exact analysis is known (item 1), then it is easy to plug in the desired failure probability ($2^{-\lambda}$) and compute the values for the different parameters. However, if only an asymptotic analysis or experimental evidence is known (items 2 and 3), then experiments must be run in order to find the parameters that set the failure probability to be smaller than $2^{-\lambda}$.

We stress that a concrete analysis is needed whenever a cryptographic protocol is to be used in practice. In that case, even an asymptotic analysis is insufficient since it does not specify any constants, which are crucial for deriving the exact parameter values.

3.2 Experimental Parameter Analysis

Verifying that the failure probability is smaller than $2^{-\lambda}$ for $\lambda = 40$ requires running many repetitions of the experiments. Furthermore, for large input sizes (large values of n), each single run of the experiment can be rather lengthy. (And one could justifiably argue that the more interesting results are for the larger values of n, since for smaller n we can use less optimal constructions and still get reasonable performance.)

[2] We note though that many probabilistic constructions are analyzed in the algorithms research literature to have a failure probability of $o(1)$, which is fine for many applications, but is typically insufficient for cryptographic applications.

Examining the failure probability for a specific choice of parameters. For a specific choice of parameters, running 2^λ repetitions of an experiment is insufficient to argue about a $2^{-\lambda}$ failure probability, since it might happen that the experiments were very unlucky and resulted in no failure even though the failure probability is somewhat larger than $2^{-\lambda}$. Instead, we can argue about a confidence interval: namely, a confidence interval of $1-\alpha$ (say, 95%, or 99.9%) states that if the failure probability is greater than $2^{-\lambda}$, then we would have *not* seen the results of the experiment, except with a probability that is smaller than α. Therefore, either the experiment was very unlucky, or the failure probability is sufficiently small. For example, an easy to remember confidence level used in statistics is the "rule of three", which states that if an event has not occurred in $3 \cdot s$ experiments, then the 95% confidence interval for its rate of occurrence in the population is $[0, 1/s]$. For our purposes this means that running $3 \cdot 2^\lambda$ experiments with no failure suffices to state that the failure probability is smaller than $2^{-\lambda}$ with 95% confidence. (We will report experiments in Sect. 6.1 which result in a 99.9% confidence interval for the failure probability.)

Examining the failure probability as a function of n. For large values of n (e.g., $n = 2^{20}$), it might be too costly to run sufficiently many (more than 2^{40}) experiments. Suppose that the experiments spend just 10 cycles on each item. This is an extremely small lower bound, which is probably optimistic by orders of magnitude compared to the actual run-time. Then the experiments take at least $10 \cdot 2^{60}$ cycles. This translates to about a million core hours on 3 GHz machines.

In order to be able to argue about the failure probability for large values of n, we can run experiments for progressively increasing values of n and identify how the failure probability behaves as a function of n. If we observe that the failure probability is decreasing, or, better still, identify the dependence on n, we can argue, given experimental results for medium-sized n values, about the failure probabilities for larger values of n.

3.3 Our Constructions

Asymptotic overhead. We present in Sect. 4 a construction of circuit-based PSI that we denote as the "mirror" construction. This construction uses four instances of standard Cuckoo hashing and therefore we know that a stash of size s guarantees a failure probability of $O(n^{-(s+1)})$ [29]. (Actually, the previously known analysis was only stated for $s = O(1)$. We show in the full version [40] that this failure probability also holds for s that is not constant.)

The bound on the failure probability implies that for any constant security parameter λ, a stash of constant size is sufficient to ensure that the failure probability is smaller than $2^{-\lambda}$ for sufficiently large n. In order to achieve a failure probability that is negligible in n, we can set the stash size s to be slightly larger than $O(1)$, e.g., $s = \log\log n$, $s = \log^* n$, or any $s = \omega(1)$. The result is a construction with an overhead of $\omega(n)$. (More accurately, the overhead is as close as desired to being linear: for any $f(n) \in \omega(n)$, the overhead is $o(f(n))$.)

Concrete overhead. In Sect. 5 we present a new variant of Cuckoo hashing that we denote as two-dimensional (or 2D) Cuckoo hashing. We analyze this construction in the full version [40] and show that when no stash is used, then the failure probability (with tables of size $O(n)$) is $O(1/n)$, as in standard Cuckoo hashing.

We only have a sketch of an analysis for the size of the stash of the construction in Sect. 5, but we observed that this construction performed much better than the asymptotic construction. Also, performance was improved with the heuristic of using half as many bins but letting each bin store two items instead of one. (This variant is known to perform much better also in the case of standard Cuckoo hashing, see [47].)

Since we do not have a theoretical analysis of this construction, we ran extensive experiments in order to examine its performance. These experiments follow the analysis paradigm given in Sect. 3.2, and are described in Sect. 6.1. For a specific ratio between the table size and n, we ran 2^{40} experiments for $n = 2^{12}$ and found that the failure probability is at most 2^{-37} with 99.9% confidence. We also ran experiments for increasing values of n, up to $n = 2^{12}$, and found that the failure probability has linear dependence on n^{-3} (an explanation of this behavior appears in the full version [40]). Therefore, we can argue that for $n \geq 2^{13} = 2 \cdot 2^{12}$ the failure probability is at most $2^{-37} \cdot 2^{-3} = 2^{-40}$.

4 An Asymptotic Construction Through Mirror Cuckoo Hashing

We show here a construction for circuit-based PSI that has an $\omega(n)$ asymptotic overhead. The analysis in this section is not intended to be tight, but rather shows the asymptotic behavior of the overhead.

The analysis is based on a construction which we denote as *mirror Cuckoo hashing* (as the placement of the hash functions that are used in one side is a mirror image of the hash functions of the other side). Hashing is computed in a single iteration. The main advantage of this construction is that it is based on four copies of standard Cuckoo hashing. Therefore, we can apply known bounds on the failure probability of Cuckoo hashing. Namely, applying the result of [29] that the failure probability when using a stash of size s is $O(n^{-(s+1)})$. Given this result, a stash of size $\omega(1)$ guarantees that the failure probability is negligible in n (while a constant stash size guarantees that for sufficiently large n the failure probability is smaller than any constant, and in particular smaller than 2^{-40}). We note that while the known results about the size of the stash are only stated for $s = O(1)$, we show in the full version [40] that the $O(n^{-(s+1)})$ bound on the failure probability also applies to a non-constant stash size.

4.1 Mirror Cuckoo Hashing

We describe a hashing scheme that uses two sets of tables. A left set including tables T_{L}, T_{R}, and a right set including tables T'_{L}, T'_{R}. Each table is also denoted

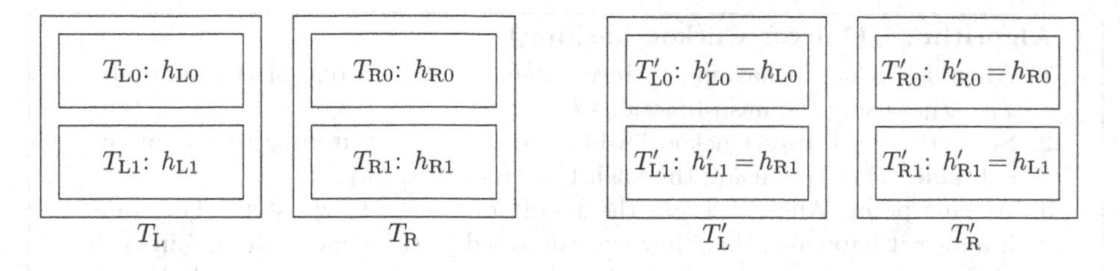

Fig. 1. The tables T_L, T_R and T'_L, T'_R. The hash functions in the upper subtables of T'_L, T'_R are the same as in T_L, T_R, and those in the lower subtables are in reverse order.

as a "column". Each table has two subtables, or "rows". So overall there are four tables (columns), each containing two subtables (rows).

Bob maps each of his items to one subtable in each table, namely to one row in each column. Alice maps each of her items to the two subtables in one of the tables, namely to both rows in just one of the columns. These mappings ensure that for any item x that is owned by both Alice and Bob, there is exactly one subtable to which it is mapped by both parties.

The tables. The construction uses two sets of tables, T_L, T_R and T'_L, T'_R. Each table is of size $2(1+\varepsilon)n$ and is composed of two subtables of size $(1+\varepsilon)n$ (T_L contains the subtables T_{L0}, T_{L1}, etc.). Each subtable is associated with a hash function that will be used by both parties. E.g., function h_{L0} will be used for subtable T_{L0}, etc. The tables and the hash functions are depicted in Fig. 1.

The hash functions. The hash functions associated with the tables are defined as follows:

- The functions for the left two tables (columns) T_L, T_R, i.e., h_{L0}, h_{L1}, h_{R0}, h_{R1}, are chosen at random. Each function maps items to the range $[0, (1+\varepsilon)n-1]$, which corresponds to the number of bins in each of T_{L0}, T_{L1}, T_{R0}, T_{R1}.
- The functions for the two right tables T'_L, $T'_R$are defined as follows:
  - The two functions of the upper subtables are equal to the functions of the upper subtables on the left. Namely, $h'_{L0} = h_{L0}$ and $h'_{R0} = h_{R0}$.
  - The two functions of the lower subtables are the *mirror image* of the functions of the lower subtables on the left. Namely, h'_{L1}, h'_{R1} are defined such that $h'_{L1} = h_{R1}$, and $h'_{R1} = h_{L1}$.

Bob's insertion algorithm. Bob needs to insert each of his items to one subtable in each of the tables T_L, T_R, T'_L, T'_R. He can do so by simply using Cuckoo hashing for each of these tables. For example, for the table T_L and its subtables T_{L0}, T_{L1}, Bob uses the functions h_{L0}, h_{L1} to insert each input x to either T_{L0} or T_{L1}. The same is applied to T_R, T'_L, and T'_R. In addition, Bob keeps a small stash of size $\omega(1)$ for each of the four tables. Overall, based on known properties of Cuckoo hashing, we can claim that the construction guarantees the following property:

Algorithm 1 (Mirror Cuckoo hashing)

1. Alice uses Cuckoo hashing to insert each item x to one of the subtables T_{L0}, T_{R0}, using the hash functions h_{L0}, h_{R0}.
2. Similarly, Alice uses Cuckoo hashing to insert each item x to one of the subtables T_{L1}, T_{R1}, using the hash functions h_{L1}, h_{R1}.
3. At this point, Alice observes the result of the first two steps. For some inputs x it happened that they were mapped to the same "column" in both of these steps. Namely, x was mapped to both T_{L0} and T_{L1}, or to both T_{R0} and T_{R1}. These are the "good" items, since they were mapped to the same column, as is required for all of Alice's inputs.
4. The other inputs of Alice, the "bad" items, were mapped to one column in Step 1 and to the other column in Step 2. Alice applies the following procedure to these items:
   (a) Each "bad" item x is removed from both locations to which it was mapped in Steps 1 and 2.
   (b) x is now inserted in either of T'_{L0}, T'_{R0} using the hash functions $h'_{L0} := h_{L0}$, $h'_{R0} := h_{R0}$ with the same mapping as in Step 1.
   (c) x is also inserted in either of T'_{L1}, T'_{R1} using the hash functions $h'_{L1} := h_{R1}$, $h'_{R1} := h_{L1}$ with the same mapping as in Step 2.

Claim. With all but negligible probability, it holds that for every input x of Bob, and for each of the four tables T_L, T_R, T'_L, T'_R, Bob inserts x to exactly one of the two subtables or to the stash.

Alice's insertion algorithm. Alice's operation is a little more complex and is described in Algorithm 1. Alice considers the two upper subtables on the left, T_{L0}, T_{R0}, as two subtables for standard Cuckoo hashing. Similarly, she considers the two lower subtables on the left, T_{L1}, T_{R1}, as two subtables for standard Cuckoo hashing. In other words, she considers the left top row and the left bottom row as standard Cuckoo hashing tables.

Alice then inserts each input item of hers to each of these two tables using standard Cuckoo hashing. (She also uses stashes of size $\omega(1)$ to store items which cannot be placed in the Cuckoo tables.) For some input items x it happens that x is inserted in the top row to T_{L0} and in the bottom row to T_{L1}; or x is inserted in the top row to T_{R0} and in the bottom row to T_{R1}. Therefore, x is inserted in two subtables in the same column. (x is denoted as "good" since this is the outcome that we want.)

Let x' be one of the other, "bad", items. Thus, x' is inserted in the top row to T_{L0} and in the bottom row to T_{R1}, or vice versa. In this case, Alice removes x' from the tables on the left and inserts it to the tables T'_L, T'_R on the right. Since the hash functions that are used in T'_L, T'_R are equal to the functions used on the left side (where in the bottom row the functions are in reverse order), Alice does not need to run a Cuckoo hash insertion algorithm on the right side: Assume that x' was stored in locations $T_{L0}[h_{L0}(x')]$ and $T_{R1}[h_{R1}(x')]$ on the left. Then Alice inserts it to locations $T'_{L0}[h'_{L0}(x')] = T'_{L0}[h_{L0}(x')]$ and $T'_{L1}[h'_{L1}(x')] = T'_{L1}[h_{R1}(x')]$ on the right.

In other words, in a global view, one can see the algorithm as composed of the following steps: (1) First, all items are placed in the left tables. (2) Each subtable is divided in two copies, where one copy contains the good items and the other copy contains the bad items. (3) The subtable copies with the good items are kept on the left, whereas the copies with the bad items are moved to the right, where in the bottom row on the right we replace the order of the subtables.

This algorithm has two important properties: First, all items that were successfully inserted in the first step to the left tables will be placed in tables on either the left or the right hand sides. Moreover, each item will be placed in two subtables in the same column — the good items happened to initially be placed in this way in the left tables; whereas the bad items were in different columns on the left side but were moved to the same column on the right side. Hence, we can state the following claim:

Claim. With all but negligible probability, Alice inserts each of her inputs either to two locations in exactly one of T_{L}, T_{R}, T'_{L}, T'_{R} and to no locations in other tables, or to a stash.

Tables size. The total size of the tables is $8(1+\varepsilon)n$.

Stash size. With regards to stashes, each party needs to keep a stash for each of the Cuckoo hashing tables that it uses. Since Alice runs the Cuckoo hashing insertion algorithm only for the left tables and re-uses the mapping for the right tables, she needs only two stashes. Bob on the other hand runs the Cuckoo hashing insertion algorithm four times and hence needs four stashes. (In order to preserve simplicity, we omitted the stashes in Fig. 1 and Algorithm 1.) Given the result of [29], and our observation in the full version [40] about its applicability to non-constant stash sizes, it holds that a total stash of size $\omega(1)$ elements suffices to successfully map all items, except with negligible probability. We note that the size of the stash can be arbitrarily close to constant, e.g., it can be set to be $O(\log\log n)$ or $O(\log^* n)$. Essentially, for any function $f(n) \in \omega(n)$, the size of the stash can be $o(f(n))$.

4.2 Circuit-Based PSI from Mirror Cuckoo Hashing

Mirror Cuckoo hashing lets the parties map their inputs to tables of size $O(n)$ and stashes of size $\omega(1)$, with negligible failure probability. It is therefore straightforward to construct a PSI protocol based on this hashing scheme:

1. The parties agree on the parameters that define the size of the tables and the stash for mirror Cuckoo hashing. They also agree on the hash functions that will be used in each table.
2. Each party maps its items to the tables using the hash functions that were agreed upon.
3. The parties evaluate a circuit that performs the following operations:
   (a) For each bin in the tables, the circuit compares the item that Alice mapped to the bin to the item that Bob mapped to the same bin.

(b) Each item that Bob mapped to his stashes is compared with all items of Alice. Similarly, each item that Alice mapped to her stashes is compared with all items of Bob.

The properties of mirror Cuckoo hashing ensure: (1) If an item x is in the intersection, then there is exactly one comparison in which x is input by both Alice and Bob. (2) The number of comparisons in Step 3 is $\omega(n)$.

5 A Concretely Efficient Construction Through 2D Cuckoo Hashing

Two-dimensional Cuckoo hashing (a.k.a. 2D Cuckoo hashing) is a new construction with the following properties:

- It uses overall $O(n)$ memory (specifically, $8(1+\varepsilon)n$ in our construction, where we set $\varepsilon = 0.2$ in our experiments).
- Both, Alice and Bob, map each of their items to $O(1)$ memory locations (specifically, to two or four memory locations in our construction).
- If x appears in the input of both parties, then there is exactly one location to which both Alice and Bob map x.

The construction uses two tables, T_{L}, T_{R}, located on the left and the right side, respectively. Each of these tables is of size $4(1+\varepsilon)n$ and is composed of two smaller subtables: T_{L} is composed of the two smaller subtables T_{L0}, T_{L1}, while T_{R} is composed of the two smaller tables T_{R0}, T_{R1}. The hash functions h_{L0}, h_{L1}, h_{R0}, h_{R1} are used to map items to T_{L0}, T_{L1}, T_{R0}, T_{R1}, respectively. The tables are depicted in Fig. 2.

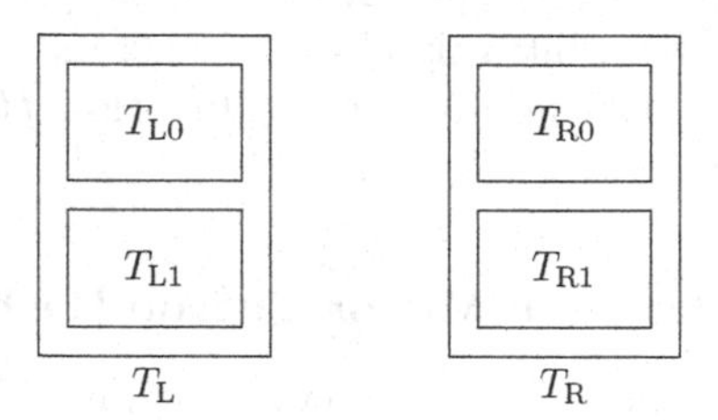

Fig. 2. The tables T_{L} and T_{R}, consisting of T_{L0}, T_{L1} and T_{R0}, T_{R1}, respectively.

Hashing is performed in the following way:

- Alice maps each of her items to all subtables on one of the two sides. Namely, each item x of Alice is either mapped to both bins $T_{\text{L0}}[h_{\text{L0}}(x)]$ and $T_{\text{L1}}[h_{\text{L1}}(x)]$ on the left side, or to bins $T_{\text{R0}}[h_{\text{R0}}(x)]$ and $T_{\text{R1}}[h_{\text{R1}}(x)]$ on the right side. In other words, ALICE maps each item to ALL subtables on one side.

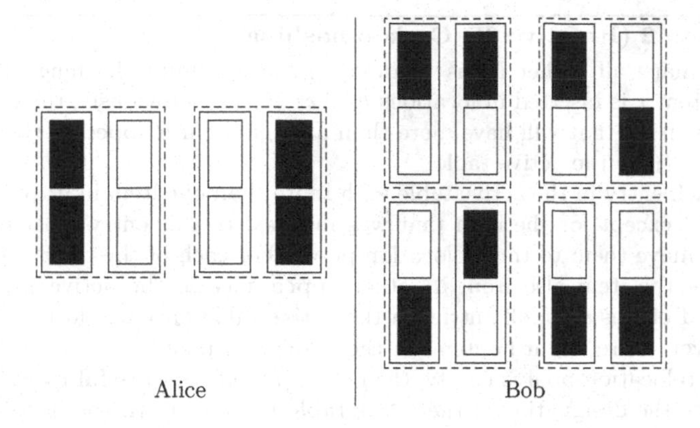

Fig. 3. The possible combinations of locations to which Alice and Bob map their inputs.

– Bob maps each of his items to one subtable on each side. This is done using standard Cuckoo hashing. Namely, each input x of Bob is mapped to one of the locations $T_{\mathrm{L}0}[h_{\mathrm{L}0}(x)]$ or $T_{\mathrm{L}1}[h_{\mathrm{L}1}(x)]$ on the left side, as well as mapped to one of the locations $T_{\mathrm{R}0}[h_{\mathrm{R}0}(x)]$ or $T_{\mathrm{R}1}[h_{\mathrm{R}1}(x)]$ on the right side. In other words, BOB maps each item to one subtable on BOTH sides.

The possible options for hashing an item x by both parties are depicted in Fig. 3. It is straightforward to see that if both parties have the same item x, there is exactly one table out of $T_{\mathrm{L}0}$, $T_{\mathrm{L}1}$, $T_{\mathrm{R}0}$, $T_{\mathrm{R}1}$ that is used by both Alice and Bob to store x.

We next describe a construction of 2D Cuckoo hashing, followed by a variant based on a heuristic optimization that stores two items in each table entry. The asymptotic behavior of the basic construction is analyzed in the full version [40]. In Sect. 6.1 we describe simulations for setting the parameters of the heuristic construction in order to reduce the hashing failure probability to below 2^{-40}.

5.1 Iterative 2D Cuckoo Hashing

This construction uses two tables, T_{L}, T_{R}, each of $4(1+\varepsilon)n$ entries. (In this construction, there is no need to assume that each table is composed of two subtables.) The parties associate two hash functions with each table, namely $h_{\mathrm{L}0}$, $h_{\mathrm{L}1}$ for T_{L}, and $h_{\mathrm{R}0}$, $h_{\mathrm{R}1}$ for T_{R}.

Bob uses Cuckoo hashing to insert each of his items into one location in each of the tables.

Alice inserts each item x either into the two locations $h_{\mathrm{L}0}(x)$ and $h_{\mathrm{L}1}(x)$ in T_{L}, or into the two locations $h_{\mathrm{R}0}(x)$ and $h_{\mathrm{R}1}(x)$ in T_{R}. This is achieved by Alice running a modified Cuckoo insertion algorithm that maps an item to two locations in one table, "kicks out" any item that is currently present in these locations and also removes the other occurrence of this item from the table, and then tries to insert this item into its two locations in the other table, and so on.

Algorithm 2 (Iterative 2D Cuckoo hashing)

1. Alice maps all of her items to table T_L, using simple hashing. That is, each item x is inserted in locations $h_{L0}(x), h_{L1}(x)$. Obviously, there will be entries in T_L that will have more than a single item mapped to them. Denote T_L as the active table.
2. For each entry in the active table with more than one item in it: remove all items – except for the item that was mapped to this entry most recently – and move them to the "relocation pool". For each of the removed items, remove the item also from its other appearance in the active table. (At the end of this step, all entries in the active table have at most one entry. However, there might be items in the relocation pool.)
3. If the relocation pool is empty, then stop (found a successful mapping).
4. Change the designation of the active table to point to the other table.
5. Move each item x from the relocation pool to locations $h_0(x), h_1(x)$ in the active table. (For example, if T_R is the active table, move x to $h_{R0}(x), h_{R1}(x)$.)
6. Go to Step 2.

Algorithm 3 (Iterative 2D Cuckoo hashing with bins of size 2)

The algorithm is identical to Algorithm 2, except for the following change in Step 2:

2. For each entry in the active table with more than *two* items in it: remove all items – except for the *two* items that were mapped to this entry most recently – and move them to the "relocation pool". For each of the removed items, remove the item also from its other appearance in the active table.

This is a new variant of Cuckoo hashing, where inserting an item into a table might result in four elements that need to be stored in the other table: storing x in $h_{L0}(x), h_{L1}(x)$ might remove two items, y_0, y_1, one from each location. These items are also removed from their other occurrences in T_L. They must now be stored in locations $h_{R0}(y_0), h_{R1}(y_0), h_{R0}(y_1), h_{R1}(y_1)$ in T_R.

It is not initially clear whether such a mapping is possible (with high probability, given random choices of the hash functions). We analyze the construction in the full version [40] and show that it only fails with probability $O(1/n)$. We ran extensive simulations, showing that the algorithm (when using a stash and a certain choice of parameters) fails with very small probability, smaller than 2^{-40}.

The insertion algorithm of Alice is described in Algorithm 2. The choice made in Step 2 of the algorithm, to first remove the oldest items that were mapped to the entry, is motivated by the intuition that it is more likely that the locations to which these items are mapped in the other table are free.

Storing two items per bin. It is known that the space utilization of Cuckoo hashing can be improved by storing more than one item per bin (cf. [15,37] or the review of multiple choice hashing in [47]). We take a similar approach and use two tables of size $2(1 + \varepsilon)n$ where each entry can store *two* items. (These tables have half as many entries as before, but each entry can store two items rather than one. The total size of the tables is therefore unchanged.) The change to the insertion algorithm is minimal and affects only Step 2. The new algorithm is defined in Algorithm 3.

Our experiments in Sect. 6.1 show that when using the same amount of space, then this variant of iterative 2D Cuckoo hashing performs better than the basic protocol with bins of size one. That is, it achieves a lower probability of hashing failure, namely of the need to use the stash, and requires less iterations to finish.

5.2 Circuit-Based PSI from 2D Cuckoo Hashing

This section describes how 2D Cuckoo hashing can be used for computing PSI. In addition, we describe two optimizations which substantially improve the efficiency of the protocol. The first optimization has the parties use permutation-based hashing [2] (as was done in [39]) in order to reduce the size of the items that are stored in each bin, and hence reduce the number of gates in the circuit. The second optimization is based on having each party use a single stash instead of using a separate stash for each Cuckoo hashing instance.
The PSI protocol is pretty straightforward given 2D Cuckoo hashing:

First, the parties agree on the hash functions to be used in each table. (These functions must be chosen at random, independently of the inputs, in order not to disclose any information about the inputs. Therefore, a participant cannot change the hash functions if some items cannot be mapped, and thus we seek parameter values that make the hashing failure probability negligible, e.g., smaller than 2^{-40}.)

Then, each party maps its items to bins using 2D Cuckoo hashing and the chosen hash functions. The important property is that if Alice and Bob have the same input item then there exists exactly one bin into which both parties map this item (or, alternatively, at least one of them places this item in a stash). Empty bins are padded with dummy elements. This ensures that no information is leaked by how empty the tables and stashes are.

Afterwards, the parties construct a circuit that compares, for each bin, the items that both parties stored in it. In addition, this circuit compares each item that Alice mapped to the stash with all of Bob's items, and vice versa. Since the number of bins is $O(n)$, the number of items in each bin is $O(1)$, and the number of items in the stash is $\omega(1)$, the total size of this circuit is $\omega(n)$. The parties can define another circuit that takes the output of this circuit and computes a desired function of it, e.g., the number of items in the intersection.

Finally, the parties run a generic MPC protocol that securely evaluates this circuit (cf. Sect. 6.3 for a concrete implementation and benchmarks).

Permutation-based Hashing. The protocol uses permutation-based hashing to reduce the bitlength of the elements that are stored in the bins and thus reduces the size of the circuit comparing them. This idea was introduced in [2] and used for PSI in [39]. It is implemented in the following way. The hash function h that is used to map an item x to one of the β bins is constructed as follows: Let $x = x_L|x_R$ where $|x_L| = \log\beta$. We first assume that β is a power of 2 and then describe the general case. Let f be a random function with range $[0, \beta - 1]$. Then h maps an element x to bin $x_L \oplus f(x_R)$ and the value stored in the bin is x_R. The important property is that the stored value has a reduced bitlength of only $|x| - \log\beta$, yet there are no collisions (since if x, y are mapped to the same bin and store the same value, then $x_R = y_R$ and $x_L \oplus f(x_R) = y_L \oplus f(y_R)$ and therefore $x = y$).

In the general case, where β is not a power of two, the output of h is reduced modulo β and a stored extra bit indicates if the output was reduced or not.

For Cuckoo hashing the protocol uses two hash functions to map the elements to the bins in one table. To avoid collisions among the two hash functions, a stored extra bit indicates which hash function was used.

Using a Combined Stash. Recall that Alice uses 2D Cuckoo hashing, for which we show experimentally in Sect. 6.1 that no stash is needed. Bob, on the other hand, uses two invocations of standard Cuckoo hashing, and therefore when he does not succeed in mapping an item to a table, he must store it in a stash and compare it with all items of Alice. In this case, the parties cannot encode their items using permutation-based hashing, and therefore these comparisons must be of the full-length original values and not of the shorter values computed using permutation-based hashing as described before. Therefore, the size of the circuits that handle the stash values have a considerable effect on the total overhead of the protocol.

We observe that, instead of keeping several stashes, Bob can collect all the values that he did not manage to map to any of the tables in a *combined* stash. Suppose that he maps items to c tables and that we have an upper bound s which holds w.h.p. on the size of each stash. A naive approach would use c stashes of that size, resulting in a total stash size of $c \cdot s$. A better approach would be to use a single stash for all these items, since it is very unlikely that all stashes will be of maximal size, and therefore we can show that with the same probability, the size s' of the combined stash is much smaller than $c \cdot s$. To do so, we determine the upper bounds for the combined stash for $c = 2$: The probability of having a combined stash of size s' is $\sum_{i=0}^{s'} P(i) \cdot P(s' - i)$, where $P(i)$ denotes the probability of having a single stash of size i. The value of $P(i)$ is $O(n^{-i}) - O(n^{-(i+1)}) \approx O(n^{-i})$ [29]. We can estimate the exact values of these probabilities based on the experiments conducted by [39]: they performed 2^{30} Cuckoo hashing experiments for each $n \in \{2^{11}, 2^{12}, 2^{13}, 2^{14}\}$ and counted the required stash sizes. Using linear regression, we extrapolated the results for larger sets of 2^{16} and 2^{20} elements. Table 1 shows the required stash sizes when binding the probability to be below 2^{-40}: it turns out that for 2^{12} and 2^{16} elements the combined stash should include only one more element compared to the upper bound for a single stash, whereas for 2^{20} even the same stash size is sufficient.

Table 1. Stash sizes required for binding the error probability to be below 2^{-40} when inserting $n \in \{2^{12}, 2^{16}, 2^{20}\}$ elements into $2.4n$ bins using Cuckoo hashing.

Number of elements n	2^{12}	2^{16}	2^{20}
Single stash size s (from [39, Table 4])	6	4	3
Stash size for two separate stashes $s' = 2s$	12	8	6
Combined stash size s'	7	5	3

All in all, when comparing to the naive solution with two separate stashes, the combined stash size is reduced by almost a factor of 2x.

5.3 Extension to a Larger Number of Parties

Computing PSI between the inputs of more than two parties has received relatively little interest. (The challenge is to compute the intersection of the inputs of all parties, without disclosing information about the intersection of the inputs of any subset of the parties.) Specific protocols for this task were given, e.g., in [20,24,32]. We note that our 2D Cuckoo hashing can be generalized to m dimensions in order to obtain a circuit-based protocol for computing the intersection of the inputs of m parties. The caveat is that the number of tables grows to 2^m and therefore the solution is only relevant for a small number of parties.

We describe the case of three parties: The hashing will be to a set of eight tables $T_{x,y,z}$, where $x, y, z \in \{0, 1\}$. Any input item of P_1 is mapped to either all tables $T_{0,0,0}, T_{0,0,1}, T_{0,1,0}, T_{0,1,1}$, or to all tables $T_{1,0,0}, T_{1,0,1}, T_{1,1,0}, T_{1,1,1}$. Namely, the index x is set to either 0 or 1, and the input item is mapped to all tables with that value of x. Every input of P_2 is mapped either to all tables whose y index is 0, or to all tables where $y = 1$. Every input of P_3 is mapped either to all tables whose z index is 0, or to all tables where $z = 1$.

It is easy to see that regardless of the choices of the values of x, y, z, the sets of tables to which all parties map an item intersect in exactly one table. Therefore, the parties can evaluate a simple circuit that checks every bin for equality of the values that were mapped to it by the three parties. It is guaranteed that if the same value is in the input sets of all parties, then there is exactly one bin to which this value is mapped by all three parties. If some items are mapped to a stash by one of the parties, they must be compared with all items of the other parties, but the overhead of this comparison is $\omega(n)$ if the stash is of size $\omega(1)$.

The remaining issue is the required size of the tables. In the full version [40] we show that inserting an item into one of two (big) tables, such that the item is mapped to k locations in that table, requires tables of size greater than $k^2(1+\varepsilon)n$. When computing PSI between three parties using the method described above, we have eight (small) tables, where each party must insert its items to four tables in one plane or to four tables in the other plane. Each such set of four small tables corresponds to a big table in the analysis and is therefore of size $16(1 + \varepsilon)n$. The total size of the tables is therefore $32(1 + \varepsilon)n$.

5.4 No Extension to Security Against Malicious Adversaries

We currently do not see how to extend our hashing-based protocols to achieve security against malicious adversaries. As pointed out by [44], it is inherently hard to extend protocols based on Cuckoo hashing to obtain security against malicious adversaries. The reason is that the placement of items depends on the exact composition of the input set, and therefore a malicious party might learn the placement used by the other party.

Coming up with a similar argument as in [44], assume that in our construction in Fig. 3, Bob maps an item x to the two upper subtables and Alice maps x to the two left subtables. Now assume Alice maliciously deviates from the protocol and places x only in the upper left subtable, but not in the lower left one. This deviation may allow Alice to learn whether Bob placed x in the upper or lower subtables: For example, in a PSI-CA protocol Alice could use only dummy elements and x as an input set and if the cardinality turns out to be 1, then she knows that Bob placed x in the upper left subtable. However, the locations in which Bob places an item cannot be simulated in the ideal world as they depend on other items in his input set. Therefore, we see no trivial way to provide security against malicious adversaries based on 2D Cuckoo hashing.

6 Evaluation

This section describes extensive experiments that set the parameters for the hashing schemes, the resulting circuit sizes, and the results of experiments evaluating PSI using these circuits.

6.1 Simulations for Setting the Parameters of 2D Cuckoo Hashing

We experimented with the iterative 2D Cuckoo hashing scheme described in Sect. 5.1, set concrete sizes for the tables, and examined the failure probabilities of hashing to the tables.

Our implementation is written in C and available online at http://encrypto.de/code/2DCuckooHashing. It repeatedly inserts a set of random elements into two tables using random hash functions. The insertion algorithm is very simple: All elements are first inserted into the two locations to which they are mapped (by the hash functions) in the first table. Obviously, many table entries will contain multiple items. Afterwards, the implementation iteratively moves items between the tables, in order to reduce the maximum bin occupancy below a certain threshold (cf. Algorithms 2 and 3 in Sect. 5.1).

Run-time. We report in Sect. 6.3 the results of experiments analyzing the run-time of the 2D Cuckoo hashing insertion algorithm. Overall, the insertion time (a few milliseconds) is negligible compared to the run-time of the entire PSI protocol.

Hashing to bins of size 1. First, we checked if it is possible to use a maximum bin occupation of 1. For this, we set the sizes of each of the two tables to be $4.8n$ (corresponding to the threshold size of $4(1+\varepsilon)n$ in the analysis available in the full version [40], as well as twice the recommended size for Cuckoo hashing, since all elements are inserted twice). We ran the experiment 100 000 times with input size $n = 2^{12}$ and bitlength 32. For all except 828 executions it was possible to reduce the maximum bin occupation to 1 after at least 7 and at most 129 iterations of the insertion algorithm. On average, 20 iterations of the insertion algorithm were necessary to achieve the desired result. In said 828 cases there remained at least one bin with more than one item even after 500 iterations of the insertion algorithm. This implies that iterative 2D Cuckoo hashing works in principle, but, as standard Cuckoo hashing, requires a stash for storing the elements of overfull bins.

Hashing to bins of size 2. For PSI protocols it would be desirable to avoid having an additional stash on Alice's side. In standard Cuckoo hashing it is possible to achieve better memory utilization and less usage of the stash by using fewer bins, where each bin can store two items [47]. Therefore, we changed the parameters as follows: the table size is halved and reduced to $2.4n$, but each bin is allowed to contain two elements. This way, while consuming the same amount of memory as before, we try to achieve better utilization. We followed the paradigm that was described in Sect. 3.2 for the experimental analysis of the failure probability. Namely, we ran massive sets of experiments to measure the number of failures for several values of n and several table sizes, and given this data we (1) found confidence intervals for the failure probability for specific values of the parameters, and (2) found how the failure probability behaves as a function of n.

Our first experiment ran 2^{40} tests within ~2 million core hours on the Lichtenberg[3] high performance computer of the TU Darmstadt for input size $n = 2^{12}$. We chose input size 2^{12} (instead of larger sizes like 2^{16} or 2^{20}) since running experiments with larger values of n would have taken even more time and would have simply been impractical. It turned out that the insertion algorithm was successful in reducing the maximum bin size to 2 (after at most 18 iterations) in all but one test.

Given this data, we calculated the confidence interval of the failure probability p. The probability of observing one failure in N experiments is $N \cdot p \cdot (1-p)^{N-1}$, where in our experiments $N = 2^{40}$. We checked the values of p for which the probability of this observation is greater than 0.001 and concluded that with 99.9% confidence, the failure probability for iterative 2D Cuckoo hashing with set size $n = 2^{12}$ and table size $2.4n$ lies within $\left[2^{-50}, 2^{-37}\right]$. (Namely, there is at most a 0.001 probability that we would have seen one failure in 2^{40} runs if p was greater than 2^{-37} or smaller than 2^{-50}.)

[3] See http://www.hhlr.tu-darmstadt.de/hhlr/index.en.jsp for details on the hardware configuration.

Measuring the dependence on the parameters. To get a better understanding on how the failure probability behaves for different input and table sizes, we performed a set of experiments that required another ~3.5 million core hours. Concretely, we ran 2^{40} tests for each set size $n \in \{2^6, 2^8, 2^{10}\}$ and each table size in the range $2.2n$, $2.4n$, and $2.6n$. We also tested the table size $3.6n$ for $n \in \{2^6, 2^8\}$ as well as table sizes $3.0n$ and $3.2n$ for $n = 2^{10}$. The results for all experiments are given in Table 2 and are depicted in Fig. 4.

The results demonstrate that, w.r.t. the dependence on n, for set sizes $n \in \{2^6, 2^8, 2^{10}\}$ it can be observed that increasing the set size by factor 4x reduces the failure probability by factor 64x. (For larger set sizes, the number of failures

Table 2. Number of observed stashes for different table sizes and set sizes n when performing 2^{40} tests of iterative 2D Cuckoo hashing.

Table size	Stash size	$n = 2^6$	$n = 2^8$	$n = 2^{10}$	$n = 2^{12}$
$2.2n$	1	64 020	1 021	16	—
	2	154	1	0	—
	3	4	0	0	—
$2.4n$	1	31 033	499	8	1
	2	65	0	0	0
$2.6n$	1	16 014	270	5	—
	2	33	0	0	—
$3.0n$	1	—	—	0	—
$3.2n$	1	—	—	0	—
$3.6n$	1	1 202	17	—	—

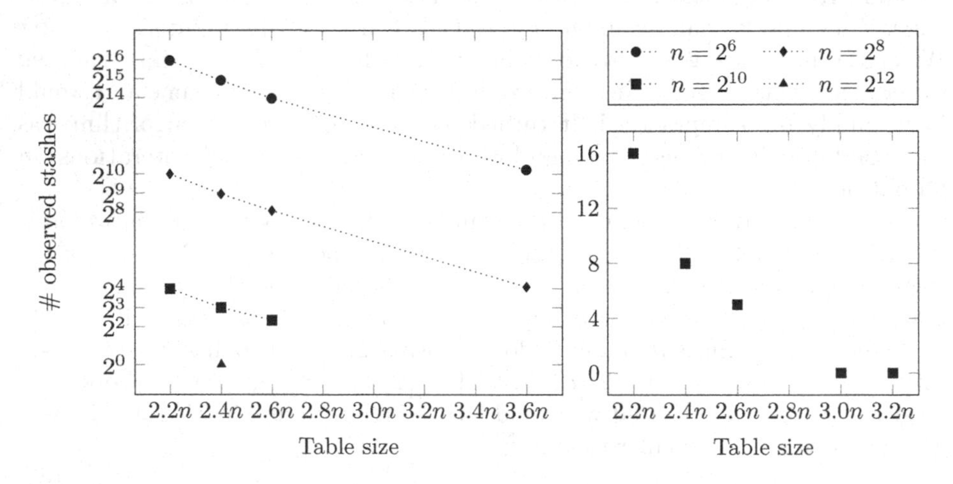

Fig. 4. Number of observed stashes for different table and set sizes when performing 2^{40} tests of iterative 2D Cuckoo hashing.

is too small to be meaningful.) These experiments also demonstrate that the dependence of the failure probability on n is $O(n^{-3})$. An intuitive theoretical explanation why the probability behaves this way is given in the full version [40]. As for the dependence on the table size, the failure probability decreases by a factor of 2x when increasing the table size in steps of $0.2n$ within the tested range $2.2n$ to $3.6n$.

From these results (a failure probability of at most 2^{-37} for $n = 2^{12}$ with table size $2.4n$ and a dependence of $O(n^{-3})$ of the failure probability on n) we conclude that the failure probability for $n \geq 2^{13}$ and table size $2.4n$ is at most 2^{-40}.

In total we spent about 5.5 million core hours on our experiments on the Lichtenberg high performance computer of the TU Darmstadt.

6.2 Circuit Complexities

We compare the complexities of the different circuit-based PSI constructions for two sets, each with n elements that have bitlength σ. We consider two possible bitlengths:

1. **Fixed bitlength:** Here, the elements have fixed bitlength $\sigma = 32$ bits (e.g., for IPv4 addresses).
2. **Arbitrary bitlength:** Here, the elements have arbitrary bitlength and are hashed to values of length $\sigma = 40 + 2\log_2(n) - 1$ bits, with a collision probability that is bounded by 2^{-40}. (See Appendix A of the full version of [41] for an analysis.) Therefore, we set the bitlength to $\sigma = 40 + 2\log_2(n) - 1$ bits.

For all protocols we report the circuit size where we count only the number of AND gates, since many secure computation protocols provide free computation of XOR gates. We compute the size of the circuits up to the step where single-bit wires indicate if a match was found for the respective element. We note that for many circuits computing functions of the intersection, this part of the circuit consumes the bulk of the total size. For example, computing the Hamming weight of these bits is equal to computing the cardinality of the intersection (PSI-CA). The size-optimal Hamming weight circuit of [5] has size $x - w_H(x)$ and depth $\log_2 x$, where x is the number of inputs and $w_H(\cdot)$ is the Hamming weight. The size of the Hamming weight circuit is negligible compared to the rest of the circuit. As another example, if the cardinality is compared with a threshold (yielding a PSI-CAT protocol), this only adds $3\log_2 n$ AND gates and depth $\log_2 \log_2 n$ using the depth-optimized construction described in [45], which is also negligible.

The size of the Sort-Compare-Shuffle circuit. The Sort-Compare-Shuffle circuit [26] has three phases. In the SORT phase, the two sorted lists of inputs are merged into one sorted list, which takes $2\sigma n \log_2(2n)$ AND gates. In the COMPARE phase, neighboring elements are compared to find the elements in the intersection, which takes $\sigma(3n - 1) - n$ AND gates. The SHUFFLE phase randomly permutes these values and takes $\sigma(n \log_2(n) - n + 1)$ AND gates. To have a fair comparison with our protocols, we remove the SHUFFLE phase and let the COMPARE phase output only a single bit that indicates if a match was found for the respective element or not; this removes n multiplexers of σ-bit values from the COMPARE phase, i.e., σn AND gates. Hence, the total size is $2\sigma n \log_2(n) + 2\sigma n - n - \sigma + 2$ AND gates.

The size of the Circuit-Phasing circuit. The Circuit-Phasing circuit [39] has $2.4nm(\sigma - \log_2(2.4n) + 1) + sn(\sigma - 1)$ AND gates where m is the maximum occupancy of a bin for simple hashing and s is the size of the stash.

The size of our iterative 2D Cuckoo hashing construction of Sect. 5.2. Each of the following operations is performed twice for the left and right side: (1) For each of the $2.4n$ bins the shortened representation (cf. Sect. 5.2) of the single item in Bob's bin is compared with the two elements in the corresponding bin of Alice. (2) Bob has a stash of size s'. Each item in the stash is compared to all of Alice's items (using the full bitlength representation). Hence, the overall complexity is $4 \cdot 2.4n(\sigma - \log_2(2.4n) + 1) + s'n(\sigma - 1)$ AND gates, where s' is the size of the combined stash.

Concrete Circuit Sizes. The Sort-Compare-Shuffle construction [26] has a circuit of size $O(\sigma n \log n)$. The Circuit-Phasing construction [39] has circuit size $O(\sigma n \log n / \log\log n)$, while the asymptotic construction we present in this paper has a size of $\omega(\sigma n)$ and the iterative 2D Cuckoo hashing construction has an even smaller size.

For a comparison of the concrete circuit sizes, we use the parameters from the analysis in [39]: For $n = 2^{12}$ elements the maximum bin size for simple hashing is $m = 18$, for $n = 2^{16}$ we set $m = 19$, and for $n = 2^{20}$ we set $m = 20$. We set the stash size s and the combined stash size s' according to Table 1 (on page 21).

On the left side of Table 3 we compare the concrete circuit sizes for *fixed* bitlength $\sigma = 32$ bit. Our best protocol ("Ours Iterative Combined") improves over the best previous protocol by factor 2.0x for $n = 2^{12}$ (over [26]), by factor 2.7x for $n = 2^{16}$ (over [39]), and by factor 3.2x for $n = 2^{20}$ (over [39]).

On the right side of Table 3 we compare the concrete circuit sizes for *arbitrary* bitlength σ. Our best protocol (Ours Iterative Combined) improves over the best previous protocol by factor 1.8x for $n = 2^{12}$ (over [26]), by factor 2.8x for $n = 2^{16}$ (over [26]), and by factor 3.8x for $n = 2^{20}$ (over [39]).

Our constructions always have smaller circuits than both former constructions, and, due to our better asymptotic size, the savings become greater as n increases.

Circuit Depths. For some protocols, the circuit depth is a relevant metric (e.g., for the GMW protocol the depth determines the round complexity of the online phase). Our constructions have the same depth as the Circuit-Phasing protocol of [39], i.e., $\log_2 \sigma$. This is much more efficient than the depth of the

Table 3. Concrete circuit sizes in #ANDs for PSI variants on n elements of fixed bitlength $\sigma = 32$ (left) and arbitrary bitlength hashed to $\sigma = 40 + 2\log_2(n) - 1$ bits (right).

Protocol	Fixed bitlength $\sigma = 32$			Arbitrary bitlength		
	$n = 2^{12}$	$n = 2^{16}$	$n = 2^{20}$	$n = 2^{12}$	$n = 2^{16}$	$n = 2^{20}$
Sort-Compare-Shuffle [26]	3 403 746	71 237 602	1 408 237 538	6 705 091	158 138 299	3 478 126 515
Circuit-Phasing [39]	4 254 256	55 155 466	688 258 388	10 501 475	181 928 305	3 201 695 060
Separate stashes $s' = 2s$						
Ours iterative separate	2 299 801	26 153 770	313 183 300	5 042 482	71 137 681	1 081 999 223
Combined stash s' (cf. Table 1)						
Ours iterative combined	1 664 921	20 058 922	215 665 732	3 772 722	57 375 121	836 632 439

Sort-Compare-Shuffle circuit of [26] which is $O(\log \sigma \cdot \log n)$ when using depth-optimized comparison circuits.

Further Optimizations. So far, we computed the comparisons with a Boolean circuit consisting of 2-input gates: For elements of bitlength ℓ, the circuit XORs the elements and afterwards computes a tree of $\ell - 1$ non-XOR gates s.t. the final output is 1 if the elements are equal or 0 otherwise. This circuit allows to use an arbitrary secure computation protocol based on Boolean gates, e.g., Yao or GMW. The recent approach of [14] shows that for security against semi-honest adversaries the communication can be improved by using multi-input lookup tables (LUTs). Their best LUT has 7 inputs and requires only 372 bits of total communication (cf. [14, Table 4]). For computing equality, 6 of the non-XOR gates in the tree can be combined into one 7-input LUT. This improves communication of the Circuit-Phasing protocol of [39] and our protocols by factor $6 \cdot 256/372 = 4.1$x.

6.3 Performance

We empirically compare the performance of our iterative 2D Cuckoo hashing PSI-CAT protocol with a combined stash described in Sect. 5.2 with the Circuit-Phasing PSI-CAT protocol of [39]. As a baseline, we also compare with the public key-based PSI-CA protocol of [9,35,46] that leaks the cardinality to one party, and the currently best specialized PSI protocol of [31] that cannot be easily modified to compute variants of the set intersection functionality.

Implementation. Pinkas et al. [39] provide the implementation of their Circuit-Phasing PSI protocol as part of the ABY framework [13]. This framework allows to securely evaluate the PSI circuit using either Yao's garbled circuit or the GMW protocol, both implemented with most recent optimizations (cf. Sect. 2). However, since the evaluation in [39] showed that using the GMW protocol yields much better run-times, we focus only on GMW. ABY also implements the LUT-based evaluation of [14] (cf. Sect. 6.2), which we compare to GMW evaluation.

For the Circuit-Phasing PSI-CAT protocol, we extended the existing codebase with the Hamming weight circuit of [5] and the depth-optimized comparison circuit of [45] to compare the Hamming weight with a threshold. Based on this, we implemented our iterative 2D Cuckoo hashing PSI-CAT protocol by duplicating the code for simple hashing and Cuckoo hashing, combining the stashes, and implementing the iterative insertion algorithm. Our implementation is available online as part of the ABY framework at http://encrypto.de/code/ABY. For the DH/ECC-based protocol of Shamir/Meadows/De Cristofaro et al. [9,35,46], we use the ECC-based implementation of [39] available online at http://encrypto.de/code/PSI that already supports computing the cardinality (PSI-CA). The implementation of the special purpose BaRK-OPRF PSI protocol of [31] is taken from https://github.com/osu-crypto/BaRK-OPRF.

Benchmarking Environment. For our benchmarks we use two machines, each equipped with an Intel Core i7-4790 CPU @ 3.6 GHz and 16 GB of RAM. The CPUs support the AES-NI instruction set for fast AES evaluations. We distinguish two network settings: a LAN setting and a WAN setting. For the LAN setting, we restrict the bandwidth of the network interfaces to 1 Gbit/s and enforce a round-trip time of 1 ms. For the WAN setting, we limit the bandwidth to 100 Mbit/s and set a round-trip time of 100 ms. We instantiate all protocols corresponding to a computational security parameter of 128 bit and a statistical security parameter of 40 bit. All reported run-times are the average of 10 executions with less than 10% variance.

Benchmarking Results. In Table 4, we give the run-times for $n \in \{2^{12}, 2^{16}, 2^{20}\}$ elements[4] of bitlength $\sigma = 32$ (suitable, e.g., for IPv4 addresses). The corresponding communication is given in Table 6. We do not use the LUT-based evaluation in the LAN setting since there is little need for better communication while the run-times are not competitive. However, to demonstrate the advantages of the LUT-based evaluation in the WAN setting, we compare the protocols when running with a single thread and four threads.[5]

Run-times (Tables 4 and 5). In comparison with the Circuit-Phasing PSI-CAT protocol of [39] in Table 4, our iterative combined PSI-CAT protocol is faster by factor 1.4x for $n = 2^{12}$ and up to factor 2.8x for $n = 2^{20}$. This holds when the circuit is evaluated with GMW in both network settings and for both 1 and 4 threads. With LUT-based evaluation [14], we observe a further improvement for the circuit-based protocols by about 13% in the WAN setting, but only for medium set sizes of $n = 2^{16}$ and 4 threads due to the higher computation complexity.

The circuit-based protocols have two steps: mapping the input items to the tables, and securely evaluating the circuit. The run-times of the hashing step are

[4] Unfortunately, the LUT-based implementation of [14] was not capable of evaluating the PSI circuits for $n = 2^{20}$ elements.

[5] We do not provide benchmarks with multiple threads for the DH/ECC PSI-CA protocol since the implementation of [39] does not support multi-threading.

shown in Table 5. The times for Cuckoo hashing into two tables in our PSI-CAT protocol are exactly twice of those for Cuckoo hashing into one table in [39]. Compared to simple hashing, our 2D Cuckoo hashing is slower by factor 1.6x up to factor 2.1x due to the additional iterations. However, all in all, the hashing procedures are by 2–3 orders of magnitude faster than the times for securely evaluating the circuit, and therefore negligible w.r.t. the overall run-time.

In comparison with the DH-based PSI-CA protocol of [9,35,46], our iterative combined PSI-CAT protocol is faster by factor 1.5x for $n = 2^{12}$ up to factor 91x for $n = 2^{20}$ in the LAN setting with a single thread. Also in the WAN setting with a single thread, our protocol is faster (except for small sets with $n = 2^{12}$), despite the substantially lower communication of the DH-based protocol described below. In both network settings even the best measured run-times of our PSI-CAT protocol are between 19x to 36x slower than the BaRK-OPRF specialized PSI protocol of [31], but our protocols are generic.

Communication (Table 6). The communication given in Table 6 is measured on the network interface, so these numbers are slightly larger than the theoretical communication (derived from the number of AND gates on the left side in Table 3) due to TCP/IP headers and padding of messages. The lowest communication is

Table 4. Total run-times in ms for PSI variants on n elements of bitlength $\sigma = 32$ bit.

Protocol	Network setting	LAN			WAN				
	Circuit evaluation protocol	GMW [21]			GMW [21]			LUT [14]	
	Set size n	2^{12}	2^{16}	2^{20}	2^{12}	2^{16}	2^{20}	2^{12}	2^{16}
DH/ECC PSI-CA [9,35,46]		3 296	49 010	7 904 054	4 082	51 866	8 008 771	4 082	51 866
BaRK-OPRF PSI [31]		113	295	3 882	540	1 247	14 604	540	1 247
1 Thread									
Circuit-Phasing PSI-CAT [39]		3 170	20 401	242 235	15 143	99 433	1 042 712	19 951	117 438
Ours iterative separate PSI-CAT		2 433	11 251	122 008	11 210	57 474	547 950	15 656	70 545
Ours iterative combined PSI-CAT		2 220	9 076	86 648	10 060	45 252	389 891	12 999	56 179
4 Threads									
Circuit-Phasing PSI-CAT [39]		2 333	10 600	123 765	12 492	97 480	987 459	15 471	76 184
Ours iterative separate PSI-CAT		1 903	6 273	64 324	9 361	56 141	541 677	11 946	46 797
Ours iterative combined PSI-CAT		1 694	5 177	49 417	8 793	44 596	376 591	9 413	39 272

Table 5. Run-times in ms for hashing n elements of bitlength $\sigma = 32$ bit.

Hashing procedure	Set size n	2^{12}	2^{16}	2^{20}
Circuit-Phasing PSI-CAT [39]				
Simple hashing		3.50	27.96	557.54
Cuckoo hashing		2.43	15.87	391.16
Ours iterative PSI-CAT				
2D Cuckoo hashing		6.23	58.90	873.19
Cuckoo hashing (for two tables with a combined stash)		4.85	31.75	782.32

Table 6. Communication in MB for PSI variants on n elements of bitlength $\sigma = 32$ bit.

Protocol / Set size n	2^{12}	2^{16}	2^{20}
DH/ECC PSI-CA [9,35,46]	0.4	6.6	106.0
BaRK-OPRF PSI [31]	0.53	8.06	127.20
GMW [21]			
Circuit-Phasing PSI-CAT [39]	121.9	1 588.9	20 028.5
Ours iterative separate PSI-CAT	72.3	826.1	9 971.4
Ours iterative combined PSI-CAT	52.7	638.8	6 950.6
LUT [14]			
Circuit-Phasing PSI-CAT [39]	32.6	418.1	—
Ours iterative separate PSI-CAT	19.4	221.3	—
Ours iterative combined PSI-CAT	14.3	171.3	—

achieved by the DH-based PSI-CA protocol of [9,35,46] which is in line with the experiments in [39]. Our best protocol for PSI-CAT has between 132x (for $n = 2^{12}$) and 66x (for $n = 2^{20}$) more communication than the DH-based PSI-CA protocol when evaluated with GMW. Recall, however, that our protocol does not leak the cardinality. Our best protocol improves the communication over the PSI-CAT protocol of [39] by factor 2.3x (for $n = 2^{12}$) to 2.9x (for $n = 2^{20}$). When using LUT-based evaluation of [14], we observe that the communication of all circuit-based PSI-CAT protocols improves over GMW by factor 3.7x which is close to the theoretical upper bound of 4.1x (cf. Sect. 6.2). Still, our best LUT-based protocol has more than 20x higher communication than the BaRK-OPRF specialized PSI protocol of [31], but it is generic.

Application to privacy-preserving ridesharing. Our PSI-CAT protocol can easily be extended for the privacy-preserving ridesharing functionality of [23], where the intersection is revealed only if the size of the intersection is larger than a threshold. The authors of [23] give a protocol that securely computes this functionality, but has quadratic computation complexity. By slightly extending our circuit for PSI-CAT to encapsulate a key that is released only if the size of the intersection is larger than the threshold and using this key to symmetrically encrypt the last message in any linear complexity PSI protocol (e.g., [31,39,41, 42]), we get a protocol with almost linear complexity. Our key encapsulation would take less than 3 s for $n = 2^{12}$ elements (cf. our results for PSI-CAT in Table 4), whereas the solution of [23] takes 5 627 s, i.e., we improve by factor 1 876x and also asymptotically.

Acknowledgments. We thank Oleksandr Tkachenko for his invaluable help with the implementation and benchmarking. We also thank Moni Naor for suggesting the application to achieve differential privacy. This work has been co-funded by the DFG as part of project E4 within the CRC 1119 CROSSING and by the German Federal

Ministry of Education and Research (BMBF), the Hessen State Ministry for Higher Education, Research and the Arts (HMWK) within CRISP, and the BIU Center for Research in Applied Cryptography and Cyber Security in conjunction with the Israel National Cyber Bureau in the Prime Minister's Office. Calculations for this research were conducted on the Lichtenberg high performance computer of the TU Darmstadt.

References

1. Amossen, R.R., Pagh, R.: A new data layout for set intersection on GPUs. In: International Symposium on Parallel and Distributed Processing (IPDPS) (2011)
2. Arbitman, Y., Naor, M., Segev, G.: Backyard cuckoo hashing: constant worst-case operations with a succinct representation. In: FOCS (2010)
3. Asharov, G., Lindell, Y., Schneider, T., Zohner, M.: More efficient oblivious transfer and extensions for faster secure computation. In: CCS (2013)
4. Asokan, N., Dmitrienko, A., Nagy, M., Reshetova, E., Sadeghi, A.-R., Schneider, T., Stelle, S.: CrowdShare: secure mobile resource sharing. In: Jacobson, M., Locasto, M., Mohassel, P., Safavi-Naini, R. (eds.) ACNS 2013. LNCS, vol. 7954, pp. 432–440. Springer, Heidelberg (2013). https://doi.org/10.1007/978-3-642-38980-1_27
5. Boyar, J., Peralta, R.: Concrete multiplicative complexity of symmetric functions. In: Královič, R., Urzyczyn, P. (eds.) MFCS 2006. LNCS, vol. 4162, pp. 179–189. Springer, Heidelberg (2006). https://doi.org/10.1007/11821069_16
6. Chen, H., Laine, K., Rindal, P.: Fast private set intersection from homomorphic encryption. In: CCS (2017)
7. Dachman-Soled, D., Malkin, T., Raykova, M., Yung, M.: Efficient robust private set intersection. In: Abdalla, M., Pointcheval, D., Fouque, P.-A., Vergnaud, D. (eds.) ACNS 2009. LNCS, vol. 5536, pp. 125–142. Springer, Heidelberg (2009). https://doi.org/10.1007/978-3-642-01957-9_8
8. Davidson, A., Cid, C.: An efficient toolkit for computing private set operations. In: Pieprzyk, J., Suriadi, S. (eds.) ACISP 2017. LNCS, vol. 10343, pp. 261–278. Springer, Cham (2017). https://doi.org/10.1007/978-3-319-59870-3_15
9. De Cristofaro, E., Gasti, P., Tsudik, G.: Fast and private computation of cardinality of set intersection and union. In: Pieprzyk, J., Sadeghi, A.-R., Manulis, M. (eds.) CANS 2012. LNCS, vol. 7712, pp. 218–231. Springer, Heidelberg (2012). https://doi.org/10.1007/978-3-642-35404-5_17
10. De Cristofaro, E., Kim, J., Tsudik, G.: Linear-complexity private set intersection protocols secure in malicious model. In: Abe, M. (ed.) ASIACRYPT 2010. LNCS, vol. 6477, pp. 213–231. Springer, Heidelberg (2010). https://doi.org/10.1007/978-3-642-17373-8_13
11. De Cristofaro, E., Tsudik, G.: Practical private set intersection protocols with linear complexity. In: Sion, R. (ed.) FC 2010. LNCS, vol. 6052, pp. 143–159. Springer, Heidelberg (2010). https://doi.org/10.1007/978-3-642-14577-3_13
12. Debnath, S.K., Dutta, R.: Secure and efficient private set intersection cardinality using bloom filter. In: Lopez, J., Mitchell, C.J. (eds.) ISC 2015. LNCS, vol. 9290, pp. 209–226. Springer, Cham (2015). https://doi.org/10.1007/978-3-319-23318-5_12
13. Demmler, D., Schneider, T., Zohner, M.: ABY – a framework for efficient mixed-protocol secure two-party computation. In: NDSS (2015)
14. Dessouky, G., Koushanfar, F., Sadeghi, A.-R., Schneider, T., Zeitouni, S., Zohner, M.: Pushing the communication barrier in secure computation using lookup tables. In: NDSS (2017)

15. Dietzfelbinger, M., Weidling, C.: Balanced allocation and dictionaries with tightly packed constant size bins. Theoret. Comput. Sci. **380**(1–2), 47–68 (2007)
16. Dong, C., Chen, L., Wen, Z.: When private set intersection meets big data: an efficient and scalable protocol. In: CCS (2013)
17. Dwork, C.: Differential privacy. In: Bugliesi, M., Preneel, B., Sassone, V., Wegener, I. (eds.) ICALP 2006. LNCS, vol. 4052, pp. 1–12. Springer, Heidelberg (2006). https://doi.org/10.1007/11787006_1
18. Eppstein, D., Goodrich, M., Mitzenmacher, M., Torres, M.: 2–3 cuckoo filters for faster triangle listing and set intersection. In: Symposium on Principles of Database Systems (PODS) (2017)
19. Freedman, M.J., Hazay, C., Nissim, K., Pinkas, B.: Efficient set intersection with simulation-based security. J. Cryptol. **29**(1), 115–155 (2016)
20. Freedman, M.J., Nissim, K., Pinkas, B.: Efficient private matching and set intersection. In: Cachin, C., Camenisch, J.L. (eds.) EUROCRYPT 2004. LNCS, vol. 3027, pp. 1–19. Springer, Heidelberg (2004). https://doi.org/10.1007/978-3-540-24676-3_1
21. Goldreich, O., Micali, S., Wigderson, A.: How to play any mental game or a completeness theorem for protocols with honest majority. In: STOC (1987)
22. Gonnet, G.H.: Expected length of the longest probe sequence in hash code searching. J. ACM **28**(2), 289–304 (1981)
23. Hallgren, P., Orlandi, C., Sabelfeld, A.: PrivatePool: privacy-preserving ridesharing. In: Computer Security Foundations Symposium (CSF) (2017)
24. Hazay, C., Venkitasubramaniam, M.: Scalable multi-party private set-intersection. In: Fehr, S. (ed.) PKC 2017. LNCS, vol. 10174, pp. 175–203. Springer, Heidelberg (2017). https://doi.org/10.1007/978-3-662-54365-8_8
25. Huang, Y., Chapman, P., Evans, D.: Privacy-preserving applications on smartphones. In: Hot Topics in Security (HotSec) (2011)
26. Huang, Y., Evans, D., Katz, J.: Private set intersection: Are garbled circuits better than custom protocols? In: NDSS (2012)
27. Huang, Y., Evans, D., Katz, J., Malka, L.: Faster secure two-party computation using garbled circuits. In: USENIX Security (2011)
28. Ion, M., Kreuter, B., Nergiz, E., Patel, S., Saxena, S., Seth, K., Shanahan, D., Yung, M.: Private intersection-sum protocol with applications to attributing aggregate ad conversions. Cryptology ePrint Archive, Report 2017/738 (2017)
29. Kirsch, A., Mitzenmacher, M., Wieder, U.: More robust hashing: cuckoo hashing with a stash. SIAM J. Comput. **39**(4), 1543–1561 (2009)
30. Kiss, Á., Liu, J., Schneider, T., Asokan, N., Pinkas, B.: Private set intersection for unequal set sizes with mobile applications. In: PoPETs, vol. 2017(4) (2017)
31. Kolesnikov, V., Kumaresan, R., Rosulek, M., Trieu, N.: Efficient batched oblivious PRF with applications to private set intersection. In: CCS (2016)
32. Kolesnikov, V., Matania, N., Pinkas, B., Rosulek, M., Trieu, N.: Practical multi-party private set intersection from symmetric-key techniques. In: CCS (2017)
33. Kolesnikov, V., Schneider, T.: Improved garbled circuit: free XOR gates and applications. In: Aceto, L., Damgård, I., Goldberg, L.A., Halldórsson, M.M., Ingólfsdóttir, A., Walukiewicz, I. (eds.) ICALP 2008. LNCS, vol. 5126, pp. 486–498. Springer, Heidelberg (2008). https://doi.org/10.1007/978-3-540-70583-3_40
34. Kreuter, B.: Secure multiparty computation at Google. In: Real World Crypto Conference (RWC) (2017)
35. Meadows, C.: A more efficient cryptographic matchmaking protocol for use in the absence of a continuously available third party. In: S&P (1986)

36. Pagh, R., Rodler, F.F.: Cuckoo hashing. In: European Symposium on Algorithms (ESA) (2001)
37. Panigrahy, R.: Efficient hashing with lookups in two memory accesses. In: ACM-SIAM Symposium on Discrete Algorithms (SODA) (2005)
38. Pettai, M., Laud, P.: Combining differential privacy and secure multiparty computation. In: ACSAC (2015)
39. Pinkas, B., Schneider, T., Segev, G., Zohner, M.: Phasing: private set intersection using permutation-based hashing. In: USENIX Security (2015)
40. Pinkas, B., Schneider, T., Weinert, C., Wieder, U.: Efficient circuit-based PSI via cuckoo hashing. In: Cryptology ePrint Archive, Report 2018/120 (2018)
41. Pinkas, B., Schneider, T., Zohner, M.: Faster private set intersection based on OT extension. In: USENIX Security (2014)
42. Pinkas, B., Schneider, T., Zohner, M.: Scalable private set intersection based on OT extension. ACM Trans. Priv. Secur. (TOPS) **21**(2) (2018)
43. Rindal, P., Rosulek, M.: Improved private set intersection against malicious adversaries. In: Coron, J.-S., Nielsen, J.B. (eds.) EUROCRYPT 2017. LNCS, vol. 10210, pp. 235–259. Springer, Cham (2017). https://doi.org/10.1007/978-3-319-56620-7_9
44. Rindal, P., Rosulek, M.: Malicious-secure private set intersection via dual execution. In: CCS (2017)
45. Schneider, T., Zohner, M.: GMW vs. Yao? Efficient secure two-party computation with low depth circuits. In: Sadeghi, A.-R. (ed.) FC 2013. LNCS, vol. 7859, pp. 275–292. Springer, Heidelberg (2013). https://doi.org/10.1007/978-3-642-39884-1_23
46. Shamir, A.: On the power of commutativity in cryptography. In: de Bakker, J., van Leeuwen, J. (eds.) ICALP 1980. LNCS, vol. 85, pp. 582–595. Springer, Heidelberg (1980). https://doi.org/10.1007/3-540-10003-2_100
47. Wieder, U.: Hashing, load balancing and multiple choice. Found. Trends Theoret. Comput. Sci. **12**(3–4), 275–379 (2017)
48. Yao, A.C.: How to generate and exchange secrets. In: FOCS (1986)
49. Yung, M.: From mental poker to core business: why and how to deploy secure computation protocols? In: CCS (2015)
50. Zahur, S., Rosulek, M., Evans, D.: Two halves make a whole: reducing data transfer in garbled circuits using half gates. In: Oswald, E., Fischlin, M. (eds.) EUROCRYPT 2015. LNCS, vol. 9057, pp. 220–250. Springer, Heidelberg (2015). https://doi.org/10.1007/978-3-662-46803-6_8

Overdrive: Making SPDZ Great Again

Marcel Keller[1(✉)], Valerio Pastro[2], and Dragos Rotaru[1,3]

[1] University of Bristol, Bristol, UK
{m.keller,dragos.rotaru}@bristol.ac.uk
[2] Yale University, New Haven, USA
vpastro86@gmail.com
[3] imec-Cosic, Department of Electrical Engineering, KU Leuven, Leuven, Belgium

Abstract. SPDZ denotes a multiparty computation scheme in the preprocessing model based on somewhat homomorphic encryption (SHE) in the form of BGV. At CCS '16, Keller et al. presented MASCOT, a replacement of the preprocessing phase using oblivious transfer instead of SHE, improving by two orders of magnitude on the SPDZ implementation by Damgård et al. (ESORICS '13). In this work, we show that using SHE is faster than MASCOT in many aspects:

1. We present a protocol that uses semi-homomorphic (addition-only) encryption. For two parties, our BGV-based implementation is six times faster than MASCOT on a LAN and 20 times faster in a WAN setting. The latter is roughly the reduction in communication.
2. We show that using the proof of knowledge in the original work by Damgård et al. (Crypto '12) is more efficient in practice than the one used in the implementation mentioned above by about one order of magnitude.
3. We present an improvement to the verification of the aforementioned proof of knowledge that increases the performance with a growing number of parties, doubling it for 16 parties.

Keywords: Multiparty computation
Somewhat homomorphic encryption · BGV
Zero-knowledge proofs of knowledge

1 Introduction

Multiparty computation (MPC) allows a set of parties to jointly compute a function over their inputs while keeping them private. In the last decade MPC

D. Rotaru—The work in this paper was partially supported by EPSRC via grants EP/M012824 and EP/N021940/1, by the European Union's H2020 Programme under grant agreement number ICT-644209, and by the Defense Advanced Research Projects Agency (DARPA) and Space and Naval Warfare Systems Center, Pacific (SSC Pacific) under contract No. N66001-15-C-4070. The second author conducted this work while working as a post-doctoral researcher at Yale University, supported by NSF grants CNS-1562888, 1565208, and DARPA SafeWare W911NF-15-C-0236. We would also like to thank the anonymous Eurocrypt reviewers as well as Ivan Damgård, Peter Scholl, and Nigel Smart for their helpful comments.

J. B. Nielsen and V. Rijmen (Eds.): EUROCRYPT 2018, LNCS 10822, pp. 158–189, 2018.
https://doi.org/10.1007/978-3-319-78372-7_6

has developed from a largely theoretical field to a practical one where many applications have been developed on top of it [DES16, GSB+17]. This is mostly due to the rise of compilers which translate high-level code to secure branching, additions and multiplications on secret data [BLW08, KSS13, DSZ15, ZE15].

A high number of applications require to evaluate an arithmetic circuit (over the integers or modulo p) due to the easiness of expressing them rather than performing bitwise operations in a binary circuit. This is especially true for linear programming of satellite collisions where fixed and floating point numbers are intensively used [DDN+16, KW15]. A recent line of work even looked at how to decrease the amount of storage needed throughout sequential computations from one MPC engine to another with symmetric key primitives evaluated as arithmetic circuits [GRR+16, RSS17].

To accomplish MPC one can select between two paradigms: garbled circuits [GLNP15, RR16, WRK17] or secret sharing [DGKN09, BDOZ11, DPSZ12]. We will concentrate on the latter because it is currently the most suitable to evaluate arithmetic circuits although there have been some recent theoretical improvements on garbling modulo p made by Ball et al. [BMR16]. Since our goal in this paper is to have secure computation within a system that scales with the number of parties as well as to provide a guarantee against malicious, players we will focus on SPDZ [DPSZ12, DKL+13].

It is no surprise that homomorphic encryption can help with multiparty computation. In the presence of malicious adversaries, however, there needs to be assurances that parties actually encrypt the information that they are supposed to Zero-knowledge proofs are the essential tool to achieve this, and there exist compilers to make passive protocols secure against an active adversary. However, these proofs are relatively expensive, and it is the aim of SPDZ to reduce this cost by using them as little as possible.

The core idea of SPDZ is that, instead of encrypting the parties' inputs, it is easier to work with random data, conduct some checks at the end of the protocol, and abort if malicious behavior is detected. In order to evaluate a function with private inputs, the computation is separated in two phases, a preprocessing or offline phase and an online phase. The latter uses information-theoretic algorithms to compute the results from the inputs and the correlated randomness produced by the offline phase.

The correlated randomness consists of secret-shared random multiplication triples, that is (a, b, ab) for random a and b. In SPDZ, the parties encrypt random additive shares of a and b under a global public key, use the homomorphic properties to sum up and multiply the shares, and then run a distributed decryption protocol to learn their share of ab. With respect to malicious parties, there are two requirements on the encrypted shares of a and b. First, they need to be independent of other parties' shares, otherwise the sum would not be random, and second, the ciphertexts have to be valid. In the context of lattice-based cryptography, this means that the noise must be limited. Both requirements are achieved by using zero-knowledge proofs of knowledge and bounds of the cleartext and encryption randomness. It turns out that this is the most expensive part of the protocol.

The original SPDZ protocol [DPSZ12] uses a relatively simple Schnorr-like protocol [CD09] to prove knowledge of cleartexts and correctness of ciphertexts, but the later implementation [DKL+13] uses more sophisticated cut-and-choose-style protocols for both covert and active security. We have found that the simpler Schnorr-like protocol, which guarantees security against active malicious parties, is actually more efficient than the cut-and-choose proof with covert security.

Intuitively, it suffices that the encryption of the sum of all shares has to be correct because only the sum is used in the protocol. We take advantage of this by replacing the per-party proof with a global proof in Sect. 4. This significantly reduces the computation because every party only has to check one proof instead of $n - 1$. However, the communication complexity stays the same because the independence requirement means that every party still has to commit to every other party in some sense. Otherwise, a rushing adversary could make its input dependent on others, resulting in a predictable triple.

Section 3 contains our largest theoretical contribution. We present a replacement for the offline phase of SPDZ based solely on the additive homomorphism of BGV. This allows to reduce the communication and computation compared to SPDZ because the ciphertext modulus can be smaller. At the core of our scheme is the two-party oblivious multiplication protocol by Bendlin et al. [BDOZ11], which is based on the multiplication of ciphertexts and constants. Unlike their work, we assume that the underlying cryptosystem achieves linear targeted malleability introduced by Bitansky et al. [BCI+13], which enables us to avoid the costliest part of their protocol, the proof of correct multiplication. Instead, we replace this check by the SPDZ sacrifice, and argue that BGV with increased entropy in the secret key is a candidate for the above-mentioned assumption.

We do not consider the restriction to BGV to be a loss. Bendlin et al. suggest two flavors for the underlying cryptosystem: lattice-based and Paillier-like. For lattice-based cryptosystems, Costache and Smart [CS16] have shown that BGV is very competitive for large enough cleartext moduli such as needed by our protocol. On the other hand, Paillier only supports simple packing techniques and makes it difficult to manipulate individual slots [NWI+13]. Another advantage of BGV over Paillier is the heavy parallelization with CRT and FFT since in the lattice-based cryptosystem the ciphertext modulus can be a product of several primes.

2 Preliminaries

In the following section we define the basic notation and give an overview of the BGV encryption scheme and the SPDZ protocol.

2.1 Security Model

We use the UC (Universally Composable) framework of Canetti [Can01] to prove the security of our schemes against malicious, static adversaries, except for proofs of knowledge where we use rewinding to extract inputs from the adversary. Previous works [BDOZ11, DPSZ12] do this by having all inputs encrypted under a public key for which the secret key is known to the simulator in the registered key model. In our Low Gear protocol, this would involve sending extra ciphertexts not used in the protocol otherwise, which is why we opt for limited UC security.

Our protocols work with n parties $\mathcal{P} = \{P_1, \ldots, P_n\}$ where up to $n-1$ corruptions can take place before the protocol starts. We say that a protocol Π implements securely a functionality $\mathcal{F}$ if any probabilistic polynomial time adversary Adv cannot distinguish between a protocol Π and a functionality $\mathcal{F}$ attached to a simulator $\mathcal{S}$ with computational security k and statistical security sec.

We require the functionality $\mathcal{F}_{\mathsf{Rand}}$ to generate public randomness. Whenever the functionality is activated by all parties, it outputs a uniformly random value $r \xleftarrow{\$} \mathbb{F}$ to all parties. $\mathcal{F}_{\mathsf{Rand}}$ can be implemented using commitments of random values, which are then added. In our experiments, we will use simple commitments based on the random oracle model.

2.2 BGV

We now give a short overview of the leveled encryption scheme developed by Brakerski et al. [BGV12] required for our pre-processing phase. Since the protocols used for generating the triples need only multiplication by scalars or ciphertext addition, the BGV scheme is instantiated with a single level. For completion we present the details required to understand our paper. The reader can consult the following papers for further details: [LPR10, BV11, GHS12a, LPR13].

Underlying Algebra. Let $R = \mathbb{Z}[X]/\langle f(x) \rangle$ be a polynomial ring with integer coefficients modulo $f(x)$. In our case $R = \mathbb{Z}[x]/\langle \Phi_m(X) \rangle$ where $\Phi_m(X) = \prod_{i \in Z_m^*}(X - \omega_m^i) \in \mathbb{Z}[X]$ and $\omega_m = \exp(2\pi/m) \in \mathbb{C}$ is a principal m'th complex root of unity and $\omega_m^i = \exp(2\pi\sqrt{-1}/m) \in \mathbb{C}$ iterates over all primitive complex mth roots of unity.

The ring R is also called the ring of algebraic integers of the m'th cyclotomic polynomial. For example when $m \geq 2$ and m is a power of two, the polynomial $\Phi_m(X) = X^{m/2} + 1$. Notice that the degree of $\Phi_m(X)$ is equal to $\phi(m)$, which makes R a field extension with degree $N = \phi(m)$. Next we define $R_q = R/qR \cong R/\langle(\Phi_m(X), q)\rangle$ where q is not necessarily a prime number. The latter will be used as the ciphertext modulus.

Plaintext Slots. Since triples are generated for arithmetic circuits modulo p, the plaintext space is the ring $R_p = R/pR$ where for technical reasons p and q are co-prime. If $p \equiv 1 \mod m$ we have that $\Phi_m(X) = F_1(X) \cdots F_l(X) \mod p$ splits into l irreducible polynomials where each $F_i(X)$ has degree $d = \phi(m)/l$ and $F_i(X) \cong \mathbb{F}_p^d$. It is useful to think of an element $a \in R_p$ as a vector of size l where each element is $(a \mod F_i(X))_{i=1}^l$. This in turn allows manipulating l plaintexts at once using SIMD (Single Instruction Multiple Data) operations.

Distributions. Throughout the definitions we will refer to a polynomial $a \in R$ as a vector of size $N = \phi(m)$. To realize the cryptosystem we need to sample at various times from different distributions to generate a vector of length N with coefficients mod p or q (which means an element from R_p or R_q). We will keep R_q throughout the following definitions:

- $\mathcal{U}(R_q)$ is the uniform distribution where each unique polynomial $a \in R_q$ has an equal chance to be drawn. This is achieved by sampling each coefficient of a uniformly at random (from the integers modulo q).
- $\mathcal{DG}(\sigma^2, R_q)$ is the discrete Gaussian with variance σ^2. Sampling proceeds as above except each coefficient $a \in R_q$ is generated by calling the normal Gaussian $\mathcal{N}(\sigma^2)$ and rounding it to the nearest integer.
- $\mathcal{ZO}(0.5)$ outputs a vector of length N where each entry has values in the set $\{-1, 0, 1\}$. Here, zero appears with a probability 1/2 whereas $\{-1, 1\}$ each appear with probability 1/4.
- $\mathcal{HWT}(h)$ outputs a random vector of length N where at least h entries are non-zero and each entry is in the $\{-1, 0, 1\}$ set.

Ring-LWE. Hardness of the BGV scheme is based on the Ring version of the Learning with Errors problem [LPR10]. For a secret $s \in R_p$, recall that a Ring-LWE sample is produced by choosing $a \in R_q$ uniformly at random and an error $e \leftarrow \chi$ from a special Gaussian distribution, and computing $b = a \cdot s + e$. It turns out that, if an adversary manages to break the BGV encryption scheme in polynomial time, one can also build a polynomial time distinguisher for Ring-LWE samples and the uniform distribution, namely $(a, b = a \cdot s + e) \cong (a', b')$ where $(a', b') \stackrel{\$}{\leftarrow} \mathcal{U}(R_q^2)$.

Key-Generation, Encryption and Decryption. The cryptosystem used in Sect. 3 is identical to the one by Damgård et al. [DKL+13] bar the augmentation data needed for modulus switching:

- $\mathsf{KeyGen}()$: Sample $s \leftarrow \mathcal{HWT}(h)$, $a \leftarrow \mathcal{U}(R_q)$, $e \leftarrow \mathcal{DG}(\sigma^2, R_q)$ and then $b \leftarrow a \cdot s + p \cdot e$. Now set the public key $\mathsf{pk} \leftarrow (a, b)$. Note that pk looks very similar to a Ring-LWE sample.
- $\mathsf{Enc}_{\mathsf{pk}}(m)$: To encrypt a message $m \in R_p$, sample a small polynomial with coefficients $v \leftarrow \mathcal{ZO}(0.5)$, and two Gaussian polynomials $e_1, e_2 \leftarrow \mathcal{DG}(\sigma^2, R_q)$. The ciphertext will be a pair $c = (c_0, c_1)$ where $c_0 = b \cdot v + p \cdot e_0 + m \in R_q$ and $c_1 = a \cdot v + p \cdot e_1 \in R_q$.
- $\mathsf{Dec}_{\mathsf{sk}}(c)$: To decrypt a ciphertext $c \in R_q^2$, one can simply compute $m' \leftarrow c_0 - s \cdot c_1 \in R_q$ and then set $m \leftarrow m' \bmod p$ to get the original plaintext. The decryption works only if the noise $\nu = (m' \bmod p) - m$ associated with c is less than $q/2$ such that the ciphertext will not wrap around the modulus R_q.

2.3 Zero-Knowledge Proofs

In a typical scenario, a zero-knowledge (ZK) proof allows a verifier to check the validity of a statement claimed by a prover without revealing anything other that the claim is true. Previous implementations have used one of two approaches: a Schnorr-like protocol [CD09, DPSZ12, DKL+12] and cut-and-choose [DKL+13]. We will call SPDZ using either of the two protocols SPDZ-1 and SPDZ-2, respectively. Analysing the communication complexity, we found that the Schnorr-like protocol is more efficient because it only involves sending two extra ciphertexts per ciphertext to be proven whereas Damgård et al. [DKL+13] suggest that, for malicious security, their protocol is most efficient with 32 extra ciphertexts. It is also worth noting that the Schnorr-like protocol seems to be easier to implement.

The Schnorr-like protocol is based on the following 3-move standard Σ-protocol. To prove knowledge of x in a field $\mathbb{F}$ such that $f(x) = y$ without revealing x:

1. The prover $\mathcal{P}$ sends a commitment $a = f(s)$ for a random s.
2. The verifier $\mathcal{V}$ then samples a random $e \xleftarrow{\$} \mathbb{F}$ and sends it to $\mathcal{P}$.
3. $\mathcal{P}$ replies with $z = s + e \cdot x$. Finally $\mathcal{V}$ checks whether $f(z) = a + e \cdot y$.

If f is homomorphic with respect to the field operations, the protocol is clearly correct. Security of the prover (honest-verifier zero-knowledge) is achieved by simulating (a, e, z) from any e by sampling $z \xleftarrow{\$} \mathbb{F}$ and computing $a = f(z) - e \cdot y$. Security for the verifier (special soundness) allows to extract the secret from two different transcripts (z, c), (z', c') with $c \neq c'$. This can be done by computing $x = (z - z') \cdot (c - c')^{-1}$, which is possible in a field.

For our setting x is an integer (or a vector thereof), and we would like to prove that $\|x\|_\infty \leq B$ for some bound. For this case, Damgård and Cramer [CD09] have presented an amortized protocol (proving several pre-images at once) where s has to be chosen in a large enough interval (to statistically hide $E \cdot x$) and the challenge E is sampled from a set of matrices such that any $(E - E')$ is invertible over $\mathbb{Z}$ for any $E \neq E'$. The preimage is now extracted as $x = (E - E')^{-1}(z - z')$, thus a bound on $\|z\|_\infty$ also implies a bound on $\|x\|_\infty$.

However, it is not possible to make these bounds tight. Namely, an honest prover using $\|x\|_\infty \leq B$ will achieve that $\|z\|_\infty \leq B'$ for some $B' > B$. The quotient between the two bounds is called slack. Damgård et al. [DPSZ12] also show that in the Fiat-Shamir setting (where the challenge is generated using a random oracle on a), a technique called rejection sampling can be used to reduce the slack. This involves sampling different s until the response z achieves the desired bound. In any case, we will see in Sect. 3.4 that the slack of this proof is too small to make it worthwhile using the cut-and-choose proof instead.

Figure 1 shows the functionality that the proofs above implement. For a simplified exposition we also assume that $\mathcal{F}^S_{\mathsf{ZKPoK}}$ generates correct keys. In previous works this has been done by separate key registration [BDOZ11] or key generation functionalities [DPSZ12, DKL+13].

2.4 Overview of SPDZ

The SPDZ protocol [DPSZ12, DKL+13] can be viewed as a two-phase protocol where inputs are shared via an additive secret sharing scheme. First there is the pre-processing phase where triples are generated independent of the inputs to the computation. The classical way to produce these triples is either by oblivious transfer or homomorphic encryption. Each has its own advantages and caveats. In this work, we are only concerned with the homomorphic encryption technique, where ciphertexts are passed around players. Since we allow parties to deviate maliciously from the protocol, they could insert too much noise in the encryption algorithm, which we mitigate by using ZK proofs.

$\mathcal{F}^{S}_{\mathsf{ZKPoK}}$

The functionality generates keys $(\mathsf{pk}, \mathsf{sk})$ and sends them both to P_A and pk to P_B. If P_A is corrupted, the adversary chooses the keys. Then, the following can happen repeatedly:

1. P_A inputs either a vector $\mathbf{a}$ or a value a. In the latter case, $\mathbf{a}$ is defined to contain a in all slots.
2. If P_A is honest, P_B receives $\mathsf{Enc}_{\mathsf{pk}}(\mathbf{a})$, otherwise $\mathsf{Enc}'_{\mathsf{pk}}(\mathbf{a})$, where Enc' has noise at most S times as much as regular encryption.
3. The adversary can abort any time.

Fig. 1. Proof of knowledge of ciphertext

These random triples are further used in the online phase where parties interact by broadcasting data whenever a value is revealed. Privacy and correctness are then guaranteed by authenticated shared values with information-theoretic MACs[1] on top of them.

More formally, an authenticated secret value $x \in \mathbb{F}$ is defined as the following:

$$[\![x]\!] = (x^{(1)}, \dots, x^{(n)}, m^{(1)}, \dots, m^{(n)}, \Delta^{(1)}, \dots, \Delta^{(n)})$$

where each player P_i holds an additive sharing tuple $(x^{(i)}, m^{(i)}, \Delta^{(i)})$ such that:

$$x = \sum_{i=1}^{n} x^{(i)}, x \cdot \Delta = \sum_{i=1}^{n} m^{(i)}, \Delta = \sum_{i=1}^{n} \Delta^{(i)}.$$

For the pre-processing phase the goal is to model a Triple command which generates a tuple $([\![a]\!], [\![b]\!], [\![c]\!])$ where $c = a \cdot b$ and a, b are uniformly random from $\mathbb{F}$.

[1] These are not be confused with the more common symmetric-key MACs.

To open a value $[\![x]\!]$, all players P_i broadcast their shares $x^{(i)}$, commit and then open $m^{(i)} - x \cdot \Delta^{(i)}$. Afterwards they check if the sum of the latter is equal to zero. One can check multiple values at once by taking a random linear combination of $m^{(i)} - x \cdot \Delta^{(i)}$ exactly as in the MAC Check protocol in Fig. 5 in Sect. 3.

In the online phase the main task is to evaluate an arbitrary circuit with secret inputs. After the parties have provided their inputs using the Input command, the next step is to perform addition and multiplication between authenticated shared values. Since the addition is linear, it can be done via local computation. However multiplying two values $[\![x]\!], [\![y]\!]$ requires some interaction between the parties. To compute $[\![x \cdot y]\!]$ a fresh random triple $[\![a]\!], [\![b]\!], [\![c]\!] = [\![ab]\!]$ has to be available for Beaver's trick [Bea92]. It works by opening $[\![x - a]\!]$ and $[\![x - b]\!]$ to get ϵ and ρ respectively. Then the authenticated product can be obtained by setting $[\![x \cdot y]\!] \leftarrow [\![c]\!] + \epsilon[\![b]\!] + \rho[\![a]\!] + \epsilon \cdot \rho$.

Offline Phase. We now outline the core ideas of the preprocessing phase of SPDZ. Assume that the parties have a global public key and a secret sharing of the secret key Δ, and that there is a distributed decryption protocol that allows the parties to decrypt an encryption such that they receive a secret sharing of the cleartext (see the Reshare procedure by Damgård et al. [DPSZ12] for details).

For passive security only, the parties can simply broadcast encryptions of randomly sampled shares a_i, b_i and their share of the MAC key Δ_i. These encryptions can be added up and multiplied to produce encryptions of $(a \cdot b, a \cdot \Delta, b \cdot \Delta, a \cdot b \cdot \Delta)$ if the encryption allows multiplicative depth two. Distributed decryption then allows the parties to receive an additive secret sharing of each of those values, which already is enough for a triple. Since achieving a higher multiplicative depth is relatively expensive, SPDZ only uses a scheme with multiplicative depth one and extends the distributed decryption to produce a fresh encryption of $a \cdot b$, which then can be multiplied with the encryption of Δ.

In the context of an active adversary there are two main issues: First, the ciphertexts input by corrupted parties have to be correct and independent of the honest parties' ciphertexts. This is where zero-knowledge is applied to prove that certain values lie within a certain bound. Second, the distributed decryption protocol actually allows the adversary to add an error - that is, the parties can end up with a triple $(a, b, ab + e)$ with e known to the adversary and where the MACs have additional errors as well. While an error on a MAC will make the MAC check fail in any case, the problem of an incorrect triple requires more attention. This is where the so-called SPDZ sacrifice comes in. Imagine two triples with potential errors $([\![a]\!], [\![b]\!], [\![ab + e]\!])$ and $([\![a']\!], [\![b']\!], [\![a'b' + e']\!])$, and let t be a random field element. Then,

$$\begin{aligned} & t \cdot (ab + e) - (a'b' + e') - (ta - a') \cdot b - a' \cdot (b - b)' \\ &= tab + te - a'b' - e' - tab - a'b - a'b + a'b' \\ &= te - e', \end{aligned}$$

which is 0 with probability negligible in sec for a field of size at least 2^{sec} if either $e \neq 0$ or $e' \neq 0$. The use of MACs means that the adversary cannot forge the result of this computation, hence any error will be caught with overwhelming probability since with the additive secret sharing of our triples the parties have to reveal $[\![ta - a']\!]$ and $[\![b - b']\!]$. Therefore, one of the triples has to be discarded in order keep the other one "fresh" for use in the online phase. For MASCOT, Keller et al. [KOS16] found that the sacrifice also works with two triples (a, b, ab) and $(a', b, a'b)$, which implies $b - b' = b - b = 0$. Such a combined triple is cheaper to produce (both in MASCOT and SPDZ), and requires less revealing for the check.

3 Low Gear: Triple Generation Using Semi-homomorphic Encryption

The multiplication of secret numbers is at the heart of many secret sharing-based multiparty computation protocols because linear secret sharing schemes make addition easy, and the two operations together are complete.[2] Both Bendlin et al. [BDOZ11] and Keller et al. [KOS16] have effectively reduced the problem of secure computation to computing an additive secret sharing of the product of two numbers known to two different parties. The former uses semi-homomorphic encryption which allows to add two ciphertexts to get an encryption of the sum of cleartexts whereas the latter uses oblivious transfer which is known to be complete for any protocol.

The semi-homomorphic solution works roughly as follows: One party sends an encryption $\mathsf{Enc}(a)$ of their input under their own public key to the other, which replies by $C := b \cdot \mathsf{Enc}(a) - \mathsf{Enc}(c_B)$, where b denotes the second party's input, and c_B is chosen at random. Any semi-homomorphic encryption scheme allows the multiplication of a known value with a ciphertext, hence the decryption of the second message is $c_A := b \cdot a - c_B$, which makes (c_A, c_B) an additive secret sharing of $a \cdot b$. Here the noise of C might reveal information about b but this can be mitigated by adding random noise from an interval that is sec larger than the maximum noise of C. This technique, sometimes called "drowning", is also used in the distributed decryption of SPDZ.

In the context of a malicious adversary there are two concerns with the above protocol: $\mathsf{Enc}(a)$ might not be a correct encryption and C might not be computed correctly. In both cases, Bendlin et al. use a zero-knowledge proof of knowledge to make sure that both parties behave correctly.

To prove the correctness of $\mathsf{Enc}(a)$, there are relatively efficient proofs based on amortized Σ-protocols (reducing the overhead per ciphertext by processing several ciphertexts at once), but for the proof of correct multiplication amortization this is not possible in our context because the underlying ciphertext $\mathsf{Enc}(a)$ is different in every instance. The main goal of our work in this section

[2] This fact is mirrored in the world of garbled circuits, where the free-XOR technique only requires to garble AND gates, which compute the product of two bits.

is therefore to avoid the proof of correct multiplication altogether and delay it to a later check in the protocol described in the previous section.

Recall that the goal in this family of protocols is to generate random multiplication triples (a, b, ab). The sacrifice will guarantee that the parties have shares of correct triples, but there is a possibility of a selective failure attack. If C was not computed correctly, just the fact that the check passed (otherwise the parties abort without using their private data) can reveal information meant to stay private in the protocol. We will show that assuming the enhanced CPA notion in Sect. 3.1 for the underlying cryptosystem suffices to achieve this.

In Sect. 3.2, we will then use our multiplication protocol a first time to compute SPDZ-style MACs, that is, additive secret sharings of the product of a value and a global MAC key, which itself is secret-shared additively. It is straightforward to compute such a global product from the two-party protocol. Consider that $\sum_i a_i \cdot \sum_i b_i = \sum_{i,j} a_i \cdot b_j$. Every summand in the right-hand side can be computed either locally or by the two-party protocol, and the additive operation is trivially commutative with the addition of shares.

Building on the authentication protocol, we present the multiplication triple generation in Sect. 3.3 using the two-party multiplication protocol once more. Note that the after-the-fact check of correct multiplication works differently in the two protocols. In the authentication protocol, we make use of the fact that changing values are always multiplied with the same share of the MAC key. In the triple generation, however, both values change from triple to triple, thus we rely on the SPDZ sacrifice there. For this, we use a trick used by Keller et al. that reduces the complexity by generating a pair of triples $((a, b, ab), (a', b, a'b))$ for the sacrifice instead of two independent triples.

Finally, we present our choice of BGV parameters in Sect. 3.4, following the considerations of Damgård et al. [DKL+13], which in turn are based on Gentry et al. [GHS12b]. We found that the ciphertext modulus is about 100 bits shorter compared to original SPDZ for fields of size 2^{64} to 2^{128}, which makes a significant contribution to the reduced complexity of our protocol because SPDZ requires a modulus of bit length about 300 for 64-bit fields and 40-bit security.

3.1 Enhanced CPA Security

We want to reduce the security of our protocol to an enhanced version of the CPA game for the encryption scheme. In other words, if the encryption scheme in use is enhanced-CPA secure, then even a selective failure caused by the adversary does not reveal private information.

We say that an encryption scheme is *enhanced-CPA secure* if, for all PPT adversaries in the game from Fig. 2, $\Pr[b = b'] - 1/2$ is negligible in k.

Achieving Enhanced-CPA Security. The game without zero-checks in step 3 clearly can be reduced to the standard CPA game. Furthermore, we have to make sure that the oracle queries cannot be used to reveal information about m. The cryptosystem is only designed to allow affine linear operations limiting the adversary to succeed only with negligible probability due to the high entropy of m.

$\mathcal{G}_{\mathsf{cpa+}}$

1. The challenger samples $(\mathsf{pk}, \mathsf{sk}) \leftarrow \mathsf{KeyGen}(k)$, sends pk to the adversary.
2. The challenger sends $c = \mathsf{Enc}_{\mathsf{pk}}(m)$ for a random message m.
3. For $j \in \mathsf{poly}(k)$:
   (a) The adversary sends c_j to the challenger.
   (b) The challenger checks if $\mathsf{Dec}_{\mathsf{sk}}(c_j) = 0$; if this is the case the challenger sends OK to the adversary; else, the challenger sends FAIL to the adversary and aborts.
4. The challenger samples $b \xleftarrow{\$} \{0,1\}$ and sends m to the adversary if $b = 0$ and a random m' otherwise.
5. The adversary sends $b' \in \{0,1\}$ to the challenger and wins the game if $b = b'$.

Fig. 2. Enhanced CPA game

However, if the cryptosystem would allow to generate an encryption of a bit of m from $\mathsf{Enc}_{\mathsf{pk}}(m)$, the adversary could test this bit for zero with success probability 1/2. Therefore, we have to assume that non-linear operations on ciphertexts are not possible. To this end, Bitansky et al. [BCI+13] have introduced the notion of linear targeted malleability. A stronger notion thereof, linear-only encryption, has been conjectured by Boneh et al. [BISW17] to apply to the cryptosystem by Peikert et al. [PVW08], which is based on the ring learning with errors problem. The definition by Bitansky et al. is as follows.

Definition 1. *An encryption scheme has the* linear targeted malleability *property if for any polynomial-size adversary A and plaintext generator $\mathcal{M}$ there is a polynomial-size simulator S such that, for any sufficiently large $\lambda \in \mathbb{N}$, and any auxiliary input $z \in \{0,1\}^{\mathsf{poly}(\lambda)}$, the following two distributions are computationally indistinguishable:*

$$\left\{ \begin{array}{l|r} \mathsf{pk}, & (\mathsf{sk}, \mathsf{pk}) \leftarrow \mathsf{Gen}(1^\lambda) \\ a_1, \ldots, a_m, & (s, a_1, \ldots, a_m) \leftarrow \mathcal{M}(\mathsf{pk}) \\ s, & (c_1, \ldots, c_m) \leftarrow (\mathsf{Enc}_{\mathsf{pk}}(a_1), \ldots, \mathsf{Enc}_{\mathsf{pk}}(a_m)) \\ \mathsf{Dec}_{\mathsf{sk}}(c'_1), \ldots, \mathsf{Dec}_{\mathsf{sk}}(c'_k) & (c'_1, \ldots, c'_k) \leftarrow A(\mathsf{pk}, c_1, \ldots, c_m; z) \\ & \textit{where} \\ & \mathsf{ImVer}_{\mathsf{sk}}(c'_1) = 1, \ldots, \mathsf{ImVer}_{\mathsf{sk}}(c'_k) = 1 \end{array} \right\}$$

and

$$\left\{ \begin{array}{l|r} \mathsf{pk}, & (\mathsf{sk}, \mathsf{pk}) \leftarrow \mathsf{Gen}(1^\lambda) \\ a_1, \ldots, a_m, & (s, a_1, \ldots, a_m) \leftarrow \mathcal{M}(\mathsf{pk}) \\ s, & (\Pi, \mathbf{b}) \leftarrow S(\mathsf{pk}; z) \\ a'_1, \ldots, a'_k & (a'_1, \ldots, a'_k)^\top \leftarrow \Pi \cdot (a_1, \ldots, a_m)^\top + \mathbf{b} \end{array} \right\}$$

where $\Pi \in \mathbb{F}^{k \times m}$, $\mathbf{b} \in \mathbb{F}^k$, and s is some arbitrary string (possibly correlated with the plaintexts).

In the context of BGV, the definition can easily be extended to vectors of field elements. Furthermore, verifying whether a ciphertext is the image of the encryption

(ImVer) can be trivially done by checking membership in $R_q \times R_q$, which is possible without the secret key.

It is straightforward to see that linear targeted malleability allows to reduce the enhanced-CPA game to a game without a zero-test oracle. We simply replace the decryption of the adversary's queries by $a'_1, \ldots, a'_k$ computed using S according to the definition, which can be tested for zero without knowing the secret key. The two games are computationally indistinguishable by definition, and the modified one can be reduced to the normal CPA game as argued above.

We now argue that BGV as used by us is a valid candidate for linear targeted malleability. First, the definition excludes computation on ciphertexts other than affine linear maps. Most notably, this excludes multiplication. Since we do not generate the key-switching material used by Damgård et al. [DKL+13], there is no obvious way of computing multiplications or operations of any higher order.

Second, the definition requires the handling of ciphertexts that were generated by the adversary without following the encryption algorithm. For example, $\mathsf{Dec}_{\mathsf{sk}}(0, 1) = s \bmod p$. The decryption of such ciphertexts can be simulated by sampling a secret key and computing the decryption accordingly. However, to avoid a security degradation due to independent consideration of standard CPA security and linear targeted malleability, we add sec bits of entropy to the secret key as follows.

The key generation of BGV generates s of length N such that s has $h = 64$ non-zero entries at randomly chosen places, which are chosen uniformly from $\{-1, 1\}$. The entropy is therefore

$$\log \binom{N}{h} + h.$$

It is easy to see that choosing $h' = h + \mathsf{sec}$ non-zero entries increases the entropy by sec bits[3] for large enough N. Because $\binom{N}{k}$ monotonously increases for $k \leq N/2$,

$$\binom{N}{h + \mathsf{sec}} \geq \binom{N}{h}$$

for $N \geq 2 \cdot (h + \mathsf{sec})$. It follows that

$$\log \binom{N}{h + \mathsf{sec}} + h + \mathsf{sec} \geq \left(\log \binom{N}{h} + h\right) + \mathsf{sec},$$

which is the desired result. We will later see that N is much bigger than $2 \cdot (h + \mathsf{sec})$ for $h = 64$ and $\mathsf{sec} = 128$.

3.2 Input Authentication

As in Keller et al. [KOS16], we want to implement a functionality (Fig. 3) that commits the parties to secret sharings and that provides the secure computation

[3] $\mathsf{sec}/2$ would suffice, but sec does not affect the efficiency and allows for a simpler analysis.

of linear combinations of inputs. However, instead of using oblivious transfer for the pairwise multiplication of secret numbers we use our building block based on semi-homomorphic encryption. See Fig. 4 for our protocol.

$\mathcal{F}_{[\![\cdot]\!]}$

Input: On input $(\mathsf{Input}, \mathsf{id}_1, \ldots, \mathsf{id}_l, x_1, \ldots, x_l, P_j)$ from party P_j and $(\mathsf{Input}, \mathsf{id}_1, \ldots, \mathsf{id}_l, P_j)$ from all players P_i where $i \neq j$, store $\mathsf{Val}[\mathsf{id}_k] \leftarrow x_k$ for all $k \in [1 \ldots l]$.

Linear Combination: On input $(\mathsf{LinComb}, \overline{\mathsf{id}}, \mathsf{id}_1, \ldots, \mathsf{id}_l, c_1, \ldots, c_l, c)$ from all parties where $\mathsf{id}_k \in \mathsf{Val.Keys}()$ store $\mathsf{Val}[\overline{\mathsf{id}}] = \sum_{k=1}^{l} \mathsf{Val}[\mathsf{id}_k] \cdot c_k + c$.

Open: On input $(\mathsf{Open}, \mathsf{id})$ from all parties, send $\mathsf{Val}[\mathsf{id}]$ to the adversary; wait for input x from the adversary and then send x to all parties.

Check: On input $(\mathsf{Check}, \mathsf{id}_1, \ldots, \mathsf{id}_l, x_1, \ldots, x_l)$ from all parties, wait for the adversary's input. If the input is OK and $\mathsf{Val}[\mathsf{id}_k] = x_k$ for all $k \in [1 \ldots l]$ then send OK to every party, otherwise send $\perp$ and terminate.

Abort: On input Abort from the adversary send $\perp$ to all parties and terminate.

Fig. 3. Functionality $\mathcal{F}_{[\![\cdot]\!]}$

In case parties P_i and P_j are honest,

$$\begin{aligned} \Delta^{(i)} \cdot \rho - \sigma^{(i)} - \sum_{k=1}^{m} t_k \cdot \mathbf{d}_k^{(i)} &= \Delta^{(i)} \cdot (\sum_{k=1}^{m} t_k \cdot x_k) - \sum_{k=1}^{m} t_k \cdot e_k^{(i)} - \sum_{k=1}^{m} t_k \cdot \mathbf{d}_k^{(i)} \\ &= \sum_{k=1}^{m} t_k \cdot (\Delta^{(i)} \cdot x_k - e_k^{(i)} - \mathbf{d}_k^{(i)}) = 0 \end{aligned}$$

for all $i \neq j$. This means that P_i's check succeeds in this case. The last equation follows from the homomorphism of the encryption scheme.

Furthermore, one can check similarly that

$$\sum_i m_k^{(i)} = x_k \cdot \sum_i \Delta^{(i)},$$

which is the desired equation underlying the MAC. If it does not hold because of P_j's behaviour, we would like the check to fail for some honest P_i. Informally, the fact that P_j cannot predict the coefficients t_k makes it impossible for P_j to provide correct $(\rho, \sigma^{(i)})$ to an honest party P_i after computing $C^{(i)}$ incorrectly. However, this opens the possibility for leakage by a selective failure attack, which is why we need the underlying cryptosystem to achieve enhanced-CPA security.

The most intricate part of the simulator $\mathcal{S}_{[\![\cdot]\!]}$ (Fig. 6) is simulating the **Input** phase for a corrupted P_j while the same phase for honest P_j is straight-forward given that Enc' statistically hides the noise of $\mathbf{x}^{(i)} \cdot \mathsf{Enc}(\boldsymbol{\Delta})$. Note that $(x_m, e_m^{(i)}, d_m^{(i)})$ are only used for the check. This maintains P_j's privacy even after sending ρ and $\{\sigma^{(i)}\}_{i \neq j}$.

$\Pi_{[\![\cdot]\!]}$

Initialize: Each party P_i does the following:

1. Sample a MAC key $\Delta^{(i)} \xleftarrow{\$} \mathbb{F}$.
2. Initialize two instances of $\mathcal{F}^{S}_{\mathsf{ZKPoK}}$ with every other party P_j (one as prover, one as verifier), receiving $(\mathsf{pk}_{ij}, \mathsf{sk}_{ij})$ and pk_{ji}.
3. Using $\mathcal{F}^{S}_{\mathsf{ZKPoK}}$, send an encryption $\mathsf{Enc}_{\mathsf{pk}_{ij}}(\boldsymbol{\Delta}^{(i)})$ to every other party where $\boldsymbol{\Delta}^{(i)}$ denotes a cleartext with all slots set to $\Delta^{(i)}$.

Input: On input $(\mathsf{Input}, \mathsf{id}_1, \ldots, \mathsf{id}_l, x_1, \ldots, x_l, P_j)$ from P_j and $(\mathsf{Input}, \mathsf{id}_1, \ldots, \mathsf{id}_l, P_j)$ from all P_i where $i \neq j$:

1. For each input x_k where $k \in [1 \ldots l]$ P_j samples randomly $x_k^{(i)} \xleftarrow{\$} \mathbb{F}$ and sends them to the designated party i. Then P_j sets its corresponding share $x_k^{(j)}$ accordingly such that $\sum_{l=1}^{n} x_k^{(l)} = x_k$.
2. We assume that $l < m$ where m is the number of ciphertext slots in the encryption scheme. Let $\mathbf{x}$ denote the vector containing x_k in the first l entries and a random number in the m-th one.
3. For every party P_i:
   (a) P_j computes $C^{(i)} = \mathbf{x} \cdot \mathsf{Enc}_{\mathsf{pk}_{ij}}(\boldsymbol{\Delta}^{(i)}) - \mathsf{Enc}'_{\mathsf{pk}_{ij}}(\mathbf{e}^{(i)})$ for random $\mathbf{e}^{(i)}$ and sends $C^{(i)}$ to P_i. Enc' denotes encryption with noise $p \cdot 2^{\mathsf{sec}}$ larger than in normal encryption.
   (b) P_i decrypts $\mathbf{d}^{(i)} = \mathsf{Dec}_{\mathsf{sk}_{ij}}(C^{(i)})$.
4. The parties use $\mathcal{F}_{\mathsf{Rand}}$ to generate random t_k for $k = 1, \ldots, m$.
5. P_j computes $\rho = \sum_{k=1}^{m} t_k \cdot x_k$ and $\sigma^{(i)} = \sum_{k=1}^{m} t_k \cdot e_k^{(i)}$, and sends $(\rho, \sigma^{(i)})$ to P_i.
6. P_i checks whether $\Delta^{(i)} \cdot \rho - \sigma^{(i)} - \sum_{k=1}^{m} t_k \cdot \mathbf{d}_k^{(i)} = 0$ and aborts if not.
7. P_j sets its MAC share associated to x_k as $m_k^{(j)} \leftarrow \sum_{i \neq j} e_k^{(i)} + x_k \cdot \Delta^{(j)}$ and each party P_i does so for $m_k^{(i)} \leftarrow d_k^{(i)}$.
8. All parties store their authenticated shares $x_k^{(i)}, m_k^{(i)}$ as $[\![x]\!]$ under the identifiers $\mathsf{id}_1, \ldots, \mathsf{id}_l$.

Linear Combination: On input $(\mathsf{LinComb}, \overline{\mathsf{id}}, \mathsf{id}_1, \ldots, \mathsf{id}_l, c_1, \ldots, c_l, c)$ from all parties, every P_i retrieves the share-MAC pairs $x_k^{(i)}, m(x_k)^{(i)}_{k \in [1 \ldots l]}$ and computes:

$$y^{(i)} = \sum_{k=1}^{l} c_k \cdot x_k^{(i)} + c \cdot s_1^{(i)}$$

$$m(y)^{(i)} = \sum_{k=1}^{l} c_k \cdot m(x_k)^{(i)} + c \cdot \Delta^{(i)}$$

$s_1^{(i)}$ denotes a fixed sharing of 1, for example, $(1, 0, \ldots, 0)$.

Open: On input $(\mathsf{Open}, \mathsf{id})$ from all parties, each P_i looks up the share $x^{(i)}$ with identifier id and broadcasts it. Then each party reconstructs $x = \sum_{i=1}^{n} x^{(i)}$.

Check: On input $(\mathsf{Check}, \mathsf{id}_1, \ldots, \mathsf{id}_l, x_1, \ldots, x_l)$ from all parties:

1. Sample public vector $r \leftarrow \mathcal{F}_{\mathsf{Rand}}(\mathbb{F}^l)$.
2. Compute $y^{(i)} = \sum_{k=1}^{m} r_k^{(i)} x_k^{(i)}$ and $m(y)^{(i)} = \sum_{k=1}^{m} r_k^{(i)} m_k(x_k)^{(i)}$.
3. Run Π_{MACCheck} with $y^{(i)}, m(y)^{(i)}$.

Fig. 4. Protocol for n-party input authentication

Π_{MACCheck}

Each party P_i uses $y^{(i)}, m(y)^{(i)}, \Delta^{(i)}$ in the following way:

1. Compute $\sigma^{(i)} \leftarrow m(y)^{(i)} - \Delta^{(i)} y^{(i)}$.
2. Call $\mathcal{F}_{\mathsf{Commit}}$ with $(\mathsf{Commit}, \sigma^{(i)})$ to receive handle τ_i.
3. Broadcast $\sigma^{(i)}$ to all parties by calling $\mathcal{F}_{\mathsf{Commit}}$ with (Open, τ_i).
4. If $\sigma^{(1)} + \cdots + \sigma^{(n)} \neq 0$ then abort and output $\perp$; otherwise continue.

Fig. 5. Protocol for MAC checking

$\mathcal{S}_{[\![\cdot]\!]}$

Let H denote the set of honest parties and A the complement thereof.

Init:
1. Emulating $\mathcal{F}^S_{\mathsf{ZKPoK}}$, generate $(\mathsf{pk}_{ij}, \mathsf{sk}_{ij})$ for all $i \in H$ and $j \in A$ and send all pk_{ij} to the adversary.
2. Emulating $\mathcal{F}^S_{\mathsf{ZKPoK}}$, send $\mathsf{Enc}_{\mathsf{pk}_{ij}}(x)$ for random x to the adversary for all $i \in H$.

Input: We assume that $j \notin H$.
1. Receive $x_k^{(i)}$ from the adversary for all $k = 1, \ldots, l$ and $i \in H$ and store them.
2. Receive the ciphertext $C^{(i)}$ from the adversary and decrypt it to get $\mathbf{d}^{(i)}$ for all $i \in H$.
3. Emulating $\mathcal{F}_{\mathsf{Rand}}$, sample random t_i for $i = 1, \ldots, m$.
4. Receive $(\rho, \sigma^{(i)})$ from the adversary for all $i \in H$.
5. Check whether $\sigma^{(i)} + \sum_{k=1}^{m} t_k \cdot d_k^{(i)} = 0$ for all $i \in H$ and abort if not.
6. Rewinding the adversary, collect enough answers $(\sigma^{(i)}, \rho)$ for random $\mathbf{t}$ in order to reconstruct $\mathbf{x}$ and $\mathbf{e}^{(i)}$. Choose $\mathbf{t}$ such that it is linearly independent from all previously used $\mathbf{t}$.
7. Compute $m_k^{(i)}$ for every $i \in A$ and store it.
8. Input $(x_1, \ldots, x_l)$ to $\mathcal{F}_{[\![\cdot]\!]}$.

Linear Combination: For every $i \in A$, compute shares and MACs as an honest party would.

Open:
1. Receive the value from $\mathcal{F}_{[\![\cdot]\!]}$.
2. If the value is a linear combination of previously opened values, compute the honest parties' shares accordingly. Otherwise, sample new shares.
3. Send the honest parties' shares to the adversary.
4. Receive the corrupted parties' shares from the adversary.
5. Input the sum to $\mathcal{F}_{[\![\cdot]\!]}$.

Check:
1. Emulating $\mathcal{F}_{\mathsf{Rand}}$, send r to all corrupted parties.
2. Emulating $\mathcal{F}_{\mathsf{Commit}}$ receive $\sigma^{(i)}$ for all $i \in A$.
3. If $\sum_{i \in A} \sigma^{(i)}$ does not match the result computed from stored shares, abort $\mathcal{F}_{[\![\cdot]\!]}$.

Fig. 6. Simulator for $\Pi_{[\![\cdot]\!]}$

Theorem 1. *$\Pi_{[\![\cdot]\!]}$ implements $\mathcal{F}_{[\![\cdot]\!]}$ in the $\mathcal{F}_{\mathsf{Commit}}$-hybrid model with rewinding in a presence of a dishonest majority if the underlying cryptosystem achieves enhanced CPA-security.*

Proof (Sketch). We focus on the case of a corrupted P_j in the **Input** phase because the adversary has a larger degree of freedom with the encryptions $C^{(i)}$. However, with rewinding in step 6 we can extract the values used by the adversary. This extraction takes time inversely dependent to the success probability, as per the soundness argument for Σ-protocols. To see this, consider that the space of all possible challenges $\{t_k\}_{k=1}^m$ has size $|\mathbb{F}|^m$. The extractor requires the responses to m linearly independent challenges $\{t_k\}_{k=1}^m$. The adversary can only prevent this by restricting the correct responses to an incomplete subspace of $S \subset \mathbb{F}^m$, that is $|S| \leq \mathbb{F}^{m-1}$. Such an adversary will succeed with probability at most $|\mathbb{F}|^{m-1}/|\mathbb{F}|^m = |\mathbb{F}|^{-1}$, which is negligible because we require the size of $\mathbb{F}$ to be exponential in the security parameter. It follows that the soundness extractor for Σ-protocols by Damgård [Dam02] can be adapted to our case.

After the extraction, it is straightforward to simulate the rest of the protocol because the **Linear Combination** phase does not involve communication, and producing a correct MAC in the **Check** phase for an incorrect output in the **Open** phase is equivalent to extracting Δ. This argument can also be extended to the random linear combination used in the **Check** phase similarly to Keller et al. [KOS16]. It is easy to see that extracting Δ is in turn equivalent to breaking the security of the underlying cryptosystem.

We therefore construct a distinguisher in the enhanced-CPA security game from an environment distinguishing between the real and the ideal world. The difference between $\mathsf{Enc}_{\mathsf{pk}_{ij}}(\Delta^{(i)})$ in the real world and $E = \mathsf{Enc}_{\mathsf{pk}_{ij}}(x)$ for random x in the simulation can trivially be reduced to our CPA security game (using the encryption as c in the game) because the adversary never receives $\Delta^{(i)}$. Furthermore, $\mathbf{x}$ and $\mathbf{e}^{(i)}$ extracted from the adversary can be used to compute $C' = C^{(i)} - \mathbf{x} \cdot E - \mathbf{e}^{(i)}$. Via the check conducted by the honest party P_i, the adversary learns whether C' decrypts to zero. We therefore forward C' to the zero test in our enhanced CPA game.

3.3 Triple Generation

Recall that the goal is to produce random authenticated triples $([\![a]\!], [\![b]\!], [\![ab]\!])$ such that a, b are randomly sampled from $\mathbb{F}$ as described in Fig. 8. Our protocol in Fig. 7 is modeled closely after MASCOT [KOS16], replacing oblivious transfer with semi-homomorphic encryption. The construction of a "global" multiplication from a two-party protocol works exactly the same way in both cases. The **Sacrifice** step is exactly the same as in SPDZ and MASCOT and essentially guarantees that corrupted parties have used the same inputs in the **Multiplication** and **Authentication** steps. This is the only freedom the adversary has because all other arithmetic is handled by $\mathcal{F}_{[\![\cdot]\!]}$ at this stage.

Theorem 2. *Π_{Triple} implements $\mathcal{F}_{\mathsf{Triple}}$ in the $(\mathcal{F}_{[\![\cdot]\!]}, \mathcal{F}_{\mathsf{Rand}})$-hybrid model with a dishonest majority of parties.*

Π_{Triple}

Multiply:

1. Each party P_i samples $\mathbf{a}^{(i)}, \mathbf{b}^{(i)}, \hat{\mathbf{b}}^{(i)} \xleftarrow{\$} \mathbb{F}$ (such that the length of every vector matches the number of slots in the encryption scheme).
2. Every unordered pair (P_i, P_j) executes the following:
   (a) P_i uses $\mathcal{F}^S_{\mathsf{ZKPoK}}$ to send P_j the encryption $\mathsf{Enc}_{\mathsf{pk}_{ij}}(\mathbf{a}^{(i)})$.
   (b) P_j computes $C^{(ij)} = \mathbf{b}^{(j)} \cdot \mathsf{Enc}_{\mathsf{pk}_{ij}}(\mathbf{a}^{(i)}) - \mathsf{Enc}'_{\mathsf{pk}_{ij}}(\mathbf{e}^{(ij)})$ for random $\mathbf{e}^{(ij)} \xleftarrow{\$} \mathbb{F}$ and sends it to P_i. $\mathsf{Enc}'_{\mathsf{pk}_{ij}}$ denotes encryption with noise $p \cdot 2^{\mathsf{sec}}$ larger than normal encryption times the slack in the zero-knowledge proof.
   (c) P_i decrypts $\mathbf{d}^{(ij)} = \mathsf{Dec}_{\mathsf{sk}_{ij}}(C^{(ij)})$.
   (d) Repeat the last two steps with $\hat{\mathbf{b}}^{(i)}$ to get $\hat{\mathbf{e}}^{(ij)}$ and $\hat{\mathbf{d}}^{(ij)}$.
3. Each party P_i computes $\mathbf{c}^{(i)} = \mathbf{a}^{(i)} \cdot \mathbf{b}^{(i)} + \sum_{j \neq i}(\mathbf{e}^{(ij)} + \mathbf{d}^{(ij)})$ and $\hat{\mathbf{c}}^{(i)}$ similarly.

Authenticate: Party P_i calls $\mathcal{F}_{[\![\cdot]\!]}.\mathsf{Input}$ with $(\mathbf{a}^{(i)}, \mathbf{b}^{(i)}, \hat{\mathbf{b}}^{(i)}, \mathbf{c}^{(i)}, \hat{\mathbf{c}}^{(i)})$ and then $\mathcal{F}_{[\![\cdot]\!]}.\mathsf{LinComb}$ to get vectors of handles of the sum of shares. E.g., we denote by $[\![\mathbf{a}]\!]$ the vector of handles for the respective sums of elements $\{\mathbf{a}^{(i)}\}_{i=1\ldots n}$.

Sacrifice: The parties do the following:

1. Call $r \leftarrow \mathcal{F}_{\mathsf{Rand}}$.
2. Call $\mathcal{F}_{[\![\cdot]\!]}.\mathsf{LinComb}$ for $r \cdot [\![\mathbf{b}]\!] - [\![\hat{\mathbf{b}}]\!]$ and store them as $[\![\boldsymbol{\rho}]\!]$.
3. Reveal $\boldsymbol{\rho} \leftarrow \mathcal{F}_{[\![\cdot]\!]}.\mathsf{Open}([\![\boldsymbol{\rho}]\!])$.
4. Call $\mathcal{F}_{[\![\cdot]\!]}.\mathsf{Open}(\cdot)$ on $\boldsymbol{\tau} \leftarrow r \cdot \mathbf{c} - \hat{\mathbf{c}} - \boldsymbol{\rho} \cdot \mathbf{a}$. If $\boldsymbol{\tau} \neq 0$ then abort; else continue.
5. Call $\mathcal{F}_{[\![\cdot]\!]}.\mathsf{Check}$ an all opened values. If any check fails then abort, otherwise continue the protocol.

Output: $([\![\mathbf{a}]\!], [\![\mathbf{b}]\!], [\![\mathbf{c}]\!])$ as a vector of valid triples.

Fig. 7. Protocol for random triple generation

$\mathcal{F}_{\mathsf{Triple}}$

$\mathcal{F}_{\mathsf{Triple}}$ offers the same interface as $\mathcal{F}_{[\![\cdot]\!]}$ and the following function:

Triple: On input $(\mathsf{Triple}, \mathsf{id}_a, \mathsf{id}_b, \mathsf{id}_c)$ from all parties sample $a, b \xleftarrow{\$} \mathbb{F}$ and store $(\mathsf{Val}[\mathsf{id}_a], \mathsf{Val}[\mathsf{id}_b], \mathsf{Val}[\mathsf{id}_c]) = (a, b, c)$ where $c = a \cdot b$.

Fig. 8. Functionality for random triple generation.

Proof (Sketch). For the proof we use $\mathcal{S}_{\mathsf{Triple}}$ in Fig. 9. The simulator is based on two important facts: First, it can decrypt $C^{(ji)}$ for a corrupted party P_j because it generates the keys emulating $\mathcal{F}^S_{\mathsf{ZKPoK}}$. Second, the adversary is committed to all shares of corrupted parties by the input to $\mathcal{F}_{[\![\cdot]\!]}$ in the **Authenticate** step. This allows the simulator to determine exactly whether the **Sacrifice** step in $\Pi_{[\![\cdot]\!]}$ will fail. Furthermore, the adversary only learns encryptions of honest parties' shares, corrupted parties' shares, $\boldsymbol{\rho}$, and the result of the check. If the check fails, the protocol aborts, $\boldsymbol{\rho}$ is independent of any output information because $\hat{\mathbf{b}}$ and $\hat{\mathbf{c}}$ are discarded at the end, and finally, an environment deducing information from the encryptions can be used to break the enhanced-CPA security of the underlying cryptosystem. In addition, the environment only learns handles to triples in the **Output** steps, from which no information can be deduced.

$\mathcal{S}_{\mathsf{Triple}}$

Let H denote the set of honest parties and A the complement thereof.

Initialize: Emulating $\mathcal{F}^{S}_{\mathsf{ZKPoK}}$, for every $i \in A$ and $j \in H$, generate all key pairs $(\mathsf{pk}_{ij}, \mathsf{sk}_{ij})$ and send the relevant parts to the corresponding party.

Multiply:

1. For every $i \in A$ and $j \in H$, emulate two instances of $\mathcal{F}^{S}_{\mathsf{ZKPoK}}$:
   (a) Send $\mathsf{Enc}_{\mathsf{pk}_{ji}}(0)$ to the adversary and receive $C^{(ji)}$.
   (b) Receive $\mathbf{a}^{(i)}$ from the adversary and reply with $\mathsf{Enc}'_{\mathsf{pk}ij}(\mathbf{e}^{(ij)})$ for random $\mathbf{e}^{(ji)}$.

Authenticate: Emulating $\mathcal{F}_{[\![\cdot]\!]}$, receive $\mathbf{a}^{(i)}, \mathbf{b}^{(i)}, \hat{\mathbf{b}}^{(i)}, \mathbf{c}^{(i)}, \hat{\mathbf{c}}^{(i)}$ for all $i \in A$ from the adversary and return the desired handles.

Sacrifice:

1. Emulating $\mathcal{F}_{\mathsf{Rand}}$, sample $r \stackrel{\$}{\leftarrow} \mathbb{F}_p$ and send it to the adversary.
2. Sample $\boldsymbol{\rho} \stackrel{\$}{\leftarrow} \mathbb{F}_p^m$ and send it to the adversary emulating $\mathcal{F}_{[\![\cdot]\!]}.\mathsf{Open}$. Set Fail if the adversary inputs a different value in response.
3. Given the adversary's inputs in **Authenticate** and $\mathsf{Dec}_{\mathsf{sk}_{ji}}(C^{(ji)})$, we can compute $\boldsymbol{\tau}$. Send it to the adversary emulating $\mathcal{F}_{[\![\cdot]\!]}.\mathsf{Open}$. If the response is different, or $\boldsymbol{\tau} \neq 0$, set Fail.
4. Emulating $\mathcal{F}_{[\![\cdot]\!]}.\mathsf{Check}$, abort if Fail is set.

Fig. 9. Simulator for Π_{Triple}

3.4 Parameter Choice

Since we do not need multiplication of ciphertexts, the list of moduli used in previous works [DKL+13, GHS12b] collapses to one q $(= q_1 = q_0 = p_0$ depending on context). The other main parameter is the number of ciphertext slots denoted by $N = \phi(m)$. Gentry et al. [GHS12b] give the following inequality for the largest modulus:

$$N \geq \frac{\log(q/\sigma)(k+110)}{7.2}$$

for a computational security k, which gives

$$N \geq \log q \cdot 33.1 \tag{1}$$

for 128-bit security. $\sigma = 3.2$ does not make a difference in this inequality.

The second constraint on q and $\phi(m)$ depends on the noise of the ciphertext to be decrypted. Damgård et al. compute the bound B_{clean} on the noise of a freshly generated ciphertext:

$$B_{\mathsf{clean}} = N \cdot p/2 + p \cdot \sigma(16 \cdot N \cdot \sqrt{n/2} + 6 \cdot \sqrt{N} + 16 \cdot \sqrt{n \cdot h \cdot N})$$

p denotes the plaintext modulus, and n denotes the number of parties, which appears because of the distributed ciphertext generation (the secret is the sum of n secret keys). Setting $n = 1$ because we do not use distributed ciphertext generation, and $h = 64 + \mathsf{sec} \leq 192$, $\sigma = 3.2$ as in the previous works, we get

$$B_{\mathsf{clean}} \leq p \cdot (37N + 685\sqrt{N}).$$

In the multiplication protocol, one party multiplies the ciphertext with a number in $\mathbb{F}_p$, adds a number in $\mathbb{F}_p$, and then "drowns" the noise with statistical security sec (adding extra noise sampling from an interval that is 2^{sec} larger than the current noise bound). Furthermore, depending on the proof of knowledge used, we can only assume that the noise of the ciphertext being sent is $S \cdot B_{\mathsf{clean}}$ for some soundness slack $S \geq 1$. Therefore, the noise before decryption is bounded by

$$p \cdot S \cdot B_{\mathsf{clean}} \cdot (1 + 2^{\mathsf{sec}}),$$

which must be smaller than $q/2$ for correct decryption. Hence,

$$2 \cdot p^2 \cdot S \cdot \left(37N + 685\sqrt{N}\right)(1 + 2^{\mathsf{sec}}) < q. \tag{2}$$

Putting things together, (2) implies that, loosely, $120 \leq \log q$ or $384 \leq \log q$ if $\mathsf{sec} = 40$ or $\mathsf{sec} = 128$ and $p \geq 2^{\mathsf{sec}}$ (the latter is a requirement of SPDZ-like sacrificing). Using this in (1) gives $N \geq 3972$ or $N \geq 12711$. For both values of N as well as a ten times larger N,

$$\log\left(37N + 685\sqrt{N}\right) \approx 20 \pm 2.$$

Hence,

$$\log q \gtrsim 21 + 2\log p + \log S + \mathsf{sec} \pm 2.$$

The proof of knowledge in the first version of SPDZ [DPSZ12] has the worst soundness slack with

$$S = N \cdot \mathsf{sec}^2 \cdot 2^{\mathsf{sec}/2+8}.$$

Thus,

$$\log S \leq \log N + 2\log \mathsf{sec} + \mathsf{sec}/2 + 8$$

and

$$\log q \gtrsim 29 + 2\log p + 3\mathsf{sec}/2 + 2\log \mathsf{sec} + \log N \pm 2.$$

Note that, even though this estimate is now five years old, we found our parameters to hold against more recent estimates [APS15] tested using the script that is available online [Alb17]. The main reason is that our parameters have a considerable margin because we require N to be a power of two.

More recently, Damgård et al. [CDXY17] presented an improved version of the cut-and-choose proof used in a previous implementation of SPDZ [DKL+13], but the reduced slack does not justify the increased complexity caused by several additional ciphertexts being computed and sent in the proof. Consider that, even for $\mathsf{sec} = 128$ and $N = 2^{16}$ (the latter being typical for our parameters), $\log S$ is about 100, increasing the ciphertext modulus length by less than 25%.

We have calculated the ciphertext modulus q's bit length for various parameters and for our protocol with semi-homomorphic encryption and SPDZ (using somewhat homomorphic encryption). Then we instantiated both protocols with several ZK proofs like the Schnorr-like protocol [CD09, DPSZ12] and the recent cut-and-choose proof [CDXY17]. Table 1 shows the results of our calculation as well as the results given by Damgård et al. [DKL+13]. One can see that using cut-and-choose instead of the Schnorr-like protocol does not make any difference for SPDZ. This is because the scaling (also called modulus switching) involves the division by a number larger than the largest possible slack of the Schnorr-like protocol (roughly 2^{100}), hence the slack will be eliminated. For our Low Gear protocol, the slack has a slight impact, increasing the size of a ciphertext by up to 25%. However, this does not justify the use of a cut-and-choose proof because it involves sending seven instead of two extra ciphertexts per proof.

Table 1 also shows Low Gear ciphertexts are about 30% shorter than SPDZ ciphertexts. Consider that Table 3 in Sect. 5 shows a reduction in the communication from SPDZ to Low Gear of up to 50%. The main reason for the additional reduction is the fact that for one guaranteed triple, SPDZ involves producing two triples $(a, b, c), (d, e, f)$, of which (a, b, d, e) require a zero-knowledge proof. In Low Gear on the other hand, we produce $(a, b, c, \hat{b}, \hat{c})$, of which only a requires a zero-knowledge proof.

4 High Gear: SPDZ with Global ZKPoK Check

In terms of computation, the most expensive part of SPDZ is anything related to the encryption scheme, encryption, decryption, and homomorphic operations. The encryption algorithm is not only used for inputs but also by both the prover and the verifier in the zero-knowledge proof. Since a non-interactive zero-knowledge protocol allows the parties to generate only one proof per input, independently of the number of parties, every party has to verify every other party's proof because every other party is assumed to be corrupted. With a growing number of parties, this is clearly the computational bottleneck of the protocol. In this section, we present a way to avoid this by summing all proofs and only checking the sum. This is similar to the threshold proofs presented by Keller et al. [KMR12]. However, this neither reduces the communication nor the asymptotic computation because every party still has to send every proof to every party and then sum all the received proofs. Nevertheless, summing up the proofs is much cheaper than verifying them individually.

Table 1. Ciphertext modulus bit length ($\log(q)$) for two parties.

Low Gear		SPDZ			sec	$\log(\lvert\mathbb{F}_p\rvert)$
[CD09]	[CDXY17]	1 [DPSZ12]	2 [CDXY17]	2 [DKL+13]		
238	199	330	330	332	40	64
367	327	526	526	526	40	128
276	224	378	378	N/A	64	64
406	352	572	572	N/A	64	128
504	418	700	700	N/A	128	128

The High Gear protocol is meant to surpass Low Gear when executed with a high number of parties. To achieve this we design a new zero-knowledge proof which scales better when increasing the number of players. One can think of the High Gear proof of knowledge as a customized interactive proof version from Damgård et al. [DPSZ12] whereas Low Gear is a protocol ran with the non-interactive proof. The latter requires knowledge of the first message of the proof (sometimes called the commitment) to compute the challenge. In the context of combining the proof with many parties, the first message is the sum of an input from each party, which means that communication is required in any case. Therefore, there is less of an advantage in using the non-interactive proof.

Figure 10 shows our adaptation of the zero-knowledge proof in Fig. 9 from Damgård et al. [DPSZ12]. The main conceptual difference is going from a two-party to a multi-party protocol. However, we have also simplified the bounds.

In the following we will prove that our protocol achieves the natural extension of the Σ-protocol properties in the multi-party setting.

Correctness. The equality in step 6 follows trivially from the linearity of the encryption. It remains to check the probability that an honest prover will fail the bounds check on $\|\mathbf{z}\|_\infty$ and $\|\mathbf{t}\|_\infty$ where the infinity norm $\|\cdot\|_\infty$ denotes the maximum of the absolute values of the components.

Remember that the honestly generated $E^{(i)}$ are (τ, ρ) ciphertexts. The bound check will succeed if the infinity norm of $\sum_{i=1}^{n}(\mathbf{y}^{(i)}+\sum_{k=1}^{\mathsf{sec}}(M_{e_{jk}}\cdot\mathbf{x}^{(i)}))$ is at most $2\cdot n\cdot B_{\mathsf{plain}}$. This is always true because $\mathbf{y}^{(i)}$ is sampled such that $\|\mathbf{y}^{(i)}\|_\infty \leq B_{\mathsf{plain}}$ and $\|M_{\mathbf{e}}\cdot\mathbf{x}^{(i)}\|_\infty \leq \mathsf{sec}\cdot\tau \leq 2^{\mathsf{sec}}\cdot\tau = B_{\mathsf{plain}}$. A similar argument holds regarding ρ and B_{rand}.

Special Soundness. To prove this property one must be able to extract the witness given responses from two different challenges. In this case consider the transcripts $(\mathbf{x}, \mathbf{a}, \mathbf{e}, (\mathbf{z}, T))$ and $(\mathbf{x}, \mathbf{a}, \mathbf{e}', (\mathbf{z}', T'))$ where $\mathbf{e} \neq \mathbf{e}'$. Recall that each party has a different secret $\mathbf{x}^{(i)}$. Because both challenges have passed the bound checks during the protocol, we get that:

$$(M_{\mathbf{e}} - M_{\mathbf{e}'})\cdot E^{\mathsf{T}} = (\mathbf{d}-\mathbf{d}')^{\mathsf{T}}$$

Π_{gZKPoK}

Let $B_{\mathsf{plain}} = 2^{\mathsf{sec}} \cdot \tau$ and $B_{\mathsf{rand}} = 2^{\mathsf{sec}} \cdot \rho$ be the bounds for plaintext and randomness used for encryption where $\rho = 2 \cdot 3.2 \cdot \sqrt{N}$ and $\tau = p/2$.
Let $V = 2 \cdot \mathsf{sec} - 1$ and $M_{\mathbf{e}} \in \{0,1\}^{V \times \mathsf{sec}}$ the be matrix associated with the challenge $\mathbf{e}$ such that $M_{kl} = e_{k-l+1}$ for $1 \leq k - l + 1 \leq \mathsf{sec}$ and 0 in all other entries. The randomness used for encryptions of $\mathbf{x}^{(i)}, \mathbf{y}^{(i)}$ is packed into matrices $\mathbf{r}^{(i)} \leftarrow (r_1^{(i)}, \ldots, r_{\mathsf{sec}}^{(i)})$ and $\mathbf{s}^{(i)} \leftarrow (s_1^{(i)}, \ldots, s_V^{(i)})$. Hence $\mathbf{r}^{(i)}, \in \mathbb{Z}^{\mathsf{sec} \times 3}$ and $\mathbf{s}^{(i)} \in \mathbb{Z}^{V \times 3}$ (each row has 3 entries accordingly to Enc defined in Section 2.2). Recall that here $\mathbf{x}^{(i)}$ is a vector with sec entries: $(\mathbf{x}_1^{(i)}, \ldots, \mathbf{x}_{\mathsf{sec}}^{(i)})$ and $\mathbf{y}^{(i)}$ has V entries: $(\mathbf{y}_1^{(i)}, \ldots, \mathbf{y}_V^{(i)})$.

1. Each party P_i broadcasts $E^{(i)} = \mathsf{Enc}_{\mathsf{pk}}(\mathbf{x}^{(i)}, \mathbf{r}^{(i)})$.
2. Each party P_i samples each entry of $\mathbf{y}^{(i)}$ and $\mathbf{s}^{(i)}$ randomly w.r.t to the bounds $\|\mathbf{y}_j^{(i)}\|_\infty \leq B_{\mathsf{plain}}$, $\|\mathbf{s}_j^{(i)}\|_\infty \leq B_{\mathsf{rand}}$ where $j \in [1 \ldots \mathrm{sec}]$. Then P_i uses the random coins $\mathbf{s}^{(i)}$ to compute $\mathbf{a}^{(i)} \leftarrow \mathsf{Enc}_{\mathsf{pk}}(\mathbf{y}^{(i)}, \mathbf{s}^{(i)})$ and broadcasts $\mathbf{a}^{(i)}$.
3. The parties use $\mathcal{F}_{\mathsf{Rand}}$ to sample $\mathbf{e} \in \{0,1\}^{\mathsf{sec}}$.
4. Each party P_i computes $\mathbf{z}^{(i)\mathsf{T}} = \mathbf{y}^{(i)\mathsf{T}} + M_{\mathbf{e}} \cdot \mathbf{x}^{(i)\mathsf{T}}$ and $T^{(i)} = \mathbf{s}^{(i)} + M_{\mathbf{e}} \cdot \mathbf{r}^{(i)}$ and broadcasts $(\mathbf{z}^{(i)}, T^{(i)})$.
5. Each party P_i computes $\mathbf{d}^{(i)} = \mathsf{Enc}_{\mathsf{pk}}(\mathbf{z}^{(i)}, \mathbf{t})$ where $\mathbf{t}$ ranges through all rows of $T^{(i)}$, then stores the sum $\mathbf{d} = \sum_{i=1}^{n} \mathbf{d}^{(i)}$.
6. The parties compute $E = \sum_i E^{(i)}$ $\mathbf{a} = \sum_i \mathbf{a}^{(i)}$, $\mathbf{z} = \sum_i \mathbf{z}^{(i)}$ and $T = \sum_i T^{(i)}$ and conduct the checks (allowing the norms to be $2n$ times bigger to accommodate the summations):
$$\mathbf{d}^{\mathsf{T}} = \mathbf{a}^{\mathsf{T}} + (M_{\mathbf{e}} \cdot E), \quad \|\mathbf{z}\|_\infty \leq 2 \cdot n \cdot B_{\mathsf{plain}}, \quad \|T\|_\infty \leq 2 \cdot n \cdot B_{\mathsf{rand}}.$$
7. If the check passes, the parties output $\sum_{i=1}^{n} E^{(i)}$.

Fig. 10. Protocol for global proof of knowledge of a ciphertext

To solve the equation for E notice that $M_{\mathbf{e}} - M_{\mathbf{e}'}$ is a matrix with entries in $\{-1, 0, 1\}$ so we must solve a linear system where $E = \mathsf{Enc}_{\mathsf{pk}}(\mathbf{x}_k, \mathbf{r}_k)$ for $k = 1, \ldots, \mathsf{sec}$. This can be done in two steps: solve the linear system for the first half: $\mathbf{c}_1, \ldots, \mathbf{c}_{\mathsf{sec}/2}$ and then for the second half: $\mathbf{c}_{\mathsf{sec}/2+1}, \ldots, \mathbf{c}_{\mathsf{sec}}$. For the first step identify a square submatrix of $\mathsf{sec} \times \mathsf{sec}$ entries in $M_{\mathbf{e}} - M_{\mathbf{e}'}$ which has a diagonal full of 1's or -1's and it is lower triangular. This can be done since there is at least one component j such that $e_j \neq e'_j$. Recall that the plaintexts $\mathbf{z}_k, \mathbf{z}'_k$ have norms less than B_{plain} and the randomness used for encrypting them, $\mathbf{t}_k, \mathbf{t}'_k$, have norms less than B_{rand} where k ranges through $1, \ldots, \mathsf{sec}$.

Solving the linear system from the top row to the middle row via substitution we obtain in the worst case: $\|\mathbf{x}_k\|_\infty \leq 2^k \cdot n \cdot B_{\mathsf{plain}}$ and $\|\mathbf{y}_k\|_\infty \leq 2^k \cdot n \cdot B_{\mathsf{rand}}$ where k ranges through $1, \ldots, \mathsf{sec}/2$. The second step is similar to the first with the exception that now we have to look for an upper triangular matrix of $\mathsf{sec} \times \mathsf{sec}$. Then solve the linear system from the last row to the middle row. In this way we extract $\mathbf{x}_k, \mathbf{r}_k$ which form $(2^{\mathsf{sec}/2+1} \cdot n \cdot B_{\mathsf{plain}}, 2^{\mathsf{sec}/2+1} \cdot n \cdot B_{\mathsf{rand}})$ or $(2^{3\mathsf{sec}/2+1} \cdot n \cdot \tau, 2^{3\mathsf{sec}/2+1} \cdot n \cdot \rho)$ ciphertexts. This means that the slack is $2^{3\mathsf{sec}/2+1}$.

Honest Verifier Zero-Knowledge. Here we give a simulator $\mathcal{S}$ for an honest verifier (each party P_i acts as one at one point during the protocol). The simula-

$\mathcal{F}^S_{\mathsf{gZKPoK}}$

Let H denote the set of honest parties. Initially, all parties input pk. Then, the following can happen repeatedly:

1. Every honest party P_i inputs $\mathbf{x}^{(i)}$.
2. Output $\mathsf{Enc}_{\mathsf{pk}}(\mathbf{x}^{(i)})$ for all $i \in H$ to the adversary.
3. The adversary inputs $\mathbf{x}'$.
4. The functionality sends $\mathsf{Enc}_{\mathsf{pk}}(\mathbf{x}' + \sum_{i \in H} \mathbf{x}^{(i)})$, all with noise increased by a factor of $n \cdot S$ to all parties.

Fig. 11. Functionality for global proof of knowledge of ciphertext

tor's purpose is to create a transcript with the verifier which is indistinguishable from the real interaction between the prover and the verifier. To achieve this, $\mathcal{S}$ samples uniformly $\mathbf{e} \xleftarrow{\$} \{0,1\}^{\mathsf{sec}}$ and then creates the transcript accordingly: sample $\mathbf{z}^{(i)}$ such that $\|\mathbf{z}^{(i)}\|_\infty \leq B_{\mathsf{plain}}$ and $T^{(i)}$ such that $\|T^{(i)}\|_\infty \leq B_{\mathsf{rand}}$ and then fix $\mathbf{a}^{(i)} = \mathsf{Enc}_{\mathsf{pk}}(\mathbf{z}^{(i)}, T^{(i)}) - (M_{\mathbf{e}} \cdot E^{(i)})$, where the encryption is applied component-wise. Clearly the produced transcript $(\mathbf{a}^{(i)}, \mathbf{e}^{(i)}, \mathbf{z}^{(i)}, T^{(i)})$ passes the final checks and the statistical distance to the real one is $2^{-\mathsf{sec}}$, which is negligible with respect to sec.

$\mathcal{S}^S_{\mathsf{gZKPoK}}$

Let A denote the set of corrupted parties, and H the set of honest ones.

1. Receive $E^{(i)}$ for all $i \in H$.
2. Sample $\mathbf{e} \xleftarrow{\$} \{0,1\}^{\mathsf{sec}}$.
3. Use the honest-verifier zero-knowledge simulator above to generate transcripts $(\mathbf{a}^{(i)}, \mathbf{e}, (\mathbf{z}^{(i)}, T^{(i)}))$ for $i \in H$.
4. Send $\{\mathbf{a}^{(i)}\}_{i \in H}$ to the adversary.
5. Receive $(E^{(i)}, \mathbf{y}^{(i)}, \mathbf{a}^{(i)})$ for every corrupted party P_i from the adversary.
6. Emulating $\mathcal{F}_{\mathsf{Rand}}$, send $\mathbf{e}$ to the adversary.
7. Receive $(\mathbf{z}^{(i)}, T^{(i)})$ for every corrupted party P_i from the adversary.
8. Check whether $\sum_{i \in A} \mathbf{z}^{(i)}$ and $\sum_{i \in A} T^{(i)}$ meets the bounds. Abort if not.
9. Rewinding the adversary, sample $\tilde{\mathbf{e}} \neq \mathbf{e}$ and conduct the same check for the adversary's responses $\{\tilde{\mathbf{z}}^{(i)}, \tilde{T}^{(i)}\}_{i \in A}$ until the check passes.
10. Use the Σ-protocol extractor on $\{(E^{(i)}, \mathbf{y}^{(i)}, \mathbf{a}^{(i)}, \mathbf{e}, \mathbf{z}^{(i)}, T^{(i)}, \tilde{\mathbf{e}}, \tilde{\mathbf{z}}^{(i)}, \tilde{T}^{(i)})\}_{i \in A}$ to compute $\{\mathbf{x}^{(i)}\}_{i \in A}$ and input $\sum_{i \in A} \mathbf{x}^{(i)}$ to $\mathcal{F}^S_{\mathsf{gZKPoK}}$.

Fig. 12. Simulator for global proof of knowledge of ciphertext

Putting Things Together. In the context of our triple generation, we model Π_{gZKPoK} as $\mathcal{F}^S_{\mathsf{gZKPoK}}$ in Fig. 11. We will argue below that Π_{gZKPoK} implements $\mathcal{F}^S_{\mathsf{gZKPoK}}$ with slack $S = 2^{3\mathsf{sec}/2+1}$.

$\mathcal{F}^S_{\mathsf{gZKPoK}}$ does not guarantee the correctness of individual corrupted parties' ciphertexts but the correctness of the resulting sum. This suffices because only the latter is used in the protocol. A rewinding simulator still can extract individual inputs, but there is no guarantee that either they are in fact pre-images of the encryptions sent by corrupted parties or they are subject to any bounds. Both properties only hold for the sum. This is modeled by $\mathcal{F}^S_{\mathsf{gZKPoK}}$ only outputting a sum, and it is easy to see that this output suffices for SPDZ.

$\mathcal{S}^S_{\mathsf{gZKPoK}}$ in Fig. 12 describes our simulator. The rewinding technique is the same as in the soundness simulator for the Σ-protocol and therefore has the same running time (roughly inverse to the success probability of a corrupted prover). See Sect. 3 of [Dam02] for details.

5 Implementation

We have implemented all three approaches to triple generation in this paper and measured the throughputs achieved by them in comparison to previous results with SPDZ [DKL+12, DKL+13] and MASCOT [KOS16]. We have used the optimized distributed decryption in the full version [KPR17] for SPDZ-1, SPDZ-2, and High Gear. Our code is written in C++ and uses MPIR [MPI17] for arithmetic with large integers.[4] We use Montgomery modular multiplication and the Chinese reminder theorem representation of polynomials wherever beneficial. See Gentry et al. [GHS12b] for more details.

Note that the parameters chosen by Damgård et al. [DKL+13][Appendix A] for the non-interactive zero-knowledge proof imply that the prover has to recompute the proof with probability 1/32 as part of a technique called rejection sampling. We have increased the parameters to reduce this probability by up to 2^{20} as long as it would not impact on the performance, i.e., the number of 64-bit words needed to represent p_o and p_1 would not change.

All previous implementations have benchmarks for two parties on a local network with 1 Gbit/s throughput on commodity hardware. We have have used i7-4790 and i7-3770S CPUs with 16 to 32 GB of RAM, and we have re-run and optimized the code by Damgård et al. [DKL+13] for a fairer comparison. Table 2 shows our results in this setting. SDPZ-1 and SPDZ-2 refer to the two different proofs for ciphertexts, the Schnorr-like protocol presented in the original paper [DPSZ12] and the cut-and-choose protocol in the follow-up work [DKL+13], the latter with either covert or active security. The c-covert security is defined as a cheating adversary being caught with probability $1/c$, and by sec-bit security we mean a statistical security parameter of sec. Throughout this section, we will round figures to the two most significant digits for a more legible presentation.

[4] We extensively use the function `mpn_addmul_1`, which we found to be 10–20% faster in MPIR compared to GMP.

To allow direct comparisons with previous works, we have benchmarked our protocols for several choices of security parameters and field size. Note that the computational security parameter is set everywhere to $k = 128$ and we highlight how the statistical parameter impacts the performance. The main difference between our implementation of SPDZ with the Schnorr-like protocol to the previous one [DKL+12], is the underlying BGV implementation because the protocol is the same.

Table 2. Triple generation for 64 and 128 bit prime fields with two parties on a 1 Gbit/s LAN.

	Triples/s	Security	BGV impl.	$\log_2(\lvert\mathbb{F}_p\rvert)$
SPDZ-1 [DKL+12]	79	40-bit active	NTL	64
SPDZ-2 [DKL+13]	158	20-covert	Specific	64
SPDZ-2 [DKL+13]	36	40-bit active	Specific	64
MASCOT [KOS16]	5,100	64-bit active	$\perp$	128
SPDZ-1 (ours)	12,000	40-bit active	Specific	64
SPDZ-1 (ours)	6,400	64-bit active	Specific	128
SPDZ-1 (ours)	4,200	128-bit active	Specific	128
SPDZ-2 (ours)	3,900	20-covert	Specific	64
SPDZ-2 (ours)	1,100	40-bit active	Specific	64
Low Gear (Sect. 3)	59,000	40-bit active	Specific	64
Low Gear (Sect. 3)	30,000	64-bit active	Specific	128
Low Gear (Sect. 3)	15,000	128-bit active	Specific	128
High Gear (Sect. 4)	11,000	40-bit active	Specific	64
High Gear (Sect. 4)	5,600	64-bit active	Specific	128
High Gear (Sect. 4)	2,300	128-bit active	Specific	128

In Table 3 we also analyze the communication per triple of some protocols with active security and compared the actual throughput to the maximum possible on a 1 Gbit/s link (network throughput divided by the communication per triple). The higher the difference between actual and maximum possible, the more time is spent on computation. The figures show that MASCOT has very low computation; the actual throughput is more than 90% of the maximum possible. On the other hand, all BGV-based implementations have a significant gap, which is to be expected. Experiments have shown that the relative gap increases in Low Gear with a growing statistical parameter. This is mostly because the ciphertexts become larger and 32 GB of memory is not enough for one triple generator thread per core, hence there is some computation capacity left unused.

Table 3. Communication per prime field triple (one way) and actual vs. maximum throughput with two parties on a 1 Gbit/s link.

	Communication	Security	$\log_2(\mathbb{F}_p|)$	Triples/s	Maximum
SPDZ-2	350 kbit	40	64	1,100	2,900
MASCOT [KOS16]	180 kbit	64	128	5,100	5,600
SPDZ-1	23 kbit	40	64	12,000	44,000
SPDZ-1	32 kbit	64	128	6,400	31,000
SPDZ-1	37 kbit	128	128	4,200	27,000
Low Gear (Sect. 3)	9 kbit	40	64	59,000	110,000
Low Gear (Sect. 3)	15 kbit	64	128	30,000	68,000
Low Gear (Sect. 3)	17 kbit	128	128	15,000	60,000
High Gear (Sect. 4)	24 kbit	40	64	11,000	42,000
High Gear (Sect. 4)	34 kbit	64	128	5,600	30,000
High Gear (Sect. 4)	42 kbit	128	128	2,300	24,000

Table 4. Communication per prime field triple (one way) and actual vs. maximum throughput with two parties on a 50 Mbit/s link.

	Communication	Security	$\log_2(\mathbb{F}_p|)$	Triples/s	Maximum
MASCOT [KOS16]	180 kbit	64	128	214	275
SPDZ-1	23 kbit	40	64	1,800	2,200
SPDZ-1	32 kbit	64	128	1,400	1,600
SPDZ-1	37 kbit	128	128	1,100	1,400
Low Gear (Sect. 3)	9 kbit	40	64	4,500	5,600
Low Gear (Sect. 3)	15 kbit	64	128	3,200	3,400
Low Gear (Sect. 3)	17 kbit	128	128	2,600	3,000
High Gear (Sect. 4)	24 kbit	40	64	1,600	2,100
High Gear (Sect. 4)	34 kbit	64	128	1,300	1,500
High Gear (Sect. 4)	42 kbit	128	128	700	1,200

WAN Setting. For a more complete picture, we have also benchmarked our protocols in the same WAN setting as Keller et al. [KOS16], restricting the bandwidth to 50 Mbit/s and imposing a delay of 50 ms to all communication. Table 4 shows our results in similar manner to Table 3. As one would expect, the gap between actual throughput and maximum possible is more narrow because the communication becomes more of a bottleneck, and the performance is closely related to the required communication.

Fields of Characteristic Two. For a more thorough comparison with MASCOT, we have also implemented our protocols for the field of size 2^{40} using

Table 5. Triple generation for characteristic two with two parties on a 1 Gbit/s LAN.

	Triples/s	Security	BGV impl.	$\mathbb{F}_{2^n}$
SPDZ-1 [DKL+12]	16	40-bit active	NTL	40
MASCOT [KOS16]	5,100	64-bit active	$\perp$	128
SPDZ-1 (ours)	67	40-bit active	Specific	40
SPDZ-2 (ours)	24	20-covert	Specific	40
SPDZ-2 (ours)	8	40-bit active	Specific	40
Low Gear (Sect. 3)	117	40-bit active	Specific	40
High Gear (Sect. 4)	67	40-bit active	Specific	40

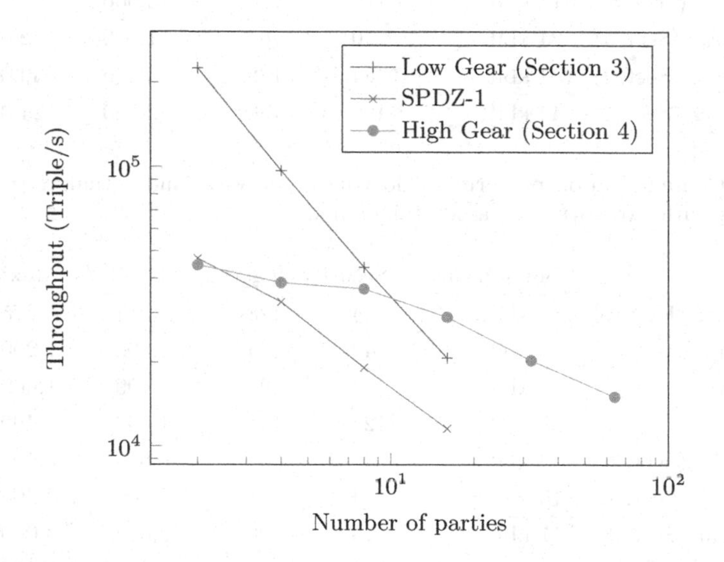

Fig. 13. Triple generation for a 128 bit prime field with 64 bit statistical security on AWS r4.16xlarge instances.

the same approach as Damgård et al. [DKL+12]. Table 5 shows the low performance of homomorphic encryption-based protocols with fields of characteristic two. This has been observed before: in the above work, the performance for $\mathbb{F}_{2^{40}}$ is an order of magnitude worse than for $\mathbb{F}_p$ with a 64-bit bit prime. The main reason is that BGV lends itself naturally to cleartexts modulo some integer p. The construction for $\mathbb{F}_{2^{40}}$ sets $p = 2$ and uses 40 slots to represent an element whereas an element of $\mathbb{F}_p$ for a prime p only requires one ciphertext slot.

More Than Two Parties. Increasing the number of parties, we have benchmarked our protocols and our implementation of SPDZ with up to 64 r4.16xlarge instances on Amazon Web Services. Figure 13 shows that both Low and High Gear improve over SPDZ-1, with High Gear taking the lead from about ten parties. Missing figures do not indicate failed experiments but rather omitted experiments due to financial constraints.

At the time of writing, one hour on an r4.16xlarge instance in US East costs $4.256. Therefore, the number of triples per dollar and party varies between 190 million (two parties with Low Gear) and 13 million (64 parties with High Gear).

5.1 Vickrey Auction for 100 Parties

As a motivation for computation with a high number of parties, we have implemented a secure Vickrey second price auction [Vic61], where 100 parties input one bid each. Table 6 shows our online phase timings for two different Amazon Web Services instances.

Table 6. Online phase of Vickrey auction with 100 parties, each inputting one bid.

AWS instance	Time	Cost per party
t2.nano	9.0 s	$0.000017
c4.8xlarge	1.4 s	$0.000741

The Vickrey auction requires 44,571 triples. In Table 7, we compare the offline cost of MASCOT and our High Gear protocol on AWS m3.2xlarge instances.

Table 7. Offline phase of Vickrey auction with 100 parties, each inputting one bid.

	Time	Cost per party
MASCOT [KOS16]	1,300 s	$0.190
High Gear (Sect. 4)	98 s	$0.014

6 Future Work

Recently, there has been an improved zero-knowledge proof of knowledge of bounded pre-images for LWE-style one-way functions [BL17]. It reduces the extra ciphertexts per proven plaintext from two (in our protocol) to any number larger than one. The technique is dependent on the number of ciphertexts that are proven simultaneously. More concretely, for $u \cdot \mathsf{sec}$ ciphertexts in one proof (and $u \geq 1$), the prover needs to send $(u+1) \cdot \mathsf{sec}$ ciphertexts in the first round, hence the amortized overhead is $(u+1)/u$. This compares to $2u \cdot \mathsf{sec} - 1$ ciphertexts amortized over $2 - 1/(u \cdot \mathsf{sec})$ in our scheme. However, we estimate that the benefits of the newer proof strongly depend on the parameters and the available memory. For some parameters, we found that our implementation would exhaust 32 GB of memory with fewer than eight generation threads.

We therefore could not exhaust the computational capacity of the CPU. Note that our implementation stores all necessary information for the proof in memory, and consider that one ciphertext takes up to $2^{16} \cdot 700 \cdot 2$ bits or ≈ 11 MBytes. This means that, for 128-bit active security, we require about $(3\mathsf{sec} - 1) \cdot 11$ MBytes or ≈ 4.4 GBytes of storage for the ciphertexts alone (not considering any cleartexts). It would be interesting to see how the newer proof fares and whether using a solid state disk for storage would improve the performance.

References

[Alb17] Albrecht, M.R.: LWE estimator (2017). https://bitbucket.org/malb/lwe-estimator. Accessed September 2017

[APS15] Albrecht, M.R., Player, R., Scott, S.: On the concrete hardness of learning with errors. J. Math. Cryptol. **9**(3), 169–203 (2015)

[BCI+13] Bitansky, N., Chiesa, A., Ishai, Y., Paneth, O., Ostrovsky, R.: Succinct non-interactive arguments via linear interactive proofs. In: Sahai, A. (ed.) TCC 2013. LNCS, vol. 7785, pp. 315–333. Springer, Heidelberg (2013). https://doi.org/10.1007/978-3-642-36594-2_18

[BDOZ11] Bendlin, R., Damgård, I., Orlandi, C., Zakarias, S.: Semi-homomorphic encryption and multiparty computation. In: Paterson, K.G. (ed.) EUROCRYPT 2011. LNCS, vol. 6632, pp. 169–188. Springer, Heidelberg (2011). https://doi.org/10.1007/978-3-642-20465-4_11

[Bea92] Beaver, D.: Efficient multiparty protocols using circuit randomization. In: Feigenbaum, J. (ed.) CRYPTO 1991. LNCS, vol. 576, pp. 420–432. Springer, Heidelberg (1992). https://doi.org/10.1007/3-540-46766-1_34

[BGV12] Brakerski, Z., Gentry, C., Vaikuntanathan, V.: (Leveled) fully homomorphic encryption without bootstrapping. In: Goldwasser, S. (ed.) ITCS 2012, pp. 309–325. ACM, January 2012

[BISW17] Boneh, D., Ishai, Y., Sahai, A., Wu, D.J.: Lattice-based SNARGs and their application to more efficient obfuscation. In: Coron, J.-S., Nielsen, J.B. (eds.) EUROCRYPT 2017. LNCS, vol. 10212, pp. 247–277. Springer, Cham (2017). https://doi.org/10.1007/978-3-319-56617-7_9

[BL17] Baum, C., Lyubashevsky, V.: Simple amortized proofs of shortness for linear relations over polynomial rings. Cryptology ePrint Archive, Report 2017/759 (2017). http://eprint.iacr.org/2017/759

[BLW08] Bogdanov, D., Laur, S., Willemson, J.: Sharemind: a framework for fast privacy-preserving computations. In: Jajodia, S., Lopez, J. (eds.) ESORICS 2008. LNCS, vol. 5283, pp. 192–206. Springer, Heidelberg (2008). https://doi.org/10.1007/978-3-540-88313-5_13

[BMR16] Ball, M., Malkin, T., Rosulek, M.: Garbling gadgets for boolean and arithmetic circuits. In: Weippl, E.R., Katzenbeisser, S., Kruegel, C., Myers, A.C., Halevi, S. (eds.) ACM CCS 2016, pp. 565–577. ACM Press, October 2016

[BV11] Brakerski, Z., Vaikuntanathan, V.: Efficient fully homomorphic encryption from (standard) LWE. In: Ostrovsky, R. (ed.) 52nd FOCS, pp. 97–106. IEEE Computer Society Press, October 2011

[Can01] Canetti, R.: Universally composable security: a new paradigm for cryptographic protocols. In: 42nd FOCS, pp. 136–145. IEEE Computer Society Press, October 2001

[CD09] Cramer, R., Damgård, I.: On the amortized complexity of zero-knowledge protocols. In: Halevi, S. (ed.) CRYPTO 2009. LNCS, vol. 5677, pp. 177–191. Springer, Heidelberg (2009). https://doi.org/10.1007/978-3-642-03356-8_11

[CDXY17] Cramer, R., Damgård, I., Xing, C., Yuan, C.: Amortized complexity of zero-knowledge proofs revisited: achieving linear soundness slack. In: Coron, J.-S., Nielsen, J.B. (eds.) EUROCRYPT 2017. LNCS, vol. 10210, pp. 479–500. Springer, Cham (2017). https://doi.org/10.1007/978-3-319-56620-7_17

[CS16] Costache, A., Smart, N.P.: Which ring based somewhat homomorphic encryption scheme is best? In: Sako, K. (ed.) CT-RSA 2016. LNCS, vol. 9610, pp. 325–340. Springer, Cham (2016). https://doi.org/10.1007/978-3-319-29485-8_19

[Dam02] Damgård, I.: On Σ-protocols (2002). http://www.daimi.au.dk/~Eivan/Sigma.pdf. Accessed September 2017

[DDN+16] Damgård, I., Damgård, K., Nielsen, K., Nordholt, P.S., Toft, T.: Confidential benchmarking based on multiparty computation. In: Grossklags, J., Preneel, B. (eds.) FC 2016. LNCS, vol. 9603, pp. 169–187. Springer, Heidelberg (2017). https://doi.org/10.1007/978-3-662-54970-4_10

[DES16] Doerner, J., Evans, D., Shelat, A.: Secure stable matching at scale. In: Weippl, E.R., Katzenbeisser, S., Kruegel, C., Myers, A.C., Halevi, S. (eds.) ACM CCS 2016, pp. 1602–1613. ACM Press, October 2016

[DGKN09] Damgård, I., Geisler, M., Krøigaard, M., Nielsen, J.B.: Asynchronous multiparty computation: theory and implementation. In: Jarecki, S., Tsudik, G. (eds.) PKC 2009. LNCS, vol. 5443, pp. 160–179. Springer, Heidelberg (2009). https://doi.org/10.1007/978-3-642-00468-1_10

[DKL+12] Damgård, I., Keller, M., Larraia, E., Miles, C., Smart, N.P.: Implementing AES via an actively/covertly secure dishonest-majority MPC protocol. In: Visconti, I., De Prisco, R. (eds.) SCN 2012. LNCS, vol. 7485, pp. 241–263. Springer, Heidelberg (2012). https://doi.org/10.1007/978-3-642-32928-9_14

[DKL+13] Damgård, I., Keller, M., Larraia, E., Pastro, V., Scholl, P., Smart, N.P.: Practical covertly secure MPC for dishonest majority - or: breaking the SPDZ limits. In: Crampton, J., Jajodia, S., Mayes, K. (eds.) ESORICS 2013. LNCS, vol. 8134, pp. 1–18. Springer, Heidelberg (2013)

[DPSZ12] Damgård, I., Pastro, V., Smart, N., Zakarias, S.: Multiparty computation from somewhat homomorphic encryption. In: Safavi-Naini, R., Canetti, R. (eds.) CRYPTO 2012. LNCS, vol. 7417, pp. 643–662. Springer, Heidelberg (2012). https://doi.org/10.1007/978-3-642-32009-5_38

[DSZ15] Demmler, D., Schneider, T., Zohner, M.: ABY - a framework for efficient mixed-protocol secure two-party computation. In: NDSS 2015. The Internet Society, February 2015

[GHS12a] Gentry, C., Halevi, S., Smart, N.P.: Fully homomorphic encryption with polylog overhead. In: Pointcheval, D., Johansson, T. (eds.) EUROCRYPT 2012. LNCS, vol. 7237, pp. 465–482. Springer, Heidelberg (2012). https://doi.org/10.1007/978-3-642-29011-4_28

[GHS12b] Gentry, C., Halevi, S., Smart, N.P.: Homomorphic evaluation of the AES circuit. In: Safavi-Naini, R., Canetti, R. (eds.) CRYPTO 2012. LNCS, vol. 7417, pp. 850–867. Springer, Heidelberg (2012). https://doi.org/10.1007/978-3-642-32009-5_49

[GLNP15] Gueron, S., Lindell, Y., Nof, A., Pinkas, B.: Fast garbling of circuits under standard assumptions. In: Ray, I., Li, N., Kruegel, C. (eds.) ACM CCS 2015, pp. 567–578. ACM Press, October 2015

[GRR+16] Grassi, L., Rechberger, C., Rotaru, D., Scholl, P., Smart, N.P.: MPC-friendly symmetric key primitives. In: Weippl, E.R., Katzenbeisser, S., Kruegel, C., Myers, A.C., Halevi, S. (eds.) ACM CCS 2016, pp. 430–443. ACM Press, October 2016

[GSB+17] Gascón, A., Schoppmann, P., Balle, B., Raykova, M., Doerner, J., Zahur, S., Evans, D.: Privacy-preserving distributed linear regression on high-dimensional data. Proc. Priv. Enhancing Technol. **4**, 248–267 (2017)

[KMR12] Keller, M., Mikkelsen, G.L., Rupp, A.: Efficient threshold zero-knowledge with applications to user-centric protocols. In: Smith, A. (ed.) ICITS 2012. LNCS, vol. 7412, pp. 147–166. Springer, Heidelberg (2012). https://doi.org/10.1007/978-3-642-32284-6_9

[KOS16] Keller, M., Orsini, E., Scholl, P.: MASCOT: faster malicious arithmetic secure computation with oblivious transfer. In: Weippl, E.R., Katzenbeisser, S., Kruegel, C., Myers, A.C., Halevi, S. (eds.) ACM CCS 2016, pp. 830–842. ACM Press, October 2016

[KPR17] Keller, M., Pastro, V., Rotaru, D.: Overdrive: making SPDZ great again. Cryptology ePrint Archive, Report 2017/1230 (2017). https://eprint.iacr.org/2017/1230

[KSS13] Keller, M., Scholl, P., Smart, N.P.: An architecture for practical actively secure MPC with dishonest majority. In: Sadeghi, A.-R., Gligor, V.D., Yung, M. (eds.) ACM CCS 2013, pp. 549–560. ACM Press, November 2013

[KW15] Kamm, L., Willemson, J.: Secure floating point arithmetic and private satellite collision analysis. Int. J. Inf. Secur. **14**(6), 531–548 (2015)

[LPR10] Lyubashevsky, V., Peikert, C., Regev, O.: On ideal lattices and learning with errors over rings. In: Gilbert, H. (ed.) EUROCRYPT 2010. LNCS, vol. 6110, pp. 1–23. Springer, Heidelberg (2010). https://doi.org/10.1007/978-3-642-13190-5_1

[LPR13] Lyubashevsky, V., Peikert, C., Regev, O.: A toolkit for ring-LWE cryptography. In: Johansson, T., Nguyen, P.Q. (eds.) EUROCRYPT 2013. LNCS, vol. 7881, pp. 35–54. Springer, Heidelberg (2013). https://doi.org/10.1007/978-3-642-38348-9_3

[MPI17] MPIR team: Multiple precision integers and rationals (2017). https://www.mpir.org. Accessed September 2017

[NWI+13] Nikolaenko, V., Weinsberg, U., Ioannidis, S., Joye, M., Boneh, D., Taft, N.: Privacy-preserving ridge regression on hundreds of millions of records. In: 2013 IEEE Symposium on Security and Privacy, pp. 334–348. IEEE Computer Society Press, May 2013

[PVW08] Peikert, C., Vaikuntanathan, V., Waters, B.: A framework for efficient and composable oblivious transfer. In: Wagner, D. (ed.) CRYPTO 2008. LNCS, vol. 5157, pp. 554–571. Springer, Heidelberg (2008). https://doi.org/10.1007/978-3-540-85174-5_31

[RR16] Rindal, P., Rosulek, M.: Faster malicious 2-party secure computation with online/offline dual execution. In: USENIX Security Symposium, pp. 297–314 (2016)

[RSS17] Rotaru, D., Smart, N.P., Stam, M.: Modes of operation suitable for computing on encrypted data. IACR Trans. Symm. Cryptol. **2017**(3), 294–324 (2017)

[Vic61] Vickrey, W.: Counterspeculation, auctions, and competitive sealed tenders. J. Finance **16**(1), 8–37 (1961)

[WRK17] Wang, X., Ranellucci, S., Katz, J.: Authenticated garbling and efficient maliciously secure two-party computation. In: Thuraisingham, B.M., Evans, D., Malkin, T., Xu, D. (eds.) ACM CCS 2017, pp. 21–37. ACM Press, October/November 2017

[ZE15] Zahur, S., Evans, D.: Obliv-C: a language for extensible data-oblivious computation. Cryptology ePrint Archive, Report 2015/1153 (2015). http://eprint.iacr.org/2015/1153

Non-interactive Zero-Knowledge

Efficient Designated-Verifier Non-interactive Zero-Knowledge Proofs of Knowledge

Pyrros Chaidos[1(✉)] and Geoffroy Couteau[2]

[1] National and Kapodistrian University of Athens, Athens, Greece
pchaidos@di.uoa.gr

[2] Karsruhe Institute of Technology, Karlsruhe, Germany

Abstract. We propose a framework for constructing efficient designated-verifier non-interactive zero-knowledge proofs (DVNIZK) for a wide class of algebraic languages over abelian groups, under standard assumptions. The proofs obtained via our framework are proofs of knowledge, enjoy statistical, and unbounded soundness (the soundness holds even when the prover receives arbitrary feedbacks on previous proofs). Previously, no efficient DVNIZK system satisfying *any* of those three properties was known. Our framework allows proving arbitrary relations between cryptographic primitives such as Pedersen commitments, ElGamal encryptions, or Paillier encryptions, in an efficient way. For the latter, we further exhibit the first non-interactive zero-knowledge proof system in the standard model that is more efficient than proofs obtained via the Fiat-Shamir transform, with still-meaningful security guarantees and under standard assumptions. Our framework has numerous applications, in particular for the design of efficient privacy-preserving non-interactive authentication.

Keywords: Zero-knowledge proofs · Non-interactive proofs

1 Introduction

Zero-knowledge proof systems allow a prover to convince someone of the truth of a statement, without revealing anything beyond the fact that the statement is true. After their introduction in the seminal work of Goldwasser, Micali, and Rackoff [34], they have proven to be a fundamental primitive in cryptography. Among them, *non-interactive zero-knowledge proofs* (NIZK proofs), where the proof consists of a single flow from the prover to the verifier, are of particular interest, in part due to their tremendous number of applications in cryptographic primitives and protocols, and in part due to the theoretical and technical challenges that they represent.

G. Couteau—Part of this work was made while the second author was at École Normale Supérieure de Paris, France.

J. B. Nielsen and V. Rijmen (Eds.): EUROCRYPT 2018, LNCS 10822, pp. 193–221, 2018.
https://doi.org/10.1007/978-3-319-78372-7_7

For almost two decades after their introduction in [10], NIZKs coexisted in two types: inefficient NIZKs secure under standard assumptions (such as doubly enhanced trapdoor permutations [30]) in the common reference string model, and practically efficient NIZKs built from the Fiat-Shamir heuristic [31,47], which are secure in the random oracle model [6] (hence only heuristically secure in the standard model). This state of affairs changed with the arrival of pairing-based cryptography, from which a fruitful line of work (starting with the work of Groth, Ostrovsky, and Sahai [37,38]) introduced increasingly more efficient NIZK proof systems in the standard model. That line of work culminated with the framework of Groth-Sahai proofs [39], which provided an efficient framework of pairing-based NIZKs for a large class of useful languages. Yet, one decade later, pairing-based NIZKs from the Groth-Sahai framework remain the only known efficient NIZK proof system in the standard model. Building efficient NIZKs in the standard model, without pairing-based assumptions, is a major open problem, and research in this direction has proven elusive.

1.1 Designated-Verifier Non-interactive Zero-Knowledge

Parallel to the research on NIZKs, an alternative promising line of research has focused on *designated-verifier* non-interactive zero-knowledge proof systems (DVNIZKs). A DVNIZK retains most of the security properties of a NIZK, but is not publicly verifiable: only the owner of some secret information (the designated verifier) can check the proof. Nevertheless, DVNIZKs can replace publicly verifiable NIZKs in a variety of applications. In addition, unlike their publicly-verifiable counterpart, it is known that efficient DVNIZKs secure in the standard model for rich classes of languages can be constructed without pairing-based assumptions [17,23,43,49]. However, to date, research in DVNIZKs has attracted less attention than NIZKs, the previously listed papers being (to our knowledge) the only existing works on this topic, and several important questions have been left open. We list the main open questions below.

Proofs Versus Arguments. A non-interactive zero-knowledge argument system is a NIZK in which the soundness property is only required to hold against computationally bounded adversaries. In a NIZK *proof* system, however, soundness is required to hold even against unbounded adversaries.

Currently, while several DVNIZK argument systems have been designed in the standard model without pairing-based assumptions, efficient DVNIZK proof systems without pairings remain an open question. In fact, to our knowledge, the only known constructions of (possibly inefficient) DVNIZK proofs rely on publicly-verifiable NIZK proofs.

Soundness Versus Knowledge Extraction. A non-interactive zero-knowledge proof (or argument) system is a NIZK *of knowledge* if it guarantees that, when the prover succeeds in convincing the verifier, he must *know* a witness for the truth of the statement. This is in constrast with the standard soundness notion, which only guarantees that the statement is true. Formally, this is ensured by requiring the existence of an efficient simulator that can extract a witness from the proof.

Non-interactive zero-knowledge proofs of knowledge are more powerful than standard NIZKs, and the knowledge-extractability property is crucial in many applications. In particular, they are necessary for the very common task of proving relations between values committed with a perfectly hiding commitment scheme, and they are a core component in privacy-preserving authentication mechanisms [4]. Currently, all known DVNIZK argument systems are not arguments of knowledge. Designing efficient DVNIZKs of knowledge without pairing-based assumptions remains an open question.

Bounded Soundness Versus Unbounded Soundness. The classical soundness security notion for non-interactive zero-knowledge proof systems states that if the statement is not true, no malicious prover can possibly convince the verifier of the truth of the statement with non-negligible probability. While this security notion is sufficient for publicly-verifiable NIZKs, it turns out to be insufficient when considering designated-verifier NIZKs, and corresponds only to a *passive* type of security notion. Indeed, the verification of a DVNIZK involves a secret value, known to the verifier. The fact that a DVNIZK satisfies the standard soundness notion does not preclude the possibility for a malicious prover to learn this secret value, e.g. by submitting a large number of proofs and receiving feedback on whether the proof was accepted or not. Intuitively, this is the same type of issue as for encryption schemes indistinguishable against chosen-plaintext attacks, which can be broken if the adversary is given access to a decryption oracle, or for signature schemes secure against key-only or known-message attacks, which can be broken if the adversary is given access to a signing oracle. Here, an adversary could possibly break the soundness of a DVNIZK if it is given access to a verification oracle.

In practice, this means that as soon as a proof system with bounded soundness is used for more than a logarithmic number of proofs, the soundness property is no longer guaranteed to hold. This calls for a stronger notion of soundness, *unbounded soundness*, which guarantees security even against adversaries that are given arbitrary access to a verification oracle.

Designing a DVNIZK with unbounded soundness has proven to be highly non-trivial. In fact, apart from publicly-verifiable NIZKs (which can be seen as particular types of DVNIZKs where the secret key of the verifier is the empty string), the only known construction of DVNIZK claiming to satisfy unbounded soundness is the construction of [23], where the claim is supported by a proof of security in an idealized model. However, we found this claim to be flawed: there is an explicit attack against the unbounded soundness of any protocol obtained using the compiler of [23], which operates by using slightly malformed proofs to extract the verification key. In the full version of this work [16], we describe our attack, and identify the flaw in the proof of Theorem 5 in [23, Appendix A]. We have notified the authors of our finding and will update future versions of this work with their reply. To our knowledge, in all current constructions, the common reference string and the public key must be refreshed after a logarithmic number of proofs.

1.2 Our Contribution

In this work, we first introduce a framework for designated-verifier NIZKs on group-dependent languages, in the spirit of the Groth-Sahai framework for NIZKs on languages related to pairing-friendly elliptic curves. Our framework only requires that the underlying abelian group on which it is instantiated has order M, where $\mathbb{Z}_M$ is the plaintext-space of an homomorphic cryptosystem with specific properties, and allows to prove a wide variety statements formulated in terms of the operation associated to this abelian group. In particular, we do not need to rely on pairings. The DVNIZKs obtained with our framework are efficient, as they only require a few group elements and ciphertexts. The zero-knowledge property of our schemes reduces to the IND-CPA security of the underlying encryption scheme. Additionally, our DVNIZKs enjoy the following properties: they are (adaptively) *knowledge-extractable*; their knowledge-extractability holds *statistically*; their knowledge-extractability is *unbounded*. We stress that previously, no efficient construction of DVNIZK in the standard model satisfying *any* of the above properties was known. The third property, unbounded soundness, was only claimed to hold for the construction of [23], and this claim was formalized with a proof in an idealized model, but as previously mentioned, we found this claim to be flawed. We also point out that in the Groth-Sahai framework, witness extraction is limited either to statements about group elements, or to statements about exponents committed in a bit-by-bit fashion (making the proof highly inefficient). In contrast, our proof system allows to efficiently extract large exponents, without harming the efficiency of the proof. In addition to the above properties, our DVNIZKs satisfy some other useful properties: they are multi-theorem [30], randomizable [3], and same-string zero-knowledge [27] (*i.e.*, the common reference string used by the prover and the simulator are the same).

Second, our framework comes with a dual variant, where the role of the encryption scheme and the abelian group are reversed, to prove statements, not about elements of the abelian group, but about the underlying homomorphic encryption scheme. This dual variant leads to DVNIZKs satisfying adaptive statistical unbounded soundness, but not knowledge-extractability (i.e. the dual variant does not give proofs of knowledge).

Third, we show that if one is willing to give up unbounded soundness for efficiency, our techniques can be used to construct extremely efficient DVNIZKs with bounded-soundness. The DVNIZKs that we obtain this way are more efficient than *any* previously known construction of non-interactive zero-knowledge proofs, even when considering NIZKs in the random oracle model using the Fiat-Shamir transform: the proofs we obtain are shorter than the proofs obtained via the Fiat-Shamir transform by almost a factor two. To our knowledge, this is the first example of a NIZK construction in the standard model which (conditionally) improves on the Fiat-Shamir paradigm.

Instantiating the Encryption Scheme. Informally, the security properties we require from the underlying scheme are the following: it must be additively homomorphic, with plaintext space $\mathbb{Z}_M$, random source $\mathbb{Z}_R$, and $\gcd(M, R) = 1$, and it must be *decodable*, which means that a plaintext m can be efficiently recovered from an encryption of m with random coin 0. A natural candidate for the above scheme is the Paillier encryption scheme [45] (and its variants, such as Damgård-Jurik [26]). This gives rise to efficient DVNIZK proofs of knowledge over abelian groups of composite order (e.g. subgroups of $\mathbb{F}_p^*$, with order a prime $p = k \cdot n + 1$ for a small k and an RSA modulus n, or composite-order elliptic curves), as well as efficient DVNIZKs for proving relations between Paillier ciphertexts (using the dual variant of our framework). Alternatively, the scheme can also be instantiated with the more recent Castagnos-Laguillaumie encryption scheme [15] to get DVNIZKs over prime-order abelian groups.

Our framework captures many useful zero-knowledge proofs of knowledge that are commonly used in cryptography. This includes DVNIZK proofs of knowledge of a discrete logarithm, of correctness of a Diffie-Hellman tuple, of multiplicative relationships between Pedersen commitments or ElGamal ciphertexts (or variants thereof), among many others. Our results show that, in the settings where a designated-verifier is sufficient, one can build efficient non-interactive zero-knowledge proofs of knowledge for most statements of interest, under well-known assumptions and with strong security properties, without having to rely on pairing-friendly groups.

1.3 Our Method

It is known that linear relations (*i.e.*, membership in linear subspaces) can be non-interactively verified, using the homomorphic properties of cryptographic primitives over abelian groups. Indeed, DVNIZK proofs for linear languages can be constructed, e.g., from hash proof systems [33,41]. In [39], pairings provide exactly the additional structure needed to evaluate degree-two relations, which can be easily generalized to arbitrary relations.

An alternative road was taken in [23] and subsequent works, to obtain non-interactive zero-knowledge proofs for a wide variety of relations, in the designated-verifier setting. To illustrate, let us consider a prover interacting with a verifier, with a common input $(g_1, g_2, h_1, h_2) \in \mathbb{G}^4$ in some group $\mathbb{G}$ of order p, where p is a λ-bit prime. The prover wants to show that (h_1, h_2) have the same discrete logarithm in the basis (g_1, g_2), *i.e.*, there exists x such that $(h_1, h_2) = (g_1^x, g_2^x)$. The standard interactive zero-knowledge proof for this statement proceeds as follows:[1]

1. The prover picks $r \xleftarrow{\$} \{0,1\}^{3\lambda}$, and sends $(a_1, a_2) \leftarrow (g_1^r, g_2^r)$.
2. The verifier picks and sends a uniformly random challenge $e \xleftarrow{\$} \mathbb{Z}_p$.
3. The prover computes and sends $d \leftarrow e \cdot x + r$. The verifier accepts the proof if and only if $(g_1^d, g_2^d) = (h_1^e a_1, h_2^e a_2)$.

[1] More formally, this proof only satisfies zero-knowledge against honest verifiers, but this property is sufficient for the construction of [23].

The idea of [23] is to squash this interactive protocol into a (designated-verifier) non-interactive proof, by giving the challenge to the prover in advance. As knowing the challenge before sending the first flow gives the prover the ability to cheat, the challenge is encrypted with an additively homomorphic encryption scheme. That way, the prover cannot see the challenge; yet, he can still compute an encryption of the value d homomorphically, using the encryption of e. The verifier, who is given the secret verification key, can decrypt the last flow and perform the above check. Thus, the proof is a tuple (a_1, a_2, c_d), where c_d is an encryption of d computed from (x, r) and an encryption c_e of the challenge e.

Although natural, this intuitive approach has proven quite tough to analyze. In [23], the authors had to rely on a new complexity-leveraging-type assumption tailored to their scheme, which (informally) states that the simulator cannot break the security of the encryption scheme, even if he is powerful enough to break the problem underlying the protocol (in the above example, the discrete logarithm problem over $\mathbb{G}$). Even in the bounded setting, analyzing the soundness guarantees of the protocols obtained by this compilation technique (and its variants) is non-trivial, and it has been the subject of several subsequent works [17,43,49]. Additionally, in the unbounded setting, where we must give an efficient simulator that can successfully answer to the proofs submitted by any malicious prover, this compilation technique breaks down. Furthermore, for DVNIZKs constructed with this method, soundness holds only computationally, and security does not guarantee that the simulator can extract a witness for the statement.

Our core idea to overcome all of the above issues is to implement the same strategy in a slightly different way: rather than encrypting the challenge e as the *plaintext* of an homomorphic encryption scheme, we encrypt it as the *random coin* of an encryption scheme which is also homomorphic over the coins. To understand how this allows us to improve over all previous constructions, suppose that we have an encryption scheme Enc which is homomorphic over both the plaintext and the random coins, with plaintext space $\mathbb{Z}_M$ and random source $\mathbb{Z}_R$, and that M is coprime to R. Consider the previously described protocol for proving equality of two discrete logarithms. Given an encryption $\mathsf{Enc}(0; e)$ of 0, where the challenge is the random coin, a prover holding (x, r) can compute and send $\mathsf{Enc}(x; \rho)$ and $\mathsf{Enc}(r; -e\rho)$, for some random ρ. This allows the verifier, who knows e, to compute $\mathsf{Enc}(x \cdot e + r; 0)$, from which she can extract $d = x \cdot e + r \bmod M$ (note that the verifier only needs to know e; unlike in previous work, she does not need to know the decryption key of Enc). Observe that the extracted value depends only on e *modulo* M. At the same time, however, the ciphertext $E(0; e)$ only leaks e *modulo* R, even to an unbounded adversary. By picking e to be sufficiently large ($e > MR$), as M is coprime to R, the verifier can ensure that this leaks *no information* (statistically) about $e \bmod M$. Therefore, we can use a statistical argument to show that the prover cannot cheat when the verification using d succeeds. To allow for efficient simulation of the verifier, we simply give to the simulator the secret key of the scheme, which will allow him to extract all encrypted values, and to check the validity of the equations, without knowing $e \bmod M$. As the simulator is able to extract the values encrypted with Enc, the scheme can be proven to be (statistically) knowledge-extractable. Contrary to previous constructions, the verification key is a random coin rather than the secret key of an encryption scheme. The secret key is only used to extract information in the simulated game.

Example: DVNIZK Proof of Knowledge of a Discrete Logarithm. We illustrate our method with the classical example of proving knowledge of a discrete logarithm. For concreteness, we describe an explicit protocol using the Paillier encryption scheme; therefore, this section assumes some basic knowledge of the Paillier encryption scheme. All necessary preliminaries can be found in Sect. 2. Let $\mathbb{G}$ be a group of order n, where $n = p \cdot q$ is an RSA modulus (*i.e.*, a product of two strong primes). Let g be a generator of $\mathbb{G}$, and let T be a group element. A prover P wishes to prove to a verifier V that he knows a value $t \in \mathbb{Z}_n$ such that $g^t = T$.

Let $h \leftarrow u^n \bmod n^2$, where u denotes an arbitrary generator of $\mathbb{J}_n$, the subgroup of elements of $\mathbb{Z}_n^*$ with Jacobi symbol 1. The Paillier encryption of a message $m \in \mathbb{Z}_n$ with randomness $r \in \mathbb{Z}_{\varphi(n)/2}$ is $\mathsf{Enc}(m; r) = (1+n)^m h^r \bmod n^2$. The public key of the DVNIZK is $E = h^e \in \mathbb{Z}_{n^2}^*$, for a random $e \gg n \cdot \varphi(n)/2$; observe that this is exactly $\mathsf{Enc}(0; e)$. The secret key is e. The DVNIZK proceeds as follows:

The prover P picks $x \xleftarrow{\$} \mathbb{Z}_n$ and a Paillier random coin r, and computes $X \leftarrow g^x$, $T' \leftarrow (1+n)^t h^r \bmod n^2$, and $X' \leftarrow (1+n)^x E^{-r} \bmod n^2$. The verifier V computes $D \leftarrow T^e X \bmod n^2$ and $D' \leftarrow (T')^e X' \bmod n^2$. Then, she checks that D' is of the form $(1+n)^d \bmod n^2$. If so, V computes $d \bmod n$ from D', and checks that $D = g^d$. V accepts iff both checks succeeded.

Let us provide an intuition of the security of this scheme. Correctness follows easily by inspection. Zero-knowledge comes from the fact that T' hides t, under the IND-CPA security of Paillier. For statistical knowledge extractability, note E only reveals $e \bmod \varphi(n)$ to an unbounded adversary, which leaks (statistically) no information on $e \bmod n$ as $\varphi(n)$ is coprime to n. This ensures the value t' encrypted in T' must be equal to t, otherwise the verification equations would uniquely define $e \bmod n$, which is statistically unknown to the prover. The simulator knows $\varphi(n)$ (but not $e \bmod n$) and gets t by decrypting T'.

1.4 Applications

A natural application of non-interactive zero-knowledge proofs of knowledge is the design of privacy-preserving non-interactive authentication schemes. This includes classical authentication protocols, but also P-signatures [4] and their many applications, such as anonymous credentials [4], group signatures [20], electronic cash [19], or anonymous authentication [48]. Our framework can lead to a variety of efficient new constructions of designated-verifier variants for the above applications without pairings, whereas all previous constructions either had to rely on the random oracle model, or use pairing-based cryptography.[2] In many scenarios of non-interactive authentication, the designated-verifier property is not an issue.

In addition, the aforementioned applications build upon the Groth-Sahai framework for NIZKs. However, Groth-Sahai NIZKs only satisfy a restricted notion of extractability, called f-extractability in [4]. As a result, constructions of privacy-preserving authentication mechanisms from Groth-Sahai NIZKs require a careful security analysis. Our framework leads to fully extractable zero-knowledge proofs, which could potentially simplify this. We note that our DVNIZKs are additionally randomizable, which has applications for delegatable anonymous credential schemes [3].

Other potential applications of our framework include round-efficient two-party computation protocols secure against malicious adversaries, electronic voting (see e.g. [17]), as well as designated-verifier variants of standard cryptographic primitives, such as verifiable encryption [13], or verifiable pseudorandom-functions [5]. Potential applications to the construction of adaptive oblivious transfers can also be envisioned: in [35], the authors mention that an adaptive oblivious transfer protocol can be designed by replacing the interactive zero-knowledge proofs of the protocol of [14] by non-interactive one. They raise two issues to this approach, namely, that Groth-Sahai proofs are only witness-indistinguishable for the required class of statements, and that they only satisfy a weak form of extractability. None of these restrictions apply to our DVNIZK constructions.

1.5 Related Work

Non-interactive zero-knowledge proofs were first introduced in [10]. Efficient publicly-verifiable non-interactive zero-knowledge proofs can be constructed in the random oracle model [31,32,47], or in the non-programmable random oracle model [42] (using a common reference string in addition). The latter construction was improved in [21]. In the standard model, the main construction of efficient publicly-verifiable NIZKs is the Groth-Sahai framework [39].

Designated-verifier non-interactive zero-knowledge arguments where first introduced in [46], where it was shown that the existence of semantically secure encryption implies the existence of DVNIZK arguments with bounded soundness; however, the construction is highly inefficient and therefore only of theoretical interest. Furthermore, even putting aside efficiency consideration, the construction is inherently limited to arguments (as opposed to proofs) with bounded soundness (as opposed to unbounded soundness).

[2] These applications typically require a proof-friendly signature scheme, but designated-verifier variants of such scheme can easily be constructed (without pairings) from algebraic MACs [18,40], by committing to the secret key of the MAC and proving knowledge of the committed value with a DVNIZK; such statements are naturally handled by our framework.

Designated-verifier NIZKs for linear languages can be constructed from hash proof systems [22,33,41]. Such NIZKs are perfectly zero-knowledge and statistically adaptively sound, but are not proofs of knowledge and are restricted to very specific statements, captured by linear equations.

Efficient designated-verifier NIZKs for more general statements were first described in [23]. The authors describe a general compiler that converts any three-round (honest-verifier) zero-knowledge protocol satisfying some (mild) requirements into a DVNIZK. However, the construction has several drawbacks: the soundness only holds under a very specific complexity-leveraging assumption, and only against adversaries making at most $O(\log \lambda)$ proofs (as already mentioned, the paper claims that the construction enjoy unbounded soundness as well, but this claim is flawed, see the full version [16]). In addition, the proofs obtained with this compiler are not proofs of knowledge.

In subsequent works [17,49], variations of the compilation technique of [23] are described, where the complexity-leveraging assumption was replaced by more standard assumptions (although achieving a more restricted type of soundness) by relying on encryption schemes with additional properties. Eventually, [43] removes some of the constraints of the constructions of [17], and provides new protocols that can be compiled using the transformation. However, all the constructions obtained in these papers are only computationally sound, do not enjoy unbounded soundness, and are not proofs of knowledge; this strongly limits their scope, and in particular, prevents them from being used in the previously discussed applications.

1.6 Organization

In Sect. 2, we introduce our notation, and necessary primitives. We refer the reader to the full version of this work [16] for classical preliminaries on commitments and cryptosystems. Section 2 also describes the notion of a DVNIZK-friendly encryption scheme, which is central to our framework. In Sect. 3, we introduce our framework for building DVNIZKs of knowledge over an abelian group, illustrate it with practical examples, and prove its security. In Sect. 4, we describe the dual variant of our framework for proving statements over plaintexts of a DVNIZK-friendly encryption scheme. In the full version of this work [16], we additionally describe optimizations on the efficiency of DVNIZKs for relations between plaintexts of a DVNIZK-friendly scheme, by eschewing unbounded soundness, as well as our attack on the unbounded soundness of [23].

2 Preliminaries

Throughout this paper, λ denotes the security parameter. A probabilistic polynomial time algorithm (PPT, also denoted *efficient* algorithm) runs in time polynomial in the (implicit) security parameter λ. A positive function f is *negligible* if for any polynomial p there exists a bound $B > 0$ such that, for any integer $k \geq B$, $f(k) \leq 1/|p(k)|$. An event depending on λ occurs with *overwhelming probability* when its probability is at least $1 - \text{negl}(\lambda)$ for a negligible function negl. Given a finite set S, the notation $x \xleftarrow{\$} S$ means a uniformly random assignment of an element of S to the variable x. We represent adversaries as interactive probabilistic Turing machines; the notation $\mathscr{A}^{\mathcal{O}}$ indicates that the machine $\mathscr{A}$ is given oracle access to $\mathcal{O}$. Adversaries will sometime output an arbitrary state st to capture stateful interactions.

Abelian Groups and Modules. We use additive notation for groups for convenience, and write $(\mathbb{G}, \boldsymbol{+})$ for an abelian group of order k. When it is clear from the context, we denote 0 its neutral element (otherwise, we denote it $0_{\mathbb{G}}$). We denote by $\bullet$ the scalar-multiplication algorithm (*i.e.* for any $(x, G) \in \mathbb{Z}_k \times \mathbb{G}$, $x \bullet G = G \boldsymbol{+} G \boldsymbol{+} \ldots \boldsymbol{+} G$, where the sum contains x terms). Observe that we can naturally view $\mathbb{G}$ as a $\mathbb{Z}_k$-module $(\mathbb{G}, \boldsymbol{+}, \bullet)$, for the ring $(\mathbb{Z}_k, +, \cdot)$. For simplicity, we write $\boldsymbol{-}G$ for $(-1) \bullet G$. We use lower case to denote elements of $\mathbb{Z}_k$, upper case to denote elements of $\mathbb{G}$, and bold notations to denote vectors. We extend the notations $(\boldsymbol{+}, \boldsymbol{-})$ to vectors and matrices in the natural way, and write $\boldsymbol{x} \bullet \boldsymbol{G}$ to denote the scalar product $x_1 \bullet G_1 \boldsymbol{+} \ldots \boldsymbol{+} x_t \bullet G_t$ (where $\boldsymbol{x}, \boldsymbol{G}$ are vectors of the same length t). For a vector $\boldsymbol{v}$, we denote by $\boldsymbol{v}^\intercal$ its transpose. By $\mathsf{GGen}(1^\lambda)$, we denote a probabilistic efficient algorithm that, given the security parameter λ, generates an abelian group $\mathbb{G}$ such that the best known algorithm for solving discrete logs in G takes time 2^λ. In the following, we write $(\mathbb{G}, k) \xleftarrow{\$} \mathsf{GGen}(1^\lambda)$. Additionally, we denote by $\mathsf{GGen}(1^\lambda, k)$ a group generation algorithm that allows us to select the order k beforehand.

RSA Groups. A *strong prime* is a prime $p = 2p' + 1$ such that p' is also a prime. We call *RSA modulus* a product $n = pq$ of two strong primes. We denote by φ Euler's totient function; it holds that $\varphi(n) = (p-1)(q-1)$. We denote by $\mathbb{J}_n$ the cyclic subgroup of $\mathbb{Z}_n^*$ of elements with Jacobi symbol 1 (the order of this group is $\varphi(n)/2$), and by QR_n the cyclic subroup of squares of $\mathbb{Z}_n^*$ (which is also a subgroup of $\mathbb{J}_n$ and has order $\varphi(n)/4$). By $\mathsf{Gen}(1^\lambda)$, we denote a probabilistic efficient algorithm that, given the security parameter λ, generates a strong RSA modulus n and secret parameters (p, q) where $n = pq$, such that the best known algorithm for factoring n takes time 2^λ. In the following, we write $(n, (p, q)) \xleftarrow{\$} \mathsf{Gen}(1^\lambda)$.

2.1 Encryption Schemes

The formal definition of an IND-CPA-secure public-key encryption scheme is recalled in the full version [16], but in short, a public-key encryption scheme S is a triple of PPT algorithms $(S.\mathsf{KeyGen}, S.\mathsf{Enc}, S.\mathsf{Dec})$, where $S.\mathsf{KeyGen}$ generates a pair $(\mathsf{ek}, \mathsf{dk})$ with an encryption key and a decryption key, decryption (with dk, deterministically) is the reverse operation of encryption (with ek, randomized), and no adversary can distinguish encryptions of one of two messages of its choice (IND-CPA security).

In this work, we will focus on additively homomorphic encryption schemes, which are homomorphic for both the message and the random coin. More formally, we require that the message space $\mathcal{M}$ and the random source $\mathcal{R}$ are integer sets $(\mathbb{Z}_M, \mathbb{Z}_R)$ for some integers (M, R), and that there exists an efficient operation $\oplus$ such that for any $(\mathsf{ek}, \mathsf{sk}) \xleftarrow{\$} \mathsf{KeyGen}(1^\lambda)$, any $(m_1, m_2) \in \mathbb{Z}_M^2$ and $(r_1, r_2) \in \mathbb{Z}_R^2$, denoting $(C_i)_{i \leq 2} \leftarrow (S.\mathsf{Enc}_{\mathsf{ek}}(m_i; r_i))_{i \leq 2}$, it holds that $C_1 \oplus C_2 = S.\mathsf{Enc}_{\mathsf{ek}}(m_1 + m_2 \bmod M; r_1 + r_2 \bmod R)$. We say an encryption scheme is *strongly additive* if it satisfies these requirements. Note that the existence of $\oplus$ implies (via a standard square-and-multiply method) the existence of an algorithm that, on input a ciphertext $C = S.\mathsf{Enc}_{\mathsf{ek}}(m; r)$ and an integer $\rho \in \mathbb{Z}$, outputs a ciphertext $C' = S.\mathsf{Enc}_{\mathsf{ek}}(\rho m \bmod M; \rho r \bmod R)$. We denote by $\rho \odot C$ the external multiplication of a ciphertext C by an integer ρ, and by $\ominus$ the operation $C \oplus (-1) \odot C'$ for two ciphertexts (C, C'). We will sometimes slightly abuse these notations, and write $C \oplus m$ (resp. $C \ominus m$) for a plaintext m to denote $C \oplus S.\mathsf{Enc}_{\mathsf{ek}}(m; 0)$ (resp. $C \ominus S.\mathsf{Enc}_{\mathsf{ek}}(m; 0)$).

A simple observation on strongly additively homomorphic encryption schemes is that IND-CPA security implies that R must either be equal to 0 mod M, or unknown given ek. Otherwise, an IND-CPA adversary would set $(m_0, m_1) = (0, 1)$ and check if $R \odot C$ equals $S.\mathsf{Enc}_{\mathsf{ek}}(0; 0)$ or $S.\mathsf{Enc}_{\mathsf{ek}}(R; 0)$.

The Paillier Encryption Scheme. The Paillier encryption scheme [45] is a well-known additively homomorphic encryption scheme over $\mathbb{Z}_n$ for an RSA modulus n. We describe here a standard variant [25, 43], where the random coin is an exponent over $\mathbb{J}_n$ rather than a group element. Note that the exponent space of $\mathbb{J}_n$ is $\mathbb{Z}_{\varphi(n)/2}$, which is a group of unknown order; however, it suffices to draw exponents at random from $\mathbb{Z}_{n/2}$ to get a distribution statistically close from uniform over $\mathbb{Z}_{\varphi(n)/2}$.

- $\mathsf{KeyGen}(1^\lambda)$: run $(n, (p, q)) \xleftarrow{\$} \mathsf{Gen}(1^\lambda)$, pick $g \xleftarrow{\$} \mathbb{J}_n$, set $h \leftarrow g^n \bmod n^2$, and compute $\delta \leftarrow n^{-1} \bmod \varphi(n)$ (n and $\varphi(n)$ are relatively prime). Return $\mathsf{ek} = (n, h)$ and $\mathsf{dk} = \delta$;
- $\mathsf{Enc}(\mathsf{ek}, m; r)$: given $m \in \mathbb{Z}_n$, for a random $r \xleftarrow{\$} \mathbb{Z}_{n/2}$, compute and output $c \leftarrow (1+n)^m \cdot h^r \bmod n^2$;
- $\mathsf{Dec}(\mathsf{dk}, c)$: compute $x \leftarrow c^{\mathsf{dk}} \bmod n$ and $c_0 \leftarrow [c \cdot x^{-n} \bmod n^2]$. Return $m \leftarrow (c_0 - 1)/n$.

Note that knowing dk is equivalent to knowing the factorization of n. The IND-CPA security of the Paillier encryption scheme reduces to the decisional composite residuosity (DCR) assumption, which states that it is computationally infeasible to distinguish random n'th powers over $\mathbb{Z}_{n^2}^*$ from random elements of $\mathbb{Z}_{n^2}^*$.[3] It is also strongly additive, where the homomorphic addition of ciphertexts is the multiplication over $\mathbb{Z}_{n^2}^*$.

[3] In the variant we consider here, we must restrict our attention to elements of $\mathbb{Z}_{n^2}^*$ which have Jacobi symbol 1 when reduced modulo n as $g \in \mathbb{J}_n$, but this can be checked in polynomial time anyway.

The ElGamal Encryption Scheme. We recall the additive variant of the famous ElGamal cryptosystem [28], over an abelian group $(\mathbb{G}, +)$ of order k.

- $\mathsf{KeyGen}(1^\lambda)$: pick $G \xleftarrow{\$} \mathbb{G}$, pick $s \xleftarrow{\$} \mathbb{Z}_k$, set $G \leftarrow s \bullet G$, and return $\mathsf{ek} = (G, H)$ and $\mathsf{dk} = s$;
- $\mathsf{Enc}(\mathsf{ek}, m; r)$: given $m \in \mathbb{Z}_k$, for a random $r \xleftarrow{\$} \mathbb{Z}_k$, output $\boldsymbol{C} \leftarrow (r \bullet G, (m \bullet G) + (r \bullet H))$;
- $\mathsf{Dec}(\mathsf{dk}, \boldsymbol{C})$: parse $\boldsymbol{C}$ as (C_0, C_1), and compute $M \leftarrow C_1 - (\mathsf{dk} \bullet C_0)$. Compute the discrete logarithm m of M in base G, and return m.

The $\mathsf{IND\text{-}CPA}$ security of the ElGamal encryption scheme reduces to the decisional Diffie-Hellman (DDH) assumption over $\mathbb{G}$, which states that it is computationally infeasible to distinguish tuples of the form $(G, H, x \bullet G, x \bullet H)$ for random x from uniformly random 4-tuples over $\mathbb{G}$. It is also strongly additive (and the homomorphic operation is the vector addition over $\mathbb{G}$). However, the decryption procedure is not efficient in general, as it requires to compute a discrete logarithm. For the decryption process to be efficient, the message m must be restricted to come from a subset of $\mathbb{Z}_k$ of polynomial size.

DVNIZK-Friendly Encryption Scheme. We say that a strongly additive encryption scheme is DVNIZK *-friendly*, when it satisfies the following additional properties:

- Coprimality Property: we require that the size M of the plaintext space and the size R of the random source are coprime[4], *i.e.*, $\gcd(M, R) = 1$;
- Decodable: for any $(\mathsf{ek}, \mathsf{sk}) \xleftarrow{\$} \mathsf{KeyGen}(1^\lambda)$, the function $f_{\mathsf{ek}} : m \mapsto \mathsf{Enc}_{\mathsf{ek}}(m; 0)$ must be efficiently invertible (*i.e.*, there is a PPT algorithm, which is given ek, computing f_{ek}^{-1} on any value from the image of f_{ek}).

One can observe that the Paillier cryptosystem is DVNIZK-friendly ($\gcd(n, \varphi(n)) = 1$, and any message m can be efficiently recovered from $\mathsf{Enc}_{\mathsf{ek}}(m; 0) = (1 + n)^m \bmod n^2$), while the ElGamal cryptosystem is not (it satisfies none of the above properties). Other DVNIZK-friendly cryptosystems include variants of the Paillier cryptosystem [12,22,24–26], and the more recent Castagnos-Laguillaumie cryptosystem [15], with prime-order plaintext space. For simplicity, we will also assume that all prime factors of the size M of the plaintext space of a DVNIZK-friendly cryptosystem are of superpolynomial size; our results can be extended to cryptosystems with a small plaintext space (or a plaintext space with small prime factors), but at a cost in efficiency. Note that by the homomorphic property, the decodability property implies that a plaintext can always be recovered from a ciphertext if the random coin is known.

2.2 Non-interactive Zero-Knowledge Proof Systems

In the definitions below, we focus on proof systems for NP-languages that admit an efficient (polynomial-time) prover. For an NP-language $\mathscr{L}$, we denote $R_{\mathscr{L}}$

[4] In view of our previous observation on $\mathsf{IND\text{-}CPA}$ security for strongly additive cryptosystems, this implies that R is secret.

its associated relation, *i.e.*, a polynomial-time algorithm which satisfies $\mathscr{L} = \{x \mid \exists w, |w| = \mathsf{poly}(|x|) \wedge R_{\mathscr{L}}(x, w) = 1\}$. It is well known that non-interactive proof systems cannot exist for non-trivial languages in the plain model [44]; our constructions will be described in the common reference string model. For conciseness, the common reference string is always implictly given as input to all algorithms. We note that all of our constructions can be readily adapted to work in the registered public-key model as well, a relaxation of the common reference string model introduced by Barak et al in [2].

While languages are naturally associated to *statements of membership*, the constructions of this paper will mainly consider *statements of knowledge*. We write $\mathsf{St}(x) = \text{ⴺ}\{w : R(x, w) = 1\}$ to denote the statement "I know a witness w such that $R(x, w) = 1$" for a word x and a polytime relation R. Similarly, we write $\mathsf{St}(x) = \exists\{w : R(x, w) = 1\}$ to denote the existential statement "there exists a witness w such that $R(x, w) = 1$".

Definition 1 *(Non-Interactive Zero-Knowledge Proof System). A non-interactive zero-knowledge (*NIZK*) proof system Π between for a family of languages $\mathscr{L} = \{\mathscr{L}_{\mathsf{crs}}\}_{\mathsf{crs}}$ is a quadruple of probabilistic polynomial-time algorithms* ($\Pi.\mathsf{Setup}, \Pi.\mathsf{KeyGen}, \Pi.\mathsf{Prove}, \Pi.\mathsf{Verify}$) *such that*

- $\Pi.\mathsf{Setup}(1^\lambda)$, *outputs a common reference string* crs *(which specifies the language $\mathscr{L}_{\mathsf{crs}}$),*
- $\Pi.\mathsf{KeyGen}(1^\lambda)$, *outputs a public key* pk *and a verification key* vk,
- $\Pi.\mathsf{Prove}(\mathsf{pk}, x, w)$, *on input the public key* pk, *a word $x \in \mathscr{L}_{\mathsf{crs}}$, and a witness w, outputs a proof π,*
- $\Pi.\mathsf{Verify}(\mathsf{pk}, \mathsf{vk}, x, \pi)$, *on input the public key* pk, *the verification key* vk, *a word x, and a proof π, outputs $b \in \{0, 1\}$,*

which satisfies the completeness, zero-knowledge, and soundness properties defined below.

We assume for simplicity that once it is generated, the common reference string crs is implicitly passed as an argument to the algorithms ($\Pi.\mathsf{KeyGen}, \Pi.\mathsf{Prove}, \Pi.\mathsf{Verify}$). In the above definition of NIZK proof systems, we let the key generation algorithm generate a verification key vk which is used by the verifier to check the proofs. We call *publicly verifiable non-interactive zero-knowledge proof system* a NIZK proof system in which vk is set to the empty string (or, equivalently, in which vk is made part of the public key). Otherwise, we call it a *designated-verifier non-interactive zero-knowledge proof system.*

Definition 2 *(Completeness). A* NIZK *proof system* $\Pi = (\Pi.\mathsf{Setup}, \Pi.\mathsf{KeyGen}, \Pi.\mathsf{Prove}, \Pi.\mathsf{Verify})$ *for a family of languages $\mathscr{L} = \{\mathscr{L}_{\mathsf{crs}}\}_{\mathsf{crs}}$ with relations R_{crs} satisfies the (perfect, statistical) completeness property if for $\mathsf{crs} \xleftarrow{\$} \Pi.\mathsf{Setup}(1^\lambda)$, for every $x \in \mathscr{L}_{\mathsf{crs}}$ and every witness w such that $R_{\mathsf{crs}}(x, w) = 1$,*

$$\Pr\begin{bmatrix}(\mathsf{pk}, \mathsf{vk}) \xleftarrow{\$} \Pi.\mathsf{KeyGen}(1^\lambda), \\ \pi \leftarrow \Pi.\mathsf{Prove}(\mathsf{pk}, x, w)\end{bmatrix} : \Pi.\mathsf{Verify}(\mathsf{pk}, \mathsf{vk}, x, \pi) = 1 \Bigg] = 1 - \mu(\lambda)$$

where $\mu(\lambda) = 0$ for perfect completeness, and $\mu(\lambda) = \mathrm{negl}(\lambda)$ for statistical completeness.

We now define the zero-knowledge property.

Definition 3 *(Composable Zero-Knowledge). A* NIZK *proof system* $\Pi = (\Pi.\mathsf{Setup}, \Pi.\mathsf{KeyGen}, \Pi.\mathsf{Prove}, \Pi.\mathsf{Verify})$ *for a family of languages* $\mathscr{L} = \{\mathscr{L}_{\mathsf{crs}}\}_{\mathsf{crs}}$ *with relations* R_{crs} *satisfies the (perfect, statistical) composable zero-knowledge property if for any* $\mathsf{crs} \xleftarrow{\$} \Pi.\mathsf{Setup}(1^\lambda)$, *there exists a probabilistic polynomial-time simulator* Sim *such that for any stateful adversary* $\mathscr{A}$,

$$\left| \Pr\left[\begin{array}{l}(\mathsf{pk},\mathsf{vk}) \xleftarrow{\$} \Pi.\mathsf{KeyGen}(1^\lambda),\\ (x,w) \leftarrow \mathscr{A}(\mathsf{pk},\mathsf{vk}),\\ \pi \leftarrow \Pi.\mathsf{Prove}(\mathsf{pk},x,w)\end{array} : (R_{\mathsf{crs}}(x,w)=1) \wedge (\mathscr{A}(\pi)=1)\right] - \Pr\left[\begin{array}{l}(\mathsf{pk},\mathsf{vk}) \xleftarrow{\$} \Pi.\mathsf{KeyGen}(1^\lambda),\\ (x,w) \leftarrow \mathscr{A}(\mathsf{pk},\mathsf{vk}),\\ \pi \leftarrow \mathsf{Sim}(\mathsf{pk},\mathsf{vk},x)\end{array} : (R_{\mathsf{crs}}(x,w)=1) \wedge (\mathscr{A}(\pi)=1)\right]\right| \le \mu(\lambda)$$

where $\mu(\lambda) = 0$ *for perfect composable zero-knowledge, and* $\mu(\lambda) = \mathrm{negl}(\lambda)$ *for statistical composable zero-knowledge. If the composable zero-knowledge property holds against efficient (PPT) verifiers, the proof system satisfies computational composable zero-knowledge.*

The composable zero-knowledge property was first introduced in [36]. It strenghtens the standard zero-knowledge definition, in that it explicitly states that the trapdoor of the simulator is exactly the verification key vk of the verifier. This strong security property guarantees that the same common reference string can be used for many different proofs, as the same trapdoor is used for simulating all proofs, which enhances the proof system with composability properties. We note that [36] additionally required indistinguishability between real and simulated common reference string; in our constructions, this will be trivially satisfied, as the simulated crs will be exactly the real one. We define below the notion of (bounded) adaptive soundness, which allows the input to be adversarially picked after the public key is fixed.

Definition 4 *(Bounded Adaptive Soundness). A* NIZK *proof system* $\Pi = (\Pi.\mathsf{Setup}, \Pi.\mathsf{KeyGen}, \Pi.\mathsf{Prove}, \Pi.\mathsf{Verify})$ *for a family of languages* $\mathscr{L} = \{\mathscr{L}_{\mathsf{crs}}\}_{\mathsf{crs}}$ *with relations* R_{crs} *satisfies the bounded adaptive soundness property if for* $\mathsf{crs} \xleftarrow{\$} \mathsf{Setup}(1^\lambda)$, *for every adversary* $\mathscr{A}$,

$$\Pr\left[\begin{array}{l}(\mathsf{pk},\mathsf{vk}) \xleftarrow{\$} \Pi.\mathsf{KeyGen}(1^\lambda),\\ (\pi,x) \leftarrow \mathscr{A}(\mathsf{pk})\end{array} : x \notin \mathscr{L}_{\mathsf{crs}} \wedge \Pi.\mathsf{Verify}(\mathsf{pk},\mathsf{vk},x,\pi)\right] = \mathrm{negl}(\lambda).$$

Definition 4 is formulated with respect to arbitrary adversaries $\mathscr{A}$, which leads to a statistical notion of soundness. A natural relaxation of this requirement is to consider only *efficient* (PPT) adversarial provers. We denote by *computational soundness* this relaxed notion of soundness. Computationally sound proof systems are called *argument systems*.

Unbounded Soundness. Definition 4 corresponds to a *bounded* notion of soundness, in the sense that soundness is only guaranteed to hold when the prover tries to forge a single proof of a wrong statement, right after the setup phase. However, if the prover is allowed to interact polynomially many times with the verifier before trying to forge a proof, sending proofs and receiving feedback on whether the proof was accepted, the previous definition provides no security guarantees.

Intuitively, in this situation, the distinction between bounded and unbounded soundness is comparable to the distinction between security against chosen plaintext attacks and security against chosen ciphertext attacks for cryptosystems. We define unbounded soundness in a similar fashion, by giving the prover access to a verification oracle $\mathcal{O}_{\mathsf{vk}}[\mathsf{pk}]$ (with crs implicitly given as parameter) which, on input (x, π), returns $b \leftarrow \mathsf{Verify}(\mathsf{pk}, \mathsf{vk}, x, \pi)$.

Definition 5 *(Q-bounded Adaptive Soundness). A* NIZK *proof system* $\Pi = (\Pi.\mathsf{Setup}, \Pi.\mathsf{KeyGen}, \Pi.\mathsf{Prove}, \Pi.\mathsf{Verify})$ *for a family of languages* $\mathscr{L} = \{\mathscr{L}_{\mathsf{crs}}\}_{\mathsf{crs}}$ *with relations* R_{crs} *satisfies the Q-bounded adaptive soundness property if for* $\mathsf{crs} \xleftarrow{\$} \Pi.\mathsf{Setup}(1^\lambda)$*, and every adversary* $\mathscr{A}$ *making at most Q queries to* $\mathcal{O}_{\mathsf{vk}}[\mathsf{pk}]$*, it holds that*

$$\Pr\left[\begin{matrix}(\mathsf{pk}, \mathsf{vk}) \xleftarrow{\$} \Pi.\mathsf{KeyGen}(1^\lambda), \\ (\pi, x) \leftarrow \mathscr{A}^{\mathcal{O}_{\mathsf{vk}}[\mathsf{pk}]}(\mathsf{pk})\end{matrix} \quad : x \notin \mathscr{L}_{\mathsf{crs}} \wedge \Pi.\mathsf{Verify}(\mathsf{pk}, \mathsf{vk}, x, \pi)\right] = \mathrm{negl}(\lambda).$$

Alternatively, the above definition can be formulated with respect to polynomial-time adversarial provers, leading to computational Q-bounded adaptive soundness. Note that the answers of the oracle are bits; therefore, if a NIZK proof system satisfies the bounded adaptive soundness property of Definition 4, it also satisfies the above Q-bounded adaptive soundness property for any $Q = O(\log \lambda)$. Indeed, if Q is logarithmic, one can always guess in advance the answers of the verification oracle with non-negligible (inverse polynomial) probability. We say that a NIZK proof system which is Q-bounded adaptively sound for any $Q = \mathsf{poly}(\lambda)$ satisfies *unbounded adaptive soundness.*

Eventually, we define (unbounded) *knowledge-extractability*, a strenghtening of the soundness property which guarantees that if the prover produces an accepting proof, then the simulator can actually *extract* a witness for the statement. To this aim, we extend the syntax of the Setup algorithm to also output a trapdoor τ, used by the extractor. The knowledge-extractibility guarantee is stronger than soundness, in that the proof guarantees not only that there exists a witness, but also that the prover must know that witness. A NIZK satisfying knowledge-extractability is called a NIZK proof of knowledge.

Definition 6 *(Q-bounded Knowledge-Extractability). A* NIZK *proof system* $\Pi = (\Pi.\mathsf{Setup}, \Pi.\mathsf{KeyGen}, \Pi.\mathsf{Prove}, \Pi.\mathsf{Verify})$ *for a family of languages* $\mathscr{L} = \{\mathscr{L}_{\mathsf{crs}}\}_{\mathsf{crs}}$ *with relations* R_{crs} *satisfies the Q-bounded knowledge-extractability property if for* $(\mathsf{crs}, \tau) \xleftarrow{\$} \Pi.\mathsf{Setup}(1^\lambda)$*, and every adversary* $\mathscr{A}$ *making at most Q queries to*

$\mathcal{O}_{\mathsf{vk}}[\mathsf{pk}]$, *there is an efficient extractor* Ext *such that*

$$\Pr\left[\begin{array}{l}(\mathsf{pk},\mathsf{vk}) \xleftarrow{\$} \Pi.\mathsf{KeyGen}(1^\lambda),\\ (\pi,x) \leftarrow \mathscr{A}^{\mathcal{O}_{\mathsf{vk}}[\mathsf{pk}]}(\mathsf{pk}),\\ w \leftarrow \mathsf{Ext}(\pi,x,\tau),\end{array} : R_{\mathsf{crs}}(x,w) \textit{ iff } \Pi.\mathsf{Verify}(\mathsf{pk},\mathsf{vk},x,\pi)\right] \approx 1.$$

3 A Framework for Designated-Verifier Non-interactive Zero-Knowledge Proofs of Knowledge

In this section, we let k be an integer, $(\mathbb{G}, +)$ be an abelian group of order k, and (α, β, γ) be three integers. We will describe a framework for proving statements of knowledge over a wide variety of algebraic relations over $\mathbb{G}$, in the spirit of the Groth-Sahai framework for NIZK proofs over bilinear groups. To describe the relations handled by our framework, we describe languages of algebraic relations via linear maps. While this system was previously used to describe membership statements [7–9], we adapt it to statements of knowledge. As previously observed in [7], this system encompasses a wider class of languages than the Groth-Sahai framework.

3.1 Statements Defined by a Linear Map over $\mathbb{G}$

Let $\boldsymbol{G} \in \mathbb{G}^\alpha$ denote a vector of *public parameters*, and let $\boldsymbol{C} \in \mathbb{G}^\beta$ denote a public *word*. We will consider statements $\mathsf{St}_\Gamma(\boldsymbol{G}, \boldsymbol{C})$ defined by a linear map $\Gamma : (\mathbb{G}^\alpha, \mathbb{G}^\beta) \mapsto \mathbb{G}^{\gamma\times\beta}$ as follows:

$$\mathsf{St}_\Gamma(\boldsymbol{G}, \boldsymbol{C}) = \text{\reflectbox{N}}\{\boldsymbol{x} \in \mathbb{Z}_k^\gamma \mid \boldsymbol{x} \bullet \Gamma(\boldsymbol{G}, \boldsymbol{C}) = \boldsymbol{C}\} \quad (1)$$

That is, the prover knows a witness-vector $\boldsymbol{x} \in \mathbb{Z}_k^\gamma$ such that the equation $\boldsymbol{x} \bullet \Gamma(\boldsymbol{G}, \boldsymbol{C}) = \boldsymbol{C}$ holds. This abstraction captures a wide class of statements. Below, we describe two examples of statements that can be handled by our framework. They aim at clarifying the way the framework can be used, illustrating its power, as well as providing useful concrete instantiations. The examples focus on the most standard primitives (Pedersen commitments, ElGamal ciphertexts), but the reader will easily recognize they can be naturaly generalized to all standard variants of these primitives (e.g., variants of ElGamal secure under t-linear assumptions [11], or under assumptions from the matrix Diffie-Hellman family of assumptions [29]).

Example 1: Knowledge of Opening to a Pedersen Commitment. We consider statements of knowledge of an opening (m, r) to a Pedersen commitment C.

- Public Parameters: $(G, H) \in \mathbb{G}^2$;
- Word: $C \in \mathbb{G}$;
- Witness: a pair $(m, r) \in \mathbb{Z}_k^2$ such that $C = m \bullet G + r \bullet H$;
- Linear Map: $\Gamma_{\mathsf{Ped}} : (G, H, C) \mapsto (G, H)^\intercal$;
- Statement: $\mathsf{St}_{\Gamma_{\mathsf{Ped}}}(G, H, C) = \text{\reflectbox{N}}\{(m, r) \in \mathbb{Z}_k^2 \mid (m, r) \bullet (G, H)^\intercal = C\}$.

Example 2: Multiplicative Relationship Between ElGamal Ciphertexts. This type of statement is of particular interest, as it can be generalized to arbitrary (polynomial) relationships between plaintexts.

- Public Parameters: $(G, H) \in \mathbb{G}^2$;
- Word: $\boldsymbol{C} = ((U_i, V_i)_{0 \le i \le 2}) \in \mathbb{G}^6$;
- Witness: a 5-tuple $\boldsymbol{x} = (m_0, r_0, m_1, r_1, r_2) \in \mathbb{Z}_k^5$ such that $U_i = r_i \bullet G$ and $V_i = m_i \bullet G + r \bullet H$ for $i = 0, 1$, and $U_2 = m_1 \bullet U_0 + r_2 \bullet G$, $V_2 = m_1 \bullet V_0 + r_2 \bullet H$;
- Linear Map:

$$\Gamma_{\mathsf{EM}} : (G, H, \boldsymbol{C}) \mapsto \begin{pmatrix} 0 & G & 0 & 0 & 0 & 0 \\ G & H & 0 & 0 & 0 & 0 \\ 0 & 0 & 0 & G & U_0 & V_0 \\ 0 & 0 & G & H & 0 & 0 \\ 0 & 0 & 0 & 0 & G & H \end{pmatrix};$$

- Statement: $\mathsf{St}_{\Gamma_{\mathsf{EM}}}(G, H, \boldsymbol{C}) = \text{Ж}\{\boldsymbol{x} \in \mathbb{Z}_k^5 \mid \boldsymbol{x} \bullet \Gamma_{\mathsf{EM}}(G, H, \boldsymbol{C}) = \boldsymbol{C}\}$.

Conjunction of Statements. The above framework naturally handles conjuctions. Consider two statements $(\mathsf{St}_{\Gamma_0}(\boldsymbol{G_0}, \boldsymbol{C_0}), \mathsf{St}_{\Gamma_1}(\boldsymbol{G_1}, \boldsymbol{C_1}))$, defined by linear maps (Γ_0, Γ_1), with public paramcters $(\boldsymbol{G_1}, \boldsymbol{G_1})$, words $(\boldsymbol{C_0}, \boldsymbol{C_1})$, and witnesses $(\boldsymbol{x_0}, \boldsymbol{x_1})$. Let $\boldsymbol{G} \leftarrow (\boldsymbol{G_1}, \boldsymbol{G_1})$, $\boldsymbol{C} \leftarrow (\boldsymbol{C_0}, \boldsymbol{C_1})$, and $\boldsymbol{x} \leftarrow (\boldsymbol{x_0}, \boldsymbol{x_1})$. We construct the linear map Γ associated to $\mathsf{St}_\Gamma(\boldsymbol{G}, \boldsymbol{C})$ as $\Gamma \leftarrow ((\Gamma_0, 0)^\intercal, (0, \Gamma_1)^\intercal)$. One can immediatly observe that $\mathsf{St}_\Gamma(\boldsymbol{G}, \boldsymbol{C}) = \mathsf{St}_{\Gamma_0}(\boldsymbol{G_0}, \boldsymbol{C_0}) \wedge \mathsf{St}_{\Gamma_1}(\boldsymbol{G_1}, \boldsymbol{C_1})$. The framework handles disjunction of statements as well, as observed in [1]; we omit the details.

3.2 A Framework for DVNIZK Proofs of Knowledge

We now introduce our framework for constructing designated-verifier non-interactive zero-knowledge proofs of knowledge for statements defined by a linear map over $\mathbb{G}$. Let $S = (S.\mathsf{KeyGen}, S.\mathsf{Enc}, S.\mathsf{Dec})$ denote a DVNIZK-friendly encryption scheme with plaintext space $\mathbb{Z}_k$. We construct a DVNIZK of knowledge $\Pi_\mathsf{K} = (\Pi_\mathsf{K}.\mathsf{Setup}, \Pi_\mathsf{K}.\mathsf{KeyGen}, \Pi_\mathsf{K}.\mathsf{Prove}, \Pi_\mathsf{K}.\mathsf{Verify})$ for a statement $\mathsf{St}_\Gamma(\boldsymbol{G}, \boldsymbol{C})$ over a word $\boldsymbol{C} \in \mathbb{G}^\beta$, with public parameters $\boldsymbol{G} \in \mathbb{G}^\alpha$, defined by a linear map $\Gamma : (\mathbb{G}^\alpha, \mathbb{G}^\beta) \mapsto \mathbb{G}^{\gamma \times \beta}$. Our construction proceeds as follows:

- $\Pi_\mathsf{K}.\mathsf{Setup}(1^\lambda)$: compute $(\mathsf{ek}, \mathsf{dk}) \xleftarrow{\$} S.\mathsf{KeyGen}(1^\lambda)$. Output $\mathsf{crs} \leftarrow \mathsf{ek}$. Note that ek defines a plaintext space $\mathbb{Z}_k$ and a random source $\mathbb{Z}_R$. As the IND-CPA and strong additive properties of S require R to be unknown, we assume that a bound B on R is publicly available. We denote $\ell \leftarrow 2^\lambda k B$.
- $\Pi_\mathsf{K}.\mathsf{KeyGen}(1^\lambda)$: pick $e \leftarrow \mathbb{Z}_\ell$, set $\mathsf{pk} \leftarrow S.\mathsf{Enc}_\mathsf{ek}(0; e)$ and $\mathsf{vk} \leftarrow e$.
- $\Pi_\mathsf{K}.\mathsf{Prove}(\mathsf{pk}, \boldsymbol{C}, \boldsymbol{x})$: on a word $\boldsymbol{C} \in \mathbb{Z}_k^\beta$, with witness $\boldsymbol{x}$ for the statement $\mathsf{St}_\Gamma(\boldsymbol{G}, \boldsymbol{C})$, pick $\boldsymbol{x'} \xleftarrow{\$} \mathbb{Z}_k^\gamma$, $\boldsymbol{r} \xleftarrow{\$} \mathbb{Z}_{2^\lambda B}^\gamma$, compute

$$\boldsymbol{X} \leftarrow S.\mathsf{Enc}_\mathsf{ek}(\boldsymbol{x}, \boldsymbol{r}), \quad \boldsymbol{X'} \leftarrow S.\mathsf{Enc}_\mathsf{ek}(\boldsymbol{x'}, 0) \ominus (\boldsymbol{r} \odot \mathsf{pk}), \quad \boldsymbol{C'} \leftarrow \boldsymbol{x'} \bullet \Gamma(\boldsymbol{G}, \boldsymbol{C}),$$

and output $\boldsymbol{\pi} \leftarrow (\boldsymbol{X}, \boldsymbol{X'}, \boldsymbol{C'})$.

- $\Pi_{\mathsf{K}}.\mathsf{Verify}(\mathsf{pk}, \mathsf{vk}, \boldsymbol{C}, \boldsymbol{\pi})$: parse $\boldsymbol{\pi}$ as $(\boldsymbol{X}, \boldsymbol{X}', \boldsymbol{C}')$. Check that $e \odot \boldsymbol{X} \oplus \boldsymbol{X}'$ is decodable, and decode it to a vector $\boldsymbol{d} \in \mathbb{Z}_k^\gamma$. Check that

$$\boldsymbol{d} \bullet \Gamma(\boldsymbol{G}, \boldsymbol{C}) = e \bullet \boldsymbol{C} \boldsymbol{+} \boldsymbol{C}'.$$

If all checks succeeded, accept. Otherwise, reject.

The proof $\boldsymbol{\pi}$ consists of 2γ ciphertexts of S, and β elements of $\mathbb{G}$. Below, we illustrate our construction of DVNIZK on the examples of statements given in the previous section. For the sake of concreteness, we instantiate the DVNIZK-friendly encryption scheme S with Paillier (hence the operation $\boldsymbol{+}$ is instantiated as the multiplication modulo n^2), so that the message space is $\mathbb{Z}_n$ and the randomizer space is $\mathbb{Z}_{\varphi(n)/2}$ for an RSA modulus n. In the examples, we use a bound $B = n$ and draw Paillier random coins from $\mathbb{Z}_{2^\lambda B}$, following our generic framework. However, observe that in the case of Paillier, we can also draw the coins from $\mathbb{Z}_{n/2}$ to get a distribution statistically close to uniform over $\mathbb{Z}_{\varphi(n)/2}$, which is more efficient.

Example 1: Knowledge of Opening to a Pedersen Commitment.

- $\Pi_{\mathsf{Ped}}.\mathsf{Setup}(1^\lambda)$: Compute $((n, h), \delta) = (\mathsf{ek}, \mathsf{dk}) \stackrel{\$}{\leftarrow} S.\mathsf{KeyGen}(1^\lambda)$. Output $\mathsf{crs} \leftarrow \mathsf{ek}$. Let $\ell \leftarrow 2^\lambda n^2$. Let $\mathbb{G} \stackrel{\$}{\leftarrow} \mathsf{GGen}(1^\lambda, n)$, $(G, H) \stackrel{\$}{\leftarrow} \mathbb{G}^2$.
- $\Pi_{\mathsf{Ped}}.\mathsf{KeyGen}(1^\lambda)$: pick $e \stackrel{\$}{\leftarrow} \mathbb{Z}_\ell$, set $\mathsf{pk} \leftarrow h^e \bmod n^2$ and $\mathsf{vk} \leftarrow e$.
- $\Pi_{\mathsf{Ped}}.\mathsf{Prove}(\mathsf{pk}, C, \boldsymbol{x})$: on a word $C \in \mathbb{G}$, with witness $\boldsymbol{x} = (m, r) \in \mathbb{Z}_n^2$ for the statement $\mathsf{St}_{\Gamma_{\mathsf{Ped}}}(\boldsymbol{G}, \boldsymbol{C})$, pick $\boldsymbol{x}' \stackrel{\$}{\leftarrow} \mathbb{Z}_n^2$, $\boldsymbol{\rho} \stackrel{\$}{\leftarrow} \mathbb{Z}_{2^\lambda B}^2$, compute $\boldsymbol{X} \leftarrow (1 + n)^{\boldsymbol{x}} h^{\boldsymbol{\rho}} \bmod n^2, \boldsymbol{X}' \leftarrow (1+n)^{\boldsymbol{x}'} \mathsf{pk}^{-\boldsymbol{\rho}} \bmod n^2, \boldsymbol{C}' \leftarrow \boldsymbol{x}' \bullet (G, H)^\intercal$, and output $\boldsymbol{\pi} \leftarrow (\boldsymbol{X}, \boldsymbol{X}', \boldsymbol{C}')$.
- $\Pi_{\mathsf{Ped}}.\mathsf{Verify}(\mathsf{pk}, \mathsf{vk}, \boldsymbol{C}, \boldsymbol{\pi})$: parse $\boldsymbol{\pi}$ as $(\boldsymbol{X}, \boldsymbol{X}', \boldsymbol{C}')$. Check that $\boldsymbol{X}^e \boldsymbol{X}'$ is of the form $(1+n)^{\boldsymbol{d}}$, and recover the vector $\boldsymbol{d} \in \mathbb{Z}_n^2$. Check that $\boldsymbol{d} \bullet (G, H)^\intercal = e \bullet \boldsymbol{C} \boldsymbol{+} \boldsymbol{C}'$.

Example 2: Multiplicative Relationship Between ElGamal Ciphertexts.

- $\Pi_{\mathsf{EM}}.\mathsf{Setup}(1^\lambda)$ as $\Pi_{\mathsf{Ped}}.\mathsf{Setup}(1^\lambda)$.
- $\Pi_{\mathsf{EM}}.\mathsf{KeyGen}(1^\lambda)$ as $\Pi_{\mathsf{Ped}}.\mathsf{KeyGen}(1^\lambda)$.
- $\Pi_{\mathsf{EM}}.\mathsf{Prove}(\mathsf{pk}, \boldsymbol{C}, \boldsymbol{x})$: on a word $\boldsymbol{C} \in \mathbb{G}^6$, with witness $\boldsymbol{x} = (m_0, r_0, m_1, r_1, r_2) \in \mathbb{Z}_n^5$ for the statement $\mathsf{St}_{\Gamma_{\mathsf{EM}}}(\boldsymbol{G}, \boldsymbol{C})$, pick $\boldsymbol{x}' \stackrel{\$}{\leftarrow} \mathbb{Z}_n^5$, $\boldsymbol{\rho} \stackrel{\$}{\leftarrow} \mathbb{Z}_{2^\lambda B}^5$, compute $\boldsymbol{X} \leftarrow (1+n)^{\boldsymbol{x}} h^{\boldsymbol{\rho}} \bmod n^2, \boldsymbol{X}' \leftarrow (1+n)^{\boldsymbol{x}} \mathsf{pk}^{-\boldsymbol{\rho}} \bmod n^2, \boldsymbol{C}' \leftarrow \boldsymbol{x}' \bullet \Gamma_{\mathsf{EM}}(\boldsymbol{G}, \boldsymbol{C})$, and output $\boldsymbol{\pi} \leftarrow (\boldsymbol{X}, \boldsymbol{X}', \boldsymbol{C}')$.
- $\Pi_{\mathsf{EM}}.\mathsf{Verify}(\mathsf{pk}, \mathsf{vk}, \boldsymbol{C}, \boldsymbol{\pi})$: parse $\boldsymbol{\pi}$ as $(\boldsymbol{X}, \boldsymbol{X}', \boldsymbol{C}')$. Check that $\boldsymbol{X}^e \boldsymbol{X}'$ is of the form $(1+n)^{\boldsymbol{d}}$, and recover the vector $\boldsymbol{d} \in \mathbb{Z}_n^5$. Check that $\boldsymbol{d} \bullet \Gamma_{\mathsf{EM}}(\boldsymbol{G}, \boldsymbol{C}) = e \bullet \boldsymbol{C} \boldsymbol{+} \boldsymbol{C}'$.

3.3 Security Proof

We now prove the generic DVNIZK construction from Sect. 3.2 is secure.

Perfect Completeness. It follows from straighforward calculations: $e \odot \boldsymbol{X} \oplus \boldsymbol{X}' = S.\mathsf{Enc}_{\mathsf{ek}}(e \cdot \boldsymbol{x} + \boldsymbol{x}'; e \cdot \boldsymbol{r} - e \cdot \boldsymbol{r}) = S.\mathsf{Enc}_{\mathsf{ek}}(e \cdot \boldsymbol{x} + \boldsymbol{x}'; 0)$ is decodable and decodes to $\boldsymbol{d} = e \cdot \boldsymbol{x} + \boldsymbol{x}' \bmod k$. Then, $\boldsymbol{d} \bullet \Gamma(\boldsymbol{G}, \boldsymbol{C}) = e \bullet (\boldsymbol{x} \bullet \Gamma(\boldsymbol{G}, \boldsymbol{C})) + \boldsymbol{x}' \bullet \Gamma(\boldsymbol{G}, \boldsymbol{C}) = e \bullet \boldsymbol{C} + \boldsymbol{C}'$ by the correctness of the statement $(\boldsymbol{x} \bullet \Gamma(\boldsymbol{G}, \boldsymbol{C}) = \boldsymbol{C})$ and by construction of $\boldsymbol{C}'$.

Composable Zero-Knowledge. We prove the following theorem:

Theorem 7 (Zero-Knowledge of Π_{K}). *If the encryption scheme S is* IND-CPA *secure, the* DVNIZK *scheme Π_{K} is composable zero-knowledge.*

We describe a simulator $\mathsf{Sim}(\boldsymbol{C}, \mathsf{pk}, \mathsf{vk})$ producing proofs computationally indistinguishable from those produced by an honest prover on true statements. The simulator operates as follows: let $\boldsymbol{d} \xleftarrow{\$} \mathbb{Z}_k^{\gamma}$, and $\boldsymbol{C}' \leftarrow \boldsymbol{d} \bullet \Gamma(\boldsymbol{G}, \boldsymbol{C}) - e \bullet \boldsymbol{C}$. Sample $\boldsymbol{x} \xleftarrow{\$} \mathbb{Z}_k^{\gamma}$, $\boldsymbol{r} \xleftarrow{\$} \mathbb{Z}_{2^{\lambda}B}^{\gamma}$, and compute $\boldsymbol{X} \leftarrow S.\mathsf{Enc}_{\mathsf{ek}}(\boldsymbol{x}, \boldsymbol{r}), \boldsymbol{X}' \leftarrow S.\mathsf{Enc}_{\mathsf{ek}}(\boldsymbol{d} - \boldsymbol{e} \cdot \boldsymbol{x}, -\boldsymbol{e} \cdot \boldsymbol{r})$. Output $\pi_s = (\boldsymbol{X}, \boldsymbol{X}', \boldsymbol{C}')$.

Let $\mathscr{A}$ be an adversary that can distinguish Sim from Prove. We will build a reduction against the IND-CPA security of S. The reduction obtains $\boldsymbol{C}, \boldsymbol{x}$ from $\mathscr{A}$, samples $\tilde{\boldsymbol{x}} \leftarrow \mathbb{Z}_k^{\gamma}$, sends $(\boldsymbol{x}, \tilde{\boldsymbol{x}})$ to the IND-CPA game and sets $\boldsymbol{X}$ to be the challenge from the IND-CPA game. Now, the reduction samples $\boldsymbol{d} \leftarrow \mathbb{Z}_k^{\gamma}$ and sets $\boldsymbol{X}' := S.\mathsf{Enc}_{\mathsf{ek}}(\boldsymbol{d}; 0) \ominus \boldsymbol{X} \odot e$. Finally, the reduction sets $\boldsymbol{C}' := \boldsymbol{d} \bullet \Gamma(\boldsymbol{G}, \boldsymbol{C}) - e \bullet \boldsymbol{C}$. Send $\pi^* = (\boldsymbol{X}, \boldsymbol{X}', \boldsymbol{C})$ to $\mathscr{A}$.

Direct calculation shows that if the IND-CPA game outputs an encryption of $\tilde{\boldsymbol{X}}$, then $\boldsymbol{X}, \boldsymbol{X}', \boldsymbol{C}$ are distributed as those produced by Sim, whereas when it outputs an encryption of $\boldsymbol{X}$ then π^* is distributed identical to a real proof. Thus, whatever advantage $\mathscr{A}$ has in distinguishing Sim from Prove is also achieved by the reduction against IND-CPA. Note that for simplicity, our proof assume that the IND-CPA game is directly played over vectors, but standard methods allow to reduce this to the classical IND-CPA game with a single challenge ciphertext.

Adaptive Unbounded Knowledge-Extractability. We start by showing that Π_{K} satisfies statistical adaptive unbounded knowledge-extractability. More precisely, we prove the following theorem:

Theorem 8 (Soundness of Π_{K}). *There is an efficient simulator* Sim *such that for any (possibly unbounded) adversary $\mathscr{A}$ that outputs an accepting proof $\boldsymbol{\pi}$ with probability ε on an arbitrary word $\boldsymbol{C}$ after making at most Q queries to the oracle $\mathcal{O}_{\mathsf{vk}}[\mathsf{pk}]$,* Sim *extracts a valid witness for the statement $\mathsf{St}_{\Gamma}(\boldsymbol{G}, \boldsymbol{C})$ with probability at least $\varepsilon - (Q+1)\beta/p_k$, where p_k is the smallest prime factor of k.*

The proof describes an efficient simulator Sim that correctly emulates the verifier, without knowing vk mod k. The simulation is done as follows:

- $\mathsf{Sim.Setup}(1^{\lambda})$: compute $(\mathsf{ek}, \mathsf{dk}) \xleftarrow{\$} S.\mathsf{KeyGen}(1^{\lambda})$. Output $\mathsf{crs} \leftarrow \mathsf{ek}$. The encryption key ek defines a plaintext space $\mathbb{Z}_k$ and a random source $\mathbb{Z}_R$ with bound B. Let $\ell \leftarrow 2^{\lambda}kB$.

- $\mathsf{Sim.KeyGen}(1^\lambda)$: compute $(\mathsf{pk}, \mathsf{vk}) \xleftarrow{\$} \Pi_\mathsf{K}.\mathsf{KeyGen}(1^\lambda)$, output pk, store $e_R \leftarrow \mathsf{vk} \bmod R$, and erase vk.
- $\mathsf{Sim.Verify}(\mathsf{pk}, \mathsf{dk}, e_R, \boldsymbol{C}, \boldsymbol{\pi})$: parse $\boldsymbol{\pi}$ as $(\boldsymbol{X}, \boldsymbol{X}', \boldsymbol{C}')$. Using the secret key dk of S, decrypt $\boldsymbol{X}$ to a vector $\boldsymbol{x}$, and $\boldsymbol{X}'$ to a vector $\boldsymbol{x}'$. Check that $(-e_R) \odot (\boldsymbol{X} \ominus \boldsymbol{x}) = \boldsymbol{X}' \ominus \boldsymbol{x}'$. Check that $\boldsymbol{x} \bullet \Gamma(\boldsymbol{G}, \boldsymbol{C}) = \boldsymbol{C}$, and that $\boldsymbol{x}' \bullet \Gamma(\boldsymbol{G}, \boldsymbol{C}) = \boldsymbol{C}'$. If all checks succeeded, accept. Otherwise, reject.

The simulator Sim first calls $\mathsf{Sim.Setup}(1^\lambda)$ to generate the common reference string (note that our simulator generates the common reference string honestly, hence the simulation of Setup cannot be distinguished from an honest run of Setup), and stores dk. Each time the adversary $\mathscr{A}$ sends a query $(\boldsymbol{C}, \boldsymbol{\pi})$ to the oracle $\mathcal{O}_{\mathsf{vk}}[\mathsf{pk}]$, Sim simulates $\mathcal{O}_{\mathsf{vk}}[\mathsf{pk}]$ (without knowing $\mathsf{vk} \bmod k$) by running $\mathsf{Sim.Verify}(\mathsf{pk}, \mathsf{dk}, e_R, \boldsymbol{C}, \boldsymbol{\pi})$, and accepts or rejects accordingly. When $\mathscr{A}$ outputs a final answer $(\boldsymbol{C}, \boldsymbol{\pi})$, Sim computes a witness $\boldsymbol{x}$ for $\mathsf{St}_\Gamma(\boldsymbol{G}, \boldsymbol{C})$ by decrypting $\boldsymbol{C}$ with dk.

Observe that the distribution $\{(\mathsf{pk}, \mathsf{vk}) \xleftarrow{\$} \Pi_\mathsf{K}.\mathsf{KeyGen}(1^\lambda), e_k \leftarrow \mathsf{vk} \bmod k : (\mathsf{pk}, e_k)\}$ is statistically indistinguishable from the distribution $\{(\mathsf{pk}, \mathsf{vk}) \xleftarrow{\$} \Pi_\mathsf{K}.\mathsf{KeyGen}(1^\lambda), e_k \xleftarrow{\$} \mathbb{Z}_k : (\mathsf{pk}, e_k)\}$. Put otherwise, the distribution of $\mathsf{vk} \bmod k$ is statistically indistinguishable from random, even given pk. Indeed, as S is a DVNIZK-friendly encryption scheme, it holds by definition that $\gcd(k, R) = 1$. As $\ell = 2^\lambda Bk \geq 2^\lambda Rk$, the distribution $\{e \xleftarrow{\$} \mathbb{Z}_\ell, e_k \leftarrow e \bmod k, e_R \leftarrow e \bmod R : (e_k, e_R)\}$ is statistically indistinguishable from the uniform distribution over $\mathbb{Z}_k \times \mathbb{Z}_R$, and the value pk only leaks e_R, even to an unbounded adversary (as $S.\mathsf{Enc}_{\mathsf{ek}}(0; e) = S.\mathsf{Enc}_{\mathsf{ek}}(0; e \bmod R)$). We now prove the following claim:

Claim. For any public parameters $\boldsymbol{G}$ and word $\boldsymbol{C}$, it holds that

$$\Pr\left[\begin{array}{l} (\mathsf{pk}, \mathsf{vk}) \xleftarrow{\$} \Pi_\mathsf{K}.\mathsf{KeyGen}(1^\lambda), \\ b \leftarrow \mathsf{Sim.Verify}(\mathsf{pk}, \mathsf{dk}, \boldsymbol{C}, \boldsymbol{\pi}), : b' = b \\ b' \leftarrow \Pi_\mathsf{K}.\mathsf{Verify}(\mathsf{pk}, \mathsf{vk}, \boldsymbol{C}, \boldsymbol{\pi}) \end{array}\right] \geq 1 - \beta/p_k,$$

where p_k is one of the prime factors of k.

Proof. First, we show that if $b = 1$, then $b' = 1$. Indeed, let us denote $(\boldsymbol{x}, \boldsymbol{x}')$ the plaintexts associated to $(\boldsymbol{X}, \boldsymbol{X}')$. Let $(\boldsymbol{r}, \boldsymbol{r}')$ be the random coins of the ciphertexts $(\boldsymbol{X}, \boldsymbol{X}')$. Observe that, by the homomorphic properties of S, the equation $(-e_R) \odot (\boldsymbol{X} \ominus \boldsymbol{x}) = \boldsymbol{X}' \ominus \boldsymbol{x}'$ is equivalent to $S.\mathsf{Enc}_{\mathsf{ek}}(0; -e_R \cdot \boldsymbol{r}) = S.\mathsf{Enc}_{\mathsf{ek}}(0; \boldsymbol{r}')$, which is equivalent to $e \odot \boldsymbol{X} \oplus \boldsymbol{X}' = S.\mathsf{Enc}(e \cdot \boldsymbol{x} + \boldsymbol{x}' \bmod k; e \cdot \boldsymbol{r} + \boldsymbol{r}' \bmod R) = S.\mathsf{Enc}(e \cdot \boldsymbol{x} + \boldsymbol{x}' \bmod k; 0)$ as $e = e_R \bmod R$. Therefore, the verifier's check that $e \odot \boldsymbol{X} \oplus \boldsymbol{X}'$ is decodable succeeds if and only if Sim's first check succeeds, and the decoded value $\boldsymbol{d} \in \mathbb{Z}_k^\gamma$ satisfies $\boldsymbol{d} = e \cdot \boldsymbol{x} + \boldsymbol{x}' \bmod k$. Moreover, if the equations $\boldsymbol{x} \bullet \Gamma(\boldsymbol{G}, \boldsymbol{C}) = \boldsymbol{C}$ and $\boldsymbol{x}' \bullet \Gamma(\boldsymbol{G}, \boldsymbol{C}) = \boldsymbol{C}'$ are both satisfied (*i.e.* Sim's other checks succeed), then it necessarily holds that $\boldsymbol{d} \bullet \Gamma(\boldsymbol{G}, \boldsymbol{C}) = (e \cdot \boldsymbol{x} + \boldsymbol{x}') \bullet \Gamma(\boldsymbol{G}, \boldsymbol{C}) = e \bullet (\boldsymbol{x} \bullet \Gamma(\boldsymbol{G}, \boldsymbol{C})) + \boldsymbol{x}' \bullet \Gamma(\boldsymbol{G}, \boldsymbol{C}) = e \bullet \boldsymbol{C} + \boldsymbol{C}'$. This concludes the proof that, conditioned on Sim's checks succeeding, the verifier's checks necessarily succeed.

Now, assume for the sake of contradiction that the converse is not true: suppose that Sim rejected the proof, while the verifier accepted. We already showed that the equation $(-e_R)\odot(\boldsymbol{X}\ominus\boldsymbol{x}) = \boldsymbol{X}'\ominus\boldsymbol{x}'$ is equivalent to the equation $e\odot\boldsymbol{X}\oplus\boldsymbol{X}' = S.\mathsf{Enc}(e\cdot\boldsymbol{x}+\boldsymbol{x}' \bmod k; 0)$; therefore, if $e\odot\boldsymbol{X}\oplus\boldsymbol{X}'$ is decodable (it has random coin 0), then Sim's check that $(-e_R)\odot(\boldsymbol{X}\ominus\boldsymbol{x}) = \boldsymbol{X}'\ominus\boldsymbol{x}'$ succeeds. As we assumed that Sim rejects the proof, this means that at least one of Sim's last checks must fail: either $\boldsymbol{x}\bullet\Gamma(\boldsymbol{G},\boldsymbol{C})\neq\boldsymbol{C}$, or $\boldsymbol{x}'\bullet\Gamma(\boldsymbol{G},\boldsymbol{C})\neq\boldsymbol{C}'$. By the first check of the verifier, it holds that $e\odot\boldsymbol{X}\oplus\boldsymbol{X}'$ is decodable; denoting $(\boldsymbol{x},\boldsymbol{x}')$ the plaintexts associated to $(\boldsymbol{X},\boldsymbol{X}')$, it therefore decodes to $\boldsymbol{d} = e\cdot\boldsymbol{x}+\boldsymbol{x}' \bmod k$. By the second check of the verifier, it holds that $\boldsymbol{d}\bullet\Gamma(\boldsymbol{G},\boldsymbol{C}) = e\bullet\boldsymbol{C}\boldsymbol{+}\boldsymbol{C}'$, which implies $e\bullet(\boldsymbol{x}\bullet\Gamma(\boldsymbol{G},\boldsymbol{C}))\boldsymbol{+}\boldsymbol{x}'\bullet\Gamma(\boldsymbol{G},\boldsymbol{C}) = e\bullet\boldsymbol{C}\boldsymbol{+}\boldsymbol{C}'$. This last equation rewrites to

$$e\bullet(\boldsymbol{x}\bullet\Gamma(\boldsymbol{G},\boldsymbol{C})\boldsymbol{-}\boldsymbol{C}) = \boldsymbol{C}'\boldsymbol{-}\boldsymbol{x}'\bullet\Gamma(\boldsymbol{G},\boldsymbol{C}) \tag{2}$$

Now, recall that by assumption, either $\boldsymbol{x}\bullet\Gamma(\boldsymbol{G},\boldsymbol{C})\neq\boldsymbol{C}$, or $\boldsymbol{x}'\bullet\Gamma(\boldsymbol{G},\boldsymbol{C})\neq\boldsymbol{C}'$. Observe that Eq. 2 further implies, as $e\neq 0$ (with overwhelming probability), that $\boldsymbol{x}'\bullet\Gamma(\boldsymbol{G},\boldsymbol{C})\boldsymbol{-}\boldsymbol{C}'\neq 0$ if and only if $\boldsymbol{x}\bullet\Gamma(\boldsymbol{G},\boldsymbol{C})\boldsymbol{-}\boldsymbol{C}\neq 0$. Therefore, conditioned on Sim rejecting the proof, it necessarily holds that $\boldsymbol{x}\bullet\Gamma(\boldsymbol{G},\boldsymbol{C})\boldsymbol{-}\boldsymbol{C}\neq 0$ and $\boldsymbol{x}'\bullet\Gamma(\boldsymbol{G},\boldsymbol{C})\boldsymbol{-}\boldsymbol{C}'\neq 0$. Let (μ_i,ν_i) be two non-zero entries of the vectors $(\boldsymbol{x}\bullet\Gamma(\boldsymbol{G},\boldsymbol{C})\boldsymbol{-}\boldsymbol{C},\boldsymbol{C}'\boldsymbol{-}\boldsymbol{x}'\bullet\Gamma(\boldsymbol{G},\boldsymbol{C}))$ at the same position $i\leq\beta$; by Eq. 2, it holds that $e = \nu_i\cdot\mu_i^{-1} \bmod p$ for at least one of the prime factors p of k. However, recall that the value $e \bmod k$ is *statistically hidden* to the prover (and therefore, so is the value $e \bmod p$), hence the probability of this event happening can be upper-bounded by $\beta/p\leq\beta/p_k$. This concludes the proof of the claim. □

Now, consider an adversary $\mathscr{A}$ that outputs an accepting proof $(\boldsymbol{C},\boldsymbol{\pi})$ with probability at least ε after a polynomial number Q of interactions with the oracle $\mathcal{O}_{\mathsf{vk}}[\mathsf{pk}]$. By the above claim and a union bound, it necessarily holds that $\mathscr{A}$ outputs an accepting proof $(\boldsymbol{C},\boldsymbol{\pi})$ with probability at least $\varepsilon - Q\beta/p_k$ after interacting Q times with $\mathsf{Sim.Verify}(\mathsf{pk},\mathsf{dk},e_R,\cdot,\cdot)$; moreover, with probability at least $1-\beta p_k$, this proof is also accepted by Sim's verification algorithm. Overall, Sim obtains a proof accepted by his verification algorithm with probability at least $\varepsilon-(Q+1)\beta/p_k$. In particular, this implies that the vector $\boldsymbol{x}$ extracted by Sim from $\boldsymbol{\pi}$ satisfies $\boldsymbol{x}\bullet\Gamma(\boldsymbol{G},\boldsymbol{C}) = \boldsymbol{C}$ with probability at least $\varepsilon-(Q+1)\beta/p_k$. Therefore, Sim extracts a valid witness for the knowledge statement $\mathsf{St}_\Gamma(\boldsymbol{G},\boldsymbol{C})$ with probability at least $\varepsilon-(Q+1)\beta/p_k$. As the size k of a DVNIZK-friendly cryptosystem has only superpolynomially large prime-factors, it holds that p_k is superpolynomially large. As $(Q+1)\beta$ is polynomial, we conclude that if $\mathscr{A}$ outputs an accepting proof with non-negligible probability, then Sim extracts a valid witness with non-negligible probability.

4 Dual Variant of the Framework

In the previous section, we described a framework for constructing efficient DVNIZKs of knowledge for relations between words defined over an abelian group

$(\mathbb{G}, +)$, using a cryptosystem with specific properties as the underlying commitment scheme for the proof system. In this section, we show that the framework can also be used in a dual way, by considering languages of relations between the plaintexts of the underlying encryption scheme – we call this variant 'dual variant' of the framework, as the roles of the underlying encryption scheme (which is used as a commitment scheme for the proof) and of the abelian group (which contains the words on which the proof is made) are partially exchanged. This allows for example to handle languages of relations between Paillier ciphertexts. To instantiate the framework, it suffices to have any perfectly binding commitment scheme defined over $\mathbb{G}$. This dual variant leads to efficient DVNIZK proofs for relations between, e.g., Paillier ciphertexts, whose zero-knowledge property reduces to the binding property of the commitment scheme over $\mathbb{G}$ (e.g. the DDH assumption, or its variants), and with statistical (unbounded, adaptive) soundness.

4.1 Perfectly Binding Commitment over $\mathbb{G}$

Suppose that we are given a perfectly binding homomorphic commitment $C = (C.\mathsf{Setup}, C.\mathsf{Com}, C.\mathsf{Verify})$, where $C.\mathsf{Com} : \mathbb{Z}_k \times \mathbb{Z}_k \mapsto \mathbb{G}^*$. Assume further that $C.\mathsf{Setup}$ generates a public vector of parameters $\boldsymbol{G} \in \mathbb{G}^*$, and that there is a linear map Γ_C associated to this commitment such that for all $(m, r) \in \mathbb{Z}_k^2$, $C.\mathsf{Com}(m, r) = (m, r) \bullet \Gamma_C(\boldsymbol{G})$. Note this implies the commitment scheme is homomorphic over $\mathbb{G}$. ElGamal (Sect. 2.1), can be used as a commitment scheme satisfying these properties, is hiding under the DDH assumption and perfectly binding. We do so by using $\mathsf{KeyGen}(1^\lambda)$ in place of $\mathsf{Setup}(1^\lambda)$ to generate group elements (G, H) (the public key of the encryption scheme), and commit (i.e encrypt) via $\Gamma_C(G, H) = ((0, G)^\intercal, (G, H)^\intercal)$. We generalize this to commitments to length-t vectors as follow: we let $\Gamma_{C,t}$ denote the extended matrix such that $C.\mathsf{Com}(\boldsymbol{m}, \boldsymbol{r}) = (\boldsymbol{m}, \boldsymbol{r}) \bullet \Gamma_{C,t}(\boldsymbol{G})$, where $(\boldsymbol{m}, \boldsymbol{r})$ are vectors of length t ($\Gamma_{C,t}$ is simply the block-diagonal matrix whose t blocks are all equal to Γ_C). Consider now the following statement, where the word is a vector $\boldsymbol{C}$ of commitments:

$$\begin{aligned} \mathsf{St}_{\Gamma_{C,t}}(\boldsymbol{G}, \boldsymbol{C}) &= \text{\reflectbox{K}}\{(\boldsymbol{m}, \boldsymbol{r}) \mid (\boldsymbol{m}, \boldsymbol{r}) \bullet \Gamma_{C,t}(\boldsymbol{G}) = \boldsymbol{C}\} \\ &= \text{\reflectbox{K}}\{(\boldsymbol{m}, \boldsymbol{r}) \mid C.\mathsf{Com}(\boldsymbol{m}, \boldsymbol{r}) = \boldsymbol{C}\}. \end{aligned}$$

One can immediatly observe that this statement (which is a proof of knowledge of openings to a vector of commitments with C) is handled by the framework of Sect. 3.

4.2 Equality of Plaintexts Between C and S

In this section, we describe a simple method to convert a DVNIZK on the statement $\mathsf{St}_{\Gamma_{C,t}}(\boldsymbol{G}, \boldsymbol{C}) = \text{\reflectbox{K}}\{(\boldsymbol{m}, \boldsymbol{r}) \mid C.\mathsf{Com}(\boldsymbol{m}, \boldsymbol{r}) = \boldsymbol{C}\}$ into a DVNIZK on the statement $\mathsf{St}'(\boldsymbol{G}, \boldsymbol{C}, \boldsymbol{X}_m) = \exists\{(\boldsymbol{m}, \boldsymbol{\rho}_m, \boldsymbol{r}) \mid \boldsymbol{X}_m = S.\mathsf{Enc}_{\mathsf{ek}}(\boldsymbol{m}, \boldsymbol{\rho}_m) \wedge \boldsymbol{C} = C.\mathsf{Com}(\boldsymbol{m}, \boldsymbol{r})\}$ for a length-t vector $\boldsymbol{C}$ of commitments with a commitment scheme over $\mathbb{G}$ satisfying the requirements defined in the previous section, and a length-t vector of DVNIZK-friendly ciphertexts $\boldsymbol{X_m}$. Instantiating the framework of Sect. 3 for the statement $\mathsf{St}_{\Gamma_{C,t}}(\boldsymbol{G}, \boldsymbol{C})$, we get the following DVNIZK Π:

- $\Pi.\mathsf{Setup}(1^\lambda)$: compute $(\mathsf{ek}, \mathsf{dk}) \xleftarrow{\$} S.\mathsf{KeyGen}(1^\lambda)$. Output $\mathsf{crs} \leftarrow \mathsf{ek}$. Note that ek defines the plaintext space $\mathbb{Z}_k$ and the random source $\mathbb{Z}_R$ with bound B. We denote $\ell \leftarrow 2^\lambda kB$.
- $\Pi.\mathsf{KeyGen}(1^\lambda)$: pick $e \leftarrow \mathbb{Z}_\ell$, set $\mathsf{pk} \leftarrow S.\mathsf{Enc}_{\mathsf{ek}}(0; e)$ and $\mathsf{vk} \leftarrow e$.
- $\Pi.\mathsf{Prove}(\mathsf{pk}, \boldsymbol{C}, (\boldsymbol{m}, \boldsymbol{r}))$: on a word $\boldsymbol{C} \in \mathbb{Z}_k^t$, with witness $(\boldsymbol{m}, \boldsymbol{r})$ for the statement $\mathsf{St}_{\Gamma_{C,t}}(\boldsymbol{G}, \boldsymbol{C})$ (where $\boldsymbol{G} \xleftarrow{\$} C.\mathsf{Setup}(1^\lambda)$), pick random $(\boldsymbol{m}', \boldsymbol{r}')$, random coins $(\boldsymbol{\rho_m}, \boldsymbol{\rho_r})$ for S, and compute

$$\begin{aligned} \boldsymbol{X_m} &\leftarrow S.\mathsf{Enc}_{\mathsf{ek}}(\boldsymbol{m}, \boldsymbol{\rho_m}), & \boldsymbol{X_r} &\leftarrow S.\mathsf{Enc}_{\mathsf{ek}}(\boldsymbol{r}, \boldsymbol{\rho_r}), \\ \boldsymbol{X'_m} &\leftarrow S.\mathsf{Enc}_{\mathsf{ek}}(\boldsymbol{m}', 0) \ominus (\boldsymbol{\rho_m} \odot \mathsf{pk}), & \boldsymbol{X'_r} &\leftarrow S.\mathsf{Enc}_{\mathsf{ek}}(\boldsymbol{r}', 0) \ominus (\boldsymbol{\rho_r} \odot \mathsf{pk}), \\ \boldsymbol{C}' &\leftarrow (\boldsymbol{m}', \boldsymbol{r}') \bullet \Gamma_{C,t}(\boldsymbol{G}, \boldsymbol{C}), \end{aligned}$$

  and output $\pi \leftarrow (\boldsymbol{X_m}, \boldsymbol{X'_m}, \boldsymbol{X_r}, \boldsymbol{X'_r}, \boldsymbol{C}')$.
- $\Pi_{\mathsf{K}}.\mathsf{Verify}(\mathsf{pk}, \mathsf{vk}, \boldsymbol{C}, \pi)$: parse π as $(\boldsymbol{X_m}, \boldsymbol{X'_m}, \boldsymbol{X_r}, \boldsymbol{X'_r}, \boldsymbol{C}')$. Check that $e \odot \boldsymbol{X_m} \oplus \boldsymbol{X'_m}$ and $e \odot \boldsymbol{X_r} \oplus \boldsymbol{X'_r}$ are decodable, and decode them to vectors $(\boldsymbol{d_m}, \boldsymbol{d_r}) \in (\mathbb{Z}_k^t)^2$. Check that $(\boldsymbol{d_m}, \boldsymbol{d_r}) \bullet \Gamma_{C,t}(\boldsymbol{G}, \boldsymbol{C}) = e \bullet \boldsymbol{C} + \boldsymbol{C}'$.

By the result of Sect. 3, this is an unbounded statistical adaptive knowledge-extractable DVNIZK of knowledge of an opening for C. Suppose now that we modify the above scheme as follow: we let $\boldsymbol{X_m}$ be part of the *word* on which the proof is executed, rather than being computed as part of the *proof* by the algorithm $\Pi.\mathsf{Prove}$. That is, we consider words of the form $(\boldsymbol{C}, \boldsymbol{X_m})$ with witness $(\boldsymbol{m}, \boldsymbol{r}, \boldsymbol{\rho_m})$ such that $(\boldsymbol{C}, \boldsymbol{X_m}) = (C.\mathsf{Com}(\boldsymbol{m}; \boldsymbol{r}), S.\mathsf{Enc}_{\mathsf{ek}}(\boldsymbol{m}, \boldsymbol{\rho_m}))$. Let Π' denote the modified proof, in which $\boldsymbol{X_m}$ is part of the word and $(\boldsymbol{X'_m}, \boldsymbol{X_r}, \boldsymbol{X'_r}, \boldsymbol{C}')$ are computed as in Π. Observe that the proof of security of our framework immediatly implies that Π' is a secure DVNIZK for *plaintext equality* between commitments with C and encryptions with S: our statistical argument shows that a (possibly unbounded) adversary has negligible probability of outputting a word $\boldsymbol{C}$ together with an accepting proof $\pi = (\boldsymbol{X_m}, \boldsymbol{X'_m}, \boldsymbol{X_r}, \boldsymbol{X'_r}, \boldsymbol{C}')$ where the plaintext extracted by the simulator from $\boldsymbol{X_m}$ is not also the plaintext of $\boldsymbol{C}$. Hence, it is trivial that the probability of outputting a word $(\boldsymbol{C}, \boldsymbol{X_m})$ and an accepting proof $\pi' = (\boldsymbol{X'_m}, \boldsymbol{X_r}, \boldsymbol{X'_r}, \boldsymbol{C}')$ where the plaintext extracted by the simulator from $\boldsymbol{X_m}$ is not also the plaintext of $\boldsymbol{C}$ is also negligible. Thus, we get:

Theorem 9. *The proof system Π' is an adaptive unbounded statistically sound proof for equality between plaintexts of C and plaintexts of S, whose composable zero-knowledge property reduces to the* IND-CPA *security of S.*

Note that the proof Π' is no longer a proof of knowledge: while the simulator can extract $(\boldsymbol{m}, \boldsymbol{r})$ from the prover, he cannot necessarily extract the random coins $\boldsymbol{\rho_m}$ of $\boldsymbol{X_m}$, which are now part of the witness. Therefore, for the protocol to make sense, it is important that C is perfectly binding.

4.3 A Framework for Relations Between Plaintexts of S

The observations of the above section suggest a very natural way for designing DVNIZKs for relations between plaintexts $\boldsymbol{m} \in \mathbb{Z}_k^*$ of the encryption scheme S, which intuitively operates in two steps: first, we create commitments to the plaintexts $\boldsymbol{m}$ over $\mathbb{G}$ using C and prove them consistent with the encrypted values using the method described in the previous section. Then, we are able to use the framework of Sect. 3 to demonstrate the desired relation holds between the commited values (this is a statement naturally captured by the framework). More formally, on input a vector of ciphertexts $\boldsymbol{X_m}$ encrypting plaintexts $\boldsymbol{m}$ with random coins $\boldsymbol{\rho_m}$,

- Pick $\boldsymbol{r}$ and compute $\boldsymbol{C} \leftarrow C.\mathsf{Com}(\boldsymbol{m}, \boldsymbol{r})$.
- Construct a DVNIZK for the statement $\mathsf{St}'(\boldsymbol{G}, \boldsymbol{C}, \boldsymbol{X_m})$ with witness $(\boldsymbol{m}, \boldsymbol{\rho_m}, \boldsymbol{r})$, using the method described in Sect. 4.2.
- Construct a DVNIZK for the statement $\mathsf{St}_\Gamma(\boldsymbol{G}, \boldsymbol{C})$ with witness $(\boldsymbol{m}, \boldsymbol{r})$, using the framework of Sect. 3.

The correctness of this approach is immediate: the second DVNIZK guarantees that the appropriate relation is satisfied between the plaintexts of the commitments, while the first one guarantees that the ciphertexts indeed encrypt the committed values. This leads to a DVNIZK proof of relation between plaintexts of S, with unbounded adaptive statistical soundness. Regarding zero-knowledge, as the proof starts by committing to $\boldsymbol{m}$ with C, we must in addition assume that the commitment scheme is hiding (the security analysis is straightforward).

Theorem 10. *The above system is an adaptive unbounded statistically sound proof for relations of plaintexts of S, whose composable zero-knowledge reduces to the* IND-CPA *security of S and the hiding property of C.*

We note that we can also obtain a variant of Theorem 10, where zero-knowledge only relies on the IND-CPA of S, and hiding of C implies the soundness property, using commitment schemes *a la* Groth-Sahai where the crs can be generated in two indistinguishable ways, one leading to a perfectly hiding scheme, and one leading to a perfectly binding scheme (such commitments are known, e.g., from the DDH assumption).

Example: Multiplicative Relationship Between Paillier Ciphertexts. We focus now on the useful case of multiplicative relationship between plaintexts of Paillier ciphertexts. We instantiate S with the Paillier encryption scheme over an RSA group $\mathbb{Z}_n$, with a public key (n, h) ($h = g^n \bmod n^2$ for a generator g of $\mathbb{J}_n$), and the commitment scheme C with the ElGamal encryption scheme over a group $\mathbb{G}$ of order n, with public key (G, H). Let $(P_0, P_1, P_2) \in (\mathbb{Z}_{n^2}^*)^3$ be three Paillier ciphertexts, and let $(m_0, m_1, m_2, \rho_0, \rho_1, \rho_2)$ be such that $m_2 = m_0 m_1 \bmod n$, and $P_0 = (1+n)^{m_0} h^{\rho_0} \bmod n^2, P_1 = (1+n)^{m_1} h^{\rho_1} \bmod n^2, P_2 = (1+n)^{m_2} h^{\rho_2} \bmod n^2$. Let $E = h^e \bmod n^2$ denote the public key of the verifier. The designated-verifier NIZK for proving that P_2 encrypts $m_0 m_1$ proceeds as follows:

- **Committing over $\mathbb{G}$:** pick (r_0, r_1, r_2) and send $(U_i, V_i)_{0 \le i \le 2} \leftarrow (r_i \bullet G, r_i \bullet H \boldsymbol{+} m_i \bullet G)_{0 \le i \le 2}$ (which are commitments with ElGamal to (m_0, m_1, m_2) over $\mathbb{G}$).
- **Proof of Plaintext Equality:** pick $(m'_i, r'_i, \rho'_i)_{0 \le i \le 2} \xleftarrow{\$} (\mathbb{Z}_n \times \mathbb{Z}_n \times \mathbb{Z}_{n/2})^3$, and send for $i = 0$ to 2, $X_i \leftarrow (1+n)^{r_i} h^{\rho'_i} \bmod n^2, X'_i \leftarrow (1+n)^{r'_i} E^{-\rho'_i} \bmod n^2, P'_i \leftarrow (1+n)^{m'_i} E^{-\rho_i} \bmod n^2$, and $(U'_i, V'_i) \leftarrow (r'_i \bullet G, r'_i \bullet H \boldsymbol{+} m'_i \bullet G)$.
- **Proof of Multiplicative Relationship Between the Committed Values:** apply the proof system of Example 2 from Sect. 3 to the word $(U_i, V_i)_{0 \le i \le 2}$, with public parameters (G, H), and the witness $\boldsymbol{x} = (m_0, r_0, m_1, r_1, r_2 - r_0 m_1)$ which satisfies $(U_0, V_0) = (r_0 \bullet G, r_0 \bullet H \boldsymbol{+} m_0 \bullet G), (U_1, V_1) = (r_1 \bullet G, r_1 \bullet H \boldsymbol{+} m_1 \bullet G)$, and $(U_2, V_2) = ((r_2 - r_0 m_1) \bullet G \boldsymbol{+} m_1 \bullet U_0, (r_2 - r_0 m_1) \bullet H \boldsymbol{+} m_1 \bullet V_0)$.
- **Proof Verification:** upon receving $(U_i, V_i, X_i, X'_i, P'_i, U'_i, V'_i)_{0 \le i \le 2}$ together with the proof of multiplicative relationship between the values committed with $(U_i, V_i)_i$, the verifier with verification key $\mathsf{vk} = e$ checks that $e \odot P_i \oplus P'_i$ and $e \odot X_i \oplus X'_i$ successfully decode (respectively) to values p_i, x_i, and that $e \bullet U_i \boldsymbol{+} U'_i = x_i \bullet G$ and $e \bullet V_i \boldsymbol{+} V'_i = x_i \bullet H \boldsymbol{+} p_i \bullet G$, for $i = 0$ to 2. The verifier additionally checks the multiplicative proof, as in Example 4 from Sect. 3. She accepts iff all checks succeed.

The proof for the multiplicative statement involves 10 Paillier ciphertexts and 3 ElGamal ciphertexts. Overall, the total proof involves 20 Paillier ciphertexts, and 9 ElGamal ciphertexts. However, this size is obtained by applying the framework naively; in this situation, it introduces a lot of redudancy. For instance, instead of computing Paillier encryptions of (m_0, r_0, m_1, r_1) in the third phase, one can simply reuse the word (P_0, P_1) and the ciphertexts (X_0, X_1), as well as reusing $(P'_i, X'_i)_i$ for the corresponding masks $(m'_i, r'_i)_i$, saving 8 Paillier ciphertexts; similar savings can be obtained for the ElGamal ciphertexts, leading to a proof of total size 12 Paillier ciphertexts + 7 ElGamal ciphertexts.

Furthermore, if we eschew unbounded soundness and accept bounds on m_i we are able to produce a much shorter proof, comprising only two Paillier ciphertexts, outperforming even Fiat-Shamir. We detail this in the full version [16].

Acknowledgements. We thank Jens Groth for insightful discussions and contributions to early versions of this work. The first author was supported by EU Horizon 2020 grant 653497 (project PANORAMIX). The second author was supported by ERC grant 339563 (project CryptoCloud) and ERC grant 724307 (project PREP-CRYPTO).

References

1. Abdalla, M., Benhamouda, F., Pointcheval, D.: Disjunctions for hash proof systems: new constructions and applications. In: Oswald, E., Fischlin, M. (eds.) EUROCRYPT 2015. LNCS, vol. 9057, pp. 69–100. Springer, Heidelberg (2015). https://doi.org/10.1007/978-3-662-46803-6_3
2. Barak, B., Canetti, R., Nielsen, J.B., Pass, R.: Universally composable protocols with relaxed set-up assumptions. In: 45th FOCS, pp. 186–195. IEEE Computer Society Press, October 2004

3. Belenkiy, M., Camenisch, J., Chase, M., Kohlweiss, M., Lysyanskaya, A., Shacham, H.: Randomizable proofs and delegatable anonymous credentials. In: Halevi, S. (ed.) CRYPTO 2009. LNCS, vol. 5677, pp. 108–125. Springer, Heidelberg (2009). https://doi.org/10.1007/978-3-642-03356-8_7
4. Belenkiy, M., Chase, M., Kohlweiss, M., Lysyanskaya, A.: P-signatures and non-interactive anonymous credentials. In: Canetti, R. (ed.) TCC 2008. LNCS, vol. 4948, pp. 356–374. Springer, Heidelberg (2008). https://doi.org/10.1007/978-3-540-78524-8_20
5. Belenkiy, M., Chase, M., Kohlweiss, M., Lysyanskaya, A.: Compact e-cash and simulatable VRFs revisited. In: Shacham, H., Waters, B. (eds.) Pairing 2009. LNCS, vol. 5671, pp. 114–131. Springer, Heidelberg (2009). https://doi.org/10.1007/978-3-642-03298-1_9
6. Bellare, M., Rogaway, P.: Random oracles are practical: a paradigm for designing efficient protocols. In: Ashby, V. (ed.) ACM CCS 1993, pp. 62–73. ACM Press, November 1993
7. Benhamouda, F., Blazy, O., Chevalier, C., Pointcheval, D., Vergnaud, D.: New techniques for SPHFs and efficient one-round PAKE protocols. In: Canetti, R., Garay, J.A. (eds.) CRYPTO 2013. LNCS, vol. 8042, pp. 449–475. Springer, Heidelberg (2013). https://doi.org/10.1007/978-3-642-40041-4_25
8. Benhamouda, F., Couteau, G., Pointcheval, D., Wee, H.: Implicit zero-knowledge arguments and applications to the malicious setting. In: Gennaro, R., Robshaw, M. (eds.) CRYPTO 2015. LNCS, vol. 9216, pp. 107–129. Springer, Heidelberg (2015). https://doi.org/10.1007/978-3-662-48000-7_6
9. Benhamouda, F., Pointcheval, D.: Trapdoor smooth projective hash functions. Cryptology ePrint Archive, Report 2013/341 (2013). http://eprint.iacr.org/2013/341
10. Blum, M., Feldman, P., Micali, S.: Non-interactive zero-knowledge and its applications (extended abstract). In: 20th ACM STOC, pp. 103–112. ACM Press, May 1988
11. Boneh, D., Boyen, X., Shacham, H.: Short group signatures. In: Franklin, M. (ed.) CRYPTO 2004. LNCS, vol. 3152, pp. 41–55. Springer, Heidelberg (2004). https://doi.org/10.1007/978-3-540-28628-8_3
12. Bresson, E., Catalano, D., Pointcheval, D.: A simple public-key cryptosystem with a double trapdoor decryption mechanism and its applications. In: Laih, C.-S. (ed.) ASIACRYPT 2003. LNCS, vol. 2894, pp. 37–54. Springer, Heidelberg (2003). https://doi.org/10.1007/978-3-540-40061-5_3
13. Camenisch, J., Damgård, I.: Verifiable encryption, group encryption, and their applications to separable group signatures and signature sharing schemes. In: Okamoto, T. (ed.) ASIACRYPT 2000. LNCS, vol. 1976, pp. 331–345. Springer, Heidelberg (2000). https://doi.org/10.1007/3-540-44448-3_25
14. Camenisch, J., Neven, G., Shelat, A.: Simulatable adaptive oblivious transfer. In: Naor, M. (ed.) EUROCRYPT 2007. LNCS, vol. 4515, pp. 573–590. Springer, Heidelberg (2007). https://doi.org/10.1007/978-3-540-72540-4_33
15. Castagnos, G., Laguillaumie, F.: Linearly homomorphic encryption from DDH. In: Nyberg, K. (ed.) CT-RSA 2015. LNCS, vol. 9048, pp. 487–505. Springer, Cham (2015). https://doi.org/10.1007/978-3-319-16715-2_26
16. Chaidos, P., Couteau, G.: Efficient designated-verifier non-interactive zero-knowledge proofs of knowledge. Cryptology ePrint Archive, Report 2017/1029 (2017). http://eprint.iacr.org/2017/1029

17. Chaidos, P., Groth, J.: Making sigma-protocols non-interactive without random oracles. In: Katz, J. (ed.) PKC 2015. LNCS, vol. 9020, pp. 650–670. Springer, Heidelberg (2015). https://doi.org/10.1007/978-3-662-46447-2_29
18. Chase, M., Meiklejohn, S., Zaverucha, G.: Algebraic MACs and keyed-verification anonymous credentials. In: ACM CCS 2014, pp. 1205–1216. ACM Press (2014)
19. Chaum, D., Fiat, A., Naor, M.: Untraceable electronic cash. In: Goldwasser, S. (ed.) CRYPTO 1988. LNCS, vol. 403, pp. 319–327. Springer, New York (1990). https://doi.org/10.1007/0-387-34799-2_25
20. Chaum, D., van Heyst, E.: Group signatures. In: Davies, D.W. (ed.) EUROCRYPT 1991. LNCS, vol. 547, pp. 257–265. Springer, Heidelberg (1991). https://doi.org/10.1007/3-540-46416-6_22
21. Ciampi, M., Persiano, G., Siniscalchi, L., Visconti, I.: A transform for NIZK almost as efficient and general as the fiat-shamir transform without programmable random oracles. In: Kushilevitz, E., Malkin, T. (eds.) TCC 2016. LNCS, vol. 9563, pp. 83–111. Springer, Heidelberg (2016). https://doi.org/10.1007/978-3-662-49099-0_4
22. Cramer, R., Shoup, V.: Universal hash proofs and a paradigm for adaptive chosen ciphertext secure public-key encryption. In: Knudsen, L.R. (ed.) EUROCRYPT 2002. LNCS, vol. 2332, pp. 45–64. Springer, Heidelberg (2002). https://doi.org/10.1007/3-540-46035-7_4
23. Damgård, I., Fazio, N., Nicolosi, A.: Non-interactive zero-knowledge from homomorphic encryption. In: Halevi, S., Rabin, T. (eds.) TCC 2006. LNCS, vol. 3876, pp. 41–59. Springer, Heidelberg (2006). https://doi.org/10.1007/11681878_3
24. Damgård, I., Jurik, M.: A length-flexible threshold cryptosystem with applications. In: Safavi-Naini, R., Seberry, J. (eds.) ACISP 2003. LNCS, vol. 2727, pp. 350–364. Springer, Heidelberg (2003). https://doi.org/10.1007/3-540-45067-X_30
25. Damgård, I., Jurik, M., Nielsen, J.B.: A generalization of paillier's public-key system with applications to electronic voting. Int. J. Inf. Secur. **9**(6), 371–385 (2010)
26. Damgård, I., Jurik, M.: A generalisation, a simpli.cation and some applications of paillier's probabilistic public-key system. In: Kim, K. (ed.) PKC 2001. LNCS, vol. 1992, pp. 119–136. Springer, Heidelberg (2001). https://doi.org/10.1007/3-540-44586-2_9
27. De Santis, A., Di Crescenzo, G., Ostrovsky, R., Persiano, G., Sahai, A.: Robust non-interactive zero knowledge. In: Kilian, J. (ed.) CRYPTO 2001. LNCS, vol. 2139, pp. 566–598. Springer, Heidelberg (2001). https://doi.org/10.1007/3-540-44647-8_33
28. ElGamal, T.: A public key cryptosystem and a signature scheme based on discrete logarithms. In: Blakley, G.R., Chaum, D. (eds.) CRYPTO 1984. LNCS, vol. 196, pp. 10–18. Springer, Heidelberg (1985). https://doi.org/10.1007/3-540-39568-7_2
29. Escala, A., Herold, G., Kiltz, E., Ràfols, C., Villar, J.: An algebraic framework for diffie-hellman assumptions. In: Canetti, R., Garay, J.A. (eds.) CRYPTO 2013. LNCS, vol. 8043, pp. 129–147. Springer, Heidelberg (2013). https://doi.org/10.1007/978-3-642-40084-1_8
30. Feige, U., Lapidot, D., Shamir, A.: Multiple non-interactive zero knowledge proofs based on a single random string (extended abstract). In: 31st FOCS, pp. 308–317. IEEE Computer Society Press, October 1990
31. Fiat, A., Shamir, A.: How to prove yourself: practical solutions to identification and signature problems. In: Odlyzko, A.M. (ed.) CRYPTO 1986. LNCS, vol. 263, pp. 186–194. Springer, Heidelberg (1987). https://doi.org/10.1007/3-540-47721-7_12

32. Fischlin, M.: Communication-efficient non-interactive proofs of knowledge with online extractors. In: Shoup, V. (ed.) CRYPTO 2005. LNCS, vol. 3621, pp. 152–168. Springer, Heidelberg (2005). https://doi.org/10.1007/11535218_10
33. Gay, R., Hofheinz, D., Kiltz, E., Wee, H.: Tightly CCA-secure encryption without pairings. In: Fischlin, M., Coron, J.-S. (eds.) EUROCRYPT 2016. LNCS, vol. 9665, pp. 1–27. Springer, Heidelberg (2016). https://doi.org/10.1007/978-3-662-49890-3_1
34. Goldwasser, S., Micali, S., Rackoff, C.: The knowledge complexity of interactive proof systems. SIAM J. Comput. **18**(1), 186–208 (1989)
35. Green, M., Hohenberger, S.: Universally composable adaptive oblivious transfer. In: Pieprzyk, J. (ed.) ASIACRYPT 2008. LNCS, vol. 5350, pp. 179–197. Springer, Heidelberg (2008). https://doi.org/10.1007/978-3-540-89255-7_12
36. Groth, J.: Simulation-sound NIZK proofs for a practical language and constant size group signatures. In: Lai, X., Chen, K. (eds.) ASIACRYPT 2006. LNCS, vol. 4284, pp. 444–459. Springer, Heidelberg (2006). https://doi.org/10.1007/11935230_29
37. Groth, J., Ostrovsky, R., Sahai, A.: Non-interactive zaps and new techniques for NIZK. In: Dwork, C. (ed.) CRYPTO 2006. LNCS, vol. 4117, pp. 97–111. Springer, Heidelberg (2006). https://doi.org/10.1007/11818175_6
38. Groth, J., Ostrovsky, R., Sahai, A.: Perfect non-interactive zero knowledge for NP. In: Vaudenay, S. (ed.) EUROCRYPT 2006. LNCS, vol. 4004, pp. 339–358. Springer, Heidelberg (2006). https://doi.org/10.1007/11761679_21
39. Groth, J., Sahai, A.: Efficient non-interactive proof systems for bilinear groups. In: Smart, N. (ed.) EUROCRYPT 2008. LNCS, vol. 4965, pp. 415–432. Springer, Heidelberg (2008). https://doi.org/10.1007/978-3-540-78967-3_24
40. Kiltz, E., Pan, J., Wee, H.: Structure-preserving signatures from standard assumptions, revisited. In: Gennaro, R., Robshaw, M. (eds.) CRYPTO 2015. LNCS, vol. 9216, pp. 275–295. Springer, Heidelberg (2015). https://doi.org/10.1007/978-3-662-48000-7_14
41. Kiltz, E., Wee, H.: Quasi-adaptive NIZK for linear subspaces revisited. In: Oswald, E., Fischlin, M. (eds.) EUROCRYPT 2015. LNCS, vol. 9057, pp. 101–128. Springer, Heidelberg (2015). https://doi.org/10.1007/978-3-662-46803-6_4
42. Lindell, Y.: An efficient transform from sigma protocols to NIZK with a CRS and non-programmable random oracle. In: Dodis, Y., Nielsen, J.B. (eds.) TCC 2015. LNCS, vol. 9014, pp. 93–109. Springer, Heidelberg (2015). https://doi.org/10.1007/978-3-662-46494-6_5
43. Lipmaa, H.: Optimally sound sigma protocols under DCRA. Cryptology ePrint Archive, Report 2017/703 (2017). http://eprint.iacr.org/2017/703
44. Oren, Y.: On the cunning power of cheating verifiers: some observations about zero knowledge proofs (extended abstract). In: 28th FOCS, pp. 462–471. IEEE Computer Society Press, October 1987
45. Paillier, P.: Public-key cryptosystems based on composite degree residuosity classes. In: Stern, J. (ed.) EUROCRYPT 1999. LNCS, vol. 1592, pp. 223–238. Springer, Heidelberg (1999). https://doi.org/10.1007/3-540-48910-X_16
46. Pass, R., Shelat, A., Vaikuntanathan, V.: Construction of a non-malleable encryption scheme from any semantically secure one. In: Dwork, C. (ed.) CRYPTO 2006. LNCS, vol. 4117, pp. 271–289. Springer, Heidelberg (2006). https://doi.org/10.1007/11818175_16
47. Pointcheval, D., Stern, J.: Security proofs for signature schemes. In: Maurer, U. (ed.) EUROCRYPT 1996. LNCS, vol. 1070, pp. 387–398. Springer, Heidelberg (1996). https://doi.org/10.1007/3-540-68339-9_33

48. Teranishi, I., Furukawa, J., Sako, K.: k-times anonymous authentication (extended abstract). In: Lee, P.J. (ed.) ASIACRYPT 2004. LNCS, vol. 3329, pp. 308–322. Springer, Heidelberg (2004). https://doi.org/10.1007/978-3-540-30539-2_22
49. Ventre, C., Visconti, I.: Co-sound zero-knowledge with public keys. In: Preneel, B. (ed.) AFRICACRYPT 2009. LNCS, vol. 5580, pp. 287–304. Springer, Heidelberg (2009). https://doi.org/10.1007/978-3-642-02384-2_18

Quasi-Optimal SNARGs via Linear Multi-Prover Interactive Proofs

Dan Boneh[1,4](✉), Yuval Ishai[2,3,4], Amit Sahai[3,4], and David J. Wu[1,4]

[1] Stanford University, Stanford, USA
dabo@cs.stanford.edu
[2] Technion, Haifa, Israel
[3] UCLA, Los Angeles, USA
[4] Center for Encrypted Functionalities, Los Angeles, USA

Abstract. Succinct non-interactive arguments (SNARGs) enable verifying NP computations with significantly less complexity than that required for classical NP verification. In this work, we focus on simultaneously minimizing the proof size and the prover complexity of SNARGs. Concretely, for a security parameter λ, we measure the asymptotic cost of achieving soundness error $2^{-\lambda}$ against provers of size 2^{λ}. We say a SNARG is *quasi-optimally succinct* if its proof length is $\widetilde{O}(\lambda)$, and that it is *quasi-optimal*, if moreover, its prover complexity is only polylogarithmically greater than the running time of the classical NP prover. We show that this definition is the best we could hope for assuming that NP does not have succinct proofs. Our definition strictly strengthens the previous notion of quasi-optimality introduced in the work of Boneh et al. (Eurocrypt 2017).

This work gives the first quasi-optimal SNARG for Boolean circuit satisfiability from a concrete cryptographic assumption. Our construction takes a two-step approach. The first is an information-theoretic construction of a quasi-optimal linear multi-prover interactive proof (linear MIP) for circuit satisfiability. Then, we describe a generic cryptographic compiler that transforms our quasi-optimal linear MIP into a quasi-optimal SNARG by relying on the notion of linear-only vector encryption over rings introduced by Boneh et al. Combining these two primitives yields the first quasi-optimal SNARG based on linear-only vector encryption. Moreover, our linear MIP construction leverages a new *robust* circuit decomposition primitive that allows us to decompose a circuit satisfiability instance into several smaller circuit satisfiability instances. This primitive may be of independent interest.

Finally, we consider (designated-verifier) SNARGs that provide *optimal* succinctness for a non-negligible soundness error. Concretely, we put forward the notion of "1-bit SNARGs" that achieve soundness error 1/2 with only one bit of proof. We first show how to build 1-bit SNARGs from indistinguishability obfuscation, and then show that 1-bit SNARGs also suffice for realizing a form of witness encryption. The latter result highlights a two-way connection between the soundness of very succinct argument systems and powerful forms of encryption.

The full version of this paper is available at https://eprint.iacr.org/2018/133.pdf.

J. B. Nielsen and V. Rijmen (Eds.): EUROCRYPT 2018, LNCS 10822, pp. 222–255, 2018.
https://doi.org/10.1007/978-3-319-78372-7_8

1 Introduction

Proof systems are fundamental to modern cryptography. Many works over the last few decades have explored different aspects of proof systems, including interactive proofs [35,48,56], zero-knowledge proofs [35], probabilistically checkable proofs [2,3,26], and computationally sound proofs [44,49]. In this work, we study one such aspect: NP proof systems where the proofs can be significantly shorter than the NP witness and can be verified much faster than the time needed to check the NP witness. We say that such proof systems are *succinct.*

In interactive proof systems for NP with statistical soundness, non-trivial savings in communication and verification time are highly unlikely [16,32,33,65]. However, if we relax the requirements and consider proof systems with computational soundness, also known as *argument systems* [17], significant efficiency improvements become possible. Kilian [44] gave the first succinct four-round interactive argument system for NP based on collision-resistant hash functions and probabilistically checkable proofs (PCPs). Subsequently, Micali [49] showed how to convert Kilian's four-round argument into a single-round argument for NP by applying the Fiat-Shamir heuristic [27] to Kilian's interactive protocol. Micali's "computationally-sound proofs" (CS proofs) represents the first candidate construction of a *succinct non-interactive argument* (that is, a "SNARG" [30]). In the standard model, single-round succinct arguments are highly unlikely for sufficiently hard languages [4,65], so we consider the weaker goal of two-message succinct arguments systems where the initial message from the verifier is independent of the statement being verified. We refer to this message as the common reference string (CRS).

In this work, we focus on simultaneously minimizing both the proof size and the prover complexity of succinct non-interactive arguments. For a security parameter λ, we measure the asymptotic cost of achieving soundness against provers of size 2^λ with $2^{-\lambda}$ error. We say that a SNARG is *quasi-optimally succinct* if its proof length is $\widetilde{O}(\lambda)$, and that it is *quasi-optimal* if in addition, the prover's runtime is only polylogarithmically greater than the the running time of the classical prover. In Sect. 5.1, we show that this notion of quasi-optimal succinctness is tight (up to polylogarithmic factors): assuming NP does not have succinct proofs, no succinct argument system can provide the same soundness guarantees with proofs of size $o(\lambda)$. Our notion of quasi-optimality is a strict strengthening of the previous notion from [14], which imposed a weaker soundness requirement on the SNARG. Notably, under the definition in [14], we show that it is possible to construct SNARGs with even shorter proofs than what they consider to be (quasi)-optimally succinct. We discuss the differences in these notions of quasi-optimality in Sect. 1.1 as well as the full version of this paper [15].

In this paper, we construct the first quasi-optimal SNARG whose security is based on a concrete cryptographic assumption similar in flavor to those of previous works [13,14]. To our knowledge, all previous candidates are either not quasi-optimal or rely on a heuristic security argument. Similar to previous works [13,14], we take a two-step approach to construct our quasi-optimal

SNARGs. First, we construct an information-theoretic proof system that provides soundness against a restricted class of provers (e.g., *linearly*-bounded provers [41]). We then leverage cryptographic tools (e.g., *linear-only* encryption [13,14]) to compile the information-theoretic primitive into a succinct argument system. In this work, the core information-theoretic primitive we use is a linear multi-prover interactive proof (linear MIP). One of the main contributions in this work is a new construction of a quasi-optimal linear MIP that can be compiled to a quasi-optimal SNARG using similar cryptographic tools as those in [14]. We give an overview of our quasi-optimal linear MIP construction in Sect. 2, and the formal construction in Sect. 4.

Background on SNARGs. We briefly introduce several properties of succinct non-interactive argument systems. In this work, we focus on constructing SNARGs for the problem of Boolean circuit satisfiability. (This suffices for building SNARGs for general RAM computations, cf. [13].) A SNARG is *publicly verifiable* if anyone can verify the proofs, and it is *designated-verifier* if only the holder of a secret verification state (generated along with the CRS) can verify proofs. In this work, we focus on constructing quasi-optimal designated-verifier SNARGs. In addition, we say a SNARG is fully succinct if the setup algorithm (i.e., the algorithm that generates the CRS, and in the designated-verifier setting, the secret verification state), is also efficient (i.e., runs in time that is only polylogarithmic in the circuit size). A weaker notion is the concept of a *preprocessing SNARG*, where the setup algorithm is allowed to run in time that is polynomial in the size of the circuit being verified. In this work, we consider preprocessing SNARGs. We provide additional background on SNARGs and other related work in Sect. 1.3.

1.1 Quasi-Optimal SNARGs

In this section, we summarize the main results of this work on defining and constructing quasi-optimal SNARGs. In Sect. 2, we provide a more technical survey of our main techniques.

Defining quasi-optimality. In this work, we are interested in minimizing the prover complexity and proof size in succinct non-interactive argument systems. To reiterate, our definition of quasi-optimality considers the prover complexity and proof size needed to ensure soundness error $2^{-\lambda}$ against provers of size 2^{λ}. We say a SNARG (for Boolean circuit satisfiability) is quasi-optimal if the proof size is $\widetilde{O}(\lambda)$ and the prover complexity is $\widetilde{O}(|C|) + \text{poly}(\lambda, \log |C|)$, where C is the Boolean circuit.[1] In Lemma 5.2, we show that this notion of quasi-optimality is the "right" one in the following sense: assuming NP does not have succinct *proofs*, the length of any succinct argument system that provides this soundness guarantee is necessarily $\Omega(\lambda)$. Thus, SNARG systems with strictly better parameters are unlikely to exist.

[1] We write $\widetilde{O}(\cdot)$ to suppress factors that are *polylogarithmic* in the circuit size $|C|$ and the security parameter λ.

Our notion is a strict strengthening of the previous notion of quasi-optimality from [14] which only required soundness error $\mathrm{negl}(\lambda)$ against provers of size 2^λ. In fact, we show in the full version [15] that the previous notion of quasi-optimality from [14] is not tight. If we only want ρ bits of soundness where $\rho = o(\lambda)$, it is possible to construct a designated-verifier SNARG where the proofs are exactly ρ bits. This means that there exists a designated-verifier SNARG which meet the soundness requirements in [14], but whose size is strictly shorter than what would be considered "optimal."

Previous SNARG constructions. Prior to this work, the only SNARG candidate that satisfies our notion of quasi-optimal prover complexity is Micali's CS proofs [49]. However, to achieve $2^{-\lambda}$ soundness, the length of a CS proof is $\Omega(\lambda^2)$, which does not satisfy our notion of quasi-optimal succinctness. Conversely, if we just consider SNARGs that provide quasi-optimal succinctness, we have many candidates [13,14,24,29,37,38,45,46]. With the exception of [14], the SNARG proof in all of these candidates contains a constant number of bilinear group elements, and so, is quasi-optimally succinct. The drawback is that to construct the proof, the prover has to perform a group operation for every gate in the underlying circuit. Since each group element is $\Omega(\lambda)$ bits, the prover overhead is at least multiplicative in λ. Consequently, none of these existing constructions satisfy our notion of quasi-optimal prover complexity. The lattice-based construction in [14] has the same limitation: the prover needs to operate on an LWE ciphertext per gate in the circuit, which introduces a multiplicative overhead $\Omega(\lambda)$ in the prover's computational cost.

Quasi-optimal linear MIPs. This work gives the first construction of a quasi-optimal SNARG for Boolean circuit satisfiability from a concrete cryptographic assumption. Following previous works on constructing SNARGs [13,14], our construction can be broken down into two components: an information-theoretic component (linear MIPs), and a cryptographic component (linear-only vector encryption). We give a brief description of the information-theoretic primitive we construct in this work: a *quasi-optimal* linear MIP. At the end of this section, we discuss why the general PCPs and linear PCPs that have featured in previous SNARG constructions do not seem sufficient for building quasi-optimal SNARGs.

We first review the notion of a linear PCP [13,41]. A linear PCP over a finite field $\mathbb{F}$ is an oracle computing a linear function $\boldsymbol{\pi}\colon \mathbb{F}^m \to \mathbb{F}$. On any query $\mathbf{q} \in \mathbb{F}^m$, the linear PCP oracle responds with the inner product $\mathbf{q}^\top \boldsymbol{\pi} = \langle \mathbf{q}, \boldsymbol{\pi} \rangle \in \mathbb{F}$. More generally, if ℓ queries are made to the linear PCP oracle, the ℓ queries can be packed into the columns of a query matrix $\mathbf{Q} \in \mathbb{F}^{m \times \ell}$. In this case, we can express the response of the linear PCP oracle as the matrix-vector product $\mathbf{Q}^\top \boldsymbol{\pi}$.

Linear MIPs are a direct generalization of linear PCPs to the setting where there are ℓ independent proof oracles $(\boldsymbol{\pi}_1, \ldots, \boldsymbol{\pi}_\ell)$, each implementing a linear function $\boldsymbol{\pi}_i\colon \mathbb{F}^m \to \mathbb{F}$. In the linear MIP model, the verifier's queries consist of a ℓ-tuple $(\mathbf{q}_1, \ldots, \mathbf{q}_\ell)$ where each $\mathbf{q}_i \in \mathbb{F}^m$. For each query $\mathbf{q}_i \in \mathbb{F}^m$ to the proof oracle $\boldsymbol{\pi}_i$, the verifier receives the response $\langle \mathbf{q}_i, \boldsymbol{\pi}_i \rangle$. We review the formal definitions of linear PCPs and linear MIPs in the full version [15].

In this work, we say that a linear MIP for Boolean circuit satisfiability is quasi-optimal if the MIP prover (for proving satisfiability of a circuit C) can be implemented by a circuit of size $\widetilde{O}(|C|) + \text{poly}(\lambda, \log |C|)$, and the linear MIP provides soundness error $2^{-\lambda}$. Existing linear PCP constructions [13,14] (which can be viewed as linear MIPs with a single prover) are not quasi-optimal: they either require embedding the Boolean circuit into an arithmetic circuit over a large field [13], or rely on making $O(\lambda)$ queries, each of length $m = O(|C|)$ [14].

Constructing quasi-optimal linear MIPs. Our work gives the first construction of a quasi-optimal linear MIP for Boolean circuit satisfiability. We refer to Sect. 2 for an overview of our construction and to Sect. 4 for the full description. At a high-level, our quasi-optimal linear MIP construction relies on two key ingredients: a robust circuit decomposition and a method for enforcing consistency.

Robust circuit decomposition. Our robust decomposition primitive takes a circuit C and produces from it a collection of constraints $f_1, \ldots, f_t$, each of which can be computed by a circuit of size roughly $|C|/t$. Each constraint reads a subset of the bits of a global witness (computed based on the statement-witness pair for C). The guarantee provided by the robust decomposition is that for any false statement $\mathbf{x}$ (that is, a statement $\mathbf{x}$ where for all witnesses $\mathbf{w}$, $C(\mathbf{x}, \mathbf{w}) = 0$), no single witness to $f_1, \ldots, f_t$ can simultaneously satisfy more than a *constant fraction* of the constraints. Now, to prove satisfiability of a circuit C, the prover instead proves that there is a consistent witness that simultaneously satisfies all of the constraints $f_1, \ldots, f_t$. Each of these proofs can be implemented by a standard linear PCP. The advantage of this approach is that for a false statement, only a constant fraction of the constraints can be satisfied (for any choice of witness), so even if each underlying linear PCP instance only provided *constant* soundness, the probability that the prover is able to satisfy *all* of the instances is amplified to $2^{-\Omega(t)} = 2^{-\Omega(\lambda)}$ if we let $t = \Theta(\lambda)$. Finally, even though the prover now has to construct t proofs for the t constraints, each of the constraints can themselves be computed by a circuit of size $\widetilde{O}(|C|/t)$. The robustness property of our decomposition is reminiscent of the relation between traditional PCPs and constraint-satisfaction problems, and one might expect that we could instantiate such a decomposition using PCPs. However, in our settings, we require that the decomposition be *input-independent*, which to the best of our knowledge, is not satisfied by existing (quasilinear) PCP constructions. We discuss this in more detail in the full version [15].

The robust decomposition can amplify soundness without introducing much additional overhead. The alternative approach of directly applying a constant-query linear PCP to check satisfiability of C has the drawback of only providing $1/\text{poly}(\lambda)$ soundness when working over a small field (i.e., as would be the case with Boolean circuit satisfiability). We state the formal requirements of our robust decomposition in Sect. 4.1, and give one instantiation in the full version by combining MPC protocols with polylogarithmic overhead [23] with the "MPC-in-the-head" paradigm [42]. Since the notion of a robust decomposition is a very natural one, we believe that our construction is of independent interest and will have applications beyond quasi-optimal linear MIP constructions.

Enforcing consistency. The second ingredient we require is a way for the verifier to check that the individual proofs the prover constructs (for showing satisfiability of each constraint $f_1, \ldots, f_t$) are self-consistent. Our construction here relies on constructing randomized permutation decompositions, and we refer to Sect. 2 for the technical overview, and Sect. 4 for the full description.

Preprocessing SNARGs from linear MIPs. To complete our construction of quasi-optimal SNARGs, we show a generic compiler from linear MIPs to preprocessing SNARGs by relying on the notion of a linear-only vector encryption scheme over rings introduced by Boneh et al. [14]. We give our construction in Sect. 5. Our primary contribution here is recasting the Boneh et al. construction, which satisfies the weaker notion of quasi-optimality, as a generic framework for compiling linear MIPs into preprocessing SNARGs. Combined with our information-theoretic construction of quasi-optimal linear MIPs, this yields the first quasi-optimal designated-verifier SNARG for Boolean circuit satisfiability in the preprocessing model (Corollaries 5.6 and 5.7).

Why linear MIPs? A natural question to ask is whether our new linear MIP to preprocessing SNARG compiler provides any advantage over the existing compilers in [13,14], which use different information-theoretic primitives as the underlying building block (namely, linear interactive proofs [13] and linear PCPs [14]). After all, any k-query, ℓ-prover linear MIP with query length m can be transformed into a $(k\ell)$-query linear PCP with query length $m\ell$ by concatenating the proofs of the different provers together, and likewise, padding the queries accordingly. While this still yields a quasi-optimal linear PCP (with sparse queries), applying the existing cryptographic compilers to this linear PCP incurs an additional prover overhead that is proportional to ℓ. In our settings, $\ell = \Theta(\lambda)$, so the resulting SNARG is no longer quasi-optimal. By directly compiling linear MIPs to preprocessing SNARGs, our compiler *preserves* the prover complexity of the underlying linear MIP, and so, combined with our quasi-optimal linear MIP construction, yields a quasi-optimal SNARG for Boolean circuit satisfiability.

Alternatively, one might ask whether a similar construction of quasi-optimal SNARGs is possible starting from standard PCPs or linear PCPs with quasi-optimal prover complexity. Existing techniques for compiling general PCPs [9,10,49] to succinct argument systems all rely on some form of cryptographic hashing to commit to the proof and then open up a small number of bits chosen by the verifier. In the random oracle model [49], this kind of construction achieves quasi-optimal prover complexity, but not quasi-optimal succinctness [14, Remark 4.16]. In the standard model [9,11], additional cryptographic tools (notably, a private information retrieval protocol) are needed in the construction, which do not preserve the prover complexity of the underlying construction.

If instead we start with linear PCPs and apply the compilers in [13,14], the challenge is in constructing a quasi-optimal linear PCP that provides soundness error $2^{-\lambda}$ over a small field $\mathbb{F}$. As noted above, existing linear PCP constructions [13,14] are not quasi-optimal for Boolean circuit satisfiability.

1.2 Optimally-Laconic Arguments and 1-Bit SNARGs

More broadly, we can view our quasi-optimal SNARGs in the preprocessing model as a quasi-optimal *interactive* argument system with a maximally *laconic* prover. Here, we allow the verifier to send an arbitrarily long string (namely, the CRS), and our goal is to minimize the prover's computational cost and the number of bits the prover communicates to the verifier. Our quasi-optimal SNARG thus gives the first interactive argument system with a *quasi-optimal* laconic prover.

Optimally-laconic arguments and 1-bit SNARGs. Independent of our results on constructing quasi-optimal SNARGs, we also ask the question of what is the minimal proof length needed to ensure ρ bits of soundness where ρ is a concrete soundness parameter. Lemma 5.2 shows that achieving $2^{-\rho}$ soundness error only requires proofs of length $\Omega(\rho)$. When $\rho = \Omega(\lambda)$, many existing SNARG candidates, including the one we construct in this paper, are quasi-optimally succinct [13,14,29,37]. More generally, this question remains interesting when $\rho = o(\lambda)$, and even independently of achieving quasi-optimal prover complexity. A natural question to ask is whether there exist SNARGs where the size of the proofs achieves the lower bound of $\Omega(\rho)$ for providing ρ bits of soundness. Taken to the extreme, we ask whether there exists a 1-bit SNARG with soundness error $1/2 + \text{negl}(\lambda)$. We note that a 1-bit SNARG immediately implies an *optimally-succinct* SNARG for all soundness parameters ρ: namely, to build a SNARG with soundness error $2^{-\rho}$, we concatenate ρ independent instances of a 1-bit SNARG.

In the full version [15], we show that the designated-verifier analog of the Sahai-Waters [53] construction of non-interactive zero-knowledge proofs from indistinguishability obfuscation and one-way functions is a 1-bit SNARG. In the *interactive* setting, we show that we can construct 1-bit laconic arguments from witness encryption. We do not know how to build 1-bit SNARGs and 1-bit laconic arguments for general languages from weaker assumptions,[2] and leave this as an open problem.

The power of optimally-laconic arguments. Finally, we show an intriguing connection between 1-bit laconic arguments and a variant of witness encryption. Briefly, a witness encryption scheme [28] allows anyone to encrypt a message m with respect to a statement x in an NP language; then, anyone who holds a witness w for x is able to decrypt the ciphertext. In the full version [15], we

[2] Note that for some special languages such as graph non-isomorphism, we do have 1-bit laconic arguments [31].

show that a 1-bit laconic argument (or SNARG) for a cryptographically-hard[3] language $\mathcal{L}$ implies a relaxed form of witness encryption for $\mathcal{L}$ where semantic security holds for messages encrypted to a *random* false instance (as opposed to an arbitrary false instance in the standard definition). While this is a relaxation of the usual notion of witness encryption, it already suffices to realize some of the powerful applications of witness encryption described in [28]. This implication thus demonstrates the power of optimally-laconic arguments, as well as some of the potential challenges in constructing them from simple assumptions.

Our construction of witness encryption from 1-bit arguments relies on the observation that for a (random) false statement $\mathbf{x}$, any computationally-bounded prover can only produce a valid proof $\pi \in \{0,1\}$ with probability that is negligibly close to 1/2. Thus, the proof π can be used to hide the message m in a witness encryption scheme (when encrypting to the statement $\mathbf{x}$). Here, we implicitly assume that a (random) statement $\mathbf{x}$ has exactly one accepting proof—this assumption holds for any cryptographically-hard language. Essentially, our construction shows how to leverage the soundness property of a proof system to obtain a secrecy property in an encryption scheme. Previously, Applebaum et al. [1] showed how to leverage secrecy to obtain soundness, so in some sense, we can view our construction as a dual of their secrecy-to-soundness construction. The recent work of Berman et al. [8] also showed how to obtain public-key encryption from laconic *zero-knowledge* arguments. While their construction relies on the additional assumption of zero-knowledge, their construction does not require the argument system be optimally laconic.

We can also view a 1-bit argument for a cryptographically-hard language as a "predictable argument" (c.f., [25]). A predictable argument is one where there is exactly one accepting proof for any statement. Faonio et al. [25] show that any predictable argument gives a witness encryption scheme. In this work, we show that soundness *alone* suffices for this transformation, provided we make suitable restrictions on the underlying language.

1.3 Additional Related Work

Gentry and Wichs [30] showed that no construction of an *adaptively-secure* SNARG (for general NP languages) can be proven secure via a black-box reduction from any falsifiable cryptographic assumption [51].[4] As a result, most existing SNARG constructions (for general NP languages) in the standard model have relied on non-falsifiable assumptions such as knowledge-of-exponent assumptions [5,21,24,29,37,39,40,45–47,50], extractable collision-resistant hashing [9,10,22], extractable homomorphic encryption [12,29], and linear-only encryption [13,14]. Other constructions have relied on showing security in idealized models such as the random oracle model [49,59] or the generic

[3] Here, we say a language is cryptographically-hard if there exists a distribution over YES instances that is computationally indistinguishable from a distribution of NO instances for the language.

[4] In the case of non-adaptive SNARGs, Sahai and Waters give a construction from indistinguishability obfuscation and one-way functions [53].

group model [38]. In many of these constructions, the underlying SNARGs also satisfy a knowledge property, which says that whenever a prover generates an accepting proof π of a statement $\mathbf{x}$, there is an efficient extractor that can extract a witness $\mathbf{w}$ from π such that $C(\mathbf{x}, \mathbf{w}) = 1$. SNARGs with this property are called SNARGs of knowledge, or more commonly, SNARKs. In many cases, SNARGs also have a zero-knowledge property [13,24,29,37,39,45–47] which says that the proof π does not reveal any additional information about the witness $\mathbf{w}$ other than the fact that $C(\mathbf{x}, \mathbf{w}) = 1$.

A compelling application of succinct argument systems is to verifiable delegation of computation. Over the last few years, there has been significant progress in leveraging SNARGs (and their variants) for implementing scalable systems for verifiable computation both in the interactive setting [19,34,54,55,57,58,60–62] as well as the non-interactive setting [6,7,18,20,52,63]. We refer to [64] and the references therein for a more comprehensive survey of this area.

2 Quasi-Optimal Linear MIP Construction Overview

In this section, we give a technical overview of our quasi-optimal linear MIP construction for arithmetic circuit satisfiability over a finite field $\mathbb{F}$. Combined with our cryptographic compiler based on linear-only vector encryption over rings, this gives the first construction of a quasi-optimal SNARG from a concrete cryptographic assumption.

Robust circuit decomposition. The first ingredient we require in our quasi-optimal linear MIP construction is a *robust* way to decompose an arithmetic circuit $C\colon \mathbb{F}^{n'} \times \mathbb{F}^{m'} \to \mathbb{F}^{h'}$ into a collection of t constraint functions $f_1, \ldots, f_t$, where each constraint $f_i\colon \mathbb{F}^n \times \mathbb{F}^m \to \{0,1\}$ takes as input a common statement $\mathbf{x} \in \mathbb{F}^n$ and witness $\mathbf{w} \in \mathbb{F}^m$. More importantly, each constraint f_i can be computed by a small arithmetic circuit C_i of size roughly $|C|/t$. This means that each arithmetic circuit C_i may only need to read some subset of the components in $\mathbf{x}$ and $\mathbf{w}$. There is a mapping $\mathsf{inp}\colon \mathbb{F}^{n'} \to \mathbb{F}^n$ that takes as input a statement $\mathbf{x}'$ for C and outputs a statement $\mathbf{x}$ for $f_1, \ldots, f_t$, and another mapping $\mathsf{wit}\colon \mathbb{F}^{n'} \times \mathbb{F}^{m'} \to \mathbb{F}^m$ that takes as input a statement-witness pair $(\mathbf{x}', \mathbf{w}')$ for C, and outputs a witness $\mathbf{w}$ for $f_1, \ldots, f_t$. The decomposition must satisfy two properties: completeness and robustness. Completeness says that whenever a statement-witness pair $(\mathbf{x}', \mathbf{w}')$ is accepted by C, then $f_i(\mathbf{x}, \mathbf{w}) = 1$ for all i if we set $\mathbf{x} = \mathsf{inp}(\mathbf{x}')$ and $\mathbf{w} = \mathsf{wit}(\mathbf{x}', \mathbf{w}')$. Robustness says that for a false statement $\mathbf{x}' \in \mathbb{F}^{n'}$, there are no valid witnesses $\mathbf{w} \in \mathbb{F}^m$ that can simultaneously satisfy more than a constant fraction of the constraints $f_1(\mathbf{x}, \cdot), \ldots, f_t(\mathbf{x}, \cdot)$, where $\mathbf{x} = \mathsf{inp}(\mathbf{x}')$.

Roughly speaking, a robust decomposition allows us to reduce checking satisfiability of a large circuit C to checking satisfiability of many smaller circuits $C_1, \ldots, C_t$. The gain in performance will be due to our ability to check satisfiability of all of the $C_1, \ldots, C_t$ in parallel. The importance of robustness will be critical for soundness amplification. We give the formal definition of a robust decomposition in Sect. 4.1.

Instantiating the robust decomposition. In the full version [15], we describe one way of instantiating the robust decomposition by applying the "MPC-in-the-head" paradigm of [42] to MPC protocols with polylogarithmic overhead [23]. We give a brief overview here. For an arithmetic circuit $C \colon \mathbb{F}^{n'} \times \mathbb{F}^{m'} \to \mathbb{F}^{h'}$, the encoding of a statement-witness pair $(\mathbf{x}, \mathbf{w})$ will be the *views* of each party in a (simulated) t-party MPC protocol computing C on $(\mathbf{x}, \mathbf{w})$, where the bits of the input and witness are evenly distributed across the parties. Each of the constraint functions f_i checks that party i outputs 1 in the protocol execution (indicating an accepting input), and that the view of party i is *consistent* with the views of the other parties. This means that the only bits of the encoded witness that each constraint f_i needs to read are those that correspond to messages that were sent or received by party i. Then, using an MPC protocol where the computation and communication overhead is polylogarithmic in the circuit size (c.f., [23]), and where the computational burden is evenly distributed across the computing parties, each $f_1, \dots, f_t$ can be implemented by a circuit of size $\widetilde{O}(|C|/t)$. Robustness of the decomposition follows from security of the underlying MPC protocol. We give the complete description and analysis in the full version [15].

Blueprint for linear MIP construction. The high-level idea behind our quasi-optimal linear MIP construction is as follows. We first apply a robust circuit decomposition to the input circuit to obtain a collection of constraints $f_1, \dots, f_t$, which can be computed by smaller arithmetic circuits $C_1, \dots, C_t$, respectively. Each arithmetic circuit takes as input a subset of the components of the statement $\mathbf{x} \in \mathbb{F}^n$ and the witness $\mathbf{w} \in \mathbb{F}^m$. In the following, we write $\mathbf{x}_i$ and $\mathbf{w}_i$ to denote the subset of the components of $\mathbf{x}$ and $\mathbf{w}$, respectively, that circuit C_i reads. We can now construct a linear MIP with t provers as follows. A proof of a true statement $\mathbf{x}'$ with witness $\mathbf{w}'$ consists of t proof vectors $(\boldsymbol{\pi}_1, \dots, \boldsymbol{\pi}_t)$, where each proof $\boldsymbol{\pi}_i$ is a linear PCP proof that $C_i(\mathbf{x}_i, \cdot)$ is satisfiable. Then, in the linear MIP model, the verifier has oracle access to the linear functions $\boldsymbol{\pi}_1, \dots, \boldsymbol{\pi}_t$, which it can use to check satisfiability of $C_i(\mathbf{x}_i, \cdot)$. Completeness of this construction is immediate from completeness of the robust decomposition.

Soundness is more challenging to argue. For any false statement $\mathbf{x}'$, robustness of the decomposition of C only ensures that for any witness $\mathbf{w} \in \mathbb{F}^m$, at least a constant fraction of the constraints $f_i(\mathbf{x}, \mathbf{w})$ will not be satisfied, where $\mathbf{x} = \mathsf{inp}(\mathbf{x}')$. However, this does *not* imply that a constant fraction of the individual circuits $C_i(\mathbf{x}_i, \cdot)$ is unsatisfiable. For instance, for all i, there could exist some witness $\mathbf{w}_i$ such that $C_i(\mathbf{x}_i, \mathbf{w}_i) = 1$. This does *not* contradict the robustness of the decomposition so long as the set of all satisfying witnesses $\{\mathbf{w}_i\}$ contain many "inconsistent" assignments. More specifically, we can view each $\mathbf{w}_i$ as assigning values to some subset of the components of the overall witness $\mathbf{w}$, and we say that a collection of witnesses $\{\mathbf{w}_i\}$ is consistent if whenever two witnesses $\mathbf{w}_i$ and $\mathbf{w}_j$ assign a value to the same component of $\mathbf{w}$, they assign the *same* value. Thus, robustness only ensures that the prover cannot find a *consistent* set of witnesses $\{\mathbf{w}_i\}$ that can simultaneously satisfy more than a fraction of the circuits C_i.

Or equivalently, if $\mathbf{x}$ is the encoding of a false statement $\mathbf{x}'$, then a constant fraction of any set of witnesses $\{\mathbf{w}_i\}$ where $C_i(\mathbf{x}_i, \mathbf{w}_i) = 1$ must be mutually inconsistent.

The above analysis shows that it is insufficient for the prover to independently argue satisfiability of each circuit $C_i(\mathbf{x}_i, \cdot)$. Instead, we need the stronger requirement that the prover uses a *consistent* set of witnesses $\{\mathbf{w}_i\}$ when constructing its proofs $\boldsymbol{\pi}_1, \ldots, \boldsymbol{\pi}_t$. Thus, we need a way to bind each proof $\boldsymbol{\pi}_i$ to a specific witness $\mathbf{w}_i$, as well as a way for the verifier to check that the complete set of witnesses $\{\mathbf{w}_i\}$ are mutually consistent. For the first requirement, we introduce the notion of a *systematic linear PCP*, which is a linear PCP where the linear PCP proof vector $\boldsymbol{\pi}_i$ contains a copy of a witness $\mathbf{w}_i$ where $C_i(\mathbf{x}_i, \mathbf{w}_i) = 1$ (Definition 4.2). Now, given a collection of systematic linear PCP proofs $\boldsymbol{\pi}_1, \ldots, \boldsymbol{\pi}_t$, the verifier's goal is to decide whether the witnesses $\mathbf{w}_1, \ldots, \mathbf{w}_t$ embedded within $\boldsymbol{\pi}_1, \ldots, \boldsymbol{\pi}_t$ are mutually consistent. Since the witnesses $\mathbf{w}_i$ are part of the proof vectors $\boldsymbol{\pi}_i$, in the remainder of this section, we will simply assume that the verifier has oracle access to the linear function $\langle \mathbf{w}_i, \cdot \rangle$ for all i since such queries can be simulated using the proof oracle $\langle \boldsymbol{\pi}_i, \cdot \rangle$.

2.1 Consistency Checking

The robust decomposition ensures that for a false statement $\mathbf{x}'$, any collection of witnesses $\{\mathbf{w}_i\}$ where $C_i(\mathbf{x}_i, \mathbf{w}_i) = 1$ for all i is guaranteed to have many inconsistencies. In fact, there must always exists $\Omega(t)$ (mutually disjoint) pairs of witnesses that contain some inconsistency in their assignments. Ensuring soundness thus reduces to developing an efficient method for testing whether $\mathbf{w}_1, \ldots, \mathbf{w}_t$ constitute a consistent assignment to the components of $\mathbf{w}$ or not. This is the main technical challenge in constructing quasi-optimal linear MIPs, and our construction proceeds in several steps, which we describe below.

Notation. We begin by introducing some notation. First, we pack the different witnesses $\mathbf{w}_1, \ldots, \mathbf{w}_t \in \mathbb{F}^q$ into the rows of an *assignment matrix* $\mathbf{W} \in \mathbb{F}^{t \times q}$. Specifically, the ith row of $\mathbf{W}$ is the witness $\mathbf{w}_i$. Next, we define the *replication structure* for the circuits $C_1, \ldots, C_t$ to be a matrix $\mathbf{A} \in [m]^{t \times q}$. Here, the $(i, j)^{\text{th}}$ entry $\mathbf{A}_{i,j}$ encodes the index in $\mathbf{w} \in \mathbb{F}^m$ to which the jth entry in $\mathbf{w}_i$ corresponds. With this notation, we say that the collection of witnesses $\mathbf{w}_1, \ldots, \mathbf{w}_t$ are consistent if for all indices (i_1, j_1) and (i_2, j_2) where $\mathbf{A}_{i_1,j_1} = \mathbf{A}_{i_2,j_2}$, the assignment matrix satisfies $\mathbf{W}_{i_1,j_1} = \mathbf{W}_{i_2,j_2}$.

Checking global consistency. To check whether an assignment matrix $\mathbf{W} \in \mathbb{F}^{t \times q}$ is consistent with respect to the replication structure $\mathbf{A} \in [m]^{t \times q}$, we can leverage an idea from Groth [36], and subsequently used in [14,43] for performing similar kinds of consistency checks. The high-level idea is as follows. Take any index $z \in [m]$ and consider the positions $(i_1, j_1), \ldots, (i_d, j_d)$ where z appears in $\mathbf{A}$. In this way, we associate a disjoint set of Hamiltonian cycles over the entries of $\mathbf{A}$, one for each of the m components of $\mathbf{w}$. Let Π be a permutation over the entries in the matrix $\mathbf{A}$ such that Π splits into a product of the Hamiltonian cycles induced by the entries of $\mathbf{A}$. In particular, this means $\mathbf{A} = \Pi(\mathbf{A})$, and moreover, $\mathbf{W}$ is consistent with respect to $\mathbf{A}$ if and only if $\mathbf{W} = \Pi(\mathbf{W})$. The insight in [36] is that the relation $\mathbf{W} = \Pi(\mathbf{W})$ can be checked using two sets of linear queries. First, the verifier draws vectors $\mathbf{r}_1, \ldots, \mathbf{r}_t \xleftarrow{\text{R}} \mathbb{F}^q$ and defines the matrix $\mathbf{R} \in \mathbb{F}^{t \times q}$ to be the matrix whose rows are $\mathbf{r}_1, \ldots, \mathbf{r}_t$. Next, the verifier computes the permuted matrix $\mathbf{R}' \leftarrow \Pi(\mathbf{R})$. Let $\mathbf{r}'_1, \ldots, \mathbf{r}'_t$ be the rows of $\mathbf{R}'$. Similarly, let $\mathbf{w}_1, \ldots, \mathbf{w}_t$ be the rows of $\mathbf{W}$. Finally, the verifier queries the linear MIP oracles $\langle \mathbf{w}_i, \cdot \rangle$ on $\mathbf{r}_i$ and $\mathbf{r}'_i$ for all i and checks the relation

$$\sum_{i \in [t]} \langle \mathbf{w}_i, \mathbf{r}_i \rangle \stackrel{?}{=} \sum_{i \in [t]} \langle \mathbf{w}_i, \mathbf{r}'_i \rangle \in \mathbb{F}. \tag{2.1}$$

By construction of Π, if $\mathbf{W} = \Pi(\mathbf{W})$, this check always succeeds. However, if $\mathbf{W} \neq \Pi(\mathbf{W})$, then by the Schwartz-Zippel lemma, this check rejects with probability $1/|\mathbb{F}|$. When working over a polynomial-size field, this consistency check achieves $1/\text{poly}(\lambda)$ soundness (where λ is a security parameter). We can use repeated queries to amplify the soundness to $\text{negl}(\lambda)$ without sacrificing quasi-optimality. However, this approach cannot give a linear MIP with $2^{-\lambda}$ soundness and still retain prover overhead that is only polylogarithmic in λ (since we would require $\Omega(\lambda)$ repetitions). This is one of the key reasons the construction in [14] only achieves $\text{negl}(\lambda)$ soundness rather than $2^{-\lambda}$ soundness. To overcome this problem, we require a more robust consistency checking procedure.

Checking pairwise consistency. The consistency check described above and used in [14, 36, 43] is designed for checking *global* consistency of all of the assignments in $\mathbf{W} \in \mathbb{F}^{t \times q}$. The main disadvantage of performing the global consistency check in Eq. (2.1) is that it only provides soundness $1/|\mathbb{F}|$, which is insufficient when $\mathbb{F}$ is small (e.g., in the case of Boolean circuit satisfiability). One way to amplify soundness is to replace the single global consistency check with $t/2$ *pairwise* consistency checks, where each pairwise consistency check affirms that the assignments in a (mutually disjoint) pair of rows of $\mathbf{W}$ are self-consistent. Specifically, each of the $t/2$ checks consists of two queries $(\mathbf{r}_i, \mathbf{r}_j)$ and $(\mathbf{r}'_i, \mathbf{r}'_j)$ to $\langle \mathbf{w}_i, \cdot \rangle$ and $\langle \mathbf{w}_j, \cdot \rangle$, constructed in exactly the same manner as in the global consistency check, except specialized to only checking for consistency in the assignments to the variables in rows i and j. Since all of the pairwise consistency checks are independent, if there are $\Omega(t)$ pairs of inconsistent rows, the probability that all $t/2$ checks pass is bounded by $2^{-\Omega(t)}$. This means that for the same cost as performing a *single* global consistency check, the verifier can perform $\Omega(t)$ pairwise consistency checks. As long as many of the pairs of rows the verifier checks contain inconsistencies, we achieve soundness amplification.

Recall from earlier that our robust decomposition guarantees that whenever $\mathbf{x}_1, \ldots, \mathbf{x}_t$ correspond to a false statement, any collection of witnesses $\{\mathbf{w}_i\}$ where $C_i(\mathbf{x}_i, \mathbf{w}_i)$ is satisfied for all i necessarily contains many pairs $\mathbf{w}_i$ and $\mathbf{w}_j$ that are inconsistent. Equivalently, many pairs of rows in the assignment matrix $\mathbf{W}$ contain inconsistencies. Now, if the verifier knew which pairs of rows of $\mathbf{W}$ are inconsistent, then the verifier can apply a pairwise consistency check to detect an inconsistent $\mathbf{W}$ with high probability. The problem, however, is that the verifier does not know *a priori* which pairs of rows in $\mathbf{W}$ are inconsistent, and so, it is unclear how to choose the rows to check in the pairwise consistency test. However, if we make the stronger assumption that not only are there many pairs of rows in $\mathbf{W}$ that contain inconsistent assignments, but also, that most of these inconsistencies appear in *adjacent* rows, then we can use a pairwise consistency test (where each test checks for consistency between an adjacent pair of rows) to decide if $\mathbf{W}$ is consistent or not. When the assignment matrix $\mathbf{W}$ has many inconsistencies in pairs of adjacent rows, we say that the inconsistency pattern of $\mathbf{W}$ is "regular," and can be checked using a pairwise consistency test.

Regularity-inducing permutations. To leverage the pairwise consistency check, we require that the assignment matrix $\mathbf{W}$ has a regular inconsistency structure that is amenable to a pairwise consistency check. To ensure this, we introduce the notion of a *regularity-inducing permutation.* Our construction relies on the observation that the assignment matrix $\mathbf{W}$ is consistent with a replication structure $\mathbf{A}$ if and only if $\Pi(\mathbf{W})$ is consistent with $\Pi(\mathbf{A})$, where Π is an arbitrary permutation over the entries of a t-by-q matrix. Thus, if we want to check consistency of $\mathbf{W}$ with respect to $\mathbf{A}$, it suffices to check consistency of $\Pi(\mathbf{W})$ with respect to $\Pi(\mathbf{A})$. Then, we say that a specific permutation Π is regularity-inducing with respect to a replication structure $\mathbf{A}$ if whenever $\mathbf{W}$ has many pairs of inconsistent rows with respect to $\mathbf{A}$ (e.g., $\mathbf{W}$ is a set of accepting witnesses to a false statement), then $\Pi(\mathbf{W})$ has many inconsistencies in pairs of *adjacent* rows with respect to $\Pi(\mathbf{A})$. In other words, a regularity-inducing permutation shuffles the entries of the assignment matrix such that any inconsistency pattern in $\mathbf{W}$ maps to a regular inconsistency pattern according to the replication structure $\Pi(\mathbf{A})$. In the construction, instead of performing the pairwise consistency test on $\mathbf{W}$, which can have an arbitrary inconsistency pattern, we perform it on $\Pi(\mathbf{W})$, which has a regular inconsistency pattern. We define the notion more formally in Sect. 4.2 and show how to construct regularity-inducing permutations in the full version.

Decomposing the permutation. Suppose Π is a regularity-inducing permutation for the replication structure $\mathbf{A}$ associated with the circuits $C_1, \ldots, C_t$ from the robust decomposition of C. Robustness ensures that for any false statement $\mathbf{x}'$, for all collections of witnesses $\{\mathbf{w}_i\}$ where $C_i(\mathbf{x}_i, \mathbf{w}_i) = 1$ for all i, and $\mathbf{x} = \mathsf{inp}(\mathbf{x}')$, the permuted assignment matrix $\Pi(\mathbf{W})$ has inconsistencies in $\Omega(t)$ pairs of adjacent rows with respect to $\Pi(\mathbf{A})$. This can be detected with probability $1 - 2^{-\Omega(t)}$ by performing a pairwise consistency test on the matrix $\mathbf{W}' = \Pi(\mathbf{W})$. The problem, however, is that the verifier only has oracle access to $\langle \mathbf{w}_i, \cdot \rangle$, and it is unclear how to *efficiently* perform the pairwise consistency test on the permuted matrix $\mathbf{W}'$ given just oracle access to the rows $\mathbf{w}_i$ of the unpermuted matrix. Our solution here is to introduce another set of t linear MIP provers for each row $\mathbf{w}_i'$ of $\mathbf{W}' = \Pi(\mathbf{W})$. Thus, the verifier has oracle access to both the rows of the original assignment matrix $\mathbf{W}$, which it uses to check satisfiability of $C_i(\mathbf{x}_i, \cdot)$, as well as the rows of the permuted assignment matrix $\mathbf{W}'$, which it uses to check consistency of the assignments in $\mathbf{W}$. The verifier accepts only if both sets of checks pass. The problem with this basic approach is that there is no reason the prover chooses the matrix $\mathbf{W}'$ so as to satisfy the relation $\mathbf{W}' = \Pi(\mathbf{W})$. Thus, to ensure soundness from this approach, the verifier needs a mechanism to also check that $\mathbf{W}' = \Pi(\mathbf{W})$, given oracle access to the rows of $\mathbf{W}$ and $\mathbf{W}'$.

To facilitate this check, we decompose the permutation Π into a sequence of α permutations $(\Pi_1, \ldots, \Pi_\alpha)$ where $\Pi = \Pi_\alpha \circ \cdots \circ \Pi_1$. Moreover, each of the intermediate permutations Π_i has the property that they themselves can be decomposed into $t/2$ independent permutations, each of which only permutes entries that appear in 2 distinct rows of the matrix. This "2-locality" property on permutations is amenable to the linear MIP model, and we show in Construction 4.8 a way for the verifier to efficiently check that two matrices $\mathbf{W}$ and $\mathbf{W}'$ (approximately) satisfy the relation $\mathbf{W} = \Pi_i(\mathbf{W}')$, where Π_i is 2-locally decomposable. To complete the construction, we have the prover provide not just the matrix $\mathbf{W}$ and its permutation $\mathbf{W}'$, but all of the intermediate matrices $\mathbf{W}_i = (\Pi_i \circ \Pi_{i-1} \circ \cdots \circ \Pi_1)(\mathbf{W})$ for all $i = 1, \ldots, \alpha$. Since each of the intermediate permutations applied are 2-locally decomposable, there is an efficient procedure for the prover to check each relation $\mathbf{W}_i = \Pi_i(\mathbf{W}_{i-1})$, where we write $\mathbf{W}_0 = \mathbf{W}$ to denote the original assignment matrix. If each of the intermediate permutations are correctly implemented, then the verifier is assured that $\mathbf{W}' = \Pi(\mathbf{W})$, and it can apply the pairwise consistency check on $\mathbf{W}'$ to complete the verification process. We use a Beneš network to implement the decomposition. This ensures that the number of intermediate permutations required is only logarithmic in t, so introducing these additional steps only incurs logarithmic overhead, and does not compromise quasi-optimality of the resulting construction.

Randomized permutation decompositions. There is one additional complication in that the intermediate consistency checks $\mathbf{W}' \stackrel{?}{=} \Pi_i(\mathbf{W})$ are imperfect. They only ensure that *most* of the rows in $\mathbf{W}'$ agree with the corresponding rows in $\Pi_i(\mathbf{W})$. What this means is that when the prover crafts its sequence of permuted assignment matrices $\mathbf{W} = \mathbf{W}_0, \mathbf{W}_1, \ldots, \mathbf{W}_\alpha$, it is able to "correct" a small number of inconsistencies that appear in $\mathbf{W}$ in each step. Thus, we must ensure that for the particular inconsistency pattern that appears in $\mathbf{W}$, the prover is not able to find a sequence of matrices $\mathbf{W}_1, \ldots, \mathbf{W}_\alpha$, where each of them approximately implements the correct permutation at each step, but at the end, is able to correct all of the inconsistencies in $\mathbf{W}$. To achieve this, we rely on a *randomized permutation decomposition*, where the verifier samples a random sequence of intermediate permutations $\Pi_1, \ldots, \Pi_\alpha$ that collectively implement the target regularity-inducing permutation Π. There are a number of technicalities that arise in the construction and its analysis, and we refer to the full version [15] for the full description.

Putting the pieces together. To summarize, our quasi-optimal linear MIP for circuit satisfiability consists of two key components. First, we apply a robust decomposition to the circuit to obtain many constraints with the property that for a false statement, a malicious prover either cannot satisfy most of the constraints, or if it does satisfy all of the constraints, it must have used an assignment with many inconsistencies. The second key ingredient we introduce is an efficient way to check if there are many inconsistencies in the prover's assignments in the linear MIP model. Our construction here relies on first constructing a regularity-inducing permutation to enable a simple method for consistency checking, and then using a randomized permutation decomposition to enforce the consistency check. We give the formal description and analysis in Sect. 4.

3 Preliminaries

We begin by defining some notation. For an integer n, we write $[n]$ to denote the set of integers $\{1, \ldots, n\}$. We use bold uppercase letters (e.g., $\mathbf{A}, \mathbf{B}$) to denote matrices and bold lowercase letters (e.g., $\mathbf{x}, \mathbf{y}$) to denote vectors. For a matrix $\mathbf{A} \in \mathbb{F}^{t \times q}$ over a finite field $\mathbb{F}$, we write $\mathbf{A}_{[i_1,i_2]}$ (where $i_1, i_2 \in [t]$) to denote the sub-matrix of $\mathbf{A}$ containing rows i_1 through i_2 of $\mathbf{A}$ (inclusive). For $i \in [t]$ and $j \in [q]$, we use $\mathbf{A}_{i,j}$ and $\mathbf{A}[i,j]$ to refer to the entry in row i and column j of $\mathbf{A}$.

For a graph $\mathcal{G}$ with n nodes, labeled with the integers $1, \ldots, n$, a matching M is a set of edges $(i, k) \in [n] \times [n]$ with no common vertices. For a finite set S, we write $x \xleftarrow{\text{R}} S$ to denote that x is drawn uniformly at random from S. For a distribution D, we write $x \leftarrow D$ to denote a draw from distribution D. Unless otherwise noted, we write λ to denote the security parameter. We say that a function $f(\lambda)$ is negligible in λ if $f(\lambda) = o(1/\lambda^c)$ for all $c \in \mathbb{N}$. We write $f(\lambda) = \text{poly}(\lambda)$ to denote that f is bounded by some (fixed) polynomial in λ, and $f = \text{polylog}(\lambda)$ if f is bounded by a (fixed) polynomial in $\log \lambda$. We say that an algorithm is efficient if it runs in probabilistic polynomial time in the length of its input.

For a Boolean circuit $C \colon \{0,1\}^n \times \{0,1\}^m \to \{0,1\}$, the Boolean circuit satisfaction problem is defined by the relation $\mathcal{R}_C = \{(\mathbf{x}, \mathbf{w}) \in \mathbb{F}^n \times \mathbb{F}^m : C(\mathbf{x}, \mathbf{w}) = 1\}$. We refer to $\mathbf{x} \in \{0,1\}^n$ as the statement and $\mathbf{w} \in \{0,1\}^m$ as the witness. We write $\mathcal{L}_C$ to denote the language associated with $\mathcal{R}_C$: namely, the set of statements $\mathbf{x} \in \{0,1\}^n$ for which there exists a witness $\mathbf{w} \in \{0,1\}^m$ such that $C(\mathbf{x}, \mathbf{w}) = 1$. In many cases in this work, it will be more natural to work with arithmetic circuits. For an arithmetic circuit $C \colon \mathbb{F}^n \times \mathbb{F}^m \to \mathbb{F}^h$ over a finite field $\mathbb{F}$, we say that C is satisfied if on an input $(\mathbf{x}, \mathbf{w}) \in \mathbb{F}^n \times \mathbb{F}^m$, all of the outputs are 0. Specifically, we define the relation for arithmetic circuit satisfiability to be $\mathcal{R}_C = \left\{(\mathbf{x}, \mathbf{w}) \in \mathbb{F}^n \times \mathbb{F}^m : C(\mathbf{x}, \mathbf{w}) = \mathbf{0}^h\right\}$. We include additional preliminaries in the full version [15].

4 Quasi-Optimal Linear MIPs

In this section, we present our core information-theoretic construction of a linear MIP with quasi-optimal prover complexity. We refer to Sect. 2 for a high-level overview of the construction. In Sects. 4.1 and 4.2, we introduce the key building blocks underlying our construction. We give the full construction of our quasi-optimal linear MIP in Sect. 4.3. We show how to instantiate our core building blocks in the full version [15].

4.1 Robust Decomposition for Circuit Satisfiability

In this section, we formally define our notion of a robust decomposition of an arithmetic circuit. We refer to the technical overview in Sect. 2 for a high-level description of how we implement our decomposition by combining the MPC-in-the-head paradigm [42] with robust MPC protocols with polylogarithmic overhead [23]. We provide the complete description in the full version [15].

Definition 4.1 (Quasi-Optimal Robust Decomposition). *Let $C\colon \mathbb{F}^{n'} \times \mathbb{F}^{m'} \to \mathbb{F}^{h'}$ be an arithmetic circuit of size s over a finite field $\mathbb{F}$, $\mathcal{R}_C$ be its associated relation, and $\mathcal{L}_C \subseteq \mathbb{F}^{n'}$ be its associated language. A (t, δ)-robust decomposition of C consists of the following components:*

- *A collection of functions $f_1, \ldots, f_t$ where each function $f_i\colon \mathbb{F}^n \times \mathbb{F}^m \to \{0,1\}$ can be computed by an arithmetic circuit C_i of size $\widetilde{O}(s/t) + \mathrm{poly}(t, \log s)$. Note that a function f_i may only depend on a (fixed) subset of its input variables; in this case, its associated arithmetic circuit C_i only needs to take the (fixed) subset of dependent variables as input.*
- *An efficiently-computable mapping $\mathsf{inp}\colon \mathbb{F}^{n'} \to \mathbb{F}^n$ that maps between a statement $\mathbf{x}' \in \mathbb{F}^{n'}$ for C to a statement $\mathbf{x} \in \mathbb{F}^n$ for $f_1, \ldots, f_t$.*
- *An efficiently-computable mapping $\mathsf{wit}\colon \mathbb{F}^{n'} \times \mathbb{F}^{m'} \to \mathbb{F}^m$ that maps between a statement-witness pair $(\mathbf{x}', \mathbf{w}') \in \mathbb{F}^{n'} \times \mathbb{F}^{m'}$ to C to a witness $\mathbf{w} \in \mathbb{F}^m$ for $f_1, \ldots, f_t$.*

Moreover, the decomposition must satisfy the following properties:

- ***Completeness:*** *For all $(\mathbf{x}', \mathbf{w}') \in \mathcal{R}_C$, if we set $\mathbf{x} = \mathsf{inp}(\mathbf{x}')$ and $\mathbf{w} = \mathsf{wit}(\mathbf{x}', \mathbf{w}')$, then $f_i(\mathbf{x}, \mathbf{w}) = 1$ for all $i \in [t]$.*
- ***δ-Robustness:*** *For all statements $\mathbf{x}' \notin \mathcal{L}_C$, if we set $\mathbf{x} = \mathsf{inp}(\mathbf{x}')$, then it holds that for all $\mathbf{w} \in \mathbb{F}^m$, the set of indices $S_{\mathbf{w}} = \{i \in [t] : f_i(\mathbf{x}, \mathbf{w}) = 1\}$ satisfies $|S_{\mathbf{w}}| < \delta t$. In other words, any single witness $\mathbf{w}$ can only simultaneously satisfy at most a δ-fraction of the constraints.*
- ***Efficiency:*** *The mappings inp and wit can be computed by an arithmetic circuit of size $\widetilde{O}(s) + \mathrm{poly}(t, \log s)$.*

Systematic linear PCPs. Recall from Sect. 2 that our linear MIP for checking satisfiability of a circuit C begins by applying a robust decomposition to the circuit C. The MIP proof is comprised of linear PCP proofs $\boldsymbol{\pi}_1, \ldots, \boldsymbol{\pi}_t$ to show that each of the circuits $C_1(\mathbf{x}_1, \cdot), \ldots, C_t(\mathbf{x}_t, \cdot)$ in the robust decomposition of C is satisfiable. Here, $\mathbf{x}_i$ denotes the bits of the statement $\mathbf{x}$ that circuit C_i reads. To provide soundness, the verifier needs to perform a sequence of consistency checks to ensure that the proofs $\boldsymbol{\pi}_1, \ldots, \boldsymbol{\pi}_t$ are *consistent* with some witness $\mathbf{w}$. To facilitate this, we require that the underlying linear PCPs are *systematic*: namely, each proof $\boldsymbol{\pi}_i$ contains a copy of some witness $\mathbf{w}_i$ where $(\mathbf{x}_i, \mathbf{w}_i) \in \mathcal{R}_{C_i}$. The consistency check then affirms that the witnesses $\mathbf{w}_1, \ldots, \mathbf{w}_t$ associated with $\boldsymbol{\pi}_1, \ldots, \boldsymbol{\pi}_t$ are mutually consistent. We give the formal definition of a systematic linear PCP below, and then describe one such instantiation by Ben-Sasson et al. [6, Appendix E].

Definition 4.2 (Systematic Linear PCPs). *Let $(\mathcal{P}, \mathcal{V})$ be an input-oblivious k-query linear PCP for a relation $\mathcal{R}_C$ where $C \colon \mathbb{F}^n \times \mathbb{F}^m \to \mathbb{F}^h$. We say that $(\mathcal{P}, \mathcal{V})$ is* systematic *if the following conditions hold:*

- *On input a statement-witness pair $(\mathbf{x}, \mathbf{w}) \in \mathbb{F}^n \times \mathbb{F}^m$ the prover's output of $\mathcal{P}(\mathbf{x}, \mathbf{w})$ has the form $\boldsymbol{\pi} = [\mathbf{w}, \mathbf{p}] \in \mathbb{F}^d$, for some $\mathbf{p} \in \mathbb{F}^{d-m}$. In other words, the witness is included as part of the linear PCP proof vector.*
- *On input a statement $\mathbf{x}$ and given oracle access to a proof $\boldsymbol{\pi}^* = [\mathbf{w}^*, \mathbf{p}^*]$, the knowledge extractor $\mathcal{E}^{\boldsymbol{\pi}^*}(\mathbf{x})$ outputs $\mathbf{w}^*$.*

Fact 4.3 (*[6,* Claim E.3]). Let $C \colon \mathbb{F}^n \times \mathbb{F}^m \to \mathbb{F}^h$ be an arithmetic circuit of size s over a finite field $\mathbb{F}$ where $|\mathbb{F}| > s$. There exists a systematic input-oblivious 5-query linear PCP $(\mathcal{P}, \mathcal{V})$ for $\mathcal{R}_C$ *over* $\mathbb{F}$ with knowledge error $O(s/|\mathbb{F}|)$ and query length $O(s)$. Moreover, letting $\mathcal{V} = (\mathcal{Q}, \mathcal{D})$, the prover and verifier algorithms satisfy the following properties:

- the prover algorithm $\mathcal{P}$ is an arithmetic circuit of size $\widetilde{O}(s)$;
- the query-generation algorithm $\mathcal{Q}$ is an arithmetic circuit of size $O(s)$;
- the decision algorithm $\mathcal{D}$ is an arithmetic circuit of size $O(n)$.

4.2 Consistency Checking

As described in Sect. 2, in our linear MIP construction, we first apply a robust decomposition to the input circuit C to obtain smaller arithmetic circuits $C_1, \ldots, C_t$, each of which depends on some subset of the components of a witness $\mathbf{w} \in \mathbb{F}^m$. The proof then consists of a collection of systematic linear PCP proofs $\boldsymbol{\pi}_1, \ldots, \boldsymbol{\pi}_t$ that $C_1, \ldots, C_t$ are individually satisfiable. The second ingredient we require is a way for the verifier to check that the prover uses a consistent witness to construct the proofs $\boldsymbol{\pi}_1, \ldots, \boldsymbol{\pi}_t$. In this section, we formally introduce the building blocks we use for the consistency check. We refer to Sect. 2.1 for an overview of our methods. We begin by defining the notion of a replication structure induced by the decomposition $C_1, \ldots, C_t$, and what it means for a collection of assignments to the circuit $C_1, \ldots, C_t$ to be consistent.

Definition 4.4 (Replication Structures and Inconsistency Matrices). *Fix integers $m, t, q \in \mathbb{N}$. A* replication structure *is a matrix $\mathbf{A} \in [m]^{t \times q}$. We say that a matrix $\mathbf{W} \in \mathbb{F}^{t \times q}$ is consistent with respect to a replication structure $\mathbf{A}$ if for all $i_1, i_2 \in [t]$ and $j_1, j_2 \in [q]$, whenever $\mathbf{A}_{i_1,j_1} = \mathbf{A}_{i_2,j_2}$, $\mathbf{W}_{i_1,j_1} = \mathbf{W}_{i_2,j_2}$. If there is a pair of indices (i_1, j_1) and (i_2, j_2) where this relation does not hold, then we say that there is an inconsistency in $\mathbf{W}$ (with respect to $\mathbf{A}$) at locations (i_1, j_1) and (i_2, j_2). For a replication structure $\mathbf{A} \in [m]^{t \times q}$ and a matrix of values*

$\mathbf{W} \in \mathbb{F}^{t \times q}$, we define the inconsistency matrix $\mathbf{B} \in \{0,1\}^{t \times q}$ where $\mathbf{B}_{i,j} = 1$ if and only if there is an inconsistency in $\mathbf{W}$ at location (i,j) with respect to the replication structure $\mathbf{A}$. In the subsequent analysis, we will sometimes refer to an arbitrary inconsistency matrix $\mathbf{B} \in \{0,1\}^{t \times q}$ (independent of any particular set of values $\mathbf{W}$ or replication structure $\mathbf{A}$).

Definition 4.5 (Consistent Inputs to Circuits). *Let $C_1, \ldots, C_t$ be a collection of circuits where each $C_i \colon \mathbb{F}^m \to \mathbb{F}^h$ only depends on at most $q \leq m$ components of an input vector $\mathbf{w} \in \mathbb{F}^m$. For each $i \in [t]$, let $a_1^{(i)}, \ldots, a_q^{(i)} \in [m]$ be the indices of the q components of the input $\mathbf{w}$ on which C_i depends. The* replication structure *of $C_1, \ldots, C_t$ is the matrix $\mathbf{A} \in [m]^{t \times q}$, where the ith row of $\mathbf{A}$ is the vector $a_1^{(i)}, \ldots, a_q^{(i)}$ (namely, the subset of indices on which C_i depends). We say that a collection of inputs $\mathbf{w}_1, \ldots, \mathbf{w}_t \in \mathbb{F}^q$ to $C_1, \ldots, C_t$ is consistent if the assignment matrix $\mathbf{W}$, where the ith row of $\mathbf{W}$ is $\mathbf{w}_i$ for $i \in [t]$, is consistent with respect to the replication structure $\mathbf{A}$.*

To simplify the analysis, we introduce the notion of an inconsistency graph for an assignment matrix $\mathbf{W} \in \mathbb{F}^{t \times q}$ with respect to a replication structure $\mathbf{A} \in [m]^{t \times q}$. At a high level, the inconsistency graph of $\mathbf{W}$ with respect to $\mathbf{A}$ is a graph with t nodes, one for each row of $\mathbf{W}$, and there is an edge between two nodes $i, j \in [t]$ if assignments $\mathbf{w}_i$ and $\mathbf{w}_j$ (in rows i and j of $\mathbf{W}$, respectively) contain an inconsistent assignment with respect to $\mathbf{A}$.

Definition 4.6 (Inconsistency Graph). *Fix positive integers $m, t, q \in \mathbb{N}$ and take a replication structure $\mathbf{A} \in [m]^{t \times q}$. For any assignment matrix $\mathbf{W} \in \mathbb{F}^{t \times q}$, we define the inconsistency graph $\mathcal{G}_{\mathbf{W},\mathbf{A}}$ of $\mathbf{W}$ with respect to $\mathbf{A}$ as follows:*

- *Graph $\mathcal{G}_{\mathbf{W},\mathbf{A}}$ is an undirected graph with t nodes, with labels in $[t]$. We associate node $i \in [t]$ with the ith row of $\mathbf{A}$.*
- *Graph $\mathcal{G}_{\mathbf{W},\mathbf{A}}$ has an edge between nodes i_1 and i_2 if there exists $j_1, j_2 \in [q]$ such that $\mathbf{A}_{i_1,j_1} = \mathbf{A}_{i_2,j_2}$ but $\mathbf{W}_{i_1,j_1} \neq \mathbf{W}_{i_2,j_2}$. In other words, there is an edge in $\mathcal{G}_{\mathbf{W},\mathbf{A}}$ whenever there is an inconsistency in the assignments to rows i_1 and i_2 in $\mathbf{W}$ (with respect to the replication structure $\mathbf{A}$).*

Definition 4.7 (Regular Matchings). *Fix integers $m, t, q \in \mathbb{N}$ where t is even, and take any replication structure $\mathbf{A} \in [m]^{t \times q}$ and assignment matrix $\mathbf{W} \in \mathbb{F}^{t \times q}$. We say that the inconsistency graph $\mathcal{G}_{\mathbf{W},\mathbf{A}}$ contains a* regular *matching of size s if $\mathcal{G}_{\mathbf{W},\mathbf{A}}$ contains a matching M of size s, where each edge $(v_1, v_2) \in M$ satisfies $(v_1, v_2) = (2i-1, 2i)$ for some $i \in [t/2]$. In other words, all matched edges are between nodes corresponding to* adjacent *rows in $\mathbf{W}$.*

Having defined these notions, we can reformulate the guarantees provided by the (t, δ)-robust decomposition (Definition 4.1). For a constant $\delta > 0$, let $(f_1, \ldots, f_t, \mathsf{inp}, \mathsf{wit})$ be a (t, δ)-robust decomposition of a circuit C. Let $\mathbf{A}$ be the replication structure of the circuits $C_1, \ldots, C_t$ computing $f_1, \ldots, f_t$. Take any statement $\mathbf{x}' \notin \mathcal{L}_C$, and consider any collection of witnesses $\mathbf{w}_1, \ldots, \mathbf{w}_t$ where $C_i(\mathbf{x}_i, \mathbf{w}_i) = 1$ for all $i \in [t]$. As usual, $\mathbf{x}_i$ denotes the bits of $\mathbf{x} = \mathsf{inp}(\mathbf{x}')$ that C_i

reads. Robustness of the decomposition ensures that no single $\mathbf{w}$ can be used to simultaneously satisfy more than a δ-fraction of the constraints. In particular, this means that there must exist $\Omega(t)$ pairs of witnesses $\mathbf{w}_i$ and $\mathbf{w}_j$ which are inconsistent. Equivalently, we say that the inconsistency graph $\mathcal{G}_{\mathbf{W},\mathbf{A}}$ contains a matching of size $\Omega(t)$. We prove this statement formally in the full version [15].

Approximate consistency check. By relying on the robust decomposition, it suffices to construct a protocol where the verifier can detect whether the inconsistency graph $\mathcal{G}_{\mathbf{W},\mathbf{A}}$ of the prover's assignments $\mathbf{W}$ with respect to a replication structure $\mathbf{A}$ contains a large matching. To facilitate this, we first describe an algorithm to check whether two assignment matrices $\mathbf{W}, \mathbf{W}' \in \mathbb{F}^{t \times q}$ (approximately) satisfy the relation $\mathbf{W}' = \Pi(\mathbf{W})$ in the linear MIP model, where Π is a 2-locally decomposable permutation. This primitive can then be used directly to detect whether an inconsistency graph $\mathcal{G}_{\mathbf{W},\mathbf{A}}$ contains a *regular* matching (Corollary 4.11). Subsequently, we show how to permute the entries in $\mathbf{W}$ according to a permutation Π' so as to convert an arbitrary matching in $\mathcal{G}_{\mathbf{W},\mathbf{A}}$ into a regular matching in $\mathcal{G}_{\Pi'(\mathbf{W}),\Pi'(\mathbf{A})}$. Our construction of the approximate consistency check is a direct generalization of the pairwise consistency check procedure described in Sect. 2.1.

Construction 4.8 (Approximate Consistency Check). Fix an even integer $t \in \mathbb{N}$, and let $P_1, \ldots, P_t, P_1', \ldots, P_t'$ be a collection of $2 \cdot t$ provers in a linear MIP system. For $i \in [t]$, let $\boldsymbol{\pi}_i \in \mathbb{F}^d$ be the proof vector associated with prover P_i and $\boldsymbol{\pi}_i' \in \mathbb{F}^d$ be the proof vector associated with prover P_i'. We can associate a matrix $\mathbf{W} \in \mathbb{F}^{t \times d}$ with provers $(P_1, \ldots, P_t)$, where the ith row of $\mathbf{W}$ is $\boldsymbol{\pi}_i$. Similarly, we associate a matrix $\mathbf{W}'$ with provers $(P_1', \ldots, P_t')$. Let Π be a 2-locally decomposable permutation on the entries of a t-by-d matrix. Then, we describe the following linear MIP verification procedure for checking that $\mathbf{W}' \approx \Pi(\mathbf{W})$.

- **Verifier's query algorithm:** The verifier chooses a random matrix $\mathbf{R} \xleftarrow{\text{R}} \mathbb{F}^{t \times d}$, and sets $\mathbf{R}' \leftarrow \Pi(\mathbf{R})$. Let $\mathbf{r}_i$ and $\mathbf{r}_i'$ denote the ith row of $\mathbf{R}$ and $\mathbf{R}'$, respectively. The query algorithm outputs the query $\mathbf{r}_i$ for prover P_i and the query $\mathbf{r}_i'$ to prover P_i'.
- **Verifier's decision algorithm:** Since Π is 2-locally decomposable, we can decompose Π into $t' = t/2$ independent permutations, $\Pi_1, \ldots, \Pi_{t'}$, where each Π_i only operates on a pair of rows (j_{2i-1}, j_{2i}), for all $i \in [t']$. Given responses $\mathbf{y}_i = \langle \boldsymbol{\pi}_i, \mathbf{r}_i \rangle \in \mathbb{F}$ and $\mathbf{y}_i' = \langle \boldsymbol{\pi}_i', \mathbf{r}_i' \rangle \in \mathbb{F}$ for $i \in [t]$, the verifier checks that the relation
$$\mathbf{y}_{j_{2i-1}} + \mathbf{y}_{j_{2i}} \stackrel{?}{=} \mathbf{y}'_{j_{2i-1}} + \mathbf{y}'_{j_{2i}},$$
for all $i \in [t']$. The verifier accepts if the relations hold for all $i \in [t']$. Otherwise, it rejects.

By construction, we see that if $\mathbf{W}' = \Pi(\mathbf{W})$, then the verifier always accepts.

Lemma 4.9 (Consistency Check Soundness). *Define t, Π, $\mathbf{W}$, and $\mathbf{W}'$ as in Construction 4.8. Then, if the matrix $\mathbf{W}'$ disagrees with $\Pi(\mathbf{W})$ on κ rows, the verifier in Construction 4.8 will reject with probability at least $1 - 2^{-\Omega(\kappa)}$.*

Proof. Consider the event where $\mathbf{W}'$ disagrees with $\hat{\mathbf{W}} = \Pi(\mathbf{W})$ on κ rows. We show that the probability of the verifier accepting in this case is bounded by $2^{-\Omega(\kappa)}$. In the linear MIP model, the verifier's decision algorithm corresponds to checking the following relation:

$$\langle \boldsymbol{\pi}_{j_{2i}}, \mathbf{r}_{j_{2i}} \rangle + \langle \boldsymbol{\pi}_{j_{2i+1}}, \mathbf{r}_{j_{2i+1}} \rangle \stackrel{?}{=} \langle \boldsymbol{\pi}'_{j_{2i}}, \mathbf{r}'_{j_{2i}} \rangle + \langle \boldsymbol{\pi}'_{j_{2i+1}}, \mathbf{r}'_{j_{2i+1}} \rangle. \quad (4.1)$$

By assumption, there are at least $\kappa/2$ indices $i \in [t]$ where $\mathbf{W}'_{[j_{2i-1},j_{2i}]} \neq \hat{\mathbf{W}}_{[j_{2i-1},j_{2i}]}$. By the Schwartz-Zippel lemma, for the indices $i \in [t]$ where $\mathbf{W}'_{[j_{2i},j_{2i+1}]} \neq \hat{\mathbf{W}}_{[j_{2i},j_{2i+1}]}$, the relation in Eq. (4.1) holds with probability at most $1/|\mathbb{F}|$ (over the randomness used to sample $\mathbf{r}_{j_{2i-1}}$ and $\mathbf{r}_{j_{2i}}$) Since there are at least $\kappa/2$ such indices, the probability that Eq. (4.1) holds for all $i \in [t']$ is at most $(1/|\mathbb{F}|)^{\kappa/2} = 2^{-\Omega(\kappa)}$. Hence, the verifier rejects with probability $1 - 2^{-\Omega(\kappa)}$. □

The approximate consistency check from Construction 4.8 immediately gives a way to check whether an inconsistency graph $\mathcal{G}_{\mathbf{W},\mathbf{A}}$ contains a regular matching of size $\Omega(t)$. To show this, it suffices to exhibit a 2-locally decomposable permutation Π where the assignment matrix $\mathbf{W}$ is consistent on adjacent pairs of rows if and only if $\mathbf{W} = \Pi(\mathbf{W})$. The construction can be viewed as composing many copies of the global consistency check permutation used in [36] (and described in Sect. 2.1), each applied to a pair of adjacent rows. We give the construction below.

Construction 4.10 (Pairwise Consistency in Adjacent Rows). Fix integers $m, t, q \in \mathbb{N}$ with t even, and let $\mathbf{A} \in [m]^{t \times q}$ be a replication structure. Let $t' = t/2$. For each $i \in [t']$, let Π_i be a permutation over 2-by-q matrices such that Π_i splits into a disjoint set of Hamiltonian cycles based on the entries of $\mathbf{A}_{[2i-1,2i]}$. Define a permutation Π on t-by-q matrices where the action of Π on rows $2i-1$ and $2i$ is given by Π_i for all $i \in [t']$. By construction, the permutation Π is 2-locally decomposable, and moreover, $\mathbf{W} \in \mathbb{F}^{t \times q}$ is pairwise consistent on adjacent rows with respect to $\mathbf{A}$ if and only if $\mathbf{W} = \Pi(\mathbf{W})$.

Corollary 4.11. *Fix integers $m, t, q \in \mathbb{N}$ with t even. Let $\mathbf{A} \in [m]^{t \times q}$ be a replication structure, and Π be the pairwise consistency test permutation for $\mathbf{A}$ from Construction 4.10. Then, for any assignment matrix $\mathbf{W} \in \mathbb{F}^{t \times q}$ where the inconsistency graph $\mathcal{G}_{\mathbf{W},\mathbf{A}}$ contains a regular matching of size $\Omega(t)$, the verifier Construction 4.8 will reject the relation $\mathbf{W} \stackrel{?}{=} \Pi(\mathbf{W})$ with probability $1 - 2^{-\Omega(t)}$.*

Proof. Since $\mathcal{G}_{\mathbf{W},\mathbf{A}}$ contains a regular matching of size $\Omega(t)$, there are inconsistencies in $\Omega(t)$ pairs of adjacent rows of $\mathbf{W}$. By construction of Π, this means that $\mathbf{W}$ and $\Pi(\mathbf{W})$ differ on $\Omega(t)$ rows. The claim then follows by Lemma 4.9. □

Regularity-inducing permutations. Recall that our objective in the consistency check is to give an algorithm that detects whether an inconsistency graph $\mathcal{G}_{\mathbf{W},\mathbf{A}}$ contains a matching of size $\Omega(t)$. Corollary 4.11 gives a way to detect if the inconsistency graph $\mathcal{G}_{\mathbf{W},\mathbf{A}}$ contains a *regular* matching of size $\Omega(t)$ with soundness error $2^{-\Omega(t)}$. Thus, to perform the consistency check, we first construct a permutation Π on $\mathbf{W}$ such that whenever $\mathcal{G}_{\mathbf{W},\mathbf{A}}$ contain a matching of size $\Omega(t)$, the inconsistency graph $\mathcal{G}_{\Pi(\mathbf{W}),\Pi(\mathbf{A})}$ contains a regular matching of similar size $\Omega(t)$. We say that such permutations are *regularity-inducing*. While we are not able to construct a single permutation Π that is regularity-inducing for all assignment matrices $\mathbf{W}$, we are able to construct a *family* of permutations $(\Pi_1, \ldots, \Pi_z)$ for a fixed replication structure $\mathbf{A}$ such that for all assignment matrices $\mathbf{W} \in \mathbb{F}^{t \times q}$, there is at least one $\beta \in [z]$ where $\mathcal{G}_{\Pi_\beta(\mathbf{W}),\Pi_\beta(\mathbf{A})}$ contains a regular matching of size $\Omega(t)$.

Definition 4.12 (Regularity-Inducing Permutations). *Fix integers $m, t, q \in \mathbb{N}$, and let $\mathbf{A} \in [m]^{t \times q}$ be a replication structure. Let Π be a permutation on t-by-q matrices and $\mathbf{W} \in \mathbb{F}^{t \times q}$ be a matrix such that the inconsistency graph $\mathcal{G}_{\mathbf{W},\mathbf{A}}$ contains a matching M of size s. We say that Π is ρ-*regularity-inducing *for $\mathbf{W}$ with respect to $\mathbf{A}$ if the inconsistency graph $\mathcal{G}_{\Pi(\mathbf{W}),\Pi(\mathbf{A})}$ contains a* regular *matching M' of size at least s/ρ. Moreover, there is a one-to-one correspondence between the edges in M' and a subset of the edges in M (as determined by Π). We say that $(\Pi_1, \ldots, \Pi_z)$ is a collection of ρ-regularity-inducing permutations with respect to a replication structure $\mathbf{A}$ if for all $\mathbf{W} \in \mathbb{F}^{t \times q}$, there exists $\beta \in [z]$ such that Π_β is ρ-regularity-inducing for $\mathbf{W}$.*

In this work, we will construct regularity-inducing permutations where $\rho = O(1)$. To simplify the following description, we will implicitly assume that $\rho = O(1)$. Given an assignment matrix $\mathbf{W}$ and a collection of ρ-regularity-inducing permutations $(\Pi_1, \ldots, \Pi_z)$ for a replication structure $\mathbf{A}$, we can affirm that the inconsistency graph $\mathcal{G}_{\mathbf{W},\mathbf{A}}$ does not contain a matching of size $\Omega(t)$ by checking that each of the graphs $\mathcal{G}_{\Pi_\beta(\mathbf{W}),\Pi_\beta(\mathbf{A})}$ does not contain a regular matching of size $\Omega(t/\rho) = \Omega(t)$ for all $\beta \in [z]$ and assuming $\rho = O(1)$. By Corollary 4.11, each of these checks can be implemented in the linear MIP model using Construction 4.8. However, to apply the protocol in Construction 4.8 to $\Pi_\beta(\mathbf{W})$, the verifier requires oracle access to the individual rows of $\Pi_\beta(\mathbf{W})$. Thus, in the linear MIP construction, in addition to providing oracle access to the rows of the assignment matrix $\mathbf{W}$, we also provide the verifier oracle access to the rows of $\Pi_\beta(\mathbf{W})$ for all $\beta \in [z]$. Of course, a malicious MIP prover may provide the rows of a different matrix $\mathbf{W}' \in \mathbb{F}^{t \times q}$ (so as to pass the consistency check). Thus, the final ingredient we require is a way for the verifier to check that two matrices $\mathbf{W}, \mathbf{W}' \in \mathbb{F}^{t \times q}$ satisfy the relation $\mathbf{W}' = \Pi_\beta(\mathbf{W})$. Note that Construction 4.8 does not directly apply because the permutation Π_β is not necessarily 2-locally decomposable.

Decomposing the permutation. To complete the description, we now describe a way for the verifier to check that two matrices $\mathbf{W}, \mathbf{W}' \in \mathbb{F}^{t \times q}$ satisfy the relation $\mathbf{W}' = \Pi(\mathbf{W})$, for an *arbitrary* permutation Π. We assume that the verifier is given oracle access to the rows of $\mathbf{W}$ and $\mathbf{W}'$ in the linear MIP model. Construction 4.8 provides a way to check the relation whenever Π is 2-locally decomposable, so a natural starting point is to decompose the permutation Π

into a sequence of 2-locally-decomposable permutations $\Pi_1, \ldots, \Pi_\alpha$, where $\Pi = \Pi_\alpha \circ \cdots \circ \Pi_1$. Then, the linear MIP proof consists of the initial and final matrices $\mathbf{W}$ and $\mathbf{W}'$, as well as the intermediate matrices $\mathbf{W}_i = (\Pi_i \circ \cdots \circ \Pi_1)(\mathbf{W})$. The linear MIP proof would consist of the rows of all of the matrices $\mathbf{W} = \mathbf{W}_0, \mathbf{W}_1, \ldots, \mathbf{W}_\alpha = \mathbf{W}'$, and the verifier would apply Construction 4.8 to check that for all $\ell \in [\alpha]$, $\mathbf{W}_i = \Pi_i(\mathbf{W}_{i-1})$.

While this general approach seems sound, there is a subtle problem. The soundness guarantee for the consistency check in Construction 4.8 only states that on input $\mathbf{W}, \mathbf{W}'$ and a permutation Π, the verifier will only reject with probability $1 - 2^{\Omega(t)}$ when $\mathbf{W}'$ and $\Pi(\mathbf{W})$ differ on $\Omega(t)$ rows. This means that a malicious prover can provide a sequence of matrices $\mathbf{W}, \mathbf{W}_1, \ldots, \mathbf{W}_\alpha$ where each $\mathbf{W}_\ell$ differs from $\Pi_\ell(\mathbf{W}_{\ell-1})$ on a small number of rows (e.g., $o(t)$ rows), and in doing so, correct all of the inconsistent assignments that appear in the final matrix $\mathbf{W}_\alpha$.

Randomizing the decomposition. Abstractly, we can view the problem as follows. Let $\mathbf{B} \in \{0,1\}^{t \times q}$ be the inconsistency matrix for $\mathbf{W}$ with respect to $\mathbf{A}$ (Definition 4.4). In other words, $\mathbf{B}_{i,j} = 1$ whenever $\mathbf{W}_{i,j}$ encodes a value that is inconsistent with another assignment elsewhere in $\mathbf{W}$. Since $\mathcal{G}_{\mathbf{W},\mathbf{A}}$ contains a matching of size $\Omega(t)$, we know that there are at least $\Omega(t)$ rows in $\mathbf{B}$ that contain a 1. The permutation Π is chosen so that $\Pi(\mathbf{W})$ has a regular matching of size $\Omega(t)$ with respect to $\Pi(\mathbf{A})$. In particular, this means that the permuted inconsistency matrix $\Pi(\mathbf{B})$ contains a 1 in $\Omega(t)$ adjacent pairs of rows.

Consider the sequence of matrices $\mathbf{W}_1, \ldots, \mathbf{W}_\alpha$ chosen by the prover. Using the approximate pairwise consistency check, we can ensure that $\mathbf{W}_i$ agrees with $\Pi_i(\mathbf{W}_{i-1})$ on all but some κ_1 rows. Now suppose that there exists some $\ell \in [\alpha]$ where $\mathbf{B}_\ell = (\Pi_\ell \circ \cdots \circ \Pi_1)(\mathbf{B})$ has the property that all of the locations with a 1 in $\mathbf{B}$ appear in just κ_1 rows of $\mathbf{B}_\ell$. If this happens, then the malicious prover can construct $\mathbf{W}_1, \ldots, \mathbf{W}_{\ell-1}$ honestly, and then choose $\mathbf{W}_\ell$ such that $\mathbf{W}_\ell = \Pi_\ell(\mathbf{W}_{\ell-1})$ on all rows where $\mathbf{B}_\ell$ does not contain a 1, and set the values in the rows where $\mathbf{B}_\ell$ does contain a 1 to be consistent with the other rows of $\mathbf{W}$. Notably, all the entries in $\mathbf{W}_\ell$ are now consistent, and moreover, $\mathbf{W}_\ell$ differs from $\Pi_\ell(\mathbf{W}_{\ell-1})$ on at most κ_1 rows (and so, will not be detected with high probability by the pairwise consistency check). This means that from the verifier's perspective, the final matrix $\Pi(\mathbf{W})$ has no inconsistencies, and thus, the verifier's final pairwise consistency check passes with probability 1 (even though the original inconsistency graph $\mathcal{G}_{\mathbf{W},\mathbf{A}}$ contains a matching of size $\Omega(t)$). Thus, we require a stronger property on the permutation decomposition. It is not sufficient that there is a matching of size $\Omega(t)$ in the starting and ending configurations $\mathbf{W}$ and $\mathbf{W}'$. Rather, we need that the size of the matching in *every* step of the decomposition cannot shrink by too much, or equivalently, the intermediate permutations $\Pi_1, \ldots, \Pi_\alpha$ cannot "concentrate" all of the inconsistencies in $\mathbf{W}$ into a small number of rows (which the malicious prover can fix without being detected). We say permutation decompositions with this property are *non-concentrating.* We now formally define the notion of a non-concentrating permutation decomposition and what it means for a collection of permutation sequences to be non-concentrating.

Definition 4.13 (Non-concentrating Permutations). *Fix positive integers* $t, q \in \mathbb{N}$*, and let* $\Gamma = (\Pi_1, \ldots, \Pi_\alpha)$ *be a sequence of permutations over t-by-q matrices. Let* $\mathbf{B} \in \{0,1\}^{t \times q}$ *be an inconsistency matrix. For* $\ell \in [\alpha]$*, define* $\mathbf{B}_\ell = (\Pi_\ell \circ \cdots \circ \Pi_1)(\mathbf{B})$*. We say that* Γ *is a sequence of* (κ_1, κ_2)-non-concentrating permutations *with respect to* $\mathbf{B}$ *if for all* $\ell \in [\alpha]$*, the inconsistency matrix* $\mathbf{B}_\ell$ *has the property that no subset of* κ_1 *rows contains more than* κ_2 *inconsistencies (indices where the value is* 1*). Next, we say a collection of permutation sequences* $\Gamma^{(1)}, \ldots, \Gamma^{(\gamma)}$ *where each* $\Gamma^{(j)} = \left(\Pi_1^{(j)}, \ldots, \Pi_\alpha^{(j)}\right)$ *is* (κ_1, κ_2)*-non-concentrating for a set* $\mathcal{B} \subseteq \{0,1\}^{t \times q}$ *of inconsistency matrices if for all* $\mathbf{B} \in \mathcal{B}$*, there is some* $j \in [\gamma]$ *such that* $\Gamma^{(j)}$ *is* (κ_1, κ_2)*-non-concentrating with respect to* $\mathbf{B}$.

Putting the pieces together. To summarize, the goal of the consistency check is to decide whether the inconsistency graph $\mathcal{G}_{\mathbf{W},\mathbf{A}}$ of some assignment matrix $\mathbf{W}$ with respect to a replication structure $\mathbf{A}$ contains a matching of size $\Omega(t)$. Our strategy relies on the following:

- Let $(\Pi_1, \ldots, \Pi_z)$ be a collection of regularity-inducing permutations with respect to $\mathbf{A}$.
- For each $\beta \in [z]$, let $\Gamma_\beta^{(1)}, \ldots, \Gamma_\beta^{(\gamma)}$ be a collection of non-concentrating permutations that implement Π_β, where $\Gamma_\beta^{(j)} = (\Pi_{\beta,1}^{(j)}, \ldots, \Pi_{\beta,\alpha}^{(j)})$ for all $j \in [\gamma]$, and each of the intermediate permutations $\Pi_{\beta,\ell}^{(j)}$ are 2-locally decomposable for all $j \in [\gamma]$, $\beta \in [z]$, and $\ell \in [\alpha]$.

The proof then consists of the initial assignment matrix $\mathbf{W}$ in addition to all of the intermediate matrices $\mathbf{W}_{\beta,\ell}^{(j)} = \Pi_{\beta,\ell}^{(j)}(\mathbf{W}_{\beta,\ell-1}^{(j)})$, where we define $\mathbf{W}_{\beta,0}^{(j)} = \mathbf{W}$ for all $j \in [\gamma]$, $\beta \in [z]$. The verifier checks consistency of all of the intermediate matrices using Construction 4.8, and applies a pairwise consistency test (Construction 4.10) to each of $\mathbf{W}_{\beta,\alpha}^{(j)}$ for all $j \in [\gamma]$ and $\beta \in [z]$. The soundness argument then proceeds roughly as follows:

- Since $(\Pi_1, \ldots, \Pi_z)$ is regularity-inducing, there is some $\beta \in [z]$ where $\mathcal{G}_{\Pi_\beta(\mathbf{W}),\Pi_\beta(\mathbf{A})}$ contains a regular matching.
- Since $\Gamma_\beta^{(1)}, \ldots, \Gamma_\beta^{(\gamma)}$ is a collection of non-concentrating permutations that implement Π_β, and all of the intermediate consistency checks pass, then there must be some $j \in [\gamma]$ such that $\mathcal{G}_{\mathbf{W}_{\beta,\alpha}^{(j)},\Pi_\beta(\mathbf{A})}$ contains a regular matching of size $\Omega(t)$. The verifier then rejects with exponentially-small probability (in t) by soundness of the pairwise consistency test.

Finally, in our concrete instantiation (described in the full version [15]), we show how to construct our collection of regularity-inducing permutations and non-concentrating permutations sequences where $z = O(1)$, $\gamma = O(\log^3 t)$, $\alpha = \Theta(\log t)$. For this setting of parameters, the overall consistency check only incurs *polylogarithmic* overhead to the prover complexity and the proof size. In Sect. 4.3, we give the formal description and analysis of our linear MIP construction.

4.3 Quasi-Optimal Linear MIP Construction

In this section, we describe our quasi-optimal linear MIP for circuit satisfiability. We give our construction (Construction 4.14) but defer the security theorem and analysis to the full version. By instantiating Construction 4.14 with the appropriate primitives, we obtain the first quasi-optimal linear MIP (Theorem 4.15).

Construction 4.14 (Linear MIP). Fix parameters $t, \delta, k, \varepsilon, d, \rho, \kappa_1, \kappa_2$, and let C be an arithmetic circuit of size s over a finite field $\mathbb{F}$. The construction relies on the following ingredients:

- Let $(f_1, \ldots, f_t, \mathsf{inp}, \mathsf{wit})$ be a quasi-optimal (t, δ)-robust decomposition of C. Let C_i be the arithmetic circuit that computes each constraint $f_i \colon \mathbb{F}^n \times \mathbb{F}^m \to \{0,1\}$.
- Let $(\mathcal{P}_1, \mathcal{V}_1), \ldots, (\mathcal{P}_t, \mathcal{V}_t)$ be k-query systematic linear PCP systems for circuits $C_1, \ldots, C_t$, respectively, with knowledge error ε and query length d.
- Let $\mathbf{A} \in [m]^{t \times q}$ be the replication structure of $C_1, \ldots, C_t$ (where q is a bound on the number of indices in a witness $\mathbf{w} \in \mathbb{F}^m$ on which each circuit depends). Let $\Pi_1, \ldots, \Pi_z$ be a collection of ρ-regularity-inducing permutations on t-by-q matrices with respect to the replication structure $\mathbf{A}$ (Definition 4.12).
- For $\beta \in [z]$, let $\mathcal{B}_\beta \subseteq \{0,1\}^{t \times q}$ be the set of inconsistency patterns where $\mathbf{B}$ and $\Pi_\beta(\mathbf{B})$ have at most one inconsistency in each row. Let $\Gamma_\beta^{(1)}, \ldots, \Gamma_\beta^{(\gamma)}$ be a collection of permutation sequences implementing Π_β that is (κ_1, κ_2)-non-concentrating for $\mathcal{B}_\beta$ (Definition 4.13). In particular, each $\Gamma_\beta^{(j)}$ is a sequence of α permutations $(\Pi_{\beta,1}^{(j)}, \ldots, \Pi_{\beta,\alpha}^{(j)})$, where each intermediate permutation $\Pi_{\beta,\ell}^{(j)}$ is 2-locally decomposable.

The linear MIP with $t \cdot (1 + \alpha\gamma z)$ provers and query length d is defined as follows:

- **Syntax:** The linear MIP consists of $t \cdot (1 + \alpha\gamma z)$ provers. We label the provers as P_i and $P_{\beta,\ell,i}^{(j)}$ for $i \in [t]$, $j \in [\gamma]$, $\beta \in [z]$, and $\ell \in [\alpha]$. To simplify the description, we will often pack the proof vectors from different provers into the rows of a matrix. To recall, when we say we associate a matrix $\hat{\mathbf{W}} \in \mathbb{F}^{t \times d}$ with provers $(P_1, \ldots, P_t)$, we mean that the ith row of $\hat{\mathbf{W}}$ is the proof vector assigned to prover P_i for all $i \in [t]$. Similarly, when we say the verifier distributes a query matrix $\mathbf{Q} \in \mathbb{F}^{t \times d}$ to provers $(P_1, \ldots, P_t)$, we mean that it submits the ith row of $\mathbf{Q}$ as a query to P_i for all $i \in [t]$.
- **Prover's algorithm:** On input the statement $\mathbf{x}' \in \mathbb{F}^{n'}$ and witness $\mathbf{w}' \in \mathbb{F}^{m'}$, the prover prepares the proof vectors as follows:
  - **Linear PCP proofs.** First, the prover computes $\mathbf{x} \leftarrow \mathsf{inp}(\mathbf{x}')$ and $\mathbf{w} \leftarrow \mathsf{wit}(\mathbf{x}', \mathbf{w}')$. For each $i \in [t]$, it computes a proof $\boldsymbol{\pi}_i \leftarrow \mathcal{P}_i(\mathbf{x}_i, \mathbf{w}_i)$, where $\mathbf{x}_i$ and $\mathbf{w}_i$ denote the bits of the statement $\mathbf{x}$ and witness $\mathbf{w}$ on which circuit C_i depends, respectively. Since $(\mathcal{P}_i, \mathcal{V}_i)$ is a systematic linear PCP, we can write $\boldsymbol{\pi}_i = [\mathbf{w}_i, \mathbf{p}_i]$ where $\mathbf{w}_i \in \mathbb{F}^q$ and $\mathbf{p}_i \in \mathbb{F}^{d-q}$. For $i \in [t]$, the prover associates the vector $\boldsymbol{\pi}_i$ with P_i.

- **Consistency proofs.** Let $\mathbf{W} \in \mathbb{F}^{t \times q}$ be the matrix where the ith row is the vector $\mathbf{w}_i$. Now, for all $j \in [\gamma]$, $\beta \in [z]$, and $\ell \in [\alpha]$, let $\mathbf{W}_{\beta,\ell}^{(j)} = (\Pi_{\beta,\ell}^{(j)} \circ \Pi_{\beta,\ell-1}^{(j)} \circ \cdots \circ \Pi_{\beta,1}^{(j)})(\mathbf{W})$. Let $\hat{\mathbf{W}}_{\beta,\ell}^{(j)} = [\mathbf{W}_{\beta,\ell}^{(j)}, \mathbf{0}^{t \times (d-q)}]$. The prover associates $\hat{\mathbf{W}}_{\beta,\ell}^{(j)}$ with provers $(P_{\beta,\ell,1}^{(j)}, \ldots, P_{\beta,\ell,t}^{(j)})$.

– **Verifier's query algorithm:** To simplify the description, we will sometimes state the query vectors the verifier submits to each prover P_i and $P_{\beta,\ell,i}^{(j)}$ rather than the explicit query matrices. The verifier's queries are constructed as follows:
  - **Linear PCP queries.** For $i \in [t]$, the verifier invokes the query generation algorithm $\mathcal{Q}_i$ for each of the underlying linear PCP instances $(\mathcal{P}_i, \mathcal{V}_i)$ to obtain a query matrix $\mathbf{Q}_i \in \mathbb{F}^{d \times k}$ and some state information st_i. The verifier gives $\mathbf{Q}_i$ to prover P_i, and saves the state $\mathsf{st} = (\mathsf{st}_1, \ldots, \mathsf{st}_t)$.
  - **Routing consistency queries.** For all $j \in [\gamma]$, $\beta \in [z]$, and $\ell \in [\alpha]$, the verifier invokes the query generation algorithm of Construction 4.8 on permutation $\Pi_{\beta,\ell}^{(j)}$ to obtain two query matrices $\mathbf{R}_{\beta,\ell}^{(j)}$ and $\mathbf{S}_{\beta,\ell}^{(j)} \in \mathbb{F}^{t \times q}$. The verifier pads the matrices to obtain $\hat{\mathbf{R}}_{\beta,\ell}^{(j)} = [\mathbf{R}_{\beta,\ell}^{(j)}, \mathbf{0}^{t \times (d-q)}]$ and $\hat{\mathbf{S}}_{\beta,\ell}^{(j)} = [\mathbf{S}_{\beta,\ell}^{(j)}, \mathbf{0}^{t \times (d-q)}]$. There are two cases:
    * If $\ell = 1$, the verifier distributes the queries $\hat{\mathbf{R}}_{\beta,\ell}^{(j)}$ to provers $(P_1, \ldots, P_t)$.
    * If $\ell > 1$, the verifier distributes the queries $\hat{\mathbf{R}}_{\beta,\ell}^{(j)}$ to provers $(P_{\beta,\ell-1,1}^{(j)}, \ldots, P_{\beta,\ell-1,t}^{(j)})$.

    In addition, the verifier distributes the queries $\hat{\mathbf{S}}_{\beta,\ell}^{(j)}$ to provers $(P_{\beta,\ell,1}^{(j)}, \ldots, P_{\beta,\ell,t}^{(j)})$. Intuitively, the verifier is applying the approximate consistency check from Construction 4.8 to every permutation $\Pi_{\beta,\ell}^{(j)}$.
  - **Pairwise consistency queries.** For each $\beta \in [z]$, let $\mathbf{A}_\beta = \Pi_\beta(\mathbf{A})$, and let Π'_β be the pairwise consistency test matrix for $\mathbf{A}_\beta$ (Construction 4.10). The verifier invokes the query generation algorithm of Construction 4.8 on permutation Π'_β to obtain two query matrices $\mathbf{R}_\beta$ and $\mathbf{S}_\beta \in \mathbb{F}^{t \times q}$. It pads the matrices to obtain $\hat{\mathbf{R}}_\beta = [\mathbf{R}_\beta, \mathbf{0}^{t \times (d-q)}]$ and $\hat{\mathbf{S}}_\beta = [\mathbf{S}_\beta, \mathbf{0}^{t \times (d-q)}]$. Next, it distributes $\hat{\mathbf{R}}_\beta$ and $\hat{\mathbf{S}}_\beta$ to $(P_{\beta,\alpha,1}^{(j)}, \ldots, P_{\beta,\alpha,t}^{(j)})$ for all $j \in [\gamma]$. In this step, the verifier is checking pairwise consistency of the permuted assignment matrices $\mathbf{W}_{\beta,\alpha}^{(j)}$ for all $j \in [\gamma]$ and $\beta \in [z]$.

  In total, the verifier makes a total of $k + \alpha\gamma z$ queries to each prover P_i for $i \in [t]$. It makes $O(1)$ queries to the other provers.

– **Verifier's decision algorithm:** First, the verifier computes the statement $\mathbf{x} \leftarrow \mathsf{inp}(\mathbf{x}')$. For $i \in [t]$, let $\mathbf{x}_i$ denote the bits of $\mathbf{x}$ on which circuit C_i depends. The verifier processes the responses from each set of queries as follows:
  - **Linear PCP queries.** For $i \in [t]$, let $\mathbf{y}_i \in \mathbb{F}^k$ be the response of prover P_i to the linear PCP queries. For $i \in [t]$, the verifier invokes the decision algorithm $\mathcal{D}_i$ for each of the underlying linear PCP instances $(\mathcal{P}_i, \mathcal{V}_i)$ on the state st_i, the statement $\mathbf{x}_i$, and the response $\mathbf{y}_i$. It rejects the proof if $\mathcal{D}_i(\mathsf{st}_i, \mathbf{x}_i, \mathbf{y}_i) = 0$ for any $i \in [t]$.

- **Consistency queries.** For each set of routing consistency query responses (for checking consistency of the intermediate permutations $\Pi_{\beta,\ell}^{(j)}$), and for each set of pairwise consistency query responses (for checking consistency of the final configurations Π'_β), the verifier applies the decision algorithm from Construction 4.8, and rejects if any check fails.

If all of the checks pass, then the verifier accepts the proof.

Instantiating the construction. We defer the security analysis of Construction 4.14 to the full version [15]. In the full version, we additionally show how to instantiate the robust decomposition, the regularity-inducing permutations, and the non-concentrating permutation sequences needed to apply Construction 4.14. Combining Construction 4.14 with our concrete instantiations, we obtain a quasi-optimal linear MIP. We state the formal theorem below, and give the proof in the full version.

Theorem 4.15 (Quasi-Optimal Linear MIP). *Fix a security parameter λ. Let $C\colon \mathbb{F}^n \times \mathbb{F}^m \to \mathbb{F}^h$ be an arithmetic circuit of size s over a* $\mathrm{poly}(\lambda)$*-size finite field $\mathbb{F}$ where $|\mathbb{F}| > s$. Then, there exists an input-oblivious k-query linear MIP $(\mathcal{P}, \mathcal{V})$ with $\ell = \widetilde{O}(\lambda)$ provers for $\mathcal{R}_C$ with soundness error $2^{-\lambda}$, query length $\widetilde{O}(s/\lambda) + \mathrm{poly}(\lambda, \log s)$, and $k = \mathrm{polylog}(\lambda)$. Moreover, letting $\mathcal{V} = (\mathcal{Q}, \mathcal{D})$, the prover and verifier algorithms satisfy the following properties:*

- *the prover algorithm $\mathcal{P}$ is an arithmetic circuit of size $\widetilde{O}(s) + \mathrm{poly}(\lambda, \log s)$;*
- *the query-generation algorithm $\mathcal{Q}$ is an arithmetic circuit of size $\widetilde{O}(s) + \mathrm{poly}(\lambda, \log s)$;*
- *the decision algorithm $\mathcal{D}$ is an arithmetic circuit of size $\widetilde{O}(\lambda n)$.*

Remark 4.16 (Soundness Against Affine Provers). To leverage our linear MIP to construct a SNARG, we often require that the linear MIP provide soundness against affine provers. We note that Construction 4.14 inherits this property as long as the underlying linear PCPs and approximate consistency check primitives provide soundness against affine strategies. It is straightforward to see that Construction 4.8 remains sound even against affine adversarial strategies, and in the full version, we show how the underlying linear PCPs can be made robust against affine strategies with minimal overhead. Importantly, these modifications do not increase the asymptotic complexity of Construction 4.14.

5 Quasi-Optimal SNARGs

In this section, we formally introduce the notion of a quasi-optimal SNARG. Next, in Sect. 5.2, we show how to compile a linear MIP into a designated-verifier SNARG in the preprocessing model using the notion of a linear-only vector encryption over rings introduced in [14]. Combined with our quasi-optimal linear MIP from Sect. 4, this yields a quasi-optimal designated-verifier SNARG for Boolean circuit satisfiability in the preprocessing model. We refer to the full version [15] for the formal definition of a succinct non-interactive argument

(SNARG) and for the definitions of a linear-only vector encryption that we use in our construction. We also introduce the notion of a 1-bit SNARG in the full version.

5.1 Defining Quasi-Optimality

In this section, we formally define our notion of a quasi-optimal SNARG. Then, in the full version, we compare our notion to the previous notion of quasi-optimality introduced in [14], as well as describe a heuristic approach for instantiating quasi-optimal SNARGs.

Definition 5.1 (Quasi-Optimal SNARG). *Let* $\Pi_{\mathsf{SNARG}} = (\mathsf{Setup}, \mathsf{Prove}, \mathsf{Verify})$ *be a SNARG for a family of Boolean circuits* $\mathcal{C} = \{C_n\}_{n \in \mathbb{N}}$. *Then,* Π_{SNARG} *is* quasi-optimal *if it achieves* $2^{-\lambda}$ *soundness error against provers of size* 2^{λ} *and satisfies the following properties:*

- ***Prover Complexity:*** *The running time of* Prove *is* $\widetilde{O}(|C_n|) + \mathrm{poly}(\lambda, \log |C_n|)$.
- ***Succinctness:*** *The length of the proof output by* Prove *is* $\widetilde{O}(\lambda)$.

Next, in Lemma 5.2, we show that our notion of quasi-optimality is tight in the following sense: assuming NP does not have succinct *proofs*, any argument system for NP that provides soundness error $2^{-\lambda}$ must have proofs of length $\Omega(\lambda)$. We state the lemma below and give the proof in the full version [15].

Lemma 5.2. *Let* $\mathcal{C} = \{C_n\}_{n \in \mathbb{N}}$ *be a family of Boolean circuits for some language* $\mathcal{L} = \bigcup_{n \in \mathbb{N}} \mathcal{L}_{C_n}$, *where* $C_n \colon \{0,1\}^n \times \{0,1\}^{m(n)} \to \{0,1\}$ *for all* $n \in \mathbb{N}$. *Fix a soundness parameter* ρ *and a security parameter* λ. *Let* $\Pi_{\mathsf{SNARG}} = (\mathsf{Setup}, \mathsf{Prove}, \mathsf{Verify})$ *be a SNARG for* $\mathcal{C}$ *with soundness* $2^{-\rho}$ *against provers of size* $\mathrm{poly}(\lambda)$. *If* $\mathcal{L}_{C_n} \not\subseteq \mathbf{DTIME}(2^{o(n)})$, *then the length* $\ell(\rho)$ *of an argument in* Π_{SNARG} *is* $\Omega(\rho)$.

5.2 Quasi-Optimal SNARGs from Quasi-Optimal Linear MIPs

In this section, we show how to combine a linear MIPs with linear-only vector encryption over rings to obtain a quasi-optimal SNARG. We refer to the full version for the definition of a linear-only vector encryption from [14]. We describe the construction and state its security theorems here, but defer the security proofs to the full version [15].

Construction 5.3 (SNARG from Linear MIP). Fix a prime p and let $\mathcal{C} = \{C_n\}_{n \in \mathbb{N}}$ be a family of arithmetic circuits over $\mathbb{F}_p$. Let $\mathcal{R}_{\mathcal{C}}$ be the relation associated with $\mathcal{C}$. Let $(\mathcal{P}, \mathcal{V})$ be a k-query linear MIP with ℓ provers and query length d for the relation $\mathcal{R}_{\mathcal{C}}$. Let $\Pi_{\mathsf{venc}} = (\mathsf{KeyGen}, \mathsf{Encrypt}, \mathsf{Decrypt})$ be a secret-key vector encryption scheme over R^k where $R \cong \mathbb{F}_p^{\ell}$. Our single-theorem, designated-verifier SNARG $\Pi_{\mathsf{SNARG}} = (\mathsf{Setup}, \mathsf{Prove}, \mathsf{Verify})$ in the preprocessing model for $\mathcal{R}_{\mathcal{C}}$ is given below:

- $\mathsf{Setup}(1^\lambda, 1^n) \to (\sigma, \tau)$: On input the security parameter λ and the circuit family parameter n, the setup algorithm does the following:
  1. Invoke the query-generation algorithm $\mathcal{Q}$ for the linear MIP to obtain a tuple of query matrices $\mathbf{Q}_1, \ldots, \mathbf{Q}_\ell \in \mathbb{F}_p^{d \times k}$ and state information st.
  2. Generate a secret key $\mathsf{sk} \leftarrow \mathsf{KeyGen}(1^\lambda, 1^\ell)$ for the vector encryption scheme.
  3. Pack the ℓ query matrices $\mathbf{Q}_1, \ldots, \mathbf{Q}_\ell$ into a single query matrix $\mathbf{Q} \in R^{d \times k}$ (recall that the ring R splits into ℓ isomorphic copies of $\mathbb{F}_p$).
  4. Encrypt each row of $\mathbf{Q}$ (an element of R^k) using the vector encryption scheme. In other words, for $i \in [d]$, let $\mathbf{q}_i \in R^d$ be the ith row of $\mathbf{Q}$. In this step, the setup algorithm computes ciphertexts $\mathsf{ct}_i \leftarrow \mathsf{Encrypt}(\mathsf{sk}, \mathbf{q}_i)$.
  5. Output the common reference string $\sigma = (\mathsf{ct}_1, \ldots, \mathsf{ct}_d)$ and the verification state $\tau = (\mathsf{sk}, \mathsf{st})$.
- $\mathsf{Prove}(\sigma, \mathbf{x}, \mathbf{w}) \to \boldsymbol{\pi}$. On input the common reference string $\sigma = (\mathsf{ct}_1, \ldots, \mathsf{ct}_d)$, a statement $\mathbf{x}$, and a witness $\mathbf{w}$, the prover's algorithm works as follows:
  1. For each $i \in [\ell]$, invoke the linear MIP prover algorithm P_i on input $\mathbf{x}$ and $\mathbf{w}$ to obtain a proof $\boldsymbol{\pi}_i \leftarrow P_i(\mathbf{x}, \mathbf{w}) \in \mathbb{F}_p^d$.
  2. Pack the ℓ proof vectors $\boldsymbol{\pi}_1, \ldots, \boldsymbol{\pi}_\ell \in \mathbb{F}_p^d$ into a single proof vector $\boldsymbol{\pi} \in R^d$. Then, viewing the ciphertexts $\mathsf{ct}_1, \ldots, \mathsf{ct}_m$ as vector encryptions of the rows of the query matrix $\mathbf{Q} \in R^{d \times k}$, homomorphically compute an encryption of the matrix-vector product $\mathbf{Q}^\top \boldsymbol{\pi} \in R^k$. In particular, the prover homomorphically computes the sum $\mathsf{ct}' = \sum_{i \in d} \boldsymbol{\pi}_i \cdot \mathsf{ct}_i$.
  3. Output the proof ct'.
- $\mathsf{Verify}(\tau, \mathbf{x}, \boldsymbol{\pi}) \to \{0, 1\}$: On input the verification state $\tau = (\mathsf{sk}, \mathsf{st})$, the statement $\mathbf{x}$, and the proof $\pi = \mathsf{ct}'$, the verifier does the following:
  1. Decrypt the proof ct' using the secret key sk to obtain the prover's responses $\mathbf{y} \leftarrow \mathsf{Decrypt}(\mathsf{sk}, \mathsf{ct}')$. If $\mathbf{y} = \perp$, the verifier terminates with output 0.
  2. The verifier decomposes $\mathbf{y} \in R^k$ into vectors $\mathbf{y}_1, \ldots, \mathbf{y}_\ell \in \mathbb{F}_p^k$. It then invokes the linear MIP decision algorithm $\mathcal{D}$ on the statement $\mathbf{x}$, the responses $\mathbf{y}_1, \ldots, \mathbf{y}_\ell$, and the verification state st and outputs $\mathcal{D}(\mathsf{st}, \mathbf{x}, \mathbf{y}_1, \ldots, \mathbf{y}_\ell)$.

Theorem 5.4. *Fix a security parameter λ and a prime p. Let $\mathcal{C} = \{C_n\}_{n \in \mathbb{N}}$ be a family of arithmetic circuits over $\mathbb{F}_p$, $\mathcal{R}_\mathcal{C}$ be the relation associated with $\mathcal{C}$, and $(\mathcal{P}, \mathcal{V})$ be a k-query linear MIP with ℓ provers, query length d, and soundness error $\varepsilon(\lambda)$ against affine provers for the relation $\mathcal{R}_\mathcal{C}$. Let $\Pi_{\mathsf{venc}} = (\mathsf{KeyGen}, \mathsf{Encrypt}, \mathsf{Decrypt})$ be a vector encryption scheme over a ring $R \cong \mathbb{F}_p^\ell$ with linear targeted malleability. Then, applying Construction 5.3 to $(\mathcal{P}, \mathcal{V})$ and Π_{venc} yields a non-adaptive designated-verifier preprocessing SNARG with soundness error $2 \cdot \varepsilon(\lambda) + \mathrm{negl}(\lambda)$.*

Theorem 5.5. *Fix a security parameter λ and a prime p. Let $\mathcal{C} = \{C_n\}_{n \in \mathbb{N}}$ be a family of arithmetic circuits over $\mathbb{F}_p$, $\mathcal{R}_{\mathcal{C}}$ be the relation associated with $\mathcal{C}$, and $(\mathcal{P}, \mathcal{V})$ be a k-query linear MIP with ℓ provers, query length d, and soundness error $\varepsilon(\lambda)$ against affine provers for the relation $\mathcal{R}_{\mathcal{C}}$. Let $\Pi_{\mathsf{venc}} = (\mathsf{KeyGen}, \mathsf{Encrypt}, \mathsf{Decrypt})$ be a linear-only vector encryption scheme. Then, applying Construction 5.3 to $(\mathcal{P}, \mathcal{V})$ and Π_{venc} yields an adaptive designated-verifier preprocessing SNARG with soundness error $\varepsilon(\lambda) + \mathrm{negl}(\lambda)$.*

Instantiating the Construction. To conclude this section, we show that combining the candidate vector encryption scheme Π_{venc} over polynomial rings R^k, where $R \cong \mathbb{F}_p^\ell$ from [14, Sect. 4.4] with our quasi-optimal linear MIP construction from Theorem 4.15 yields a quasi-optimal SNARG from linear-only vector encryption. We first recall from [14, Sect. 4.4] that the candidate vector encryption scheme Π_{venc} has the following properties:

- When $k = \mathrm{polylog}(\lambda)$, $\ell = \widetilde{O}(\lambda)$, and $|\mathbb{F}| = \mathrm{poly}(\lambda)$, each ciphertext encrypting an element of R^k has length $\widetilde{O}(\lambda)$.
- Scalar multiplication and homomorphic addition of two ciphertexts can be performed in time $\widetilde{O}(\lambda)$.

When we apply Construction 5.3 to the linear MIP from Theorem 4.15 and Π_{venc}, the prover complexity and proof sizes are then as follows (targeting soundness error $2^{-\lambda}$):

- **Prover complexity:** The SNARG prover first invokes the underlying linear MIP prover to obtain proofs $\boldsymbol{\pi}_1, \ldots, \boldsymbol{\pi}_\ell$ for each of the $\ell = \widetilde{O}(\lambda)$ provers. From Theorem 4.15, this step requires time $\widetilde{O}(s) + \mathrm{poly}(\lambda, \log s)$, where s is the size of the circuit. To construct the proof, the prover has to perform d homomorphic operations, where $d = \widetilde{O}(s/\lambda) + \mathrm{poly}(\lambda, \log s)$ is the query length of the construction from Theorem 4.15. Since each homomorphic operation can be computed in $\widetilde{O}(\lambda)$ time, the overall prover complexity is $\widetilde{O}(s) + \mathrm{poly}(\lambda, \log s)$.
- **Proof size:** The proof in Construction 5.3 consists of a single ciphertext, which for our parameter settings, have length $\widetilde{O}(\lambda)$.

From this analysis, we obtain the following quasi-optimal SNARG instantiations:

Corollary 5.6. *Assuming the vector encryption scheme Π_{venc} from [14, Sect. 4.4] satisfies linear targeted malleability (with exponential security), then applying Construction 5.3 to the quasi-optimal linear MIP from Theorem 4.15 and Π_{venc} yields a non-adaptive designated-verifier quasi-optimal SNARG for Boolean circuit satisfiability in the preprocessing model.*

Corollary 5.7. *Assuming the vector encryption scheme Π_{venc} from [14, Sect. 4.4] (with the "double-encryption" transformation described in [14, Remark C.4]) is linear-only (with exponential security), then applying Construction 5.3 to the quasi-optimal linear MIP from Theorem 4.15 and Π_{venc} yield an adaptive designated-verifier quasi-optimal SNARG for Boolean circuit satisfiability in the preprocessing model.*

Construction 5.3 gives a construction of a *single-theorem* SNARG from any linear MIP system. In the full version [15], we discuss some of the challenges in extending our construction to provide multi-theorem security.

Remark 5.8 (Multi-theorem SNARGs). Construction 5.3 gives a construction of a *single-theorem* SNARG from any linear MIP system. The works of [13,14] show how to construct multi-theorem designated-verifier SNARGs by relying on a stronger notion of soundness at the linear PCP level coupled with a stronger interactive linear-only encryption assumption. While we could rely on the same type of cryptographic assumption as in [14], our linear MIP from Sect. 4 does not satisfy the notion of "reusable" or "strong" soundness from [13]. Strong soundness essentially says that for all proofs, the probability that the verifier accepts or that it rejects is negligible close to 1 (where the probability is taken over the randomness used to generate the queries). In particular, whether the verifier decides to accept or reject should be *uncorrelated* with the randomness associated with its secret verification state. In our linear MIP model, we operate over a polynomial-size field, so a prover making a local change will cause the verifier's decision procedure to change with noticeable probability. This reveals information about the secret verification state, which can enable the malicious prover to break soundness. We leave it as an open problem to construct a quasi-optimal linear MIP that provides strong soundness. Such a primitive would be useful in constructing a quasi-optimal multi-theorem SNARGs.

Acknowledgments. We thank the anonymous reviewers for helpful feedback on the presentation. D. Boneh and D. J. Wu are supported by NSF, DARPA, a grant from ONR, and the Simons Foundation. Y. Ishai and A. Sahai are supported in part from a DARPA/ARL SAFEWARE award, NSF Frontier Award 1413955, NSF grants 1619348, 1228984, 1136174, and 1065276, BSF grant 2012378, NSF-BSF grant 2015782, a Xerox Faculty Research Award, a Google Faculty Research Award, an equipment grant from Intel, and an Okawa Foundation Research Grant. Y. Ishai is additionally supported by ISF grant 1709/14 and ERC grant 742754. This material is based upon work supported by the Defense Advanced Research Projects Agency through the ARL under Contract W911NF-15-C-0205. The views expressed are those of the authors and do not reflect the official policy or position of the Department of Defense, the National Science Foundation, or the U.S. Government.

References

1. Applebaum, B., Ishai, Y., Kushilevitz, E.: From secrecy to soundness: efficient verification via secure computation. In: Abramsky, S., Gavoille, C., Kirchner, C., Meyer auf der Heide, F., Spirakis, P.G. (eds.) ICALP 2010. LNCS, vol. 6198, pp. 152–163. Springer, Heidelberg (2010). https://doi.org/10.1007/978-3-642-14165-2_14
2. Arora, S., Lund, C., Motwani, R., Sudan, M., Szegedy, M.: Proof verification and the hardness of approximation problems. J. ACM **45**(3), 501–555 (1998)
3. Babai, L., Fortnow, L., Levin, L.A., Szegedy, M.: Checking computations in polylogarithmic time. In: STOC (1991)

4. Barak, B., Pass, R.: On the possibility of one-message weak zero-knowledge. In: Naor, M. (ed.) TCC 2004. LNCS, vol. 2951, pp. 121–132. Springer, Heidelberg (2004). https://doi.org/10.1007/978-3-540-24638-1_7
5. Bellare, M., Palacio, A.: The knowledge-of-exponent assumptions and 3-round zero-knowledge protocols. In: Franklin, M. (ed.) CRYPTO 2004. LNCS, vol. 3152, pp. 273–289. Springer, Heidelberg (2004). https://doi.org/10.1007/978-3-540-28628-8_17
6. Ben-Sasson, E., Chiesa, A., Genkin, D., Tromer, E., Virza, M.: SNARKs for C: verifying program executions succinctly and in zero knowledge. In: Canetti, R., Garay, J.A. (eds.) CRYPTO 2013. LNCS, vol. 8043, pp. 90–108. Springer, Heidelberg (2013). https://doi.org/10.1007/978-3-642-40084-1_6
7. Ben-Sasson, E., Chiesa, A., Tromer, E., Virza, M.: Succinct non-interactive zero knowledge for a Von Neumann architecture. In: USENIX Security Symposium (2014)
8. Berman, I., Degwekar, A., Rothblum, R., Vasudevan, P.N.: From laconic zero-knowledge to public-key cryptography. In: Electronic Colloquium on Computational Complexity (ECCC) (2017)
9. Bitansky, N., Canetti, R., Chiesa, A., Goldwasser, S., Lin, H., Rubinstein, A., Tromer, E.: The hunting of the SNARK. J. Cryptol. **30**(4), 989–1066 (2017)
10. Bitansky, N., Canetti, R., Chiesa, A., Tromer, E.: From extractable collision resistance to succinct non-interactive arguments of knowledge, and back again. In: ITCS (2012)
11. Bitansky, N., Canetti, R., Chiesa, A., Tromer, E.: Recursive composition and bootstrapping for SNARKS and proof-carrying data. In: STOC (2013)
12. Bitansky, N., Chiesa, A.: Succinct arguments from multi-prover interactive proofs and their efficiency benefits. In: Safavi-Naini, R., Canetti, R. (eds.) CRYPTO 2012. LNCS, vol. 7417, pp. 255–272. Springer, Heidelberg (2012). https://doi.org/10.1007/978-3-642-32009-5_16
13. Bitansky, N., Chiesa, A., Ishai, Y., Paneth, O., Ostrovsky, R.: Succinct non-interactive arguments via linear interactive proofs. In: Sahai, A. (ed.) TCC 2013. LNCS, vol. 7785, pp. 315–333. Springer, Heidelberg (2013). https://doi.org/10.1007/978-3-642-36594-2_18
14. Boneh, D., Ishai, Y., Sahai, A., Wu, D.J.: Lattice-based SNARGs and their application to more efficient obfuscation. In: Coron, J.-S., Nielsen, J.B. (eds.) EUROCRYPT 2017. LNCS, vol. 10212, pp. 247–277. Springer, Cham (2017). https://doi.org/10.1007/978-3-319-56617-7_9
15. Boneh, D., Ishai, Y., Sahai, A., Wu, D.J.: Quasi-optimal SNARGs via linear multi-prover interactive proofs. IACR Cryptology ePrint Archive (2018). https://eprint.iacr.org/2018/133.pdf
16. Boppana, R.B., Håstad, J., Zachos, S.: Does co-NP have short interactive proofs? Inf. Process. Lett. **25**(2), 127–132 (1987)
17. Brassard, G., Chaum, D., Crépeau, C.: Minimum disclosure proofs of knowledge. J. Comput. Syst. Sci. **37**(2), 156–189 (1988)
18. Braun, B., Feldman, A.J., Ren, Z., Setty, S.T.V., Blumberg, A.J., Walfish, M.: Verifying computations with state. In: SOSP (2013)
19. Cormode, G., Mitzenmacher, M., Thaler, J.: Practical verified computation with streaming interactive proofs. In: ITCS (2012)
20. Costello, C., Fournet, C., Howell, J., Kohlweiss, M., Kreuter, B., Naehrig, M., Parno, B., Zahur, S.: Geppetto: versatile verifiable computation. In: IEEE SP (2015)

21. Damgård, I.: Towards practical public key systems secure against chosen ciphertext attacks. In: Feigenbaum, J. (ed.) CRYPTO 1991. LNCS, vol. 576, pp. 445–456. Springer, Heidelberg (1992). https://doi.org/10.1007/3-540-46766-1_36
22. Damgård, I., Faust, S., Hazay, C.: Secure two-party computation with low communication. In: Cramer, R. (ed.) TCC 2012. LNCS, vol. 7194, pp. 54–74. Springer, Heidelberg (2012). https://doi.org/10.1007/978-3-642-28914-9_4
23. Damgård, I., Ishai, Y., Krøigaard, M.: Perfectly secure multiparty computation and the computational overhead of cryptography. In: Gilbert, H. (ed.) EUROCRYPT 2010. LNCS, vol. 6110, pp. 445–465. Springer, Heidelberg (2010). https://doi.org/10.1007/978-3-642-13190-5_23
24. Danezis, G., Fournet, C., Groth, J., Kohlweiss, M.: Square span programs with applications to Succinct NIZK arguments. In: Sarkar, P., Iwata, T. (eds.) ASIACRYPT 2014. LNCS, vol. 8873, pp. 532–550. Springer, Heidelberg (2014). https://doi.org/10.1007/978-3-662-45611-8_28
25. Faonio, A., Nielsen, J.B., Venturi, D.: Predictable arguments of knowledge. In: Fehr, S. (ed.) PKC 2017. LNCS, vol. 10174, pp. 121–150. Springer, Heidelberg (2017). https://doi.org/10.1007/978-3-662-54365-8_6
26. Feige, U., Goldwasser, S., Lovász, L., Safra, S., Szegedy, M.: Approximating clique is almost NP-complete (preliminary version). In: FOCS (1991)
27. Fiat, A., Shamir, A.: How to prove yourself: practical solutions to identification and signature problems. In: Odlyzko, A.M. (ed.) CRYPTO 1986. LNCS, vol. 263, pp. 186–194. Springer, Heidelberg (1987). https://doi.org/10.1007/3-540-47721-7_12
28. Garg, S., Gentry, C., Sahai, A., Waters, B.: Witness encryption and its applications. In: STOC (2013)
29. Gennaro, R., Gentry, C., Parno, B., Raykova, M.: Quadratic span programs and Succinct NIZKs without PCPs. In: Johansson, T., Nguyen, P.Q. (eds.) EUROCRYPT 2013. LNCS, vol. 7881, pp. 626–645. Springer, Heidelberg (2013). https://doi.org/10.1007/978-3-642-38348-9_37
30. Gentry, C., Wichs, D.: Separating Succinct non-interactive arguments from all falsifiable assumptions. In: STOC (2011)
31. Goldreich, O.: The Foundations of Cryptography, Basic Techniques, vol. 1. Cambridge University Press, Cambridge (2001)
32. Goldreich, O., Håstad, J.: On the complexity of interactive proofs with bounded communication. Inf. Process. Lett. **67**(4), 205–214 (1998)
33. Goldreich, O., Vadhan, S., Wigderson, A.: On interactive proofs with a Laconic prover. In: Orejas, F., Spirakis, P.G., van Leeuwen, J. (eds.) ICALP 2001. LNCS, vol. 2076, pp. 334–345. Springer, Heidelberg (2001). https://doi.org/10.1007/3-540-48224-5_28
34. Goldwasser, S., Kalai, Y.T., Rothblum, G.N.: Delegating computation: interactive proofs for muggles. In: STOC (2008)
35. Goldwasser, S., Micali, S., Rackoff, C.: The knowledge complexity of interactive proof-systems. In: STOC (1985)
36. Groth, J.: Linear algebra with sub-linear zero-knowledge arguments. In: Halevi, S. (ed.) CRYPTO 2009. LNCS, vol. 5677, pp. 192–208. Springer, Heidelberg (2009). https://doi.org/10.1007/978-3-642-03356-8_12
37. Groth, J.: Short pairing-based non-interactive zero-knowledge arguments. In: ASIACRYPT (2010)
38. Groth, J.: On the size of pairing-based non-interactive arguments. In: EUROCRYPT (2016)

39. Groth, J., Maller, M.: Snarky signatures: minimal signatures of knowledge from simulation-extractable SNARKs. In: Katz, J., Shacham, H. (eds.) CRYPTO 2017. LNCS, vol. 10402, pp. 581–612. Springer, Cham (2017). https://doi.org/10.1007/978-3-319-63715-0_20
40. Hada, S., Tanaka, T.: On the existence of 3-round zero-knowledge protocols. In: Krawczyk, H. (ed.) CRYPTO 1998. LNCS, vol. 1462, pp. 408–423. Springer, Heidelberg (1998). https://doi.org/10.1007/BFb0055744
41. Ishai, Y., Kushilevitz, E., Ostrovsky, R.: Efficient arguments without short PCPs. In: CCC (2007)
42. Ishai, Y., Kushilevitz, E., Ostrovsky, R., Sahai, A.: Zero-knowledge from secure multiparty computation. In: STOC (2007)
43. Ishai, Y., Prabhakaran, M., Sahai, A.: Secure arithmetic computation with no honest majority. In: TCC (2009)
44. Kilian, J.: A note on efficient zero-knowledge proofs and arguments. In: STOC (1992)
45. Lipmaa, H.: Progression-free sets and sublinear pairing-based non-interactive zero-knowledge arguments. In: Cramer, R. (ed.) TCC 2012. LNCS, vol. 7194, pp. 169–189. Springer, Heidelberg (2012). https://doi.org/10.1007/978-3-642-28914-9_10
46. Lipmaa, H.: Succinct non-interactive zero knowledge arguments from span programs and linear error-correcting codes. In: Sako, K., Sarkar, P. (eds.) ASIACRYPT 2013. LNCS, vol. 8269, pp. 41–60. Springer, Heidelberg (2013). https://doi.org/10.1007/978-3-642-42033-7_3
47. Lipmaa, H.: Prover-efficient commit-and-prove zero-knowledge SNARKs. In: Pointcheval, D., Nitaj, A., Rachidi, T. (eds.) AFRICACRYPT 2016. LNCS, vol. 9646, pp. 185–206. Springer, Cham (2016). https://doi.org/10.1007/978-3-319-31517-1_10
48. Lund, C., Fortnow, L., Karloff, H.J., Nisan, N.: Algebraic methods for interactive proof systems. In: FOCS (1990)
49. Micali, S.: Computationally sound proofs. SIAM J. Comput. **30**(4), 1253–1298 (2000)
50. Mie, T.: Polylogarithmic two-round argument systems. J. Math. Cryptology **2**(4), 343–363 (2008)
51. Naor, M.: On cryptographic assumptions and challenges. In: Boneh, D. (ed.) CRYPTO 2003. LNCS, vol. 2729, pp. 96–109. Springer, Heidelberg (2003). https://doi.org/10.1007/978-3-540-45146-4_6
52. Parno, B., Howell, J., Gentry, C., Raykova, M.: Pinocchio: nearly practical verifiable computation. In: IEEE Symposium on Security and Privacy (2013)
53. Sahai, A., Waters, B.: How to use indistinguishability obfuscation: deniable encryption, and more. In: STOC (2014)
54. Setty, S.T.V., McPherson, R., Blumberg, A.J., Walfish, M.: Making argument systems for outsourced computation practical (sometimes). In: NDSS (2012)
55. Setty, S.T.V., Vu, V., Panpalia, N., Braun, B., Blumberg, A.J., Walfish, M.: Taking proof-based verified computation a few steps closer to practicality. In: USENIX Security Symposium (2012)
56. Shamir, A.: IP=PSPACE. In: FOCS (1990)
57. Thaler, J.: Time-optimal interactive proofs for circuit evaluation. In: Canetti, R., Garay, J.A. (eds.) CRYPTO 2013. LNCS, vol. 8043, pp. 71–89. Springer, Heidelberg (2013). https://doi.org/10.1007/978-3-642-40084-1_5
58. Thaler, J., Roberts, M., Mitzenmacher, M., Pfister, H.: Verifiable computation with massively parallel interactive proofs. In: HotCloud (2012)

59. Valiant, P.: Incrementally verifiable computation or proofs of knowledge imply time/space efficiency. In: TCC (2008)
60. Vu, V., Setty, S.T.V., Blumberg, A.J., Walfish, M.: A hybrid architecture for interactive verifiable computation. In: IEEE SP (2013)
61. Wahby, R.S., Howald, M., Garg, S.J., Shelat, A., Walfish, M.: Verifiable ASICs. In: IEEE Symposium on Security and Privacy (2016)
62. Wahby, R.S., Ji, Y., Blumberg, A.J., Shelat, A., Thaler, J., Walfish, M., Wies, T.: Full accounting for verifiable outsourcing. In: ACM CCS (2017)
63. Wahby, R.S., Setty, S.T.V., Ren, Z., Blumberg, A.J., Walfish, M.: Efficient RAM and control flow in verifiable outsourced computation. In: NDSS (2015)
64. Walfish, M., Blumberg, A.J.: Verifying computations without reexecuting them. Commun. ACM **58**(2), 74–84 (2015)
65. Wee, H.: On round-efficient argument systems. In: Caires, L., Italiano, G.F., Monteiro, L., Palamidessi, C., Yung, M. (eds.) ICALP 2005. LNCS, vol. 3580, pp. 140–152. Springer, Heidelberg (2005). https://doi.org/10.1007/11523468_12

Anonymous Communication

Untagging Tor: A Formal Treatment of Onion Encryption

Jean Paul Degabriele[1(✉)] and Martijn Stam[2]

[1] Department of Computer Science, TU Darmstadt, Darmstadt, Germany
jeanpaul.degabriele@cryptoplexity.de
[2] Department of Computer Science, University of Bristol, Bristol, UK
martijn.stam@bristol.ac.uk

Abstract. Tor is a primary tool for maintaining anonymity online. It provides a low-latency, circuit-based, bidirectional secure channel between two parties through a network of onion routers, with the aim of obscuring exactly who is talking to whom, even to adversaries controlling part of the network. Tor relies heavily on cryptographic techniques, yet its onion encryption scheme is susceptible to *tagging attacks* (Fu and Ling 2009), which allow an active adversary controlling the first and last node of a circuit to deanonymize with near-certainty. This contrasts with less active traffic correlation attacks, where the same adversary can at best deanonymize with high probability. The Tor project has been actively looking to defend against tagging attacks and its most concrete alternative is proposal 261, which specifies a new onion encryption scheme based on a variable-input-length tweakable cipher.

We provide a formal treatment of low-latency, circuit-based onion encryption, relaxed to the unidirectional setting, by expanding existing secure channel notions to the new setting and introducing *circuit hiding* to capture the anonymity aspect of Tor. We demonstrate that circuit hiding prevents tagging attacks and show proposal 261's *relay* protocol is circuit hiding and thus resistant against tagging attacks.

Keywords: Anonymity · Onion routing · Secure channels · Tor
Tagging attacks

1 Introduction

Anonymity as a separate security goal to confidentiality and integrity was recognized early on. Chaum [14] provided a number of suggestions for anonymous communication, of which his mix-nets later evolved into onion routing. Onion routing protocols come in a variety of flavours, depending on whether they are low-latency or not, whether they are circuit-oriented or ciphertext-oriented, the TCP/IP layer at which they operate, and a number of other factors. Examples include I2P [27], Mixminion [16], MorphMix [38] and Tarzan [23], but the best known and most widely used onion routing solution is Tor [20]. Tor is a

J. B. Nielsen and V. Rijmen (Eds.): EUROCRYPT 2018, LNCS 10822, pp. 259–293, 2018.
https://doi.org/10.1007/978-3-319-78372-7_9

low-latency, circuit-oriented onion routing protocol operating at the transport layer. Its original architecture was laid out in a quick succession of articles by Goldschlag et al. [25,37,42]. The extent to which Tor and its brethren defend against (mostly) passive traffic correlation analysis has been an active research area [13,28,29,33,40], yet the impact of active attacks against the core cryptographic components on anonymity remains relatively unexplored. Indeed, it is not even clear what the formal design desiderata would be to provide any meaningful form of provable anonymity.

When talking about anonymity, it is worth bearing in mind the original goal set for Tor [25]: "The goal of onion routing is *not* to provide anonymous communication. Parties are free to (and usually should) identify themselves within a message. But the use of a public network should not automatically give away the identity and locations of the communication parties" (emphasis ours). In practice, an onion routing network (see Fig. 1) enables two parties Anna (A) and Xavier (X) to route their communication through various intermediate nodes. As a result, there is no longer a direct link of communication between Anna and Xavier to observe and the hope is that their traffic gets lost in the masses. Ideally, even the intermediate nodes in direct contact with Anna and Xavier cannot link Anna and Xavier together.

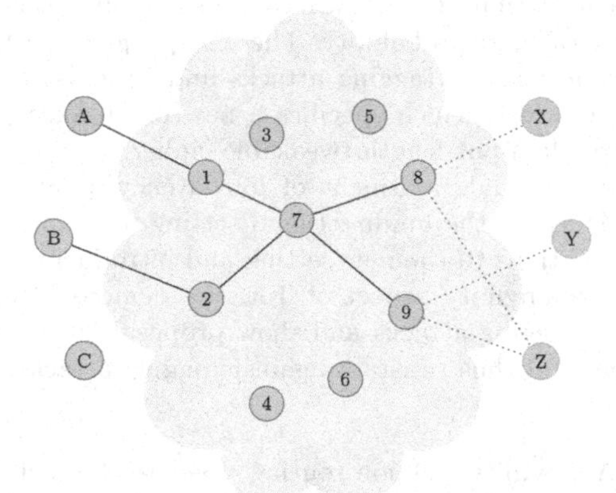

Fig. 1. Onion routing in a nutshell. Nodes A, B, and C are the onion proxies; nodes 1 to 9 are the onion routers making up the network; finally nodes X, Y, and Z are the destinations. User A created the dark blue circuit 1–7–8, using exit node 8 to communicate with destinations X and Z, whereas user B created the dark red circuit 2–7–9 and using its exit node 9 to communicate with destinations Y and Z. (Color figure online)

The core components of Tor are the *link* protocol, the *circuit extend* protocol, the *relay* protocol, and the *stream* protocol. Any communication between any pair of interacting parties is secured by the link layer, which uses TLS, and all communication occurs on top of it. The circuit extend protocol establishes multi-hop tunnels called circuits between the sender and the receiver. In essence it uses public-key cryptography to exchange key material between nodes for onion encryption. The relay protocol is the component that actually handles the onion encryption and will be our main focus. The stream protocol operates over the relay protocol and is used to establish TCP connections, send data, etc.

At a very high level, the relay protocol operates as follows. The sender, which shares a symmetric key with every other node on the circuit, encrypts a message by applying multiple layers of encryption in succession, one for each node along the circuit. Specifically, a message is first encoded with a two-byte field of zeros and a SHA1 digest truncated to four bytes, and each layer of encryption then consists of 128-bit AES in counter mode. The resulting ciphertext, or *cell* in Tor's terminology, is then passed by the sender to the first node in the circuit. Each node in turn strips off one layer of encryption and either forwards the cell to the next node in the circuit or acts on that cell itself if it determines that it is the intended recipient. Note that only the final node in the circuit checks, and can check, the integrity by considering the redundancy introduced by the sender's encoding.

The combination of the final-node integrity check and the high level of malleability of counter-mode encryption leave Tor's relay protocol susceptible to the following *tagging attack* [24]. Assume an adversary controls the first and last nodes in a circuit. It can then 'tag' a cell c during the first hop by xoring it with some pattern δ, i.e. it sends $c \oplus \delta$ instead of c. If an honest exit node receives the corresponding cell, the integrity check will very likely fail, and the honest exit node will reject it. However, if the *adversary* controls the exit node, it can check for an invalid received cell c' that $c' \oplus \delta$ *does* pass the integrity check. Thus, the adversary has established that the two nodes are on the same circuit, and thereby linked the user (known to the first node) to its activities, as seen by the last node.

Superficially, tagging attacks expose a similar vulnerability as traffic correlation attacks by adversaries controlling both the first and last node of a circuit. Moreover, as Tor is a low-latency system, it cannot adequately protect against these passive traffic correlation attacks, which begs the question what active tagging attacks add to an adversary's arsenal. Indeed, back in 2004 when Tor was conceived [20], its authors already "accepted that our design is vulnerable to end-to-end timing attacks; so tagging attacks performed within the circuit provide no additional information to the attacker". Thus the choice for low latency—a compromise trading stronger security for usability—appeared to render tagging attacks redundant or even irrelevant, as seemingly equally powerful traffic correlation attacks are possible.

This perception changed around 2012, following an anonymous post on the Tor developers' mailing list by The23rd Raccoon, pointing out that tagging attacks *are* considerably more potent than traffic correlation attacks [36]: a successful tagging attack gives an adversary certainty when linking a circuit's entry and exit node, whereas for traffic correlation attacks a degree of uncertainty remains, including false positives where two nodes are incorrectly assumed to be on the same circuit. Consequently, tagging attacks scale better and with increased severity compared to traffic correlation attacks. We expand on these observations in Sect. 2.2, where we also address why detection (of the active tagging) does not lead to a satisfactory defense mechanism.

All in all, the Tor project has reversed its position and is currently seeking alternative onion encryption schemes that do protect against tagging attacks [31]. Whereas traffic correlation attacks cannot be prevented by cryptographic means (without sacrificing low latency), conceivably protection against tagging attacks without significant performance penalty is achievable.

Taking a broader perspective, we observe that while there has been ample work focusing on anonymity, circuit-based onion encryption *as a cryptographic primitive* has been largely overlooked. Yet in onion routing networks, anonymity is achieved through a combination of factors, such as the number of users in the system, the amount of traffic, and cryptographic mechanisms like onion encryption. The aforementioned tagging attacks against Tor clearly indicate that the latter is not well understood: it is unclear what properties the cryptographic component should provide, let alone how to do so.

Our Contribution. Our aim is exactly this, to characterise what security properties *should* and *can* be expected from an onion encryption scheme. While we try to maintain as much generality as possible, our attention is focused on the setting typified by Tor. That is, we consider onion encryption for the case of low-latency and circuit-oriented systems, operating on top of a link protocol, like TLS, that secures communication between adjacent nodes. Tarzan and Morphmix, like Tor, fall within this category and are also captured by our models.

Our three design choices change the landscape quite significantly compared to high-latency mix-nets or public key ciphertext-oriented onion routing. In particular, requiring low-latency precludes the possibility of shuffling cells as in the case of mix-nets. A circuit-oriented architecture assumes the existence of a complementary protocol (in the case of Tor the circuit extend protocol) that sets up circuits across the network on which cells can be transmitted. Thus cells must follow predefined paths. In contrast, in a ciphertext-oriented architecture (as in I2P and Mixminion), each ciphertext can specify and follow a distinct path. This dichotomy corresponds quite closely to the distinction between symmetric and public-key onion encryption. One benefit of a circuit-oriented architecture is that, being stateful, it can protect against replay and reordering attacks. Finally, because the onion encryption operates on top of a link protocol, the adversary can only access the onion encryption if it controls some subset of the nodes in the network. We exploit this fact in our circuit hiding definition, but we don't make use of it in other security definitions as we can achieve the stronger definition with little effort or added complexity.

Clearly we would like an onion encryption scheme to provide confidentiality, integrity, and anonymity, ideally even if a subset of the nodes are under adversarial control. After establishing a syntax of circuit-based onion routing in Sect. 3, we adapt the end-to-end security notions for confidentiality and integrity of a secure unidirectional channel to the context of onion routing in Sect. 4, before tackling the most challenging and novel part, namely anonymity, in Sect. 5.

Anonymity in low-latency onion routing is never absolute and relies crucially on non-cryptographic factors—beyond what can be guaranteed by a suitable choice of onion encryption—such as the network size, its number of active users and the amount of traffic flowing through the network. Accordingly we do not aim for a full-blown definition of anonymity, but instead aim for a more refined security goal that can be achieved purely by cryptographic means (assuming ideal traffic conditions). Our proposed notion is that of circuit hiding, which roughly states that an adversary should not be able to learn any information about the circuits' topology in the network beyond what is inevitably leaked through the nodes that it controls. In particular this should hold even when the adversary is allowed to choose the messages that get encrypted and is able to re-order, inject, and manipulate ciphertexts on the network. Indeed the latter is exactly how tagging attacks operate, and consequently these are captured by our model.

Following on from the two potential directions to thwart tagging attacks [31], the only concrete proposal to date is Tor proposal 261 [32]. Our second contribution is a security analysis of the onion encryption scheme specified therein. Proposal 261 is based on AEZ [26] (see Sect. 6 for more details), but it could be instantiated with any other variable-input-length (VIL) tweakable cipher, such as Farfalle [9] or HHFHFH [8] which have both been suggested as alternatives to AEZ. Indeed, our analysis is general enough to apply to any such instantiation and thus its scope surpasses proposal 261.

Security in our framework guarantees protection from tagging attacks, but also ensures that it is not done at the detriment of some other security aspect. For instance, a naive solution to stop tagging attacks would be to extend Tor's counter mode AES with a MAC in an encrypt-then-MAC configuration. While this fix might foil tagging attacks, it would not suffice to guarantee circuit hiding, as the length of a circuit and a node's relative position within a circuit can now be inferred from the size of a cell. This is another instance of leakage on the circuits' topology that is captured by our notion.

We emphasize however that our analysis is limited in scope. In particular we only consider static node corruptions that are chosen by the adversary but fixed before it can interact with the network. Furthermore, we assume that circuits have already been established in a secure and anonymous way and we do not study how circuits should be chosen either.

Related Work. Camenisch and Lysyanskaya [11] gave a formal security definition of public-key onion routing in the Universal Composability (UC) framework, as well as an alternative, compound game-based definition. For the latter they identify three core properties: correctness, integrity, and security. Combined with secure point-to-point channels these three imply the UC security notion. However their security model focuses on ciphertext-oriented architectures, where onion routers are stateless, and is therefore not applicable to Tor. For instance, their security definition does not and cannot capture circuits or provide protection against replay and reordering attacks.

Feigenbaum et al. [21] provide a black-box probabilistic analysis of onion routing based on a very high-level idealised functionality. Here an adversary is allowed to statically corrupt a fraction of the routers, so when a user selects a destination (for a circuit), the functionality randomly selects a path from the user to the destination and thereby determines whether the nodes adjacent to the user, resp. the destination, are corrupt or not. For corrupt adjacent nodes, the adversary will learn the user, resp. the destination, connected to it. In particular, if both adjacent nodes are corrupt, then the adversary will learn that a circuit has been established between the user and its destination. This model is useful for analysing the overall effect of traffic correlation attacks under the assumption that an adversary is capable to link traffic flowing in and out of the honest routers [43]. However, when considering how the cryptographic component affects this traffic linking assumption, their model is unsuitable.

Motivated by Tor, Backes et al. [3] propose another security definition for onion encryption in the UC framework. They consider a combined ideal functionality closely mirroring Tor's syntax, incorporating both the circuit establishment protocol and the onion encryption component. Meeting their security notion relies on the onion encryption being *predictably malleable*. Unfortunately, this predictable malleability is exactly what enables tagging attacks. In other words, schemes secure in their framework are *guaranteed* to be insecure against tagging attacks.

Danezis and Goldberg [17] propose Sphinx, a cryptographic packet format for relaying anonymized messages within high latency mix networks. It improves over prior constructions by being more space efficient, in part by replacing RSA encryption with elliptic curve cryptography. Sphinx is designed to protect against tagging attacks, but as it follows a ciphertext oriented architecture, is inapplicable to Tor.

In concurrent work, Rogaway and Zhang provide an independent treatment of onion encryption [39], see our full version [18] for a comparison.

2 Background and Preliminaries

2.1 An Overview of Tor

We now give a brief overview of how Tor works. A more detailed description can be found in Tor's introduction [20] or its protocol specification [19]; for a comparison of Tor with other anonymous communication systems see Danezis et al. [15].

In Tor, an end-to-end overlay network is formed wherein participating nodes, called onion routers (OR), relay messages across the overlay network. Users can run an onion proxy (OP) to access the Tor network (though we will use the terms sender, user, and onion proxy interchangeably). Onion routers maintain TLS connections with each other and onion proxies join the Tor network by establishing a TLS connection to one or more onion routers (the *link* protocol).

All peer-to-peer communication occurs over these TLS connections. The aim of the network is to prevent outsiders or participating nodes from linking the recipient of a message to its source.

A user's application data is routed over fixed paths called *circuits*. A circuit is a path within the Tor network consisting of two or more ORs; the default is three. At the start of the circuit is an onion proxy which transmits data over the circuit, though Tor does not consider the OP to be part of the circuit. The exit node is the onion router responsible for delivering application data to the intended recipient (who may well reside outside the network). By default, the last node in a circuit acts as an exit node, but other nodes in the circuit may also act as exit nodes—a feature sometimes referred to as "leaky pipes". These variable exit nodes are supported in Tor by allowing multiple streams to run over the same circuit; for the most part we will ignore this complication and assume the exit node will be the last node of the circuit.

An onion proxy is responsible for establishing its circuits. To this end, it will select a sequence of onion routers in the circuit, where each node can appear only once. The proxy establishes a symmetric key with each of the routers in sequence using the *circuit extend* protocol: at each step the circuit is extended by one hop in a telescopic fashion, so the key agreement with the $i+1^{\text{th}}$ node in the circuit runs over the current, partial circuit with the ith node temporarily taking on the role of exit node. Circuit establishment in Tor enables a bidirectional channel between the onion proxy and the exit node.

Once a circuit is established, the OP shares a symmetric key with each node in that circuit. Furthermore, each OR shares a distinct circuit identifier with each node that is adjacent to it in the circuit. The circuit is then used by the OP to instruct the exit node to establish a TCP connection to a specific address and port (the *stream* protocol). Data intended for this stream is then encapsulated in relay cells, and the *relay* protocol protects each cell with a checksum and multiple layers of encryption: the OP adds a layer of encryption (128-bit AES in counter mode) for each OR in the circuit. Upon receiving a relay cell, an OR looks up the cell's circuit identifier and uses the corresponding key to remove a layer of encryption. If the cell is headed away from the OP, the OR then checks whether the resulting cell has a valid checksum. The checksum is composed of two all-zero bytes and a four-byte digest computed through a seeded running hash over the data. If valid, the OR interprets the relay cell to be intended for itself (any node in the circuit can act as an exit node). Otherwise it looks up the circuit identifier and the OR for the next hop in the circuit, replaces the circuit identifier, and forwards the relay cell to the next OR in the circuit. If the OR at the end of the circuit received an unrecognised relay cell, an error has occurred and the circuit is torn down. An OP treats incoming relay cells similarly: it iteratively removes an encryption layer for each OR on the circuit from closest to farthest. If at any point the checksum is valid, the cell must have originated at the OR whose layer has just been removed.

In Tor, all data exchanged between nodes is encapsulated in *cells*. In the majority of cases, cells are of a fixed size. In version 4 and higher, fixed-size cells are 514 bytes long and consist of a header and a payload portion. The header is composed of a four-byte circuit identifier *id*, and a single-byte command field *cmd* indicating what to do with the cell's payload. Circuit identifiers are connection-specific, so as a cell travels along a circuit it will have a different circuit identifier on each OP/OR and OR/OR connection that it traverses. The cell's payload is protected using onion encryption, where each cell is additionally TLS encrypted on its OP/OR, resp. OR/OR connection. Based on their command field, cells are either control cells to be interpreted by the node that receives them, or relay cells which carry end-to-end stream data. Control cells serve to create, maintain and tear down circuits. Relay cells have an additional relay header located at the front of the payload composed of a two-byte stream identifier, a six-byte checksum, a two-byte length field, and a single-byte relay command field. The stream identifier allows multiple stream traffic to be multiplexed over the same circuit. The checksum is used by ORs to determine whether they are the intended recipient of the cell, while the length field specifies the size of the relay payload in bytes. Relay commands are exchanged between the Onion Proxy and the exit node to manage TCP streams, such as for instance to instruct an exit node to open a TCP connection to some destination specified in the relay payload.

Our focus is on the onion encryption component of Tor, which means we will abstract away most of the details of how an onion proxy initially connects to the Tor network, how it chooses which circuit to create, and how the telescopic key agreement operates. Instead, we will assume that all necessary keys have already been established in a secure manner and that secure channels between nodes (e.g. based on TLS) are readily available. We collapse the *stream* and the *relay* protocols and only directly consider sending arbitrary length messages, thus ignoring complications arising from treating data as a stream [22].

2.2 On the Relative Severity of Tagging Attacks

We have already alluded to the similarity between tagging attacks and traffic correlation attacks, which in turn raises the question as to why should we bother with tagging attacks at all when seemingly equally powerful attacks are possible. There is indication however that tagging attacks can be significantly more damaging than traffic correlation attacks. The arguments in support of this claim stem from the analysis in two anonymous posts on the Tor developers' mailing list by The23rd Raccoon from 2008 and 2012 [35,36]. In turn these observations prompted the Tor project to reverse its decision and seek to protect against tagging attacks [31,32]. We here attempt to give some insight into this rationale but refer the reader to the actual posts for further details.

The main distinctive advantage of tagging attacks over traffic correlation attacks is that a circuit can be confirmed with a zero chance of a false positive (i.e. two end points being categorised as belonging to the same circuit when in reality they do not). In contrast, traffic correlation techniques inevitably incur false positives with non-zero probability. Moroever, the base rate fallacy implies that even a relatively small false positive rate severely reduces the overall efficacy of traffic correlation attacks as the network size increases [35]. While the original post [35] did not make any mention of tagging attacks it is easy to see that tagging attacks are immune to this phenomenon and therefore scale much better than traffic correlation attacks.

Another argument in support of the severity of tagging attacks is their inherent "amplification" effect as described by The23rd Raccoon in 2012 [36]. (Perhaps amplification is not the most appropriate term but to avoid confusion we stick with The23rd Raccoon's choice.) The amplification relies on the tear down of circuits as soon as a tagged cell is not untagged at the exit of the network (and similarly, whenever cells that were not previously tagged are "untagged"). The immediate effect is that uncompromised circuits will be automatically filtered out and the adversary does not have to dedicate further resources to them. A secondary effect is that, when the OP attempts to re-establish a circuit using a new path, with some probability both entry and exit routers will be under adversarial control. Thus tagging attacks bias the creation of more compromised circuits. In principle, this amplification could be simulated using a traffic correlation attack by actively tearing down uncorrelated circuits, though again, false positives limit the efficacy of this approach [36].

Regrettably, tagging attacks cannot easily be prevented by detection and subsequent eviction of dishonest routers. For instance, although tagging attacks have been known since at least 2004, in 2014 they were successfully deployed against Tor without being noticed until months later [2]. Secondly, a client can only detect modification as a circuit failure but the natural failure rate in the Tor network is high enough to complicate timely detection of an attack. Moreover, even if a circuit failure is correctly classified as an attack, identifying the malicious onion routers is far from obvious. It requires independent onion routers to collaborate, including a mechanism to resolve disputes as misbehaving routers could manipulate the evidence in order to shift the blame on other routers. Such collaboration is further hampered as the required exchange of information should not allow the reconstruction of the affected circuits as it would deanonymise their users in the process—precisely what we are trying to prevent in the first place. Finally, an attacker in full control of the exit nodes through which the tagged traffic flows can avoid detection altogether. Using tagging attack in conjunction with preliminary traffic analysis could realistically lead to such a scenario.

2.3 Notation

If $\mathcal{S}$ is a finite set then $|\mathcal{S}|$ denotes its size, and $y \leftarrow_{\$} \mathcal{S}$ denotes the process of selecting an element from $\mathcal{S}$ uniformly at random and assigning it to y. An oracle may return the special symbol $\lightning$ to suppress output; in contrast $\perp$ denotes an error message that is output by some scheme.

We denote vectors in bold letters or explicitly by listing their components in between []. For any vector $\mathbf{v}$, we denote its ith component by $\mathbf{v}[i]$, its size by $|\mathbf{v}|$ and we endow it with a function $\mathbf{v}.\mathsf{append}(e)$ that extends $\mathbf{v}$ with a new component of value e. We use $[e]_1^n$ to denote a vector of size n whose entries are all set to e. We also make use of queue structures, where for any queue Q

the function calls Q.enqueue() and Q.dequeue() bear their usual meaning. Unless otherwise specified, all vectors and queues are initially empty.

3 Modelling Onion Routing Networks

Ultimately, our goal is to quantify how well the cryptologic component of cell creation and processing provides security and anonymity, even against adaptive adversaries. Our model abstracts away certain aspects that are highly relevant in practice (e.g. key management and traffic analysis), but that are to a large extent orthogonal to the cryptographic channel security. To ensure our formalism reasonably matches intuition, we embedded some Tor specific design choices into our syntax, yet our syntax is considerably more general in order to capture alternative cryptographic solutions as well. As a result, we strike a balance to avoid needless complexity as much as possible.

We consider two types of roles, corresponding to Tor's onion proxies and onion routers, respectively. The onion routers and proxies are modelled by nodes in a graph, with the (directed) edges representing possible direct communication. Our assumption is that each directed edge corresponds to an independent, unidirectional secure channel, and that the graph is a complete directed graph, allowing any party to communicate securely and directly (but not anonymously!) with any other party. If desired, one could consider other graphs to represent topological restrictions. All parties have a unique, publicly known identifier, and can take on the role of both proxy and router—even though we often write as if these are completely different entities.

As in Tor, the onion proxies are responsible for initializing circuits and for encrypting messages to a circuit, which we assume is used for *unidirectional* anonymous communication. A circuit consists of an onion proxy and a path $\mathbf{p}$ through the graph of onion routers. The path should be acyclic and its length is denoted ℓ. The circuit is then represented as a vector of nodes $\mathbf{p}[1], \ldots, \mathbf{p}[\ell]$ where we abuse $\mathbf{p}[0]$ to refer to the circuit's onion proxy so $\mathbf{p}$ allows the identification of not only the ℓ routers, but also the proxy. By convention, we set $\mathbf{p}[\ell+1] = \oslash$ to indicate the end of a circuit, where the symbol $\oslash$ is reserved solely for this purpose and cannot be assigned to any node. For any circuit, we use the terms sending node, receiving node and forwarding nodes to refer respectively to the OP, the path's last node $\mathbf{p}[\ell]$ and intermediate nodes $\mathbf{p}[1], \ldots, \mathbf{p}[\ell-1]$ in the path. Note that our receiving node corresponds to Tor's exit node, whereas the party outside the network with which the exit node communicates is beyond the scope of our formalism.

Both onion proxies and onion routers maintain a vector of states, containing a state for each of the circuits they are involved with. There is a notable difference in their use, as proxies use their state for encryption and will know which circuit it is for (and therefore which 'state' to use), whereas for routers, upon receipt of a cell they will first have to figure out which of its circuits it is intended for, if any. Accordingly, we split decryption into two separate stages, D and $\bar{\mathsf{D}}$, where D is responsible for figuring out the relevant circuit and $\bar{\mathsf{D}}$ for the proper processing of the cell. Note that the identity of the cell's sender can be used to help identify the circuit, though not necessarily uniquely as multiple circuits could be routed along the same edge.

We require that a node maintains its individual decryption states in two separate state vectors $\boldsymbol{\tau}$ and $\bar{\boldsymbol{\tau}}$, relevant for D and $\bar{\mathsf{D}}$, respectively. Each time a new circuit is created, every forwarding and receiving node in that path will append a new component to its decryption state vectors $\boldsymbol{\tau}$ and $\bar{\boldsymbol{\tau}}$.

Our two-stage model $(\mathsf{D}, \bar{\mathsf{D}})$ for the processing of cells, is very much a choice for which level of generality to strive for. On the one hand, it reflects practical protocol designs such as Tor, without being overly prescriptive on quite how routing has to work. On the other hand, our model is evidently less general than a single stage model, that might allow arbitrary changes to its state. Our split in two stages, coupled with the restrictions on how the state looks and can be affected, guarantees that the processing of cells for one circuit cannot unduly influence the later processing of a cell associated to a different circuit. This guaranteed *robustness* significantly simplifies the definition of security games later on, as all circuits which the adversary has not interacted with, will still behave correctly. For a more general syntax, robustness does not follow automatically and would have to be modelled separately.

3.1 Onion Encryption

A (symmetric) *onion encryption scheme* $\mathsf{OE} = (\mathsf{G}, \mathsf{E}, \mathsf{D}, \bar{\mathsf{D}})$ is a quadruple of algorithms (see Fig. 2) to which we associate a message space $\mathsf{MsgSp} \subseteq \{0,1\}^*$ and a cell space $\mathsf{CelSp} \subseteq \{0,1\}^*$.

- The stateful circuit creation algorithm G is an abstraction of how circuits are created (which in reality is more likely to be an interactive process). It takes as input a path $\mathbf{p}$ that is not allowed to loop (by using the same router multiple times) and includes the proxy $\mathbf{p}[0]$. It updates its own state ϱ (initially $\varrho = \varepsilon$) and returns an initial encryption state

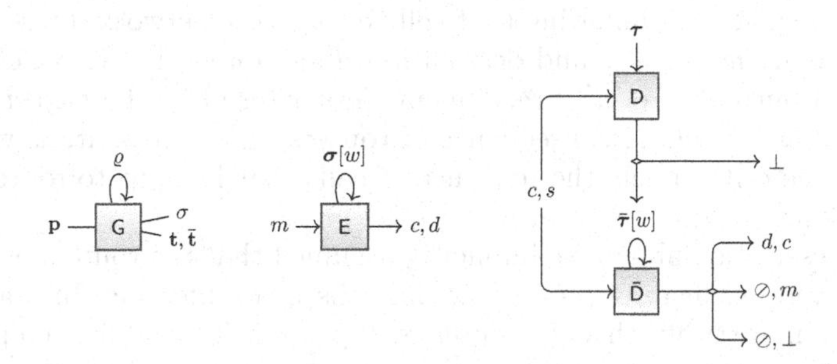

Fig. 2. Our syntax illustrated, showing the various possible outcomes during decryption. The end-of-circuit symbol $\oslash$ indicates that the current node is the intended recipient; the loops above algorithms indicate a state update.

σ (given to $\mathbf{p}[0]$) and two vectors, $\mathbf{t}$ and $\bar{\mathbf{t}}$, of initial decryption state components, one for each router in the path, so $|\mathbf{t}| = |\bar{\mathbf{t}}| = |\mathbf{p}|$. Upon receipt of their respective entries of $\mathbf{t}$ and $\bar{\mathbf{t}}$, the routers append these entries to their decryption state pair $(\boldsymbol{\tau}, \bar{\boldsymbol{\tau}})$. That is, if $\mathbf{p} = [a, b, c, d, e]$, we update the individual decryption states by the following sequence of operations: $\boldsymbol{\tau}_b.\mathsf{append}(\mathbf{t}[1]), \bar{\boldsymbol{\tau}}_b.\mathsf{append}(\bar{\mathbf{t}}[1]), \ldots, \boldsymbol{\tau}_e.\mathsf{append}(\mathbf{t}[4]), \bar{\boldsymbol{\tau}}_e.\mathsf{append}(\bar{\mathbf{t}}[4])$. Similarly the proxy's encryption state vector is updated by $\boldsymbol{\sigma}_a.\mathsf{append}(\sigma)$. As shown above, we will indicate the identity of the node to which a state vector or state variable belongs through its subscript. See also ADD($\mathbf{p}$) (Fig. 3) for G in action.
- The algorithm E is used by a proxy to send messages to one of its circuits. Given the current encryption state $\boldsymbol{\sigma}[w]$ for a circuit indexed locally by w and a message $m \in \mathsf{MsgSp}$, the algorithm E updates the encryption state and returns an initial cell $c \in \mathsf{CelSp}$ as well as the identity d of the router to which the cell has to be forwarded to.
- The deterministic algorithms D and $\bar{\mathsf{D}}$ are jointly responsible for processing an incoming cell c by a router. In the first stage, D associates the cell c to one of its circuits, where it can also use the identity of the source node s from which it received the cell. Importantly, D takes as additional input the node's entire first decryption state $\boldsymbol{\tau}$, but *without* the possibility to change this state. It returns a 'local' index w indicating to which circuit it has associated the cell and hence which component of the second decryption state $\bar{\boldsymbol{\tau}}$ should be used by $\bar{\mathsf{D}}$ to process the cell. The symbol $\perp$ indicates that the cell could not be associated to a circuit.
  In the second stage, $\bar{\mathsf{D}}$ takes the state component $\bar{\boldsymbol{\tau}}[w]$, as well as the source node and cell. It can update the decryption state component (though not any other part of the state) and return an output string x and a destination node d. The value d indicates the node to which the string x is to be forwarded, where $x \in \mathsf{CelSp}$. Alternatively, if $d = \oslash$, the router knows it is the intended recipient, in which case $x \in \mathsf{MsgSp} \cup \{\perp\}$.

A Cell's Trajectory and Lifecycle. Once E has output a cell c and an initial router d, we could start following that cell through the network: present the cell c to d, receiving new cell c' and destination d', so forward c' to d', etc. until a router either outputs $\perp$ or $d' = \oslash$. This process determines the *trajectory* of the cell, namely the chronological sequence of routers that process it, as well as the *lifecycle* of the cell, namely the sequence of cells that is input to routers during this processing.

In the description above, we implicitly assumed that the routers on the cell's trajectory were exclusively processing the cells in its lifecycle. In reality there will be much more traffic that the routers will process. This additional processing can affect the routers' states and consequently change the real-life trajectory and lifecycle of a cell. Our syntactical choices, such as deterministic processing by $\bar{\mathsf{D}}$, ensure that the lifecycle of a cell is fixed, as long as the real-life trajectory matches the path corresponding to the cell's intended circuit (cf. the security notion trajectory integrity, see the full version [18]). The ability to effectively predetermine a cell's lifecycle will turn out crucial when defining the security notion circuit hiding (Definition 4); it was exploited using a slightly different formalism in the context of public key, circuitless onion routing [11].

Local Versus Global Perspective. A key goal of onion routing is to ensure that routers are unable to link the recipient of a message to the proxy from which it originated, unless all the routers on a circuit collude. This necessitates that the router's view of a circuit is local: it knows which of its own circuits a cell belongs to (D's output), but otherwise a router should only be aware of the nodes that are directly adjacent to it.

Yet, when formalizing security notions (or correctness), we will need a global view and a way to move effortlessly from a router's local perspective to a more global view. To this end, we associate a global circuit index to each circuit upon creation and define the function map that takes a global circuit index and router index on the corresponding circuit and maps it to the node identifier and local circuit index. We allow the router index to be 0, so for instance $(v, w) = \mathsf{map}(i, 0)$ indicates that $v = \mathbf{p}_i[0]$ is the proxy for the circuit with global index i. The partial inverse map map^{-1} takes a node's identifier (which cannot be $\oslash$) and its local circuit index, and maps it back to the global view: which circuit is this and how far along the circuit does v occur. Both map and map^{-1} are dynamically defined (as new circuits can be created) and both are only ever called on their proper domains (values for which the functions are by then well-defined), with the convention that $\mathsf{map}(i, |\mathbf{p}_i| + 1)$ and $\mathsf{map}(i, -1)$ are set to $(\oslash, 0)$.

3.2 Correctness

Correctness guarantees that honestly generated cells are routed correctly and decrypt to the original messages at their intended destination in the same order as they were sent. Correctness should hold regardless of which circuits are created when, or the order in which cells are processed, as long as the order of cells belonging to the same circuit is preserved.

We formulate correctness through a game (see Fig. 3) whereby a scheduler is allowed to create circuits, choose the messages to be sent by the sending nodes, and determine the order in which individual routers process cells across different circuits. Reordering of cells belonging to different circuits models unpredictability of delays across the physical network, as well as the router's (limited) liberty to mix up the processing of cells to hamper traffic analysis. The scheduler however is *not* allowed to tamper with cells.

Concretely, for each circuit i we maintain a list $\mathbf{m}_i$ of the messages being sent through that circuit (using ENC), and check at the router's end (PASS) whether the messages arrive in order (the counter ctr_i indicates the next message on $\mathbf{m}_i$ that should be received). Moreover, for each circuit i and each of the $|\mathbf{p}_i|$ routers on its path (counted using j), we maintain a queue Q_j^i to keep track of the cells that are waiting to be processed by that router. Thus processing a cell by a forwarding router results in dequeuing for the current router and circuit, and enqueuing for its successor router.

Definition 1 (Correctness). *An onion encryption scheme* OE *is said to be correct if for all scheduling algorithms* $\mathcal{S}$ *(including computationally unbounded ones) it holds that:*

$$\Pr\left[\,\text{TRANSMIT}^{\mathcal{S}}_{\mathsf{OE}} \Rightarrow \mathsf{true}\,\right] = 0,$$

where the game TRANSMIT *is given in Fig. 3.*

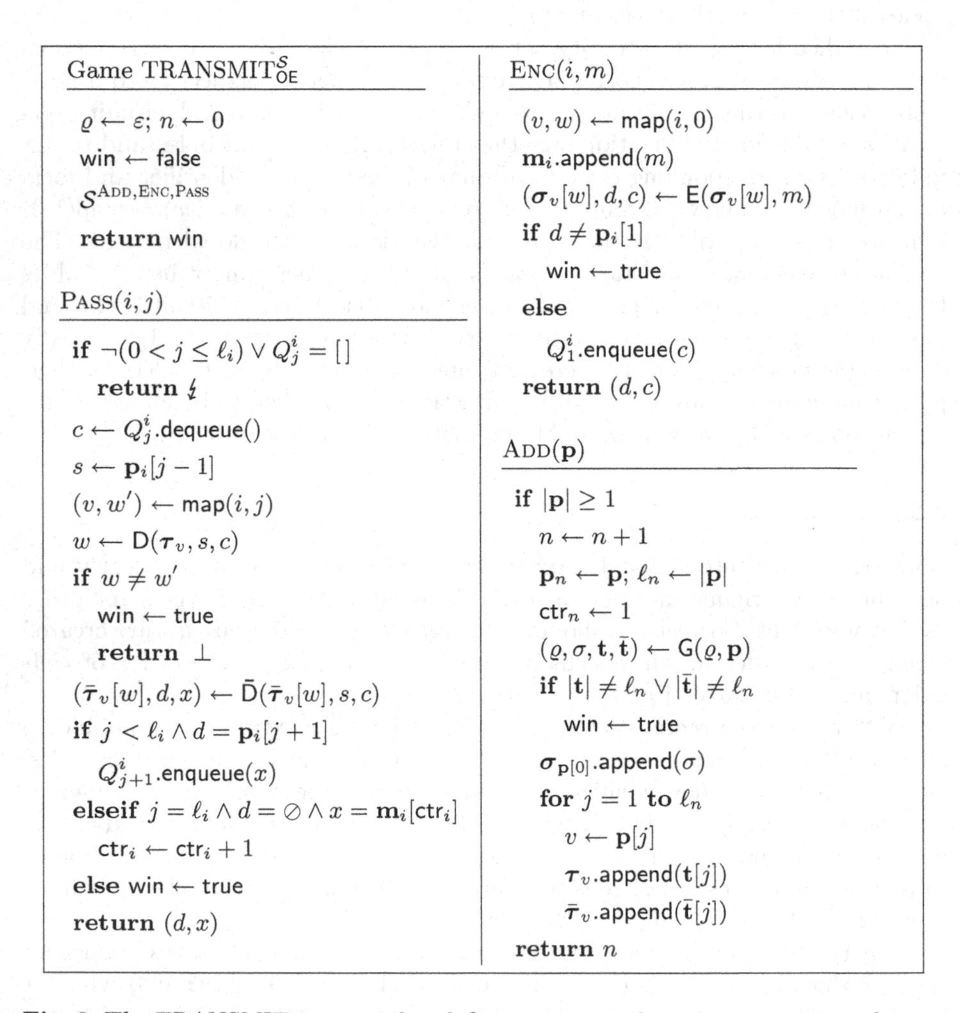

Fig. 3. The TRANSMIT game used to define correctness for onion encryption schemes.

3.3 Security

Onion routing networks should satisfy a range of security notions. In Sect. 5 we will deal with anonymity (in the form of circuit-hiding), and in Sect. 4 we concentrate on integrity and confidentiality, where the goal is that every circuit should implement a secure channel, even if the adversary has full control of the intermediate routers. Though we cannot give full control to an adversary in the anonymity setting, our security definitions in both sections share a number of modelling choices, as explained below.

Firstly, all our security notions are game-based where we simply define an adversary's advantage, without making an explicit and precise statement of what constitutes "secure". This concrete security approach is gaining traction for real world cryptosystems and would be harder to achieve in for instance an asymptotic UC framework. Secondly, all our formal definitions are multi-user definitions in the context of the entire routing network. For simplicity and wherever possible, our intuitive explanations only address what happens for a single circuit.

The customary threat model is to protect against adversaries who can "observe some fraction of network traffic; who can generate, modify, delete, or delay traffic; who can operate onion routers of their own; and who can compromise some fraction of the onion routers." [20]. When we map this threat model to our formal model, we first need to factor in the effect of the secure, unidirectional node-to-node communication. On a single edge, a passive outside adversary will be able to see the timing and volume of traffic. While this is extremely potent information to perform traffic analysis, our focus on the core cryptographic component renders this information largely out of scope. An active outside adversary can delay traffic on an edge (or delete all future traffic), but the edge's channel security will prevent it from inserting, modifying (including reordering and replaying), or deleting any of these cells. However, if a router is set to receive two cells from different routers, the adversary could control which one will arrive first. Fortunately, our two-stage approach to decryption with the router's state update restricted to a single circuit, makes the order in which cells associated to different circuits are processed irrelevant. Thus, for circuit hiding (Sect. 5) we restrict the adversary to the network interface it obtains from the compromised onion routers. For channel security we expand the adversary's power slightly (see Sect. 4).

The operation and compromise of routers is modelled by *selective* corruptions, where the adversary has to specify the set $\mathcal{C}$ of nodes it wishes to corrupt at the outset. For corrupted nodes, an adversary will learn the state of the router (incl. future updates), have access to all incoming cells to that router, and have full control over the cells being sent out to other routers. Recalling that circuit creation G outputs the triple $(\sigma, \mathbf{t}, \bar{\mathbf{t}})$ encoding the state updates of the proxy and routers on the circuit's path, we denote with $(\sigma, \mathbf{t}, \bar{\mathbf{t}})|_{\mathcal{C}}$ those state updates that are associated with corrupted nodes.

Our choice for selective corruptions only is informed by the often unforeseen complications that adaptive corruptions bring with them (see for instance selective opening attacks [4] and non-committing encryption [34]). Moreover, formalizing secure channels in a multi-user setting is relatively uncommon (cf. [30]) and introducing adaptive corruptions is, as far as we are aware of, unexplored.

4 Channel Security

We model channel security by considering both integrity and confidentiality, where we concentrate on the end-to-end effect (so plaintext integrity instead of ciphertext integrity and left-or-right indistinguishability instead of ciphertext indistinguishability). Moreover, we consider a slightly stronger threat model as the one alluded to above: even if an adversary has not corrupted a node, we will allow the adversary full control over its incoming and outgoing edges. Thus the end-to-end channel security of a circuit established by Tor should rely purely on its two end points not being compromised. Consequently, the unidirectional node-to-node security provided by TLS is of no use to establish channel security.

Plaintext Integrity. Plaintext integrity guarantees that, even in the presence of an adversary with *almost* full knowledge of all states and full control of the network, an honest receiving node can be reassured that the messages it outputs correspond to those being sent (assuming the sending node is uncompromised). This captures the inability for an adversary to inject, modify, reorder, or replay messages.

The game PINT (Fig. 4) models plaintext integrity. For each circuit with an honest proxy, we maintain a list $\mathbf{m}_i$ to check whether messages arrive unmodified and in the correct order at the honest receiver. The oracle $\textsc{Proc}(s, v, c)$ models node v's processing of cell c received from s. We do not insist that s is corrupt, thus we allow an adversary to inject cells even on edges for which it does not control the sending node, notwithstanding our assumption on secure node-to-node communication. This modest strengthening of the notion results in a slightly cleaner game.

Mirroring the correctness game, the counter ctr_n indicates the next message on $\mathbf{m}_n$ that should be received. As one would when defining plaintext integrity for ordinary channels, if the honest receiver accepts a message that wasn't sent (in that order), the adversary wins. Additionally—and this concept appears unique for routing networks—if an intermediate forwarding router believes a cell contains a valid message intended for it, the adversary wins. This win is a consequence of our choice *not* to allow "leaky pipes" [20].

Also note that indexing a vector component that does not exist is assumed to return a special symbol outside the set $\{0,1\}^*$. That is, for any vector $\mathbf{m}$ and any $k > 0$, if $|\mathbf{m}| = t$ then $\mathbf{m}[t+k] \neq \varepsilon$. In particular if the onion encryption scheme allows 'dummy' cells that decrypt to the empty string, an adversary that is able to forge such a dummy cell is deemed successful in the game PINT.

Definition 2. *The plaintext integrity advantage of adversary $\mathcal{A}$ against* OE *is defined by*

$$\mathbf{Adv}^{\mathrm{PINT}}_{\mathsf{OE}}(\mathcal{A}) = \Pr\left[\, \mathrm{PINT}^{\mathcal{A}}_{\mathsf{OE}} \Rightarrow \mathsf{true} \,\right],$$

where the game PINT *is given in Fig. 4.*

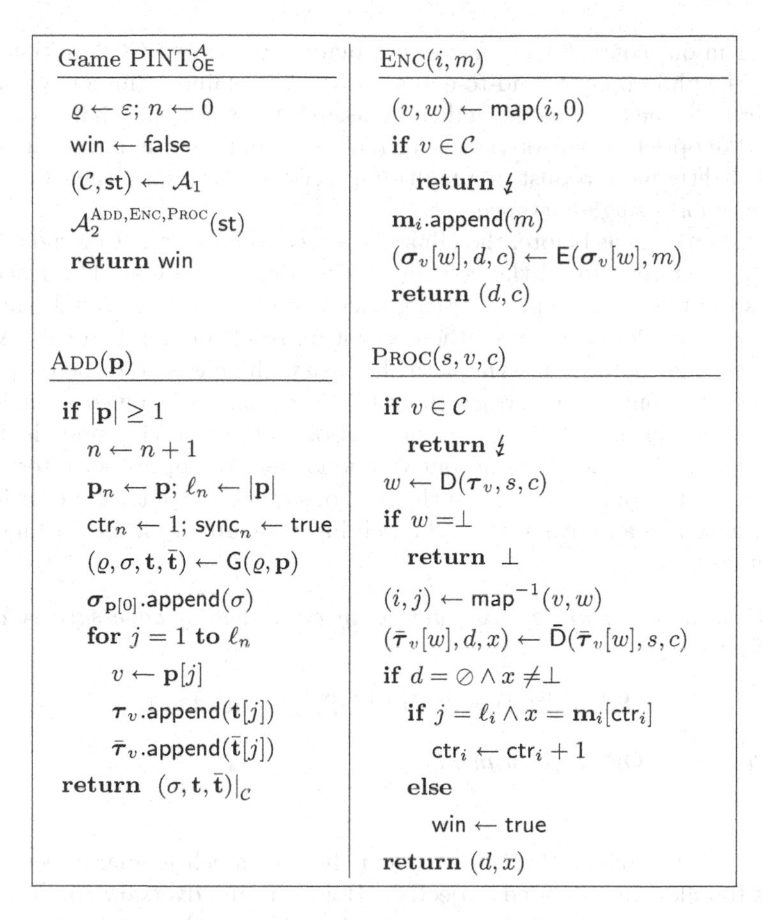

Fig. 4. The PINT game used to define plaintext integrity for onion encryption schemes. The sync_n flags in the ADD(**p**) oracle are used in later games.

Confidentiality. Confidentiality guarantees that an adversary gains no knowledge about the content of messages being sent on a circuit, as long as both the receiving and sending nodes are uncompromised. Otherwise, the adversary may have full knowledge of all states (except ϱ) and full control of the network. As usual for confidentiality, it is possible to provide both a passive 'CPA' and an active 'CCA' variant. Our game (Fig. 5) captures 'chosen cell attacks'; for the weaker chosen plaintext attack variant simply remove adversarial access to PROC.

The mechanism to define chosen cell attacks is an adaptation of left-or-right CCA indistinguishability for stateful encryption [6], combined with the plaintext-oriented suppression of 'decryption' queries as introduced in the context of RCCA security [12]. (The 'R' from RCCA for replayable has become a misnomer in our context.) In our view, a plaintext-oriented CCA notion better matches the philosophy of end-to-end security (like plaintext integrity), making it cleaner to define and less dependent on assumptions how the channel is implemented. We opted for left-or-right over real-or-random as the former in general appears slightly more robust in a multi-user setting [7], despite their fairly tight equivalence for a single-instance.

The lists $\mathbf{m}_i$ are as before, though as we are considering a left-or-right notion, they will either contain all the 'left' or all the 'right' messages. The PROC oracle plays the role of decryption oracle, which would result in trivial wins if an adversary were allowed to learn the decryption result of the final cell. As long as the receiving router is in-sync with the proxy, the message to be output will be suppressed from the adversary (by returning $\lightning$ instead). Once a single message deviates (which includes the error symbol $\perp$) the receiving node is deemed out-of-sync and henceforth its output will no longer be suppressed. Intermediate nodes are deemed out-of-sync from the get go, so their output will never be suppressed; of course an adversary might well have corrupted all forwarding nodes in the circuit.

Definition 3. *The plaintext confidentiality advantage of adversary $\mathcal{A}$ against* OE *is defined by*

$$\mathbf{Adv}_{\mathsf{OE}}^{\mathrm{LOR}}(\mathcal{A}) = 2 \cdot \Pr\left[\, \mathrm{LOR}_{\mathsf{OE}}^{\mathcal{A}} \Rightarrow \mathsf{true} \,\right] - 1,$$

where the game LOR *is given in Fig. 5.*

Trajectory Integrity. If all behaviour is honest, a cell is guaranteed by correctness to follow its intended trajectory. However, an adversary could interfere by injecting and modifying traffic potentially affecting the routing of cells. We provide a formal definition of inconsistent routing in the full version [18] under the name (cell) trajectory integrity. We assume an adversary has full control over the network and it knows all the parties' secrets, notwithstanding we assume all parties will still honestly process cells using the PROC(s, v, c) oracle. Both D and $\bar{\mathsf{D}}$ could lead to inconsistent routing, where a cell's trajectory does not match the path of a single circuit, or does not match the path of the circuit originally used by the proxy to create the cell (modelled by the ENC(i, m) oracle).

5 Anonymity

Overview. In an onion routing network, anonymity relies on a number of factors, such as the size of the network, the amount of traffic, the length of the anonymous channels (circuits), etc. Anonymity services such as sender anonymity and unlinkability can only be attained if the topology of the network of circuits remains hidden. We investigate how the *cryptographic* properties of an onion encryption scheme can contribute towards hiding the network's topology from an adversary.

Game $\text{LOR}_{\text{OE}}^{\mathcal{A}}$

$\varrho \leftarrow \varepsilon;\ n \leftarrow 0$
$\mathsf{win} \leftarrow \mathsf{false}$
$b \leftarrow_{\$} \{0,1\}$
$(\mathcal{C}, \mathsf{st}) \leftarrow \mathcal{A}_1$
$b' \leftarrow \mathcal{A}_2^{\text{ADD,ENC,PROC}}(\mathsf{st})$
return $b = b'$

$\text{ENC}(i, m_0, m_1)$

$(v, w) \leftarrow \mathsf{map}(i, 0)$
if $v \in \mathcal{C} \vee \mathbf{p}_i[\ell_i] \in \mathcal{C} \vee |m_0| \neq |m_1|$
  return ^{3}
$\mathbf{m}_i.\mathsf{append}(m_b)$
$(\boldsymbol{\sigma}_v[w], d, c) \leftarrow \mathsf{E}(\boldsymbol{\sigma}_v[w], m_b)$
return (d, c)

$\text{PROC}(s, v, c)$

if $v \in \mathcal{C}$
  return $\lightning$
$w \leftarrow \mathsf{D}(\boldsymbol{\tau}_v, s, c)$
if $w = \perp$
  return $\perp$
$(i, j) \leftarrow \mathsf{map}^{-1}(v, w)$
$(\bar{\boldsymbol{\tau}}_v[w], d, x) \leftarrow \bar{\mathsf{D}}(\bar{\boldsymbol{\tau}}_v[w], s, c)$
if $j = \ell_i \wedge d = \oslash$
  if $c = \mathbf{m}_i[\mathsf{ctr}_i] \wedge \mathsf{sync}_i = \mathsf{true}$
    $\mathsf{ctr}_i \leftarrow \mathsf{ctr}_i + 1$
    return $\lightning$
  else
    $\mathsf{sync}_i \leftarrow \mathsf{false}$
return (d, x)

Fig. 5. The LOR game used to define left-or-right indistinguishability for onion encryption schemes. For the $\text{ADD}(\mathbf{p})$ oracle refer to Fig. 4.

Our starting point is an indistinguishability game where the adversary gets to interact with one of two possible networks of his choice, and is required to guess which network it is interacting with. The adversary's interaction with and view of the network is facilitated by the nodes of the network it has corrupted and thus controls. An adversary controlling part of the network will inevitably gain partial information about that network, in particular about the topology of its circuits. For instance, for a corrupted node, an adversary will always be able to learn the previous and subsequent nodes of each of the corrupted node's circuits. For a circuit being routed through a contiguous sequence of corrupted nodes, the adversary can piece together the *directed subcircuit* as formed by those corrupted nodes and their adjacent honest nodes. The restrictions on an adversary's behaviour to avoid 'trivial' wins (e.g. if these observable, directed subcircuits differ between the two worlds) form a critical component of our circuit hiding game C-HIDE.

In the first stage of this game (see Fig. 7), the adversary $\mathcal{A}_1$ specifies a pair of vectors of circuits $\mathcal{W}_0$ and $\mathcal{W}_1$, and a set of corrupted nodes $\mathcal{C}$. Subject to a number of checks to avoid trivial wins (implemented by the predicate VALID as explained below), the game uses the procedure INIT-CIRC to initialize either the $\mathcal{W}_0$ or $\mathcal{W}_1$ network. The adversary is given $\tau_{\mathcal{C}}$ containing the states of corrupted nodes, but with a twist: after all circuits have been created by INIT-CIRC, the router's decryption-state vectors are all shuffled (and the map function will refer to the state post-shuffle). The shuffling reflects a secure implementation that avoids "order" correlation attacks by linking traffic through the order in which circuits were set up. For further justification on why this shuffling is necessary, we refer the reader to the full version of this paper [18]. Without this shuffling, security against active attacks appears a lot harder to achieve in the absence of very strong cell integrity.

In addition, the adversary is given access to the network by means of two oracles. The encryption oracle ENC can be used to trigger honest proxies to encrypt any message for one of its circuits. The network oracle NET provides collective and suitably restricted access to the honest routers in the network, as explained below. The goal of the adversary it to guess which of the two networks ($\mathcal{W}_0$ or $\mathcal{W}_1$) it is interacting with.

Below we will often refer to *segments* of a circuit. A circuit segment is defined to be a maximal subpath of a circuit such that its constituent nodes are either all honest or all corrupt. Thus any circuit uniquely decomposes into multiple segments (alternating honest and corrupt) and we can refer to, say, the first honest segment or the second corrupted segment in a circuit. Here the order of segments is understood to start from and include the proxy.

Challenge Validity $(\mathcal{W}_0, \mathcal{W}_1, \mathcal{C})$. The predicate $\text{VALID}(\mathcal{W}_0, \mathcal{W}_1, \mathcal{C})$ checks that the adversary's choice of networks does not allow a trivial win. A fair number of conditions are checked for this purpose, where we additionally disallow some settings where, without loss of generality, an adversary could achieve the same advantage while adhering to our restrictions (if corruptions were adaptive, these simplifications would be less clean). We list the conditions and their justifications below.

1. **The two circuit vectors $\mathcal{W}_0$ and $\mathcal{W}_1$ contain the same number of circuits, i.e. $|\mathcal{W}_0| = |\mathcal{W}_1|$.**

The interface which we provide to the adversary for interacting with the network allows it to easily infer the number n of circuits present in the network; mainly through its oracles and by inspecting the states of the nodes that it controls. While for sufficiently large networks this may be hard to determine in practice, we do not aim to conceal this information through cryptographic means.

2. **Every circuit in $\mathcal{W}_0$ and $\mathcal{W}_1$ contains at most two corrupted segments.**

This restriction keeps the complexity of the security definition manageable. The consequence of this assumption is that the most complicated honest–corrupt configuration for a circuit will be two corrupted segments and up to three honest segments, with one of these honest segments sandwiched between the corrupt nodes. This middle honest segment will play an import role. For circuits consisting of a proxy and three routers—the default circuit length in Tor and sufficient for a minimal working example of the tagging attack [24]—the restriction is without loss of generality.

3. **For each** i, **circuits** $\mathcal{W}_0[i]$ **and** $\mathcal{W}_1[i]$ **share a subpath** $[v_1, v_2, \ldots, v_{m-1}, v_m]$ **where** v_2 **is the first corrupted node in either circuit, nodes** $v_2 \ldots, v_{m-1}$ **are corrupted, and either** v_m **is honest or it is the last node in both** $\mathcal{W}_0[i]$ **and** $\mathcal{W}_1[i]$.

When we introduce the Enc oracle, an adversary will be able to select a circuit with an honest proxy and ask for a message to be encrypted. The resulting ciphertext will be processed by the honest routers before a cell is handed to one of the routers under adversarial control. While the specific path a circuit index points to depends on which network the adversary is interacting with, this condition ensures that a message encrypted for circuit i will reach the adversary on the same edge and, where applicable, reenters the honest component identically, irrespective of the challenge bit b.

4. **For a given circuit vector consider the multiset of subpaths** $[v_1, v_2, \ldots, v_{m-1}, v_m]$ **where nodes** v_1 **and** v_m **are honest and nodes** $v_2 \ldots, v_{m-1}$ **are corrupted. Then the corresponding multisets for** $\mathcal{W}_0$ **and** $\mathcal{W}_1$ **should be identical.**

An adversary can always infer the directed subcircuits overlapping with the nodes it has corrupted, by observing to which state component (w) a cell gets associated to at each corrupted node. Thus the two networks are required to match on these directed subcircuits, including the adjacent honest nodes.

5. **For all** i, **if either** $\mathcal{W}_0[i][0] \in \mathcal{C}$ **or** $\mathcal{W}_1[i][0] \in \mathcal{C}$ **then** $\mathcal{W}_0[i] = \mathcal{W}_1[i]$.

If a circuit's proxy is corrupted in either of the two worlds, then the corresponding circuit must be the same in both networks. The rationale is that we assume that a proxy's state reveals the entire path of routers for each of the circuits it is involved in.

Altogether the conditions so far ensure that any information that is inevitably leaked through the corrupted nodes is identical in both worlds; the final two conditions are simplifying conditions.

6. **For all** i **there exists** $j > 0$ **such that** $\mathcal{W}_0[i][j] \notin \mathcal{C}$.

Every circuit must contain at least one honest router. If all routers in a circuit were corrupted (possibly with an honest proxy), by condition 3 the circuit (including the proxy) must be identical in both networks. The inclusion of such circuits does not benefit the adversary, as can be shown by a straightforward reduction, so for simplicity we assume that every circuit includes at least one honest router.

7. **Every circuit in** $\mathcal{W}_0$ **and** $\mathcal{W}_1$ **contains at least one router in** $\mathcal{C}$.

An adversary has very little control over a circuit consisting entirely of honest nodes: while it could trigger the encryption oracle, it wouldn't actually be able to observe any of the cells travelling on that circuit (as all the connections between honest nodes are protected). Moreover, for schemes that satisfy trajectory integrity, the creation and operation of honest circuits has no influence on the rest of the network (cryptographically speaking). Therefore, without loss of generality, we assume that the network does not contain all-honest circuits.

As each circuit has to contain at least one honest node and one corrupt node, for all circuits $\mathbf{p}$ in the C-HIDE game we will have that $|\mathbf{p}| \geq 2$.

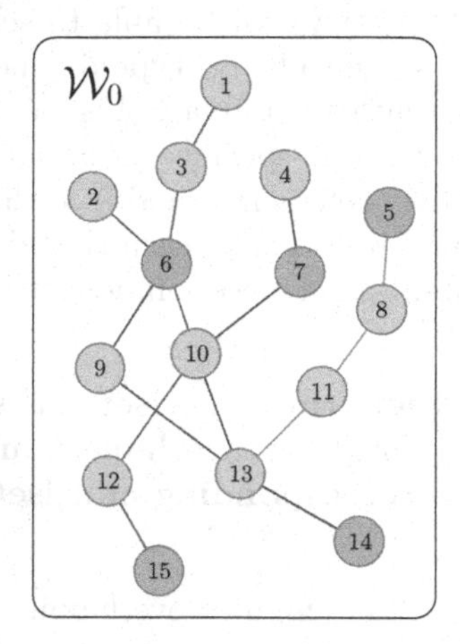

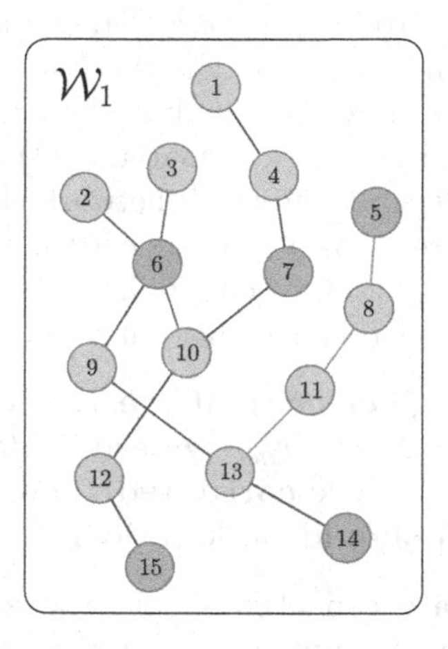

Fig. 6. An example of a valid challenge, where $\mathcal{W}_0 = [[2,6,9,13],[1,3,6,10,12,15],[4,7,10,13,14],[5,8,11,13]]$, $\mathcal{W}_1 = [[2,6,9,13,14],[3,6,10],[1,4,7,10,12,15],[5,8,11,13]]$, and $\mathcal{C} = \{5,6,7,14,15\}$ (marked in red). (Color figure online)

An example of a valid challenge for the circuit-hiding game is depicted in Figure 6. Both circuit vectors contain 4 circuits, each of which contain at most two corrupted segments. In this particular case the proxy of circuit $[5,8,11,13]$ is corrupt and accordingly it is identical in both worlds. The encryption oracle can be queried on any of the first three circuits (i.e. $i \in \{1,2,3\}$), as their proxies are honest. In $\mathcal{W}_0$ the onion proxies corresponding to the three indices are 2, 1, and 4, whereas in $\mathcal{W}_1$ they are 2, 3, and 1. However in either case the corresponding cells are returned on the same edges, i.e. $(2,6),(3,6),(4,7)$. The NET oracle consists of the subgraph containing the nodes $8,9,10,11,12,13$ and can be accessed through four input edges $(6,9),(6,10),(7,10),(5,8)$ and two output edges $(12,15),(13,14)$. Note that while the internal structure of this subgraph differs between the two cases, the interface that the adversary sees, i.e. the set of input and output edges, is identical in both worlds, as required.

The Init-Circ Procedure. For each circuit in $\mathcal{W}_b$ this procedure calls G to create initial states for the proxy and routers involved. Additionally some bookkeeping is performed similar to prior games. Novel are the two sets of circuit indices, $\mathcal{I}_{\text{EN}}$ and $\mathcal{I}_{\text{NOP}}$, that INIT-CIRC keeps track of, for later use by the NET oracle.

The set $\mathcal{I}_{\text{EN}}$ contains the indices of all circuits which have an honest proxy and contain an *entry edge* (the predicate EN returns true), namely an edge from a corrupt node to an honest node.

Game C-HIDE$^{\mathcal{A}}_{\mathrm{OE}}$

$(\mathcal{W}_0, \mathcal{W}_1, \mathcal{C}, \mathsf{st}) \leftarrow \mathcal{A}_1$
if $\neg$ VALID$(\mathcal{W}_0, \mathcal{W}_1, \mathcal{C})$
  return false
$\forall i$ $\mathsf{sync}_i \leftarrow$ true
$\varrho \leftarrow \varepsilon$; $n \leftarrow 0$; $b \leftarrow_\$ \{0,1\}$
INIT-CIRC$(\mathcal{W}_b)$
$\tau_\mathcal{C} \leftarrow \{(v, \boldsymbol{\sigma}_v, \boldsymbol{\tau}_v, \bar{\boldsymbol{\tau}}_v) \mid v \in \mathcal{C}\}$
$b' \leftarrow \mathcal{A}_2^{\mathrm{ENC,NET}}(\mathsf{st}, \tau_\mathcal{C})$
return $b = b'$

NET(z)

$\forall i$ $\mathsf{assc}_i \leftarrow 0$; $\mathbf{x} \leftarrow [\,]$
for $i' = 1$ **to** $|\mathbf{z}|$
  $(s, v, c) \leftarrow \mathbf{z}[i']$
  $w \leftarrow \mathsf{D}(\boldsymbol{\tau}_v, s, c)$
  if $s \notin \mathcal{C} \vee v \in \mathcal{C} \vee w = \perp$
    return $\lightning$
for $i' = 1$ **to** $|\mathbf{z}|$
  $(s, v, c) \leftarrow \mathbf{z}[i']$; $c^* \leftarrow c$
  $w \leftarrow \mathsf{D}(\boldsymbol{\tau}_v, s, c)$
  $(\bar{\boldsymbol{\tau}}_v[w], d, c) \leftarrow \bar{\mathsf{D}}(\bar{\boldsymbol{\tau}}_v[w], s, c)$
  $(i, j) \leftarrow \mathsf{map}(v, w)$
  while $d \notin \mathcal{C} \wedge d \neq \oslash$
    $s \leftarrow v$; $v \leftarrow d$
    $w \leftarrow \mathsf{D}(\boldsymbol{\tau}_v, s, c)$
    $(\bar{\boldsymbol{\tau}}_v[w], d, c) \leftarrow \bar{\mathsf{D}}(\bar{\boldsymbol{\tau}}_v[w], s, c)$
  if $d \in \mathcal{C}$
    $\mathbf{x}$.append(v, d, c)
  if $d \in \mathcal{C} \vee i \in \mathcal{I}_{\mathrm{NOP}}$
    $\mathsf{assc}_i \leftarrow \mathsf{assc}_i + 1$
    if $c^* \neq Q^i$.dequeue()
      $\mathsf{sync}_i \leftarrow$ false
if $\bigvee_{i \in \mathcal{I}_{\mathrm{EN}}} (\mathsf{sync}_i \vee \mathsf{assc}_i \neq 1)$
  return $\lightning$
return sort$(\mathbf{x})$

INIT-CIRC$(\mathcal{W})$

for $i = 1$ **to** $|\mathcal{W}|$
  $n \leftarrow n + 1$; $\mathbf{p}_n \leftarrow \mathcal{W}[i]$
  $(\varrho, \sigma, \mathbf{t}, \bar{\mathbf{t}}) \leftarrow \mathsf{G}(\varrho, \mathbf{p}_n)$
  $\ell_n \leftarrow |\mathbf{p}_n|$
  $\mathsf{sync}_n \leftarrow$ true
  $\boldsymbol{\sigma}_{\mathbf{p}_n[0]}$.append$(\sigma)$
  for $j = 1$ **to** ℓ_n
    $v \leftarrow \mathbf{p}_n[j]$
    $\boldsymbol{\tau}_v$.append$(\mathbf{t}[j])$
    $\bar{\boldsymbol{\tau}}_v$.append$(\bar{\mathbf{t}}[j])$
  if EN$(\mathbf{p}_n, \mathcal{C}) \wedge \mathbf{p}_n[0] \notin \mathcal{C}$
    $\mathcal{I}_{\mathrm{EN}} \leftarrow \mathcal{I}_{\mathrm{EN}} \cup \{i\}$
    if NOP$(\mathbf{p}_n, \mathcal{C})$
      $\mathcal{I}_{\mathrm{NOP}} \leftarrow \mathcal{I}_{\mathrm{NOP}} \cup \{i\}$
foreach v
  Shuffle$(\boldsymbol{\sigma}_v, \boldsymbol{\tau}_v, \bar{\boldsymbol{\tau}}_v)$

ENC(i, m)

$(v, w) \leftarrow \mathsf{map}(i, 0)$
if $v \in \mathcal{C}$
  return $\lightning$
$(\boldsymbol{\sigma}_v[w], d, c) \leftarrow \mathsf{E}(\boldsymbol{\sigma}_v[w], m)$
while $d \notin \mathcal{C}$
  $s \leftarrow v$; $v \leftarrow d$
  $w \leftarrow \mathsf{D}(\boldsymbol{\tau}_v, s, c)$
  $(\bar{\boldsymbol{\tau}}_v[w], d, c) \leftarrow \bar{\mathsf{D}}(\bar{\boldsymbol{\tau}}_v[w], s, c)$
$(v^*, d^*, c^*) \leftarrow (v, d, c)$
while $d \in \mathcal{C}$
  $s \leftarrow v$; $v \leftarrow d$
  $w \leftarrow \mathsf{D}(\boldsymbol{\tau}_v, s, c)$
  $(\bar{\boldsymbol{\tau}}_v[w], d, c) \leftarrow \bar{\mathsf{D}}(\bar{\boldsymbol{\tau}}_v[w], s, c)$
if $d \neq \oslash$
  $(i, j) \leftarrow \mathsf{map}^{-1}(v, w)$
  Q^i.enqueue(c)
return (v^*, d^*, c^*)

Fig. 7. The C-HIDE game used to define circuit-hiding security for onion encryption.

The set $\mathcal{I}_{\text{NOP}}$ is a subset of $\mathcal{I}_{\text{EN}}$. It contains those circuits in $\mathcal{I}_{\text{EN}}$ for which the adversary is unable to observe an output. Specifically, the predicate NOP returns true iff after the first entry edge the circuit contains no corrupted nodes, i.e. the circuit contains only one corrupted segment. Thus an adversary may inject cells into these circuits through the entry edge, but lacking a later corrupted segment, it is unable to 'catch' the processed cells. Note however that such circuits are still highly relevant in the C-HIDE game as the adversary should not be able to infer which cells produce no output.

The Enc Oracle. This oracle allows the adversary to encrypt a message m under any circuit i whose proxy is honest. As the adversary can only observe edges where one of the constituent nodes is corrupted, it will only get the ciphertext as output by the proxy if it happens to control the node which the proxy forwards it to. Otherwise we need to progress the ciphertext through the honest part of the circuit, until hitting a corrupted router. The first **while** loop takes care of this progression, resulting in a cell c^* and edge (v^*, d^*) that will be returned to the adversary.

However, before doing so, there is some further bookkeeping to be done on behest of the NET oracle. Recall that we allow up to two corrupt segments per circuit, so presumably after the corrupt segment starting with d^* the circuit can turn honest, and then corrupt again. In other words, there will be an honest middle segment. The NET oracle will allow the adversary to query these honest middle segments for all circuits simultaneously. Clearly, given ENC's interface, the adversary will know what messages are concealed across the various cells. Then it can always forward these to the NET oracle, where the cells it returns (at the interface between the honest segments and the second corrupt component) will correspond to a subset of the original set of messages. However, the adversary does not know which messages reside in which cells, though figuring this out would trivially identify the circuit over which the cell was sent (e.g. by embedding i in the message). To prevent trivial wins when an adversary also controls exit nodes (and is therefore able to recover plaintext) the NET oracle will suppress certain queries, based on the bookkeeping that ENC is about to do.

In the second **while** loop, the cell c^* is progressed further along the corrupted segment until the first honest node is encountered. Here the premise is that the nodes in the corrupted segment behave honestly (where the cells are processed using the routers' original state variables $\bar{\tau}_v[w]$ as the adversary has its own separate states).

This process allows us to predetermine the cell c that the first honest node d will receive from the corrupted segment. Unless we already reached the end of the circuit (in which case $d = \oslash$), we add this cell to a queue Q^i corresponding to the circuit i. The queue will be used by the NET oracle to detect when cells sent by corrupt routers digress from these stored values. Once this happens, an adversary has become active with respect to that circuit.

The Net Oracle. Finally, we give the adversary oracle access to the honest component of the network that it can inject traffic into. These are the honest circuit segments that are preceded by corrupted nodes. In turn these honest circuit segments may lead into a second corrupted segment or remain honest until the end. Hence, the NET oracle may return less cells than it receives in its input. Circuits with a corrupted proxy are also accessible through the NET oracle, but since they must be identical in both worlds, their corresponding output edges will be known to the adversary.

We will impose a number of restrictions on the NET oracle. After all, if the adversary could query each honest circuit segment individually, it could distinguishing the two networks simply by observing on which edge the corresponding cell is received. For this reason, the adversary can only query the honest circuit segments in parallel, so the sandwiched honest component behaves a little like a mix net.

Moreover, as already mentioned when discussing ENC, if the adversary is able to forward the cells obtained from ENC straight into NET (without modification), decryption of the resulting cells (using corrupted routers) would again allow to distinguish the two networks. Accordingly we restrict the oracle to only return an output when all honest circuit segments are queried in parallel and are all *out of sync*. The flag sync_i keeps track of whether circuit i is still *in sync* or whether it has gone *out of sync*. The exact meaning of a circuit being in or out of sync will be explained shortly.

Throughout, the adversary may query an honest circuit segment individually, but the output will be suppressed. This allows the adversary to progress the states of the routers along a particular circuit, prior to making the next 'parallel' query.

After intercepting cells through the ENC oracle, the adversary can manipulate, replay and re-order these cells and re-inject them into the honest part of the network through the NET oracle. As its input it takes a vector $\mathbf{z}$ of triples, each identifying a cell together with the edge on which it is incident. The first **for** loop verifies that this input satisfies two conditions. The first is that all cells be incident on an edge which the adversary has access to, that is, an entry edge. The other condition is that all cells must be associated to some circuit (i.e. $w \neq \perp$) by the honest node of the entry edge on which they are incident. If both checks are successful, the second **for** loop progresses each cell through the honest segments of their respective trajectories. In every iteration it stores the initial cell c^* and the circuit index i to which c^* was associated by the first processing node v. Every cell is progressed along its trajectory until it reaches a corrupted node or its destination. If the cell has reached a corrupted node, the cell and the corresponding output edge are stored in the output vector $\mathbf{x}$. In addition if the cell has reached a corrupted node or the circuit produces no output ($i \in \mathcal{I}_{\text{NOP}}$), the assc and sync variables for the associated circuit are updated. The variable assc_i keeps track of how many ciphertexts are associated to circuit i in a single oracle call. On the other hand, sync_i keeps track of whether the adversary has become active with respect to circuit i. This is determined by comparing c^* with

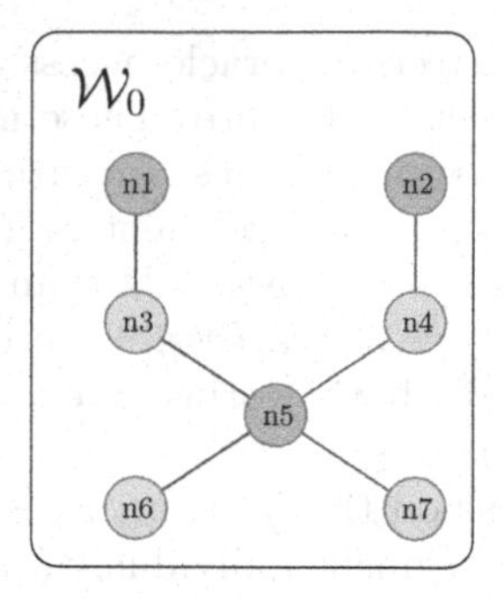

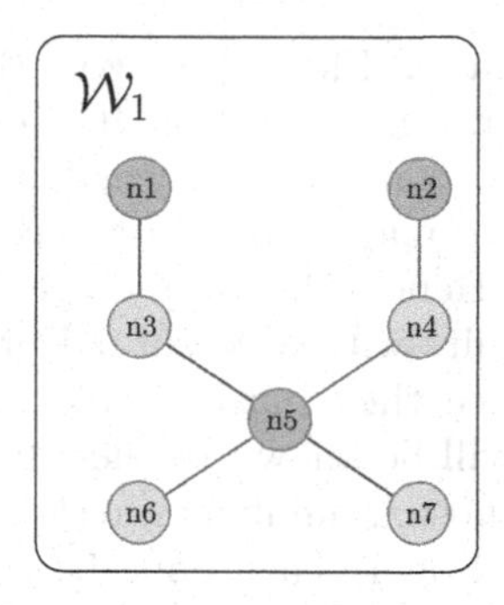

Fig. 8. The challenge used in the tagging attack example of Sect. 5.1, where $\mathcal{W}_0 = [[n1, n3, n5, n6], [n2, n4, n5, n7]]$, $\mathcal{W}_1 = [[n1, n3, n5, n7], [n2, n4, n5, n6]]$, and $\mathcal{C} = \{n3, n4, n6, n7\}$ (marked in red). (Color figure online)

the next available ciphertext in the queue corresponding to circuit i. If these don't match, sync_i is set to false, indicating that the circuit went out of sync. Once a circuit goes out of sync, it stays out of sync.

It is important to note the conditions under which we update these variables. In particular, if a circuit contains two entry edges, a cell will only affect these variables if it has been injected through the first entry edge of that circuit. Clearly cells injected through the second entry edge will produce no output either.

An output is returned to the adversary only if *every* circuit in the set $\mathcal{I}_{\text{EN}}$ has exactly one cell in $\mathbf{z}$ associated to it and is out of sync. The first condition stops the adversary from correlating the endpoints merely through the number of cells that are input and output at each end. The latter condition is analogous to the suppression of output in stateful security definitions [1,5,10,22]. On the other hand, circuits that have a corrupted proxy or whose routers are either all corrupted or all honest are excluded from this requirement (since we quantify over $\mathcal{I}_{\text{EN}}$).

Finally, before returning the output to the adversary we sort its components lexicographically to prevent the adversary from correlating the outputs with the inputs based on the ordering in which they have been processed by NET.

Definition 4. *The circuit hiding advantage of adversary $\mathcal{A}$ against* OE *is defined by*

$$\mathbf{Adv}_{\mathsf{OE}}^{\text{C-HIDE}}(\mathcal{A}) = 2 \cdot \Pr\left[\, \text{C-HIDE}_{\mathsf{OE}}^{\mathcal{A}} \Rightarrow \mathsf{true} \,\right] - 1,$$

where the game C-HIDE *is given in Fig. 7.*

5.1 Capturing Tagging Attacks

One of our main goals was to arrive to an anonymity definition that captures tagging attacks, we now confirm that this is indeed the case. Consider Tor's current onion encryption scheme, based on counter-mode AES, described in Sect. 2.1. This scheme is not C-HIDE secure as evidenced by the following attack. The adversary outputs the challenge described in Fig. 8. For any arbitrary message m it makes two encryption queries, $(1, m)$ and $(2, m)$, obtaining in return the respective replies, $(n1, n3, c_1)$ and $(n2, n4, c_2)$. The adversary progresses these cells past nodes $n3$ and $n4$ respectively, using its own copy of the nodes' states, to obtain the respective cells c_3 and c_4. It then tags both cells by flipping the last bit and last two bits of each cell respectively, that is, it queries $[(n3, n5, c_3 \oplus 1), (n4, n5, c_4 \oplus 11)]$ to the NET oracle. Since the cells' headers are unchanged each circuit will have exactly one cell associated to it, in addition both circuits will be out of sync because both cells have been modified by the adversary. Thus the oracle's output will not be suppressed and it will be of the form $[(n5, n6, c_5), (n5, n7, c'_5)]$. At this point the adversary attempts to untag the cells and process them at their respective exit nodes, i.e. $c_5 \oplus 1$ at node $n6$ and $c'_5 \oplus 11$ at node $n7$. If both decrypt correctly the adversary outputs 0 as its guess and 1 otherwise. It is easy to see that the adversary's advantage is very close to 1; the only time it fails is when both $c_5 \oplus 10$ and $c'_5 \oplus 10$ decrypt correctly, which happens with low probability.

6 Preventing Tagging Attacks

On an intuitive level it is evident that tagging attacks in Tor are enabled by the inherent malleability of counter-mode encryption which carries on across multiple layers of encryption. Proposal 202 [31] identified two potential ways of addressing this. One approach would be to borrow from mix-net designs by appending a MAC tag to each encryption layer, where after tag verification each node would re-pad the cell to its original length [16,17]. The other was to replace counter mode encryption with a Variable-Input-Length (VIL) tweakable cipher, as used in disk encryption, which impedes malleability without incurring any ciphertext expansion. Clearly the increased space efficiency of this latter approach is a huge bonus, but back in 2012 all known VIL tweakable cipher constructions were significantly slower than counter mode encryption. This changed however with the advent of AEZ [26] whose efficiency is comparable to that of counter mode AES, albeit at the expense of a more heuristic security analysis. Now having a viable instantiation, the Tor project put forward a concrete design for a new onion encryption scheme in proposal 261 [32]. We will refer to the onion encryption scheme described therein as Tor261. We emphasize that a VIL tweakable cipher is only a building block, and constructing a secure onion encryption scheme from it is substantially non-trival. The rest of this section is devoted to put the security of Tor261 on firm grounds, but we first describe Tor261 in more detail.

6.1 VIL Tweakable Ciphers and AEZ

As the name suggests, a VIL tweakable cipher is a tweakable cipher that can operate over inputs of varying length. More precisely it is a pair of deterministic algorithms (Π, Π^{-1}) each of which takes a key K, a tweak tw and a string x, respectively y, to return a string y, respectively x, where $|x| = |y|$ and for all K and tw, $\Pi(K, tw, \cdot)$ is a permutation and $\Pi^{-1}(K, tw, \cdot)$ is its inverse. Recall that in Tor the cell size is fixed to 509 bytes and it would therefore suffice to have a tweakable cipher that can handle inputs of this length. Accordingly the term wide-block tweakable cipher is often used instead, but in reality all known constructions admit inputs of varying length. In terms of security, a tweakable cipher is expected to be a (strong) tweakable pseudorandom permutation. We refer the reader to [41] for an up-to-date introduction to VIL tweakable ciphers.

Technically, AEZ embodies a different primitive called Robust Authenticated Encryption (RAE) [26]. An RAE is a pair of deterministic algorithms (Π, Π^{-1}) where Π takes a key K, a nonce no, associated data ad, a message x and a stretch τ to return a ciphertext y of length $|x| + \tau$. The decryption algorithm Π^{-1} inverts this operation, taking a K, a nonce no, associated data ad and a ciphertext y to return either a message x, if y was generated honestly, or the special symbol $\perp$ indicating that y is invalid. When the key is chosen uniformly at random security requires that for any nonce and associated data, (Π, Π^{-1}) should behave as a pseudorandom injection, and its inverse, from binary strings to τ-bit longer ones. It is easy to see that if we set $\tau = 0$ RAE collapses to a VIL tweakable cipher, where the nonce and the associated data, collectively play the role of the tweak. Indeed in Tor proposal 261 τ is set to zero and we will therefore treat AEZ as a VIL tweakable cipher where the tweak is represented by the pair (no, ad).

6.2 Tor261: The Onion Encryption Scheme in Tor Proposal 261

The onion encryption scheme Tor261 is obtained by instantiating the Tor relay protocol [19] with the layer encryption described in [32]. Note that proposal 261 only affects the relay protocol and in particular the cryptography used in the circuit extend protocol is unaffected. A pseudocode description of Tor261 is displayed in Fig. 9.

In addition to a VIL tweakable cipher, the scheme also makes use of a block cipher BC (instantiated with AES) in a Davies-Meyer-type configuration to compute a chain value h (by means of a separate chain key L) that is included in the tweak of every layer of encryption, i.e. Π evaluation. It is intended to provide forward security rather than anonymity or standard channel security, consequently it does not surface in our analysis. In addition, the tweak also contains the xor of the input and output strings from the *previous* layer encryption call. Intuitively this serves to create a domino effect whereby the corruption of any cell will corrupt all subsequent cells. The no component of the tweak is composed of a counter ctr and two binary flags, fwd and $early$, encoded as single-byte strings. These indicate respectively the direction of travel of the cell with respect to the direction in which the circuit was established, and whether the cell is of the type RELAY or RELAY_EARLY (the two cell types handled by the relay protocol). Whether a cell is of type RELAY or RELAY_EARLY is indicated in the command field (cmd) in the cell header, through byte values 3 and 9 respectively. Thus during decryption the $early$ flag is set according to the value described in

algorithm $\mathsf{G}(\varrho, \mathbf{p})$

$\ell \leftarrow |\mathbf{p}|, ctr \leftarrow 0, \mathbf{c}_0 \leftarrow [\varepsilon]_1^\ell$
$(v^*, d^*) \leftarrow (\mathbf{p}[0], \mathbf{p}[1])$
do $id_o \leftarrow_{\$} \{0,1\}^{32}$ **until** $(v^*, d^*, id_o) \notin \varrho$
$\varrho \leftarrow \varrho \cup \{(v^*, d^*, id_o)\}$
$id_o^* \leftarrow id_o$
for $j = 1$ **to** ℓ
  $(s, v, d) \leftarrow (\mathbf{p}[j-1], \mathbf{p}[j], \mathbf{p}[j+1])$
  $\mathbf{K}[j] \leftarrow_{\$} \{0,1\}^k$
  $\mathbf{L}[j] \leftarrow_{\$} \{0,1\}^{256}$
  $\mathbf{h}[j] \leftarrow_{\$} \{0,1\}^{128}$
  $id_i \leftarrow id_o$
  do $id_o \leftarrow_{\$} \{0,1\}^{32}$ **until** $(v, d, id_o) \notin \varrho$
  $\varrho \leftarrow \varrho \cup \{(v, d, id_o)\}$
  $\mathbf{t}[j] \leftarrow (s, id_i)$
  $\bar{\mathbf{t}}[j] \leftarrow (v, \mathbf{K}[j], \mathbf{L}[j], \mathbf{h}[j], ctr, \varepsilon, \varepsilon, d, id_o)$
$\sigma \leftarrow (\ell, \mathbf{K}, \mathbf{L}, \mathbf{h}, ctr, \mathbf{c}_0, d^*, id_o^*)$
return $(\varrho, \sigma, \mathbf{t}, \bar{\mathbf{t}})$

algorithm $\mathsf{E}(\sigma, m)$

$early \leftarrow 0$
parse σ **as** $(\ell, \mathbf{K}, \mathbf{L}, \mathbf{h}, ctr, \mathbf{c}_0, d, id_o)$
$\mathbf{c}_1[\ell] \leftarrow \mathsf{encode}(m)$
for $j = \ell$ **to** 1
  $no \leftarrow \langle ctr \rangle_{64} \parallel \langle fwd \rangle_8 \parallel \langle early \rangle_8$
  $\mathbf{h}[n] \leftarrow \mathsf{BC}(\mathbf{L}[j], \mathbf{h}[j]) \oplus \mathbf{h}[j]$
  $ad \leftarrow (\mathbf{c}_0[j] \oplus \mathbf{c}_0[j-1]) \parallel \mathbf{h}[j]$
  $\mathbf{c}_1[j-1] \leftarrow \Pi(\mathbf{K}[j], (no, ad), \mathbf{c}_1[j])$
if $early = 1 : cmd \leftarrow \langle 9 \rangle_8$
else : $cmd \leftarrow \langle 3 \rangle_8$
$\hat{c} \leftarrow (id_o, cmd, \mathbf{c}_1[0])$
$ctr \leftarrow ctr + 1$
$\sigma \leftarrow (\ell, \mathbf{K}, \mathbf{L}, \mathbf{h}, ctr, \mathbf{c}_1, d, id_o)$
return $(\sigma, d, \hat{c})$

algorithm $\mathsf{D}(\boldsymbol{\tau}, s, \hat{c})$

parse $\hat{c}$ **as** (id_i, cmd, c)
if $cmd \neq \langle 3 \rangle_8$
  return $\perp$
for $w = 1$ **to** $|\boldsymbol{\tau}|$
  if $(s, id_i) = \boldsymbol{\tau}[w]$
    return w
return $\perp$

algorithm $\bar{\mathsf{D}}(\bar{\boldsymbol{\tau}}[w], s, \hat{c})$

parse $\bar{\boldsymbol{\tau}}[w]$
  as $(v, K, L, h, ctr, c_0, c_0', d, id_o)$
parse $\hat{c}$ **as** (id_i, cmd, c_1)
if $cmd = \langle 9 \rangle_8 : early \leftarrow 1$
else : $early \leftarrow 0$
$no \leftarrow \langle ctr \rangle_{64} \parallel \langle fwd \rangle_8 \parallel \langle early \rangle_8$
$h \leftarrow \mathsf{BC}(L, h) \oplus h$
$ad \leftarrow (c_0 \oplus c_0') \parallel h$
$c_1' \leftarrow \Pi^{-1}(K, (no, ad), c_1)$
$ctr \leftarrow ctr + 1$
if $\mathsf{chkzeros}(c_1') \wedge d \neq \perp$
  $x \leftarrow \mathsf{decode}(c_1')$; $d^* \leftarrow \oslash$
else
  if $d \in \{\perp, \oslash\}$
    $d \leftarrow \perp$; $d^* \leftarrow \oslash$; $x \leftarrow \perp$
  else
    $d^* \leftarrow d$
    $x \leftarrow (id_o, cmd, c_1')$
$\bar{\boldsymbol{\tau}}[w] \leftarrow (K, L, h, ctr, c_1, c_1', d, id_o)$
return $(\bar{\boldsymbol{\tau}}[w], d^*, x)$

Fig. 9. The Onion Encryption Scheme Tor261 and its variant $\overline{\text{Tor261}}$. Scheme $\overline{\text{Tor261}}$ includes the shaded code but Tor261 does not.

the cell's command field. In all other cases we treat fwd and $early$ as internal variables set in accordance with the context in which encryption and decryption are operating. Since we only consider unidirectional anonymous channels in the forward direction, $fwd = 1$, always.

In our modelling of the Tor relay protocol we make the following assumptions and simplifications. We assume a version 4 or higher cell format with a total cell size of 514 bytes. In addition to the cell header relay cells include a payload header and proposal 261 alters the format of the payload header. However the only cryptographic processing of this header is limited to checking that the redundancy in certain fields is correctly formatted. Specifically it identifies 55 bits that should be verified to contain zeros upon decryption, but suggests that this verification could be extended to other fields for added security. We model the processing of a message by padding it and prepending the relay header through an encode function. Similarly during the decryption we will employ a decode function to reverse this process and a function $\mathsf{chkzeros}$ to verify that the relevant fields contain zeros. For generality, in our analysis we assume the number of bits set to zero by encode and later verified by $\mathsf{chkzeros}$ is r.

As before, a node determines to which circuit it should associate a cell from the circuit identifier id in the cell header as described in Sect. 2.1. Circuit identifiers are chosen during circuit establishment by the various nodes involved, which we abstracted in the circuit creation algorithm G. Except for a mechanism to avoid collisions, the Tor specification does not specify how circuit identifiers are to be chosen. Since it is not particularly relevant for our analysis, for simplicity we assume these are sampled by G uniformly at random without replacement with respect to every edge. In particular, G maintains a state ϱ comprised of a set of triples (a, b, id) to keep track that the circuit identifier id is in use on edge (a, b) and thereby avoid collisions.

The Tor specification allows for certain messages to be delivered to nodes in the circuit other than the last one. As far as we are aware, this functionality is not actually used in the relay protocol in practice. Our syntax does not allow an onion proxy to specify cells for an intermediate node, but intermediate nodes may nonetheless recognise a cell as being intended for them in Tor261. Thus from the perspective of an intermediate node a cell is either recognised or else it is forwarded along the circuit, but it is never deemed invalid. On the other hand, if the last node in a circuit receives a cell which it does not recognise, it declares the cell as invalid and the circuit is torn down [20]. To model this behaviour we overload the semantics of the variable d in the node's decryption state component. For any circuit, it is intended to store the next hop in the circuit with respect to the current node and is set to $\oslash$ if the node is last in the circuit. When the last node detects an invalid cell it sets $d \leftarrow \perp$ to indicate such an event, and returns an error for that cell and any subsequent ones. However the other nodes in the circuit are unaffected since the adversary may be able to block the cells instructing the circuit teardown.

6.3 Circuit Hiding

As described in Sect. 6.2 the relay protocol supports two types of cells, RELAY and RELAY_EARLY. The sole purpose of RELAY_EARLY cells is to enable a mechanism for limiting the length of circuits in Tor, see [19]. While the details of this mechanism are beyond our scope, partly because it extends over to the circuit extend protocol, the support of RELAY_EARLY cells in Tor261 exposes it to tagging attacks. In essence an adversary can tag a cell by flipping the *cmd* field in its header which will then propagate along the circuit unaltered. As an example consider the challenge depicted in Fig. 8. After making two ENC queries, one on each circuit, the adversary forges a RELAY cell on $(n3, n5)$ and a RELAY _EARLY cell on $(n4, n5)$, and submits both as a single NET query. If the cell output on $(n5, n7)$ is RELAY_EARLY it knows it is interacting with $\mathcal{W}_1$.

Tagging attacks manipulating the *cmd* field were already exploited in the infamous 2014 incident [2], and as such this vulnerability is a real concern and not just an artifact of our security definition. Interestingly, Tor261 appears to attempt to protect against this by including the *early* flag in the nonce *no* but as we just pointed out this does not prevent the attack. It could be argued however that this attack is somewhat limited in practice since it only admits one type of tag. On the other hand, the current onion encryption scheme allows an adversary to tag each cell with a unique mark allowing it to de-anonymise multiple circuits in parallel. This limitation could potentially be overcome by instead tagging a unique mark in a *sequence* of cells. However this possibility is limited by the fact that in Tor honest nodes are required to tear down the circuit if they observe more than eight RELAY_EARLY cells. Moreover, in a typical setting three RELAY_EARLY cells would already have been used up during circuit establishment. Thus in practice an adversary would have at most 2^5 unique tags at its disposal. On the one hand, this improves significantly over Tor's current state of affairs but it still falls short of the best possible security.

Unable to prove Tor261 secure we consider its variant $\overline{\mathsf{Tor261}}$, also described in Fig. 9, which supports only RELAY cells and prove that it meets our circuit hiding notion. This serves to show that the above attack is the only way of mounting tagging attacks on Tor261, which could possibly be prevented by adopting an alternative mechanism, not involving RELAY_EARLY cells, for limiting circuit size or mitigated further by reducing the maximum circuit size. Informally, Theorem 1 states that $\overline{\mathsf{Tor261}}$ is circuit hiding as long as (Π, Π^{-1}) is a secure VIL tweakable cipher in the $\pm\widetilde{\mathrm{prp}}$ sense [26,41], and m and r are sufficiently large.

Theorem 1. ($\overline{\mathsf{Tor261}}$ is Circuit Hiding). *Let* $\overline{\mathsf{Tor261}}$ *be the scheme described in Fig. 9 composed of a VIL tweakable cipher* (Π, Π^{-1}) *and an encoding scheme* encode *that prepends messages with r zeros. Then for any circuit hiding adversary* $\mathcal{A} = (\mathcal{A}_1, \mathcal{A}_2)$ *running in time t, making* q_e *queries to* ENC *and* q_n *queries to* NET*, there exists a* $\pm\widetilde{\mathrm{prp}}$ *adversary* $\mathcal{B}$ *running in time* t' *and making* q_f *and* q_i *queries to its forward and inverse oracles, such that:*

$$\mathbf{Adv}^{\text{C-HIDE}}_{\overline{\mathsf{Tor261}}}(\mathcal{A}) \leq |\mathcal{I}_{\text{EN}}| \left(2\,\mathbf{Adv}^{\pm\widetilde{\mathrm{prp}}}_{\Pi,\Pi^{-1}}(\mathcal{B}) + 2^{-r+1} + \min(q_e, q_n)\, 2^{-m+1} \right),$$

where $\mathcal{I}_{\text{EN}}$ is the set of circuits with an honest proxy that the adversary can inject cells into (i.e. circuits containing an entry edge). Furthermore we have that $q_f \leq q_e$, $q_i \leq q_n$, and $t' \leq t + q_e \ell_{max} T + q_n |\mathcal{I}_{\text{EN}}|(\ell_{max} - 1)T$ where ℓ_{max} is the maximum length of a circuit and T is the maximum time needed to evaluate Π or Π^{-1}.

The full proof of Theorem 1 can be found in the full version [18]. Below we outline the main intuition.

Proof outline. We prove the theorem through a standard game-hopping technique. We start with the C-HIDE game instantiated with $\overline{\text{Tor261}}$ and focus on the honest node in the entry edge (if any) of each circuit. We then replace the VIL tweakable cipher instance corresponding to that node with a truly random permutation. We gradually chop parts of the circuits until we eventually end up with a game that depends solely the state information pertaining to those subpaths that are required (by VALID) to be common to both worlds (corresponding to the two possible values of b). It can then be shown that for $\overline{\text{Tor261}}$ the state information corresponding to these subpaths is identically distributed in either world. Thus we eventually end up with a game that is independent of the bit b in which case the adversary can do no better than guess the bit b.

7 Conclusion

Motivated by Tor's susceptibility to tagging attacks and the ongoing effort in the Tor community to thwart them, we initiated a formal treatment of circuit-based onion encryption. In our treatment, we opted for a level of abstraction that is closer to practice than previous works. For instance, we explicitly included the routing functionality that characterizes onion routing. While this choice arguably complicates security definitions and analysis, we believe it provides for a more informative model, allowing us to expose certain hitherto unsuspected conflicts between routing and security. One illustration is the potential for correlating traffic between onion router based on the seniority of circuit, which has implications for the data structure used to store a router's full decryption state.

We analysed Tor's new proposal Tor261, intended to prevent tagging attacks, using our new framework. Our analysis confirms that its overall design is sound, yet also exposes that its support of RELAY_EARLY cells still enables tagging attacks. Presently, we focused on the circuit-hiding property of the proposed scheme, leaving open the analysis of Tor261's end-to-end channel security.

Finally, unidirectional channels are only a first step in the formal analysis of Tor and related onion routing protocols. Marson and Poettering [30] recently exposed important challenges when composing ordinary unidirectional secure channels and it is unclear how those challenges affect anonymous channels. (This cautionary remark does not refer to any specific issue that we identified, rather it delineates our analysis.) Furthermore, extending our analysis to include dynamic circuit establishment and deal with adaptive corruptions, are challenging open problems.

Acknowledgments. We would like to thank Matthew Green for suggesting this problem to us and Jonathan Katz for helpful initial discussions. We are indebted to Nick Matthewson for clarifying certain historical and practical aspects of Tor. We also thank the anonymous reviewers for their constructive feedback.

Degabriele was supported in part by EPSRC grant EP/M013472/1 (UK Quantum Technology Hub for Quantum Communications Technologies) and in part by the German Federal Ministry of Education and Research (BMBF) within CRISP.

References

1. Albrecht, M.R., Degabriele, J.P., Hansen, T.B., Paterson, K.G.: A surfeit of SSH cipher suites. In: Weippl, E.R., Katzenbeisser, S., Kruegel, C., Myers, A.C., Halevi, S., (eds.) ACM CCS 2016, pp. 1480–1491. ACM Press, October 2016
2. Dingledine (arma), R.: Tor security advisory: "relay early" traffic confirmation attack, July 2014. https://blog.torproject.org/blog/tor-security-advisory-relay-early-traffic-confirmation-attack
3. Backes, M., Goldberg, I., Kate, A., Mohammadi, E.: Provably secure and practical onion routing. In: CSF, pp. 369–385. IEEE Computer Society (2012)
4. Bellare, M., Hofheinz, D., Yilek, S.: Possibility and impossibility results for encryption and commitment secure under selective opening. In: Joux, A. (ed.) EUROCRYPT 2009. LNCS, vol. 5479, pp. 1–35. Springer, Heidelberg (2009). https://doi.org/10.1007/978-3-642-01001-9_1
5. Bellare, M., Kohno, T., Namprempre, C.: Authenticated encryption in SSH: provably fixing the SSH binary packet protocol. In: Atluri, V. (ed.) ACM CCS 2002, pp. 1–11. ACM Press, November 2002
6. Bellare, M., Namprempre, C.: Authenticated encryption: relations among notions and analysis of the generic composition paradigm. In: Okamoto, T. (ed.) ASIACRYPT 2000. LNCS, vol. 1976, pp. 531–545. Springer, Heidelberg (2000). https://doi.org/10.1007/3-540-44448-3_41
7. Bellare, M., Ristenpart, T., Tessaro, S.: Multi-instance security and its application to password-based cryptography. In: Safavi-Naini, R., Canetti, R. (eds.) CRYPTO 2012. LNCS, vol. 7417, pp. 312–329. Springer, Heidelberg (2012). https://doi.org/10.1007/978-3-642-32009-5_19
8. Daniel, J.: Bernstein, Mridul Nandi, and Palash Sarkar. HHFHFH, Dagstuhl (2016)
9. Bertoni, G., Daemen, J., Peeters, M., Van Assche, G., Van Keer, R.: Farfalle: parallel permutation-based cryptography. Cryptology ePrint Archive, Report 2016/1188 (2016). http://eprint.iacr.org/2016/1188
10. Boldyreva, A., Degabriele, J.P., Paterson, K.G., Stam, M.: Security of symmetric encryption in the presence of ciphertext fragmentation. In: Pointcheval, D., Johansson, T. (eds.) EUROCRYPT 2012. LNCS, vol. 7237, pp. 682–699. Springer, Heidelberg (2012). https://doi.org/10.1007/978-3-642-29011-4_40
11. Camenisch, J., Lysyanskaya, A.: A formal treatment of onion routing. In: Shoup, V. (ed.) CRYPTO 2005. LNCS, vol. 3621, pp. 169–187. Springer, Heidelberg (2005). https://doi.org/10.1007/11535218_11
12. Canetti, R., Krawczyk, H., Nielsen, J.B.: Relaxing chosen-ciphertext security. In: Boneh, D. (ed.) CRYPTO 2003. LNCS, vol. 2729, pp. 565–582. Springer, Heidelberg (2003). https://doi.org/10.1007/978-3-540-45146-4_33

13. Chakravarty, S., Barbera, M.V., Portokalidis, G., Polychronakis, M., Keromytis, A.D.: On the effectiveness of traffic analysis against anonymity networks using flow records. In: Faloutsos, M., Kuzmanovic, A. (eds.) PAM 2014. LNCS, vol. 8362, pp. 247–257. Springer, Cham (2014). https://doi.org/10.1007/978-3-319-04918-2_24
14. Chaum, D.: Untraceable electronic mail, return addresses, and digital pseudonyms. Commun. ACM **24**(2), 84–88 (1981)
15. Danezis, G., Diaz, C., Syverson, P.: Systems for anonymous communication. In: CRC Handbook of Financial Cryptography and Security, p. 61 (2009)
16. Danezis, G., Dingledine, R., Mathewson, N.: Mixminion: design of a type III anonymous remailer protocol. In: 2003 IEEE Symposium on Security and Privacy, pp. 2–15. IEEE Computer Society Press, May 2003
17. Danezis, G., Goldberg, I.: Sphinx: a compact and provably secure mix format. In: 2009 IEEE Symposium on Security and Privacy, pp. 269–282. IEEE Computer Society Press, May 2009
18. Degabriele, J.P., Stam, M.: Untagging Tor: a formal treatment of onion encryption. Cryptology ePrint Archive, Report 2018/162 (2018). https://eprint.iacr.org/2018/162
19. Dingledine, R., Mathewson, N.: Tor protocol specification. https://gitweb.torproject.org/torspec.git/plain/tor-spec.txt
20. Dingledine, R., Mathewson, N., Syverson, P.F.: Tor: the second-generation onion router. In: USENIX Security Symposium, pp. 303–320. USENIX (2004)
21. Feigenbaum, J., Johnson, A., Syverson, P.F.: Probabilistic analysis of onion routing in a black-box model. ACM Trans. Inf. Syst. Secur. **15**(3), 14:1–14:28 (2012)
22. Fischlin, M., Günther, F., Marson, G.A., Paterson, K.G.: Data is a stream: security of stream-based channels. In: Gennaro, R., Robshaw, M. (eds.) CRYPTO 2015. LNCS, vol. 9216, pp. 545–564. Springer, Heidelberg (2015). https://doi.org/10.1007/978-3-662-48000-7_27
23. Freedman, M.J., Morris, R.: Tarzan: a peer-to-peer anonymizing network layer. In: Atluri, V. (ed.) ACM CCS 2002, pp. 193–206. ACM Press, November 2002
24. Fu, X., Ling, Z.: One cell is enough to break Tor's anonymity. In: Proceedings of Black Hat DC 2009, p. 10 (2009)
25. Goldschlag, D.M., Reed, M.G., Syverson, P.F.: Hiding routing information. In: Anderson, R. (ed.) IH 1996. LNCS, vol. 1174, pp. 137–150. Springer, Heidelberg (1996). https://doi.org/10.1007/3-540-61996-8_37
26. Hoang, V.T., Krovetz, T., Rogaway, P.: Robust authenticated-encryption AEZ and the problem that it solves. In: Oswald, E., Fischlin, M. (eds.) EUROCRYPT 2015. LNCS, vol. 9056, pp. 15–44. Springer, Heidelberg (2015). https://doi.org/10.1007/978-3-662-46800-5_2
27. The invisible internet project (I2P). https://geti2p.net
28. Johnson, A., Wacek, C., Jansen, R., Sherr, M., Syverson, P.F.: Users get routed: traffic correlation on Tor by realistic adversaries. In: Sadeghi, A.-R., Gligor, V.D., Yung, M. (eds.) ACM CCS 2013, pp. 337–348. ACM Press, November 2013
29. Levine, B.N., Reiter, M.K., Wang, C., Wright, M.: Timing attacks in low-latency mix systems. In: Juels, A. (ed.) FC 2004. LNCS, vol. 3110, pp. 251–265. Springer, Heidelberg (2004). https://doi.org/10.1007/978-3-540-27809-2_25
30. Marson, G.A., Poettering, B.: Security notions for bidirectional channels. IACR Trans. Symm. Cryptol. **2017**(1), 405–426 (2017)
31. Mathewson, N.: Proposal 202: two improved relay encryption protocols for Tor cells, June 2012. https://lists.torproject.org/pipermail/tor-dev/2012-June/003649.html

32. Mathewson, N.: Proposal 261: AEZ for relay cryptography, December 2015. https://lists.torproject.org/pipermail/tor-dev/2015-December/010080.html
33. Murdoch, S.J., Zieliński, P.: Sampled traffic analysis by internet-exchange-level adversaries. In: Borisov, N., Golle, P. (eds.) PET 2007. LNCS, vol. 4776, pp. 167–183. Springer, Heidelberg (2007). https://doi.org/10.1007/978-3-540-75551-7_11
34. Nielsen, J.B.: Separating random Oracle proofs from complexity theoretic proofs: the non-committing encryption case. In: Yung, M. (ed.) CRYPTO 2002. LNCS, vol. 2442, pp. 111–126. Springer, Heidelberg (2002). https://doi.org/10.1007/3-540-45708-9_8
35. The23rd Raccoon. How I learned to stop ph34ring NSA and love the base rate fallacy, September 2008. http://archives.seul.org/or/dev/Sep-2008/msg00016.html
36. The23rd Raccoon. Analysis of the relative severity of tagging attacks, March 2012. http://archives.seul.org/or/dev/Mar-2012/msg00019.html
37. Reed, M.G., Syverson, P.F., Goldschlag, D.M.: Proxies for anonymous routing. In: ACSAC 1996, pp. 95–104. IEEE Computer Society (1996)
38. Rennhard, M., Plattner, B.: Practical anonymity for the masses with MorphMix. In: Juels, A. (ed.) FC 2004. LNCS, vol. 3110, pp. 233–250. Springer, Heidelberg (2004). https://doi.org/10.1007/978-3-540-27809-2_24
39. Rogaway, P., Zhang, Y.: Onion-AE: foundations of nested encryption. Cryptology ePrint Archive, Report 2018/126 (2018). https://eprint.iacr.org/2018/126
40. Serjantov, A., Sewell, P.: Passive attack analysis for connection-based anonymity systems. In: Snekkenes, E., Gollmann, D. (eds.) ESORICS 2003. LNCS, vol. 2808, pp. 116–131. Springer, Heidelberg (2003). https://doi.org/10.1007/978-3-540-39650-5_7
41. Shrimpton, T., Terashima, R.S.: A modular framework for building variable-input-length Tweakable ciphers. In: Sako, K., Sarkar, P. (eds.) ASIACRYPT 2013. LNCS, vol. 8269, pp. 405–423. Springer, Heidelberg (2013). https://doi.org/10.1007/978-3-642-42033-7_21
42. Syverson, P.F., Goldschlag, D.M., Reed, M.G.: Anonymous connections and onion routing. In: 1997 IEEE Symposium on Security and Privacy, pp. 44–54. IEEE Computer Society Press (1997)
43. Syverson, P., Tsudik, G., Reed, M., Landwehr, C.: Towards an analysis of onion routing security. In: Federrath, H. (ed.) Designing Privacy Enhancing Technologies. LNCS, vol. 2009, pp. 96–114. Springer, Heidelberg (2001). https://doi.org/10.1007/3-540-44702-4_6

Exploring the Boundaries of Topology-Hiding Computation

Marshall Ball[1,2](✉), Elette Boyle[2], Tal Malkin[1], and Tal Moran[2]

[1] Columbia University, New York, USA
{marshall,tal}@cs.columbia.edu
[2] IDC Herzliya, Herzliya, Israel
{elette.boyle,talm}@idc.ac.il

Abstract. Topology-hiding computation (THC) is a form of multi-party computation over an incomplete communication graph that maintains the privacy of the underlying graph topology. In a line of recent works [Moran, Orlov & Richelson TCC'15, Hirt et al. CRYPTO'16, Akavia & Moran EUROCRYPT'17, Akavia et al. CRYPTO'17], THC protocols for securely computing any function in the *semi-honest* setting have been constructed. In addition, it was shown by Moran et al. that in the fail-stop setting THC with negligible leakage on the topology is impossible.

In this paper, we further explore the feasibility boundaries of THC.

- We show that even against semi-honest adversaries, topology-hiding broadcast on a small (4-node) graph implies oblivious transfer; in contrast, trivial broadcast protocols exist unconditionally if topology can be revealed.
- We strengthen the lower bound of Moran et al. identifying and extending a relation between the *amount* of leakage on the underlying graph topology that must be revealed in the fail-stop setting, as a function of the number of parties and communication round complexity: Any n-party protocol leaking δ bits for $\delta \in (0, 1]$ must have $\Omega(n/\delta)$ rounds.

We then present THC protocols providing close-to-optimal leakage rates, for *unrestricted graphs* on n nodes against a fail-stop adversary controlling a dishonest majority of the n players. These constitute the first general fail-stop THC protocols. Specifically, for this setting we show:

- A THC protocol that leaks at most one bit and requires $O(n^2)$ rounds.

M. Ball—Supported in part by the Defense Advanced Research Project Agency (DARPA) and Army Research Office (ARO) under Contract #W911NF-15-C-0236, NSF grants #CNS1445424 and #CCF-1423306, ISF grant no. 1790/13, and the Check Point Institute for Information Security.

E. Boyle—Supported in part by ISF grant 1861/16, AFOSR Award FA9550-17-1-0069, and ERC Grant no. 307952.

T.Malkin—Supported in part by the Defense Advanced Research Project Agency (DARPA) and Army Research Office (ARO) under Contract #W911NF-15-C-0236, NSF grants #CNS1445424 and #CCF-1423306, and the Leona M. & Harry B. Helmsley Charitable Trust. Any opinions, findings and conclusions or recommendations expressed are those of the authors and do not necessarily reflect the views of the Defense Advanced Research Projects Agency, Army Research Office, the National Science Foundation, or the U.S. Government.

T. Moran—Supported in part by ISF grant no. 1790/13 and by the Bar-Ilan Cyber-center.

J. B. Nielsen and V. Rijmen (Eds.): EUROCRYPT 2018, LNCS 10822, pp. 294–325, 2018.
https://doi.org/10.1007/978-3-319-78372-7_10

- A THC protocol that leaks at most δ bits for arbitrarily small non-negligible δ, and requires $O(n^3/\delta)$ rounds.

These protocols also achieve full security (with no leakage) for the semi-honest setting. Our protocols are based on one-way functions and a (stateless) *secure hardware box* primitive. This provides a theoretical feasibility result, a heuristic solution in the plain model using general-purpose obfuscation candidates, and a potentially practical approach to THC via commodity hardware such as Intel SGX. Interestingly, even with such hardware, proving security requires sophisticated simulation techniques.

1 Introduction

Secure multiparty computation (MPC) is a fundamental research area in cryptography. Seminal results, initiated in the 1980s [8,18,33,56], and leading to a rich field of research which is still flourishing, proved that mutually distrustful parties can compute arbitrary functions of their input securely in many settings. Various adversarial models, computational assumptions, complexity measures, and execution environments have been studied in the literature. However, until recently, almost the entire MPC literature assumed the participants are connected via a *complete graph*, allowing any two players to communicate with each other.

Recently, Moran et al. [52] initiated the study of *topology-hiding computation (THC)*. THC addresses settings where the network communication graph may be partial, and the network topology itself is sensitive information to keep hidden. Here, the goal is to allow parties who see only their immediate neighborhood, to securely compute arbitrary functions (that may depend on their secret inputs and/or on the secret underlying communication graph). In particular, the computation should not reveal any information about the graph topology beyond what is implied by the output.

Topology-hiding computation is of theoretical interest, but is also motivated by real-world settings where it is desired to keep the underlying communication graph private. These include social networks, ISP networks, vehicle-to-vehicle communications, wireless and ad-hoc sensor networks, and other Internet of Things networks. Examples indicating interest in privacy of the network graph in these application domains include the project diaspora* [1], which aims to provide a distributed social network with privacy as an important goal; works such as [16,54] which try to understand the internal ISP network topology despite the ISP's wish to hide them; and works such as [25,45] that try to protect location privacy in sensor network routing, among others.

There are only a few existing THC constructions, and they focus mostly on the *semi-honest* adversarial setting, where the adversary follows the prescribed protocol. In particular, for the semi-honest setting, the work of Moran et al. [52] achieves THC for network graphs with a *logarithmic diameter* in the number of players, from the assumptions of oblivious transfer (OT) and PKE.[1] Hirt et al. [42] improve these results, relying on the DDH assumption, but still requiring the graph to have logarithmic diameter. Akavia and Moran [3] achieve THC for other classes of graphs, in particular graphs with small circumference. Recently, this was extended by Akavia et al. [2] to DDH-based THC for *general* graphs.

In the *fail-stop* setting, where an adversary may abort at any point but otherwise follows the protocol, the only known construction is one from [52], where they achieve THC for a very limited corruption and abort pattern: the adversary is not allowed to corrupt a full neighborhood (even a small one) of any honest party, and not allowed an abort pattern that disconnects the graph. This result is matched with a lower bound, proving that THC in the fail-stop model is *impossible* (the proof utilizes an adversary who disconnects the graph using aborts).

In this paper, we further explore the feasibility boundaries of THC. In the semi-honest model, we study the minimal required computational assumptions for THC, and in the fail-stop model we study lower and upper bounds on the necessary leakage. All our upper bounds focus on THC for arbitrary graphs with arbitrary corruption patterns (including dishonest majority). The security notion in the fail-stop setting is one of "security with abort", in which the adversary is allowed to abort honest parties after receiving the protocol's output.

1.1 Our Results

We will often describe our results in terms of the special case of *Topology-Hiding Broadcast* (THB), where one party is broadcasting an input to all other parties. We note that all our results apply both to THB and to THC (for arbitrary functionalities). In general, THB can be used to achieve THC for arbitrary functions using standard techniques, and for our upper bound protocols in particular the protocols can be easily changed to directly give THC of any functionality instead of broadcast.

Lower Bounds. We first ask what is the minimal assumption required to achieve THB in the semi-honest model. Our answer is that at the very least, OT is required (and this holds even for small graphs). Specifically, we prove:

Theorem (informal): If there exists a 4-party protocol realizing topology-hiding broadcast against a semi-honest adversary, then there exists a protocol for Oblivious Transfer.

Note that without the topology-hiding requirement, it is trivial to achieve broadcast *unconditionally* in the semi honest case, as well as the fail-stop case with security with abort. Indeed, the trivial protocol (sometimes referred to as "flooding") consists of propagating everything you received from your neighbors in the previous round, and then aborting if there is any inconsistency, for sufficiently many rounds (as many as the diameter of the graph).

[1] Alternatively, the [52] results can be interpreted as results for arbitrary graphs, but where the adversary is limited in its corruption pattern, and not allowed to corrupt any k-neighborhoods where k depends on the graph.

We mentioned above the result of [52], who prove that THC in the fail-stop model is impossible to achieve, since any protocol in the fail-stop model must have some *non-negligible* leakage. We next refine their attack to characterize (and amplify) the *amount* of leakage required, as a function of the number of parties n and communication rounds r of the protocol. We model the leakage of a protocol by means of a leakage oracle $\mathcal{L}$ evaluated on the parties' inputs (including graph topology) made available to the ideal-world simulator, and say that a protocol has $(\delta, \mathcal{L})$-leakage if the simulator only accesses $\mathcal{L}$ with probability δ over its randomness.[2]

In particular, we demonstrate the following:

Theorem (informal): For an arbitrary leakage oracle $\mathcal{L}$ (even one which completely reveals all inputs), the existence of r-round, n-party THB with $(\delta, \mathcal{L})$-leakage implies $\delta \in \Omega(n/r)$.

The theorem holds even if all parties are given oracle access to an arbitrary functionality, as is the case with the secure hardware box assumption mentioned below. This improves over the bound of [52], which corresponds to $\delta \in \Omega(1/r)$ when analyzed in this fashion.

Upper Bounds. We start by noting that a modification of the construction in [52] gives a scheme achieving TH computation in the *semi-honest* setting, for *log-diameter graphs* from OT alone (rather than OT + PKE as in the original work). This matches our lower bound above, showing THC if and only if OT, in the case of low diameter graphs.

Our main upper bound result is a THC construction for arbitrary graph structures and corruptions, in the *fail-stop, dishonest majority* setting, and (since leakage is necessary), with *almost no leakage*.

We have two versions of our scheme. The first is a scheme in $O(n^2)$ rounds (where n is a bound on the number of parties), which leaks at most one bit about the graph topology (i.e., simulatable given a single-output-bit leakage oracle $\mathcal{L}$). The leaked bit is information about whether or not one given party has aborted at a given time in the computation. This information may depend on the graph topology.

We then extend the above to a randomized scheme with arbitrarily small inverse polynomial leakage δ, in $O(n^3/\delta)$ rounds; more specifically, $(\delta, \mathcal{L})$-leakage for single-bit-output oracle $\mathcal{L}$. Here the leakage from $\mathcal{L}$ also consists of information about whether or not one given party has aborted at a given time in the computation. However, roughly speaking, the protocol is designed so that this bit depends on the graph topology only if the adversary has chosen to obtain this information in a specific "lucky" round, chosen at random (and kept hidden during the protocol), and thus happens with low probability.

We also point out that a simpler version of our scheme achieves full security (with no leakage) in the semi-honest model (for arbitrary graphs and arbitrary corruption pattern). Moreover, we leverage our stronger assumption to achieve essentially optimal round complexity in the semi-honest model—the protocol runs in $O(\text{diam}(G))$ rounds (where $\text{diam}(G)$ is a bound on the diameter of the communication graph G) and can directly compute any functionality (any broadcast protocol must have at least $\text{diam}(G)$

[2] Interestingly, this formalization is not equivalent to (and slightly weaker than) $\mathcal{L}_\delta$-leakage for respective functionality $\mathcal{L}_\delta$ that provides the output of $\mathcal{L}$ only with probability δ; see the full version for details(Note that ruling out a weaker notion means a stronger lower bound.)

rounds, otherwise the information might not reach all of the nodes in the graph). In contrast, the only previous THC protocol for general graphs [2] requires $\Omega(n^3)$ rounds for a single broadcast; computing more complex functionalities requires composing this with another layer of MPC on top.

Our schemes relies on the existence of one-way functions (OWF), as well as a *secure hardware box*, which is a stateless "black box", or oracle, with a fixed secret program, given to each participant before the protocol begins. We next discuss the meaning and implications of this underlying assumption, but first we summarize our main upper bound results:

Theorem (informal): If OWF exist and given a secure hardware box, for any n-node graph G and poly-time computable function f,

- *There exists an efficient topology-hiding computation protocol for f against poly-time fail-stop adversaries, which leaks at most one bit of information about G, and requires $O(n^2)$ rounds.*
- *For any inverse polynomial δ, there exists an efficient topology-hiding computation protocol for f against poly-time fail-stop adversaries, which leaks at most δ bits of information about G, and requires $O(n^3/\delta)$ rounds.*

We remark that the first result gives an n-party, r-round protocol with $(O(n^2/r), \mathcal{L})$-leakage, in comparison to our lower bound that shows impossibility of $(o(n/r), \mathcal{L})$-leakage. Closing this gap is left as an intriguing open problem.

On Secure Hardware Box Assumption. A secure hardware box is an oracle with a fixed, stateless secret program. This bears similarity to the notion of *tamper-proof hardware tokens*, introduced by Katz [46] to achieve UC secure MPC, and used in many followup works in various contexts, both with stateful and stateless tokens (cf. [15,20,37] and references within).

A hardware box is similar to a stateless token, but is incomparable in terms of the strength of the assumption. On one hand, a hardware box is worse, as we assume an honest setup of the box (by a party who does not need to know the topology of the graph, but needs to generate a secret key and embed the right program), while hardware tokens are typically allowed to be generated maliciously (although other notions of secure hardware generated honestly have been considered before, e.g. [19,43]). On the other hand, a hardware box is better, in that, unlike protocols utilizing tokens, it does not need to be passed around during the protocol, and the players do not need to embed their own program in the box: there is a single program that is written to all the boxes before the start of the protocol.

Unlike previous uses of secure hardware in the UC settings (where we know some setup assumption is necessary for security), we do not have reason to believe that strong setup (much less a hardware oracle) is necessary to achieve THC. However, we believe many of the core problems of designing a THC protocol remain even given a secure hardware oracle. For example, the lower bounds on leakage hold even in this setting. In particular, our hardware assumption does not make the solution trivial (in fact, in some senses the proofs become harder, since even a semi-honest adversary may query the oracle "maliciously"). Our hope is that the novel techniques we use in constructing the protocol, and in proving its correctness, will be useful in eventually constructing a protocol in the standard model. We note that this paradigm is a common one in cryptography: protocols are first constructed using a helpful "hardware' oracle", and then ways are found to replace the hardware assumption with a more standard one. Examples include the ubiquitous "random oracle", but also hardware assumptions much more similar to ours, such as the Signature Card assumption first used to construct Proof-Carrying Data (PCD) schemes [19]. (Signature cards contain a fixed program and a secret key, and can be viewed as a specific instance of our secure hardware assumption.)

Thus, one way to think of our upper bound result is as a step towards a protocol in the standard model.

At the same time, our phrasing of the assumption as "secure hardware" is intentional, and physical hardware may turn out to be the most practical approach to actually implementing a THC protocol. Because our functionality is fixed, stateless, and identical for all parties, our secure hardware box can be instantiated by a wide range of physical implementations, including general-purpose "trusted execution environments", that are becoming widespread in commodity hardware (for example, both ARM (TrustZone) and Intel (SGX) have their own flavors implemented in their CPUs). We discuss a potentially practical approach to THC through the use of SGX secure hardware in the full version of this paper.

Future directions. Our work leaves open many interesting directions to further pursue, such as the following.

- Obtain better constructions for the case of *honest majority*.
- Obtain THC in the fail-stop model from standard cryptographic assumptions. In particular, can THC be achieved from OT alone, matching our lower bound?
- The results for THC in the fail-stop setting are in some ways reminiscent of the results for optimally fair coin tossing. In particular, in both cases there is an impossibility result if no leakage or bias is allowed, and there are lower bounds and upper bounds trading off the amount of leakage with the number of rounds (cf. [22–24,51]). It would be interesting to explore whether there is a formal connection between THC and fair coin tossing, and whether such a connection can yield tighter bounds for THC.
- THC with security against malicious adversaries is an obvious open problem, with no prior work addressing it (to the best of our knowledge). Could our results be extended to achieve security in the malicious settings? More generally, could a secure hardware box be useful towards maliciously secure THC?

1.2 Technical Overview

A starting point for our upper bound protocol is the same starting point underlying the previous THB constructions [3,42,52]: Consider the trivial flooding protocol that achieves broadcast with no topology hiding, by propagating the broadcasted bit to all the neighbors repeatedly until it reaches everyone. One problem with this protocol is that the messages received by a node leak a lot of information about the topology of the graph (e.g., the distance to the broadcaster). Previous works mitigate that by encrypting the communication, and also requiring all nodes to send a bit in every round: the broadcaster sends its bit, and the other nodes send a 0; each node then ORs all the incoming bits, and forwards to its neighbors in the next round. However, this leaves the question of how the bit will be decrypted to obtain the final result. This is where previous works differ in their techniques to address this issue (using nested MPC, homomorphic encryption, ideas inspired by mix-nets or onion routing to allow gradual decryption, etc.), and the different techniques imply different limitations on the allowed graph topology (or corruption patterns) that support the solution.

We also begin with the same starting point of trying to implement flooding. We then use the secure hardware box, which will contain a relevant secret key that allows it to process encrypted inputs (partial transcripts propagated from different parts of the graph) and produce an encrypted output to propagate further in the next round, as well as decrypting the output at the end. However, we have several new technical challenges that arise, both because of the fail-stop setting, and because of the existence of the box itself.

First, the fail-stop setting presents a significant challenge (indeed, provably necessitating some leakage). Intuitively, abort behavior by the adversary will influence the behavior of honest parties (e.g., if an honest party is isolated by aborts in their immediate neighborhood, they would not be able to communicate and will have to abort rather than output something; aborting behavior of honest parties can in turn provide information to the adversary about the graph topology). The hardware box will help in checking consistency of partial transcripts, and helping honest parties manage when and how they disclose their plan to output abort at the end of the protocol.

A second source of difficulty stems from having the secure hardware box at the disposal of the adversary. This allows to inject a malicious aspect to the adversary's behavior, even in a fail-stop (or even semi-honest) setting. Indeed, since each player has their own box, and the box is stateless, the adversary can run the boxes with arbitrary inputs, providing different partial transcripts, abort or non-abort behavior, etc., in order to try and learn information about the graph topology. This presents challenges that make the proof of security much more involved and quite subtle.

Overview of our solution. Recall the core source of information leakage is from the abort-or-not values of various parties, as a function of fail-stop aborts caused by the adversary. The first idea of our construction is to limit the amount of leakage to a *single* bit by ensuring that for any fail-stop abort strategy, the abort-or-not value of only a single party will be topology dependent. This is achieved by designating a special "threshold" round T_i for each party: if the party P_i learns of an abort somewhere in the graph before round T_i, he will output abort at the conclusion of the protocol, and if he only learns of an abort after this round he will output the correct bit value. By sufficiently separating these threshold rounds, and leveraging the fact that an abort will travel to all nodes within n rounds (independent of the graph topology), we can guarantee that any given abort structure will either reach before or after the threshold round of a *single* party in a manner dependent on the topology.

Note that in the above, if the threshold rounds T_i are known, then there exists an adversarial strategy which indeed leaks a full bit on the topology. To obtain arbitrarily small leakage δ, we modify the above protocol by expanding the "zone" of each party into a collection of $O(n/\delta)$ possible threshold rounds. The value of the true threshold for each party is determined (pseudo-)randomly during protocol execution and is hidden from the parties themselves (who see only encrypted state vectors from their respective secure hardware boxes). Because of this, the probability that an adversary will be able to successfully launch a leakage attack on any single party's threshold round will drop by a factor of δ/n; because this attack can be amplified by attacking across several parties' zones, the overall winning probability becomes comparable to δ. Note that such an increase in rounds to gain smaller leakage is to be expected, based on our leakage lower bound.

The more subtle and complex portion of our solution comes in the simulation strategy, in particular for simulating the output of the hardware box on arbitrary local queries by adversarial parties. At a high level, the simulator will maintain a collection of graph structures corresponding to query sequences to the boxes (where outputs from previous box queries are part of input to a later box query), and will identify a specific set of conditions in which a query to the leakage oracle must be made. See below for a more detailed description.

Overview of simulation strategy. Simulation consists almost entirely of answering queries to the hardware box. As intermediate outputs from the box are encrypted, the chief difficulty lies in determining what output to give to queries corresponding to the *final round* of the protocol: either $\perp$ (abort); the broadcast bit; or something else, given partial leakage information about the graph and only the local neighborhood of the corrupted parties.

The simulator uses a data structure to keep track of the relationship between queries to the hardware box and outputs from previous queries. In the real world, this relationship is enforced by the unforgeability of the authenticated encryption scheme. The simulator can use this data structure to determine whether a query is "derived," in part, from 'honest' (simulated) messages, and additionally, what initializations were used for the non-honest parties connected to the node expecting output.

One of the major difficulties is that even a semi-honest adversary can locally query his hardware box in malicious ways: combining new initializations in novel ways with pieces of the honest transcript, or aborting in multiple different patterns. The bulk of the proof is devoted to showing that all of these cases can be simulated.

One key fact utilized in this process is that if the protocol gives any output at all, then all honest nodes must have encrypted states at round n (the maximum diameter of the graph) that contain a complete picture of the graph, inputs, session keys, etc. Therefore, the real hardware box will not give plaintext output if such an honest state is mixed with states in a manner that deviates from the real protocol evaluation significantly.

An added wrinkle is that the hardware box, by virtue of the model, is required to handle a variety of abort sequences. Moreover, the kind of output received after certain abort timing inherently leaks information about the topology. Yet, the simulator must decide output behavior without additional leakage queries. Here again the honest messages will essentially "lock" an adversary into aborts that are "consistent" with the aborts in the real protocol evaluation. (For example, an adversary can "fast-forward" a node after the nodes output is guaranteed by pretending all of its neighbors aborted.)

The honest messages also aid in replay attacks as they allow the simulator to only consider connected groups of corrupt nodes. If two nodes are separated by honest nodes in the real world, then in their replay attack no new abort information will be transmitted from one to the other if the protocol is replayed in a locally consistent manner (modulo aborts). (If the protocol is not locally consistent no descendent of that query will yield plaintext output.)

Finally, if a query doesn't have any honest ancestors, the simulator can simulate output trivially as it knows all of the initialization information.

In short, the difficulties in the proof come from the fact that output depends on topology and abort structure, and a fail-stop adversary can use his box to essentially simulate malicious runs of the protocol after its completion to attempt to gain more leakage on the topology. However, the simulator can only query the leakage oracle at most once. Accordingly, the specific timing of its query in protocol evaluation is very delicate: if it is too early the adversary can abort other nodes to change output behavior in an unsimulatable manner, if it is too late then the adversary can fast-forward to get output in an unsimulatable manner. Moreover, output behavior must be known for all replay attacks where the simulator has incomplete initialization information (pieces of the honest transcript are used). As a consequence, we are forced to consider elaborate consistency conditions to bind the adversary to a specific evaluation (modulo aborts), and prove that these conditions achieve bind the adversary while still allowing him the freedom to *actively* attack the protocol using the hardware.

1.3 Related Work

We have already discussed above the prior works on topology-hiding computation in the computational setting [3,42,52], which are the most relevant to our work.

Topology-hiding computation was also considered earlier in the *information-theoretic setting*, by Hinkelmann and Jakoby [41]. They provide an impossibility result, proving that any information-theoretic THC protocol leaks information to an adversary (roughly, when two nodes who are not neighbors communicate across the graph, some party will be able to learn that it is on the path between them). They also provide an upper bound, achieving information theoretic THC that leaks a routing table of the network, but no other information about the graph.

There are several other lines of work that are related to communication over incomplete networks, but in different contexts, not with the goal of hiding the topology. For example, a line of work studied the feasibility of reliable communication over (known) incomplete networks (cf. [4,5,7,10,14,26–28,48]). More recent lines of work study secure computation with restricted interaction patterns in a few settings, motivated by improving efficiency, latency, scalability, usability, or security. Examples include [6,11,13,36,38,39]. Some of these works utilize a secret communication subgraph of the complete graph that is available to the parties as a tool to achieve their goal (e.g. [11,13] use this idea in order to achieve communication locality).

An early use of a hidden communication graph which is selected as a subgraph of an available known larger graph, is in the context of anonymous communication and protection against traffic analysis. Particularly noteworthy are the mix-net and onion routing techniques ([17,53,55] and many follow up works), which also inspired some of the recent THC techniques.

There is a long line of work related to the use of secure hardware in cryptography, in various flavors with or without assuming honest generation, state, complete tamper proofness, etc. This could be dated back to the notion of oblivious RAM ([34] and many subsequent works). Katz [46] introduced the notion of a hardware token in the context of UC-secure computation, and this notion has been used in many followup works (e.g. [15,20,37] and many others). Variations on the hardware token, where the hardware is generated honestly by a trusted setup, include signature cards [43], trusted smartcards [40], and so called non-local-boxes [9]. The latter are similar to global hardware boxes that are generated honestly and take inputs and output from multiple parties (in contrast to our notion of a hardware box, which is local). Other variations and relaxations include tamper-evident seals [50], one time programs [35], and various works allowing some limited tampering ([31,44] and subsequent works). Finally, there is a line of works using other physical tools to perform cryptographic tasks securely, including [29,30,32].

2 Preliminaries

2.1 Secure Hardware

We model our secure hardware box as an ideal oracle, parameterized by a stateless program Π. The oracle query $\mathcal{O}(\Pi)(x)$ returns the value $\Pi(x)$. Our definition is much simpler than the standard secure hardware token definitions, since all parties have access to the same program, and it is stateless—there is no need for a more complex functionality that keeps track of the "physical location" of the token or its internal state.

2.2 Topology Hiding Computation

The work of [52] put forth two formal notions of topology hiding: a simulation-based definition, and a weaker indistinguishability-based definition. In this work, we primarily focus on the simulation-based definition, given below. However, some of our lower bounds apply also to the indistinguishability-based notion.

The definition of [52] works in the $\mathcal{F}_{\text{graph}}$-hybrid model, for $\mathcal{F}_{\text{graph}}$ functionality (shown in Fig. 1) that takes as input the network graph from a special "graph party" P_{graph} and returns to each other party a description of their neighbors. It then handles communication between parties, acting as an "ideal channel" functionality allowing neighbors to communicate with each other without this communication going through the environment.

In a real-world implementation, $\mathcal{F}_{\text{graph}}$ models the actual communication network; i.e., whenever a protocol specifies a party should send a message to one of its neighbors using $\mathcal{F}_{\text{graph}}$, this corresponds to the real-world party directly sending the message over the underlying communication network.

Participants/Notation:
This functionality involves all the parties $P_1, \ldots, P_m$ and a special graph party P_{graph}.
Initialization Phase:
Inputs: $\mathcal{F}_{\text{graph}}$ waits to receive the graph $G = (V, E)$ from P_{graph}.
Outputs: $\mathcal{F}_{\text{graph}}$ outputs $\mathrm{N}_G[v]$ to each P_v.
Communication Phase:
Inputs: $\mathcal{F}_{\text{graph}}$ receives from a party P_v a destination/data pair (w, m) where $w \in \mathrm{N}(v)$ and m is the message P_v wants to send to P_w. (If w is not a neighbor of v, $\mathcal{F}_{\text{graph}}$ does nothing.)
Output: $\mathcal{F}_{\text{graph}}$ gives output (v, m) to P_w indicating that P_v sent the message m to P_v.

Fig. 1. The functionality $\mathcal{F}_{\text{graph}}$.

Since $\mathcal{F}_{\text{graph}}$ provides local information about the graph to all corrupted parties, *any* ideal-world adversary must have access to this information as well (regardless of the functionality we are attempting to implement). To capture this, we define the functionality $\mathcal{F}_{\text{graphInfo}}$, that is identical to $\mathcal{F}_{\text{graph}}$ but contains only the initialization phase. For any functionality $\mathcal{F}$, we define a "composed" functionality $(\mathcal{F}_{\text{graphInfo}} \| \mathcal{F})$ that adds the initialization phase of $\mathcal{F}_{\text{graph}}$ to $\mathcal{F}$. We can now define topology-hiding MPC in the UC framework:

Definition 1 (Topology Hiding (Simulation-Based)). *We say that a protocol Π securely realizes a functionality $\mathcal{F}$* hiding topology *if it securely realizes $(\mathcal{F}_{graphInfo} \| \mathcal{F})$ in the $\mathcal{F}_{graph}$-hybrid model.*

Note that this definition can also capture protocols that realize functionalities depending on the graph (e.g., find a shortest path between two nodes with the same input, or count the number of triangles in the graph).

2.3 Extended Definitions of THC

We extend the simulation definition of Topology-Hiding Computation beyond the semi-honest model, capturing fail-stop corruptions, and formalizing a measure of leakage of a protocol.

Topology Hiding with Leakage

We consider a weakened notion of topology hiding with partial information leakage. This is modeled by giving the ideal-world simulator access to a reactive functionality leakage oracle $\mathcal{L}$, where the type/amount of leakage revealed by the protocol is captured by the choice of the leakage oracle $\mathcal{L}$. For example, we will say a protocol "leaks a single bit" about the topology if it is topology hiding for some oracle $\mathcal{L}$ which outputs at most 1 bit throughout the simulation.[3]

Definition 2 (Topology Hiding with $\mathcal{L}$-Leakage). *We say that a protocol Π securely realizes a functionality $\mathcal{F}$* hiding topology with $\mathcal{L}$-leakage *if it realizes* $(\mathcal{F}_{graphInfo}||\mathcal{F}||\mathcal{L})$ *in the $\mathcal{F}_{graph}$-hybrid model, where $\mathcal{L}$ is treated as an ideal (possibly reactive) functionality which outputs only to corrupt parties.*

Note that the above functionality $(\mathcal{F}_{\text{graphInfo}}||\mathcal{F}||\mathcal{L})$ is not a "well-formed" functionality in the sense of [12], as the output of the functionality depends on the set of corrupt parties. However, this is limited to additional information given to corrupt parties, which does not run into the simple impossibilities mentioned in [12] (indeed, it is easier to securely realize than $(\mathcal{F}_{\text{graphInfo}}||\mathcal{F})$). The definition also extends directly to topology hiding within different adversarial models, by replacing $\mathcal{F}_{\text{graph}}$ with the corresponding functionality (such as $\mathcal{F}_{\text{graph-failstop}}$ for fail-stop adversaries; see below).

It will sometimes be convenient when analyzing lower bounds and considering fractional bits of leakage to consider the following restricted notion of $(\delta, \mathcal{L})$-leakage, for probability $\delta \in [0, 1]$. Loosely, a $(\delta, \mathcal{L})$-leakage simulator is restricted to only utilizing the leakage oracle $\mathcal{L}$ with probability δ over the choice of its random coins. Note that this notion is closely related to $\mathcal{L}_\delta$-leakage for the oracle $\mathcal{L}_\delta$ which internally tosses coins and decides with probability δ to respond with the output of $\mathcal{L}$. Interestingly, however, the two notions are *not* equivalent: in the full version of this paper we show that there exist choices of $\mathcal{F}$, $\delta \in [0, 1]$, oracle $\mathcal{L}$, and protocols Π for which Π is a $(\delta, \mathcal{L})$-leakage secure protocol, but not $\mathcal{L}_\delta$-leakage secure. For our purposes, $(\delta, \mathcal{L})$-leakage will be more convenient.

Definition 3 (Topology Hiding with $(\delta, \mathcal{L})$-Leakage). *Let $\delta \in [0, 1]$ and $\mathcal{L}$ a leakage oracle functionality. We say that a protocol Π securely realizes a functionality $\mathcal{F}$* hiding topology with $(\delta, \mathcal{L})$-leakage *if it realizes* $(\mathcal{F}_{graphInfo}||\mathcal{F}||\mathcal{L})$ *in the $\mathcal{F}_{graph}$-hybrid model with the following property: For any adversarial environment $\mathcal{Z}$, it holds with probability $(1 - \delta)$ over the random coins of the simulator $\mathcal{S}$, that $\mathcal{S}$ does not make any call to $\mathcal{L}$.*

In the full version of this paper we show that this notion of $(\delta, \mathcal{L})$-leakage provides a natural form of composability.

Topology Hiding in the Fail-Stop Model

We now define security for the case that the adversary must follow the protocol (as in the semi-honest case), but may fail nodes. Consider the functionality $\mathcal{F}_{\text{graph-failstop}}$ given in Fig. 2, which serves as the analog of $\mathcal{F}_{\text{graph}}$ in the semi-honest model.

As the initialization phase (and ideal-world-counterpart) of $\mathcal{F}_{\text{graph-failstop}}$ is identical to that of $\mathcal{F}_{\text{graph}}$, we denote it the same: $\mathcal{F}_{\text{graphInfo}}$. As before, the communication phase consists of repeated invocation of $\mathcal{F}_{\text{graph-failstop}}$. The fail input in the communication phase represents failing a node, as such, it should only be invoked adversarially (not part of normal protocol operation).

[3] Note that this is related to, but a different setting than leakage-resilient protocols, where the model considers leakage information to the adversary in the *real-world* execution.

Participants/Notation:

This functionality involves all the parties $P_1, \ldots, P_m$ and a special graph party P_{graph}.

Initialization Phase:

Inputs: $\mathcal{F}_{\text{graph-failstop}}$ waits to receive the graph $G = (V, E)$ from P_{graph}.

Outputs: $\mathcal{F}_{\text{graph-failstop}}$ outputs $\mathrm{N}_G[v]$ to each P_v.

Communication Phase:

Inputs: $\mathcal{F}_{\text{graph-failstop}}$ may receive one of the following from a party P_v:

1. a destination/data pair (w, m) where $w \in \mathrm{N}(v)$ and m is the message P_v wants to send to P_w. (If w is not a neighbor of v, or if either v or w has previously sent (Fail), $\mathcal{F}_{\text{graph-failstop}}$ ignores this message.)
2. Fail

Output: If the input is of the form:

1. (destination/data pair (w, m)) $\mathcal{F}_{\text{graph-failstop}}$ gives output (v, m) to P_w indicating that P_v sent the message m to P_v.
2. (Fail) $\mathcal{F}_{\text{graph-failstop}}$ gives output Fail to all neighbors of P_v.

Fig. 2. The functionality $\mathcal{F}_{\text{graph-failstop}}$.

Topology-hiding security-with-abort. As is the case for standard (non-topology-hiding) MPC, when we allow active adversaries we relax the security definition to security-with-abort. However, there are wrinkles specific to the topology-hiding setting that make our security-with-abort definition slightly different.

In the standard extension of simulation-based security to security-with-abort, we add a special **abort** command to the ideal functionality; when invoked by the ideal-world simulator, all the honest parties' outputs are replaced by $\perp$. When the communication graph is complete, this extra functionality is trivial to add to any protocol: an honest party will output $\perp$ if it receives an **abort** message from any party (since honest parties will never send **abort**, this allows the adversary to abort any honest party, but does not otherwise change the protocol).

In the topology-hiding setting, this extra functionality—by itself—might already be too strong to realize, since, depending on when the abort occurs, the "signal" might not have time to reach all honest parties. (In fact, this is essentially the crux of the fail-stop impossibility result of [52] and of our leakage lower bound in Sect. 3.2).

Thus, when we define security-with-abort for topology-hiding computation, we augment the ideal functionality with a slightly more complex **abort** command: it now receives list of parties as input (the "abort vector"); only the outputs of those parties will be replaced with $\perp$, while the rest of the parties will output as usual.

Note that in the UC model, the environment sees the outputs of *all* parties, including the honest parties. Hence, to securely realize a functionality-with-abort, the simulator must ensure that the simulation transcript, together with the honest parties' output, is indistinguishable in the real and ideal worlds. In the topology-hiding case, this means that the set of aborting parties must also be indistinguishable. Since whether or not a party aborts during protocol execution depends on the topology of the graph, in order to determine the abort vector the simulator may require the aid of the leakage oracle (in our case, this is actually the only use of the leakage oracle).

Definition 4 (Fail-stop Topology Hiding). *We say that a protocol Π securely realizes a functionality $\mathcal{F}$* hiding topology against fail-stop adversaries *if it realizes* $(\mathcal{F}_{graphInfo}||\mathcal{F})$ with abort *in the $\mathcal{F}_{graph\text{-}failstop}$-hybrid model.*

Recall that general topology hiding computation against fail-stop adversaries is impossible [52]; we thus consider the notion of topology hiding against fail-stop *with $(\delta, \mathcal{L})$-leakage.*

Definition 5 (Fail-stop Topology Hiding with Leakage). *We say that a protocol Π securely realizes a functionality $\mathcal{F}$* hiding topology against fail-stop adversaries with $(\delta, \mathcal{L})$-leakage *if it realizes $(\mathcal{F}_{graphInfo}||\mathcal{F}||\mathcal{L})$ with abort in the $\mathcal{F}_{graph\text{-}failstop}$-hybrid model, with the following property: For any adversarial environment $\mathcal{Z}$, it holds with probability $(1-\delta)$ over the random coins of the simulator $\mathcal{S}$, that $\mathcal{S}$ does not make any call to $\mathcal{L}$.*

3 Lower Bounds

We begin by exploring lower bounds on the feasibility of topology-hiding computation protocols. In this direction, we present two results.

First, we demonstrate that topology hiding is inherently a non-trivial cryptographic notion, in the sense that even for semi-honest adversaries and the simple goal of broadcast (achievable trivially when topology hiding is not a concern), topology-hiding protocols imply the existence of oblivious transfer.

We then shift to the fail-stop model, and provide a lower bound on the *amount* of leakage that must be revealed by any protocol achieving broadcast, as a function of the number of rounds and number of parties. This refines the lower bound of [52], which shows only that non-negligible leakage must occur.

Both results rely only on the correctness guarantee of the broadcast protocol in the "legal" setting, where a single broadcaster sends a valid message. We make no assumptions as to what occurs in the protocol if parties supply an invalid set of inputs. (In particular, this behavior will not be need to be encountered within our lower bounds.)

More formally, our lower bounds apply to THC protocols achieving any functionality $\mathcal{F}$ that satisfies the following *single-broadcaster-correctness* property:

Definition 6 (Single-Broadcaster Correctness). *An ideal n-party functionality* $\mathcal{F}$: $\{0,1,\bot\}^n \to \{0,1,\bot\}^n$ *will be said to satisfy* single-broadcaster correctness *if for any input vector $(b_1,\dots,b_n) \in \{0,1,\bot\}^n$ in which a* single *input $b := b_i$ is non-$\bot$, the functionality $\mathcal{F}$ outputs b to all parties within the connected component of P_i (and no output to all other parties).*

3.1 Semi-honest Topology-Hiding Broadcast Implies OT

Consider the task of broadcast on a given communication graph. If parties are semi-honest, and no topology hiding is required, then such a protocol is trivial: In each round, every party simply passes the broadcast value to each of his neighbors; within n rounds, all parties are guaranteed to learn the value. However, such a protocol leaks information about the graph structure. For instance, the round in which a party receives the broadcast bit is precisely the distance of this party to the broadcaster. It is not clear at first glance whether this approach could be adapted unconditionally, or perhaps enhanced by tools such as symmetric-key encryption, in order to hide the topology.

We demonstrate that such an approach will not be possible. Namely, we show that even semi-honest topology-hiding broadcast (THB) implies the existence of oblivious transfer. This holds even for the weaker notion of *indistinguishability-based* (IND-CTA) topology-hiding security [52] which directly implies the same lower bound for the simulation-based definition. As described above, our bound applies to protocols for any functionality which satisfies single-broadcaster correctness.

Theorem 1 (THB implies OT). *If there exists an n-party protocol for $n \geq 4$ achieving IND-CTA topology hiding against a semi-honest adversary, for any functionality $\mathcal{F}$ with single-broadcaster correctness, then there exists a protocol for oblivious transfer.*

We note that, because both the following protocol and proof are black box with respect to the IND-CTA topology-hiding broadcast protocol, the proof holds in the presence of secure hardware.

Proof. We present a protocol for semi-honest secure 2-party computation of the OR functionality given such a semi-honest topology-hiding broadcast protocol for $n = 4$ parties. This implies existence of oblivious transfer [21,47,49].

First, observe that in the semi-honest setting, topology-hiding broadcast of messages of any length (even of a single bit) directly implies topology-hiding broadcast of arbitrary-length messages, by sequential repetition.

In a secure OR computation protocol, two parties A, B begin with inputs $x_A, x_B \in \{0,1\}$, and must output $(x_A \vee x_B)$. In our construction, each party A, B will emulate *two* parties in an execution of the 4-party topology-hiding broadcast protocol $\mathcal{P}_{\text{sh-broadcast}}$ for messages of length λ: namely, A emulates P_0^A, P_1^A, and B emulates P_0^B, P_1^B, where P_0^A, P_0^B are connected as neighbors and P_1^A, P_1^B are similarly neighbors. Each of the parties A, B will emulate an edge between its own pair of parties if and only if its protocol input bit $x_A, x_B \in \{0,1\}$ is 1. More formally, the secure 2-party OR protocol is given in Fig. 3.

We now demonstrate a simulator for the secure 2-party computation protocol. The simulator receives as input the security parameter 1^λ, the corrupted party C's input x_C (where $C \in \{A, B\}$), the final output $b \in \{0,1\}$, equal to the OR of x_C with the (secret) honest party input bit, and auxiliary input z. As its output, $\mathcal{S}_{\mathcal{A}}(1^\lambda, x_C, b, z)$ simulates an execution of $\mathcal{P}_{\text{OR}}$ interacting with the adversary $\mathcal{A}$ while emulates the role of the uncorrupted party $C' \neq C \in \{A, B\}$, *but using input b in the place of the (unknown) input $x_{C'}$.*

Denote by $\mathsf{view}_{\mathcal{A}}^{\mathcal{P}_{\text{OR}}}(1^\lambda, (x_A, x_B), z)$ the (real) view of the adversary $\mathcal{A}$ within the protocol $\mathcal{P}_{\text{OR}}$ on inputs x_A, x_B, when given auxiliary input z.

Claim. For every $x_A, x_B \in \{0,1\}$, non-uniform polynomial-time adversary $\mathcal{A}$, and auxiliary input z, it holds that

$$(x_A, x_B, b, \mathsf{view}_{\mathcal{A}}^{\mathcal{P}_{\text{OR}}}(1^\lambda, (x_A, x_B), z)) \stackrel{c}{\cong} (x_A, x_B, b, \mathcal{S}_{\mathcal{A}}(1^\lambda, x_C, b, z)).$$

Inputs: Parties A, B have inputs $x_A, x_B \in \{0, 1\}$.
Outputs: Parties A, B output $(x_A \vee x_B)$.
Protocol $\mathcal{P}_{\text{OR}}$:

1. A chooses a string $R \leftarrow \{0, 1\}^\lambda$ at random.
2. A, B honestly emulate an execution of $\mathcal{P}_{\text{sh-broadcast}}$, as follows:
   - A emulates parties P_0^A, P_1^A with respective neighborhoods $\{P_0^B\}, \{P_1^B\}$ if $x_A = 0$ and with neighborhoods $\{P_1^A, P_0^B\}, \{P_0^A, P_1^B\}$ if $x_A = 1$.
   - B emulates parties P_0^B, P_1^B with respective neighborhoods $\{P_0^A\}, \{P_1^A\}$ if $x_B = 1$ and with neighborhoods $\{P_1^B, P_0^A\}, \{P_0^B, P_1^A\}$ if $x_B = 1$.

   Party P_0^A runs as the designated broadcaster, with input R. All other parties run with input $\perp$, indicating they are not broadcasting.
3. For each $X \in \{A, B\}$, $b \in \{0, 1\}$, denote by out_b^X the output of emulated party P_b^X at the conclusion of $\mathcal{P}_{\text{sh-broadcast}}$. A, B output as follows:
   - A outputs 1 iff $\mathsf{out}_0^A = \mathsf{out}_1^A = R$.
   - B outputs 1 iff $\mathsf{out}_0^B = \mathsf{out}_1^B$.

Fig. 3. Secure 2-party OR protocol $\mathcal{P}_{\text{OR}}$ from semi-honest THB

Proof. First observe that output correctness of $\mathcal{P}_{\text{OR}}$ holds, as follows. By single-broadcaster correctness of $\mathcal{P}_{\text{sh-broadcast}}$ (note that indeed there is a single broadcaster), all parties in the connected component of the broadcaster P_0^A within the emulated execution will output the string R. In particular, this includes P_0^B: i.e., $\mathsf{out}_0^B = R$. In contrast, any party outside the connected component of P_0^A will have a view in the emulated THB protocol that is information theoretically independent of the choice of R, and thus will output R with negligible probability. This means out_1^A and out_1^B will equal R precisely when there exists an edge between x_D^0 and x_D^1 for at least one $D \in \{A, B\}$: that is, iff $(x_A \vee x_B) = 1$.

In the case of $b = 0$, the simulation is perfect. In the case of $b = 1$, indistinguishability of the above real-world and ideal-world distributions follows directly by the indistinguishability under chosen topology attack (IND-CTA) security of $\mathcal{P}_{\text{sh-broadcast}}$. Namely, the simulation corresponds to execution of $\mathcal{P}_{\text{sh-broadcast}}$ on the graph G with an edge between the two uncorrupted parties $P_0^{C'}, P_1^{C'}$, whereas depending on the value of the honest input $x_{C'}$, the real distribution is an execution on either this graph G or the graph G' with this edge removed. A successful distinguisher thus breaks IND-CTA for the challenge graphs G, G'.

3.2 Lower Bound on Information Leakage in Fail-Stop Model

The work of [52] demonstrated that non-negligible leakage on the graph topology must occur in any broadcast protocol in the presence of fail-stop corruptions. In what follows, we extend this lower bound, quantifying and amplifying the amount of information revealed.

Roughly, we prove that any protocol realizing broadcast with abort must leak $\Omega(n/R)$ bits of information on the graph topology, where n is the number of parties, and R is the number of rounds of interaction of the protocol.[4] More formally, we demonstrate an attack that successfully distinguishes between two different honest party graph structures with advantage $\Omega(n/R)$. This, in particular, rules out the existence of $(\delta, \mathcal{L})$-leakage topology hiding for $\delta \in o(n/R)$, for any leakage oracle $\mathcal{L}$. We compare this to our protocol construction in Sect. 4.2, which *achieves* $(\delta, \mathcal{L})$-leakage in this model for a single-output-bit $\mathcal{L}$ and $\delta \in O(n^2/R)$. We leave open the intriguing question of closing this gap.

The proof follows an enhanced version of the attack approach of [52], requiring the adversary to control only 4 parties, and perform only 2 fail-stop aborts. At a high level, parties are arranged in a chain with a broadcaster at one end, 2 aborting parties in the middle, and an additional corrupted party who is either on the same side or opposite side of the chain as the broadcaster. In the attack, one of the 2 middle parties aborts in round i, and the second aborts in round $i+d$ as soon as the first's abort message reaches him. Parties on one end of the chain thus see a single abort at round i, whereas parties on the other end see only an abort at round $i+d$. In [52] it is shown that the view of a party given an abort in round i versus $i+1$ can be distinguished with advantage $\Omega(1/R)$, where R is the number of rounds.

We improve over [52] by separating the two aborting parties by a distance of $\Theta(n)$, instead of distance 1. Roughly, the corrupted party's view in the two positions will be consistent with either an abort in round i or round $i+\Theta(n)$ of the protocol, (versus i and $i+1$ in [52]), which can be shown to yield distinguishing advantage $\Theta(n)$ better than in [52].

As in Sect. 3.1, our attack does not leverage any behavior outside the scope of a single broadcaster, and thus applies to any functionality $\mathcal{F}$ satisfying single-broadcaster correctness. Further, the proof only requires that the protocol is correct and that information is required to travel over the network topology: that is, each node can only transmit information to adjacent nodes in any given round. Therefore the theorem holds in the presence of secure hardware (which is only held locally and cannot be jointly accessed by different parties).

Theorem 2. *Let $\mathcal{L}$ be an arbitrary leakage oracle. Then no R-round n-party protocol can securely emulate broadcast (with abort) in the fail-stop model while hiding topology with $(\delta, \mathcal{L})$-leakage for any $\delta \in o(n/R)$.*

Proof. Let $\mathcal{P}$ be an arbitrary protocol which achieves broadcast with abort as above. We demonstrate a pair of graphs G_0, G_1 and an attack strategy $\mathcal{A}$ such that $\mathcal{A}$ can distinguish with advantage $\Omega(n/R)$ the executions of $\mathcal{P}$ within G_0 versus G_1. We then prove this suffices to imply the theorem.

Both graphs G_0, G_1 are line graphs on n nodes. In graph G_0, the parties appear in order (i.e., the neighbors of P_i are P_{i-1} and P_{i+1}). In graph G_1, the parties appear in order, except with the following change: The location of parties P_3, P_4, P_5 (in nodes

[4] For simplicity we consider a fixed number of rounds R; however, the techniques can be extended to probabilistic R as well.

$3, 4, 5$ of G_0) are now in nodes $n-2, n-1$, and n, respectively; in turn, parties P_{n-2}, P_{n-1}, and P_n (in nodes $n-2, n-1$, and n of G_0) are now in nodes $3, 4, 5$.

The adversary $\mathcal{A}$ will corrupt: party P_1 (always at position 1 in both graphs), which we will denote as B the broadcaster; party P_4 (who is in position 4 of G_0 and position $n-1$ in G_1), which we will denote as D the "detective" party; and parties P_7 and P_{n-4} (in fixed positions $7, n-4$) who we will denote as A_1 and A_2 the aborting parties. For simplicity of notation, in the following analysis, we will denote the two nodes $4, n-1$ in which D can be located as v, v'. We will further denote the distance $(n-4)-7$ between the aborting parties A_1 and A_2 as m; note that $m \in \Theta(n)$.

Note that the neighbors of all corrupted parties are the same across G_0 and G_1 (this is the purpose of moving the uncorrupted parties P_3, P_5 in addition to P_3, as well as maintaining a gap between the collections of relevant corrupted parties).

We define two events:

$$E_i := \text{Event that the first abort occurs in round } i, \text{ by either } A_1 \text{ or } A_2$$
$$H_{v,b} := \text{Event that the party at node } v \text{ outputs the correct broadcast bit } b$$
$$\text{(note that this depends on the protocol and the graph on which it is run)}$$

By (single-broadcaster) correctness of $\mathcal{P}$, it must hold for every broadcast bit b that $\Pr[H_{v,b}|E_R] = 1$: that is, if node A_1 or A_2 aborts in the final round, then the news of the abort will not reach node v, in which case the corresponding party must output in the same (correct) fashion as if no abort occurred.

By an information argument, it must be the case for v' that for some choice of $b \in \{0, 1\}$, it holds that $\Pr[H_{v'_\ell,b}|E_1] \leq 1/2$. Recall that v'_ℓ lies on the opposite end of the aborting parties compared to the broadcasting node B.

Combining the above two statements, it holds that $\exists b \in \{0, 1\}$ such that we have $\Pr[H_{v,b}|E_R] - \Pr[H_{v',b}|E_1] \geq 1/2$.

By telescoping and the pigeonhole principle, there must exist some m-step of rounds between R and 1 which contains at least (m/R) of this mass:

$$\exists b \in \{0,1\}, \exists j^* \in \left[\left\lfloor \frac{R}{m} \right\rfloor\right] \text{ s.t. } \begin{cases} \Pr[H_{v_\ell,b}|E_{j^*m}] - \Pr[H_{v'_\ell,b}|E_{(j^*-1)m}] \geq \frac{m}{2R}, \text{ or} \\ \Pr[H_{v'_\ell,b}|E_{j^*m}] - \Pr[H_{v_\ell,b}|E_{(j^*-1)m}] \geq \frac{m}{2R}. \end{cases} \quad (1)$$

We now leverage these facts to describe an attack.

The Attack. Consider a non-uniform adversary $\mathcal{A}$ hardcoded with: the $b \in \{0, 1\}$ and $j^* \in [\lfloor R/m \rfloor]$ from Eq. (1), and whether we are in the top or the bottom of the two cases (in which the roles of v and v' are reversed). Suppose temporarily that we are the first case. $\mathcal{A}$ proceeds as follows:

1. Corrupt the set of parties $\{B, A_1, A_2, D\}$ from the set of n parties.
2. Execute an honest execution of protocol $\mathcal{P}$ up to round $(j^* - 1)m$. In the execution, party B is initialized with input broadcast bit b, and all other emulated parties with input $\perp$ (i.e., not broadcasting).
3. At round $(j^* - 1)m$, abort party A_2. Continue honestly simulating all other corrupted parties for m rounds.
4. At round j^*m, abort party A_1.

5. Continue honestly simulating all other corrupted parties until the conclusion of the protocol. Denote by out_D the protocol output of the "detective" party D.
6. $\mathcal{A}$ outputs $\mathsf{out}_D \oplus b$.

If we are instead in the setting of case 2 in Eq. (1), then the attack is identical, except that the roles of A_1 and A_2 in Step 2 are swapped.

Claim. $\mathcal{A}$ distinguishes between the execution of $\mathcal{P}$ on graphs G_0 and G_1 with advantage $\Omega(n/R)$.

Proof. This argument follows a similar structure as that of [52]. Suppose wlog we are in Case 1 of Eq. (1). (Case 2 is handled in an identical symmetric manner.) Recall that the aborting parties A_1, A_2 are distance m from one another. This means that A_1 aborts at some round $(j^* - 1)m$, then the view of parties to his left (in particular, the party at node $v = 4$) is as in the event $E_{(j^*-1)m}$. However, information of this abort must take at least m rounds in order to reach any parties to the *right* of A_2; thus, since A_2 aborts already at this round j^*m, A_1's abort is never seen by parties to the right of A_2 (in particular, the party at node $v' = m - 1$), who will have view consistent with event E_{j^*m}.

If the execution took place on G_0, then D is at node v, otherwise it is at node v'. Thus, the advantage of the adversary $\mathcal{A}$ is precisely $\Pr[H_{v_\ell,b}|E_{j^*m}] - \Pr[H_{v'_\ell,b}|E_{(j^*-1)m}] \geq \frac{m}{2R}$.

Now, suppose that $\mathcal{P}$ securely realizes $\mathcal{F}_{BC}^{sh}$ hiding topology with $(\delta, \mathcal{L})$-leakage, for some leakage oracle $\mathcal{L}$. Consider the distribution $\mathcal{S}'$ generated by running the $(\delta, \mathcal{L})$-leakage simulator $\mathcal{S}$, but aborting and outputting $\perp$ in the event that the randomness of $\mathcal{S}$ indicates it will query the leakage oracle. By construction, the statistical distance between $\mathcal{S}'$ and the properly simulated distribution $\mathcal{S}$ with access to leakage $\mathcal{L}$ for any fixed choice of real inputs (i.e., honest graph) is bounded by δ. In particular, $\mathcal{S}'$ is within δ statistical distance from both $\mathcal{S}^{\mathcal{L}(G_0)}$ and $\mathcal{S}^{\mathcal{L}(G_1)}$ (denoting oracle access to leakage on the respective graphs G_0, G_1). By the assumed $(\delta, \mathcal{L})$-leakage simulation security of the protocol, for both $b = 0, 1$ it holds that $\mathcal{S}^{\mathcal{L}(G_b)}$ is computationally indistinguishable from the adversarial view of execution of $\mathcal{P}$ on G_b. Combining these steps, we see that no efficient adversary can distinguish between the executions of $\mathcal{P}$ on graphs G_0 and G_1 with advantage non-negligibly better than δ.

Therefore, combined with Claim 3.2 it follows that δ must be bounded below by $\delta \in \Omega(n/R)$.

4 Upper Bounds

In this section, we observe that oblivious transfer implies semi-honest topology-hiding computation on small diameter graphs, and then present two constructions of topology-hiding broadcast with security against fail-stop adversaries from secure hardware.

The construction of semi-honest THC for graphs with small diameter follows is a modified variant of the protocol given in [52].

The first fail-stop secure topology-hiding broadcast protocol leaks at most one bit in the presence of aborts, by exploiting a stratified structure where the protocol is broken into epochs corresponding to the parties playing. If at the end of an epoch the communication network is still intact, the corresponding party will receive output at the end of the protocol. If the network is not intact, the party will not receive the broadcast bit. Aborting during a given an epoch may leak a bit about the distance from some aborting party to the one corresponding to the epoch. But by the next epoch all parties (or rather, their secure hardware) will be aware that the network has been disrupted and no future epochs will yield output to their corresponding parties.

The second protocol is a simple modification of the first which extends each epoch into many smaller eras. The era that actually determines the party's output is randomly (and secretly) chosen by the secure hardware. So, unless the first abort occurs in this era (leaking a single bit), all parties (namely, their secure hardware) will reach consensus about the network being disrupted and what their output is, independently of the network topology. Thus a bit is only leaked with probability that degrades inversely with the number of eras.

4.1 OT Implies Semi-honest THC (for Small-Diameter Graphs)

Semi-Honest THC for small-diameter graphs is, in fact, *equivalent* to OT. This follows from a minor modification to the MPC-based protocol of [52]. Recall, the high-level approach of [52] is a recursive construction:

- At the base level, nodes run the (insecure) OR-and-forward protocol, except that every node has a key pair for a PKE scheme, and every message to node i from one of its neighbors is encrypted under pk_i.
- The recursion step is to replace every node by an MPC protocol in its local neighborhood (the node and all its immediate neighbors), such that its internal state is revealed only if the entire neighborhood colludes.
  The communication pattern for each of these MPCs is a star. Since leaf nodes can't communicate directly, they must pass messages through the center. In order to simulate private channels the leaf nodes first exchange PKE public keys (we are in the semi-honest model, so man-in-the-middle attacks are not relevant) and then use the PKE scheme to encrypt messages between them.
  Note that the MPC simulates the node's next-message function. All nodes receive as input a secret-share of the state from the previous round, and output a secret share of the updated state. In addition, the input of the central node contains the list of messages received from its neighbors in the previous round, and its output contains the list of messages to send to its neighbors in this round. At the end of the MPC execution, the central node sends the messages to its neighbors (who will then use them as part of their input in MPC executions in the next round).
  This structure is the reason for requiring the messages to a node at the base level to be encrypted—the MPC doesn't hide the messages themselves from the central node, hence privacy would be lost if they were unencrypted.

We will replace the PKE scheme with a key-agreement protocol and a symmetric encryption scheme. Since the existence of OT implies both of these primitives, the resulting protocol can be build from OT (we note that, unlike the construction of OT from THB, this construction is *not* black-box in the OT primitive—the recursion step will require non-black-box access to the OT).

For the base step, instead of using a PKE scheme, every node will perform a key-agreement protocol with all of its neighbors. Henceforth, messages from p_i to a direct neighbor p_j will be encrypted under their shared key (using the symmetric encryption scheme). This ensures that to an adversary that does not have access to the state of either p_i or p_j messages between the two are indistinguishable from random.

For the recursion step, we do the same thing except with the leaf nodes. That is, every pair of leaf nodes p_i, p_j will execute the key agreement protocol, using the central node to pass messages. Henceforth, the private channel between them in the MPC protocol will be simulated by encrypting messages using their shared key and passing the messages via the central node.

4.2 Constructions for Fail-Stop Adversaries

Both fail-stop protocols presented in this section achieve a standard notion of broadcast. The broadcast functionalities considered previously were defined with respect to the network topology, particularly its connected components. Forthwith, we will assume the network (before failures) is fully connected.

Definition 7 ($\mathcal{F}_{BC}$)**.** *The ideal n-party broadcast functionality* $\mathcal{F}_{BC}$ *is defined by the following output behavior on input* $b_i \in \{0, 1, \bot\}$ *from every party* P_i*: if a exactly one* $b := b_i$ *is non-*$\bot$*, then* $\mathcal{F}_{BC}$ *outputs b to all parties; otherwise, all outputs are* $\bot$.

Protocol with One Bit of Leakage

We present a topology-hiding broadcast protocol secure against fail-stop adversaries making static corruptions given one bit of leakage. We assume a secure hardware box and one-way functions. We also note that the protocol presented here is secure against semi-honest adversaries without any leakage.

In what follows we consider parties to correspond to their node in the set of all network nodes, $[n]$.

Our protocol has two major phases:

1. **Graph Collection:** Collect a description of the graph, inputs, and aliases. This phase runs for a number of rounds proportional to the network diameter. Any abort seen during this phase by any party will cause that part to abort in the final round.
2. **Consistency Checking and Abort Segregation:** Checking consistency and outputting. This phase has a number of subphases corresponding to the number of parties, n.
   Each subphase runs for a number of rounds proportional to the size of the network. During subphase i, party i will no longer abort if the first abort it sees takes place during that subphase or later. However, an abort seen by any party in $\{i + 1, i + 2, \ldots, n\}$ will still cause that party to abort in the final round. The intuition is that if any party aborts in a subphase, all non-aborted parties are guaranteed to know that an abort has occurred by the end of next subphase.

By subphase n, if no abort has been seen, all honest parties will output correctly, regardless of subsequent abort behaviors.

The hardware box, aside from initialization and final output, will take as input and output authenticated-encryptions of the player's current "state." The plaintext state of a party P_u with session key k_u after round i, denoted s_u^i, contains the following information:

- The party's alias: id_u, a random λ-length string chosen at outset.
- The round number: i_u.
- Current "knowledge" of the graph: G_u^i.
- Current "knowledge" of inputs: $\mathbf{m}^{u,i} = (m_1^{u,i}, \ldots, m_n^{u,i})$.
- Current "knowledge" of session keys: $\mathbf{sk}^{u,i} = (\mathrm{sk}_1^{u,i}, \ldots, \mathrm{sk}_n^{u,i})$.
- First round an abort was seen: a_u^i.
- Indicator of whether or not a neighbor has aborted in a previous round: $\mathbf{b}_u^i := (b_{v_1}^i, \ldots, b_{v_d}^i)$. (This information is not strictly necessary, but convenient when proving security.)
- abort flag: α.

We now define the C_{fs} functionality that is embedded in the hardware box. As stated above, we take network locations (typically denoted u or v) to be elements of $[n]$. State information is represented as vectors over the alphabet $\Sigma = \{0, 1, ?, \perp\}$. We take $\boldsymbol{e}_i(x)$ to be the vector of all ?s with $x \in \{0, 1, \perp\}$ in the ith position (the length of the vector should be clear from context, unless otherwise specified). If the vector is in fact an $m \times n$-matrix, we take $\boldsymbol{e}_i(\boldsymbol{x})$ to denote the vector with $?^n$ in all rows, except i which contains $\boldsymbol{x} \in \Sigma^n$. The network (graph) is represented as an adjacency matrix, with ?s denoting what is unknown and $\perp$s representing errors, or inconsistencies. In particular, the closed neighborhood of u, $N[u]$, is the $n \times n$ matrix with the adjacency vector of u in the uth row and ?s elsewhere. Let $H : 2^{\Sigma^m} \to \Sigma^m$ denote the component-wise "accumulation" operator where the ith output symbol, for $i \in [m]$, is as defined as follows:

$$H_i X := \begin{cases} 1 \text{ if } \exists \boldsymbol{x} \in \boldsymbol{X} \in [k] : \boldsymbol{x}_i = 1, \text{ and } \forall \boldsymbol{y} \in \boldsymbol{X} : \boldsymbol{y}_i \in \{1, ?\} \\ 0 \text{ if } \exists \boldsymbol{x} \in \boldsymbol{X} : \boldsymbol{x}_i = 0, \text{ and } \forall \boldsymbol{y} \in \boldsymbol{X} : \boldsymbol{x}_i \in \{0, ?\} \\ ? \text{ if } \forall \boldsymbol{x} \in \boldsymbol{X} : \boldsymbol{x}_i = ? \\ \perp \text{ otherwise} \end{cases}$$

Finally, let $R = n(n + 2) + 1$ (final number of rounds) (Fig. 4).

Let $\mathcal{L}$ denote the class of efficient leakage functions that leak one bit about the topology of the network.

Theorem 3. *The protocol $\mathcal{F}_{\text{fs-broadcast}}$ topology hiding realizes broadcast with $\mathcal{L}$ leakage with respect to static corruptions.*

Remark 1. By observing that the leakage oracle is only used by the simulator in the event of an abort, the protocol $\mathcal{F}_{\text{fs-broadcast}}$ is secure against probabilistic polynomial time semi-honest adversaries without leakage.

Notation:

Let (G, E, D) denote a symmetric key authenticated encryption scheme. Let $\mathrm{PRF} \leftarrow \{\mathrm{PRF}_k\}_k$ denote a pseudorandom function with security parameter λ, chosen randomly from some such family during setup. Additionally, let $\mathrm{msk} \leftarrow \mathrm{G}(1^\lambda)$, again chosen during setup.

We take u to denote the location in the network associated with the party, P_u, using this instance of C_{fs}.

As above, we use the following to denote state information:

$$s_v^i = (v; \mathrm{sk}_v; i; G_v^i; \mathbf{m}^{v,i} = m_1^{v,i}, \ldots, m_n^{v,i}; \mathbf{sk}^{v,i} = \mathrm{sk}_1^{v,i}, \ldots, \mathrm{sk}_n^{v,i}; a_v^i; \mathbf{b}_v^i := b_{v_1}^i, \ldots, b_{v_d}^i, \alpha).$$

Input:

- (Initialization input) $x = (m, u, N[u], \mathrm{sk}_u)$, where m should be the broadcast message if u is broadcaster, and $\perp$ otherwise. $N[u]$ denotes the closed neighborhood of u. sk is a random session key.
- (Round input) d authenticated encryptions of the form:

$$x = \left(x_u, x_{v_1}, \ldots, x_{v_d}, \mathrm{sk}_u\right)$$

where $v_1, \ldots, v_d$ are in $N(u)$, additionally in numerical order, and ignored if $\deg(u) < d$.
For all $v \in N[v]$, x_v should either an encrypted state, $\mathrm{E}_{\mathrm{msk}}(s_v^i)$, or Abort.
Moreover, we

Fig. 4. The functionality C_{fs} (Part 1: continued in Fig. 5).

Correctness. Assuming there are no aborts, correctness follows by induction on the rounds. By inspection, local consistency checks will pass under honest evaluation if there are no aborts. Clearly, at the end of round i, the encrypted broadcast message will have reached all parties whose distance is at most i from the broadcaster. So by round $n + 1$, all parties will have the message. Similarly, by round $n + 1$, all local descriptions of the network and aliases will have reached all parties. Moreover, no abort flags will be triggered. Thus, the global consistency checks will pass in the final round, R, and all parties will receive the broadcast message.

By inspection, it is easy to see that if evaluation is semi-honest with possible aborts each party will either output the unique non-$\perp$ input, or abort (Fig. 6).

Security. We start with a rough overview of simulation and why it works, with the full proof of security given in the full version.

Crucially, the authenticated encryption of internal states makes it infeasible for an adversary to either forge states, or glean any information about their contents. As the simulator may have incomplete information about the graph topology, this allows it to send fake states and simply output consistently with the real protocol. Moreover, the unforgeability gives the simulator full knowledge of any initial information used when querying the box, and, importantly, how these queries relate to one another (especially whether or not they are consistent).

Computation:

If input is of "Initialization" format: I.e., $x = (m, u, N[u], \mathrm{sk}_u)$.
Let $s_u^1 = (u; \mathrm{sk}; 1; G_u^1 = N[u]; e_u(m_u); e_u(\mathbf{sk}_u); \perp; 0, \ldots, 0; 0)$, and output: $\mathrm{E}_{\mathrm{msk}}(s_u^1; \mathrm{PRF}(x))$. If $N[u]$ is not a closed neighborhood (i.e. if there is more than one node with more than one edge), output $\perp$.

If input is of "Round" format: I.e., $x = (x_u, x_{v_1}, \ldots, x_{v_d}, \mathrm{sk})$ where $x_v = \mathrm{E}_{\mathrm{msk}}(\hat{s}_v)$ or Abort, and $\hat{s}_v = (\hat{v}; \hat{\mathrm{sk}}_v; \hat{i}_v; G_v; \hat{\mathbf{m}}^v; \hat{a}_v; \hat{\mathbf{b}}_v; \hat{\alpha}_v)$, for $v \in N[u]$

Perform local consistency checks:

- If x_u = Abort, output $\perp$.
- Authenticate/decrypt all non-Abort inputs, to get $\hat{\mathbf{s}}_u, \hat{\mathbf{s}}_{v_1}, \ldots, \hat{\mathbf{s}}_{v_\ell}$. If authentication or encryptions fails, halt and output $\perp$.
- If round counters $(\hat{i}_u, \hat{i}_{v_1}, \ldots, \hat{i}_{v_\ell})$ are not all equal, halt and output $\perp$.
- For each network location $v \in N[u]$, where $N[u]$ is extracted from s_u, if v does not correspond to plaintext state input (in correct position) or Abort, set $\alpha' = 1$.
  Additionally, if some $v \notin N[u]$ is associated with any input state, output $\perp$.
- If there exists $x_v \neq$ Abort but $\hat{b}_v^u = 1$, output $\perp$.
- If the plain text session key input sk does not match the session key extracted from $\hat{\mathbf{s}}_u$, output $\perp$.

Perform global consistency checks and accumulation:

- Initialize abort flag to be the same as in $\hat{\mathbf{s}}_u$: $\alpha' = \hat{\alpha}_u$.
- If sender keys don't match "stored" keys, in other words $\hat{\mathbf{sk}}_{\hat{v}} = (\hat{\mathbf{sk}}^{\hat{u}})_{\hat{v}}$ for all input non-Abort-ing locations $\hat{v}$, then set $\alpha' := 1$ (encrypted abort).
- $i' := \hat{i}_u + 1$,
- Generate $G' := H\{\hat{G}^v : v \in N[u]\}$, If any component of G' is $\perp$, let $\alpha' := 1$.
- $\mathbf{m}' := H\{\hat{\mathbf{m}}^v : v \in N[u]\}$, If any component of $\mathbf{m}'$ is $\perp$, let $\alpha' := 1$.
- $\mathbf{sk}' := H\{\hat{\mathbf{sk}}_v : v \in N[u]\}$. If any component of $\mathbf{sk}'$ is $\perp$, let $\alpha' := 1$.
- For all $v \in \hat{N}(u)$, if x_v = Abort then set $b'_v := 1$. Otherwise, set $b'_v = \hat{b}_v^u$. Let $\mathbf{b}' = (b'_{v_1}, \ldots, b'_{v_d})$.
- If if $\hat{a}_u \neq \perp$ and either $\hat{a}_v^i \neq \perp$ or x_v = Abort for any $v \in \hat{N}[u]$,then set $a' := \hat{i}_u$. Otherwise, $a' := \hat{a}_u$.

Check to enforce commitment:
If $\hat{i}_u = n + 1$ and $\hat{G}_u$ contains any '?', set $\alpha' := 1$.

Output:

- if $\hat{i}_u < R$, then output: $\mathrm{E}_{\mathrm{msk}}(\boldsymbol{s}'; \mathrm{PRF}(x))$, where $\boldsymbol{s}' := (\hat{u}_u; \hat{\mathrm{sk}}_u; i'; G'; \mathbf{m}'; \mathbf{sk}'; a'; \mathbf{b}', \alpha')$.
- Else, if $\hat{i}_u = R$: If there exists more than one v such that $(\hat{\mathbf{m}}^u)_v \neq \perp$, output $\perp$. If $a' \leq (\hat{u} + 1) \cdot n$ or $\alpha' = 1$, output $\perp$. Otherwise, output the message corresponding to the unique non-$\perp$ location: $(\hat{\mathbf{m}}^u)_{v^*}$.
- Else, output $\perp$.

Fig. 5. The functionality C_{fs} (Part 2)

Notation: Let $O(C_{\text{fs}})$ denote a secure hardware box (oracle) running C_{fs}.
Input: Each party P_x receives $O(C_{\text{fs}})$. If P_x is broadcaster, P_x also has input $m = m_x$, the message to be broadcast. If P_x is not broadcaster, m_x is the all zero vector.
Protocol for Broadcast:

- (Initialization) P_u chooses a random id $\leftarrow \{0,1\}^\lambda$. P_u computes $c_u^1 = O(C_{\text{fs}})(m_u; u; N(x); \text{sk}_u)$. P_u sends $(v; c_u^1)$ to $\mathcal{F}_{\text{graph}}$ for all $v \in N(u)$. Each $v \in Nu$ receives $(u; c_u^1)$. P_u receives $(v; c_v^1)$ for each $v \in N(u)$.
- (Round $i = 2$ to $R-1$) P_u computes $c_u^i = O(C_{\text{fs}})(c_u^{i-1}, c_{v_1}^{i-1}, \ldots, c_{v_d}^{i-1})$. P_u sends $(v; c_u^i)$ to $\mathcal{F}_{\text{graph}}$ for all $v \in N(u)$. Each $v \in N(u)$ receives $(u; c_u^i)$. P_u receives $(v; c_v^i)$ for each $v \in N(u)$.
- (Round R) P_u computes $\hat{m} = O(C_{\text{fs}})(c_u^R, c_{v_1}^R, \ldots, c_{v_d}^R; \text{sk}_u)$, and outputs $\hat{m}$.

Fig. 6. The protocol $\mathcal{F}_{\text{fs-broadcast}}$ in $\mathcal{F}_{\text{graph}}$-hybrid model.

The difficulty in the proof is in dealing with "replay" attacks, where the adversary combines information from the honest nodes in malicious ways with other initializations. The session keys aid in this by rendering it infeasible for the adversary to replay as an uncorrupted party with a modified local topology. Additionally, the collection phase implies that honest nodes have complete information about topology and initialization information used in execution by round n. Thus, when this information is later combined with initializations that do not match the execution exactly, the only plaintext output such a malicious adversary will receive is $\perp$. Together this means the simulator only has to provide output when query structure matches execution almost exactly (up to somewhat local aborts). The upshot being that the simulator can provide output identical to a real execution, even though it has incomplete knowledge of the network topology.

One of the dangers in simulation is if an adversary corrupts all the nodes within a distance r of a given node, it has enough information to "fast forward" the node to get its program outputs for the next $r+1$-rounds. Additionally, after its threshold round has occurred, an adversary can abort all the neighbors of a node, and iterate the remaining rounds by itself to get output. However, we show that the simulator will always have enough information to fool such adversaries.

- For each party, P_u, generate $(n+2)n$ random encryptions of 0. These will constitute the messages sent by honest parties to corrupted parties, and the output of the oracle for queries consistent with semi-honest evaluation.
- The simulated oracle will remember all queries from the adversary in a data structure, the outputs given, and how they relate to one another (as we explain in the full version). The idea is that "valid" inputs will return more encryptions of 0 until the last "round." Then, the simulator will use the data structure to determine the appropriate output given the initialization queries used in conjunction with the single bit of leakage (supposing there was an Abort in the execution, described below). We describe the simulation of the hardware box program in more detail below in the full

version.
Any query which isn't an initialization input or a concatenation of previous queries will immediately return $\bot$. Likewise, any combination of previous queries that correspond to locally inconsistent topologies or round numbers. Moreover, after n rounds any queries that yield an inconsistent topology (the combined the base initializations of all queries, and the previous queries they depend on, does not yield a single consistent graph) will be recognized in the real world. Thus the simulator need only give output in the final round if all queries, including their ancestors, correspond to consistent graph initialization.
If the first abort occurred in round i, the simulator will query to determine if the real encrypted state of party $j = \lfloor i/n \rfloor + 2$ contained information "witnessed" the abort after round $(j + 2)n$. If so, the queries corresponding to the final round of execution for parties $P_1, P_2, \ldots, P_j$ will return the broadcast message, and ABORT to all other parties. If P_j "witnessed" the abort on or before round $(j + 2)n$, then the query corresponding to P_j's final input to the black box program will return ABORT as well (all other outputs for "final" queries are unchanged from the previous case). The simulator uses the single bit of leakage to determine if an abort reach P_j in time.

- When the adversary corrupts a party once the protocol is underway, first choose a random sk. Then, fix the oracle to yield pre-determined ciphertexts corresponding to the honest initialization and pre-determined messages from its neighborhood.

We refer the reader to the full version of this paper for a complete description of the simulator and hybrids.

Protocol with Arbitrarily Low Leakage
This protocol is only a slight modification of the previous one. To achieve δ leakage, each party is not associated with a single subphase, but instead a sequence of $n\lceil/\delta\rceil$ subphases of a *zone*. At the outset, parties provide randomness (which can be drawn from the session keys), which will assist in selecting one of these subphases to be the true one. Thus, the probability of an aborting adversary successfully hitting any sub-subphase with its first abort is dependent on the graph structure is $< 1/\delta$.

The state is identical to the previous, with one additional parameter, t, encoding the threshold round.

The protocol here is the same as before, except now $R = n(n^2\lceil 1/\delta\rceil + 2) + 1$.

We now define $C_{\text{rand-fs}}$ functionality. We take notation to be consistent with the previous construction, C_{fs}, where not otherwise specified (Fig. 7).

Theorem 4. *For any $\delta = 1/poly(\lambda, n)$, the protocol $\mathcal{F}_{fs\text{-}broadcast}$, when C_{fs} is replaced with $C_{rand\text{-}fs}(\delta)$, topology hiding realizes broadcast with $(\delta, \mathcal{L})$ leakage with respect to static corruptions.*

Correctness. The proof here is nearly identical to the preceding one.

Security. Here the simulator is nearly identical to the previous, except it chooses each location's random threshold itself, and only queries the leakage oracle if when the first real abort occurs in chosen block. Here, it queries more-or-less identically to before. For all other nodes it outputs according to whether the threshold has already occurred

Notation:

Let (G, E, D) denote a symmetric key authenticated encryption scheme. Let PRF denote a pseudorandom function with security parameter λ, chosen randomly from some such family during setup. Additionally, let msk $\leftarrow$ G(1^λ), again chosen during setup.

Let Threshold$_{\delta,n}$ be a pseudorandom function such that Threshold$_{\delta,n}(u; \mathrm{sk}_u)$ that outputs a a pseudorandom integer in $\{3n + (u-1)n^2\lceil 1/\delta\rceil, 4n + (u-1)n^2\lceil 1/\delta\rceil, \ldots, 2n + un^2\lceil 1/\delta\rceil\}$.

As above, we use the following to denote state information:

$$s_v^i = (v; \mathrm{sk}_v; i; G_v^i; \mathbf{m}^{v,i} = m_1^{v,i}, \ldots, m_n^{v,i}; \mathbf{sk}^{v,i} = \mathrm{sk}_1^{v,i}, \ldots, \mathrm{sk}_n^{v,i}; a_v^i; \mathbf{b}_v^i := b_{v_1}^i, \ldots, b_{v_d}^i; \alpha_v; t_v).$$

Input:

- (Initialization input) $x = (m, u, N[u], \mathrm{sk}_u)$, where m should be the broadcast message if u is broadcaster, and $\perp$ otherwise. $N[u]$ denotes the closed neighborhood of u, a binary vector representing all adjacencies (or lack thereof). sk is a random session key.
- (Round input) d authenticated encryptions of the form:

$$x = \left(x_u, x_{v_1}, \ldots, x_{v_d}, \mathrm{sk}_u\right)$$

where $v_1, \ldots, v_d$ are in $N(u)$ and ignored if $\deg(u) < d$.
For all $v \in N[v]$, x_v is either an encrypted state, $\mathrm{E}_{\mathrm{msk}}(s_v^i)$, or ABORT.

Fig. 7. The functionality $C_{\text{rand-fs}}$ (Part 1: continued in Fig. 8).

or not. The leakage oracle itself will represent an identical functionality to the previous case.

Because the distribution of thresholds is computationally indistinguishable in simulated case from the real one, an adversary will be unable to distinguish. As many of the lemmas from the previous construction hold here, we will simply bound the probability that the leakage oracle is called (and hence, the leakage itself).

The Simulator. As before the first the simulator first generates a non-aborting execution for each corrupt component, which will form the basis of the messages from honest parties. Additionally, in this case, the simulator selects a threshold block uniformly for each party's zone. Having chosen thresholds and an execution, simulation of $C_{\text{rand-fs}}$ proceeds identically to the previous protocol.

Lemma 1. *For any probabilistic poly-time adversary, the simulator only needs to query the leakage oracle with probability at most* δ.

Proof. Recall that the simulator only needs to query the leakage oracle with respect to at most one party.

For any fixed network location, we will bound the probability that the simulator needs to call the leakage oracle for that location by δ/n. Then by a union bound, the probability that the simulator needs to call the leakage oracle for *any* node.

Computation:

If input is of "Initialization" format: I.e., $x = (m, u, N[u], \text{sk}_u)$.
Let $\boldsymbol{s}_u^1 = (u; \text{sk}; 1; G_u^1 = N[u]; \boldsymbol{e}_u(m_u); \boldsymbol{e}_u(\textbf{sk}_u); \bot; 0, \ldots, 0; 0; \text{Threshold}(u; \text{sk}))$, and output: $\text{E}_{\text{msk}}(s_u^1; \text{PRF}(x))$. If $N[u]$ is not a closed neighborhood (i.e. if there is more than one node with more than one edge), output $\bot$.

If input is of "Round" format: I.e., $x = (x_u, x_{v_1}, \ldots, x_{v_d})$ where $x_v = \text{E}_{\text{msk}}(\hat{s}_v^i)$ or Abort, and $\hat{s}_v^i = (\hat{v}; \hat{\text{sk}}_v; \hat{i}_v; G_v; \hat{\mathbf{m}}^v; \hat{a}_v; \hat{\mathbf{b}}_v; \hat{\alpha}_v; \hat{t}_v)$, for $v \in N[u]$.

Perform local consistency checks:

* If x_u = Abort, output $\bot$.
* Authenticate/decrypt all non-Abort inputs, to get $\hat{\boldsymbol{s}}_u, \hat{\boldsymbol{s}}_{v_1}, \ldots, \hat{\boldsymbol{s}}_{v_\ell}$.
  If authentication or encryptions fails, halt and output $\bot$.
* If round counters $(\hat{i}_u, \hat{i}_{v_1}, \ldots, \hat{i}_{v_\ell})$ are not all equal, halt and output $\bot$.
* For each network location $v \in N[u]$, where $N[u]$ is extracted from $\boldsymbol{s}_u$, if v does not correspond to plaintext state input (in correct position) or Abort, output $\bot$.
  Additionally, if some $v \notin N[u]$ is associated with any input state, output $\bot$.
* If there exists $x_v \neq$ Abort but $\hat{b}_v^u = 1$, output $\bot$.
* If the plain text local session key input sk does not match the session key extracted from $\hat{\boldsymbol{s}}_u$, output $\bot$.

Perform global consistency checks and accumulation:

* Initialize abort flag to be the same as in $\hat{\boldsymbol{s}}_u$: $\alpha' = \hat{\alpha}_u$.
* If sender keys don't match "stored" keys, in other words $\hat{\mathbf{sk}}_{\hat{v}} = (\hat{\mathbf{sk}}^{\hat{u}})_{\hat{v}}$ for all input non-Abort-ing locations $\hat{v}$, then set $\alpha' := 1$ (encrypted abort).
* $i' := \hat{i}_u + 1$,
* Generate $G' := H\{\hat{G}^v : v \in N[u]\}$, If any component of G' is $\bot$, let $\alpha' := 1$.
* $\mathbf{m}' := H\{\hat{\mathbf{m}}^v : v \in N[u]\}$, If any component of $\mathbf{m}'$ is $\bot$, let $\alpha' := 1$.
* $\mathbf{sk}' := H\{\hat{\mathbf{sk}}^v : v \in N[u]\}$. If any component of $\mathbf{sk}'$ is $\bot$, let $\alpha' := 1$.
* For all $v \in \hat{N}(u)$, if x_v = Abort then set $b_v' := 1$. Otherwise, set $b_v' = \hat{b}_v^u$, for all $v \in N(u)$. Let $\mathbf{b}' = (b_{v_1}', \ldots, b_{v_d}')$.
* If $\hat{a}_u \neq \bot$ and either $\hat{a}_v \neq \bot$ or x_v = Abort for any $v \in \hat{N}[u]$,then set $a' := \hat{i}_u$. Otherwise, $a' := \hat{a}_u$.

Check to enforce commitment:
If $\hat{i}_u = n + 1$ and $\hat{G}_u$ contains any '?', set $\alpha' = 1$.

Output:

- if $\hat{i}_u < R$, then output: $\text{E}_{\text{msk}}(\boldsymbol{s}'; \text{PRF}(x))$, where $\boldsymbol{s}' := (\hat{u}_u; \hat{\text{sk}}_u; i'; G'; \mathbf{m}'; \mathbf{sk}'; a'; \mathbf{b}'; \alpha'; \hat{t}_u)$.
- Else, if $\hat{i}_u = R$: If there does not exist exactly one v such that $(\hat{\mathbf{m}}^u)_v \neq \bot$, output $\bot$. If $a' \leq t$ or $\alpha' = 1$, output $\bot$. Otherwise, output the message corresponding to the unique non-$\bot$ location: $(\hat{\mathbf{m}}^u)_{v^*}$.
- Else, output $\bot$.

Fig. 8. The functionality $C_{\text{rand-fs}}$ (Part 2)

In the full version we show that any non-aborting query graph induced by a threshold node must match the non-aborting execution exactly. As a consequence to get non-aborting output, the adversary must have run the protocol up to at least $r + 1$ rounds before the node's threshold. Thus, anything that happens before such a time will give no information about the threshold round (beyond whether it has or has not occurred yet). Additionally, by the time an adversary can learn whether the threshold follows its current block, it will be to late to execute a non-simulatable abort (outside of the corruption radius of the node).

If the first abort occurs in the i-th block of a location's zone, then the probability the adversary hits the chosen block is:

$$\underbrace{\left(1 - \frac{i-1}{n/\delta}\right)}_{\text{prob. threshold hasn't occured}} \cdot \underbrace{\left(\frac{1}{n/\delta - (i-1)}\right)}_{\text{cond. prob. of hitting relevant block}} = \frac{1}{n/\delta} = \frac{\delta}{n}$$

References

1. diaspora*: The online social world where you are in control
2. Akavia, A., LaVigne, R., Moran, T.: Topology-hiding computation on all graphs. In: Katz, J., Shacham, H. (eds.) CRYPTO 2017. LNCS, vol. 10401, pp. 447–467. Springer, Cham (2017). https://doi.org/10.1007/978-3-319-63688-7_15
3. Akavia, A., Moran, T.: Topology-hiding computation beyond logarithmic diameter. In: Coron, J.-S., Nielsen, J.B. (eds.) EUROCRYPT 2017. LNCS, vol. 10212, pp. 609–637. Springer, Cham (2017). https://doi.org/10.1007/978-3-319-56617-7_21
4. Beimel, A.: On private computation in incomplete networks. Distrib. Comput. **19**(3), 237–252 (2007)
5. Beimel, A., Franklin, M.K.: Reliable communication over partially authenticated networks. Theor. Comput. Sci. **220**(1), 185–210 (1999)
6. Beimel, A., Gabizon, A., Ishai, Y., Kushilevitz, E., Meldgaard, S., Paskin-Cherniavsky, A.: Non-interactive secure multiparty computation. In: Garay, J.A., Gennaro, R. (eds.) CRYPTO 2014. LNCS, vol. 8617, pp. 387–404. Springer, Heidelberg (2014). https://doi.org/10.1007/978-3-662-44381-1_22
7. Beimel, A., Malka, L.: Efficient reliable communication over partially authenticated networks. Distrib. Comput. **18**(1), 1–19 (2005)
8. Ben-Or, M., Goldwasser, S., Wigderson, A.: Completeness theorems for noncryptographic fault-tolerant distributed computations. In: Proceedings of the 20th Annual ACM Symposium on Theory of Computing (STOC), pp. 1–10 (1988)
9. Bhurman, H., Christandl, M., Unger, F., Wehner, S., Winter, A.: Implications of superstrong nonlocality for cryptography. Proc. R. Soc. A **462**(2071), 1919–1932 (2006)
10. Bläser, M., Jakoby, A., Liskiewicz, M., Manthey, B.: Private computation: k-connected versus 1-connected networks. J. Cryptol. **19**(3), 341–357 (2006)
11. Boyle, E., Goldwasser, S., Tessaro, S.: Communication locality in secure multi-party computation: how to run sublinear algorithms in a distributed setting. In: Sahai, A. (ed.) TCC 2013. LNCS, vol. 7785, pp. 356–376. Springer, Heidelberg (2013). https://doi.org/10.1007/978-3-642-36594-2_21

12. Canetti, R., Lindell, Y., Ostrovsky, R., Sahai, A.: Universally composable two-party and multi-party secure computation. In: Reif, J.H. (ed.) Proceedings on 34th Annual ACM Symposium on Theory of Computing, 19–21 May 2002, Montréal, Québec, Canada, pp. 494–503. ACM (2002)
13. Chandran, N., Chongchitmate, W., Garay, J.A., Goldwasser, S., Ostrovsky, R., Zikas, V.: The hidden graph model: communication locality and optimal resiliency with adaptive faults. In: Proceedings of the 2015 Conference on Innovations in Theoretical Computer Science, ITCS 2015, pp. 153–162. ACM, New York (2015)
14. Chandran, N., Garay, J.A., Ostrovsky, R.: Edge fault tolerance on sparse networks. In: Czumaj, A., Mehlhorn, K., Pitts, A., Wattenhofer, R. (eds.) ICALP 2012. LNCS, vol. 7392, pp. 452–463. Springer, Heidelberg (2012). https://doi.org/10.1007/978-3-642-31585-5_41
15. Chandran, N., Goyal, V., Sahai, A.: New constructions for UC secure computation using tamper-proof hardware. In: Smart, N. (ed.) EUROCRYPT 2008. LNCS, vol. 4965, pp. 545–562. Springer, Heidelberg (2008). https://doi.org/10.1007/978-3-540-78967-3_31
16. Chang, H., Govindan, R., Jamin, S., Shenker, S.J., Willinger, W.: Towards capturing representative AS-level Internet topologies. Comput. Netw. **44**(6), 737–755 (2004)
17. Chaum, D.: Untraceable electronic mail, return addresses, and digital pseudonyms. Commun. ACM **24**(2), 84–88 (1981)
18. Chaum, D., Crepeau, C., Damgard, I.: Multiparty unconditionally secure protocols. In: Proceedings of the 20th Annual ACM Symposium on Theory of Computing (STOC), pp. 11–19 (1988)
19. Chiesa, A., Tromer, E.: Proof-carrying data and hearsay arguments from signature cards. In: Yao, A.C. (ed.) Proceedings of the Innovations in Computer Science – ICS 2010, Tsinghua University, Beijing, China, 5–7 January 2010, pp. 310–331. Tsinghua University Press (2010)
20. Choi, S.G., Katz, J., Schröder, D., Yerukhimovich, A., Zhou, H.-S.: (Efficient) universally composable oblivious transfer using a minimal number of stateless tokens. In: Lindell, Y. (ed.) TCC 2014. LNCS, vol. 8349, pp. 638–662. Springer, Heidelberg (2014). https://doi.org/10.1007/978-3-642-54242-8_27
21. Chor, B., Kushilevitz, E.: A zero-one law for boolean privacy. SIAM J. Discrete Math. **4**(1), 36–47 (1991)
22. Cleve, R.: Limits on the security of coin flips when half the processors are faulty (extended abstract). In: STOC, pp. 364–369 (1986)
23. Cleve, R., Impagliazzo, R.: Martingales, collective coin flipping and discrete control processes (1993, unpublished)
24. Dachman-Soled, D., Lindell, Y., Mahmoody, M., Malkin, T.: On the black-box complexity of optimally-fair coin tossing. In: Ishai, Y. (ed.) TCC 2011. LNCS, vol. 6597, pp. 450–467. Springer, Heidelberg (2011). https://doi.org/10.1007/978-3-642-19571-6_27
25. Deng, J., Han, R., Mishra, S.: Decorrelating wireless sensor network traffic to inhibit traffic analysis attacks. Pervasive Mob. Comput. **2**(2), 159–186 (2006)
26. Dolev, D.: The Byzantine generals strike again. J. Algorithms **3**(1), 14–30 (1982)
27. Dolev, D., Dwork, C., Waarts, O., Yung, M.: Perfectly secure message transmission. J. ACM **40**(1), 17–47 (1993)
28. Dwork, C., Peleg, D., Pippenger, N., Upfal, E.: Fault tolerance in networks of bounded degree. SIAM J. Comput. **17**(5), 975–988 (1988)
29. Fisch, B., Freund, D., Naor, M.: Physical zero-knowledge proofs of physical properties. In: Garay, J.A., Gennaro, R. (eds.) CRYPTO 2014. LNCS, vol. 8617, pp. 313–336. Springer, Heidelberg (2014). https://doi.org/10.1007/978-3-662-44381-1_18
30. Fisch, B.A., Freund, D., Naor, M.: Secure physical computation using disposable circuits. In: Dodis, Y., Nielsen, J.B. (eds.) TCC 2015. LNCS, vol. 9014, pp. 182–198. Springer, Heidelberg (2015). https://doi.org/10.1007/978-3-662-46494-6_9

31. Gennaro, R., Lysyanskaya, A., Malkin, T., Micali, S., Rabin, T.: Algorithmic Tamper-Proof (ATP) security: theoretical foundations for security against hardware tampering. In: Naor, M. (ed.) TCC 2004. LNCS, vol. 2951, pp. 258–277. Springer, Heidelberg (2004). https://doi.org/10.1007/978-3-540-24638-1_15
32. Glaser, A., Barak, B., Goldston, R.: A zero-knowledge protocol for nuclear warhead verification. Nature **510**, 497–502 (2004)
33. Goldreich, O., Micali, S., Wigderson, A.: How to play any mental game or a completeness theorem for protocols with honest majority. In: STOC, pp. 218–229. ACM (1987)
34. Goldreich, O., Ostrovsky, R.: Software protection and simulation on oblivious rams. J. ACM **43**(3), 431–473 (1996)
35. Goldwasser, S., Kalai, Y.T., Rothblum, G.N.: One-time programs. In: Wagner, D. (ed.) CRYPTO 2008. LNCS, vol. 5157, pp. 39–56. Springer, Heidelberg (2008). https://doi.org/10.1007/978-3-540-85174-5_3
36. Gordon, S.D., Malkin, T., Rosulek, M., Wee, H.: Multi-party computation of polynomials and branching programs without simultaneous interaction. In: Johansson, T., Nguyen, P.Q. (eds.) EUROCRYPT 2013. LNCS, vol. 7881, pp. 575–591. Springer, Heidelberg (2013). https://doi.org/10.1007/978-3-642-38348-9_34
37. Goyal, V., Ishai, Y., Sahai, A., Venkatesan, R., Wadia, A.: Founding cryptography on tamper-proof hardware tokens. In: Micciancio, D. (ed.) TCC 2010. LNCS, vol. 5978, pp. 308–326. Springer, Heidelberg (2010). https://doi.org/10.1007/978-3-642-11799-2_19
38. Halevi, S., Ishai, Y., Jain, A., Kushilevitz, E., Rabin, T.: Secure multiparty computation with general interaction patterns. In: Proceedings of the 2016 ACM Conference on Innovations in Theoretical Computer Science, ITCS 2016, pp. 157–168. ACM, New York (2016)
39. Halevi, S., Lindell, Y., Pinkas, B.: Secure computation on the web: computing without simultaneous interaction. In: Rogaway, P. (ed.) CRYPTO 2011. LNCS, vol. 6841, pp. 132–150. Springer, Heidelberg (2011). https://doi.org/10.1007/978-3-642-22792-9_8
40. Hazay, C., Lindell, Y.: Constructions of truly practical secure protocols using standardsmartcards. In: Ning, P., Syverson, P.F., Jha, S. (eds.) Proceedings of the 2008 ACM Conference on Computer and Communications Security, CCS 2008, Alexandria, Virginia, USA, 27–31 October 2008, pp. 491–500. ACM (2008)
41. Hinkelmann, M., Jakoby, A.: Communications in unknown networks: preserving the secret of topology. Theoret. Comput. Sci. **384**(2–3), 184–200 (2007). Structural Information and Communication Complexity (SIROCCO 2005)
42. Hirt, M., Maurer, U., Tschudi, D., Zikas, V.: Network-hiding communication and applications to multi-party protocols. In: Robshaw, M., Katz, J. (eds.) CRYPTO 2016. LNCS, vol. 9815, pp. 335–365. Springer, Heidelberg (2016). https://doi.org/10.1007/978-3-662-53008-5_12
43. Hofheinz, D., Muller-Quade, J., Unruh, D.: Universally composable zero-knowledge arguments and commitments from signature cards. In: 5th Central European Conference on Cryptology (2005)
44. Ishai, Y., Sahai, A., Wagner, D.: Private circuits: securing hardware against probing attacks. In: Boneh, D. (ed.) CRYPTO 2003. LNCS, vol. 2729, pp. 463–481. Springer, Heidelberg (2003). https://doi.org/10.1007/978-3-540-45146-4_27
45. Kamat, P., Zhang, Y., Trappe, W., Ozturk, C.: Enhancing source-location privacy in sensor network routing. In: 25th International Conference on Distributed Computing Systems (ICDCS 2005), 6–10 June 2005, Columbus, OH, USA, pp. 599–608 (2005)
46. Katz, J.: Universally composable multi-party computation using tamper-proof hardware. In: Naor, M. (ed.) EUROCRYPT 2007. LNCS, vol. 4515, pp. 115–128. Springer, Heidelberg (2007). https://doi.org/10.1007/978-3-540-72540-4_7

47. Kilian, J.: A general completeness theorem for two-party games. In: Koutsougeras, C., Vitter, J.S. (eds.) Proceedings of the 23rd Annual ACM Symposium on Theory of Computing, 5–8 May 1991, New Orleans, Louisiana, USA, pp. 553–560. ACM (1991)
48. Kumar, M.V.N.A., Goundan, P.R., Srinathan, K., Rangan, C.P.: On perfectly secure communication over arbitrary networks. In: PODC, pp. 193–202 (2002)
49. Kushilevitz, E.: Privacy and communication complexity. SIAM J. Discrete Math. **5**(2), 273–284 (1992)
50. Moran, T., Naor, M.: Basing cryptographic protocols on tamper-evident seals. Theor. Comput. Sci. **411**(10), 1283–1310 (2010)
51. Moran, T., Naor, M., Segev, G.: An optimally fair coin toss. J. Cryptol. **29**(3), 491–513 (2016)
52. Moran, T., Orlov, I., Richelson, S.: Topology-hiding computation. In: Dodis, Y., Nielsen, J.B. (eds.) TCC 2015. LNCS, vol. 9014, pp. 159–181. Springer, Heidelberg (2015). https://doi.org/10.1007/978-3-662-46494-6_8
53. Reiter, M.K., Rubin, A.D.: Anonymous web transactions with crowds. Commun. ACM **42**(2), 32–38 (1999)
54. Spring, N.T., Mahajan, R., Wetherall, D.: Measuring ISP topologies with Rocketfuel. In: Proceedings of SIGCOMM 2002 (2002)
55. Syverson, P.F., Goldschlag, D.M., Reed, M.G.: Anonymous connections and onion routing. In: 1997 IEEE Symposium on Security and Privacy, 4–7 May 1997, Oakland, CA, USA, pp. 44–54 (1997)
56. Yao, A.C.: Protocols for secure computations. In: Proceedings of the 23rd IEEE Symposium on Foundations of Computer Science (FOCS), pp. 160–164 (1982)

Isogeny

Supersingular Isogeny Graphs and Endomorphism Rings: Reductions and Solutions

Kirsten Eisenträger[1(✉)], Sean Hallgren[2], Kristin Lauter[3], Travis Morrison[1], and Christophe Petit[4]

[1] Department of Mathematics, The Pennsylvania State University, University Park, USA
eisentra@math.psu.edu

[2] Department of Computer Science and Engineering, The Pennsylvania State University, University Park, USA

[3] Microsoft Research, Redmond, USA

[4] University of Birmingham, Birmingham, UK

Abstract. In this paper, we study several related computational problems for supersingular elliptic curves, their isogeny graphs, and their endomorphism rings. We prove reductions between the problem of path finding in the ℓ-isogeny graph, computing maximal orders isomorphic to the endomorphism ring of a supersingular elliptic curve, and computing the endomorphism ring itself. We also give constructive versions of Deuring's correspondence, which associates to a maximal order in a certain quaternion algebra an isomorphism class of supersingular elliptic curves. The reductions are based on heuristics regarding the distribution of norms of elements in quaternion algebras.

We show that conjugacy classes of maximal orders have a representative of polynomial size, and we define a way to represent endomorphism ring generators in a way that allows for efficient evaluation at points on the curve. We relate these problems to the security of the Charles-Goren-Lauter hash function. We provide a collision attack for special but natural parameters of the hash function and prove that for general parameters its preimage and collision resistance are also equivalent to the endomorphism ring computation problem.

This paper is the result of a merge of [EHM17,PL17].

The first author was partially supported by National Science Foundation awards DMS-1056703 and CNS-1617802, and by the National Security Agency (NSA) under Army Research Office (ARO) contract number W911NF-12-1-0541.

The second author was partially supported by National Science Foundation awards CNS-1617802 and CCF-1618287, and by the National Security Agency (NSA) under Army Research Office (ARO) contract number W911NF-12-1-0541.

The fourth author was partially supported by National Science Foundation grants DMS-1056703 and CNS-1617802.

J. B. Nielsen and V. Rijmen (Eds.): EUROCRYPT 2018, LNCS 10822, pp. 329–368, 2018.
https://doi.org/10.1007/978-3-319-78372-7_11

1 Introduction

The recent search for new "post-quantum" cryptographic primitives and the ongoing international PQC competition sponsored by NIST has motivated a new era of research in the mathematics of cryptography. Ideas for cryptographic primitives based on hard mathematical problems are being actively proposed and examined. This paper focuses on supersingular isogeny-based cryptography, and in particular on the hardness of computing endomorphism rings of supersingular elliptic curves and its possible applications in cryptography.

In 2006, Charles et al. [CGL06, CGL09] introduced the hardness of finding paths in Supersingular Isogeny Graphs into cryptography and used it for constructing cryptographic hash functions. In the CGL hash function, preimage resistance relies on the hardness of computing certain ℓ-power isogenies (for ℓ a small prime) between supersingular elliptic curves. Since then, this problem and related hard problems have been used as the basis for key exchange protocols [JDF11], signature schemes [YAJ+17, GPS17], and public key encryption [DFJP14]. There is also a submission [ACC+17] to the PQC standardization competition based on supersingular isogeny problems. While polynomial-time quantum algorithms are known for attacking widely deployed public key cryptosystems such as RSA and Elliptic Curve Cryptography (ECC), there are currently no known subexponential quantum attacks against these supersingular isogeny graph-based schemes.

In the supersingular case three problems have emerged as potential computational hardness assumptions related to the above systems. The first is computing isogenies between supersingular elliptic curves, the second one is computing the endomorphism ring of a supersingular elliptic curve, and the third is to compute a maximal order isomorphic to the endomorphism ring of a supersingular elliptic curve. In order to develop confidence that these new systems are secure against quantum computers, it is important to understand these problems, their relationships, and how they relate to the cryptosystems. The natural way to do this is to give polynomial-time reductions between the problems when possible, and there are heuristics for doing this [Koh96, KLPT14]. However, one quickly runs into problems when attempting to find efficient reductions. For example, the main parameter for these problems is a large prime p, and it is not obvious that the endomorphism ring of an elliptic curve even has a basis with a representation size that is polynomial in $\log p$. The same problem exists for maximal orders.

The computational hardness assumption introduced in [CGL09] which underlies the security of Supersingular Isogeny Graph-based cryptography can be equivalently described as finding paths in the isogeny graph or as producing an ℓ-power isogeny (for ℓ a small prime) between two given supersingular elliptic curves. However, there exists another language to describe this problem, thanks to Deuring's correspondence [Deu41], which establishes (non-constructively) a one-to-one correspondence between supersingular j-invariants and maximal orders in a quaternion algebra, up to some equivalence relations. Following this correspondence, path-finding in the Supersingular Isogeny Graph can be translated, in theory, into a problem involving maximal orders in quaternion algebras which was solved in [KLPT14]. So this motivates the problem of finding explicit versions of Deuring's correspondence, namely constructive, efficient algorithms to translate j-invariants into maximal orders in the quaternion algebra and conversely.

1.1 Contributions

Section 2 introduces preliminary material on supersingular elliptic curves and the arithmetic of quaternion algebras, and we recall some well-known facts from [Mes86, Piz80, Wat69], with an emphasis on explicit computations and representations. We state several problems for supersingular elliptic curves in Sect. 3. In Sect. 4, we show that an isomorphism class of maximal orders in a quaternion algebra has at least one representative of polynomial size. Since computing maximal orders is one of the central problems we consider, such a theorem is necessary to have meaningful polynomial-time reductions. The results in Sect. 4 are conditional on GRH but do not use any heuristics. In Sect. 6.4, we construct the quaternion algebra analogue of a factorization of an isogeny of ℓ-power degree into degree ℓ isogenies. The results in that section do not use any heuristics and are unconditional. The construction of Sect. 6.4 is used in our reductions between algorithms involving maximal orders and paths in the ℓ-isogeny graph in Sects. 5 and 6.

Section 5 reduces three hard problems in supersingular graphs to each other: a constructive version of Deuring's correspondence from j-invariants to maximal orders in $B_{p,\infty}$ (Problem 2); the endomorphism ring computation problem (Problem 3); and the preimage and collision resistance of the Charles-Goren-Lauter hash function, for a randomly chosen initial vertex. These reductions rely on various heuristic assumptions underlying the quaternion ℓ-isogeny algorithm of [KLPT14] and its powersmooth version described explicitly in [GPS17], along with new heuristics about using loops in the isogeny graph to generate endomorphism rings.

Section 6 shows that constructing paths in the ℓ-isogeny graph reduces to a different type of endomorphism ring computation. However, instead of just requiring an algorithm for computing the maximal order, one also needs to know how the generators of the order act on the ℓ-torsion of the curve. Thus this section contains a reduction to a harder problem. On the other hand, this section removes some of the heuristics used in Sect. 5. More precisely, the reductions in Sect. 5 use both the quaternion ℓ-isogeny algorithm and its powersmooth version, whereas the reductions in Sect. 6 only use the quaternion ℓ-isogeny algorithm [KLPT14].

Intuitively these heuristics say that numbers generated by the norm form of a quaternion algebra in the algorithm behave in the same way as random numbers of the same size, with respect to their factorization patterns.

Section 7 provides a (heuristic) probabilistic polynomial-time algorithm for computing the Deuring correspondence in one direction, and a partial attack on a special case of the Charles-Goren-Lauter hash function. In Sect. 8, we start by defining the notion of a compact representation of an endomorphism, which has as a requirement that it has size polynomial in $\log p$. We prove that every endomorphism ring has a basis specified by compact representations, and that we can evaluate the endomorphism at points using the representation. We then show that the endomorphism problem reduces to computing a maximal order and the Action-on-ℓ-Torsion problem.

1.2 Related Work

The endomorphism ring computation problem and constructive versions of Deuring's correspondence have been studied in the past independently of their cryptographic applications, and all known algorithms for these problems have required exponential time. Computing the endomorphism ring of a supersingular elliptic curve was first studied by Kohel [Koh96, Theorem 75], who gave an approach for finding four linearly independent endomorphisms, generating a finite-index subring of $\mathrm{End}(E)$. The algorithm was based on finding loops in the ℓ-isogeny graph of supersingular elliptic curves, and the running time of the probabilistic algorithm is $O(p^{1+\varepsilon})$. Another problem that has been considered is to list all isomorphism classes of supersingular elliptic curves together with a description of the maximal order in a quaternion algebra that is isomorphic to $\mathrm{End}(E)$. This was done in [Cer04,LM04] and improved in [CG14, Sect. 5.2]. However, this approach is necessarily exponential in $\log p$ because there are roughly $\lfloor p/12 \rfloor$ isomorphism classes of supersingular elliptic curves.

The problem of computing isogenies between supersingular elliptic curves has also been studied, both in the classical setting [DG16, Sect. 4] where the complexity of the algorithm is $\tilde{O}(p^{1/2})$, and in the quantum setting [BJS14], where the complexity is $\tilde{O}(p^{1/4})$.

A signature scheme based on endomorphism ring computation is given in [GPS17, Sect. 4], where the secret key is a maximal order isomorphic to the endomorphism ring of a supersingular elliptic curve. While the scheme in [DFJP14] had to reveal auxiliary points, this is not necessary in this scheme.

Recently there have been several partial attacks on isogeny-based protocols (see [GPST16,Ti17,GW17]). These attacks target the key exchange protocol of Jao-De Feo [JDF11] in specific attack models, such as fault attacks, and are complementary to our work.

2 Preliminaries

2.1 Background on Elliptic Curves

Elliptic Curves and Isogenies. By an elliptic curve E over a field k of characteristic $p > 3$ we mean a curve with equation $E : y^2 = x^3 + Ax + B$ for some $A, B \in k$ satisfying $4A^3 + 27B^2 \neq 0$. The points of E are the points (x, y) satisfying the curve equation, together with the point at infinity. These points form an abelian group. The j-invariant of an elliptic curve given as above is $j(E) = \frac{256 \cdot 27 \cdot A^3}{4A^3 + 27B^2}$. Two elliptic curves E, E' defined over a field k have the same j-invariant if and only if they are isomorphic over the algebraic closure of k.

We write $j(E)$ for the j-invariant of E. Given a j-invariant $j \neq 0, 1728$, we write $E(j)$ for the curve defined by the equation

$$y^2 + xy = x^3 - \frac{36}{j - 1728}x - \frac{1}{j - 1728}.$$

Such a curve can be put into a short Weierstrass equation $y^2 = x^3 + Ax + B$. We also write $E(0)$ and $E(1728)$ for the curves with equations $y^2 = x^3 + 1$ and $y^2 = x^3 + x$ respectively.

Let E_1 and E_2 be elliptic curves defined over a field k of positive characteristic p. An *isogeny* $\varphi : E_1 \to E_2$ defined over k is a non-constant rational map defined over k which is also a group homomorphism from $E_1(k)$ to $E_2(k)$ [Sil09, III.4]. The degree of an isogeny is its degree as a rational map. When the degree d of the isogeny φ is coprime to p, then φ is separable and the kernel of φ is a subgroup of the points on E_1 of size d. Every isogeny of degree n greater than one can be factored into a composition of isogenies of prime degrees such that the product of the degrees equals n. If $\psi : E_1 \to E_2$ is an isogeny of degree d, the *dual isogeny* of ψ is the unique isogeny $\widehat{\psi} : E_2 \to E_1$ satisfying $\psi\widehat{\psi} = [d]$, where $[d] : E_1 \to E_1$ is the multiplication-by-d map.

We can describe an isogeny via its kernel. Given an elliptic curve E and a finite subgroup H of E, there is, up to isomorphism a unique isogeny $\varphi : E \to E'$ having kernel H (see [Sil09, III.4.12]). Hence we can describe an isogeny of E to some other elliptic curve by giving its kernel. We can compute equations for the isogeny from its kernel by using Vélu's formula [Vél71].

Endomorphisms and Supersingular Versus Ordinary Curves. An isogeny of an elliptic curve E to itself is called an endomorphism of E. If E is defined over some finite field $\mathbb{F}_q$, then an endomorphism of E will be defined over a finite extension of $\mathbb{F}_q$. The set of endomorphisms of E defined over $\overline{\mathbb{F}_q}$ together with the zero map form a ring under the operations addition and composition. It is called the endomorphism ring of E, and is denoted by $\mathrm{End}(E)$. When E is defined over a finite field, then $\mathrm{End}(E)$ is isomorphic either to an order in a quadratic imaginary field or to an order in a quaternion algebra. In the first case we call E an *ordinary elliptic curve*. An elliptic curve whose endomorphism ring is isomorphic to an order in a quaternion algebra is called a *supersingular elliptic curve*. Every supersingular elliptic curve over a field of characteristic p has a model that is defined over $\mathbb{F}_{p^2}$ because the j-invariant of such a curve is in $\mathbb{F}_{p^2}$.

ℓ-Power Isogenies Between Supersingular Elliptic Curves. Let E, E' be two supersingular elliptic curves defined over $\mathbb{F}_{p^2}$. It is a fact that for each prime $\ell \neq p$, E and E' are connected by a chain of isogenies of degree ℓ [Mes86]. By [Koh96, Theorem 79], E and E' can be connected by m isogenies of degree ℓ, where $m = O(\log p)$. So any two supersingular elliptic curves can be connected by an isogeny of degree ℓ^m with $m = O(\log p)$. If $\ell = O(\log p)$ is a fixed prime, then any ℓ-isogeny in the chain above can either be specified by rational maps or by giving the kernel of the isogeny, and both of these representations will have polynomial size in $\log p$. By Vélu's formula, and since $\ell = O(\log p)$, there is an efficient way to go back and forth between these two representations.

2.2 Quaternion Algebras, $B_{p,\infty}$ and the Deuring Correspondence

Quaternion Algebras. For $a, b \in \mathbb{Q}^\times$, let $H(a, b)$ denote the quaternion algebra over $\mathbb{Q}$ with basis $1, i, j, ij$ such that $i^2 = a$, $j^2 = b$ and $ij = -ji$. That is,

$$H(a, b) = \mathbb{Q} + \mathbb{Q}i + \mathbb{Q}j + \mathbb{Q}ij.$$

It is a fact that any quaternion algebra over $\mathbb{Q}$ can be written in this form. Now let $B_{p,\infty}$ be the unique quaternion algebra over $\mathbb{Q}$ that is ramified exactly at p and ∞. Then $B_{p,\infty}$ is a definite quaternion algebra, so $B_{p,\infty} = H(a, b)$ for some $a, b \in \mathbb{Q}^\times$, and one can show a and b can be chosen to be negative integers. For example, when $p \equiv 3 \pmod 4$, then $B_{p,\infty} = H(-p, -1)$.

There is a *canonical involution* on $B_{p,\infty}$ which sends an element $\alpha = a_1 + a_2 i + a_3 j + a_4 ij$ to $\overline{\alpha} := a_1 - a_2 i - a_3 j - a_4 ij$. Define the *reduced trace* of an element α as above to be

$$\mathrm{Trd}(\alpha) = \alpha + \overline{\alpha} = 2a_1,$$

and the *reduced norm* to be

$$\mathrm{Nrd}(\alpha) = \alpha\overline{\alpha} = a_1^2 - a a_2^2 - b a_3^2 + ab a_4^2.$$

We say that Λ is a *lattice* in $B_{p,\infty}$ if $\Lambda = \mathbb{Z}x_1 + \cdots + \mathbb{Z}x_4$ and the elements $x_1, \ldots, x_4$ are a vector space basis for $B_{p,\infty}$.

If $I \subseteq B_{p,\infty}$ is a lattice, the reduced norm of I, $\mathrm{Nrd}(I)$, is the positive generator of the fractional $\mathbb{Z}$-ideal generated by $\{\mathrm{Nrd}(\alpha) : \alpha \in I\}$. The quaternion algebra $B_{p,\infty}$ is an inner product space with respect to the bilinear form

$$\langle x, y \rangle = \frac{\mathrm{Nrd}(x + y) - \mathrm{Nrd}(x) - \mathrm{Nrd}(y)}{2}.$$

The basis $\{1, i, j, ij\}$ is an orthogonal basis with respect to this inner product.

Orders in $B_{p,\infty}$ and Representation of Elements in $B_{p,\infty}$. An *order* $\mathcal{O}$ of $B_{p,\infty}$ is a subring of $B_{p,\infty}$ which is also a lattice, and if $\mathcal{O}$ is not properly contained in any other order, we call it a *maximal order*. For a lattice $I \subseteq B_{p,\infty}$ we define

$$\mathcal{O}_R(I) := \{x \in B_{p,\infty} : Ix \subseteq I\}$$

to be the *right order of the lattice* I, and we similarly define its left order $\mathcal{O}_L(I)$. If $\mathcal{O}$ is a maximal order in $B_{p,\infty}$ and $I \subseteq \mathcal{O}$ is a left ideal of $\mathcal{O}$, then $\mathcal{O}_R(I)$ is also a maximal order. Given any two maximal orders $\mathcal{O}, \mathcal{O}'$, there is a lattice $I \subseteq B_{p,\infty}$ such that $\mathcal{O}_L(I) = \mathcal{O}$ and $\mathcal{O}_R(I) = \mathcal{O}'$; we say that I connects $\mathcal{O}$ and $\mathcal{O}'$.

An element $\beta \in B_{p,\infty}$ is represented as a coefficient vector (a_1, a_2, a_3, a_4) in $\mathbb{Q}^4$ such that $\beta = a_1 + a_2 i + a_3 j + a_4 ij$ in terms of the basis $\{1, i, j, ij\}$ for $B_{p,\infty}$. This will be used for specifying basis elements of maximal orders $\mathcal{O}$ and elements of left ideals I of $\mathcal{O}$.

The Deuring Correspondence and Describing Isogenies via Kernel Ideals. For a detailed overview of the information in this section, see Chap. 42 in [Voi]. Let E be a supersingular elliptic curve defined over $\mathbb{F}_{p^2}$. In [Deu41] Deuring proved that the endomorphism ring of E is isomorphic to a maximal order in $B_{p,\infty}$. Under this isomorphism, degrees and traces of endomorphisms correspond to norms and traces of quaternions. The correspondence between isomorphism classes of supersingular elliptic curves and maximal orders is often referred to as Deuring's correspondence.

Fix E, a supersingular elliptic curve over $\mathbb{F}_{p^2}$. We can associate to each pair (E', ϕ) with ϕ an isogeny $E \to E'$ of degree n a left $\mathrm{End}(E)$-ideal $I = \mathrm{Hom}(E', E)\phi$ of norm n, and it was shown in [Koh96, Sect. 5.3] that every left $\mathrm{End}(E)$-ideal arises in this way. We now describe how to construct an isogeny from a left $\mathrm{End}(E)$-ideal.

Let I be a nonzero integral left ideal of $\mathrm{End}(E)$. Define $E[I]$ to be the scheme-theoretic intersection

$$E[I] = \bigcap_{\alpha \in I} \ker(\alpha).$$

Thus to each left ideal I of $\mathrm{End}(E)$ there is an associated isogeny $\phi_I : E \to E/E[I]$. If $\mathrm{Nrd}(I)$ is coprime to p, then

$$E[I] = \{P \in E(\overline{\mathbb{F}}_{p^2}) : \alpha(P) = 0 \quad \forall \alpha \in I\}.$$

2.3 Supersingular Isogeny Graphs

For any prime $\ell \neq p$, one can construct a so-called *ℓ-isogeny graph*, where each vertex is associated to a supersingular j-invariant, and an edge between two vertices is associated to a degree ℓ isogeny between the corresponding curves. Isogeny graphs are regular with regularity degree $\ell + 1$; they are directed graphs (unless $p \equiv 1 \pmod{12}$). Isogeny graphs are Ramanujan, i.e. they are optimal *expander graphs*, with the consequence that random walks on the graph quickly reach the uniform distribution [HLW06].

2.4 The Charles-Goren-Lauter Hash Function

The first cryptographic construction based on supersingular isogeny problems is a hash function proposed by Charles, Goren and Lauter [CGL09]. The security of this construction relies on the hardness of computing some isogenies of special degrees between two supersingular elliptic curves.

More precisely, consider an ℓ-isogeny graph over $\mathbb{F}_{p^2}$, where p is a "large" prime and ℓ is a "small" prime. The authors suggest to take $p \equiv 1 \pmod{12}$ to avoid some annoying backtracking issues. The message is first mapped into $\{0,\ldots,\ell-1\}^*$, with some padding if necessary. At each vertex, a deterministic ordering of the edges is fixed (this can be done by sorting the j-invariants of the $\ell+1$ neighbors). An initial vertex j_0 is also fixed, as well as an initial incoming direction.

Given a message $(m_1, m_2, \ldots, m_N) \in \{0,\ldots,\ell-1\}^*$, an edge adjacent to j_0 (excluding the incoming edge) is first chosen according to the value of m_1, and the corresponding neighbor E_1 is computed. Then an edge of j_1 (excluding the edge between j_0 and j_1) is chosen according to the value of m_2, and the corresponding neighbor j_2 is computed, etc. The final invariant j_N reached by this computation is mapped to $\{0,1\}^n$ in some deterministic way (here $n \approx \log p$) and the value obtained is returned as the output of the hash function.

Clearly the function is preimage resistant if and only if, given two supersingular j-invariants j_1 and j_2, it is computationally hard to compute a positive integer e and an isogeny $\varphi : E(j_1) \to E(j_2)$ of degree ℓ^e.

In this paper we give two new results on the security of this construction. On the one hand (Sect. 5.5), we show that for a randomly chosen starting point j_0 the function is preimage and collision resistant if and only if the endomorphism ring computation problem is hard: loosely speaking this means computing some endomorphisms of $E(j)$ but not necessarily of the correct norms. The interest of this result lies in that computing endomorphisms of elliptic curves is a natural problem to consider from an algorithmic number theory point of view, and it has indeed been studied since Kohel's thesis in 1996. On the other hand (Sect. 7.2), we also show that the collision resistance problem is easy for some particular starting points.

2.5 Isogeny-Based Cryptography

A few years after Charles, Goren and Lauter designed their hash function, Jao and De Feo proposed a variant of the Diffie-Hellman protocol based on supersingular isogeny problems, which is now known as the supersingular isogeny key exchange protocol [JDF11]. We briefly describe it here in a way to encompass both the original parameters and the generalization recently suggested by Petit [Pet17].

The parameters include a large prime p, a supersingular curve E, and two coprime integers N_A and N_B. Alice and Bob select cyclic subgroups of E of order N_A and N_B, respectively; they compute the corresponding isogenies and they exchange the values of the end vertices, which are E/G_A and E/G_B, respectively. The shared key is the value $j(E/\langle G_A, G_B\rangle)$. This shared key could a priori not be computed by any party from E/G_A, E/G_B and their respective secret keys only, so Alice (resp. Bob) additionally sends the images of a basis of $E[N_B]$ by ϕ_A (resp. a basis of $E[N_A]$ by ϕ_B).

Jao-De Feo suggested to use $N_A = 2^{e_B} \approx p^{1/2} \approx N_B = 3^{e_B}$ such that $(p-1)/N_A N_B$ is a small integer for efficiency reasons; in [Pet17] Petit argued that choosing $N_A \approx N_B \approx p^2$ both powersmooth numbers is a priori better from a security point of view while preserving polynomial-time complexity for the protocol execution. It was shown by Gabraith-Petit-Shani-Ti [GPST16] that computing the endomorphism ring of E and E_A is sufficient to break the key exchange for the parameters suggested by Jao-De Feo. The argument uses the fact that isogenies generated for Jao-De Feo's parameters are of relatively small degree, and this does not seem to apply to Petit's parameters.

The security of Jao-De Feo's protocol relies on the hardness of computing isogenies of a given degree between two given curves, when provided in addition with the action of the isogeny on a large torsion group. This problem is not known to be equivalent to the endomorphism ring computation problem. Recent results by Petit [Pet17] show that revealing the action of isogenies on a torsion group does make some isogeny problems easier to solve, though at the moment his techniques do not apply to Jao-De Feo's original parameters. We *believe* that the security of the key exchange protocol lies between these hard and easy problems, but leave its study to future work.

The interest in isogeny-based cryptography has recently increased in the context of NIST's call for post-quantum cryptography algorithms [NIS16], and a submitted proposal was based on isogeny-based cryptography [ACC+17]. At the moment the best algorithms to solve supersingular isogeny problems all require exponential time in the security parameter, even when including quantum algorithms. Besides the hash function and the key exchange protocols, there are now constructions based on isogeny problems for public key encryption, identification protocols and signatures [DFJP14, YAJ+17, GPS17]. Constructions in the first two papers build on the key exchange protocol and rely on similar assumptions. The second signature scheme in [GPS17], however, only relies on the endomorphism computation problem.

3 Problem Statements and Heuristics

3.1 The Deuring Correspondence

The Deuring correspondence states that

$$\{\mathcal{O} \subseteq B_{p,\infty} \text{ maximal}\} / \simeq \quad \leftrightarrow \{j \in \mathbb{F}_{p^2} : E(j) \text{ supersingular}\} / \mathrm{Gal}(\mathbb{F}_{p^2}/\mathbb{F}_p)$$

is a bijective correspondence, given by associating a supersingular j-invariant to a maximal order in $B_{p,\infty}$ isomorphic to $\mathrm{End}(E(j))$.

In this paper we will be interested in *constructing* Deuring's correspondence for arbitrary maximal orders and supersingular j-invariants. This could a priori have different meanings, given by Problems 1 and 2 below.

Problem 1 (Constructive Deuring Correspondence). *Given a maximal order $\mathcal{O} \subset B_{p,\infty}$, return a supersingular j-invariant such that the endomorphism ring of $E(j)$ is isomorphic to $\mathcal{O}$.*

We refer to the problem of computing a maximal order isomorphic to $\text{End}(E(j))$ for given a supersingular j-invariant as Problem MaxOrder or the "Inverse Deuring Correspondence."

Problem 2 (MaxOrder). *Given p, the standard basis for $B_{p,\infty}$, and a supersingular elliptic curve E defined over $\mathbb{F}_{p^2}$, output vectors β_1, β_2, β_3, $\beta_4 \in B_{p,\infty}$ that form a $\mathbb{Z}$-basis of a maximal order $\mathcal{O}$ in $B_{p,\infty}$ such that $\text{End}(E) \cong \mathcal{O}$. In addition, the output basis is required to have representation size polynomial in $\log p$.*

The j-invariant is naturally represented as an element of $\mathbb{F}_{p^2}$, and it is unique up to Galois conjugation. The maximal order is unique up to conjugation by an invertible quaternion element, and it can be described by a $\mathbb{Z}$-basis, namely four elements $1, \omega_2, \omega_3, \omega_4 \in B_{p,\infty}$ such that $\mathcal{O} = \mathbb{Z} + \omega_2\mathbb{Z} + \omega_3\mathbb{Z} + \omega_4\mathbb{Z}$. Choosing a Hermite basis makes this description unique.

In this paper we will provide a polynomial-time algorithm for Problem 1 (Sect. 7.1). We will also provide explicit connections between Problem 1 and the endomorphism ring computation problem, where instead of a maximal order in $B_{p,\infty}$ one needs to output a basis for $\text{End}(E(j))$.

3.2 The Endomorphism Ring Computation Problem

Given an elliptic curve, it is natural to ask to compute its endomorphism ring.

Problem 3 (Endomorphism ring computation problem). *Given p and a supersingular j-invariant j, compute the endomorphism ring of $E(j)$.*

The endomorphism ring can be returned as four rational maps that form a $\mathbb{Z}$-basis with respect to scalar multiplication (in fact 3 maps, since one of these maps can always be chosen equal to the identity map). The maps themselves can usually not be returned in their canonical expression as rational maps, as in general this representation will require a space larger than the degree, and the degrees can be as big as p.

Various representations of the maps are a priori possible. We believe that any valid representation should be *concise* and *useful*, in the sense that it must require a space polynomial in $\log p$ to store, and it must allow the evaluation of the maps at arbitrary elliptic curve points in a time polynomial in both $\log p$ and the space required to store those points. To the best of our knowledge these two conditions are sufficient for all potential applications of Problem 3. When its degree is a smooth number, an endomorphism can be efficiently represented as a composition of small degree isogenies. In Sect. 5.1 we will consider a more general representation.

A first approximation to a solution to Problem 3 was provided by Kohel in his PhD thesis [Koh96], and later improved by Galbraith [Gal99] using a birthday argument. The resulting algorithm explores a tree in an ℓ-isogeny graph (for some small integer ℓ) until a collision is found, corresponding to an endomorphism. The expected cost of this procedure is $O(\sqrt{p})$ times a polynomial in $\log p$. Repeating this procedure a few times, possibly with different values of ℓ, we obtain a set of endomorphisms which generate a subring of the whole endomorphism ring. The endomorphism ring computation problem was also considered in [DG16] for curves defined over $\mathbb{F}_p$. The identification protocol and signature schemes developed in [GPS17] explicitly rely on its potential hardness for security.

We observe that Problems 2 and 3 take the same input, and their outputs are also "equal" in the sense they are isomorphic. For this reason the two problems have sometimes been referred to interchangeably. In particular, a solution to Problem 2 does not a priori provide a useful description of the endomorphism ring so that one can evaluate endomorphisms at given points. Similarly, a solution to Problem 2 does not a priori provide a $\mathbb{Z}$-basis for an order in $B_{p,\infty}$, and this is necessary to apply the algorithms of [KLPT14].

It turns out that the two problems are equivalent: in Sects. 5.1 and 5.4, we provide efficient algorithms to go from a representation of the endomorphism ring as a $\mathbb{Z}$ basis over $\mathbb{Q}$ to a representation as rational maps and conversely.

In Sects. 6 and 8, our reductions will involve the following problem.

Problem 4 (Action-on-ℓ-Torsion). *Given p, a supersingular elliptic curve E defined over $\mathbb{F}_{p^2}$, and four elements $\{\beta_1, \beta_2, \beta_3, \beta_4\}$ in a maximal order $\mathcal{O}$ of $B_{p,\infty}$ such that there exists an isomorphism $\iota : \mathrm{End}(E) \to \mathcal{O}$, output eight pairs of points on E, (P_1, Q_{1r}), (P_2, Q_{2r}) ($r = 1, \ldots, 4$) such that P_1, P_2 form a basis for the ℓ-torsion $E[\ell]$ of E, and such that $Q_{1r} = \iota^{-1}(\beta_r)(P_1)$ and $Q_{2r} = \iota^{-1}(\beta_r)(P_2)$ for $r = 1, \ldots, 4$.*

The combination of this problem with Problem MaxOrder is, intuitively, to ask for both the algebraic structure of $\mathrm{End}(E)$ (by asking for generators in $B_{p,\infty}$ for a maximal order $\mathcal{O} \simeq \mathrm{End}(E)$, along with a small amount of geometric information, meaning asking for how those generators act as endomorphisms on $E[\ell]$.

Finally, we will be relating these various endomorphism ring problems to pathfinding in the ℓ-isogeny graph, which we often refer to as preimage resistance for the Charles-Goren-Lauter has function or Problem ℓ-PowerIsogeny.

Problem 5 (ℓ-PowerIsogeny). *Given a prime p, along with two supersingular elliptic curves E and E' over $\mathbb{F}_{p^2}$, output an isogeny from E to E' represented as a chain of k isogenies whose degrees are ℓ.*

Since E is given as $y^2 = x^3 + ax + b$ with $a, b \in \mathbb{F}_{p^2}$, the input size for this problem is $O(\log p)$. By Sect. 2.1, the representation size of the output is also polynomial in $\log p$, if $\ell \in O(\log p)$ and the isogenies are represented by rational maps.

Below we map out the various reductions in this paper. An arrow represents the reduction from one problem to another, and its label indicates the algorithm or theorem giving that reduction.

Pathfinding in ℓ-isogeny graph $\xrightarrow{\text{Algorithm 7}}$ Endomorphism Ring $\xrightarrow{\text{Algorithm 6}}$ Max Order

Pathfinding in ℓ-isogeny graph $\xleftarrow{\text{Algorithm 8}}$ Endomorphism Ring $\xleftarrow{\text{Algorithm 4}}$ Max Order

Pathfinding in ℓ-isogeny graph $\xrightarrow{\text{Algorithm 9}}$ Max Order and Action on ℓ-Torsion

Endomorphism Ring $\xrightarrow{\text{Theorem 16}}$ Max Order and Action on ℓ-Torsion

3.3 Heuristics

Our reductions require several heuristics related to the distribution of numbers represented by certain quadratic forms and on isogeny graphs. When we refer to plausible heuristic assumptions, we mean one or more of the following:

1. We assume the heuristics used in [KLPT14], which can be summarized as saying that the distribution of outputs of quadratic forms arising from the norm form of a maximal order in $B_{p,\infty}$ is approximately like the uniform distribution on numbers of the same size.
2. We also assume the heuristics used in [GPS17] on representing powersmooth numbers by these quadratic forms.
3. We assume that the endomorphism ring of an elliptic curve can be generated by endomorphisms arising from loops in the ℓ-isogeny graph. In particular, we assume that given a suborder $\mathcal{O}'$ of a maximal order $\mathcal{O}$ such that $\mathcal{O}'$ is generated by loops in an ℓ-isogeny graph, the probability that a randomly generated loop in the graph is in $\mathcal{O}'$ is inversely proportional to $[\mathcal{O} : \mathcal{O}']$.

4 Efficient Computations with Maximal Orders and Their Ideals

One of the main problems we consider in this paper is computing a maximal order associated to an elliptic curve E. The following sections will show that computing isogenies and computing endomorphisms reduces to computing maximal orders, together with a problem about ℓ-torsion action. In this section we show that maximal orders have polynomial-representation size, so that the reductions are meaningful. We will also show that the representation size of ideals inside these orders is related to their norms. Maximal orders are inside the algebra $B_{p,\infty}$, so we start with that.

Let p be a prime. In Proposition 5.1 of [Piz80] it is shown that $B_{p,\infty} = H(-1,-1)$ if $p = 2$, $B_{p,\infty} = H(-1,-p)$ if $p \equiv 3 \pmod 4$, $B_{p,\infty} = H(-2,-p)$ if $p \equiv 5 \pmod 8$, and $B_{p,\infty} = H(-q,-p)$ if $p \equiv 1 \pmod 8$, where $q \equiv 3 \pmod 4$ is prime and p is not a square modulo q.

So given p, we choose a and b as above (depending on the congruence class of p) such that $B_{p,\infty} = H(a,b)$. We obtain a basis $1, i, j, ij$ for $B_{p,\infty}$ such that $i^2 = a$ and $j^2 = b$. We refer to this as the *standard basis* of $B_{p,\infty}$. As stated in Sect. 2.2, we represent elements of $B_{p,\infty}$ as their coefficient vectors in $\mathbb{Q}^4$ with respect to the standard basis.

To reduce problems to Problem MaxOrder in polynomial time, one requirement is that in every conjugacy class there is a maximal order that has a basis with representation size that is polynomial in $\log p$. Since a prime p is given, and E is given as $y^2 = x^3 + ax + b$ with $a, b \in \mathbb{F}_{p^2}$, the input size for this problem is $O(\log p)$.

To show that there is a maximal order that has a polynomial representation size, we first show this is true for a special maximal order $\mathcal{O}_0$ and then express all other classes of maximal orders as right orders $\mathcal{O}_R(I)$ for a left ideal I of $\mathcal{O}_0$. Since every left ideal class of $\mathcal{O}_0$ contains an ideal whose reduced norm is $O(p^2)$, it will follow that in each conjugacy class of maximal orders, there is one with polynomial representation size.

As mentioned above, Pizer [Piz80] gave the following explicit description of $B_{p,\infty}$ for all p along with a basis for one maximal order.

Proposition 1. *Let $p > 2$ be a prime. Then we can define $B_{p,\infty}$ and a maximal order $\mathcal{O}_0$ as follows:*

p	(a, b)	$\mathcal{O}_0$
$3 \pmod 4$	$(-p, -1)$	$\langle 1, j, \frac{j+k}{2}, \frac{1+i}{2} \rangle$
$5 \pmod 8$	$(-p, -2)$	$\langle 1, j, \frac{2-j+k}{4}, \frac{-1+i+j}{2} \rangle$
$1 \pmod 8$	$(-p, -q)$	$\langle \frac{1+j}{2}, \frac{i+k}{2}, \frac{j+ck}{q}, k \rangle$

where in the last row $q \equiv 3 \pmod 4$, $(p/q) = -1$ and c is some integer with $q | c^2p+1$. Assuming that the generalized Riemann hypothesis is true, there exists $q = O(\log^2 p)$ satisfying these conditions.

Proof. The information in the table follows from [Piz80, pp. 368–369]. The only thing we need to prove is the statement that when $p \equiv 1 \pmod 8$ there exists a prime $q \equiv 3 \pmod 4$ such that $\left(\frac{p}{q}\right) = -1$. Equivalently, we require that q be an unramified prime which does not split in either $K_1 = \mathbb{Q}(\sqrt{p})$ or $K_2 = \mathbb{Q}(\sqrt{-1})$.

This is equivalent to the condition that the Frobenius of q in $\mathrm{Gal}(K_1K_2/\mathbb{Q})$ is the unique automorphism which restricts to the nontrivial automorphisms of $\mathrm{Gal}(K_1/\mathbb{Q})$ and $\mathrm{Gal}(K_2/\mathbb{Q})$. By [LO77], there is a prime q of size $O((\log |D|)^2)$ whose Frobenius is this element, where D is the absolute discriminant of the compositum $K_1K_2/\mathbb{Q}$. The absolute discriminant of $K_1/\mathbb{Q}$ is p since $p \equiv 1 \pmod 4$, and the absolute discriminant of $K_2/\mathbb{Q}$ is -4. Because $(4, p) = 1$, we have that $\mathcal{O}_{K_1K_2} = \mathcal{O}_{K_1}\mathcal{O}_{K_2}$, and using this, a computation shows that $D = \mathrm{Disc}(K_1K_2/\mathbb{Q}) = 4^2p^2$. Hence $q = O(\log^2 p)$, as desired. □

We stress that in all cases the maximal orders $\mathcal{O}_0$ given by Proposition 1 contain $\langle 1, i, j, k \rangle$ as a small index subring.

For the remainder of this section, fix such an order $\mathcal{O}_0$ together with the small basis $\{b_1, \ldots, b_4\}$ as in Proposition 1. We will now show that ideals of $\mathcal{O}_0$ of norm N have representations of size polynomial in $\log(N)$ in terms of the basis $\{b_1, \ldots, b_4\}$.

Lemma 1. *Let I be a left ideal of $\mathcal{O}_0$. Then there is a $\mathbb{Z}$-basis $\{\alpha_1, \ldots, \alpha_4\}$ of I, consisting of elements $\alpha_i \in \mathcal{O}_0$, such that the coefficients of the α_i expressed, in terms of the basis $\{b_1, b_2, b_3, b_4\}$ of $\mathcal{O}_0$, are bounded by* $\mathrm{Nrd}(I)^2$.

Proof. Let $\{\gamma_1, \ldots, \gamma_4\}$ be a $\mathbb{Z}$-basis of I and write γ_i as $\gamma_i = \sum_j a_{ij} b_j$. Let $A = (a_{ij})$ be the matrix whose rows are the coefficients of γ_i. Let $H = UA$ where H is the (row-)Hermite normal form of A and $U \in \mathrm{SL}_4(\mathbb{Z})$. Then the rows of H correspond to elements of $\mathcal{O}_0$ which generate I as a $\mathbb{Z}$-basis. Additionally, H is upper triangular, its diagonal elements satisfy $0 < h_{ii}$, and $h_{ij} < h_{jj}$ for $i < j$. We have $\mathrm{Nrd}(I)^2 = \det(A) = \prod h_{ii}$ and hence all $h_{ij} < \mathrm{Nrd}(I)^2$. This gives us the desired basis $\{\alpha_1, \ldots, \alpha_4\}$. □

We will now prove that every conjugacy class of maximal orders has a representative whose basis has representation size $O(\log p)$ when written in terms of the standard basis $1, i, j, ij$ for $B_{p,\infty}$.

For this, we will show that the reduced norm Nrd is the Euclidean norm on $B_{p,\infty} = H(-q, -p)$ considered as a lattice in $\mathbb{R}^4$. (Here $q = 1, 2$ or a prime $\equiv 3 \pmod 4$ that is not a square modulo p, depending on the congruence class of p.) We can view orders $\mathcal{O}$ in $B_{p,\infty}$ as lattices in $\mathbb{R}^4$, and we will relate the covolume of a lattice to its discriminant. This is similar to the number field case. Together with Minkowski's Theorem, this will give us the desired result.

Note that $B_{p,\infty} \otimes \mathbb{R}$ is isomorphic to $\mathbb{H}$, the Hamiltonians. Let $1, i', j', i'j'$ be the basis of $\mathbb{H}$ with $i'^2 = j'^2 = -1$. Let

$$f : B_{p,\infty} \otimes \mathbb{R} \xrightarrow{\simeq} \mathbb{H},$$

and let the isomorphism be given by $i \mapsto \sqrt{q}i'$, $j \mapsto \sqrt{p}j'$. Then the norm on $\mathbb{H}$, which is the (square of) the standard Euclidean norm on $\mathbb{R}^4$, is just the reduced norm on the image of $B_{p,\infty}$ in $\mathbb{H}$ under the isomorphism f. Let $\Lambda \subseteq \mathbb{R}^n$ be a lattice. Define its *covolume*, denoted $\mathrm{Covol}(\Lambda)$, to be $\sqrt{\det(L^T L)}$ for any matrix L consisting of a basis for Λ. If $\mathcal{O} \subseteq B_{p,\infty}$ is a lattice, define its covolume to be $\mathrm{Covol}(f(\mathcal{O}))$.

If a lattice $\mathcal{O} \subseteq B_{p,\infty}$ has generators $\beta_1, \ldots, \beta_4$, its *discriminant*, denoted $\mathrm{Disc}(\mathcal{O})$, is $\det((\mathrm{Trd}(\beta_i \overline{\beta_j})))$. If a lattice $\mathcal{O}$ is a maximal order in $B_{p,\infty}$, then $\mathrm{Disc}(\mathcal{O}) = p^2$.

Proposition 2. *Let $\mathcal{O}$ be a lattice in $B_{p,\infty}$. Then* $\mathrm{Covol}(\mathcal{O})^2 = \frac{1}{16}\mathrm{Disc}(\mathcal{O})$.

Proof. This is Eq. 2.2 of [CG14]. □

We need the notion of a Minkowski-reduced basis. A basis $\{v_1, \ldots, v_n\}$ of a lattice $\Lambda \subseteq \mathbb{R}^n$ is *Minkowski-reduced* if for $1 \leq k \leq n$,

$$||v_k||_2 \leq \left| \sum_{i=1}^{n} x_i ||v_i||_2 \right|,$$

whenever $x_1, \ldots, x_n$ are coprime integers. Here $||\cdot||_2$ denotes the Euclidean norm. Given a lattice Λ in $\mathbb{R}^n$, define the *ith successive minimum of* Λ, $\lambda_i(\Lambda)$, to be the smallest nonnegative, real number r such that there are i linearly independent lattice vectors of Λ contained in the closed ball of radius r centered at the origin. So $\lambda_1(\Lambda)$ is the length of a shortest nonzero vector of Λ. For $n \leq 4$, there is a basis $v_1, \ldots, v_n$ of Λ such that $||v_i||_2 = \lambda_i(\Lambda)$; see [NS09]. Such a basis is Minkowski-reduced. When we refer to a Minkowski-reduced basis, we will always assume we choose such a basis.

Theorem 1 (Minkowski's second theorem). *Let V denote the volume of the n-dimensional unit ball of $\mathbb{R}^n$. Then*

$$\frac{2^n}{n!}\frac{\mathrm{Covol}(\Lambda)}{V} \leq \prod_{i=1}^{n} \lambda_i(\Lambda) \leq \frac{2^n}{V}\mathrm{Covol}(\Lambda).$$

Corollary 1. *Let p be a prime, and let $\mathcal{O}_0$ be the maximal order of $B_{p,\infty}$ as above. Let $I \subseteq \mathcal{O}_0$ be a left ideal and let $\mathcal{O} := \mathcal{O}_R(I)$. Let $\alpha_1, \ldots, \alpha_4$ be a basis of $\mathcal{O}$ such that $||\alpha_i||_2 = \lambda_i(\mathcal{O})$ for $i = 1, \ldots, 4$. Then*

$$\prod_{i=1}^{4} \mathrm{Nrd}(\alpha_i) \leq \mathrm{Disc}(\mathcal{O}) = p^2.$$

Proof. We use Minkowski's second theorem applied to $\mathcal{O}$, and the fact that by Proposition 2, $\mathrm{Covol}(\mathcal{O})^2 = \mathrm{Disc}(\mathcal{O})/16$. These two facts, together with $\mathrm{Nrd}(\alpha) = ||f(\alpha)||_2^2$ give us that

$$\prod \mathrm{Nrd}(\alpha_i) = \prod \lambda_i(\mathcal{O})^2 \leq \frac{16}{\pi^4/4}\mathrm{Disc}(\mathcal{O}) \leq p^2.$$

□

Now we prove the main theorem on representation sizes of maximal orders:

Theorem 2. *Every conjugacy class of maximal orders in $B_{p,\infty}$ has a $\mathbb{Z}$-basis $x_1, \ldots, x_4$ with $\mathrm{Nrd}(x_i) \in O(p^2)$. If we express x_r (for $1 \leq r \leq 4$) as a coefficient vector in terms of $1, i, j, ij$, then the rational numbers appearing have numerators and denominators whose representation size are polynomial in $\log p$.*

Proof. The map $[I] \to [\mathcal{O}_R(I)]$ is a surjection from left ideal classes of $\mathcal{O}_0$ to isomorphism classes of maximal orders of $B_{p,\infty}$; see [Gro87], page 116. Every left ideal class of $\mathcal{O}_0$ contains an ideal I with $\mathrm{Nrd}(I) \in O(p^2)$; see [Vig80, Proposition 17.5.6]. Set $\mathcal{O} = \mathcal{O}_R(I)$ and let $\langle 1, x_2, x_3, x_4 \rangle$ be a Minkowski-reduced $\mathbb{Z}$-basis of $\mathcal{O}$. By Corollary 1, $\mathrm{Nrd}(x_i) \leq p^2$, since each x_i is integral. Since $\mathcal{O} = \mathcal{O}_R(I)$, it follows that $x_i \mathrm{Nrd}(I) \in I$. This implies that if we express x_i as a $\mathbb{Q}$-linear combination of the elements $1, i, j, ij$, then the denominators of the coefficients are divisors of $\mathrm{Nrd}(I) \cdot 4q$ where $q = \mathrm{Nrd}(j)$. The numerator of each coefficient is then bounded by $8pq \mathrm{Nrd}(I)$: indeed, if a/b is a coefficient of x_r, $(1 \leq r \leq 4)$, then $(a/b)^2 \leq \mathrm{Nrd}(x_r) \leq p^2$. Then

$$|a| \leq pb \leq 4pq \mathrm{Nrd}(I).$$

□

5 Equivalent Hard Problems in Supersingular Isogeny Graphs

In this section we consider the following problems:

- A constructive version of Deuring's correspondence, from j-invariants to maximal orders in $B_{p,\infty}$ (Problem 2).
- The endomorphism ring computation problem (Problem 3).
- The preimage and collision resistance of the Charles-Goren-Lauter hash function, for a randomly chosen initial vertex.

We show that all these problems are heuristically equivalent, in the sense that there exist efficient reductions from one problem to another under plausible heuristics assumptions.

The first two problems have the same inputs and in a sense their outputs are also equal, so it is perhaps no surprise to the reader that they are equivalent. However, the two problems differ in the way the output should be represented: as a maximal order in $B_{p,\infty}$ for Problem 2, and as four rational maps for Problem 3. Sections 5.1 and 5.4 below clarify the steps from one representation to the other.

It should also be clear intuitively that (heuristically at least) an algorithm to find preimages or collisions for the hash function can be used to compute endomorphism rings. The other implication is perhaps not as intuitive, and our solution crucially requires the tools developed in [KLPT14]. These reductions are discussed in Sect. 5.5 below.

5.1 Endomorphism Ring Computation Is not Harder than Inverse Deuring Correspondence

When $p \equiv 3 \pmod 4$ the curve $y^2 = x^3 + x$ is supersingular with invariant $j = 1728$. This curve corresponds to a maximal order $\mathcal{O}_0$ with $\mathbb{Z}$-basis $\{1, i, \frac{1+k}{2}, \frac{i+j}{2}\}$ under Deuring's correspondence, and there is an isomorphism of quaternion algebras $\theta : B_{p,\infty} \to \mathrm{End}(E_0) \otimes \mathbb{Q}$ sending $(1, i, j, k)$ to $(1, \phi, \pi, \pi\phi)$ where $\pi : (x, y) \to (x^p, y^p)$ is the Frobenius endomorphism, and $\phi : (x, y) \to (-x, \iota y)$ with $\iota^2 = -1$. More generally, it is easy to compute j-invariants corresponding to the maximal orders given by Proposition 1.

Proposition 3. *There is a polynomial-time algorithm that given a prime $p > 2$, computes a supersingular j-invariant $j_0 \in \mathbb{F}_p$ such that $\mathrm{End}(E(j_0)) \cong \mathcal{O}_0$ (where $\mathcal{O}_0$ is as given by Proposition 1 together with a map $\phi \in \mathrm{End}(E(j_0))$) such that $\theta : B_{p,\infty} \to \mathrm{End}(E(j_0)) \otimes \mathbb{Q} : (1, i, j, k) \to (1, \phi, \pi, \pi\phi)$ is an isomorphism of quaternion algebras.*

Proof. Let q be chosen such that $B_{p,\infty} = H(-q, -p)$ as in Proposition 1 and let R be the ring of integers of $\mathbb{Q}(\sqrt{-q})$. Consider Algorithm 3 below. Step 1 can be executed in time polynomial in $\log p$ using a modification of Bröker's Algorithm 2.4 in [Brö09]: the cardinality of $\mathcal{J} := \{j \in \mathbb{F}_{p^2} : R \subseteq \mathrm{End}(E(j))\}$ is equal to the class number h_{-q} of R, and this is bounded by q. To see this requires a surjectivity and injectivity argument. Suppose $j \in \mathbb{F}_{p^2}$ is a supersingular j-invariant such that R embeds into $\mathrm{End}(E(j))$. Then if $R = \mathbb{Z}[\alpha]$, by Deuring's Lifting Theorem [Lan87, Theorem 14, p. 184] applied to $E(j)$ and α, there is an elliptic curve $\tilde{E}/\mathbb{C}$ such that $\mathrm{End}(\tilde{E}) \simeq R$ and a prime $\mathfrak{p}$ of R dividing p such that $\tilde{E} \pmod{\mathfrak{p}} = E(j)$. Since $\tilde{E}$ has complex multiplication by R, $j(\tilde{E})$ is a root of the Hilbert class polynomial of $\mathbb{Q}(\sqrt{-q})$. Because $E(j)$ is supersingular, p is inert in R and $\mathfrak{p} = pR$. We see that the map is injective because principal prime ideals of R split completely in H, and so the Hilbert class polynomial will have h_{-q} distinct roots modulo p. To compute ϕ in Step 3 one can simply compute all isogenies of degree q using Vélu's formulae and identify the one corresponding to an endomorphism. The map ϕ defines an isomorphism of quaternion algebras $\theta : B_{p,\infty} \to \mathrm{End}(E(j_0)) \otimes \mathbb{Q} : (1, i, j, k) \to (1, \phi, \pi, \pi\phi)$. To perform the check in Step 4, one applies θ to the numerators of $\mathcal{O}_0$ basis elements, and check whether the resulting maps annihilate the D torsion, where D is the denominator. □

Algorithm 3. *Computing the Deuring correspondence for special orders*
Input: A prime p.
Output: A supersingular j-invariant $j_0 \in \mathbb{F}_p$ such that $\mathcal{O}_0 \cong \mathrm{End}(E(j_0))$, and an endomorphism $\phi \in \mathrm{End}(E(j_0))$ such that $\mathrm{Nrd}(\phi) = q$ and $\mathrm{Trd}(\phi) = 0$.

1. *Compute $\mathcal{J}$, a set of supersingular j-invariants such that for $j \in \mathcal{J}$, R_{-q} embeds into $\mathrm{End}(E(j))$, where R_{-q} is the integer ring of $\mathbb{Q}(\sqrt{-q})$.*
2. *For $j \in \mathcal{J}$:*
   (a) *Compute ϕ, an endomorphism of degree q of $E(j)$.*
   (b) *If $\mathrm{End}(E(j)) \cong \mathcal{O}_0$:*
      i. *Return j and ϕ.*

5.2 Quaternion ℓ-Isogeny Algorithm

The quaternion ℓ-isogeny problem was introduced and solved in [KLPT14] as a step forward in the cryptanalysis of the Charles-Goren-Lauter hash function.

We refer to [KLPT14, GPS17] for a full description of the algorithm and its powersmooth version as well as their analysis. For our purposes the following proposition will be sufficient.

Lemma 2 *[KLPT14, GPS17]. Under various heuristic assumptions, there exist two polynomial-time algorithms that given I a left ideal of $\mathcal{O}_0$, returns J another left ideal of $\mathcal{O}_0$ in the same class as I of norm N such that $N \approx p^{7/2}$. Moreover for the first algorithm we have $N = \prod p_i^{e_i}$ with $p_i^{e_i} < \log p$ and for the second algorithm we have $N = \ell^e$ for some integer e and some small prime ℓ.*

Interestingly, [GPS17] also proves that (after a minor tweak) the outputs of these algorithms only depend on the ideal class of their inputs and not on the particular ideal class representative.

Many of our algorithms and reductions below will use these algorithms as black boxes. Their correctness will therefore rely on the same heuristics, and possibly some more.

5.3 Translating $\mathcal{O}_0$-Ideals to Isogenies

Let $\mathcal{O}_0$ be the maximal order given by Proposition 1, let E_0 be a corresponding supersingular elliptic curve, and let I be a left $\mathcal{O}_0$-ideal of norm N such that I is not contained in $\mathcal{O}_0 m$ for any $m \in \mathbb{N}$. This ideal corresponds to an isogeny $\phi : E_0 \to E_1$ of degree N. This isogeny is uniquely defined by its kernel, which is a cyclic subgroup of order N in E_0 by Proposition 10. Following Waterhouse [Wat69] one can identify the correct subgroup by evaluating the maps corresponding to an $\mathcal{O}_0$-basis at a generator of each subgroup. Moreover when N is composite, the kernel can be represented more efficiently as a product of cyclic subgroups whose orders are powers of primes, and similarly the isogenies are represented more efficiently as a composition of prime degree isogenies. The details of such an algorithm can be found in [GPS17], which also analyzes its complexity. The following proposition will be sufficient for our purposes.

Proposition 4. *There exists an algorithm which, given an $\mathcal{O}_0$ left ideal I of norm $N = \prod_i p_i^{e_i}$, returns an isogeny $\phi : E_0 \to E_1$ corresponding to this ideal through Deuring's correspondence. Moreover the complexity of this algorithm is polynomial in* $\max_i p_i^{e_i}$.

We stress that this translation algorithm requires us to know the endomorphism ring of E_0, and that it is only efficient when $\max_i p_i^{e_i}$ is small.

Let us first assume that we have an efficient algorithm for Problem 2, returning a $\mathbb{Z}$ basis for a maximal order as discussed above. Algorithm 4 below uses this algorithm to solve Problem 3.

Algorithm 4. *Reduction from Problem 3 to Problem 2*
Input: A supersingular j-invariant j.
Output: Four maps that generate $\mathrm{End}(E(j))$.

1. *Use an algorithm for Problem 2 to obtain a maximal order* $\mathcal{O} \simeq \mathrm{End}(E(j))$.
2. *Compute an ideal I connecting $\mathcal{O}_0$ and $\mathcal{O}$.*
3. *Compute an ideal J with powersmooth norm in the same class as I.*
4. *Translate the ideal J into an isogeny $\varphi : E_0 \to E$.*
5. *Let N be the norm of J.*
6. *Let $1, \phi_2, \phi_3, \phi_4$ generate* $\mathrm{End}(E(j_0))$.
7. *Let $1, \omega_2, \omega_3, \omega_4$ generate $\mathcal{O}$, and let $1, \omega_{2,0}, \omega_{3,0}, \omega_{4,0} \in \mathcal{O}_0$ correspond to $1, \phi_2, \phi_3, \phi_4$.*
8. *Find integers c_{ij} such that $\omega_i = \frac{\sum_j c_{ij}\omega_{j,0}}{N}$.*
9. *Return N, φ, c_{ij} implicitly representing the maps $\frac{\sum_{i=1}^4 c_{ij}\widehat{\varphi}\phi_i\varphi}{N}$ for each i.*

The maps returned by Algorithm 4 are of the form $\phi = \frac{\sum_{i=1}^4 c_{ij}\widehat{\varphi}\phi_i\varphi}{N}$ where N is a smooth number, $c_{ij} \in \mathbb{Z}$, $\{\phi_i\}_{i=1,2,3,4}$ form a basis for the endomorphism ring of a special curve E_0, and $\varphi : E_0 \to E(j)$ is an isogeny of degree N, given as a composition of isogenies of low degree. In Sect. 8 we define compact representations of endomorphisms, and the data given by Algorithm 4 define four compact representations. This is arguably not the most natural representation of endomorphisms, but it still allows to efficiently evaluate them at arbitrary points, as shown by Algorithm 5 and Lemma 3 below. See Sect. 8 for a detailed definition of how to represent the output of this algorithm.

Algorithm 5. *Endomorphism evaluation*
Input: A curve E, an isogeny $\varphi : E_0 \to E$ with powersmooth degree N, and integers a, b, c, d defining an endomorphism $\phi = \frac{\varphi(a + b\phi_2 + c\phi_3 + d\phi_4)\widehat{\varphi}}{N} \in \mathrm{End}(E)$.
Input: A point $P \in E$.
Output: $\phi(P)$.

1. *Let $N = \prod_i p_i^{e_i}$ and let $m_i = N/p_i^{e_i}$.*
2. *For all i:*
   (a) *Compute Q_i such that $p_i^{e_i} Q_i = P$.*
   (b) *Compute $S_i = \varphi(a + b\phi_2 + c\phi_3 + d\phi_4)\widehat{\varphi}(Q_i)$*
3. *Compute S such that $S_i = m_i S$ for all i.*
4. *Return S.*

Lemma 3. *Let $P \in E(K)$ with K an extension of $\mathbb{F}_{p^2}$. Assume that* $\log N$ *and* $\max_i p_i^{e_i}$ *are polynomial in* $\log p$. *Then Algorithm 5 computes $\phi(P)$ and can be implemented to run in time polynomial in* $\log |K|$.

Proof. We will first prove the correctness of the above algorithm. Let $\gamma := \varphi(a + b\phi_2 + c\phi_3 + d\phi_4)\widehat{\varphi}$, so $[N] \circ \phi = \gamma$. While the choice of Q_i in Step 2a is not unique, in Step 2b the point S_i is independent of the choice of Q_i, because of the calculation

$$S_i = \gamma(Q_i) = ([N] \circ \phi)(Q_i) = ([m_i] \circ \phi)(P).$$

We now show that the S in Step 3 exists, is unique, and equals $\phi(P)$. The above calculation showed $\phi(P)$ satisfies $m_i\phi(P) = S_i$. On the other hand, the point S also satisfies $m_i S = S_i$ for all i, so $\phi(P) - S \in E[m_i]$ for all i. Since $\gcd(\{m_1, \ldots, m_k\}) = 1$, we have $\bigcap_{i=1}^k E[m_i] = \{0\}$. This implies that $S = \phi(P)$.

We can efficiently compute S in Step 3 as follows. Since the greatest common divisor of $\{m_1, \ldots, m_k\}$ is 1, there are integers $a_1, \ldots, a_k$ such that $\sum_{j=1}^k a_j m_j = 1$. These integers can be efficiently computed with the extended Euclidean algorithm since $k = O(\log p)$. Define $S := \sum_{i=1}^k a_i S_i$. Observe that for $i \neq j$, we have

$$m_i S_j = \frac{N}{p_i^{e_i} p_j^{e_j}} p_j^{e_j} S_j = \frac{N}{p_i^{e_i} p_j^{e_j}} p_j^{e_j} \gamma(Q_j) = \frac{N}{p_i^{e_i} p_j^{e_j}} \gamma(P) = \frac{N}{p_i^{e_i} p_j^{e_j}} \gamma(p_i^{e_i} Q_i) = m_j S.$$

This implies that $m_i S_j = m_j S_i$. Now we calculate

$$m_i S = m_i \sum_{j=1}^{k} a_j S_j = S_i - \left(\sum_{j \neq i} a_j m_j S_i \right) + \sum_{j \neq i} m_i a_j S_j = S_i.$$

Although Q may lie in a very large extension of $\mathbb{F}_{p^2}$, each of the Q_i lies in a reasonably small extension, namely the extension degree is polynomial in $\log p$.

Note that S lies in an extension of K of degree at most 6 by Theorem 4.1 of [Wat69], so Step 3 is efficient. Step 2a involves some univariate polynomial factorization, a task that is polynomial in both the degree of the polynomial and the logarithm of the field size. In Step 2b the isogeny φ and its dual can be evaluated stepwise, and evaluating the map $a + b\phi_2 + c\phi_3 + d\phi_4$ at an arbitrary point involves 4 scalar multiplications, three additions and the evaluation of the maps $\phi_i \in \mathrm{End}(E(j_0))$ at certain points. □

Proposition 5. *Under plausible heuristic assumptions, the reduction in Algorithm 4 from Problem 3 to Problem 2 can be implemented to run in time polynomial in* $\log p$.

Proof. By Theorem 2, we may assume that the maximal order isomorphic to $\mathrm{End}(E(j))$ has size polynomial in $\log p$. In Step 2, the ideal I can be computed with Algorithm 3.5 of [KV10]. This can be done in time polynomial in $\log p$ since $\mathcal{O}_0$ and $\mathcal{O}$ have size polynomial in $\log p$. By Lemma 2 the output of Step 3 is an ideal of norm $N = \prod p_i^{e_i}$ such that $S = \max_i p_i^{e_i} = O(\log p)$. The translation algorithm runs in a time polynomial in S, hence in $\log p$. The other steps also run in polynomial time. □

5.4 Inverse Deuring Correspondence Is not Harder than Endomorphism Ring Computation

Let us now assume that we have an efficient algorithm for Problem 3, returning four maps generating the endomorphism ring, in some format that allows efficient evaluation of the maps at arbitrary points. Algorithm 6 below uses this algorithm and then constructs a sequence of linear transformations that map $1, \alpha, \beta, \gamma$ to four orthogonal maps $1, \iota, \lambda, \iota\lambda$ corresponding to $1, i, j, k \in B_{p,\infty}$. Composing the inverses of these maps then gives a $\mathbb{Z}$-basis for $\mathcal{O}$.

Algorithm 6. *Reduction from Problem 2 to Problem 3*
Input: A supersingular j-invariant j.
Output: A maximal order $\mathcal{O} \subset B_{p,\infty}$ such that $\mathrm{End}(E(j)) \simeq \mathcal{O}$.

1. *Use an algorithm for Problem 3 to obtain four maps $1, \alpha, \beta, \gamma$ which generate* $\mathrm{End}(E(j))$*, in a format that allows efficient evaluation at elliptic curve points.*
2. *Compute the Gram matrix associated to the sequence $(1, \alpha, \beta, \gamma)$.*
3. *Find a rational invertible linear transformation sending $(1, \alpha, \beta, \gamma)$ to some $(1, \alpha', \beta', \alpha'\beta')$, where $1, \alpha', \beta', \alpha'\beta'$ generate an orthogonal basis for $B_{p,\infty}$ over $\mathbb{Q}$.*
4. *If the numerators and denominators of* $\mathrm{Nrd}(\alpha')$ *and* $\mathrm{Nrd}(\beta')$ *are not easy to factor:*
   (a) *Apply a random invertible linear transformation to (α, β, γ).*
   (b) *Go to Step 3.*
5. *Find $a, b, c \in \mathbb{Q}$ such that $\mathrm{Nrd}(\iota) = q$, where $\iota = a\alpha' + b\beta' + c\alpha'\beta'$.*

6. *Find a rational invertible linear transformation sending* $(1, \alpha', \beta', \alpha'\beta')$ *to* $(1, \iota, \delta, \iota\delta)$ *for some* $\delta \in B_{p,\infty}$ *where* $1, \iota, \delta, \iota\delta$ *generate an orthogonal basis for* $B_{p,\infty}$ *over* $\mathbb{Q}$.
7. *If the numerator and denominator of* $\mathrm{Nrd}(\delta)$ *is not easy to factor:*
   (a) *Apply a random invertible linear transformation to* (α, β, γ).
   (b) *Go to Step 3.*
8. *Find* $a, b \in \mathbb{Q}$ *such that* $\mathrm{Nrd}(\delta)(a^2 + b^2 q) = p$. *Let* $\lambda = a\delta + b\iota\delta$.
9. *Compute a rational invertible linear transformation sending* $(1, \iota, \delta, \iota\delta)$ *to* $(1, \iota, \lambda, \iota\lambda)$.
10. *Invert and compose all linear transformations to express* $1, \alpha, \beta, \gamma$ *in the basis* $(1, \iota, \lambda, \iota\lambda)$, *and deduce a basis of* $\mathcal{O}$ *in* $B_{p,\infty}$.
11. *Return the basis of* $\mathcal{O}$.

Let B be a bound on the degrees of the maps α, β, γ returned in Step 1 of Algorithm 6. We analyze the complexity of the algorithm through the following lemmas and proposition.

Lemma 4. *There exists an algorithm for Step 2 that runs in time polynomial in* $\log p$ *and* $\log B$.

Proof. Given two endomorphisms α, β, one can compute their inner product $\langle \alpha, \beta \rangle = \alpha\bar{\beta} + \beta\bar{\alpha} \in \mathbb{Z}$ by evaluating it on an appropriate set of torsion points of small prime order, and then applying the Chinese Remainder Theorem, following a strategy similar to Schoof's point counting algorithm (see [Koh96, Theorem 81]). Applying this algorithm to every pair of maps from $(1, \alpha, \beta, \gamma)$ gives the result. □

Lemma 5. *There exists an algorithm for Steps 3 and 6 that runs in time polynomial in* $\log p$ *and* $\log B$.

Proof. We focus on Step 3, and Step 6 is similar. Given the Gram matrix one can apply the Gram-Schmidt orthogonalization process to obtain a new basis $(1, \alpha', \beta', \gamma')$. It remains to show that $\alpha'\beta'$ is a scalar multiple of γ' so that we can normalize γ' to obtain the result. It suffices to show that $\alpha'\beta'$ is orthogonal to 1, α' and β'. Indeed we have $\langle \alpha'\beta', 1 \rangle = \alpha'\beta' + \bar{\beta}'\bar{\alpha}' = \langle \alpha', \bar{\beta}' \rangle = -\langle \alpha', \beta' \rangle = 0$; we have $\langle \alpha'\beta', \alpha' \rangle = \alpha'\beta'\bar{\alpha}' + \alpha'\bar{\beta}'\bar{\alpha}' = \mathrm{Nrd}(\alpha')\,\mathrm{Trd}(\beta') = 0$; and similarly $\langle \alpha'\beta', \beta' \rangle = \alpha'\beta'\bar{\beta}' + \beta'\bar{\beta}'\bar{\alpha}' = \mathrm{Nrd}(\beta')\,\mathrm{Trd}(\alpha') = 0$. □

Lemma 6. *Given the factorizations of the numerators and denominators of both* $\mathrm{Nrd}(\alpha')$ *and* $\mathrm{Nrd}(\beta')$, *there exists an algorithm for Step 5 that runs in time polynomial in* $\log p$ *and* $\log B$.

Proof. Finding such $a, b, c \in \mathbb{Q}$ satisfying the condition amounts to finding $a', b', c', d \in \mathbb{Z}$ such that $a'^2\,\mathrm{Nrd}(\alpha') + b'^2\,\mathrm{Nrd}(\beta') + c'^2\,\mathrm{Nrd}(\alpha')\,\mathrm{Nrd}(\beta') = d^2 q$. According to Simon [Sim05, Sect. 8] there is an algorithm to solve this Diophantine equation in polynomial time. □

Lemma 7. *Given the factorizations of the numerator and of the denominator of* $\mathrm{Nrd}(\delta)$, *there exists an algorithm for Step 8 that runs in time polynomial in* $\log p$ *and* $\log B$.

Proof. Note that $\langle \delta, \iota\delta \rangle$ is by construction the orthogonal space of $\langle 1, \iota \rangle$, and this space must contain an element of norm p, so the equation has a solution. Given factorizations for both the numerator and the denominator of δ one can use Cornacchia's algorithm [Cor08] to solve Step 8. □

Proposition 6. *Under plausible heuristic assumptions, the reduction provided by Algorithm 6 can be implemented to run in polynomial time.*

Proof. In Steps 4 and 7 the algorithm requires that some numbers are easy to factor. In Step 4 we may expect these numbers to behave like random numbers of the same sizes. In Step 7, p must divide the numerator of $\mathrm{Nrd}(\delta)$. We may expect that both the numerator and the denominator factor like random numbers of the same size. One can require all those numbers to be large primes, or a product of large primes and small cofactors, two properties that will be satisfied with a probability inversely proportional to a polynomial function of $\log p$. Steps 4a and 7a randomize α, β, γ so that we expect the conditions to be satisfied after a number of steps that is polynomial in $\log p$. By the four lemmas before we then expect that the whole reduction runs in a time polynomial in $\log p$. □

The reduction provided by Algorithm 6 and its runtime analysis relies on several heuristics, namely the probability to obtain suitable norms in Steps 4 and 7 as discussed in the above proposition, and the runtime assumption of Simon's algorithm for Step 5.

5.5 Preimage and Collision Resistance of the CGL Hash Function

In this section we show that the hardness of the endomorphism ring computation problem is equivalent to the security of the Charles-Goren-Lauter hash function.

Proposition 7. *Assume there exists an efficient algorithm for the endomorphism ring computation problem. Then there is an efficient algorithm to solve the preimage and collision problems for the Charles-Goren-Lauter hash function.*

Proof. By standard arguments on hash functions it is enough to focus on preimage resistance. Our reduction of this problem to the endomorphism ring computation problem is given in Algorithm 7. Besides two black box calls to an algorithm for the endomorphism ring computation problem, it uses other efficient algorithms described in this paper, including Algorithm 4 to translate a description of an endomorphism ring as rational maps into a description of a maximal order in $B_{p,\infty}$, both the ℓ-power and the powersmooth versions of the quaternion isogeny algorithm, and the translation algorithm from ideals to isogenies. All these routines are efficient by the lemmas and propositions of this paper. By the results in Sect. 6.4, the algorithm is correct. □

Algorithm 7. *Reduction from preimage resistance to endomorphism ring computation*
Input: Two supersingular j-invariants $j_s, j_t \in \mathbb{F}_{p^2}$.
Output: A sequence of j-invariants $j_s = j_0, j_1, \ldots, j_e = j_t$ such that for any i there exists an isogeny of degree ℓ from $E(j_i)$ to $E(j_{i+1})$.

1. *Compute* $\mathrm{End}(E(j_s))$ *and* $\mathrm{End}(E(j_t))$.
2. *Compute* $\mathcal{O}_s \simeq \mathrm{End}(E(j_s))$ *and* $\mathcal{O}_t \simeq \mathrm{End}(E(j_t))$ *with Algorithm 4.*
3. *Compute ideals* I_s *and* I_t *connecting* $\mathcal{O}_0$ *respectively to* $\mathcal{O}_s$ *and* $\mathcal{O}_t$.
4. *Compute ideals* $J_s = \mathcal{O}_0\alpha_s + \mathcal{O}_0\ell^{e_s}$ *and* $J_t = \mathcal{O}_0\alpha_t + \mathcal{O}_0\ell^{e_t}$ *with norm* ℓ^{e_s}, ℓ^{e_t} *for some* e_s, e_t, *in the same classes as* I_s *and* I_t *respectively.*
5. *For* $r = s, t$ *and corresponding* $E = E(j_r)$:
   (a) *Compute a sequence of ideals* $J_{r,i} = \mathcal{O}_0\alpha_r + \mathcal{O}_0\ell^i$ *for* $i = 0, \ldots, e_r$
   (b) *For* $0 \leq i \leq e_r$:
   (c) *Compute* $K_{r,i}$ *with powersmooth norm in the same class as* $J_{r,i}$.
   (d) *Translate* $K_{r,i}$ *into an isogeny* $\varphi_{r,i} : E_0 \to E_{r,i}$.
   (e) *Deduce a sequence* $(j_0, j(E_{r,1}), j(E_{r,2}), \ldots, j(E_{r,e}) = j(E))$.
6. *Return* $(j(E_s), \ldots, j_0, \ldots, j(E_t))$ *the concatenation of both paths.*

The reverse direction may a priori look easier. By standard arguments on hash functions it is sufficient to prove the claim with respect to a collision algorithm. A collision for the Charles-Goren-Lauter hash function gives a non-scalar endomorphism of the curve; four linearly independent endomorphisms give a full rank subring of the endomorphism ring; and heuristically one expects that a few such maps will be sufficient to generate the whole ring. To compute the endomorphism ring one would therefore call the collision finding algorithms multiple times until the resulting maps generate the full endomorphism ring. This strategy, however, has a potential caveat: the collision algorithm might be such that it always returns the same endomorphism. In Algorithm 8 we get around this problem by performing a random walk from the input invariant j, calling the collision algorithm on the end-vertex of the random walk, and concatenating paths to form endomorphisms of $E(j)$.

Proposition 8. *Assume there exists an efficient preimage or collision algorithm for the Charles-Goren-Lauter hash function. Then under plausible heuristic assumptions there is an efficient algorithm to solve the endomorphism ring computation problem.*

Proof. The reduction algorithm for collision resistance is given by Algorithm 8 below. Note that in Step 7 the discriminant can be computed from the Gram matrix, which by Lemma 4 can be efficiently computed. Heuristically, one expects that the loop will be executed at most $O(\log p)$ times. Indeed let us assume that after adding some elements to the subring we have a subring of index N. Then we can heuristically expect any new randomly generated endomorphism to lie in this subring with probability only $1/N$. Moreover when it does not lie in the subring, the element will decrease the index by a non trivial integer factor of N. □

Algorithm 8. *Reduction from endomorphism ring computation to collision resistance*
Input: A supersingular j*-invariant* $j \in \mathbb{F}_{p^2}$.
Output: The endomorphism ring of $E(j)$.

1. *Let* $\mathcal{R} = \langle 1 \rangle \subset \mathrm{End}(E(j))$.
2. *While* $disc(\mathcal{R}) \neq 4p^2$:
   (a) *Perform a random walk in the graph, leading to a new vertex* j'.
   (b) *Apply a collision finding algorithm on* j', *leading to an endomorphism of* $E(j')$.
   (c) *Deduce an endomorphism* ϕ *of* $E(j)$ *by concatenating paths.*
   (d) *Set* $\mathcal{R} \leftarrow \langle \mathcal{R}, \phi \rangle$.
   (e) *Compute the discriminant of* $\mathcal{R}$.
3. *Return a* $\mathbb{Z}$-*basis for* $\mathcal{R}$.

6 ℓ-PowerIsogeny Reduces to MaxOrder and Action-on-ℓ-Torsion

In this section we show that computing an ℓ-isogeny between two supersingular elliptic curves reduces to computing maximal orders of elliptic curves and solving the Action-on-ℓ-Torsion Problem.

6.1 Outline of Reduction

Given two supersingular elliptic curves E, E' over $\mathbb{F}_{p^2}$, and oracles for the problems Action-on-ℓ-Torsion and MaxOrder, we will construct an ℓ-power isogeny $E \to E'$ by constructing a chain of ℓ-isogenies through intermediate curves. First, the oracle will give us two maximal orders $\mathcal{O}, \mathcal{O}' \subseteq B_{p,\infty}$ with $\mathcal{O} \simeq \mathrm{End}(E)$ and $\mathcal{O}' \simeq \mathrm{End}(E')$. We then compute a connecting ideal, meaning a left ideal of $\mathcal{O}$, whose left order is $\mathcal{O}$ and right order is $\mathcal{O}'$. Next we use the main algorithm of [KLPT14] to compute an equivalent ideal I whose norm is ℓ^e for some $e = O(\log p)$. The isogeny $\phi_I : E \to E'$ corresponding to I has degree ℓ^e, so the representation size of the isogeny is exponential. To remedy this we will, given I, compute a chain of ℓ-isogenies $\psi_1, \ldots, \psi_e$ such that $\phi_I = \psi_e \circ \cdots \circ \psi_1$. Since $\psi_1, \ldots, \psi_e$ have degree ℓ, they are of polynomial representation size as rational maps. To obtain the ψ_i we will first show that there is a factorization of the ideal I. The proper notion here is that of a *filtration* of ideals, namely a sequence

$$I = I_e \subseteq I_{e-1} \subseteq \cdots \subseteq I_1 \subseteq I_0 = \mathcal{O}$$

such that the isogeny corresponding to I_k is a map ϕ_k from E to some intermediate curve E_k. The factorization of ϕ_I gives us a path starting at E and ending at E' of length e in the graph of isogenies of degree ℓ, and the filtration of I leads to a corresponding "path" between maximal orders in $B_{p,\infty}$. The maximal orders that appear in this path are $\mathcal{O}_R(I_k)$ and the ideal connecting $\mathcal{O}_R(I_k)$ to $\mathcal{O}_R(I_{k+1})$ is $J_k := I_{k-1}^{-1} I_k$. These paths are given in the following diagrams:

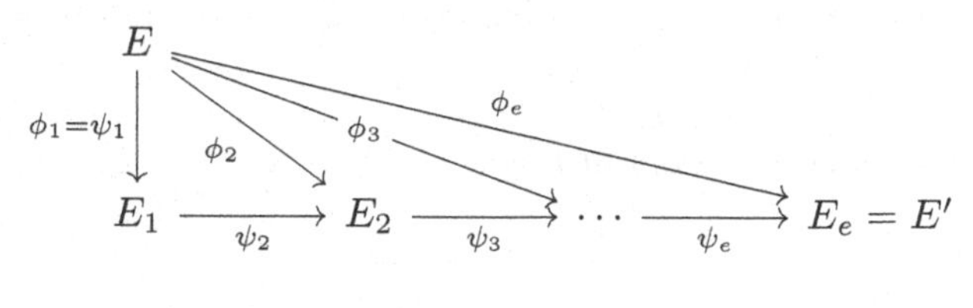

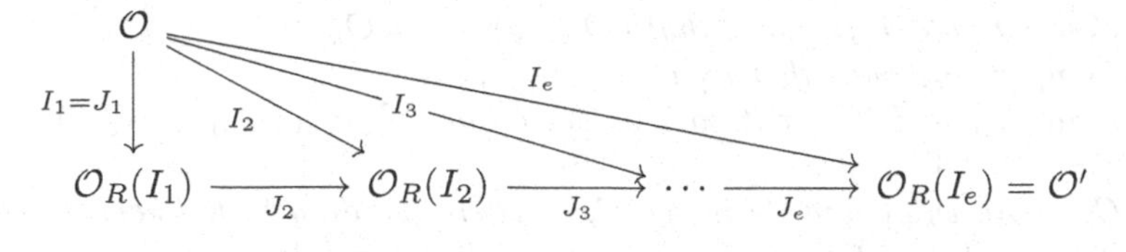

For each k, the isogeny $\phi_k : E_0 \to E_k$ has degree ℓ^k, and so corresponds to a left $\mathcal{O}$-ideal I_k of norm ℓ^k. We will show that $I_k = I + \mathcal{O}\ell^k$ is the desired ideal. As k grows, these ideals will have norms which are too big to find the corresponding isogenies, so we will compute the maps $\psi_k : E_{k-1} \to E_k$ which correspond to left ideals J_k of $\mathcal{O}_R(I_{k-1})$ of norm ℓ. Suppose we have computed ψ_k, the curve E_k, and J_{k+1} as above. We can use the oracle for MaxOrder to identify generators of J_{k+1} with endomorphisms of E_k. On the other hand, J_{k+1} corresponds to the isogeny ψ_{k+1}, whose kernel we compute using the information from the oracle Action-on-ℓ-Torsion. Using Vélu's formula, we can compute ψ_{k+1} from its kernel. This procedure iteratively computes the desired maps $\psi_1, \psi_2, \ldots, \psi_e$.

6.2 Reduction from ℓ-PowerIsogeny to MaxOrder and Action-on-ℓ-Torsion

In this section, we give the reduction from ℓ-Power Isogeny to the problems MaxOrder and Action-on-ℓ-Torsion.

Algorithm 9. *Reduction from ℓ-PowerIsogeny to MaxOrder and Action-on-ℓ-Torsion*
Input: E, E' supersingular elliptic curves over $\mathbb{F}_{p^2}$, a prime $\ell \neq p$.
Output: a chain of ℓ-isogenies connecting E and E'.

1. *Compute a basis $\langle 1, i, j, ij \rangle$ for $B_{p,\infty}$.*
2. *Call oracle MaxOrder on $p, \langle 1, i, j, ij \rangle, E$, resulting in $\alpha_1, \alpha_2, \alpha_3, \alpha_4$ where* $\mathrm{End}(E) \simeq \mathcal{O} := \langle \alpha_1, \alpha_2, \alpha_3, \alpha_4 \rangle \subseteq B_{p,\infty}$.
3. *Call oracle MaxOrder on $p, \langle 1, i, j, ij \rangle, E'$, resulting in $\alpha'_1, \alpha'_2, \alpha'_3, \alpha'_4$ where* $\mathrm{End}(E') \simeq \mathcal{O}' := \langle \alpha'_1, \alpha'_2, \alpha'_3, \alpha'_4 \rangle \subseteq B_{p,\infty}$.
4. *Compute connecting ideal: use $\alpha_1, \ldots, \alpha_4$ and $\alpha'_1, \ldots, \alpha'_4$ to compute a left ideal I of $\mathcal{O}$ such that $\mathcal{O}_R(I) = \mathcal{O}'$ and $\mathrm{Nrd}(I) = \ell^e$ with $e = O(\log p)$. Adjust I so that $I \not\subseteq \ell^k \cdot \mathcal{O}$ for any positive integer k.*
5. *For $0 \leq k \leq e$:*
   (a) *Compute $I_k := I + \mathcal{O}\ell^k$. This is a left ideal of $\mathcal{O}$ of norm ℓ^k. Also compute its right order $\mathcal{O}_R(I_k)$.*
   (b) *Compute a $\mathbb{Z}$-basis $\gamma_1, \gamma_2, \gamma_3, \gamma_4$ for the ideal $J_{k+1} := I_k^{-1} I_{k+1}$ of $\mathcal{O}_R(I_k)$.*
6. *Set $E_0 := E$.*
7. *For $0 \leq k \leq e - 1$:*
   (a) *Compute a basis $\{P_1, P_2\}$ for $E_k[\ell]$.*
   (b) *Call oracle MaxOrder with $p, \langle 1, i, j, ij \rangle$, E_k, resulting in β_1, β_2, β_3, β_4 that generate $\mathcal{O}_k \subseteq B_{p,\infty}$.*
   (c) *Call oracle Action-on-ℓ-Torsion with parameters p, P_1, P_2, $\langle 1, i, j, ij \rangle$, E_k, β_1, β_2, β_3, β_4 resulting in $Q_{st} = \iota_k^{-1}(\beta_s)(P_t)$ for $s = 1, \ldots, 4$, $t = 1, 2$. Here, $\iota_k : \mathrm{End}(E_k) \to \langle \beta_1, \ldots, \beta_4 \rangle$ is an isomorphism.*

(d) *Compute* $v \in B_{p,\infty}$ *such that* $v\mathcal{O}_R(I_k)v^{-1} = \mathcal{O}_k$.
(e) *Compute* c_{rs} *such that* $v\gamma_r v^{-1} = \sum_s c_{rs}\beta_s$.
(f) *Find* $x, y \in \mathbb{Z}/\ell\mathbb{Z}$, *not both* 0, *such that* $\sum_s c_{rs}(xQ_{s1} + yQ_{s2}) = 0$ *for* $r = 1, \ldots, 4$.
(g) *Compute* ψ_{k+1} *and its image* E_{k+1} *corresponding to the kernel subgroup* $\langle xP_1 + yP_2\rangle = E_k[\iota_k^{-1}(J_{k+1})]$ *using Vélu's formula*

8. *Return* $\psi_1, \psi_2, \ldots, \psi_e$.

Theorem 10. *ℓ-PowerIsogeny efficiently reduces to MaxOrder and Action-on-ℓ-Torsion. In particular, given a prime p, a prime $\ell \neq p$, and supersingular elliptic curves E, E' over $\mathbb{F}_{p^2}$, Algorithm 9 returns isogenies $\psi_1, \ldots, \psi_e$ of degree ℓ whose composition is an isogeny $\psi := \psi_e \circ \cdots \circ \psi_1$ of degree ℓ^e from E to E'. Assuming ℓ is of size $O(\log p)$, Algorithm 9 runs in time polynomial in* $\log p$ *and makes $O(\log p)$ queries of MaxOrder and Action-on-ℓ-Torsion.*

Proof. By Theorem 2, the oracle returns a basis for $\mathcal{O}$ and for $\mathcal{O}'$ of polynomial size. To do Step 4, we first compute an arbitrary connecting ideal for $\mathcal{O}$ and $\mathcal{O}'$ in polynomial time using Algorithm 3.5 of [KV10]. An equivalent connecting ideal of norm ℓ^e, where $e = O(\log p)$, can be computed in polynomial time as claimed in [KLPT14].

Define $E_k := E/E[I_k]$ (here by $E[I_k]$ we mean the subgroup $E[\iota^{-1}(I_k)]$, where $\iota : \mathrm{End}(E) \to \mathcal{O}$ is an isomorphism). We need to show that I_k has norm ℓ^k and that the left $\mathcal{O}_R(I_k)$-ideal J_{k+1} corresponds to the isogeny $\psi_{k+1} : E_k \to E_{k+1}$ in the factorization $\phi_k = \psi_k \circ \phi_{k-1}$; this is proved in Theorem 11. Right orders and products of ideals can be computed efficiently with linear algebra over $\mathbb{Z}$, hence Step 4 is efficient; see [Rón92], Theorem 3.2 for the statement on right orders. Inverses can be computed from the formula $I^{-1} = \frac{1}{\mathrm{Nrd}(I)}\overline{I}$. We make e calls to the oracle for generators of $\mathrm{End}(E_k)$ and their action on ℓ-torsion. If $\mathcal{O} \simeq \mathcal{O}_k$, we can compute v such that $v\mathcal{O}_k v^{-1} = \mathcal{O}$ in polynomial time by Lemma 2.5, Corollary 3.6, and Proposition 6.9 of [KV10]. By Theorem 11, the isogeny corresponding to I factors as the product of the isogenies corresponding to J_k, $k = 1, \ldots, e$, all of which have degree ℓ. Now compute the kernel of ψ_k using J_k and the action of $\mathrm{End}(E_{k-1})$ on the ℓ-torsion of E_{k-1}; see Proposition 9. Since ℓ is $O(\log p)$, rational maps for ψ_k from its kernel can be efficiently computed. □

6.3 Going from an Ideal of Norm ℓ to a Corresponding Subgroup of Order ℓ

At the beginning of Step 7 of the algorithm, we have an isogeny $E_{k-1} \to E_k$ represented by a left $\mathcal{O}_R(I_{k-1})$-ideal J_k. We wish to specify the subgroup of E_{k-1} which is the kernel of this isogeny. If $\widetilde{J_k} \subseteq \mathrm{End}(E_{k-1})$ is the ideal isomorphic to J_k, recall from Sect. 2.2 that

$$E_{k-1}[\widetilde{J_k}] = \bigcap_{\gamma \in \tilde{J}_k} \ker(\gamma_k),$$

and it suffices to compute $\ker(\gamma_1) \cap \cdots \cap \ker(\gamma_4)$, where $\gamma_1, \ldots, \gamma_4$ are a $\mathbb{Z}$-basis of $\widetilde{J_k}$. Once we have $E_{k-1}[\widetilde{J_k}]$, we can use Vélu's formula to compute ψ_k.

Step 7 in our algorithm computes $E_{k-1}[\tilde{J}_k]$ and is similar to Algorithm 2 in [GPS17]. In our version, we are working with ideals in consecutive endomorphism rings, rather than in the endomorphism ring of the starting curve, and we give proofs of correctness along with analysis of input size of left ideals of a maximal order.

Proposition 9. *Let E be a supersingular elliptic curve over $\mathbb{F}_{p^2}$, and assume $\iota : \mathrm{End}(E) \to \mathcal{O} \subseteq B_{p,\infty}$ is an isomorphism, where $\mathcal{O}$ has a basis of size polynomial in $\log p$. Let $I \subseteq \mathcal{O}$ be an ideal of norm ℓ^e for a prime $\ell \neq p$ with $\ell = O(\log p)$. For $k = 1, \ldots, e$, define $I_k := I + \mathcal{O} \cdot \ell^k$ and $J_k = I_{k-1}^{-1} I_k \subseteq \mathcal{O}_R(I_{k-1})$ and $E_k := E/E[\iota^{-1}(I_k)]$ as in Theorem 11. Then if we are given $\iota_{k-1}(\mathrm{End}(E_{k-1}))$ in $B_{p,\infty}$ where $\iota_{k-1} : \mathrm{End}(E_{k-1}) \otimes \mathbb{Q} \to B_{p,\infty}$ is an isomorphism of quaternion algebras, along with the action of $\mathrm{End}(E_{k-1})$ on $E_{k-1}[\ell]$, we can compute the kernel of the isogeny corresponding to $\iota_{k-1}^{-1}(J_k)$ in time polynomial in $\log p$.*

Proof. We wish to determine $E_{k-1}[\iota_{k-1}^{-1}(J_k)]$ so that we can compute the corresponding isogeny $\psi_k : E_{k-1} \to E_k$. If J_k has a $\mathbb{Z}$-basis $\gamma_1, \ldots, \gamma_4 \in \mathcal{O}_R(I_{k-1})$, we need to understand how the γ_i act as endomorphisms of E_{k-1}. Suppose we are given the action of generators $\phi_1, \ldots, \phi_4$ of $\mathrm{End}(E_{k-1})$ on $E_{k-1}[\ell]$ and the image of an embedding $\iota_{k-1} : \mathrm{End}(E_{k-1}) \to B_{p,\infty}$. Set $\mathcal{O}_{k-1} := \iota_{k-1}(\mathrm{End}(E_{k-1}))$; then we can compute $v \in B_{p,\infty}^{\times}$ such that $\mathcal{O}_{k-1} = v\mathcal{O}_R(I_{k-1})v^{-1}$ in polynomial time by [KV10]. By expressing $v\gamma_i v^{-1}$ in terms of $\iota_{k-1}(\phi_j)$, say

$$v\gamma_r v^{-1} = \sum_s c_{rs}\iota_{k-1}(\phi_s),$$

we discern the kernel of the isogeny corresponding to J_k as follows. We require a nonzero point $P \in E_{k-1}[\ell]$ such that for all $r = 1, \ldots, 4$,

$$\sum_s c_{rs}\phi_s(P) = 0.$$

Because we assume that we are given $\phi_s(P)$ for $s = 1, \ldots, 4$ and $P \in E_{k-1}[\ell]$, we can find such a P by just calculating the sum for all $r = 1, \ldots, 4$ and $P \neq 0 \in E_{k-1}[\ell]$. □

6.4 Isogeny Paths and Corresponding Filtrations of Left Ideals

Let $E, E'/\mathbb{F}_{p^2}$ be supersingular elliptic curves. We now prove the correctness of our earlier claims on how an ℓ-isogeny path between E and E' corresponds to a sequence of ideals of norm ℓ in $\mathrm{End}(E) \otimes \mathbb{Q}$. In particular, suppose $\phi : E \to E'$ has degree ℓ^e for some prime $\ell \neq p$. Then the kernel ideal I of ϕ in $\mathrm{End}(E)$ has degree ℓ^e. There is a factorization $\phi = \psi_e \circ \cdots \circ \psi_1$ with $\deg(\psi_k) = \ell$, and by setting $\phi_k := \psi_k \circ \cdots \circ \psi_1$, there is a corresponding ideal I_k of $\mathrm{End}(E)$ of norm ℓ^k. Additionally, there is an ideal J_k of $\mathcal{O}_R(I_{k-1})$ which corresponds to the factorization of the isogeny $\phi_k = \psi_k \circ \psi_{k-1}$; in this section, we construct I_k and J_k from I. Let I be a left ideal of $\mathrm{End}(E)$ of norm ℓ^e such that $I \not\subseteq \mathrm{End}(E) \cdot \ell^m$ for any positive integer m. In this section, we prove that for $k = 0, \ldots, e$, $I_k = I + \mathrm{End}(E) \cdot \ell^k$ is an ideal of norm ℓ^k and that

$$I = I_e \subseteq I_{e-1} \subseteq \cdots \subseteq I_1 \subseteq I_0 = \mathrm{End}(E).$$

We first establish when an ideal corresponds to an isogeny with cyclic kernel.

Proposition 10. *Suppose $I \subseteq \mathrm{End}(E)$ is a left ideal with $\mathrm{Nrd}(I)$ coprime to p. Then I is not contained in $\mathrm{End}(E) \cdot m$ for any $m \in \mathbb{N}$ if and only if $E[I]$ is cyclic.*

Proof. Suppose that $I \subseteq \mathrm{End}(E) \cdot m$. Then $E[I] \supset E[\mathrm{End}(E) \cdot m] = E[m]$ and thus $m | \deg(\phi_I)$. Since p does not divide $\deg(\phi_I)$, it also does not divide m, so $E[m] \neq 0$ and has rank two as a $\mathbb{Z}/m\mathbb{Z}$-module. Hence $E[I]$ is not cyclic. For the other direction, suppose that $E[I]$ is not cyclic. Then, by the structure theorem of abelian groups,

$$E[I] \simeq \bigoplus_{i=1}^{j} \mathbb{Z}/k_i\mathbb{Z}$$

and we can choose the k_i uniquely such that $k_i | k_{i+1}$. Since $E[I]$ is not cyclic, $j \neq 1$ and hence $E[I]$ has two elements of order k_1 which are linearly independent. Thus $E[k_1] \subseteq E[I]$ and hence $I \supset \mathrm{End}(E) \cdot k_1$. □

Proposition 11. *Suppose $I \subseteq \mathrm{End}(E)$ and $N := \mathrm{Nrd}(I)$ is coprime to p. Also suppose $M|N$, and that I is not contained in $\mathrm{End}(E) \cdot m$ for any $m \in \mathbb{N}$. Then $I + \mathrm{End}(E) \cdot M$ has norm M.*

Proof. We claim that

$$E[I + M\mathcal{O}] = E[I] \cap E[M].$$

Indeed, for an arbitrary left ideal J of $\mathrm{End}(E)$ with $\mathrm{Nrd}(J)$ coprime to p, $E[J]$ is the intersection of the kernels of a generating set of J, and for two left $\mathrm{End}(E)$-ideals J, J', $J + J'$ is generated by $J \cup J'$. Since $E[I]$ is cyclic by Proposition 10, there is some $Q \in E[N]$ so that $E[I] = \langle Q \rangle$. Then $E[I] \cap E[M] = \langle [N/M]Q \rangle$, a group of order M as desired. □

6.5 Matching up a Filtration of an Ideal with a Factorization of an Isogeny

In this section, we show that the definition of J_k in Algorithm 9 gives us the ideal which corresponds to the isogeny $E_{k-1} \to E_k$ of degree ℓ. To do this, it suffices to understand the horizontal isogeny and corresponding ideal in the following diagram:

$$\begin{array}{ccc} E & & \\ {\scriptstyle I_{k-1}}\downarrow & \searrow^{I_k} & \\ E_{k-1} := E/E[I_{k-1}] & \xrightarrow[J_k]{} & E_k := E/E[I_k] \end{array}$$

We will describe the relationship between the horizontal isogeny and its kernel ideal for two arbitrary left ideals I, I' of $\text{End}(E)$ satisfying $I' \subseteq I$, so in the above picture, we replace I_{k-1} with I and I_k with I'. The goal is to find, given $I' \subseteq I$, the horizontal isogeny $E_I \to E_{I'}$ by first computing its corresponding ideal $\tilde{J}$ in the following diagram:

$$\begin{array}{ccc} E & & \\ {\scriptstyle I}\downarrow & \searrow^{I'} & \\ E_I := E/E[I] & \xrightarrow[\tilde{J}]{} & E_{I'} := E/E[I'] \end{array}$$

Let $\phi_I : E \to E_I := E/E[I]$ and $\phi_{I'} : E \to E_{I'} := E/E[I']$ be the corresponding isogenies; then $E[I] \subseteq E[I']$ and hence $\phi_{I'}$ factors as $\phi_{I'} = \psi\phi_I$ for some isogeny $\psi : E_I \to E_{I'}$. We wish to view the kernel of ψ as $E_I[\tilde{J}]$ for some left ideal $\tilde{J}$ of $\text{End}(E_I)$. We make this idea precise in the following proposition.

Proposition 12. *Let $I' \subseteq I$ be two left $\text{End}(E)$-ideals whose norms are coprime to p. Then there exists a separable isogeny $\psi : E_I \to E_{I'}$ such that $\phi_I = \psi \circ \phi_{I'}$, and a left ideal $\tilde{J}$ of $\text{End}(E_I)$ with $E_I[\tilde{J}] = \ker(\psi)$ such that $J = \iota(\tilde{J}) = I^{-1}I'$, where $\iota : \text{End}(E_I) \to \text{End}(E) \otimes \mathbb{Q}$ is the map in Lemma 9 below.*

To prove this, we need the following three lemmas:

Lemma 8. *For a left ideal I of $\text{End}(E)$, the map*

$$\phi_I^* : \text{Hom}(E_I, E) \to I$$
$$\psi \mapsto \psi\phi_I$$

is an isomorphism of left $\text{End}(E)$-modules.

Proof. This is Lemma 42.2.6 of [Voi]. It also follows from Proposition 48 of [Koh96]. □

Lemma 9. *Set $B = \text{End}(E) \otimes \mathbb{Q}$. The map*

$$\iota : \text{End}(E_I) \to B$$
$$\beta \mapsto \frac{1}{\deg(\phi_I)}\widehat{\phi_I}\beta\phi_I$$

is injective, and its image is $\mathcal{O}_R(I)$.

Proof. This is Lemma 42.2.8 of [Voi] or Proposition 3.9 of [Wat69]. □

Lemma 10. *We have a bijection*

$$g : \mathrm{Hom}(E_{I'}, E_I) \to I^{-1}I'$$
$$\psi \mapsto \frac{1}{\deg(\phi_I)}\widehat{\phi_I}\psi\phi_{I'}.$$

Proof. This is Lemma 42.2.19 of [Voi]. □

Now we can prove the proposition.

Proof (Proof of Proposition 12). We have that $I^{-1} = \frac{1}{\mathrm{Nrd}(I)}\overline{I}$. Consider an element $x \in I^{-1}I'$ of the form

$$x = \frac{1}{\deg(\phi_I)}\widehat{\alpha}'\beta',$$

where $\alpha' \in I$, $\beta' \in I'$. Then by Lemma 8, there exists $\alpha \in \mathrm{Hom}(E_I, E)$ and $\beta \in \mathrm{Hom}(E_{I'}, E)$ with

$$\alpha' = \alpha\phi_I, \beta' = \beta\phi_{I'}.$$

Thus

$$x = \frac{1}{\deg(\phi_I)}\widehat{\phi_I}\widehat{\alpha}\beta\phi_{I'} = g(\widehat{\alpha}\beta),$$

where $g : \mathrm{Hom}(E_{I'}, E_I) \to I^{-1}I'$ is the map in Lemma 10. Since $E[I] \subseteq E[I']$, and $\phi_I, \phi_{I'}$ are separable, by Corollary III.4.11 of [Sil09] there exists a unique separable isogeny $\psi : E_I \to E_{I'}$ such that $\phi_{I'} = \psi \circ \phi_I$. Then define

$$\tilde{J} := \{\alpha \in \mathrm{End}(E_1) : \alpha(P) = 0 \quad \forall P \in \ker(\psi)\}.$$

Now map $g^{-1}(x) = \widehat{\alpha}\beta \in \mathrm{Hom}(E_{I'}, E_I)$ to an element of $\tilde{J}$ using ψ^*: $\widehat{\alpha}\beta\psi = \psi^*(\widehat{\alpha}\beta) \in \tilde{J}$. Finally, compute

$$\begin{aligned} x &= \frac{1}{\deg(\phi_I)}\widehat{\phi_I}\widehat{\alpha}\beta\phi_{I'} \\ &= \frac{1}{\deg(\phi_I)}\widehat{\phi_I}\widehat{\alpha}\beta\psi\phi_I \\ &= \iota(\widehat{\alpha}\beta\psi) \\ &= \iota(\psi^*(\widehat{\alpha}\beta)) \\ &= (\iota \circ \psi^* \circ g^{-1})(x). \end{aligned}$$

In other words, we have

$$g = \iota \circ \psi^*.$$

From this, we conclude that the left ideal of $\mathcal{O}_R(I_1)$ corresponding to $\tilde{J}$ indeed is $I^{-1}I'$. □

Combining the above results, we have our main theorem on matching up filtrations of ideals with factorizations of isogenies:

Theorem 11. *Suppose that* $I \subseteq \text{End}(E)$ *satisfies* $\text{Nrd}(I) = \ell^e$ *where* $\ell \neq p$ *is a prime and* $I \not\subseteq \text{End}(E) \cdot \ell^k$ *for any* $k \in \mathbb{N}$. *Then there exists a filtration*

$$I = I_e \subsetneq I_{e-1} \subsetneq \cdots \subsetneq I_1 \subsetneq I_0 = \text{End}(E)$$

and a chain of isogenies

$$E = E_0 \xrightarrow{\psi_1} E_1 \xrightarrow{\psi_2} \cdots \xrightarrow{\psi_{e-2}} E_{e-1} \xrightarrow{\psi_e} E_e = E'$$

such that if we set $\phi_k : E \to E/E[I_k]$, *then* $\phi_{k+1} = \psi_k \phi_k$. *Moreover, for* $k = 0, \ldots, e-1$, *the map* $\psi_{k+1} : E_k \to E_{k+1}$ *has degree* ℓ, *and its kernel ideal in* $\text{End}(E_k)$ *is isomorphic to* $I_k^{-1} I_{k+1} \subseteq \mathcal{O}_R(I_k)$ *under the map*

$$\iota_k : \text{End}(E_k) \to \mathcal{O}_R(I_k)$$
$$\rho \mapsto \frac{1}{\deg(\phi_k)} \hat{\phi}_k \rho \phi.$$

Proof. For $k = 0, 1, \ldots, e$, define $I_k := I + \text{End}(E) \cdot \ell^k$. By Proposition 11, $\text{Nrd}(I_k) = \ell^k$. Let $\phi_I : E \to E_e := E/E[I_e] = E/E[I]$ be the isogeny corresponding to $I = I_e$. Set $\mathcal{O}_k := \mathcal{O}_R(I_k) \subseteq \text{End}(E) \otimes \mathbb{Q}$, and $J_k := I_{k-1}^{-1} I_k$. Then $\text{Nrd}(J_k) = \ell$. Let $E_k := E/E[I_k]$. From the ideals J_k, we have isogenies $\psi_k : E_{k-1} \to E_k$ such that

$$\phi = \psi_e \circ \cdots \circ \psi_1$$

by Proposition 12 applied inductively to the ideals $I_{k+1} \subsetneq I_k$. □

7 Some Easy Problems in Supersingular Isogeny Graphs

The previous sections relied heavily on the quaternion ℓ-isogeny algorithm of [KLPT14] to derive the computational equivalence of several problems. In this section, we provide two additional applications of this algorithm. First, we give an algorithm for constructing the Deuring correspondence from maximal orders in $B_{p,\infty}$ to supersingular j-invariants. Second, we give a polynomial-time collision algorithm against the Charles-Goren-Lauter hash function when a special curve is chosen as the initial point.

7.1 Constructive Deuring Correspondence, from Quaternion Orders to j-invariants

In this section we provide an efficient algorithm to solve Problem 1. Algorithm 12 first computes an ideal connecting $\mathcal{O}_0$ to $\mathcal{O}$. Then it uses the quaternion ℓ-isogeny algorithm from [KLPT14] (or rather, its powersmooth version) to compute another ideal in the same class but with a norm $N = \prod p_i^{e_i}$ such that $\max_i p_i^{e_i}$ is small. It finally translates that ideal into an isogeny $\phi : E_0 \to E_1$ that corresponds to it via Deuring's correspondence.

Algorithm 12. *Constructive Deuring correspondence, from maximal orders to j-invariants.*
Input: Maximal order $\mathcal{O} \subset B_{p,\infty}$.
Output: Supersingular j*-invariant* j *such that* $\mathrm{End}(E(j)) \simeq \mathcal{O}$.

1. *Compute an ideal* I *that is a left ideal of* $\mathcal{O}_0$ *and a right ideal of* $\mathcal{O}$.
2. *Compute an ideal* J *in the same class as* I *but with powersmooth norm.*
3. *Compute an isogeny* $\phi : E_0 \to E_I$ *that corresponds to* J *via Deuring's correspondence.*
4. *Return* $j(E_I)$.

Let $\langle 1, \omega_2, \omega_2, \omega_3 \rangle$ be a basis for $\mathcal{O}$, and let $M \in GL(4, \mathbb{Q})$ be such that $(1, \omega_2, \omega_2, \omega_3) = M(1, i, j, k)$. Let B be a bound on the numerators and denominators of all the coefficients of M.

Proposition 13. (Constructive Deuring Correspondence). *Under plausible heuristic assumptions, Algorithm 12 can be implemented to run in time polynomial in both* $\log B$ *and* $\log p$.

Proof. The analysis is similar to the proof of Proposition 5. □

We remark that this algorithm is implicitly used in the recent identification protocol of Galbraith, Silva and Petit [GPS17].

7.2 An Attack on the CGL Hash Function

It was shown in [CGL09] that computing collisions or preimages for the Charles-Goren-Lauter hash function amounts to computing large ℓ-power degree isogenies between two (possibly isomorphic) elliptic curves. The hardness arguments for these problems then essentially relied on the following arguments:

1. In general, these isogenies must have a degree so large that they cannot be efficiently computed with current algorithms.
2. The best known algorithms for these problems were variants that used birthday arguments, with an exponential complexity in the parameter's size [Gal99].

Paradoxically, the quaternion ℓ-isogeny algorithm [KLPT14] leads to both the security arguments of Sect. 5.5 and to a partial attack against the hash function. More precisely, in this section we present a collision attack for the hash function when the initial point used in the random walk is the special elliptic curve E_0 as constructed in Algorithm 3.

Our attack is summarized by Algorithm 13 below. We first compute $\alpha \in \langle 1, i, j, k \rangle \subset \mathcal{O}_0$ with $\mathrm{Nrd}(\alpha) = \ell^e$ for some e, which defines a sequence of ideals I_i corresponding to a loop starting and ending at $\mathcal{O}_0$. To ensure there is no backtracking in the loop (and moreover, that $\alpha \neq \ell^{e/2}$), we require that for any natural number k, $\ell^{-k}\alpha \notin \mathcal{O}_0$. Applying the translation algorithm directly to this sequence of ideals would have a prohibitive cost because ℓ^e is larger than p. As in Algorithm 7, we first replace each ideal in the sequence by another ideal in the same class but with powersmooth norm, and we apply the translation algorithm to each of them individually to obtain corresponding isogenies. The end vertices of these isogenies form a sequence of j-invariants that define a collision for the original elliptic curve version of the Charles-Goren-Lauter hash function.

Algorithm 13. *Collision attack on CGL hash function for special initial points*
Input: Special j_0 and $\mathcal{O}_0$ from Algorithm 3.
Output: A sequence of j-invariants $j_0, j_1, \ldots, j_e = j_0$ such that for any i there exists an isogeny of degree ℓ from $E(j_i)$ to $E(j_{i+1})$.

1. *Compute $e \in \mathbb{N}$ and $\alpha \in \langle 1, i, j, k \rangle \subset \mathcal{O}_0$ with $\mathrm{Nrd}(\alpha) = \ell^e$.*
2. *Compute a sequence of ideals $I_i = \mathcal{O}_0 q + \mathcal{O}_0 \ell^i$.*
3. *For all i:*
   (a) *Compute J_i with powersmooth norm in the same class as I_i.*
   (b) *Translate J_i into an isogeny $\varphi_i : E_0 \to E_i$.*
4. *Return $(j_0, j(E_1), j(E_2), \ldots, j(E_e) = j_0)$.*

To obtain an element whose norm is a power of ℓ in Step 1, we fix e large enough, then pick random values of y and z until the equation $w^2 + qx^2 = \ell^e - p(y^2 + qz^2)$ can be solved with Cornacchia's algorithm. This solution is described in Algorithm 14.

Algorithm 14. *ℓ-power norm element in $\mathcal{O}_0$*
Input: Maximal order $\mathcal{O}_0 \subset B_{p,\infty}$ as defined in Proposition 1.
Output: $e \in \mathbb{N}$ and $\alpha \in \mathcal{O}_0$ with $\mathrm{Nrd}(\alpha) = \ell^e$.

1. *Let $e = \lceil 2 \log p \rceil$.*
2. *Choose random y, z smaller than $\sqrt{p/q}$.*
3. *Let $N \leftarrow \ell^e - p(y^2 + qz^2)$.*
4. *Find $w, x \in \mathbb{Z}$ such that $w^2 + qx^2 = N$ if there are some, otherwise go to Step 2.*
5. *Return $\alpha = w + xi + yj + zk$.*

Proposition 14. *There exists an algorithm that computes a collision for the Charles-Goren-Lauter hash function when the initial vertex is a special curve in time polynomial in $\log p$.*

Proof. In Algorithm 14 we expect that the equation in Step 4 will have a solution for $1/2q \log p$ of the random choices (y, z), so we expect this algorithm to run in time polynomial in $\log p$. Note that $e = \lceil 2 \log p \rceil$, and that Steps 4 and 5 in Algorithm 13 both run in time polynomial in $\log p$. We conclude that the runtime of Algorithm 13 is also polynomial in $\log p$. To ensure there is no backtracking in the loop in the isogeny graph, we require that the ideal $\mathcal{O}_0 \alpha$ satisfies $\mathcal{O}_0 \alpha \not\subset \mathcal{O}_0 \ell^k$ for any k. □

We remark that we described our attack only for the maximal orders $\mathcal{O}_0$ defined in Proposition 1, but it can be extended to other maximal orders as long as the corresponding curve is known or can be computed, and as long as elements of norm a power of ℓ can be found in the order. This is the case for "special" orders, as defined in [KLPT14].

The attack provided by Algorithm 13 can be extended into a "backdoor attack" where an entity in charge of deciding the initial vertex for the hash function plays the role of the attacker. This entity could take a random walk from j_0 to another curve E and publish this $j(E)$ as the initial vertex for the hash function. Due to the random walk the vertex $j(E)$ will be uniformly distributed, hence the function will be collision resistant based on the assumption that the endomorphism ring computation problem is hard (see Proposition 8). However, the entity can concatenate the path from j_0 to j and the collision which begins and ends at j_0 to obtain a collision which begins and ends at j.

To the best of our knowledge, there exists no efficient algorithm to sample supersingular j-invariants that does not involve this random walk procedure, so the backdoor attack cannot really be avoided. On the other hand, by inspecting such a collision, it is easy to recover a path to $\mathcal{O}_0$ and that will reveal that a backdoor was inserted. In that sense, the backdoor mechanism may not be too much of an issue in practice.

8 The EndomorphismRing Problem

In this section we provide an alternative study of the computational hardness of computing endomorphism rings of supersingular elliptic curves. The inputs are p and the curve, and so the running time must be polynomial in $\log p$. This brings up two important questions: (1) Does the endomorphism ring of an elliptic curve have a polynomial representation size? And (2) If it does, can the endomorphisms be evaluated in polynomial time? To have any meaningful efficient reduction, or to analyze how hard it is to compute the endomorphism ring, we need to know what the representation size of an endomorphism ring is. In particular, we need to discuss what we mean by *computing the endomorphism ring*.

We will define a compact representation of endomorphisms which has polynomial size, and show that the endomorphism ring of any supersingular elliptic curve has a basis of such representations. This answers question 1. We also show that these representations can be evaluated efficiently at arbitrary points, answering question 2. We then define the problem EndomorphismRing in terms of this new definition, and show that it efficiently reduces to MaximalOrder and Action-on-ℓ-Torsion for $\ell = 2, 3$. Our definition of compact representations is implicitly used in Algorithm 4. We also identify another problem that it reduces to, which is related to computing isogenies.

8.1 Representation Size of Endomorphism Rings

There are two typical ways to represent the endomorphism ring of E. The first is to give rational functions $F_1(x, y), \ldots, F_4(x, y)$ and $G_1(x, y), \ldots, G_4(x, y)$ such that $\phi_i : (x, y) \mapsto (F_i(x, y), G_i(x, y))$ $(i = 1, \ldots, 4)$ are endomorphisms of E that form a basis for $\mathrm{End}(E)$. The second is to give the kernel of the maps ϕ_i, which in general is not good enough for computations. However, it is not known if a basis for $\mathrm{End}(E)$ exists in either representation that is of polynomial size. For example, the basis may contain an endomorphism of exponential degree, where exponentially many coefficients would be needed to describe it in general. For the case of using the kernel, the generators may lie in a finite field of exponential degree over the base field, and there will be exponentially many points in the kernel.

8.2 Compact Representations of Endomorphisms

We will now show that the endomorphism ring $\mathrm{End}(E)$ of any supersingular elliptic curve $E/\mathbb{F}_{p^2}$ has compact representations if $p \equiv 3 \pmod 4$. The proof will require a special curve E_0 for which a basis of the endomorphism ring is known; such a curve exists if $p \not\equiv 1 \pmod{12}$.

For simplicity, we will focus on the case where $p \equiv 3 \pmod 4$ is a prime and let $E_0 : y^2 = x^3 + x$. Let $\pi : E_0 \to E_0$ denote the Frobenius map, and let $\phi : E_0 \to E_0$ be the map $(x, y) \mapsto (-x, \sqrt{-1}y)$. The maps $1 + \phi\pi$ and $\phi + \pi$ both have kernels containing $E[2]$, so they factor through the map $[2] : E_0 \to E_0$. Let $(1 + \phi\pi)/2$ and $(\phi + \pi)/2$ represent the maps in these factorizations. It can be shown that $1, \phi, (1 + \phi\pi)/2, (\phi + \pi)/2$ form a basis for $\mathrm{End}(E_0)$, see [GPS17]. As rational maps, the size of this basis may not be polynomial in $\log p$, but the description as rational linear combinations of $1, \phi, \pi, \phi\pi$ uniquely identifies them, and so it is enough that ϕ and π have polynomial size. This representation allows for efficient evaluation at points P of E_0 by writing $P = [2]Q$ and then evaluating linear combinations of $1, \phi, \pi, \phi\pi$ at Q. Define $[\beta_1, \beta_2, \beta_3, \beta_4] := [1, \phi, (1 + \phi\pi)/2, (\phi + \pi)/2]$. We will use $\beta_1, \beta_2, \beta_3, \beta_4$ in our definition of compact representatives of endomorphisms for all other supersingular elliptic curves $E/\mathbb{F}_{p^2}$.

Definition 1 (Compact representation of an endomorphism). Let $p \equiv 3 \pmod 4$ be a prime, let $E_0 : y^2 = x^3 + x$, and $\beta_1, \ldots, \beta_4 := 1, \phi, (1 + \phi\pi)/2, (\phi + \pi)/2$ be the endomorphisms of E_0 as above. Let $E/\mathbb{F}_{p^2}$ be another supersingular elliptic curve, and let $\rho \in \mathrm{End}(E)$. Define a *compact representation* of ρ to be a list

$$[d, [c_1, \ldots, c_4], [\phi_1, \ldots, \phi_m], [\widehat{\phi_1}, \ldots, \widehat{\phi_m}]],$$

where $c_1, \ldots, c_4, d \in \mathbb{Z}$, ϕ_i are isogenies on a path from E_0 to E, the total size of the list

$$\log(|d|) + \log(|c_1|) + \cdots + \log(|c_4|) + \sum_{i=1}^{m} \log(\deg(\phi_m))$$

is at most polynomial in $\log p$, and

$$\rho = \frac{1}{d}\left(\phi_m \circ \cdots \circ \phi_1 \circ \left(\sum_{i=1}^{4} c_i\beta_i\right) \circ \widehat{\phi_1} \circ \cdots \circ \widehat{\phi_m}\right).$$

Theorem 15. *Let $p \equiv 3 \pmod 4$ and let $E/\mathbb{F}_{p^2}$ be a supersingular elliptic curve. Then there exist two lists of four compact representatives of endomorphisms of E, such that each list represents a $\mathbb{Z}$-basis of* $\mathrm{End}(E)$.

Moreover, assume $\rho \in \mathrm{End}(E)$ is a linear combination of the endomorphisms corresponding to one such basis, and assume that its coefficient vector in terms of this basis is of size polynomial in $\log p$. Using the two lists, we can evaluate ρ at arbitrary points of E in time polynomial in $\log p$ and the size of the point P.

Proof. Let $\mathcal{O}_0$ be the maximal order in $B_{p,\infty}$ with basis

$$b_1, \ldots, b_4 := 1, i, (1+ij)/2, (i+j)/2.$$

Then $\mathcal{O}_0 \cong \mathrm{End}(E_0)$ and $b_1, \ldots, b_4$ correspond to $\beta_1, \ldots, \beta_4$ under an isomorphism. There exist chains of isogenies $\phi_1, \ldots, \phi_m$ and $\psi_1, \ldots, \psi_n$ between E_0 and E with $\deg(\phi_k) = 2$ and $\deg(\psi_k) = 3$, and with $m, n = O(\log p)$. Set $\phi = \phi_m \circ \cdots \circ \phi_1$ and $\psi = \psi_n \circ \cdots \circ \psi_1$. Let $I \subseteq \mathcal{O}_0$ and $J \subseteq \mathcal{O}_0$ be the left $\mathcal{O}_0$-ideals corresponding to ϕ and ψ respectively.

There exist rational numbers c^I_{rs} whose denominators are divisors of $2\,\mathrm{Nrd}(I)$ and rational numbers c^J_{rs} whose denominators are divisors of $2\,\mathrm{Nrd}(J)$ such that

$$\gamma^I_r := \sum_s c^I_{rs} b_s, 1 \le r \le 4$$

is a a Minkowski-reduced basis of $\mathcal{O}_R(I)$, and

$$\gamma^J_r := \sum_s c^J_{rs} b_s, 1 \le r \le 4$$

is a Minkowski-reduced basis of $\mathcal{O}_R(J)$. This follows from Theorem 2 and its proof. We can also efficiently find $v \in B_{p,\infty}$ such that $v\mathcal{O}_R(I)v^{-1} = \mathcal{O}_R(J)$, see [KV10].

Then $\rho^J_r := \frac{1}{2^m}\phi\gamma^I_r\widehat{\phi}$ and $\rho^I_r := \frac{1}{3^n}\psi\gamma^J_r\widehat{\psi}$ $(r = 1, \ldots, 4)$ each form a basis for $\mathrm{End}(E)$. Then our compact representations are, for $r = 1, \ldots, 4$,

$$[\mathrm{Nrd}(I), c^I_{r1}, \ldots, c^I_{r4}, [\phi_1, \ldots, \phi_m,], [\widehat{\phi_1}, \ldots, \widehat{\phi_m}]],$$
$$[\mathrm{Nrd}(J), c^J_{r1}, \ldots, c^J_{r4}, [\psi_1, \ldots, \psi_n], [\widehat{\psi_1}, \ldots, \widehat{\psi_n}]].$$

Observe that we can efficiently evaluate ρ^J_r at any point P of E whose order is coprime to 2. This is because $[2^m]\rho^I_r$ can be evaluated at P as it is a composition of the $\widehat{\phi_k}$, an integer linear combination of the β_k and then ϕ_k, all of which we can efficiently evaluate in terms of the size of P. Set $Q = [2^m]\rho^I_r(P)$. Let N be the inverse of 2^m modulo the order of P. Then $[N]Q = \rho^I_r(P)$.

If we want to evaluate ρ^I_r at a point P with $P \in E[2^f]$, we will instead express $v\rho^I_r v^{-1}$ as an integral linear combination of $\rho^J_1, \ldots \rho^J_4$. We can evaluate each $\rho^J_1, \ldots, \rho^J_4$ at any point of order coprime to 3 by the same argument.

Thus we can evaluate at arbitrary points P: if P has order $2^f M$ with $(2, M) = 1$, then we can write P as a sum of a point P_2 of order 2^f and P_M of order M. We can then evaluate at P by evaluating it at each summand with the two above strategies. □

Computing compact representations of endomorphisms which can be evaluated at points of E and which generate $\mathrm{End}(E)$ is a natural interpretation of the problem of computing endomorphism rings, so we formally state it here before relating it to other isogeny problems.

Problem 6 (EndomorphismRing). *Given a prime p and a supersingular elliptic curve $E/\mathbb{F}_{p^2}$, find a list of total length bounded by $O(\log p)$ of compact representations of endomorphisms of E such that using this list, we can evaluate the corresponding endomorphisms at points of E, and such that the corresponding endomorphisms generate* $\mathrm{End}(E)$ *as a $\mathbb{Z}$-module.*

In the next section, we will discuss two reductions from EndomorphismRing.

8.3 EndomorphismRing Reduces to MaxOrder and Action-on-2-Torsion and Action-on-3-Torsion

In Algorithm 9, we used embeddings of endomorphism rings in $B_{p,\infty}$, together with their action on ℓ-torsion, to construct an ℓ-isogeny.

Theorem 16. *If $p \equiv 3 \pmod 4$, EndomorphismRing reduces to MaxOrder and Action-on-ℓ-Torsion for $\ell = 2$ and 3.*

Proof. Let E be a supersingular elliptic curve. Let E_0 be the curve $y^2 = x^3 + x$ and let $\mathcal{O}_0$ be the order isomorphic to $\mathrm{End}(E_0)$. By Theorem 15, the necessary data to give compact representations of generators of $\mathrm{End}(E)$ is a 2-power and 3-power isogeny from E_0 to E, and a basis for the right orders of the ideals which correspond to these isogenies in $B_{p,\infty}$. In the proof of Theorem 10, note that all of this data is constructed using the oracles for MaxOrder, and Problems Action-on-2-Torsion and Action-on-3-Torsion. □

8.4 EndomorphismRing Reduces to an Isogeny Problem

We can also reduce the problem EndomorphismRing to a variant of the ℓ-Isogeny Problem, where we require the ℓ-power isogeny to be represented both by a chain of ℓ-isogenies and by a left ideal in a maximal order.

Problem 7 (FindKernelIdeal). *Given a prime p and a sequence of supersingular elliptic curves $E_0, \ldots, E_{m-1}$ and ℓ-isogenies $\phi_k : E_{k-1} \to E_k$, $k = 1, \ldots, m$, with $m = O(\log p)$, along with a maximal order $\mathcal{O}_0 \subseteq B_{p,\infty}$ isomorphic to* $\mathrm{End}(E_0)$*, compute the ideal I of $\mathcal{O}_0 \subseteq B_{p,\infty}$ corresponding to $\phi_m \circ \cdots \circ \phi_1 : E_0 \to E_m$.*

Theorem 17. *Problem EndomorphismRing reduces in polynomial time to Problems ℓ-PowerIsogeny and FindKernelIdeal.*

Proof. Let E be a supersingular elliptic curve. Assume we are given $\phi_1, \dots, \phi_m$ and $\psi_1, \dots, \psi_n$ whose compositions are 2^m- and 3^n-isogenies $E_0 \to E$ and m, n are $O(\log p)$. Also assume we are given ideals A and B of $\mathcal{O}_0$ such that A is the kernel ideal of $\phi := \phi_m \circ \cdots \phi_1 : E_0 \to E$ and B is the kernel ideal of $\psi := \psi_m \circ \cdots \circ \psi_1$. Then we can compute $\mathbb{Z}$-bases of $\mathcal{O}_R(A)$ and $\mathcal{O}_R(B)$. The sequences $\{\phi_r\}$ and $\{\psi_s\}$ for $r = 1, \dots, m$ and $s = 1, \dots, n$, along with $\mathbb{Z}$-bases of $\mathcal{O}_R(A)$ and $\mathcal{O}_R(B)$, give us the compact representations of generators of $\mathrm{End}(E)$ constructed in the proof of Theorem 15. □

Acknowledgments. We thank John Voight for many helpful discussions regarding orders in quaternion algebras and their connection with supersingular elliptic curves. We would also like to thank the anonymous referees for their helpful suggestions and corrections.

References

[ACC+17] Azarderakhsh, R., Campagna, M., Costello, C., De Feo, L., Hess, B., Jalali, A., Jao, D., Koziel, B., LaMacchia, B., Longa, P., Naehrig, M., Renes, J., Soukharev, V., Urbanik, D.: Supersingular isogeny key encapsulation. Submission to the NIST Post-Quantum Standardization Project (2017). https://csrc.nist.gov/Projects/Post-Quantum-Cryptography/Round-1-Submissions

[BJS14] Biasse, J.-F., Jao, D., Sankar, A.: A quantum algorithm for computing isogenies between supersingular elliptic curves. In: Meier, W., Mukhopadhyay, D. (eds.) INDOCRYPT 2014. LNCS, vol. 8885, pp. 428–442. Springer, Cham (2014). https://doi.org/10.1007/978-3-319-13039-2_25

[Brö09] Bröker, R.: Constructing supersingular elliptic curves. J. Comb. Number Theory **1**(3), 269–273 (2009)

[Cer04] Cerviño, J.M.: Supersingular elliptic curves and maximal quaternionic orders. Mathematisches Institut. Georg-August-Universität Göttingen: Seminars Summer Term 2004, pp. 53–60. Universitätsdrucke Göttingen, Göttingen (2004)

[CG14] Chevyrev, I., Galbraith, S.D.: Constructing supersingular elliptic curves with a given endomorphism ring. LMS J. Comput. Math. **1**(suppl. A), 71–91 (2014)

[CGL06] Charles, D., Goren, E., Lauter, K.: Cryptographic hash functions from expander graphs. Cryptology ePrint Archive, Report 2006/021 (2006). https://eprint.iacr.org/2006/021

[CGL09] Charles, D.X., Goren, E.Z., Lauter, K.: Cryptographic hash functions from expander graphs. J. Cryptol. **22**(1), 93–113 (2009)

[Cor08] Cornacchia, G.: Su di un metodo per la risoluzione in numeri interi dell' equazione $\sum_{h=0}^{n} c_h x^{n-h} y^h = p$. Giornale di Matematiche di Battaglini **46**, 33–90 (1908)

[Deu41] Deuring, M.: Die Typen der Multiplikatorenringe elliptischer Funktionenkörper. Abh. Math. Sem. Univ. Hambg. **14**(1), 197–272 (1941)

[DFJP14] De Feo, L., Jao, D., Plût, J.: Towards quantum-resistant cryptosystems from supersingular elliptic curve isogenies. J. Math. Cryptol. **3**(3), 209–247 (2014)

[DG16] Delfs, C., Galbraith, S.D.: Computing isogenies between supersingular elliptic curves over $\mathbb{F}_p$. Des. Codes Cryptogr. **78**(2), 425–440 (2016)

[EHM17] Eisenträger, K., Hallgren, S., Morrison, T.: On the hardness of computing endomorphism rings of supersingular elliptic curves. Cryptology ePrint Archive, Report 2017/986 (2017). https://eprint.iacr.org/2017/986

[Gal99] Galbraith, S.D.: Constructing isogenies between elliptic curves over finite fields. LMS J. Comput. Math. **2**, 118–138 (1999)

[GPS17] Galbraith, S.D., Petit, C., Silva, J.: Identification protocols and signature schemes based on supersingular isogeny problems. In: Takagi, T., Peyrin, T. (eds.) ASIACRYPT 2017. LNCS, vol. 10624, pp. 3–33. Springer, Cham (2017). https://doi.org/10.1007/978-3-319-70694-8_1

[GPST16] Galbraith, S.D., Petit, C., Shani, B., Ti, Y.B.: On the security of supersingular isogeny cryptosystems. In: Cheon, J.H., Takagi, T. (eds.) ASIACRYPT 2016, Part I. LNCS, vol. 10031, pp. 63–91. Springer, Heidelberg (2016). https://doi.org/10.1007/978-3-662-53887-6_3

[Gro87] Gross, B.H.: Heights and the special values of L-series. In: Number Theory, Montreal, QC, 1985. CMS Conference Proceedings, vol. 7, pp. 115–187. American Mathematical Society, Providence (1987)

[GW17] Gélin, A., Wesolowski, B.: Loop-abort faults on supersingular isogeny cryptosystems. In: Lange, T., Takagi, T. (eds.) PQCrypto 2017. LNCS, vol. 10346, pp. 93–106. Springer, Cham (2017). https://doi.org/10.1007/978-3-319-59879-6_6

[HLW06] Hoory, S., Linial, N., Wigderson, A.: Expander graphs and their applications. Bull. Amer. Math. Soc. (N.S.) **43**(4), 439–561 (2006)

[JDF11] Jao, D., De Feo, L.: Towards quantum-resistant cryptosystems from supersingular elliptic curve isogenies. In: Yang, B.-Y. (ed.) PQCrypto 2011. LNCS, vol. 7071, pp. 19–34. Springer, Heidelberg (2011). https://doi.org/10.1007/978-3-642-25405-5_2

[KLPT14] Kohel, D., Lauter, K., Petit, C., Tignol, J.-P.: On the quaternion ℓ-isogeny path problem. LMS J. Comput. Math. **17**, 418–432 (2014)

[Koh96] Kohel, D.: Endomorphism rings of elliptic curves over finite fields. Ph.D. thesis, University of California, Berkeley (1996)

[KV10] Kirschmer, M., Voight, J.: Algorithmic enumeration of ideal classes for quaternion orders. SIAM J. Comput. **39**(5), 1714–1747 (2010)

[Lan87] Lang, S.: Elliptic Functions. Graduate Texts in Mathematics, vol. 112, 2nd edn. Springer, New York (1987). https://doi.org/10.1007/978-1-4612-4752-4. With an appendix by J. Tate

[LM04] Lauter, K., McMurdy, K.: Explicit generators of endomorphism rings of supersingular elliptic curves. Preprint (2004)

[LO77] Lagarias, J.C., Odlyzko, A.M.: Effective versions of the Chebotarev density theorem. In: Algebraic Number Fields: L-functions and Galois Properties: Proceedings of Symposium, Durham University, Durham, 1975, pp. 409–464. Academic Press, London (1977)

[Mes86] Mestre, J.-F.: La méthode des graphes. Exemples et applications. In: Proceedings of the International Conference on Class Numbers and Fundamental Units of Algebraic Number Fields, Katata, 1986, pp. 217–242. Nagoya University, Nagoya (1986)

[NIS16] NIST: Post-quantum cryptography (2016). http://csrc.nist.gov/Projects/Post-Quantum-Cryptography. Accessed 30 Sept 2017

[NS09] Nguyen, P.Q., Stehlé, D.: Low-dimensional lattice basis reduction revisited. ACM Trans. Algorithms **5**(4), 48 (2009). Article No. 46

[Pet17] Petit, C.: Faster algorithms for isogeny problems using torsion point images. In: Takagi, T., Peyrin, T. (eds.) ASIACRYPT 2017. LNCS, vol. 10625, pp. 330–353. Springer, Cham (2017). https://doi.org/10.1007/978-3-319-70697-9_12

[Piz80] Pizer, A.: An algorithm for computing modular forms on $\Gamma_0(N)$. J. Algebra **64**(2), 340–390 (1980)

[PL17] Petit, C., Lauter, K.: Hard and easy problems for supersingular isogeny graphs. Cryptology ePrint Archive, Report 2017/962 (2017). https://eprint.iacr.org/2017/962

[Rón92] Rónyai, L.: Algorithmic properties of maximal orders in simple algebras over Q. Comput. Complex. **2**(3), 225–243 (1992)

[Sil09] Silverman, J.H.: The Arithmetic of Elliptic Curves. Springer, New York (2009). https://doi.org/10.1007/978-0-387-09494-6

[Sim05] Simon, D.: Quadratic equations in dimensions 4, 5 and more. Preprint (2005)

[Ti17] Ti, Y.B.: Fault attack on supersingular isogeny cryptosystems. In: Lange, T., Takagi, T. (eds.) PQCrypto 2017. LNCS, vol. 10346, pp. 107–122. Springer, Cham (2017). https://doi.org/10.1007/978-3-319-59879-6_7

[Vél71] Vélu, J.: Isogénies entre courbes elliptiques. C. R. Acad. Sci. Paris Sér. A-B **273**, A238–A241 (1971)

[Vig80] Vignéras, M.-F.: Arithmétique des Algèbres de Quaternions. LNM, vol. 800. Springer, Heidelberg (1980). https://doi.org/10.1007/BFb0091027

[Voi] Voight, J.: Quaternion Algebras. Version v0.9.7, 3 September 2017

[Wat69] Waterhouse, W.C.: Abelian varieties over finite fields. Ann. Sci. École Norm. Sup. **4**(2), 521–560 (1969)

[YAJ+17] Yoo, Y., Azarderakhsh, R., Jalali, A., Jao, D., Soukharev, V.: A post-quantum digital signature scheme based on supersingular isogenies. In: Kiayias, A. (ed.) FC 2017. LNCS, vol. 10322, pp. 163–181. Springer, Cham (2017). https://doi.org/10.1007/978-3-319-70972-7_9

Leakage

On the Complexity of Simulating Auxiliary Input

Yi-Hsiu Chen[1(✉)], Kai-Min Chung[2(✉)], and Jyun-Jie Liao[2(✉)]

[1] Harvard John A. Paulson School of Engineering and Applied Sciences, Harvard University, Cambridge, USA
yihsiuchen@g.harvard.edu

[2] Institute of Information Science, Academia Sinica, Taipei, Taiwan
{kmchung,jjliao}@iis.sinica.edu.tw

Abstract. We construct a simulator for the simulating auxiliary input problem with complexity better than all previous results and prove the optimality up to logarithmic factors by establishing a black-box lower bound. Specifically, let ℓ be the length of the auxiliary input and ϵ be the indistinguishability parameter. Our simulator is $\tilde{O}(2^\ell \epsilon^{-2})$ more complicated than the distinguisher family. For the lower bound, we show the relative complexity to the distinguisher of a simulator is at least $\Omega(2^\ell \epsilon^{-2})$ assuming the simulator is restricted to use the distinguishers in a black-box way and satisfy a mild restriction.

1 Introduction

In the *simulating auxiliary inputs* problem [JP14], a joint distribution (X, Z) over $\{0,1\}^n \times \{0,1\}^\ell$ is given. the goal is to find a "low complexity" simulator function $h : \{0,1\}^n \to \{0,1\}^\ell$ such that (X, Z) and $(X, h(X))$ are indistinguishable by a family of distinguishers. The non-triviality comes from the "low complexity" requirement. Otherwise, one can simply hardcode the distributions $Z|_{X=x}$ for each x to approximate Z. We call the lemma that addresses this problem *Leakage Simulation Lemma.*

Theorem 1 (Leakage Simulation Lemma, informal). *Let $\mathcal{F}$ be a family of deterministic distinguishers from $\{0,1\}^n \times \{0,1\}^\ell$. For every joint distribution (X, Z) over $\{0,1\}^n \times \{0,1\}^\ell$, There exists a simulator function $h : \{0,1\}^n \to \{0,1\}^\ell$ with complexity* $\mathrm{poly}(2^\ell, \epsilon^{-1})$ *relative to $\mathcal{F}$ such that for all $f \in \mathcal{F}$,*

$$\left|\Pr\left[f(X,Z)=1\right] - \Pr\left[f(X,h(X))\right] = 1\right| \leq \epsilon.$$

The "relative complexity" means if we have oracle gates that compute functions in $\mathcal{F}$, then what is the circuit complexity of h when considering those oracle

Y.-H. Chen—Supported by NSF grant CCF-1749750.
K.-M. Chung—This research is partially supported by the 2016 Academia Sinica Career Development Award under Grant no. 23-17 and Ministry of Science and Technology, Taiwan, under Grant no. MOST 106-2628-E-001-002-MY3.

J. B. Nielsen and V. Rijmen (Eds.): EUROCRYPT 2018, LNCS 10822, pp. 371–390, 2018.
https://doi.org/10.1007/978-3-319-78372-7_12

gates [JP14]. A typical choice of a family of distinguishers is a set of all circuits of size s. In that case, we can get a simulator of size $s \cdot \mathrm{poly}(2^\ell, \epsilon^{-1})$.

The Leakage Simulation Lemma implies many theorems in computational complexity and cryptography. For instance, Jetchev and Pietrzak [JP14] used the lemma to give a simpler and quantitatively better proof for the leakage-resilient stream-cipher [DP08]. Also, Chung et al. [CLP15] apply the lemma[1] to study connections between various notions of Zero-Knowledge. Moreover, the leakage simulation lemma can be used to deduce the technical lemma of Gentry and Wichs [GW11] (for establishing lower bounds for succinct arguments) and the Leakage Chain Rule [JP14] for *relaxed-HILL pseudoentropy* [HILL99, GW11].

Before Jetchev and Pietrzak described the Leakage Simulation Lemma as in Theorem 1, Trevisan, Tulsiani and Vadhan proved a similar lemma called *Regularity Lemma* [TTV09], which can be viewed as a special case of the Leakage Simulation Lemma by restricting the family of distinguishers in certain forms. In [TTV09], they also showed that all Dense Model Theorem [RTTV08], Impagliazzo Hardcore Lemma [Imp95] and Weak Szemerédi Regularity Lemma [FK99] can be derived from the Regularity Lemma. That means the Leakage Simulation Lemma also implies all those theorems.

As the Leakage Simulation Lemma has many implications, achieving the better complexity bound in $\mathrm{poly}(\epsilon^{-1}, 2^\ell)$ is desirable. Notably, in certain parameter settings, the provable security level of a leakage-resilient stream-cipher can be improved significantly if we can prove the better bound for the Leakage Simulation Lemma with better complexity bound. (See the next section for a concrete example). Therefore, an interesting question is what is the optimal parameter complexity bound we can get for the Leakage Simulation Lemma? In this paper, we provide an improved upper bound and also show the bound is "almost" optimal.

1.1 Upper Bound Results

Previous Results. In [TTV09], they provided two different approaches for proving the Regularity Lemma. One is by the min-max theorem, and another one is via boosting-type of proof. Although it is not known whether the Regularity Lemma implies the Leakage Simulation Lemma directly, [JP14] adopted both techniques and used them to show the Leakage Chain Rule with complexity bound $\tilde{O}(2^{4\ell}\epsilon^{-4})$.[2] On the other hand, Vadhan and Zheng derived the Leakage Simulation Lemma [VZ13, Lemma 6.8] using so-called "uniform min-max theorem", which is proved via multiplicative weight update (MWU) method incorporating with KL-projections. The circuit complexity of the simulator they got is $\tilde{O}(s \cdot 2^\ell\epsilon^{-2} + 2^\ell\epsilon^{-4})$ where s is the size of the distinguisher circuits. Recently, Skórski also used the boosting-type method to achieve the bound $\tilde{O}(2^{5\ell}\epsilon^{-2})$ [Skó16a], then later improved it to $\tilde{O}(2^{3\ell}\epsilon^{-2})$ by incorporating the

[1] They also consider the interactive version.

[2] In the original paper, they claimed to achieve the bound $\tilde{O}(2^{3\ell}\epsilon^{-2})$. However, Skórski pointed out some analysis flaws [Skó16a].

subgradient method [Skó16b]. Note that the complexity bound in [VZ13] has an additive term, so their result is incomparable to the others.

Our Results. In this paper, we achieve the bound $\tilde{O}(2^\ell \epsilon^{-2})$ for relative complexity, which contains the best components out of three complexity bounds mentioned above. The algorithm we use is also of multiplicative weight update (MWU) method as in [VZ13] but without going through the uniform min-max theorem argument. The additive term $2^\ell \epsilon^{-4}$ in [VZ13] is due to the precision issue when performing multiplication of "real numbers". The saving of the additive term is based on the observation mentioned in [VZ13] – the KL-projection step in their MWU algorithm is not needed when proving the Leakage Simulation Lemma. Thus we can potentially simplify the circuit construction. Indeed, we prove that certain level of truncation on weights does not effect the accuracy too much but helps us reducing the circuit complexity. In Table 1, we list out and compare all previous results to ours.

Table 1. Summary of existing upper bound results and our results.

Paper	Techinque	Complexity of simulator
[JP14]	Min-max/Boosting	$\tilde{O}(s \cdot 2^{4\ell} \epsilon^{-4})$
[VZ13]	Boosting with KL-projection	$\tilde{O}(s \cdot 2^\ell \epsilon^{-2} + 2^\ell \epsilon^{-4})$
[Skó16a]	Boosting with self-defined projection	$\tilde{O}(s \cdot 2^{5\ell} \epsilon^{-2})$
[Skó16b]	Boosting with subgradient method	$\tilde{O}(s \cdot 2^{3\ell} \epsilon^{-2})$
This work	Boosting	$O(s \cdot \ell 2^\ell \epsilon^{-2})$
	Black-box lower bound	$\Omega(s \cdot 2^\ell \epsilon^{-2})$

Implication of Our Results. As mentioned before, our result yields a proof of better security in leakage-resilient stream-cipher. All previous results suffer from the term ϵ^{-4} [JP14, VZ13][3] or the $2^{3\ell}$ multiplicative factor [Skó16b] in the complexity bound. In particular, Skórski's gave legitimate examples [Skó16a] where the bounds in [JP14] and [VZ13] only guarantee trivial security bounds when ϵ is set to be 2^{-40}. On the other hand, the factor $2^{3\ell}$ (or even $2^{5\ell}$) is significant and makes the guaranteed security bound trivial when the leakage is more than few bits. Therefore, in some reasonable parameter settings, our bound is the only one that can achieve a useful security. Here is a concrete example. If we consider the stream cipher in [JP14] and follow the settings in [Skó16a, Sect. 1.6]: The underlying weak PRF has 256 bits security, the target cipher security is $\epsilon' = 2^{-40}$ and the round is 16. If the leakage is $\lambda = 17$ per rounds, then using our bound, we can guarantee the security against 2^{50}-size circuit but all the analyses in [JP14, VZ13, Skó16a] guarantee nothing.

[3] It appears as an additive complexity in [VZ13] and/or a multiplicative term in [JP14].

1.2 Lower Bound Results

Our Results. We show that the simulator must have a "relative complexity" $\Omega(2^\ell \epsilon^{-2})$ to the distinguisher family by establishing a black-box lower bound, where a simulator can only use the distinguishers in a black-box way. Our lower bound requires an additional mild assumption that the simulator on a given input x, does not make a query an $x' \neq x$ to distinguishers.[4] Querying at points different from the input seems not helpful, but that makes the behaviors on different inputs not completely independent, which causes a problem in analysis. Indeed, all the known upper bound algorithms (including the one in this work) satisfy the assumptions we made. Still, we leave it as an open problem to close this gap completely.

Comparison to Related Results. In [JP14], they proved a $\Omega(2^\ell)$ lower bound for relative complexity under a hardness assumption for one-way functions. Besides, there are also lower bound results on the theorems that implied by the Leakage Simulation Lemma, including Regularity Lemma [TTV09], Hardcore Lemma [LTW11], Dense Model Theorem [Zha11], Leakage Chain Rule [PS16] and Hardness Amplification [SV10, AS11]. The best lower bound one can obtain before this work is $\Omega(\epsilon^{-2})$ (from [LTW11, SV10, Zha11]) or $\Omega(2^\ell \epsilon^{-1})$ (from [PS16]). Thus our lower bound is the first tight lower bound $\Omega(2^\ell \epsilon^{-2})$ for Leakage Simulation Lemma. See Sect. 4.2 for more detailed comparison.

Proof Overview. We define an oracle and a joint distribution $(X, Z) \in \{0,1\}^n \times \{0,1\}^\ell$. Considering a family of the distinguishers that each of them makes a single query to the oracle, the simulator has to query the oracle at least $\Omega(2^\ell \epsilon^{-2})$ times to fool all the distinguishers in the family. Therefore, if the only way to access the oracle is through the distinguishers, the simulator must use at least $\Omega(2^\ell \epsilon^{-2})$ distinguishers.

We can treat Z as a randomized function of X. That is, if can define $g : \{0,1\}^n \to \{0,1\}^\ell$ such that $\Pr[g(x) = z] = \Pr[Z = z | X = x]$, then $(X, Z) = (X, g(X))$. The distribution we consider is that the function g is deterministic, but the images are "hidden" from the simulator. Note that it is impossible for a simulator to hardwire all 2^n images. If the oracle receives a query $(x, z) \in \{0,1\}^n \times \{0,1\}^\ell$ with $z = g(x)$, it returns an answer based on the distribution $\mathrm{Bern}(1/2 + \epsilon)$. Otherwise, use the distribution $\mathrm{Bern}(1/2)$. Intuitively, the goal of the simulator is to find $g(x)$ for a given input x. For each z, due to the anti-concentration bound, it has to make $\Omega(\epsilon^{-2})$ many queries to check if $g(x) = z$. And if it has to check a constant fraction of all $z \in \{0,1\}^n$, then the total query complexity is $\Omega(2^\ell \epsilon^{-2})$.

[4] Many black-box lower bounds in related contexts [LTW11, Zha11, PS16] (implicitly) make the same mild assumption. See Sect. 4.2 for more details.

2 Preliminaries

2.1 Basic Definitions

Notations. For a natural number n, $[n]$ denotes the set $\{1, 2, \ldots, n\}$ and U_n denotes the uniform distribution over $\{0,1\}^n$. For a finite set $\mathcal{X}$, $|\mathcal{X}|$ denotes its cardinality, and $U_{\mathcal{X}}$ denotes the uniform distribution over $\mathcal{X}$. For a distribution X over $\mathcal{X}$, $x \leftarrow X$ means x is a random sample drawn from X. Bern(p) denotes the Bernoulli distribution with parameter $0 \leq p \leq 1$. For any function f, $\tilde{O}(f)$ means $O(f \log^k f)$ and $\tilde{\Omega}(f)$ means $\Omega(f/\log^k f)$ for some constant $k > 0$.

Definition 1 (Statistical Distance). *Let X and Y be two random variables. The statistical distance (or total variation) between X and Y is denoted as*

$$\Delta(X,Y) = \sum_x \frac{1}{2} \left| \Pr[X = x] - \Pr[Y = x] \right|.$$

Also, we say X and Y are ϵ-close *if $\Delta(X,Y) \leq \epsilon$.*

Definition 2 (Indistinguishability). *Let X, Y be distributions over $\{0,1\}^n$. We say X and Y are* (s,ϵ)-indistinguishable *if for every circuit $f : \{0,1\}^n \to \{0,1\}$ of size s,*

$$\left| \mathop{\mathbb{E}}_{x \leftarrow X}[f(x)] - \mathop{\mathbb{E}}_{y \leftarrow Y}[f(y)] \right| \leq \epsilon.$$

2.2 Multiplicative Weight Update

Consider the following prediction game. In each round, a predictor makes a prediction and receive a payoff. There are N experts that the predictor can refer to. That is, the predictor can (randomly) choose an expert to follow. The goal of the predictor is to minimize total payoff in many rounds. We called the difference between the total payoff of predictor and of the best expert *regret*, which is the criterion we use to measure the performance of the predictor. The Multiplicative weight update (MWU) algorithm provides a good probabilistic strategy for prediction. The overview of the algorithm is as follows. In the first round, the predictor simply chooses an expert uniformly at random. In the following rounds, the predictor updates the probabilities of choosing experts "multiplicatively" according to their performances in the previous round. The formal algorithm and the guarantees by the MWU algorithm is stated below.

Lemma 1 (Multiplicative weight update [AHK12]**).** *Consider a T-round game such that in t-th round, the predictor chooses a distribution D_t over $[N]$, and obtains a payoff according to the function $f_t : [N] \to [0,1]$. Let $0 < \eta \leq 1/2$ be an update rate. If player 1 chooses D_t as in Algorithm 1, then for every $i \in [N]$,*

$$\sum_{t=1}^{T} \mathop{\mathbb{E}}_{j \leftarrow D_t}[f_t(j)] \leq \sum_{t=1}^{T} f_t(i) + \frac{\log N}{\eta} + T\eta.$$

In particular, if we set $\eta = \sqrt{\log N/T}$*, we have*

$$\sum_{t=1}^{T} \mathop{\mathbb{E}}_{j \leftarrow D_t} [f_t(j)] \leq \sum_{t=1}^{T} f_t(i) + O\left(\sqrt{T \log N}\right)$$

Algorithm 1. Multiplicative weight update

```
1 For all i ∈ [N] set w_i := 1.
2 for t := 1 to T do
3     Choose D_t such that D_t(i) ∝ w_i.
4     for i := 1 to N do
5         w_i := w_i · (1 − η)^{f_t(i)};
```

As the regret grows sub-linearly to T, the predictor can achieve δ average regret when T is large enough.

Corollary 1. *There exists* $T = O\left(\frac{\ln N}{\epsilon^2}\right)$ *such that for all* $i \in [N]$,

$$\frac{1}{T} \sum_t \mathop{\mathbb{E}}_{j \leftarrow D_t} [f_t(j)] \leq \frac{1}{T} \sum_t f_t(i) + \epsilon.$$

Freund and Schapire discovered the connection between MWU algorithm and zero sum game [FS96] by treating the best response of Player 2 as the payoff function. MWU algorithm not only gives a new proof of von Neumanns Min-Max Theorem, but also provides a way to "approximate" the universal strategy obtained by the Min-Max Theorem[5].

Lemma 2 ([FS96]). *Consider a zero-sum game between Player 1 and Player 2 whose (pure) strategy spaces are* $\mathcal{P}$ *and* $\mathcal{Q}$*, respectively, and* $|\mathcal{P}| = N$*. The payoff to Player 2 is defined by the function* $u : \mathcal{P} \times \mathcal{Q} \to [0, 1]$*. We apply the MWU algorithm (Algorithm 1) in the following way to get the mixed strategy* P^* *and* Q^**.*

1. *Treat each pure strategy in* $|\mathcal{P}|$ *as an expert. Let* P_t *denote the mixed strategy described by* D_t *(the i-th pure strategy is chosen with probability* $D_t(i)$*).*
2. *Let* Q_t *denote the best response of Player 2 to* P_t*. Namely*

$$Q_t = \min_{Q} \mathop{\mathbb{E}}_{p \leftarrow P_t, q \leftarrow Q} f(p, q)$$

3. *Set the payoff function in the MWU algorithm as* $f_t(\cdot) = M(\cdot, Q_t)$*.*
4. *Let* $P^* = \frac{1}{T} \sum_t P_t$ *and* $Q^* = \frac{1}{T} \sum_t Q_t$*.*

[5] It is called Non-uniform Min-Max Theorem in [VZ13].

If we conduct the above procedure for $T = O(\log N/\epsilon^2)$ rounds, the mixed strategies P^*, Q^* are almost the equilibrium strategies. That is

$$\max_q \mathop{\mathbb{E}}_{p \leftarrow P^*}[u(p,q)] - \epsilon \leq \max_Q \min_P \mathop{\mathbb{E}}_{p \leftarrow P, q \leftarrow Q}[u(p,q)]$$
$$= \min_P \max_Q \mathop{\mathbb{E}}_{p \leftarrow P, q \leftarrow Q}[u(p,q)] \leq \min_p \mathop{\mathbb{E}}_{q \leftarrow Q^*}[u(p,q)] + \epsilon.$$

3 Simulating Auxiliary Inputs

The formal description of Leakage Simulation Lemma with our improved parameters is as follows.

Theorem 2 (Leakage Simulation Lemma). *Let $n, \ell \in \mathbb{N}$, $\epsilon > 0$ and $\mathcal{F}$ be a collection of deterministic distinguishers $f : \{0,1\}^n \times \{0,1\}^\ell \to \{0,1\}$. For every distribution (X, Z) over $\{0,1\}^n \times \{0,1\}^\ell$, there exists a simulator circuit $h : \{0,1\}^n \to \{0,1\}^\ell$ such that*

1. *h has complexity $\tilde{O}(2^\ell \epsilon^{-2})$ relative to $\mathcal{F}$. i.e., h can be computed by an oracle-aided circuit of size $\tilde{O}(2^\ell \epsilon^{-2})$ with oracle gates are functions in $\mathcal{F}$.*
2. *(X, Z) and $(X, h(X))$ are indistinguishable by $\mathcal{F}$. That is, for every $f \in \mathcal{F}$,*

$$\left| \mathop{\mathbb{E}}_{(x,z) \leftarrow (X,Z)}[f(x,z)] - \mathop{\mathbb{E}}_{h, x \leftarrow X}[f(x, h(x))] \right| \leq \epsilon.$$

Set $\mathcal{F}$ to be a set of Boolean circuits of size at most s, we immediate have the following corollary.

Corollary 2. *Let $s, n, \ell \in \mathbb{N}$ and $\epsilon > 0$. For every distribution (X, Z) over $\{0,1\}^n \times \{0,1\}^\ell$, there exists a simulator circuit of size $s' = \tilde{O}(s \cdot 2^\ell \epsilon^{-2})$ such that (X, Z) and $(X, h(X))$ are (s, ϵ)-indistinguishable.*

3.1 Boosting

There are numbers of proof of Leakage Simulation Lemma as discussed in the introduction. We focus on the "boosting" type of proof as it usually gives us better circuit complexity. The boosting framework has the following structure:

1. Choose a proper initial simulator h.
2. If h satisfies the constraint above, return h. Otherwise, find $f \in \mathcal{F}'$ which violates the constraint.
3. Update h with f and repeat.

Previous proofs in the framework are different in how they update h and correspondingly how they prove the convergence. If the algorithm converges fast and each update does not take too much time, we can get an efficient simulator. Starting from [TTV09], then followed [JP14] and [Skó16a], they use *additive* update on the probability mass function of each $h(x)$. However, additive update may cause negative weights, so they need an extra efforts (Both algorithm-wise and complexity-wise) to fix it. Vadhan and Zheng use multiplicative weight update instead [VZ13], which not only avoids the issue above but also converges faster. However, the number of bits to represent weights increases drastically after multiplications, and that causes the $O(2^\ell \epsilon^{-4})$ additive term in the complexity. Since the backbone of our algorithm is same as in [VZ13], we review their idea first in the next section, and then show how the additive term can be eliminated in Sect. 3.3.

3.2 Simulate Leakage with MWU

In this section, we show how MWU algorithm helps in simulating auxiliary inputs and why we can achieve the low round complexity. It is convenient to think Z as a randomized function of X. That is, we can define $g : \{0,1\}^n \to \{0,1\}^\ell$ such that $\Pr[g(x) = z] = \Pr[Z = z | X = x]$, then $(X, Z) = (X, g(X))$. Essentially, the goal is to find an "efficient function" h to simulate g.

Now we show that how the simulation problem problem is related to a zero-sum game, thus can be solved via MWU algorithm. The first step is to remove the one-sided error constraint. Let $\mathcal{F}'$ denote the closure of $\mathcal{F}$ under complement, namely, $\mathcal{F}' = \{f, 1-f : f \in \mathcal{F}\}$. Then the indistinguishability constraint is equivalent to

$$\forall f \in \mathcal{F}', \quad \mathop{\mathbb{E}}_{h, x \leftarrow X}[f(x, h(x))] - \mathop{\mathbb{E}}_{g, x \leftarrow X}[f(x, g(x))] \leq \epsilon.$$

Then consider the following zero-sum game: Player 1 choose a simulator h, Player 2 choose a distinguisher f, and the payoff to Player 2 is

$$\mathop{\mathbb{E}}_{h, x \leftarrow X}[f(x, h(x))] - \mathop{\mathbb{E}}_{g, x \leftarrow X}[f(x, g(x))].$$

One can get a bounded relative complexity of g by simply applying Lemma 2 with treating all functions from $\{0,1\}^n$ to $\{0,1\}^\ell$ as pure strategies of Player 1. However, relative complexity is $O(s \cdot 2^n \ell \epsilon^{-2})$ and hence is inefficient. To solve the above issue, Vadhan and Zheng observed that the marginal distribution of X-part is fixed. Thus we can consider the MWU algorithm for every $X = x$, where in each run of MWU, the Player 1 strategy space is simply a distribution over $\{0,1\}^\ell$, hence the round complexity is merely $O(\ell/\epsilon^2)$.

While the framework Vadhan and Zheng's considered is more general, the proof is also more complicated. Below we give a simpler proof which only uses the no-regret property of MWU.[6] Note that any no-regret algorithms for expert learning will work for this proof. Indeed, by applying online gradient descent instead of MWU we will get an additive boosting simulator. Nevertheless, multiplicative weight update is optimal in expert learning, which explains why MWU converges faster than additive boosting proofs.

[6] We say an online decision-making algorithm is *no-regret* if the average regret tends to zero as T approaches infinity. See, e.g., [Rou16].

Algorithm 2. Construction of Simulator h

1 **Input**: $x \in \{0,1\}^n$
2 **Parameter**: $\epsilon > 0$
3 Let $T = O(n/\epsilon^2)$, $\eta = \sqrt{\log N/T}$.
4 For all $z \in \{0,1\}^\ell$, set $w_x(z) = 1$.
5 Let h_0 be a randomized function such that $\Pr[h_0(x) = z] \propto w_x(z)$.
6 **for** $t = 1 \to T$ **do**
7 $\quad$ $Let f_t \in \mathcal{F}' = arg\,max_{f \in \mathcal{F}'} \mathbb{E}_{h_{t-1}, x \leftarrow X}[f(x, h_{t-1}(x))] - \mathbb{E}_{g, x \leftarrow X}[f(x, g(x))]$.
8 $\quad$ **if** $\mathbb{E}_{h_{t-1}, x \leftarrow X}[f(x, h_{t-1}(x))] - \mathbb{E}_{g, x \leftarrow X}[f(x, g(x))] \leq \epsilon$ **then**
9 $\quad\quad$ **Return** $h_{t-1}(x)$ as the output $h(x)$
10 $\quad$ For all $z \in \{0,1\}^\ell$, set $w_x(z) = w_x(z) \cdot (1 - \eta)^{f_t(x,z)}$
11 $\quad$ Let h_t be a randomized function such that $\Pr[h_t(x) = z] \propto w_x(z)$.
12 **Return** $h_T(x)$ as the output $h(x)$

Lemma 3. *Let X be a distribution over $\{0,1\}^n$ and $g : \{0,1\}^n \to \{0,1\}^\ell$ be a randomized function. For a given error parameter ϵ, the function h defined by Algorithm 2 satisfies*

$$\forall f \in \mathcal{F}' , \quad \mathop{\mathbb{E}}_{x \leftarrow X}[f(x, h(x))] - \mathop{\mathbb{E}}_{x \leftarrow X}[f(x, g(x))] \leq \epsilon.$$

Proof. For a fixed x, if there exists $f \in \mathcal{F}'$ such that

$$\mathop{\mathbb{E}}_{h}[f(x, h(x))] - \mathop{\mathbb{E}}_{g}[f(x, g(x))] > \epsilon,$$

then the algorithm returns at the line 12. That means for all $t \in [T]$, we have

$$\mathop{\mathbb{E}}_{h_{t-1}}[f_t(x, h_{t-1}(x))] - \mathop{\mathbb{E}}_{g}[f_t(x, g(x))] > \epsilon, \tag{1}$$

and so

$$\frac{1}{T}\sum_{t=1}^{T} \mathop{\mathbb{E}}_{h_{t-1}}[f_t(x, h_{t-1}(x))] - \frac{1}{T}\sum_{t=1}^{T} \mathop{\mathbb{E}}_{g}[f_t(x, g(x))] > \epsilon, \tag{2}$$

However, by Corollary 1, for every $z \in \{0,1\}^\ell$,

$$\frac{1}{T}\sum_{t=1}^{T} \mathop{\mathbb{E}}_{h_{t-1}}[f_t(x, h_{t-1}(x))] \leq \frac{1}{T}\sum_{t=1}^{T} f_t(x, z) + \epsilon.$$

By taking z over $g(x)$, we get a contradiction. Therefore, for all $f \in \mathcal{F}$,

$$\mathop{\mathbb{E}}_{h}[f(x, h(x))] - \mathop{\mathbb{E}}_{g}[f(x, g(x))] > \epsilon.$$

Take the expectation of x over X, we conclude the lemma.

3.3 Efficient Approximation

Algorithm 2 provides a simulator which fools all distinguishers in $\mathcal{F}$ by error up to ϵ. However, we have only proved a bound for the number of iterations, but not for the complexity of h_T itself. Actually, the circuit complexity of a naive implementation of Algorithm 2 is not better than using additive boosting. Nevertheless, we will show that there exists an efficient way to implement h_T approximately, of which the complexity is not much larger than evaluating the distinguishers T times.

In below, we assume all functions $f \in \mathcal{F}$ has circuit complexity at most s. From Algorithm 2, we can see $h_T(x)$ returns z with probability proportional to $(1-\eta)^{\sum_i f_i(x,z)}$. A natural way to approximate h_T is to compute $(1-\eta)^{\sum_i f_i(x,z)}$ for each z and apply a rejection sampling. Without loss of generality, we can assume that $(1-\eta)$ can be represented in $O(\log \frac{1}{\eta})$ bits, and thus, it takes at most $O(k \log \frac{1}{\eta})$ to represent $(1-\eta)^k$ for $k \in \mathbb{N}$. Since $\sum_i f_i(x,z)$ is at most T, it takes $O(Ts + T^2 \log^2 \frac{1}{\eta})$ complexity to compute $(1-\eta)^{\sum_i f_i(x,z)}$ by naive multiplication, or $O(Ts + T^2 \log T \log \frac{1}{\eta})$ via lookup table. Therefore there exists an approximation of h_T of size $O((T^2 \log^2 \frac{1}{\eta} + Ts) \cdot 2^\ell)$, which is $\tilde{O}(s \cdot 2^\ell \epsilon^{-2} + 2^\ell \epsilon^{-4}))$ after expanding T and η. This is the complexity claimed in [VZ13]. As mentioned in [Skó16a], the $\tilde{O}(2^\ell \epsilon^{-4})$ term may dominate in some settings, so the bound in [VZ13] is not always better.

Now we state the idea of approximating normalized weights efficiently. Observe that weights are of the form $(1-\eta)^{\sum_i f_i(x,z)}$. If the total weight is guaranteed to be at least 1, then intuitively, truncating the weight at each $z \in \{0,1\}^\ell$ a little amount does not influence the result distribution too much. Hopefully, if the truncated values can be stored with a small number of bits, a lookup table which maps $\sum_i f_i(x,z)$ to the truncated value of $(1-\eta)^{\sum_i f_i(x,z)}$ is affordable. In the lemma below we formalize the above intuition.

Lemma 4. *Suppose there are two sequences of positive real numbers* $\{\gamma_i\}_{i\in[n]}$, $\{w_i\}_{i\in[n]}$ *such that* $\forall i \in [n]$, $\gamma_i \leq w_i$. *Let* $r = \sum_i \gamma_i / \sum_i w_i$ *and* X, X' *be a distribution over* $[n]$ *such that* $\Pr[X = i] \propto w_i$ *and* $\Pr[X' = i] \propto (w_i - \gamma_i)$, *respectively. Then* $\Delta(X, X') \leq \frac{r}{1-r}$.

Proof.

$$\begin{aligned}
\Delta(X, X') &= \frac{1}{2} \sum_z \left| \frac{w_z}{\sum_i w_i} - \frac{w_z - \gamma_z}{\sum_i (w_i - \gamma_i)} \right| \\
&= \frac{1}{2} \sum_z \left| \frac{\gamma_z \sum_i w_i - w_z \sum_i \gamma_i}{(\sum_i w_i)^2 (1-r)} \right| \\
&\leq \frac{1}{2} \sum_z \frac{w_z \sum_i \gamma_i + \gamma_z \sum_i w_i}{(\sum_i w_i)^2 (1-r)} \\
&= \frac{\sum_i w_i \sum_i \gamma_i}{(\sum_i w_i)^2 (1-r)} = \frac{r}{1-r}
\end{aligned}$$

where the inequality follows from the triangle inequality.

Corollary 3. *Let $h' : \{0,1\}^n \to \{0,1\}^\ell$ be a function which satisfies*

$$\Pr[h'(x) = z] = \frac{(1-\eta)^{\sum_i f_i(x,z)} - \gamma_{x,z}}{\sum_{z'} \left((1-\eta)^{\sum_i f_i(x,z')} - \gamma_{x,z'}\right)}$$

where

$$\gamma_{x,z} \leq \min\left\{(1-\eta)^{\sum_i f_i(x,z)}, \frac{\eta}{2^\ell(1+\eta)} \cdot \sum_{z'}(1-\eta)^{\sum_i f_i(x,z')}\right\}.$$

Then for any $x \in \mathcal{X}$, $h'(x)$ is η-close to $h_T(x)$.

By the above corollary, the following procedure gives a good approximation of h_T.

1. For every $z \in \{0,1\}^\ell$, compute $Adv(x,z) = \sum_i f_i(x,z) - \min_{z'}(\sum_i f_i(x,z'))$. This can be done by a circuit of size $O(2^\ell \cdot (sT + T\log T))$.
2. Because there is z_0 such that $Adv(x,z_0) = 0$, we have $\sum_z (1-\eta)^{Adv(x,z)} \geq 1$. Let $k = O(\ell \log(1/\delta))$ be the smallest integer which satisfies $2^{-k} \leq \frac{\eta}{2^\ell(1+\eta)}$. By Corollary 3, if we truncate $(1-\eta)^{Adv(x,z)}$ down to the closest multiple of 2^{-k}, the corresponding distribution is still η-close to $h_T(x)$. Let $h'(x)$ denote the truncated distribution.
3. Observe that the truncated value is positive only if $Adv(x,z)$ is less than some threshold $t = O(k/\eta)$. Therefore we can build a lookup table consists of the truncated value of $(1-\eta)^j$ for $j \in [t]$. Such table is of size $O(t \log t \cdot k)$. With this table we can query truncated value of $(1-\eta)^{Adv(x,z)}$ for each z.
4. By rejection sampling, we can sample a η-approximation of $h'(x)$ in at most $O(2^\ell \log(1/\delta))$ rounds, and each round takes only $O(k)$ time.

Let h^* be the circuit which uses above steps to approximate h_T. Since $\eta = O(\epsilon)$ and $h(x)$ is 2η-close to $h_T(x)$, we have

$$\mathop{\mathbb{E}}_{h,x \leftarrow X}[f(x,h(x))] - \mathop{\mathbb{E}}_{g,x \leftarrow X}[f(x,g(x))] \leq \epsilon + 2\eta = O(\epsilon)$$

for any $f \in \mathcal{F}'$. (Note that we can always rescale ϵ to make the final gap is at most ϵ.) Since the complexity of the first step dominates all other steps, h is of complexity $O(2^\ell \cdot (sT + T\log T)) = \tilde{O}(s \cdot 2^\ell \epsilon^{-2})$.

4 Lower Bound for Leakage Simulation

We have seen that there exists an MWU algorithm which combines only $O(\ell\epsilon^{-2})$ distinguishers to make a good simulator h. Besides, for every chosen distinguisher f the algorithm queries $f(x,z)$ for every $z \in \{0,1\}^\ell$ when computing $h(x)$. Therefore the algorithm makes $O(\ell 2^\ell \epsilon^{-2})$ queries in total. In the previous section, we also showed that evaluating the $O(\ell\epsilon^{-2})$ chosen distinguishers is the bottleneck of the simulation. Then a natural question arises: can we construct a simulator which makes fewer queries? It might be possible to find a boosting procedure using fewer distinguishers, or maybe we can skip some $z \in \{0,1\}^\ell$ when querying $f(x,z)$ for some f. However, in this section we will show that the MWU approach is almost optimal: any *black-box* simulator which satisfies an independence restriction has to make $\Omega(2^\ell \epsilon^{-2})$ queries to fool the distinguishers.

4.1 Black-Box Model

To show the optimality of the MWU approach, we consider black-box simulation, which means we only use only the distinguishers as black-box and does not rely on how they are implemented. Note that all known results of leakage simulation ([JP14, Skó16a, VZ13]) are black-box. Indeed, all the leakage simulation results are in the following form: first learn a set of distinguishers $\{f_1, \ldots, f_{q'}\}$ which is common for each x, then query $f_i(x,z)$ for each $z \in \{0,1\}^\ell$ and $i \in [q']$, and finally combine them to obtain the distribution of $h(x)$. The model we consider is more general than this form, so it also rules out some other possible black-box approaches.

Definition 3 (Simulator). *Given a function $g : \{0,1\}^n \to \{0,1\}^\ell$, a distribution X over $\{0,1\}^n$ and a set $\mathcal{F}$ of functions $\{0,1\}^{n+\ell} \to \{0,1\}$, we say function $h : \{0,1\}^n \to \{0,1\}^\ell$ is an $(\epsilon, X, \mathcal{F})$-*simulator of g *if*

$$\forall f \in \mathcal{F}\,, \left| \mathop{\mathbb{E}}_{g,x \leftarrow X} [f(x, g(x))] - \mathop{\mathbb{E}}_{h,x \leftarrow X} [f(x, h(x))] \right| \le \epsilon.$$

Definition 4 (Black-Box Simulator). *Let $\ell, m, a \in \mathbb{N}$ and $\epsilon > 0$. We say an oracle-aid simulation circuit $D^{(\cdot)}$ which takes two inputs $x \in \{0,1\}^n$ and $\alpha \in \{0,1\}^a$ is a black-box (ϵ, ℓ, m, a)-simulator with query complexity q if it satisfies the follows. For every function $g : \{0,1\} \to \{0,1\}^\ell$, distribution X over $\{0,1\}^n$ and a set of distinguishers $\mathcal{F}$ with $|\mathcal{F}| \le m$, there exists $\alpha \in \{0,1\}^a$ (which we call "advice string") such that $D^{\mathcal{F}}(\cdot, \alpha)$ is an $(\epsilon, X, \mathcal{F})$-simulator for g and D uses at most q oracle gates.*

We say a black-box simulator is a same-input *black-box simulator if for every $f \in \mathcal{F}$, D only queries $f(x, \cdot)$ when computing on input x. We say a black-box simulator is* non-adaptive *if the choice of the oracle queries (including the choice of f and query input) does not depend on any response of the oracle.*

Remark 1. A reasonable range of parameters are $\epsilon^{-1}, 2^\ell, \log |\mathcal{F}| < 2^{o(n)}$ since all the simulations we know is of complexity $\text{poly}(\epsilon^{-1}, 2^\ell, \log |\mathcal{F}|)$. Note that when we consider $\mathcal{F}$ to be the set of every distinguisher of size at most s, $\log |\mathcal{F}| = O(s \log s)$. Besides, we also assume $a = 2^{o(n)}$ so that the simulator cannot trivially take α as an expression of g.

The lower bound we prove in this paper is for *same-input black-box simulator.* The same-input assumption is also made in related works including [LTW11, Zha11, PS16]. See the next section for more discussions about the black-box models in related results.

It is not hard to see that all the boosting approaches we mentioned above are in this model: the advice α is of length $O(q \log |\mathcal{F}|)$ and stands for "which distinguishers should be chosen", and D queries every chosen distinguisher f with input (x, z) for every $z \in \{0,1\}^\ell$ when computing $D^{\mathcal{F},\alpha}(x)$. Moreover, these simulation algorithms are non-adaptive. We can write the MWU approach as the following corollary:

Corollary 4. *For every $0 < \epsilon < \frac{1}{2}$, $\ell, m \in \mathbb{N}$, there exists an non-adaptive same-input black-box (ϵ, ℓ, m, a)-simulator with query complexity $q = O(\ell 2^\ell \epsilon^{-2})$ and $a = \tilde{O}(q \log |\mathcal{F}|)$.*

Besides capturing all known simulators, our lower bound also rules out the adaptive approaches. Whether there exists a faster simulation not satisfying the same-input restriction is left open, but it is hard to imagine how querying different input is useful.

4.2 Main Theorem and Related Results

Theorem 3. *For every $2^{-o(n)} < \epsilon < 0.001$, $\ell = o(n)$, $\omega(2^\ell/\epsilon^3) < m < 2^{2^{o(n)}}$ and $a = 2^{o(n)}$, a same-input black-box (ϵ, ℓ, m, a)-simulator must have query complexity $q = \Omega(2^\ell \epsilon^{-2})$.*

Remark 2. For ϵ we require it to be smaller than some constant so that $\mathrm{Bern}(\frac{1}{2} + \Theta(\epsilon))$ is well defined. Besides, we also require the size of distinguisher set m to be large enough to guarantee that the simulator must "simulate" the function instead of fooling distinguishers one by one. As we saw in Remark 1, the range of parameters here is reasonable.

Before this paper, there were some lower bounds either for Leakage Simulation Lemma itself or for its implications. We classify these results by their models as follows.

- **Non-Adaptive Same-Input Black-Box Lower Bounds.** Recall that Leakage Simulation implies Hardcore Theorem and Dense Model Theorem. Lu et al. [LTW11] proved an $\Omega(\log(\frac{1}{\delta})/\epsilon^2)$ lower bound for query complexity in Hardcore Lemma proof where δ denotes the density of the hardcore set. By taking $\delta = \Theta(1)$ we can obtain an $\Omega(1/\epsilon^2)$ lower bound for query complexity of Leakage Simulation. Similarly, Zhang [Zha11] proved a lower bound for query complexity in Dense Model Theorem proof which implies the same $\Omega(1/\epsilon^2)$ lower bound.[7] Besides, Pietrzak and Skórski [PS16] proved a $\Omega(2^\ell/\epsilon)$ lower bound for leakage chain rule, which also implies a $\Omega(2^\ell/\epsilon)$ lower bound for Leakage Simulation. These lower bounds assume both the non-adaptivity and the independence of inputs.[8]

[7] The black-box model these results considered is more restricted. Actually, the black-box model in [LTW11] does not contain Holenstein's proof [Hol05]. Nevertheless, their proof for query lower bound also works for the model we define here.

[8] Interestingly, in the reduction from Leakage Chain Rule to Leakage Simulation, there exists a distinguisher in the reduction which only need to be queried on one adaptively chosen input. In this case non-adaptivity causes a 2^ℓ additive loss. This can be viewed as an evidence that adaptivity might be useful.

- **Non-Adaptive Black-Box Lower Bounds.** Impagliazzo [Imp95] proved that the Hardcore Lemma implies Yao's XOR Lemma [GNW95, Yao82], which is an important example of hardness amplification. Since the reduction is black-box, it is not hard to see that the $\Omega(\log(\frac{1}{\delta})/\epsilon^2)$ lower bound for hardness amplification proved by Shaltiel and Viola [SV10] is also applicable to Hardcore Lemma. Similarly, by setting $\delta = \Theta(1)$ we get a $\Omega(1/\epsilon^2)$ lower bound for Leakage Simulation. Moreover, this lower bound does not require the same-input assumption.[9] Nevertheless, the proof highly relies on non-adaptivity.
- **General Black-Box Lower Bounds.** Artemenko and Shaltiel [AS11] proved an $\Omega(1/\epsilon)$ lower bound for a simpler type of hardness amplification, and removed the non-adaptivity. Their result implies a general black-box lower bound for Leakage Simulation, but the lower bound is far from optimal.
- **Non-Black-Box Lower Bounds.** Trevisan, Tulsiani and Vadhan show that the simulator cannot be much more efficient than the distinguishers [TTV09, Remark 1.6]. Indeed, for any large enough $s \in \mathbb{N}$ they construct a function g such that any simulator h of complexity s can be distinguished from g by a distinguisher of size $\tilde{O}(ns)$. Jetchev and Pietrzak [JP14] also show an $\Omega(2^\ell \cdot s)$ lower bound under some hardness assumptions for one-way functions.

None of the existing results imply an optimal lower bound for Leakage Simulation. However, proving a lower bound for Leakage Simulation might be a simpler task, and it turns out that we can prove a lower bound of $\Omega(2^\ell \epsilon^{-2})$. The basic ideas is as follows, and would be further explained in the proof. To capture the 2^ℓ factor, for each distinguisher f and input x we hide information at $f(x, z)$ for a random z, similar to the proof in [PS16]. Then checking all z over $\{0, 1\}^\ell$ is necessary. Although the claim seems trivial, the analysis would be more complicated in our adaptive model. To capture the ϵ^{-2} factor, we utilize the anti-concentration of almost uniform Bernoulli distribution $\text{Bern}(\frac{1}{2} + \Theta(\epsilon))$, so that $\Omega(1/\epsilon^2)$ samples are needed to distinguish it from uniform distribution with constant probability. A similar concept can be found for example in [Fre95, LTW11, PS16]. Note that in [PS16] they only require an advantage of ϵ when distinguishing such Bernoulli distribution from uniform, which causes an $O(1/\epsilon)$ loss in complexity.

4.3 Proof of Theorem 3

Overview. We would like to show that there exists a function g and a set of distinguisher $\mathcal{F}$ such that any simulator h with limited queries to $\mathcal{F}$ cannot approximate g well. Since $|\mathcal{F}|$ is much larger than the number of queries, there exist some distinguishers which can distinguish g and any bad simulator h "fairly", i.e. these distinguishers are independent of h. Therefore more queries are required to successfully simulate g and fool $\mathcal{F}$. We will prove the existence of g and $\mathcal{F}$ by probabilistic argument.

[9] Actually, such assumption is not even natural in hardness amplification.

To make the simulation task as hard as possible, let g be a random function. Besides, for any distinguisher $f \in \mathcal{F}$, let $f(x,z)$ be a random bit drawn from $\text{Bern}(\frac{1}{2}+c_1\epsilon)$ for some constant c_1 if $z = g(x)$, or from $\text{Bern}(\frac{1}{2})$ otherwise, so that a query to f provides least possible information.[10] To understand such setting, we can imagine that there exists a random oracle $\mathcal{O}$ which takes input (x,z) and only return biased bit at $z = g(x)$ for each x. Then $g(x)$ is considered as the *key* to the oracle, and our goal is to find out the correct key. Each $f \in \mathcal{F}$ can be viewed as a collection of samples from the oracle with certain randomness. Intuitively, since $f(x, g(x))$ is only $\Theta(\epsilon)$ away from uniform, f can distinguish g and any bad simulator h which does not approximate g with constant probability. To approximate g well, we need to test all 2^ℓ keys to find the correct one. Besides, it requires $\Omega(1/\epsilon^2)$ samples to distinguish $\text{Bern}(\frac{1}{2}+\Theta(\epsilon))$ and $\text{Bern}(\frac{1}{2})$ with constant probability, so $\Omega(1/\epsilon^2)$ queries are required for each key to make sure we can distinguish the real key from other fake keys. Therefore a successful simulator h should make at least $\Omega(\epsilon^{-2}2^\ell)$ queries.

Now we proceed to the formal proof. Assume for contradiction that D is a black-box (ϵ, ℓ, m, a)-simulator with query complexity $q \le c_2(2^\ell\epsilon^{-2})$, where $c_2 = \frac{1}{360000}$. Let $g : \{0,1\}^n \to \{0,1\}^\ell$ be a random function such that for every $x \in \{0,1\}^n$, $g(x)$ is chosen uniformly at random from $\{0,1\}^\ell$. Let $\mathcal{F}$ be a set of random function defined in previous paragraph, and we specify that $c_1 = 30$. First we prove that given any fixed advice string α, the decision function $D^{\mathcal{F}}(,\alpha)$ cannot guess g correctly with high enough probability over the choice of $\mathcal{F}$ and g.

Lemma 5. *Fix α and let $h = D^{\mathcal{F},\alpha}$. For any $x \in \{0,1\}^n$, we have $\Pr[h(x) = g(x)] \le 1 - \frac{3}{c_1}$, where the probability is taken over the choice of $g(x)$, $f(x,\cdot)$ for every $f \in \mathcal{F}$ (abbreviated as $\mathcal{F}(x)$), and the randomness of h.*

Proof. Without loss of generality, assume that h has no randomness other than oracle queries. (We can obtain the same bound for probabilistic h by taking average over deterministic circuits.) We also assume that h always make q different queries by adding dummy queries.

Consider h as a decision tree where queries are the nodes and different answers represent different branches. For every fixed $g(x)$ and $\mathcal{F}(x)$, the computation of $h(x)$ corresponds to a root-to-leaf path denoted as $t = \{a_1, \ldots, a_q\}$ where a_i is the answer to the i-th query, and we call t *transcript*. Let T be a random variable over $\{0,1\}^q$ which represents such transcript. Note that the output of $h(x)$ is uniquely determined by its transcript. Let $Dec : \{0,1\}^q \to \{0,1\}^\ell$ denote the corresponding decision function from transcript to output. Then we have

$$\Pr[h(x) = g(x)] = \Pr[Dec(T) = g(x)] = \sum_{t,k} \Pr[T = t, g(x) = k, Dec(t) = k].$$

[10] Note that $\mathcal{F}$ should be able to distinguish g from easy functions with advantage ϵ, otherwise the simulation is trivial.

To prove the upper bound for $\Pr[h(x) = g(x)]$, first we consider an ideal case such that each function in $\mathcal{F}$ is an uniform random function. In this case, for every $(t, k) \in \{0,1\}^q \times \{0,1\}^\ell$, $\Pr[T^* = t, g(x) = k] = 2^{-(q+\ell)}$ where T^* is the ideal transcript, i.e., uniform distribution over $\{0,1\}^q$. Since for each t there exists a unique k where $Dec(t) = k$, only 2^q pairs (t, k) are *correct* (i.e. $Dec(t) = k$). In such ideal case, we have $\Pr[h^*(x) = g(x)] = 2^{-\ell}$ where h^* denotes the ideal variant of h. In the real case, $\Pr[T = t, g(x) = k]$ can be at most $2^{-\ell}(\frac{1}{2} + c_1\epsilon)^q$, in the case that h queries with correct key in every query and all the responses are 1. However, there does not exist too many extreme cases like this. Besides, we have seen that most of the pairs (t, k) over $\{0,1\}^q \times \{0,1\}^\ell$ do not satisfy $Dec(t) = k$. Therefore we can expect that a large fraction of pairs are *normal* (i.e. chosen with probability $\Theta(2^{-(q+\ell)})$) and *wrong* (i.e. $Dec(t) \neq k$). Such statement implies a lower bound for $\Pr[h(x) \neq g(x)]$.

Next we formally prove the statement above. Consider any transcript $t = \{a_1, a_2, \ldots, a_q\}$. Recall that the queries made by h are uniquely determined by t: the first query is fixed, the second query is determined by the first bit of t, and so on. Let $\{z_1, z_2, \ldots, z_q\}$ be the sequence of key such that the i-th query is $f_i(x, z_i)$ for some $f_i \in \mathcal{F}$. For any $k \in \{0,1\}^\ell, t \in \{0,1\}^q$, let u_i denote the index of the i-th *useful query*, which means the i-th index satisfying $z_{u_i} = k$. Then we define $N_b(t, k) = \sum_i [a_{u_i} = b]$ for $b \in \{0,1\}$, which represents the number of useful queries with response b. Besides, let $N(t, k) = N_0(t, k) + N_1(t, k)$ and $N_\Delta(t, k) = N_0(t, k) - N_1(t, k)$. Similarly, for $j \le N(t, k)$, we define $N_b(t, k, j) = \sum_{i=1}^{j} [a_{u_i} = b]$ for $b \in \{0,1\}$ and $N_\Delta(t, k, j) = N_0(t, k, j) - N_1(t, k, j)$, which only consider the first j useful queries. Recall that for any $f \in \mathcal{F}$, $f(x, z)$ is uniform when $z \neq g(x)$ and biased when $z = g(x)$. For any fixed (t, k),

$$\begin{aligned}
\Pr[g(x) = k, T = t] &= \left(\frac{1}{2}\right)^{(\ell+q-N(t,k))} \left(\frac{1}{2} - c_1\epsilon\right)^{N_0(t,k)} \left(\frac{1}{2} + c_1\epsilon\right)^{N_1(t,k)} \\
&= \left(\frac{1}{2}\right)^{(\ell+q)} (1 - 2c_1\epsilon)^{N_\Delta(t,k)} \left(1 - 4c_1^2\epsilon^2\right)^{N_1(t,k)} \\
&\ge \left(\frac{1}{2}\right)^{(\ell+q)} (1 - 2c_1\epsilon)^{N_\Delta(t,k)} \left(1 - 4c_1^2\epsilon^2\right)^{N(t,k)} \qquad (3)
\end{aligned}$$

Therefore a pair (t, k) is normal if $N_\Delta(t, k) = O(1/\epsilon)$ and $N(t, k) = O(1/\epsilon^2)$. We claim that a large enough fraction of pairs over $\{0,1\}^q \times \{0,1\}^\ell$ are wrong and normal as following:

Claim. Let $q' = 5q/2^\ell \le 5c_2\epsilon^{-2}$. Then for at least $\frac{1}{5}$ fraction of pairs (t, k) over $\{0,1\}^q \times \{0,1\}^\ell$ satisfies the following conditions:

1. $Dec(t) \neq k$.
2. $N(t, k) < q'$.
3. $N_\Delta(t, k) < \sqrt{5q'}$.

Proof. We will prove upper bounds for correct pairs and extreme cases to make sure a large fraction of normal and wrong pairs are left. More precisely, we prove upper bound for the contrary of each condition one by one.

1. Only $2^{-\ell}$ of pairs are correct:
   This obviously holds because (t, k) is correct only when $Dec(t) = k$.
2. At most $\frac{1}{5}$ of pairs (t, k) satisfy $N(t, k) \geq q'$:
   For any t we have $\mathbb{E}_{k \leftarrow U_\ell}[N(t, k)] = \frac{q}{2^\ell}$. By Markov's inequality, at most $\frac{q}{2^\ell q'} = \frac{1}{5}$ of pairs satisfy $N(t, k) \geq q'$.
3. For at most $\frac{1}{10}$ of pairs (t, k), $N(t, k) < q'$ and $N_\Delta(t, k) > \sqrt{5q'}$:
   Fix k. Let T^* be a random transcript which is uniform over $\{0, 1\}^q$. Consider a sequence of random variable $\{Y_j\}$ depending on T^* such that

$$Y_j = \begin{cases} N_\Delta(T^*, k, j), & \text{if } j < N(T^*, k) \\ N_\Delta(T^*, k), & \text{otherwise.} \end{cases}$$

   It's not hard to see that $\{Y_i\}$ is a martingale with difference at most 1. By Azuma's inequality, we have $\Pr[Y_{q'} \geq \sqrt{5q'}] \leq e^{-5q'/2q'} < 0.1$. Since T^* is uniform, the statement above is the same as saying for at most 0.1 fraction of $t \in \{0, 1\}^q$, $Y_{q'}(t) \geq \sqrt{5q'}$. When restricted to t satisfying $N(t, k) < q'$ we have $N_\Delta(t, k) = Y_{q'}(t) \geq \sqrt{5q'}$.

By union bound, all three conditions in the claim hold simultaneously for at least $\frac{1}{5}$ of pairs over $\{0, 1\}^q \times \{0, 1\}^\ell$.

Now consider any pair (t, k) which satisfies condition 2 and 3 in the claim above, in other word a *normal* pair. By inequality (3), we have

$$\begin{aligned} \Pr[g(x) = k, T = t] &\geq (1/2)^{(\ell+q)} (1 - 2c_1\epsilon)^{N_\Delta(t,k)} (1 - 4c_1^2\epsilon^2)^{N(t,k)} \\ &\geq (1/2)^{(\ell+q)} (1 - 2c_1\epsilon)^{\sqrt{5q'}} (1 - 4c_1^2\epsilon^2)^{q'} \qquad (4) \\ &= (1/2)^{(\ell+q)} (1 - 2c_1\epsilon)^{5\sqrt{c_2}\epsilon^{-1}} (1 - 4c_1^2\epsilon^2)^{5c_2\epsilon^{-2}} \\ &\geq (1/2)^{(\ell+q)} (0.3)^{10c_1\sqrt{c_2}} (0.3)^{20c_1^2c_2} \qquad (5) \\ &\geq (1/2)^{(\ell+q)} \cdot 0.5 \qquad (6) \end{aligned}$$

The inequality (5) holds because $(1 - \delta)^{1/\delta} \geq 0.3$ for any $0 < \delta \leq 0.1$. Since $\frac{1}{5}$ of pairs satisfy the conditions above, we have

$$\Pr[h(x) \neq g(x)] = \sum_{k,t} [g(x) = k, T = t, Dec(t) \neq k] \geq 0.1. \qquad (7)$$

Therefore $\Pr[h(x) = g(x)] \leq 0.9 = 1 - \frac{3}{c_1}$.

With the lemma above, we can finish the proof simply with a concentration bound and probabilistic method. Consider the probabilistic distinguisher f_R which is a uniform distribution over all distinguishers in $\mathcal{F}$. Fix any advice α and consider $h(\cdot) = D^{\mathcal{F}}(\cdot, \alpha)$. For any $x \in \{0, 1\}^n, f \in \mathcal{F}$ such that f is not queried by $h(x)$, we have $\mathbb{E}\left[f(x, h(x))\right] = \frac{1}{2} + \Pr[h(x) = g(x)] \cdot c_1\epsilon$ by definition of f. Since h makes at most q query when computing $h(x)$, f_R chooses a query coincident with queries in h with probability $\frac{q}{m}$. Even in the worst case that f_R returns 1 in all these cases, we still have

$$\mathbb{E}\left[f_R(x, g(x))\right] \leq \frac{1}{2} + \Pr[g(x) = h(x)] \cdot c_1\epsilon + \frac{q}{m} \tag{8}$$

$$\leq \frac{1}{2} + (c_1 - 2)\epsilon \tag{9}$$

because m is large enough. Also we have $\mathbb{E}\left[f_R(x, g(x))\right] = \frac{1}{2} + c_1\epsilon$ by definition. Therefore, $\mathbb{E}\left[f_R(x, g(x)) - f_R(x, h(x))\right] \geq 2\epsilon$. Let X be the uniform distribution. Note that for different x, $g(x)$ and $\mathcal{F}(x)$ are chosen independently. Therefore $\mathbb{E}_h\left[f_R(x, g(x)) - f_R(x, h(x))\right]$[11] for each x are independent random variables since it is only influenced by randomness of $g(x)$ and $\mathcal{F}(x)$. By Chernoff-Hoeffding bound, $\mathbb{E}_{x \leftarrow X}\left[f_R(x, g(x)) - f_R(x, h(x))\right] < \epsilon$ holds with probability $2^{-\Omega(\epsilon^2 2^n)}$ over the choice of $\mathcal{F}$ and g. By taking union bound over α, we have

$$\forall \alpha \in \{0,1\}^{2^{o(n)}}, \quad \mathbb{E}_{x \leftarrow X}\left[f_R(x, g(x)) - f_R(x, D^{\mathcal{F}}(x, \alpha))\right] \leq \epsilon \tag{10}$$

with probability $2^{-\Omega(\epsilon^2 2^n) + 2^{o(n)}}$, which is less than 1 for large enough n. By the probabilistic argument there exists a function g and a set $\mathcal{F}$ such that

$$\mathbb{E}_{x \leftarrow X}\left[f_R(x, g(x)) - f_R(x, D^{\mathcal{F}}(x, \alpha))\right] > \epsilon. \tag{11}$$

By averaging argument, for any α, there exists $f \in \mathcal{F}$ such that f can distinguish $(X, D^{\mathcal{F}}(X, \alpha))$ and $(X, g(X))$. Therefore the simulation fails no matter what α is, which contradicts to our assumption. Thus there is no simulator with query complexity $c_2(2^\ell \epsilon^{-2})$.

To summarize, we proved an $\Omega(2^\ell \epsilon^{-2})$ lower bound for black-box (ϵ, ℓ, k, a)-simulator, while the upper bound is only $O(\ell 2^\ell \epsilon^{-2})$. Note that in order to apply Chernoff bound, we need the same-input assumption (i.e. $D(x)$ cannot query $\mathcal{F}(x')$ for $x' \neq x$) to guarantee the independence of different x, even though querying with different input seems useless. A general black-box tight lower bound is left for future work.

References

[AHK12] Arora, S., Hazan, E., Kale, S.: The multiplicative weights update method: a meta-algorithm and applications. Theory Comput. **8**(1), 121–164 (2012)

[AS11] Artemenko, S., Shaltiel, R.: Lower bounds on the query complexity of non-uniform and adaptive reductions showing hardness amplification. In: Goldberg, L.A., Jansen, K., Ravi, R., Rolim, J.D.P. (eds.) APPROX/RANDOM -2011. LNCS, vol. 6845, pp. 377–388. Springer, Heidelberg (2011). https://doi.org/10.1007/978-3-642-22935-0_32

[11] The expectation is taken over the local randomness of h, which does not need to be considered in the probabilistic argument.

[CLP15] Chung, K.-M., Lui, E., Pass, R.: From weak to strong zero-knowledge and applications. In: Dodis, Y., Nielsen, J.B. (eds.) TCC 2015. LNCS, vol. 9014, pp. 66–92. Springer, Heidelberg (2015). https://doi.org/10.1007/978-3-662-46494-6_4

[DBL08] 49th Annual IEEE Symposium on Foundations of Computer Science, FOCS 2008, Philadelphia, PA, USA, 25–28 October 2008. IEEE Computer Society (2008)

[DP08] Dziembowski, S., Pietrzak, K.: Leakage-resilient cryptography. In: 49th Annual IEEE Symposium on Foundations of Computer Science, FOCS 2008, Philadelphia, PA, USA, 25–28 October 2008 [DBL08], pp. 293–302

[FK99] Frieze, A.M., Kannan, R.: Quick approximation to matrices and applications. Combinatorica **19**(2), 175–220 (1999)

[Fre95] Freund, Y.: Boosting a weak learning algorithm by majority. Inf. Comput. **121**(2), 256–285 (1995)

[FS96] Freund, Y., Schapire, R.E.: Game theory, on-line prediction and boosting. In: Blum, A., Kearns, M. (eds.) Proceedings of the Ninth Annual Conference on Computational Learning Theory, COLT 1996, Desenzano del Garda, Italy, 28 June–1 July 1996, pp. 325–332. ACM (1996)

[GNW95] Goldreich, O., Nisan, N., Wigderson, A.: On Yao's XOR-lemma. In: Electronic Colloquium on Computational Complexity (ECCC), vol. 2, no. 50 (1995)

[GW11] Gentry, C., Wichs, D.: Separating succinct non-interactive arguments from all falsifiable assumptions. In: Fortnow, L., Vadhan, S.P. (eds.) Proceedings of the 43rd ACM Symposium on Theory of Computing, STOC 2011, San Jose, CA, USA, 6–8 June 2011, pp. 99–108. ACM (2011)

[HILL99] Håstad, J., Impagliazzo, R., Levin, L.A., Luby, M.: A pseudorandom generator from any one-way function. SIAM J. Comput. **28**(4), 1364–1396 (1999)

[Hol05] Holenstein, T.: Key agreement from weak bit agreement. In: Gabow, H.N., Fagin, R. (eds.) Proceedings of the 37th Annual ACM Symposium on Theory of Computing, Baltimore, MD, USA, 22–24 May 2005, pp. 664–673. ACM (2005)

[HS16] Hirt, M., Smith, A. (eds.): TCC 2016-B. LNCS, vol. 9985. Springer, Heidelberg (2016). https://doi.org/10.1007/978-3-662-53641-4

[Imp95] Impagliazzo, R.: Hard-core distributions for somewhat hard problems. In: 36th Annual Symposium on Foundations of Computer Science, Milwaukee, Wisconsin, 23–25 October 1995, pp. 538–545. IEEE Computer Society (1995)

[JP14] Jetchev, D., Pietrzak, K.: How to fake auxiliary input. In: Lindell, Y. (ed.) TCC 2014. LNCS, vol. 8349, pp. 566–590. Springer, Heidelberg (2014). https://doi.org/10.1007/978-3-642-54242-8_24

[LTW11] Lu, C.-J., Tsai, S.-C., Wu, H.-L.: Complexity of hard-core set proofs. Comput. Complex. **20**(1), 145–171 (2011)

[PS16] Pietrzak, K., Skórski, M.: Pseudoentropy: lower-bounds for chain rules and transformations. In: Hirt and Smith [HS16], pp. 183–203

[Rou16] Roughgarden, T.: No-Regret Dynamics, pp. 230–246. Cambridge University Press, Cambridge (2016)

[RTTV08] Reingold, O., Trevisan, L., Tulsiani, M., Vadhan, S.P.: Dense subsets of pseudorandom sets. In: 49th Annual IEEE Symposium on Foundations of Computer Science, FOCS 2008, Philadelphia, PA, USA, 25–28 October 2008 [DBL08], pp. 76–85

[Skó16a] Skórski, M.: Simulating auxiliary inputs, revisited. In: Hirt and Smith [HS16], pp. 159–179

[Skó16b] Skórski, M.: A subgradient algorithm for computational distances and applications to cryptography. IACR Cryptology ePrint Archive, 2016:158 (2016)

[SV10] Shaltiel, R., Viola, E.: Hardness amplification proofs require majority. SIAM J. Comput. **39**(7), 3122–3154 (2010)

[TTV09] Trevisan, L., Tulsiani, M., Vadhan, S.P.: Regularity, boosting, and efficiently simulating every high-entropy distribution. In: Proceedings of the 24th Annual IEEE Conference on Computational Complexity, CCC 2009, Paris, France, 15–18 July 2009, pp. 126–136. IEEE Computer Society (2009)

[VZ13] Vadhan, S., Zheng, C.J.: A uniform min-max theorem with applications in cryptography. In: Canetti, R., Garay, J.A. (eds.) CRYPTO 2013. LNCS, vol. 8042, pp. 93–110. Springer, Heidelberg (2013). https://doi.org/10.1007/978-3-642-40041-4_6

[Yao82] Yao, A.C.-C.: Theory and applications of trapdoor functions (extended abstract). In 23rd Annual Symposium on Foundations of Computer Science, Chicago, Illinois, USA, 3–5 November 1982, pp. 80–91. IEEE Computer Society (1982)

[Zha11] Zhang, J.: On the query complexity for showing dense model. In: Electronic Colloquium on Computational Complexity (ECCC), vol. 18, p. 38 (2011)

Key Exchange

Fuzzy Password-Authenticated Key Exchange

Pierre-Alain Dupont[1,2,3(✉)], Julia Hesse[4], David Pointcheval[2,3], Leonid Reyzin[5], and Sophia Yakoubov[5]

[1] DGA, Paris, France
[2] DIENS, École Normale Supérieure, CNRS, PSL University, Paris, France
{pierre-alain.dupont,david.pointcheval}@ens.fr
[3] INRIA, Paris, France
[4] Technische Universität Darmstadt, Darmstadt, Germany
julia.hesse@ens.fr
[5] Boston University, Boston, USA
reyzin@cs.bu.edu, sonka@bu.edu

Abstract. Consider key agreement by two parties who start out knowing a common secret (which we refer to as "pass-string", a generalization of "password"), but face two complications: (1) the pass-string may come from a low-entropy distribution, and (2) the two parties' copies of the pass-string may have some noise, and thus not match exactly. We provide the first efficient and general solutions to this problem that enable, for example, key agreement based on commonly used biometrics such as iris scans.

The problem of key agreement with each of these complications individually has been well studied in literature. Key agreement from low-entropy shared pass-strings is achieved by *password-authenticated key exchange* (PAKE), and key agreement from noisy but high-entropy shared pass-strings is achieved by information-reconciliation protocols as long as the two secrets are "close enough." However, the problem of key agreement from noisy low-entropy pass-strings has never been studied.

We introduce (universally composable) *fuzzy password-authenticated key exchange* (fPAKE), which solves exactly this problem. fPAKE does not have any entropy requirements for the pass-strings, and enables secure key agreement as long as the two pass-strings are "close" for some notion of closeness. We also give two constructions. The first construction achieves our fPAKE definition for any (efficiently computable) notion of closeness, including those that could not be handled before even in the high-entropy setting. It uses Yao's garbled circuits in a way that is only two times more costly than their use against semi-honest adversaries, but that guarantees security against malicious adversaries. The second construction is more efficient, but achieves our fPAKE definition only for pass-strings with low Hamming distance. It builds on very simple primitives: robust secret sharing and PAKE.

J. Hesse—Work done while at École Normale Supérieure.

J. B. Nielsen and V. Rijmen (Eds.): EUROCRYPT 2018, LNCS 10822, pp. 393–424, 2018.
https://doi.org/10.1007/978-3-319-78372-7_13

Keywords: Authenticated key exchange · PAKE
Hamming distance · Error correcting codes · Yao's garbled circuits

1 Introduction

Consider key agreement by two parties who start out knowing a common secret (which we refer to as "pass-string", a generalization of "password"). These parties may face several complications: (1) the pass-string may come from a non-uniform, low-entropy distribution, and (2) the two parties' copies of the pass-string may have some noise, and thus not match exactly. The use of such pass-strings for security has been extensively studied; examples include biometrics and other human-generated data [15,23,29,39,46,49,66], physically unclonable functions (PUFs) [30,52,57,58,64], noisy channels [61], quantum information [9], and sensor readings of a common environment [32,33].

The Noiseless Case. When the starting secret is not noisy (i.e., the same for both parties), existing approaches work quite well. The case of low-entropy secrets is covered by *password-authenticated key exchange* (PAKE) (a long line of work, with first formal models introduced in [7,14]). A PAKE protocol allows two parties to agree on a shared high-entropy key if and only if they hold the same short password. Even though the password may have low entropy, PAKE ensures that off-line dictionary attacks are impossible. Roughly speaking, an adversary has to participate in one on-line interaction for every attempted guess at the password. Because key agreement is not usually the final goal, PAKE protocols need to be composed with whatever protocols (such as authenticated encryption) use the output key. This composability has been achieved by universally composable (UC) PAKE defined by Canetti *et al.* [20] and implemented in several follow-up works.

In the case of high-entropy secrets, off-line dictionary attacks are not a concern, which enables more efficient protocols. If the adversary is passive, randomness extractors [51] do the job. The case of active adversaries is covered by the literature on so-called robust extractors defined by Boyen *et al.* [13] and, more generally, by many papers on privacy amplification protocols secure against active adversaries, starting with the work of Maurer [45]. Composability for these protocols is less studied; in particular, most protocols leak information about the pass-string itself, in which case reusing the pass-string over multiple protocol executions may present problems [12] (with the exception of [19]).

The Noisy Case. When the pass-string is noisy (i.e., the two parties have slightly different versions of it), this problem has been studied only for the case of high-entropy pass-strings. A long series of works on information-reconciliation protocols started by Bennett *et al.* [9] and their one-message variants called fuzzy extractors (defined by Dodis *et al.* [26], further enhanced for active security starting by Renner and Wolf [54]) achieves key agreement when the pass-string has a lot of entropy and not too much noise. Unfortunately, these approaches do not extend to the low-entropy setting and are not designed to prevent off-line dictionary attacks.

Constructions for the noisy case depend on the specific noise model. The case of binary Hamming distance—when the n pass-string characters held by the two parties are the same at all but δ locations—is the best studied. Most existing constructions require, at a minimum, that the pass-string should have at least δ bits of entropy. This requirement rules out using most kinds of biometric data as the pass-string—for example, estimates of entropy for iris scans (transformed into binary strings via wavelet transforms and projections) are considerably lower than the amount of errors that need to be tolerated [11, Sect. 5]. Even the PAKE-based construction of Boyen *et al.* [13] suffers from the same problem.

One notable exception is the construction of Canetti *et al.* [19], which does not have such a requirement, but places other stringent limitations on the probability distribution of pass-strings. In particular, because it is a one-message protocol, it cannot be secure against off-line dictionary attacks.

1.1 Our Contributions

We provide definitions and constant-round protocols for key agreement from noisy pass-strings that:

- Resist off-line dictionary attacks and thus can handle low-entropy pass-strings,
- Can handle a variety of noise types and have high error-tolerance, and
- Have well specified composition properties via the UC framework [17].

Instead of imposing entropy requirements or other requirements on the distribution of pass-strings, our protocols are secure as long as the adversary cannot guess a pass-string value that is sufficiently close. There is no requirement, for example, that the amount of pass-string entropy is greater than the number of errors; in fact, one of our protocols is suitable for iris scans. Moreover, our protocols prevent off-line attacks, so each adversarial attempt to get close to the correct pass-string requires an on-line interaction by the adversary. Thus, for example, our protocols can be meaningfully run with pass-strings whose entropy is only 30 bits—something not possible with any prior protocols for the noisy case.

New Models. Our security model is in the Universal Composability (UC) Framework of Canetti [17]. The advantage of this framework is that it comes with a composability theorem that ensures that the protocol stays secure even running in arbitrary environments, including arbitrary parallel executions. Composability is particularly important for key agreement protocols, because key agreement is rarely the ultimate goal. The agreed-upon key is typically used for some subsequent protocol—for example, a secure channel. Further, this framework allows to us to give a definition that is agnostic to how the initial pass-strings are generated. We have no entropy requirements or constraints on the pass-string distribution; rather, security is guaranteed as long as the adversary's input to the protocol is not close enough to the correct pass-string.

As a starting point, we use the definition of UC security for PAKE from Canetti *et al.* [20]. The PAKE ideal functionality is defined as follows: the secret pass-strings (called "passwords" in PAKE) of the two parties are the inputs to the functionality, and two random keys, which are equal if and only if the two inputs are equal, are the outputs. The main change we make to PAKE is enhancing the functionality to give equal keys even if the two inputs are not equal, as long as they are close enough. We also relax the security requirement to allow one party to find out some information about the other party's input—perhaps even the entire input—if the two inputs are close. This relaxation makes sense in our application: if the two parties are honest, then the differences between their inputs are a problem rather than a feature, and we would not mind if the inputs were in fact the same. The benefit of this relaxation is that it permits us to construct more efficient protocols. (We also make a few other minor changes which will be described in Sect. 2.) We call our new UC functionality "Fuzzy Password-Authenticated Key Exchange" or fPAKE.

New Protocols. The only prior PAKE-based protocol for the noisy setting by Boyen *et al.* [13], although more efficient than ours, does not satisfy our goal. In particular, it is not composable, because it reveals information about the secret pass-strings (we demonstrate this formally in the full version of this paper [28]). Because some information about the pass-strings is unconditionally revealed, high-entropy pass-strings are required. Thus, in order to realize our definition for arbitrary low-entropy pass-strings, we need to construct new protocols.

Realizing our fPAKE definition is easy using general two-party computation techniques for protocols with malicious adversaries and without authenticated channels [4]. However, we develop protocols that are considerably more efficient: our definitional relaxation allows us to build protocols that achieve security against malicious adversaries but cost just a little more than the generic two-party computation protocols that achieve security only against honest-but-curious adversaries (i.e., adversaries who do not deviate from the protocol, but merely try to infer information they are not supposed to know).

Our first construction uses Yao's garbled circuits [6,63] and oblivious transfer (see [21] and references therein). The use of these techniques is standard in two-party computation. However, by themselves they give protocols secure only against honest-but-curious adversaries. In order to prevent malicious behavior of the players, one usually applies the cut-and-choose technique [42], which is quite costly: to achieve an error probability of $2^{-\lambda}$, the number of circuits that need to be garbled increases by a factor of λ, and the number of oblivious transfers that need to be performed increases by a factor of $\lambda/2$. We show that for our special case, to achieve malicious security, it suffices to repeat the honest-but-curious protocol twice (once in each direction), incurring only a factor of 2 overhead over the semi-honest case.[1] Mohassel and Franklin [48] and Huang *et al.* [34] suggest

[1] Gasti *et al.* [31] similarly use Yao's garbled circuits for continuous biometric user authentication on a smartphone. Our approach can eliminate the third party in their application, at the cost of requiring two garbled circuits instead of one. As far as we know, ours is the first use of garbled circuits in the two-party fully malicious setting without calling on an expensive transformation.

a similar technique (known as "dual execution"), but at the cost of leaking a bit of the adversary's choice to the adversary. In contrast, our construction leaks nothing to the adversary at all (as long as the pass-strings are not close). This construction works regardless of what it means for the two inputs to be "close," as long as the question of closeness can be evaluated by an efficient circuit.

Our second construction is for the Hamming case: the two n-character pass-strings have low Hamming distance if not too many characters of one party's pass-string are different from the corresponding characters of the other's pass-string. The two parties execute a PAKE protocol for each position in the string, obtaining n values each that agree or disagree depending on whether the characters of the pass-string agree or disagree in the corresponding positions. It is important that at this stage, agreement or disagreement at individual positions remains unknown to everyone; we therefore make use of a special variant of PAKE which we call *implicit-only PAKE* (we give a formal UC security definition of implicit-only PAKE and show that it is realized by the PAKE protocol from [1,8]). This first step upgrades Hamming distance over a potentially small alphabet to Hamming distance over an exponentially large alphabet. We then secret-share the ultimate output key into n shares using a robust secret sharing scheme, and encrypt each share using the output of the corresponding PAKE protocol.

The second construction is more efficient than the first in the number of rounds, communication, and computation. However, it works only for Hamming distance. Moreover, it has an intrinsic gap between functionality and security: if the honest parties need to be within distance δ to agree, then the adversary may break security by guessing a secret within distance 2δ. See Fig. 10 for a comparison between the two constructions.

The advantages of our protocols are similar to the advantages of UC PAKE: They provide composability, protection against off-line attacks, the ability to use low-entropy inputs, and handle any distribution of secrets. And, of course, because we construct *fuzzy* PAKE, our protocols can handle noisy inputs—including many types of noisy inputs that could not be handled before. Our first protocol can handle any type of noise as long as the notion of "closeness" can be efficiently computed, whereas most prior work was for Hamming distance only. However, these advantages come at the price of efficiency. Our protocols require 2–5 rounds of interaction, as opposed to many single-message protocols in the literature [19,25,60]. They are also more computationally demanding than most existing protocols for the noisy case, requiring one public-key operation per input character. We emphasize, however, that our protocols are much less computationally demanding than the protocols based on general two-party computation, as already discussed above, or general-purpose obfuscation, as discussed in [10, Sect. 4.3.4].

2 Security Model

We now present a security definition for fuzzy password-authenticated key exchange (fPAKE). We adapt the definition of PAKE from Canetti *et al.* [20]

to work for pass-strings (a generalization of "passwords") that are similar, but not necessarily equal. Our definition uses measures of the distance $d(\mathsf{pw}, \mathsf{pw}')$ between pass-strings $\mathsf{pw}, \mathsf{pw}' \in \mathbb{F}_p^n$. In Sects. 3.3 and 4, Hamming distance is used, but in the generic construction of Sect. 3, any other notion of distance can be used instead. We say that pw and pw' are "similar enough" if $d(\mathsf{pw}, \mathsf{pw}') \leq \delta$ for a distance notion d and a threshold δ that is hard-coded into the functionality.

To model the possibility of dictionary attacks, the functionality allows the adversary to make one pass-string guess against each player ($\mathcal{P}_0$ and $\mathcal{P}_1$). In the real world, if the adversary succeeds in guessing (a pass-string similar enough to) party $\mathcal{P}_i$'s pass-string, it can often choose (or at least bias) the session key computed by $\mathcal{P}_i$. To model this, the functionality then allows the adversary to set the session key for $\mathcal{P}_i$.

As usual in security notions for key exchange, the adversary also sets the session keys for corrupted players. In the definition of Canetti *et al.* [20], the adversary additionally sets $\mathcal{P}_i$'s key if $\mathcal{P}_{1-i}$ is corrupted. However, contrarily to the original definition, we do not allow the adversary to set $\mathcal{P}_i$'s key if $\mathcal{P}_{1-i}$ is corrupted but did not guess $\mathcal{P}_i$'s pass-string. We make this change in order to protect an honest $\mathcal{P}_i$ from, for instance, revealing sensitive information to an adversary who did not successfully guess her pass-string, but did corrupt her partner.

Another minor change we make is considering only two parties—$\mathcal{P}_0$ and $\mathcal{P}_1$—in the functionality, instead of considering arbitrarily many parties and enforcing that only two of them engage the functionality. This is because universal composability takes care of ensuring that a two-party functionality remains secure in a multi-party world.

As in the definition of Canetti *et al.* [20], we consider only static corruptions in the standard corruption model of Canetti [17]. Also as in their definition, we chose not to provide the players with confirmation that key agreement was successful. The players might obtain such confirmation from subsequent use of the key.

By default, in the fPAKE functionality the `TestPwd` interface provides the adversary with one bit of information—whether the pass-string guess was correct or not. This definition can be strengthened by providing the adversary with no information at all, as in implicit-only PAKE ($\mathcal{F}_{\mathsf{iPAKE}}$, Fig. 7), or weakened by providing the adversary with extra information when the adversary's guess is close enough.

To capture the diversity of possibilities, we introduce a more general `TestPwd` interface, described in Fig. 2. It includes three leakage functions that we will instantiate in different ways below—L_c if the guess is close-enough to succeed, L_f if it is too far. Moreover, a third leakage function—L_m for medium distance—allows the adversary to get some information even if the adversary's guess is only somewhat close (closer than some parameter $\gamma \geq \delta$), but not close enough for successful key agreement. We thus decouple the distance needed for functionality from the (possibly larger) distance needed to guarantee security; the smaller the gap between these two distances, the better, of course.

The functionality fPAKE is parameterized by a security parameter λ and tolerances $\delta \leq \gamma$. It interacts with an adversary $\mathcal{S}$ and two parties $\mathcal{P}_0$ and $\mathcal{P}_1$ via the following queries:

- **Upon receiving a query (`NewSession`, sid, pw_i) from party $\mathcal{P}_i$:**
  - Send (`NewSession`, sid, $\mathcal{P}_i$) to $\mathcal{S}$;
  - If this is the first `NewSession` query, or if this is the second `NewSession` query and there is a record $(\mathcal{P}_{1-i}, \mathsf{pw}_{1-i})$, then record $(\mathcal{P}_i, \mathsf{pw}_i)$ and mark this record `fresh`.
- **Upon receiving a query (`TestPwd`, sid, $\mathcal{P}_i$, pw'_i) from the adversary $\mathcal{S}$:** If there is a `fresh` record $(\mathcal{P}_i, \mathsf{pw}_i)$, then set $d \leftarrow d(\mathsf{pw}_i, \mathsf{pw}'_i)$ and do:
  - If $d \leq \delta$, mark the record `compromised` and reply to $\mathcal{S}$ with "correct guess";
  - If $d > \delta$, mark the record `interrupted` and reply to $\mathcal{S}$ with "wrong guess".
- **Upon receiving a query (`NewKey`, sid, $\mathcal{P}_i$, sk) from the adversary $\mathcal{S}$:** If there is no record of the form $(\mathcal{P}_i, \mathsf{pw}_i)$, or if this is not the first `NewKey` query for $\mathcal{P}_i$, then ignore this query. Otherwise:
  - If at least one of the following is true, then output (sid, sk) to player $\mathcal{P}_i$:
    * The record is `compromised`
    * $\mathcal{P}_i$ is corrupted
    * The record is `fresh`, $\mathcal{P}_{1-i}$ is corrupted, and there is a record $(\mathcal{P}_{1-i}, \mathsf{pw}_{1-i})$ with $d(\mathsf{pw}_i, \mathsf{pw}_{1-i}) \leq \delta$
  - If this record is `fresh`, both parties are honest, there is a record $(\mathcal{P}_{1-i}, \mathsf{pw}_{1-i})$ with $d(\mathsf{pw}_i, \mathsf{pw}_{1-i}) \leq \delta$, a key sk′ was sent to $\mathcal{P}_{1-i}$, and $(\mathcal{P}_{1-i}, \mathsf{pw}_{1-i})$ was `fresh` at the time, then output (sid, sk′) to $\mathcal{P}_i$;
  - In any other case, pick a new random key sk′ of length λ and send (sid, sk′) to $\mathcal{P}_i$.
  - Mark the record $(\mathcal{P}_i, \mathsf{pw}_i)$ as `completed`.

Fig. 1. Ideal functionality fPAKE

- **Upon receiving a query (`TestPwd`, sid, $\mathcal{P}_i$, pw'_i) from the adversary $\mathcal{S}$:** If there is a `fresh` record $(\mathcal{P}_i, \mathsf{pw}_i)$, then set $d \leftarrow d(\mathsf{pw}_i, \mathsf{pw}'_i)$ and do:
  - If $d \leq \delta$, mark the record `compromised` and reply to $\mathcal{S}$ with $L_c(\mathsf{pw}_i, \mathsf{pw}'_i)$;
  - If $\delta < d \leq \gamma$, mark the record `compromised` and reply to $\mathcal{S}$ with $L_m(\mathsf{pw}_i, \mathsf{pw}'_i)$;
  - If $\gamma < d$, mark the record `interrupted` and reply to $\mathcal{S}$ with $L_f(\mathsf{pw}_i, \mathsf{pw}'_i)$.

Fig. 2. A modified `TestPwd` interface to allow for different leakage

Below, we list the specific leakage functions L_c, L_m and L_f that we consider in this work, in order of decreasing strength (or increasing leakage):

1. The strongest option is to provide no feedback at all to the adversary. We define fPAKE^N to be the functionality described in Fig. 1, except that `TestPwd` is from Fig. 2 with

$$L_c^N(\mathsf{pw}_i, \mathsf{pw}'_i) = L_m^N(\mathsf{pw}_i, \mathsf{pw}'_i) = L_f^N(\mathsf{pw}_i, \mathsf{pw}'_i) = \bot\,.$$

2. The basic functionality fPAKE, described in Fig. 1, leaks the correctness of the adversary's guess. That is, in the language of Fig. 2,

$$L_c(\mathsf{pw}_i, \mathsf{pw}_i') = \text{"correct guess"},$$
$$\text{and} \quad L_m(\mathsf{pw}_i, \mathsf{pw}_i') = L_f(\mathsf{pw}_i, \mathsf{pw}_i') = \text{"wrong guess"}.$$

   The classical PAKE functionality from [20] has such a leakage.
3. Assume the two pass-strings are strings of length n over some finite alphabet, with the jth character of the string pw denoted by $\mathsf{pw}[j]$. We define fPAKE^M to be the functionality described in Fig. 1, except that `TestPwd` is from Fig. 2, with L_c and L_m that leak the indices at which the guessed pass-string differs from the actual one when the guess is close enough (we will call this leakage the *mask* of the pass-strings). That is,

$$L_c^M(\mathsf{pw}_i, \mathsf{pw}_i') = (\{j \text{ s.t.} \mathsf{pw}_i[j] = \mathsf{pw}_i'[j]\}, \text{"correct guess"}),$$
$$L_m^M(\mathsf{pw}_i, \mathsf{pw}_i') = (\{j \text{ s.t.} \mathsf{pw}_i[j] = \mathsf{pw}_i'[j]\}, \text{"wrong guess"})$$
$$\text{and} \quad L_f^M(\mathsf{pw}_i, \mathsf{pw}_i') = \text{"wrong guess"}.$$

4. The weakest definition—or the strongest leakage—reveals the entire actual pass-string to the adversary if the pass-string guess is close enough. We define fPAKE^P to be the functionality described in Fig. 1, except that `TestPwd` is from Fig. 2, with

$$L_c^P(\mathsf{pw}_i, \mathsf{pw}_i') = L_m^P(\mathsf{pw}_i, \mathsf{pw}_i') = \mathsf{pw}_i \text{ and } L_f^P(\mathsf{pw}_i, \mathsf{pw}_i') = \text{"wrong guess"}.$$

   Here, L_c^P and L_m^P do not need to include "correct guess" and "wrong guess", respectively, because this is information that can be easily derived from pw_i itself.

The first two functionalities are the strongest, but there are no known constructions that realize them, other than through generic two-party computation secure against malicious adversaries, which is an inefficient solution. The last two functionalities, though weaker, still provide meaningful security, especially when $\gamma = \delta$. Intuitively, this is because strong leakage only occurs when an adversary guesses a "close" pass-string, which enables him to authenticate as though he knows the real pass-string anyway.

In Sect. 3, we present a construction satisfying fPAKE^P for any efficiently computable notion of distance, with $\gamma = \delta$ (which is the best possible). We present a construction for Hamming distance satisfying fPAKE^M in Sect. 4, with $\gamma = 2\delta$.

3 General Construction Using Garbled Circuits

In this section, we describe a protocol realizing fPAKE^P that uses Yao's garbled circuits [63]. We briefly introduce this primitive in Sect. 3.1 and refer to Yakoubov [62] for a more thorough introduction.

The Yao's garbled circuit-based fPAKE construction has two advantages:

1. It is more flexible than other approaches; any notion of distance that can be efficiently computed by a circuit can be used. In Sect. 3.3, we describe a suitable circuit for Hamming distance. The total size of this circuit is $O(n)$, where n is the length of the pass-strings used. Edit distance is slightly less efficient, and uses a circuit whose total size is $O(n^2)$.
2. There is no gap between the distances required for functionality and security—that is, there is no leakage about the pass-strings used unless they are similar enough to agree on a key. In other words, $\delta = \gamma$.

Informally, the construction involves the garbled evaluation of a circuit that takes in two pass-strings as input, and computes whether their distance is less than δ. Because Yao's garbled circuits are only secure against semi-honest garblers, we cannot simply have one party do the garbling and the other party do the evaluation. A malicious garbler could provide a garbling of the wrong function—maybe even a constant function—which would result in successful key agreement even if the two pass-strings are very different. However, as suggested by Mohassel and Franklin [48] and Huang *et al.* [34], since a malicious evaluator (unlike a malicious garbler) cannot compromise the computation, by performing the protocol twice with each party playing each role once, we can protect against malicious behavior. They call this the *dual execution* protocol.

The dual execution protocol has the downside of allowing the adversary to specify and receive a single additional bit of leakage. It is important to note that because of this, dual execution cannot directly be used to instantiate fPAKE, because a single bit of leakage can be too much when the entropy of the pass-strings is low to begin with—a few adversarial attempts will uncover the entire pass-string. Our construction is as efficient as that of Mohassel *et al.* and Huang *et al.*, while guaranteeing no leakage to a malicious adversary in the case that the pass-strings used are not close. We describe how we achieve this in Sect. 3.1.3.

3.1 Building Blocks

In Sect. 3.1.1, we briefly review oblivious transfer. In Sect. 3.1.2, we review Yao's Garbled Circuits. In Sect. 3.1.3, we describe in more detail our take on the dual execution protocol, and how we avoid leakage to the adversary when the pass-strings used are dissimilar.

3.1.1 Oblivious Transfer (OT)

Informally, 1-out-of-2 Oblivious Transfer (see [21] and citations therein) enables one party (the sender) to transfer exactly one of two secrets to another party (the receiver). The receiver chooses (by index 0 or 1) which secret she wants. The security of the OT protocol guarantees that the sender does not learn this choice bit, and the receiver does not learn anything about the other secret.

3.1.2 Yao's Garbled Circuits (YGC)

Next, we give a brief introduction to Yao's garbled circuits [63]. We refer to Yakoubov [62] for a more detailed description, as well as a summary of some of the Yao's garbled circuits optimizations [3,5,38,40,53,65]. Informally, Yao's garbled circuits are an asymmetric secure two-party computation scheme. They enable two parties with sensitive inputs (in our case, pass-strings) to compute a joint function of their inputs (in our case, an augmented version of similarity) without revealing any additional information about their inputs. One party "garbles" the function they wish to evaluate, and the other evaluates it in its garbled form.

Below, we summarize the garbling scheme formalization of Bellare *et al.* [6], which is a generalization of YGC.

Functionality. A garbling scheme $\mathcal{G}$ consists of four polynomial-time algorithms $(\mathsf{Gb}, \mathsf{En}, \mathsf{Ev}, \mathsf{De})$:

1. $\mathsf{Gb}(1^\lambda, f) \rightarrow (F, e, d)$. The garbling algorithm Gb takes in the security parameter λ and a circuit f, and returns a garbled circuit F, encoding information e, and decoding information d.
2. $\mathsf{En}(e, x) \rightarrow X$. The encoding algorithm En takes in the encoding information e and an input x, and returns a garbled input X.
3. $\mathsf{Ev}(F, X) \rightarrow Y$. The evaluation algorithm Ev takes in the garbled circuit F and the garbled input X, and returns a garbled output Y.
4. $\mathsf{De}(d, Y) \rightarrow y$. The decoding algorithm De takes in the decoding information d and the garbled output Y, and returns the plaintext output y.

A garbling scheme $\mathcal{G} = (\mathsf{Gb}, \mathsf{En}, \mathsf{Ev}, \mathsf{De})$ is *projective* if encoding information e consists of $2n$ *wire labels* (each of which is essentially a random string), where n is the number of input bits. Two wire labels are associated with each bit of the input; one wire label corresponds to the event of that bit being 0, and the other corresponds to the event of that bit being 1. The garbled input includes only the wire labels corresponding to the actual values of the input bits. In projective schemes, in order to give the evaluator the garbled input she needs for evaluation, the garbler can send her all of the wire labels corresponding to the garbler's input. The evaluator can then use OT to retrieve the wire labels corresponding to her own input.

Similarly, we call a garbling scheme *output-projective* if decoding information d consists of two labels for each output bit, one corresponding to each possible value of that bit. The garbling schemes used in this paper are both projective and output-projective.

Correctness. Informally, a garbling scheme $(\mathsf{Gb}, \mathsf{En}, \mathsf{Ev}, \mathsf{De})$ is *correct* if it always holds that $\mathsf{De}(d, \mathsf{Ev}(F, \mathsf{En}(e, x))) = f(x)$.

Security. Bellare *et al.* [6] describe three security notions for garbling schemes: *obliviousness*, *privacy* and *authenticity*. Informally, a garbling scheme $\mathcal{G} = (\mathsf{Gb}, \mathsf{En}, \mathsf{Ev}, \mathsf{De})$ is *oblivious* if a garbled function F and a garbled input X do not reveal anything about the input x. It is *private* if additionally knowing the decoding information d reveals the output y, but does not reveal anything more about the input x. It is *authentic* if an adversary, given F and X, cannot find a garbled output $Y' \neq \mathsf{Ev}(F, X)$ which decodes without error.

In the full version of this paper [28], we define a new property of output-projective garbling schemes called *garbled output randomness.* Informally, it states that even given one of the output labels, the other should be indistinguishable from random.

3.1.3 Malicious Security: A New Take on Dual Execution with Privacy-Correctness Tradeoffs

While Yao's garbled circuits are naturally secure against a malicious evaluator, they have the drawback of being insecure against a malicious garbler. A garbler can "mis-garble" the function, either replacing it with a different function entirely or causing an error to occur in an informative way (this is known as "selective failure").

Typically, malicious security is introduced to Yao's garbled circuits by using the cut-and-choose transformation [35,41,43]. To achieve a $2^{-\lambda}$ probability of cheating without detection, the parties need to exchange λ garbled circuits [41].[2] Some of the garbled circuits are "checked", and the rest of them are evaluated, their outputs checked against one another for consistency. Because of the factor of λ computational overhead, though, cut-and-choose is expensive, and too heavy a tool for fPAKE. Other, more efficient transformations such as LEGO [50] and authenticated garbling [59] exist as well, but those rely heavily on pre-processing, which cannot be used in fPAKE since it requires advance interaction between the parties.

Mohassel and Franklin [48] and Huang *et al.* [34] suggest an efficient transformation known as "dual execution": each party plays each role (garbler and evaluator) once, and then the two perform a comparison step on their outputs in a secure fashion. Dual execution incurs only a factor of 2 overhead over semi-honest garbled circuits. However, it does not achieve fully malicious security. It guarantees correctness, but reduces the privacy guarantee by allowing a malicious garbler to learn one bit of information of her choice. Specifically, if a malicious garbler garbles a wrong circuit, she can use the comparison step to learn one bit about the output of this wrong circuit on the other party's input. This one extra bit of information could be crucially important, violating the privacy of the evaluator's input in a significant way.

We introduce a tradeoff between correctness and privacy for boolean functions. For one of the two possible outputs (without loss of generality, '0'), we restore full privacy at the cost of correctness. The new privacy guarantee is that if the correct output is '0', then a malicious adversary cannot learn anything beyond this output, but if the correct output is '1', then she can learn a single bit of her choice. The new correctness guarantee is that a malicious adversary

[2] There are techniques [44] that improve this number in the amortized case when many computations are done—however, this does not fit our setting.

can cause the computation that should output '1' to output '0' instead, but not the other way around.

The main idea of dual execution is to have the two parties independently evaluate one another's circuits, learn the output values, and compare the output labels using a secure comparison protocol. In our construction, however, the parties need not learn the output values before the comparison. Instead, the parties can compare output labels *assuming* an output of '1', and if the comparison fails, the output is determined to be '0'.

More formally, let $d_0[0]$, $d_0[1]$ be the two output labels corresponding to $\mathcal{P}_0$'s garbled circuit, and $d_1[0]$, $d_1[1]$ be the two output labels corresponding to $\mathcal{P}_1$'s circuit. Let Y_0 be the output label learned by $\mathcal{P}_1$ as a result of evaluation, and Y_1 be the label learned by $\mathcal{P}_0$. The two parties securely compare $(d_0[1], Y_1)$ to $(Y_0, d_1[1])$; if the comparison succeeds, the output is "1".

Our privacy–correctness tradeoff is perfect for fPAKE. If the parties' inputs are similar, learning a bit of information about each other's inputs is not problematic, since arguably the small amount of noise in the inputs is a bug, not a feature. If the parties' inputs are not similar, however, we are guaranteed to have no leakage at all. We pay for the lack of leakage by allowing a malicious party to force an authentication failure even when authentication should succeed. However, either party can do so anyway by providing an incorrect input.

In Sect. 3.2.2, we describe our Yao's garbled circuit-based fPAKE protocol. Note that in this protocol, we omit the final comparison step; instead, we use the output lables $((d_0[1], Y_1)$ and $(Y_0, d_1[1]))$ to compute the agreed-upon key directly.

3.2 Construction

Building a fPAKE from YGC and OT is not straightforward, since all constructions of OT assume authenticated channels, and fPAKE (or PAKE) is designed with unauthenticated channels in mind. We therefore follow the framework of Canetti *et al.* [18], who build a UC secure PAKE protocol using OT. We first build our protocol assuming authenticated channels, and then apply the generic transformation of Barak *et al.* [4] to adapt it to the unauthenticated channel setting. More formally, we proceed in three steps:

1. First, in Sect. 3.2.1, we define a randomized fuzzy equality-testing functionality $\mathcal{F}_{\mathsf{RFE}}$, which is analogous to the randomized equality-testing functionality of Canetti *et al.*
2. In Sect. 3.2.2, we build a protocol that securely realizes $\mathcal{F}_{\mathsf{RFE}}$ in the OT-hybrid model, assuming authenticated channels.
3. In Sect. 3.2.3, we apply the transformation of Barak *et al.* to our protocol. This results in a protocol that realizes the "split" version of functionality $\mathcal{F}^{P}_{\mathsf{RFE}}$, which we show to be enough to implement to fPAKE^{P}. Split functionalities, which were introduced by Barak *et al.*, adapt functionalities which assume authenticated channels to an unauthenticated channels setting. The only additional ability an adversary has in a split functionality is the ability to execute the protocol separately with the participating parties.

The functionality $\mathcal{F}_{\mathsf{RFE}}$ is parameterized by a security parameter λ and a tolerance δ. It interacts with an adversary $\mathcal{S}$ and two parties $\mathcal{P}_0$ and $\mathcal{P}_1$ via the following queries:

- **Upon receiving a query** $(\texttt{NewSession}, \mathsf{sid}, \mathsf{pw}_i)$ **from party** $\mathcal{P}_i \in \{\mathcal{P}_0, \mathcal{P}_1\}$**:**
  - Send $(\texttt{NewSession}, \mathsf{sid}, \mathcal{P}_i)$ to $\mathcal{S}$;
  - If this is the first `NewSession` query, or if this is the second `NewSession` query and there is a record $(\mathcal{P}_{1-i}, \mathsf{pw}_{1-i})$, then record $(\mathcal{P}_i, \mathsf{pw}_i)$.
- **Upon receiving a query** $(\texttt{TestPwd}, \mathsf{sid}, \mathcal{P}_i)$ **from the adversary** $\mathcal{S}$, $\mathcal{P}_i \in \{\mathcal{P}_0, \mathcal{P}_1\}$**:**
  If records of the form $(\mathcal{P}_0, \mathsf{pw}_0)$ and $(\mathcal{P}_1, \mathsf{pw}_1)$ do not exist, if $\mathcal{P}_{1-i}$ is not corrupted, or this is not the first `TestPwd` query for $\mathcal{P}_i$, ignore this query. Otherwise, if $d(\mathsf{pw}_0, \mathsf{pw}_1) \leq \delta$, send pw_i to the adversary $\mathcal{S}$.
- **Upon receiving a query** $(\texttt{NewKey}, \mathsf{sid}, \mathcal{P}_i, \mathsf{sk})$ **from the adversary** $\mathcal{S}$, $\mathcal{P}_i \in \{\mathcal{P}_0, \mathcal{P}_1\}$**:**
  If there are no records of the form $(\mathcal{P}_i, \mathsf{pw}_i)$ and $(\mathcal{P}_{1-i}, \mathsf{pw}_{1-i})$, or if this is not the first `NewKey` query for $\mathcal{P}_i$, then ignore this query. Otherwise:
  - If at least one of the following is true, then output $(\mathsf{sid}, \mathsf{sk})$ to party $\mathcal{P}_i$.
    - $\mathcal{P}_i$ is corrupted
    - $\mathcal{P}_{1-i}$ is corrupted and $d(\mathsf{pw}_0, \mathsf{pw}_1) \leq \delta$
  - If both parties are honest, $d(\mathsf{pw}_0, \mathsf{pw}_1) \leq \delta$, and a key k_{1-i} was sent to $\mathcal{P}_{1-i}$, then output $(\mathsf{sid}, \mathsf{k}_{1-i})$ to $\mathcal{P}_i$.
  - In any other case, pick a new random key k_i of length λ and send $(\mathsf{sid}, \mathsf{k}_i)$ to $\mathcal{P}_i$.

Fig. 3. Ideal functionality $\mathcal{F}^P_{\mathsf{RFE}}$ for randomized fuzzy equality

3.2.1 The Randomized Fuzzy Equality Functionality

Figure 3 shows the randomized fuzzy equality functionality $\mathcal{F}^P_{\mathsf{RFE}}$, which is essentially what $\mathcal{F}^P_{\mathsf{fPAKE}}$ would look like assuming authenticated channels. The primary difference between $\mathcal{F}^P_{\mathsf{RFE}}$ and $\mathcal{F}^P_{\mathsf{fPAKE}}$ is that the only pass-string guesses allowed by $\mathcal{F}^P_{\mathsf{RFE}}$ are the ones actually used as protocol inputs; this limits the adversary to guessing by corrupting one of the participating parties, not through man in the middle attacks. Like $\mathcal{F}^P_{\mathsf{fPAKE}}$, if a pass-string guess is "similar enough", the entire pass-string is leaked. This leakage could be replaced with any other leakage from Sect. 2; $\mathcal{F}_{\mathsf{RFE}}$ would leak the correctness of the guess, $\mathcal{F}^M_{\mathsf{RFE}}$ would leak which characters are the same between the two pass-strings, etc.

Note that, unlike the randomized equality functionality in the work of Canetti *et al.* [18], $\mathcal{F}^P_{\mathsf{fPAKE}}$ has a `TestPwd` interface. This is because `NewKey` does not return the necessary leakage to an honest user. So, an interface enabling the adversary to retrieve additional information is necessary.

3.2.2 A Randomized Fuzzy Equality Protocol

In Fig. 4 we introduce a protocol Π_{RFE} that securely realizes $\mathcal{F}^P_{\mathsf{RFE}}$ using Yao's garbled circuits. Garbled circuits are secure against a malicious evaluator, but only a semi-honest garbler; however, we obtain security against malicious

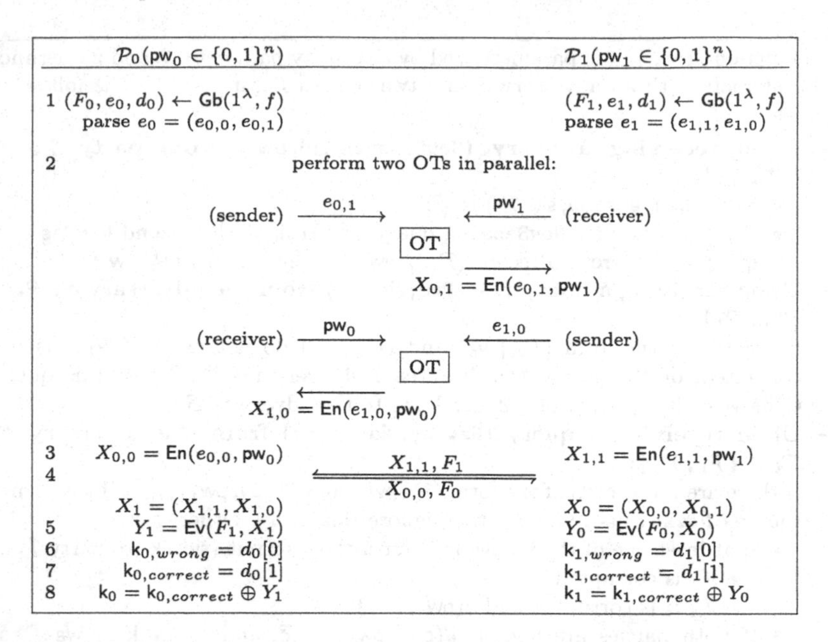

Fig. 4. A protocol Π_{RFE} realizing $\mathcal{F}^{P}_{\mathsf{RFE}}$ using Yao's garbled circuits and an Ideal OT Functionality. If at any point an expected message fails to arrive (or arrives malformed), the parties output a random key. Subscripts are used to indicate who produced the object in question. If a double subscript is present, the second subscript indicates whose data the object is meant for use with. For instance, a double subscript 0, 1 denotes that the object was produced by party $\mathcal{P}_0$ for use with $\mathcal{P}_1$'s data; $e_{0,1}$ is encoding information produced by $\mathcal{P}_0$ to encode $\mathcal{P}_1$'s pass-string. Note that we abuse notation by encoding inputs to a single circuit separately; the input to $\mathcal{P}_0$'s circuit corresponding to pw_0 is encoded by $\mathcal{P}_0$ locally, and the input corresponding to pw_1 is encoded via OT. For any projective garbling scheme, this is not a problem.

adversaries by having each party play each role once, as describe in Sect. 3.1.3. In more detail, both parties $\mathcal{P}_i \in \{\mathcal{P}_0, \mathcal{P}_1\}$ proceed as follows:

1. $\mathcal{P}_i$ garbles the circuit f that takes in two pass-strings pw_0 and pw_1, and returns '1' if $d(\mathsf{pw}_0, \mathsf{pw}_1) \leq \delta$ and '0' otherwise. Section 3.3 describes how f can be designed efficiently for Hamming distance. Instead of using the output of f ('0' or '1'), we will use the garbled output, also referred to as an *output label* in an output-projective garbling scheme. The possible output labels are two random strings—one corresponding to a '1' output (we call this label $\mathsf{k}_{i,correct}$), and one corresponding to a '0' output (we call this label $\mathsf{k}_{i,wrong}$).
2. $\mathcal{P}_i$ uses OT to retrieve the input labels from $\mathcal{P}_{1-i}$'s garbling that correspond to $\mathcal{P}_i$'s pass-string.
3. $\mathcal{P}_i$ sends $\mathcal{P}_{1-i}$ her garbled circuit, together with the input labels from her garbling that correspond to her own pass-string. After this step, $\mathcal{P}_i$ should have $\mathcal{P}_{1-i}$'s garbled circuit and a garbled input consisting of input labels corresponding to the bits of the two pass-strings.

4. $\mathcal{P}_i$ evaluates $\mathcal{P}_{1-i}$'s garbled circuit, and obtains an output label Y_{1-i}.
5. $\mathcal{P}_i$ outputs $\mathsf{k}_i = \mathsf{k}_{i,correct} \oplus Y_{1-i}$.

The natural question to ask is why Π_{RFE} only realizes $\mathcal{F}^{P}_{\mathsf{RFE}}$, and not a stronger functionality with less leakage. We argue this assuming (without loss of generality) that $\mathcal{P}_1$ is corrupted. Π_{RFE} cannot realize a functionality that leaks less than the full pass-string pw_0 to $\mathcal{P}_1$ if $d(\mathsf{pw}_0, \mathsf{pw}_1) \leq \delta$; intuitively, this is because if $\mathcal{P}_1$ knows a pass-string pw_1 such that $d(\mathsf{pw}_0, \mathsf{pw}_1) \leq \delta$, $\mathcal{P}_1$ can extract the actual pass-string pw_0, as follows. If $\mathcal{P}_1$ plays the role of OT receiver and garbled circuit evaluator honestly, $\mathcal{P}_0$ and $\mathcal{P}_1$ will agree on $\mathsf{k}_{0,correct}$. $\mathcal{P}_1$ can then mis-garble a circuit that returns $\mathsf{k}_{1,correct}$ if the first bit of pw_0 is 0, and $\mathsf{k}_{1,wrong}$ if the first bit of pw_0 is 1. By testing whether the resulting keys k_0 and k_1 match (which $\mathcal{P}_1$ can do in subsequent protocols where the key is used), $\mathcal{P}_1$ will be able to determine the actual first bit of pw_0. $\mathcal{P}_1$ can then repeat this for the second bit, and so on, extracting the entire pass-string pw_0. Of course, if $\mathcal{P}_1$ does *not* know a sufficiently close pw_1, $\mathcal{P}_1$ will not be able to perform these tests, because the keys will not match no matter what circuit $\mathcal{P}_1$ garbles.

More formally, if $\mathcal{P}_1$ knows a pass-string pw_1 such that $d(\mathsf{pw}_0, \mathsf{pw}_1) \leq \delta$ and carries out the mis-garbling attack described above, then in the real world, the keys produced by $\mathcal{P}_0$ and $\mathcal{P}_1$ either will or will not match based on some predicate p of $\mathcal{P}_1$'s choosing on the two pass-strings pw_0 and pw_1. Therefore, in the ideal world, the keys should also match or not match based on $p(\mathsf{pw}_0, \mathsf{pw}_1)$; otherwise, the environment will be able to distinguish between the two worlds. In order to make that happen, since the simulator does not know the predicate p in question, the simulator must be able to recover the entire pass-string pw_0 (given a sufficiently close pw_1) through the `TestPwd` interface.

Theorem 1. *If* $(\mathsf{Gb}, \mathsf{En}, \mathsf{Ev}, \mathsf{De})$ *is a projective, output-projective and garbled-output random secure garbling scheme, then protocol* Π_{RFE} *with authenticated channels in the* $\mathcal{F}_{\mathsf{OT}}$*-hybrid model securely realizes* $\mathcal{F}^{P}_{\mathsf{RFE}}$ *with respect to static corruptions for any threshold* δ*, as long as the pass-string space and notion of distance are such that for any pass-string* pw*, it is easy to compute another pass-string* pw' *such that* $d(\mathsf{pw}, \mathsf{pw}') > \delta$.

Proof (Sketch). For every efficient adversary $\mathcal{A}$, we describe a simulator $\mathcal{S}_{\mathsf{RFE}}$ such that no efficient environment can distinguish an execution with the real protocol Π_{RFE} and $\mathcal{A}$ from an execution with the ideal functionality $\mathcal{F}^{P}_{\mathsf{RFE}}$ and $\mathcal{S}_{\mathsf{RFE}}$. $\mathcal{S}_{\mathsf{RFE}}$ is described in the full version of this paper. We prove indistinguishability in a series of hybrid steps. First, we introduce the ideal functionality as a dummy node. Next, we allow the functionality to choose the parties' keys, and we prove the indistinguishability of this step from the previous using the garbled output randomness property of our garbling scheme Next, we simulate an honest party's interaction with another honest party without using their pass-string, and prove the indistinguishability of this step from the previous using the obliviousness property of our garbling scheme. Finally, we simulate an honest party's interaction with a corrupted party without using the honest party's pass-string, and prove the indistinguishability of this step from the previous using the privacy property of our garbling scheme.

We give a more formal proof of Theorem 1 in the full version of this paper [28].

3.2.3 From Split Randomized Fuzzy Equality to fPAKE

The Randomized Fuzzy Equality (RFE) functionality $\mathcal{F}_{\mathsf{RFE}}^{P}$ assumes authenticated channels, which an fPAKE protocol cannot do. In order to adapt RFE to our setting, we use the split functionality transformation defined by Barak *et al.* [4]. Barak *et al.* provide a generic transformation from protocols which require authenticated channels to protocols which do not. In the "transformed" protocol, an adversary can engage in two separate instances of the protocol with the sender and receiver, and they will not realize that they are not talking to one another. However, it does guarantee that the adversary cannot do anything beyond this attack. In other words, it provides "session authentication", meaning that each party is guaranteed to carry out the entire protocol with the same partner, but not "entity authentication", meaning that the identity of the partner is not guaranteed.

Barak *et al.* achieve this transformation in three steps. First, the parties generate signing and verification keys, and send one another their verification keys. Next, the parties sign the list of all keys they have received (which, in a two-party protocol, consists of only one key), sign that list, and send both list and signature to all other parties. Finally, they verify all of the signatures they have received. After this process—called "link initialization"—has been completed, the parties use those public keys they have exchanged to authenticate subsequent communication.

We describe the Randomized Fuzzy Equality Split Functionality in Fig. 5. It is simplified from Fig. 1 in Barak *et al.* [4] because we only need to consider two parties and static corruptions.

It turns out that $s\mathcal{F}_{\mathsf{RFE}}^{P}$ is enough to realize $\mathcal{F}_{\mathsf{fPAKE}}^{P}$. In fact, the protocol Π_{RFE} with the split functionality transformation directly realizes $\mathcal{F}_{\mathsf{fPAKE}}^{P}$. In the full version of this paper [28], we prove that this is the case.

3.3 An Efficient Circuit f for Hamming Distance

The Hamming distance of two pass-strings $\mathsf{pw}, \mathsf{pw}' \in \mathbb{F}_p^n$ is equal to the number of locations at which the two pass-strings have the same character. More formally,

$$d(\mathsf{pw}, \mathsf{pw}') := |\{j \mid \mathsf{pw}[j] \neq \mathsf{pw}'[j], j \in [n]\}|.$$

We design f for Hamming distance as follows:

1. First, f XORs corresponding (binary) pass-string characters, resulting in a list of bits indicating the (in)equality of those characters.
2. Then, f feeds those bits into a threshold gate, which returns 1 if at least $n - \delta$ of its inputs are 0, and returns 0 otherwise. f returns the output of that threshold gate, which is 1 if and only if at least $n - \delta$ pass-string characters match.

The functionality $s\mathcal{F}^{P}_{\mathsf{RFE}}$ is parameterized by a security parameter λ. It interacts with an adversary $\mathcal{S}$ and two parties $\mathcal{P}_0$ and $\mathcal{P}_1$ via the following queries:

- **Initialization**
  - **Upon receiving a query** $(\texttt{Init}, \mathsf{sid})$ **from a party** $\mathcal{P}_i \in \{\mathcal{P}_0, \mathcal{P}_1\}$, send $(\texttt{Init}, \mathsf{sid}, \mathcal{P}_i)$ to the adversary $\mathcal{S}$.
  - **Upon receiving a query** $(\texttt{Init}, \mathsf{sid}, \mathcal{P}_i, H, \mathsf{sid}_H)$ **from the adversary** $\mathcal{S}$:
    * Verify that $H \subseteq \{\mathcal{P}_0, \mathcal{P}_1\}$, that $\mathcal{P}_i \in H$, and that if a previous set H' was recorded, either (1) $H \cap H'$ contains only corrupted parties and $\mathsf{sid}_H \neq \mathsf{sid}_{H'}$, or (2) $H = H'$ and $\mathsf{sid}_H = \mathsf{sid}_{H'}$.
    * If verification fails, do nothing.
    * Otherwise, record the pair (H, sid_H) (if it was not already recorded), output $(\texttt{Init}, \mathsf{sid}, \mathsf{sid}_H)$ to $\mathcal{P}_i$, and locally initialize a new instance of the original RFE functionality $\mathcal{F}_{\mathsf{RFE}}$ denoted $H\mathcal{F}^{P}_{\mathsf{RFE}}$, letting the adversary play the role of $\{\mathcal{P}_0, \mathcal{P}_1\} - H$ in $H\mathcal{F}^{P}_{\mathsf{RFE}}$.
- **RFE**
  - **Upon receiving a query from a party** $\mathcal{P}_i \in \{\mathcal{P}_0, \mathcal{P}_1\}$, find the set H such that $\mathcal{P}_i \in H$, and forward the query to $H\mathcal{F}^{P}_{\mathsf{RFE}}$. Otherwise, ignore the query.
  - **Upon receiving a query from the adversary** $\mathcal{S}$ **on behalf of** $\mathcal{P}_i$ **corresponding to set** H, if $H\mathcal{F}^{P}_{\mathsf{RFE}}$ is initialized and $\mathcal{P}_i \notin H$, then forward the query to $H\mathcal{F}^{P}_{\mathsf{RFE}}$. Otherwise, ignore the query.

Fig. 5. Functionality $s\mathcal{F}^{P}_{\mathsf{RFE}}$

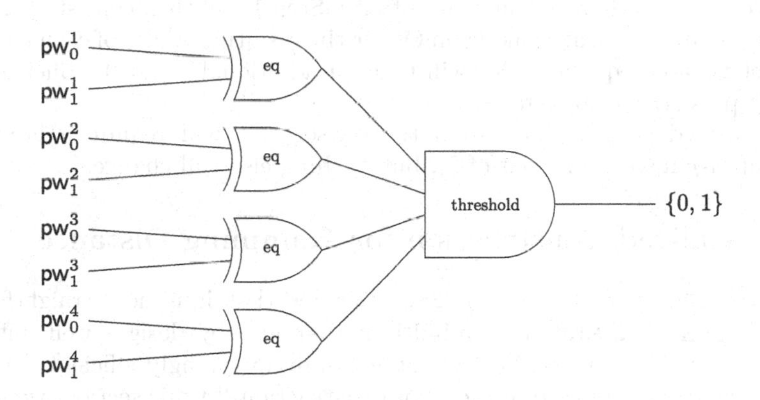

Fig. 6. The f circuit

This circuit, illustrated in Fig. 6, is very efficient to garble; it only requires n ciphertexts. Below, we briefly explain this garbling. Our explanation assumes familiarity with YGC literature [62, and references therein]. Briefly, garbled gadget labels [3] enable the evaluation of modular addition gates for free (there is no need to include any information in the garbled circuit to enable this addition). However, for a small modulus m, converting the output of that addition to a binary decision requires $m - 1$ ciphertexts. We utilize garbled gadgets with a modulus of $n + 1$ in our efficient garbling as follows:

1. The input wire labels encode 0 or 1 modulo $n+1$. However, instead of having those input wire labels encode the characters of the two pass-strings directly, they encode the outputs of the comparisons of corresponding characters. If the jth character of $\mathcal{P}_i$'s pass-string is 0, then $\mathcal{P}_i$ puts the 0 label first; however, if the jth character of $\mathcal{P}_i$'s pass-string is 1, then $\mathcal{P}_i$ flips the labels. Then, when $\mathcal{P}_{1-i}$ is using oblivious transfer to retrieve the label corresponding to her jth pass-string character, she will retrieve the 0 label if the two characters are equal, and the 1 label otherwise. (Note that this pre-processing on the garbler's side eliminates the need to send $X_{0,0}$ and $X_{1,1}$ in Fig. 4.)
2. Compute a n-input threshold gate, as illustrated in Fig. 6 of Yakoubov [62]. This gate returns 0 if the sum of the inputs is above a certain threshold (that is, if at least $n - \delta$ pass-string characters differ), and 1 otherwise. This will require n ciphertexts.

Thus, a garbling of f consists of n ciphertexts. Since fPAKE requires two such garbled circuits (Fig. 4), $2n$ ciphertexts will be exchanged.

Larger Pass-string Characters. If larger pass-string characters are used, then Step 1 above needs to change to check (in)equality of the larger characters instead of bits. Step 2 will remain the same. There are several ways to perform an (in)equality check on characters in $\mathbb{F}_p$ for $p \geq 2$:

1. Represent each character in terms of bits. Step 1 will then consist of XORing corresponding bits, and taking an OR or the resulting XORs of each character to get negated equality. This will take an additional $n \log(p)$ ciphertexts for every pass-string character.
2. Use garbled gadget labels from the outset. We will require a larger OT (1-out-of-p instead of 1-out-of-2), but nothing else will change.

4 Specialized Construction for Hamming Distance

In the full version of this paper [28], we show that it is not straightforward to build a secure fPAKE from primitives that are, by design, well-suited for correcting errors. However, PAKE protocols are appealingly efficient compared to the garbled circuits used in the prior construction. In this section, we will see whether the failed approach can be rescued in an efficient way, and we answer this question in the affirmative.

4.1 Building Blocks

4.1.1 Robust Secret Sharing

We recall the definition of a robust secret sharing scheme, slightly simplified for our purposes from Cramer *et al.* [22]. For a vector $c \in \mathbb{F}_q^n$ and a set $A \subseteq [n]$, we denote with c_A the projection $\mathbb{F}_q^n \rightarrow \mathbb{F}_q^{|A|}$, i.e., the sub-vector $(c_i)_{i \in A}$.

Definition 2. *Let $\mathbb{F}_q$ be a finite field and $n, t, r \in \mathbb{N}$ with $t < r \leq n$. An (n, t, r) robust secret sharing scheme (RSS) consists of two probabilistic algorithms* $\mathsf{Share} : \mathbb{F}_q \rightarrow \mathbb{F}_q^n$ *and* $\mathsf{Reconstruct} : \mathbb{F}_q^n \rightarrow \mathbb{F}_q$ *with the following properties:*

- *t-privacy: for any $s, s' \in \mathbb{F}_q, A \subset [n]$ with $|A| \leq t$, the projections c_A of $c \stackrel{\$}{\leftarrow} \mathsf{Share}(s)$ and c'_A of $c' \stackrel{\$}{\leftarrow} \mathsf{Share}(s')$ are identically distributed.*
- *r-robustness: for any $s \in \mathbb{F}_q, A \subset [n]$ with $|A| \geq r$, any c output by $\mathsf{Share}(s)$, and any $\tilde{c}$ such that $c_A = \tilde{c}_A$, it holds that $\mathsf{Reconstruct}(\tilde{c}) = s$.*

In other words, an (n, t, r)-RSS is able to reconstruct the shared secret even if the adversary tampered with up to $n - r$ shares, while each set of t shares is distributed independently of the shared secret s and thus reveals nothing about it. We note that we allow for a gap, i.e., $r \geq t + 1$. Schemes with $r > t + 1$ are called *ramp* RSS.

4.1.2 Linear Codes

A linear q-ary code of length n and rank k is a subspace C with dimension k of the vector space $\mathbb{F}_q^n$. The vectors in C are called codewords. The size of a code is the number of codewords it contains, and is thus equal to q^k. The weight of a word $w \in \mathbb{F}_q^n$ is the number of its non-zero components, and the distance between two words is the Hamming distance between them (equivalently, the weight of their difference). The minimal distance d of a linear code C is the minimum weight of its non-zero codewords, or equivalently, the minimum distance between any two distinct codewords.

A code for an alphabet of size q, of length n, rank k, and minimal distance d is called an $(n, k, d)_q$-code. Such a code can be used to detect up to $d - 1$ errors (because if a codeword is sent and fewer than $d - 1$ errors occur, it will not get transformed to another codeword), and correct up to $\lfloor (d-1)/2 \rfloor$ errors (because for any received word, there is a unique codeword within distance $\lfloor (d-1)/2 \rfloor$). For linear codes, the encoding of a (row vector) word $W \in \mathbb{F}_q^k$ is performed by an algorithm $C.\mathsf{Encode} : \mathbb{F}_q^k \rightarrow \mathbb{F}_q^n$, which is the multiplication of W by a so-called "generating matrix" $G \in \mathbb{F}_q^{k \times n}$ (which defines an injective linear map). This leads to a row-vector codeword $c \in C \subset \mathbb{F}_q^n$.

The Singleton bound states that for any linear code, $k + d \leq n + 1$, and a *maximum distance separable* (or MDS) code satisfies $k + d = n + 1$. Hence, $d = n - k + 1$ and MDS codes are fully described by the parameters (q, n, k). Such an $(n, k)_q$-MDS code can correct up to $\lfloor (n-k)/2 \rfloor$ errors; it can detect if there are errors whenever there are no more than $n - k$ of them.

For a thorough introduction to linear codes and proof of all statements in this short overview we refer the reader to [55].

Observe that a linear code, due to the linearity of its encoding algorithm, is not a primitive designed to hide anything about the encoded message. However, we show in the following lemma how to turn an MDS code into a RSS scheme.

Lemma 3. *Let C be a $(n+1, k)_q$-MDS code. We set L to be the last column of the generating matrix G of the code C and we denote by C' the $(n, k)_q$-MDS* *code whose generating matrix G' is G without the last column. Let* Share *and* Reconstruct *work as follows:*

- Share(s) *for $s \in \mathbb{F}_q$ first chooses a random row vector $W \in \mathbb{F}_q^k$ such that $W \cdot L = s$, and outputs $c \leftarrow C'$*.Encode(W) *(equivalently, we can say that* Share(s) *chooses a uniformly random codeword of C whose last coordinate is s, and outputs the first n coordinates as c).*
- Reconstruct(w) *for $w \in \mathbb{F}_q^n$ first runs C'*.Decode(w). *If it gets a vector W', then output $s = W' \cdot L$, otherwise output $s \overset{\$}{\leftarrow} \mathbb{F}_q$.*

Then Share *and* Reconstruct *form a (n, t, r)-RSS for $t = k-1$ and $r = \lceil (n+k)/2 \rceil$.*

Proof. Let us consider the two properties from Definition 2.

- t-privacy: Assume $|A| = t$ (privacy for smaller A will follow immediately by adding arbitrary coordinates to it to get to size t). Let $J = A \cup \{n+1\}$; note that $|J| = t + 1 = k$. Note that for the code C, any k coordinates of a codeword determine uniquely the input to Encode that produces this codeword (otherwise, there would be two codewords that agreed on k elements and thus had distance $n - k + 1$, which is less than the minimum distance of C). Therefore, the mapping given by Encode$_J$: $\mathbb{F}_q^k \rightarrow \mathbb{F}_q^{|J|}$ is bijective; thus coordinates in J are uniform when the input to Encode is uniform. The algorithm Share chooses the input to Encode uniformly subject to fixing the coordinate $n+1$ of the output. Therefore, the remaining coordinates (i.e., the coordinates in A) are uniform.
- r-robustness: Note that C has minimum distance $n - k + 2$, and therefore C' has minimum distance $n - k + 1$ (because dropping one coordinate reduces the distance by at most 1). Therefore, C' can correct $\lfloor (n-k)/2 \rfloor = n - r$ errors. Since $c_A = \tilde{c}_A$ and $|A| \geq r$, there are at most $n - r$ errors in $\tilde{c}$, so the call to C'.Decode(c') made by Reconstruct(c') will output $W' = W$. Then Reconstruct(c') will output $s = W' \cdot L = W \cdot L$.

Note that the Shamir's secret sharing scheme is exactly the above construction with Reed-Solomon codes [47].

4.1.3 Implicit-Only PAKE

PAKE protocols can have two types of authentication: implicit authentication, where at the end of the protocol the two parties share the same key if they used the same pass-string and random independent keys otherwise; or explicit authentication where, in addition, they actually know which of the two situations they are in. A PAKE protocol that only achieves implicit authentication can provide explicit authentication by adding key-confirmation flows [7].

The standard PAKE functionality $\mathcal{F}_{\mathsf{pwKE}}$ from [20]is designed with explicit authentication in mind, or at least considers that success or failure will later be detected by the adversary when he will try to use the key. Thus, it reveals to the adversary whether a pass-string guess attempt was successful or not. However, some applications could require a PAKE that does not provide any feedback, and so does not reveal the situation before the keys are actually used. Observe that, regarding honest players, already $\mathcal{F}_{\mathsf{pwKE}}$ features implicit authentication since the players do not learn anything but their own session key.

Definition of implicit-only PAKE. Hence, we introduce a new notion, called implicit-only PAKE or iPAKE (see Fig. 7). This ideal functionality is designed to implement implicit authentication also with respect to an adversary, namely by not providing him with any feedback upon a dictionary attack. Of course, in many cases, the parties as well as the adversary can later check whether their session keys match or not, and so whether the pass-strings were the same or not. We stress that this is not a leakage from the PAKE protocol itself, but from the global system.

In terms of functionalities, there are two differences from $\mathcal{F}_{\mathsf{pwKE}}$ to $\mathcal{F}_{\mathsf{iPAKE}}$. First, the `TestPwd` query only silently updates the internal state of the record (from `fresh` to either `compromised` or `interrupted`), meaning that its outcome is not given to the adversary $\mathcal{S}$. Second, the `NewKey` query is modified so that the adversary gets to choose the key for a non-corrupted party only if it uses the

The functionality $\mathcal{F}_{\mathsf{iPAKE}}$ is parameterized by a security parameter λ. It interacts with an adversary $\mathcal{S}$ and the (dummy) parties $\mathcal{P}_0$ and $\mathcal{P}_1$ via the following queries:

- **Upon receiving a query (`NewSession`, sid, pw_i) from party $\mathcal{P}_i$:**
  - Send (`NewSession`, sid, $\mathcal{P}_i$) to $\mathcal{S}$;
  - If this is the first `NewSession` query, or if this is the second `NewSession` query and there is a record $(\mathcal{P}_{1-i}, \mathsf{pw}_{1-i})$, then record $(\mathcal{P}_i, \mathsf{pw}_i)$ and mark this record `fresh`.
- **Upon receiving a query (`TestPwd`, sid, $\mathcal{P}_i$, pw'_i) from $\mathcal{S}$:**
  If there is a `fresh` record $(\mathcal{P}_i, \mathsf{pw}_i)$, then:
  - If $\mathsf{pw}_i = \mathsf{pw}'_i$, mark the record `compromised`;
  - If $\mathsf{pw}_i \neq \mathsf{pw}'_i$, mark the record `interrupted`.
- **Upon receiving a query (`NewKey`, sid, $\mathcal{P}_i$, sk) from $\mathcal{S}$, where $|\mathsf{sk}| = \lambda$:**
  If there is no record of the form $(\mathcal{P}_i, \mathsf{pw}_i)$, or if this is not the first `NewKey` query for $\mathcal{P}_i$, then ignore this query. Otherwise:
  - If at least one of the following is true, then output (sid, sk) to player $\mathcal{P}_i$:
    * The record is `compromised`
    * $\mathcal{P}_i$ is corrupted
    * The record is `fresh`, $\mathcal{P}_{1-i}$ is corrupted, and there is a record $(\mathcal{P}_{1-i}, \mathsf{pw}_{1-i})$ with $\mathsf{pw}_i = \mathsf{pw}_{1-i}$
  - If this record is `fresh`, both parties are honest, there is a record $(\mathcal{P}_{1-i}, \mathsf{pw}_{1-i})$ with $\mathsf{pw}_i = \mathsf{pw}_{1-i}$, a key sk' was sent to $\mathcal{P}_{1-i}$, and $(\mathcal{P}_{1-i}, \mathsf{pw}_{1-i})$ was `fresh` at the time, then output (sid, sk′) to $\mathcal{P}_i$;
  - In any other case, pick a new random key sk' of length λ and send (sid, sk′) to $\mathcal{P}_i$.
  - Mark the record $(\mathcal{P}_i, \mathsf{pw}_i)$ as `completed`.

Fig. 7. Functionality $\mathcal{F}_{\mathsf{iPAKE}}$

correct pass-string (corruption of the other party is no longer enough), as already discussed earlier. Without going too much into the details, it is intuitively clear that simulation of an honest party is hard if the simulator does not know whether it should proceed the simulation with a pass-string extracted from a dictionary attack or not. Regarding the output, i.e., the question whether the session keys computed by both parties should match or look random, the simulator thus gets help from our functionality by modifying the `NewKey` queries.

We further alter this functionality to allow for public labels, as shown in the full version of this paper [28]. The resulting functionality $\mathcal{F}_{\ell\text{-iPAKE}}$ idealizes what we call *labeled implicit-only PAKE* (or ℓ-iPAKE for short), resembling the notion of labeled public key encryption as formalized in [56]. In a nutshell, labels are public authenticated strings that are chosen by each user individually for each execution of the protocol. Authenticated here means that tampering with the label can be efficiently detected. Such labels can be used to, e.g., distribute public information such as public keys reliably over unauthenticated channels.

A UC-Secure ℓ-iPAKE Protocol. In the seminal paper by Bellovin and Merritt [8], the Encrypted Key Exchange protocol (EKE) is proposed, which is essentially a Diffie-Hellman [24] key exchange. The two flows of the protocol are encrypted using the pass-string as key with an appropriate symmetric encryption scheme. The EKE protocol has been further formalized by Bellare *et al.* [7] under the name EKE2. We present its labeled variant in Fig. 8. The idea of appending the label to the symmetric key is taken from [1]. We prove security of this protocol in the $\mathcal{F}_{\mathsf{RO}}, \mathcal{F}_{\mathsf{IC}}, \mathcal{F}_{\mathsf{CRS}}$-hybrid model. That is, we use an ideal random oracle functionality $\mathcal{F}_{\mathsf{RO}}$ to model the hash function, and ideal cipher functionality $\mathcal{F}_{\mathsf{IC}}$ to model the encryption scheme and assume a publicly available common reference string modeled by $\mathcal{F}_{\mathsf{CRS}}$. Formal definitions of these functionalities are given in the full version of this paper [28].

$A(\mathsf{pw} \in \mathbb{F}_p)$		$B(\mathsf{pw}' \in \mathbb{F}_p)$
$x \xleftarrow{\$} \mathbb{F}_P,\ \ell \leftarrow \mathcal{L},\ X \leftarrow g^x$		
$X^* \leftarrow \mathcal{E}_{\mathsf{pw}\|\|\ell}(X)$	$\xrightarrow{\ell, X^*}$	$y \xleftarrow{\$} \mathbb{F}_P,\ \ell' \leftarrow \mathcal{L},\ Y \leftarrow g^y$
	$\xleftarrow{\ell', Y^*}$	$Y^* \leftarrow \mathcal{E}_{\mathsf{pw}'\|\|\ell'}(Y)$
$Z \leftarrow \mathcal{D}_{\mathsf{pw}\|\|\ell'}(Y^*)^x$		$Z' \leftarrow \mathcal{D}_{\mathsf{pw}'\|\|\ell}(X^*)^y$
$k \leftarrow H(X^*, Y^*, Z)$		$k' \leftarrow H(X^*, Y^*, Z')$
output (ℓ', k)		output (ℓ, k')

Fig. 8. Protocol EKE2, in a group $\mathbb{G} = \langle g \rangle$ of prime order P, with a hash function $H : \mathbb{G}^3 \rightarrow \{0,1\}^k$ and a symmetric cipher $\mathcal{E}, \mathcal{D}$ onto $\mathbb{G}$ for keys in $\mathbb{F}_p \times \mathcal{L}$, where $\mathcal{L}$ is the label space.

Theorem 4. *If the CDH assumption holds in* $\mathbb{G}$, *the protocol EKE2 depicted in Fig. 8 securely realizes* $\mathcal{F}_{\ell\text{-iPAKE}}$ *in the* $\mathcal{F}_{RO}, \mathcal{F}_{IC}, \mathcal{F}_{CRS}$*-hybrid model with respect to static corruptions.*

We note that this result is not surprising, given that other variants of EKE2 have already been proven to UC-emulate $\mathcal{F}_{\mathsf{pwKE}}$. Intuitively, a protocol with only two flows not depending on each other does not leak the outcome to the adversary via the transcript, which explains why EKE2 is implicit-only. Hashing of the transcript keeps the adversary from biasing the key unless he knows the correct pass-string or breaks the ideal cipher. For completeness, we include the full proof in the full version of this paper [28].

4.2 Construction

We show how to combine an RSS with a signature scheme and an ℓ-iPAKE to obtain an fPAKE. The high-level idea is to fix the issue that arose in the protocol from the full version of this paper [28] due to pass-strings being used as one-time pads. Instead, we first expand the pass-string characters to session keys with large entropy using ℓ-iPAKE. The resulting session keys are then used as a one-time pad on the entirety of shares of a nonce. We also apply known techniques from the literature, such as executing the protocol twice with reversed roles to protect against malicious parties, and adding signatures and labels to prevent man-in-the-middle attacks. Our full protocol is depicted in Fig. 9. It works as follows:

1. In the first phase, the two parties aim at enhancing their pass-strings to a vector of session keys with good entropy. For this, pass-strings are viewed as vectors of characters. The parties repeatedly execute a PAKE on each of these characters separately. The PAKE will ensure that the key vectors held by the two parties match in all positions where their pass-strings matched, and are uniformly random in all other positions.
2. In the second phase, the two parties exchange nonces of their choice, in such a way that the nonce reaches the other party only if enough of the key vector matches. This is done by applying an RSS to the nonce, and sending it to the other party using the key vector as a one time pad. Both parties do this symmetrically, each using half of the bits of the key vector. The robustness property of the RSS ensures that a few non-matching pass-string characters do not prevent both parties from recovering the other party's nonce. The final key is then obtained by adding the nonces (again, as a one-time pad): this is a scalar in $\mathbb{F}_q$.

When using the RSS from MDS codes described in Lemma 3, the one-time pad encryption of the shares (which form a codeword) can be viewed as the code-offset construction for information reconciliation (aka secure sketch) [27,36] applied to the key vectors. While our presentation goes through RSS as a separate object, we could instead present this construction using information reconciliation. The syndrome construction of secure sketches Lemma 3 can also be used here instead of the code-offset construction.

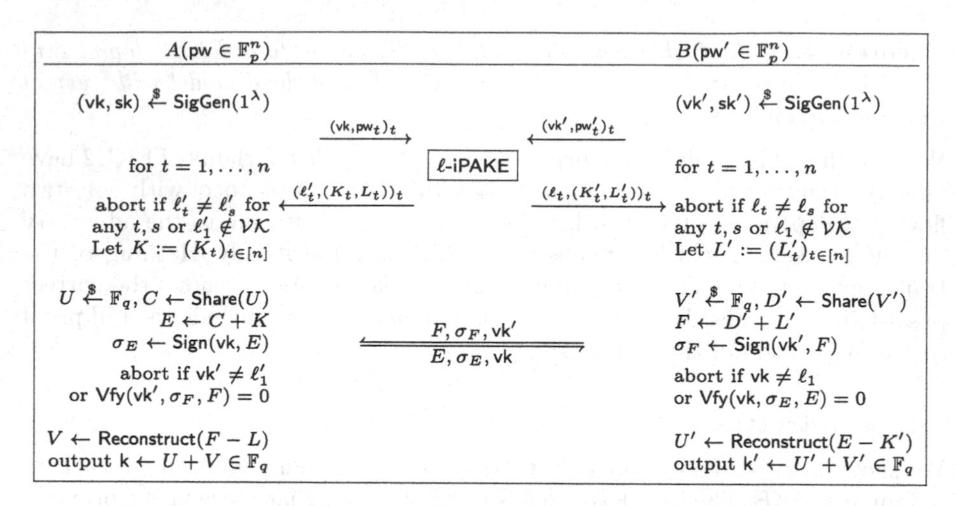

Fig. 9. Protocol $\mathsf{fPAKE_{RSS}}$ where $q \approx 2^\lambda$ is a prime number and $+$ denotes the group operation in $\mathbb{F}_q^n$. $(\mathsf{Share}, \mathsf{Reconstruct})$ is a Robust Secret Sharing scheme with $\mathsf{Share} : \mathbb{F}_q \to \mathbb{F}_q^n$, and $(\mathsf{SigGen} \to \mathcal{VK} \times \mathcal{SK}, \mathsf{Sign}, \mathsf{Vfy})$ is a signature scheme. The parties repeatedly execute a labeled implicit-only PAKE protocol with label space $\mathcal{VK}$ and key space $\mathbb{F}_q^2$, which takes inputs from $\mathbb{F}_p$. If at any point an expected message fails to arrive (or arrives malformed), the parties output a random key.

4.3 Security of $\mathsf{fPAKE_{RSS}}$

We show that our protocol realizes functionality $\mathcal{F}_{\mathsf{fPAKE}}^M$ in the $\mathcal{F}_{\ell\text{-}\mathsf{iPAKE}}$-hybrid model. In a nutshell, the idea is to simulate without the pass-strings by adjusting the keys outputted by $\mathcal{F}_{\ell\text{-}\mathsf{iPAKE}}$ to the mask of the pass-strings, which is leaked by $\mathcal{F}_{\mathsf{fPAKE}}^M$.

Theorem 5. *If* $(\mathsf{Share} : \mathbb{F}_q \to \mathbb{F}_q^n, \mathsf{Reconstruct} : \mathbb{F}_q^n \to \mathbb{F}_q)$ *is an* (n, t, r) *RSS and* $(\mathsf{SigGen}, \mathsf{Sign}, \mathsf{Vfy})$ *is an EUF-CMA secure one-time signature scheme, protocol* fPAKE_{RSS} *securely realizes* $\mathcal{F}_{\mathsf{fPAKE}}^M$ *with* $\gamma = n - t - 1$ *and* $\delta = n - r$ *in the* $\mathcal{F}_{\ell\text{-}\mathsf{iPAKE}}$*-hybrid model with respect to static corruptions.*

In particular, if we wish key agreement to succeed as long as there are fewer than δ errors, we instantiate RSS using the construction of Lemma 3 based on a $(n+1, k)_q$ MDS code, with $k = n - 2\delta$. This will give $r = \lceil (n+k)/2 \rceil = n - \delta$, so δ will be equal to $n - r$, as required. It will also give $\gamma = n - t - 1 = 2\delta$.

We thus obtain the following corollary:

Corollary 6. *For any* δ *and* $\gamma = 2\delta$*, given an* $(n+1, k)_q$*-MDS code for* $k = n - 2\delta$ *(with minimal distance* $d = n - k + 2$*) and an EUF-CMA secure one-time signature scheme, protocol* fPAKE_{RSS} *securely realizes* $\mathcal{F}_{\mathsf{fPAKE}}^M$ *in the* $\mathcal{F}_{\ell\text{-}\mathsf{iPAKE}}$*-hybrid model with respect to static corruptions.*

Proof sketch of Theorem 5. We start with the real execution of the protocol and indistinguishably switch to an ideal execution with dummy parties relaying their inputs to and obtaining their outputs from $\mathcal{F}_{\mathsf{fPAKE}}^{M}$. To preserve the view of the distinguisher, the environment $\mathcal{Z}$, a simulator $\mathcal{S}$ plays the role of the real world adversary by controlling the communication between $\mathcal{F}_{\mathsf{fPAKE}}^{M}$ and $\mathcal{Z}$. During the proof, we built $\mathcal{F}_{\mathsf{fPAKE}}^{M}$ and $\mathcal{S}$ by subsequently randomizing pass-strings (since the final simulation has to work without them) and session keys (since $\mathcal{F}_{\mathsf{fPAKE}}^{M}$ hands out random session keys in certain cases). We have to tackle the following difficulties, which we will describe in terms of attacks.

- Passive attack: in this attack, $\mathcal{Z}$ picks two pass-strings and then observes the transcript and outputs of the protocol, without having access to any internal state of the parties. We show that $\mathcal{Z}$ cannot distinguish between transcript and outputs that were either produced using $\mathcal{Z}$'s pass-strings or random pass-strings. Regarding the outputs, we argue that even in the real execution the session keys were chosen uniformly at random (with $\mathcal{Z}$ not knowing the coins consumed by this choice) as long as the distance check is reliable. Using properties of the RSS, we show that this is the case with overwhelming probability. Regarding the transcript, randomization is straightforward using properties of the one-time pad.
- Man-in-the-middle attack: in this attack, $\mathcal{Z}$ injects a malicious message into a session of two honest parties. There are several ways to secure protocols that have to run in unauthenticated channels and are prone to this attack. Basically, all of them introduce methods to bind messages together to prevent the adversary from injecting malicious messages. To do this, we need the *labeled* version of our iPAKE and a one-time signature scheme[3]. Unless $\mathcal{Z}$ is able to break a one-time-signature scheme, this attack always results in an abort.
- Active attack: in this attack, $\mathcal{Z}$ injects a malicious message into a session with one corrupted party, thereby knowing the internal state of this party. We show how to produce transcript and outputs looking like in a real execution, but without using the pass-strings of the honest party. Since $\mathcal{Z}$ can now actually decrypt the one-time pad and therefore the transcript reveals the positions of the errors in the pass-strings, $\mathcal{S}$ has to rely on $\mathcal{F}_{\mathsf{fPAKE}}^{M}$ revealing the mask of the pass-strings used in the real execution. If, on the other hand, the pass-strings are too far away from each other, we show that the privacy property of the RSS actually hides the number and positions of the errors. This way, $\mathcal{S}$ can use a random pass-string to produce the transcript in that case.

One interesting subtlety that arises is the usage of the iPAKE. Observe that the UC security notion for a regular PAKE as defined in [20] and recalled in the full version of this paper [28] provides an interface to the adversary to test

[3] Instead of labels and one-time signature, one could just sign all the messages, as would be done using the split-functionality [4], but this would be less efficient. This trade-off, with labels, is especially useful when we use a PAKE that admits adding labels basically for free, as it is the case with the special PAKE protocol we use.

a pass-string once and learn whether it is right or wrong. Using this notion, our simulator would have to answer to such queries from $\mathcal{Z}$. Since this is not possible without $\mathcal{F}_{\text{fPAKE}}^{M}$ leaking the mask all the time, it is crucial to use the iPAKE variant that we introduced in Sect. 4.1.3. Using this stronger notion, the adversary is still allowed one pass-string guess which may affect the output, but the adversary learns nothing more about the outcome of his guess than he can infer from whatever access he has to the outputs alone. Since our protocol uses the outputs of the PAKE as one-time pad keys, it is intuitively clear that by preventing $\mathcal{Z}$ from getting additional leakage about these keys, we protect the secrets of honest parties.

4.4 Further Discussion

4.4.1 Adaptive Corruptions

Adaptive security of our protocol is not achievable without relying on additional assumptions. To see this, consider the following attack: $\mathcal{Z}$ starts the protocol with two equal pass-strings and, without corrupting anyone, silently observes the transcript produced by $\mathcal{S}$ using random pass-strings. Afterwards, $\mathcal{Z}$ corrupts both players to learn their internal state. $\mathcal{S}$ may now choose a value K. This also fixes $L' = K$ since the pass-strings were equal. Now note that $\mathcal{S}$ is committed to E, F since signatures are not equivocable. Since perfect shares are sparse in $\mathbb{F}_q^n$, the probability that there exists a K such that $E - K$ and $F - K$ are both perfect shares is negligible. Thus, there do not exist plausible values U, V' that explain the transcript[4].

4.4.2 Removing Modeling Assumptions

All modeling assumptions of our protocol come from the realization of the ideal $\mathcal{F}_{\ell\text{-iPAKE}}$ functionality. E.g., the ℓ-iPAKE protocol from Sect. 4.1.3 requires a random oracle, an ideal cipher and a CRS. We note that we can remove everything up to the CRS by, e.g., taking the PAKE protocol introduced in [37]. This protocol also securely realizes our $\mathcal{F}_{\ell\text{-iPAKE}}$ functionality[5]. However, it is more costly than our ℓ-iPAKE protocol since both messages each contain one non-interactive zero knowledge proof.

[4] We note that additional assumptions like assuming erasures can enable an adaptive security proof.

[5] In a nutshell, their protocol is implicit-only for the same reason as the ℓ-iPAKE protocol we use here: there are only two flows that do not depend on each other, so the transcript cannot reveal the outcome of a guess unless it reveals the pass-string to anyone. Regarding the session keys, usage of a hash function takes care of randomizing the session key in case of a failed dictionary attack. Furthermore, the protocol already implements labels. A little more detailed, looking at the proof in [37], the simulator does not make use of the answer of `TestPwd` to simulate any messages. Regarding the session key that an honest player receives in an corrupted session, they are chosen to be random in the simulation (in Expt_3). Letting this happen already in the functionality makes the simulation independent of the answer of `TestPwd` also regarding the computation of the session keys.

Since fPAKE implies a regular PAKE (simply set $\delta = 0$), [20] gives strong evidence that we cannot hope to realize $\mathcal{F}_{\mathsf{fPAKE}}$ without a CRS.

5 Comparison of fPAKE Protocols

In this section, we give a brief comparison of our fPAKE protocols. First, in Fig. 10, we describe the assumptions necessary for the two constructions, and the security parameters that they can achieve.

Then, in Fig. 11, we describe the efficiency of the constructions when concrete primitives (OT/ℓ-iPAKE) are used to instantiate them. $\mathsf{fPAKE}_{\mathsf{RSS}}$ is instantiated as the construction in Fig. 9 with the ℓ-iPAKE in Fig. 8 and an RSS. $\mathsf{fPAKE}_{\mathsf{YGC}}$ is instantiated as the construction in Fig. 4 with the UC-secure oblivious transfer protocol of Chou and Orlandi [21], with the garbling scheme of Bal *et al.* [3], and with the split functionality transformation of Barak *et al.* [4]. Though $\mathsf{fPAKE}_{\mathsf{YGC}}$ can handle any efficiently computable notion of distance, Fig. 11 assumes that both constructions use Hamming distance (and that, specifically, $\mathsf{fPAKE}_{\mathsf{YGC}}$ uses the circuit described in Fig. 6). We describe efficiency in terms of sub-operations (per-party, not in aggregate).

	Assumptions	Threshold δ	Gap $\gamma - \delta$
$\mathsf{fPAKE}_{\mathsf{RSS}}$	UC-secure ℓ-iPAKE	$< n/2$	δ
$\mathsf{fPAKE}_{\mathsf{YGC}}$	(1) UC-secure OT (2) projective, output-projective and garbled-output random secure garbling scheme	Any	None

Fig. 10. Assumptions, distance thresholds and functionality/security gaps achieved by the two schemes. $\mathsf{fPAKE}_{\mathsf{RSS}}$ is the construction in Fig. 9. $\mathsf{fPAKE}_{\mathsf{YGC}}$ is the construction in Fig. 4 with the split functionality transformation of Barak *et al.* [4].

	Output Key Format	# (Bidirectional) Communication Flows	# Exponentiations	# Hashes	# Encryptions	# Decryptions	# Share	# Reconstruct	# SigKeyGens	# Signs	# SigVerifies
$\mathsf{fPAKE}_{\mathsf{RSS}}$	$\mathbb{F}_q$	2	$2n$	n	n	n	1	1	1	1	1
$\mathsf{fPAKE}_{\mathsf{YGC}}$	$\{0,1\}^\lambda$	5	$3n+2$	$4n+7$	$2n$	n	–	–	1	5	5

Fig. 11. Efficiency (in terms of sub-operations) of the two constructions. $\mathsf{fPAKE}_{\mathsf{RSS}}$ is the construction in Fig. 9 instantiated with the ℓ-iPAKE in Fig. 8. $\mathsf{fPAKE}_{\mathsf{YGC}}$ is the construction in Fig. 4 instantiated with the UC-secure oblivious transfer protocol of Chou and Orlandi [21], the garbling scheme of Bal *et al.* [3], and with the split functionality transformation of Barak *et al.* [4].

Note that these concrete primitives each have their own set of required assumptions. Specifically, the ℓ-iPAKE in Fig. 8 requires a random oracle (RO), ideal cipher (IC) and common reference string (CRS). The oblivious transfer protocol of Chou and Orlandi [21] requires a random oracle. The garbling scheme of Bal *et al.* [3] requires a mixed modulus circular correlation robust hash function, which is a weakening of the random oracle assumption.

For $\mathsf{fPAKE}_{\mathsf{RSS}}$, the factor of n arises from the n times EKE2 is executed. For $\mathsf{fPAKE}_{\mathsf{YGC}}$, the factor of n comes from the garbled circuit. Additionally, in $\mathsf{fPAKE}_{\mathsf{YGC}}$, three rounds of communication come from OT. The last of these is combined with sending the garbled circuits. Two additional rounds of communication come from the split functionality transformation. The need for signatures also arises from the split functionality transformation.

Efficiency Optimizations to $\mathsf{fPAKE}_{\mathsf{YGC}}$. We can make several small efficiency improvements to the $\mathsf{fPAKE}_{\mathsf{YGC}}$ construction which are not reflected in Fig. 11. First, instead of using the split functionality transformation of Barak *et al.* [4], we can use the split functionality of Camenisch *et al.* [16]. It uses a split key exchange functionality to establish symmetric keys, and then uses those to symmetrically encrypt and authenticate each flow. While this does not save any rounds, it does reduce the number of public key operations needed. Second, if the pass-strings are more than λ bits long (where λ is the security parameter), OT extensions that are secure against malicious adversaries [2] can be used. If the pass-strings are fewer than λ bits long, then nothing is to be gained from using OT extensions, since OT extensions require λ "base OTs". However, if the pass-strings are longer—say, if they are some biometric measurement that is thousands of bits long—then OT extensions would save on the number of public key operations, at the cost of an extra round of communication.

Acknowledgments. We thank Ran Canetti for guidance on the details of UC key agreement definitions, and Adam Smith for discussions on coding and information reconciliation.

This work was supported in part by the European Research Council under the European Community's Seventh Framework Programme (FP7/2007-2013 Grant Agreement no. 339563 – CryptoCloud). Leonid Reyzin gratefully acknowledges the hospitality of École Normale Supérieure, where some of this work was performed. He was supported, in part, by US NSF grants 1012910, 1012798, and 1422965.

References

1. Abdalla, M., Catalano, D., Chevalier, C., Pointcheval, D.: Efficient two-party password-based key exchange protocols in the UC framework. In: Malkin, T. (ed.) CT-RSA 2008. LNCS, vol. 4964, pp. 335–351. Springer, Heidelberg (2008). https://doi.org/10.1007/978-3-540-79263-5_22
2. Afshar, A., Hu, Z., Mohassel, P., Rosulek, M.: How to efficiently evaluate RAM programs with malicious security. In: Oswald, E., Fischlin, M. (eds.) EUROCRYPT 2015, Part I. LNCS, vol. 9056, pp. 702–729. Springer, Heidelberg (2015). https://doi.org/10.1007/978-3-662-46800-5_27

3. Ball, M., Malkin, T., Rosulek, M.: Garbling gadgets for boolean and arithmetic circuits. In: Weippl, E.R., Katzenbeisser, S., Kruegel, C., Myers, A.C., Halevi, S. (eds.) ACM CCS 2016, pp. 565–577. ACM Press, New York (2016)
4. Barak, B., Canetti, R., Lindell, Y., Pass, R., Rabin, T.: Secure computation without authentication. In: Shoup, V. (ed.) CRYPTO 2005. LNCS, vol. 3621, pp. 361–377. Springer, Heidelberg (2005). https://doi.org/10.1007/11535218_22
5. Beaver, D., Micali, S., Rogaway, P.: The round complexity of secure protocols (extended abstract). In: 22nd ACM STOC, pp. 503–513. ACM Press, May 1990
6. Bellare, M., Hoang, V.T., Rogaway, P.: Foundations of garbled circuits. In: Yu, T., Danezis, G., Gligor, V.D. (eds.) ACM CCS 2012, pp. 784–796. ACM Press, New York (2012)
7. Bellare, M., Pointcheval, D., Rogaway, P.: Authenticated key exchange secure against dictionary attacks. In: Preneel, B. (ed.) EUROCRYPT 2000. LNCS, vol. 1807, pp. 139–155. Springer, Heidelberg (2000). https://doi.org/10.1007/3-540-45539-6_11
8. Bellovin, S.M., Merritt, M.: Encrypted key exchange: password-based protocols secure against dictionary attacks. In: 1992 IEEE Symposium on Security and Privacy, pp. 72–84. IEEE Computer Society Press, May 1992
9. Bennett, C.H., Brassard, G., Robert, J.M.: Privacy amplification by public discussion. SIAM J. Comput. **17**(2), 210–229 (1988)
10. Bitansky, N., Canetti, R., Kalai, Y.T., Paneth, O.: On virtual grey box obfuscation for general circuits. In: Garay, J.A., Gennaro, R. (eds.) CRYPTO 2014, Part II. LNCS, vol. 8617, pp. 108–125. Springer, Heidelberg (2014). https://doi.org/10.1007/978-3-662-44381-1_7
11. Blanton, M., Hudelson, W.M.P.: Biometric-based non-transferable anonymous credentials. In: Qing, S., Mitchell, C.J., Wang, G. (eds.) ICICS 2009. LNCS, vol. 5927, pp. 165–180. Springer, Heidelberg (2009). https://doi.org/10.1007/978-3-642-11145-7_14
12. Boyen, X.: Reusable cryptographic fuzzy extractors. In: Atluri, V., Pfitzmann, B., McDaniel, P. (eds.) ACM CCS 2004, pp. 82–91. ACM Press, New York (2004)
13. Boyen, X., Dodis, Y., Katz, J., Ostrovsky, R., Smith, A.: Secure remote authentication using biometric data. In: Cramer, R. (ed.) EUROCRYPT 2005. LNCS, vol. 3494, pp. 147–163. Springer, Heidelberg (2005). https://doi.org/10.1007/11426639_9
14. Boyko, V., MacKenzie, P., Patel, S.: Provably secure password-authenticated key exchange using Diffie-Hellman. In: Preneel, B. (ed.) EUROCRYPT 2000. LNCS, vol. 1807, pp. 156–171. Springer, Heidelberg (2000). https://doi.org/10.1007/3-540-45539-6_12
15. Brostoff, S., Sasse, M.A.: Are passfaces more usable than passwords? A field trial investigation. In: McDonald, S., Waern, Y., Cockton, G. (eds.) People and Computers XIV – Usability or Else!, pp. 405–424. Springer, London (2000). https://doi.org/10.1007/978-1-4471-0515-2_27
16. Camenisch, J., Casati, N., Gross, T., Shoup, V.: Credential authenticated identification and key exchange. In: Rabin, T. (ed.) CRYPTO 2010. LNCS, vol. 6223, pp. 255–276. Springer, Heidelberg (2010). https://doi.org/10.1007/978-3-642-14623-7_14
17. Canetti, R.: Universally composable security: a new paradigm for cryptographic protocols. In: 42nd FOCS, pp. 136–145. IEEE Computer Society Press, October 2001

18. Canetti, R., Dachman-Soled, D., Vaikuntanathan, V., Wee, H.: Efficient password authenticated key exchange via oblivious transfer. In: Fischlin, M., Buchmann, J., Manulis, M. (eds.) PKC 2012. LNCS, vol. 7293, pp. 449–466. Springer, Heidelberg (2012). https://doi.org/10.1007/978-3-642-30057-8_27
19. Canetti, R., Fuller, B., Paneth, O., Reyzin, L., Smith, A.: Reusable fuzzy extractors for low-entropy distributions. In: Fischlin, M., Coron, J.-S. (eds.) EUROCRYPT 2016, Part I. LNCS, vol. 9665, pp. 117–146. Springer, Heidelberg (2016). https://doi.org/10.1007/978-3-662-49890-3_5
20. Canetti, R., Halevi, S., Katz, J., Lindell, Y., MacKenzie, P.: Universally composable password-based key exchange. In: Cramer, R. (ed.) EUROCRYPT 2005. LNCS, vol. 3494, pp. 404–421. Springer, Heidelberg (2005). https://doi.org/10.1007/11426639_24
21. Chou, T., Orlandi, C.: The simplest protocol for oblivious transfer. In: Lauter, K., Rodríguez-Henríquez, F. (eds.) LATINCRYPT 2015. LNCS, vol. 9230, pp. 40–58. Springer, Cham (2015). https://doi.org/10.1007/978-3-319-22174-8_3
22. Cramer, R., Damgård, I.B., Döttling, N., Fehr, S., Spini, G.: Linear secret sharing schemes from error correcting codes and universal hash functions. In: Oswald, E., Fischlin, M. (eds.) EUROCRYPT 2015, Part II. LNCS, vol. 9057, pp. 313–336. Springer, Heidelberg (2015). https://doi.org/10.1007/978-3-662-46803-6_11
23. Daugman, J.: How iris recognition works. IEEE Trans. Circuits Syst. Video Technol. **14**(1), 21–30 (2004)
24. Diffie, W., Hellman, M.E.: New directions in cryptography. IEEE Trans. Inf. Theory **22**(6), 644–654 (1976)
25. Dodis, Y., Kanukurthi, B., Katz, J., Reyzin, L., Smith, A.: Robust fuzzy extractors and authenticated key agreement from close secrets. IEEE Trans. Inf. Theory **58**(9), 6207–6222 (2012). https://doi.org/10.1109/TIT.2012.2200290
26. Dodis, Y., Ostrovsky, R., Reyzin, L., Smith, A.: Fuzzy extractors: how to generate strong keys from biometrics and other noisy data. SIAM J. Comput. **38**(1), 97–139 (2008)
27. Dodis, Y., Reyzin, L., Smith, A.: Fuzzy extractors: how to generate strong keys from biometrics and other noisy data. In: Cachin, C., Camenisch, J.L. (eds.) EUROCRYPT 2004. LNCS, vol. 3027, pp. 523–540. Springer, Heidelberg (2004). https://doi.org/10.1007/978-3-540-24676-3_31
28. Dupont, P.A., Hesse, J., Pointcheval, D., Reyzin, L., Yakoubov, S.: Fuzzy authenticated key exchange. Cryptology ePrint Archive, Report 2017/1111 (2017). https://eprint.iacr.org/2017/1111
29. Ellison, C., Hall, C., Milbert, R., Schneier, B.: Protecting secret keys with personal entropy. Future Gener. Comput. Syst. **16**(4), 311–318 (2000)
30. Gassend, B., Clarke, D.E., van Dijk, M., Devadas, S.: Silicon physical random functions. In: Atluri, V. (ed.) ACM CCS 2002, pp. 148–160. ACM Press, New York (2002)
31. Gasti, P., Sedenka, J., Yang, Q., Zhou, G., Balagani, K.S.: Secure, fast, and energy-efficient outsourced authentication for smartphones. Trans. Info. For. Sec. **11**(11), 2556–2571 (2016). https://doi.org/10.1109/TIFS.2016.2585093
32. Han, J., Chung, A., Sinha, M.K., Harishankar, M., Pan, S., Noh, H.Y., Zhang, P., Tague, P.: Do you feel what I hear? Enabling autonomous IoT device pairing using different sensor types. In: IEEE Symposium on Security and Privacy (2018)
33. Han, J., Harishankar, M., Wang, X., Chung, A.J., Tague, P.: Convoy: physical context verification for vehicle platoon admission. In: 18th ACM International Workshop on Mobile Computing Systems and Applications (HotMobile) (2017)

34. Huang, Y., Katz, J., Evans, D.: Quid-Pro-Quo-tocols: strengthening semi-honest protocols with dual execution. In: 2012 IEEE Symposium on Security and Privacy, pp. 272–284. IEEE Computer Society Press, May 2012
35. Huang, Y., Katz, J., Evans, D.: Efficient secure two-party computation using symmetric cut-and-choose. In: Canetti, R., Garay, J.A. (eds.) CRYPTO 2013, Part II. LNCS, vol. 8043, pp. 18–35. Springer, Heidelberg (2013). https://doi.org/10.1007/978-3-642-40084-1_2
36. Juels, A., Wattenberg, M.: A fuzzy commitment scheme. In: ACM CCS 1999, pp. 28–36. ACM Press, November 1999
37. Katz, J., Vaikuntanathan, V.: Round-optimal password-based authenticated key exchange. In: Ishai, Y. (ed.) TCC 2011. LNCS, vol. 6597, pp. 293–310. Springer, Heidelberg (2011). https://doi.org/10.1007/978-3-642-19571-6_18
38. Kolesnikov, V., Mohassel, P., Rosulek, M.: FleXOR: flexible garbling for XOR gates that beats free-XOR. In: Garay, J.A., Gennaro, R. (eds.) CRYPTO 2014, Part II. LNCS, vol. 8617, pp. 440–457. Springer, Heidelberg (2014). https://doi.org/10.1007/978-3-662-44381-1_25
39. Kolesnikov, V., Rackoff, C.: Password mistyping in two-factor-authenticated key exchange. In: Aceto, L., Damgård, I., Goldberg, L.A., Halldórsson, M.M., Ingólfsdóttir, A., Walukiewicz, I. (eds.) ICALP 2008, Part II. LNCS, vol. 5126, pp. 702–714. Springer, Heidelberg (2008). https://doi.org/10.1007/978-3-540-70583-3_57
40. Kolesnikov, V., Schneider, T.: Improved garbled circuit: free XOR gates and applications. In: Aceto, L., Damgård, I., Goldberg, L.A., Halldórsson, M.M., Ingólfsdóttir, A., Walukiewicz, I. (eds.) ICALP 2008, Part II. LNCS, vol. 5126, pp. 486–498. Springer, Heidelberg (2008). https://doi.org/10.1007/978-3-540-70583-3_40
41. Lindell, Y.: Fast cut-and-choose based protocols for malicious and covert adversaries. In: Canetti, R., Garay, J.A. (eds.) CRYPTO 2013, Part II. LNCS, vol. 8043, pp. 1–17. Springer, Heidelberg (2013). https://doi.org/10.1007/978-3-642-40084-1_1
42. Lindell, Y., Pinkas, B.: Secure two-party computation via cut-and-choose oblivious transfer. In: Ishai, Y. (ed.) TCC 2011. LNCS, vol. 6597, pp. 329–346. Springer, Heidelberg (2011). https://doi.org/10.1007/978-3-642-19571-6_20
43. Lindell, Y., Pinkas, B.: An efficient protocol for secure two-party computation in the presence of malicious adversaries. J. Cryptol. **28**(2), 312–350 (2015)
44. Lindell, Y., Riva, B.: Cut-and-choose Yao-based secure computation in the online/offline and batch settings. In: Garay, J.A., Gennaro, R. (eds.) CRYPTO 2014, Part II. LNCS, vol. 8617, pp. 476–494. Springer, Heidelberg (2014). https://doi.org/10.1007/978-3-662-44381-1_27
45. Maurer, U.: Information-theoretically secure secret-key agreement by NOT authenticated public discussion. In: Fumy, W. (ed.) EUROCRYPT 1997. LNCS, vol. 1233, pp. 209–225. Springer, Heidelberg (1997). https://doi.org/10.1007/3-540-69053-0_15
46. Mayrhofer, R., Gellersen, H.: Shake well before use: intuitive and secure pairing of mobile devices. IEEE Trans. Mob. Comput. **8**(6), 792–806 (2009)
47. McEliece, R.J., Sarwate, D.V.: On sharing secrets and Reed-Solomon codes. Commun. ACM **24**(9), 583–584 (1981). http://doi.acm.org/10.1145/358746.358762
48. Mohassel, P., Franklin, M.: Efficiency tradeoffs for malicious two-party computation. In: Yung, M., Dodis, Y., Kiayias, A., Malkin, T. (eds.) PKC 2006. LNCS, vol. 3958, pp. 458–473. Springer, Heidelberg (2006). https://doi.org/10.1007/11745853_30

49. Monrose, F., Reiter, M.K., Wetzel, S.: Password hardening based on keystroke dynamics. Int. J. Inf. Secur. **1**(2), 69–83 (2002)
50. Nielsen, J.B., Orlandi, C.: LEGO for two-party secure computation. In: Reingold, O. (ed.) TCC 2009. LNCS, vol. 5444, pp. 368–386. Springer, Heidelberg (2009). https://doi.org/10.1007/978-3-642-00457-5_22
51. Nisan, N., Zuckerman, D.: More deterministic simulation in logspace. In: 25th ACM STOC, pp. 235–244. ACM Press, May 1993
52. Pappu, R., Recht, B., Taylor, J., Gershenfeld, N.: Physical one-way functions. Science **297**(5589), 2026–2030 (2002)
53. Pinkas, B., Schneider, T., Smart, N.P., Williams, S.C.: Secure two-party computation is practical. In: Matsui, M. (ed.) ASIACRYPT 2009. LNCS, vol. 5912, pp. 250–267. Springer, Heidelberg (2009). https://doi.org/10.1007/978-3-642-10366-7_15
54. Renner, R., Wolf, S.: The exact price for unconditionally secure asymmetric cryptography. In: Cachin, C., Camenisch, J.L. (eds.) EUROCRYPT 2004. LNCS, vol. 3027, pp. 109–125. Springer, Heidelberg (2004). https://doi.org/10.1007/978-3-540-24676-3_7
55. Roth, R.: Introduction to Coding Theory. Cambridge University Press, New York (2006)
56. Shoup, V.: A proposal for an ISO standard for public key encryption. Cryptology ePrint Archive, Report 2001/112 (2001). http://eprint.iacr.org/2001/112
57. Suh, G.E., Devadas, S.: Physical unclonable functions for device authentication and secret key generation. In: Proceedings of the 44th Annual Design Automation Conference, pp. 9–14. ACM (2007)
58. Tuyls, P., Schrijen, G.-J., Škorić, B., van Geloven, J., Verhaegh, N., Wolters, R.: Read-proof hardware from protective coatings. In: Goubin, L., Matsui, M. (eds.) CHES 2006. LNCS, vol. 4249, pp. 369–383. Springer, Heidelberg (2006). https://doi.org/10.1007/11894063_29
59. Wang, X., Ranellucci, S., Katz, J.: Authenticated garbling and efficient maliciously secure two-party computation. In: Thuraisingham, B.M., Evans, D., Malkin, T., Xu, D. (eds.) ACM CCS 2017, pp. 21–37. ACM Press, New York (2017)
60. Woodage, J., Chatterjee, R., Dodis, Y., Juels, A., Ristenpart, T.: A new distribution-sensitive secure sketch and popularity-proportional hashing. In: Katz, J., Shacham, H. (eds.) CRYPTO 2017, Part III. LNCS, vol. 10403, pp. 682–710. Springer, Cham (2017). https://doi.org/10.1007/978-3-319-63697-9_23
61. Wyner, A.D.: The wire-tap channel. Bell Syst. Tech. J. **54**, 1355–1387 (1975)
62. Yakoubov, S.: A gentle introduction to Yao's garbled circuits (2017). http://web.mit.edu/sonka89/www/papers/2017ygc.pdf
63. Yao, A.C.C.: How to generate and exchange secrets (extended abstract). In: 27th FOCS, pp. 162–167. IEEE Computer Society Press (Oct 1986)
64. Yu, M.D.M., Devadas, S.: Secure and robust error correction for physical unclonable functions. IEEE Des. Test **27**(1), 48–65 (2010)
65. Zahur, S., Rosulek, M., Evans, D.: Two halves make a whole: reducing data transfer in garbled circuits using half gates. In: Oswald, E., Fischlin, M. (eds.) EUROCRYPT 2015, Part II. LNCS, vol. 9057, pp. 220–250. Springer, Heidelberg (2015). https://doi.org/10.1007/978-3-662-46803-6_8
66. Zviran, M., Haga, W.J.: A comparison of password techniques for multilevel authentication mechanisms. Comput. J. **36**(3), 227–237 (1993)

Bloom Filter Encryption and Applications to Efficient Forward-Secret 0-RTT Key Exchange

David Derler[1(✉)], Tibor Jager[2], Daniel Slamanig[3], and Christoph Striecks[3]

[1] Graz University of Technology, Graz, Austria
david.derler@tugraz.at
[2] Paderborn University, Paderborn, Germany
tibor.jager@upb.de
[3] AIT Austrian Institute of Technology, Vienna, Austria
{daniel.slamanig,christoph.striecks}@ait.ac.at

Abstract. Forward secrecy is considered an essential design goal of modern key establishment (KE) protocols, such as TLS 1.3, for example. Furthermore, efficiency considerations such as zero round-trip time (0-RTT), where a client is able to send cryptographically protected payload data along with the very first KE message, are motivated by the practical demand for secure low-latency communication.

For a long time, it was unclear whether protocols that simultaneously achieve 0-RTT and full forward secrecy exist. Only recently, the first forward-secret 0-RTT protocol was described by Günther et al. (Eurocrypt 2017). It is based on Puncturable Encryption. Forward secrecy is achieved by "puncturing" the secret key after each decryption operation, such that a given ciphertext can only be decrypted once (cf. also Green and Miers, S&P 2015). Unfortunately, their scheme is completely impractical, since one puncturing operation takes between 30 s and several minutes for reasonable security and deployment parameters, such that this solution is only a first feasibility result, but not efficient enough to be deployed in practice.

In this paper, we introduce a new primitive that we term Bloom Filter Encryption (BFE), which is derived from the probabilistic Bloom filter data structure. We describe different constructions of BFE schemes, and show how these yield new puncturable encryption mechanisms with extremely efficient puncturing. Most importantly, a puncturing operation only involves a small number of very efficient computations, plus the deletion of certain parts of the secret key, which outperforms previous constructions by orders of magnitude. This gives rise to the first forward-secret 0-RTT protocols that are efficient enough to be deployed in practice. We believe that BFE will find applications beyond forward-secret 0-RTT protocols.

Keywords: Bloom Filter Encryption · Bloom filter · 0-RTT
Forward secrecy · Key exchange · Puncturable encryption

The full version of this paper is available in the IACR Cryptology ePrint archive.

J. B. Nielsen and V. Rijmen (Eds.): EUROCRYPT 2018, LNCS 10822, pp. 425–455, 2018.
https://doi.org/10.1007/978-3-319-78372-7_14

1 Introduction

One central ingredient to secure today's Internet are key exchange (KE) protocols with the most prominent and widely deployed instantiations thereof in the Transport Layer Security (TLS) protocol [15]. Using a KE protocol, two parties (e.g., a server and a client) are able to establish a shared secret (session key) which afterwards can be used to cryptographically protect data to be exchanged between those parties. The process of arriving at a shared secret requires the exchange of messages between client and server, which adds latency overhead to the protocol. The time required to establish a key is usually measured in round-trip times (RTTs). A novel design goal, which was introduced by Google's QUIC protocol and also adopted in the upcoming version of TLS 1.3, aims at developing zero round-trip time (0-RTT) protocols with strong security guarantees. So far, quite some effort was made in the cryptographic literature, e.g. [21,31], and, indeed, 0-RTT protocols are probably going to be used heavily in the future Internet as TLS version 1.3 [28] is approaching fast. Already today, Google's QUIC protocol [29] is used on Google webservers and within the Chrome and Opera browsers to support 0-RTT. Unfortunately, none of the above mentioned protocols are enjoying 0-RTT and full forward secrecy at the same time. Only recently, Günther, Hale, Jager, and Lauer (GHJL henceforth) [20] made progress and proposed the first 0-RTT key exchange protocol with full forward secrecy for all transmitted payload messages. However, although their 0-RTT protocol offers the desired features, their construction is not yet practical.

In more detail, GHJL's forward-secure 0-RTT key-exchange solution is based on puncturable encryption (PE), which they showed can be constructed in a black-box way from any selectively secure hierarchical identity-based encryption (HIBE) scheme. Loosely speaking, PE is a public-key encryption primitive which provides a Puncture algorithm that, given a secret key and ciphertext, produces an updated secret key that is able to decrypt all ciphertexts except the one it has been punctured on. PE has been introduced by Green and Miers [19] (GM henceforth) who provide an instantiation relying on a binary-tree encryption (BTE) scheme—or selectively secure HIBE—together with a key-policy attribute-based encryption (KP-ABE) [18] scheme for non-monotonic (NM) formulas with specific properties. In particular, the KP-ABE needs to provide a non-standard property to enhance existing secret keys with additional NOT gates, which is satisfied by the NM KP-ABE in [27]. Since then, PE has proved to be a valuable building block to construct public-key watermarking schemes [13], forward-secret proxy re-encryption [14], or for achieving chosen-ciphertext security for fully-homomorphic encryption [11]. However, the mentioned PE instantiations from [11,13] are based on indistinguishability obfuscation and, thus, do not yield practical schemes at all while [14] uses the same techniques as in GHJL.

When looking at the two most efficient PE schemes available, i.e., GM and GHJL, they still come with severe drawbacks. In particular, puncturing in GHJL is highly inefficient and takes several seconds to minutes on decent hardware for reasonable deployment parameters. In the GM scheme, puncturing is more efficient, but the cost of decryption is very significant and increases with the number of puncturings. More precisely, cost of decryption requires a number of pairing evaluations that depends on the number of puncturings, and can be in the order of 2^{10} to 2^{20} for realistic deployment parameters. These issues make both of them especially unsuitable for the application in forward-secret 0-RTT key exchange in a practical setting.

Contributions. In this paper, we introduce Bloom filter encryption (BFE), which can be considered as a variant of PE [11,13,19,20]. The main difference to other existing PE constructions is that in case of BFE, we tolerate a non-negligible correctness error.[1] This allows us to construct PE and in particular puncturable key encapsulation (PKEM) schemes with highly efficient puncturing and in particular where puncturing only requires a few very efficient operations, i.e., to *delete* parts of the secret key, but no further expensive cryptographic operations. Altogether, this makes BFE a very suitable building block to construct practical forward-secret 0-RTT key exchange. In more detail, our contributions are as follows:

- We formalize the notion of BFE by presenting a suitable security model. The intuition behind BFE is to provide a highly efficient decryption and puncturing. Interestingly, puncturing mainly consists of *deleting* parts of the secret key. This approach is in contrast to existing puncturable encryption schemes, where puncturing and/or decryption is a very expensive operation.
- We propose efficient constructions of BFE. First, we present a direct construction which uses ideas from the Boneh-Franklin identity-based encryption (IBE) scheme [9]. Additionally, we present a black-box construction from a ciphertext-policy attibute-based encryption (CP-ABE) scheme that only needs to be small-universe (i.e., bounded) and support threshold policies, which allows us to achieve compact ciphertexts. To improve efficiency, we finally provide a time-based BFE (TB-BFE) from selectively-secure HIBEs.
- To achieve CCA security, we adopt the Fujisaki-Okamoto (FO) transformation [16] to the BFE setting. This is technically non-trivial, and therefore we consider it as another interesting aspect of this work. In particular, the original FO transformation [16] works only for schemes with *perfect* correctness. Recently, Hofheinz et al. [23] described a variant which works also for schemes with *negligible* correctness error. We adopt the FO transformation to BFE and PKEMs with *non-negligible* correctness error respectively. To this end, we formalize additional properties of the PKEM that are required to apply the FO transform to BFE schemes, and show that our CPA-secure constructions satisfy them. This serves as a template that allows an easy application of the FO transform in a black-box manner to BFE schemes.
- We provide a construction of a forward-secret 0-RTT key exchange protocol (in the sense of GHJL) from TB-BFE. Furthermore, we give a detailed comparison of (TB-)BFE with other PE schemes and discuss the efficiency in the context of the proposed application to forward-secret 0-RTT key exchange.

[1] We discuss below why this is not only tolerable, but actually a very reasonable approach for applications like 0-RTT key exchange.

In particular, our construction of forward-secret 0-RTT key-exchange from TB-BFE has none of the drawbacks mentioned in the introduction (at the cost of a somewhat larger secret key, that, however, shrinks with the number of puncturings). Consequently, our forward-secret 0-RTT key exchange can be seen as a significant step forward to construct very *practical* forward-secret 0-RTT key exchange protocols.

On tolerating a non-negligible correctness error for 0-RTT. The huge efficiency gain of our construction stems partially from the relaxation of allowing a non-negligible correctness error, which, in turn, stems from the potentially non-negligible false-positive probability of a Bloom filter. While this is unusual for classical public-key encryption schemes, we consider it as a reasonable approach to accept a small, but non-negligible correctness error for the 0-RTT mode of a key exchange protocol, in exchange for the huge efficiency gain.

For example, a $1/10000$ chance that the key establishment fails allows to use 0-RTT in 9999 out of 10000 cases on average, which is a significant practical efficiency improvement. Furthermore, the communicating parties can implement a fallback mechanism which immediately continues with running a standard 1-RTT key exchange protocol with perfect correctness, if the 0-RTT exchange fails. Thus, the resulting protocol can have the same worst-case efficiency as a 1-RTT protocol, while most of the time 0-RTT is already sufficient to establish a key and full forward secrecy is *always* achieved.

Compared to other practical 0-RTT solutions, note that both TLS 1.3 [28] and QUIC [29] have similar fallback mechanisms. Furthermore, in order to achieve at least a very weak form of forward secrecy, they define so called *tickets* [28] or *server configuration (SCFG)* messages [29], which expire after a certain time. Forward secrecy is only achieved after the ticket/SCFG message has expired and the associated secrets have been erased. Therefore the lifetime should be kept short. If a client connects to a server after the ticket/SCFG message has expired, then the fallback mechanism is invoked and a full 1-RTT handshake is performed. In particular, for settings where a client connects only occasionally to a server, and for reasonably chosen parameters and a moderate life time of the ticket/SCFG message, which at least guarantees some weak form of forward secrecy, this requires a full handshake more often than with our approach.

Finally, note that puncturable encryption with perfect (or negligible) correctness error inherently seems to require secret keys whose size at least grows linearly with the number of puncturings. This is because any such scheme inherently must (implicitly or explicitly) encode information about the list of punctured ciphertexts into the secret key, which lower-bounds the size of the secret key. By tolerating a non-negligible correctness error, we are also able to restrict the growth of the secret key to a limit which seems tolerable in practice.

2 Bloom Filter Encryption

The key idea behind Bloom Filter Encryption (BFE) is that the key pair of such a scheme is associated to a Bloom filter (BF) [7], a probabilistic data structure for the approximate set membership problem with a non-negligible false-positive probability in answering membership queries. The initial secret key sk output by the key generation algorithm of a BFE scheme corresponds to an empty BF where all bits are set to 0. Encryption takes a message M and the public key pk, samples a random element s (acting as a tag for the ciphertext) corresponding to the universe $\mathcal{U}$ of the BF and encrypts a message using pk with respect to the k positions set in the BF by s. A ciphertext is then basically identified by s and decryption works as long as at least one index pointed to by s in the BF is still set to 0. Puncturing the secret key with respect to a ciphertext (i.e., the tag s of the ciphertext) corresponds to inserting s in the BF (i.e., updating the corresponding indices to 1 and deleting the corresponding parts of the secret key). This basically means updating sk such that it no longer can decrypt any position indexed by s.

2.1 Formal Definition of Bloom Filters

A Bloom filter (BF) [7] is a probabilistic data structure for the approximate set membership problem. It allows a succinct representation T of a set $\mathcal{S}$ of elements from a large universe $\mathcal{U}$. For elements $s \in \mathcal{S}$ a query to the BF always answers 1 ("yes"). Ideally, a BF would always return 0 ("no") for elements $s \notin \mathcal{S}$, but the succinctness of the BF comes at the cost that for any query to $s \notin \mathcal{S}$ the answer can be 1, too, but only with small probability (called the *false-positive probability*).

We will only be interested in the original construction of Bloom filters by Bloom [7], and omit a general abstract definition. Instead we describe the construction from [7] directly. For a general definition refer to [26].

Definition 1 (Bloom Filter). *A* Bloom filter B *for set* $\mathcal{U}$ *consists of algorithms* $\mathsf{B} = (\mathsf{BFGen}, \mathsf{BFUpdate}, \mathsf{BFCheck})$, *which are defined as follows.*

$\mathsf{BFGen}(m, k)$: *This algorithm takes as input two integers* $m, k \in \mathbb{N}$. *It first samples k universal hash functions* $H_1, \ldots, H_k$, *where* $H_j : \mathcal{U} \to [m]$, *defines* $H := (H_j)_{j \in [k]}$ *and* $T := 0^m$ *(that is, T is an m-bit array with all bits set to* 0*), and outputs* (H, T).

$\mathsf{BFUpdate}(H, T, u)$: *Given* $H = (H_j)_{j \in [k]}$, $T \in \{0,1\}^m$, *and* $u \in \mathcal{U}$, *this algorithm defines the updated state* T' *by first assigning* $T' := T$. *Then, writing* $T'[i]$ *to denote the i-th bit of* T', *it sets* $T'[H_j(u)] := 1$ *for all* $j \in [k]$, *and finally returns* T'.

$\mathsf{BFCheck}(H, T, u)$: *Given* $H = (H_j)_{j \in [k]}$, $T \in \{0,1\}^m$ *where we write* $T[i]$ *to denote the i-th bit of* T, *and* $u \in \mathcal{U}$, *this algorithm returns a bit* $b := \bigwedge_{j \in [k]} T[H_j(u)]$

Relevant properties of Bloom filters. Let us summarize the properties of Bloom filters relevant to our work.

Perfect completeness. A Bloom filter always "recognizes" elements that have been added with probability 1. More precisely, let $\mathcal{S} = (s_1, \ldots, s_n) \in \mathcal{U}^n$ be any vector of n elements of $\mathcal{U}$. Let $(H, T_0) \xleftarrow{\$} \mathsf{BFGen}(m, k)$ and define

$$T_i = \mathsf{BFUpdate}(H, T_{i-1}, s_i) \text{ for } i \in [n].$$

Then for all $s^* \in \mathcal{S}$ and all $(H, T_0) \xleftarrow{\$} \mathsf{BFGen}(m, k)$ with $m, k \in \mathbb{N}$, it holds that

$$\Pr[\mathsf{BFCheck}(H, T_n, s^*) = 1] = 1.$$

Compact representation of $\mathcal{S}$. Independent of the size of the set $\mathcal{S} \subset \mathcal{U}$ and the representation of individual elements of $\mathcal{U}$, the size of representation T is a constant number of m bits. A larger size of $\mathcal{S}$ increases only the false-positive probability, as discussed below, but not the size of the representation.

Bounded false-positive probability. The probability that an element which has not yet been added to the Bloom filter is erroneously "recognized" as being contained in the filter can be made arbitrarily small, by choosing m and k adequately, given (an upper bound on) the size of $\mathcal{S}$.
More precisely, let $\mathcal{S} = (s_1, \ldots, s_n) \in \mathcal{U}^n$ be any vector of n elements of $\mathcal{U}$. Then for any $s^* \in \mathcal{U} \setminus \mathcal{S}$, we have

$$\Pr[\mathsf{BFCheck}(H, T_n, s^*) = 1] \approx (1 - e^{-kn/m})^k,$$

where $(H, T_0) \xleftarrow{\$} \mathsf{BFGen}(m, k)$, $T_i = \mathsf{BFUpdate}(H, T_{i-1}, s_i)$ for $i \in [n]$, and the probability is taken over the random coins of BFGen.

Discussion on the choice of parameters. In order to provide a first intuition on the choice of parameters n, m and k for the use of BFs within BFE, we subsequently discuss some reasonable choices. Let us assume that we want to have $n = 2^{20}$, which amounts to adding for a full year every day about 2^{12} elements to the BF. Then, assuming the optimal number of hash functions k, and tolerating a false-positive probability of $p = 10^{-3}$, we obtain a size of the BF given by $m = -{}^{n \ln p}/_{(\ln 2)^2}$, as $m \approx 15\,\text{Mb} \approx 2\,\text{MB}$. The optimal number of hash functions k is given by $k = {}^m/_n \ln 2$, and we will instantiate Bloom filters with

$$k := \lceil {}^m/_n \ln 2 \rceil .$$

This yields a correctness error $p \approx (1 - e^{-kn/m})^k = (1 - e^{-n/m \cdot \lceil \frac{m}{n} \rceil \ln 2})^k \leq 2^{-k}$. For above parameters n, m and p we obtain $k = 10$.

Looking ahead to the BFE construction in Sect. 2.5, at a 120-bit security level (using the pairing-friendly `BLS12-381` curve), this choice of parameters would yield ciphertexts of size $<$ 720 B and public as well as secret keys of size $<$ 100 B and $\approx$ 700 MB respectively. Thereby, we need to emphasize that initially the secret key (representing the empty BF) has its maximum size, but every puncturing (i.e., addition of an element to the BF), reduces the size of the secret key. Moreover, we stress that the false-positive probability represents an upper bound as it assumes that all $n = 2^{20}$ elements are already added to the BF, i.e., the secret key has already been punctured with respect to 2^{20} ciphertexts. Finally, when we use our time-based BFE approach (TB-BFE) from Sect. 2.7, we can even reduce the secret key size by reducing the maximum number of puncturings at the cost of switching the time intervals more frequently.

2.2 Formal Model of BFE

Subsequently, we introduce the formal model for BFE which essentially is a variant of puncturable encryption (PE) [11,13,19,20] with the only difference that with BFE we tolerate a non-negligible correctness error. Thus, although we are speaking of BFE, we choose to introduce a formal model for PE with a relaxed correctness definition[2] and treat BFE as an instantiation of PE. Consequently, our Definition 2 below is a variant of the one in [20], with the only difference that we allow the key generation to take the additional parameters m and k (of the BF) as input, which specify the correctness error.

For 0-RTT key establishment, our prime application in this paper, we do not need a full-blown encryption scheme, but only a key-encapsulation mechanisms (KEM) to transport a symmetric encryption key. Consequently, we chose to present our definitions by means of a puncturable KEM (PKEM). We stress that defining PKEM instead of PE does not represent any limitation, as any KEM can generically be converted into a secure full-blown encryption scheme [16]. Conversely, any secure encryption scheme trivially yields a secure KEM. Nonetheless, for completeness, we give stand-alone definitions of PE tolerating a non-negligible correctness error in the full version.

Definition 2 (PKEM). *A puncturable key encapsulation (PKEM) scheme with key space $\mathcal{K}$ is a tuple* $(\mathsf{KGen}, \mathsf{Enc}, \mathsf{Punc}, \mathsf{Dec})$ *of* PPT *algorithms:*

$\mathsf{KGen}(1^\lambda, m, k)$: *Takes as input a security parameter λ, parameters m and k and outputs a secret and public key* $(\mathsf{sk}, \mathsf{pk})$ *(we assume that $\mathcal{K}$ is implicit in* pk*).*

$\mathsf{Enc}(\mathsf{pk})$: *Takes as input a public key* pk *and outputs a ciphertext C and a symmetric key* K.

$\mathsf{Punc}(\mathsf{sk}, C)$: *Takes as input a secret key* sk, *a ciphertext C and outputs an updated secret key* sk'.

$\mathsf{Dec}(\mathsf{sk}, C)$: *Takes as input a secret key* sk, *a ciphertext C and outputs a symmetric key* K *or $\perp$ if decapsulation fails.*

Correctness. We start by defining correctness of a PKEM scheme. Basically, here one requires that a ciphertext can always be decapsulated with unpunctured secret keys. However, we allow that if punctured secret keys are used for decapsulation then the probability that the decapsulation fails is bounded by some non-negligible function in the scheme's parameters m, k.

Definition 3 (Correctness). *For all $\lambda, m, k, \in \mathbb{N}$, any* $(\mathsf{sk}, \mathsf{pk}) \xleftarrow{\$} \mathsf{KGen}(1^\lambda, m, k)$ *and* $(C, \mathsf{K}) \xleftarrow{\$} \mathsf{Enc}(\mathsf{pk})$, *we have that* $\mathsf{Dec}(\mathsf{sk}, C) = \mathsf{K}$. *Moreover, for any (arbitrary interleaved) sequence $i = 1, \ldots, \ell$ (where ℓ is determined by m, k) of invocations of* $\mathsf{sk}' \xleftarrow{\$} \mathsf{Punc}(\mathsf{sk}, C')$ *for any $C' \neq C$ it holds that* $\Pr\left[\mathsf{Dec}(\mathsf{sk}', C) = \perp\right] \leq \mu(m, k)$, *where $\mu(\cdot)$ is some (possibly non-negligible) bound.*

---

[2] This moreover allows to compactly present our construction of forward-secret 0-RTT key exchange as this then essentially follows the argumentation in [20].

2.3 Additional Properties of a PKEM

In this section, we will define additional properties of a PKEM. One will be necessary for the application to 0-RTT key exchange from [20]. The others are required to construct a CCA-secure PKEM via the Fujisaki-Okamoto (FO) transformation, as described in Sect. 2.6. We will show below that our constructions of CPA-secure PKEMs satisfy these additional properties, and thus are suitable for our variant of the FO transformation, and to construct 0-RTT key exchange.

Extended correctness. Intuitively, we first require an extended variant of correctness which demands that (1) decapsulation yields a failure when attempting to decapsulate under a secret key previously punctured for that ciphertext. This is analogous to [20]. Second, we additionally demand that (2) decapsulating an honest ciphertext with the unpuctured key does always succeed and (3) if decryption does *not* fail, then the decapsulated value must match the key returned by the Enc algorithm, for any key sk' obtained from applying any sequence of puncturing operations to the initial secret key sk.

Definition 4 (Extended Correctness). *For all* $\lambda, m, k, \ell \in \mathbb{N}$, *any* $(\mathsf{sk}, \mathsf{pk}) \xleftarrow{\$} \mathsf{KGen}(1^\lambda, m, k)$ *and* $(C, \mathsf{K}) \xleftarrow{\$} \mathsf{Enc}(\mathsf{pk})$ *and any (arbitrary interleaved and possibly empty) sequence* $C_1, \ldots, C_\ell$ *of invocations of* $\mathsf{sk}' \xleftarrow{\$} \mathsf{Punc}(\mathsf{sk}, C_i)$ *it holds that:*

1. ***Impossibility of false-negatives:***
   $\mathsf{Dec}(\mathsf{sk}', C_i) = \bot$ *for all* $i \in [\ell]$.
2. ***Perfect correctness of the initial, non-punctured secret key:***
   If $(C, \mathsf{K}) \xleftarrow{\$} \mathsf{Enc}(\mathsf{pk})$ *then* $\mathsf{Dec}(\mathsf{sk}, C) = \mathsf{K}$, *where* sk *is the initial, non-punctured secret key.*
3. ***Semi-correctness of punctured secret keys:***
   If $\mathsf{Dec}(\mathsf{sk}', C) \neq \bot$ *then* $\mathsf{Dec}(\mathsf{sk}', C) = \mathsf{Dec}(\mathsf{sk}, C)$.

Separable randomness. We require that the encapsulation algorithm Enc essentially reads the key K in $(C, \mathsf{K}) \xleftarrow{\$} \mathsf{Enc}(\mathsf{pk})$ directly from its random input tape. Intuitively, this will later enable us to make the randomness r used by the encapsulation algorithm Enc dependent on the key K computed by Enc.

Definition 5 (Separable Randomness). *Let* $\mathsf{PKEM} = (\mathsf{KGen}, \mathsf{Enc}, \mathsf{Punc}, \mathsf{Dec})$ *be a PKEM. We say that* PKEM *has* separable randomness, *if one can equivalently write the encapsulation algorithm* Enc *as*

$$(C, \mathsf{K}) \xleftarrow{\$} \mathsf{Enc}(\mathsf{pk}) = \mathsf{Enc}(\mathsf{pk}; (r, \mathsf{K})),$$

for uniformly random $(r, \mathsf{K}) \in \{0,1\}^{\rho+\lambda}$, *where* $\mathsf{Enc}(\cdot;\cdot)$ *is a deterministic algorithm whose output is uniquely determined by* pk *and the randomness* $(r, \mathsf{K}) \in \{0,1\}^{\rho+\lambda}$.

Remark. We note that one can generically construct a separable PKEM from any non-separable PKEM. Given a non-separable PKEM with encapsulation algorithm Enc, a separable PKEM with encryption algorithm Enc' can be obtained as follows:

$\mathsf{Enc}'(\mathsf{pk}; (r, \mathsf{K}'))$: Run $(C, \mathsf{K}) \xleftarrow{\$} \mathsf{Enc}(\mathsf{pk}; r)$, set $C' := (C, \mathsf{K} \oplus \mathsf{K}')$ return (C', K').

We need separability in order to apply our variant of the FO transformation, which is the reason why we have to make it explicit. Alternatively, we could have started from a non-separable PKEM and applied the above construction. However, this adds an additional component to the ciphertext, while the construction given in Sect. 2.5 will already be separable, such that we can avoid this overhead.

Publicly-checkable puncturing. Finally, we need that it is efficiently checkable whether the decapsulation algorithm outputs $\bot = \mathsf{Dec}(\mathsf{sk}, C)$, given *not* the secret key sk, but only the public key pk, the ciphertext C to be decrypted, and the sequence $C_1, \ldots, C_w$ at which the secret key sk has been punctured.

Definition 6 (Publicly-Checkable Puncturing). *Let $\mathcal{Q} = (C_1, \ldots, C_w)$ be any list of ciphertexts. We say that* PKEM *allows* publicly-checkable puncturing, *if there exists an efficient algorithm* CheckPunct *with the following correctness property.*

1. *Run* $(\mathsf{sk}, \mathsf{pk}) \xleftarrow{\$} \mathsf{KGen}(1^\lambda, m, k)$.
2. *Compute* $C_i \xleftarrow{\$} \mathsf{Enc}(\mathsf{pk})$ *and* $\mathsf{sk} = \mathsf{Punc}(\mathsf{sk}, C_i)$ *for* $i \in [w]$.
3. *Let C be any string. We require that*

$$\bot = \mathsf{Dec}(\mathsf{sk}, C) \iff \bot = \mathsf{CheckPunct}(\mathsf{pk}, \mathcal{Q}, C).$$

From a high-level perspective, this additional property will be necessary to simulate the decryption oracle properly in the CCA security experiment when our variant of the FO transformation is applied. Together with the second and third property of Definition 4, it replaces the perfect correctness property required in the original FO transformation.

Min-entropy of ciphertexts. Following [23], we require that ciphertexts of a randomness-separable PKEM have sufficient min-entropy, even if K is fixed:

Definition 7 (γ-Spreadness). *Let* $\mathsf{PKEM} = (\mathsf{KGen}, \mathsf{Enc}, \mathsf{Punc}, \mathsf{Dec})$ *be a* randomness-separable *PKEM with ciphertext space $\mathcal{C}$. We say that* PKEM *is* γ-spread, *if for any honestly generated* pk, *any key* K *and any* $C \in \mathcal{C}$

$$\Pr_{r \xleftarrow{\$} \{0,1\}^\rho} [C = \mathsf{Enc}(\mathsf{pk}; (r, \mathsf{K}))] \leq 2^{-\gamma}.$$

2.4 Security Definitions

We define three notions of security for PKEMs. The two "standard" security notions are indistinguishability under chosen-plaintext (IND-CPA) and chosen-ciphertext (IND-CCA) attacks. We also consider one-wayness under chosen-plaintext attacks (OW-CPA). The latter is the weakest notion among the ones considered in this paper, and implied by both IND-CPA and IND-CCA, but sufficient for our generic construction of IND-CCA-secure PKEMs.

Indistinguishability-based security. Figure 1 defines the IND-CPA and IND-CCA experiments for PKEMs. The experiments are similar to the security notions for conventional KEMs, but the adversary can arbitrarily puncture the secret key via the Punc oracle and retrieve the punctured secret key via the Corr oracle, once it has been punctured on the challenge ciphertext C^*.

$\mathbf{Exp}^{\mathsf{T}}_{\mathcal{A},\mathsf{PKEM}}(\lambda, m, k)$:
$(\mathsf{sk}, \mathsf{pk}) \overset{\$}{\leftarrow} \mathsf{KGen}(1^\lambda, m, k)$, $(C^*, \mathsf{K}_0) \overset{\$}{\leftarrow} \mathsf{Enc}(\mathsf{pk})$, $\mathcal{Q} \leftarrow \emptyset$
$\mathsf{K}_1 \overset{\$}{\leftarrow} \mathcal{K}$, $b \overset{\$}{\leftarrow} \{0, 1\}$
$b^* \overset{\$}{\leftarrow} \mathcal{A}^{\mathcal{O},\mathsf{Punc}(\mathsf{sk},\cdot),\mathsf{Corr}}(\mathsf{pk}, C^*, \mathsf{K}_b)$
where $\mathcal{O} \leftarrow \{\mathsf{Dec}'(\mathsf{sk}, \cdot)\}$ if $\mathsf{T} = \mathsf{IND\text{-}CCA}$ and $\mathcal{O} \leftarrow \emptyset$ otherwise.
$\mathsf{Dec}'(\mathsf{sk}, C)$ behaves as Dec but returns $\perp$ if $C = C^*$
$\mathsf{Punc}(\mathsf{sk}, C)$ runs $\mathsf{sk} \overset{\$}{\leftarrow} \mathsf{Punc}(\mathsf{sk}, C)$ and $\mathcal{Q} \leftarrow \mathcal{Q} \cup \{C\}$
Corr returns sk if $C^* \in \mathcal{Q}$ and $\perp$ otherwise
If $b^* = b$ then return 1
return 0

Fig. 1. Indistinguishability-based security for PKEMs.

Definition 8 (Indistinguishability-Based Security of PKEM). *For* $\mathsf{T} \in \{\mathsf{IND\text{-}CPA}, \mathsf{IND\text{-}CCA}\}$, *we define the advantage of an adversary* $\mathcal{A}$ *in the* T *experiment* $\mathbf{Exp}^{\mathsf{T}}_{\mathcal{A},\mathsf{PKEM}}(\lambda, m, k)$ *as*

$$\mathbf{Adv}^{\mathsf{T}}_{\mathcal{A},\mathsf{PKEM}}(\lambda, m, k) := \left| \Pr\left[\mathbf{Exp}^{\mathsf{T}}_{\mathcal{A},\mathsf{PKEM}}(\lambda, m, k) = 1\right] - \frac{1}{2} \right|.$$

A puncturable key-encapsulation scheme PKEM is $\mathsf{T} \in \{\mathsf{IND\text{-}CPA}, \mathsf{IND\text{-}CCA}\}$ secure, if $\mathbf{Adv}^{\mathsf{T}}_{\mathcal{A},\mathsf{PKEM}}(\lambda, m, k)$ is a negligible function in λ for all $m, k > 0$ and all PPT adversaries $\mathcal{A}$.

One-wayness under chosen-plaintext attack. Figure 2 defines the OW-CPA experiment. The experiment is similar to the IND-CPA experiment, except that the goal of the adversary is to recover the encapsulated key, given a random challenge ciphertext.

$\mathbf{Exp}^{\mathsf{OW\text{-}CPA}}_{\mathcal{A},\mathsf{PKEM}}(\lambda, m, k)$:
$(\mathsf{sk}, \mathsf{pk}) \overset{\$}{\leftarrow} \mathsf{KGen}(1^\lambda, m, k)$, $(C^*, \mathsf{K}_0) \overset{\$}{\leftarrow} \mathsf{Enc}(\mathsf{pk})$, $\mathcal{Q} \leftarrow \emptyset$
$\mathsf{K}_0^* \overset{\$}{\leftarrow} \mathcal{A}^{\mathsf{Punc}(\mathsf{sk},\cdot),\mathsf{Corr}}(\mathsf{pk}, C^*)$
where $\mathsf{Punc}(\mathsf{sk}, C)$ runs $\mathsf{sk} \overset{\$}{\leftarrow} \mathsf{Punc}(\mathsf{sk}, C)$ and $\mathcal{Q} \leftarrow \mathcal{Q} \cup \{C\}$
Corr returns sk if $C^* \in \mathcal{Q}$ and $\perp$ otherwise
If $\mathsf{K}_0^* = \mathsf{K}_0$ then return 1
return 0

Fig. 2. OW-CPA security for PKEMs.

Definition 9 (One-Wayness Under Chosen-Plaintext Attack). *We define the advantage of an adversary* $\mathcal{A}$ *in experiment* $\mathbf{Exp}^{\mathsf{OW\text{-}CPA}}_{\mathcal{A},\mathsf{PKEM}}(\lambda, m, k)$ *as*

$$\mathbf{Adv}^{\mathsf{OW\text{-}CPA}}_{\mathcal{A},\mathsf{PKEM}}(\lambda, m, k) := \Pr\left[\mathbf{Exp}^{\mathsf{OW\text{-}CPA}}_{\mathcal{A},\mathsf{PKEM}}(\lambda, m, k) = 1\right].$$

A PKEM is OW-CPA secure, if $\mathbf{Adv}^{\mathsf{OW\text{-}CPA}}_{\mathcal{A},\mathsf{PKEM}}(\lambda, m, k)$ is a negligible function in λ for all $m, k > 0$ and all PPT adversaries $\mathcal{A}$.

2.5 Basic Bloom Filter Encryption

Bilinear maps and notation. In the sequel, let BilGen be an algorithm that, on input a security parameter 1^λ, outputs $(p, e, \mathbb{G}_1, \mathbb{G}_2, \mathbb{G}_T, g_1, g_2) \xleftarrow{\$} \mathsf{BilGen}(1^\lambda)$, where $\mathbb{G}_1$, $\mathbb{G}_2$, $\mathbb{G}_T$ are groups of prime order p with bilinear map $e : \mathbb{G}_1 \times \mathbb{G}_2 \to \mathbb{G}_T$ and generators $g_i \in \mathbb{G}_i$ for $i \in \{1, 2\}$.

Construction. In the sequel, let $\mathsf{Params} := (p, e, \mathbb{G}_1, \mathbb{G}_2, \mathbb{G}_T, g_1, g_2) \xleftarrow{\$} \mathsf{BilGen}(1^\lambda)$, and $g_T = e(g_1, g_2)$. We will always assume that all algorithms described below implicitly receive these parameters as additional input. Let $\mathsf{B} = (\mathsf{BFGen}, \mathsf{BFUpdate}, \mathsf{BFCheck})$ be a Bloom filter for set $\mathbb{G}_1$. Furthermore, let $G : \mathbb{N} \to \mathbb{G}_2$ and $G' : \mathbb{G}_T \to \{0,1\}^\lambda$ be cryptographic hash functions (which will be modelled as random oracles [5] in the security proof).

Let $\mathsf{PKEM} = (\mathsf{KGen}, \mathsf{Enc}, \mathsf{Punc}, \mathsf{Dec})$ be defined as follows.

$\underline{\mathsf{KGen}(1^\lambda, m, k)}$: This algorithm first generates a Bloom filter instance by running $(H, T) \xleftarrow{\$} \mathsf{BFGen}(m, k)$. Then it chooses $\alpha \xleftarrow{\$} \mathbb{Z}_p$, and computes and returns

$$\mathsf{sk} := (T, (G(i)^\alpha)_{i \in [m]}) \text{ and } \mathsf{pk} := (g_1^\alpha, H).$$

Remark. The reader familiar with the Boneh-Franklin IBE scheme [9] may note that the secret key contains m elements of $\mathbb{G}_2$, each essentially being a secret key of the Boneh-Franklin scheme for "identity" i, $i \in [m]$, with respect to "master public-key" g_1^α.

$\underline{\mathsf{Enc}(\mathsf{pk})}$: This algorithm takes as input a public key pk of the above form. It samples a uniformly random key $\mathsf{K} \xleftarrow{\$} \{0,1\}^\lambda$ and exponent $r \xleftarrow{\$} \mathbb{Z}_p$. Then it computes $i_j := H_j(g_1^r)$ for $(H_j)_{j \in [k]} := H$, then $y_j = e(g_1^\alpha, G(i_j))^r$ for $j \in [k]$, and finally

$$C := \left(g_1^r, (G'(y_j) \oplus \mathsf{K})_{j \in [k]}\right).$$

It outputs $(C, \mathsf{K}) \in (\mathbb{G}_1 \times \{0,1\}^{k\lambda}) \times \{0,1\}^\lambda$.

Remark. Note that for each $j \in [k]$, the tuple $(g_1^r, G'(y_j) \oplus \mathsf{K})$ is essentially a "hashed Boneh-Franklin IBE" ciphertext, encrypting K for "identity" $i_j = H_j(g_1^r)$ and with respect to master public key g_1^α, where the identity is derived deterministically from a "unique" (with overwhelming probability) ciphertext component g_1^r. Thus, the ciphertext C essentially consists of k Boneh-Franklin ciphertexts that share the same randomness r, each encrypting the same key K for an "identity" derived deterministically from g_1^r.

Note also that this construction of Enc satisfies the requirement of separable randomness from Definition 5. Furthermore, ciphertexts are γ-spread according to Definition 7 with $\gamma = \log_2 p$, because g_1^r is uniformly distributed over $\mathbb{G}_1$.

$\underline{\mathsf{Punc}(\mathsf{sk}, C)}$: Given a ciphertext $C := \left(g_1^r, (G'(y_j) \oplus \mathsf{K})_{j \in [k]}\right)$ and secret key $\mathsf{sk} = (T, (\mathsf{sk}[i])_{i \in [m]})$, the puncturing algorithm first computes $T' = \mathsf{BFUpdate}(H, T, g_1^r)$. Then, for each $i \in [m]$ it defines

$$\mathsf{sk}'[i] := \begin{cases} \mathsf{sk}[i] & \text{if } T'[i] = 0, \text{ and} \\ \bot & \text{if } T'[i] = 1, \end{cases}$$

where $T'[i]$ denotes the i-th bit of T'. Finally, this algorithm returns

$$\mathsf{sk}' := (T', (\mathsf{sk}'[i])_{i \in [m]}).$$

Remark. Note that the above procedure is correct even if the procedure is applied repeatedly with different ciphertexts C, since the $\mathsf{BFUpdate}$ algorithm only changes bits of T from 0 to 1, but never from 1 to 0. So we can delete a secret key element $\mathsf{sk}[i]$ once $T'[i]$ has been set to 1. Furthermore, we have $\mathsf{sk}'[i] = \bot \iff T'[i] = 1$. Intuitively, this will ensure that we can use this key to decrypt a ciphertext $C := \left(g_1^r, (G'(y_j) \oplus \mathsf{K})_{j \in [k]}\right)$ if and only if $\mathsf{BFCheck}(H, T, g_1^r) = 0$, where (H, T) is the Bloom filter instance contained in the public key. Note also that the puncturing algorithm essentially only evaluates k universal hash functions $H = (H_j)_{j \in [k]}$ and then deletes a few secret keys, which makes this procedure extremely efficient. Finally, observe that the filter state T can be efficiently re-computed given only public information, namely the list of hash functions H contained in pk and the sequence of ciphertexts $C_1, \ldots, C_w$ on which a secret key has been punctured. This yields the existence of an efficient $\mathsf{CheckPunct}$ according to Definition 6.

$\underline{\mathsf{Dec}(\mathsf{sk}, C)}$: Given a secret key $\mathsf{sk} = (T, (\mathsf{sk}[i])_{i \in [m]})$ and a ciphertext $C := (C[0], C[i_1], \ldots, C[i_k])$ it first checks whether $\mathsf{BFCheck}(H, T, C[0]) = 1$, and outputs $\bot$ in this case. Otherwise, note that $\mathsf{BFCheck}(H, T, C[0]) = 0$ implies that there exists at least one index i^* with $\mathsf{sk}[i^*] \neq \bot$. It picks the smallest index $i^* \in \{i_1, \ldots, i_k\}$ such that $\mathsf{sk}[i^*] = G(i^*)^\alpha \neq \bot$, computes

$$y_{i^*} := e(g_1^r, G(i^*)^\alpha),$$

and returns $\mathsf{K} := C[i^*] \oplus G'(y_{i^*})$.

Remark. If $\mathsf{BFCheck}(H, T_n, C[0]) = 0$, then the decryption algorithm performs a "hashed Boneh-Franklin" decryption with a secret key for one of the identities. Note that $\mathsf{Dec}(\mathsf{sk}_n, C) \neq \bot \iff \mathsf{BFCheck}(H, T, C[0]) = 0$, which guarantees the first extended correctness property required by Definition 4. It is straightforward to verify that the other two extended correctness properties of Definition 4 hold as well.

Design choices. We note that we have chosen to base our Bloom filter encryption scheme on *hashed* Boneh-Franklin IBE instead of standard Boneh-Franklin for two reasons. First, it allows us to keep ciphertexts short and independent of the size of the binary representation of elements of $\mathbb{G}_T$. This is useful, because the recent advances for computing discrete logarithms in finite extension fields [24] apply to the target group of state-of-the-art pairing-friendly elliptic curve groups. Recent assessments of the impact of these advances by Menezes et al. [25] as well as Barbulescu and Duquesne [2] suggest that for currently used efficient curve families such as BN [4] or BLS [3] curves a conservative choice of parameters for the 128 bit security level yields sizes of $\mathbb{G}_T$ elements of $\approx$4600–5500 bits. The hash function allows us to "compress" these group elements in the ciphertext to 128 bits. Even if future research enables the construction of bilinear maps where elements of $\mathbb{G}_T$ can be represented by 2λ bits for λ-bit security (which is optimal), it is still preferable to hash group elements to λ bits to reduce the ciphertext by a factor of about 2. Second, by modelling G' as a random oracle, we can reduce security to a weaker complexity assumption.

Correctness error of this scheme. We will now explain that the correctness error of this scheme is essentially identical to the false-positive probability of the Bloom filter, up to a statistically small distance which corresponds to the probability that two independent ciphertexts share the same randomness r.

For $m, k \in \mathbb{N}$, let $(\mathsf{sk}_0, \mathsf{pk}) \xleftarrow{\$} \mathsf{KGen}(1^\lambda, m, k)$, let $\mathcal{U} := \{C : (C, \mathsf{K}) \xleftarrow{\$} \mathsf{Enc}(\mathsf{pk})\}$ denote the set of all valid ciphertext with respect to pk. Let $\mathcal{S} = (C_1, \ldots, C_n)$ be a list of n ciphertexts, where $(C_i, \mathsf{K}_i) \xleftarrow{\$} \mathsf{Enc}(\mathsf{pk})$, and run $\mathsf{sk}_i = \mathsf{Punc}(\mathsf{sk}_{i-1}, C_i)$ for $i \in [n]$ to determine the secret key sk_n obtained from puncturing sk_0 iteratively on all ciphertexts $C_i \in \mathcal{S}$.

Now let us consider the probability

$$\Pr\left[\mathsf{Dec}(\mathsf{sk}_n, C^*) \neq \mathsf{K}^* : (C^*, \mathsf{K}^*) \xleftarrow{\$} \mathsf{Enc}(\mathsf{pk}), C^* \notin \mathcal{S}\right]$$

that a newly generated ciphertext $C^* \notin \mathcal{S}$ is not correctly decrypted by sk_n. To this end, let $C^*[0] = g_1^{r^*}$ denote the first component of ciphertext $C^* = (g_1^{r^*}, C_1^*, \ldots, C_k^*)$, and likewise let $C_i[0]$ denote the first component of ciphertext C_i for all $C_i \in \mathcal{S}$. Writing $\mathsf{sk}_n = (T_n, (\mathsf{sk}_n[i])_{i \in [m]})$ and $\mathsf{pk} = (g_1^\alpha, H)$, one can now verify that we have $\mathsf{Dec}(\mathsf{sk}_n, C^*) \neq \mathsf{K}^* \iff \mathsf{BFCheck}(H, T_n, C^*[0]) = 1$, because $\mathsf{BFCheck}(H, T_n, C^*[0]) = 0$ guarantees that there exists at least one index j such that $\mathsf{sk}_n[H_j(C^*[0])] \neq \bot$, so correctness of decryption follows essentially from correctness of the Boneh-Franklin scheme. Thus, we have to consider the probability that $\mathsf{BFCheck}(H, T_n, C^*[0]) = 1$. We distinguish between two cases:

1. There exists an index $i \in [n]$ such that $C^*[0] = C_i[0]$. Note that this implies immediately that $\mathsf{BFCheck}(H, T_n, C^*[0]) = 1$. However, recall that $C^*[0] = g_1^{r^*}$ is a uniformly random element of $\mathbb{G}_1$. Therefore the probability that this happens is upper bounded by n/p, which is negligibly small.
2. $C^*[0] \neq C_i[0]$ for all $i \in [n]$. In this case, as explained in Sect. 2.1, the soundness of the Bloom filter guarantees that $\Pr[\mathsf{BFCheck}(H, T_n, C^*[0]) = 1] \approx 2^{-k}$.

In summary, the correctness error of this scheme is approximately $2^{-k} + n/p$. Since n/p is negligibly small, this essentially amounts to the correctness error of the Bloom filter, which in turn depends on the number of ciphertexts n, and the choice of parameters m, k.

Flexible instantiability of this scheme. Our scheme is highly parameterizable in the sense that we can adjust the size of keys and ciphertexts by adjusting the correctness error (determined by the choice of parameters m, k that in turn determine the false-positive probability of the Bloom filter) of our scheme.

Additional properties. As already explained in the remarks after the description of the individual algorithms of PKEM, the scheme satisfies the requirements of Definitions 4, 5, 6, and 7.

IND-CPA-security. We base IND-CPA-security on a bilinear computational Diffie-Hellman variant in the bilinear groups generated by BilGen.

Definition 10 (BCDH). *We define the advantage of adversary $\mathcal{A}$ in solving the BCDH problem with respect to* BilGen *as*

$$\mathbf{Adv}^{\mathsf{BCDH}}_{\mathcal{A},\mathsf{BilGen}}(\lambda) := \Pr\left[e(g_1, h_2)^{r\alpha} \xleftarrow{\$} \mathcal{A}(\mathsf{Params}, g_1^r, g_1^\alpha, g_2^\alpha, h_2)\right],$$

where $\mathsf{Params} = (p, e, \mathbb{G}_1, \mathbb{G}_2, \mathbb{G}_T, g_1, g_2) \xleftarrow{\$} \mathsf{BilGen}(1^\lambda)$, and $(g_1^r, g_1^\alpha, g_2^\alpha, h_2) \xleftarrow{\$} \mathbb{G}_1^2 \times \mathbb{G}_2$.

Theorem 1. *From each efficient adversary $\mathcal{B}$ that issues q queries to random oracle G' we can construct an efficient adversary $\mathcal{A}$ with*

$$\mathbf{Adv}^{\mathsf{BCDH}}_{\mathcal{A},\mathsf{BilGen}}(\lambda) \geq \frac{\mathbf{Adv}^{\mathsf{IND\text{-}CPA}}_{\mathcal{B},\mathsf{PKEM}}(\lambda, m, k)}{kq}.$$

Proof. Algorithm $\mathcal{A}$ receives as input a BCDH-challenge tuple $(g_1^r, g_1^\alpha, g_2^\alpha, h_2)$. It runs adversary $\mathcal{B}$ as a subroutine by simulating the $\mathbf{Exp}^{\mathsf{IND\text{-}CPA}}_{\mathcal{B},\mathsf{PKEM}}(\lambda, m, k)$ experiment, including random oracles G and G', as follows.

First, it defines $\mathcal{Q} := \emptyset$, runs $(H, T) \xleftarrow{\$} \mathsf{BFGen}(m, k)$, and defines the public key as $\mathsf{pk} := (g_1^\alpha, H)$. Note that this public key is identically distributed to a public key output by $\mathsf{KGen}(1^\lambda, m, k)$. In order to simulate the challenge ciphertext, the adversary chooses a random key $\mathsf{K} \xleftarrow{\$} \{0,1\}^\lambda$ and k uniformly random values $Y_j \xleftarrow{\$} \{0,1\}^\lambda$, $j \in [k]$, and defines the challenge ciphertext as $C^* := (g_1^r, (Y_j)_{j \in [k]})$. Finally, it outputs $(\mathsf{pk}, C^*, \mathsf{K})$ to $\mathcal{B}$.

Whenever $\mathcal{B}$ queries $\mathsf{Punc}(\mathsf{sk}, \cdot)$ on input $C = (C[0], \ldots)$, then $\mathcal{A}$ updates T by running $T = \mathsf{BFUpdate}(H, T, C[0])$, and $\mathcal{Q} \leftarrow \mathcal{Q} \cup \{C\}$.

Whenever a random oracle query to $G : \mathbb{N} \to \mathbb{G}_2$ is made (either by $\mathcal{A}$ or $\mathcal{B}$), with input $\ell \in \mathbb{N}$, then $\mathcal{A}$ responds with $G(\ell)$, if $G(\ell)$ has already been defined. If not, then $\mathcal{A}$ chooses a random integer $r_\ell \xleftarrow{\$} \mathbb{Z}_p$, and returns $G(\ell)$, where

$$G(\ell) := \begin{cases} h_2 \cdot g_2^{r_\ell} & \text{if } \ell \in \{H_j(g_1^r) : j \in [k]\}, \text{ and} \\ g_2^{r_\ell} & \text{otherwise.} \end{cases}$$

This definition of G allows $\mathcal{A}$ to simulate the Corr oracle as follows. When $\mathcal{B}$ queries Corr, then it first checks whether $C^* \in \mathcal{Q}$, and returns $\perp$ if this does not hold. Otherwise, note that we must have $\forall j \in [k] : T[H_j(g_1^r)] = 0$, where $H = (H_j)_{j \in [k]}$ and $T[\ell]$ denotes the ℓ-th bit of T. Thus, by the simulation of G described above, $\mathcal{A}$ is able to compute and return $G(\ell)^\alpha = (g_2^{r_\ell})^\alpha = (g_2^\alpha)^{r_\ell}$ for all ℓ with $\ell \notin \{H_j(g_1^r) : j \in [k]\}$, and therefore in particular for all ℓ with $T[\ell] = 1$. This enables the perfect simulation of Corr.

Finally, whenever $\mathcal{B}$ queries random oracle $G' : \mathbb{G}_T \to \{0,1\}^\lambda$ on input y, then $\mathcal{A}$ responds with $G'(y)$, if $G'(y)$ has already been defined. If not, then $\mathcal{A}$ chooses a random string $Y \xleftarrow{\$} \{0,1\}^\lambda$, assigns $G'(y) := Y$, and returns $G'(y)$. Now we have to distinguish between two types of adversaries.

1. A Type-1 adversary $\mathcal{B}$ never queries G' on input of a value y, such that there exists $j \in [k]$ such that $y = e(g_1^\alpha, G(H_j(g_1^r)))^r$. Note that in this case the value $Y_j' := G'(e(g_1^\alpha, G(H_j(g_1^r))))$ remains undefined for all $j \in [k]$ throughout the entire experiment. Thus, information-theoretically, a Type-1 adversary receives no information about the key encrypted in the challenge ciphertext C^*, and thus can only have advantage $\mathbf{Adv}_{\mathcal{B},\mathsf{PKEM}}^{\mathsf{IND\text{-}CPA}}(\lambda, m, k) = 0$, in which case the theorem holds trivially.
2. A Type-2 adversary queries $G'(y)$ such that there exists $j \in [k]$ with $y = e(g_1^\alpha, G(H_j(g_1^r)))^r$. $\mathcal{A}$ uses a Type-2 adversary to solve the BCDH challenge as follows. At the beginning of the game, it picks two indices $(q^*, j^*) \xleftarrow{\$} [q] \times [k]$ uniformly random. When $\mathcal{B}$ outputs y in its q^*-th query to G', then $\mathcal{A}$ computes and outputs $W := y \cdot e(g_1^\alpha, g_2^r)^{-r_\ell}$. Since $\mathcal{B}$ is a Type-2 adversary, we know that at some point it will query $G'(y)$ with $y = e(g_1^\alpha, G(H_j(g_1^r)))^r$ for some $j \in [k]$. If this is the q^*-th query and we have $j = j^*$, which happens with probability $1/(qk)$, then we have
$$\begin{aligned} W &= y \cdot e(g_1^\alpha, g_2^r)^{-r_\ell} = e(g_1^\alpha, G(H_j(g_1^r)))^r \cdot e(g_1^\alpha, g_2^r)^{-r_\ell} \\ &= e(g_1^\alpha, h_2 \cdot g_2^{r_\ell})^r \cdot e(g_1^\alpha, g_2^r)^{-r_\ell} = e(g_1^\alpha, h_2)^r \cdot e(g_1^\alpha, g_2^{r_\ell})^r \cdot e(g_1^\alpha, g_2^r)^{-r_\ell} \end{aligned}$$
   and thus W is a solution to the given BCDH instance. □

OW-CPA-Security. The following theorem can either be proven analogous to Theorem 1, or based on the fact that IND-CPA-security implies OW-CPA-security. Therefore we give it without proof.

Theorem 2. *From each efficient adversary $\mathcal{B}$ that issues q queries to random oracle G' we can construct an efficient adversary $\mathcal{A}$ with*
$$\mathbf{Adv}_{\mathcal{A},\mathsf{BilGen}}^{\mathsf{BCDH}}(\lambda) \geq \frac{\mathbf{Adv}_{\mathcal{B},\mathsf{PKEM}}^{\mathsf{OW\text{-}CPA}}(\lambda, m, k)}{kq}.$$

Remark 1. The construction presented above allows to switch the roles of $\mathbb{G}_1$ and $\mathbb{G}_2$, i.e., to switch all elements in $\mathbb{G}_1$ to $\mathbb{G}_2$ and vice versa. This might be beneficial regarding the size of the secret key when instantiating our construction using a bilinear group where the representation of elements in $\mathbb{G}_2$ requires more space than the representation of elements in $\mathbb{G}_1$.

2.6 CCA-Security via Fujisaki-Okamoto

We obtain a CCA-secure PKEM by adopting the Fujisaki-Okamoto (FO) transformation [16] to the PKEM setting. Since the FO transformation does not work generically for any KEM, we have to use the additional requirements on the underlying PKEM that have been defined in Sect. 2.3. These additional properties enable us to overcome the difficulty that the original Fujisaki-Okamoto transformation from [16] requires *perfect* correctness, what no puncturable KEM can provide. We note that Hofheinz *et al.* [23] give a new, modular analysis of the FO transformation, which also works for public key *encryption* schemes with *negligible* correctness error, however, it is not applicable to PKEMs with non-negligible correctness error because the bounds given in [23] provide insufficient security in this case.

Construction. Let $\mathsf{PKEM} = (\mathsf{KGen}, \mathsf{Enc}, \mathsf{Punc}, \mathsf{Dec})$ be a PKEM with *separable randomness* according to Definition 5. Recall that this means that we can write Enc equivalently as $(C, \mathsf{K}) \xleftarrow{\$} \mathsf{Enc}(\mathsf{pk}) = \mathsf{Enc}(\mathsf{pk}; (r, \mathsf{K}))$ for uniformly random $(r, \mathsf{K}) \xleftarrow{\$} \{0,1\}^{\rho+\lambda}$. In the sequel, let R be a hash function (modeled as a random oracle in the security proof), mapping $R : \{0,1\}^* \rightarrow \{0,1\}^{\rho+\lambda}$. We construct a new scheme $\mathsf{PKEM}' = (\mathsf{KGen}', \mathsf{Enc}', \mathsf{Punc}', \mathsf{Dec}')$ as follows.

$\underline{\mathsf{KGen}'(1^\lambda, m, k)}$: This algorithm is identical to KGen.

$\underline{\mathsf{Enc}'(\mathsf{pk})}$: Algorithm Enc' samples $\mathsf{K} \xleftarrow{\$} \{0,1\}^\lambda$. Then it computes $(r, \mathsf{K}') := R(\mathsf{K}) \in \{0,1\}^{\rho+\lambda}$, runs $(C, \mathsf{K}) \xleftarrow{\$} \mathsf{Enc}(\mathsf{pk}; (r, \mathsf{K}))$, and returns (C, K').

$\underline{\mathsf{Punc}'(\mathsf{sk}, C)}$: This algorithm is identical to Punc.

$\underline{\mathsf{Dec}'(\mathsf{sk}, C)}$: This algorithm first runs $\mathsf{K} \xleftarrow{\$} \mathsf{Dec}(\mathsf{sk}, C)$, and returns $\perp$ if $\mathsf{K} = \perp$. Otherwise, it computes $(r, \mathsf{K}') = R(\mathsf{K})$, and checks consistency of the ciphertext by verifying that $(C, \mathsf{K}) = \mathsf{Enc}(\mathsf{pk}; (r, \mathsf{K}))$. If this does not hold, then it outputs $\perp$. Otherwise it outputs K'.

Correctness error and extended correctness. Both the correctness error and the extended correctness according to Definition 4 are not affected by the Fujisaki-Okamoto transform. Therefore these properties are inherited from the underlying scheme. The fact that the first property of Definition 4 is satisfied makes the scheme suitable for the application to 0-RTT key establishment.

IND-CCA-security. The security proof reduces security of our modified scheme to the OW-CPA-security of the scheme from Sect. 2.5.

Theorem 3. *Let* $\mathsf{PKEM} = (\mathsf{KGen}, \mathsf{Enc}, \mathsf{Punc}, \mathsf{Dec})$ *be a BFKEM scheme that satisfies the additional properties of Definitions 4 and 6, and which is γ-spread according to Definition 7. Let* $\mathsf{PKEM}' = (\mathsf{KGen}', \mathsf{Enc}', \mathsf{Punc}', \mathsf{Dec}')$ *be the scheme described in Sect. 2.6. From each efficient adversary $\mathcal{A}$ that issues at most $q_{\mathcal{O}}$ queries to oracle $\mathcal{O}$ and q_R queries to random oracle R, we can construct an efficient adversary $\mathcal{B}$ with*

$$\mathbf{Adv}^{\mathsf{OW\text{-}CPA}}_{\mathcal{B},\mathsf{PKEM}}(\lambda, m, k) \geq \frac{\mathbf{Adv}^{\mathsf{IND\text{-}CCA}}_{\mathcal{A},\mathsf{PKEM}'}(\lambda, m, k) - q_{\mathcal{O}}/2^\gamma}{q_R}.$$

Proof. We proceed in a sequence of games. In the sequel, $\mathcal{O}_i$ is the implementation of the decryption oracle in Game i.

Game 0. This is the original IND-CCA security experiment from Definition 8, played with the scheme described above. In particular, the decryption oracle $\mathcal{O}_0$ is implemented as follows:

$\mathcal{O}_0(C)$

$\mathsf{K} \stackrel{\$}{\leftarrow} \mathsf{Dec}(\mathsf{sk}, C)$
If $\mathsf{K} = \bot$ **then return** $\bot$
$(r, \mathsf{K}') = R(\mathsf{K})$
If $(C, \mathsf{K}) \neq \mathsf{Enc}(\mathsf{pk}; (r, \mathsf{K}))$ **then return** $\bot$
Return K'

Recall that K_0 denotes the encapsulated key computed by the IND-CCA experiment. K_0 is uniquely defined by the challenge ciphertext C^* via $\mathsf{K}_0 := \mathsf{Dec}(\mathsf{sk}_0, C^*)$, where sk_0 is the initial (non-punctured) secret key, since the scheme satisfies extended correctness (Definition 4, second property). Let Q_0 denote the event that $\mathcal{A}$ ever queries K_0 to random oracle R. Note that $\mathcal{A}$ has zero advantage in distinguishing K' from random, until Q_0 occurs, because R is a random function. Thus, we have $\Pr[Q_0] \geq \mathbf{Adv}^{\mathsf{IND\text{-}CCA}}_{\mathcal{A},\mathsf{PKEM}'}(\lambda, m, k)$. In the sequel, we denote with Q_i the event that $\mathcal{A}$ ever queries K_0 to random oracle R in Game i.

Game 1. This game is identical to Game 0, except that after computing $\mathsf{K} \stackrel{\$}{\leftarrow} \mathsf{Dec}(\mathsf{sk}, C)$ and checking whether $\mathsf{K} \neq \bot$, the experiment additionally checks whether the adversary has ever queried random oracle R on input K, and returns $\bot$ if not. More precisely, the experiment maintains a list

$$L_R = \{(\mathsf{K}, (r, \mathsf{K}')) : \mathcal{A} \text{ queried } R(\mathsf{K}) = (r, \mathsf{K}')\}$$

to record all queries K made by the adversary to random oracle R, along with the corresponding response $(r, \mathsf{K}') = R(\mathsf{K})$. The decryption oracle $\mathcal{O}_1$ uses this list as follows (boxed statements highlight changes to $\mathcal{O}_0$):

$\mathcal{O}_1(C)$

$\mathsf{K} \stackrel{\$}{\leftarrow} \mathsf{Dec}(\mathsf{sk}, C)$
If $\nexists (r, \mathsf{K}') : (\mathsf{K}, (r, \mathsf{K}')) \in L_R$ **then return** $\bot$
$(r, \mathsf{K}') = R(\mathsf{K})$
If $(C, \mathsf{K}) \neq \mathsf{Enc}(\mathsf{pk}; (r, \mathsf{K}))$ **then return** $\bot$
Return K'

Note that Games 0 and 1 are perfectly indistinguishable, unless $\mathcal{A}$ ever outputs a ciphertext C with $\mathcal{O}_1(C) = \bot$, but $\mathcal{O}_0(C) \neq \bot$. Note that this happens if and only if $\mathcal{A}$ outputs C such that $C = \mathsf{Enc}(\mathsf{pk}; (r, \mathsf{K}))$, where r is the randomness defined by $(r, \mathsf{K}') = R(\mathsf{K})$, but without prior query of $R(\mathsf{K})$.

The random oracle R assigns a uniformly random value $r \in \{0,1\}^\rho$ to each query, so, by the γ-spreadness of PKEM, the probability that the ciphertext C output by the adversary "matches" the ciphertext produced by $\mathsf{Enc}(\mathsf{pk}; (r, \mathsf{K}))$ is $2^{-\gamma}$. Since $\mathcal{A}$ issues at most $q_\mathcal{O}$ queries to $\mathcal{O}_1$, this yields $\Pr[Q_1] \geq \Pr[Q_0] - q_\mathcal{O}/2^\gamma$.

Game 2. We make a minor conceptual modification. Instead of computing $(r, \mathsf{K}') = R(\mathsf{K})$ by evaluating R, $\mathcal{O}_2$ reads (r, K') from list L_R. More precisely:

$\mathcal{O}_2(C)$

$\mathsf{K} \xleftarrow{\$} \mathsf{Dec}(\mathsf{sk}, C)$
If $\nexists (r, \mathsf{K}') : (\mathsf{K}, (r, \mathsf{K}')) \in L_R$ **then return** $\perp$
Define (r, K') to be the unique tuple such that $(\mathsf{K}, (r, \mathsf{K}')) \in L_R$.
If $(C, \mathsf{K}) \neq \mathsf{Enc}(\mathsf{pk}; (r, \mathsf{K}))$ **then return** $\perp$
Return K'

By definition of L_R it always holds that $(r, \mathsf{K}') = R(\mathsf{K})$ for all $(\mathsf{K}, (r, \mathsf{K}')) \in L_R$. Indeed (r, K'), is uniquely determined by K, because $(r, \mathsf{K}') = R(\mathsf{K})$ is a function. Since R is only evaluated by $\mathcal{O}_1$ if there exists a corresponding tuple $(\mathsf{K}, (r, \mathsf{K}')) \in L_R$ anyway, due to the changes introduced in Game 1, oracle $\mathcal{O}_2$ is equivalent to $\mathcal{O}_1$ and we have $\Pr[Q_2] = \Pr[Q_1]$.

Game 3. This game is identical to Game 2, except that whenever $\mathcal{A}$ queries a ciphertext C to oracle $\mathcal{O}_3$, then $\mathcal{O}_3$ first runs the $\mathsf{CheckPunct}$ algorithm associated to PKEM (cf. Definition 6). If $\mathsf{CheckPunct}(\mathsf{pk}, \mathcal{Q}, C) = \perp$, then it immediately returns $\perp$. Otherwise, it proceeds exactly like $\mathcal{O}_2$. More precisely:

$\mathcal{O}_3(C)$

If $\mathsf{CheckPunct}(\mathsf{pk}, \mathcal{Q}, C) = \perp$ **then return** $\perp$
$\mathsf{K} \xleftarrow{\$} \mathsf{Dec}(\mathsf{sk}, C)$
If $\nexists (r, \mathsf{K}') : (\mathsf{K}, (r, \mathsf{K}')) \in L_R$ **then return** $\perp$
Define (r, K') to be the unique tuple such that $(\mathsf{K}, (r, \mathsf{K}')) \in L_R$.
If $(C, \mathsf{K}) \neq \mathsf{Enc}(\mathsf{pk}; (r, \mathsf{K}))$ **then return** $\perp$
Return K'

Recall that by public checkability (Definition 6) we have $\perp = \mathsf{Dec}(\mathsf{sk}, C) \iff \perp = \mathsf{CheckPunct}(\mathsf{pk}, \mathcal{Q}, C)$. Therefore the introduced changes are conceptual, and $\Pr[Q_3] = \Pr[Q_2]$.

Game 4. We modify the secret key used to decrypt the ciphertext. Let sk_0 denote the initial secret key generated by the experiment (that is, before any puncturing operation was performed). $\mathcal{O}_4$ uses sk_0 to compute $\mathsf{K} \xleftarrow{\$} \mathsf{Dec}(\mathsf{sk}_0, C)$ instead of $\mathsf{K} \xleftarrow{\$} \mathsf{Dec}(\mathsf{sk}, C)$, where sk is a possibly punctured secret key. More precisely:

$\mathcal{O}_4(C)$

If $\mathsf{CheckPunct}(\mathsf{pk}, \mathcal{Q}, C) = \bot$ **then return** $\bot$
$\mathsf{K} \xleftarrow{\$} \mathsf{Dec}(\mathsf{sk}_0, C)$
If $\nexists(r, \mathsf{K}') : (\mathsf{K}, (r, \mathsf{K}')) \in L_R$ **then return** $\bot$
Define (r, K') to be the unique tuple such that $(\mathsf{K}, (r, \mathsf{K}')) \in L_R$.
If $(C, \mathsf{K}) \neq \mathsf{Enc}(\mathsf{pk}; (r, \mathsf{K}))$ **then return** $\bot$
Return K'

For indistinguishability from Game 3, we show that $\mathcal{O}_4(C) = \mathcal{O}_3(C)$ for all ciphertexts C. Let us first consider the case $\mathsf{Dec}(\mathsf{sk}, C) = \bot$. Then public checkability guarantees that $\mathcal{O}_4(C) = \mathcal{O}_3(C) = \bot$, due to the fact that $\mathsf{Dec}(\mathsf{sk}, C) = \bot \iff \mathsf{CheckPunct}(\mathsf{pk}, \mathcal{Q}, C) = \bot$.

Now let us consider the case $\mathsf{Dec}(\mathsf{sk}, C) \neq \bot$. In this case, the semi-correctness of punctured keys (3rd requirement of Definition 4) guarantees that $\mathsf{Dec}(\mathsf{sk}, C) = \mathsf{Dec}(\mathsf{sk}_0, C) = \mathsf{K} \neq \bot$.

After computing $\mathsf{Dec}(\mathsf{sk}_0, C)$, $\mathcal{O}_4$ performs exactly the same operations as $\mathcal{O}_3$ after computing $\mathsf{Dec}(\mathsf{sk}, C)$. Thus, in this case both oracles are perfectly indistinguishable, too. This yields that the changes introduced in Game 4 are purely conceptual, and we have $\Pr[Q_4] = \Pr[Q_3]$.

Remark. Due to the fact that we are now using the initial secret key to decrypt C, we have reached a setting where, due to the perfect correctness of the initial secret key sk_0, essentially a perfectly-correct encryption scheme is used – except that the decryption oracle implements a few additional abort conditions. Thus, we can now basically apply the standard Fujisaki-Okamoto transformation, but we must show that we are also able to simulate the additional abort imposed by the additional consistency checks properly. To this end, we first replace these checks with equivalent checks before applying the FO transformation.

Game 5. We replace the consistency checks performed by $\mathcal{O}_4$ with an equivalent check. More precisely, $\mathcal{O}_5$ works as follows:

$\mathcal{O}_5(C)$

If $\mathsf{CheckPunct}(\mathsf{pk}, \mathcal{Q}, C) = \bot$ **then return** $\bot$
$\mathsf{K} \xleftarrow{\$} \mathsf{Dec}(\mathsf{sk}_0, C)$
If $\nexists(r, \mathsf{K}') : ((\mathsf{K}, (r, \mathsf{K}')) \in L_R \wedge (C, \mathsf{K}) = \mathsf{Enc}(\mathsf{pk}; (r, \mathsf{K})))$ **then return** $\bot$
Return K' such that $(\mathsf{K}, (r, \mathsf{K}')) \in L_R \wedge (C, \mathsf{K}) = \mathsf{Enc}(\mathsf{pk}; (r, \mathsf{K}))$

This is equivalent, so that we have $\Pr[Q_5] = \Pr[Q_4]$.

Game 6. Observe that in Game 5 we check whether there exists a tuple (r, K') with $(\mathsf{K}, (r, \mathsf{K}')) \in L_R$ and $(C, \mathsf{K}) = \mathsf{Enc}(\mathsf{pk}; (r, \mathsf{K}))$, where K must match the secret key computed by $\mathsf{K} \xleftarrow{\$} \mathsf{Dec}(\mathsf{sk}_0, C)$.

In Game 6, we relax this check. We test only whether there exists any tuple $(\tilde{\mathsf{K}}, (\tilde{r}, \tilde{\mathsf{K}'})) \in L_R$ such that $(C, \tilde{\mathsf{K}}) = \mathsf{Enc}(\mathsf{pk}; (\tilde{r}, \tilde{\mathsf{K}}))$ holds. Thus, it is not explicitly checked whether $\tilde{\mathsf{K}}$ matches the value $\mathsf{K} \xleftarrow{\$} \mathsf{Dec}(\mathsf{sk}_0, C)$. Furthermore, the corresponding value $\tilde{\mathsf{K}'}$ is returned. More precisely:

$\mathcal{O}_6(C)$

If $\mathsf{CheckPunct}(\mathsf{pk}, \mathcal{Q}, C) = \bot$ **then return** $\bot$
$\mathsf{K} \xleftarrow{\$} \mathsf{Dec}(\mathsf{sk}_0, C)$
If $\nexists(\tilde{r}, \tilde{\mathsf{K}'}) : ((\tilde{\mathsf{K}}, (\tilde{r}, \tilde{\mathsf{K}'})) \in L_R \wedge (C, \tilde{\mathsf{K}}) = \mathsf{Enc}(\mathsf{pk}; (\tilde{r}, \tilde{\mathsf{K}})))$ **then return** $\bot$
Return $\tilde{\mathsf{K}'}$ such that $(\tilde{\mathsf{K}}, (\tilde{r}, \tilde{\mathsf{K}'})) \in L_R \wedge (C, \tilde{\mathsf{K}}) = \mathsf{Enc}(\mathsf{pk}; (\tilde{r}, \tilde{\mathsf{K}}))$

By the perfect correctness of the initial secret key sk_0, we have

$$(C, \tilde{\mathsf{K}}) = \mathsf{Enc}(\mathsf{pk}; (\tilde{r}, \tilde{\mathsf{K}})) \implies \mathsf{Dec}(\mathsf{sk}_0, C) = \tilde{\mathsf{K}},$$

so that we must have $\mathsf{K} = \tilde{\mathsf{K}}$. $\mathcal{O}_6$ is equivalent to $\mathcal{O}_5$, and $\Pr[Q_6] = \Pr[Q_5]$.

Game 7. This game is identical to Game 6, except that we change the decryption oracle again. Observe that the value K computed by $\mathsf{K} \xleftarrow{\$} \mathsf{Dec}(\mathsf{sk}_0, C)$ is never used by $\mathcal{O}_6$. Therefore the computation of $\mathsf{K} \xleftarrow{\$} \mathsf{Dec}(\mathsf{sk}_0, C)$ is obsolete, and we can remove it. More precisely, $\mathcal{O}_7$ works as follows.

$\mathcal{O}_7(C)$

If $\mathsf{CheckPunct}(\mathsf{pk}, \mathcal{Q}, C) = \bot$ **then return** $\bot$
If $\nexists(\tilde{r}, \tilde{\mathsf{K}'}) : ((\tilde{\mathsf{K}}, (\tilde{r}, \tilde{\mathsf{K}'})) \in L_R \wedge (C, \tilde{\mathsf{K}}) = \mathsf{Enc}(\mathsf{pk}; (\tilde{r}, \tilde{\mathsf{K}})))$ **then return** $\bot$
Return $\tilde{\mathsf{K}'}$ such that $(\tilde{\mathsf{K}}, (\tilde{r}, \tilde{\mathsf{K}'})) \in L_R \wedge (C, \tilde{\mathsf{K}}) = \mathsf{Enc}(\mathsf{pk}; (\tilde{r}, \tilde{\mathsf{K}}))$

We have only removed an obsolete instruction, which does not change the output distribution of the decryption oracle. Therefore $\mathcal{O}_7$ simulates $\mathcal{O}_6$ perfectly, and we have $\Pr[Q_7] = \Pr[Q_6]$.

Reduction to OW-CPA-security. Now we are ready to describe the OW-CPA-adversary $\mathcal{B}$. $\mathcal{B}$ receives (pk, C^*). It samples a uniformly random key $\mathsf{K}' \xleftarrow{\$} \{0,1\}^\lambda$ and runs the IND-CCA-adversary $\mathcal{A}$ as a subroutine on input $(\mathsf{pk}, C^*, \mathsf{K}')$. Whenever $\mathcal{A}$ issues a Punc- or Corr-query, then $\mathcal{B}$ forwards this query to the OW-CPA-experiment and returns the response. In order to simulate the decryption oracle $\mathcal{O}$, adversary $\mathcal{B}$ implements the simulated oracle $\mathcal{O}_7$ from Game 7 described above. When $\mathcal{A}$ terminates, then $\mathcal{B}$ picks a uniformly random entry $(\hat{\mathsf{K}}, (\hat{r}, \hat{\mathsf{K}'})) \xleftarrow{\$} L_R$, and outputs $\hat{\mathsf{K}}$.

Analysis of the reduction. Let $\hat{Q}$ denote the event that $\mathcal{A}$ ever queries K_0 to random oracle R. Note that $\mathcal{B}$ simulates Game 7 perfectly until Q_7 occurs, thus we have $\mathsf{Pr}[\hat{Q}] \geq \mathsf{Pr}[Q_7]$. Summing up, the probability that the value $\hat{\mathsf{K}}$ output by $\mathcal{B}$ matches the key encapsulated in C^* is therefore at least

$$\frac{\mathsf{Pr}[\hat{Q}]}{q_R} \geq \frac{\mathbf{Adv}^{\mathsf{IND\text{-}CCA}}_{\mathcal{A},\mathsf{PKEM}'}(\lambda, m, k) - q_{\mathcal{O}}/2^{\gamma}}{q_R}.$$

□

Remark on the tightness. Alternatively, we could have based the security of our IND-CCA-secure scheme on the IND-CPA (rather than OW-CPA) security of PKEM′. In this case, we would have achieved a tighter reduction, as we would have been able to avoid guessing the index $(\hat{\mathsf{K}}, (\hat{r}, \hat{\mathsf{K}}')) \xleftarrow{\$} L_R$, at the cost of requiring stronger security of the underlying scheme.

From IND-CCA-secure KEMs to IND-CCA-secure encryption. It is well-known that one can generically transform an IND-CCA-secure KEM into an IND-CCA-secure encryption scheme, by combining it with a CCA-secure symmetric encryption scheme [16]. This construction applies to PKEMs as well.

2.7 Time-Based Bloom Filter Encryption

For a standard BFE scheme we have to update the public key after the secret key has been punctured n-times, because otherwise the false-positive probability would exceed an acceptable bound. In this section, we describe a construction of a scheme where the lifetime of the public key is split into *time slots*. Ciphertexts are associated with time slots, which assumes loosely synchronized clocks between sender and receiver of a ciphertext. The main advantage is that for a given bound on the correctness error, we are able to handle about the same number of puncturings *per time slot* as the basic scheme during the entire life time of the public key. We call this approach *time-based* Bloom filter encryption. It is inspired by the time-based approach used to construct puncturable encryption in [19,20], which in turn is inspired by the construction of forward-secret public-key encryption by Canetti et al. [10].

Note that a time-based BFE scheme can trivially be obtained from any BFE scheme, by assigning an individual public/secret key pair for each time slot. However, if we want to split the life time of the public key into, say, 2^t time slots, then this would of course increase the size of keys by a factor 2^t. Since we want to enable a fine-grained use of time slots, to enable a very large number of puncturings over the entire lifetime of the public key without increasing the false positive probability beyond an unacceptable bound, we want to have 2^t as large as possible, but without increasing the size of the public key beyond an acceptable bound. To this end, we give a direct construction which increases the size of secret keys only by an *additive* amount of additional group elements, which is only *logarithmic* in the number of time slots. Thus, for 2^t time slots we have to add merely about t elements to the secret key, while the size of public keys remains even *constant*.

Formal definition. Likewise to considering our Bloom filter KEMs as an instantiation of a puncturable KEM with non-negligible correctness error, we can view the time-based approach analogously as an instantiation of a puncturable forward-secret KEM (PFSKEM) [20] with non-negligible correctness error. Consequently, we also chose to stick with the existing formal framework for PFSKEM, which we present subsequently. It is essentially our BFKEM Definition 2, augmented by time slots and an additional algorithm PuncInt that allows to puncture a secret key not with respect to a given ciphertext in a given time slot, but with respect to an entire time slot.

Definition 11 (PFSKEM [20]**).** *A puncturable forward-secret key encapsulation (PFSKEM) scheme is a tuple of the following* PPT *algorithms:*

$\mathsf{KGen}(1^\lambda, m, k, t)$: *Takes as input a security parameter* λ*, parameters* m *and* k *for the Bloom filter, and a parameter* t *specifying the number of time slots. It outputs a secret and public key* $(\mathsf{sk}, \mathsf{pk})$*, where we assume that the key-space* $\mathcal{K}$ *is implicit in* pk*.*

$\mathsf{Enc}(\mathsf{pk}, \tau)$: *Takes as input a public key* pk *and a time slot* τ *and outputs a ciphertext* C *and a symmetric key* K*.*

$\mathsf{PuncCtx}(\mathsf{sk}, \tau, C)$: *Takes as input a secret key* sk*, a time slot* τ*, a ciphertext* C *and outputs an updated secret key* sk'*.*

$\mathsf{Dec}(\mathsf{sk}, \tau, C)$: *Takes as input a secret key* sk*, a time slot* τ*, a ciphertext* C *and outputs a symmetric key* K *or* $\perp$ *if decapsulation fails.*

$\mathsf{PuncInt}(\mathsf{sk}, \tau)$: *Takes as input a secret key* sk*, a time slot* τ *and outputs an updated secret key* sk' *for the next slot* $\tau + 1$*.*

Due to the lack of space, we postpone the presentation of correctness, the additional properties (which are rather straightforward adaptions of the ones of a PKEM introduced in Sect. 2.3), as well as the IND-CPA/IND-CCA security notions to the full version.

Hierarchical IB-KEMs. We recall the basic definition of hierarchical identity-based key encapsulation schemes (HIB-KEMs) and their security.

Definition 12. *A* $(t+1)$*-level hierarchical identity-based key encapsulation scheme (*HIB-KEM*) with identity space* $\mathcal{D} = \mathcal{D}_1 \times \cdots \times \mathcal{D}_{t+1}$*, ciphertext space* $\mathcal{C}$*, and key space* $\mathcal{K}$ *consists of the following four algorithms:*

$\mathsf{HIBGen}(1^\lambda)$: *Takes as input a security parameter and outputs a key pair* $(\mathsf{mpk}, \mathsf{sk}_0)$*. We say that* mpk *is the master public key, and* sk_0 *is the* level-0 *secret key.*

$\mathsf{HIBDel}(\mathsf{sk}_{i-1}, d)$: *Takes as input a level-*$i-1$ *secret key* sk_{i-1} *with* $i \in [t]$ *and an element* $d \in \mathcal{D}_i$ *and outputs a level-*i *secret key* sk_i*.*

$\mathsf{HIBEnc}(\mathsf{mpk}, \boldsymbol{d})$: *Takes as input the master public key* mpk *and an identity* $\boldsymbol{d} \in \mathcal{D}$ *and outputs a ciphertext* $C \in \mathcal{C}$ *and a key* $\mathsf{K} \in \mathcal{K}$*.*

$\mathsf{HIBDec}(\mathsf{sk}_\ell, C)$: *Takes as input a level-*t *secret key* sk_t *and a ciphertext* C*, and outputs a value* $\mathsf{K} \in \mathcal{K} \cup \{\perp\}$*, where* $\perp$ *is a distinguished error symbol.*

$\mathbf{Exp}^{\mathsf{OW\text{-}sID\text{-}CPA}}_{\mathcal{A},\mathsf{HIB\text{-}KEM}}(\lambda)$
$(\boldsymbol{d}^*, \mathtt{state}_{\mathcal{A}}) \xleftarrow{\$} \mathcal{A}(1^\lambda)$
if $\boldsymbol{d}^* \notin \mathcal{D}$ return 0
$(\mathsf{mpk}, \mathsf{sk}_0) \xleftarrow{\$} \mathsf{HIBGen}(1^\lambda)$, $(C, \mathsf{K}) \xleftarrow{\$} \mathsf{HIBEnc}(\mathsf{mpk}, \boldsymbol{d}^*)$
$\mathsf{K}^* \xleftarrow{\$} \mathcal{A}(\mathsf{mpk}, C, \mathtt{state}_{\mathcal{A}})$
return 1, if $\mathsf{K}^* = \mathsf{K}$
return 0

Fig. 3. OW-sID-CPA security for HIB-KEMs.

Security definition. We will require only the very weak notion of one-wayness under selective-ID and chosen-plaintext attacks (OW-sID-CPA) where the corresponding experiment is defined in Fig. 3.

Definition 13 (OW-sID-CPA Security of HIB-KEM). *We define the advantage of an adversary $\mathcal{A}$ in the OW-sID-CPA experiment* $\mathbf{Exp}^{\mathsf{OW\text{-}sID\text{-}CPA}}_{\mathcal{A},\mathsf{HIB\text{-}KEM}}(\lambda)$ *as*

$$\mathbf{Adv}^{\mathsf{OW\text{-}sID\text{-}CPA}}_{\mathcal{A},\mathsf{HIB\text{-}KEM}}(\lambda) := \Pr\left[\mathbf{Exp}^{\mathsf{OW\text{-}sID\text{-}CPA}}_{\mathcal{A},\mathsf{HIB\text{-}KEM}}(\lambda) = 1\right].$$

We call a HIB-KEM *OW-sID-CPA secure, if* $\mathbf{Adv}^{\mathsf{OW\text{-}sID\text{-}CPA}}_{\mathcal{A},\mathsf{HIB\text{-}KEM}}(\lambda)$ *is a negligible function in λ for all PPT adversaries $\mathcal{A}$.*

Time slots. We will construct a Bloom filter encryption scheme that allows to use 2^t time slots. We associate the i-th time slot with the string in $\{0,1\}^t$ that corresponds to the canonical t-bit binary representation of integer i.

Following [10,19,20], each time slot forms a leaf of an ordered binary tree of depth t. The root of the tree is associated with the empty string ϵ. We associate the left-hand descendants of the root with bit string 0, and the right-hand descendant with 1. Continuing this way, we associate the left descendant of node 0 with 00 and the right descendant with 01, and so on. We continue this procedure for all nodes, until we have constructed a complete binary tree of depth t. Note that two nodes at level t' of the tree are siblings if and only if their first $t'-1$ bits are equal, and that each bit string in $\{0,1\}^t$ is associated with a leaf of the tree. Note also that the tree is ordered, in the sense that the leftmost leaf is associated with 0^t, its right neighbour with $0^{t-1}1$, and so on.

Intuition of the construction. The basic idea behind the construction combines the binary tree approach of [10,19,20] with the Bloom filter encryption construction described in Sect. 2.5. We use a HIB-KEM with identity space

$$\mathcal{D} = \mathcal{D}_1 \times \cdots \times \mathcal{D}_{t+1} = \underbrace{\{0,1\} \times \cdots \times \{0,1\}}_{t \text{ times}} \times [m].$$

Each bit vector $\tau \in \mathcal{D}_1 \times \cdots \times \mathcal{D}_t = \{0,1\}^t$ corresponds to one time slot, and we set $\mathcal{D}_{t+1} = [m]$, where m is the size of the Bloom filter. The hierarchical key delegation property of the HIB-KEM enables the following features:

First, given a HIB-KEM key sk_τ for some "identity" (= time slot) $\tau \in \{0,1\}^t$, we can derive keys for all Bloom filter bits from sk_τ by computing

$$\mathsf{sk}_{\tau|d} \xleftarrow{\$} \mathsf{HIBDel}(\mathsf{sk}_\tau, d) \quad \text{for all} \quad d \in [m].$$

Second, in order to advance from time slot $\tau - 1$ to τ, we first compute

$$\mathsf{sk}_{\tau|d} \xleftarrow{\$} \mathsf{HIBDel}(\mathsf{sk}_\tau, d) \quad \text{for all} \quad d \in [m].$$

As soon as we have computed all Bloom filter keys for time slot τ, we "puncture" the tree "from left to right", such that we are able to compute all $\mathsf{sk}_{\tau'}$ with $\tau' > \tau$, but not any $sk_{\tau'}$ with $\tau' \leq \tau$. Here, we proceed exactly as in [10,19,20]. That is, in order to puncture at time slot τ, we first compute the HIB-KEM secret keys associated to all *right-hand siblings* of nodes that lie on the path from node τ to the root, and then we delete all secret keys associated to nodes that lie on the path from node τ to the root, including sk_τ itself. This yields a new secret key, which contains m level-$(t+1)$ HIB-KEM secret keys plus at most t HIB-KEM secret keys for levels $\leq t$, even though we allow for 2^t time slots.

Construction. Let (HIBGen, HIBDel, HIBEnc, HIBDec) be a HIB-KEM with key space $\mathcal{K}$ and identity space $\mathcal{D} = \mathcal{D}_1 \times \cdots \times \mathcal{D}_{t+1}$, where $\mathcal{D}_1 = \cdots = \mathcal{D}_t = \{0,1\}$, $\mathcal{D}_{t+1} = [m]$, and m is the size of the Bloom filter. Since we will only need selective security, one can instantiate such a HIB-KEM very efficiently, for example in bilinear groups based on the Boneh-Boyen-Goh [8] scheme, or based on lattices [1]. In the sequel, we will write $\{0,1\}^t$ shorthand for $\mathcal{D}_1 \times \cdots \times \mathcal{D}_t$, but keep in mind that the HIB-KEM supports more fine-grained key delegation. Let $\mathsf{B} = (\mathsf{BFGen}, \mathsf{BFUpdate}, \mathsf{BFCheck})$ be a Bloom filter for set $\{0,1\}^\lambda$. Furthermore, let $G' : \mathcal{K} \to \{0,1\}^\lambda$ be a hash function (which will be modeled as a random oracle [5] in the security proof).

We define $\mathsf{PKEM} = (\mathsf{KGen}, \mathsf{Enc}, \mathsf{PuncCtx}, \mathsf{Dec}, \mathsf{PuncInt})$ as follows.

$\underline{\mathsf{KGen}(1^\lambda, m, k, 2^t)}$: This algorithm first runs $((H_j)_{j\in[k]}, T) \xleftarrow{\$} \mathsf{BFGen}(m, k)$ to generate a Bloom filter, and $(\mathsf{mpk}, \mathsf{sk}_\epsilon) \xleftarrow{\$} \mathsf{HIBGen}(1^\lambda)$ to generate a key pair. Finally, the algorithm generates the keys for the first time slot. To this end, it first computes the HIB-KEM key for identity 0^t by recursively computing

$$\mathsf{sk}_{0^d} \xleftarrow{\$} \mathsf{HIBDel}(\mathsf{sk}_{0^{d-1}}, 0) \quad \text{for all} \quad d \in [t].$$

Then it computes the m Bloom filter keys for time slot 0^t by computing

$$\mathsf{sk}_{0^t|d} \xleftarrow{\$} \mathsf{HIBDel}(\mathsf{sk}_{0^t}, d) \quad \text{for all} \quad d \in [m],$$

and setting $\mathsf{sk}_{\mathsf{Bloom}} := (\mathsf{sk}_{0^t|d})_{d\in[m]}$. Finally, it punctures the secret key sk_ϵ at position 0^t, by computing

$$\mathsf{sk}_{0^{d-1}1} \xleftarrow{\$} \mathsf{HIBDel}(\mathsf{sk}_{0^{d-1}}, 1) \quad \text{for all} \quad d \in [t],$$

and setting $\mathsf{sk}_{\mathsf{time}} := (\mathsf{sk}_{0^{d-1}|1})_{d\in[t]}$. The algorithm outputs

$$\mathsf{sk} := (T, \mathsf{sk}_{\mathsf{Bloom}}, \mathsf{sk}_{\mathsf{time}}) \text{ and } \mathsf{pk} := (\mathsf{mpk}, (H_j)_{j\in[k]}).$$

$\underline{\mathsf{Enc}(\mathsf{mpk}, \tau)}$: On input mpk and time slot identifier $\tau \in \{0,1\}^t$, this algorithm first samples a random string $c \xleftarrow{\$} \{0,1\}^\lambda$ and a random key $\mathsf{K} \xleftarrow{\$} \{0,1\}^\lambda$. Then it defines k HIB-KEM identities as $\boldsymbol{d}_j := (\tau, H_j(c)) \in \mathcal{D}$ for $j \in [k]$, and generates k HIB-KEM key encapsulations as

$$(C_j, \mathsf{K}_j) \xleftarrow{\$} \mathsf{HIBEnc}(\mathsf{mpk}, \boldsymbol{d}_j) \quad \text{for} \quad j \in [k].$$

Finally, it outputs the ciphertext $C := (c, (C_j, G'(\mathsf{K}_j) \oplus \mathsf{K})_{j\in[k]})$.

Note that the ciphertexts essentially consists of $k+1$ elements of $\{0,1\}^\lambda$, plus k elements of $\mathcal{C}$, where k is the Bloom filter parameter.

$\underline{\mathsf{PuncCtx}(\mathsf{sk}, C)}$: Given a ciphertext $C := (c, (C_j, G'(\mathsf{K}_j) \oplus \mathsf{K})_{j\in[k]})$, and secret key $\mathsf{sk} = (T, \mathsf{sk}_{\mathsf{Bloom}}, \mathsf{sk}_{\mathsf{time}})$ where $\mathsf{sk}_{\mathsf{Bloom}} = (\mathsf{sk}_{\tau|d})_{d\in[m]}$, the puncturing algorithm first computes $T' = \mathsf{BFUpdate}((H_j)_{j\in[k]}, T, c)$. Then, for each $i \in [m]$, it defines

$$\mathsf{sk}'_{\tau|i} := \begin{cases} \mathsf{sk}_{\tau|i} & \text{if } T'[i] = 0, \text{ and} \\ \bot & \text{if } T'[i] = 1, \end{cases}$$

where $T'[i]$ denotes the i-th bit of T'. Finally, this algorithm sets $\mathsf{sk}'_{\mathsf{Bloom}} = (\mathsf{sk}'_{\tau|d})_{d\in[m]}$ and returns $\mathsf{sk}' = (T', \mathsf{sk}'_{\mathsf{Bloom}}, \mathsf{sk}_{\mathsf{time}})$.

Remark. We note again that the above procedure is correct even if the procedure is applied repeatedly, with the same arguments as for the construction from Sect. 2.5. Also, the puncturing algorithm essentially only evaluates k universal hash functions and then deletes a few secret keys, which makes this procedure extremely efficient.

$\underline{\mathsf{Dec}(\mathsf{sk}, C)}$: Given $\mathsf{sk} = (T, \mathsf{sk}_{\mathsf{Bloom}}, \mathsf{sk}_{\mathsf{time}})$ where $\mathsf{sk}_{\mathsf{Bloom}} = (\mathsf{sk}_{\tau|d})_{d\in[m]}$ and ciphertext $C := (c, (C_j, G_j)_{j\in[k]})$. If $\mathsf{sk}_{\tau|H_j(c)} = \bot$ for all $j \in [k]$, then it outputs $\bot$. Otherwise, it picks the smallest index j such that $\mathsf{sk}_{\tau|H_j(c)} \neq \bot$, computes

$$\mathsf{K}_j = \mathsf{HIBDec}(\mathsf{sk}_{\tau|H_j(c)}, C_j),$$

and returns $\mathsf{K} = G_j \oplus G'(\mathsf{K}_j)$.

Remark. Again we have $\mathsf{Dec}(\mathsf{sk}, C) \neq \bot \iff \mathsf{BFCheck}(H, T, c) = 0$, which guarantees extended correctness in the sense of Definition 4.

$\underline{\mathsf{PuncInt}(\mathsf{sk}, \tau)}$: Given a secret key $\mathsf{sk} = (T, \mathsf{sk}_{\mathsf{Bloom}}, \mathsf{sk}_{\mathsf{time}})$ for time interval $\tau' < \tau$, the time puncturing algorithm proceeds as follows. First, it resets the Bloom filter by setting $T := 0^m$. Then it uses the key delegation algorithm to first compute sk_τ. This key can be computed from the keys contained in $\mathsf{sk}_{\mathsf{time}}$, because sk is a key for time interval $\tau' < \tau$. Then it computes

$$\mathsf{sk}_{\tau|d} \xleftarrow{\$} \mathsf{HIBDel}(\mathsf{sk}_\tau, d) \quad \text{for all} \quad d \in [m],$$

and redefines $\mathsf{sk}_{\mathsf{Bloom}} := (\mathsf{sk}_{\tau|d})_{d\in[m]}$. Finally, it updates $\mathsf{sk}_{\mathsf{time}}$ by computing the HIB-KEM secret keys associated to all *right-hand siblings* of nodes that lie on the path from node τ to the root and adds the corresponding keys to $\mathsf{sk}_{\mathsf{time}}$. Then it deletes all keys from $\mathsf{sk}_{\mathsf{time}}$ that lie on the path from τ to the root.

Remark. Note that puncturing between time intervals may become relatively expensive. Depending on the choice of Bloom filter parameters, in particular on m, this may range between 2^{15} and 2^{25} HIBE key delegations. However, the main advantage of Bloom filter encryption over previous constructions of puncturable encryption is that these computations must not be performed "online", during puncturing, but can actually be computed separately (for instance, parallel on a different computer, or when a server has low workload, etc.).

Correctness error of this scheme. With exactly the same arguments as for the scheme from Sect. 2.5, one can verify that the correctness error of this scheme is essentially identical to the false positive probability of the Bloom filter, unless a given ciphertext $C = (c, (C_j, G_j)_{j \in [k]})$ has a value of c which is identical to the value of c of any previous ciphertext. Since c is uniformly random in $\{0,1\}^\lambda$, this probability is approximately $2^{-k} + n \cdot 2^{-\lambda}$.

Extended correctness. It is straightforward to verify that the scheme satisfies extended correctness in the sense of Definition 4.

CPA Security. Below we state theorem for CPA security of our scheme.

Theorem 4. *From each efficient adversary $\mathcal{B}$ that issues q queries to random oracle G' we can construct an efficient adversary $\mathcal{A}$ with*

$$\mathbf{Adv}^{\mathsf{OW\text{-}sID\text{-}CPA}}_{\mathcal{A},\mathsf{HIB\text{-}KEM}}(\lambda) \geq \frac{\mathbf{Adv}^{\mathsf{s\text{-}CPA}}_{\mathcal{B},\mathsf{PFSKEM}}(\lambda, m, k)}{qk}.$$

The proof is almost identical to the proof of Theorem 1 and a straightforward reduction to the security of the underlying HIB-KEM. We sketch it in the full version.

CCA Security. In order to apply the Fujisaki-Okamoto [16] transform in the same way as done in Sect. 2.6 to achieve CCA security, we need to show that the time based variants of the properties presented in Sect. 2.3 are satisfied (for the formal definitions of those properties we refer the reader to the full version). First, using a full-blown HIBE as a starting point yields a separable HIB-KEM as discussed in Sect. 2.3. Hence, the separable randomness is satisfied. Moreover, the publicly-checkable puncturing is given by construction (as in Sect. 2.5). Regarding extended correctness, the impossibility of false-negatives is given by construction, the perfect correctness of the non-punctured secret key is given by the perfect correctness of the HIBE and the semi-correctness of punctured secret keys is given by construction. Finally, γ-spreadness is also given by construction: the ciphertext component c is chosen uniformly at random from $\{0,1\}^\lambda$. Consequently, all properties are satisfied. We note that one could omit c in the ciphertext if the concretely used HIBE ciphertexts are already sufficiently random. Considering the HIBE of Boneh-Boyen-Goh [8], HIBE ciphertexts are of the form $(g^r, (h_1^{I_1} \cdots h_t^{I_t} \cdot h_0)^r, H(e(g_1, g_2)^r) \oplus \mathsf{K})$, for honestly generated fixed group elements $g, g_1, g_2, h_0, \ldots, h_t$, universal hash function H, fixed K and fixed integers $I_1, \ldots, I_t$. Consequently, we have that the ciphertext has at least min-entropy $\log_2 p$ with p being the order of the groups. We want to mention that also many other HIBE construction satisfy the required properties, including, for example [12,17,30].

Remark on CCA Security. Alternatively to applying the FO transform to a PFSKEM satisfying the additional properties of extended correctness, separable randomness, publicly checkable puncturing and γ-spreadness to obtain CCA security, we can add another HIBE level to obtain IND-CCA security via the CHK transform [10] in the standard model, and thus to avoid random oracles if required.

3 Forward-Secret 0-RTT Key Exchange

In [20], GHJL provide a formal model for forward-secret one-pass key exchange (FSOPKE) by extending the one-pass key exchange [22] by Halevi and Krawczyk. They provide a security model for FSOPKE which requires both forward secrecy and replay protection from the FSOPKE protocol and captures unilateral authentication of the server and mutual authentication simultaneously. We recap the definition of FSOPKE with a slightly adapted correctness notion in the full version.

Construction. The construction in [20] builds on puncturable forward-secret key encapsulation (PFSKEM), and we can now directly plug our construction of time-based BFE (PFSKEM) as defined in Definition 11 into the construction of [20, Definition 12], yielding a forward-secret 0-RTT key exchange protocols with non-negligible correctness error:

$\underline{\mathsf{FSOPKE.KGen}(1^\lambda, r, \tau_{max})}$: Outputs $(\mathsf{pk}, \mathsf{sk})$ as follows: if $r = \mathsf{server}$, then obtain $(PK, SK) \leftarrow \mathsf{KGen}(1^\lambda, m, k, t)$ (for suitable choices of m, k and t) and set $\mathsf{pk} := (PK, \tau_{max})$ and $\mathsf{sk} := (SK, \tau, \tau_{max})$, for $\tau := 1$. If $r = \mathsf{client}$, then set $(\mathsf{pk}, \mathsf{sk}) := (\perp, \tau)$, for $\tau := 1$.

$\underline{\mathsf{FSOPKE.RunC}(\mathsf{sk}, \mathsf{pk})}$: Outputs $(\mathsf{sk}', \mathsf{K}, M)$ as follows: for $\mathsf{sk} = \tau$ and $\mathsf{pk} = (PK, \tau_{max})$, if $\tau > \tau_{max}$, then set $(\mathsf{sk}', \mathsf{K}, M) := (\mathsf{sk}, \perp, \perp)$, otherwise obtain $(C, \mathsf{K}) \leftarrow \mathsf{Enc}(\mathsf{pk}, \tau)$ and set $(\mathsf{sk}', \mathsf{K}, M) := (\tau + 1, \mathsf{K}, C)$.

$\underline{\mathsf{FSOPKE.RunS}(\mathsf{sk}, \mathsf{pk}, M)}$: Outputs $(\mathsf{sk}', \mathsf{K})$ as follows: for $\mathsf{sk} = (SK, \tau, \tau_{max})$ and $\mathsf{pk} = \perp$, if $SK = \perp$ or $\tau > \tau_{max}$, then set $(\mathsf{sk}', \mathsf{K}) := (\mathsf{sk}, \perp)$ and abort. Obtain $\mathsf{K} \leftarrow \mathsf{Dec}(SK, \tau, M)$. If $\mathsf{K} = \perp$, then set $(\mathsf{sk}', \mathsf{K}) = (\mathsf{sk}, \perp)$, otherwise obtain $SK' \leftarrow \mathsf{PuncCtx}(SK, \tau, M)$ and set $(\mathsf{sk}', \mathsf{K}) = ((SK', \tau, \tau_{max}), \mathsf{K})$.

$\underline{\mathsf{FSOPKE.TimeStep}(\mathsf{sk}, r)}$: Outputs sk' as follows: if $r = \mathsf{server}$, then for $\mathsf{sk} = (SK, \tau, \tau_{max})$: if $\tau \geq \tau_{max}$, then set $\mathsf{sk}' := (\perp, \tau + 1, \tau_{max})$ and abort, otherwise obtain $SK' \leftarrow \mathsf{PuncInt}(SK, \tau)$ and set $\mathsf{sk}' := (SK', \tau + 1, \tau_{max})$ and abort. If $r = \mathsf{client}$, then for $sk = \tau$, set $\mathsf{sk}' := \tau + 1$.

Correctness of the FSOPKE follows from the (extended) correctness property of the underlying PFSKEM and security guarantees hold due to [20, Theorem 2]. We state the following corollary:

Corollary 1. *When instantiated with the PFSKEM from Sect. 2.7, the above FSOPKE construction is a correct and secure FSOPKE protocol (with unilateral authentication).*

3.1 Analysis

In Table 1, we provide an overview of all existing practically instantiable approaches to construct forward-secret (time-based) PKEM with the one proposed in this paper.[3] We compare all schemes for an arbitrary number ℓ of time slots, where for sake of simplicity we assume $\ell = 2^w$ for some integer w, (corresponding to our time-based BFE/BFKEM) and only count the expensive cryptographic operations, i.e., such as group exponentiations and pairings.

Table 1. Overview of the existing approaches to PFSKEM. We denote by p the number a secret key is already punctured, and ℓ the maximum number of time slots. We consider the GHJL [20] instantiation with the BKP-HIBE of [6], the GM [19] and our instantiations with the BBG-HIBE [8], though other HIBE schemes may lead to different parameters. Finally, note that $p \leq 2^{20}$, k and m refer to the parameters in the Bloom filter, where k is some orders of magnitude smaller than λ, i.e., $k = 10$ vs. $\lambda = 128$, and $|\mathbb{G}_i|$ denotes the bitlength of an element from $\mathbb{G}_i$.

Scheme	|pk|	|sk|	$\|C\|$	Dec	PuncCtx	PuncInt
$\ell = 2^w$ time slots (PFSKEM)						
GM	$(w+5)\|\mathbb{G}_1\|$	$(2w+3p+5)\|\mathbb{G}_2\|$	$3\|\mathbb{G}_1\| + \|\mathbb{G}_T\|$	$O(p)$	$O(1)$	$O(w^2)$
GHJL	$(w+35)\|\mathbb{G}_2\|$	$\leq 3(p \cdot 2\lambda + w)\|\mathbb{G}_2\|$	$6\|\mathbb{G}_1\| + 2\|\mathbb{Z}_p\|$	$O(\lambda^2)$	$O(\lambda^2)$	$O(w^2)$
Ours	$(w+4)\|\mathbb{G}_2\|$	$(2me^{-kp/m} + w(2+w))\|\mathbb{G}_2\|$	$2\|\mathbb{G}_1\| + (4k+2)\lambda$	$O(k)$	$O(k)$	$O(w^2+m)$

To quickly summarize the schemes: The most interesting characteristic of our approach compared to previous approaches is that our scheme allows to offload all expensive operation to an offline phase, i.e., to the puncturing of time intervals. Here, in addition to the $O(w^2)$ operations which are common to all existing approaches, we have to generate a number of keys, linear in the size m of the Bloom filter. We believe that accepting this additional overhead in favor of blazing fast online puncturing and decryption operations is a viable tradeoff. For the online phase, our approach has a ciphertext size depending on k (where $k = 10$ is a reasonable choice), decryption depends on k, the secret key shrinks with increasing amount of puncturings and one does only require to securely delete secret keys during puncturing (note that all constructions have to implement a secure-delete functionality for secret keys within puncturing anyways). In contrast, decryption and puncturing in GHJL is highly inefficient and takes several seconds to minutes on decent hardware for reasonable deployment parameters as it involves a large amount of $O(\lambda^2)$ HIBE delegations and consequently expensive group operations. In the GM scheme[4], puncturing is efficient, but the size of

[3] We consider all but the PE schemes from indistinguishability obfuscation [11,13].

[4] Although GM supports an arbitrary number d of tags in a ciphertext, we consider the scheme with only using a single tag (which is actually favourable for the scheme) to be comparable to GHJL as well as our approach.

the secret key and thus cost of decryption grows in the number of puncturings p. Hence, it gets impractical very soon. More precisely, cost of decryption requires a number of pairing evaluations that depends on the number of puncturings, and can be in the order of 2^{20} for realistic deployment parameters.

4 Conclusion

In this paper we introduced the new notion of Bloom filter encryption (BFE) as a variant of puncturable encryption which tolerates a non-negligible correctness error. We presented various BFKEM constructions. The first one is a simple and very efficient construction which builds upon ideas known from the Boneh-Franklin IBE. The second one, which is presented in the full version, is a generic construction from CP-ABEs which achieves constant size ciphertexts. Furthermore, we extended the notion of BFE to the forward-secrecy setting and also presented a construction of what we call a time-based BFE (TB-BFE). This construction is based on HIBEs and in particular can be instantiated very efficiently using the Boneh-Boyen-Goh Tiny HIBE [8]. Our time-based BFKEM can directly be used to instantiate forward-secret 0-RTT key exchange (fs 0-RTT KE) as in [20].

From a practical viewpoint, our motivation stems from the observation that forward-secret 0-RTT KE requires very efficient decryption and puncturing. Our framework—for the first time—allows to realize practical forward-secret 0-RTT KE, even for larger server loads: while we only require to delete secret keys upon puncturing, puncturing in [20] requires, besides deleting secret-key components, additional computations in the order of seconds to minutes on decent hardware. Likewise, when using [19] in the forward-secret 0-RTT KE protocol given in [20], one requires computations in the order of the current number of puncturings upon decryption, while we achieve decryption to be independent of this number. Finally, we believe that BFE will find applications beyond forward-secret 0-RTT KE protocols.

Acknowledgments. This research was supported by H2020 project PRISMACLOUD, grant agreement n°644962, H2020 project CREDENTIAL, grant agreement n°653454, and the German Research Foundation (DFG), project JA 2445/2-1. We thank Kai Gellert and all anonymous reviewers for their valuable comments.

References

1. Agrawal, S., Boneh, D., Boyen, X.: Efficient lattice (H)IBE in the standard model. In: Gilbert, H. (ed.) EUROCRYPT 2010. LNCS, vol. 6110, pp. 553–572. Springer, Heidelberg (2010). https://doi.org/10.1007/978-3-642-13190-5_28
2. Barbulescu, R., Duquesne, S.: Updating key size estimations for pairings. Cryptology ePrint Archive, Report 2017/334 (2017). http://eprint.iacr.org/2017/334
3. Barreto, P.S.L.M., Lynn, B., Scott, M.: Constructing elliptic curves with prescribed embedding degrees. In: Cimato, S., Persiano, G., Galdi, C. (eds.) SCN 2002. LNCS, vol. 2576, pp. 257–267. Springer, Heidelberg (2003). https://doi.org/10.1007/3-540-36413-7_19

4. Barreto, P.S.L.M., Naehrig, M.: Pairing-friendly elliptic curves of prime order. In: Preneel, B., Tavares, S. (eds.) SAC 2005. LNCS, vol. 3897, pp. 319–331. Springer, Heidelberg (2006). https://doi.org/10.1007/11693383_22
5. Bellare, M., Rogaway, P.: Random oracles are practical: a paradigm for designing efficient protocols. In: Ashby, V. (ed.) ACM CCS 1993, Fairfax, Virginia, USA, pp. 62–73. ACM Press, 3–5 November 1993
6. Blazy, O., Kiltz, E., Pan, J.: (Hierarchical) identity-based encryption from affine message authentication. In: Garay, J.A., Gennaro, R. (eds.) CRYPTO 2014. LNCS, vol. 8616, pp. 408–425. Springer, Heidelberg (2014). https://doi.org/10.1007/978-3-662-44371-2_23
7. Bloom, B.H.: Space/time trade-offs in hash coding with allowable errors. Commun. ACM **13**(7), 422–426 (1970)
8. Boneh, D., Boyen, X., Goh, E.-J.: Hierarchical identity based encryption with constant size ciphertext. In: Cramer, R. (ed.) EUROCRYPT 2005. LNCS, vol. 3494, pp. 440–456. Springer, Heidelberg (2005). https://doi.org/10.1007/11426639_26
9. Boneh, D., Franklin, M.: Identity-based encryption from the weil pairing. In: Kilian, J. (ed.) CRYPTO 2001. LNCS, vol. 2139, pp. 213–229. Springer, Heidelberg (2001). https://doi.org/10.1007/3-540-44647-8_13
10. Canetti, R., Halevi, S., Katz, J.: A forward-secure public-key encryption scheme. In: Biham, E. (ed.) EUROCRYPT 2003. LNCS, vol. 2656, pp. 255–271. Springer, Heidelberg (2003). https://doi.org/10.1007/3-540-39200-9_16
11. Canetti, R., Raghuraman, S., Richelson, S., Vaikuntanathan, V.: Chosen-ciphertext secure fully homomorphic encryption. In: Fehr, S. (ed.) PKC 2017. LNCS, vol. 10175, pp. 213–240. Springer, Heidelberg (2017). https://doi.org/10.1007/978-3-662-54388-7_8
12. Chen, J., Wee, H.: Fully, (almost) tightly secure IBE and dual system groups. In: Canetti, R., Garay, J.A. (eds.) CRYPTO 2013. LNCS, vol. 8043, pp. 435–460. Springer, Heidelberg (2013). https://doi.org/10.1007/978-3-642-40084-1_25
13. Cohen, A., Holmgren, J., Nishimaki, R., Vaikuntanathan, V., Wichs, D.: Watermarking cryptographic capabilities. In: Wichs, D., Mansour, Y. (eds.) 48th ACM STOC, Cambridge, MA, USA, pp. 1115–1127. ACM Press, 18–21 June 2016
14. Derler, D., Krenn, S., Lorünser, T., Ramacher, S., Slamanig, D., Striecks, C.: Revisiting proxy re-encryption: forward secrecy, improved security, and applications. In: Abdalla, M., Dahab, R. (eds.) PKC 2018. LNCS, vol. 10769, pp. 219–250. Springer, Cham (2018). https://doi.org/10.1007/978-3-319-76578-5_8
15. Dierks, T.: The Transport Layer Security (TLS) Protocol Version 1.2. RFC 5246, August 2008. https://rfc-editor.org/rfc/rfc5246.txt
16. Fujisaki, E., Okamoto, T.: Secure integration of asymmetric and symmetric encryption schemes. In: Wiener, M. (ed.) CRYPTO 1999. LNCS, vol. 1666, pp. 537–554. Springer, Heidelberg (1999). https://doi.org/10.1007/3-540-48405-1_34
17. Gentry, C., Silverberg, A.: Hierarchical ID-based cryptography. In: Zheng, Y. (ed.) ASIACRYPT 2002. LNCS, vol. 2501, pp. 548–566. Springer, Heidelberg (2002). https://doi.org/10.1007/3-540-36178-2_34
18. Goyal, V., Pandey, O., Sahai, A., Waters, B.: Attribute-based encryption for fine-grained access control of encrypted data. In: Juels, A., Wright, R.N., Vimercati, S. (eds.) ACM CCS 2006, Alexandria, Virginia, USA, pp. 89–98. ACM Press, 30 October –3 November 2006. Cryptology ePrint Archive Report 2006/309
19. Green, M.D., Miers, I.: Forward secure asynchronous messaging from puncturable encryption. In: 2015 IEEE Symposium on Security and Privacy, San Jose, CA, USA, pp. 305–320. IEEE Computer Society Press, 17–21 May 2015

20. Günther, F., Hale, B., Jager, T., Lauer, S.: 0-RTT key exchange with full forward secrecy. In: Coron, J.-S., Nielsen, J.B. (eds.) EUROCRYPT 2017. LNCS, vol. 10212, pp. 519–548. Springer, Cham (2017). https://doi.org/10.1007/978-3-319-56617-7_18
21. Hale, B., Jager, T., Lauer, S., Schwenk, J.: Simple security definitions for and constructions of 0-RTT key exchange. In: Gollmann, D., Miyaji, A., Kikuchi, H. (eds.) ACNS 2017. LNCS, vol. 10355, pp. 20–38. Springer, Cham (2017). https://doi.org/10.1007/978-3-319-61204-1_2
22. Halevi, S., Krawczyk, H.: One-pass HMQV and asymmetric key-wrapping. In: Catalano, D., Fazio, N., Gennaro, R., Nicolosi, A. (eds.) PKC 2011. LNCS, vol. 6571, pp. 317–334. Springer, Heidelberg (2011). https://doi.org/10.1007/978-3-642-19379-8_20
23. Hofheinz, D., Hövelmanns, K., Kiltz, E.: A modular analysis of the fujisaki-okamoto transformation. Cryptology ePrint Archive, Report 2017/604 (2017). http://eprint.iacr.org/2017/604
24. Kim, T., Barbulescu, R.: Extended tower number field sieve: a new complexity for the medium prime case. In: Robshaw, M., Katz, J. (eds.) CRYPTO 2016. LNCS, vol. 9814, pp. 543–571. Springer, Heidelberg (2016). https://doi.org/10.1007/978-3-662-53018-4_20
25. Menezes, A., Sarkar, P., Singh, S.: Challenges with assessing the impact of NFS advances on the security of pairing-based cryptography. IACR Cryptology ePrint Archive 2016, 1102 (2016). http://eprint.iacr.org/2016/1102
26. Naor, M., Yogev, E.: Bloom filters in adversarial environments. In: Gennaro, R., Robshaw, M. (eds.) CRYPTO 2015. LNCS, vol. 9216, pp. 565–584. Springer, Heidelberg (2015). https://doi.org/10.1007/978-3-662-48000-7_28
27. Ostrovsky, R., Sahai, A., Waters, B.: Attribute-based encryption with non-monotonic access structures. In: Ning, P., di Vimercati, S.D.C., Syverson, P.F. (eds.) ACM CCS 2007, Alexandria, Virginia, USA, pp. 195–203. ACM Press, 28–31 October 2007
28. Rescorla, E.: The Transport Layer Security (TLS) Protocol Version 1.3. Internet-Draft draft-ietf-tls-tls13-20, Internet Engineering Task Force, April 2017. Work in Progress. https://datatracker.ietf.org/doc/html/draft-ietf-tls-tls13-20
29. Thomson, M., Iyengar, J.: QUIC: A UDP-Based Multiplexed and Secure Transport. Internet-Draft draft-ietf-quic-transport-02, Internet Engineering Task Force, March 2017. Work in Progress. https://datatracker.ietf.org/doc/html/draft-ietf-quic-transport-02
30. Waters, B.: Dual system encryption: realizing fully secure IBE and HIBE under simple assumptions. In: Halevi, S. (ed.) CRYPTO 2009. LNCS, vol. 5677, pp. 619–636. Springer, Heidelberg (2009). https://doi.org/10.1007/978-3-642-03356-8_36
31. Wu, D.J., Taly, A., Shankar, A., Boneh, D.: Privacy, discovery, and authentication for the internet of things. In: Askoxylakis, I., Ioannidis, S., Katsikas, S., Meadows, C. (eds.) ESORICS 2016. LNCS, vol. 9879, pp. 301–319. Springer, Cham (2016). https://doi.org/10.1007/978-3-319-45741-3_16

OPAQUE: An Asymmetric PAKE Protocol Secure Against Pre-computation Attacks

Stanislaw Jarecki[1(✉)], Hugo Krawczyk[2], and Jiayu Xu[1]

[1] University of California, Irvine, USA
stasio@ics.uci.edu, jiayux@uci.edu
[2] IBM Research, Yorktown Heights, USA
hugo@ee.technion.ac.il

Abstract. Password-Authenticated Key Exchange (PAKE) protocols allow two parties that only share a password to establish a shared key in a way that is immune to offline attacks. *Asymmetric* PAKE (aPAKE) strengthens this notion for the more common client-server setting where the server stores a mapping of the password and security is required even upon server compromise, that is, the only allowed attack in this case is an (inevitable) offline exhaustive dictionary attack against individual user passwords. Unfortunately, most suggested aPAKE protocols (that dispense with the use of servers' public keys) allow for *pre-computation attacks* that lead to the *instantaneous compromise* of user passwords upon server compromise, thus forgoing much of the intended aPAKE security. Indeed, these protocols use – in essential ways – deterministic password mappings or use random "salt" transmitted *in the clear* from servers to users, and thus are vulnerable to pre-computation attacks.

We initiate the study of *Strong aPAKE* protocols that are secure as aPAKE's but *are also secure against pre-computation attacks.* We formalize this notion in the Universally Composable (UC) settings and present two modular constructions using an Oblivious PRF as a main tool. The first builds a Strong aPAKE from *any* aPAKE (which in turn can be constructed from any PAKE [18]) while the second builds a Strong aPAKE from *any* authenticated key-exchange protocol secure against reverse impersonation (a.k.a. KCI). Using the latter transformation, we show *a practical instantiation of a UC-secure Strong aPAKE* in the Random Oracle model. The protocol ("OPAQUE") consists of 2 messages (3 with mutual authentication), requires 3 and 4 exponentiations for server and client, respectively (2 to 4 of which can be fixed-base depending on optimizations), provides forward secrecy, is PKI-free, supports user-side hash iterations, and allows a user-transparent server-side threshold implementation.

1 Introduction

Passwords constitute the most ubiquitous form of authentication in the Internet, from the most mundane to the most sensitive applications. The almost universal

Extended version available from IACR Cryptology ePrint Archive 2018:163.

J. B. Nielsen and V. Rijmen (Eds.): EUROCRYPT 2018, LNCS 10822, pp. 456–486, 2018.
https://doi.org/10.1007/978-3-319-78372-7_15

password authentication method in practice relies on TLS/SSL and consists of the user sending its password to the server under the protection of a client-to-server confidential TLS channel. At the server, the password is decrypted and verified against a one-way image typically computed via hash iterations applied to the password and a random "salt" value. Both the password image and salt are stored for each user in a so-called "password file." In this way, an attacker who succeeds in stealing the password file is forced to run an exhaustive *offline dictionary attack* to find users' passwords given a set ("dictionary") of candidate passwords. The two obvious disadvantages of this approach are: (i) the password appears in cleartext at the server during login; and (ii) security breaks if the TLS channel is established with a compromised server's public key (a major concern given today's too-common PKI failures[1]).

Password protocols have been extensively studied in the crypto literature – including in the above client-server setting where the user is assumed to possess an authentic copy of the server's public key [19,20], but the main focus has been on *password-only* protocols where the user does not need to rely on any outside keying material (such as public keys). The basic setting considers two parties that share the same low-entropy password with the goal of establishing shared session keys secure against *offline dictionary attacks*, namely, against an active attacker that possesses a small dictionary from which the password has been chosen. The only viable option for the attacker should be the inevitable *online* impersonation attack with guessed passwords. Such model, known as *password-authenticated key exchange (PAKE)*, was first studied by Bellovin and Merritt [5] and later formalized by Bellare et al. [4] in the game-based indistinguishability approach. Canetti et al. [12] formalized PAKE in the Universally Composable (UC) framework [11], which better captures PAKE security issues such as the use of arbitrary password distributions, the inputting of wrong passwords by the user, and the common use in practice of related passwords for different services.

Whereas the cryptographic literature on PAKE's focuses on the above basic setting, in practice the much more common application of password protocols is in the client-server setting. However, sharing the same password between user and server would mean that a break to the server leaks plaintext passwords for all its users. Thus, what's needed is that upon a server compromise, and the stealing of the password file, an attacker is forced to perform an exhaustive offline dictionary attack as in the above TLS scenario. No other attack, except for an inevitable online guessing attack, should be feasible. In particular, *the two main shortcomings of password-over-TLS* mentioned earlier - reliance on public keys and exposure of the password to the server - *need to be eliminated.* This setting, known as *aPAKE*, for *asymmetric PAKE* (also called *augmented* or *verifier-based*), was introduced by Bellovin and Merrit [6], later formalized in the simulation-based approach by Boyko et al. [10], and in the UC framework by

[1] PKI failures include stealing of server private keys, software that does not verify certificates correctly, users that accept invalid or suspicious certificates, certificates issued by rogue CAs, rogue CAs accepted as roots of trust, servers that share their TLS keys with others, e.g. CDN providers or security monitoring software; and more.

Gentry et al. [18]. Early protocols proven in the simulation-based model include [10,28,29]. Later, Gentry et al. [18] presented a compiler that transforms any UC-PAKE protocol into a UC-aPAKE (adding an extra round of communication and a client's signature). This was followed by [24] who show the first simultaneous one-round adaptive UC-aPAKE protocol. In addition, several aPAKE protocols targeting practicality have been proposed, most with ad-hoc security arguments, and some have been (and are being) considered for standardization (see below).

A common unfortunate property of all these aPAKE protocols, including those being proposed for practical use and regardless of their underlying formalism, is that they are *all vulnerable to pre-computation attacks.* Namely, the attacker $\mathcal{A}$ can *pre-compute* a table of values based on a passwords dictionary D, so as soon as $\mathcal{A}$ succeeds in compromising a server it can *instantly* find a user's password. This significantly weakens the benefits of security against server compromise that motivate the aPAKE notion in the first place. Moreover, while current definitions require that the attacker cannot exploit a server compromise without incurring a workload proportional to the dictionary size $|D|$, these definitions allow all this workload to be spent *before* the actual server compromise happens. Indeed, this weakening in the existing aPAKE security definition [18] is needed to accommodate aPAKE protocols that store a one-way *deterministic* mapping of the user's password at the server, say $H(\mathsf{pw})$. Such protocols trivially fall to a pre-computation attack as the attacker $\mathcal{A}$ can build a table of $(H(\mathsf{pw}), \mathsf{pw})$ pairs for all $\mathsf{pw} \in D$, and once it compromises the server, it finds the value $H(\mathsf{pw})$ associated with a user and immediately, in $\log(|D|)$ time, finds that user's password. Such devastating attack can be mitigated by "personalizing" the password map, e.g., hashing the password together with the user id. This forces $\mathcal{A}$ to pre-compute separate tables for individual users, yet all this effort can still be spent before the actual server compromise. Note that in the case of passwords transmitted over TLS, pre-computation is prevented since password are hashed with a random salt visible to the server only. In contrast, existing aPAKE protocols that do not rely on PKI, either don't use salt or if they do, the salt is transmitted from server to user during login *in the clear*[2]. Given that password stealing via server compromise is the main avenue for collecting billions of passwords by attackers, the above vulnerability of existing aPAKE protocols to pre-computation attacks is a serious flaw, and in this aspect password-over-TLS is more secure than all known aPAKE schemes.

Our Contributions

We initiate the study of *Strong aPAKE (SaPAKE)* protocols that strengthen the aPAKE security notion by *disallowing pre-computation attacks.* We formalize this notion in the Universally Composable (UC) model by modifying the aPAKE functionality from [18] to eliminate an adversarial action which allowed such pre-computation attacks. As we explain above, allowing pre-computation attacks was indeed necessary to model the security of existing aPAKE protocols.

[2] While aPAKE protocols are not intended to run over TLS, we point out that even in such a case, the transmitted salt would be open to a straightforward active attack.

The next contribution is building Strong aPAKE (SaPAKE) protocols. For this we present two generic constructions. The first builds the SaPAKE protocol from any aPAKE protocol (namely one that satisfies the original definition from [18]) so that one can "salvage" existing aPAKE protocols. To do so we resort to Oblivious PRF (OPRF) functions [17,22], namely, a PRF with an associated two-party protocol that in our case is run between a server S that stores a PRF key k and a user U with a password pw. At the end of the interaction, U learns the PRF output $F_k(\mathsf{pw})$ and S learns nothing (in particular, nothing about pw). We show that by preceding any aPAKE protocol with an OPRF interaction in which U computes the value $\mathsf{rw} = F_k(\mathsf{pw})$ with the help of S and uses rw as the password in the aPAKE protocol, one obtains a Strong aPAKE protocol. We show that if the OPRF and the given aPAKE protocol are, respectively, UC realizations of the OPRF functionality (defined in [22]) and the original aPAKE functionality from [18], the resultant scheme realizes our UC functionality $\mathcal{F}_{\mathsf{SaPAKE}}$.

Our second transformation consists of the composition of an OPRF as above with a regular authenticated key exchange protocol AKE. We require UC security for the AKE protocol as well as a property known as resistance to KCI attacks. The latter means that an attacker that learns the secret keys of one party P, but does not actively control P, cannot use this information to impersonate another party P' to P. KCI resistance is a common property of most AKE protocols. In our SaPAKE construction, U first runs the OPRF with S to compute $\mathsf{rw} = F_k(\mathsf{pw})$; then it runs the AKE protocol with S using a private key stored, encrypted using an *authenticated* encryption under rw, at S who sends it to U. Crucial to the security of the protocol is the use of authenticated encryption with a "random-key robustness" property, which is achieved naturally by some schemes or otherwise can be easily ensured, e.g., by adding an HMAC to a symmetric encryption scheme. Under these conditions we show that the composed scheme realizes our UC functionality $\mathcal{F}_{\mathsf{SaPAKE}}$.

Next, we use the above second transformation to instantiate a Strong aPAKE protocol with a very efficient OPRF and any efficient AKE with the KCI property. The OPRF scheme we use, essentially a Chaum-type blinded DH computation, has been proven UC-secure by Jarecki et al. [21,22]. We show that this OPRF scheme, which we call DH-OPRF (called 2HashDH in [21,22]), remains secure in spite of changes to the OPRF functionality that we introduce for supporting a stronger OPRF notion needed in our setting. We call the result of this instantiation, the *OPAQUE protocol.*

OPAQUE combines the best properties of existing aPAKE protocols and of the standard password-over-TLS. As any aPAKE-secure protocol, it offers two fundamental advantages over the TLS-based solution: It does not rely on PKI and the plaintext password is never in the clear at the server. The only way for an attacker that observes (or actively controls) a session at a server to learn the password is via an exhaustive offline dictionary attack. Watching or participating in a session with the user does not help the attacker. At the same time, OPAQUE resolves the major flaw of existing aPAKE protocols relative to password-over-TLS, namely, their vulnerability to pre-computation attacks.

In addition to the above fundamental properties, OPAQUE enjoys important properties for use in practice. Its modularity allows for its use with different key-exchange schemes that can provide different features and performance tradeoffs. When implemented with a 2-message implicit-authentication KE protocol (e.g., HMQV [27]), OPAQUE takes only 2 messages (or 3 with mutual explicit authentication). The computational cost (using the DH-OPRF scheme from Appendix A) is one exponentiation for the server and two for the client[3] in addition to the KE protocol cost (with HMQV, this cost is 2.17 exponentiations per party). OPAQUE offers forward secrecy (a particularly crucial property for password protocols) if the KE does. OPAQUE further supports password hardening for increasing the cost of offline dictionary attacks (upon server compromise) through user-side iterated hashing without the need to transmit salt from S to U. In Fig. 7 in Sect. 6 we show an instantiation of OPAQUE in the RO model with HMQV as the AKE.

Compared to the practical aPAKE protocols that have been and are being considered for standardization (cf., [1,32]), OPAQUE fares clearly better on the security side as the only scheme that offers resistance to pre-computation attacks while all others are vulnerable. Performance-wise, OPAQUE is competitive with the more efficient among these protocols (see Sect. 6). Additional advantages of OPAQUE include its ability to store and retrieve user's secrets such as a bitcoin wallet, authentication credentials, encrypted backup keys, etc., and to support a user-transparent server-side threshold implementation [23] (where the only exposure of the user password - or any stored secrets - is in case a threshold of servers is compromised and even then a full dictionary attack is required). Finally, we comment that while OPAQUE can completely replace password authentication in TLS, it can also be used in conjunction with TLS, either for bootstrapping client authentication (via an OPAQUE-retrieved client signing key) or as an hedge against PKI failures. In other words, while we are accustomed to use TLS to protect passwords, OPAQUE can be used to protect TLS.

We stress that variants of OPAQUE have been studied in prior work in several settings but none of these works presents a formal analysis of the protocol as an aPAKE, let alone as a Strong aPAKE, a notion that we introduce here for the first time. While our treatment frames OPAQUE in the context of Oblivious PRFs [21,22], its design can be seen as an instantiation of the Ford-Kaliski paradigm for password hardening and credential retrieval using Chaum's blinded exponentiation. Boyen [9] specifies and studies the protocol (called HPAKE) in the setting of client-side *halting KDF* [8]. Jarecki et al. [21,22] study a threshold version (also using the OPRF abstraction) in the context of *password-protected secret sharing (PPSS)* protocols. Because of the relation between PPSS and Threshold PAKE protocols [21], this analysis implies security of OPAQUE as a PAKE protocol in the BPR model [4] but not as an aPAKE (let alone as a strong aPAKE).

[3] A variant of the protocol discussed in Sect. 6.2 allows one or both of the client's exponentiations to be fixed-base and offline.

2 The Strong aPAKE Functionality

We present the ideal UC Strong aPAKE functionality, $\mathcal{F}_{\mathsf{SaPAKE}}$, that will serve as our definition of Strong aPAKE security; namely, we call a protocol a secure *Strong aPAKE* if it realizes $\mathcal{F}_{\mathsf{SaPAKE}}$. Functionality $\mathcal{F}_{\mathsf{SaPAKE}}$ is a simple but significant variant of the UC aPAKE functionality $\mathcal{F}_{\mathsf{aPAKE}}$ from [18] (it was denoted $\mathcal{F}_{\mathsf{apwKE}}$ in [18]) which we recall in Fig. 1.

The aPAKE functionality of [18] is based on the UC PAKE functionality from [12], and it includes extensions needed for taking care of the asymmetric nature of the aPAKE setting. First, in an aPAKE scheme the server and the user run different programs: The user runs an aPAKE session on a password (via command UsrSession) while the server runs it on a "password file" $\mathsf{file}[sid]$ that represents server's user-specific state corresponding to the user's password, e.g., a password hash, which the server creates on input the user's password during aPAKE initialization, via command StorePwdFile. Furthermore, $\mathcal{F}_{\mathsf{aPAKE}}$ models a possible compromise of a server, via command StealPwdFile, from which the attacker obtains $\mathsf{file}[sid]$. Such compromise subsequently allows the attacker to (1) impersonate the server to the user, via command Impersonate, and (2) find the password via an offline dictionary attack, via command OfflineTestPwd. The way functionality $\mathcal{F}_{\mathsf{aPAKE}}$ of [18] handles the offline dictionary attack is the focus of the Strong aPAKE functionality we propose, and we discuss them below.

Strong aPAKE vs. aPAKE. Our functionality $\mathcal{F}_{\mathsf{SaPAKE}}$ is almost identical to $\mathcal{F}_{\mathsf{aPAKE}}$ except that the text with the gray background in Fig. 1 is omitted. That is, the only difference between $\mathcal{F}_{\mathsf{SaPAKE}}$ and $\mathcal{F}_{\mathsf{aPAKE}}$ are in the actions upon the stealing of the password file; specifically, $\mathcal{F}_{\mathsf{SaPAKE}}$ omits recording the (offline, pw) pairs and does not allow for OfflineTestPwd queries made before the StealPwdFile query. Let us explain. Let's consider first the definition of $\mathcal{F}_{\mathsf{SaPAKE}}$, i.e., with the gray text omitted. In this case, the actions upon server compromise, i.e., StealPwdFile, are simple. First, a flag is defined to mark that the password file has been compromised. Second, once this event happens, the adversary is allowed to submit password guesses and be informed if a guess was correct. Note that each guess "costs" the attacker one OfflineTestPwd query. This together with the restriction that these queries can only be made after the password file is compromised ensure that shortcuts in finding the password after such compromise are not possible, namely that the attacker needs to pay with one OfflineTestPwd query for each password it wants to test. Thus, pre-computation attacks are made infeasible.

Now, consider the $\mathcal{F}_{\mathsf{aPAKE}}$ functionality from [18] which includes the text in gray too. This functionality allows the attacker, via (offline, pw) records, to make guess queries against the password even before the password file is compromised. The restriction is that the responses to whether a guess was correct or not are provided to the attacker only after a StealPwdFile event. But note that if one of these guesses was correct, the attacker learns it *immediately* upon server compromise. This provision was necessary in [18] because the $\mathsf{file}[sid]$ in their aPAKE construction contains a deterministic publicly-computable hash of the

In the description below, we assume $\mathsf{P} \in \{\mathsf{U}, \mathsf{S}\}$.

Password Registration

- On (StorePwdFile, *sid*, U, pw) from S, if this is the first StorePwdFile message, record $\langle$file, U, S, pw$\rangle$ and mark it uncompromised.

Stealing Password Data

- On (StealPwdFile, *sid*) from $\mathcal{A}^*$, if there is no record $\langle$file, U, S, pw$\rangle$, return "no password file" to $\mathcal{A}^*$. Otherwise, if the record is marked uncompromised, mark it compromised; regardless,
  - If there is a record $\langle$offline, pw$\rangle$, send pw to $\mathcal{A}^*$.
  - Else Return "password file stolen" to $\mathcal{A}^*$.
- On (OfflineTestPwd, *sid*, pw*) from $\mathcal{A}^*$, do:
  - If there is a record $\langle$file, U, S, pw$\rangle$ marked compromised, do: if pw* = pw, return "correct guess" to $\mathcal{A}^*$; else return "wrong guess."
  - Else record $\langle$offline, pw$\rangle$.

Password Authentication

- On (UsrSession, *sid*, *ssid*, S, pw′) from U, send (UsrSession, *sid*, *ssid*, U, S) to $\mathcal{A}^*$. Also, if this is the first UsrSession message for *ssid*, record $\langle$*ssid*, U, S, pw′$\rangle$ and mark it fresh.
- On (SvrSession, *sid*, *ssid*) from S, retrieve $\langle$file, U, S, pw$\rangle$, and send (SvrSession, *sid*, *ssid*, U, S) to $\mathcal{A}^*$. Also, if this is the first SvrSession message for *ssid*, record $\langle$*ssid*, S, U, pw$\rangle$ and mark it fresh.

Active Session Attacks

- On (TestPwd, *sid*, *ssid*, P, pw*) from $\mathcal{A}^*$, if there is a record $\langle$*ssid*, P, P', pw′$\rangle$ marked fresh, do: if pw* = pw′, mark it compromised and return "correct guess" to $\mathcal{A}^*$; else mark it interrupted and return "wrong guess."
- On (Impersonate, *sid*, *ssid*) from $\mathcal{A}^*$, if there is a record $\langle$*ssid*, U, S, pw′$\rangle$ marked fresh, do: if there is a record $\langle$file, U, S, pw$\rangle$ marked compromised and pw′ = pw, mark $\langle$*ssid*, U, S, pw′$\rangle$ compromised and return "correct guess" to $\mathcal{A}^*$; else mark it interrupted and return "wrong guess."

Key Generation and Authentication

- On (NewKey, *sid*, *ssid*, P, SK) from $\mathcal{A}^*$ where $|SK| = \tau$, if there is a record $\langle$*ssid*, P, P', pw′$\rangle$ not marked completed, do:
  - If the record is marked compromised, or either P or P' is corrupted, send (*sid*, *ssid*, SK) to P.
  - If the record is marked fresh, a (*sid*, *ssid*, SK') tuple was sent to P', and at that time there was a record $\langle$*ssid*, P', P, pw′$\rangle$ marked fresh, send (*sid*, *ssid*, SK') to P.
  - Else pick $SK'' \leftarrow_{\mathrm{R}} \{0,1\}^\tau$ and send (*sid*, *ssid*, SK'') to P.

  Finally, mark $\langle$*ssid*, P, P', pw′$\rangle$ completed.
- On (TestAbort, *sid*, *ssid*, P) from $\mathcal{A}^*$, if there is a record $\langle$*ssid*, P, P', pw′$\rangle$ not marked completed, do:
  - If it is marked fresh and record $\langle$*ssid*, P', P, pw′$\rangle$ exists, send Succ to $\mathcal{A}^*$.
  - Else send Fail to $\mathcal{A}^*$ and (abort, *sid*, *ssid*) to P, and mark $\langle$*ssid*, P, P', pw′$\rangle$ completed.

7

Fig. 1. Functionalities $\mathcal{F}_{\mathsf{aPAKE}}$ (full text) and $\mathcal{F}_{\mathsf{SaPAKE}}$ (shadowed text omitted)

password, thus allowing for a pre-computation attack which lets the adversary instantaneously identify the password with a single table lookup upon server compromise. Indeed, one can think of the pairs (OFFLINE, pw) in the original $\mathcal{F}_{\mathsf{aPAKE}}$ functionality as a pre-computed table that the attacker builds overtime and which it can use to identify the password as soon as the server is compromised. By eliminating the ability to get guesses (OFFLINE, pw) answered before server compromise in our $\mathcal{F}_{\mathsf{SaPAKE}}$ functionality, we make such pre-computation attacks infeasible in the case of a Strong aPAKE.

Modeling Server Compromise and Offline Dictionary Queries. As in [18], we specify that STEALPWDFILE and OFFLINETESTPWD messages from $\mathcal{A}^*$ to $\mathcal{F}_{\mathsf{SaPAKE}}$ are accounted for by the environment. This is consistent with the UC treatment of adaptive corruption queries and is crucial to our modeling. Note that if the environment does not observe adaptive corruption queries then the ideal model adversary, i.e., the simulator, could immediately corrupt all parties at the beginning of the protocol, learning their private inputs and thus making the work of simulation easier. By making the player-corruption queries, modeled by STEALPWDFILE command in our context, observable by the environment, we ensure that the environment's view of both the ideal and the real execution includes the same player-corruption events. This way we keep the simulator "honest," because it can only corrupt a party if the environment accounts for it.

The same concern pertains to offline dictionary queries OFFLINETESTPWD, because if they were not observable by the environment, the ideal adversary could make such queries even if the real adversary does not. In particular, without environmental accounting for these queries the $\mathcal{F}_{\mathsf{aPAKE}}$ and $\mathcal{F}_{\mathsf{SaPAKE}}$ functionalities would be equivalent because the simulator could internally gather all the offline dictionary attack queries made by the real-world adversary before server corruption, and it would send them all via the OFFLINETESTPWD query to $\mathcal{F}_{\mathsf{SaPAKE}}$ after server corruption via the STEALPWDFILE query. Such simulator would make the ideal-world view indistinguishable from the real-world view to the environment *if* the environment does not observe the sequence of OFFLINETESTPWD and STEALPWDFILE queries.

Finally, we note that the functionality $\mathcal{F}_{\mathsf{SaPAKE}}$, like $\mathcal{F}_{\mathsf{aPAKE}}$, has effectively two separate notions of a server corruption. Formally, it considers a *static* adversarial model where all entities, including users and servers, are either honest or corrupt throughout the life-time of the scheme. In addition, it allows for an *adaptive* server compromise of an honest server, via the STEALPWDFILE, which leaks to the adversary the server's private state corresponding to a particular password file, but it does not give the adversary full control over the server's entity. In particular, the accounts on the same server for which the adversary does not explicitly issue the STEALPWDFILE command must remain unaffected. We adopt this convention from [18] and we call a server "corrupted" if it is (statically) corrupt and adversarially controlled, and we call an aPAKE instance "compromised" if the adversary steals its password file from the server.

Non-black-box Assumptions. Note that the aPAKE functionality requires the simulator, playing the role of the ideal-model adversary, to detect offline password guesses made by the real-world adversary. As pointed out by [18], this seems to require a non-black-box hardness assumption on some cryptographic primitive, e.g., the Random Oracle Model (ROM), which would allow the simulator to extract a password guess from adversary's *local* computation, e.g., a local execution of aPAKE interaction on a password guess and a stolen password file.

Server Initialization. We note that while $\mathcal{F}_{\mathsf{aPAKE}}$ defines password registration as an internal action of server S, with the user's password as a local input, one can modify it to support an interactive procedure between user and server, e.g., to prevent S from ever learning the plaintext password. To that end one needs to assume that during the Password Registration phase there is an *authenticated channel* from server to user, so the user can verify that it is registering the password with the correct server. (Functionality $\mathcal{F}_{\mathsf{aPAKE}}$ effectively also assumes such authenticated channel because otherwise the user's password cannot be safely transported to S.) In practice, the server also needs to verify the user's identity, and the password file could be created by the user and transported to the server. However, this is beyond the scope of the formal aPAKE functionality.

Public Parameters: PRF output-length ℓ, polynomial in security parameter τ.

Convention on F and tx: For every (sid, S) value $\mathsf{tx}(sid, \mathsf{S})$ is initially set to 0, and for every x value $F_{sid,\mathsf{S}}(x)$ is initially undefined. Whenever $F_{sid,\mathsf{S}}(x)$ is referenced below and it is undefined, pick $F_{sid,\mathsf{S}}(x) \leftarrow_{\mathrm{R}} \{0,1\}^{\ell}$.

Initialization

- On $(\textsc{Init}, sid, x)$ from server S, send $(\textsc{Init}, sid, F_{sid,\mathsf{S}}(x))$ to S.

Server Compromise

- On $(\textsc{Compromise}, sid)$ from $\mathcal{A}^*$ for a server S, mark S COMPROMISED.
- On $(\textsc{OfflineEval}, sid, \mathsf{S}, x)$ from $\mathcal{A}^*$, if S is corrupted or marked COMPROMISED, send $(\textsc{OfflineEval}, sid, F_{sid,\mathsf{S}}(x))$ to $\mathcal{A}^*$.

Evaluation

- On $(\textsc{Eval}, sid, ssid, \mathsf{S}, x)$ from party $\mathsf{P} \in \{\mathsf{U}, \mathcal{A}^*\}$, record $\langle ssid, \mathsf{P}, \mathsf{S}, x\rangle$ and send $(\textsc{Eval}, sid, ssid, \mathsf{P}, \mathsf{S})$ to $\mathcal{A}^*$.
- On $(\textsc{SndrComplete}, sid, ssid)$ from S, send $(\textsc{SndrComplete}, sid, ssid, \mathsf{S})$ to $\mathcal{A}^*$. If this is the first SNDRCOMPLETE message for $ssid$, set $\mathsf{tx}(sid, \mathsf{S})$++.
- On $(\textsc{RcvComplete}, sid, ssid, \mathsf{S}^*)$ from $\mathcal{A}^*$, retrieve $\langle ssid, \mathsf{P}, \mathsf{S}, x\rangle$ (where $\mathsf{P} \in \{\mathsf{U}, \mathcal{A}^*\}$); abort if (i) such record does not exist, or (ii) S is honest and not marked COMPROMISED and $\mathsf{S}^* \neq \mathsf{S}$, or (iii) $\mathsf{tx}(sid, \mathsf{S}^*) = 0$. Otherwise set $\mathsf{tx}(sid, \mathsf{S}^*)$−− and send $(\textsc{Eval}, sid, ssid, F_{sid,\mathsf{S}^*}(x))$ to P.

Fig. 2. Functionality $\mathcal{F}_{\mathsf{OPRF}}$ with adaptive compromise

3 Oblivious Pseudorandom Function

Oblivious Pseudorandom Functions (OPRF) are a central tool in all our constructions. An OPRF consists of a pseudorandom function family F with an associated two-party protocol run between a server that holds a key k for F and a user with an input x. At the end of the interaction, the user learns the PRF output $F_k(x)$ and nothing else, and the server learns nothing (in particular, nothing about x). The notion of OPRF was introduced in [17]. The first UC formulation of it was given in [21], including a verifiability property that lets the user check the correct behavior of the server during the OPRF execution. Later [22] gave an alternative UC definition of OPRF which dispensed with the verifiability property, allowing for more efficient instantiations. The main idea in the OPRF formulations of [21,22] is the use of a *ticketing mechanism* $\mathsf{tx}(\cdot)$ that ensures that the number of input values on which anyone can compute the OPRF on a key held by an honest server S is no more than the number of executions of the OPRF recorded by S. This mechanism dispenses with the need to extract users' inputs as is typically needed in UC simulations and it leads to much more efficient OPRF instantiations.

Here we adopt the formulation from [22] as the basis for our definition of functionality $\mathcal{F}_{\mathsf{OPRF}}$ presented in Fig. 2. We refer to [22] for detailed rationale, but we note that it requires PRF outputs to be pseudorandom even to the owner of the PRF key k. This does not seem achievable under non-black-box assumptions, but it is achievable, indeed very efficiently, in the Random Oracle Model (ROM). Note that the reliance on non-black-box assumptions like ROM is called for in the aPAKE context, see Sect. 2.

Changes from OPRF Functionality of [22]**.** To use UC OPRF in our application(s) we need to make some changes to the way functionality $\mathcal{F}_{\mathsf{OPRF}}$ was defined in [22], as described below. Changes (2), (3) and (4) are essentially syntactic and require only cosmetic changes in the security argument. Change (1) is the only one which influences the security argument in a more essential way. Fortunately, the DH-OPRF protocol that we use for OPRF instantiation in our protocols, shown in [22] to realize their version of the OPRF functionality $\mathcal{F}_{\mathsf{OPRF}}$, also realizes our modified $\mathcal{F}_{\mathsf{OPRF}}$ functionality. We recall the DH-OPRF protocol in Fig. 9 in Appendix A, adapting its syntax to our changes in $\mathcal{F}_{\mathsf{OPRF}}$, and we argue that the security proof of [22] which shows that it realizes $\mathcal{F}_{\mathsf{OPRF}}$ defined by [22] extends to the modified functionality $\mathcal{F}_{\mathsf{OPRF}}$ presented here.

(1) We extend the OPRF functionality to allow the *adaptive compromise* of a server holding the PRF key via a COMPROMISE message. Such action is needed in the aPAKE setting where the attacker $\mathcal{A}^*$ can compromise a server's password file that contains the server's OPRF key. After the compromise, $\mathcal{A}^*$ is allowed to compute that server's PRF function by itself on any value of its choice using OFFLINEEVAL and without the restrictions of the ticketing mechanism. We note that functionality $\mathcal{F}_{\mathsf{OPRF}}$ distinguishes between (statically) *corrupted servers* and (adaptively) *compromised sessions* (the latter representing different OPRF keys at the same server).

This distinction allows for a granular separation between compromised and uncompromised OPRF keys held by the same server. We adopt this distinction for consistency with the aPAKE functionality from Fig. 1 that distinguishes between an entirely corrupted server and particular aPAKE instances that can be adaptively compromised by an adversary.

(2) We change the SNDRCOMPLETE message such that it is sent from S instead of $\mathcal{A}$, thus restricting the number of OPRF invocations per *ssid* to one. This enforces a single password guess per aPAKE sub-session which is crucial for aPAKE security.

(3) We change the session-id syntax used in [22] to model the use of multiple OPRF keys by the same server. In the formulation of [22] each PRF key was identified with a server identity making a one-to-one correspondence between OPRF keys and servers. Here, we allow multiple OPRF keys to be associated with one server. Each such key is identified with a tag *sid* and a server can be associated with multiple such tags. In the context of our application to aPAKE protocols, each aPAKE session is associated with a unique OPRF key used by the server for a particular user, so the session-id *sid* corresponds to a user account at that server. Any *sid* can include *sub-sessions*, denoted by *ssid*, corresponding to different runs of the OPRF protocol between a user and a server.

(4) We add an Initialization phase to the functionality, which models a server picking an OPRF key and, in addition, computing the OPRF value on any input. This interface simplifies the usage of OPRF in our applications to aPAKE, where the server will pick an OPRF key for a new user and evaluate the OPRF on the user's password (for generating an encryption key). This modeling differs from [22] who framed OPRF initialization as an interactive procedure through an EVAL call while here it is performed locally by the server.

4 A Compiler from aPAKE to Strong aPAKE via OPRF

In Fig. 3 we specify a compiler that transforms any OPRF and any aPAKE into a Strong aPAKE protocol. In UC terms the Strong aPAKE protocol is defined in the $(\mathcal{F}_{\mathsf{OPRF}}, \mathcal{F}_{\mathsf{aPAKE}})$-hybrid world, for $\mathcal{F}_{\mathsf{OPRF}}$ with the output length parameter $\ell = 2\tau$. The compiler is simple. First, the user transforms its password pw into a randomized value rw by interacting with the server in an OPRF protocol where the user inputs pw and the server inputs the OPRF key. Nothing is learned at the server about pw (i.e., rw is indistinguishable from random as long as the input pw is not queried as input to the OPRF). Next, the user sets rw as its password in the given aPAKE protocol. Note that since the password rw is taken from a pseudorandom set, then even if the size of this set is the same as the original dictionary D from which pw was taken, the pseudorandom set is unknown to the attacker (the attacker can only learn this set via OPRF queries which require an online dictionary attack). Thus, any previous ability to run a pre-computation attack against the aPAKE protocol based on dictionary D is now lost.

We assume that $\mathcal{A}$ always simultaneously sends queries (COMPROMISE, *sid*) and (STEALPWDFILE, *sid*) for the same *sid*, resp. to $\mathcal{F}_{\mathsf{OPRF}}$ to $\mathcal{F}_{\mathsf{aPAKE}}$, because in any instantiation of this scheme the server's OPRF-related state and aPAKE-related state would be part of the same file[*sid*]. Consequently, for a single *sid*, S's status (COMPROMISED or not) in $\mathcal{F}_{\mathsf{OPRF}}$ and $\mathcal{F}_{\mathsf{aPAKE}}$ is always the same.

Password Registration

1. On input (STOREPWDFILE, *sid*, U, pw), S sends (INIT, *sid*, pw) to $\mathcal{F}_{\mathsf{OPRF}}$. On $\mathcal{F}_{\mathsf{OPRF}}$'s response (INIT, *sid*, rw), S sends (STOREPWDFILE, *sid*, U, rw) to $\mathcal{F}_{\mathsf{aPAKE}}$.

Password Authentication and Key Generation

1. On input (USRSESSION, *sid*, *ssid*, S, pw′), U sends (EVAL, *sid*, *ssid*, S, pw′) to $\mathcal{F}_{\mathsf{OPRF}}$. On $\mathcal{F}_{\mathsf{OPRF}}$'s response (EVAL, *sid*, *ssid*, rw′), U sends (USRSESSION, *sid*, *ssid*, S, rw′) to $\mathcal{F}_{\mathsf{aPAKE}}$.
2. On input (SVRSESSION, *sid*, *ssid*), S sends (SNDRCOMPLETE, *sid*, *ssid*) to $\mathcal{F}_{\mathsf{OPRF}}$ and (SVRSESSION, *sid*, *ssid*) to $\mathcal{F}_{\mathsf{aPAKE}}$.
3. On (*sid*, *ssid*, *SK*) or (ABORT, *sid*, *ssid*) from $\mathcal{F}_{\mathsf{aPAKE}}$, the recipient, either U or S, outputs this message.

Fig. 3. Strong aPAKE protocol in the ($\mathcal{F}_{\mathsf{OPRF}}, \mathcal{F}_{\mathsf{aPAKE}}$)-hybrid world

4.1 Proof of Security

Theorem 1. *The protocol in Fig. 3 UC-realizes the* $\mathcal{F}_{\mathsf{SaPAKE}}$ *functionality assuming access to the OPRF functionality* $\mathcal{F}_{\mathsf{OPRF}}$ *and aPAKE functionality* $\mathcal{F}_{\mathsf{aPAKE}}$*.*

Concretely, for any adversary $\mathcal{A}$ *against the protocol, there is a simulator* SIM *that produces a view in the simulated ideal world (henceforth simulated world) such that the advantage that an environment has in distinguishing between this view and the view in the* ($\mathcal{F}_{\mathsf{OPRF}}, \mathcal{F}_{\mathsf{aPAKE}}$)*-hybrid real world (henceforth real world) is no more than* $(q_F^2 + 2q_O + 6)/2^{2\tau+1}$*, where* τ *is the security parameter,* q_F *is the number of* EVAL *and* OFFLINEEVAL *messages aimed at* $\mathcal{F}_{\mathsf{OPRF}}$ *from* $\mathcal{A}$*, and* q_O *is the number of* OFFLINETESTPWD *messages aimed at* $\mathcal{F}_{\mathsf{aPAKE}}$ *from* $\mathcal{A}$*. (In the real world,* $\mathcal{A}$ *sends the messages to* $\mathcal{F}_{\mathsf{OPRF}}$ *and* $\mathcal{F}_{\mathsf{aPAKE}}$*. In the simulated world,* $\mathcal{A}$ *sends the messages to* SIM *acting as both* $\mathcal{F}_{\mathsf{OPRF}}$ *and* $\mathcal{F}_{\mathsf{aPAKE}}$*.)*

Due to lack of space, we leave the proof to the full version of this paper.

5 A Compiler from AKE-KCI to Strong aPAKE via OPRF

Our second transformation for building a Strong aPAKE protocol composes an OPRF with an Authenticated Key Exchange (AKE) protocol, "glued" together using authenticated encryption. We require the AKE to be secure in the UC model, namely, to realize the UC KE functionality of [14], but we also require it to be "KCI secure," a property which we call here "security against reverse impersonation." The notion of AKE-KCI security has been formalized with a game-based approach in [27], but to the best of our knowledge it was not formalized in UC setting, and we present such formalization in Sect. 5.1.

5.1 UC Definition of AKE-KCI

The KCI notion for KE protocols, which stands for "key-compromise impersonation," captures the property we call "security against reverse impersonation," which concerns an attacker $\mathcal{A}$ who learns party P's long-term keys but otherwise does not actively control P. Resistance to KCI attacks, or "KCI security" for short, postulates that even though $\mathcal{A}$ can impersonate P to other parties, sessions which P itself runs with honest peers need to remain secure. A game-based definition of this notion appears in [27], and here we formalize it in the UC model through functionality $\mathcal{F}_{\mathsf{AKE-KCI}}$ presented in Fig. 4. We specialize functionality $\mathcal{F}_{\mathsf{AKE-KCI}}$ to our user-server setting where only servers can be compromised, but it can be extended to allow for compromise of any protocol party.

Functionality $\mathcal{F}_{\mathsf{AKE-KCI}}$ extends the standard KE functionality of [14] with two adversarial actions. The first, COMPROMISE, is targeted at a server and captures the compromise of the server's keys. The second is IMPERSONATE which is borrowed from the aPAKE functionality of [18] shown in Fig. 1. This action can only be targeted at users' sessions, and only for sessions with servers compromised via the COMPROMISE action, and it marks such session as COMPROMISED, which implies that the attacker can determine the session key this session outputs, via the NEWKEY action. This models the fact that user's sessions with a compromised S as a peer cannot be assumed to be secure since they could have been run with the adversary who has stolen S's keys. However, sessions at S itself must not be affected by the IMPERSONATE action, and they remain secure. All other elements in $\mathcal{F}_{\mathsf{AKE-KCI}}$ are the same as in the basic UC KE functionality, except of some syntactic specialization to the user-server setting.

AKE-KCI Security of HMQV. A concrete instantiation of protocol OPAQUE shown in Fig. 7 in Sect. 6, which instantiates the generic Strong aPAKE protocol shown in Sect. 5.2 below, using HMQV [27] as the AKE-KCI protocol. The KCI property of HMQV was proved in [27] in the game-based Canetti-Krawczyk model [13] extended to include KCI security. Here we require UC security, namely, a protocol that realizes functionality $\mathcal{F}_{\mathsf{AKE-KCI}}$. Fortunately, [14] proves the equivalence of the game-based definition of [13] and their UC AKE formulation. Thanks to this equivalence, HMQV, as a basic KE, is secure in the

In the description below, we assume $\mathsf{P} \in \{\mathsf{U}, \mathsf{S}\}$.

- On (UsrSession, sid, $ssid$, S) from U, send (UsrSession, sid, $ssid$, U, S) to $\mathcal{A}^*$. If there is no record $(ssid, \mathsf{U}, \cdot)$, record $(ssid, \mathsf{U}, \mathsf{S})$ and mark it FRESH.
- On (SvrSession, sid, $ssid$, U) from S, send (SvrSession, sid, $ssid$, U, S) to $\mathcal{A}^*$. If there is no record $(ssid, \mathsf{S}, \cdot)$, record $(ssid, \mathsf{S}, \mathsf{U})$ and mark it FRESH.
- On (Compromise, sid) from $\mathcal{A}^*$, mark S COMPROMISED.
- On (Impersonate, sid, $ssid$) from $\mathcal{A}^*$, if S is marked COMPROMISED and there is a record $(ssid, \mathsf{U}, \mathsf{S})$ marked FRESH, mark the record COMPROMISED.
- On (NewKey, sid, $ssid$, P, SK) from $\mathcal{A}^*$ where $|SK| = \tau$, if there is a record $(ssid, \mathsf{P}, \mathsf{P}')$ not marked COMPLETED, do:
  - If the record is marked COMPROMISED, or either P or P' is corrupted, send $(sid, ssid, SK)$ to P.
  - If the record is marked FRESH, a $(sid, ssid, SK')$ tuple was sent to P', and at that time there was a record $(ssid, \mathsf{P}', \mathsf{P})$ marked FRESH, send $(sid, ssid, SK')$ to P.
  - Else pick $SK'' \leftarrow_{\mathrm{R}} \{0,1\}^\tau$ and send $(sid, ssid, SK'')$ to P.

  Finally, mark $(ssid, \mathsf{P}, \mathsf{P}')$ COMPLETED.

Fig. 4. Functionality $\mathcal{F}_{\mathsf{AKE-KCI}}$

UC model. More precisely, this applies to the three-message HMQV with client authentication (which satisfies the "ACK" property required for the equivalence in [14]). For the 2-message version of HMQV, the equivalence still holds using the notion of *non-information oracle* [14] that holds for HMQV under Computational Diffie-Hellman (CDH) assumption in the RO model. For our purposes, however, we need HMQV to realize the extended AKE-KCI functionality of Fig. 4. Luckily, the equivalence with the game-based definition extends to this case. Indeed, since the original equivalence from [14] holds even in the case of adaptive party corruptions, the Compromise and Impersonate actions introduced here – which constitute a *limited* form of adaptive corruptions – follow as a special case. Finally, we note that the equivalence between the above models also preserves forward secrecy, so this property (proved in the game-based Canetti-Krawczyk model in [27]) holds in the UC too. We note that by the results in [27], the 3-message HMQV enjoys full PFS while the 2-message only weak PFS (against passive attackers only). The above security of HMQV (without including security against the leakage of ephemeral exponents) is based on the CDH assumption in the RO model [27].

5.2 Strong aPAKE Construction from OPRF and AKE-KCI

Our Strong aPAKE protocol based on OPRF and AKE-KCI is shown in Fig. 5. The protocol uses the same OPRF tool as the Strong aPAKE construction of Sect. 4, for length parameter $\ell = 2\tau$, which defines the "randomized password"

Public Components:

- KCI-secure AKE protocol Π with private/public keys denoted p_s, P_s, p_u, P_u;
- Random-key robust authenticated encryption $\mathsf{AE} = (\mathsf{AuthEnc}, \mathsf{AuthDec})$ with (2τ)-bit keys;
- Functionality $\mathcal{F}_{\mathsf{OPRF}}$ with output length parameter $\ell = 2\tau$.

Password Registration

1. On input (StorePwdFile, sid, U, pw), S generates pairs (p_s, P_s) and (p_u, P_u), and sends (Init, sid, pw) to $\mathcal{F}_{\mathsf{OPRF}}$.
   On $\mathcal{F}_{\mathsf{OPRF}}$'s response (Init, sid, rw), S computes $c \leftarrow \mathsf{AuthEnc}_{\mathsf{rw}}(p_u, P_u, P_s)$ and records $\mathsf{file}[sid] := \langle p_s, P_s, P_u, c \rangle$.

Server Compromise

1. On (StealPwdFile, sid) from $\mathcal{A}$, S retrieves $\mathsf{file}[sid]$ and sends it to $\mathcal{A}$.

Login

1. On (UsrSession, sid, $ssid$, S, pw′), U sends (Eval, sid, $ssid$, S, pw′) to $\mathcal{F}_{\mathsf{OPRF}}$.
2. On (SvrSession, sid, $ssid$), S retrieves $\mathsf{file}[sid] = \langle p_s, P_s, P_u, c \rangle$, sends c to U, sends (SndrComplete, sid, $ssid$) to $\mathcal{F}_{\mathsf{OPRF}}$ and runs Π on input (p_s, P_s, P_u).
3. On (Eval, sid, $ssid$, rw′) from $\mathcal{F}_{\mathsf{OPRF}}$ and c from S, U computes $\mathsf{AuthDec}_{\mathsf{rw}'}(c)$. If the result is $\perp$, U outputs (abort, sid, $ssid$) and halts. Otherwise U parses $(p'_u, P'_u, P'_s) := \mathsf{AuthDec}_{\mathsf{rw}'}(c)$ and runs Π on input (p'_u, P'_u, P'_s).
4. Given Π's local output SK, the corresponding party, either U or S, outputs $(sid, ssid, SK)$.

Fig. 5. Strong aPAKE based on AKE-KCI in the $\mathcal{F}_{\mathsf{OPRF}}$-hybrid world

value $\mathsf{rw} = F_k(\mathsf{pw})$ for user U's password pw and OPRF key k held by server S. We assume that in the AKE-KCI protocol Π each party holds a (private, public) key pair, and that the each party runs the Login subprotocol using its key pair and the public key of the counterparty as inputs. In Password Registration phase, server S generates the user U's keys, and S's password file contains S's key pair p_s, P_s; U's public key P_u; and a ciphertext c of U's private key p_u, and the public keys P_u and P_s created using an Authenticated Encryption scheme using $\mathsf{rw} = F_k(\mathsf{pw})$ as the key. After creating the password file, value p_u is erased at S. In Login phase, S runs OPRF with U, which lets U compute $\mathsf{rw} = F_k(\mathsf{pw})$, it sends c to U, who can decrypt it under rw and retrieves its key-pair p_u, P_u together with the server's key P_s, at which point both parties have appropriate inputs to the AKE-KCI protocol Π to compute the session key.

Role of Authenticated Encryption. The Strong aPAKE protocol of Fig. 5 utilizes an *Authenticated Encryption* scheme $\mathsf{AE} = (\mathsf{AuthEnc}, \mathsf{AuthDec})$ to encrypt and authenticate U's AKE "credential" $m = (p_u, P_u, P_s)$. We encrypt the whole payload m for simplicity, because unlike U's private key p_u, values P_u, P_s

could be public and need to be only authenticated, not encrypted. However, the authentication property of AE must apply to the whole payload. Intuitively, U must authenticate S's public key P_s, but if U derived even its key pair (p_u, P_u) using just the secrecy of $\mathsf{rw} = F_k(\mathsf{pw})$, e.g., using rw as randomness in a key generation, and U then executed AKE on such (p_u, P_u) pair, the resulting protocol would already be insecure. To see an example, if an AKE leaks U's public key input P_u (note that AKE does not guarantee privacy of the public key) then an adversary $\mathcal{A}$ who engages U in a single protocol instance can find U's password pw via an offline dictionary attack by running the OPRF with U on some key k^*, and then given P_u leaked in the subsequent AKE it finds pw s.t. the key generation outputs P_u as a public key on randomness $\mathsf{rw} = F_{k^*}(\mathsf{pw})$.

Thus the role of the authentication property in authenticated encryption is to commit $\mathcal{A}$ to a single guess of rw and consequently, given the OPRF key k^*, to a single guess pw. (Note that our UC OPRF notion implies that F is collision-resistant.) To that end we need the authenticated encryption to satisfy the following property which we call *random-key robustness*:[4] For any efficient algorithm $\mathcal{A}$ there is a negligible probability that $\mathcal{A}$ on input (k_1, k_2) for two random keys k_1, k_2 outputs c s.t. $\mathsf{AuthDec}_{k_1}(c) \neq \bot$ and $\mathsf{AuthDec}_{k_2}(c) \neq \bot$. In other words, it must be infeasible to create an authenticated ciphertext that successfully decrypts under two different randomly generated keys. This property can be achieved in the standard model using e.g. encrypt-then-MAC with a MAC that is collision resistant with respect to the message and key, a property enjoyed by HMAC with full hash output. In the RO model used by our aPAKE application one can also enforce it for any authenticated encryption scheme by attaching to its ciphertext c a hash $H(k, c)$ for a RO hash H with 2τ-bit outputs.

Note on Not Utilizing $\mathcal{F}_{\mathsf{AKE-KCI}}$. In Fig. 5 we abstract the OPRF protocol as functionality $\mathcal{F}_{\mathsf{OPRF}}$, but we use the real-world AKE-KCI protocol Π, rather than functionality $\mathcal{F}_{\mathsf{AKE-KCI}}$. The reason for this presentation is that in the KE functionality of [14], of which $\mathcal{F}_{\mathsf{AKE-KCI}}$ is an extension, it is not clear how to support a usage of the KE protocol on keys which are computed via some other mechanism than the intended KE key generation. The KE functionality of [14] assumes that each entity keeps its private key as a permanent state, authenticates to a counterparty given its identity, and a KE party cannot specify any bitstring as one's own private key and a counterparty's public key. This is not how we use AKE in our Strong aPAKE of Fig. 5 precisely because U does not keep state and has to reconstruct its keys from a password (via OPRF). However, we can still use the real-world protocol Π, which UC-realizes $\mathcal{F}_{\mathsf{AKE-KCI}}$, giving it the OPRF-computed information as input. In the proof of security we utilize the simulator $\mathsf{SIM}_{\mathsf{AKE}}$, which shows that Π UC-realizes $\mathcal{F}_{\mathsf{AKE-KCI}}$, in our simulator construction, but we rely on its correctness only if U runs Π on the correctly reconstructed (p_u, P_s, P_s), and if the adversary causes U to reconstruct a different string we interpret this as a successful attack on U's login session.

[4] This notion is a weakening of *full robustness (FROB)* from [16] where the attacker is allowed to choose k_1, k_2 (in our case these keys are random). An even weaker notion, Semi-FROB, is defined in [16] where k_1, k_2 are random but only k_1 is provided to $\mathcal{A}$.

5.3 Proof of Security

In Theorem 2 below we state security of the Strong aPAKE protocol of Fig. 5.

Theorem 2. *If protocol Π UC-realizes functionality $\mathcal{F}_{\mathsf{AKE-KCI}}$ then protocol in Fig. 5 UC-realizes functionality $\mathcal{F}_{\mathsf{SaPAKE}}$ in the $\mathcal{F}_{\mathsf{OPRF}}$-hybrid model.*

Concretely, suppose that there is a simulator $\mathsf{SIM}_{\mathsf{AKE}}$ such that the distinguishing advantage of an environment $\mathcal{Z}$ between the real execution of Π and $\mathcal{Z}$'s interaction with $\mathsf{SIM}_{\mathsf{AKE}}$ is at most $\mathbf{Adv}^{\mathsf{DIST}}_{\mathsf{SIM}_{\mathsf{AKE}},\mathcal{Z}}(\tau)$, where τ is the security parameter. Then for any adversary $\mathcal{A}$ with running time T against the protocol, there is a simulator SIM that produces a view in the simulated world such that the advantage that $\mathcal{Z}$ has in distinguishing between this view and the view in the real world is no more than $\mathbf{Adv}^{\mathsf{AUTH}}_{\mathsf{AE},T}(\tau) + q_F^2 \cdot \mathbf{Adv}^{\mathsf{RK-RBST}}_{\mathsf{AE},T}(\tau) + 2\mathbf{Adv}^{\mathsf{DIST}}_{\mathsf{SIM}_{\mathsf{AKE}},\mathcal{Z}}(\tau)$, where q_F is the number of Eval *and* OfflineEval *messages aimed at $\mathcal{F}_{\mathsf{OPRF}}$ from $\mathcal{A}$, and $\mathbf{Adv}^{\mathsf{AUTH}}_{\mathsf{AE},T}(\tau)$ and $\mathbf{Adv}^{\mathsf{RK-RBST}}_{\mathsf{AE},T}(\tau)$ are the probabilities that any algorithm in running time T breaks the authenticity of* AE *and the random-key robustness of* AE*, respectively.*

Proof. For any adversary $\mathcal{A}$, we construct a simulator SIM as in Fig. 6. While interacting with $\mathsf{SIM}_{\mathsf{AKE}}$, SIM plays the role of both $\mathcal{F}_{\mathsf{AKE-KCI}}$ and $\mathcal{A}$.

Following [11], without loss of generality, we may assume that $\mathcal{A}$ is a "dummy" adversary that merely passes all its messages and computations to the environment $\mathcal{Z}$. We omit all interactions with corrupted U and S where SIM acts as $\mathcal{F}_{\mathsf{OPRF}}$, since the simulation is trivial (SIM gains all information needed and simply follows the code of $\mathcal{F}_{\mathsf{OPRF}}$). To keep notation brief we denote functionality $\mathcal{F}_{\mathsf{SaPAKE}}$ as $\mathcal{F}$.

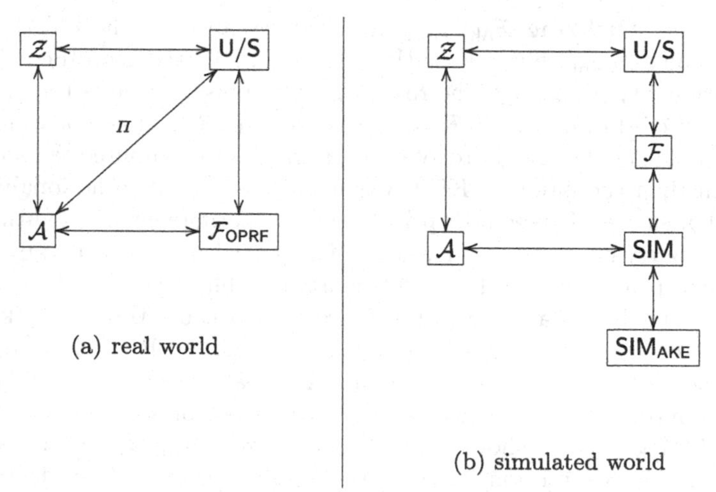

(a) real world

(b) simulated world

In order to account for the advantage of the environment $\mathcal{Z}$ in distinguishing between its views in the real world and the simulated world, we compare between these two settings in the different simulator actions and derive the distinguishing advantages in cases where the simulation is not perfect. Below we assume that

For every sid and every server S, initialize $\mathsf{tx}(sid, \mathsf{S})$ to 0.

Stealing Password Data and Offline Queries

1. On (COMPROMISE, sid) from $\mathcal{A}$ aimed at $\mathcal{F}_{\mathsf{OPRF}}$ and (STEALPWDFILE, sid) from $\mathcal{A}$ aimed at S, send (STEALPWDFILE, sid) to $\mathcal{F}$.
   If $\mathcal{F}$ returns "password file stolen," mark S COMPROMISED, generate two key pairs (p_s, P_s) and (p_u, P_u), pick $\mathsf{rw} \leftarrow_{\mathrm{R}} \{0,1\}^{2\tau}$, compute $c \leftarrow \mathsf{AuthEnc}_{\mathsf{rw}}(p_u, P_u, P_s)$, record $\mathsf{file}[sid] := \langle p_s, P_s, P_u, c\rangle$, and send $\mathsf{file}[sid]$ to $\mathcal{A}$ as a message from S.
2. On (OFFLINEEVAL, sid, S, x) from $\mathcal{A}$ aimed at $\mathcal{F}_{\mathsf{OPRF}}$, if S is marked COMPROMISED or corrupted, send (OFFLINETESTPWD, sid, x) to $\mathcal{F}$. If $\mathcal{F}$ returns "correct guess," set $F_{sid,\mathsf{S}}(x) := \mathsf{rw}$. Regardless, send (OFFLINEEVAL, $sid, F_{sid,\mathsf{S}}(x)$) to $\mathcal{A}$ as a message from $\mathcal{F}_{\mathsf{OPRF}}$ (if $F_{sid,\mathsf{S}}(x)$ is undefined, pick $\rho \leftarrow_{\mathrm{R}} \{0,1\}^{2\tau}$ and set $F_{sid,\mathsf{S}}(x) := \rho$).

Password Authentication

1. On (USRSESSION, $sid, ssid, \mathsf{U}, \mathsf{S}$) from $\mathcal{F}$, send (EVAL, $sid, ssid, \mathsf{U}, \mathsf{S}$) to $\mathcal{A}$ as a message from $\mathcal{F}_{\mathsf{OPRF}}$. Also, if this is the first USRSESSION message for $ssid$, record $\langle ssid, \mathsf{U}, \mathsf{S}, \cdot\rangle$.
2. On (SVRSESSION, $sid, ssid, \mathsf{U}, \mathsf{S}$) from $\mathcal{F}$, retrieve $\mathsf{file}[sid] = \langle p_s, P_s, P_u, c\rangle$, send c and (SNDRCOMPLETE, $sid, ssid, \mathsf{S}$) to $\mathcal{A}$ as a message from S to U and from $\mathcal{F}_{\mathsf{OPRF}}$, respectively, and send (SVRSESSION, $sid, ssid, \mathsf{U}, \mathsf{S}$) to $\mathsf{SIM}_{\mathsf{AKE}}$ as a message from $\mathcal{F}_{\mathsf{AKE-KCI}}$. Also, if this is the first SVRSESSION message for $ssid$, set $\mathsf{tx}(sid, \mathsf{S})$++.
3. On (RCVCOMPLETE, $sid, ssid, \mathsf{S}^*$) from $\mathcal{A}$ aimed at $\mathcal{F}_{\mathsf{OPRF}}$, retrieve $\langle ssid, \mathsf{U}, \mathsf{S}, \cdot\rangle$; ignore this message if (i) such record does not exist, or (ii) S is honest and not marked COMPROMISED and $\mathsf{S}^* \neq \mathsf{S}$, or (iii) $\mathsf{tx}(sid, \mathsf{S}^*) = 0$. Else set $\mathsf{tx}(sid, \mathsf{S}^*)$−−, augment $\langle ssid, \mathsf{U}, \mathsf{S}, \cdot\rangle$ to $\langle ssid, \mathsf{U}, \mathsf{S}, \mathsf{S}^*, \cdot\rangle$ and mark $(ssid, \mathsf{U})$ COMPLETED.

Key Generation and Authentication

1. As soon as $(ssid, \mathsf{U})$ is marked COMPLETED and a c' is sent from $\mathcal{A}$ aimed at U, retrieve $\mathsf{file}[sid] = \langle p_s, P_s, P_u, c\rangle$ and $\langle ssid, \mathsf{U}, \mathsf{S}, \mathsf{S}^*, \cdot\rangle$, and proceed as follows:
   - If $c' = c$ and $\mathsf{S}^* = \mathsf{S}$, send (TESTABORT, $sid, ssid, \mathsf{U}$) to $\mathcal{F}$.
     If $\mathcal{F}$ returns SUCC, send (USRSESSION, $sid, ssid, \mathsf{U}, \mathsf{S}$) to $\mathsf{SIM}_{\mathsf{AKE}}$ as a message from $\mathcal{F}_{\mathsf{AKE-KCI}}$. Mark this case (1).
   - Else for every x such that $F_{sid,\mathsf{S}^*}(x)$ is defined (denote it y), check whether $\mathsf{AuthDec}_y(c') \neq \perp$.
     - If there are more than one such x's, output HALT and abort.
     - If there is a unique such x, send (TESTPWD, $sid, ssid, \mathsf{U}, x$) to $\mathcal{F}$.
       If $\mathcal{F}$ returns "correct guess," parse $(p'_u, P'_u, P'_s) := \mathsf{AuthDec}_y(c')$. Mark this case (2).
       If $\mathcal{F}$ returns "wrong guess," send (TESTABORT, $sid, ssid, \mathsf{U}$) to $\mathcal{F}$ and halt.
     - If there is no such x, send (TESTABORT, $sid, ssid, \mathsf{U}$) to $\mathcal{F}$ and halt.
2. In case (1): (i) On (IMPERSONATE, $sid, ssid$) from $\mathsf{SIM}_{\mathsf{AKE}}$, if S is marked COMPROMISED, pass this message to $\mathcal{F}$; (ii) While $\mathsf{SIM}_{\mathsf{AKE}}$ simulates the execution of Π, pass messages between it and $\mathcal{A}$; (iii) On (NEWKEY, $sid, ssid, \mathsf{P}, SK$) from $\mathsf{SIM}_{\mathsf{AKE}}$, pass this message to $\mathcal{F}$.
3. In case (2): (i) On $\mathcal{A}$'s message as from S to U, run U's algorithm in Π (henceforth Π_u) on (p'_u, P'_u, P'_s); (ii) On $\mathcal{A}$'s message as from U to S, pass it to $\mathsf{SIM}_{\mathsf{AKE}}$ as a message from $\mathcal{A}$, and pass $\mathsf{SIM}_{\mathsf{AKE}}$'s response to $\mathcal{A}$ as from S to U; (iii) When Π_u is completed with output SK, send (NEWKEY, $sid, ssid, \mathsf{U}, SK$) to $\mathcal{F}$; (iv) On (NEWKEY, $sid, ssid, \mathsf{S}, SK$) from $\mathsf{SIM}_{\mathsf{AKE}}$, send (NEWKEY, $sid, ssid, \mathsf{S}, 0^\tau$) to $\mathcal{F}$.

Fig. 6. The simulator SIM

$\mathcal{Z}$ issues the (STOREPWDFILE, sid, U, pw) command to S for some pw; otherwise any subsequent server-side commands of $\mathcal{Z}$ will not have any effect.

- $\mathsf{file}[sid] = \langle p_s, P_s, P_u, c\rangle$ (from $\mathcal{A}$): In both worlds, $\mathcal{Z}$ receives this message after $\mathcal{A}$ sends (COMPROMISE, sid) aimed at $\mathcal{F}_{\mathsf{OPRF}}$ and (STEALPWDFILE, sid) to S, provided that $\mathcal{Z}$ input (STOREPWDFILE, sid, U, pw) to S previously.
  In both worlds, p_s, P_s and P_u are generated in the same way, and c is computed as $\mathsf{AuthEnc}_{\mathsf{rw}}(p_u, P_u, P_s)$. The only difference is that rw is $F_{sid,\mathsf{S}}(\mathsf{pw})$ in the real world, while it is chosen from random in the simulated world. There is no way for $\mathcal{Z}$ to distinguish unless and until it queries $F_{sid,\mathsf{S}}(\mathsf{pw})$ by letting $\mathcal{A}$ send (OFFLINEEVAL, sid, S, pw) aimed at $\mathcal{F}_{\mathsf{OPRF}}$. However, once $\mathcal{A}$ sends such message, SIM sets $F_{sid,\mathsf{S}}(\mathsf{pw})$ to rw. Therefore, in both worlds, $F_{sid,\mathsf{S}}(\mathsf{pw}) = \mathsf{rw}$ and $\mathcal{Z}$ cannot distinguish.
- (OFFLINEEVAL, sid, ρ) (from $\mathcal{A}$): In both worlds, $\mathcal{Z}$ receives this message after $\mathcal{A}$ sends (OFFLINEEVAL, sid, S, x) to $\mathcal{F}_{\mathsf{OPRF}}$, provided that S is corrupted or marked COMPROMISED. The selection of ρ is the same in the two worlds, except that in the simulated world, if $x = \mathsf{pw}$, ρ is set to rw which was chosen from random in advance, while in the real world, ρ is always chosen from random directly. There is no way to distinguish between these two cases.
- (EVAL, sid, $ssid$, U, S) (from $\mathcal{A}$): In both worlds, $\mathcal{Z}$ receives this message after inputting (USRSESSION, sid, $ssid$, S, pw$'$) to U.
- c and (SNDRCOMPLETE, sid, $ssid$, S) (from $\mathcal{A}$): In both worlds, $\mathcal{Z}$ receives these two messages after inputting (SVRSESSION, sid, $ssid$) to S. As argued above, $\mathcal{Z}$ cannot distinguish the two c's in the two worlds.
- (ABORT, sid, $ssid$) (from U): In both worlds, $\mathcal{Z}$ may receive this message after $\mathcal{A}$ sends (RCVCOMPLETE, sid, $ssid$, S^*) aimed at $\mathcal{F}_{\mathsf{OPRF}}$ and c' aimed at U, provided that (i) there is a record $\langle ssid, \mathsf{U}, \mathsf{S}, \mathsf{pw}'\rangle$ in $\mathcal{F}_{\mathsf{OPRF}}$ (or a record $\langle ssid, \mathsf{U}, \mathsf{S}, \cdot\rangle$ in SIM), (ii) if S is honest and not marked COMPROMISED, then $\mathsf{S}^* = \mathsf{S}$, and (iii) $\mathsf{tx}(sid, \mathsf{S}^*) > 0$.
  Note that $\mathcal{Z}$ may see a HALT message from SIM at this time. HALT occurs when there exists $x_1 \neq x_2$ such that $\mathsf{AuthDec}_{y_1}(c') \neq \perp$ and $\mathsf{AuthDec}_{y_2}(c') \neq \perp$, where $y_1 = F_{sid,\mathsf{S}^*}(x_1)$ and $y_2 = F_{sid,\mathsf{S}^*}(x_2)$. Since $F_{sid,\mathsf{S}^*}(\cdot)$ is a random function onto $\{0,1\}^{2\tau}$, y_1 and y_2 are independent random strings in $\{0,1\}^{2\tau}$; thus, for fixed y_1 and y_2, the probability that $\mathcal{A}$ finds c' such that $\mathsf{AuthDec}_{y_1}(c') \neq \perp$ and $\mathsf{AuthDec}_{y_2}(c') \neq \perp$ is at most $\mathbf{Adv}^{\mathsf{RK-RBST}}_{\mathsf{AE},T}(\tau)$ due to the random-key robustness of AE. Since $\mathcal{A}$ queries F q_F times, there are q_F independent y's; using a polynomial reduction, we have $\Pr[\text{HALT}] \leq q_F^2 \cdot \mathbf{Adv}^{\mathsf{RK-RBST}}_{\mathsf{AE},T}(\tau)$.
  Next we assume that HALT does not occur. In the real world, $\mathcal{Z}$ receives (ABORT, sid, $ssid$) from U if and only if $\mathsf{AuthDec}_{\mathsf{rw}'}(c') = \perp$; that is, $\mathcal{Z}$ does not receive this message if and only if $\mathsf{AuthDec}_{\mathsf{rw}'}(c') \neq \perp$. There are only three possibilities:
  (1) $(\mathsf{pw}', \mathsf{S}^*, c') = (\mathsf{pw}, \mathsf{S}, c)$: Then $\mathsf{rw}' = \mathsf{rw} = F_{sid,\mathsf{S}}(\mathsf{pw})$, thus $\mathsf{AuthDec}_{\mathsf{rw}'}(c') = \mathsf{AuthDec}_{\mathsf{rw}}(c) = (p_u, P_u, P_s)$.
  (2) $\mathcal{A}$ queries $\mathsf{rw}' = F_{sid,\mathsf{S}^*}(\mathsf{pw}')$ previously, and $\mathsf{AuthDec}_{\mathsf{rw}'}(c') \neq \perp$: If $\mathcal{A}$ learns rw', then it can compute an AuthEnc instance on rw' and any message to find a c' such that $\mathsf{AuthDec}_{\mathsf{rw}'}(c') \neq \perp$.

(3) Other cases where $\mathcal{A}$ finds a c' such that $\mathsf{AuthDec}_{\mathsf{rw}'}(c') \neq \bot$, while rw' is independently random of everything else in $\mathcal{Z}$'s view (since $\mathcal{A}$ does not query $F_{sid,\mathsf{S}^*}(\mathsf{pw}')$), and $\mathcal{Z}$ does not query $\mathsf{AuthEnc}_{\mathsf{rw}'}(p'_u, P'_u, P'_s)$ ($\mathcal{Z}$ queries $\mathsf{AuthEnc}_{\mathsf{rw}'}(p'_u, P'_u, P'_s)$ by setting $\mathsf{pw}' = \mathsf{pw}$ [thus making $\mathsf{rw}' = \mathsf{rw}$] and receiving $c = \mathsf{AuthEnc}_{\mathsf{rw}}(p_u, P_u, P_s)$ from S). Since AE is an authenticated encryption, the probability of (3) is at most $\mathbf{Adv}^{\mathsf{AUTH}}_{\mathsf{AE},T}(\tau)$.

In the simulated world, $\mathcal{Z}$ does not receive this message if and only if either of the following two conditions holds:

(1) $c' = c$, $\mathsf{S}^* = \mathsf{S}$ and $\mathcal{F}$ returns SUCC on (TESTABORT, sid, $ssid$, U) from SIM. The last condition holds if and only if there are two records $\langle ssid, \mathsf{U}, \mathsf{S}, \mathsf{pw}' \rangle$ and $\langle ssid, \mathsf{S}, \mathsf{U}, \mathsf{pw}'' \rangle$, the former marked FRESH and $\mathsf{pw}' = \mathsf{pw}''$. Note that no TESTPWD, IMPERSONATE or NEWKEY message has been issued yet, so the record must be FRESH. According to the syntax of SVRSESSION, we have $\mathsf{pw}'' = \mathsf{pw}$. Therefore, the last condition is equivalent to $\mathsf{pw}' = \mathsf{pw}$, thus this case is equivalent to case (1) in the real world.

(2) There exists x s.t. $y = F_{sid,\mathsf{S}^*}(x)$ is defined in SIM, $\mathsf{AuthDec}_y(c') \neq \bot$ and $\mathcal{F}$ returns "correct guess" on (TESTPWD, sid, $ssid$, x) from SIM. The last condition is equivalent to $x = \mathsf{pw}'$; thus, the three conditions combined are equivalent to $\mathsf{rw}' = F_{sid,\mathsf{S}^*}(\mathsf{pw}')$ is defined in SIM and $\mathsf{AuthDec}_{\mathsf{rw}'}(c') \neq \bot$. SIM defines $F_{sid,\mathsf{S}^*}(\mathsf{pw}')$ only when receiving (OFFLINEEVAL, sid, S^*, pw') from $\mathcal{A}$. Therefore, this case is equivalent to case (2) in the real world.

Hence, $\mathcal{Z}$ receives this message in the two worlds under the same conditions, except for case (3) in the real world.

- Messages sent from U and S while executing Π (in the real world), or messages sent from SIM (in the simulated world) (from $\mathcal{A}$): In case (1) and messages sent from S in case (2), they are simulated by SIM who in turn receives them from $\mathsf{SIM_{AKE}}$. Since $\mathsf{SIM_{AKE}}$ generates $\mathcal{A}$'s view indistinguishable from $\mathcal{A}$'s view in the real world, SIM, who merely passes messages between $\mathsf{SIM_{AKE}}$ and $\mathcal{A}$, can also achieve that; the distinguishing advantage of $\mathcal{Z}$ is at most $\mathbf{Adv}^{\mathsf{DIST}}_{\mathsf{SIM_{AKE}},\mathcal{Z}}(\tau)$. For messages sent from U in case (2), they are the results of Π_u on (p'_u, P'_u, P'_s), and are simulated perfectly.
- $(sid, ssid, SK')$ (from U): In both worlds, $\mathcal{Z}$ receives this message when Π is completed and sends output to U. In the real world, there are two cases:
  - $(p'_u, P'_u, P'_s) = (p_u, P_u, P_s)$, i.e., the input of U to Π is correct. This corresponds to case (1) above. There are two subcases regarding Π:
    * S is not compromised. Then according to the syntax of $\mathcal{F}_{\mathsf{AKE-KCI}}$, SK' is a random string in $\{0,1\}^\tau$ (independent of everything else, or the same with S's output if S already output previously). In the simulated world, the record $\langle ssid, \mathsf{U}, \mathsf{S}, \mathsf{pw}' \rangle$ in $\mathcal{F}$ is marked FRESH, so SK' is also a random string in $\{0,1\}^\tau$.
    * S is compromised (then $\mathcal{A}$ may impersonate S while interacting with U in the execution of Π and set U's output). In the simulated world, $\mathsf{SIM_{AKE}}$ sends (IMPERSONATE, sid, $ssid$) to SIM, who transfers this message to $\mathcal{F}$, which makes the record $\langle ssid, \mathsf{U}, \mathsf{S}, \mathsf{pw}' \rangle$ marked

COMPROMISED (note that we have $\mathsf{pw}' = \mathsf{pw}$ here since this is a condition of case (1)). Therefore, SK' chosen by $\mathsf{SIM_{AKE}}$ (which is the same with the SK' output by Π in the real world except for probability at most $\mathbf{Adv}^{\mathsf{DIST}}_{\mathsf{SIM_{AKE}},\mathcal{Z}}(\tau)$) is the value output to U.

- $(p'_u, P'_u, P'_s) \neq (p_u, P_u, P_s)$, i.e., the input of U to Π is incorrect. This may occur only in cases (2) and (3) above. As argued above, the probability of (3) is at most $\mathbf{Adv}^{\mathsf{AUTH}}_{\mathsf{AE},T}(\tau)$.

  (2) is equivalent to case (2) in the simulated world, where SIM sends $(\text{TESTPWD}, sid, ssid, \mathsf{U}, x)$ to $\mathcal{F}$ and $\mathcal{F}$ returns "correct guess" (meaning that $x = \mathsf{pw}'$). After this, the record $\langle ssid, \mathsf{U}, \mathsf{S}, \mathsf{pw}' \rangle$ is marked COMPROMISED. Therefore, SK', which is computed by SIM as Π_u's output on (p'_u, P'_u, P'_s), is the value output to U. In the real world, U also outputs SK'.

– $(sid, ssid, SK)$ (from S): In both worlds, $\mathcal{Z}$ receives this message when Π is completed and sends output to S.

  In the real world, SK is always a random string in $\{0,1\}^\tau$ (independent of everything else, or the same with U's output if U already output previously). Note that in the simulated world, the record $\langle ssid, \mathsf{S}, \mathsf{U}, \mathsf{pw}' \rangle$ is always marked FRESH. Therefore, SK is also random string in $\{0,1\}^\tau$.

It remains to show that $\mathsf{SIM_{AKE}}$'s view while interacting with SIM is the same as interacting with $\mathcal{F}_{\mathsf{AKE-KCI}}$ and $\mathcal{A}$. When SIM acts as $\mathcal{A}$, the interaction is trivial since SIM merely passes messages between $\mathsf{SIM_{AKE}}$ and the real $\mathcal{A}$. Consider when SIM acts as $\mathcal{F}_{\mathsf{AKE-KCI}}$, and note that SIM engages with $\mathsf{SIM_{AKE}}$ only in cases (1) and (2):

(1) U's input is correct: Same effect as honest U and S executing Π;
(2) U's input is incorrect: Same effect as corrupted U and honest S executing Π. Note that SIM engages with $\mathsf{SIM_{AKE}}$ on the side of S only, so $\mathsf{SIM_{AKE}}$'s view is again the same.

We conclude that $\mathcal{Z}$'s view in the real world and the simulated world is the same, except for (1) $(\text{ABORT}, sid, ssid)$ or HALT after $\mathcal{A}$ sends $(\text{RCVCOMPLETE}, sid, ssid, \mathsf{S}^*)$ and c', (2) messages sent during the execution of Π, and (3) $(sid, ssid, SK')$ output from U. The probabilities that (1), (2) and (3) are different in the two worlds are no more than $\mathbf{Adv}^{\mathsf{AUTH}}_{\mathsf{AE},T}(\tau) + q_F^2 \cdot \mathbf{Adv}^{\mathsf{RK-RBST}}_{\mathsf{AE},T}(\tau)$, $\mathbf{Adv}^{\mathsf{DIST}}_{\mathsf{SIM_{AKE}},\mathcal{Z}}(\tau)$ and $\mathbf{Adv}^{\mathsf{DIST}}_{\mathsf{SIM_{AKE}},\mathcal{Z}}(\tau)$, respectively. Using a hybrid argument, we can see that $\mathcal{Z}$'s advantage is no more than $\mathbf{Adv}^{\mathsf{AUTH}}_{\mathsf{AE},T}(\tau) + q_F^2 \cdot \mathbf{Adv}^{\mathsf{RK-RBST}}_{\mathsf{AE},T}(\tau) + 2\mathbf{Adv}^{\mathsf{DIST}}_{\mathsf{SIM_{AKE}},\mathcal{Z}}(\tau)$.

6 OPAQUE: A Strong Asymmetric PAKE Instantiation

Figure 7 shows OPAQUE, a concrete instantiation of the generic OPRF+AKE protocol from Fig. 5. An illustration is presented in Fig. 8.

The OPRF is instantiated with the DH-OPRF scheme from [22] recalled in Appendix A, while the AKE protocol can be instantiated with any UC-secure

2-message implicitly-authenticated AKE-KCI; in Fig. 7 this is illustrated with HMQV [27]. Fortunately, the two messages of DH-OPRF and the two messages from HMQV (or a similar protocol) can be run "in parallel" hence obtaining a 2-message SaPAKE.

By Theorem 2 on the security of the generic OPRF+AKE construction, by Lemma 1 in Appendix A on the security of DH-OPRF, and by security of HMQV (see below), we get that protocol OPAQUE realizes functionality $\mathcal{F}_{\mathsf{SaPAKE}}$, hence it is a provably-secure Strong aPAKE, under the One-More Diffie-Hellman assumption [3,22] in ROM.

6.1 Protocol Details and Properties

We expand on the specification of OPAQUE and the protocol's properties.

• *Password registration.* Password registration is the only part of the protocol assumed to run over secure channels where parties can authenticate each other. We note that while OPAQUE is presented with S doing all the registration operations, in practice one may want to avoid that. Instead, we can let S choose an OPRF key k_s and U choose pw, and then run the OPRF protocol between U and S so only U learns its secrets $(\mathsf{pw}, \mathsf{rw}, p_u)$ and only S learns p_s. A problem arises with this approach if S's policy is to check the user's password for compliance with some rules. A possible workaround is to adapt techniques from [26] that present zero-knowledge proofs for proving compliance without disclosing the password.

• *Authenticated encryption.* As specified in Sect. 5.2, the scheme $\mathsf{AuthEnc}$ used in the protocol needs to satisfy the key-committing property defined there. In practice, using an encrypt-then-mac scheme with HMAC-256 (or larger) as the MAC provides this property (if a scheme does not have this property then adding on top of it such a HMAC computed on the scheme's ciphertext will ensure this property).

• *Key exchange.* The generic AKE representation via the KE formula applies to any protocol whose session key is computed as a function of the long-term private-public key pair of each party and ephemeral session-specific private-public values. These values are represented as (p_s, P_s, x_s, X_s) for the server and (p_u, P_u, x_u, X_u) for the user. We note that while more general key-exchange protocols can be used with OPAQUE, this representation applies to many such protocols and, in particular, to HMQV [27] which we use here as our main instantiation.

• *Explicit mutual authentication.* The protocol as illustrated takes just two messages but does not provide explicit user authentication. With a third message the protocol achieves mutual authentication by simply adding the value $f_K(1)$ to the server's message and adding a third message where U sends $f_K(2)$ to S. Each party verifies that the value received from the other is computed correctly and if not it aborts.

• *Use of HMQV.* Recall that the security of OPAQUE depends on the KE protocol being AKE-secure in the UC model with the additional KCI property;

Public Parameters and Components

- Security parameter τ
- Group G of prime order q, $|q| = 2\tau$ and generator g (G^* denotes $G \setminus \{1\}$).
- Hash functions $H(\cdot,\cdot)$, $H'(\cdot)$ with ranges $\{0,1\}^{2\tau}$ and G, respectively.
- Pseudorandom function (PRF) $f(\cdot)$ with range $\{0,1\}^{2\tau}$.
- OPRF function defined as $F_k(x) = H(x, (H'(x))^k)$ for key $k \in \mathbb{Z}_q$.
- Key-committing authenticated encryption scheme ($\mathsf{AuthEnc}, \mathsf{AuthDec}$).
- Key exchange formula KE defined below.

Password Registration

1. ($\textsc{StorePwdFile}$, sid, U, pw): S computes $k_s \leftarrow_{\mathrm{R}} \mathbb{Z}_q$, $\mathsf{rw} := F_{k_s}(\mathsf{pw})$, $p_s \leftarrow_{\mathrm{R}} \mathbb{Z}_q$, $p_u \leftarrow_{\mathrm{R}} \mathbb{Z}_q$, $P_s := g^{p_s}$, $P_u := g^{p_u}$, $c \leftarrow \mathsf{AuthEnc}_{\mathsf{rw}}(p_u, P_u, P_s)$; it records $\mathsf{file}[sid] := \langle k_s, p_s, P_s, P_u, c\rangle$.

Login

1. ($\textsc{UsrSession}$, sid, $ssid$, S, pw): U picks $r, x_u \leftarrow_{\mathrm{R}} \mathbb{Z}_q$; sets $\alpha := (H'(\mathsf{pw}))^r$ and $X_u := g^{x_u}$; sends α and X_u to S.
2. ($\textsc{SvrSession}$, sid, $ssid$): On input α from U, S proceeds as follows:
   (a) Checks that $\alpha \in G^*$. If not, outputs ($\textsc{abort}$, sid, $ssid$) and halts;
   (b) Retrieves $\mathsf{file}[sid] = \langle k_s, p_s, P_s, P_u, c\rangle$;
   (c) Picks $x_s \leftarrow_{\mathrm{R}} \mathbb{Z}_q$ and computes $\beta := \alpha^{k_s}$ and $X_s := g^{x_s}$;
   (d) Computes $K := \mathsf{KE}(p_s, x_s, P_u, X_u)$ and $SK := f_K(0)$;
   (e) Sends β, X_s and c to U;
   (f) Outputs (sid, $ssid$, SK).
3. On β, X_s and c from S, U proceeds as follows:
   (a) Checks that $\beta \in G^*$. If not, outputs ($\textsc{abort}$, sid, $ssid$) and halts;
   (b) Computes $\mathsf{rw} := H(\mathsf{pw}, \beta^{1/r})$;
   (c) Computes $\mathsf{AuthDec}_{\mathsf{rw}}(c)$. If the result is $\perp$, outputs ($\textsc{abort}$, sid, $ssid$) and halts. Otherwise sets $(p_u, P_u, P_s) := \mathsf{AuthDec}_{\mathsf{rw}}(c)$;
   (d) Computes $K := \mathsf{KE}(p_u, x_u, P_s, X_s)$ and $SK := f_K(0)$;
   (e) Outputs (sid, $ssid$, SK).

Key exchange formula KE with HMQV instantiation (if any of $X_u, P_u, X_s, P_s \notin G^*$ the receiving party outputs ($\textsc{abort}$, sid, $ssid$) and halts)

$$\text{For } \mathsf{S}\text{: } \mathsf{KE}(p_s, x_s, P_u, X_u) = H\left((X_u P_u^{e_u})^{x_s + e_s p_s}\right)$$
$$\text{For } \mathsf{U}\text{: } \mathsf{KE}(p_u, x_u, P_S, X_S) = H\left((X_s P_s^{e_s})^{x_u + e_u p_u}\right)$$

where $e_u = H(X_u, \mathsf{S}) \bmod q$, $e_s = H(X_s, \mathsf{U}) \bmod q$.

Fig. 7. Protocol OPAQUE

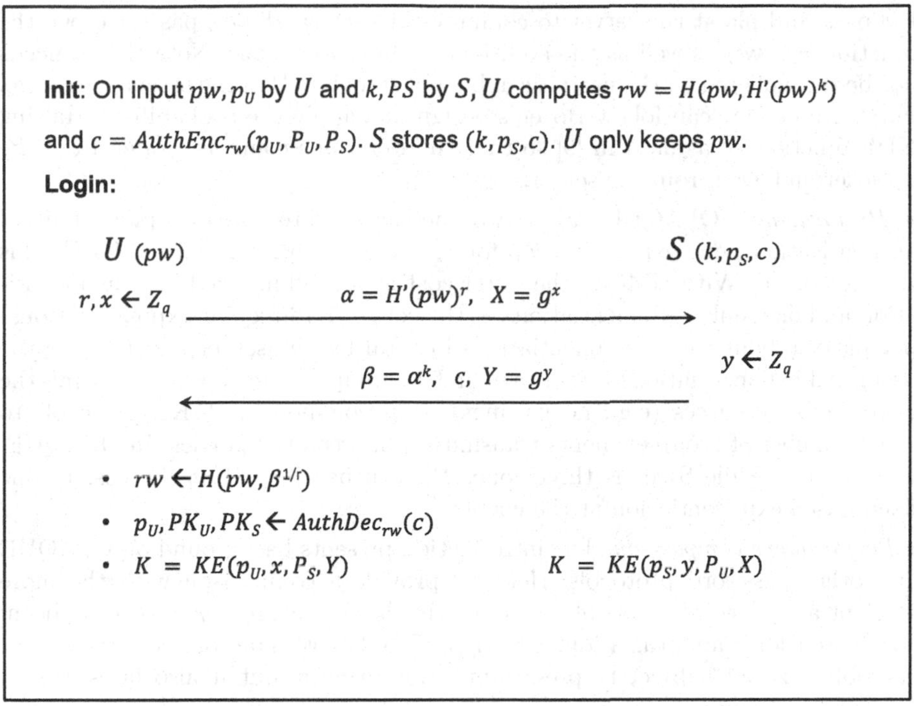

Init: On input pw, p_U by U and k, PS by S, U computes $rw = H(pw, H'(pw)^k)$ and $c = AuthEnc_{rw}(p_U, P_U, P_S)$. S stores (k, p_S, c). U only keeps pw.

Login:

U (pw)		S (k, p_S, c)
$r, x \leftarrow Z_q$	$\alpha = H'(pw)^r,\ X = g^x$ $\longrightarrow$	
	$\longleftarrow$ $\beta = \alpha^k,\ c,\ Y = g^y$	$y \leftarrow Z_q$

- $rw \leftarrow H(pw, \beta^{1/r})$
- $p_U, PK_U, PK_S \leftarrow AuthDec_{rw}(c)$
- $K = KE(p_U, x, P_S, Y)$ $\qquad\qquad$ $K = KE(p_S, y, P_U, X)$

Fig. 8. Schematic representation of OPAQUE (see Fig. 7 for the details)

namely, it should realize the AKE-KCI UC functionality from Fig. 4. As argued in Sect. 5.1, HMQV indeed realizes this functionality (under the CDH assumption in the RO model), hence it is appropriate for use in OPAQUE. Moreover, HMQV enjoys forward secrecy. Specifically, the 2-message protocol provides weak forward secrecy (i.e., forward secrecy is guaranteed for sessions where the user's message delivered to the server came from the real U) while the 3-message variant with explicit client authentication provides full forward secrecy, namely, against arbitrary active attacks [27].

• *Forward secrecy.* This property (or lack of it) is inherited by OPAQUE from the key exchange component KE. In the case of HMQV, forward secrecy is achieved as stated above. *One cannot overstate the importance of forward secrecy in password protocols: it guarantees that past session keys remain secure upon the compromise of a user's password (or server's information).*

• *User iterated hashing.* OPAQUE can be strengthened by increasing the cost of a dictionary attack in case of server compromise. This is done by changing the computation of rw to $\mathsf{rw} = H^n(F_k(\mathsf{pw}))$, that is, the client applies n iterations of the function H on top of the result of the OPRF value $F_k(\mathsf{pw})$. In practice, the iterations H^n would be replaced with one of the standard password-based KDFs, such as PBKDF2 [25] or bcrypt [31]. This forces an attacker that compromises the password file at the server to compute for *each* candidate password pw' the function $F_k(\mathsf{pw}')$ as well as the additional n hash iterations. Note that n needs not be remembered by the user; it can be sent from S to U in the server's message. Furthermore, one can follow Boyen's design and apply the probabilistic Halting KDF function [8] as used in [9] so that the iterations count is hidden from the attacker and even from the server.

• *Performance.* OPAQUE takes two messages (three with explicit mutual authentication); one exponentiation for S, two and a hashing-into-G for U, plus the cost of KE. With HMQV, the latter cost is one offline fixed-base exponentiation and one multi-exponentiation (at the cost of 1.16 regular exponentiations) per party (about three exponentiations in total for the server and four for the user). All exponentiations are in regular DH groups, hence accommodating the fastest elliptic curves (e.g., no pairings). It is common in PAKE protocols to count number of group elements transmitted between the parties. In OPAQUE, U sends two while S sends three (one, P_u, can be omitted at the cost of one fixed-based exponentiation at the client).

• *Performance comparison.* The introduction presents background on OPAQUE and other password protocols. Here we provide a comparison with the more efficient among these protocols, particularly those that are being, or have been, considered for standardization. Clearly, OPAQUE is superior security-wise as the only one not subject to pre-computation attacks, but it also fares well in terms of performance.

AugPAKE [33,34], is computationally very efficient with only 2.17 exponentiations per party; however, it uses 4 messages and does not provide forward secrecy. In addition, the protocol has only been analyzed as a PAKE protocol, not aPAKE [34]. Another proposed aPAKE protocol, SPAKE2+ [2,15], uses two messages only and 3 multi-exponentiations (or about 3.5 exponentiations) per party which is similar to OPAQUE cost. The security of the protocol has only been informally argued in [15] and to the best of our knowledge no formal analysis has appeared. We also mention SRP which has been included in TLS ciphersuites in the past but is considered outdated as it does not have an instantiation that works over elliptic curves (the protocol is defined over rings and uses both addition and multiplication). Its implementations over RSA moduli is therefore less efficient than those over elliptic curve; it also takes 4 messages.

We also mention two very recent schemes that have been formally analyzed as aPAKE protocols but, as the rest, are vulnerable to pre-computation. The protocol VTBPEKE in [30] uses 3 messages and 4 exponentiations per party and was proven secure in the non-UC aPAKE model of [7], while [24] shows a *simultaneous* one-round scheme that they prove secure in the UC aPAKE model of [18] augmented with adaptive security. The protocol works over bilinear groups and its computational cost includes 4 exponentiations and 3 pairing per party. We note that all of the above protocols require an initial message from server to user in order to transmit salt, which results in one or two added messages to the above message counts (except for VTBPEKE which already includes the salt transmission in its 3 messages). Also, all these protocols, like OPAQUE, work in the RO model.

• *Threshold implementation.* We comment on a simple extension of OPAQUE that can be very valuable in large deployments, namely, the ability to implement the OPRF phase as a Threshold OPRF [23]. In this case, an attacker needs to break into a threshold of servers to be able to impersonate the servers to the user or to run an offline dictionary attack. Such an implementation requires no user-side changes, i.e., the user does not need to know if the system is implemented with one or multiple servers.

• *Secret retrieval and hedging TLS.* Additional features of OPAQUE include the ability to store and retrieve user's secrets (such as a bitcoin wallet, authentication credentials, encrypted backup keys, etc.) as part of the information encrypted and authenticated at the server under ciphertext c. In one particular use case such secret can be a client signature key for TLS. In this case, the key exchange part of OPAQUE can reuse that of TLS and a server's certificate can be replaced with the server's public key stored under the client-authenticated ciphertext c.

6.2 An OPAQUE Variant: Multiplicative Blinding

A variant of OPAQUE is obtained by replacing the user's exponential blinding operation $\alpha := (H'(\mathsf{pw}))^r$ with $\alpha := (H'(\mathsf{pw})) \cdot g^r$. The server responds as before with $\beta = \alpha^{k_s}$. Assuming that U knows the value $y = g^{k_s}$ (previously stored or received from S), it can compute the same "hashed Diffie-Hellman" value $(H'(\mathsf{pw}))^{k_s}$ as β / y^r. The advantage of this variant is that while the number of client exponentiations remains the same, one is fixed-base (g^r) and the other (y^r) can also be fixed-base if U caches y, a realistic possibility for accounts where the user logs in frequently (e.g., a personal email or social network). Computing y^r can also be done while waiting for the server's response to reduce latency. Moreover, both exponentiations can be done offline although only short-term storage is recommended as the leakage of r exposes $H'(\mathsf{pw})$. If U does not store y, it needs to be transmitted to U by S together with the response β. This still allows for fixed-base optimization for computing g^r but not for y^r.

However, it turns out that this multiplicative mechanism results in an OPRF protocol that does *not* realize our OPRF functionality $\mathcal{F}_{\mathsf{OPRF}}$. Thus, our analysis here does not imply the security of the multiplicative OPAQUE variant in general. If rw is redefined as $\mathsf{rw} := H(\mathsf{pw}, y, H'(\mathsf{pw})^{k_s})$, i.e. if y is included under the hash, then the resulting OPRF does realize our functionality, and OPAQUE remains secure as SaPAKE under both blinding variants. This change, however, introduces a (slight) overhead of having to transmit y even if it is not strictly needed, e.g. if the client implements the exponential blinding operation. An alternative approach would be to replace the OPRF functionality $\mathcal{F}_{\mathsf{OPRF}}$ with a weaker form $\mathcal{F}'_{\mathsf{OPRF}}$ and to show that (i) $\mathcal{F}'_{\mathsf{OPRF}}$ is realized by the multiplicative variant (even without hashing y) and (ii) $\mathcal{F}'_{\mathsf{OPRF}}$ is sufficient for proving Theorem 2 hence implying the security of OPAQUE as SaPAKE. We intend to investigate this weakening of $\mathcal{F}_{\mathsf{OPRF}}$.

A The DH-OPRF Protocol Realizing Revised $\mathcal{F}_{\mathsf{OPRF}}$

Figure 9 shows the DH-OPRF protocol of [22] (who calls it 2HashDH), syntactically modified to realize functionality $\mathcal{F}_{\mathsf{OPRF}}$, see Fig. 2 in Sect. 3. Recall that the $\mathcal{F}_{\mathsf{OPRF}}$ functionality we show in Sect. 3 is a revision of the OPRF functionality defined in [22], with the most important difference being modeling adaptive corruptions. The protocol shown below is essentially the same as in [22], and requires the same One-More Diffie-Hellman assumption [3,22] for security.

Components: Hash functions $H(\cdot,\cdot)$, $H'(\cdot)$ with ranges $\{0,1\}^\ell$ and G, respectively.

Initialization

- On input (INIT, sid, x), S picks $k \leftarrow_{\mathrm{R}} \mathbb{Z}_q$ and outputs (INIT, $sid, H(x, H'(x)^k)$).

Evaluation

- On input (EVAL, $sid, ssid$, S, x), U proceeds as follows:
  - If there is a record $\langle \mathsf{S}, x, r, y\rangle$, outputs (EVAL, $sid, ssid, y$) to $\mathcal{Z}$.
  - Else if there is a record $\langle \mathsf{S}', x, r, y\rangle$ (where $\mathsf{S}' \neq \mathsf{S}$), sends $a := H'(x)^r$ to S.
  - Else picks $r \leftarrow_{\mathrm{R}} \mathbb{Z}_q$, records $\langle \mathsf{S}, x, r, \cdot\rangle$ and sends $a := H'(x)^r$ to S.
- On input (SNDRCOMPLETE, $sid, ssid$) and a from U, S sends $b := a^k$ to U.
- On b from S, if this is the first such message for $ssid$, U retrieves record $\langle \mathsf{S}, x, r, \cdot\rangle$, replaces $\cdot$ with $y := H_2(x, b^{1/r})$ and outputs (EVAL, $sid, ssid, y$).

Fig. 9. Protocol DH-OPRF (for PRF output length ℓ)

We defer the proof of the following Lemma 1 to the full version because it is very similar to the proof of security given in [22].

Lemma 1. *The DH-OPRF protocol shown in Fig. 9 UC-realizes the OPRF functionality $\mathcal{F}_{\mathsf{OPRF}}$ under the One-More Diffie Hellman assumption in ROM.*

Modifications in the Proof of [22]**.** We briefly discuss how our modifications to $\mathcal{F}_{\mathsf{OPRF}}$ influence the security proof, and leave the detailed proof to the full version of this paper.

Since no message is sent to $\mathcal{A}^*$ in the Initialization phase, adding Initialization has no impact on simulation. Allowing for sub-sessions (identified by *ssid*) results in adding *ssid* in the simulator whenever appropriate. The impact of changing SNDRCOMPLETE messages as sent from $\mathcal{Z}$, instead of from $\mathcal{A}^*$, is that no such messages are sent from SIM any more in steps 6 and 7; however, this does not influence the reduction that Pr[HALT] is negligible, since the only SNDRCOMPLETE messages which count are those in step 5, which are still there

1. Pick $r_1, \ldots, r_N \leftarrow_R \mathbb{Z}_m$, and compute $g_1 := g^{r_1}, \ldots, g_N := g^{r_N}$. Record $(r_1, g_1), \ldots, (r_N, g_N)$. Set counter $J := 1$.
2. Every time when there is a fresh query x to $H'(\cdot)$, answer it with g_J and record (x, r_J, g_J). Finally, set J++.
3. On (COMPROMISE, *sid*) from $\mathcal{A}$ as a message to S, mark S COMPROMISED, send (COMPROMISE, *sid*) to $\mathcal{F}$ and do:
   - If there is no record $\langle \mathsf{S}, k, z \rangle$, pick $k \leftarrow_R \mathbb{Z}_q$, compute $z := g^k$, record $\langle \mathsf{S}, k, z \rangle$ and send k to $\mathcal{A}$ as S's response.
   - Else retrieve $\langle \mathsf{S}, k, z \rangle$ and send k to $\mathcal{A}$ as S's response.
4. On (EVAL, *sid*, *ssid*, U, S) from $\mathcal{F}$, send g_J to $\mathcal{A}$ as U's message to S and record $\langle ssid, \mathsf{U}, \mathsf{S}, r_J, g_J \rangle$. Finally, set J++.
5. On (SNDRCOMPLETE, *sid*, *ssid*, S) from $\mathcal{F}$ and a from $\mathcal{A}$ as some user U's message to S, do:
   - If there is no record $\langle \mathsf{S}, k, z \rangle$, pick $k \leftarrow_R \mathbb{Z}_q$, compute $z := g^k$, record $\langle \mathsf{S}, k, z \rangle$ and send a^k to $\mathcal{A}$ as S's response to U.
   - Else retrieve $\langle \mathsf{S}, k, z \rangle$ and send a^k to $\mathcal{A}$ as S's response to U.
6. On b from $\mathcal{A}$ as some server S's message to a user U, retrieve record $\langle ssid, \mathsf{U}, \cdot, r_j, g_j \rangle$ and do:
   - If there is a record $\langle \mathsf{S}', \cdot, z \rangle$ such that $b^{1/r_j} = z$, send (RCVCOMPLETE, *sid*, *ssid*, S′) to $\mathcal{F}$.
   - Else create a new server S′, record $\left\langle \mathsf{S}', \cdot, b^{1/r_j} \right\rangle$ and send (RCVCOMPLETE, *sid*, *ssid*, S′) to $\mathcal{F}$.
7. Every time when there is a fresh query (x, u) to $H(\cdot, \cdot)$, do:
   (a) If there is a record (x, r_j, g_j), do:
      (1) If there is a record $\langle \mathsf{S}, k, z \rangle$ such that $u^{1/r_j} = z$ and S is marked COMPROMISED, send (OFFLINEEVAL, *sid*, S, x) to $\mathcal{F}$.
      On $\mathcal{F}$'s response (OFFLINEEVAL, *sid*, y), set $H(x, u) := y$.
      (2) Else if there is a record $\langle \mathsf{S}, \cdot, z \rangle$ such that $u = z^{r_j}$ and S is not marked COMPROMISED, send (EVAL, *sid*, *ssid*, S, x) and then (RCVCOMPLETE, *sid*, *ssid*, S) to $\mathcal{F}$.
      If $\mathcal{F}$ ignores this message, output HALT and abort.
      Otherwise on $\mathcal{F}$'s response (EVAL, *sid*, *ssid*, y), set $H(x, u) := y$.
   (b) In any other case, set $H(x, u) \leftarrow_R \{0, 1\}^l$.

Fig. 10. The simulator SIM for the DH-OPRF protocol ($\mathcal{F}_{\mathsf{OPRF}}$ denoted $\mathcal{F}$)

(the only difference is that their issuers become $\mathcal{Z}$ instead of SIM, but they still have the effect of increasing the tx value).

The remaining change is that $\mathcal{A}$ may compromise a server (for a specific *sid*) at any time; after that, $\mathcal{A}$ can compute the server's function value on any valid input. SIM is able to simulate this by sending OFFLINEEVAL messages to $\mathcal{F}$. Furthermore, note that HALT may only occur on servers who is not marked COMPROMISED at that time; therefore, the argument upper-bounding Pr[HALT] (in the setting where a server cannot be compromised) is not influenced.

References

1. CFRG: Crypto Forum Research Group. https://datatracker.ietf.org/rg/cfrg/documents/
2. Abdalla, M., Pointcheval, D.: Simple password-based encrypted key exchange protocols. In: Menezes, A. (ed.) CT-RSA 2005. LNCS, vol. 3376, pp. 191–208. Springer, Heidelberg (2005). https://doi.org/10.1007/978-3-540-30574-3_14
3. Bellare, M., Namprempre, C., Pointcheval, D., Semanko, M.: The one-more-RSA-inversion problems and the security of Chaum's blind signature scheme. J. Cryptol. **16**(3), 185–215 (2003)
4. Bellare, M., Pointcheval, D., Rogaway, P.: Authenticated key exchange secure against dictionary attacks. In: Preneel, B. (ed.) EUROCRYPT 2000. LNCS, vol. 1807, pp. 139–155. Springer, Heidelberg (2000). https://doi.org/10.1007/3-540-45539-6_11
5. Bellovin, S.M., Merritt, M.: Encrypted key exchange: Password-based protocols secure against dictionary attacks. In: IEEE Computer Society Symposium on Research in Security and Privacy – S&P 1992, pp. 72–84. IEEE (1992)
6. Bellovin, S.M., Merritt, M.: Augmented encrypted key exchange: a password-based protocol secure against dictionary attacks and password file compromise. In: ACM Conference on Computer and Communications Security - CCS 1993, pp. 244–250. ACM (1993)
7. Benhamouda, F., Pointcheval, D.: Verifier-based password-authenticated key exchange: New models and constructions. IACR Cryptology ePrint Archive, 2013:833 (2013)
8. Boyen, X.: Halting password puzzles. In: USENIX Security Symposium – SECURITY 2007, pp. 119–134. The USENIX Association (2007)
9. Boyen, X.: HPAKE: password authentication secure against cross-site user impersonation. In: Garay, J.A., Miyaji, A., Otsuka, A. (eds.) CANS 2009. LNCS, vol. 5888, pp. 279–298. Springer, Heidelberg (2009). https://doi.org/10.1007/978-3-642-10433-6_19
10. Boyko, V., MacKenzie, P., Patel, S.: Provably secure password-authenticated key exchange using Diffie-Hellman. In: Preneel, B. (ed.) EUROCRYPT 2000. LNCS, vol. 1807, pp. 156–171. Springer, Heidelberg (2000). https://doi.org/10.1007/3-540-45539-6_12
11. Canetti, R.: Universally composable security: a new paradigm for cryptographic protocols. In: IEEE Symposium on Foundations of Computer Science – FOCS 2001, pp. 136–145. IEEE (2001)
12. Canetti, R., Halevi, S., Katz, J., Lindell, Y., MacKenzie, P.: Universally composable password-based key exchange. In: Cramer, R. (ed.) EUROCRYPT 2005. LNCS, vol. 3494, pp. 404–421. Springer, Heidelberg (2005). https://doi.org/10.1007/11426639_24
13. Canetti, R., Krawczyk, H.: Analysis of key-exchange protocols and their use for building secure channels. In: Pfitzmann, B. (ed.) EUROCRYPT 2001. LNCS, vol. 2045, pp. 453–474. Springer, Heidelberg (2001). https://doi.org/10.1007/3-540-44987-6_28
14. Canetti, R., Krawczyk, H.: Universally composable notions of key exchange and secure channels. In: Knudsen, L.R. (ed.) EUROCRYPT 2002. LNCS, vol. 2332, pp. 337–351. Springer, Heidelberg (2002). https://doi.org/10.1007/3-540-46035-7_22

15. Cash, D., Kiltz, E., Shoup, V.: The twin Diffie-Hellman problem and applications. In: Smart, N. (ed.) EUROCRYPT 2008. LNCS, vol. 4965, pp. 127–145. Springer, Heidelberg (2008). https://doi.org/10.1007/978-3-540-78967-3_8
16. Farshim, P., Orlandi, C., Rosie, R.: Security of symmetric primitives under incorrect usage of keys. IACR Trans. Symmetric Cryptol. **2017**(1), 449–473 (2017)
17. Freedman, M.J., Ishai, Y., Pinkas, B., Reingold, O.: Keyword search and oblivious pseudorandom functions. In: Kilian, J. (ed.) TCC 2005. LNCS, vol. 3378, pp. 303–324. Springer, Heidelberg (2005). https://doi.org/10.1007/978-3-540-30576-7_17
18. Gentry, C., MacKenzie, P., Ramzan, Z.: A method for making password-based key exchange resilient to server compromise. In: Dwork, C. (ed.) CRYPTO 2006. LNCS, vol. 4117, pp. 142–159. Springer, Heidelberg (2006). https://doi.org/10.1007/11818175_9
19. Gong, L., Lomas, M.A., Needham, R.M., Saltzer, J.H.: Protecting poorly chosen secrets from guessing attacks. IEEE J. Sel. Areas Commun. **11**(5), 648–656 (1993)
20. Halevi, S., Krawczyk, H.: Public-key cryptography and password protocols. ACM Trans. Inf. Syst. Secur. (TISSEC) **2**(3), 230–268 (1999)
21. Jarecki, S., Kiayias, A., Krawczyk, H.: Round-optimal password-protected secret sharing and T-PAKE in the password-only model. In: Sarkar, P., Iwata, T. (eds.) ASIACRYPT 2014. LNCS, vol. 8874, pp. 233–253. Springer, Heidelberg (2014). https://doi.org/10.1007/978-3-662-45608-8_13
22. Jarecki, S., Kiayias, A., Krawczyk, H., Xu, J.: Highly-efficient and composable password-protected secret sharing (or: how to protect your bitcoin wallet online). In: IEEE European Symposium on Security and Privacy - EuroS&P 2016, pp. 276–291. IEEE (2016)
23. Jarecki, S., Kiayias, A., Krawczyk, H., Xu, J.: TOPPSS: cost-minimal password-protected secret sharing based on threshold OPRF. In: Gollmann, D., Miyaji, A., Kikuchi, H. (eds.) ACNS 2017. LNCS, vol. 10355, pp. 39–58. Springer, Cham (2017). https://doi.org/10.1007/978-3-319-61204-1_3
24. Jutla, C.S., Roy, A.: Smooth NIZK arguments with applications to asymmetric UC-PAKE. IACR Cryptology ePrint Archive 2016:233 (2016)
25. Kaliski, B.: PKCS #5: password-based cryptography specification version 2.0 (2000)
26. Kiefer, F., Manulis, M.: Zero-knowledge password policy checks and verifier-based PAKE. In: Kutyłowski, M., Vaidya, J. (eds.) ESORICS 2014. LNCS, vol. 8713, pp. 295–312. Springer, Cham (2014). https://doi.org/10.1007/978-3-319-11212-1_17
27. Krawczyk, H.: HMQV: a high-performance secure Diffie-Hellman protocol. In: Shoup, V. (ed.) CRYPTO 2005. LNCS, vol. 3621, pp. 546–566. Springer, Heidelberg (2005). https://doi.org/10.1007/11535218_33
28. MacKenzie, P.: More efficient password-authenticated key exchange. In: Naccache, D. (ed.) CT-RSA 2001. LNCS, vol. 2020, pp. 361–377. Springer, Heidelberg (2001). https://doi.org/10.1007/3-540-45353-9_27
29. MacKenzie, P., Patel, S., Swaminathan, R.: Password-authenticated key exchange based on RSA. In: Okamoto, T. (ed.) ASIACRYPT 2000. LNCS, vol. 1976, pp. 599–613. Springer, Heidelberg (2000). https://doi.org/10.1007/3-540-44448-3_46
30. Pointcheval, D., Wang, G.: VTBPEKE: verifier-based two-basis password exponential key exchange. In: ACM Asia Conference on Computer and Communications Security – AsiaCCS 2017, pp. 301–312. ACM (2017)
31. Provos, N., Mazieres, D.: A future-adaptable password scheme. In: USENIX Annual Technical Conference, FREENIX Track, pp. 81–91 (1999)

32. Schmidt, J.: Requirements for password-authenticated key agreement (PAKE) schemes. Technical report (2017)
33. Shin, S., Kobara, K.: Augmented password-authenticated key exchange (AugPAKE). draft-irtf-cfrg-augpake-08
34. Shin, S., Kobara, K., Imai, H.: Security proof of AugPAKE. IACR Cryptology ePrint Archive 2010:334 (2010)

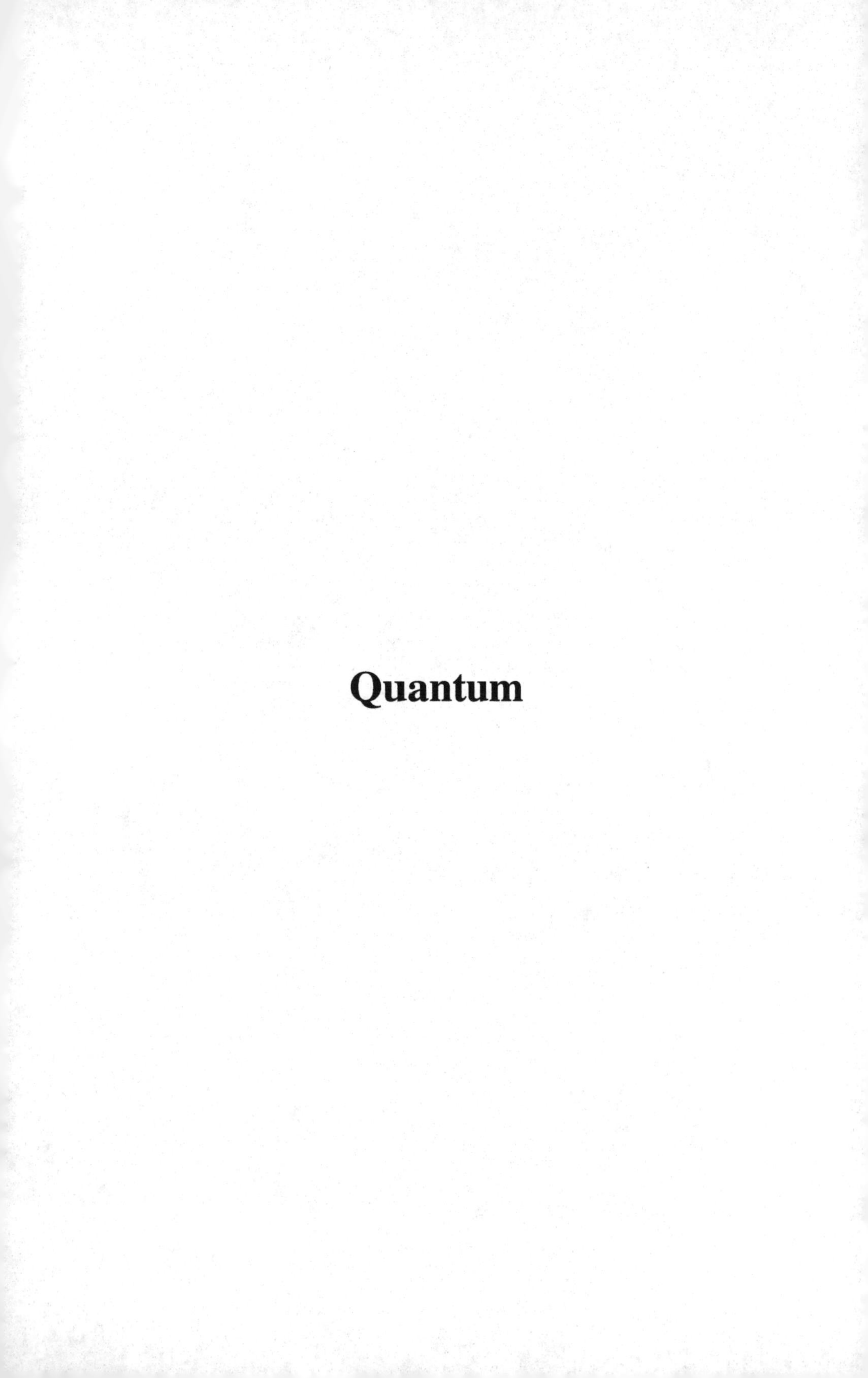

Quantum

Unforgeable Quantum Encryption

Gorjan Alagic[1,2(✉)], Tommaso Gagliardoni[3], and Christian Majenz[4,5]

[1] Joint Center for Quantum Information and Computer Science, University of Maryland, College Park, MD, USA
galagic@umd.edu
[2] National Institute of Standards and Technology, Gaithersburg, MD, USA
[3] IBM Research, Zurich, Switzerland
tog@zurich.ibm.com
[4] Institute for Logic, Language and Computation, University of Amsterdam, Amsterdam, Netherlands
c.majenz@uva.nl
[5] Centrum for Wiskunde en Informatica, Amsterdam, Netherlands

Abstract. We study the problem of encrypting and authenticating quantum data in the presence of adversaries making adaptive chosen plaintext and chosen ciphertext queries. Classically, security games use string copying and comparison to detect adversarial cheating in such scenarios. Quantumly, this approach would violate no-cloning. We develop new techniques to overcome this problem: we use entanglement to detect cheating, and rely on recent results for characterizing quantum encryption schemes. We give definitions for (i) ciphertext unforgeability, (ii) indistinguishability under adaptive chosen-ciphertext attack, and (iii) authenticated encryption. The restriction of each definition to the classical setting is at least as strong as the corresponding classical notion: (i) implies INT-CTXT, (ii) implies IND-CCA2, and (iii) implies AE. All of our new notions also imply QIND-CPA privacy. Combining one-time authentication and classical pseudorandomness, we construct symmetric-key quantum encryption schemes for each of these new security notions, and provide several separation examples. Along the way, we also give a new definition of one-time quantum authentication which, unlike all previous approaches, authenticates ciphertexts rather than plaintexts.

1 Introduction

Given the rapid development of quantum information processing, it is reasonable to conjecture that future communication networks will include at least some large-scale quantum computers and high-capacity quantum channels. What will secure communication look like on the resulting "quantum Internet"? For instance, how will we transmit quantum messages securely over a completely insecure channel? One approach is via interactive and information-theoretically secure methods, e.g., combining entanglement distillation with teleportation.

J. B. Nielsen and V. Rijmen (Eds.): EUROCRYPT 2018, LNCS 10822, pp. 489–519, 2018.
https://doi.org/10.1007/978-3-319-78372-7_16

In this work, we will instead consider the non-interactive, highly efficient approach which dominates the current classical Internet. A natural goal here is to achieve, in the quantum setting, all the basic features that are enjoyed by classical encryption: (i) a single small key suffices for transmitting an essentially unlimited amount of data, (ii) these keys can be exchanged over public channels, and (iii) the security guarantees are as strong as possible. Previous work has shown how to achieve both (i) and (ii), but only for secrecy against chosen-plaintext and non-adaptive chosen-ciphertext attacks [3,14]. Authentication or adaptive chosen-ciphertext security for such schemes has, as yet, not been considered. In fact, at the time of writing, there is not even a definition for *two-time quantum authentication*, much less for quantum analogues of EUF-CMA or IND-CCA2. The aim of this work is to address this problem.

The security definitions we seek do not yet exist due to a number of technical obstacles, all of which can be traced to quantum no-cloning and the destructiveness of quantum measurements. These obstacles make it difficult even just to formulate the basic security notion, much less to prove reductions or to construct secure schemes. In unforgeability, for example, no-cloning makes it impossible to record the adversary's queries and check whether the final output is a fresh forgery. In adaptive chosen-ciphertext security, no-cloning makes it impossible to record the challenge ciphertext and ensure that the adversary doesn't "cheat" by simply decrypting it (and thus win against any scheme). Moreover, due to the destructiveness of quantum measurement, it is unclear if one can *both* perform cheat-detection *and* answer non-cheating queries correctly.

In this work, we overcome these obstacles, and present the first definitions of multiple-query unforgeability and adaptive chosen-ciphertext indistinguishability for quantum encryption schemes, thereby solving a longstanding open problem [3,12,20]. While our definitions are inherently quantum in nature, we are able to show that they are in fact natural analogues of well-known classical security definitions, such as INT-CTXT and IND-CCA2. The strongest security notion we define is called *quantum authenticated encryption* (or QAE) and corresponds to the strongest form of security normally studied in the classical setting. A secret-key scheme satisfying QAE is unforgeable and indistinguishable even against adversaries that can make adaptive encryption and decryption queries.

In an effort to explore this new landscape, we prove several theorems which relate our new notions to each other and to established quantum and classical security definitions. We also show how to satisfy each of our new security notions with explicit, efficient constructions. In particular, we show that combining a post-quantum secure pseudorandom function with a unitary 2-design yields the strongest form of secret-key quantum encryption defined thus far, i.e., QAE.

Related Work. Computationally-secure quantum encryption has garnered significant interest in the past few years, beginning with basic security notions like QIND-CPA and QIND-CCA1 [3,14], and then with more advanced concepts such as quantum fully-homomorphic encryption (QFHE) [14,17]. For authentication, uncloneability, and non-malleability, the one-time setting has received considerable attention (see, e.g., [5,6,15,19,21,23,24,27].) We will make use of the authentication definition of [19], a characterization lemma of [5], and a simulation adversary of [15]. For classical notions of unforgeability and chosen-ciphertext security, see e.g. [25].

1.1 Our Approach

The Problem. We begin by outlining the technical difficulties in some further detail. Let us consider many-time authentication for symmetric-key encryption schemes first. In the classical setting, secure many-time authentication is defined in terms of *unforgeability.* A scheme is unforgeable if no adversary, even if granted the black-box power to authenticate with our secret key, can generate a fresh and properly authenticated message (i.e., a forgery). Translating this idea to the quantum setting presents immediate technical difficulties. First, no-cloning prevents us from recording the adversary's previous queries. Second, even if the first problem is surmounted, the nature of measurement might make it difficult to reliably identify whether the adversary's output is indeed fresh. For example, we might need many copies of the adversary's query, as well as many copies of their final output.

A similar problem occurs for secrecy. The current state-of-the-art is the so-called QIND-CCA1 model. In this model, the transmitted state (the "challenge") remains secret even to adversaries with the black-box power to both encrypt and non-adaptively decrypt with our secret key. Our experience in the classical world tells us that this model is too weak, because real-world adversaries can sometimes gain *adaptive* access to decryption (e.g., in WEP and early versions of SSL [8]). Classically, this is addressed using the so-called IND-CCA2 model, where the adversary is allowed adaptive decryption queries *but cannot use them on the challenge* (without this caveat, security becomes impossible). Here again, the quantum setting presents numerous technical difficulties: no-cloning prevents us from recording the challenge, and the nature of measurement makes it difficult to tell if the adversary is attempting to decrypt the challenge.

Recall that the strongest form of classical security, so-called "authenticated encryption" (or AE) is defined to be IND-CCA2 together with unforgeability of ciphertexts [25]. Achieving a comparable quantum notion thus seems to require solving all of the above problems.

Using classical intuition, one might attempt a solution as follows: consider only pure-state plaintexts, and demand that the final forgery is orthogonal to the previous queries (or, in CCA2, that decryption queries are orthogonal to the challenge). This may seem promising at first, but a closer look reveals numerous issues; for example: (i) quantum states are in general not pure, and may include side registers kept by the adversary, (ii) this idea charges the adversary with adhering to very strict demands, contrary to good theory practice, (iii) checking whether a particular adversary satisfies the demands cannot be done efficiently.

A Promising Approach. We now describe a more promising solution, beginning with unforgeability. We will express security in terms of the performance of adversaries $\mathcal{A}$ in two games: (1) F-Real, where $\mathcal{A}$ gets oracle access to Enc_k and wins if he outputs *any* valid ciphertext, and (2) F-Cheat, where we attempt to ascertain if $\mathcal{A}$ is cheating by feeding us an output of the oracle. How do we detect this kind of cheating? Recall that, even in the one-time setting, quantum authentication implies indistinguishability of ciphertexts. A consequence of this is that, whenever $\mathcal{A}$ performs an encryption query on a certain plaintext state, we are free to respond with an encryption *of a different state* – for example, half of a maximally-entangled state. This will be our approach: we prepare an entangled pair $|\phi^+\rangle_{MM'}$, apply Enc_k to register M, give the resulting ciphertext register to $\mathcal{A}$, and keep M'. When the game ends, we decrypt the output of $\mathcal{A}$ into a register O, and then perform the measurement $\{\Pi_{\phi^+}, \mathbb{1} - \Pi_{\phi^+}\}$ on OM'. We then declare that $\mathcal{A}$ is cheating if and only if the first outcome is recorded.

This idea can also be applied to the multiple-query setting. There, we respond to the jth query with an encryption of register M of $|\phi^+\rangle_{MM_j}$, and save M_j; at the end of the game, we perform the aforementioned measurement on OM_j for all j and declare that $\mathcal{A}$ cheated if any of them return the first outcome.

To define a quantum analogue of IND-CCA2, we can try a similar strategy. We again compare the performance of $\mathcal{A}$ in two games: (1) C-Real, which is just like the classical IND-CCA2 game, except with no restrictions on $\mathcal{A}$'s use of the Dec_k oracle, and (2) C-Cheat, where we again attempt to detect cheating. In C-Cheat, when the adversary sends us the challenge plaintext, we discard it and respond with the ciphertext register of $(\mathsf{Enc}_k \otimes \mathbb{1}_{M'})|\phi^+\rangle_{MM'}$ instead, while keeping M' to ourselves. Whenever $\mathcal{A}$ queries the decryption oracle, we first apply Dec_k and place the resulting plaintext in a register O. Then we apply the measurement $\{\Pi_{\phi^+}, \mathbb{1} - \Pi_{\phi^+}\}$ to OM' to see if the adversary is cheating. If we get the first outcome, we declare that $\mathcal{A}$ cheated.

The above ideas do lead to reasonable security definitions, which (at least partly) fulfill our original goals. However, they suffer from a number of drawbacks. First, repeated measurement of the plaintext requires the use of a so-called "gentle measurement lemma" [29], and thus can only apply to large plaintext spaces (e.g., n^c qubits for $c > 0$). Second, they only offer *plaintext authentication* and a kind-of *plaintext CCA security*; modification of ciphertexts (that does not also modify the underlying plaintext) cannot be detected. Our classical experience tells us that this is insufficient, and that we should demand impossibility of any ciphertext manipulation whatsoever. Addressing these problems is where many of our new technical contributions (in addition to the above ideas) are needed. While our actual approach will be different, and more sophisticated techniques are required, we will still follow the spirit of the idea outlined above.

1.2 Summary of Results

Recall that, in the setting of quantum data, copying is impossible and authentication implies encryption [9]. In particular, there is no direct quantum analogue of a MAC. As a result, the central objects of study in our work will be symmetric-key quantum encryption schemes, or SKQES for short, but our results on quantum CCA2 security carry over to the public-key setting as well.

Quantum Ciphertext Authentication. All previous definitions of authentication for quantum data allow manipulation of the ciphertext (see Sect. 2), thus only authenticating the plaintext state. In our first main contribution, we solve this problem, laying the necessary groundwork for our remaining results.

- We give a new definition: information-theoretic *quantum one-time ciphertext authentication* (QCA), inspired by ideas of [5,15].
- We prove that QCA is a strengthening of "DNS"-authentication [19].

Theorem 1 (informal). *If a SKQES authenticates ciphertexts* (QCA)*, then it also authenticates plaintexts* (DNS)*; in particular, it satisfies secrecy* (QIND).

- *We define computational-security (one-time) analogues:* cQCA *and* cDNS.

Quantum Unforgeability. In this setting, the adversary is granted access to an encryption oracle, and must generate a valid "fresh" ciphertext.

- We give a new definition: *quantum unforgeability* (QUF), combining ideas of Sect. 1.1 and [5]. We also define a bounded-query analogue (t-QUF).
- We show that UF, the classical analogue of QUF, is remarkably strong.

Theorem 2 (informal). *For classical schemes,* UF $\iff$ AE.

Quantum Chosen-Ciphertext Security. We address the longstanding problem of defining quantum security under adaptive chosen-ciphertext attack [3,12,20]; the state of the art was previously the non-adaptive QIND-CCA1 [3].

- We give a new definition: *quantum indistinguishability under adaptive chosen-ciphertext attack*
  (QIND-CCA2), using all of the aforementioned ideas.
- We relate QIND-CCA2 to existing security notions.

Theorem 3 (informal).

1. *For quantum schemes,* QIND-CCA2 $\implies$ QIND-CCA1.
2. *The classical analogue of* QIND-CCA2 *is equivalent to classical* IND-CCA2.

Quantum Authenticated Encryption. In our main contribution, we define a natural quantum analogue of the classical concept of *authenticated encryption* (AE). All previous quantum security notions lacked both unforgeability and adaptive chosen-ciphertext security.

- We give a new definition: *quantum authenticated encryption* (QAE), combining the ideas of Sect. 1.1, the notion of QCA, and a real/ideal approach [28].
- We give evidence that QAE is indeed the correct quantum analogue of AE.

Theorem 4 (informal).

1. *Unforgeability and secure authentication:* QAE $\implies$ QUF $\wedge$ cQCA.
2. *Chosen-ciphertext security:* QAE $\implies$ QIND-CCA2.
3. *The classical analogue of* QAE *is equivalent to classical* AE.

The new notions and connections we develop are summarized in Fig. 1.

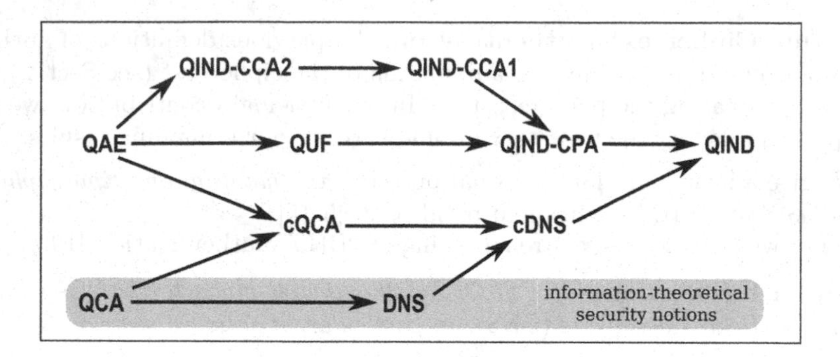

Fig. 1. Implications between quantum security notions

Constructions and Separations. Our new constructions combine a SKQES Π with a classical keyed function family f to build a new SKQES Π^f, as follows. In Π^f, key generation outputs a key for f; to encrypt a state ϱ, we generate a random r and output $(r, \mathsf{Enc}^{\Pi}_{f_k(r)}(\varrho))$. For example, if Π is the quantum one-time pad and f is a pqPRF (i.e., a post-quantum-secure pseudo-random function), then Π^f is the IND-CCA1-secure scheme from [3]. We will also need the standard one-time authentication scheme 2desTag, defined by $\mathsf{Enc}_k : \varrho \mapsto C_k(\varrho \otimes |0^n\rangle\langle 0^n|)C_k^\dagger$ where C is an (exact or approximate) unitary two-design.

Theorem 5 (informal). *Let Π be a* 2desTag *scheme, let f be a* pqPRF*, and let g be a t-wise independent classical function family. Then*

1. *Π is one-time ciphertext authenticating (*QCA*).*
2. *Π^g is t-time quantum unforgeable (*t-QUF*).*
3. *Π^f satisfies quantum authenticated encryption (*QAE*); in particular, it is quantum unforgeable (*QUF*) and chosen-ciphertext secure (*QIND-CCA2*).*

Theorem 6 (informal).

1. *There exists an SKQES which is* QIND-CCA1 *but not* QIND-CCA2.
2. *There exists an SKQES which is* QIND-CCA2 *but not* QAE.

Our Choice of Primitives. The reader may wonder why our constructions do not need "quantum-oracle-secure" primitives (e.g., QPRFs for unforgeability and $2t$-wise independence for t-time security, as in the quantum-secure classical setting of [11]). In our work, the classical portion of the ciphertext is generated by honest parties during encryption, and measured during decryption. As a result, oracle access to Enc_k and Dec_k (as CPTP maps) never grants quantum oracle access to the underlying classical primitive. Of course, one could grant the adversary more powerful oracles that do grant this kind of access, and then quantum-oracle-secure primitives (such as QPRFs) would indeed be required.

A Remark on Applicability. While all of our definitions apply to arbitrary quantum encryption schemes, security reductions sometimes require the following additional condition. As discussed in Sect. 3, all quantum encryption algorithms can be characterized as (1) drawing a random pure state from a probability distribution, (2) attaching it to the plaintext, and (3) applying a unitary operator. For the implication $\mathsf{QAE} \Rightarrow \mathsf{cQCA}$ of Theorem 4 to hold, it is required that (1), (2) and (3) are efficiently implementable. This condition holds for all schemes known to us. However, it is *in principle* possible that there are schemes for which Enc_k is efficiently implementable, but the particular implementation "(1), then (2), then (3)" is not. We leave this as an open problem.

2 Preliminaries

Basic Notation and Conventions. In the rest of this work, we use "classical" to denote "non-quantum", "iff" for "if and only if", and n to denote the security parameter. A function $\varepsilon(n)$ is negligible (denoted $\varepsilon(n) \leq \mathrm{negl}(n)$) if it is asymptotically smaller than $1/p(n)$ for every polynomial function p. The notation $x \xleftarrow{\$} X$ means that x is a sample from the uniform distribution over the set X. By "PPT" we mean a polynomial-time uniform family of probabilistic circuits, and by "QPT" we mean a polynomial-time uniform family of quantum circuits. We will frequently give such algorithms names like "adversary" or "challenger," but this is only to help remember the role of the algorithm.

For notation and conventions regarding quantum information, we refer the reader to [26]. We recall a few basics here. We denote by $\mathcal{H}_M$ a complex Hilbert space with label M and finite dimension $\dim M$. We use the standard bra-ket notation to work with pure states $|\varphi\rangle \in \mathcal{H}_M$. The class of positive, Hermitian, trace-one linear operators on $\mathcal{H}_M$ is denoted by $\mathfrak{D}(\mathcal{H}_M)$. A *quantum register* is a physical system whose set of valid states is $\mathfrak{D}(\mathcal{H}_M)$; in this case we label by M the register itself. We reserve the notation τ_M for the maximally mixed state (i.e., uniform classical distribution) $\mathbb{1}/\dim M$ on M.

In a typical cryptographic scenario, a "quantum register M" is in fact an infinite family of registers $\{M_n\}_{n\in\mathbb{N}}$ consisting of $p(n)$ qubits, where p is some fixed polynomial. This family is parameterized by n, which is typically also the security parameter. We will consider completely positive (CP), trace-preserving (TP) maps (i.e., quantum channels) when describing quantum algorithms. To indicate that Φ is a channel from register A to B, we will write $\Phi_{A\to B}$. When it helps to clarify notation, we will use $\circ$ to denote composition of operators. We will also often drop tensor products with the identity, e.g., given a map $\Psi_{BC\to D}$, we will write $\Psi \circ \Phi$ to denote the map $\Psi \circ (\Phi \otimes \mathbb{1}_C)$ from AC to D.

The support of a quantum state ϱ is its cokernel (as a linear operator). Equivalently, this is the span of the pure states making up any decomposition of ϱ as a convex combination of pure states. We will denote the orthogonal projection operator onto this subspace by P_ϱ. The two-outcome projective measurement (to test if a state has the same or different support as ϱ) is then $\{P_\varrho, \mathbb{1} - P_\varrho\}$.

Next, we single out some unitary operators that will appear frequently. First, the group of n-qubit operators generated by Paulis I, X, Y, Z (applied to individual qubits) is a well-known *unitary one-design*. The Clifford group on n qubits is defined to be the normalizer of the Pauli group inside the unitary group. It can also be seen as the group generated by the gate set $(H, P, CNOT)$ [22]; it is also a *unitary two-design* [16].

A *unitary t-design* (for a fixed t) is an infinite collection $\mathcal{U} = \{\mathcal{U}^{(n)} : n \in \mathbb{N}\}$, where $\mathcal{U}^{(n)}$ forms an n-qubit unitary t-design in the standard sense, i.e.,

$$\frac{1}{|\mathcal{U}^{(n)}|} \sum_{U \in \mathcal{U}^{(n)}} U^{\otimes t} X \left(U^{\dagger}\right)^{\otimes t} = \int U^{\otimes t} X \left(U^{\dagger}\right)^{\otimes t} dU \,. \tag{1}$$

In the above, the integral is taken over the n-qubit unitary group according to the Haar measure. We assume that there is an explicit polynomial function $m(n)$ and a deterministic polynomial-time algorithm which, given 1^n and $k \xleftarrow{\$} \{0,1\}^{m(n)}$, produces a circuit for a unitary operator $U_{k,n}$ which is distributed uniformly at random in $\mathcal{U}^{(n)}$. We will not refer to this algorithm explicitly and will simply write $\{U_{k,n} : k \in \{0,1\}^{m(n)}\}$ for the resulting distribution on unitary operators; we will also frequently suppress one index and write U_k when n is clear from context. We refer to the polynomial m as the *key length* of the t-design. Standard examples are: (i) the Pauli one-design (where we apply $X^a Z^b$ to each qubit for random $a, b \in \{0,1\}$) is a unitary one-design on n qubits with key length $2n$; (ii) the Clifford group (where we apply a uniformly random element of the n-qubit Clifford group, efficiently generated via the Gottesman-Knill theorem [1]) is a unitary 3-design, and therefore in particular a unitary 2-design, on n qubits with key length $O(n^2)$; (iii) random poly(t, n)-size quantum circuits, randomly generated from a universal gate set, are approximate t-designs on n qubits [13].

In this work, we will only require one-designs and two-designs, and we will assume for simplicity that the designs are exact. While approximate designs would also suffice, some additional (but straightforward) analysis would be required.

Quantum Encryption. We will follow the conventions set in [3]; the exception is that decryption can reject by outputting a special symbol $\perp$.

Definition 1. *A* symmetric-key quantum encryption scheme (or SKQES) *is a triple of QPT algorithms:*

1. *(key generation)*[1] KeyGen : *on input* 1^n*, outputs* $k \xleftarrow{\$} \mathcal{K}$
2. *(encryption)* Enc : $\mathcal{K} \times \mathfrak{D}(\mathcal{H}_M) \to \mathfrak{D}(\mathcal{H}_C)$
3. *(decryption)* Dec : $\mathcal{K} \times \mathfrak{D}(\mathcal{H}_C) \to \mathfrak{D}(\mathcal{H}_M \oplus |\perp\rangle\langle\perp|)$

such that $\|\mathsf{Dec}_k \circ \mathsf{Enc}_k - \mathbb{1}_M \oplus 0_\perp\|_\diamond \leq \mathrm{negl}(n)$ *for all* $k \in \mathbf{supp}$ $\mathsf{KeyGen}(1^n)$.

[1] A more general definition uses arbitrary key generation algorithms. We assume a uniform key in this paper for technical and notational convenience.

It is implicit that the key space $\mathcal{K}$ is classical and of size poly(n); likewise, the registers C and M are quantum registers of at most poly(n) qubits. We will only consider SKQES of *fixed-length*, meaning that the number of qubits in M is a fixed function of the security parameter n. We assume that honest parties will apply the measurement $\{\Pi_\perp, \mathbb{1} - \Pi_\perp\}$ (where $\Pi_\perp = |\perp\rangle\langle\perp|$) immediately after decryption. This allows us to write, e.g., $\mathsf{Dec}_k(\varrho) \neq \perp$ to mean that decryption (followed by this measurement) successfully produced a valid plaintext.

We will often combine quantum schemes with classical (keyed) function families. A keyed function family consists of functions $f : \{0,1\}^{p(n)} \times \{0,1\}^{q(n)} \to \{0,1\}^{s(n)}$ where p, q, s are polynomials in n. In typical usage, we sample a key $k \xleftarrow{\$} \{0,1\}^{p(n)}$ and then consider the restricted function $f_k : \{0,1\}^{q(n)} \to \{0,1\}^{s(n)}$ defined by $f_k(x) = f(k, x)$. All keyed function families are assumed to be computable by a deterministic polynomial-time uniform classical algorithm.

Definition 2. *Let* $\Pi = (\mathsf{KeyGen}^\Pi, \mathsf{Enc}^\Pi, \mathsf{Dec}^\Pi)$ *be a SKQES, and* $f : \{0,1\}^{p(n)} \times \{0,1\}^{q(n)} \to \{0,1\}^{s(n)}$ *a classical keyed function family. Define a new SKQES* $\Pi^f = (\mathsf{KeyGen}, \mathsf{Enc}, \mathsf{Dec})$ *as follows:*

1. KeyGen : *on input* 1^n, *outputs* $k \xleftarrow{\$} \{0,1\}^{p(n)}$;
2. Enc_k : *on input* ϱ, *outputs* $|r\rangle\langle r| \otimes \mathsf{Enc}^{\Pi}_{f_k(r)}(\varrho)$, *where* $r \xleftarrow{\$} \{0,1\}^{q(n)}$;
3. $\mathsf{Dec}_k : |s\rangle\langle s| \otimes \sigma \mapsto \mathsf{Dec}^{\Pi}_{f_k(s)}(\sigma)$.

We extend Dec_k to arbitrary inputs by postulating that it begins by measuring the first register in the computational basis. Note that Π^f has plaintext length $t(s(n))$ where $t(.)$ is the plaintext length of Π as a function of Π's key length. This construction can be extended to schemes Π with a non-uniform key by using the output of the keyed function family as a random tape for KeyGen^Π.

Quantum Secrecy. The literature contains a number of information-theoretic definitions of quantum secrecy (see, e.g., [3,6,7,14]). It is well-known that a unitary one-design (e.g., the Pauli group) is an information-theoretically secret scheme. In this work, however, we focus on the computational setting [3,14].

Definition 3 (QIND). *A SKQES* $\Pi = (\mathsf{KeyGen}, \mathsf{Enc}, \mathsf{Dec})$ *has indistinguishable encryptions (or is* QIND*) if for every QPT adversary* $\mathcal{A} = (\mathcal{M}, \mathcal{D})$ *we have:*

$$\Big|\Pr\big[\mathcal{D}\{(\mathsf{Enc}_k \otimes \mathbb{1}_E)\varrho_{ME}\} = 1\big] - \Pr\big[\mathcal{D}\{(\mathsf{Enc}_k \otimes \mathbb{1}_E)(|0\rangle\langle 0|_M \otimes \varrho_E)\} = 1\big]\Big| \leq \text{negl}(n),$$

where $\varrho_{ME} \leftarrow \mathcal{M}(1^n)$, $\varrho_E = \mathrm{Tr}_M(\varrho_{ME})$, *and the probabilities are taken over* $k \leftarrow \mathsf{KeyGen}(1^n)$ *and the coins and measurements of* Enc, $\mathcal{M}$, $\mathcal{D}$. *We also define:*

- QIND-CPA*: In addition to the above,* $\mathcal{M}$ *and* $\mathcal{D}$ *have oracle access to* Enc_k.
- QIND-CCA1*: In addition to* QIND-CPA, $\mathcal{M}$ *has oracle access to* Dec_k.

Recall that a pqPRF (post-quantum pseudorandom function) is a classical, deterministic, efficiently computable keyed function family $\{f_k\}_k$ which appears random to QPT algorithms with classical oracle access to f_k for uniformly random k. The strongest notion (QIND-CCA1) is satisfied by Π^f where Π is a one-design and f is a pqPRF [3]. We let $\mathsf{1des}^{\mathsf{PRF}}$ denote such schemes.

One-Time Authentication. We recall quantum authentication as defined by Dupuis et al. [19], and adapt it to our conventions. Given an attack map $\Lambda_{CB\to C\tilde{B}}$ on a scheme $\Pi = (\mathsf{KeyGen}, \mathsf{Enc}, \mathsf{Dec})$ (where the adversary holds B and $\tilde{B}$), we define the "averaged effective plaintext map" (or just "effective map") as follows.

$$\Lambda^{\Pi}_{MB\to M\tilde{B}} := \mathbb{E}_{k\leftarrow \mathsf{KeyGen}(1^n)}\left[\mathsf{Dec}_k \circ \Lambda \circ \mathsf{Enc}_k\right].$$

We then require that, conditioned on acceptance, this map is the identity on M.

Definition 4 ([19]). *A SKQES* $\Pi = (\mathsf{KeyGen}, \mathsf{Enc}, \mathsf{Dec})$ *is* DNS*-authenticating if, for any CP-map* $\Lambda_{CB\to C\tilde{B}}$*, there exist CP-maps* $\Lambda^{\mathsf{acc}}_{B\to\tilde{B}}$ *and* $\Lambda^{\mathsf{rej}}_{B\to\tilde{B}}$ *that sum to a TP map, such that*

$$\left\|\Lambda^{\Pi}_{MB\to M\tilde{B}} - \left(\mathrm{id}_M \otimes \Lambda^{\mathsf{acc}}_{B\to\tilde{B}} + |\bot\rangle\langle\bot|_M \otimes \Lambda^{\mathsf{rej}}_{B\to\tilde{B}}\right)\right\|_{\diamond} \leq \mathrm{negl}(n)\,. \qquad (2)$$

An important observation is that this definition only provides for authentication of the plaintext state. To see that this cannot be "ciphertext authentication," simply take a scheme which is DNS and change it so that (i) an extra bit is added to the ciphertext during encryption, and (ii) that same bit is ignored during decryption. The resulting scheme still satisfies DNS, but the adversary can clearly forge ciphertexts by flipping the extra bit. A perhaps more compelling example just adds encoding (in some QEC code) after encryption, and decoding prior to decryption. The adversary is then free to modify ciphertexts with correctable errors without violating DNS. We remark that, in this respect, the recent strengthening of DNS due to Garg et al. [21] is no different: a scheme secure according to this stronger notion of authentication can be modified in the same way without losing security.

Next, we recall a standard one-time authentication scheme. We encrypt by appending n "tag" qubits in the fixed state $|0\rangle$ and then applying a random element of a 2-design. Decryption first undoes the 2-design, then outputs the plaintext iff all tag qubits measure to 0; otherwise it outputs $\bot$.

Scheme 1. The scheme family 2desTag is defined as follows. Select a unitary 2-design $\mathcal{U}$ with key length $m(\cdot)$, and define algorithms:

1. KeyGen: on input 1^n, output $k \xleftarrow{\$} \{0,1\}^{m(2n)}$;
2. Enc_k: on input ϱ_M, output $U_k(\varrho_M \otimes |0^n\rangle\langle 0^n|_T)U_k^\dagger$
3. Dec_k: on input σ_{MT}, output

$$\langle 0^n|_T U_k^\dagger \sigma_{MT} U_k |0^n\rangle_T + \mathrm{Tr}\left[(\mathbb{1} - |0^n\rangle\langle 0^n|_T)U_k^\dagger \sigma_{MT} U_k\right] |\bot\rangle\langle\bot|_M\,.$$

We chose 2desTag to have plaintext and tag length n. It is well-known that, for plaintexts of at most polynomial length and tags of length at least n^c, these schemes are DNS-authenticating [2,19].

3 One-Time Ciphertext Authentication

One-time quantum authentication has been extensively studied [5,9,15,18,19, 21]. As we observed above, all of these works concern *plaintext authentication*, which ensures that manipulated ciphertexts decrypt to either the original plaintext or the reject symbol. Classical MACs, on the other hand, provide *ciphertext authentication*, which ensures that any ciphertext manipulation whatsoever will result in rejection. This distinction is important; for instance, in classical IND-CCA2, the adversary can defeat plaintext-authenticating schemes by invoking the decryption oracle on a modified challenge ciphertext.

In this section we show how to define and construct ciphertext authentication in the quantum setting. These ideas will be crucial to defining more advanced notions (such as ciphertext unforgeability and adaptive chosen-ciphertext security) later in the paper. We start with the information-theoretical security setting, and then we discuss how to apply these notions to the computational setting.

A Characterization of Encryption Schemes. We recall a lemma from [5] stating that all SKQES encrypt by (i) attaching some (possibly key-dependent) auxiliary state, and (ii) applying a unitary[2] operator. Decryption undoes the unitary, and then checks if the support of the state in the auxiliary register has changed. We emphasize that this characterization follows from correctness only, and thus applies to all schemes.

Lemma 1 (Lemma B.9 in [5], restated). *Let $\Pi = (\mathsf{KeyGen}, \mathsf{Enc}, \mathsf{Dec})$ be a SKQES. Then* Enc *and* Dec *have the following form:*

$$\mathsf{Enc}_k(X_M) = V_k \left(X_M \otimes (\sigma_k)_T\right) V_k^\dagger$$
$$\mathsf{Dec}_k(Y_C) = \mathrm{Tr}_T \left[P_T^{\sigma_k} \left(V_k^\dagger Y_C V_k\right) P_T^{\sigma_k}\right] + \hat{D}_k \left[\bar{P}_T^{\sigma_k} \left(V_k^\dagger Y_C V_k\right) \bar{P}_T^{\sigma_k}\right].$$

Here, σ_k is a state on register T, $P_T^{\sigma_k}$ and $\bar{P}_T^{\sigma_k}$ are the orthogonal projectors onto the support of $\sigma^{(k)}$ (see Sect. 2) and its complement (respectively), V_k is a unitary operator, and $\hat{D}_k$ is a channel.

In practice, $\hat{D}_k$ (i.e., the map that is applied to any ciphertext outside of the range of Enc_k) will just discard the state and replace it with $\perp$. Let us explain how the schemes we have seen so far fit into this characterization. For 2desTag, σ_k is simply the (key-independent) pure state $|0^n\rangle\langle 0^n|_T$, V_k is the unitary operator of the two-design corresponding to key k, $P^{\sigma_k} = |0^n\rangle\langle 0^n|$, and $\hat{D}_k$ replaces the state with $\perp$. For $\mathsf{1des}^{\mathsf{PRF}}$, σ_k is the maximally mixed state τ (i.e., the classical randomness r from Definition 2), and V_k is the controlled-unitary which applies a quantum one-time pad on the first register, controlled on the contents of the second register (using the pqPRF f), i.e., $|x\rangle|r\rangle \mapsto P_{f_k(r)}|x\rangle|r\rangle$. Decryption

[2] If the dimension of the plaintext space does not divide the dimension of the ciphertext space, then we may need an isometry. In our case, all spaces are made up of qubits.

undoes the controlled unitary and never rejects, i.e., $P^{\sigma_k} = \mathbb{1}$. This corresponds to the fact that τ has full support.

By considering the spectral decomposition of the state σ_k from Lemma 1, it is straightforward to show that encryption can always be implemented using unitary operators and only classical randomness. We state this fact as follows.

Corollary 1. *Let $\Pi = (\mathsf{KeyGen}, \mathsf{Enc}, \mathsf{Dec})$ be a SKQES. Then for every k, there exists a probability distribution $p_k : \{0,1\}^t \to [0,1]$ and a family of quantum states $|\psi^{(k,r)}\rangle_T$ such that Enc_k is equivalent to the following algorithm:*

1. *sample $r \in \{0,1\}^t$ according to p_k;*
2. *apply the following map: $\mathsf{Enc}_{k;r}(X_M) = V_k \left(X_M \otimes |\psi^{(k,r)}\rangle\langle\psi^{(k,r)}|_T\right) V_k^\dagger$.*

Here V_k and T are defined as in Lemma 1, and t is the number of qubits in T.

For example, in the case of 2desTag, the distribution is a point distribution and $|\psi^{(k,r)}\rangle = |0^t\rangle$. In $\mathsf{1des}^{\mathsf{PRF}}$, the distribution is uniform and $|\psi^{(k,r)}\rangle = |r\rangle$.

It is important to remark here that, even if Enc_k is a polynomial-time algorithm, the functionally-equivalent algorithm provided by Corollary 1 may not be. We thus define the following.

Condition 1. *Let Π be a SKQES, and let p_k, $|\psi^{(k,r)}\rangle$ and V_k be as given in Corollary 1. We say that Π* satisfies Condition *1 if there exist efficient quantum algorithms for (i) sampling from p_k, (ii) preparing $|\psi^{(k,r)}\rangle$, and (iii) implementing V_k, and this holds for all but a negligible fraction of k and r.*

We are not aware of any examples of SKQES that violate Condition 1. In fact, in all schemes we will consider (including all schemes constructed via Definition 2), the distribution p_k and the states $|\psi^{(k,r)}\rangle$ are trivial to prepare, and the unitaries V_k are implementable by poly-size quantum circuits. In any case, when Condition 1 is required for a particular result, we will state this explicitly.

Defining Ciphertext Authentication. We begin by outlining our approach. Fix an encryption scheme Π with plaintext register M and ciphertext register C. Let $\Lambda_{CB\to C\tilde{B}}$ be an attack map. Intuitively, we would like to decide whether to accept or reject conditioned on whether Λ has changed the ciphertext. A possible approach would be to use the simulator from Theorem 5.1 in [15]: in the case of acceptance, this simulator[3] ensures that Λ is equivalent to $\mathbb{1}_C \otimes \Phi$ for some side-information map $\Phi_{B\to\tilde{B}}$. While this approach is on the right track, it is unnecessarily strong as a definition of security: it prevents the adversary from even looking at (or copying) classical parts of the ciphertext! This would place strange requirements on encryption. It would disallow constant classical messages (e.g., "begin PGP message") accompanying ciphertexts. It would also disallow a large class of natural schemes, including all schemes Π^f from Sect. 2.

[3] In [15], this simulator was used to prove DNS security of the 2desTag scheme. Here, we consider whether that simulator can be used to *define* secure authentication.

This class has many schemes that (intuitively speaking) should be adequate for authenticating poly-many quantum ciphertexts, such as the case where Π applies a random unitary and f is a random function.

The key to finding the middle ground lies in Corollary 1: any scheme can be decomposed in a way that enables us to check separately whether the identity has been applied to the quantum part, and whether the classical register has changed. In effect, this will amount to an additional constraint over DNS-authentication[4] (Definition 4), demanding extra structure from the simulator.

Recall that an attack $\Lambda_{CB\to C\tilde{B}}$ on the scheme Π defines the averaged effective plaintext map $\Lambda^{\Pi}_{MB\to M\tilde{B}} = \mathbb{E}_k[\mathsf{Dec}_k \circ \Lambda \circ \mathsf{Enc}_k]$. We define ciphertext authentication as follows, using notation from Lemma 1 and Corollary 1.

Definition 5. *A SKQES $\Pi = (\mathsf{KeyGen}, \mathsf{Enc}, \mathsf{Dec})$ is* ciphertext authenticating, *or* QCA, *if for all CP-maps $\Lambda_{CB\to C\tilde{B}}$, there exists a CP-map $\Lambda^{\mathsf{rej}}_{B\to\tilde{B}}$ such that:*

$$\left\|\Lambda^{\Pi}_{MB\to M\tilde{B}} - \left(\mathrm{id}_M \otimes \Lambda^{\mathsf{acc}}_{B\to\tilde{B}} + |\bot\rangle\langle\bot|_M \otimes \Lambda^{\mathsf{rej}}_{B\to\tilde{B}}\right)\right\|_\diamond \leq \mathrm{negl}(n), \qquad (3)$$

and $\Lambda^{\mathsf{acc}}_{B\to\tilde{B}} + \Lambda^{\mathsf{rej}}_{B\to\tilde{B}}$ is TP. Here $\Lambda^{\mathsf{acc}}_{B\to\tilde{B}}$ is given by:

$$\Lambda^{\mathsf{acc}}_{B\to\tilde{B}}(Z_B) = \mathbb{E}_{k,r}\left[\langle\Phi_{k,r}|V_k^\dagger \Lambda\left(\mathsf{Enc}_{k;r}\left(\phi^+_{MM'} \otimes Z_B\right)\right) V_k|\Phi_{k,r}\rangle\right] \qquad (4)$$

where $|\Phi_{k,r}\rangle = |\phi^+\rangle_{MM'} \otimes |\psi^{(k,r)}\rangle_T$.

Condition (3) is simply DNS. It ensures that, in the accept case, the adversary performs the identity on the plaintext. Condition (4) demands that the rest of the action (i.e., on the side-information) is well-simulated by the following:

1. prepare a maximally entangled state $\phi^+_{MM'}$ and attach it to the input B;
2. run encryption, saving the classical randomness r used (meaning that the tag register T was prepared in the state $|\psi^{(k,r)}\rangle$);
3. apply decryption while conditioning on (i) the plaintext still being maximally entangled with M', and (ii) register T still containing $|\psi^{(k,r)}\rangle$;
4. output the contents of $\tilde{B}$.

Note that this definition only adds further constraints to DNS. Recalling that DNS implies QIND [9,21], we thus have the following.

Theorem 7. *If a SKQES is* QCA, *then it is also* DNS; *in particular, it is* QIND.

It is not difficult to see that the security proof in Theorem 5.1 of [15] (for establishing DNS of the Clifford scheme) actually applies to arbitrary 2-designs, and in fact proves QCA and not only DNS. We thus have that the scheme 2desTag fulfills ciphertext authentication. For details on the separation between QCA and DNS, see the appendix of the full version of this paper [4].

[4] One might also start from the authentication definitions of [21,27] rather than DNS. However, this is not necessary: these definitions' advantage over DNS is in key recycling; our setting is non-interactive and has no back-channel for key recycling.

Computational-Security Variant. We now briefly record a computational-security variant of one-time ciphertext authentication, which simply requires that all elements in Definition 5 are efficient.

Definition 6. *A* SKQES $\Pi = (\mathsf{KeyGen}, \mathsf{Enc}, \mathsf{Dec})$ *is* computationally ciphertext authenticating (cQCA) *if, for any efficiently implementable attack map* $\Lambda_{CB\to C\tilde{B}}$*, the effective attack* $\tilde{\Lambda}_{MB\to M\tilde{B}}$ *is computationally indistinguishable from the simulator:*

$$\Lambda^{\mathrm{sim}}_{MB\to M\tilde{B}} = \mathrm{id}_M \otimes \Lambda^{\mathsf{acc}}_{B\to\tilde{B}} + |\bot\rangle\langle\bot|_M \otimes \Lambda^{\mathsf{reject}}_{B\to\tilde{B}}. \tag{5}$$

Here the simulator is given by:

$$\Lambda^{\mathsf{acc}}_{B\to\tilde{B}} = \mathbb{E}_{k,r}\left[\langle\Phi_{k,r}|V_k^\dagger \Lambda\left(\mathsf{Enc}_{k;r}\left(\phi^+_{MM'}\otimes(\cdot)_B\right)\right)V_k|\Phi_{k,r}\rangle\right] \text{ and}$$
$$\Lambda^{\mathsf{reject}}_{B\to\tilde{B}} = \mathbb{E}_{k,r}\left[\mathrm{Tr}\left(\mathbb{1}-|\Phi_{k,r}\rangle\langle\Phi_{k,r}|\right)V_k^\dagger\Lambda\left(\mathsf{Enc}_{k;r}\left(\phi^+_{MM'}\otimes(\cdot)_B\right)\right)V_k\right], \tag{6}$$

where: $|\Phi_{k,r}\rangle = |\phi^+\rangle_{MM'}\otimes|\psi^{(k,r)}\rangle_T$.

Because we fix the form of the simulator in the reject case, the simulator is efficiently implementable just as in [15] for schemes that satisfy Condition 1. It is straightforward to define a computational variant of DNS [15], which we denote by cDNS. Given that Theorem 7 only talks about computationally bounded quantum adversaries, it also applies to cDNS. In particular we have the following.

Proposition 1. *If a* SKQES *is* cQCA*, then it is also* cDNS*; in particular, it satisfies* QIND.

4 Quantum Unforgeability

Translating the standard classical intuition of ciphertext unforgeability to the quantum setting appears nontrivial. As we develop our approach, it will be useful to keep in mind a "prototype" scheme that should (intuitively) satisfy quantum unforgeability against a polynomial-time adversary making an arbitrary number of queries. This is the scheme $\mathsf{2desTag}^{\mathsf{PRF}}$, which encrypts via:

$$\mathsf{Enc}_k(\varrho) = U_{f_k(r)}\left(\varrho\otimes|0^n\rangle\langle0^n|\right)U^\dagger_{f_k(r)}\otimes|r\rangle\langle r|$$

where k is a key for the pqPRF f and r is randomness selected freshly for each encryption. This scheme is characterized (via Lemma 1) by the key-independent "tag state" $|0^n\rangle\langle0^n|\otimes\tau$ (where τ is the maximally mixed state) and the unitary V_k which applies $U_{f_k(\cdot)}$ on the first two registers, controlled on the third register (i.e., the randomness r).

To see why this scheme should be unforgeable, assume for the moment that U_s is a Haar-random unitary and f_k is a perfectly random function. Intuitively, from the point of view of the adversary, each plaintext is mapped into a subspace which is fresh, independent, random, and exponentially-small as a fraction of the total dimension (of the ciphertext space). Security should then reduce to the security of multiple uses of a QCA one-time scheme, each time with a freshly generated key. We will carefully formalize this intuition in a later section.

Formal Definitions. Our definition will compare the performance of an adversary in two games: an unrestricted forgery game, and a cheat-detecting game. Fix an SKQES $\Pi = (\mathsf{KeyGen}, \mathsf{Enc}, \mathsf{Dec})$ and let $\mathcal{A}$ be an adversary in the following.

Experiment 1. The $\mathsf{QUF\text{-}Forge}(\Pi, \mathcal{A}, n)$ experiment:

1: $k \leftarrow \mathsf{KeyGen}(1^n)$;
2: **if** $\mathsf{Dec}_k(\mathcal{A}^{\mathsf{Enc}_k}(1^n)) \neq \bot$, **output** win; otherwise **output** reject.

We will think about this experiment as taking place between the adversary $\mathcal{A}$ and a challenger $\mathcal{C}$, who generates the key k, answers the queries of $\mathcal{A}$, and then decrypts to see the outcome of the game.

We now consider a different experiment where $\mathcal{C}$ attempts to check $\mathcal{A}$ for cheating. We will make use of the maximally entangled state $|\phi^+\rangle_{M'M''}$ on two copies (M' and M'') of the plaintext register, and the corresponding measurement $\{\Pi^+_{M'M''}, \mathbb{1} - \Pi^+_{M'M''}\}$. We will also need a measurement that will help $\mathcal{C}$ identify previously generated ciphertexts. Recall from Sect. 3 that correctness implies that Enc can be written in the form $\mathsf{Enc}_k(X) = V_k\big(X_M \otimes \sigma_k\big)V_k^\dagger$ where $\sigma_T^{(k)} = \sum_r p_k(r)\Pi_{k,r}$ and $\Pi_{k,r} = |\psi^{(k,r)}\rangle\langle\psi^{(k,r)}|_T$. This also defines, for each (k, r), the two-outcome measurement $\{\Pi_{k,r}, \mathbb{1} - \Pi_{k,r}\}$. In all these two-outcome measurements, we denote the first outcome by 0 and the second outcome by 1. Notice that these projectors commute, as $|\psi^{(k,r)}\rangle_T$ are elements of an orthonormal basis of eigenvectors.

Experiment 2. The $\mathsf{QUF\text{-}Cheat}(\Pi, \mathcal{A}, n)$ experiment:

1: $\mathcal{C}$ runs $k \leftarrow \mathsf{KeyGen}(1^n)$;
2: $\mathcal{A}$ receives 1^n and oracle access to E_k (controlled by $\mathcal{C}$), defined as follows:
  (1) $\mathcal{A}$ sends plaintext register M to $\mathcal{C}$;
  (2) $\mathcal{C}$ discards M and prepares $|\phi^+\rangle_{M'M''}$;
  (3) $\mathcal{C}$ applies Enc_k to M' using fresh randomness r, sends result C to $\mathcal{A}$;
  (4) $\mathcal{C}$ stores (M'', r) in a set $\mathcal{M}$.
3: $\mathcal{A}$ sends final output register C_{out} to $\mathcal{C}$;
4: $\mathcal{C}$ applies $V_k^\dagger$ to C_{out}, places results in MT;
5: **for each** $(M'', r) \in \mathcal{M}$ **do**
6:   $\mathcal{C}$ applies $\{\Pi_{k,r}, \mathbb{1} - \Pi_{k,r}\}$ to T;
7:   **if** outcome is 0 **then**:
8:     $\mathcal{C}$ applies $\{\Pi^+, \mathbb{1} - \Pi^+\}$ to MM'';
9:     **if** outcome is 0: **output** cheat; **end if**
10:   **end if**
11: **end for**
12: **output** reject.

Note that the experiment always outputs reject if $\mathcal{A}$ makes no queries. We emphasize that $\mathcal{C}$ is a fixed algorithm defined by the security game and the properties of Π. The challenger is efficient if the states $|\psi^{(k,r)}\rangle\langle\psi^{(k,r)}|$ and the unitary V_k are efficiently implementable and the probability distribution p_k is efficiently sampleable. We believe this is not a significant constraint. It is easily satisfied in all schemes we are aware of. Moreover, in light of Lemma 1, it seems unlikely that any reasonable form of ciphertext unforgeability can be defined without this requirement. We are now ready to define security.

Definition 7. *A SKQES Π has unforgeable ciphertexts (or is* QUF*) if, for all QPT adversaries $\mathcal{A}$, it holds:*

$$|\Pr[\mathsf{QUF\text{-}Forge}(\Pi, \mathcal{A}, n) \to \mathsf{win}] - \Pr[\mathsf{QUF\text{-}Cheat}(\Pi, \mathcal{A}, n) \to \mathsf{cheat}]| \leq \mathrm{negl}(n)\,.$$

It is straightforward to adapt the above definition to the bounded-query setting, where we fix some positive integer t (at scheme design time) and demand that adversaries can make no more than t queries. We call the resulting notion QUF_t. One then has the obvious implications $\mathsf{QUF} \Rightarrow \mathsf{QUF}_t \Rightarrow \mathsf{QUF}_{t-1} \forall t \in \mathbb{N}$.

Let us briefly discuss a potential concern with these definitions. Consider the repeated measurements applied to the adversary's final output C_{out} (Line 6 and Line 8) in QUF-Cheat. The first measurement simply compares the randomness of C_{out} to that of previously generated ciphertexts. Such measurements will not disturb properly-formed ciphertexts at all, and malformed ones will not affect our security definition. The second measurement actually measures the plaintext register M, and thus might (a priori) appear to be concerning. Indeed, if multiple such measurements are applied to M, this might open up a vulnerability to attacks. As it turns out, this is not a problem. We will shortly show (see Theorem 8 below) that QUF implies QIND-CPA. For QIND-CPA schemes, any given random string r is only chosen with negligible probability at encryption time (if not, querying the encryption oracle a polynomial number of times with the challenge plaintext would be enough to compromise security). It follows that, with overwhelming probability, the random strings chosen in the different oracle calls in QUF-Cheat are pairwise distinct. This, in turn, implies that the measurement in Line 8 is applied at most once in a given run of the experiment.

Relationship to Other Security Notions. It is well-known that even one-time quantum authentication implies QIND secrecy [9]. As we now show, QUF implies an even stronger notion of secrecy, QIND-CPA. This is a significant departure from classical unforgeability, which is completely independent of secrecy.

Theorem 8. *If a SKQES satisfies* QUF*, then it also satisfies* QIND-CPA.

Proof. Let Π be a SKQES, and let $\mathcal{A}$ be an adversary winning QIND-CPA with non-negligible advantage ν over guessing, with pre-challenge algorithm $\mathcal{A}_1$ and post-challenge algorithm $\mathcal{A}_2$. We will build an adversary $\mathcal{B}$ with black-box oracle access to $\mathcal{A}$, able to distinguish between the QUF-Forge game and the QUF-Cheat game with non-negligible advantage over guessing, as follows:

1. $\mathcal{B}$ runs $\mathcal{A}_1(1^n)$, answering its queries using his own oracle $\mathcal{O}$;
2. get registers M (challenge plaintext) and B (side information) from $\mathcal{A}_1$;
3. choose a random bit $b \xleftarrow{\$} \{0,1\}$; if $b = 1$, then replace contents of M with a maximally-mixed state;

4. invoke oracle $\mathcal{O}$ on M and place result in register C;
5. run $\mathcal{A}_2$ on registers C and B, receiving output $b' \in \{0,1\}$;
6. if $b = b'$, then output real; else output real or ideal with equal probability.

Note that, if $\mathcal{B}$ is playing QUF-Forge, then $\mathcal{O} = \mathsf{Enc}_k$ and we are faithfully simulating the QIND-CPA game for $\mathcal{A}$. It follows that $b = b'$ with probability at least $1/2 + \nu$. If $\mathcal{B}$ is playing QUF-Cheat instead, $\mathcal{O}$ discards its input (and replaces it with half of a maximally-entangled state) on every call. In that case, all inputs to $\mathcal{A}_1$ and $\mathcal{A}_2$ are completely uncorrelated with b, so that $b' = b$ with probability $1/2$. Therefore, $\mathcal{A}'$ will correctly guess the game it is playing in with non-negligible advantage.

Now it is easy to see how to use $\mathcal{B}$ to violate the main condition in the definition of QUF with the same distinguishing advantage. First, query the oracle once and store the output in register C. Next, run $\mathcal{B}$. If $\mathcal{B}$ outputs real, then output the contents of C (achieving win in QUF-Forge). Otherwise, output a random state in the ciphertext register (achieving reject in QUF-Cheat). □

We also study the restriction of the quantum notion QUF to the classical case, i.e., classical symmetric-key encryption schemes (SKES) vs classical adversaries. We denote this classical restriction by UF. In this notion, the classical unrestricted forgery game UF-Forge is defined precisely as in Experiment 1. Regarding the quantum game QUF-Cheat, notice that, in any classical scheme, one can apply ciphertext verification to a string c as follows: (i) make a copy c' of c, (ii) decrypt c, (iii) if decryption rejected, output reject, and otherwise output c'. In other words, all classical encryption schemes automatically satisfy Condition 1. The appropriate classical restriction UF-Cheat of this game thus proceeds as Experiment 2, with two modifications: (i) in step 2:, $\mathcal{C}$ replaces the plaintext in register M_j by a random plaintext, encrypts it, and stores a copy of the resulting ciphertext in C_j; and (ii) in step 4:, without decrypting, the game outputs cheat if the challenge ciphertext C equals any one of the saved C_j's. We then have the following.

Definition 8. *A SKES Π has unforgeable ciphertexts (or is* UF*) if, for all PPT adversaries $\mathcal{A}$,*

$$|\Pr[\mathsf{UF\text{-}Forge}(\Pi, \mathcal{A}, n) \to \mathsf{win}] - \Pr[\mathsf{UF\text{-}Cheat}(\Pi, \mathcal{A}, n) \to \mathsf{cheat}]| \leq \mathrm{negl}(n) .$$

The proof of Theorem 8 carries over easily to the classical case. Moreover, one can show how UF implies the classical security notion of *integrity of ciphertexts* INT-CTXT [10], which states that no bounded adversary with oracle access to an encryption oracle can produce a ciphertext which is at the same time (i) valid, and (ii) fresh, i.e., never output by the oracle. Recall that, classically, it is known [10] that INT-CTXT plus IND-CPA defines authenticated encryption AE. Therefore, the notion of unforgeability of ciphertexts, when restricted to the classical case, is at least as strong as authenticated encryption. However, one can also show the converse, i.e., AE implies UF.

Theorem 9. $\mathsf{UF} \iff \mathsf{AE}$.

Proof. The first non-trivial part to prove is $\mathsf{UF} \implies \mathsf{INT\text{-}CTXT}$. Let Π be an INT-CTXT insecure SKES. Then there exists an adversary $\mathcal{A}$ with oracle access to Enc_k which, with non-negligible probability ν, outputs a ciphertext c which was never output by the encryption oracle. Define a PPT algorithm $\mathcal{B}$ with oracle access to Enc_k, as follows. First, $\mathcal{B}$ executes $\mathcal{A}$ and records a list L of all Enc_k's answers c_j output to $\mathcal{A}$. When $\mathcal{A}$ outputs a ciphertext c, if $c \in L$, $\mathcal{B}$ outputs a random ciphertext c'; else it outputs c. For $\mathcal{B}$, the success probabilities in the games defining UF are as follows:

- in the UF-Forge experiment, since c is a fresh ciphertext with non-negligible probability ν, $\mathcal{B}$ wins UF-Forge with probability at least ν.
- In UF-Cheat instead, whenever the ciphertext is not fresh, $\mathcal{B}$ replaces it with a random one, and hence only wins UF-Cheat with negligible probability.

The fact that a random ciphertext is invalid with overwhelming probability follows by considering an adversary that does not make any queries. So we have:

$$|\Pr[\mathsf{UF\text{-}Forge}(\Pi, \mathcal{A}', n) \to \mathsf{win}] - \Pr[\mathsf{UF\text{-}Cheat}(\Pi, \mathcal{A}', n) \to \mathsf{cheat}]| \geq \nu,$$

and hence Π cannot be UF.

The other direction to prove is $\mathsf{AE} \implies \mathsf{UF}$. For this, we will use an equivalent characterization of AE, also known in the literature as IND-CCA3 [28]. In this definition, the adversary's goal is to distinguish whether he's playing in the AE-Real world, or in the AE-Ideal world. In the AE-Real world, the adversary can interact freely with an encryption oracle Enc_k, and with a restricted decryption oracle Dec_k which always rejects ($\perp$) decryption queries over any ciphertext which was output by Enc_k. In the AE-Ideal world, instead, the adversary is interacting with an oracle $\mathsf{Enc}_k(\$)$ (which ignores the input query, and always returns the encryption of a fresh random plaintext), and a constant $\perp$ oracle (which simulates the decryption oracle but always rejects any query). A scheme Π is AE secure iff, for any adversary $\mathcal{A}$ it holds:

$$|\Pr[\mathsf{AE\text{-}Real}(\Pi, \mathcal{A}, n) \to 1] - \Pr[\mathsf{AE\text{-}Ideal}(\Pi, \mathcal{A}, n) \to 1]| \leq \mathrm{negl}(n).$$

Now, let $\mathcal{A}$ be a PPT adversary breaking UF for a scheme Π. This means that there exists a non-negligible function ν such that:

$$|\Pr[\mathsf{UF\text{-}Forge}(\Pi, \mathcal{A}, n) \to \mathsf{win}] - \Pr[\mathsf{UF\text{-}Cheat}(\Pi, \mathcal{A}, n) \to \mathsf{cheat}]| \geq \nu(n).$$

We use $\mathcal{A}$ to build an adversary $\mathcal{B}$ able to distinguish AE-Real from AE-Ideal. The new adversary $\mathcal{B}$ runs $\mathcal{A}$ and forwards all of $\mathcal{A}$'s encryption queries to his own encryption oracle. Finally, when $\mathcal{A}$ outputs a ciphertext c, $\mathcal{B}$ queries his own decryption oracle on c, and looks at the oracle's response. If the response is *not* $\perp$, then $\mathcal{B}$ returns real, otherwise returns real or ideal with equal chance.

It is easy to see that $\mathcal{B}$ distinguishes AE-Ideal from AE-Real with non-negligible advantage at least $\nu/2$ over guessing. The reason is as follows. If $\mathcal{B}$ is in the AE-Real world (probability 1/2), then he is correctly simulating for $\mathcal{A}$ the UF-Forge game. Since $\mathcal{A}$ breaks UF by assumption, it means that, with probability at least ν, his output c will be a fresh valid ciphertext; in that case, also $\mathcal{B}$ wins. On the other hand, if the world is AE-Ideal, $\mathcal{B}$ still wins with probability 1/2. □

This means that UF is actually *another characterization of authenticated encryption.* This is an interesting observation, given that UF comes from the classical restriction of a quantum notion "merely" concerning the unforgeability of ciphertexts. However, we stress that this equivalence only holds at the classical level, and that this is insufficient evidence to declare that UF serves the same purpose quantumly as AE does classically. In fact, in Sect. 6 we introduce a quantum analogue of AE which we call QAE, and provide stronger evidence that the latter is in fact the correct analogue.

5 Quantum IND-CCA2

Next, we move to the problem of defining adaptive chosen-ciphertext security for quantum encryption. In the usual classical formulation (IND-CCA2), the adversary $\mathcal{A}$ receives both an encryption oracle and a decryption oracle for the entire duration of the indistinguishability game. To eliminate the trivial strategy, we do not permit $\mathcal{A}$ to query the decryption oracle on the challenge ciphertext. This last condition does not make sense in the quantum setting, for a number of reasons we've seen before: no-cloning prevents us from storing a copy of the challenge, measurement may destroy the states involved, and so on. However, our approach to defining unforgeability can be adapted to this case. The resulting notion of *quantum indistinguishability under adaptive chosen-ciphertext attacks* (QIND-CCA2) can also be recast in the public-key quantum encryption setting.

Formal Definition. As before, we will compare the performance of the adversary in two games. In each case, the adversary $\mathcal{A} = (\mathcal{A}_1, \mathcal{A}_2)$ consists of two parts (pre-challenge and post-challenge), and is playing against the challenger $\mathcal{C}$, which is a fixed algorithm determined only by the security game and the scheme.

Experiment 3. The QCCA2-Test$(\Pi, \mathcal{A}, n)$ experiment:

1: $\mathcal{C}$ runs $k \leftarrow \mathsf{KeyGen}(1^n)$ and flips a coin $b \xleftarrow{\$} \{0,1\}$;
2: $\mathcal{A}_1$ receives 1^n and access to oracles Enc_k and Dec_k;
3: $\mathcal{A}_1$ prepares a side register S, and sends $\mathcal{C}$ a challenge register M;
4: $\mathcal{C}$ puts into C either $\mathsf{Enc}_k(M)$ (if $b = 0$) or $\mathsf{Enc}_k(\tau_M)$ (if $b = 1$);
5: $\mathcal{A}_2$ receives registers C and S and oracles Enc_k and Dec_k;
6: $\mathcal{A}_2$ outputs a bit b'. **If** $b' = b$, **output** win; otherwise **output** fail.

Notice that in this game there are no restrictions on the use of Dec_k by $\mathcal{A}_2$. In particular, $\mathcal{A}_2$ is free to decrypt the challenge. In the second game, the challenge plaintext is replaced by half of a maximally entangled state, and $\mathcal{A}$ only gains an advantage over guessing if he cheats, i.e., if he tries to decrypt the challenge.

Experiment 4. The QCCA2-Fake($\Pi, \mathcal{A}, n$) experiment:

1: $\mathcal{C}$ runs $k \leftarrow \mathsf{KeyGen}(1^n)$;
2: $\mathcal{A}_1$ receives 1^n and access to oracles Enc_k and Dec_k;
3: $\mathcal{A}_1$ prepares a side register S, and sends $\mathcal{C}$ a challenge register M;
4: $\mathcal{C}$ discards M, prepares $|\phi^+\rangle_{M'M''}$ and fresh randomness r, and stores (M'', r); then $\mathcal{C}$ encrypts the M' register and sends the resulting ciphertext C' to $\mathcal{A}_2$;
5: $\mathcal{A}_2$ receives registers C' and S and oracles Enc_k and D_k, where D_k is defined as follows. On input a register C:
 (1) $\mathcal{C}$ applies $V_k^\dagger$ to C, places results in MT;
 (2) $\mathcal{C}$ applies $\{\Pi_{k,r}, \mathbb{1} - \Pi_{k,r}\}$ to T;
 (3) **if** outcome is 0 **then**:
 (4) $\mathcal{C}$ applies $\{\Pi^+, \mathbb{1} - \Pi^+\}$ to MM'';
 (5) **if** outcome is 0: **output** cheat;
 (6) **end if**
 (7) **return** M;
6: $\mathcal{C}$ draws a bit b at random. **If** $b = 1$, **output** cheat; if $b = 0$ **output** reject.

We now define quantum IND-CCA2 in terms of the advantage gap of adversaries between the above two games.[5]

Definition 9. *A SKQES Π is* QIND-CCA2 *if, for all QPT adversaries $\mathcal{A}$,*

$$\Pr[\mathsf{QCCA2\text{-}Test}(\Pi, \mathcal{A}, n) \to \mathsf{win}] - \Pr[\mathsf{QCCA2\text{-}Fake}(\Pi, \mathcal{A}, n) \to \mathsf{cheat}] \leq \mathrm{negl}(n).$$

The omission of absolute values in the above is intentional. Indeed, an adversary can artificially inflate his cheating probability by querying the decryption oracle on the challenge and then ignoring the result. What he should not be able to do (against a secure scheme) is make his win probability larger than his cheating probability. We note that QIND-CCA2 clearly implies QIND-CCA1.

Proposition 2. QIND-CCA2 $\implies$ QIND-CCA1.

Proof. Suppose we have a scheme Π which is not QIND-CCA1, i.e., there exists an adversary $\mathcal{A}$ which wins the usual QIND-CCA1 game with non-negligible advantage ν over guessing. Clearly $\mathcal{A}$ can also play the games QCCA2-Test and QCCA2-Fake, but will not query the decryption oracle post-challenge. Note that $\mathcal{A}$ wins QCCA2-Test with probability $1/2 + \nu$, but is declared as cheating in QCCA2-Fake with probability exactly $1/2$. Hence Π is not QIND-CCA2. □

Next, we show that the classical restriction of QIND-CCA2 is equivalent to the classical security notion IND-CCA2. We denote the classical restriction of QIND-CCA2 by IND-CCA2′. This is defined by adapting the replacement and verification procedure of the challenger in QCCA2-Test in the same way as when defining UF. We denote the classical versions of the games QCCA2-Test and QCCA2-Fake by CCA2-Test and CCA2-Fake, respectively.

[5] The interface that the two games provide to the adversary differ slightly in that the adversary is not asked to output a bit in the end of the QCCA2-Fake game. This is not a problem as the games have the same interface until the second one terminates.

Theorem 10. *A SKES Π is* IND-CCA2′ *iff it is* IND-CCA2.

Proof. Suppose first that $\mathcal{A}$ is an adversary breaking IND-CCA2′, i.e., winning CCA2-Test with non-negligible advantage ν over the probability of winning CCA2-Fake. We construct an adversary $\mathcal{A}'$, that runs $\mathcal{A}$, keeps a copy of the challenge ciphertext and aborts by giving a random answer whenever $\mathcal{A}$ is about to query the decryption oracle with the challenge ciphertext. Note that $\mathcal{A}'$ wins CCA2-Fake with probability exactly 1/2. We call $\mathcal{A}'$ the *self-checking version of* $\mathcal{A}$. It is easy to show that $\mathcal{A}'$ wins the CCA2-Test game with probability at least $1/2 + \nu$:

$$\begin{aligned}
&\Pr\left[\mathcal{A}' \text{ wins } \mathsf{CCA2\text{-}Test}\right] \\
&= \Pr\left[\mathcal{A} \text{ wins } \mathsf{CCA2\text{-}Test} \wedge \mathcal{A} \text{ does not cheat}\right] + \frac{1}{2}\Pr\left[\mathcal{A} \text{ cheats}\right] \\
&\geq \Pr\left[\mathcal{A} \text{ wins } \mathsf{CCA2\text{-}Test}\right] - \frac{1}{2}\Pr\left[\mathcal{A} \text{ cheats}\right] \\
&\geq \Pr\left[\mathcal{A} \text{ wins } \mathsf{CCA2\text{-}Fake}\right] + \frac{1}{\nu} - \frac{1}{2}\Pr\left[\mathcal{A} \text{ cheats}\right] = \frac{1}{2} + \nu\,.
\end{aligned}$$

The first inequality is $\Pr[A \wedge B] \geq \Pr[A] - \Pr[\neg B]$ and the second inequality is the assumption. But the CCA2-Test and IND-CCA2 games are identical for adversaries that do not query the challenge, and $\mathcal{A}'$ has been constructed not to, i.e., $\mathcal{A}'$ wins the IND-CCA2 game with probability $1/2 + \nu$.

For the other direction, let $\mathcal{A}$ be an adversary that wins the IND-CCA2 game with non-negligible advantage. Let $\mathcal{A}'$ be the self-checking version of $\mathcal{A}$. Note that $\mathcal{A}$ and $\mathcal{A}'$ behave the same in both the IND-CCA2 and CCA2-Test games, as $\mathcal{A}$ never submits the challenge ciphertext there by assumption. In the CCA2-Fake game, however, $\mathcal{A}$ could, in principle, query the oracle with the challenge ciphertext, which is why we have to resort to the use of $\mathcal{A}'$. The latter is a successful adversary for IND-CCA2′: It wins the CCA2-Test game with non-negligible advantage over random guessing by assumption, but it wins the CCA2-Fake game with probability exactly $\frac{1}{2}$. □

6 Quantum Authenticated Encryption

In the classical setting, authenticated encryption (AE) is defined as IND-CCA2 and unforgeability of ciphertexts (see Definition 4.17 in [25]) or, equivalently, IND-CPA and unforgeability of ciphertexts [10]. A third equivalent formulation due to Shrimpton [28] defines AE in terms of a real vs ideal scenario. According to this definition, a classical scheme $\Pi = (\mathsf{KeyGen}, \mathsf{Enc}, \mathsf{Dec})$ is AE if no adversary, given oracles E and D, can distinguish these two scenarios:

- AE-Real: (E, D) is $(\mathsf{Enc}_k, \mathsf{Dec}_k)$ with $k \leftarrow \mathsf{KeyGen}$;
- AE-Ideal: E discards the input and returns $\mathsf{Enc}_k(m)$ for random m, and D always rejects; here again $k \leftarrow \mathsf{KeyGen}$;

This is not yet enough, because the adversary $\mathcal{A}$ can always distinguish real from ideal by composing E with D. To patch this problem, we can (i) demand that

$\mathcal{A}$ cannot do that, as in [28], or (ii) add the condition $D \circ E = \mathbb{1}$ to the ideal case[6]. We will take the latter approach.

Motivated by this formulation of AE and our general strategy so far, we will define quantum authenticated encryption by comparing the performance of the adversary in a real world and an ideal world. In the real world, the adversary gets unrestricted access to Enc_k and Dec_k. In the ideal world, the challenger $\mathcal{C}$ stores the Enc_k queries, replacing them with halves of maximally-entangled states; when a Dec_k query is detected as corresponding to a particular earlier Enc_k query, $\mathcal{C}$ replies with the contents of the stored register; otherwise Dec_k rejects. Cheat detection is performed just as in the unforgeability game QUF-Cheat.

Formal Definition. We now formally define the two worlds: the real world QAE-Real, and the ideal (or cheat-detecting) world QAE-Ideal. In both cases, the adversary $\mathcal{A}$ receives two oracles and then outputs a single bit.

Experiment 5. The QAE-Real$(\Pi, \mathcal{A}, n)$ experiment:

1: $k \leftarrow \mathsf{KeyGen}(1^n)$;
2: **output** $\mathcal{A}^{\mathsf{Enc}_k, \mathsf{Dec}_k}(1^n)$.

In the ideal setting, it will be convenient to describe the experiment in terms of an interaction between $\mathcal{A}$ and the challenger $\mathcal{C}$, a fixed algorithm determined only by the security game and the properties of Π.

Experiment 6. The QAE-Ideal$(\Pi, \mathcal{A}, n)$ experiment:

1: $\mathcal{C}$ runs $k \leftarrow \mathsf{KeyGen}(1^n)$;
2: initialize oracles $E_{M \to C}$ and $D_{C \to M}$:
- E is defined as follows. On input a register M:
  (1) $\mathcal{C}$ prepares $|\phi^+\rangle_{M'M''}$, and generates fresh randomness r;
  (2) $\mathcal{C}$ stores (r, M'', M) in a set $\mathcal{M}$;
  (3) $\mathcal{C}$ applies Enc_k to M' using randomness r; **return** result to $\mathcal{A}$.
- D is defined as follows. On input a register C:
  (1) $\mathcal{C}$ applies $V_k^\dagger$ to C, places results in $M'T$;
  (2) **for each** $(r, M'', M) \in \mathcal{M}$ **do**:
  (3) $\quad$ $\mathcal{C}$ applies $\{\Pi_{k,r}, \mathbb{1} - \Pi_{k,r}\}$ to T;
  (4) $\quad$ **if** outcome is 0 **then**:
  (5) $\quad\quad$ $\mathcal{C}$ applies $\{\Pi^+, \mathbb{1} - \Pi^+\}$ to $M'M''$;
  (6) $\quad\quad$ **if** outcome is 0: **return** M;
  (7) $\quad$ **end if**
  (8) **end for**
  (9) **return** $|\perp\rangle\langle\perp|$;

3: **output** $\mathcal{A}^{E,D}(1^n)$.

Note that, as before, we number the measurement outcomes by 0 (the first outcome) and 1 (the second outcome). With the above games defined, we can now set down our definition of quantum authenticated encryption.

[6] More precisely, the ideal world maintains a list of all queries that $\mathcal{A}$ makes to E, and ensures that D will respond correctly if queried on an output of E.

Definition 10. *A SKQES* $\Pi = (\mathsf{KeyGen}, \mathsf{Enc}, \mathsf{Dec})$ *is an* authenticated quantum encryption scheme *(or is* QAE*) if, for all QPT adversaries* $\mathcal{A}$*:*

$$|\Pr[\mathsf{QAE\text{-}Real}(\Pi, \mathcal{A}, n) \to \mathsf{real}] - \Pr[\mathsf{QAE\text{-}Ideal}(\Pi, \mathcal{A}, n) \to \mathsf{real}]| \leq \mathrm{negl}(n).$$

Relationship to Other Security Notions. Next, we give evidence that QAE is indeed the correct formalization of a quantum analogue of AE, by showing that it implies all of the quantum security notions defined thus far. We begin with adaptive chosen-ciphertext security.

Theorem 11. QAE $\Longrightarrow$ QIND-CCA2.

Proof. The proof is similar to that of Theorem 8. For a scheme Π, let $\mathcal{A}$ be an adversary against QIND-CCA2, e.g., let us say that:

$$\Pr[\mathsf{QCCA2\text{-}Test}(\Pi, \mathcal{A}, n) \to \mathsf{win}] = \Pr[\mathsf{QCCA2\text{-}Fake}(\Pi, \mathcal{A}, n) \to \mathsf{cheat}] + \nu(n)\,,$$

for non-negligible ν. We then show how to build another adversary $\mathcal{B}$ with black-box access to $\mathcal{A}$, able to distinguish QAE-Real from QAE-Ideal.

$\mathcal{B}$ runs $\mathcal{A}$, and forwards all of $\mathcal{A}$'s queries to his own oracles. When eventually $\mathcal{A}$ outputs a challenge plaintext state, $\mathcal{B}$ flips a random bit b. If $b = 0$, then $\mathcal{B}$ forwards the challenge plaintext to his encryption oracle as usual. Otherwise, if $b = 1$, $\mathcal{B}$ replaces the challenge with a totally mixed plaintext state before relaying it to the oracle. After that, $\mathcal{B}$ continues to answer $\mathcal{A}$'s queries during the second quantum CCA phase as before, by forwarding all the queries to his oracles, until $\mathcal{A}$ produces an output bit b'. Finally, if $b = b'$, then $\mathcal{B}$ outputs real, otherwise he outputs ideal.

Now notice the following: If we are in the QAE-Real environment (that is, $\mathcal{B}$ has unrestricted Enc and Dec oracles), then $\mathcal{B}$ is faithfully simulating for $\mathcal{A}$ the QCCA2-Test game, which means that the probability of $\mathcal{B}$ correctly outputting real is exactly the same probability of $\mathcal{A}$ of winning QCCA2-Test.

If we are in the QAE-Ideal world, instead, $\mathcal{B}$ is playing in a "malformed" game, where all his encryption queries are replaced by random plaintexts before encryption. This means that the best $\mathcal{A}$ could do in order to guess the secret bit b is guessing at random, *unless* $\mathcal{A}$ uses a "cheating decryption query" on the challenge ciphertext (in this case the modified decryption oracle of the game QAE-Ideal would actually return the encrypted plaintext). Looking at the description of the QCCA2-Fake game, it is clear that this is exactly the same as $\Pr[\mathsf{QCCA2\text{-}Fake}(\Pi, \mathcal{A}, n) \to \mathsf{cheat}]$. So, summing up:

$$\begin{aligned} &\Big| \Pr[\mathsf{QAE\text{-}Real}(\Pi, \mathcal{B}, n) \to \mathsf{Real}] - \Pr[\mathsf{QAE\text{-}Ideal}(\Pi, \mathcal{B}, n) \to \mathsf{Real}] \Big| \\ =& \Big| \Pr[\mathsf{QCCA2\text{-}Test}(\Pi, \mathcal{A}, n) \to \mathsf{win}] - \Pr[\mathsf{QCCA2\text{-}Fake}(\Pi, \mathcal{A}, n) \to \mathsf{cheat}] \Big| = \nu\,, \end{aligned}$$

which concludes the proof. □

In terms of authentication security, we can show that QAE implies cQCA (computational one-time ciphertext authentication), and hence also cDNS.

Theorem 12. *Let $\Pi = (\mathsf{KeyGen}, \mathsf{Enc}, \mathsf{Dec})$ be a SKQES that is* QAE *secure and satisfies Condition 1. Then it is* cQCA.

Proof. Assume Π is not cQCA. Then there exists an algorithm $\mathcal{A} = (\mathcal{A}_1, \mathcal{A}_2, \mathcal{A}_3)$ that achieves the following. $\mathcal{A}_1$ gets an input 1^n and outputs registers M (the plaintext register) and B. $\mathcal{A}_2$ implements a map $\Lambda_{CB \to C\tilde{B}}$ on two registers C (the ciphertext register) and B. $\mathcal{A}_3$ is a distinguisher between the two states resulting from applying $\tilde{\Lambda}_{CB \to C\tilde{B}}$ or the corresponding simulator according to Eqs. (5) and (6) to the output of $\mathcal{A}_1$.

The crucial observation is, that the map on registers MB resulting from sending M to the challenger $\mathcal{C}'_{\mathsf{ideal}}$ as an encryption query in the ideal QAE game, applying $\Lambda_{CB \to C\tilde{B}}$ to the output and sending the resulting C-register to $\mathcal{C}'_{\mathsf{ideal}}$ as a decryption query, is exactly the simulator defined in Eqs. (5) and (6). Thus, the adversary that runs $\mathcal{A}_1$, queries the encryption oracle, runs $\mathcal{A}_2$, queries the decryption oracle and finally runs $\mathcal{A}_3$ is a successful QAE adversary. □

In addition, QAE implies quantum unforgeability.

Theorem 13. QAE $\Longrightarrow$ QUF.

Proof. For a scheme Π, let $\mathcal{A}$ be an adversary against QUF, e.g., let us say that:

$$\Pr\left[\mathsf{QUF\text{-}Forge}(\Pi, \mathcal{A}, n) \to \mathsf{win}\right] = \Pr\left[\mathsf{QUF\text{-}Cheat}(\Pi, \mathcal{A}, n) \to \mathsf{cheat}\right] + \nu,$$

where ν is non-negligible. We then build another adversary $\mathcal{B}$ with black-box access to $\mathcal{A}$, able to distinguish QAE-Real from QAE-Ideal with non-negligible advantage. $\mathcal{B}$ runs $\mathcal{A}$, and forwards all of $\mathcal{A}$'s queries to his own encryption oracle. When eventually $\mathcal{A}$ outputs a forgery, $\mathcal{B}$ sends it for decryption to his own decryption oracle. If the decryption succeeds (that is, the oracle does not return $|\bot\rangle\langle\bot|$), then $\mathcal{B}$ outputs real, otherwise he outputs ideal.

The idea is the following: suppose the decryption of the forgery state succeeds (i.e., it does not decrypt to $|\bot\rangle\langle\bot|$). This can happen in two cases:

1. we are in the QAE-Real game, and $\mathcal{A}$ produced a valid forgery (i.e., he won the QUF-Forge game); or
2. we are in the QAE-Ideal game, and $\mathcal{A}$ cheated by replaying an output of the encryption oracle (i.e., he won the QUF-Cheat game).

Recall that, by assumption, $\mathcal{A}$ produces a valid forgery with probability at least ν over cheating. Therefore the case 2. above happens with noticeable less probability than case 1., which is in fact the one $\mathcal{B}$ "bets" on. Analogously, suppose the decryption fails. This can happen in two cases:

1. we are in the QAE-Real game, but $\mathcal{A}$ produced an invalid forgery (i.e., he lost the QUF-Forge game); or
2. we are in the QAE-Ideal game, and $\mathcal{A}$ did not cheat (i.e., he lost QUF-Cheat).

For the same reasoning as above, 2. is noticeably more likely than 1., which is in fact $\mathcal{B}$'s bet. More in detail, we have:

$$\begin{aligned}
&\Big| \Pr\left[\mathcal{B}(\mathsf{QAE\text{-}Real}) \to \mathsf{Real}\right] - \Pr\left[\mathcal{B}(\mathsf{QAE\text{-}Ideal}) \to \mathsf{Real}\right] \Big| \\
=& \Big| \Pr\left[\mathsf{QAE\text{-}Real}\right] \cdot \Pr\left[\mathcal{A}(\mathsf{QUF\text{-}Forge}) \to \mathsf{win}\right] \\
&- \Pr\left[\mathsf{QAE\text{-}Ideal}\right] \cdot \Pr\left[\mathcal{A}(\mathsf{QUF\text{-}Cheat}) \to \mathsf{cheat}\right] \Big| \\
=& \frac{1}{2} \Big| \Pr\left[\mathcal{A}(\mathsf{QUF\text{-}Forge}) \to \mathsf{win}\right] - \left(\Pr\left[\mathcal{A}(\mathsf{QUF\text{-}Forge}) \to \mathsf{win}\right] - \nu \right) \Big| = \frac{\nu}{2},
\end{aligned}$$

which is non-negligible. □

Finally, we consider the classical restriction AE' of QAE.

Proposition 3. $\mathsf{AE}' \iff \mathsf{AE}$.

Proof. The security notion AE' is given in terms of two experiments which are like the AE-Real and AE-Ideal experiments in Shrimpton's formulation of AE security, with the following difference:

1. in the modified AE-Real experiment, the decryption oracle does not reject non-fresh ciphertexts, i.e. it is unrestricted; and
2. in the modified AE-Ideal experiment, the decryption oracle does not always return $\perp$: in case it is queried on a non-fresh ciphertext, it decrypts correctly.

Since classically we can store and compare plaintexts and ciphertexts, it is easy to construct an efficient simulator able to switch between the experiments of AE and AE', by inspecting $\mathcal{A}$'s decryption queries and reacting accordingly. Namely:

1. to switch from AE to AE', record $\mathcal{A}$'s plaintexts and ciphertexts during encryption queries, and reply with the right plaintext whenever $\mathcal{A}$ asks to decrypt a non-fresh ciphertext (otherwise, just send the query to the decryption oracle); and
2. to switch from AE' to AE, record $\mathcal{A}$'s received ciphertexts during encryption queries, and reply with $\perp$ whenever $\mathcal{A}$ asks to decrypt a non-fresh ciphertext (otherwise, just send the query to the decryption oracle).

This concludes the proof, as it shows the two cases to be equivalent. □

In particular, AE' is equivalent to UF. We provide evidence that a quantum analogue of this statement does not hold in the next section.

7 Constructions and Separations

In this section we exhibit constructions of SKQES that fulfill and separate the different security notions presented in the preceding sections. We begin by showing that augmenting a one-time scheme by a (perfectly) random function family using the construction in Definition 2 turns a QCA secure scheme into a QAE secure scheme. Then we will move on to show how to satisfy QAE with an efficiently implementable scheme. Recall that efficient QCA-secure SKQES can be constructed, e.g., from unitary two-designs like the Clifford group.

Theorem 14. *Let Π be a* QCA*-secure SKQES, and let $f : \mathcal{K}\times\{0,1\}^n \to \{0,1\}^m$ be a random function family. Then the scheme Π^f in Definition 2 is* QAE *secure.*

Proof. We let $\Pi = (\mathsf{KeyGen}, \mathsf{Enc}, \mathsf{Dec})$ and $\Pi^{\mathcal{F}} = (\mathsf{KeyGen}', \mathsf{Enc}', \mathsf{Dec}')$ where

1. $\mathsf{KeyGen}'(1^n)$ outputs a random function F from $\{0,1\}^n$ to $\{0,1\}^m$;
2. $\mathsf{Enc}'_F(X_M)$ outputs $|s\rangle\langle s|_R \otimes \mathsf{Enc}_{F(s)}(X)_C$, where $s \xleftarrow{\$} \{0,1\}^n$;
3. $\mathsf{Dec}'_F(Y_{RC})$ first measures the R register to get outcome s'; then it runs $\mathsf{Dec}_{F(s')}$ on register C and outputs the result.

Suppose $\mathcal{A}$ is a QAE adversary against $\Pi^{\mathcal{F}}$, i.e., a QPT algorithm with oracle access to Enc'_k and Dec'_k. Suppose $\mathcal{A}$ makes $\ell(n)$ queries to the oracle, where ℓ is some polynomial function of n. We assume that the randomnesses s_i and the keys $F(s_i)$ used for the scheme Π in the different encryption queries (for $i = 1, \ldots, \ell(n)$) are all distinct; this is true except with negligible probability.

Let us first analyze what happens in the QAE-Real experiment. Consider the i-th decryption oracle call. The decryption begins with a measurement of the R register, yielding some outcome s and thereby a key $\bar{k} = F(s)$. We can analyze the situation for each outcome s that occurs with non-negligible probability, separately. This is because if an adversary is successful, it is easy to see that there is also a modified successful adversary, that submits only decryption queries with a fixed string s in the randomness register.

Suppose first that $\bar{k} = F(s) \neq F(s_i)$ for all i. In this case, the Π-encrypted part of the forgery candidate gets decrypted with a key different from all the ones used for encryption. We analyze the attack map $\Lambda = \tilde{\mathcal{A}}(1^n)\mathrm{Tr}_C$ against the QCA scheme Π, where $\tilde{\mathcal{A}}$ is defined to first run $\mathcal{A}$ until the ith decryption query, while answering each encryption query by sampling a fresh key for the scheme Π. Note that Λ does not use initial side information, therefore $\sigma^{\mathsf{acc}} := \Lambda^{\mathsf{acc}}$ and $\sigma^{\mathsf{rej}} := \Lambda^{\mathsf{rej}}$ are just positive semidefinite matrices whose trace sums to one.

According to Eq. (4) in the definition of QCA, the trace of σ^{acc} is the probability that the simulator applies the identity to the plaintext. The output of the attack map Λ does not depend on it's input, i.e. the same holds for the effective map Λ^{Π} and hence for $(\mathbb{1} - |\bot\rangle\langle\bot|)\Lambda^{\Pi}(\cdot)(\mathbb{1} - |\bot\rangle\langle\bot|)$. Any such map is far from any non-negligible multiple of the identity channel so the trace of σ^{acc} is negligible according to Eq. 3. We have hence shown that the decryption oracle returns $\bot$ with overwhelming probability, so we can take $\sigma^{crej} = \mathrm{Tr}_C\tilde{\mathcal{A}}(1^n)$.

Let now $s' = r_j$, and write $\mathcal{A} = \mathcal{A}_1\mathsf{Enc}_{\tilde{k}}\mathcal{A}_0$, splitting the adversary into two parts before and after the j-th encryption query. Let $(\tilde{\mathcal{A}}_1)_{CE_1\to CE_2}$ be defined analogous to $\tilde{\mathcal{A}}$. E_1 and E_2 are the internal memory registers of $\mathcal{A}$ at the time of the j-th encryption query and the i-th decryption query, respectively. Π is QCA secure, implying that $\tilde{\mathcal{A}}_1^{\Pi} = \mathbb{E}_{\tilde{k}}\left[\mathsf{Dec}_k \circ \tilde{\mathcal{A}}_1 \circ \mathsf{Enc}_{\tilde{k}}\right]$ fulfills:

$$\|(\tilde{\mathcal{A}}_1^{\Pi})_{ME_1\to ME_2} - \mathrm{id}_M \otimes (\tilde{\mathcal{A}}_1^{\mathsf{acc}})_{E_1\to E_2} - \bot \otimes (\tilde{\mathcal{A}}_1^{\mathsf{rej}})_{E_1\to E_2}\|_\diamond \leq \mathrm{negl}(n), \quad (7)$$

where (using $P_{\mathrm{inv}} = \mathbb{1} - |\Phi_{\bar{k},\bar{r}}\rangle\langle\Phi_{\bar{k},\bar{r}}|$):

$$\tilde{\mathcal{A}}_1^{\mathsf{acc}} = \mathbb{E}_{\bar{k},\bar{r}}\left[\langle \Phi_{\bar{k},\bar{r}}|V_{\bar{k}}^{\dagger}\tilde{\mathcal{A}}_1^{\mathsf{acc}}\left(\mathsf{Enc}_{\bar{k};\bar{r}}\left(\phi_{MM'}^{+}\right)\otimes(\cdot)_{E_1}\right)V_{\bar{k}}|\Phi_{\bar{k},\bar{r}}\rangle\right] \text{ and}$$
$$\tilde{\mathcal{A}}_1^{\mathsf{rej}} = \mathbb{E}_{\bar{k},\bar{r}}\left[\mathrm{Tr}_{MM'T}P_{\mathrm{inv}}V_{\bar{k}}^{\dagger}\tilde{\mathcal{A}}_1^{\mathsf{acc}}\left(\mathsf{Enc}_{\bar{k};\bar{r}}\left(\phi_{MM'}^{+}\right)\otimes(\cdot)_{E_1}\right)V_{\bar{k}}\right]. \quad (8)$$

The form of the simulator in the reject case follows by using that the maximally entangled state is a point in the optimization defining the diamond norm in (3) and using the monotonicity of the trace norm under partial trace.

We now show indistinguishability of the real and ideal experiments by induction over the decryption queries. Since QCA implies IND, the two are indistinguishable before the first decryption query. Assume now that the two experiments cannot be distinguished using an algorithm that makes at most $i-1$ decryption queries. Consider $\mathcal{A}$ running in the ideal experiment until right before the $(i+1)$-th decryption query (or until the end, if $i=\ell$). We make the same case distinction as before. In the first case the measurement in line (3) in the ideal decryption oracle in Experiment 5 never returns 0, i.e. the output is always reject. Therefore we can replace the i-th decryption oracle by the constant reject function, thereby reducing the number of decryption oracle calls of to $i-1$. By the induction hypothesis, the contents of the internal register are therefore indistinguishable whether in the QAE-Real or in the QAE-Ideal experiment.

Turning to the second case, we make a very similar argument. We have $s=s_j$, i.e. the only encryption query where the measurement from line (3) in the definition of the ideal decryption oracle in Experiment 5 can possibly return 0 is the j-th. Here it is left to observe that the rest of the ideal decryption oracle implements exactly the same map as in the ideal world, i.e. the ones from Eqs. (7) and (8). Replacing the j-th encryption and the i-th decryption oracle call by this map, and using the induction hypothesis, we get that $\mathcal{A}$ run until before the $i+1$-th decryption oracle call cannot distinguish QAE-Real from QAE-Ideal. This ends the proof by induction. □

We now show how to satisfy QAE efficiently, by means of a post-quantum-secure pseudorandom function.

Corollary 2. *Let Π be a* QCA*-secure SKQES that satisfies Condition 1, and let f be a* pqPRF*. Then the scheme Π^f (from Definition 2) satisfies* QAE.

Proof. As a contradiction, suppose there exists a QPT algorithm $\mathcal{A}$ that distinguishes QAE-Real from QAE-Ideal. We claim that this also holds if f is replaced with a completely random function family $\mathcal{F}$. If $\mathcal{A}$ cannot break the random scheme $\Pi^{\mathcal{F}}$, then we can build a distinguisher for f versus $\mathcal{F}$, as follows. What we would like to do is the following. Given an oracle $\mathcal{O}$, we:

1. choose a random bit $b \xleftarrow{\$} \{0,1\}$;
2. if $b=0$, we simulate the QAE-Real$(\Pi^{\mathcal{O}}, \mathcal{A}, n)$ experiment using our oracle;
3. if $b=1$, we simulate the QAE-Ideal$(\Pi^{\mathcal{O}}, \mathcal{A}, n)$ experiment using our oracle;
4. output $b \oplus s$ where s is the output of $\mathcal{A}$.

This may at first not seem possible using the classical oracle we are provided with, as the ideal decryption oracle has to implement the unitary $V_k^{\dagger}$, which seems to require superposition access to the random/pseudorandom function. However, observe that steps 5–11 of Experiment 2 commute with a measurement of the randomness register R in the computational basis, and afterwards this register is discarded. Therefore the outcome of the experiment is not changed by first measuring the register R, which yields an outcome r. Then the modified challenger can use classical oracle access to the random/pseudorandom function to implement $V_k^\dagger$ on the measured input state.

Note that, if $\Pi^{\mathcal{O}}$ is secure, then b and s are independent (up to negligible terms) and $b \oplus s$ is a fair coin. If $\Pi^{\mathcal{O}}$ is insecure, then it deviates from uniform by the QUF distinguishing advantage of $\mathcal{A}$. This yields a distinguisher between the case $\mathcal{O} = f$ and $\mathcal{O} = \mathcal{F}$. The claim then follows from Theorem 14. □

In particular, the scheme family $\mathsf{2desTag}^{\mathsf{pqPRF}}$ is sufficient for QAE. We remark that the proof uses the fact that, given classical oracle access to f, the scheme Π^f is efficiently implementable in the sense of Condition 1 – regardless of the nature of the family f. Of course, in the special case where f is a pqPRF, then Π^f simply satisfies Condition 1 without any need for oracles.

As QAE implies both QUF and QIND-CCA2 (see Theorems 11 and 13), we have the following corollary.

Corollary 3. *Let Π be a* QCA*-secure* SKQES *that satisfies Condition 1, and let f be a* pqPRF*. Then the scheme Π^f (from Definition 2) satisfies* QUF *and* QIND-CCA2.

We can also show how to satisfy bounded-query unforgeability, i.e., QUF_t. Recall that a t-wise independent function is a deterministic, efficiently computable keyed function family $\{f_k\}_k$ which appears random to any algorithm (of unbounded computational power) which gets classical oracle access to f_k for uniformly random k, and can make at most t queries. One can apply the proof technique of Corollary 2 and Theorem 14 to obtain the following.

Corollary 4. *Let Π be a* QCA*-secure* SKQES*, and let f be a t-wise independent function family. Then the scheme Π^f (as defined in Definition 2) satisfies* QUF_t.

Proof (Sketch). If there exists a QPT $\mathcal{A}$ which can break QUF_t for Π^f using t-many queries, then it also breaks $\Pi^{\mathcal{F}}$ where $\mathcal{F}$ is a random function. If not, we construct an oracle distinguisher for $\mathcal{O} = f$ versus $\mathcal{O} = \mathcal{F}$ which simulates $\mathcal{A}$ in one of the two games (each with probability 1/2) and outputs a bit which is biased depending on $\mathcal{O}$. Note that we only need t queries to do this, since we only run one of the games (and not both). It then remains to invoke Theorem 14, and observe that Theorem 13 holds in the case of a bounded number of queries. □

Separations. While QAE implies QIND-CCA2 according to Theorem 11, the converse does not hold. In fact, consider any QAE secure scheme and modify the decryption function by replacing the reject symbol by a fixed plaintext, e.g. the all zero state. Such a scheme is certainly still QIND-CCA2 secure, as any adversary against it can be used against the original scheme by simulating the modified one. The modified scheme is, however, manifestly not QAE as it never outputs $\perp$. The same reasoning works for QUF in place of QAE.

Proposition 4. QIND-CCA2 $\not\Rightarrow$ QUF, *and therefore* QIND-CCA2 $\not\Rightarrow$ QAE.

Finally, we turn to the relationship of QAE and QUF, and propose a separation as follows. Let Π be a scheme that fulfills cQCA (Definition 6) for trivial register $\tilde{B}$, but can be broken using an efficient attack with nontrivial $\tilde{B}$. For any PRF f, Π^f is clearly QUF, as the security notion ignores side information. It can however not be QAE, as QAE implies cQCA.

8 Discussion

In this work, we presented four new security notions for symmetric key quantum encryption: QCA, QUF, QIND-CCA2 and QAE. While we have already made significant progress on understanding these notions, a number of open questions remain. A few are as follows. Does an encryption scheme as discussed below Proposition 4 exist, proving QUF $\not\Rightarrow$ QAE? If so, does QUF imply QIND-CCA2 or QIND-CCA1? Classically, unforgeability and IND-CCA2 imply AE; does this hold quantumly as well? Finally, is there a scheme that satisfies QIND-CCA2 but cannot be upgraded to QAE by simply modifying the decryption function?

Acknowledgements. The authors would like to thank Anne Broadbent, Frédéric Dupuis, Yfke Dulek, Alex Russell, Christian Schaffner, and Fang Song for insightful discussions about the problems solved in this work. Part of this work was done while T.G. was supported by the TU Darmstadt. Part of this work was done while G.A. and C.M. were at QMATH, University of Copenhagen. Part of this work was sponsored by the COST CryptoAction IC1306. T.G. acknowledges financial support from the European Commissions PERCY grant (agreement 321310). G.A. and C.M. acknowledge financial support from the European Research Council (ERC Grant Agreement no 337603), the Danish Council for Independent Research (Sapere Aude) and VILLUM FONDEN via the QMATH Centre of Excellence (Grant No. 10059). This work is part of the research programme "Cryptography in the Quantum Age" with project number 639.022.519, which is financed by the Netherlands Organisation for Scientific Research (NWO).

References

1. Aaronson, S., Gottesman, D.: Improved simulation of stabilizer circuits. CoRR, quant-ph/0406196 (2004)
2. Aharonov, D., Ben-Or, M., Eban, E.: Interactive proofs for quantum computations. In: Proceedings of the Innovations in Computer Science - ICS 2010, Tsinghua University, Beijing, China, 5–7 January 2010, pp. 453–469 (2010)
3. Alagic, G., Broadbent, A., Fefferman, B., Gagliardoni, T., Schaffner, C., St. Jules, M.: Computational security of quantum encryption. In: Nascimento, A.C.A., Barreto, P. (eds.) ICITS 2016. LNCS, vol. 10015, pp. 47–71. Springer, Cham (2016). https://doi.org/10.1007/978-3-319-49175-2_3
4. Alagic, G., Gagliardoni, T., Majenz, C.: Unforgeable quantum encryption. Cryptology ePrint Archive, Report 2017/960 (2017). https://eprint.iacr.org/2017/960

5. Alagic, G., Majenz, C.: Quantum non-malleability and authentication. In: Katz, J., Shacham, H. (eds.) CRYPTO 2017. LNCS, vol. 10402, pp. 310–341. Springer, Cham (2017). https://doi.org/10.1007/978-3-319-63715-0_11
6. Ambainis, A., Bouda, J., Winter, A.: Non-malleable encryption of quantum information. J. Math. Phys. **50**(4), 042106 (2009)
7. Ambainis, A., Mosca, M., Tapp, A., de Wolf, R.: Private quantum channels. In: 41st Annual Symposium on Foundations of Computer Science, FOCS 2000, Redondo Beach, California, USA, 12–14 November 2000, pp. 547–553 (2000)
8. Barak, B.: Cs127 course notes, Chap. 6. http://www.boazbarak.org/cs127/chap06_CCA.pdf. Accessed 7 Sept 2017
9. Barnum, H., Crépeau, C., Gottesman, D., Smith, A.D., Tapp, A.: Authentication of quantum messages. In: Proceedings of the 43rd Symposium on Foundations of Computer Science (FOCS 2002), Vancouver, BC, Canada, 16–19 November 2002, pp. 449–458 (2002)
10. Bellare, M., Namprempre, C.: Authenticated encryption: relations among notions and analysis of the generic composition paradigm. In: Okamoto, T. (ed.) ASIACRYPT 2000. LNCS, vol. 1976, pp. 531–545. Springer, Heidelberg (2000). https://doi.org/10.1007/3-540-44448-3_41
11. Boneh, D., Zhandry, M.: Quantum-secure message authentication codes. In: Johansson, T., Nguyen, P.Q. (eds.) EUROCRYPT 2013. LNCS, vol. 7881, pp. 592–608. Springer, Heidelberg (2013). https://doi.org/10.1007/978-3-642-38348-9_35
12. Boneh, D., Zhandry, M.: Secure signatures and chosen ciphertext security in a quantum computing world. In: Canetti, R., Garay, J.A. (eds.) CRYPTO 2013. LNCS, vol. 8043, pp. 361–379. Springer, Heidelberg (2013). https://doi.org/10.1007/978-3-642-40084-1_21
13. Brandão, F.G.S.L., Harrow, A.W., Horodecki, M.: Local random quantum circuits are approximate polynomial-designs. Commun. Math. Phys. **346**(2), 397–434 (2016)
14. Broadbent, A., Jeffery, S.: Quantum homomorphic encryption for circuits of low T-gate complexity. In: Gennaro, R., Robshaw, M. (eds.) CRYPTO 2015. LNCS, vol. 9216, pp. 609–629. Springer, Heidelberg (2015). https://doi.org/10.1007/978-3-662-48000-7_30
15. Broadbent, A., Wainewright, E.: Efficient simulation for quantum message authentication. In: Nascimento, A.C.A., Barreto, P. (eds.) ICITS 2016. LNCS, vol. 10015, pp. 72–91. Springer, Cham (2016). https://doi.org/10.1007/978-3-319-49175-2_4
16. DiVincenzo, D.P., Leung, D.W., Terhal, B.M.: Quantum data hiding. IEEE Trans. Inf. Theory **48**(3), 580–598 (2002)
17. Dulek, Y., Schaffner, C., Speelman, F.: Quantum homomorphic encryption for polynomial-sized circuits. In: Robshaw, M., Katz, J. (eds.) CRYPTO 2016. LNCS, vol. 9816, pp. 3–32. Springer, Heidelberg (2016). https://doi.org/10.1007/978-3-662-53015-3_1
18. Dupuis, F., Nielsen, J.B., Salvail, L.: Secure two-party quantum evaluation of unitaries against specious adversaries. In: Rabin, T. (ed.) CRYPTO 2010. LNCS, vol. 6223, pp. 685–706. Springer, Heidelberg (2010). https://doi.org/10.1007/978-3-642-14623-7_37
19. Dupuis, F., Nielsen, J.B., Salvail, L.: Actively secure two-party evaluation of any quantum operation. In: Safavi-Naini, R., Canetti, R. (eds.) CRYPTO 2012. LNCS, vol. 7417, pp. 794–811. Springer, Heidelberg (2012). https://doi.org/10.1007/978-3-642-32009-5_46

20. Gagliardoni, T., Hülsing, A., Schaffner, C.: Semantic security and indistinguishability in the quantum world. In: Robshaw, M., Katz, J. (eds.) CRYPTO 2016. LNCS, vol. 9816, pp. 60–89. Springer, Heidelberg (2016). https://doi.org/10.1007/978-3-662-53015-3_3
21. Garg, S., Yuen, H., Zhandry, M.: New security notions and feasibility results for authentication of quantum data. In: Katz, J., Shacham, H. (eds.) CRYPTO 2017. LNCS, vol. 10402, pp. 342–371. Springer, Cham (2017). https://doi.org/10.1007/978-3-319-63715-0_12
22. Gottesman, D.: The Heisenberg representation of quantum computers. arXiv quant-ph/9807006 (1998)
23. Gottesman, D.: Uncloneable encryption. Quantum Inf. Comput. **3**(6), 581–602 (2003)
24. Hayden, P., Leung, D.W., Mayers, D.W.: The universal composable security of quantum message authentication with key recyling. arXiv quant-ph/1610.09434 (2016)
25. Katz, J., Lindell, Y.: Introduction to Modern Cryptography, 2nd edn. CRC Press, Boca Raton (2014)
26. Nielsen, M.A., Chuang, I.L.: Quantum Computation and Quantum Information: 10th Anniversary Edition, 10th edn. Cambridge University Press, New York (2011)
27. Portmann, C.: Quantum authentication with key recycling. In: Coron, J.-S., Nielsen, J.B. (eds.) EUROCRYPT 2017. LNCS, vol. 10212, pp. 339–368. Springer, Cham (2017). https://doi.org/10.1007/978-3-319-56617-7_12
28. Shrimpton, T.: A characterization of authenticated-encryption as a form of chosen-ciphertext security. IACR Cryptology ePrint Archive 2004:272 (2004)
29. Winter, A.J.: Coding theorem and strong converse for quantum channels. IEEE Trans. Inf. Theory **45**(7), 2481–2485 (1999)

Tightly-Secure Key-Encapsulation Mechanism in the Quantum Random Oracle Model

Tsunekazu Saito(✉), Keita Xagawa(✉), and Takashi Yamakawa(✉)

NTT Secure Platform Laboratories, 3-9-11, Midori-cho, Musashino-shi, Tokyo 180-8585, Japan
{saito.tsunekazu,xagawa.keita,yamakawa.takashi}@lab.ntt.co.jp

Abstract. Key-encapsulation mechanisms secure against chosen ciphertext attacks (IND-CCA-secure KEMs) in the quantum random oracle model have been proposed by Boneh, Dagdelen, Fischlin, Lehmann, Schafner, and Zhandry (CRYPTO 2012), Targhi and Unruh (TCC 2016-B), and Hofheinz, Hövelmanns, and Kiltz (TCC 2017). However, all are non-tight and, in particular, security levels of the schemes obtained by these constructions are less than half of original security levels of their building blocks.

In this paper, we give a conversion that tightly converts a weakly secure public-key encryption scheme into an IND-CCA-secure KEM in the quantum random oracle model. More precisely, we define a new security notion for deterministic public key encryption (DPKE) called the disjoint simulatability, and we propose a way to convert a disjoint simulatable DPKE scheme into an IND-CCA-secure key-encapsulation mechanism scheme without incurring a significant security degradation. In addition, we give DPKE schemes whose disjoint simulatability is tightly reduced to post-quantum assumptions. As a result, we obtain IND-CCA-secure KEMs tightly reduced to various post-quantum assumptions in the quantum random oracle model.

Keywords: Tight security · Chosen-ciphertext security
Post-quantum cryptography · KEM

1 Introduction

1.1 Background

Indistinguishability against chosen ciphertext attacks (IND-CCA-security) is considered to be a *de facto* standard security notion of a public key encryption (PKE) and a key encapsulation mechanism (KEM). For constructing efficient IND-CCA-secure PKEs and KEMs, an idealized model called the random oracle model (ROM) [BR93] is often used. In the ROM, a hash function is idealized to be a publicly accessible oracle that simulates a truly random function. There are many known generic constructions of efficient IND-CCA-secure

J. B. Nielsen and V. Rijmen (Eds.): EUROCRYPT 2018, LNCS 10822, pp. 520–551, 2018.
https://doi.org/10.1007/978-3-319-78372-7_17

PKE/KEM in the ROM; Bellare-Rogaway (BR) [BR93], OAEP [BR95,FOPS04], REACT [OP01], GEM [CHJ+02], Fujisaki-Okamoto (FO) [FO99,FO13], etc. KEM variants of these constructions were studied by Dent [Den03], which is summarized in Fig. 10 in Sect. B.

Quantum Random Oracle Model. Though the ROM has been widely used to heuristically analyze security of cryptographic primitives, Boneh et al. [BDF+11] pointed out that the ROM is rather problematic when considering a *quantum* adversary. The problem is that in the ROM, an adversary is only given a classical access to a random oracle. Since a random oracle is an idealization of a real hash function, a quantum adversary should be able to quantumly compute it. On the basis of this observation, they proposed a new model called the quantum random oracle model (QROM) where an adversary can quantumly access a random oracle. Since many techniques used in the ROM including adaptive programmability or extractability cannot be directly translated into the ones in the QROM, proving security in the QROM often requires different techniques from proofs in the ROM (see [BDF+11] for more details). Nonetheless, some above mentioned IND-CCA-secure PKE/KEMs in the ROM (and their variants) can be shown to also be secure in the QROM: Boneh et al. [BDF+11] proved that a variant of Bellare-Rogaway is IND-CCA-secure in the QROM. Targhi and Unruh [TU16] proposed variants of the Fujisaki-Okamoto and OAEP and proved that they are IND-CCA-secure in the QROM.

Tight Security. When proving the security of a primitive P under the hardness of a problem S, we usually construct a reduction algorithm $\mathcal{R}$ that uses an adversary $\mathcal{A}$ against the security of P as a subroutine and solves the problem S. Let (T, ϵ) and (T', ϵ') denote running times and success probabilities of $\mathcal{A}$ and $\mathcal{R}$, respectively. We say that a reduction is tight if we have $T' \approx T$ and $\epsilon' \approx \epsilon$. Tight security is desirable since it ensures that breaking the security of P is as hard as solving an underlying hard problem S. Conversely, if a security reduction is non-tight, we cannot immediately conclude that breaking the security of a primitive P is hard even if an underlying problem S is hard. For example, Menezes [Men12] shows an example of a provably secure primitive with non-tight security that is insecure with a realistic parameter setting. Therefore, tight security is important to ensure the real security of a primitive.

From that perspective, the above mentioned IND-CCA-secure PKE/KEMs in the QROM do not serve as satisfactory solutions for constructing post-quantum IND-CCA-secure PKE/KEMs because they are non-tight. To clarify this, we give more details on these results below, where (T, ϵ) and (T', ϵ') denote running times and success probabilities of an adversary and a reduction algorithm, respectively, q_{H} denotes the number of random oracle queries, and t_{RO} denotes the time needed to simulate one evaluation of a random oracle (for further explanation of t_{RO}, see Subsect. 2.2).

- Boneh et al. [BDF+11] proved that a KEM variant of Bellare-Rogaway based on a one-way trapdoor function is IND-CCA-secure in the QROM.[1] According to their security proof, we have $T' \approx T + q_{\mathsf{H}} \cdot t_F + (q_{\mathsf{H}} + q_{\mathsf{Dec}}) \cdot t_{\mathsf{RO}}$ and $\epsilon' \approx \epsilon^2/q_{\mathsf{H}}^2$ where t_F denotes the time needed for evaluating an underlying one-way trapdoor function and q_{Dec} denotes the number of decryption queries.
- Targhi and Unruh [TU16] proposed a variant of Fujisaki-Okamoto and proved that their construction is secure in the QROM assuming OW-CPA security of an underlying PKE scheme. According to their security proof, we have $T' \geq T + O(q_{\mathsf{H}}^2)$ and $\epsilon' \approx \epsilon^4/q_{\mathsf{H}}^6$. We note that Hofheinz et al. [HHK17] subsequently gave a modular analysis for the conversion but did not improve the tightness.
- Targhi and Unruh [TU16] proposed a variant of OAEP and proved that their construction is secure in the QROM assuming a partial domain one-way function. According to their security proof, we have $T' \geq T + O(q_{\mathsf{H}}^2)$ and $\epsilon' \approx \epsilon^8/\mathsf{poly}(q_{\mathsf{H}})$.

As seen above, known constructions of IND-CCA-secure PKE/KEMs in the QROM incur at least quadratic security loss, and their security degrades rapidly as q_{H} grows. For example, in the Bellare-Rogaway KEM, if we start from a trapdoor function with 128-bit security (i.e., $\epsilon' = 2^{-128}$) and set $q_{\mathsf{H}} = 2^{60}$, then the bound given by Boneh et al. [BDF+11] only ensures 4-bit security (i.e., $\epsilon = 2^{-4}$) for a resulting KEM. Conversely, if we want to ensure 128-bit security (i.e., $\epsilon = 2^{-128}$) for a resulting KEM, we have to start from a trapdoor function with 376-bit security ($\epsilon' = 2^{-376}$) which incurs significant blowup of parameters. The other two constructions are even worse in regard to tightness. Therefore, to obtain an efficient construction of post-quantum IND-CCA-secure PKE/KEM, we need a construction with tighter security reduction that does not incur a quadratic security loss.

1.2 Our Contributions

In this paper, we give a construction of an IND-CCA-secure KEM based on a deterministic PKE (DPKE) scheme that satisfies a newly introduced security notion that we call the disjoint simulatability. Our security reduction is much tighter than those of existing constructions of IND-CCA-secure PKE schemes and does not incur quadratic security loss. By using the same notations as in the previous subsection, we have $T' \approx T + q_{\mathsf{H}} \cdot t_{\mathsf{Enc}} + (q_{\mathsf{H}} + q_{\mathsf{Dec}}) \cdot t_{\mathsf{RO}}$ and $\epsilon' \approx \epsilon$ where t_{Enc} denotes a time needed for encryption of an underlying DPKE scheme. We note that t_{Enc} is a fixed polynomial of the security parameter, and thus we believe that this blowup is much less significant than the quadratic (or quartic/octic) blowup for ϵ as in the previous constructions.

[1] More precisely, they proved that a hybrid encryption variant of the Bellare-Rogaway PKE scheme based on a one-way trapdoor function plus a CCA-secure symmetric-key encryption scheme is IND-CCA-secure in the QROM. Their proof is easily turned into the proof for the KEM variant of the Bellare-Rogaway conversion.

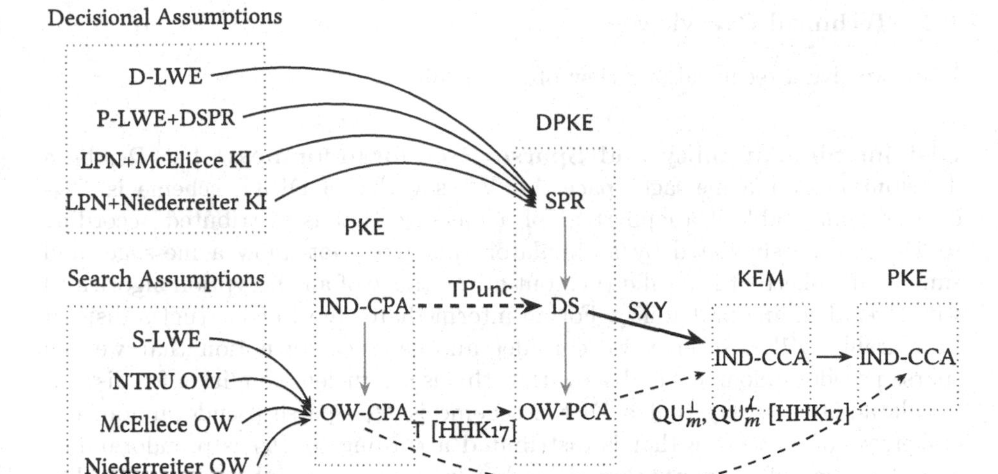

Fig. 1. Transformations among PKE, DPKE and KEM in the QROM: D-LWE and S-LWE denote the decisional and search learning-with-errors assumptions; P-LWE denotes the polynomial-LWE assumption; DSPR denotes the decisional small polynomial ratio assumption; LPN denotes the learning-parity-with-noise assumption; McEliece KI and Niederreiter KI denote the McEliece-key-indistinguishability and Niederreiter-key-indistinguishability assumptions, respectively; NTRU OW, McEliece OW, and Niederreiter OW denote onewayness of the NTRU, McEliece encryption, and Niederreiter encryption, respectively; OW-CPA, OW-PCA, IND-CPA, and IND-CCA denote onewayness under chosen-plaintext attacks, onewayness under plaintext-checking attacks, indistinguishability under chosen-plaintext attacks, and indistinguishability under chosen-ciphertext attacks, respectively; SPR denotes the sparse pseudorandomness; and DS denotes the disjoint simulatability. Solid arrows indicate quantum tight reductions, dashed arrows indicate quantum non-tight reductions, thin arrows indicate existing reductions, thick arrows indicate our new reductions, and gray arrows indicate trivial implications.

Moreover, we construct some DPKE schemes whose disjoint simulatabilities are tightly reduced to some post-quantum assumptions like learning with errors (LWE) and some other assumptions related to NTRU, the McEliece PKE, and the Niederreiter PKE. As a result, we obtain the first IND-CCA-secure KEMs that do not incur a quadratic security loss in the QROM based on these assumptions. We also construct a disjoint simulatable DPKE scheme from any IND-CPA-secure PKE scheme on an exponentially large message space with quadratic security loss. This gives a construction of an IND-CCA-secure KEM based on an IND-CPA-secure PKE scheme on an exponentially large message space with quadratic (rather than quartic as in previous works) security loss. Our results are summarized in Fig. 1.

We implement an instantiation based on NTRU-HRSS [HRSS17] on a desktop PC and a RasPi. Assuming that NTRU-HRSS is disjoint simulatable, the obtained KEM is CCA secure in the QROM. See Sect. 5.

1.3 Technical Overview

Here, we give a technical overview of our results.

Disjoint Simulatability and Sparse Pseudorandomness. Let $\mathcal{D}_{\mathcal{M}}$ be a distribution over a message space $\mathcal{M}$. We say that a DPKE scheme is $\mathcal{D}_{\mathcal{M}}$-disjoint simulatable if a ciphertext of a message that is distributed according to $\mathcal{D}_{\mathcal{M}}$ can be simulated by a simulator that does not know a message, and simulated ciphertext is invalid (i.e., out of the range of an encryption algorithm) with overwhelming probability. For an intermediate step to construct a disjoint simulatable DPKE scheme, we consider another security notion that we call sparse pseudorandomness and show that this is a sufficient condition for disjoint simulatability. We say that a DPKE scheme is $\mathcal{D}_{\mathcal{M}}$-sparse pseudorandom if a ciphertext of a message that is distributed according to $\mathcal{D}_{\mathcal{M}}$ is pseudorandom and the range of an encryption algorithm is sparse in a ciphertext space. The $\mathcal{D}_{\mathcal{M}}$-sparse pseudorandomness implies the $\mathcal{D}_{\mathcal{M}}$-disjoint simulatability because if the sparse pseudorandomness is satisfied, then a simulator that simply outputs a random element of a ciphertext space suffices for the disjoint simulatability[2].

Instantiations of Disjoint Simulatable DPKE. We construct DPKE schemes based on the concepts of the Gentry–Peikert–Vaikuntanathan (GPV) trapdoor function for LWE [GPV08], NTRU [HPS98], the McEliece PKE [McE78], and the Niederreiter PKE [Nie86] and prove that they are sparse pseudorandom (and thus disjoint simulatable) w.r.t. a certain message distribution under the LWE assumption, or other related assumptions to an underlying PKE scheme. Moreover, the reductions are tight. See Subsect. 3.3 for details of instantiations from concrete assumptions

We also construct a disjoint simulatable DPKE scheme based on any IND-CPA-secure PKE scheme with an exponentially large message space in the QROM. Unfortunately, this reduction is not tight and incurs a square security loss. See Subsect. 3.4 for details.

Previous Construction: BR-KEM. Before describing our construction, we review the construction and security proof of the Bellare-Rogaway KEM (BR-KEM), which was proven IND-CCA-secure in the QROM by Boneh et al. [BDF+11] because our construction is based on their idea. BR-KEM is a construction of an IND-CCA-secure KEM based on a one-way trapdoor function with an efficiently recognizable range[3]. For compatibility with ours, we treat a one-way trapdoor function as a perfectly correct OW-CPA-secure DPKE scheme by considering a function and an inversion to be an encryption and a

[2] In fact, we have to additionally assume that a ciphertext space is efficiently sampleable.

[3] The efficient recognizability of a range was not explicitly assumed in [BDF+11] but is actually needed for their proof.

decryption, respectively. Let $(\mathsf{Gen}, \mathsf{Enc}, \mathsf{Dec})$ denote algorithms of an underlying DPKE scheme. Then BR-KEM $= (\mathsf{Gen_{BR}}, \mathsf{Enc_{BR}}, \mathsf{Dec_{BR}})$ is described as follows:

- $\mathsf{Gen_{BR}}$ is exactly the same as Gen.
- $\mathsf{Enc_{BR}}$, given a public key ek as an input, chooses a randomness m from a message space uniformly at random, computes a ciphertext $C := \mathsf{Enc}(ek, m)$ and a key $K := \mathsf{H}(m)$ where H is a hash function modeled as a random oracle, and outputs (C, K).
- $\mathsf{Dec_{BR}}$, given a ciphertext C and a decryption key dk as an input, checks if C is in the valid ciphertext space and returns $\perp$ if not. Otherwise it computes $K := \mathsf{H}(\mathsf{Dec}(dk, C))$ and returns K.

In the security proof in the QROM, we first replace a random oracle H with $\mathsf{H}_q \circ \mathsf{Enc}(ek,)$ where H_q is another random oracle that is not given to an adversary. Since $\mathsf{Enc}(ek, \cdot)$ is injective due to its perfect correctness, $\mathsf{H}_q \circ \mathsf{Enc}(ek, \cdot)$ still works as a random oracle from the view of an adversary. After this replacement, we notice that a decryption oracle can be simulated by using H_q without the help of a decryption key because we have $\mathsf{H}(\mathsf{Dec}(dk, c)) = \mathsf{H}_q \circ \mathsf{Enc}(ek, \mathsf{Dec}(dk, c)) = \mathsf{H}_q(c)$. For proving IND-CCA security, we have to prove that $\mathsf{H}_q(c^*)$ is pseudorandom from the view of an adversary. If we were in a classical world, then this could be proven quite easily: the only way for an adversary to obtain any information of $\mathsf{H}_q(c^*)$ is to query m^* such that $c^* = \mathsf{Enc}(ek, m^*)$, in which case the adversary breaks the OW-CPA security of an underlying DPKE scheme. In a quantum world, things do not go as easily because even if an adversary queries a quantum state whose magnitude on m^* is large, a reduction algorithm cannot notice that immediately. Nonetheless, by using the One-Way to Hiding (OW2H) lemma proven by Unruh [Unr15] (Lemma 2.1), we can show that the advantage for an adversary to distinguish $\mathsf{H}_q(c^*)$ from a truly random string is at most a square root of the probability that measurement of a randomly chosen adversary's query to H is equal to m^*. Hence, we can reduce the IND-CCA security of BR-KEM to the OW-CPA security of the underlying DPKE scheme with a quadratic security loss. On the other hand, to avoid the quadratic security loss, it seems that we have to avoid the usage of the OW2H lemma because the lemma inherently incurs a quadratic security loss.

Our Conversion, SXY. In the above proof, we used the fact that the only way for an adversary to obtain any information of $\mathsf{H}_q(c^*)$ is to query m^* to H such that $c^* = \mathsf{Enc}(ek, m^*)$. Our key idea is based on the observation that if such m^* does not exist, i.e., c^* is out of the range of $\mathsf{Enc}(ek, \cdot)$, then it is information-theoretically impossible for an adversary to obtain any information of $\mathsf{H}_q(c^*)$. Indeed, though c^* is in the range of $\mathsf{Enc}(ek, \cdot)$ in the real game, if we choose an encryption randomness m according to a distribution $\mathcal{D}_{\mathcal{M}}$, then we can replace c^* with a simulated ciphertext that is out of the range of $\mathsf{Enc}(ek, \cdot)$ by using the $\mathcal{D}_{\mathcal{M}}$-disjoint simulatability. After replacing c^* with a simulated one, we can information-theoretically bound an adversary's advantage and need not use the OW2H lemma. This seems to simply resolve the problem, and we obtain an IND-CCA-secure KEM without a quadratic security loss. However, another problem arises here: a valid ciphertext space of a disjoint simulatable DPKE scheme is inherently not efficiently recognizable (otherwise real and simulated ciphertexts are easy to distinguish), whereas the simulation of decryption algorithm has to first verify if a given ciphertext is valid or not. To resolve the problem, we modify the decryption algorithm so that if a ciphertext is invalid, then it returns a random value rather than $\perp$. In the security proof of BR-KEM, a decryption oracle is simulated just by evaluating a random oracle H_q for a ciphertext, and this enables a reduction algorithm to simulate a decryption oracle for both valid and invalid ciphertexts even though it cannot determine if a given ciphertext is valid. Hence, we can reduce the IND-CCA-security of the resulting KEM without using the OW2H lemma and thus without a quadratic security loss.

Curiously, this conversion is essentially the same as $\mathsf{U}_m^{\not\perp}$ in [HHK17]. This means that we can remove an "additional" hash from $\mathsf{QU}_m^{\not\perp}$ assuming a stronger underlying DPKE in the QROM. In addition, this means that the obtained KEM is tightly secure assuming that the underlying DPKE is OW-CPA secure in the ROM as shown in [HHK17].

1.4 Related Work

In a concurrent and independent work, Jiang, Zhang, Chen, Wang, and Ma [JZC+17] proposed two new constructions of an IND-CCA-secure KEM based on a OW-CPA-secure PKE scheme with quadratic security loss. However, both constructions incur quadratic security loss.

2 Preliminaries

2.1 Notation

A security parameter is denoted by κ. We use the standard O-notations: O, Θ, Ω, and ω. DPT and PPT stand for deterministic polynomial time and probabilistic polynomial time. A function $f(\kappa)$ is said to be *negligible* if $f(\kappa) = \kappa^{-\omega(1)}$. We denote a set of negligible functions by $\mathsf{negl}(\kappa)$. For two finite sets $\mathcal{X}$ and $\mathcal{Y}$, $\mathrm{Map}(\mathcal{X}, \mathcal{Y})$ denote a set of all functions whose domain is $\mathcal{X}$ and codomain is $\mathcal{Y}$.

For a distribution χ, we often write "$x \leftarrow \chi$," which indicates that we take a sample x from χ. For a finite set S, $U(S)$ denotes the uniform distribution over S. We often write "$x \leftarrow S$" instead of "$x \leftarrow U(S)$." For a set S and a deterministic algorithm A, $\mathsf{A}(S)$ denotes the set $\{\mathsf{A}(x) \mid x \in S\}$.

If inp is a string, then "$\mathsf{out} \leftarrow \mathsf{A}(\mathsf{inp})$" denotes the output of algorithm A when run on input inp. If A is deterministic, then out is a fixed value and we write "$\mathsf{out} := \mathsf{A}(\mathsf{inp})$." We also use the notation "$\mathsf{out} := \mathsf{A}(\mathsf{inp}; r)$" to make the randomness r explicit.

For the Boolean statement P, $\mathsf{boole}(P)$ denotes the bit that is 1 if P is true, and 0 otherwise. For example, $\mathsf{boole}(b' \stackrel{?}{=} b)$ is 1 if and only if $b' = b$.

2.2 Quantum Computation

We refer to [NC00] for basic of quantum computation.

Quantum Random Oracle Model. Roughly speaking, the quantum random oracle model (QROM) is an idealized model where a hash function is modeled as a publicly and quantumly accessible random oracle. See [BDF+11] for a more detailed description of the model.

Lemmas. We review some useful lemmas regarding the quantum random oracles. The first one is called the oneway-to-hiding (OW2H) lemma, which is proven by Unruh [Unr15, Lemma 6.2]. Roughly speaking, the lemma states that if any quantum adversary issuing at most q queries to a quantum random oracle H can distinguish $(x, \mathsf{H}(x))$ from (x, y), where y is chosen uniformly at random, then we can find x by measuring one of the adversary's queries even it causes a quadratic security loss. The lemma of the following form is taken from [HHK17].

Lemma 2.1 (Algorithmic Oneway to Hiding [Unr15, HHK17]). *Let $\mathsf{H} : \mathcal{X} \to \mathcal{Y}$ be a quantum random oracle, and let $\mathcal{A}$ be an adversary issuing at most q queries to H that on input $(x, y) \in \mathcal{X} \times \mathcal{Y}$ outputs either $0/1$. For all (probabilistic) algorithms F whose input space is $\mathcal{X} \times \mathcal{Y}$ and which do not make any hash queries to H, we have*

$$\begin{aligned} &\left| \begin{array}{l} \Pr[\mathcal{A}^{\mathsf{H}}(\mathrm{inp}) \to 1 \mid x \leftarrow \mathcal{X}; \mathrm{inp} \leftarrow \mathsf{F}(x, \mathsf{H}(x))] \\ \quad - \Pr[\mathcal{A}^{\mathsf{H}}(\mathrm{inp}) \to 1 \mid (x, y) \leftarrow \mathcal{X} \times \mathcal{Y}; \mathrm{inp} \leftarrow \mathsf{F}(x, y)] \end{array} \right| \\ &\quad \leq 2q \cdot \sqrt{\Pr[\mathsf{EXT}^{\mathcal{A},\mathsf{H}}(\mathrm{inp}) \to x \mid (x, y) \leftarrow \mathcal{X} \times \mathcal{Y}; \mathrm{inp} \leftarrow \mathsf{F}(x, y)]}, \end{aligned}$$

where EXT picks $i \leftarrow \{1, \ldots, q\}$, runs $\mathcal{A}^{\mathsf{H}}(\mathrm{inp})$ until i-th query $|\hat{x}\rangle$ to H, and returns $x' := \mathsf{Measure}(|\hat{x}\rangle)$ (when $\mathcal{A}$ makes fewer than i queries, EXT outputs $\bot \notin \mathcal{X}$).

(Unruh's original statement is recovered by letting F be an identity function.)

The second one claims that a random oracle can be used as a pseudorandom function even in the quantum setting.

Lemma 2.2. *Let ℓ be an integer. Let $\mathsf{H}\colon \{0,1\}^\ell \times \mathcal{X} \to \mathcal{Y}$ and $\mathsf{H}'\colon \mathcal{X} \to \mathcal{Y}$ be two independent random oracles. If an unbounded time quantum adversary $\mathcal{A}$ makes a query to H at most q_H times, then we have*

$$\left| \Pr[\mathcal{A}^{\mathsf{H}, \mathsf{H}(s, \cdot)}() \to 1 \mid s \leftarrow \{0,1\}^\ell] - \Pr[\mathcal{A}^{\mathsf{H}, \mathsf{H}'}() \to 1] \right| \leq q_\mathsf{H} \cdot 2^{\frac{-\ell+1}{2}}$$

where all oracle accesses of $\mathcal{A}$ can be quantum.

Though this seems to be a folklore, we give a proof of this lemma in Sect. C for completeness.[4]

Simulation of Random Oracle. In the original quantum random oracle model introduced by Boneh et al. [BDF+11], they do not allow a reduction algorithm to access a random oracle, so it has to simulate a random oracle by itself. In contrast, in this paper, we give a random oracle access to a reduction algorithm. We remark that this is just a convention and not a modification of the model since we can simulate a random oracle against quantum adversaries in several ways.

[4] Jiang et al. [JZC+17] also gave a proof of an essentially identical lemma.

1. The first way is a simulation by a $2q$-wise independent hash function, where q denotes the number of random oracle queries by an adversary, as introduced by Zhandry [Zha12b]. The simulation is perfect, that is, no adversary can distinguish the real QRO from the simulated one. A drawback of this simulation is a $O(q^2)$ blowup for a running time of a reduction algorithm since it has to compute a $2q$-wise independent hash function for each random oracle query.
2. The second way is a simulation by a quantumly secure PRF as used in [BDF+11]. If we use this simulation, then the blowup of a running time of a reduction algorithm is $O(q \cdot t_{\mathsf{PRF}})$ where t_{PRF} is the time needed for evaluating a PRF, which is usually much smaller than $O(q^2)$. However, we have to additionally assume the existence of a quantumly secure PRF, which is known to exist if a quantumly secure one-way function exists [Zha12a].
3. The third way is a simulation by a real hash function like SHA-2 and to think that this is a "random oracle." Since we adopt the QROM, we idealize a real hash function as a random oracle in the construction of primitives. Thus, it may be natural to assume the same thing even in *a reduction*, that is, the reduction algorithm implements the random oracle by a concrete hash function. If we use this simulation, then the blowup of a running time of a reduction algorithm is $O(q \cdot t_{\mathsf{hash}})$ where t_{hash} denotes a time to evaluate a hash function. This gives a tightest reduction at the expense of additional idealization of a hash function. We note that a similar convention is also used by Kiltz et al. [KLS17].
   We finally note that this way strengthens the assumption, that is, we need to assume that some problem is hard *in the QROM*.

We use t_{RO} to denote a time needed to simulate a random oracle. We have $t_{\mathsf{RO}} = O(q)$, t_{PRF}, or t_{hash}, if we use the first, second, or third way, respectively. We note that in the proof of quantum variants of Fujisaki-Okamoto and OAEP [TU16, HHK17], we have to simulate a random oracle in the 1st way, because a simulator has to "invert" a random oracle in a simulation.

2.3 Public-Key Encryption

The model for PKE schemes is summarized as follows:

Definition 2.1. *A PKE scheme* PKE *consists of the following triple of polynomial-time algorithms* (Gen, Enc, Dec)*.*

- $\mathsf{Gen}(1^\kappa; r_g) \to (ek, dk)$*: a key-generation algorithm that on input* 1^κ*, where* κ *is the security parameter, outputs a pair of keys* (ek, dk)*.* ek *and* dk *are called the encryption key and decryption key, respectively.*
- $\mathsf{Enc}(ek, m; r_e) \to c$*: an encryption algorithm that takes as input encryption key* ek *and message* $m \in \mathcal{M}$ *and outputs ciphertext* $c \in \mathcal{C}$*.*
- $\mathsf{Dec}(dk, c) \to m/\bot$*: a decryption algorithm that takes as input decryption key* dk *and ciphertext* c *and outputs message* $m \in \mathcal{M}$ *or a rejection symbol* $\bot \notin \mathcal{M}$*.*

Definition 2.2. *We say a PKE scheme* PKE *is deterministic if* Enc *is deterministic. DPKE stands for deterministic public key encryption.*

Definition 2.3 (Correctness). *We say* $\mathsf{PKE} = (\mathsf{Gen}, \mathsf{Enc}, \mathsf{Dec})$ *has* perfect correctness *if for any* (ek, dk) *generated by* Gen *and for any* $m \in \mathcal{M}$*, we have that*

$$\Pr[\mathsf{Dec}(dk, c) = m \mid c \leftarrow \mathsf{Enc}(ek, m)] = 1.$$

An additional property, γ-spread, is in Sect. A

Security: Here, we define onewayness under chosen-plaintext attacks (OW-CPA), indistinguishability under chosen-plaintext attacks (IND-CPA), and indistinguishability under chosen-ciphertext attacks (IND-CCA) for a PKE.

Definition 2.4 (Security notions for PKE). *For any adversary* $\mathcal{A}$*, we define its* OW-CPA*,* IND-CPA*, and* IND-CCA *advantages against a PKE scheme* $\mathsf{PKE} = (\mathsf{Gen}, \mathsf{Enc}, \mathsf{Dec})$ *as follows:*

$$\begin{aligned}
\mathsf{Adv}^{\text{ow-cpa}}_{\mathcal{A},\mathsf{PKE}}(\kappa) &:= \Pr[\mathsf{Expt}^{\text{ow-cpa}}_{\mathsf{PKE},\mathcal{A}}(\kappa) = 1], \\
\mathsf{Adv}^{\text{ind-cpa}}_{\mathsf{PKE},\mathcal{A}}(\kappa) &:= \left|2\Pr[\mathsf{Expt}^{\text{ind-cpa}}_{\mathsf{PKE},\mathcal{A}}(\kappa) = 1] - 1\right|, \\
\mathsf{Adv}^{\text{ind-cca}}_{\mathsf{PKE},\mathcal{A}}(\kappa) &:= \left|2\Pr[\mathsf{Expt}^{\text{ind-cca}}_{\mathsf{PKE},\mathcal{A}}(\kappa) = 1] - 1\right|,
\end{aligned}$$

where $\mathsf{Expt}^{\text{ow-cpa}}_{\mathsf{PKE},\mathcal{A}}(\kappa)$*,* $\mathsf{Expt}^{\text{ind-cpa}}_{\mathsf{PKE},\mathcal{A}}(\kappa)$*, and* $\mathsf{Expt}^{\text{ind-cca}}_{\mathsf{PKE},\mathcal{A}}(\kappa)$ *are experiments described in Fig. 2. For* GOAL-ATK $\in$ {OW-CPA, IND-CPA, IND-CCA}*, we say that* PKE *is GOAL-ATK-secure if* $\mathsf{Adv}^{\text{goal-atk}}_{\mathcal{A},\mathsf{PKE}}(\kappa)$ *is negligible for any PPT adversary* $\mathcal{A}$*.*

Additional definitions are in Sect. A

$\mathsf{Expt}^{\text{ow-cpa}}_{\mathsf{PKE},\mathcal{A}}(\kappa)$

$(ek, dk) \leftarrow \mathsf{Gen}(1^\kappa)$
$m^* \leftarrow \mathcal{M}$
$c^* \leftarrow \mathsf{Enc}(ek, m^*)$
$m' \leftarrow \mathcal{A}(ek, c^*)$
return $\mathsf{boole}(m' \overset{?}{=} \mathsf{Dec}(dk, c^*))$

$\mathsf{Expt}^{\text{ind-cpa}}_{\mathsf{PKE},\mathcal{A}}(\kappa)$

$b \leftarrow \{0, 1\}$
$(ek, dk) \leftarrow \mathsf{Gen}(1^\kappa)$
$(m_0, m_1, st) \leftarrow \mathcal{A}_1(ek)$
$c^* \leftarrow \mathsf{Enc}(ek, m_b)$
$b' \leftarrow \mathcal{A}_2(c^*, st)$
return $\mathsf{boole}(b' \overset{?}{=} b)$

$\mathsf{Expt}^{\text{ind-cca}}_{\mathsf{PKE},\mathcal{A}}(\kappa)$

$b \leftarrow \{0, 1\}$
$(ek, dk) \leftarrow \mathsf{Gen}(1^\kappa)$
$(m_0, m_1, st) \leftarrow \mathcal{A}_1^{\textsc{Dec}_\perp(\cdot)}(ek)$
$c^* \leftarrow \mathsf{Enc}(ek, m_b)$
$b' \leftarrow \mathcal{A}_2^{\textsc{Dec}_{c^*}(\cdot)}(c^*, st)$
return $\mathsf{boole}(b' \overset{?}{=} b)$

$\textsc{Dec}_a(c)$

if $c = a$, **return** $\perp$
$m := \mathsf{Dec}(dk, c)$
return m

Fig. 2. Games for PKE schemes

2.4 Key Encapsulation

The model for KEM schemes is summarized as follows:

Definition 2.5. *A KEM scheme* KEM *consists of the following triple of polynomial-time algorithms* (Gen, Encaps, Decaps)*:*

- $\mathsf{Gen}(1^\kappa; r_g) \to (ek, dk)$*: a key-generation algorithm that on input* 1^κ*, where* κ *is the security parameter, outputs a pair of keys* (ek, dk)*.* ek *and* dk *are called the encapsulation key and decapsulation key, respectively.*
- $\mathsf{Encaps}(ek; r_e) \to (c, K)$*: an encapsulation algorithm that takes as input encapsulation key* ek *and outputs ciphertext* $c \in \mathcal{C}$ *and key* $K \in \mathcal{K}$*.*
- $\mathsf{Decaps}(dk, c) \to K/\bot$*: a decapsulation algorithm that takes as input decapsulation key* dk *and ciphertext* c *and outputs key* K *or a rejection symbol* $\bot \notin \mathcal{K}$*.*

Definition 2.6 (Correctness). *We say* $\mathsf{KEM} = (\mathsf{Gen}, \mathsf{Encaps}, \mathsf{Decaps})$ *has* perfect correctness *if for any* (ek, dk) *generated by* Gen*, we have that*

$$\Pr[\mathsf{Decaps}(dk, c) = K : (c, K) \leftarrow \mathsf{Encaps}(ek)] = 1.$$

Security: We define indistinguishability under chosen-plaintext and chosen-ciphertext attacks (denoted by IND-CPA and IND-CCA) for KEM, respectively.

Definition 2.7. *For any adversary* $\mathcal{A}$*, we define its* IND-CPA *and* IND-CCA *advantages against a KEM scheme* $\mathsf{KEM} = (\mathsf{Gen}, \mathsf{Encaps}, \mathsf{Decaps})$ *as follows:*

$$\mathsf{Adv}^{\text{ind-cpa}}_{\mathsf{KEM},\mathcal{A}}(\kappa) := \left| 2\Pr[\mathsf{Expt}^{\text{ind-cpa}}_{\mathsf{KEM},\mathcal{A}}(\kappa) = 1] - 1 \right|,$$
$$\mathsf{Adv}^{\text{ind-cca}}_{\mathsf{KEM},\mathcal{A}}(\kappa) := \left| 2\Pr[\mathsf{Expt}^{\text{ind-cca}}_{\mathsf{KEM},\mathcal{A}}(\kappa) = 1] - 1 \right|,$$

where $\mathsf{Expt}^{\text{ind-cpa}}_{\mathsf{KEM},\mathcal{A}}(\kappa)$ *and* $\mathsf{Expt}^{\text{ind-cca}}_{\mathsf{KEM},\mathcal{A}}(\kappa)$ *are experiments described in Fig. 3.*

$\mathsf{Expt}^{\text{ind-cpa}}_{\mathsf{KEM},\mathcal{A}}(\kappa)$	$\mathsf{Expt}^{\text{ind-cca}}_{\mathsf{KEM},\mathcal{A}}(\kappa)$	$\mathrm{DEC}_{c^*}(c)$
$b \leftarrow \{0,1\}$	$b \leftarrow \{0,1\}$	if $c = c^*$, **return** $\bot$
$(ek, dk) \leftarrow \mathsf{Gen}(1^\kappa)$	$(ek, dk) \leftarrow \mathsf{Gen}(1^\kappa)$	$K := \mathsf{Decaps}(dk, c)$
$(c^*, K_0^*) \leftarrow \mathsf{Encaps}(ek)$;	$(c^*, K_0^*) \leftarrow \mathsf{Encaps}(ek)$;	**return** K
$K_1^* \leftarrow \mathcal{K}$	$K_1^* \leftarrow \mathcal{K}$	
$b' \leftarrow \mathcal{A}(ek, c^*, K_b^*)$	$b' \leftarrow \mathcal{A}^{\mathrm{DEC}_{c^*}(\cdot)}(ek, c^*, K_b^*)$	
return $\mathsf{boole}(b' \stackrel{?}{=} b)$	**return** $\mathsf{boole}(b' \stackrel{?}{=} b)$	

Fig. 3. Games for KEM schemes

For ATK $\in$ {CPA, CCA}, *we say that* KEM *is IND-ATK-secure if* $\mathsf{Adv}^{\text{ind-atk}}_{\mathcal{A},\mathsf{PKE}}(\kappa)$ *is negligible for any PPT adversary* $\mathcal{A}$.

2.5 eXtendable-Output Functions

An eXtendable-Output Function (XOF) is a function on input bit strings in which the output can be extended to an arbitrary desired length. An XOF is denoted by $\mathsf{XOF}(X, L)$, where X is the input bit string and L is the desired output length. We modeled the XOF as a quantumly-accessible random oracle. We employ SHAKE256, standardized as an XOF by NIST [NIS15].

2.6 Assumptions

Preliminaries: Let $\rho_s(x) = \exp(-\pi\|x\|^2/s^2)$ for $x \in \mathbb{R}^n$ be a Gaussian function scaled by a factor s. For any real $s > 0$ and lattice Λ, we define the discrete Gaussian distribution $D_{\Lambda,s}$ over Λ with parameter s by

$$D_{\Lambda,s}(x) = \rho_s(x)/\rho_s(\Lambda) \text{ for } x \in \Lambda,$$

where $\rho_s(\Lambda) = \sum_{x\in\Lambda} \rho_s(x)$. The following norm bound is useful.

Lemma 2.3 (Adapted version of [MR07, Lemma 4.4]**).** *For* $\sigma = \omega(\sqrt{\log(n)})$, *it holds that*

$$\Pr_{e \leftarrow D_{\mathbb{Z}^n,\sigma}}[\|e\| > \sigma\sqrt{n}] \leq 2^{-n+1}.$$

LWE and its variants: We review the assumptions for lattice-based PKEs. The most basic one is the learning-with-errors (LWE) assumption [Reg09], which is a generalized version of the learning-parity-with-noise assumption [BFKL93, KSS10].

Definition 2.8 (LWE assumption in matrix form). *For all* κ, *let* $n = n(\kappa)$ *and* $q = q(\kappa)$ *be integers and let* χ *be a distribution over* $\mathbb{Z}$.

The decisional learning-with-errors (LWE) assumption $\mathsf{LWE}_{n,q}$ *states that, for any* $m = \mathsf{poly}(\kappa)$,

the following two distributions are computationally hard to distinguish:

- $A, sA + e$, *where* $A \leftarrow \mathbb{Z}_q^{n\times m}$, $s \leftarrow \mathbb{Z}_q^n$, *and* $e \leftarrow \chi^m$
- A, u, *where* $A \leftarrow \mathbb{Z}_q^{n\times m}$ *and* $u \leftarrow \mathbb{Z}_q^m$.

We also review its polynomial version [LPR10, BV11]. We here use the Hermite-normal form of the assumption [ACPS09, LPR10, BV11], where secret s is chosen from the noise distribution.

Definition 2.9 (Poly-LWE assumption – Hermite normal form). *For all* κ, *let* $\Phi(x) = \Phi_\kappa(x) \in \mathbb{Z}[x]$ *be a polynomial of degree* $n = n(\kappa)$, *let* $q = q(\kappa)$ *be an integer, let* $R := \mathbb{Z}[x]/(\Phi(x))$ *and* $R_q := \mathbb{Z}_q[x]/(\Phi(x))$, *and let* χ *denote a distribution over the ring* R.

The decisional polynomial learning-with-errors (Poly-LWE) assumption $\mathsf{PolyLWE}_{\Phi,q,\chi}$ *states that, for any* $\ell = \mathsf{poly}(\kappa)$, *the following two distributions are hard to distinguish:*

- $\{(a_i, a_i s + e_i)\}_{i=1,\ldots,\ell}$, *where* $a_i \leftarrow R_q$, $s, e_i \leftarrow \chi$
- $\{(a_i, u_i)\}_{i=1,\ldots,\ell}$, *where* $a_i, u_i \leftarrow R_q$.

Next, we recall the decisional small polynomial ratio (DSPR) assumption defined by López-Alt, Tromer, and Vaikuntanathan [LTV12]. We here employ an adapted version of the DSPR assumption.

Definition 2.10 (DSPR assumption). *For all* κ, *let* $\Phi(x) = \Phi_\kappa(x) \in \mathbb{Z}[x]$ *be a polynomial of degree* $n = n(\kappa)$, *let* $q = q(\kappa)$ *be a positive integer, let* $R := \mathbb{Z}[x]/(\Phi(x))$ *and* $R_q := \mathbb{Z}_q[x]/(\Phi(x))$, *and let* χ *denote a distribution over the ring* R.

The decisional small polynomial ratio (DSPR) assumption $\mathsf{DSPR}_{\Phi,q,\chi_g,\chi_f}$ *says that the following two distributions are hard to distinguish:*

- *a polynomial* $h := g \cdot f^{-1} \in R_q$, *where* $g \leftarrow \chi_g$ *and* $f \leftarrow \chi_f$.
- *a polynomial* $u \leftarrow R_q$.

Remark 2.1. Stehlé and Steinfeld [SS11] showed that $\mathsf{DSPR}_{\Phi,q,\chi}$ is statistically hard if n is a power of two, $\Phi(x) = x^n + 1$, and $\chi_g = \chi_f = D_{\mathbb{Z}^n,r}$ for $r > \sqrt{q} \cdot \mathsf{poly}(\kappa)$.

3 Disjoint Simulatability of Deterministic PKE

Here, we define a new security notion, *disjoint simulatability*, for DPKE. We also define another security notion called *sparse pseudorandomness* and prove that it implies the disjoint simulatability. Then we give some instantiations of sparse pseudorandom (and thus disjoint simulatable) deterministic PKE schemes based on the LWE assumption or various assumptions related to NTRU, the McEliece PKE, and the Niederreiter PKE with tight reductions. We also construct a disjoint simulatable DPKE scheme from any IND-CPA-secure PKE scheme with a sufficiently large message space in the QROM, though the reduction is non-tight.

3.1 Definition

We define a new security notion, *disjoint simulatability*, for DPKE. Intuitively, a deterministic PKE scheme is disjoint simulatable if there exists a simulator that is only given a public key and generates a "fake ciphertext" that is indistinguishable from a real ciphertext of a random message. Moreover, we require that a fake ciphertext falls in a valid ciphertext space with negligible probability. The formal definition is as follows.

Definition 3.1 (Disjoint simulatability). *Let $\mathcal{D}_\mathcal{M}$ denote an efficiently sampleable distribution on a set $\mathcal{M}$. A deterministic PKE scheme* $\mathsf{PKE} = (\mathsf{Gen}, \mathsf{Enc}, \mathsf{Dec})$ *with plaintext and ciphertext spaces $\mathcal{M}$ and $\mathcal{C}$ is* $\mathcal{D}_\mathcal{M}$-disjoint simulatable *if there exists a PPT algorithm $\mathcal{S}$ that satisfies the following.*

- *(Statistical disjointness:)*

$$\mathsf{Disj}_{\mathsf{PKE},\mathcal{S}}(\kappa) := \max_{(ek,dk)\in\mathsf{Gen}(1^\kappa;\mathcal{R})} \Pr[c \in \mathsf{Enc}(ek, \mathcal{M}) \mid c \leftarrow \mathcal{S}(ek)]$$

  is negligible, where $\mathcal{R}$ denotes a randomness space for Gen.
- *(Ciphertext-indistinguishability:) For any PPT adversary $\mathcal{A}$,*

$$\mathsf{Adv}^{\text{ds-ind}}_{\mathsf{PKE},\mathcal{D}_\mathcal{M},\mathcal{A},\mathcal{S}}(\kappa) := \left| \begin{array}{l} \Pr\left[\mathcal{A}(ek, c^*) \to 1 \;\middle|\; \begin{array}{c}(ek, dk) \leftarrow \mathsf{Gen}(1^\kappa); m^* \leftarrow \mathcal{D}_\mathcal{M}; \\ c^* := \mathsf{Enc}(ek, m^*)\end{array}\right] \\ -\Pr\left[\mathcal{A}(ek, c^*) \to 1 \mid (ek, dk) \leftarrow \mathsf{Gen}(1^\kappa); c^* \leftarrow \mathcal{S}(ek)\right] \end{array} \right|$$

  is negligible.

3.2 Sufficient Condition: Sparse Pseudorandomness

Here, we define another security notion for DPKE called *sparse pseudorandomness*, which is a sufficient condition to be disjoint simulatable. Intuitively, a deterministic PKE scheme is sparse pseudorandom if valid ciphertexts are sparse in a ciphertext sparse and pseudorandom when a message is randomly chosen. In other words, an encryption algorithm can be seen as a pseudorandom generator (PRG). The formal definition is as follows.

Definition 3.2 (Sparse pseudorandomness). *Let $\mathcal{D}_\mathcal{M}$ denote an efficiently sampleable distribution on a set $\mathcal{M}$. A deterministic PKE scheme* $\mathsf{PKE} = (\mathsf{Gen}, \mathsf{Enc}, \mathsf{Dec})$ *with plaintext and ciphertext spaces $\mathcal{M}$ and $\mathcal{C}$ is* $\mathcal{D}_\mathcal{M}$-sparse pseudorandom *if the following two properties are satisfied.*

- *(Sparseness:)*

$$\mathsf{Sparse}_{\mathsf{PKE}}(\kappa) := \max_{(ek,dk)\in\mathsf{Gen}(1^\kappa;\mathcal{R})} \frac{|\mathsf{Enc}(ek, \mathcal{M})|}{|\mathcal{C}|}$$

  is negligible where $\mathcal{R}$ denotes a randomness space for Gen.
- *(Pseudorandomness:) For any PPT adversary $\mathcal{A}$,*

$$\mathsf{Adv}^{\text{pr}}_{\mathsf{PKE},\mathcal{D}_\mathcal{M},\mathcal{A}}(\kappa) := \left| \begin{array}{l} \Pr\left[\mathcal{A}(ek, c^*) \to 1 \;\middle|\; \begin{array}{c}(ek, dk) \leftarrow \mathsf{Gen}(1^\kappa); m^* \leftarrow \mathcal{D}_\mathcal{M}; \\ c^* := \mathsf{Enc}(ek, m^*)\end{array}\right] \\ -\Pr\left[\mathcal{A}(ek, c^*) \to 1 \mid (ek, dk) \leftarrow \mathsf{Gen}(1^\kappa), c^* \leftarrow \mathcal{C}\right] \end{array} \right|$$

  is negligible.

Then we prove that the sparse pseudorandomness implies the disjoint simulatability if a ciphertext space is efficiently sampleable.

Lemma 3.1. *If a deterministic PKE scheme* $\mathsf{PKE} = (\mathsf{Gen}, \mathsf{Enc}, \mathsf{Dec})$ *with plaintext and ciphertext spaces* $\mathcal{M}$ *and* $\mathcal{C}$ *is* $\mathcal{D}_{\mathcal{M}}$*-sparse pseudorandom and* $\mathcal{C}$ *is efficiently sampleable, then* PKE *is also* $\mathcal{D}_{\mathcal{M}}$*-disjoint simulatable. In particular, there exists a PPT simulator* $\mathcal{S}$ *such that* $\mathsf{Disj}_{\mathsf{PKE},\mathcal{S}}(\kappa) = \mathsf{Sparse}_{\mathsf{PKE}}(\kappa)$ *and* $\mathsf{Adv}^{\text{ds-ind}}_{\mathsf{PKE},\mathcal{D}_{\mathcal{M}},\mathcal{A},\mathcal{S}}(\kappa) = \mathsf{Adv}^{\text{pr}}_{\mathsf{PKE},\mathcal{D}_{\mathcal{M}},\mathcal{A}}(\kappa)$.

Proof. Let $\mathcal{S}$ be an algorithm that outputs a random element of $\mathcal{C}$. Then we clearly have $\mathsf{Disj}_{\mathsf{PKE},\mathcal{S}}(\kappa) = \mathsf{Sparse}_{\mathsf{PKE}}(\kappa)$ and $\mathsf{Adv}^{\text{ds-ind}}_{\mathsf{PKE},\mathcal{D}_{\mathcal{M}},\mathcal{A},\mathcal{S}}(\kappa) = \mathsf{Adv}^{\text{pr}}_{\mathsf{PKE},\mathcal{D}_{\mathcal{M}},\mathcal{A}}(\kappa)$. □

3.3 Instantiations

Here, we give examples of a DPKE scheme that is disjoint simulatable. In particular, we construct a DPKE scheme that has the sparse pseudorandomness based on the LWE assumption or some other assumptions related to NTRU. (We further construct them based on the McEliece PKE and the Niederreiter PKE in the full version.) We remark that the reductions are tight. By combining those with Lemma 3.1, we obtain disjoint simulatable DPKE schemes based on any of these assumptions with tight security.

LWE-based DPKE. We review the GPV trapdoor function for LWE [GPV08, Pei09, MP12]. The LWE assumption (in matrix form) states that $(A, sA + e)$ and (A, u) are computationally indistinguishable, where $A \leftarrow \mathbb{Z}_q^{n\times m}$, $s \leftarrow \mathbb{Z}_q^n$, $e \leftarrow \chi^m$, and $u \leftarrow \mathbb{Z}_q^m$. The GPV trapdoor function for LWE exploited that if we have a "short" matrix T satisfying $AT \equiv O \bmod q$, we can retrieve s and e from $c = sA+e$. The trapdoor T for A is generated by an algorithm $\mathsf{TrapGen}$:

Theorem 3.1 ([Ajt99, AP11]). *For any positive integers* n *and* $q \geq 3$, *any* $\delta > 0$ *and* $m \geq (2+\delta)n \lg q$, *there is a probabilistic polynomial-time algorithm* $\mathsf{TrapGen}$ *that outputs a pair* $T \in \mathbb{Z}^{m\times m}$ *and* $A \in \mathbb{Z}_q^{n\times m}$ *such that: the distribution of* A *is within a negligible statistical distance of uniform over* $\mathbb{Z}_q^{n\times m}$, T *is non-singular (over the rationals),* $\|t_i\| \leq L = O(m \lg m)$ *for every column vector* t_i *of* T, *and* $AT \equiv O \pmod q$.

Let us construct a DPKE scheme $\mathsf{PKE} = (\mathsf{Gen}, \mathsf{Enc}, \mathsf{Dec})$ as follows:

Parameters: We require several parameters: the dimension $n = n(\kappa)$, the modulus $q = q(\kappa)$, and $m = m(\kappa)$. We also employ $L = O(m \lg m)$, $\sigma = \omega(\sqrt{\lg n})$, $\beta = \sigma\sqrt{n}$.
We require that $\beta L < q/2$ and $q^m \gg q^n \cdot (2\beta+1)^m$.
- The plaintext space $\mathcal{M} := \mathbb{Z}_q^n \times B_m(\beta)$, where $B_m(\beta) := \{e \in \mathbb{Z}^m \mid \|e\| \leq \beta\}$.
- The sampler $\mathcal{D}_{\mathcal{M}}$ samples $s \leftarrow \mathbb{Z}_q^n$ and $e \leftarrow D_{\mathbb{Z}^m,\sigma}$ conditioned on $\|e\| \leq \beta$.
- The ciphertext space $\mathcal{C} := \mathbb{Z}_q^m$

Key Generation: $\mathsf{Gen}(1^\kappa)$ invokes $\mathsf{TrapGen}(1^n, 1^m, q)$ and obtains $A \in \mathbb{Z}_q^{n\times m}$ and $T \in \mathbb{Z}^{m\times m}$. It outputs $ek = A$ and $dk = (A, T)$.

Encryption: $\mathsf{Enc}(ek, (s, e))$ outputs $c = sA + e \bmod q$.

Decryption: $\mathsf{Dec}(dk, c)$ computes $e = (c \cdot T \bmod q) \cdot T^{-1}$ and $s = (c - e) \cdot A^{+} \bmod q$, where $A^{+} := A^{\top} \cdot (A \cdot A^{\top}) \in \mathbb{Z}_q^{m \times n}$, the left inverse of A.

The properties of PKE are summarized as follows:

Perfect Correctness: We know $c \cdot T \equiv sAT + eT \equiv eT \pmod{q}$. If $\|eT\|_{\infty} < q/2$, then $c \cdot T \bmod q = eT \in \mathbb{Z}^m$ holds and e is recovered by $e = (c \cdot T \bmod q) \cdot T^{-1}$. Once correct e is obtained, s is recovered by $(c - e) \cdot A^{+} \in \mathbb{Z}_q^n$. The condition $\|eT\|_{\infty} < q/2$ is satisfied because $\|eT\|_{\infty} \leq \max_i \|e\| \cdot \|t_i\| \leq \beta L < q/2$, where t_i is the column vectors of T.

Sparseness: $|\mathcal{C}| = q^m$ and $|\mathsf{Enc}(ek, \mathcal{M})| \leq \mathcal{M} = |\mathbb{Z}_q^n \times B_m(\beta)| \leq q^n \cdot (2\beta + 1)^m$. Sparseness follows from the fact $q^m \gg q^n \cdot (2\beta + 1)^m$.

Pseudorandomness: We consider the following hybrid games:

- (Original game 1:) The adversary is given (A, c^*), where $(A, T) \leftarrow \mathsf{TrapGen}(1^n, 1^m, q)$, $(s, e) \leftarrow \mathcal{D}_{\mathcal{M}}$, and $c^* \leftarrow \mathbb{Z}_q^m$.
- (Hybrid game 1:) Let us replace the public key A. We consider (A, c^*), where $A \leftarrow \mathbb{Z}_q^{n \times m}$, $(s, e) \leftarrow \mathcal{D}_{\mathcal{M}}$, and $c^* := sA + e \bmod q$. This change is justified by Theorem 3.1.
- (Hybrid game 2:) Let us replace the sampler $\mathcal{D}_{\mathcal{M}}$. We consider (A, c^*), where $A \leftarrow \mathbb{Z}_q^{n \times m}$, $(s, e) \leftarrow U(\mathbb{Z}_q^n) \times D_{\mathbb{Z}^m, \sigma}$, and $c^* := sA + e \bmod q$. This replacement is justified by Lemma 2.3.
- (Hybrid game 3:) We next replace the ciphertext c^*. We consider (A, c^*), where $A \leftarrow \mathbb{Z}_q^{n \times m}$ and $c^* \leftarrow \mathbb{Z}_q^m$. This game is computationally indistinguishable from the previous game under the LWE assumption $\mathsf{LWE}_{n,q,D_{\mathbb{Z},\sigma}}$.
- (Original game 2:) We replace the public key A. We consider (A, c^*), where $(A, T) \leftarrow \mathsf{TrapGen}(1^n, 1^m, q)$ and $c^* := sA + e \bmod q$. This change is justified by Theorem 3.1.

Remark 3.1. For simplicity, we employ the simple version of the GPV trapdoor function for LWE. Further improvements are available, e.g., [MP12, Section 5].

NTRU-based DPKE. We next review the original version of NTRUEncrypt [HPS98]. Let $\Phi(x) = x^n - 1 \in \mathbb{Z}[x]$, let $p < q$ be positive integers with $\gcd(p, q) = 1$, and let $R := \mathbb{Z}[x]/(\Phi(x))$ and $R_q := \mathbb{Z}_q[x]/(\Phi(x))$. We often set $p = 3$ and $q = 2^k$ for some k. Let $\mathcal{T}$ be a set of ternary-coefficient polynomials in R, that is, $\mathcal{T} := \{t = \sum_{i=0}^{n-1} t_i x^i \in R \mid t_i \in \{-1, 0, +1\}\}$. Let $\mathcal{L}_f, \mathcal{L}_g, \mathcal{L}_r, \mathcal{L}_m \subseteq \mathcal{T}$. The public key is $h = g/f$, where $f \leftarrow \mathcal{L}_f, g \leftarrow \mathcal{L}_g$ with f has inverses in R_p and R_q. The the ciphertext of $m \in \mathcal{L}_m$ with randomness $r \in \mathcal{L}_r$ is $c = prh + m$. Roughly speaking, we can retrieve m if we know f; $cf = prg + mf \in R_q$ and it holds in R.

Parameters: We require that $\|prg + mf \bmod q\|_{\infty} < q/2$ for any g, f, m, r in their domains, where, for $t = \sum_{i=0}^{n-1} t_i x^i \in R$, we define $\|t\|_{\infty} := \max_i |t_i|$. For simplicity, we assume that $\mathcal{L}_m = \mathcal{L}_r$.

- The plaintext space is $\mathcal{M} := \mathcal{L}_m \times \mathcal{L}_r$.

- The sampler $\mathcal{D}_{\mathcal{M}}$ samples $(m, r) \leftarrow \mathcal{L}_m \times \mathcal{L}_r$.
- The ciphertext space is $\mathcal{C} := R_q$.

Key Generation: $\mathsf{Gen}()$ chooses $g \leftarrow \mathcal{L}_g$ and $f \leftarrow \mathcal{L}_f$ until f is invertible in R_q and R_p. It outputs $ek = h = g/f \in R_q$ and $dk = (h, f)$.

Encryption: $\mathsf{Enc}(ek, (m, r))$ outputs $c = prh + m \in R_q$.

Decryption: $\mathsf{Dec}(sk, c)$ computes $m := (fc \bmod q) \cdot f^{-1} \bmod p$ and $r := (c - m) \cdot (ph)^{-1} \bmod q$.

The properties of this DPKE are summarized as follows:

Perfect correctness: Note that $fc \equiv prg + mf \pmod q$. Since $\|prg + mf \bmod q\|_\infty < q/2$ from our requirement, we have $(fc \bmod q) = prg + mf \in R$. Hence, we have $(fc \bmod q) \cdot f^{-1} \equiv (prg + mf) \cdot f^{-1} \equiv m \pmod p$ as we wanted. r is also recovered because $(c - m) \cdot (ph)^{-1} \equiv prh \cdot (ph)^{-1} \equiv r \pmod q$.

Sparseness: Sparseness follows from $|\mathcal{C}| = q^n \gg 3^{2n} = |\mathcal{T}^2| \geq |\mathcal{L}_m \times \mathcal{L}_r| = |\mathsf{Enc}(ek, \mathcal{M})|$.

Pseudorandomness: What we want to show is

$$(h, c = prh + m) \approx_c (h, u),$$

where $h = g/f$ is a public key with $f \leftarrow \mathcal{L}_f, g \leftarrow \mathcal{L}_g$ with condition f has inverses R_p and R_q, $(m, r) \leftarrow \mathcal{L}_m \times \mathcal{L}_r$, and $u \leftarrow R_q$. Let $\chi_g := U(\mathcal{L}_g)$ and $\chi_f := U(\mathcal{L}_f \cap R_p^* \cap R_q^*)$, where R_k^* for $k \in \{p, q\}$ denotes $\{f \in R \mid f \text{ has an inverse in } R_k\}$. Let $\chi := U(\mathcal{L}_m) = U(\mathcal{L}_r)$.

- We first replace $h = g/f$ with random h', which is justified by the DSPR assumption $\mathsf{DSPR}_{\Phi,q,\chi_f,\chi_g}$.
- We next replace $c = prh' + m$ with random c', which is justified by the Poly-LWE assumption $\mathsf{PolyLWE}_{\Phi,q,\chi}$; Given $\tilde{h}$ and $c = r\tilde{h} + m$ or random, we convert them into $h' = p^{-1}\tilde{h}$ and c. Since p is co-prime to q, h' is truly random. If $c = r\tilde{h} + e$, then $c = pr \cdot p^{-1}\tilde{h} + e = prh' + e$ as we wanted.
- We then go backward by replacing random h' with $h = g/f$, which is justified by the DSPR assumption $\mathsf{DSPR}_{\Phi,q,\chi_f,\chi_g}$ again.

3.4 Generic Conversion from IND-CPA-Secure PKE

Here, we show that any perfectly-correct IND-CPA-secure PKE whose plaintext space is sufficiently large can be converted into a disjoint-simulatable DPKE scheme in the quantum random oracle model. We note that the conversion is *non-tight.*

Intuitively, we replace randomness of an underlying IND-CPA-secure PKE scheme with a hash value of a message similarly to the conversion T given in [HHK17] (which is in turn based on the Fujisaki-Okamoto conversion). The difference from the conversion T is that we "puncture" a message space by 0[5]. That is, if a message space of an underlying IND-CPA-secure PKE scheme is $\mathcal{M}$, then

[5] We assume that $0 \in \mathcal{M}$. In fact, we can replace 0 with an arbitrary message in $\mathcal{M}$. We assume that $0 \in \mathcal{M}$ for notational simplicity.

$\mathsf{Gen}_1(1^\kappa)$	$\mathsf{Enc}_1(ek, m)$, where $m \in \mathcal{M}'$	$\mathsf{Dec}_1(dk, c)$
$(ek, dk) \leftarrow \mathsf{Gen}(1^\kappa)$	$r := \mathsf{G}(m)$	$m := \mathsf{Dec}(dk, c)$
return (ek, dk)	$c := \mathsf{Enc}(ek, m; r)$	**if** $m \notin \mathcal{M}'$ **return** $\bot$
	return c	**else return** m

$\mathcal{S}(ek)$
$r \leftarrow \mathcal{R}$
$c := \mathsf{Enc}(ek, 0; r)$
return c

Fig. 4. $\mathsf{PKE}_1 = (\mathsf{Gen}_1, \mathsf{Enc}_1, \mathsf{Dec}_1) = \mathsf{TPunc}[\mathsf{PKE}, \mathsf{G}]$ with simulator $\mathcal{S}$.

a message space of the resulting scheme is $\mathcal{M}' := \mathcal{M} \setminus \{0\}$. In this meaning, we call our conversion TPunc. We give the concrete description of the conversion TPunc below.

Let $\mathcal{M}$ and $\mathcal{R}$ be the message and randomness spaces of PKE, respectively, and let $\mathcal{M}' := \mathcal{M} \setminus \{0\}$. Then the resulting DPKE scheme $\mathsf{PKE}_1 = \mathsf{TPunc}[\mathsf{PKE}, \mathsf{G}]$ is described in Fig. 4 where $\mathsf{G}\colon \mathcal{M} \to \mathcal{R}$ denotes a random oracle. Here, we remark that the message space of PKE_1 is restricted to $\mathcal{M}' := \mathcal{M} \setminus \{0\}$. The security of PKE_1 is stated as follows.

Theorem 3.2 (Security of TPunc). *Let $\mathcal{S}$ be the algorithm described in Fig. 4. If* PKE *is perfectly correct, then we have* $\mathsf{Disj}_{\mathsf{PKE}_1,\mathcal{S}}(\kappa) = 0$. *Moreover, for any quantum adversary $\mathcal{A}$ against* PKE_1 *issuing at most q_G quantum queries to* G, *there exist quantum adversaries $\mathcal{B}$ and $\mathcal{C}$ against* IND-CPA *security of* PKE *such that*

$$\mathsf{Adv}^{\text{ds-ind}}_{\mathsf{PKE}_1,U_{\mathcal{M}'},\mathcal{A},\mathcal{S}}(\kappa) \leq 2q_\mathsf{G}\sqrt{\mathsf{Adv}^{\text{ind-cpa}}_{\mathsf{PKE},\mathcal{B}}(\kappa) + \frac{2}{|\mathcal{M}|}} + \mathsf{Adv}^{\text{ind-cpa}}_{\mathsf{PKE},\mathcal{C}}(\kappa)$$

where $U_{\mathcal{M}'}$ denotes the uniform distribution on $\mathcal{M}'$, and $\mathsf{Time}(\mathcal{B}) \approx \mathsf{Time}(\mathcal{C}) \approx \mathsf{Time}(\mathcal{A}) + q_\mathsf{G} \cdot t_\mathsf{RO}$.

Security Proof. We obviously have $\mathsf{Disj}_{\mathsf{PKE}_1,\mathcal{S}}(\kappa) = 0$ since PKE is perfectly correct.

To prove the rest of the theorem, we consider the following sequence of games. See Table 1 for the summary of games and justifications.

Game_0: This game is defined as follows:

$$(ek, dk) \leftarrow \mathsf{Gen}(1^\kappa); m^* \leftarrow \mathcal{M}'; r^* \leftarrow \mathsf{G}(m^*); c^* := \mathsf{Enc}(ek, m^*; r^*); \\ b' \leftarrow \mathcal{A}^{\mathsf{G}(\cdot)}(ek, c^*); \textbf{return } b'.$$

Game_1: This game is the same as Game_0 except that a randomness to generate a challenge ciphertext is freshly generated:

$$(ek, dk) \leftarrow \mathsf{Gen}(1^\kappa); m^* \leftarrow \mathcal{M}'; r^* \leftarrow \mathcal{R}; c^* := \mathsf{Enc}(ek, m^*; r^*); \\ b' \leftarrow \mathcal{A}^{\mathsf{G}(\cdot)}(ek, c^*); \textbf{return } b'.$$

Table 1. Summary of games for the security proof of Theorem 3.2

Game	m^*	r^*	c^*	Justification
Game_0	$\mathcal{M}'$	$\mathsf{G}(m^*)$	$\mathsf{Enc}(ek, m^*; r^*) = \mathsf{Enc}_1(ek, m^*)$	
Game_1	$\mathcal{M}'$	r^*	$\mathsf{Enc}(ek, m^*; r^*)$	OW-CPA security of PKE and the OW2H lemma
Game_2	0	r^*	$\mathsf{Enc}(ek, 0; r^*) = \mathcal{S}(ek)$	IND-CPA security of PKE

$\mathcal{B}^{\mathsf{G}}(ek, c^*)$:	$\mathsf{F}(m^*, r^*)$
$\mathsf{inp} := (ek, c^*)$	$(ek, dk) \leftarrow \mathsf{Gen}(1^\kappa)$
$i \leftarrow [q_{\mathsf{H}}]$	$c^* := \mathsf{Enc}(ek, m^*; r^*)$
Run $\mathcal{A}^{\mathsf{G}}(\mathsf{inp})$ until i-th query $\lvert\hat{x}\rangle$ to G	$\mathsf{inp} := (ek, c^*)$
if $i >$ number of queries to G, **return** $\perp$	**return** inp
else return $x' := \mathsf{Measure}(\lvert\hat{x}\rangle)$	

Fig. 5. Adversary $\mathcal{B}$ and Algorithm F

Game_2: This game is the same as Game_1 except that a challenge ciphertext is generated by $\mathsf{Enc}(ek, m^*; r^*)$, where $m^* := 0$ rather than $m^* \leftarrow \mathcal{M}'$:

$$(ek, dk) \leftarrow \mathsf{Gen}(1^\kappa); r^* \leftarrow \mathcal{R}; c^* := \mathsf{Enc}(ek, 0; r^*); b' \leftarrow \mathcal{A}^{\mathsf{G}(\cdot)}(ek, c^*); \textbf{return } b'.$$

This completes the descriptions of games. It is easy to see that we have

$$\mathsf{Adv}^{\text{ds-ind}}_{\mathsf{PKE}_1, U_{\mathcal{M}'}, \mathcal{A}, \mathcal{S}}(\kappa) = \lvert \Pr[\mathsf{Game}_0 = 1] - \Pr[\mathsf{Game}_2 = 1] \rvert .$$

We give an upperbound for this by the following lemmas.

Lemma 3.2. *There exists an adversary $\mathcal{B}$ such that*

$$\lvert \Pr[\mathsf{Game}_0 = 1] - \Pr[\mathsf{Game}_1 = 1] \rvert \leq 2 q_{\mathsf{G}} \sqrt{\mathsf{Adv}^{\text{ind-cpa}}_{\mathsf{PKE}, \mathcal{B}}(\kappa) + \frac{2}{\lvert \mathcal{M} \rvert}}$$

and $\mathsf{Time}(\mathcal{B}) \approx \mathsf{Time}(\mathcal{A}) + q_{\mathsf{G}} \cdot t_{\mathsf{RO}}$.

Proof. Let F be an algorithm described in Fig. 5. It is easy to see that Game_0 can be restated as

$$m^* \leftarrow \mathcal{M}'; r^* \leftarrow \mathsf{G}(m^*); \mathsf{inp} := \mathsf{F}(ek, m^*; r^*); b' \leftarrow \mathcal{A}^{\mathsf{G}(\cdot)}(\mathsf{inp}); \textbf{return } b'.$$

and Game_1 can be restated as

$$m^* \leftarrow \mathcal{M}'; r^* \leftarrow \mathcal{R}; \mathsf{inp} := \mathsf{F}(ek, m^*; r^*); b' \leftarrow \mathcal{A}^{\mathsf{G}(\cdot)}(\mathsf{inp}); \textbf{return } b'.$$

Then applying the Algorithmic-OW2H lemma (Lemma 2.1) with $\mathcal{X} = \mathcal{M}'$, $\mathcal{Y} = \mathcal{R}$, $x = m^*$, $y = r^*$, and algorithms $\mathcal{A}$ and F, we have

$$|\Pr[\mathsf{Game}_0 = 1] - \Pr[\mathsf{Game}_1 = 1]| \leq 2q_{\mathsf{G}}\sqrt{\Pr[m^* \leftarrow \mathcal{B}^{\mathsf{G}}(ek, c^*)]}.$$

where $\mathcal{B}^{\mathsf{G}}$ is an algorithm described in Fig. 5, $(ek, dk) \leftarrow \mathsf{Gen}(1^\kappa)$, $m^* \leftarrow \mathcal{M}'$, $r^* \leftarrow \mathcal{R}$, and $c^* := \mathsf{Enc}(ek, m^*, r^*)$. Since the statistical distance between uniform distributions on $\mathcal{M}$ and $\mathcal{M}'$ is $\frac{1}{|\mathcal{M}|}$, we have $\Pr[m^* \leftarrow \mathcal{B}^{\mathsf{G}}(ek, c^*)] \leq \mathsf{Adv}^{\text{ow-cpa}}_{\mathsf{PKE},\mathcal{B}}(\kappa) + \frac{1}{|\mathcal{M}|}$ where the probability in the left-hand side is taken as in the above. (Note that additional $\frac{1}{|\mathcal{M}|}$ appears because m^* is taken from $\mathcal{M}' = \mathcal{M} \setminus \{0\}$ in the left-hand side probability.) Moreover, we have $\mathsf{Adv}^{\text{ow-cpa}}_{\mathsf{PKE},\mathcal{B}}(\kappa) \leq \mathsf{Adv}^{\text{ind-cpa}}_{\mathsf{PKE},\mathcal{B}}(\kappa) + \frac{1}{|\mathcal{M}|}$ in general. By combining these inequalities, the lemma is proven. □

Lemma 3.3. *There exists an adversary* $\mathcal{C}$ *such that* $|\Pr[\mathsf{Game}_1 = 1] - \Pr[\mathsf{Game}_2 = 1]| \leq \mathsf{Adv}^{\text{ind-cpa}}_{\mathsf{PKE},\mathcal{C}}(\kappa)$ *and* $\mathsf{Time}(\mathcal{C}) \approx \mathsf{Time}(\mathcal{A}) + q_{\mathsf{G}} \cdot t_{\mathsf{RO}}$.

Proof. We construct an adversary $\mathcal{C}$ against the IND-CPA security of PKE as follows.

$\mathcal{C}^{\mathsf{G}}(ek)$: It chooses $m_0 \leftarrow \mathcal{M}'$ and sets $m_1 := 0$. Then it queries (m_0, m_1) to its challenge oracle and obtains $c^* \leftarrow \mathsf{Enc}(ek, m^*; r^*)$, where m^* is m_b for a random bit b chosen by the challenger. It invokes $b' \leftarrow \mathcal{A}^{\mathsf{G}}(ek, c^*)$ and outputs b'.

This completes the description of $\mathcal{C}$. It is obvious that $\mathcal{C}$ perfectly simulates Game_{b+1} depending on the challenge bit $b \in \{0, 1\}$. Therefore, we have

$$\begin{aligned}
\mathsf{Adv}^{\text{ind-cpa}}_{\mathsf{PKE},\mathcal{C}}(\kappa) &= |2\Pr[b' = b] - 1| \\
&= |(1 - \Pr[b' = 1 \mid b = 0]) + \Pr[b' = 1 \mid b = 1] - 1| \\
&= |1 - \Pr[\mathsf{Game}_1 = 1] + \Pr[\mathsf{Game}_2 = 1] - 1| \\
&= |\Pr[\mathsf{Game}_2 = 1] - \Pr[\mathsf{Game}_1 = 1]|
\end{aligned}$$

as we wanted. □

4 Conversion from Disjoint Simulatability to IND-CCA

In this section, we convert a disjoint simulatable DPKE scheme into an IND-CCA-secure KEM. Let $\mathsf{PKE}_1 = (\mathsf{Gen}_1, \mathsf{Enc}_1, \mathsf{Dec}_1)$ be a deterministic PKE scheme and let $\mathsf{H}\colon \mathcal{M} \to \mathcal{K}$ and $\mathsf{H}'\colon \{0,1\}^\ell \times \mathcal{C} \to \mathcal{K}$ be random oracles. Our conversion SXY is described in Fig. 6. The securities of our conversion can be stated as follows.

Theorem 4.1 (Security of SXY in the ROM (an adapted version of [HHK17, Theorem 3.6])). *Let* PKE_1 *be a perfectly correct DPKE scheme.*

$\overline{\mathsf{Gen}}(1^\kappa)$	$\overline{\mathsf{Enc}}(ek')$	$\overline{\mathsf{Dec}}(dk, c)$, where $dk = (dk', ek', s)$
$(ek', dk') \leftarrow \mathsf{Gen}_1(1^\kappa)$	$m \leftarrow \mathcal{D}_\mathcal{M}$	$m := \mathsf{Dec}_1(dk', c)$
$s \leftarrow \{0,1\}^\ell$	$c := \mathsf{Enc}_1(ek', m)$	**if** $m = \bot$, **return** $K := \mathsf{H}'(s, c)$
$dk \leftarrow (dk', ek', s)$	$K := \mathsf{H}(m)$	**if** $c \neq \mathsf{Enc}_1(ek', m)$, **return** $K := \mathsf{H}'(s, c)$
return (ek', dk)	**return** (K, c)	**else return** $K := \mathsf{H}(m)$

Fig. 6. $\mathsf{KEM} := \mathsf{SXY}[\mathsf{PKE}_1, \mathsf{H}, \mathsf{H}']$.

For any IND-CCA adversary $\mathcal{A}$ against KEM *issuing q_H and $q_{\mathsf{H}'}$ quantum random oracle queries to* H *and* H′ *and $q_{\overline{\mathsf{Dec}}}$ decryption queries, there exists an* OW-CPA *adversary $\mathcal{B}$ against* PKE_1, *such that*

$$\mathsf{Adv}^{\text{ind-cca}}_{\mathsf{KEM},\mathcal{A}}(\kappa) \leq \mathsf{Adv}^{\text{ow-cpa}}_{\mathsf{PKE}_1,\mathcal{B}}(\kappa) + q_{\mathsf{H}'} \cdot 2^{-\ell}$$

and $\mathsf{Time}(\mathcal{B}) \approx \mathsf{Time}(\mathcal{A}) + q_\mathsf{H} \cdot \mathsf{Time}(\mathsf{Enc}_1) + (q_\mathsf{H} + q_{\mathsf{H}'} + q_{\overline{\mathsf{Dec}}}) \cdot t_\mathsf{CRO}$, *where t_CRO is the running time to simulate the classical random oracle.*

Theorem 4.2 (Security of SXY in the QROM). *Let* PKE_1 *be a perfectly correct DPKE scheme that satisfies the $\mathcal{D}_\mathcal{M}$-disjoint simulatability with a simulator $\mathcal{S}$. For any IND-CCA quantum adversary $\mathcal{A}$ against* KEM *issuing q_H and $q_{\mathsf{H}'}$ quantum random oracle queries to* H *and* H′ *and $q_{\overline{\mathsf{Dec}}}$ decryption queries, there exists an adversary $\mathcal{B}$ against the disjoint simulatability of* PKE_1 *such that*

$$\mathsf{Adv}^{\text{ind-cca}}_{\mathsf{KEM},\mathcal{A}}(\kappa) \leq \mathsf{Adv}^{\text{ds-ind}}_{\mathsf{PKE}_1,\mathcal{D}_\mathcal{M},\mathcal{S},\mathcal{B}}(\kappa) + \mathsf{Disj}_{\mathsf{PKE}_1,\mathcal{S}}(\kappa) + q_{\mathsf{H}'} \cdot 2^{\frac{-\ell+1}{2}}$$

and $\mathsf{Time}(\mathcal{B}) \approx \mathsf{Time}(\mathcal{A}) + q_\mathsf{H} \cdot \mathsf{Time}(\mathsf{Enc}_1) + (q_\mathsf{H} + q_{\mathsf{H}'} + q_{\overline{\mathsf{Dec}}}) \cdot t_\mathsf{RO}$.

The proof of Theorem 4.2 follows.

Remark 4.1. We also note that our reduction enables the decapsulation oracle $\overline{\mathsf{Dec}}$ to quantumly queried.

Security Proof. We use game-hopping proof. The overview of all games is given in Table 2.

Game_0: This is the original game, $\mathsf{Expt}^{\text{ind-cca}}_{\mathsf{KEM},\mathcal{A}}(\kappa)$.
Game_1: This game is the same as Game_0 except that $\mathsf{H}'(s,c)$ in the decryption oracle is replaced with $\mathsf{H}_q(c)$ where $\mathsf{H}_q : \mathcal{C} \rightarrow \mathcal{K}$ is another random oracle. We remark that $\mathcal{A}$ is not given direct access to H_q.
$\mathsf{Game}_{1.5}$: This game is the same as Game_1 except that the random oracle $\mathsf{H}(\cdot)$ is simulated by $\mathsf{H}'_q(\mathsf{Enc}_1(ek,\cdot))$ where H'_q is yet another random oracle. We remark that a decryption oracle and generation of K_0^* also use $\mathsf{H}'_q(\mathsf{Enc}_1(ek,\cdot))$ as $\mathsf{H}(\cdot)$ and that $\mathcal{A}$ is not given direct access to H'_q.
Game_2: This game is the same as $\mathsf{Game}_{1.5}$ except that the random oracle $\mathsf{H}(\cdot)$ is simulated by $\mathsf{H}_q(\mathsf{Enc}_1(ek,\cdot))$ instead of $\mathsf{H}'_q(\mathsf{Enc}_1(ek,\cdot))$. We remark that a decryption oracle and generation of K_0^* also use $\mathsf{H}_q(\mathsf{Enc}_1(ek,\cdot))$ as $\mathsf{H}(\cdot)$.

Table 2. Summary of games for the proof of Theorem 4.2

Game	H	c^*	K_0^*	K_1^*	Decryption of		Justification
					valid c	invalid c	
Game_0	$\mathsf{H}(\cdot)$	$\mathsf{Enc}_1(ek', m^*)$	$\mathsf{H}(m^*)$	random	$\mathsf{H}(m)$	$\mathsf{H}'(s, c)$	
Game_1	$\mathsf{H}(\cdot)$	$\mathsf{Enc}_1(ek', m^*)$	$\mathsf{H}(m^*)$	random	$\mathsf{H}(m)$	$\mathsf{H}_q(c)$	Lemma 2.2
$\mathsf{Game}_{1.5}$	$\mathsf{H}'_q(\mathsf{Enc}_1(ek', \cdot))$	$\mathsf{Enc}_1(ek', m^*)$	$\mathsf{H}(m^*)$	random	$\mathsf{H}(m)$	$\mathsf{H}_q(c)$	Perfect correctness
Game_2	$\mathsf{H}_q(\mathsf{Enc}_1(ek', \cdot))$	$\mathsf{Enc}_1(ek', m^*)$	$\mathsf{H}(m^*)$	random	$\mathsf{H}(m)$	$\mathsf{H}_q(c)$	Conceptual
Game_3	$\mathsf{H}_q(\mathsf{Enc}_1(ek', \cdot))$	$\mathsf{Enc}_1(ek', m^*)$	$\mathsf{H}_q(c^*)$	random	$\mathsf{H}_q(c)$	$\mathsf{H}_q(c)$	Perfect correctness
Game_4	$\mathsf{H}_q(\mathsf{Enc}_1(ek', \cdot))$	$\mathcal{S}(ek')$	$\mathsf{H}_q(c^*)$	random	$\mathsf{H}_q(c)$	$\mathsf{H}_q(c)$	DS-IND

Game_3: This game is the same as Game_2 except that K_0^* is set as $\mathsf{H}_q(c^*)$ and the decryption oracle always returns $\mathsf{H}_q(c)$ as long as $c \neq c^*$. We denote the modified decryption oracle by $\overline{\mathsf{Dec}}'$.
Game_4: This game is the same as Game_3 except that c^* is set as $\mathcal{S}(ek')$.

The above completes the descriptions of games. We clearly have

$$\mathsf{Adv}^{\mathsf{ind\text{-}cca}}_{\mathsf{KEM},\mathcal{A}}(\kappa) = |2\Pr[\mathsf{Game}_0 = 1] - 1|$$

by the definition. We upperbound this by the following lemmas.

Lemma 4.1. *We have*

$$|\Pr[\mathsf{Game}_0 = 1] - \Pr[\mathsf{Game}_1 = 1]| \leq q_{\mathsf{H}'} \cdot 2^{\frac{-\ell+1}{2}}.$$

Proof. This is obvious from Lemma 2.2. □

Lemma 4.2. *We have*

$$\Pr[\mathsf{Game}_1 = 1] = \Pr[\mathsf{Game}_{1.5} = 1].$$

Proof. Since we assume that PKE_1 has a perfect correctness, $\mathsf{Enc}_1(ek', \cdot)$ is injective. Therefore, if $\mathsf{H}'_q(\cdot)$ is a random function, then $\mathsf{H}'_q(\mathsf{Enc}_1(ek, \cdot))$ is also a random function. Remarking that access to H'_q is not given to $\mathcal{A}$, it causes no difference from the view of $\mathcal{A}$ if we replace $\mathsf{H}(\cdot)$ with $\mathsf{H}'_q(\mathsf{Enc}_1(ek, \cdot))$. □

Lemma 4.3. *We have*

$$\Pr[\mathsf{Game}_{1.5} = 1] = \Pr[\mathsf{Game}_2 = 1].$$

Proof. We call a ciphertext c valid if we have $\mathsf{Enc}_1(ek', \mathsf{Dec}_1(dk', c)) = c$ and invalid otherwise. We remark that H_q is used only for decrypting an invalid ciphertext c as $\mathsf{H}_q(c)$ in $\mathsf{Game}_{1.5}$. This means that a value of $\mathsf{H}_q(c)$ for a valid c is not used at all in $\mathsf{Game}_{1.5}$. On the other hand, any output of $\mathsf{Enc}_1(ek', \cdot)$ is valid due to the perfect correctness of PKE_1. Since H'_q is only used for evaluating an output of $\mathsf{Enc}(ek', \cdot)$, a value of $\mathsf{H}_q(c)$ for a valid c is not used at all in $\mathsf{Game}_{1.5}$. Hence, it causes no difference from the view of $\mathcal{A}$ if we use the same random oracle H_q instead of two independent random oracles H_q and H'_q. □

Lemma 4.4. *We have*

$$\Pr[\mathsf{Game}_2 = 1] = \Pr[\mathsf{Game}_3 = 1].$$

Proof. Since we set $\mathsf{H}(\cdot) := \mathsf{H}_q(\mathsf{Enc}_1(ek', \cdot))$, for any valid c and $m := \mathsf{Dec}_1(dk', c)$, we have $\mathsf{H}(m) = \mathsf{H}_q(\mathsf{Enc}_1(ek', m)) = \mathsf{H}_q(c)$. Therefore, responses of the decryption oracle are unchanged. We also have $\mathsf{H}(m^*) = \mathsf{H}_q(c^*)$ for a similar reason. □

Lemma 4.5. *There exists an adversary $\mathcal{B}$ such that*

$$|\Pr[\mathsf{Game}_3 = 1] - \Pr[\mathsf{Game}_4 = 1]| \leq \mathsf{Adv}^{\text{ds-ind}}_{\mathsf{PKE}_1, \mathcal{D}_\mathcal{M}, \mathcal{S}, \mathcal{B}}(\kappa).$$

and $\mathsf{Time}(\mathcal{B}) \approx \mathsf{Time}(\mathcal{A}) + q_\mathsf{H} \cdot \mathsf{Time}(\mathsf{Enc}_1) + (q_\mathsf{H} + q_{\mathsf{H}'} + q_{\overline{\mathsf{Dec}}}) \cdot t_\mathsf{RO}$.

Proof. We construct an adversary $\mathcal{B}$, which is allowed to access two random oracles H_q and H', against the disjoint simulatability as follows[6].

$\mathcal{B}^{\mathsf{H}_q, \mathsf{H}'}(ek', c^*)$: It picks $b \leftarrow \{0,1\}$, sets $K_0^* := \mathsf{H}_q(c^*)$ and $K_1^* \leftarrow \mathcal{K}$, and invokes $b' \leftarrow \mathcal{A}^{\mathsf{H}, \mathsf{H}', \overline{\mathsf{Dec}}'}(ek', c^*, K_b^*)$ where $\mathcal{A}$'s oracles are simulated as follows.
- $\mathsf{H}(\cdot)$ is simulated by $\mathsf{H}_q(\mathsf{Enc}_1(ek', \cdot))$.
- H' can be simulated because $\mathcal{B}$ has access to an oracle H'.
- $\overline{\mathsf{Dec}}'(\cdot)$ is simulated by forwarding to $\mathsf{H}_q(\cdot)$.

Then $\mathcal{B}$ returns $\mathsf{boole}(b \stackrel{?}{=} b')$.

This completes the description of $\mathcal{B}$. It is easy to see that $\mathcal{B}$ perfectly simulates Game_3 if $c^* = \mathsf{Enc}_1(ek, m^*)$ and Game_4 if $c^* = \mathcal{S}(ek')$. Therefore, we have

$$|\Pr[\mathsf{Game}_3 = 1] - \Pr[\mathsf{Game}_4 = 1]| \leq \mathsf{Adv}^{\text{ds-ind}}_{\mathsf{PKE}_1, \mathcal{D}_\mathcal{M}, \mathcal{S}, \mathcal{B}}(\kappa)$$

as wanted. Since $\mathcal{B}$ invokes $\mathcal{A}$ once, H is simulated by one evaluation of Enc_1 plus one evaluation of a random oracle, and H' and $\overline{\mathsf{Dec}}'$ are simulated by one evaluation of random oracles, we have $\mathsf{Time}(\mathcal{B}) \approx \mathsf{Time}(\mathcal{A}) + q_\mathsf{H} \cdot \mathsf{Time}(\mathsf{Enc}_1) + (q_\mathsf{H} + q_{\mathsf{H}'} + q_{\overline{\mathsf{Dec}}}) \cdot t_\mathsf{RO}$. □

Lemma 4.6. *We have*

$$|2\Pr[\mathsf{Game}_4 = 1] - 1| \leq \mathsf{Disj}_{\mathsf{PKE}_1, \mathcal{S}}(\kappa).$$

Proof. Let Bad denote an event in which $c^* \in \mathsf{Enc}_1(ek', \mathcal{M})$ in Game_4. It is easy to see that we have

$$\Pr[\mathsf{Bad}] \leq \mathsf{Disj}_{\mathsf{PKE}_1, \mathcal{S}}(\kappa).$$

When Bad does not occur, i.e., $c^* \notin \mathsf{Enc}_1(ek', \mathcal{M})$, $\mathcal{A}$ obtains no information about $K_0^* = \mathsf{H}_q(c^*)$. This is because queries to H only reveal $\mathsf{H}_q(c)$ for $c \in \mathsf{Enc}_1(ek', \mathcal{M})$, and $\overline{\mathsf{Dec}}'(c)$ returns $\perp$ if $c = c^*$. Therefore, we have

$$\Pr[\mathsf{Game}_4 = 1 \mid \overline{\mathsf{Bad}}] = 1/2.$$

---

[6] We allow a reduction algorithm to access the random oracles. See Subsect. 2.2 for details.

Combining the above, we have

$$\begin{aligned}
&|2\Pr[\mathsf{Game}_4 = 1] - 1| \\
&= \left|\Pr[\mathsf{Bad}] \cdot (2\Pr[\mathsf{Game}_4 = 1 \mid \mathsf{Bad}] - 1) + \Pr[\overline{\mathsf{Bad}}] \right. \\
&\quad \left. \cdot (2\Pr[\mathsf{Game}_4 = 1 \mid \overline{\mathsf{Bad}}] - 1)\right| \\
&\leq \Pr[\mathsf{Bad}] + \left|2\Pr[\mathsf{Game}_4 = 1 \mid \overline{\mathsf{Bad}}] - 1\right| \\
&\leq \mathsf{Disj}_{\mathsf{PKE}_1,\mathcal{S}}(\kappa)
\end{aligned}$$

as we wanted. □

5 Implementation

We report the implementation results on a desktop PC and on a RasPi, which are based on the previous implementation of a variant of NTRU [HRSS17].

5.1 NTRU-HRSS

We review a variant of NTRU, which we call $\mathsf{NTRU}_{\mathsf{HRSS17}}$, developed by Hülsing, Rijneveld, Schanck, and Schwabe [HRSS17].

Let $\Phi_m(x) \in \mathbb{Z}[x]$ be the m-th cyclotomic polynomial. We have $\Phi_1 = x - 1$. If m is prime, then we have $\Phi_m = 1 + x + \cdots + x^{m-1}$. Define $S_n := \mathbb{Z}[x]/(\Phi_n)$ and $R_n := \mathbb{Z}[x]/(x^n - 1)$. For prime n, we have $x^n - 1 = \Phi_1\Phi_n$ and $R_n \simeq S_1 \times S_n$. We define $\mathrm{Lift}_p\colon S_n/(p) \to R_n$ as

$$\mathrm{Lift}_p(v) := \left[\Phi_1 [v/\Phi_1]_{(p,\Phi_n)}\right]_{(x^n-1)}.$$

By definition, we have $\mathrm{Lift}_p(v) \equiv 0 \pmod{\Phi_1}$ and $\mathrm{Lift}_p(v) \equiv v \pmod{(p, \Phi_n)}$. Let $\mathfrak{p} = (p, \Phi_n)$ and $\mathfrak{q} = (q, x^n - 1)$. Let

$$\begin{aligned}
\mathcal{T} &:= \{a \in \mathbb{Z}[x] : a = [a]_{\mathfrak{p}}\} = \{a \in \mathbb{Z}[x] : a_i \in (p) \text{ and } \deg(a) < \deg(\Phi_n)\}, \\
\mathcal{T}_+ &:= \{a \in \mathcal{T} : \langle xa, a\rangle \geq 0\}.
\end{aligned}$$

The definition of $\mathsf{NTRU}_{\mathsf{HRSS17}}$ is in Fig. 7. Note that all ciphertexts are equivalent to 0 modulo (q, Φ_1), which prevents a trivial distinguishing attack.

$\mathsf{Gen}(1^\kappa)$	$\mathsf{Enc}(h, m), m \in \mathcal{T}$	$\mathsf{Dec}(f, c)$
$g, f \leftarrow \mathcal{T}_+$	$r \leftarrow \mathcal{T}$	$m' := [[cf]_{\mathfrak{q}} f^{-1}]_{\mathfrak{p}}$
$f_q := [1/f]_{(q,\Phi_n)}$	$c := [prh + \mathrm{Lift}_p(m)]_{\mathfrak{q}}$	**return** m'
$h := [\Phi_1 g f_q]_{\mathfrak{q}}$	**return** c	
return $dk = f, ek = h$		

Fig. 7. $\mathsf{NTRU}_{\mathsf{HRSS17}}$

$\mathsf{Gen}'(1^\kappa) = \mathsf{Gen}$	$\mathsf{Enc}'(h, (m, r)), (m, r) \in \mathcal{T}^2$	$\mathsf{Dec}'(f, c)$
$g, f \leftarrow \mathcal{T}_+$	$c := [prh + \mathrm{Lift}_p(m)]_{\mathfrak{q}}$	$m' := [[cf]_{\mathfrak{q}} f^{-1}]_{\mathfrak{p}}$
$f_q := [1/f]_{(q,\Phi_n)}$	**return** c	$r' := \left[[(c - \mathrm{Lift}_p(m')) \cdot (ph)^{-1}]_{\mathfrak{q}}\right]_{\mathfrak{p}}$
$h := [\Phi_1 g f_q]_{\mathfrak{q}}$		**return** (m', r')
return $dk = f, ek = h$		

Fig. 8. Our modification $\mathsf{NTRU_{HRSS17}}'$

Hülsing et al. choose $(n, p, q) = (701, 3, 8192)$: The scheme is perfectly correct, and they claimed 128-bit post-quantum security of this parameter set. The implementation of $\mathsf{NTRU_{HRSS17}}$ and $\mathsf{QFO}^{\perp}[\mathsf{NTRU_{HRSS17}}, \mathsf{G}, \mathsf{H}, \mathsf{H}']$ is reported in [HRSS17].

Our Modification: We want PKE_1 to be *deterministic.* Hence, we consider a pair of (m, r) as a plaintext and make the decryption algorithm output (m, r) rather than m. The modification $\mathsf{NTRU_{HRSS17}}'$ is summarized in Fig. 8.

The properties of this DPKE are summarized as follows:

Perfect Correctness: This follows from the perfect correctness of the original PKE.

Sparseness: This follows from the parameter setting of the original PKE.

Pseudorandomness: We assume that the modified PKE $\mathsf{NTRU_{HRSS17}}'$ satisfies pseudorandomness.

We also implement $\mathsf{SXY}[\mathsf{NTRU_{HRSS17}}', \mathsf{H}, \mathsf{H}']$, where H and H' are implemented by SHAKE256. We define

$$\mathsf{H}(m, r) := \mathsf{XOF}\big((r, m, 0), 256\big) \text{ and } \mathsf{H}'(s, c) := \mathsf{XOF}\big((c, (s \| 00 \cdots 00), 1), 256\big),$$

where we treat $r \in R_n/(q)$ and the last bit is the context string.

To avoid the inversion of polynomials in decapsulation, we add f^{-1} modulo $\mathfrak{p}$ to dk as Hüsling et al. did [HRSS17]. This requires 139 extra bytes. In addition, we put $(ph)^{-1}$ modulo $\mathfrak{q}$ in dk, which requires 1140 extra bytes. Thus, our decapsulation key is 2557 bytes long.

5.2 Experimental Results

We preform the experiment with

- one core of an Intel Core i7-6700 at 3.40 GHz on a desktop PC with 8 GB memory and Ubuntu16.04 and
- a RasPi3 with 32-bit Rasbian.

We use gcc to compile the programs with option -O3. We generate 200 keys and ciphertexts to estimate the running time of key generation, encryption, and decryption. The experimental results are summarized

Table 3. Experimental results: We have $|ek| = 1140$ bytes, $|dk| = 2557$ bytes, and $|c| = 1140$ bytes.

(a) Our Experiments on a PC

	min	med.	avg.	max
Gen_1	1 767	1 778	1 815	2 592
Enc_1	327	329	328	331
Dec_1	958	959	959	1 021
	min	med.	avg.	max
$\overline{\mathsf{Gen}}$	2 565	2 580	2 579	2 601
$\overline{\mathsf{Enc}}$	332	334	333	336
$\overline{\mathsf{Dec}}$	1 280	1 282	1 282	1 286

(b) Our Experiments on a RasPi

	min	med.	avg.	max
Gen_1	33 675	33 685	33 687	45 460
Enc_1	3 085	3 089	3 091	3 121
Dec_1	8 839	8 851	8 850	8 880
	min	med.	avg.	max
$\overline{\mathsf{Gen}}$	49 151	49 169	49 174	49 263
$\overline{\mathsf{Enc}}$	3 200	3 205	3 207	3 232
$\overline{\mathsf{Dec}}$	11 837	11 841	11 843	11 888

in Table 3. $(\mathsf{Gen}_1, \mathsf{Enc}_1, \mathsf{Dec}_1)$ and $(\overline{\mathsf{Gen}}, \overline{\mathsf{Enc}}, \overline{\mathsf{Dec}})$ indicate $\mathsf{NTRU}_{\mathsf{HRSS17}}'$ and $\mathsf{SXY}[\mathsf{NTRU}_{\mathsf{HRSS17}}']$. The results reflect Hüsling et al.'s constant-time implementation and ours. Our conversion adds only small extra costs for hashing in encryption and adds about T_{Enc_1} for re-encrypting in decryption.

Note that our implementations are for reference and we did not optimize them. Further optimizations will speed up the algorithms as Hüsling et al. did [HRSS17]. The source code is available at https://info.isl.ntt.co.jp/crypt/eng/archive/contents.html#sxy.

Acknolwedgements. We would like to thank anonymous reviewers of Eurocrypt 2018, Eike Kiltz, Daniel J. Bernstein, Edoardo Persichetti, and Joost Rijneveld for their insightful comments.

A Missing Definitions

Definition A.1 (γ-spread). *Let* $\mathsf{PKE} = (\mathsf{Gen}, \mathsf{Enc}, \mathsf{Dec})$ *be a PKE scheme. We say* PKE *is* γ-spread *if for every* (ek, dk) *generated by* $\mathsf{Gen}(1^\kappa)$ *and for any* $m \in \mathcal{M}$*, we have that*

$$-\lg\left(\max_{c \in \mathcal{C}} \Pr_{r \leftarrow \mathcal{R}}[c = \mathsf{Enc}(ek, m; r)]\right) \geq \gamma.$$

(In other words, the min entropy of $\mathsf{Enc}(ek, m; U(\mathcal{R}))$ *is at least* γ*.) We say* PKE *is* well-spread *in* κ *if* $\gamma = \gamma(\kappa) = \omega(\lg \kappa)$.

We additionally review the definitions of onewayness under validity-checking attacks (OW-VA), onewayness under plaintext-checking attacks (OW-PCA), and onewayness under plaintext and validity checking attacks (OW-PCVA) for PKE.

Definition A.2 (Security notions for PKE). *Let* $\mathsf{PKE} = (\mathsf{Gen}, \mathsf{Enc}, \mathsf{Dec})$ *be a PKE scheme with message space* $\mathcal{M}$*. For any adversary* $\mathcal{A}$ *and for* $\mathrm{ATK} \in$

$\mathsf{Expt}^{\text{ow-atk}}_{\mathsf{PKE},\mathcal{A}}(\kappa)$	$\mathsf{Pco}(m \in \mathcal{M}, c)$	$\mathsf{Cvo}(c)$
$(ek, dk) \leftarrow \mathsf{Gen}(1^\kappa)$	**return** $\mathsf{boole}(m \stackrel{?}{=} \mathsf{Dec}(dk, c))$	**if** $c = c^*$, **return** $\perp$
$m^* \leftarrow \mathcal{M}$		$m := \mathsf{Dec}(dk, c)$
$c^* \leftarrow \mathsf{Enc}(ek, m^*)$		**return** $\mathsf{boole}(m \in \mathcal{M})$
$m' \leftarrow \mathcal{A}^{O_{\text{ATK}}}(ek, c^*)$		
return $\mathsf{boole}(m' \stackrel{?}{=} \mathsf{Dec}(dk, c^*))$		

Fig. 9. Games for PKE schemes

{VA, PCA, PCVA}, *we define the experiments* $\mathsf{Expt}^{\text{ow-va}}_{\mathsf{PKE},\mathcal{A}}(\kappa)$, $\mathsf{Expt}^{\text{ow-pca}}_{\mathsf{PKE},\mathcal{A}}(\kappa)$, *and* $\mathsf{Expt}^{\text{ow-pcva}}_{\mathsf{PKE},\mathcal{A}}(\kappa)$ *as in Fig. 9, where*

$$O_{\text{ATK}} := \begin{cases} \text{Cvo}(\cdot) & (\text{ATK} = \text{VA}) \\ \text{Pco}(\cdot, \cdot) & (\text{ATK} = \text{PCA}) \\ \text{Cvo}(\cdot), \text{Pco}(\cdot, \cdot) & (\text{ATK} = \text{PCVA}). \end{cases}$$

For any adversary $\mathcal{A}$, we define its OW-VA, OW-PCA, and OW-PCVA advantages as follows:

$$\mathsf{Adv}^{\text{ow-va}}_{\mathcal{A},\mathsf{PKE}}(\kappa) := \Pr[\mathsf{Expt}^{\text{ow-va}}_{\mathsf{PKE},\mathcal{A}}(\kappa) = 1],$$
$$\mathsf{Adv}^{\text{ow-pca}}_{\mathcal{A},\mathsf{PKE}}(\kappa) := \Pr[\mathsf{Expt}^{\text{ow-pca}}_{\mathsf{PKE},\mathcal{A}}(\kappa) = 1],$$
$$\mathsf{Adv}^{\text{ow-pcva}}_{\mathcal{A},\mathsf{PKE}}(\kappa) := \Pr[\mathsf{Expt}^{\text{ow-pcva}}_{\mathsf{PKE},\mathcal{A}}(\kappa) = 1].$$

For ATK $\in$ {VA, PCA, PCVA}, *we say that* PKE *is OW-ATK-secure if* $\mathsf{Adv}^{\text{ow-atk}}_{\mathcal{A},\mathsf{PKE}}(\kappa)$ *is negligible for any PPT adversary $\mathcal{A}$.*

B Transformations in the Random Oracle Model

We summarize transformations among PKE, DPKE and KEM in the ROM in Fig. 10.

GOAL-ATTACKg indicate the class of PKEs that is GOAL-ATTACK-secure and $2^{-\omega(\lg \kappa)}$-uniformity [FO00,FO99], or equivalently $\omega(\lg \kappa)$-spreading [FO13]. Solid arrows indicate tight reductions, dashed arrows indicate non-tight reductions, thin arrows indicate trivial reductions, thick black arrows indicate reductions in [FO00], thick green arrows indicate reductions in [Den03], and thick blue arrows indicate reductions in [HHK17].

- The transformation R is in [FO00, Remark 5.5]; R converts PKE = (Gen, Enc, Dec) with randomness space $\mathcal{R}$ into PKE$'$ = (Gen$'$, Enc$'$, Dec$'$) with randomness space $\mathcal{R} \times \mathcal{R}'$. They defined $\mathsf{Gen}' := \mathsf{Gen}$, $\mathsf{Enc}'(ek, x; (r, r')) := (\mathsf{Enc}(ek, x; r), r')$ and $\mathsf{Dec}'(dk, (c, r')) := \mathsf{Dec}(dk, c)$. This change amplifies γ-uniformity of PKE into $(\gamma / |\mathcal{R}'|)$-uniformity.

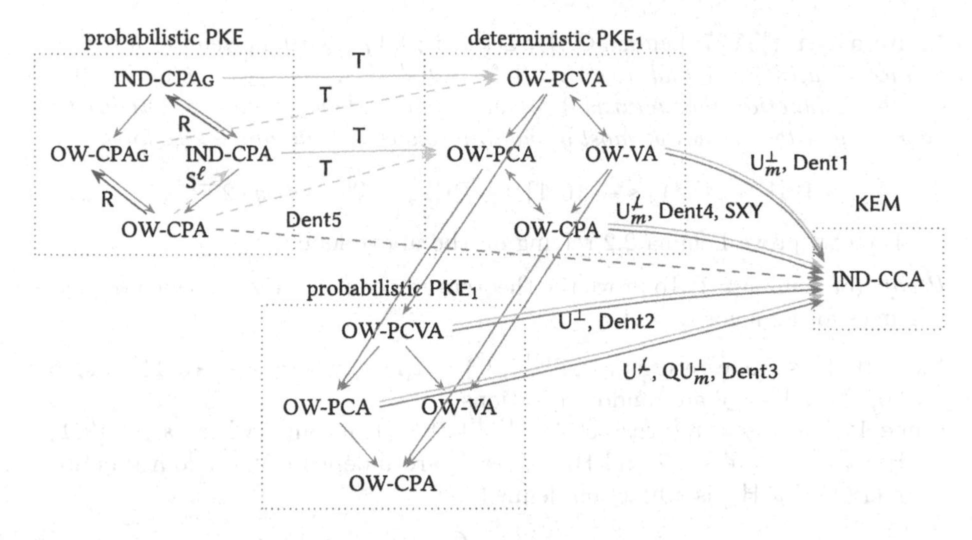

Fig. 10. Transformations in the ROM. GOAL-ATTACKg indicates the class of PKEs that is GOAL-ATTACK-secure and $2^{-\omega(\lg\kappa)}$-uniformity [FO00,FO99], or equivalently $\omega(\lg\kappa)$-spreading [FO13]. Solid arrows indicate tight reductions, dashed arrows indicate non-tight reductions, thin arrows indicate trivial reductions, thick black arrows indicate reductions in [FO00], thick green arrows indicate reductions in [Den03], and thick blue arrows indicate reductions in [HHK17]. The transformation R is in [FO00, Remark 5.5]. The transformations Dent1, Dent2, Dent3, Dent4, and Dent5 are given in [Den03]. The transformations S^{ℓ}, T, $\mathsf{U}^{\perp}$, $\mathsf{U}^{\not\perp}$, $\mathsf{U}_m^{\perp}$, $\mathsf{U}_m^{\not\perp}$, and $\mathsf{QU}_m^{\perp}$ are given in [HHK17]. (Color figure online)

- The transformations Dent1, Dent2, Dent3, Dent4, and Dent5 are given in [Den03].
- The transformations S^{ℓ}, T, $\mathsf{U}^{\perp}$, $\mathsf{U}^{\not\perp}$, $\mathsf{U}_m^{\perp}$, $\mathsf{U}_m^{\not\perp}$, and $\mathsf{QU}_m^{\perp}$ are given in [HHK17].

Note that $\mathsf{Dent1} \approx \mathsf{U}_m^{\perp}$, which is a KEM variant of BR93; $\mathsf{Dent2} \approx \mathsf{U}^{\perp}$, which is a KEM variant of REACT/GEM; $\mathsf{Dent4} \approx \mathsf{QU}_m^{\perp}$; $\mathsf{Dent5} \approx \mathsf{FO}_m^{\perp} = \mathsf{U}_m^{\perp} \circ \mathsf{T}$, which is a KEM variant of FO.

Albrecht, Orsini, Paterson, Peer, and Smart [AOP+17] gave the tight security proof for Dent5 when the underlying PKE is a certain Ring-LWE-based PKE scheme. We also observe that Dent5 is decomposed into $\mathsf{U}_m^{\perp} \circ \mathsf{T}$. Thus, starting from IND-CPAg-secure PKE, we obtain the similar proof by combining reductions in [HHK17].

C Omitted Proofs

C.1 Proof of Lemma 2.2

Here, we prove Lemma 2.2. Before proving the lemma, we introduce another lemma, which gives a lower bound for a decisional variant of Grover's search problem.

Lemma C.1 ([SY17, Lemma C.1]). *Let $g_s : \{0,1\}^\ell \to \{0,1\}$ denotes a function defined as $g_s(s) := 1$ and $g_s(s') := 0$ for all $s' \neq s$, and $g_\perp : \{0,1\}^\ell \to \{0,1\}$ denotes a function that returns 0 for all inputs. Then for any unbounded time adversary $\mathcal{A}$ that issues at most q quantum queries to its oracle, we have*

$$\Pr[1 \leftarrow \mathcal{A}^{g_s}() \mid s \leftarrow \{0,1\}^\ell] - \Pr[1 \leftarrow \mathcal{A}^{g_\perp}()] \leq q \cdot 2^{\frac{-\ell+1}{2}}.$$

Then we prove Lemma 2.2 relying on the above lemma.

Proof. (of Lemma 2.2) To prove the theorem, we consider the following sequence of games for an algorithm $\mathcal{A}$.

Game 0: This game returns as $\mathcal{A}^{\mathsf{H},\mathsf{H}(s,\cdot)}()$ outputs, where $s \leftarrow \{0,1\}^\ell$ and $\mathsf{H} : \{0,1\}^\ell \times \mathcal{X} \to \mathcal{Y}$ are random functions.

Game 1: This game returns as $\mathcal{A}^{O[s,\mathsf{H}_0,\mathsf{H}_1],\mathsf{H}_1(\cdot)}()$ outputs, where $s \leftarrow \{0,1\}^\ell$, $\mathsf{H}_0 \colon \{0,1\}^\ell \times \mathcal{X} \to \mathcal{Y}$ and $\mathsf{H}_1 \colon \mathcal{X} \to \mathcal{Y}$ are independent random functions, and $O[s,\mathsf{H}_0,\mathsf{H}_1]$ is a function defined as

$$O[s,\mathsf{H}_0,\mathsf{H}_1](s',x) := \begin{cases} \mathsf{H}_0(s',x) & \text{if } s' \neq s, \\ \mathsf{H}_1(x) & \text{if } s' = s. \end{cases} \tag{1}$$

Game 2: This game returns as $\mathcal{A}^{\mathsf{H}_0,\mathsf{H}_1}()$ outputs, where $\mathsf{H}_0 \colon \{0,1\}^\ell \times \mathcal{X} \to \mathcal{Y}$ and $\mathsf{H}_1 \colon \mathcal{X} \to \mathcal{Y}$ are independent random functions.

This completes the descriptions of games. We want to prove that $|\Pr[\mathsf{Game}_2 = 1] - \Pr[\mathsf{Game}_0 = 1]| \leq q_{\mathsf{H}} \cdot 2^{\frac{-\ell+1}{2}}$. It is easy to see that we have $\Pr[\mathsf{Game0} = 1] = \Pr[\mathsf{Game1} = 1]$. What is left is to prove that $|\Pr[\mathsf{Game}_2 = 1] - \Pr[\mathsf{Game}_1 = 1]| \leq q_{\mathsf{H}} \cdot 2^{\frac{-\ell+1}{2}}$. We prove this by a reduction to Lemma C.1. We consider the following algorithm $\mathcal{B}$ that has access to g that is g_s for randomly chosen $s \leftarrow \{0,1\}^\ell$ or $g_\perp$ where g_s and $g_\perp$ are as defined in Lemma C.1.

$\mathcal{B}^g$: It picks two random functions $\mathsf{H}_0 \colon \{0,1\}^\ell \times \mathcal{X} \to \mathcal{Y}$ and $\mathsf{H}_1 \colon \mathcal{X} \to \mathcal{Y}$, and runs $\mathcal{A}^{O,\mathsf{H}_1}$ where $\mathcal{B}$ simulates O as follows: If $\mathcal{A}$ queries (s',x) to O, $\mathcal{B}$ queries s' to its own oracle g to obtain a bit b. If $b = 0$, then $\mathcal{B}$ returns $\mathsf{H}_0(s',x)$ to $\mathcal{A}$ and if $b = 1$, then $\mathcal{B}$ returns $\mathsf{H}_1(x')$ to $\mathcal{A}$.

This completes the description of $\mathcal{B}$. It is easy to see that if $g = g_s$ for randomly chosen $s \leftarrow \{0,1\}^\ell$, then $\mathcal{B}$ perfectly simulates Game_1, and if $g = g_\perp$, then $\mathcal{B}$ perfectly simulates Game_2. Therefore, we have

$$|\Pr[\mathsf{Game}_1 = 1] - \Pr[\mathsf{Game}_2 = 1]| = \left|\Pr[1 \leftarrow \mathcal{B}^{g_s}() \mid s \leftarrow \{0,1\}^\ell] - \Pr[1 \leftarrow \mathcal{B}^{g_\perp}()]\right|.$$

On the other hand, by Lemma C.1, we have

$$\left|\Pr[1 \leftarrow \mathcal{B}^{g_s}() \mid s \leftarrow \{0,1\}^\ell] - \Pr[1 \leftarrow \mathcal{B}^{g_\perp}()]\right| \leq q_{\mathsf{H}} \cdot 2^{\frac{-\ell+1}{2}},$$

since the number of $\mathcal{B}$'s queries to its own oracle is exactly the same as the number of $\mathcal{A}$'s queries to O, which is equal to q_{H}. This completes the proof of Lemma 2.2. $\square$

References

[ACPS09] Applebaum, B., Cash, D., Peikert, C., Sahai, A.: Fast cryptographic primitives and circular-secure encryption based on hard learning problems. In: Halevi, S. (ed.) CRYPTO 2009. LNCS, vol. 5677, pp. 595–618. Springer, Heidelberg (2009). https://doi.org/10.1007/978-3-642-03356-8_35

[Ajt99] Ajtai, M.: Generating hard instances of the short basis problem. In: Wiedermann, J., van Emde Boas, P., Nielsen, M. (eds.) ICALP 1999. LNCS, vol. 1644, pp. 1–9. Springer, Heidelberg (1999). https://doi.org/10.1007/3-540-48523-6_1

[AOP+17] Albrecht, M.R., Orsini, E., Paterson, K.G., Peer, G., Smart, N.P.: Tightly secure ring-LWE based key encapsulation with short ciphertexts. In: Foley, S.N., Gollmann, D., Snekkenes, E. (eds.) ESORICS 2017. LNCS, vol. 10492, pp. 29–46. Springer, Cham (2017). https://doi.org/10.1007/978-3-319-66402-6_4

[AP11] Alwen, J., Peikert, C.: Generating shorter bases for hard random lattices. Theory Comput. Syst. **48**(3), 535–553 (2011). A preliminary versions appeared in STACS 2009 (2009)

[BDF+11] Boneh, D., Dagdelen, Ö., Fischlin, M., Lehmann, A., Schaffner, C., Zhandry, M.: Random oracles in a quantum world. In: Lee, D.H., Wang, X. (eds.) ASIACRYPT 2011. LNCS, vol. 7073, pp. 41–69. Springer, Heidelberg (2011). https://doi.org/10.1007/978-3-642-25385-0_3

[BFKL93] Blum, A., Furst, M., Kearns, M., Lipton, R.J.: Cryptographic primitives based on hard learning problems. In: Stinson, D.R. (ed.) CRYPTO 1993. LNCS, vol. 773, pp. 278–291. Springer, Heidelberg (1994). https://doi.org/10.1007/3-540-48329-2_24

[BR93] Bellare, M., Rogaway, P.: Random oracle are practical: a paradigm for designing efficient protocols. In: CCS 1993, pp. 62–73. ACM (1993)

[BR95] Bellare, M., Rogaway, P.: Optimal asymmetric encryption. In: De Santis, A. (ed.) EUROCRYPT 1994. LNCS, vol. 950, pp. 92–111. Springer, Heidelberg (1995). https://doi.org/10.1007/BFb0053428

[BV11] Brakerski, Z., Vaikuntanathan, V.: Fully homomorphic encryption from ring-LWE and security for key dependent messages. In: Rogaway, P. (ed.) CRYPTO 2011. LNCS, vol. 6841, pp. 505–524. Springer, Heidelberg (2011). https://doi.org/10.1007/978-3-642-22792-9_29

[CHJ+02] Jean-Sébastien, C., Handschuh, H., Joye, M., Paillier, P., Pointcheval, D., Tymen, C.: GEM: a generic chosen-ciphertext secure encryption method. In: Preneel, B. (ed.) CT-RSA 2002. LNCS, vol. 2271, pp. 263–276. Springer, Heidelberg (2002). https://doi.org/10.1007/3-540-45760-7_18

[Den03] Dent, A.W.: A designer's guide to KEMs. In: Paterson, K.G. (ed.) Cryptography and Coding 2003. LNCS, vol. 2898, pp. 133–151. Springer, Heidelberg (2003). https://doi.org/10.1007/978-3-540-40974-8_12

[FO99] Fujisaki, E., Okamoto, T.: Secure integration of asymmetric and symmetric encryption schemes. In: Wiener, M. (ed.) CRYPTO 1999. LNCS, vol. 1666, pp. 537–554. Springer, Heidelberg (1999). https://doi.org/10.1007/3-540-48405-1_34

[FO00] Fujisaki, E., Okamoto, T.: How to enhance the security of public-key encryption at minimum cost. IEICE Trans. Fundam. Electron. Commun. Comput. Sci. **83**(1), 24–32 (2000). A preliminary version appeared in PKC 1999 (1999)

[FO13] Fujisaki, E., Okamoto, T.: Secure integration of asymmetric and symmetric encryption schemes. J. Cryptol. **26**(1), 80–101 (2013)

[FOPS04] Fujisaki, E., Okamoto, T., Pointcheval, D., Stern, J.: RSA-OAEP is secure under the RSA assumption. J. Cryptol. **17**(2), 81–104 (2004)

[GPV08] Gentry, C., Peikert, C., Vaikuntanathan, V.: Trapdoors for hard lattices and new cryptographic constructions. In: Dwork, C. (ed.) STOC 2008, pp. 197–206. ACM (2008). https://eprint.iacr.org/2007/432

[HHK17] Hofheinz, D., Hövelmanns, K., Kiltz, E.: A modular analysis of the fujisaki-okamoto transformation. In: Kalai, Y., Reyzin, L. (eds.) TCC 2017, Part I. LNCS, vol. 10677, pp. 341–371. Springer, Cham (2017). https://doi.org/10.1007/978-3-319-70500-2_12

[HPS98] Hoffstein, J., Pipher, J., Silverman, J.H.: NTRU: a ring-based public key cryptosystem. In: Buhler, J.P. (ed.) ANTS 1998. LNCS, vol. 1423, pp. 267–288. Springer, Heidelberg (1998). https://doi.org/10.1007/BFb0054868

[HRSS17] Hülsing, A., Rijneveld, J., Schanck, J., Schwabe, P.: High-speed key encapsulation from NTRU. In: Fischer, W., Homma, N. (eds.) CHES 2017. LNCS, vol. 10529, pp. 232–252. Springer, Cham (2017). https://doi.org/10.1007/978-3-319-66787-4_12

[JZC+17] Jiang, H., Zhang, Z., Chen, L., Wang, H., Ma, Z.: Post-quantum IND-CCA-secure KEM without additional hash. IACR Cryptology ePrint Archive 2017/1096 (2017)

[KLS17] Kiltz, E., Lyubashevsky, V., Schaffner, C.: A concrete treatment of fiat-shamir signatures in the quantum random-oracle model. IACR Cryptology ePrint Archive 2017/916 (2017)

[KSS10] Katz, J., Shin, J.S., Smith, A.: Parallel and concurrent security of the HB and HB^+ protocols. J. Cryptology **23**(3), 402–421 (2010)

[LPR10] Lyubashevsky, V., Peikert, C., Regev, O.: On ideal lattices and learning with errors over rings. In: Gilbert, H. (ed.) EUROCRYPT 2010. LNCS, vol. 6110, pp. 1–23. Springer, Heidelberg (2010). https://doi.org/10.1007/978-3-642-13190-5_1

[LTV12] López-Alt, A., Tromer, E., Vaikuntanathan, V.: On-the-fly multiparty computation on the cloud via multikey fully homomorphic encryption. In: Karloff, H.J., Pitassi, T. (eds.) STOC 2012, pp. 1219–1234. ACM (2012)

[McE78] McEliece, R.J.: A public key cryptosystem based on algebraic coding theory. Technical report, DSN progress report (1978)

[Men12] Menezes, A.: Another look at provable security. In: Pointcheval, D., Johansson, T. (eds.) EUROCRYPT 2012. LNCS, vol. 7237, p. 8. Springer, Heidelberg (2012). https://doi.org/10.1007/978-3-642-29011-4_2

[MP12] Micciancio, D., Peikert, C.: Trapdoors for lattices: simpler, tighter, faster, smaller. In: Pointcheval, D., Johansson, T. (eds.) EUROCRYPT 2012. LNCS, vol. 7237, pp. 700–718. Springer, Heidelberg (2012). https://doi.org/10.1007/978-3-642-29011-4_41

[MR07] Micciancio, D., Regev, O.: Worst-case to average-case reductions based on gaussian measures. SIAM J. Comput. **37**(1), 267–302 (2007). A preliminary version appeared in FOCS 2004 (2004)

[NC00] Nielsen, M.A., Chuang, I.L.: Quantum Computation and Quantum Information. Cambridge University Press, Cambridge (2000)

[Nie86] Niederreiter, H.: Knapsack-type cryptosystems and algebraic coding theory. Probl. Control Inf. Theory **15**, 159–166 (1986)

[NIS15] Fips 202: Sha-3 standard: Permutation-based hash and extendable-output functions. U.S.Department of Commerce/National Institute of Standards and Technology (2015)

[OP01] Okamoto, T., Pointcheval, D.: REACT: rapid enhanced-security asymmetric cryptosystem transform. In: Naccache, D. (ed.) CT-RSA 2001. LNCS, vol. 2020, pp. 159–174. Springer, Heidelberg (2000). https://doi.org/10.1007/3-540-45353-9_13

[Pei09] Peikert, C.: Public-key cryptosystems from the worst-case shortest vector problem: extended abstract. In: Mitzenmacher, M. (ed.) STOC 2009, pp. 333–342. ACM (2009)

[Reg09] Regev, O.: On lattices, learning with errors, random linear codes, and cryptography. J. ACM **56**(6), Article 34 (2009). A preliminary version appeared in STOC 2005 (2005)

[SS11] Stehlé, D., Steinfeld, R.: Making NTRU as secure as worst-case problems over ideal lattices. In: Paterson, K.G. (ed.) EUROCRYPT 2011. LNCS, vol. 6632, pp. 27–47. Springer, Heidelberg (2011)

[SY17] Song, F., Yun, A.: Quantum security of NMAC and related constructions. In: Katz, J., Shacham, H. (eds.) CRYPTO 2017. LNCS, vol. 10402, pp. 283–309. Springer, Cham (2017). https://doi.org/10.1007/978-3-319-63715-0_10

[TU16] Targhi, E.E., Unruh, D.: Post-quantum security of the fujisaki-okamoto and OAEP transforms. In: Hirt, M., Smith, A. (eds.) TCC 2016. LNCS, vol. 9986, pp. 192–216. Springer, Heidelberg (2016). https://doi.org/10.1007/978-3-662-53644-5_8

[Unr15] Unruh, D.: Revocable quantum timed-release encryption. J. ACM **62**(6), No. 49 (2015). The preliminary version appeared in EUROCRYPT 2014. https://eprint.iacr.org/2013/606

[Zha12a] Zhandry, M.: How to construct quantum random functions. In: 53rd Annual IEEE Symposium on Foundations of Computer Science, FOCS 2012, New Brunswick, pp. 679–687, 20–23 October 2012

[Zha12b] Zhandry, M.: Secure identity-based encryption in the quantum random oracle model. In: Safavi-Naini, R., Canetti, R. (eds.) CRYPTO 2012. LNCS, vol. 7417, pp. 758–775. Springer, Heidelberg (2012). https://doi.org/10.1007/978-3-642-32009-5_44

A Concrete Treatment of Fiat-Shamir Signatures in the Quantum Random-Oracle Model

Eike Kiltz[1(✉)], Vadim Lyubashevsky[2], and Christian Schaffner[3]

[1] Ruhr Universität Bochum, Bochum, Germany
eike.kiltz@rub.de
[2] IBM Research – Zurich, Zurich, Switzerland
vad@zurich.ibm.com
[3] QuSoft and ILLC, University of Amsterdam, Amsterdam, The Netherlands
c.schaffner@uva.nl
http://www.qusoft.org/

Abstract. The Fiat-Shamir transform is a technique for combining a hash function and an identification scheme to produce a digital signature scheme. The resulting scheme is known to be secure in the random oracle model (ROM), which does not, however, imply security in the scenario where the adversary also has quantum access to the oracle. The goal of this current paper is to create a generic framework for constructing tight reductions in the QROM from underlying hard problems to Fiat-Shamir signatures.

Our generic reduction is composed of two results whose proofs, we believe, are simple and natural. We first consider a security notion (UF-NMA) in which the adversary obtains the public key and attempts to create a valid signature without accessing a signing oracle. We give a tight reduction showing that deterministic signatures (i.e., ones in which the randomness is derived from the message and the secret key) that are UF-NMA secure are also secure under the standard chosen message attack (UF-CMA) security definition. Our second result is showing that if the identification scheme is "lossy", as defined in (Abdalla et al. Eurocrypt 2012), then the security of the UF-NMA scheme is tightly based on the hardness of distinguishing regular and lossy public keys of the identification scheme. This latter distinguishing problem is normally exactly the definition of some presumably-hard mathematical problem. The combination of these components gives our main result.

As a concrete instantiation of our framework, we modify the recent lattice-based Dilithium digital signature scheme (Ducas et al., TCHES 2018) so that its underlying identification scheme admits lossy public keys. The original Dilithium scheme, which is proven secure in the classical ROM based on standard lattice assumptions, has 1.5 KB public keys and 2.7 KB signatures. The new scheme, which is tightly based on the hardness of the Module-LWE problem in the QROM using our generic reductions, has 7.7 KB public keys and 5.7 KB signatures for the same security level. Furthermore, due to our proof of equivalence between the UF-NMA and UF-CMA security notions of deterministic signature schemes, we can formu-

J. B. Nielsen and V. Rijmen (Eds.): EUROCRYPT 2018, LNCS 10822, pp. 552–586, 2018.
https://doi.org/10.1007/978-3-319-78372-7_18

late a new non-interactive assumption under which the original Dilithium signature scheme is also tightly secure in the QROM.

1 Introduction

FIAT-SHAMIR SIGNATURES FROM IDENTIFICATION PROTOCOLS. A canonical identification scheme [2] is a three-move authentication protocol ID of a specific form. The prover (holding the secret-key) sends a commitment W to the verifier. The verifier (holding the public-key) returns a random challenge c. The prover sends a response Z. Finally, using the verification algorithm, the verifier accepts if the transcript (W, c, Z) is correct. The Fiat-Shamir transformation [2,20] combines a canonical identification scheme ID and a hash function H to obtain a digital signature scheme $\mathsf{FS} = \mathsf{FS}[\mathsf{ID}, \mathsf{H}]$. The signing algorithm first iteratively generates a transcript (W, c, Z), where the challenge c is derived via $c := \mathsf{H}(W \parallel M)$. Signature $\sigma = (W, Z)$ is valid if the transcript $(W, c := \mathsf{H}(W \parallel M), Z)$ makes the verification algorithm accept. Lyubashevsky [26] further generalized this to the "Fiat-Shamir with aborts" transformation to account for aborting provers.

SECURITY OF FIAT-SHAMIR SIGNATURES IN THE ROM. Security of $\mathsf{FS}[\mathsf{ID}, \mathsf{H}]$ in the ROM can be proved in two steps. Firstly, if the underlying identification scheme has statistical Honest-Verifier Zero-Knowledge (HVZK), then UnForgeability against Chosen Message Attack (UF-CMA) and UnForgeability against No Message Attack (UF-NMA) are tightly equivalent (UF-NMA security means that the adversary is not allowed to make any signing queries). Secondly, the Forking Lemma [9,34] (based on a technique called "rewinding") is used to prove UF-NMA security in the random-oracle model (ROM) [11] from computational Special Soundness (SS). The latter part of the security reduction is non-tight and the loss in tightness is known to be inherent (e.g., [24,32]).

LOSSY IDENTIFICATION SCHEMES. With the goal of constructing signature schemes with a tight security reduction and generalizing a signature scheme by Katz and Wang [22], AFLT [3] introduced the new concept of lossy identification schemes and proved that Fiat-Shamir transformed signatures have a tight security reduction in the ROM. A lossy identification scheme comes with an additional lossy key generator that produces a lossy public key, computationally indistinguishable from a honestly generated public key. Further, relative to

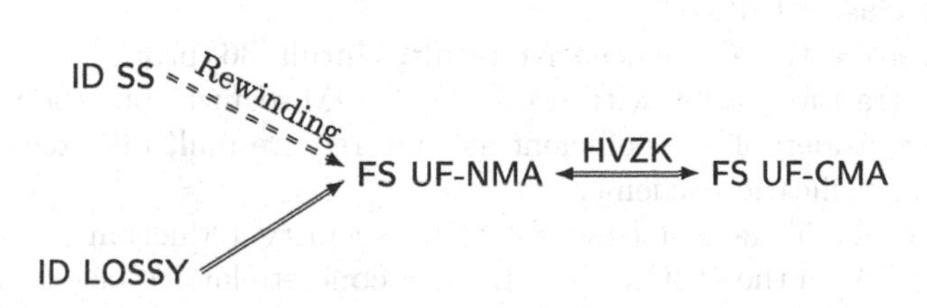

Fig. 1. Known security results of Fiat-Shamir signatures $\mathsf{FS} = \mathsf{FS}[\mathsf{ID}, \mathsf{H}]$ in the ROM. Solid arrows denote tight reductions, dashed arrows non-tight reductions.

a lossy public key the identification scheme has *statistical soundness*, i.e., not even an unbounded adversary can successfully impersonate a prover. Figure 1 summarizes the known security results of Fiat-Shamir signatures in the ROM.

Quantum Random-Oracle Model. Recently, NIST announced a competition with the goal to standardize new asymmetric encryption and signature schemes [1] with security against quantum adversaries, i.e., adversaries equipped with a quantum computer. There exists a number of (sometimes only implicitly defined) canonical identification schemes (e.g., [3,5,7,16,23,26]) whose security relies on the hardness of certain problems over lattices and codes, which are generally believed to resist quantum adversaries. Quantum computers may execute all "offline primitives" such as the hash function on arbitrary superpositions, which motivated the introduction of the quantum (accessible) random-oracle model (QROM) [13]. That is, in the UF-CMA security experiment for signatures in the QROM, an adversary has quantum access to a perfect hash function H and classical access to the signing oracle. Aiding in the construction of UF-CMA secure signatures with provable (post-quantum) security in the QROM is the main motivation of this paper.

Security of Fiat-Shamir signatures in the QROM. A number of recent works considered the security of Fiat-Shamir transformed signatures in the QROM. [13] proved a general result showing that if a reduction in the classical ROM is *history-free*, then it can also be carried out in the QROM. History-free reductions basically determine random oracle answers independently of the history of previous queries. For reductions that are not history-free, adaptive re-programming of the quantum random oracle is required which is problematic in the QROM: with one single quantum query to all inputs in superposition, an adversary might learn a superposition of all possible random oracle values which essentially means the reduction has to provide plausible values for the whole random oracle at this point. Hence, adaptive reprogramming in the QROM is difficult (but not impossible e.g., [12,18,36]).

Unfortunately, the known random-oracle proofs of Fiat-Shamir signatures [3,24,34] are not history-free. Beyond the general problem of adaptive reprogramming, the classical proof [34] uses rewinding and the Forking Lemma, a technique that we currently do not know how to extend to the quantum setting. Even worse, Ambanis et al. [6] proved that Fiat-Shamir signatures cannot be proven secure in a black-box way by just assuming computational special soundness and HVZK (these two conditions are, on the other hand, sufficient for a proof in the classical ROM).

To circumvent the above negative result, Unruh [36] proposed an alternative Fiat-Shamir transformation with provable QROM security but the resulting signatures are considerably less efficient as they require multiple executions of the underlying identification scheme.

Alkim et al. [5] gave a concrete tight security reduction for a signature scheme, TESLA, in the QROM. TESLA is a concrete lattice-based digital signature scheme implicitly derived via the Fiat-Shamir transformation. Their QROM proof from the learning with errors (LWE) assumption adaptively re-programs the quantum random oracle using a technique from [12] and seems tailored to their particular identification protocol. As described in [5], the intuition behind the QROM security proof for TESLA comes from the fact that the underlying identification scheme is lossy. They leave it as an open problem to prove Fiat-Shamir signatures generically secure from lossy identification schemes.

Unruh [37] could prove (among other things) that identification schemes with HVZK and statistical soundness yield UF-CMA secure Fiat-Shamir signatures in the QROM when additionally assuming a "dual-mode hard instance generator" for generating key pairs of the identification scheme. The latter dual mode hard instance generator is very similar to lossy identification schemes. Whereas the original publication [37] only contains asymptotic proofs, a recently updated version of the full version [38] also provides concrete security bounds. Below, in Sect. 1.2, we will compare them with our bounds.

1.1 Our Results

This work contains a simple and modular security analysis in the QROM of signatures FS[ID, H] obtained via the Fiat-Shamir transform with aborts [26] from any lossy identification scheme ID. We also consider the security of a deterministic variant DFS[ID, H, PRF] with better tightness. DFS derives the randomness for signing deterministically using a pseudo-random function PRF. Our main security statements are summarized in Fig. 2. Most importantly, if ID is a lossy identification scheme and has HVZK, then DFS[ID, H, PRF] is tightly UF-CMA secure and FS[ID, H] is (non-tightly) UF-CMA secure in the QROM. Our results suggest to prefer DFS[ID, H, PRF] over FS[ID, H].

The main component of our proof is a tweak to the AFLT Fiat-Shamir proof [3] that makes it history-free. Together with the general result of [13], one can immediately obtain *asymptotic* (i.e., non-concrete) versions of our QROM proof as a simple corollary. In this work, we instead give direct proofs with concrete, tight security bounds.

To demonstrate the efficacy of our generic framework, we construct a lattice-based signature scheme. The most compact lattice-based schemes, in terms of public key and signature sizes, crucially require sampling from a discrete Gaussian distribution [15,17]. Such schemes, however, have been shown to be particularly vulnerable to side-channel attacks (c.f. [14,19]), and it therefore seems prudent to consider schemes that only require simple uniform sampling over the integers. Of those, the most currently efficient one is the Dilithium signature scheme [16]. This signature scheme is proved secure based on the MSIS (Module-SIS) and the MLWE (Module-LWE) assumptions in the ROM implicitly using the framework from Fig. 1.

In this paper, we provide a practical instantiation of a lossy identification scheme to obtain a new digital signature scheme, Dilithium-QROM, with a tight security reduction in the QROM from the MLWE problem, derived using our new framework from Fig. 2. Dilithium-QROM is essentially a less compact variant ($\approx$3X larger) of Dilithium with modified parameters to allow the underlying

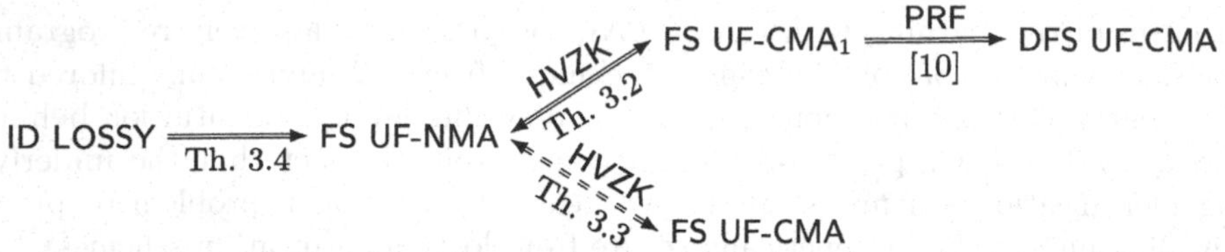

ID LOSSY —Th. 3.4→ FS UF-NMA —HVZK, Th. 3.2→ FS UF-CMA$_1$ —PRF, [10]→ DFS UF-CMA

FS UF-NMA ⇢HVZK, Th. 3.3⇢ FS UF-CMA

Fig. 2. Security of standard Fiat-Shamir signatures $\mathsf{FS} = \mathsf{FS}[\mathsf{ID}, \mathsf{H}]$ and deterministic Fiat-Shamir signatures $\mathsf{DFS} = \mathsf{DFS}[\mathsf{ID}, \mathsf{H}, \mathsf{PRF}]$ in the QROM. Solid arrows denote tight reductions, dashed arrows non-tight reductions. The considered security notions are: UF-CMA (unforgeability against chosen-message attack), UF-CMA$_1$ (unforgeability against one-query-per-message chosen-message attack), and UF-NMA (unforgeability against no-message attack).

identification scheme to admit a lossy mode. We additionally prove the security of the original Dilithium scheme in the QROM based on MLWE and another non-interactive assumption.

Security of Fiat-Shamir Signatures. Security of deterministic Fiat-Shamir signatures $\mathsf{DFS}[\mathsf{ID}, \mathsf{H}, \mathsf{PRF}]$ in the QROM is proved in two independent steps, see Fig. 2.

STEP 1: LOSSY $\Longrightarrow$ UF-NMA. We sketch an adaptation of the standard history-free proof implicitly contained in [3]. By the security properties of the lossy identification scheme, the public key can be set in lossy mode which remains unnoticed by a computationally bounded quantum adversary. Further, breaking the signature scheme in lossy mode with at most Q_{H} queries to the quantum random oracle essentially requires to solve the generic quantum search problem, whose complexity is $\Theta(Q_{\mathsf{H}}^2 \cdot \varepsilon_{\mathsf{ls}})$ [21,39], where $\varepsilon_{\mathsf{ls}}$ is the statistical soundness parameter of ID in lossy mode. A similar argument is implicitly contained in [5,37].

STEP 2: UF-NMA $\Longrightarrow$ UF-CMA. We will now sketch a history-free proof of UF-NMA $\Rightarrow$ UF-CMA$_1$, where (compared to UF-CMA security) UF-CMA$_1$ security limits the number of queried signatures per message M to one. We then apply a standard (history-free, tight) reduction to show that UF-CMA$_1$ secure signatures de-randomized with a PRF yield UF-CMA secure signatures with deterministic signing [10].

The standard ROM proof of UF-NMA $\Rightarrow$ UF-CMA (implicitly contained in [3]) works as follows: one uses the HVZK property of ID to show that the signing oracle can be efficiently simulated only knowing the public-key. Concretely, the HVZK simulator generates a transcript (W, c, Z) and later "patches" the random oracle by defining $\mathsf{H}(W \parallel M) := c$ to make (W, Z) a valid signature. The problem is that the random oracle patching (i.e., defining $\mathsf{H}(W \parallel M) := c$) can only be done *after* the signing query on M because only then W and c are known. This renders the AFLT standard reduction non history-free. In our history-free UF-NMA $\Rightarrow$ UF-CMA$_1$ proof, we resolve this problem as follows. We use the HVZK property to generate the transcript (W_M, c_M, Z_M) *deterministically* using message-dependent randomness. Hence, for each message M, the transcript (W_M, c_M, Z_M) is unique and can be computed at any time. This uniqueness allows us to patch the random oracle $\mathsf{H}(W \parallel M)$ to c_M at any time of the proof (i.e., iff $W = W_M$), even before the adversary has established a signing query on message M. This trick makes the proof history-free, see Theorem 3.2. Clearly, this only works if the adversary receives at most one signature for each messages M, which is guaranteed by the UF-CMA$_1$ experiment.

In order to deal with (full) UF-CMA security of probabilistic Fiat-Shamir signatures FS[ID, H], the above trick can be adapted to also obtain a history-free reduction, see Theorem 3.3. However, the proof is less tight as the reduction suffers from a quadratic blow-up in its running time.

Our results furthermore prove *strong* unforgeability if the identification scheme satisfies an additional property called computational unique response (CUR). CUR essentially says that it is hard to come up with two accepting transcripts with the same commitment and challenge but different responses.

Dilithium-QROM*: A signature scheme with provable security in the QROM.* The digital signature scheme Dilithium [16] is constructed from a canonical identification scheme using the Fiat-Shamir with aborts approach [26]. In the ROM, its security is based (via non-tight reductions) on the hardness of the MSIS and MLWE problems. We show that by increasing the size of the modulus and the dimension of the public key matrix, the resulting identification scheme admits a lossy mode such that distinguishing real from lossy keys is based on the hardness of MLWE. We can then apply our main reduction to conclude that the resulting digital signature scheme is based on the hardness of the MLWE problem.

In order to construct an identification scheme with a lossy mode, in addition to increasing the size of the modulus and the overall dimension, we also choose our prime modulus q so that the underlying ring $\mathbb{Z}_q[X]/(X^n+1)$ has the property that all elements with coefficients less than $\sqrt{q/2}$ have an inverse [29] – having all small elements be invertible is crucial to having lossiness.[1] For the same security levels as Dilithium, the total size of the public key and signature is increased by a factor of a little over 3.

Revisiting the Security of Dilithium. Due to the way the parameters are set, the underlying identification scheme of the original Dilithium scheme does not have a lossy mode, and so we cannot apply Theorem 3.4 in the reduction sequence in Fig. 2. Nevertheless, the reduction from Theorem 3.2 is still applicable. In the classical ROM, one then obtains a reduction from MSIS to the UF-NMA scheme via the forking lemma (see Fig. 1).

The main downside of this last step is that the reduction is inherently non-tight. In practice, however, parameters are set based on the hardness of the underlying MSIS problem and the non-tightness of the reduction is ignored.

[1] There do not exist q for which $\mathbb{Z}_q[X]/(X^n+1)$ is a field.

This is not just the case in lattice-based schemes, but is the prevalent practice for every signature scheme built via the Fiat-Shamir transform. The implicit assumption is, therefore, that the UF-NMA scheme is *exactly* as secure as MSIS (assuming that H is secure). We point out that the assumption that the UF-NMA scheme is secure is a non-interactive assumption that is reasonably simple to state, and so the fact that several decades of cryptanalysis haven't produced any improved attacks against schemes whose parameters ignore the non-tightness of the reduction, gives us confidence that equating the hardness of the UF-NMA scheme with the hardness of the underlying problem is very reasonable.

In Sect. 4.5, we formulate the security of the UF-NMA scheme as a "convolution" of a lattice/hash function problem, which we call SelfTargetMSIS, and then show that based on the hardness of MLWE and SelfTargetMSIS, the deterministic version of the Dilithium scheme is (tightly) UF-CMA secure in the QROM. In other words, we show that the security of the *tight* version of the signature scheme is based on exactly the same assumptions in the ROM and the QROM.

Other Instantiations. Our framework can be applied to obtain a security proof in the QROM for a number of existing Fiat-Shamir signature schemes that are similar to Dilithium (e.g., [3,5,7,26]) and those that have a somewhat different structure and possibly based on different assumptions (e.g., [23]). Our rationale for setting the parameters in Dilithium-QROM was to minimize the total sum of the public key and the signature. If one, on the other hand, wished to only minimize the signature size, one could create a public key whose "height" is larger than its "width" (e.g., as in [5]). For optimal efficiency, this may possibly require working over polynomial rings $\mathbb{Z}_q[X]/(f(x))$ which are finite fields.

1.2 Concrete Bounds and Comparison with Unruh [37,38]

Ignoring all constants and the computational term accounting for the pseudo-random function, our concrete bound for the UF-CMA security of deterministic Fiat-Shamir signatures DFS in the QROM is

$$\mathrm{Adv}_{\mathsf{DFS}}^{\mathsf{UF\text{-}CMA}}(\mathsf{A}) \leq \mathrm{Adv}_{\mathsf{ID}}^{\mathsf{LOSS}}(\mathsf{B}) + Q_{\mathsf{H}}^2 \cdot \varepsilon_{\mathsf{ls}} + Q_S \cdot \varepsilon_{\mathsf{zk}} + 2^{-\alpha}, \quad \mathrm{Time}(\mathsf{B}) \approx \mathrm{Time}(\mathsf{A}) \tag{1}$$

where $\mathrm{Adv}_{\mathsf{ID}}^{\mathsf{LOSS}}(\mathsf{B})$ is the lossyness advantage of ID, $\varepsilon_{\mathsf{ls}}$ is the statistical soundness parameter of ID in lossy mode, α is the min-entropy of ID's commitments, and $\varepsilon_{\mathsf{zk}}$ is the HVZK parameter of ID.

From Unruh [38] one can derive the following concrete bound which even holds for (standard) probabilistic Fiat-Shamir signatures FS.

$$\begin{aligned}\mathrm{Adv}_{\mathsf{FS}}^{\mathsf{UF\text{-}CMA}}(\mathsf{A}) \leq \mathrm{Adv}_{\mathsf{ID}}^{\mathsf{LOSS}}(\mathsf{B}) + Q_{\mathsf{H}}^2 \cdot \varepsilon_{\mathsf{ls}} + Q_S \cdot \varepsilon_{\mathsf{zk}} + Q_S Q_H^{1/2} \cdot 2^{-\alpha/4},\\ \mathrm{Time}(\mathsf{B}) \approx \mathrm{Time}(\mathsf{A}) + Q_{\mathsf{H}} Q_S.\end{aligned} \tag{2}$$

Compared to (1), bound (2) has two sources of non-tightness.

The first source of non-tightness in (2) is the term $Q_S Q_H^{1/2} \cdot 2^{-\alpha/4}$ which stems from a generic re-programming technique from [36]. In most practical lattice-based schemes the commitment's min-entropy α is large enough not to make a big impact on the worse bounds. However, this term puts a lower bound on the min-entropy of commitments which translates to an unnatural lower bound on the size of quantum-resistant Fiat-Shamir signatures. Furthermore, it is sometimes not that easy to exactly compute the min-entropy α. Further, simple techniques to get a "good-enough" bound (as we did for regular Dilithium when we obtained $\alpha = 255$) would no longer result in something meaningful when used with (2).

The second and more important sources of non-tightness in (2) is the quadratic (in the number of queries) blow-up in the running time $\mathrm{Time}(\mathsf{B}) \approx \mathrm{Time}(\mathsf{A}) + Q_{\mathsf{H}} Q_S$ which renders the reduction non-tight in all practical aspects. Interestingly, our proof for the security of *probabilistic* Fiat-Shamir signatures (Theorem 3.3) introduces the same source of non-tightness. However, under the assumption that superposition queries to classical data can be performed in a single time step (denoted by QRAM in [38]), the running time in (2) drops to $\mathrm{Time}(\mathsf{B}) \approx \mathrm{Time}(\mathsf{A})$ and hence the reduction is tight again. We leave it as an open problem to come up with a tight reduction for probabilistic Fiat-Shamir signatures in the QROM without using QRAM.

2 Preliminaries

For $n \in \mathbb{N}$, let $[n] := \{1, \ldots, n\}$. For a set S, $|S|$ denotes the cardinality of S. For a finite set S, we denote the sampling of a uniform random element x by $x \leftarrow S$, while we denote the sampling according to some distribution $\mathfrak{D}$ by $x \leftarrow \mathfrak{D}$. By $[\![B]\!]$ we denote the bit that is 1 if the Boolean Statement B is true, and 0 otherwise.

Algorithms. Let A be an algorithm. Unless stated otherwise, we assume all our algorithms to be probabilistic. We denote by $y \leftarrow \mathsf{A}(x)$ the probabilistic computation of algorithm A on input x. If A is deterministic, we write $y := \mathsf{A}(x)$. The notation $y \in \mathsf{A}(x)$ is used to indicate all possible outcomes y of the probabilistic algorithm A on input x. We can make any probabilistic A deterministic by running it with fixed randomness. We write $y := \mathsf{A}(x; r)$ to indicate that A is run on input x with randomness r. Finally, the notation $\mathsf{A}(x) \Rightarrow y$ denotes the event that A on input x returns y.

Games. We use code-based games. We implicitly assume boolean flags to be initialized to false, numerical types to 0, sets to $\varnothing$, and strings to the empty string ϵ. We make the convention that a procedure terminates once it has returned an output.

2.1 Quantum Computation

Quantum States. The state of a qubit $|\phi\rangle$ is described by a two-dimensional complex vector $|\phi\rangle = \alpha|0\rangle + \beta|1\rangle$ where $\{|0\rangle, |1\rangle\}$ form an orthonormal basis of $\mathbb{C}^2$ and $\alpha, \beta \in \mathbb{C}$ with $|\alpha|^2 + |\beta|^2 = 1$ are called the complex *amplitudes* of $|\phi\rangle$. The qbit $|\phi\rangle$ is said to be *in superposition* if $0 < |\alpha| < 1$. A classical bit $b \in \{0, 1\}$ is naturally encoded as state $|b\rangle$ of a qubit.

The state $|\psi\rangle$ of n qubits can be expressed as $|\psi\rangle = \sum_{x\in\{0,1\}^n} \alpha_x|x\rangle \in \mathbb{C}^{2^n}$ where $\{\alpha_x\}_{x\in\{0,1\}^n}$ is a set of 2^n complex amplitudes such that $\sum_{x\in\{0,1\}^n} |\alpha_x|^2 = 1$. As for one qubit, the standard orthonormal or *computational basis* is given by $\{|x\rangle\}_{x\in\{0,1\}^n}$. When the quantum state $|\psi\rangle$ is *measured* in the computational basis, the outcome is the classical string $x \in \{0,1\}^n$ with probability $|\alpha_x|^2$ and the quantum state collapses to what is observed, namely $|x\rangle$.

The evolution of a quantum system in state $|\psi\rangle$ can be described by a linear length-preserving transformation $U : \mathbb{C}^{2^n} \to \mathbb{C}^{2^n}$. Such transformations correspond to *unitary* matrices U of size 2^n by 2^n, i.e. U has the property that $UU^\dagger = \mathbb{1}$, where $U^\dagger$ is the complex-conjugate transpose of U.

For further details about basic concepts and notation of quantum computing, we refer to the standard text book by Nielsen and Chuang [31].

Quantum oracles and quantum adversaries. For a classical oracle function $\mathrm{O} : \{0,1\}^n \to \{0,1\}^m$, we follow the standard approach as in [8,13] to make the execution of the classical function O a reversible unitary transformation. We model quantum access to O by

$$U_{\mathrm{O}} : |x\rangle|y\rangle \mapsto |x\rangle|y \oplus \mathrm{O}(x)\rangle,$$

where $x \in \{0,1\}^n$ and $y \in \{0,1\}^m$. Note that due to the XOR function in the second register, U_{O} is its own inverse, i.e. executing U_{O} twice results in the identity for any function O.[2] Quantum oracle adversaries $\mathsf{A}^{|\mathrm{O}\rangle}$ can access O in superposition by applying U_{O}. The quantum time it takes to apply U_{O} is linear in the time it takes to evaluate O classically. We write $\mathsf{A}^{|\mathrm{O}\rangle}$ to indicate that an oracle is quantum-accessible, contrary to oracles which can only be accessed classically which are denoted by A^{O}. We also abuse notation and use $|O\rangle$ to denote the oracle that is quantumly accessible.

Quantum random-oracle model. We consider security games in the quantum random-oracle model (QROM) [13] like their counterparts in the classical random-oracle model [11], with the difference that we consider quantum adversaries that are given **quantum** access to the random oracles involved, and **classical** access to all other oracles (e.g., the signing oracle). Zhandry [40] proved that no quantum algorithm $\mathsf{A}^{|\mathsf{H}\rangle}$, issuing at most Q quantum queries to $|\mathsf{H}\rangle$, can distinguish between a random function $\mathsf{H} : \{0,1\}^m \to \{0,1\}^n$ and a $2Q$-wise independent function f_{2Q}. For concreteness, we view $f_{2Q} : \{0,1\}^m \to \{0,1\}^n$ as a random polynomial of degree $2Q$ over the finite field $\mathbb{F}_{2^n}$. The running time to evaluate f_{2Q} is linear in Q.

In this article, we will use this observation in the context of security reductions, where quantum adversary B simulates quantum adversary $\mathsf{A}^{|\mathsf{H}\rangle}$ which

[2] Together with the observation that taking the conjugate-complex and transposing U_{O} do not change U_{O}, we obtain $U_{\mathrm{O}}^\dagger = U_{\mathrm{O}}$, and hence, $U_{\mathrm{O}}U_{\mathrm{O}}^\dagger = U_{\mathrm{O}}^2 = \mathbb{1}$, showing that U_{O} is indeed a unitary transformation.

GAME GSPB_λ
01 $(\lambda(x))_{x\in X} \leftarrow \mathsf{A}_1$
02 If $\exists x \in X$ s.t. $\lambda(x) > \lambda$ then **return** 0
03 For all $x \in X$: $g(x) \leftarrow \mathcal{B}_{\lambda(x)}$
04 $x \leftarrow \mathsf{A}_2^{|g\rangle}$
05 **return** $g(x)$

Fig. 3. The generic search game GSPB_λ with bounded maximal Bernoulli parameter $\lambda \in [0, 1]$.

makes at most Q queries to $|\mathsf{H}\rangle$. Hence, the running time of B is $\text{Time}(\mathsf{B}) = \text{Time}(\mathsf{A}) + q \cdot \text{Time}(\mathsf{H})$, where $\text{Time}(\mathsf{H})$ is the time it takes to simulate $|\mathsf{H}\rangle$. Using the observation above, B can use a $2Q$-wise independent function in order to (information-theoretically) simulate $|\mathsf{H}\rangle$ and we obtain that the running time of B is $\text{Time}(\mathsf{B}) = \text{Time}(\mathsf{A}) + Q \cdot \text{Time}(f_{2Q})$, and the time $\text{Time}(f_{2Q})$ to evaluate f_{2Q} is linear in Q. The second term of this running time (quadratic in Q) can be further reduced to linear in Q in the quantum random-oracle model where B can simply use another random oracle to simulate $|\mathsf{H}\rangle$. Assuming evaluating the random oracle takes one time unit, we write $\text{Time}(\mathsf{B}) = \text{Time}(\mathsf{A}) + Q$ which is approximately $\text{Time}(\mathsf{A})$.

Generic Quantum Search. For $\lambda \in [0, 1]$ let $\mathcal{B}_\lambda$ be the Bernoulli distribution, i.e., $\Pr[b = 1] = \lambda$ for the bit $b \leftarrow \mathcal{B}_\lambda$. Let X be some finite set. The generic quantum search problem GSP [21,39] is to find an $x \in X$ satisfying $g(x) = 1$ given quantum access to an oracle $g : X \to \{0, 1\}$, such that for each $x \in X$, $g(x)$ is distributed according to $\mathcal{B}_\lambda$. We will need the following slight variation of GSP. The Generic quantum Search Problem with Bounded probabilities GSPB is like the quantum search problem with the difference that the Bernoulli parameter $\lambda(x)$ may (adversarially) depend on x but it is upper bounded by a global λ.

Lemma 2.1. *(Generic Search Problem with Bounded Probabilities). Let $\lambda \in [0, 1]$. For any (unbounded, quantum) algorithm* A *issuing at most Q quantum queries to $|g\rangle$, $\Pr[\mathsf{GSPB}_\lambda^{\mathsf{A}} \Rightarrow 1] \leq 8 \cdot \lambda \cdot (Q + 1)^2$, where Game* GSPB_λ *is defined in Fig. 3.*

The bound on GSPB can be reduced to the known bound on GSP [21,39] by artificially increasing the Bernoulli parameter to obtain the dependence on each $x \in X$.

2.2 Pseudorandom Functions

A pseudorandom function PRF is a mapping $\mathsf{PRF} : \mathcal{K} \times \{0, 1\}^n \to \{0, 1\}^k$, where $\mathcal{K}$ is a finite key space and n, k are integers. To a quantum adversary A and PRF we associate the advantage function

$$\text{Adv}_{\mathsf{PRF}}^{\mathsf{PR}}(\mathsf{A}) := \left| \Pr[\mathsf{A}^{\mathsf{PRF}(K,\cdot)} \Rightarrow 1 \mid K \leftarrow \mathcal{K}] - \Pr[\mathsf{A}^{\mathsf{RF}(\cdot)} \Rightarrow 1] \right|,$$

where $\mathsf{RF} : \{0,1\}^n \to \{0,1\}^k$ is a perfect random function. We note that while adversary A is quantum, it only gets classical access to the oracles $\mathsf{PRF}(K, \cdot)$ and $\mathsf{RF}(\cdot)$.

2.3 Canonical Identification Schemes

A canonical identification scheme ID is a three-move protocol of the form depicted in Fig. 4. The prover's first message W is called *commitment*, the verifier selects a uniform *challenge* c from set ChSet, and, upon receiving a *response* Z from the prover, makes a deterministic decision.

Definition 2.2 (Canonical Identification Scheme). *A canonical identification scheme* ID *is defined as a tuple of algorithms* $\mathsf{ID} := (\mathsf{IGen}, \mathsf{P}, \mathsf{ChSet}, \mathsf{V})$.

- *The key generation algorithm* IGen *takes system parameters* par *as input and returns public and secret key* (pk, sk). *We assume that* pk *defines* ChSet *(the set of challenges),* WSet *(the set of commitments), and* ZSet *(the set of responses).*
- *The prover algorithm* $\mathsf{P} = (\mathsf{P}_1, \mathsf{P}_2)$ *is split into two algorithms.* P_1 *takes as input the secret key* sk *and returns a commitment* $W \in \mathsf{WSet}$ *and a state* St*;* P_2 *takes as input the secret key* sk*, a commitment* W*, a challenge* c*, and a state* St *and returns a response* $Z \in \mathsf{ZSet} \cup \{\bot\}$*, where* $\bot \notin \mathsf{ZSet}$ *is a special symbol indicating failure.*
- *The verifier algorithm* V *takes the public key* pk *and the conversation transcript as input and outputs a* deterministic decision, *1 (acceptance) or 0 (rejection).*

We make a couple of useful definitions. A *transcript* is a three-tuple $(W, c, Z) \in \mathsf{WSet} \times \mathsf{ChSet} \times \mathsf{ZSet} \cup \{\bot, \bot, \bot\}$. It is called *valid* (with respect to public-key pk) if $\mathsf{V}(pk, W, c, Z) = 1$. In Fig. 5 we also define a transcript oracle Trans that returns a real interaction (W, c, Z) between prover and verifier as depicted in Fig. 4, with the important convention that the transcript is defined as $(\bot, \bot, \bot)$ if $Z = \bot$.

Definition 2.3 (Correctness Error). *Identification scheme* ID *has correctness error* δ *if for all* $(pk, sk) \in \mathsf{IGen}(\mathsf{par})$ *the following holds:*

- *All possible transcripts* (W, c, Z) *satisfying* $Z \neq \bot$ *are valid, i.e., for all* $(W, St) \in \mathsf{P}_1(sk)$*, all* $c \in \mathsf{ChSet}$ *and all* $Z \in \mathsf{P}_2(sk, W, c, St)$ *with* $Z \neq \bot$*, we have* $\mathsf{V}(pk, W, c, Z) = 1$.

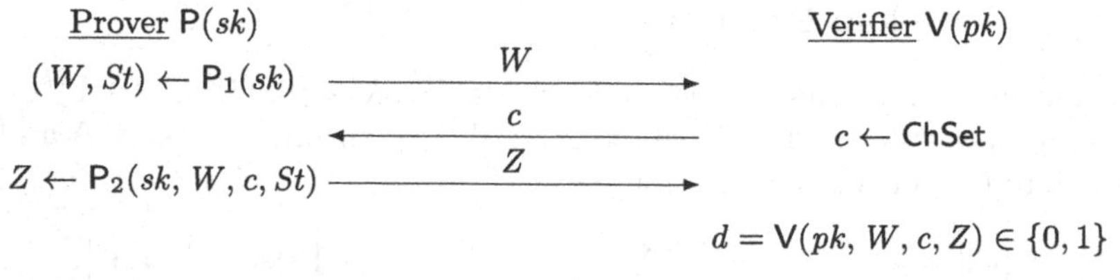

Prover $\mathsf{P}(sk)$		Verifier $\mathsf{V}(pk)$
$(W, St) \leftarrow \mathsf{P}_1(sk)$	$\xrightarrow{\quad W \quad}$	
	$\xleftarrow{\quad c \quad}$	$c \leftarrow \mathsf{ChSet}$
$Z \leftarrow \mathsf{P}_2(sk, W, c, St)$	$\xrightarrow{\quad Z \quad}$	
		$d = \mathsf{V}(pk, W, c, Z) \in \{0,1\}$

Fig. 4. A canonical identification scheme and its transcript (W, c, Z).

```
Algorithm Trans(sk):
01 (W, St) ← P1(sk)
02 c ← ChSet
03 Z ← P2(sk, W, c, St)
04 if Z = ⊥ then return (⊥, ⊥, ⊥)
05 return (W, c, Z)
```

Fig. 5. An honestly generated transcript (W, c, Z) output by the transcript oracle $\mathsf{Trans}(sk)$.

- *The probability that an honestly generated transcript* (W, c, Z) *contains* $Z = \bot$ *is bounded by* δ, *i.e.,* $\Pr[Z = \bot \mid (W, c, Z) \leftarrow \mathsf{Trans}(sk)] \leq \delta$.

Definition 2.4. *We call* ID commitment-recoverable, *if for any* $(pk, sk) \in \mathsf{IGen}(\mathsf{par})$, $c \in \mathsf{ChSet}$, *and* $Z \in \mathsf{ZSet}$, *there exists a unique* $W \in \mathsf{WSet}$ *such that* $\mathsf{V}(pk, W, c, Z) = 1$. *This unique* W *can be publicly computed using a commitment recovery algorithm as* $W := \mathsf{Rec}(pk, c, Z)$.

We define no-abort honest-verifier zero-knowledge, a weak variant of honest-verifier zero-knowledge that requires the transcript (as generated by $\mathsf{Trans}(sk)$) to be publicly simulatable, conditioned on $Z \neq \bot$.

Definition 2.5 (No-Abort Honest-verifier Zero-knowledge). *A canonical identification scheme* ID *is said to be* $\varepsilon_{\mathsf{zk}}$*-perfect* naHVZK *(no-abort honest-verifier zero-knowledge) if there exists an algorithm* Sim *that, given only the public key* pk, *outputs* (W, c, Z) *such that the following conditions hold:*

- *The distribution of* $(W, c, Z) \leftarrow \mathsf{Sim}(pk)$ *has statistical distance at most* $\varepsilon_{\mathsf{zk}}$ *from* $(W', c', Z') \leftarrow \mathsf{Trans}(sk)$, *where* Trans *is defined in Fig. 5.*
- *The distribution of* c *from* $(W, c, Z) \leftarrow \mathsf{Sim}(pk)$ *conditioned on* $c \neq \bot$ *is uniform random in* ChSet.

Note that if ID is commitment-recoverable, then we can abandon the W in the output of Trans and Sim since W can be publicly computed from (c, Z).

Definition 2.6 (Min-Entropy). *If the most likely value of a random variable* W *that is chosen from a discrete distribution* D *occurs with probability* $2^{-\alpha}$, *then we say that min-entropy*$(W \mid W \leftarrow D) = \alpha$. *We will say that a canonical identification scheme* ID *has* α bits of min-entropy, *if*

$$\Pr_{(pk,sk)\leftarrow \mathsf{IGen}(\mathsf{par})}[\textit{min-entropy}(W \mid (W, St) \leftarrow \mathsf{P}_1(sk)) \geq \alpha] \geq 1 - 2^{-\alpha}.$$

In other words, except with probability $2^{-\alpha}$ *over the choice of* (pk, sk), *the min-entropy of* W *will be at least* α.

An identification scheme has unique responses if for all W and c there exists at most one Z to make the verifier accept, i.e., $\mathsf{V}(pk, W, c, Z) = 1$. We relax this property to computational unique response (CUR) for which we require it to be computationally difficult to come up with (W, c, Z, Z') with $\mathsf{V}(pk, W, c, Z) = \mathsf{V}(pk, W, c, Z') = 1$ and $Z' \neq Z$.

Definition 2.7 (Computational Unique Response). *To an adversary* A *we associate the advantage function*

$$\mathrm{Adv}_{\mathsf{ID}}^{\mathsf{CUR}}(\mathsf{A}) := \Pr\left[\begin{array}{l}\mathsf{V}(pk, W, c, Z) = 1 \\ \mathsf{V}(pk, W, c, Z') = 1 \wedge Z \neq Z'\end{array} \middle| \begin{array}{l}(pk, sk) \leftarrow \mathsf{IGen}(\mathsf{par}); \\ (W, c, Z, Z') \leftarrow \mathsf{A}(pk)\end{array}\right].$$

LOSSY IDENTIFICATION SCHEMES. We now recall lossy identification schemes [3].

Definition 2.8. *An identification scheme* $\mathsf{ID} = (\mathsf{IGen}, \mathsf{P}, \mathsf{ChSet}, \mathsf{V})$ *is lossy if there exists a lossy key generation algorithm* $\mathsf{LossyIGen}$ *that takes system parameters* par *as input and returns public key* pk_{ls} *(and no secret key sk).*

We refer to $\mathsf{LID} = (\mathsf{IGen}, \mathsf{LossyIGen}, \mathsf{P}, \mathsf{ChSet}, \mathsf{V})$ as a lossy identification scheme.

We now define two security properties of a lossy identification scheme LID. The first property says that public keys generated with the real key generator IGen are indistinguishable from ones generated by the lossy key generator $\mathsf{LossyIGen}$. Concretely, we define the LOSS *advantage function of a quantum adversary* A *against* ID as

$$\begin{aligned}\mathrm{Adv}_{\mathsf{LID}}^{\mathsf{LOSS}}(\mathsf{A}) := \big| &\Pr[\mathsf{A}(pk_{\mathsf{ls}}) \Rightarrow 1 \mid pk_{\mathsf{ls}} \leftarrow \mathsf{LossyIGen}(\mathsf{par})] \\ &- \Pr[\mathsf{A}(pk) \Rightarrow 1 \mid (pk, sk) \leftarrow \mathsf{IGen}(\mathsf{par})]\big|.\end{aligned}$$

The second security property is statistical and says that relative to a lossy key pk_{ls}, not even an unbounded quantum adversary can impersonate the prover. We say that ID has $\varepsilon_{\mathsf{ls}}$-lossy soundness if for every (possibly unbounded, quantum) adversary C, $\Pr[\mathsf{LOSSY\text{-}IMP}^{\mathsf{C}} \Rightarrow 1] \leq \varepsilon_{\mathsf{ls}}$, where game $\mathsf{LOSSY\text{-}IMP}$ is defined in Fig. 6.

Since C is unbounded, we can upper bound $\Pr[\mathsf{LOSSY\text{-}IMP}^{\mathsf{C}} \Rightarrow 1]$ as

$$\begin{aligned}&\Pr[\mathsf{LOSSY\text{-}IMP}^{\mathsf{C}} \Rightarrow 1] \\ &\quad \leq \mathbf{E}\left[\max_{W \in \mathsf{WSet}} \left(\Pr_{c \leftarrow \mathsf{ChSet}}[\exists Z \in \mathsf{ZSet} : \mathsf{V}(pk_{\mathsf{ls}}, W, c, Z) = 1]\right)\right],\end{aligned} \tag{3}$$

where the expectation is taken over $pk_{\mathsf{ls}} \leftarrow \mathsf{LossyIGen}(\mathsf{par})$. Note that equality in Eq. (3) is achieved for the "optimal" adversary C which on the "easiest" commitment $W \in \mathsf{WSet}$ and a random challenge $c \leftarrow \mathsf{ChSet}$ finds a response $Z \in \mathsf{ZSet}$ that the verifier accepts.

GAME LOSSY-IMP:
01 $pk_{\mathsf{ls}} \leftarrow \mathsf{LossyIGen}(\mathsf{par})$
02 $(W^*, St) \leftarrow \mathsf{C}(pk_{\mathsf{ls}})$
03 $c^* \leftarrow \mathsf{ChSet}$
04 $Z^* \leftarrow \mathsf{C}(St, c^*)$
05 **return** $[\![\mathsf{V}(pk_{\mathsf{ls}}, W^*, c^*, Z^*)]\!]$

Fig. 6. The lossy impersonation game LOSSY-IMP.

2.4 Digital Signatures

We now define syntax and security of a digital signature scheme. Let par be common system parameters shared among all participants.

Definition 2.9 (Digital Signature). *A digital signature scheme* SIG *is defined as a triple of algorithms* $\mathsf{SIG} = (\mathsf{Gen}, \mathsf{Sign}, \mathsf{Ver})$.

- *The key generation algorithm* $\mathsf{Gen}(\mathsf{par})$ *returns the public and secret keys* (pk, sk)*. We assume that* pk *defines the message space* MSet.
- *The signing algorithm* $\mathsf{Sign}(sk, M)$ *returns a signature* σ.
- *The deterministic verification algorithm* $\mathsf{Ver}(pk, M, \sigma)$ *returns 1 (accept) or 0 (reject).*

Signature scheme SIG has correctness error γ if for all $(pk, sk) \in \mathsf{Gen}(\mathsf{par})$, all messages $M \in \mathsf{MSet}$, we have $\Pr[\mathsf{Ver}(pk, M, \mathsf{Sign}(sk, M)) = 0] \leq \gamma$.

Security. We define the $\mathsf{UF\text{-}CMA}$ (unforgeability against chosen-message attack), $\mathsf{UF\text{-}CMA_1}$ (unforgeability against one-per-message chosen-message attack), and $\mathsf{UF\text{-}NMA}$ (unforgeability against no-message attack) advantage functions of a quantum adversary A against SIG as $\mathrm{Adv}_{\mathsf{SIG}}^{\mathsf{UF\text{-}CMA}}(\mathsf{A}) := \Pr[\mathsf{UF\text{-}CMA}^{\mathsf{A}} \Rightarrow 1]$, $\mathrm{Adv}_{\mathsf{SIG}}^{\mathsf{UF\text{-}CMA_1}}(\mathsf{A}) := \Pr[\mathsf{UF\text{-}CMA_1}^{\mathsf{A}} \Rightarrow 1]$, and $\mathrm{Adv}_{\mathsf{SIG}}^{\mathsf{UF\text{-}NMA}}(\mathsf{A}) := \Pr[\mathsf{UF\text{-}NMA}^{\mathsf{A}} \Rightarrow 1]$, where the games $\mathsf{UF\text{-}CMA}$, $\mathsf{UF\text{-}CMA_1}$, and $\mathsf{UF\text{-}NMA}$ are given in Fig. 7. We also consider *strong* unforgeability where the adversary may return a forgery on a message previously queried to the signing oracle, but with a different signature. In the corresponding experiments $\mathsf{sUF\text{-}CMA}$ and $\mathsf{sUF\text{-}CMA_1}$, the set $\mathcal{M}$ contains tuples (M, σ) and for the winning condition it is checked that $(M^*, \sigma^*) \notin \mathcal{M}$.

Any $\mathsf{UF\text{-}CMA_1}$ ($\mathsf{sUF\text{-}CMA_1}$) secure signature scheme can be combined with a pseudo-random function PRF to obtain an $\mathsf{UF\text{-}CMA}$ ($\mathsf{sUF\text{-}CMA}$) secure signature scheme by defining $\mathsf{Sign}'((sk, K), M) := \mathsf{Sign}(sk, M; \mathsf{PRF}_K(M))$, where K is a secret PRF key which is part of the secret key. This construction is well known in the classical setting [10], and the same proof works in the quantum setting. Here PRF only has to provide security against quantum adversaries where the access to PRF is classical.

GAMES $\mathsf{UF\text{-}CMA}/\mathsf{UF\text{-}CMA_1}/\mathsf{UF\text{-}NMA}$:		$\textsc{Sign}(M)$	$\textsc{Sign}_1(M)$
01 $(pk, sk) \leftarrow \mathsf{Gen}(\mathsf{par})$		06 $\mathcal{M} = \mathcal{M} \cup \{M\}$	09 **if** $M \in \mathcal{M}$ then return $\perp$
02 $(M^*, \sigma^*) \leftarrow \mathsf{A}^{\textsc{Sign}}(pk)$	//$\mathsf{UF\text{-}CMA}$	07 $\sigma \leftarrow \mathsf{Sign}(sk, M)$	10 $\mathcal{M} = \mathcal{M} \cup \{M\}$
03 $(M^*, \sigma^*) \leftarrow \mathsf{A}^{\textsc{Sign}_1}(pk)$	//$\mathsf{UF\text{-}CMA_1}$	08 **return** σ	11 $\sigma \leftarrow \mathsf{Sign}(sk, M)$
04 $(M^*, \sigma^*) \leftarrow \mathsf{A}(pk)$	//$\mathsf{UF\text{-}NMA}$		12 **return** σ
05 **return** $[\![M^* \notin \mathcal{M}]\!] \wedge \mathsf{Ver}(pk, M^*, \sigma^*)$			

Fig. 7. Games $\mathsf{UF\text{-}CMA}$, $\mathsf{UF\text{-}CMA_1}$, and $\mathsf{UF\text{-}NMA}$.

3 Fiat-Shamir in the Quantum Random-Oracle Model

3.1 Signatures from Identification Schemes

Let $\mathsf{ID} := (\mathsf{IGen}, \mathsf{P}, \mathsf{ChSet}, \mathsf{V})$ be a canonical identification scheme, let κ_m be a positive integer, and let $\mathsf{H} : \{0,1\}^* \to \mathsf{ChSet}$ be a hash function. The following signature scheme $\mathsf{SIG} := (\mathsf{Gen} = \mathsf{IGen}, \mathsf{Sign}, \mathsf{Ver})$ is obtained by the Fiat-Shamir transformation with aborts $\mathsf{FS}[\mathsf{ID}, \mathsf{H}, \kappa_m]$ [26].

```
Sign(sk, M)
01 κ := 0
02 while Z = ⊥ and κ ≤ κ_m do
03    κ := κ + 1
04    (W, St) ← P_1(sk)
05    c = H(W || M)
06    Z ← P_2(sk, W, c, St)
07 if Z = ⊥ return σ = ⊥
08 return σ = (W, Z)

Ver(pk, M, σ)
09 Parse σ = (W, Z) ∈ WSet × ZSet
10 c = H(W || M)
11 return V(pk, W, c, Z) ∈ {0, 1}
```

We make the convention that if $\sigma = (W, Z)$ is not in $\mathsf{WSet} \times \mathsf{ZSet}$, then $\mathsf{Ver}(pk, M, \sigma)$ returns 0 (reject). Clearly, if ID has correctness error δ, then SIG has correctness error $\gamma = \delta^{\kappa_m}$.

Fiat-Shamir for Commitment-Recoverable Identification. For commitment-recoverable ID (see Definition 2.4), we can define an alternative Fiat-Shamir transformation $\mathsf{SIG}' = \mathsf{FS}'[\mathsf{ID}, \mathsf{H}, \kappa_m] := (\mathsf{Gen} = \mathsf{IGen}, \mathsf{Sign}', \mathsf{Ver}')$. Algorithm $\mathsf{Sign}'(sk, M)$ is defined as $\mathsf{Sign}(sk, M)$ with the modified output $\sigma' = (c, Z)$. Algorithm $\mathsf{Ver}'(pk, M, \sigma')$ first parses $\sigma' = (c, Z)$, then recomputes the commitment as $W' := \mathsf{Rec}(pk, c, Z)$, and finally returns 1 iff $\mathsf{H}(W' \parallel M) = c$.

```
Sign'(sk, M)
01 κ := 0
02 while Z = ⊥ and κ ≤ κ_m do
03    κ := κ + 1
04    (W, St) ← P_1(sk)
05    c = H(W || M)
06    Z ← P_2(sk, W, c, St)
07 if Z = ⊥ return σ' = ⊥
08 return σ' = (c, Z)

Ver'(pk, M, σ')
09 Parse σ' = (c, Z) ∈ ChSet × ZSet
10 W' := Rec(pk, c, Z)
11 return [[H(W' || M) = c]]
```

Since $\sigma = (W, Z)$ can be publicly transformed into $\sigma' = (c, Z)$ and vice versa, SIG and SIG' are equivalent in terms of security. The alternative Fiat-Shamir transform yields shorter signatures if $c \in \mathsf{ChSet}$ has a smaller representation size than the commitment $W \in \mathsf{WSet}$.

Main Security Statement. The following is our main security statement for $\mathsf{SIG} := \mathsf{FS}[\mathsf{ID}, \mathsf{H}, \kappa_m]$ in the QROM.

Theorem 3.1. *Assume the identification scheme* ID *is lossy,* $\varepsilon_{\mathsf{zk}}$*-perfect* naHVZK*, has* α *bits of min entropy, and is* $\varepsilon_{\mathsf{ls}}$*-lossy sound. For any quantum adversary* A *against* $\mathsf{UF\text{-}CMA_1}$ *(*$\mathsf{sUF\text{-}CMA_1}$*) security that issues at most* Q_{H} *queries to the quantum random oracle* $|\mathsf{H}\rangle$ *and* Q_S *classical queries to the signing oracle* SIGN_1*, there exists a quantum adversary* B *(and a quantum adversary* C *against* CUR*)such that*

$$\begin{aligned}
\mathrm{Adv}^{\mathsf{UF\text{-}CMA_1}}_{\mathsf{SIG}}(\mathsf{A}) &\le \mathrm{Adv}^{\mathsf{LOSS}}_{\mathsf{ID}}(\mathsf{B}) + 8(Q_{\mathsf{H}}+1)^2 \cdot \varepsilon_{\mathsf{ls}} + \kappa_m Q_S \cdot \varepsilon_{\mathsf{zk}} + 2^{-\alpha+1},\\
\mathrm{Adv}^{\mathsf{sUF\text{-}CMA_1}}_{\mathsf{SIG}}(\mathsf{A}) &\le \mathrm{Adv}^{\mathsf{LOSS}}_{\mathsf{ID}}(\mathsf{B}) + 8(Q_{\mathsf{H}}+1)^2 \cdot \varepsilon_{\mathsf{ls}} + \kappa_m Q_S \cdot \varepsilon_{\mathsf{zk}} + 2^{-\alpha+1}\\
&\quad + \mathrm{Adv}^{\mathsf{CUR}}_{\mathsf{ID}}(\mathsf{C}),
\end{aligned}$$

and $\mathrm{Time}(\mathsf{B}) = \mathrm{Time}(\mathsf{C}) = \mathrm{Time}(\mathsf{A}) + \kappa_m Q_{\mathsf{H}} \approx \mathrm{Time}(\mathsf{A})$.

Note that with this observation the bound of Theorem 3.1 is tight, i.e., the computational advantages appear with a constant factor (one). In the classical ROM setting, the only difference is that the bound depends linearly on Q_{H}, instead of quadratic.

DETERMINISTIC FIAT-SHAMIR. Let PRF be a pseudo-random function. Consider a deterministic variant $\mathsf{DSIG} := \mathsf{DFS}[\mathsf{ID}, \mathsf{H}, \mathsf{PRF}, \kappa_m] = (\mathsf{Gen}, \mathsf{DSign}, \mathsf{Ver})$ of FS where lines 04 and 06 of Sign is derandomized using the PRF, where the random key K is part of the secret key.

```
DSign((sk, K), M)
01 κ := 0
02 while Z = ⊥ and κ ≤ κ_m do
03     κ := κ + 1
04     (W, St) := P_1(sk; PRF_K(0 || m || κ))
05     c = H(W || M)
06     Z := P_2(sk, W, c, St; PRF_K(1 || m || κ))
07 if Z = ⊥ return σ = ⊥
08 return σ = (W, Z)
```

As discussed at the end of Sect. 2.4, the $\mathsf{UF\text{-}CMA}$ ($\mathsf{sUF\text{-}CMA}$) security of DSIG is implied by the $\mathsf{UF\text{-}CMA_1}$ ($\mathsf{sUF\text{-}CMA_1}$) security of FS. Concretely the advantages are upper bounded by the same terms as in Theorem 3.1 plus an additional term $\mathrm{Adv}^{\mathsf{PR}}_{\mathsf{PRF}}(\mathsf{D})$ accounting for the quantum security of the PRF.

3.2 Security Proof

The proof of Theorem 3.1 is modular. First, in Theorem 3.2 we prove that $\mathsf{UF\text{-}NMA}$ security plus naHVZK implies $\mathsf{UF\text{-}CMA_1}$ security. Second, in Theorem 3.4 we prove that a lossy identification scheme is always $\mathsf{UF\text{-}NMA}$ secure.

Theorem 3.2. *Assume the identification scheme* ID *is* $\varepsilon_{\mathsf{zk}}$*-perfect* naHVZK *and has* α *bits of min entropy. For any* $\mathsf{UF\text{-}CMA_1}$ *(*$\mathsf{sUF\text{-}CMA_1}$*) quantum adversary* A *that issues at most* Q_{H} *queries to the quantum random oracle* $|\mathsf{H}\rangle$ *and* Q_S

(classical) queries to the signing oracle SIGN_1*, there exists a quantum adversary* B *against* UF-NMA *security making* Q_{H} *queries to its own quantum random oracle (and a quantum adversary* C *against* CUR*) such that*

$$\text{Adv}_{\mathsf{SIG}}^{\mathsf{UF\text{-}CMA_1}}(\mathsf{A}) \leq \text{Adv}_{\mathsf{SIG}}^{\mathsf{UF\text{-}NMA}}(\mathsf{B}) + 2^{-\alpha+1} + \kappa_m Q_S \cdot \varepsilon_{\mathsf{zk}}$$
$$\text{Adv}_{\mathsf{SIG}}^{\mathsf{sUF\text{-}CMA_1}}(\mathsf{A}) \leq \text{Adv}_{\mathsf{SIG}}^{\mathsf{UF\text{-}NMA}}(\mathsf{B}) + 2^{-\alpha+1} + \text{Adv}_{\mathsf{ID}}^{\mathsf{CUR}}(\mathsf{C}) + \kappa_m Q_S \cdot \varepsilon_{\mathsf{zk}},$$

and $\text{Time}(\mathsf{B}) = \text{Time}(\mathsf{C}) = \text{Time}(\mathsf{A}) + \kappa_m(Q_{\mathsf{H}} + Q_S) \approx \text{Time}(\mathsf{A})$.

Proof (of Theorem 3.2*).* We first prove standard unforgeability ($\mathsf{UF\text{-}CMA_1}$ security) and then show how the proof can be modified to obtain strong unforgeability ($\mathsf{sUF\text{-}CMA_1}$ security). Let A be a quantum adversary against the $\mathsf{UF\text{-}CMA_1}$ security of SIG, issuing at most Q_{H} queries to $|\mathsf{H}\rangle$ and at most Q_S queries to SIGN_1. Consider the games given in Fig. 8. Recall that A has classical access to the signing oracle SIGN_1 and quantum access to the random oracle H. The quantum random oracle H is called with $|W \parallel M\rangle$ and returns $|\mathsf{H}(|W \parallel M\rangle)\rangle$. The games in Fig. 8 describe the computation that is performed for any $W \parallel M$ that has a non-zero amplitude in $|W \parallel M\rangle$.

GAME G_0. Note that game G_0 is the original $\mathsf{UF\text{-}CMA_1}$ game. The signing oracle SIGN_1 produces a signature using internal deterministic algorithm GetTrans which, in lines 10 and 12, derives the randomness of P_1 and P_2 using a perfect random function RF that cannot be accessed by A. Since in the $\mathsf{UF\text{-}CMA_1}$ game only one single signing query is allowed per message,

$$\Pr[G_0^{\mathsf{A}} \Rightarrow 1] = \text{Adv}_{\mathsf{SIG}}^{\mathsf{UF\text{-}CMA_1}}(\mathsf{A}).$$

```
GAME G0-G2
01 (pk, sk) <- IGen(par)
02 (M*, σ*) <- A^{|H>,SIGN1}(pk)
03 Parse σ* = (W*, Z*)
04 c* := H(W* || M*)
05 if c* ≠ H'(W* || M*) then return 0                //G2
06 return [[M* ∉ 𝓜]] ∧ V(pk, W*, c*, Z*)

GetTrans(M)                                          //G0
07 κ := 0
08 while Z_M = ⊥ and κ ≤ κ_m do
09    κ := κ + 1
10    (W_M, St) := P1(sk; RF(0 || M || κ))
11    c_M := H(W_M || M)
12    Z_M := P2(sk, W_M, c_M, St; RF(1 || M || κ))
13 if Z_M = ⊥ then (W_M, c_M, Z_M) = (⊥, ⊥, ⊥)
14 return (W_M, c_M, Z_M)

SIGN1(M)
15 if M ∈ 𝓜 then return ⊥
16 𝓜 = 𝓜 ∪ {M}
17 (W_M, c_M, Z_M) := GetTrans(M)
18 return σ_M := (W_M, Z_M)

H(W || M)                                            //quantum access
19 (W_M, c_M, Z_M) := GetTrans(M)                    //G1-G2
20 if W = W_M then return c := c_M                   //G1-G2
21 return c := H'(W || M)

GetTrans(M)                                          //G1-G2
22 κ := 0
23 while Z_M = ⊥ and κ ≤ κ_m do
24    κ := κ + 1
25    (W_M, c_M, Z_M) := Sim(pk; RF(M || κ))
26 if Z_M = ⊥ then (W_M, c_M, Z_M) = (⊥, ⊥, ⊥)
27 return (W_M, c_M, Z_M)
```

Fig. 8. Games G_0, G_1, G_2 for the proof of Theorem 3.2. Here RF and H′ are perfect random function that cannot be accessed by A. Deterministic algorithm GetTrans(M) is only used internally and cannot be accessed by A.

GAME G_1. This game computes the signatures on M using the naHVZK simulation algorithm Sim and patches the quantum random oracle H accordingly.

Concretely, consider a classical query $\textsc{Sign}_1(M)$ and let κ_M be the smallest integer $1 \leq \kappa \leq \kappa_m$ satisfying $(W, c, Z) := \mathsf{Sim}(pk; \mathsf{RF}(M \parallel \kappa))$ and $Z \neq \perp$. If no such integer exists, then we define $\kappa_M := \perp$. It deterministically computes

$$(W_M, c_M, Z_M) := \mathsf{GetTrans}(M) = \begin{cases} \mathsf{Sim}(pk; \mathsf{RF}(M \parallel \kappa_M)) & 1 \leq \kappa_M \leq \kappa_m \\ (\perp, \perp, \perp) & \kappa_M = \perp \end{cases} \quad (4)$$

The signature on M is returned as

$$\sigma_M := (W_M, Z_M).$$

By the naHVZK property and the union bound, the distribution of each σ_M has statistical distance at most $\kappa_m \varepsilon_{\mathsf{zk}}$ from one computed in game G_0. To ensure that σ_M is a valid signature on M, in line 20 the random oracle is patched such that $\mathsf{H}(W_M \parallel M) = c_M$ holds. Concretely, a query $W \parallel M$ to quantum random oracle H with non-zero amplitude is patched with $\mathsf{H}(W \parallel M) := c_M$ iff $W = W_M$, where c_M and W_M are computed by $\mathsf{GetTrans}(M)$, see Eq. (4). Note that the output distribution of the random oracle H in this game remains unchanged since c_M generated by the naHVZK simulator Sim is required to be uniformly distributed.

Overall, by a union bound we obtain

$$|\Pr[G_1^{\mathsf{A}} \Rightarrow 1] - \Pr[G_0^{\mathsf{A}} \Rightarrow 1]| \leq \kappa_m Q_S \cdot \varepsilon_{\mathsf{zk}}.$$

GAME G_2. This game returns 0 in line 05 if $c^* \neq \mathsf{H}'(W^* \parallel M^*)$. Games G_1 and G_2 can only differ if $W_{M^*} = W^*$ and $M^* \notin \mathcal{M}$. (In that case G_2 returns 0 and G_1 returns 1.) Since $M^* \notin \mathcal{M}$, the random variable W_{M^*} was not yet revealed as part of an established signature and is completely hidden from the view of the adversary. It has α bits of min-entropy, meaning we have $\Pr[W_{M^*} = W^*] \leq 2^{-\alpha}$. We obtain

$$|\Pr[G_2^{\mathsf{A}} \Rightarrow 1] - \Pr[G_1^{\mathsf{A}} \Rightarrow 1]| \leq 2^{-\alpha+1}.$$

Consider adversary B against the UF-NMA game from Fig. 9 having quantum access to random oracle H′. It perfectly simulates A's view in game G_2, using its own random oracle H′ to simulate H′ and perfectly simulating the random function RF with a $2\kappa_m Q_{\mathsf{H}}$-wise independent hash function. Assume A's forgery (M^*, σ^*) is valid in game G_2, i.e., $M^* \notin \mathcal{M}$ and $\mathsf{V}(pk, W^*, c^*, Z^*) = 1$, where $c^* = \mathsf{H}(W^* \parallel M^*)$. If $\mathsf{H}(W^* \parallel M^*) = \mathsf{H}'(W^* \parallel M^*)$, then (M^*, σ^*) is also a valid forgery in the UF-NMA game, i.e., $\mathsf{V}(pk, W^*, c^*, Z^*) = 1$, where $c^* = \mathsf{H}'(W^* \parallel M^*)$. Hence,

$$\Pr[G_2^{\mathsf{A}} \Rightarrow 1] = \mathrm{Adv}_{\mathsf{SIG}}^{\mathsf{UF\text{-}NMA}}(\mathsf{B}).$$

The proof of $\mathsf{UF\text{-}CMA}_1$ security follows by collecting the probabilities. The running time Time(B) of adversary B is given by the time Time(A) to run A as a

Adversary $\mathsf{B}^{|\mathsf{H}'\rangle}(pk)$
01 $(M^*, \sigma^*) \leftarrow \mathsf{A}^{|\mathsf{H}\rangle, \textsc{Sign}_1}(pk)$
02 Parse $\sigma^* = (W^*, Z^*)$
03 $c^* := \mathsf{H}(W^* \parallel M^*)$
04 **if** $c^* \neq \mathsf{H}'(W^* \parallel M^*)$ then **abort**
05 **if** $[\![M^* \notin \mathcal{M}]\!] \wedge \mathsf{V}(pk, W^*, c^*, Z^*)$ then **return** (M^*, σ^*)
06 **abort**

Fig. 9. Adversary B against UF-NMA security of SIG with quantum access to random oracle H′. The oracles $\textsc{Sign}_1$ and H simulated by B are defined as in game G_2 of Fig. 8.

blackbox in game G_2 where in every of the Q_{H} oracle- and Q_S signature-queries, at most $O(\kappa_m)$ computations need to be performed.

Strong unforgeability. For $\mathsf{sUF\text{-}CMA}_1$ security we consider exactly the same games with the difference that in all games the winning condition in line 06 is changed to $[\![(M^*, \sigma^*) \notin \mathcal{M}]\!] \wedge \mathsf{V}(pk, W^*, c^*, Z^*)$ to account for strong unforgerability, where $\mathcal{M}$ now records all tuples (M, σ_M) of previously established messages/signature pairs.

The difference between games G_1 and G_2 is that game G_2 returns 0 in line 05 if $c^* \neq \mathsf{H}'(W^* \parallel M^*)$, i.e., if $\mathsf{H}(W^* \parallel M^*)$ was previously patched in line 20 with $\mathsf{H}(W^* \parallel M^*) := c_{M^*}$. Games G_1 and G_2 can only differ if $W_{M^*} = W^*$, $(M^*, \sigma^*) \notin \mathcal{M}$, and $\mathsf{V}(pk, W^*, c^*, Z^*) = 1$. (In that case G_2 returns 0 and G_1 returns 1.)

We distinguish two cases. If $(M^*, \cdot) \notin \mathcal{M}$ then we are in the situation that the adversary did not query a signature on M^* and we can use the same argument as in standard unforgeability to argue $|\Pr[G_2^{\mathsf{A}} \Rightarrow 1] - \Pr[G_1^{\mathsf{A}} \Rightarrow 1]| \leq 2^{-\alpha+1}$. It leaves to handle the case $(M^*, \cdot) \in \mathcal{M}$, i.e., the adversary obtained a signatures $\sigma_{M^*} = (W_{M^*}, Z_{M^*})$ on message M^* and submits a correct forgery $\sigma^* = (W^*, Z^*)$ satisfying $W^* = W_{M^*}$ and $Z^* \neq Z_{M^*}$. The problem of finding values (W^*, c^*, Z_{M^*}, Z^*) with two accepting transcripts (W^*, c^*, Z^*) and (W^*, c^*, Z_{M^*}) is exactly bounded by the advantage of an adversary C against the CUR experiment, i.e., $|\Pr[G_2^{\mathsf{A}} \Rightarrow 1] - \Pr[G_1^{\mathsf{A}} \Rightarrow 1]| \leq \mathrm{Adv}_{\mathsf{ID}}^{\mathsf{CUR}}(\mathsf{C})$.

In combination this proves

$$|\Pr[G_2^{\mathsf{A}} \Rightarrow 1] - \Pr[G_1^{\mathsf{A}} \Rightarrow 1]| \leq 2^{-\alpha+1} + \mathrm{Adv}_{\mathsf{ID}}^{\mathsf{CUR}}(\mathsf{C}).$$

Finally, a straightforward modification of adversary B against UF-NMA security to account for the strong unforgerability check proves

$$\Pr[G_2^{\mathsf{A}} \Rightarrow 1] = \mathrm{Adv}_{\mathsf{SIG}}^{\mathsf{UF\text{-}NMA}}(\mathsf{B})$$

and completes proof of $\mathsf{sUF\text{-}CMA}_1$ security.

The running times Time(B) and Time(C) can be derived as above. □

The following theorem shows that we can also prove directly UF-CMA security of SIG, but (in terms of the running time) the reduction is less tight than the one of Theorem 3.2.

Theorem 3.3. *Assume the identification scheme* ID *is* $\varepsilon_{\mathsf{zk}}$*-perfect* naHVZK *and has* α *bits of min entropy. For any* UF-CMA *(*sUF-CMA*) quantum adversary* A *that issues at most* Q_{H} *queries to the quantum random oracle* $|\mathsf{H}\rangle$ *and* Q_S *classical queries to the signing oracle* SIGN, *there exists a quantum adversary* B *against* UF-NMA *security making* Q_{H} *queries to its own quantum random oracle (and a quantum adversary* C *against* CUR*) such that*

$$\mathrm{Adv}_{\mathsf{SIG}}^{\mathsf{UF\text{-}CMA}}(\mathsf{A}) \leq \mathrm{Adv}_{\mathsf{SIG}}^{\mathsf{UF\text{-}NMA}}(\mathsf{B}) + Q_S \cdot 2^{-\alpha+1} + \kappa_m Q_S \cdot \varepsilon_{\mathsf{zk}},$$
$$\mathrm{Adv}_{\mathsf{SIG}}^{\mathsf{sUF\text{-}CMA}}(\mathsf{A}) \leq \mathrm{Adv}_{\mathsf{SIG}}^{\mathsf{UF\text{-}NMA}}(\mathsf{B}) + Q_S \cdot 2^{-\alpha+1} + \kappa_m Q_S \cdot \varepsilon_{\mathsf{zk}} + \mathrm{Adv}_{\mathsf{ID}}^{\mathsf{CUR}}(\mathsf{C}),$$

and $\mathrm{Time}(\mathsf{B}) = \mathrm{Time}(\mathsf{C}) = \mathrm{Time}(\mathsf{A}) + \kappa_m Q_{\mathsf{H}} Q_S$.

The proof of Theorem 3.3 is similar to the one of Theorem 3.2 and appears in the full version.

Theorem 3.4. *Assume the identification scheme is lossy and* $\varepsilon_{\mathsf{ls}}$*-lossy sound. For any* UF-NMA *quantum adversary* A *that issues at most* Q_{H} *queries to the quantum random oracle* $|\mathsf{H}\rangle$, *there exists a quantum adversary* B *against* LOSS *such that*

$$\mathrm{Adv}_{\mathsf{SIG}}^{\mathsf{UF\text{-}NMA}}(\mathsf{A}) \leq \mathrm{Adv}_{\mathsf{ID}}^{\mathsf{LOSS}}(\mathsf{B}) + 8(Q_{\mathsf{H}}+1)^2 \cdot \varepsilon_{\mathsf{ls}},$$

and $\mathrm{Time}(\mathsf{B}) = \mathrm{Time}(\mathsf{A}) + Q_{\mathsf{H}} \approx \mathrm{Time}(\mathsf{A})$.

Proof. Let A be an adversary against the UF-NMA security of SIG, issuing at most Q_{H} quantum queries to $|\mathsf{H}\rangle$. Consider the games given in Fig. 10.

GAME G_0. Since game G_0 is the original UF-NMA game,

$$\Pr[G_0^{\mathsf{A}} \Rightarrow 1] = \mathrm{Adv}_{\mathsf{SIG}}^{\mathsf{UF\text{-}NMA}}(\mathsf{A}).$$

GAME G_1. In this game, the public key pk is changed to lossy mode. Clearly, there exists an adversary B simulating H by a $2Q_{\mathsf{H}}$-wise independent hash function such that

$$|\Pr[G_1^{\mathsf{A}} \Rightarrow 1] - \Pr[G_0^{\mathsf{A}} \Rightarrow 1]| \leq \mathrm{Adv}_{\mathsf{ID}}^{\mathsf{LOSS}}(\mathsf{B}).$$

Finally, we will reduce a successful A in game G_1 to the generic search problem GSPB to show

$$\Pr[G_1^{\mathsf{A}} \Rightarrow 1] \leq 8(Q_{\mathsf{H}}+1)^2 \varepsilon_{\mathsf{ls}}. \tag{5}$$

GAME G_0-G_1

01	$(pk, sk) \leftarrow \mathsf{IGen}(\mathsf{par})$	// G_0
02	$pk \leftarrow \mathsf{LossyIGen}(\mathsf{par})$	// G_1
03	$(M^*, \sigma^*) \leftarrow \mathsf{A}^{|\mathsf{H}\rangle}(pk)$	
04	Parse $\sigma^* = (W^*, Z^*)$	
05	$c^* := \mathsf{H}(W^* \parallel M^*)$	
06	**return** $\mathsf{V}(pk, W^*, c^*, Z^*)$	

Fig. 10. Games G_0-G_1 for the proof of Theorem 3.4.

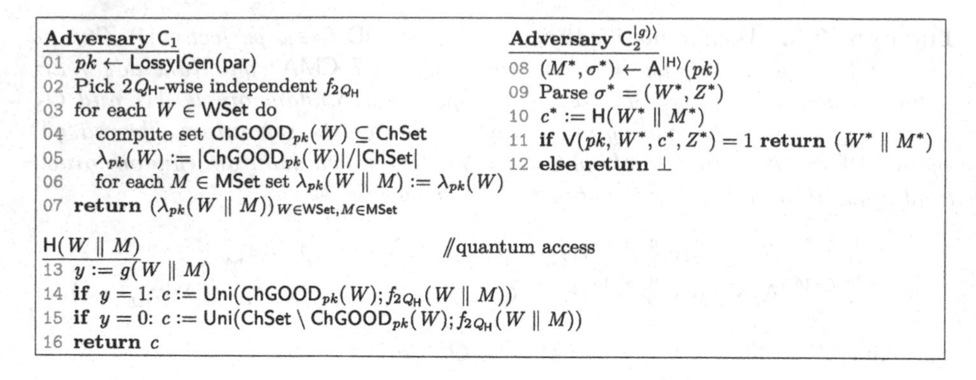

Fig. 11. Adversary $\mathsf{C} = (\mathsf{C}_1, \mathsf{C}_2)$ in game GSPB for the proof of Theorem 3.4. The set of good challenges $\mathsf{ChGOOD}_{pk}(W)$ is defined in Eq. (6).

For a finite set S, let $\mathsf{Uni}(S)$ be a probabilistic algorithm that returns uniform $x \leftarrow S$ and recall that $x := \mathsf{Uni}(S; r)$ denotes the deterministic execution of $\mathsf{Uni}(S)$ using explicitly given random tape r. To prove Eq. (5), consider the unbounded adversary $\mathsf{C} = (\mathsf{C}_1, \mathsf{C}_2)$ defined in Fig. 11 that is executed in the generic search game GSPB, making at most Q_H quantum queries to the oracle $|g(\cdot)\rangle$. First note that computing the probabilities $\lambda_{pk}(W \parallel M) = \lambda_{pk}(W)$ in line 05 for all $W \in \mathsf{WSet}$ and $M \in \mathsf{MSet}$ may take exponential time but since C is computationally unbounded it does not matter.

To analyze C's success probability in game GSPB, we first fix a public-key pk. Now consider some $W \parallel M$ with non-zero amplitude as part of a query to quantum random oracle H. Set $\mathsf{ChGOOD}_{pk}(W)$ of "good challenges" is defined as

$$\mathsf{ChGOOD}_{pk}(W) := \{c \in \mathsf{ChSet} \mid \exists Z \in \mathsf{ZSet} : \mathsf{V}(pk, W, c, Z) = 1\}. \qquad (6)$$

That is, the set $\mathsf{ChGOOD}_{pk}(W)$ contains all challenges c for which there exists a possible response Z to make (W, c, Z) a valid transcript (with respect to pk). By definition of GSPB, each query to oracle $g(W \parallel M)$ returns $y = 1$ with probability $\lambda_{pk}(W \parallel M) = |\mathsf{ChGOOD}_{pk}(W)|/|\mathsf{ChSet}|$. Hence, the output distribution of $\mathsf{H}(W \parallel M)$ sampled in lines 14 and 15 is uniform over ChSet, as in game G_1. Consistency of H is assured by deriving the randomness to sample c in case $y = 0$ (lines 14 and 15) using fixed random coins $f_{2Q_\mathsf{H}}(W \parallel M)$, derived by a $2Q_\mathsf{H}$-wise independent hash function f_{2Q_H} (which looks like a perfectly random function to A).

Now consider A's forgery $\sigma^* = (W^*, Z^*)$ on message M^* and define $c^* := \mathsf{H}(W^* \parallel M^*)$. If the signature is valid (i.e., $\mathsf{V}(pk, W^*, c^*, Z^*) = 1$), then clearly c^* is a good challenge from set $\mathsf{ChGOOD}_{pk}(W^*)$ which implies $g(W^* \parallel M^*) = 1$. This proves

$$\Pr[G_1 \Rightarrow 1 \mid pk] = \Pr[\mathsf{GSPB}^{\mathsf{C}}_{\lambda_{pk}} \Rightarrow 1 \mid pk] \leq 8(Q_\mathsf{H} + 1)^2 \lambda_{pk}, \qquad (7)$$

where

$$\lambda_{pk} = \max_{W \in \mathsf{WSet}, M \in \mathsf{MSet}} \lambda_{pk}(W \parallel M)$$

Averaging Eq. (7) over $pk \leftarrow \mathsf{LossyIGen}$ we finally obtain

$$\Pr[G_1 \Rightarrow 1] \leq 8(Q_{\mathsf{H}}+1)^2 \cdot \mathbf{E}_{pk}[\lambda_{pk}] \leq 8(Q_{\mathsf{H}}+1)^2 \varepsilon_{\mathsf{ls}},$$

where the last inequality uses Eq. (3) for the optimal adversary. □

4 Dilithium-QROM

In this section, we present a modification of the Dilithium digital signature scheme [16] whose security is based on MLWE in the QROM. We also present a new security proof of the original Dilithium that shows it to be tightly-secure in the QROM based on a different non-interactive assumption. Since Dilithium is a highly-optimized version of a scheme constructed via the "Fiat-Shamir with Aborts" framework [26], its details may be somewhat overwhelming to readers who are not already comfortable with such constructions. For this reason, we present a much simpler version of the signature scheme without any optimizations in the full version of this paper.

4.1 Preliminaries

RINGS AND DISTRIBUTIONS. We let R and R_q respectively denote the rings $\mathbb{Z}[X]/(X^n+1)$ and $\mathbb{Z}_q[X]/(X^n+1)$, for an integer q. We will assume that $q \equiv 5 (\bmod 8)$, as such a choice of q ensures that all polynomials in R_q with coefficients less than $\sqrt{q/2}$ have an inverse in the ring [29, Lemma 2.2]. This property is crucial to our security proof. Regular font letters denote elements in R or R_q (which includes elements in $\mathbb{Z}$ and $\mathbb{Z}_q$) and bold lower-case letters represent column vectors with coefficients in R or R_q. By default, all vectors will be column vectors. Bold upper-case letters are matrices.

MODULAR REDUCTIONS. For an even (resp. odd) positive integer α, we define $r' = r \bmod^{\pm} \alpha$ to be the unique element r' in the range $-\frac{\alpha}{2} < r' \leq \frac{\alpha}{2}$ (resp. $-\frac{\alpha-1}{2} \leq r' \leq \frac{\alpha-1}{2}$) such that $r' = r \bmod \alpha$. We will sometimes refer to this as a *centered* reduction modulo q. For any positive integer α, we define $r' = r \bmod^{+} \alpha$ to be the unique element r' in the range $0 \leq r' < \alpha$ such that $r' = r \bmod \alpha$. When the exact representation is not important, we simply write $r \bmod \alpha$.

SIZES OF ELEMENTS. For an element $w \in \mathbb{Z}_q$, we write $\|w\|_\infty$ to mean $|w \bmod^{\pm} q|$. We now define the ℓ_∞ and ℓ_2 norms for $w = w_0 + w_1X + \ldots + w_{n-1}X^{n-1} \in R$:

$$\|w\|_\infty = \max_i \|w_i\|_\infty, \quad \|w\| = \sqrt{\|w_0\|_\infty^2 + \ldots + \|w_{n-1}\|_\infty^2}.$$

Similarly, for $\mathbf{w} = (w_1, \ldots, w_k) \in R^k$, we define

$$\|\mathbf{w}\|_\infty = \max_i \|w_i\|_\infty, \quad \|\mathbf{w}\| = \sqrt{\|w_1\|^2 + \ldots + \|w_k\|^2}.$$

We will write S_η to denote all elements $w \in R$ such that $\|w\|_\infty \leq \eta$.

Extendable output function. Suppose that Sam is an extendable output function, that is a function on bit strings in which the output can be extended to any desired length. If we would like Sam to take as input x and then produce a value y that is distributed according to distribution S (or uniformly over a set S), we write $y \sim S := \mathsf{Sam}(x)$. It is important to note that this procedure is completely deterministic: a given x will always produce the same y. For simplicity we assume that the output distribution of Sam is perfect, whereas in practice Sam will be implemented using random oracles and produce an output that is statistically close to the perfect distribution. If K is a secret key, then $\mathsf{Sam}(K\|x)$ is a pseudo-random function from $\{0,1\}^* \to \{0,1\}^*$.

The Challenge Space. The challenge space in our identification and signature schemes needs to be a subset of the ring R, have size a little larger than 2^{256}, and consist of polynomials with small norms. In this paper, the dimension n of the ring R will be taken to be 512,[3] and so we will define the challenge space accordingly as

$$\mathsf{ChSet} := \{c \in R \,|\, \|c\|_\infty = 1 \text{ and } \|c\| = \sqrt{46}\}. \tag{8}$$

In other words, ChSet consists of elements in R with $-1/0/1$ coefficients that have exactly 46 non-zero coefficients. The size of this set is $\binom{n}{46} \cdot 2^{46}$, which for $n = 512$ is greater than 2^{265}.

The MLWE Assumption. For integers m, k, and a probability distribution $D : R_q \to [0,1]$, we say that the advantage of algorithm A in solving the decisional $\mathsf{MLWE}_{m,k,D}$ problem over the ring R_q is

$$\begin{aligned}\mathrm{Adv}^{\mathsf{MLWE}}_{m,k,D} := \big|&\Pr[\mathsf{A}(\mathbf{A},\mathbf{t}) \Rightarrow 1 \,|\, \mathbf{A} \leftarrow R_q^{m\times k}; \mathbf{t} \leftarrow R_q^m] \\ &- \Pr[\mathsf{A}(\mathbf{A}, \mathbf{A}\mathbf{s}_1 + \mathbf{s}_2) \Rightarrow 1 \,|\, \mathbf{A} \leftarrow R_q^{m\times k}; \mathbf{s}_1 \leftarrow D^k; \mathbf{s}_2 \leftarrow D^m]\big|.\end{aligned}$$

The MLWE assumption states that the above advantage is negligible for all polynomial-time algorithms A. This assumption was introduced in [25], and is generalization of the LWE assumption from [35]. The $\mathsf{Ring\text{-}LWE}$ assumption [30] is a special case of MLWE where $k = 1$. Analogously to LWE and $\mathsf{Ring\text{-}LWE}$, it was shown in [25] that solving the MLWE problem for certain parameters is as hard as solving certain worst-case problems in certain algebraic lattices.

Summary of Supporting Algorithms. To reduce the size of the public key, we will need some simple algorithms that extract "higher-order" and "lower-order" bits of elements in $\mathbb{Z}_q$. The goal is that when given an arbitrary element

[3] In Sect. 4.5, we will also discuss a scheme where $n = 256$. For that scheme the challenge space consists of 60 ±1's.

```
Power2Round_q(r, d)
01 r := r mod+ q
02 r0 := r mod± 2^d
03 return (r − r0)/2^d

UseHint_q(h, r, α)
04 m := (q − 1)/α
05 (r1, r0) := Decompose_q(r, α)
06 if h = 1 and r0 > 0 return (r1 + 1) mod+ m
07 if h = 1 and r0 ≤ 0 return (r1 − 1) mod+ m
08 return r1

MakeHint_q(z, r, α)
09 r1 := HighBits_q(r, α)
10 v1 := HighBits_q(r + z, α)
11 return [[r1 ≠ v1]]

Decompose_q(r, α)
12 r := r mod+ q
13 r0 := r mod± α
14 if r − r0 = q − 1
15   then r1 := 0; r0 := r0 − 1
16 else r1 := (r − r0)/α
17 return (r1, r0)

HighBits_q(r, α)
18 (r1, r0) := Decompose_q(r, α)
19 return r1

LowBits_q(r, α)
20 (r1, r0) := Decompose_q(r, α)
21 return r0
```

Fig. 12. Supporting algorithms for Dilithium and Dilithium-QROM.

$r \in \mathbb{Z}_q$ and another small element $z \in \mathbb{Z}_q$, we would like to be able to recover the higher order bits of $r + z$ without needing to store z. We therefore define algorithms that take r, z and produce a 1-bit hint h that allows one to compute the higher order bits of $r + z$ just using r and h. This hint is essentially the "carry" caused by z in the addition. The algorithms are exactly as in [16], and we repeat them for convenience in Fig. 12. The algorithms are described as working on integers modulo q, but are extended to polynomials in R_q by simply being applied individually to each coefficient.

The below Lemmas recall the crucial properties of these supporting algorithms that are necessary for the correctness and security of our scheme.

Lemma 4.1. *Suppose that q and α are positive integers satisfying $q > 2\alpha$, $q \equiv 1 \pmod{\alpha}$ and α even. Let $\mathbf{r}$ and $\mathbf{z}$ be vectors of elements in R_q where $\|\mathbf{z}\|_\infty \leq \alpha/2$, and let $\mathbf{h}, \mathbf{h}'$ be vectors of bits. Then the $\mathsf{HighBits}_q$, $\mathsf{MakeHint}_q$, and $\mathsf{UseHint}_q$ algorithms satisfy the following properties:*

1. $\mathsf{UseHint}_q(\mathsf{MakeHint}_q(\mathbf{z}, \mathbf{r}, \alpha), \mathbf{r}, \alpha) = \mathsf{HighBits}_q(\mathbf{r} + \mathbf{z}, \alpha)$.
2. Let $\mathbf{v}_1 = \mathsf{UseHint}_q(\mathbf{h}, \mathbf{r}, \alpha)$. *Then* $\|\mathbf{r} - \mathbf{v}_1 \cdot \alpha\|_\infty \leq \alpha + 1$.
3. For any $\mathbf{h}, \mathbf{h}'$, *if* $\mathsf{UseHint}_q(\mathbf{h}, \mathbf{r}, \alpha) = \mathsf{UseHint}_q(\mathbf{h}', \mathbf{r}, \alpha)$, *then* $\mathbf{h} = \mathbf{h}'$.

Lemma 4.2. *If $\|\mathbf{s}\|_\infty \leq \beta$ and $\|\mathsf{LowBits}_q(\mathbf{r}, \alpha)\|_\infty < \alpha/2 - \beta$, then*

$$\mathsf{HighBits}_q(\mathbf{r}, \alpha) = \mathsf{HighBits}_q(\mathbf{r} + \mathbf{s}, \alpha).$$

4.2 The Identification Protocol

The constituting algorithms of our identification protocol $\mathsf{ID} = (\mathsf{IGen}, \mathsf{P}_1, \mathsf{P}_2, \mathsf{V})$ are described in Fig. 13 with the concrete parameters $\mathsf{par} = (q, n, k, \ell, d, \gamma, \gamma', \eta, \beta)$ given later in Table 1.

$\underline{\mathsf{IGen}(\mathsf{par})}$

01 $\rho \leftarrow \{0,1\}^{256}$
02 $\mathbf{A} \leftarrow R_q^{k\times\ell} := \mathsf{Sam}(\rho)$
03 $(\mathbf{s}_1, \mathbf{s}_2) \leftarrow S_\eta^\ell \times S_\eta^k$
04 $\mathbf{t} := \mathbf{A}\mathbf{s}_1 + \mathbf{s}_2$
05 $\mathbf{t}_1 := \mathsf{Power2Round}_q(\mathbf{t}, d)$
06 $\mathbf{t}_0 := \mathbf{t} - \mathbf{t}_1 \cdot 2^d$
07 $pk = (\rho, \mathbf{t}_1, \boxed{\mathbf{t}_0})$
08 $sk = (\rho, \mathbf{s}_1, \mathbf{s}_2, \mathbf{t}_0)$
09 **return** (pk, sk)

$\underline{\mathsf{P}_1(sk)}$

10 $\mathbf{A} \leftarrow R_q^{k\times\ell} := \mathsf{Sam}(\rho)$
11 $\mathbf{y} \leftarrow S_{\gamma'-1}^\ell$
12 $\mathbf{w} := \mathbf{A}\mathbf{y}$
13 $\mathbf{w}_1 := \mathsf{HighBits}_q(\mathbf{w}, 2\gamma)$
14 **return** $(W = \mathbf{w}_1, St = (\mathbf{w}, \mathbf{y}))$

$\underline{\mathsf{P}_2(sk, W = \mathbf{w}_1, c, St = (\mathbf{w}, \mathbf{y}))}$

15 $\mathbf{z} := \mathbf{y} + c\mathbf{s}_1$
16 **if** $\|\mathbf{z}\|_\infty \geq \gamma' - \beta$ **or** $\|\mathsf{LowBits}_q(\mathbf{w} - c\mathbf{s}_2, 2\gamma)\|_\infty \geq \gamma - \beta$
17 then $(\mathbf{z}, \mathbf{h}) := \bot$
18 **else** $\mathbf{h} := \mathsf{MakeHint}_q(-c\mathbf{t}_0, \mathbf{w} - c\mathbf{s}_2 + c\mathbf{t}_0, 2\gamma)$
19 **return** $Z = (\mathbf{z}, \mathbf{h})$

$\underline{\mathsf{V}(pk, W = \mathbf{w}_1, c, Z = (\mathbf{z}, \mathbf{h}))}$

20 **return** $[\![\|\mathbf{z}\|_\infty < \gamma' - \beta]\!]$ **and** $[\![\mathbf{w}_1 = \mathsf{UseHint}_q(\mathbf{h}, \mathbf{A}\mathbf{z} - c\mathbf{t}_1 \cdot 2^d, 2\gamma)]\!]$

Fig. 13. Our ID scheme – a concrete instantiation based on the hardness of the MLWE problem of the commitment-recoverable (Definition 2.4) canonical identification scheme in Fig. 4. The $\mathbf{t}_0$ part of the public key is assumed to be known by the adversary in the security proofs, but is not needed by the verifier for verification. Thus in the real scheme, $\mathbf{t}_0$ would not be included as part of the public key.

Key Generation. The key generation proceeds by choosing a random 256-bit seed ρ and expanding into a matrix $\mathbf{A} \in R_q^{k\times\ell}$ by an extendable output function Sam modeled as a random oracle. The secret keys $(\mathbf{s}_1, \mathbf{s}_2) \in S_\eta^\ell \times S_\eta^k$ have uniformly random coefficients between $-\eta$ and η (inclusively). The value $\mathbf{t} = \mathbf{A}\mathbf{s}_1 + \mathbf{s}_2$ is then computed. The public key that is needed for verification is $(\rho, \mathbf{t}_1)$ with $\mathbf{t}_1$ output by the $\mathsf{Power2Round}_q(\mathbf{t}, d)$ algorithm in Fig. 12 (we have $\mathbf{t} = \mathbf{t}_1 \cdot 2^d + \mathbf{t}_0$ for some small $\mathbf{t}_0$), while the secret key is $(\rho, \mathbf{s}_1, \mathbf{s}_2, \mathbf{t}_0)$.

While the verifier never needs the value $\mathbf{t}_0$ (and thus it does not need to be included in the public key of the actual scheme), we do need this value in order to simulate transcripts (see Sect. 4.3). Thus the security of our scheme is based on the fact that the adversary gets $\mathbf{t}_1$ and $\mathbf{t}_0$, whereas in reality he only gets $\mathbf{t}_1$.

The set ChSet is defined as in Eq. (8), and $\mathsf{ZSet} = S_{\gamma'-\beta-1}^\ell \times \{0,1\}^k$. The set of commitments WSet is defined as $\mathsf{WSet} = \{\mathbf{w}_1 : \exists \mathbf{y} \in S_{\gamma'-1}^\ell \text{ s.t. } \mathbf{w}_1 = \mathsf{HighBits}_q(\mathbf{A}\mathbf{y}, 2\gamma)\}$.

Protocol Execution. The prover starts the identification protocol by reconstructing $\mathbf{A}$ from the random seed ρ. The next step has the prover sample $\mathbf{y} \leftarrow S_{\gamma'-1}^\ell$ and then compute $\mathbf{w} = \mathbf{A}\mathbf{y}$. He then writes $\mathbf{w} = 2\gamma \cdot \mathbf{w}_1 + \mathbf{w}_0$, with $\mathbf{w}_0$ between $-\gamma$ and γ (inclusively), and then sends $\mathbf{w}_1$ to the verifier. The verifier generates a random challenge $c \leftarrow \mathsf{ChSet}$ and sends it to the prover. The prover computes $\mathbf{z} = \mathbf{y} + c\mathbf{s}$. If $\mathbf{z} \notin S_{\gamma'-\beta-1}^\ell$, then the prover sets his response to $\bot$. He also replies with $\bot$ if $\mathsf{LowBits}_q(\mathbf{w} - c\mathbf{s}_2, 2\gamma) \notin S_{\gamma-\beta-1}^k$. This part of the protocol is necessary for security – it makes sure that $\mathbf{z}$ does not leak anything about the secret key $\mathbf{s}_1, \mathbf{s}_2$.

If the checks pass and a $\perp$ is not sent, then it can be shown (see Sect. 4.3) that $\mathsf{HighBits}_q(\mathbf{Az} - c\mathbf{t}, 2\gamma) = \mathbf{w}_1$. At this point, if the verifier knew the entire element $\mathbf{t}$ and $(\mathbf{z}, c)$, he could have recovered $\mathbf{w}_1$ and checked that $\|\mathbf{z}\|_\infty < \gamma' - \beta$ and that the high-order bits of $\mathbf{Az} - c\mathbf{t}$ are indeed $\mathbf{w}_1$. However, since we want to compress the size of the public key, the verifier only knows $\mathbf{t}_1$. Hence, the signer needs to provide a "hint" $\mathbf{h}$ which will allow the verifier to compute $\mathsf{HighBits}_q(\mathbf{Az} - c\mathbf{t}, 2\gamma)$.

The verifier checks whether $\|\mathbf{z}\|_\infty < \gamma' - \beta$ and that $\mathbf{Az} - c\mathbf{t}_1 \cdot 2^d$ together with the hint $\mathbf{h}$ allow him to reconstruct $\mathbf{w}_1$. We should point out that in the identification scheme it is actually not necessary for the verifier to be able to recover exactly $\mathbf{w}_1$. He could have simply checked that $\mathbf{Az} - c\mathbf{t}_1 \cdot 2^d \approx \mathbf{w}_1$ and this would be good enough for security. The reason that we want the verifier to be able to exactly recover $\mathbf{w}_1$ is to make the ID scheme *commitment-recoverable* and be able to reduce the communication size in the Fiat-Shamir transform (see Sect. 3.1).

4.3 Security Properties

In this section we analyze the security of ID. Most of the proofs are postponed to the full version.

NON ABORT HONEST VERIFIER ZERO-KNOWLEDGE. In this section, we will show that ID is perfectly naHVZK, i.e., the distribution of the output of the Trans algorithm (Fig. 14, left) that uses the secret key as input is exactly that of the Sim algorithm (Fig. 14, right) that uses only the public key as input.

Lemma 4.3. *If* $\beta \geq \max_{s \in S_\eta, c \in \mathsf{ChSet}} \|cs\|_\infty$, *then* ID *is perfectly* naHVZK.

CORRECTNESS. In this section, we compute the probability that the Prover does not send $\perp$ and then show that the verification procedure will always accept a transcript when the Prover does not send $\perp$.

Lemma 4.4. *If* $\beta \geq \max_{s \in S_\eta, c \in \mathsf{ChSet}} \|cs\|_\infty$ *then* ID *has correctness error* $\delta \approx 1 - \exp(-\beta n \cdot (k/\gamma + \ell/\gamma'))$.

Algorithm $\mathsf{Trans}(sk)$:	**Algorithm** $\mathsf{Sim}(pk)$:
01 $\mathbf{A} \leftarrow R_q^{k\times\ell} := \mathsf{Sam}(\rho)$	11 $\mathbf{A} \leftarrow R_q^{k\times\ell} := \mathsf{Sam}(\rho)$
02 $\mathbf{y} \leftarrow S_{\gamma'-1}^\ell$	12 with probability $1 - \frac{\lvert S_{\gamma'-\beta-1}^\ell\rvert}{\lvert S_{\gamma'-1}^\ell\rvert}$, **return** $\perp$
03 $\mathbf{w} := \mathbf{Ay}$	
04 $\mathbf{w}_1 := \mathsf{HighBits}_q(\mathbf{w}, 2\gamma)$	13 $\mathbf{z} \leftarrow S_{\gamma'-\beta-1}^\ell$
05 $c \leftarrow \mathsf{ChSet}$	14 $c \leftarrow \mathsf{ChSet}$
06 $\mathbf{z} := \mathbf{y} + c\mathbf{s}_1$	15 **if** $\|\mathsf{LowBits}_q(\mathbf{Az} - c\mathbf{t}, 2\gamma)\|_\infty \geq \gamma - \beta$
07 **if** $\|\mathbf{z}\|_\infty \geq \gamma' - \beta$ **then return** $\perp$	16 **then return** $\perp$
08 **if** $\|\mathsf{LowBits}_q(\mathbf{w} - c\mathbf{s}_2, 2\gamma)\|_\infty \geq \gamma - \beta$ **then return** $\perp$	17 $\mathbf{h} := \mathsf{MakeHint}_q(-c\mathbf{t}_0, \mathbf{Az} - c\mathbf{t} + c\mathbf{t}_0, 2\gamma)$
09 $\mathbf{h} := \mathsf{MakeHint}_q(-c\mathbf{t}_0, \mathbf{w} - c\mathbf{s}_2 + c\mathbf{t}_0, 2\gamma)$	18 **return** $(c, (\mathbf{z}, \mathbf{h}))$
10 **return** $(c, (\mathbf{z}, \mathbf{h}))$	

Fig. 14. Left: a real transcript output by the transcript algorithm $\mathsf{Trans}(sk)$; Right: a simulated transcript output by the $\mathsf{Sim}(pk)$ algorithm.

```
LossyIGen(par)
01 ρ ← {0,1}^256; A ← R_q^{k×ℓ} := Sam(ρ)
02 t ← R_q^k
03 t_1 := Power2Round_q(t, d)
04 t_0 := t − t_1 · 2^d
05 return pk = (ρ, t_1, t_0)
```

Fig. 15. The lossy instance generator LossyIGen.

LOSSYNESS. In this section, we analyze the scheme in which the public key is generated uniformly at random, as in algorithm LossyIGen of Fig. 15, rather than as in IGen of Fig. 13. Our goal is to show that even if the prover is computationally unbounded, he only has approximately a $1/|\mathsf{ChSet}|$ probability of making the verifier accept during each run of the identification scheme. This will show that the probability in Eq. (3) is upper-bounded by approximately $1/|\mathsf{ChSet}|$.

By observing that the output of LossyIGen is uniformly random over $R_q^{k\times\ell} \times R_q^k$ and the output of IGen in Fig. 13 is $(\mathbf{A}, \mathbf{A}\mathbf{s}_1 + \mathbf{s}_2)$ where $\mathbf{A} \leftarrow R_q^{k\times\ell}$ and $(\mathbf{s}_1, \mathbf{s}_2) \leftarrow S_\eta^\ell \times S_\eta^k$, we have that

$$\mathrm{Adv}_{\mathsf{ID}}^{\mathsf{LOSS}}(\mathsf{A}) = \mathrm{Adv}_{k,\ell,D}^{\mathsf{MLWE}}(\mathsf{A}),$$

where D is the uniform distribution over S_η.

Lemma 4.5. *If $4\gamma + 2$, $2\gamma' < \sqrt{q/2}$ and $\gamma' < \gamma\beta$, and $\ell \leq k$, then* ID *has $\varepsilon_{\mathsf{ls}}$-lossy soundness for*

$$\varepsilon_{\mathsf{ls}} \leq \frac{1}{|\mathsf{ChSet}|} + 2 \cdot |\mathsf{ChSet}|^2 \cdot \left(\frac{32\gamma\gamma'}{q}\right)^{nk}.$$

Our proof follows the framework from [3,22]. Then to prove Lemma 4.5, we show that if C, who outputs the first message $(\mathbf{w}_1, St)$ in the LOSSY-IMP game (see Fig. 16) is able to correctly respond to more than one random challenge c, then the previously mentioned linear equation will have a solution, which with high probability is not possible. Therefore we conclude that for virtually all $\mathbf{A}, \mathbf{t}$ output by LossyIGen, there exists (at most) only one challenge for which the prover can respond to, and therefore his success probability is at most $1/|\mathsf{ChSet}|$.

MIN ENTROPY. In Lemma 4.6 we will prove that the $\mathbf{w}_1$ sent by the honest prover in the first step is extremely likely to be distinct for every run of the protocol.

Lemma 4.6. *If 2γ, $2\gamma' < \sqrt{q/2}$ and $\ell \leq k$, then the identification scheme* ID *in Fig. 13 has*

$$\alpha > n\ell \cdot \log\left(\min\left\{\frac{q}{(4\gamma+1)(4\gamma'+1)}, 2\gamma' - 1\right\}\right)$$

bits of min-entropy (as in Definition 2.6).

GAME LOSSY-IMP:
01 $pk_{\mathsf{ls}} := (\rho, \mathbf{t}_1, \mathbf{t}_0) \leftarrow \mathsf{LossyIGen}(\mathsf{par})$
02 $(\mathbf{w}_1, St) \leftarrow \mathsf{C}(pk_{\mathsf{ls}})$
03 $c \leftarrow \mathsf{ChSet}$
04 $(\mathbf{z}, \mathbf{h}) \leftarrow \mathsf{C}(St, c)$
05 **return** $[\![\mathbf{w}_1 = \mathsf{UseHint}_q(\mathbf{h}, \mathbf{Az} - c\mathbf{t}_1 \cdot 2^d, 2\gamma)]\!]$ **and** $[\![\|\mathbf{z}\|_\infty < \gamma' - \beta]\!]$

Fig. 16. The lossy impersonation game LOSSY-IMP in case of Dilithium.

Computational Unique Response. In this section we state that our scheme satisfies the Computational Unique Response (CUR) property required for strong-unforgeability of the signature scheme.

Lemma 4.7. *If $4\gamma + 2, 2\gamma' < \sqrt{q/2}$ and $\gamma' < \gamma\beta$, and $\ell \leq k$ (i.e. the same conditions as in Lemma 4.5), then $\mathrm{Adv}_{\mathsf{ID}}^{\mathsf{CUR}}(\mathsf{A}) < \left(\frac{32\gamma\gamma'}{q}\right)^{nk}$ for every (even unbounded) adversary* A.

4.4 The Dilithium-QROM Signature Scheme and Concrete Parameters

In this section, we describe the signature scheme Dilithium-QROM (Fig. 17) which is obtained via the Fiat-Shamir transform from the scheme ID of Fig. 13 and using $\mathsf{Sam}(K \parallel \cdot)$ as a pseudorandom function. We then instantiate it with concrete parameters (Table 1) and compare them for the same security level with those in [16].

The parameters for our scheme are dictated by the requirements for the scheme to be strongly-unforgeable in Theorem 3.1 which gives an upper bound on $\mathrm{Adv}_{\mathsf{Dilithium\text{-}QROM}}^{\mathsf{sUF\text{-}CMA}}(\mathsf{A})$. Following [24], for "$\kappa$ bits of quantum security" for Dilithium-QROM we require that for all quantum adversaries A running in time at most 2^κ,

$$\mathrm{Adv}_{\mathsf{Dilithium\text{-}QROM}}^{\mathsf{UF\text{-}CMA}}(\mathsf{A})/\mathrm{Time}(\mathsf{A}) \leq 2^{-\kappa}. \quad (9)$$

To this end, we need to put bounds on the parameters $\varepsilon_{\mathsf{ls}}, \varepsilon_{\mathsf{zk}}$, and α. Lemma 4.3 tells us that

$$\varepsilon_{\mathsf{zk}} = 0.$$

To lower-bound α, note that in the parameters, we always have $2\gamma = 2\gamma' < \sqrt{q/2}$, and using a lemma in the full version of the paper, we can conclude that α is greater than 2900. Thus the $2^{-\alpha}$ term has absolutely no practical effect in Theorem 3.1 for the parameters in Sect. 4.4.

Lemma 4.7 states that as long as $4\gamma + 2$ and $2\gamma' < \sqrt{q/2}$, we will have $\mathrm{Adv}_{\mathsf{ID}}^{\mathsf{CUR}}(\mathsf{C}) < \left(\frac{32\gamma\gamma'}{q}\right)^{nk}$. The parameters in Table 1 indeed satisfy the preconditions, and so $\mathrm{Adv}_{\mathsf{ID}}^{\mathsf{CUR}}(\mathsf{C}) < \left(\frac{32\gamma\gamma'}{q}\right)^{nk} < 2^{-865}$.

We finally turn to bounding $\varepsilon_{\mathsf{ls}}$. Notice that Lemma 4.5 directly implies that

$$\varepsilon_{\mathsf{ls}} \leq \frac{1}{|\mathsf{ChSet}|} + 2 \cdot |\mathsf{ChSet}|^2 \cdot \left(\frac{32\gamma\gamma'}{q}\right)^{nk}.$$

The size of the challenge set ChSet defined in Eq. (8) is larger than 2^{265}, and so the above is at most

$$\varepsilon_{\mathsf{ls}} \leq 2^{-265} + 2^{-334} \leq 2^{-264}.$$

Plugging everything into the equation at the end of Sect. 3.1, we obtain

$$\begin{aligned} \mathrm{Adv}^{\mathsf{UF\text{-}CMA}}_{\mathsf{Dilithium\text{-}QROM}}(\mathsf{A}) &\leq \mathrm{Adv}^{\mathsf{LOSS}}_{\mathsf{ID}}(\mathsf{B}) + \mathrm{Adv}^{\mathsf{CUR}}_{\mathsf{ID}}(\mathsf{C}) + 8 \cdot (Q_{\mathsf{H}}+1)^2 \cdot \varepsilon_{\mathsf{ls}} \\ &\quad + \mathrm{Adv}^{\mathsf{PR}}_{\mathsf{Sam}}(\mathsf{D}) + \frac{200}{(1-\delta)} \cdot Q_S \cdot \varepsilon_{\mathsf{zk}} + 2^{-\alpha} \\ &< \mathrm{Adv}^{\mathsf{MLWE}}_{\mathsf{ID}}(\mathsf{B}) + Q_{\mathsf{H}}^2 \cdot 2^{-261} + \mathrm{Adv}^{\mathsf{PR}}_{\mathsf{Sam}}(\mathsf{D}). \end{aligned}$$

Table 1 also shows that the parameters of the MLWE problem are chosen such that it provides 128 bits of quantum security (using the same metric as was used in the original Dilithium scheme [16].) Assuming Sam provides 128 bits security when used as a pseudorandom function, we conclude that for all

$\underline{\mathsf{Sign}((sk, K), M)}$
01 $\kappa := 0$
02 $\mathbf{A} \leftarrow R_q^{k\times\ell} := \mathsf{Sam}(\rho)$
03 **while** $(\mathbf{z}, \mathbf{h}) = \perp$ **and** $\kappa \leq 200/(1-\delta)$ **do**
04 $\quad\kappa := \kappa + 1$
05 $\quad\mathbf{y} \leftarrow S^{\ell}_{\gamma'-1} := \mathsf{Sam}(K \parallel M \parallel \kappa)$
06 $\quad\mathbf{w} := \mathbf{A}\mathbf{y}$
07 $\quad\mathbf{w}_1 := \mathsf{HighBits}_q(\mathbf{w}, 2\gamma)$
08 $\quad c := \mathsf{H}(\mathbf{w}_1 \parallel M)$
09 $\quad\mathbf{z} := \mathbf{y} + c\mathbf{s}_1$
10 $\quad$**if** $\|\mathbf{z}\|_\infty \geq \gamma' - \beta$ **or** $\|\mathsf{LowBits}_q(\mathbf{w} - c\mathbf{s}_2, 2\gamma)\|_\infty \geq \gamma - \beta$ **then** $(\mathbf{z}, \mathbf{h}) := \perp$
11 $\quad$**else** $\mathbf{h} := \mathsf{MakeHint}_q(-c\mathbf{t}_0, \mathbf{w} - c\mathbf{s}_2 + c\mathbf{t}_0, 2\gamma)$
12 **return** $\sigma = (\mathbf{z}, \mathbf{h}, c)$

$\underline{\mathsf{Ver}(pk, M, \sigma = (\mathbf{z}, \mathbf{h}, c))}$
13 $\mathbf{A} \leftarrow R_q^{k\times\ell} := \mathsf{Sam}(\rho)$
14 $\mathbf{w}_1' := \mathsf{UseHint}_q(\mathbf{h}, \mathbf{A}\mathbf{z} - c\mathbf{t}_1 \cdot 2^d, 2\gamma)$
15 **return** $[\![\|\mathbf{z}\|_\infty < \gamma' - \beta]\!]$ **and** $[\![c = \mathsf{H}(\mathbf{w}_1' \parallel M)]\!]$

Fig. 17. Our signature scheme Dilithium-QROM := DFS[ID]. The key generation algorithm is IGen from Fig. 13, where the secret key also contains a random key K for the pseudorandom function $\mathsf{Sam}(K \parallel \cdot)$. The bound $200/(1-\delta)$ on κ can be ignored as there is only a $\delta^{200/(1-\delta)} < \exp(-200)$ chance that it will be reached in any call to the signing procedure. Its presence is for consistency with the generic signing algorithm in Sect. 3.1.

Table 1. Parameters for Dilithium-QROM and Dilithium. The security analysis for the MLWE and MSIS problems is as described in [16].

	Dilithium-QROM		Dilithium [16]	
	Recomm.	Very high	Recomm.	Very high
q (ring modulus)	$2^{45}-21283$	$2^{45}-21283$	$2^{23}-8191$	$2^{23}-8191$
n (ring dimension)	512	512	256	256
(k,ℓ) (dimension of matrix $\mathbf{A}$)	(4, 4)	(5, 5)	(5, 4)	(6, 5)
d (dropped bits from $\mathbf{t}$)	15	15	14[a]	14
# of ± 1's in $c \in$ ChSet	46	46	60	60
γ s.t $2\gamma \mid q-1$	905679	905679	261888	261888
γ' ($\approx$ max. sig. coefficient)	905679	905679	523776	523776
η (maximum coefficient of $\mathbf{s}_1, \mathbf{s}_2$)	7	3	5	3
β ($= \eta\cdot$(# of ± 1's in c))	322	138	275[b]	175
pk size (bytes)	7712	9632	1472	1760
Sig size (bytes)	5690	7098	2701	3366
Exp. repeats ($1/(1-\delta)$ from Lemma 4.4)	4.3	2.2	6.6	4.3
BKZ block-size to break LWE	480	600	485	595
Best known classical bit-cost	140	175	141	174
Best known quantum bit-cost	127	159	128	158
BKZ block-size to break SIS	NA	NA	475	605
Best known classical bit-cost	NA	NA	138	176
Best known quantum bit-cost	NA	NA	125	160

[a]For added compactness of the public key, the size of d (i.e. the amount of bits that one can drop from $\mathbf{t}$) can be such that the necessary condition $\|c\mathbf{t}_0\|_\infty < \gamma$ is not always satisfied. This would invalidate the correctness of the scheme – in particular the proof of Lemma 4.4. Nevertheless, if this condition is satisfied most of the time and the signer simply checks whether $\|c\mathbf{t}_0\|_\infty < \gamma$ before sending the signature (and aborts the signing attempt otherwise), then the scheme retains its correctness property. Since for security, we assumed that $\mathbf{t}_0$ is known to the adversary, this check does not affect security. In the Dilithium scheme, this check is performed at the end of the while loop of the signing algorithm.

[b]The β values for Dilithium were chosen such that $\Pr_{s\leftarrow S_\eta, c\leftarrow \mathsf{ChSet}}[\|sc\|_\infty > \beta]$ is very close to 0. Increasing/decreasing the value of β changes the value δ, which has an effect on the run-time of the scheme.

quantum adversaries running in time at most 2^{128} and making $1 \le Q_\mathsf{H} \le 2^{128}$ (quantum) queries to H, and we have

$$\frac{\mathrm{Adv}^{\mathsf{UF\text{-}CMA}}_{\mathsf{Dilithium\text{-}QROM}}(\mathsf{A})}{\mathrm{Time}(\mathsf{A})} \le \frac{\mathrm{Adv}^{\mathsf{MLWE}}_{\mathsf{ID}}(\mathsf{B})}{\mathrm{Time}(\mathsf{B})} + \frac{\mathrm{Adv}^{\mathsf{PR}}_{\mathsf{Sam}}(\mathsf{D})}{\mathrm{Time}(\mathsf{D})} + Q_\mathsf{H}\cdot 2^{-261} \le 2^{-128}$$

The signature size in Dilithium-QROM is $(n\cdot\ell\cdot(\lceil\log(2\gamma)\rceil)+nk+46\cdot(\log(n)+1))/8$ bytes, while the public key is $(n\cdot k\cdot(\lceil\log(q)\rceil-d)+256)/8$ bytes.

In Table 1, we compare the parameters from the current scheme, which can be proved secure based on the hardness of MLWE in the QROM, to those of the original Dilithium scheme from [16], which only has a classical security reduction from the combination of MLWE and MSIS (we introduce this latter problem in the next section). One can see that the sum of the public key and signature sizes are approximately 3.2 times larger in Dilithium-QROM than in Dilithium.

4.5 Security Assumptions for Non-lossy Schemes

The reduction from the MLWE problem to the hardness of the Dilithium-QROM scheme was a direct consequence of Theorem 3.1, which is itself a combination of Theorems 3.2 and 3.4. In this section, we consider the security of schemes for which Theorem 3.4 is inapplicable. In particular, in these schemes it is no longer true that a computationally-unbounded adversary cannot win the LOSSY-IMP game. The reason that one would like to use schemes constructed in such a manner is because they turn out to be more efficient. In particular, the original Dilithium scheme[4] [16], which is virtually identical to the Dilithium-QROM presented in this paper except for the parameter sizes, has outputs (of the public key plus signature) that are smaller by a factor of a little over 3 (see Table 1).

But while the Dilithium scheme has a security reduction from standard lattice problems in the *classical* random-oracle model, there is no such reduction in the quantum random-oracle model. Nevertheless, it is unclear whether this lack of reduction implies any weakness against quantum attacks. It would therefore be useful to understand exactly what assumptions the more efficient scheme is relying on in the quantum random-oracle model.

Let us suppose that the parameters for the Dilithium scheme are set such that Theorem 3.2 is still applicable. That is, suppose that $\varepsilon_{\mathsf{zk}} = 0$, α is very large, and the scheme is commitment-recoverable. In this case, ignoring the $2^{-\alpha+1}$ term, Theorem 3.2 states that the security of the full signature scheme is exactly the security of the UF-NMA signature scheme in the quantum random-oracle model. Since the adversary does not obtain any valid signatures in the UF-NMA security game, the security assumption of such signatures is non-interactive.

Below, we recall the standard MSIS assumption and then define a new assumption, SelfTargetMSIS, upon which the security of Dilithium is based. We also point out that in the *classical* random-oracle model, there is a (non-tight) reduction from the MSIS to the SelfTargetMSIS problem. Then we show that the Dilithium scheme for which Theorem 3.4 is not necessarily applicable, still has a security reduction from the combination of MLWE and SelfTargetMSIS problems.

THE MSIS AND SelfTargetMSIS PROBLEMS. The MSIS problem [25] is a generalization of the SIS [4] and Ring-SIS [28,33] problems in the same way that MLWE is a generalization of LWE and Ring-LWE. To an algorithm A we associate the advantage function $\mathrm{Adv}^{\mathsf{MSIS}}_{m,k,\gamma}(\mathsf{A})$ to solve the (Hermite Normal Form) $\mathsf{MSIS}_{m,k,\gamma}$ problem over the ring R_q as

$$\mathrm{Adv}^{\mathsf{MSIS}}_{m,k,\gamma}(\mathsf{A}) := \Pr\left[0 < \|\mathbf{y}\|_\infty \leq \gamma \wedge [\mathbf{I} \mid \mathbf{A}] \cdot \mathbf{y} = \mathbf{0} \mid \mathbf{A} \leftarrow R_q^{m\times k}; \mathbf{y} \leftarrow \mathsf{A}(\mathbf{A})\right].$$

As for SIS and Ring-SIS, it was shown that solving MSIS for certain parameters is as hard as worst-case instances of lattice problems over algebraic lattices of a certain form [25].

[4] We refer to the deterministic version of the scheme.

Suppose that $\mathsf{H} : \{0,1\}^* \to \mathsf{ChSet}$ is a cryptographic hash function. To an algorithm A we associate the advantage function $\mathrm{Adv}^{\mathsf{SelfTargetMSIS}}_{\mathsf{H},m,k,\gamma}(\mathsf{A})$ to solve the $\mathsf{SelfTargetMSIS}_{\mathsf{H},m,k,\gamma}$ problem over the ring R_q as

$$\mathrm{Adv}^{\mathsf{SelfTargetMSIS}}_{\mathsf{H},m,k,\gamma}(\mathsf{A}) := \Pr\left[\begin{array}{l}\|\mathbf{y}\|_\infty \leq \gamma \\ \wedge \mathsf{H}([\mathbf{I} \mid \mathbf{A}] \cdot \mathbf{y} \parallel M) = c\end{array} \middle| \mathbf{A} \leftarrow R_q^{m \times k}; \left(\mathbf{y} := \begin{bmatrix}\mathbf{r} \\ c\end{bmatrix}, M\right) \leftarrow \mathsf{A}^{|\mathsf{H}\rangle}(\mathbf{A})\right].$$

If A only has classical access to H, then there is a reduction, using the forking lemma [9,34], to prove that $\mathrm{Adv}^{\mathsf{SelfTargetMSIS}}_{\mathsf{H},m,k,\gamma}(\mathsf{B}) \approx \sqrt{\mathrm{Adv}^{\mathsf{MSIS}}_{m,k,2\gamma}(\mathsf{A})/Q_{\mathsf{H}}}$, where Q_{H} is the number of classical queries to H.[5] This reduction is standard and is implicit in the (classical) security proofs of digital signatures based on the hardness of the SIS problem (cf. [16,27]).

Security based on MLWE, MSIS, and SelfTargetMSIS in the QROM. The QROM security of (deterministic) Dilithium can be expressed as

$$\mathrm{Adv}^{\mathsf{sUF\text{-}CMA}}_{\mathsf{Dilithium}}(\mathsf{A}) \leq \mathrm{Adv}^{\mathsf{MLWE}}_{k,\ell,D}(\mathsf{B}) + \mathrm{Adv}^{\mathsf{SelfTargetMSIS}}_{\mathsf{H},k,\ell+1,\zeta}(\mathsf{C}) \quad (10)$$

$$+ \mathrm{Adv}^{\mathsf{PR}}_{\mathsf{Sam}}(\mathsf{D}) + \mathrm{Adv}^{\mathsf{MSIS}}_{k,\ell,\zeta'}(\mathsf{E}) + 2^{-\alpha+1}, \quad (11)$$

for D a uniform distribution over S_η,

$$\zeta = \max\{\gamma' - \beta, 2\gamma + 1 + 2^{d-1} \cdot \rho\}, \quad (12)$$

where ρ is the number of ± 1's in the challenge set ChSet, and

$$\zeta' = \max\{2(\gamma' - \beta), 4\gamma + 2\}. \quad (13)$$

The proof that the min-entropy α is greater than 255, and the proof for strong unforgeability appears in the full version of the paper. The bound in Eq. (10) is then obtained by combining Theorem 3.2 with results from Sect. 4.3.

Acknowledgments. Eike Kiltz was supported in part by ERC Project ERCC (FP7/615074) and by DFG SPP 1736 Big Data. Vadim Lyubashevsky was supported by the SNSF ERC Transfer Starting Grant CRETP2-166734-FELICITY and the H2020 Project SAFEcrypto. Christian Schaffner was supported by a NWO VIDI grant (639.022.519). The authors are grateful to Dominique Unruh and the anonymous reviewers for comments and discussions.

[5] This can be improved to $Q_{\mathsf{H}}\mathrm{Adv}^{\mathsf{SelfTargetMSIS}}_{\mathsf{H},m,k,\gamma}(\mathsf{B})/\mathrm{Time}(\mathsf{B}) \approx \mathrm{Adv}^{\mathsf{MSIS}}_{m,k,2\gamma}(\mathsf{A})/\mathrm{Time}(\mathsf{A})$.

References

1. NIST Special Publication 800–165 Computer Security Division 2012 Annual Report, p. 39, June 2013. https://csrc.nist.gov/Projects/Post-Quantum-Cryptography. Accessed 30 Jan 2014. 554
2. Abdalla, M., An, J.H., Bellare, M., Namprempre, C.: From identification to signatures via the Fiat-Shamir transform: minimizing assumptions for security and forward-security. In: Knudsen, L.R. (ed.) EUROCRYPT 2002. LNCS, vol. 2332, pp. 418–433. Springer, Heidelberg (2002). https://doi.org/10.1007/3-540-46035-7_28. 553
3. Abdalla, M., Fouque, P.-A., Lyubashevsky, V., Tibouchi, M.: Tightly-secure signatures from lossy identification schemes. In: Pointcheval, D., Johansson, T. (eds.) EUROCRYPT 2012. LNCS, vol. 7237, pp. 572–590. Springer, Heidelberg (2012). https://doi.org/10.1007/978-3-642-29011-4_34. 553, 554, 555, 556, 558, 564, 578
4. Ajtai, M.: Generating hard instances of lattice problems (extended abstract). In: 28th ACM STOC, pp. 99–108. ACM Press, May 1996. 582
5. Alkim, E., Bindel, N., Buchmann, J., Dagdelen, Ö., Eaton, E., Gutoski, G., Krämer, J., Pawlega, F.: Revisiting TESLA in the quantum random oracle model. In: Lange, T., Takagi, T. (eds.) PQCrypto 2017. LNCS, vol. 10346, pp. 143–162. Springer, Cham (2017). https://doi.org/10.1007/978-3-319-59879-6_9. 554, 555, 556, 558
6. Ambainis, A., Rosmanis, A., Unruh, D.: Quantum attacks on classical proof systems: the hardness of quantum rewinding. In: 55th FOCS, pp. 474–483. IEEE Computer Society Press, October 2014. 554
7. Bai, S., Galbraith, S.D.: An improved compression technique for signatures based on learning with errors. In: Benaloh, J. (ed.) CT-RSA 2014. LNCS, vol. 8366, pp. 28–47. Springer, Cham (2014). https://doi.org/10.1007/978-3-319-04852-9_2. 554, 558
8. Beals, R., Buhrman, H., Cleve, R., Mosca, M., Wolf, R.: Quantum lower bounds by polynomials. In: 39th FOCS, pp. 352–361. IEEE Computer Society Press, November 1998. 560
9. Bellare, M., Neven, G.: Multi-signatures in the plain public-key model and a general forking lemma. In: Juels, A., Wright, R.N., Vimercati, S. (eds.) ACM CCS 2006, pp. 390–399. ACM Press, October/November 2006. 553, 583
10. Bellare, M., Poettering, B., Stebila, D.: From identification to signatures, tightly: a framework and generic transforms. In: Cheon, J.H., Takagi, T. (eds.) ASIACRYPT 2016, Part II. LNCS, vol. 10032, pp. 435–464. Springer, Heidelberg (2016). https://doi.org/10.1007/978-3-662-53890-6_15. 556, 565
11. Bellare, M., Rogaway, P.: Random oracles are practical: a paradigm for designing efficient protocols. In: Ashby, V. (ed.) ACM CCS 1993, pp. 62–73. ACM Press, November 1993. 553, 560
12. Bennett, C.H., Bernstein, E., Brassard, G., Vazirani, U.V.: Strengths and weaknesses of quantum computing. SIAM J. Comput. **26**(5), 1510–1523 (1997). 554, 555
13. Boneh, D., Dagdelen, Ö., Fischlin, M., Lehmann, A., Schaffner, C., Zhandry, M.: Random oracles in a quantum world. In: Lee, D.H., Wang, X. (eds.) ASIACRYPT 2011. LNCS, vol. 7073, pp. 41–69. Springer, Heidelberg (2011). https://doi.org/10.1007/978-3-642-25385-0_3. 554, 555, 560

14. Groot Bruinderink, L., Hülsing, A., Lange, T., Yarom, Y.: Flush, gauss, and reload – a cache attack on the BLISS lattice-based signature scheme. In: Gierlichs, B., Poschmann, A.Y. (eds.) CHES 2016. LNCS, vol. 9813, pp. 323–345. Springer, Heidelberg (2016). https://doi.org/10.1007/978-3-662-53140-2_16. 555
15. Ducas, L., Durmus, A., Lepoint, T., Lyubashevsky, V.: Lattice signatures and bimodal Gaussians. In: Canetti, R., Garay, J.A. (eds.) CRYPTO 2013, Part I. LNCS, vol. 8042, pp. 40–56. Springer, Heidelberg (2013). https://doi.org/10.1007/978-3-642-40041-4_3. 555
16. Ducas, L., Kiltz, E., Lepoint, T., Lyubashevsky, V., Schwabe, P., Seiler, G., Stehlé, D.: CRYSTALS-Dilithium: a lattice-based digital signature scheme. IACR Trans. Cryptogr. Hardw. Embed. Syst. **2018**(1), 238–268 (2018). 554, 555, 557, 573, 575, 579, 580, 581, 582, 583
17. Ducas, L., Lyubashevsky, V., Prest, T.: Efficient identity-based encryption over NTRU lattices. In: Sarkar, P., Iwata, T. (eds.) ASIACRYPT 2014, Part II. LNCS, vol. 8874, pp. 22–41. Springer, Heidelberg (2014). https://doi.org/10.1007/978-3-662-45608-8_2. 555, 554
18. Eaton, E., Song, F.: Making existential-unforgeable signatures strongly unforgeable in the quantum random-oracle model. In: 10th Conference on the Theory of Quantum Computation, Communication and Cryptography, TQC 2015, Brussels, Belgium, pp. 147–162, 20–22 May 2015. 554
19. Espitau, T., Fouque, P., Gérard, B., Tibouchi, M.: Side-channel attacks on BLISS lattice-based signatures - exploiting branch tracing against strongSwan and electromagnetic emanations in microcontrollers. IACR Cryptology ePrint Archive 2017, 505 (2017). http://eprint.iacr.org/2017/505. 555
20. Fiat, A., Shamir, A.: How to prove yourself: practical solutions to identification and signature problems. In: Odlyzko, A.M. (ed.) CRYPTO 1986. LNCS, vol. 263, pp. 186–194. Springer, Heidelberg (1987). https://doi.org/10.1007/3-540-47721-7_12. 553
21. Hülsing, A., Rijneveld, J., Song, F.: Mitigating multi-target attacks in hash-based signatures. In: Cheng, C.-M., Chung, K.-M., Persiano, G., Yang, B.-Y. (eds.) PKC 2016, Part I. LNCS, vol. 9614, pp. 387–416. Springer, Heidelberg (2016). https://doi.org/10.1007/978-3-662-49384-7_15. 556, 561
22. Katz, J., Wang, N.: Efficiency improvements for signature schemes with tight security reductions. In: Jajodia, S., Atluri, V., Jaeger, T. (eds.) ACM CCS 2003, pp. 155–164. ACM Press, October 2003. 553, 578
23. Kawachi, A., Tanaka, K., Xagawa, K.: Concurrently secure identification schemes based on the worst-case hardness of lattice problems. In: Pieprzyk, J. (ed.) ASIACRYPT 2008. LNCS, vol. 5350, pp. 372–389. Springer, Heidelberg (2008). https://doi.org/10.1007/978-3-540-89255-7_23. 554, 558
24. Kiltz, E., Masny, D., Pan, J.: Optimal security proofs for signatures from identification schemes. In: Robshaw, M., Katz, J. (eds.) CRYPTO 2016, Part II. LNCS, vol. 9815, pp. 33–61. Springer, Heidelberg (2016). https://doi.org/10.1007/978-3-662-53008-5_2. 554, 579
25. Langlois, A., Stehlé, D.: Worst-case to average-case reductions for module lattices. Des. Codes Cryptogr. **75**(3), 565–599 (2015). 574, 582
26. Lyubashevsky, V.: Fiat-Shamir with aborts: applications to lattice and factoring-based signatures. In: Matsui, M. (ed.) ASIACRYPT 2009. LNCS, vol. 5912, pp. 598–616. Springer, Heidelberg (2009). https://doi.org/10.1007/978-3-642-10366-7_35. 553, 554, 555, 557, 558, 566, 573

27. Lyubashevsky, V.: Lattice signatures without trapdoors. In: Pointcheval, D., Johansson, T. (eds.) EUROCRYPT 2012. LNCS, vol. 7237, pp. 738–755. Springer, Heidelberg (2012). https://doi.org/10.1007/978-3-642-29011-4_43. 583
28. Lyubashevsky, V., Micciancio, D.: Generalized compact knapsacks are collision resistant. In: Bugliesi, M., Preneel, B., Sassone, V., Wegener, I. (eds.) ICALP 2006, Part II. LNCS, vol. 4052, pp. 144–155. Springer, Heidelberg (2006). https://doi.org/10.1007/11787006_13. 582
29. Lyubashevsky, V., Neven, G.: One-shot verifiable encryption from lattices. In: Coron, J.-S., Nielsen, J.B. (eds.) EUROCRYPT 2017, Part I. LNCS, vol. 10210, pp. 293–323. Springer, Cham (2017). https://doi.org/10.1007/978-3-319-56620-7_11. 557, 573
30. Lyubashevsky, V., Peikert, C., Regev, O.: On ideal lattices and learning with errors over rings. In: Gilbert, H. (ed.) EUROCRYPT 2010. LNCS, vol. 6110, pp. 1–23. Springer, Heidelberg (2010). https://doi.org/10.1007/978-3-642-13190-5_1. 574
31. Nielsen, M.A., Chuang, I.L.: Quantum Computation and Quantum Information. Cambridge University Press, Cambridge (2000). 560
32. Paillier, P., Vergnaud, D.: Discrete-log-based signatures may not be equivalent to discrete log. In: Roy, B. (ed.) ASIACRYPT 2005. LNCS, vol. 3788, pp. 1–20. Springer, Heidelberg (2005). https://doi.org/10.1007/11593447_1. 553
33. Peikert, C., Rosen, A.: Efficient collision-resistant hashing from worst-case assumptions on cyclic lattices. In: Halevi, S., Rabin, T. (eds.) TCC 2006. LNCS, vol. 3876, pp. 145–166. Springer, Heidelberg (2006). https://doi.org/10.1007/11681878_8. 582
34. Pointcheval, D., Stern, J.: Security arguments for digital signatures and blind signatures. J. Cryptol. **13**(3), 361–396 (2000). 553, 554, 583
35. Regev, O.: On lattices, learning with errors, random linear codes, and cryptography. In: Gabow, H.N., Fagin, R. (eds.) 37th ACM STOC, pp. 84–93. ACM Press, May 2005. 574
36. Unruh, D.: Non-interactive zero-knowledge proofs in the quantum random oracle model. In: Oswald, E., Fischlin, M. (eds.) EUROCRYPT 2015, Part II. LNCS, vol. 9057, pp. 755–784. Springer, Heidelberg (2015). https://doi.org/10.1007/978-3-662-46803-6_25. 554, 558
37. Unruh, D.: Post-quantum security of Fiat-Shamir. In: Takagi, T., Peyrin, T. (eds.) ASIACRYPT 2017, Part I. LNCS, vol. 10624, pp. 65–95. Springer, Cham (2017). https://doi.org/10.1007/978-3-319-70694-8_3. 555, 556, 558
38. Unruh, D.: Post-quantum security of fiat-shamir. Cryptology ePrint Archive, Report 2017/398 (2017). http://eprint.iacr.org/2017/398. 555, 558, 559
39. Zhandry, M.: How to construct quantum random functions. In: 53rd FOCS, pp. 679–687. IEEE Computer Society Press, October 2012. 561, 556
40. Zhandry, M.: Secure identity-based encryption in the quantum random oracle model. In: Safavi-Naini, R., Canetti, R. (eds.) CRYPTO 2012. LNCS, vol. 7417, pp. 758–775. Springer, Heidelberg (2012). https://doi.org/10.1007/978-3-642-32009-5_44. 560

Non-maleable Codes

Non-malleable Randomness Encoders and Their Applications

Bhavana Kanukurthi[1](✉), Sai Lakshmi Bhavana Obbattu[1](✉), and Sruthi Sekar[2](✉)

[1] Department of Computer Science and Automation, Indian Institute of Science, Bangalore, India
bhavana@iisc.ac.in, oslbhavana@gmail.com

[2] Department of Mathematics, Indian Institute of Science, Bangalore, India
sruthi.sekar1@gmail.com

Abstract. Non-malleable Codes (NMCs), introduced by Dziembowski, Peitrzak and Wichs (ITCS 2010), serve the purpose of preventing "related tampering" of encoded messages. The most popular tampering model considered is the 2-split-state model where a codeword consists of 2 states, each of which can be tampered independently. While NMCs in the 2-split state model provide the strongest security guarantee, despite much research in the area we only know how to build them with poor rate ($\Omega(\frac{1}{logn})$, where n is the codeword length). However, in many applications of NMCs one only needs to be able to encode randomness i.e., security is not required to hold for arbitrary, adversarially chosen messages. For example, in applications of NMCs to tamper-resilient security, the messages that are encoded are typically randomly generated secret keys. To exploit this, in this work, we introduce the notion of "*Non-malleable Randomness Encoders*" (NMREs) as a relaxation of NMCs in the following sense: NMREs output a random message along with its corresponding non-malleable encoding.

Our main result is the construction of a 2-split state, rate $\frac{1}{2}$ NMRE. While NMREs are interesting in their own right and can be directly used in applications such as in the construction of tamper-resilient cryptographic primitives, we also show how to use them, in a black-box manner, to build a 3-split-state (standard) NMCs with rate $\frac{1}{3}$. This improves both the number of states, as well as the rate, of existing constant-rate NMCs.

1 Introduction

How do we protect sensitive data from being tampered? Can we ensure that tampering of the data is detected? These are precisely the kind of questions answered in the rich area of Coding Theory. Encoding data using an error correcting code ensures that data stays the same so long as the errors introduced are appropriately limited. Dziembowski et al. [DPW10], introduced an important variant of

B. Kanukurthi—Research supported in part by Department of Science and Technology Inspire Faculty Award.

J. B. Nielsen and V. Rijmen (Eds.): EUROCRYPT 2018, LNCS 10822, pp. 589–617, 2018.
https://doi.org/10.1007/978-3-319-78372-7_19

ECCs based on the well-established intuition in cryptography that, often times, tampering data into something independent doesn't threaten the security of the underlying cryptosystem. (For example, an adversary who obtains signatures on an independently generated signing key, will not be able to forge signatures with respect to the original secret key.) Specifically, they introduced Non-malleable Codes which provide a guarantee that an adversary cannot tamper the codeword of message m into the codeword of a related message m'.

As observed in [DPW10], it is impossible to build NMCs secure against all functions. Therefore, NMCs are defined with respect to a family of tampering functions. A natural class of tampering functions that have been considered is the t-split state model where a codeword consists of t states, each of which is tampered independently by the adversary. An important parameter of interest for NMCs is its $\mathsf{Rate} = \frac{k}{n}$ where $k = \mathsf{message\ length}$ and $n = \mathsf{codeword\ length}$.

Prior to this work, the best known results for various t-state tampering models were given in Table 1.

Table 1. Prior work

Result	States	Rate
[CG14b]	n	1
[KOS17]	4	1/3
[Li17]	2	$\Omega(\frac{1}{logn})$

As we can see, while 2-split-state NMCs provide the strongest security guarantee, despite significant effort in this direction, we only know how to build them with poor rate of $\Omega(\frac{1}{logn})$. An important observation about the definition of non-malleable codes is that they ensure non-malleability of the codeword of *any* message, even adversarially chosen ones. However, in most applications of non-malleable codes, such as tamper-resilient security, the message is not adversarially controlled. In fact, it is typically a randomly chosen secret key. With that in mind, in this work, we ask the following question:

Is there any advantage in non-malleably encoding randomness?

With this question in mind, we introduce "*Non-malleable Randomness Encoders*" (NMRE) as objects which allow you to generate randomness along with its corresponding non-malleable encoding. We then go on to show that NMREs can, infact, be built efficiently and with interesting parameters: Specifically, we build a **2-state, rate-$\frac{1}{2}$ Non-malleable Randomness Encoder**. Given the major open problem in this area of NMCs i.e., of building an explicit 2-state, constant-rate non-malleable code, we propose NMREs as a useful alternative to NMCs in applications where they suffice, of which we give some examples.

Application of NMREs. Consider the key generation process of symmetric key cryptosystems. These processes typically use uniform randomness r to generate a secret key k. Using NMREs we can generate r along with its non-malleable encoding C. Instead of storing the secret key k directly, we store C in the secret state. The advantage is that this secret state is now resilient to tampering attacks. Of course, this will require us to decode C and regenerate the secret key k whenever we need to use it. Therefore, the applicability of NMREs is for scenarios where key generation is an efficient process.

As another application of NMREs, we show that NMREs can be used, in a black-box, to improve the current state of the art of standard non-malleable codes. Specifically, we build 3-state Non-malleable Codes with a rate of $\frac{1}{3}$.

1.1 Prior Work

We now survey the main results in the area of Non-malleable Codes. For the sake of completeness, we may revisit some of the terminology introduced in the previous section. Informally, a non-malleable code (NMC) [DPW10] provides the following guarantee – a codeword of message m, if tampered, will decode to one of the following:

- $\perp$ i.e., it detects tampering.
- the original message m itself i.e., the tampering did not change the message.
- something independent of m.

While each of these cases may occur with varying probabilities (for example, a tampering function that maps codeword to identity always results in Case 2), the probability with which these cases occur need to be independent of the underlying message. In [DPW10], the authors observe that it is impossible to build NMCs which are secure against unrestricted tampering. Specifically, a function $f(c) \stackrel{\text{def}}{=} \mathsf{Enc}(\mathsf{Dec}(c) + 1)$ clearly tampers $m = \mathsf{Dec}(c)$ into a related $m + 1$. This necessitates the need to define non-malleable codes with respect to the class of functions they protect against. ([DPW10] show the existence of non-malleable codes w.r.t tampering families of size less than 2^{2^n}, where n is the codeword length.)

Tampering Families and Rate. One family that has been considered in several works is that of t-state tampering families: here, a codeword consists of t blocks or states and the adversary tampers each of these independently. The family of functions $\mathcal{F}$ therefore consists of t-functions $f_1, \ldots, f_t$. For $t = n$, the model is referred to as the *bit-wise tampering model.* Dziembowski et al. [DPW10] constructed non-malleable codes resilient against this family. In addition to the class of tampering functions, another important parameter is that of $\mathsf{Rate} = \frac{\mathsf{message\ length}}{\mathsf{codeword\ length}}$ they achieve. Cheraghchi and Guruswami [CG14b] built an explicit construction of an *optimal rate* NMC in the bit-wise tampering model. While building NMCs for this model is technically challenging, a disadvantage is that, from a practical stand point, requiring each bit of the codeword to be stored in an independent state makes the model less desirable. Indeed, the best possible t-split state model would be where $t = 2$. On this front, the first efficient solution was obtained for 1-bit messages by Dziembowski et al. [DKO13]. The first construction for encoding arbitrary-length messages, was an $\Omega(n^{-6/7})$-rate construction due to Aggarwal et al. [ADL14]. At the same time, in [CG14a], Cheraghchi and Guruswami show a $1 - 1/t$ upper bound on best achievable rate for the t-split state family (and, specifically, $1/2$ when $t = 2$). The first constant rate construction for any $t = o(n)$, was due to Chattopadhyay and Zuckerman [CZ14]. Specifically, they build a constant rate, 10-state NMC. Recently, Kanukurthi et al. [KOS17] obtained a 4–state construction (i.e., $t = 4$) with rate $\frac{1}{3}$. For $t = 2$, the current best known construction is due to Li [Li17] with a rate of $\Omega(1/\log n)$. In other results, Aggarwal et al. [ADKO15] demonstrated connections between various split-state models and Agrawal et al. [AGM+15] build optimal NMCs which are simultaneously resilient to permutation attacks as well as bit-wise tampering attacks. On the computational front, there are constructions in the 2-split-state model such as [LL12] and the optimal construction of [AAG+16].

Variants of Non-malleable Codes. Since the introduction of Non-malleable codes several variants of Non-malleable codes have been considered. Some of them are Continuous NMCs [FMNV14, JW15, AKO15, DNO17], Locally updatable and decodable NMCs [DLSZ14, DKS17, CKR16].

1.2 Our Results

In this work, we introduce *Non-malleable Randomness Encoders.* Informally, NMREs allow for the generation of randomness r along with its corresponding non-malleable encoding C. The non-malleability is, as for standard NMCs, defined with respect to $\mathcal{F}$, a family of tampering functions. Note that any non-malleable code NMC is, by default, a secure NMRE (simply generate randomness r at random and let the codeword be the output of NMC). The main challenge is in building a **rate-optimal, state-optimal NMRE**. We give an overview of our construction which uses Information-theoretic *one-time message authentication codes (MACs)* as well as *Randomness Extractors.*

Randomness extractors Ext are objects that allow us to generate randomness from a source W with a Min-entropy guarantee using a short *seed* (s) of true randomness. Message authentication codes $\mathsf{MAC} = (\mathsf{Tag}, \mathsf{Vrfy})$ are secret key primitives which guarantee that even given $\mathsf{Tag}(m; k)$, an adversary cannot generate m', t' such that $m' \neq m$ and $\mathsf{Vrfy}(m', t') = 1$. Our construction makes a black-box use of a 2-split-state non-malleable code NMEnc.

Recall that our goal is to construct a 2-state NMRE with constant rate. For now, consider a 3-state codeword $C = W||L||R$ where $(L, R) \leftarrow NMC(s)$ where W is the source of the extractor and s is a randomly chosen seed. We can see that this is a three-state NMRE resilient to $f_{\mathsf{ID}}, f_2, f_3$ where f_{ID} is the identity function, f_2 and f_3 are arbitrary functions. The idea is that since L, R is the output of an NMC, any independent tampering of L, R respectively renders a tampered s', if not $\perp$, to be independent of s. From here, extractor security can be used– recall that W remains unchanged by our choice of the function family – to argue non-malleability. (This argument isn't trivial. Particularly, to complete it, we must show how $\mathsf{Ext}(W; s')$ can be simulated to complete the proof of non-malleability. While we don't go into the details, it can be done.) Note also that this argument crucially relies on f_1 being f_{ID}. Indeed, if we let W to be tampered to W', then there is no extractor security. (One can come up with concocted constructions of randomness extractors such that tampering w' to a related w and keeping s the same, can result in a related extractor output.) To prevent tampering of W, we use a one-time message authentication code: we let $(L, R) \leftarrow NMC(s, k, \mathsf{Tag}_k(W))$. This gives us a 3-split-state construction ($C = W||L||R$), i.e., one that is resilient to (f_1, f_2, f_3) where each f_i acts independently on each state.

We note that our techniques are similar in spirit to those of [KOS17]'s 4-state NMC. However, our goal here is to build 2-state NMREs. So, on the one hand, we can leverage the fact that the security we are trying to achieve is weaker. On the other hand, the task of bringing down the number of states to 3 while retaining good rate is challenging. To bring down the number of states in our current proposed 3-state NMRE, we wish to explore possibility of combining two of the states. Can we combine W with, say, L? Without going into too much detail regarding the definition of a NMC, an adversary breaking non-malleability can be viewed as consisting of two parts: one that specifies the tampering functions and the other that actually distinguishes the output of the tampering experiment from the simulated experiment.

When we combine W with L, to use the underlying NMC, we would need to be able to do two things: (a) specify the tampering functions that act on L and R and (b) use the distinguisher of the NMRE to build a distinguisher for the NMC. Indeed, the former can be done by merely hardwiring the value of W. Unfortunately, we will not be able to use the distinguisher for the NMRE for the simple reason that we won't know how W was tampered. It is for this reason that we require our NMCs to satisfy a stronger property of "augmented non-malleability". An augmented nonmalleable code is one that remains non-malleable even when the adversary, after specifying the tampering function, additionally obtains one of the states along with the decoded (tampered) message. In our proof, we carefully use the augmented non-malleability of the underlying NMC to argue non-malleability of 2-split state NMRE.

The question still remains of how to instantiate the underlying augmented NMC. We note that the Augmented Non-malleable Codes due to [ADL14] would, asymptotically, indeed give us a constant-rate solution. However, the parameters would be less desirable in terms of tradeoffs between the error and the rate. (Given that this isn't our final construction, a more detailed parameter calculation would be tedious.) To overcome these shortcomings, we instead resort to Li's 2-state construction which has the so-far best-known rate. Since Li only proves the standard non-malleability of his scheme, in Appendix A, we give a proof that it is indeed augmented non-malleable. (This follows by revisiting the connection between seedless non-malleable extractors and non-malleable codes due to [CG14b] and reproving it to achieve augmented non-malleability from strong NME.) Combining this with the outline laid out above, we get our final NMRE construction.

Building NMCs from NMRE as a Black-Box. Our next goal is to use NMREs in a black-box to build NMCs for arbitrary messages m. To do so, we use the "random message" encoded as a part of the NMRE to both compute the ciphertext (using a one-time pad) $c = Enc_{k_e}(m)$ as well as authenticate the ciphertext i.e., compute $t = \mathsf{Tag}_{\mathsf{k}_2}(\mathsf{Enc}_{\mathsf{k}_e}(\mathsf{m}))$. In order to build it in a black-box using the NMRE, it is important that we do not use anything pertaining to the message m in our underlying NMRE. The codeword now needs to have the codeword of NMRE and, additionally, c, t. In the proof, we show that the non-malleability of k_a, k_e essentially suffices to argue the over-all non-malleability and achieve constant rate. Further we show that c, t can stored jointly in a single state giving us a 3-state NMC for arbitrary messages with rate 1/3.

1.3 Organization of the Paper

We write preliminaries and building blocks in Sects. 2 and 3. We give definition of NMRE in Sect. 4.1, an explicit construction of NMRE in Subsect. 4.3, security proof of the construction in Sect. 4.4, instantiate it and analyze rate and error in rest of the Sect. 4. We show how to build a 3-state augmented non-malleable code from an NMRE, prove security, instantiate and analyze in Sects. 5.1, 5.2 and 5.3 respectively. We add concluding remarks in Sect. 6. Appendix B gives details about [Li17]'s 2-state NMC being augmented.

2 Preliminaries

Notation. κ denotes security parameter throughout. $s \in_R S$ denotes uniform sampling from set S. $x \leftarrow X$ denotes sampling from a probability distribution X. $x||y$ represents concatenation of two binary strings x and y. $|x|$ denotes length of binary string x. U_l denotes the uniform distribution on $\{0,1\}^l$. All logarithms are base 2.

Statistical Distance and Entropy. Let X_1, X_2 be two probability distributions over some set S. Their *statistical distance* is

$$\mathbf{SD}\left(X_1, X_2\right) \stackrel{\text{def}}{=} \max_{T \subseteq S}\{\Pr[X_1 \in T] - \Pr[X_2 \in T]\} = \frac{1}{2}\sum_{s\in S}\left|\Pr_{X_1}[s] - \Pr_{X_2}[s]\right|$$

(they are said to be ε-close if $\mathbf{SD}\left(X_1, X_2\right) \leq \varepsilon$ and denoted by $X_1 \approx_\varepsilon X_2$). The *min-entropy* of a random variable W is $\mathbf{H}_\infty(W) = -\log(\max_w \Pr[W = w])$. For a joint distribution (W, E), define the (average) conditional min-entropy of W given E [DORS08] as

$$\widetilde{\mathbf{H}}_\infty(W \mid E) = -\log(\mathop{\mathbf{E}}_{e \leftarrow E}(2^{-\mathbf{H}_\infty(W|E=e)}))$$

(here the expectation is taken over e for which $\Pr[E = e]$ is nonzero). For a random variable W over $\{0,1\}^n$, $W|E$ is said to be an (n, t) - source if $\widetilde{\mathbf{H}}_\infty(W|E) \geq t$.

We now state some Lemmata about statistical distance and average entropy loss.

Proposition 1. *Let $A_1, ..., A_n$ be mutually exclusive and exhaustive events. Then, for probability distributions X_1, X_2 over some set S, we have:*

$$\mathbf{SD}\left(X_1, X_2\right) \leq \sum_{i=1}^{n} \Pr[A_i].\mathbf{SD}\left(X_1|A_i, X_2|A_i\right)$$

where $X_j|A_i$ is the distribution of X_j conditioned on the event A_i.

Lemma 1. *For any random variables A, B, C if $(A, B) \approx_\epsilon (A, C)$, then $B \approx_\epsilon C$.*

Lemma 2. *For any random variables A, B if $A \approx_\epsilon B$, then for any function f, $f(A) \approx_\epsilon f(B)$.*

Lemma 3 *[KOS17]. Let A, B be correlated random variables over $\mathcal{A}, \mathcal{B}$. For randomized functions $F : \mathcal{A} \rightarrow \mathcal{X}$, $G : \mathcal{A} \rightarrow \mathcal{X}$ (randomness used is independent of B) if $\forall\, a \in \mathcal{A}, F(a) \approx_\epsilon G(a)$, then $(B, A, F(A)) \approx_\epsilon (B, A, G(A))$.*

Lemma 4 *[DORS08]. If B has at most 2^λ possible values, then $\widetilde{\mathbf{H}}_\infty(A \mid B) \geq \mathbf{H}_\infty(A, B) - \lambda \geq \mathbf{H}_\infty(A) - \lambda$. and, more generally, $\widetilde{\mathbf{H}}_\infty(A \mid B, C) \geq \widetilde{\mathbf{H}}_\infty(A, B \mid C) - \lambda \geq \widetilde{\mathbf{H}}_\infty(A \mid C) - \lambda$.*

2.1 Definitions

Definition 1. *A (possibly randomized) function* $\mathsf{Enc} : \{0,1\}^l \rightarrow \{0,1\}^n$ *and a deterministic function* $\mathsf{Dec} : \{0,1\}^n \rightarrow \{0,1\}^l \cup \{\perp\}$ *is said to be a* coding scheme *if $\forall\, m \in \{0,1\}^l$, $\Pr[\mathsf{Dec}(\mathsf{Enc}(m)) = m] = 1$. l is called the message length and n is called the block length or the codeword length. Rate of a coding scheme is given by $\frac{l}{n}$.*

We now state the definition of non-malleable codes, as given in [CG14b].

Definition 2. *A coding scheme* $(\mathsf{Enc}, \mathsf{Dec})$ *with message and codeword spaces as $\{0,1\}^l, \{0,1\}^n$ respectively, is ϵ- non-malleable with respect to a function family $\mathcal{F} \subseteq \{f : \{0,1\}^n \rightarrow \{0,1\}^n\}$ if $\forall\, f \in \mathcal{F}$, $\exists$ a distribution Sim_f over $\{0,1\}^l \cup \{same^*, \perp\}$ such that $\forall\, m \in \{0,1\}^l$*

$$\mathsf{Tamper}_f^m \approx_\epsilon \mathsf{Copy}_{Sim_f}^m$$

where Tamper_f^m *denotes the distribution* $\mathsf{Dec}(f(\mathsf{Enc}(m)))$ *and* $\mathsf{Copy}_{Sim_f}^m$ *is defined as*

$$\tilde{m} \leftarrow Sim_f$$

$$\mathsf{Copy}_{Sim_f}^m = \begin{cases} m \; if \; \tilde{m} = same^* \\ \tilde{m} \text{ otherwise} \end{cases}$$

Sim_f should be efficiently samplable given oracle access to $f(.)$.

We now generalize the definition of 2-state augmented-NMC as defined in [AAG+16], to a j-augmented NMC for t-split state family, i.e., j of the t-states is also simulatable independent of the message (where $j < t$).

Definition 3. *A coding scheme* $(\mathsf{Enc}, \mathsf{Dec})$ *with message and codeword spaces as* $\{0,1\}^\alpha, (\{0,1\}^\beta)^t$ *respectively, is* $[\epsilon, j]$*-augmented-non-malleable (where* $j < t$*) with respect to the function family* $\mathcal{F} = \{(f_1, \cdots, f_t) : f_i : \{0,1\}^\beta \to \{0,1\}^\beta\}$ *if* $\forall\ (f_1, \cdots, f_t) \in \mathcal{F}$, $\exists$ *a distribution* $Sim_{f_1,\cdots,f_t}$ *over* $(\{0,1\}^\beta)^j \times (\{0,1\}^\alpha \cup \{same^*, \perp\})$ *such that* $\forall\ m \in \{0,1\}^\alpha$

$$\mathsf{Tamper}^m_{f_1,\cdots,f_t} \approx_\epsilon \mathsf{Copy}^m_{Sim_{f_1,\cdots,f_t}}$$

where $\mathsf{Tamper}^m_{f,g}$ *denotes the distribution* $(X_{i_1}, \cdots, X_{i_j}, \mathsf{Dec}(f_1(X_1), \cdots, f_t(X_t)))$*, where* $\mathsf{Enc}(m) = (X_1, \cdots, X_t)$ *and* $(X_{i_1}, \cdots, X_{i_j})$ *represents some* j *states of the total* t *states.* $\mathsf{Copy}^m_{Sim_{f_1,\cdots,f_t}}$ *is defined as*

$$(X_{i_1}, \cdots, X_{i_j}, \tilde{m}) \leftarrow Sim_{f_1,\cdots,f_t}$$

$$\mathsf{Copy}^m_{Sim_{f_1,\cdots,f_t}} = \begin{cases} (X_{i_1}, \cdots, X_{i_j}, m) \text{ if } (X_{i_1}, \cdots, X_{i_j}, \tilde{m}) = (X_{i_1}, \cdots, X_{i_j}, same^*) \\ (X_{i_1}, \cdots, X_{i_j}, \tilde{m}) \text{ otherwise} \end{cases}$$

$Sim_{f_1,\cdots,f_t}$ *should be efficiently samplable given oracle access to* $(f_1, \cdots, f_t)(.)$.

3 Building Blocks

We use information-theoretic message authentication codes, strong average case extractor and an augmented non-malleable code for 2-split-state family, as building blocks to our construction. We define these building blocks below.

3.1 One-Time Message Authentication Codes

A family of pair of functions $\{\mathsf{Tag}_{k_a} : \{0,1\}^\gamma \to \{0,1\}^\delta,\ \mathsf{Vrfy}_{k_a} : \{0,1\}^\gamma \times \{0,1\}^\delta \to \{0,1\}\}_{k_a \in \{0,1\}^\tau}$ is said to be a $\mu -$ secure one time MAC if

1. For $k_a \in_R \{0,1\}^\tau$, $\forall\ m \in \{0,1\}^\gamma$, $\Pr[\mathsf{Vrfy}_{k_a}(m, \mathsf{Tag}_{k_a}(m)) = 1] = 1$
2. For any $m \neq m', t, t'$, $\Pr_{k_a}[\mathsf{Tag}_{k_a}(m) = t | \mathsf{Tag}_{k_a}(m') = t'] \leq \mu$ for $k_a \in_R \{0,1\}^\tau$

3.2 Average-Case Extractors

Definition 4 *[DORS08, Sect. 2.5]. Let* $Ext : \{0,1\}^n \times \{0,1\}^d \to \{0,1\}^l$ *be a polynomial time computable function. We say that* Ext *is an efficient average-case* (n, t, d, l, ϵ)*-strong extractor if for all pairs of random variables* (W, I) *such that* W *is an* n*-bit string satisfying* $\widetilde{\mathbf{H}}_\infty(W|I) \geq t$*, we have* $\mathbf{SD}((Ext(W; X), X, I), (U, X, I))$*, where* X *is uniform on* $\{0,1\}^d$.

4 Non-malleable Randomness Encoders

We now formally define non-malleable randomness encoding and give a construction for the same.

4.1 Definition

We first formalize the definition of a non-malleable randomness encoder. The goal is to argue that the original message looks random, even given the modified message. But, here the message and the codeword are both generated within the tampering experiment and the experiment outputs the message along with the modified message. This is where the non-malleability definition will defer from the regular NMC Definition 2. We capture the goal by saying that, we are able to simulate the modified message, such that its joint distribution with a message chosen independently uniformly at random is statistically close to the tampering experiment's output. The case where the simulator outputs *same** is a technicality, which we address in the definition below.

Definition 5. *Let* $(\mathsf{NMREnc}, \mathsf{NMRDec})$ *be s.t.* $\mathsf{NMREnc} : \{0,1\}^r \to \{0,1\}^k \times (\{0,1\}^{n_1} \times \{0,1\}^{n_2})$ *is defined as* $\mathsf{NMREnc}(r) = (\mathsf{NMREnc}_1(r), \mathsf{NMREnc}_2(r)) = (m, (x, y))$ *and* $\mathsf{NMRDec} : \{0,1\}^{n_1} \times \{0,1\}^{n_2} \to \{0,1\}^k$.

We say that $(\mathsf{NMREnc}, \mathsf{NMRDec})$ *is a* ϵ*-non-malleable randomness encoder with message space* $\{0,1\}^k$ *and codeword space* $\{0,1\}^{n_1} \times \{0,1\}^{n_2}$*, for the distribution* $\mathcal{R}$ *on* $\{0,1\}^r$ *with respect to the 2-split-state family* $\mathcal{F}$ *if the following is satisfied:*

- ***Correctness:***

$$\Pr_{r \leftarrow \mathcal{R}}[\mathsf{NMRDec}(\mathsf{NMREnc}_2(r)) = \mathsf{NMREnc}_1(r)] = 1$$

- ***Non-malleability:*** *For each* $(f, g) \in \mathcal{F}$, $\exists$ *a distribution* $\mathsf{NMRSim}_{f,g}$ *over* $\{0,1\}^k \cup \{same^*, \perp\}$ *such that*

$$\mathsf{NMRTamper}_{f,g} \approx_\epsilon Copy(U_k, \mathsf{NMRSim}_{f,g})$$

where $\mathsf{NMRTamper}_{f,g}$ *denotes the distribution* $(\mathsf{NMREnc}_1(\mathcal{R}), \mathsf{NMRDec}((f,g)(\mathsf{NMREnc}_2(\mathcal{R}))))$[1] *and* $Copy(U_k, \mathsf{NMRSim}_{f,g})$ *is defined as:*

$$u \leftarrow U_k;\ \tilde{m} \leftarrow \mathsf{NMRSim}_{f,g}$$

$$Copy(u, \tilde{m}) = \begin{cases} (u, u), & \text{if } \tilde{m} = same^* \\ (u, \tilde{m}), & \text{otherwise} \end{cases}$$

$\mathsf{NMRSim}_{f,g}$ *should be efficiently samplable given oracle access to* $(f, g)(.)$.

Further, the rate of this code is defined as $k/(n_1 + n_2)$.

[1] Here $(f, g)(\mathsf{NMREnc}_2(\mathcal{R}))$ just denotes the tampering by the split-state tampering functions f and g on the corresponding states.

While the non-malleability condition above, in flavor, resembles the seedless non-malleable extractors (the decoder function in the above protocol behaves like a seedless non-malleable extractor), the key difference is that, here the two states being tampered are correlated (through the encoder), while in a 2-source seedless NME, the sources need to be independent.

4.2 Notation

- NMEnc, NMDec be an $[\varepsilon_1, 1]$-augmented-non-malleable code for 2-split state family over message and codeword spaces as $\{0,1\}^\alpha$, $\{0,1\}^{\beta_1} \times \{0,1\}^{\beta_2}$ respectively (as in Definition 3), with the message length α and the length of the 2 states, β_1, β_2, respectively. $\mathsf{NMTamper}^{\mathsf{m}}_{\mathsf{f,g}}$, $\mathsf{NMSim}_{\mathsf{f,g}}$ denote the tampered message distribution of m and the simulator of NMEnc, NMDec with respect to tampering functions (f, g).
- Tag′, Vrfy′ be an information theoretic ε_2-secure one time MAC over key, message, tag spaces as $\{0,1\}^{\tau_1}, \{0,1\}^n, \{0,1\}^{\delta_1}$ respectively.
- Ext be an $(n, t, d, l + \tau, \varepsilon_3)$ average case strong extractor.

The parameters will be chosen such that $\alpha = \tau_1 + \delta_1 + d$ and $n > 2 + l + \tau + t$. (Refer to Sect. 4.5 for details).

4.3 Construction Overview

We now build a non-malleable randomness encoder, where the randomness is generated as the output of an extractor. To encode the seed, we use a regular 2-state aug-NMC. As mentioned in the introduction, in order to ensure that the source is not modified, when the seed is the same, we authenticate it using a MAC and encode the MAC key and tag along with the seed. In addition, to obtain a 2-state code, we combine the source with one of the states of the underlying aug-NMC.

$\mathsf{NMREnc}(r)$:	$\mathsf{NMRDec}(L\|\|w, R)$:
– Parse r as $s\|\|w\|\|k_{a_1}$ – $k_e\|\|k_{a_2} = \mathsf{Ext}(w; s)$ – $t_1 = \mathsf{Tag'}_{k_{a_1}}(w)$ – $(L, R) \leftarrow \mathsf{NMEnc}(k_{a_1}\|\|t_1\|\|s)$ – O/P: $(k_e\|\|k_{a_2}, (L\|\|w, R))$	– $k_{a_1}\|\|t_1\|\|s = \mathsf{NMDec}(L, R)$ – If $k_{a_1}\|\|t_1\|\|s = \perp$ output $\perp$ – else if $\mathsf{Vrfy'}_{k_{a_1}}(w, t_1) = 1$ Output $\mathsf{Ext}(w, s)$ else Output $\perp$

Theorem 1. *Let* NMEnc, NMDec *be an* $[\varepsilon_1, 1]$*-augmented-non-malleable code for the 2-split state family,* Tag′, Vrfy′ *be an information theoretic* ε_2*-secure one time* MAC *given above. Let* Ext *be an* $(n, t, d, l + \tau, \varepsilon_3)$ *average case strong extractor. Let* $\alpha = \tau_1 + \delta_1 + d$ *and* $n > 2 + l + \tau + t$.

Then (NMREnc, NMRDec) *is a non-malleable randomness encoding for the uniform distribution on* $\{0,1\}^{d+n+\tau_1}$*, with respect to the 2-split-state family.*

Further, the above construction can be instantiated, as in Sect. 4.5, to achieve a constant rate of $\frac{1}{2+\zeta}$*, for any* $\zeta > 0$ *and an error of* $2^{-\Omega(l/\log^{\rho+1} l)}$*, for any* $\rho > 0$.

Proof. We give the proof in two steps. Firstly, we prove that the proposed construction is a non-malleable randomness encoding scheme (Sect. 4.4). Secondly, we set the parameters to achieve the desired rate and error (Sect. 4.5).

4.4 Security Proof

Define the 2-split-state tampering family for the above construction as

$$\mathcal{F} = \{(f,g) : f : \{0,1\}^{\beta_1} \times \{0,1\}^n \to \{0,1\}^{\beta_1} \times \{0,1\}^n, g : \{0,1\}^{\beta_2} \to \{0,1\}^{\beta_2}\}$$

Correctness of the construction follows by its definition.

To show that (NMREnc, NMRDec) satisfies non-malleability, we need to show that $\forall\ (f,g) \in \mathcal{F}$, $\exists$ $\mathsf{NMRSim}_{f,g}$ such that

$$\mathsf{NMRTamper}_{f,g} \approx_\varepsilon \mathit{Copy}(U_k, \mathsf{NMRSim}_{f,g}).$$

Let $f, g \in \mathcal{F}$. We define the simulator $\mathsf{NMRSim}_{f,g}$ as follows:

$\mathsf{NMRSim}_{f,g}$:

1. $w \in_R \{0,1\}^n$
2. $(L, \tilde{k_{a_1}}||\tilde{t_1}||\tilde{s}) \leftarrow \mathsf{NMSim}_{f_w,g}$
   // where f_w is the function f with w hardcoded.//
3. $\tilde{w} = f_L(w)$
   // where f_L is the function f with L hardcoded.//
4. If $\tilde{k_{a_1}}||\tilde{t_1}||\tilde{s} = \mathit{same}^*$:
   - If $\tilde{w} = w$ output same^*
   - else output $\perp$
5. Else if $\mathsf{Vrfy'}_{\tilde{k_{a_1}}}(\tilde{w}, \tilde{t_1}) = 1$ output $\mathsf{Ext}(\tilde{w}; \tilde{s})$
6. Else output $\perp$

We now prove the closeness of $\mathsf{NMRTamper}_{f,g}$ and $\mathit{Copy}(U_k, \mathsf{NMRSim}_{f,g})$ through a sequence of hybrids:

$\mathsf{NMRTamper}_{f,g}$:

1. $r \in_R \{0,1\}^{d+n+\tau_1}$;
   Parse r as $s||w||k_{a_1}$
2. $t_1 = \mathsf{Tag'}_{k_{a_1}}(w)$
3. $(L, \tilde{k_{a_1}}||\tilde{t_1}||\tilde{s}) \leftarrow \mathsf{NMTamper}_{f_w,g}^{k_{a_1}||t_1||s}$
4. $\tilde{w} = f_L(w)$
5. $k_e||k_{a_2} = \mathsf{Ext}(w; s)$
6. If $\mathsf{Vrfy'}_{\tilde{k_{a_1}}}(\tilde{w}, \tilde{t_1}) = 1$
   output $k_e||k_{a_2}, \mathsf{Ext}(\tilde{w}; \tilde{s})$
7. Else output $k_e||k_{a_2}, \perp$.

$\mathsf{Hybrid1}_{f,g}$:

1. $r \in_R \{0,1\}^{d+n+\tau_1}$
   Parse r as $s||w||k_{a_1}$
2. $t_1 = \mathsf{Tag'}_{k_{a_1}}(w)$
3. $(L, \tilde{k_{a_1}}||\tilde{t_1}||\tilde{s}) \leftarrow \mathsf{NMSim}_{f_w,g}$
   If $\tilde{k_{a_1}}||\tilde{t_1}||\tilde{s} = \mathit{same}^*$,
   set $\tilde{k_{a_1}}||\tilde{t_1}||\tilde{s} = k_{a_1}||t_1||s$
4. $\tilde{w} = f_L(w)$
5. $k_e||k_{a_2} = \mathsf{Ext}(w; s)$
6. If $\mathsf{Vrfy'}_{\tilde{k_{a_1}}}(\tilde{w}, \tilde{t_1}) = 1$
   output $k_e||k_{a_2}, \mathsf{Ext}(\tilde{w}; \tilde{s})$
   Else output $k_e||k_{a_2}, \perp$.

Hybrid2$_{f,g}$:	Hybrid3$_{f,g}$:
1. $s\|\|w \in_R \{0,1\}^{d+n}$	1. $w \in_R \{0,1\}^n$
2. $(L, \tilde{k_{a_1}}\|\|\tilde{t_1}\|\|\tilde{s}) \leftarrow \mathsf{NMSim}_{f_w,g}$	2. $(L, \tilde{k_{a_1}}\|\|\tilde{t_1}\|\|\tilde{s}) \leftarrow \mathsf{NMSim}_{f_w,g}$
3. $\tilde{w} = f_L(w)$	3. $\tilde{w} = f_L(w)$
4. $k_e\|\|k_{a_2} = \mathsf{Ext}(w; s)$	4. $k_e\|\|k_{a_2} \in_R \{0,1\}^{l+\tau}$
5. If $\tilde{k_{a_1}}\|\|\tilde{t_1}\|\|\tilde{s} = same^*$:	5. If $\tilde{k_{a_1}}\|\|\tilde{t_1}\|\|\tilde{s} = same^*$:
• If $\tilde{w} = w$ output $k_e\|\|k_{a_2}, k_e\|\|k_{a_2}$	• If $\tilde{w} = w$ output $k_e\|\|k_{a_2}, k_e\|\|k_{a_2}$
• else output $k_e\|\|k_{a_2}, \perp$	• else output $k_e\|\|k_{a_2}, \perp$
Else if $\mathsf{Vrfy}'_{\tilde{k_{a_1}}}(\tilde{w}, \tilde{t_1}) = 1$	Else if $\mathsf{Vrfy}'_{\tilde{k_{a_1}}}(\tilde{w}, \tilde{t_1}) = 1$
output $k_e\|\|k_{a_2}, \mathsf{Ext}(\tilde{w}; \tilde{s})$	output $k_e\|\|k_{a_2}, \mathsf{Ext}(\tilde{w}; \tilde{s})$
Else output $k_e\|\|k_{a_2}, \perp$	Else output $k_e\|\|k_{a_2}, \perp$

Claim 1. If $(\mathsf{NMEnc}, \mathsf{NMDec})$ is a ε_1-augmented-non-malleable code, then $\mathsf{NMRTamper}_{f,g} \approx_{\varepsilon_1} \mathsf{Hybrid1}_{f,g}$.

Proof. By augmented non-malleability of $(\mathsf{NMEnc}, \mathsf{NMDec})$, we get

$$\mathsf{NMTamper}_{f_w,g}^{k_{a_1}||t_1||s} \approx_{\varepsilon_1} Copy_{\mathsf{NMSim}_{f_w,g}}^{k_{a_1}||t_1||s}$$

By using Lemma 3, we get

$$w, k_{a_1}||t_1||s, \mathsf{NMTamper}_{f_w,g}^{k_{a_1}||t_1||s} \approx_{\varepsilon_1} w, k_{a_1}||t_1||s, Copy_{\mathsf{NMSim}_{f_w,g}}^{k_{a_1}||t_1||s}$$

Now, the outputs of $\mathsf{NMRTamper}_{f,g}$ and $\mathsf{Hybrid1}_{f,g}$ are deterministic functions of above random variables. Hence, by Lemma 2, we get

$$\mathsf{NMRTamper}_{f,g} \approx_{\varepsilon_1} \mathsf{Hybrid1}_{f,g}$$

Claim 2. If $(\mathsf{Tag}', \mathsf{Vrfy}')$ is an information theoretic ε_2-secure one time MAC, then $\mathsf{Hybrid1}_{f,g} \approx_{\varepsilon_2} \mathsf{Hybrid2}_{f,g}$.

Proof. If $same^*$ is not the value sampled from NMSim_{h_1,h_2}, then the output of the two hybrids are identical. Therefore, the statistical distance is zero in this case. When $same^*$ is sampled, the key difference between $\mathsf{Hybrid1}_{f,g}$ and $\mathsf{Hybrid2}_{f,g}$ is that, corresponding to this case, we remove the two verify checks in $\mathsf{Hybrid2}_{f,g}$ and simply replace it with the equality checks. Intuitively, in this case, the statistical closeness would hold due to unforgeability of MAC. The full proof can be found in Appendix A.1.

Claim 3. If Ext is an $(n, t, d, l+\tau, \varepsilon_3)$ average case extractor, then $\mathsf{Hybrid2}_{f,g} \approx_{\varepsilon_3} \mathsf{Hybrid3}_{f,g}$.[2]

[2] We refer the reader to Appendix A.2 for an alternate proof of this claim.

Proof. We first consider the following random variables, which capture the auxiliary information. We then use extractor security and Lemma 2 to prove the closeness of the two hybrids.

We consider the output of $\mathsf{NMSim}_{f_w,g}$, which is $(L, \tilde{k_{a_1}}||\tilde{t_1}||\tilde{s})$ and define the following random variables, dependent on this:

We start with b_{same^*}, which indicates whether $\mathsf{NMSim}_{f_w,g}$ has output *same** or not

$$b_{same^*} = \begin{cases} 1 & \text{if } \tilde{k_{a_1}}||\tilde{t_1}||\tilde{s} = same^* \\ 0 & \text{otherwise} \end{cases}$$

Further, $b_\perp$ is an indicator of whether $\mathsf{NMSim}_{f_w,g}$ output $\perp$ or not.

$$b_\perp = \begin{cases} 1 & \text{if } \tilde{k_{a_1}}||\tilde{t_1}||\tilde{s} = \perp \\ 0 & \text{otherwise} \end{cases}$$

We also have:

$$eq(w) = \begin{cases} 0 \text{ if } & f_L(w) \neq w \\ 1 \text{ if } & f_L(w) = w \end{cases}$$

which is an indicator of whether $\tilde{w}$ is modified or not. And,

$$Verify(w) = \mathsf{Vrfy'}_{\tilde{k_{a_1}}}(f_L(w), \tilde{t_1})$$

which is the indicator of the MAC verification bit.
Further define:

$$Y(w, b_1, b_2) := \begin{cases} eq(w) & \text{if } b_1 = 1 \\ (Verify(w), \mathsf{Ext}(\tilde{w}; \tilde{s})) & \text{if } b_1 = 0 \wedge b_2 = 0 \\ \perp & \text{otherwise} \end{cases}$$

We now define the auxiliary information by $\hat{E} = (b_{same^*}, b_\perp, Y(W, b_{same^*}, b_\perp))$.

We now define the following function
$G(e, k)$:

- Parse $e = (b_{same^*}, b_\perp, y = Y(w, b_{same^*}, b_\perp))$[3].
- If $b_{same^*} = 1$:
  - If $y = 1$, output (k, k)
  - Else output $(k, \perp)$
- Else:
  - If $b_\perp = 1$, output $(k, \perp)$.
  - Else parse $y = (Verify(w), \mathsf{Ext}(\tilde{w}; \tilde{s}))$.
    * if $Verify(w) = 1$ output $(k, \mathsf{Ext}(\tilde{w}; \tilde{s}))$
    * else output $(k, \perp)$

[3] Here, we abuse the notation: b_{same^*} and $b_\perp$ represent the particular values taken by the corresponding random variables.

The outputs of $\mathsf{Hybrid2}_{f,g}$ and $\mathsf{Hybrid3}_{f,g}$ are $G(\hat{E}, \mathsf{Ext}(W; S))$ and $G(\hat{E}, U_{l+\tau})$ respectively, where G is deterministic. So, to prove this claim it suffices to show

$$\hat{E}, \mathsf{Ext}(W; S) \approx_{\varepsilon_3} \hat{E}, U_{l+\tau} \tag{1}$$

Observe that $\hat{E}$ depends on $\mathsf{NMSim}_{f_w,g}$ and w, which are independent of the seed s. Therefore it can be captured as auxiliary information. $\hat{E}$ takes at most $2^{3+l+\tau}$ possible values. Hence, $\widetilde{\mathbf{H}}_\infty(W|\hat{E}) \geq \mathbf{H}_\infty(W) - (3+l+\tau) = n-(3+l+\tau)$, by Lemma 4. As $n-(3+l+\tau) > t$ (due to the way we set parameters in Sect. 4.5), by security of average case extractor, Eq. 1 holds. This proves the claim.

From above Claims 1, 2 and 3, we get:

$$\mathsf{NMRTamper}_{f,g} \approx_{\varepsilon_1} \mathsf{Hybrid1}_{f,g} \approx_{\varepsilon_2} \mathsf{Hybrid2}_{f,g} \approx_{\varepsilon_3} \mathsf{Hybrid3}_{f,g} \equiv \mathit{Copy}(U_k, \mathsf{NMRSim}_{f,g})$$
$$\text{i.e., } \mathsf{NMRTamper}_{f,g} \approx_{\varepsilon_1+\varepsilon_2+\varepsilon_3} \mathit{Copy}(U_k, \mathsf{NMRSim}_{f,g})$$

4.5 Rate and Error Analysis

We now present the details of the rate of the code as well as the error it achieves. We instantiate the above construction using specific MAC construction, average case extractor Ext and non-malleable code (NMEnc, NMDec), as given in the lemmata below.

As we are encoding the seed of the extractor using the underlying non-malleable code, it is important that the strong extractor we use has short seed length. This is guaranteed by the following lemma.

Lemma 5 *[GUV07]. For every constant $\nu > 0$ all integers $n \geq t$ and all $\epsilon \geq 0$, there is an explicit (efficient) (n, t, d, l, ϵ)–strong extractor with $l = (1-\nu)t - \mathcal{O}(\log(n) + \log(\frac{1}{\epsilon}))$ and $d = \mathcal{O}(\log(n) + \log(\frac{1}{\epsilon}))$.*

Now, as we give some auxiliary information about the source, we require the security of the extractor to hold, even given this information. Hence, we use average case extractors, given in the following lemma.

Lemma 6 *[DORS08]. For any $\mu > 0$, if* Ext *is a (worst case)(n, t, d, l, ϵ)–strong extractor, then* Ext *is also an average-case $(n, t+\log(\frac{1}{\mu}), d, l, \epsilon+\mu)$ strong extractor.*

We now combine the Lemmata 5 and 6 to get an average case extractor with optimal seed length.

Corollary 1. *For any $\mu > 0$ and every constant $\nu > 0$ all integers $n \geq t$ and all $\epsilon \geq 0$, there is an explicit (efficient) $(n, t+\log(\frac{1}{\mu}), d, l, \epsilon+\mu)$– average case strong extractor with $l = (1-\nu)t - \mathcal{O}(\log(n) + \log(\frac{1}{\epsilon}))$ and $d = \mathcal{O}(\log(n) + \log(\frac{1}{\epsilon}))$.*

Now, we also encode the authentication keys and tags using the underlying non-malleable code. Hence, we require them to have short lengths. This is guaranteed by the following lemma [JKS93]:

Lemma 7. *For any $n', \varepsilon_2 > 0$ there is an efficient ε_2–secure one time* MAC *with $\delta \leq (\log(n') + \log(\frac{1}{\varepsilon_2}))$, $\tau \leq 2\delta$, where τ, n', δ are key, message, tag length respectively.*

Further, we use the 2-split-state non-malleable code by [Li17] to instantiate our construction.

Lemma 8 *[Li17, Theorem 7.12]. For any $\beta \in \mathbb{N}$ there exists an explicit non-malleable code with efficient encoder/decoder in 2-split state model with block length 2β, rate $\Omega\left(\frac{1}{\log\ \beta}\right)$ and error $= 2^{-\Omega\left(\frac{\beta}{\log\ \beta}\right)}$.*

Further, we show in Appendix B (Corollary 2) that the construction corresponding to Lemma 8 is in fact an $[2^{-\Omega(\beta/\log\ \beta)}, 1]$-augmented-non-malleable code for the two split-state family with the same rate as above.

4.5.1 Setting parameters

We instantiate our construction using (NMEnc, NMDec) as in Corollary 2, strong average case extractors, as in Corollary 1 and one time information theoretic MAC, as in Lemma 7.

- We set the error parameters as $\epsilon, \mu, \varepsilon_1, \varepsilon_2 = 2^{-\lambda}$ and $\varepsilon_3 = \epsilon + \mu$.
- The message length and codeword length in the construction of (NMREnc, NMRDec) above, are $l + \tau$ and $2\beta + n$ respectively. Here we take k_{a_2} to be of size $\tau = \mathcal{O}(\log l + \lambda)$.
- We estimate the length of the source (n). As we saw in the Claim 3 of the proof (Sect. 4.4), we leak auxiliary information of length at most $3 + l + \tau$. Hence, by Lemma 4, the average entropy of the source, given auxiliary information is $\geq n - (3 + l + \tau)$.

  To use extractor security, we require that the average entropy is at least the entropy threshold $t + log(\frac{1}{\mu})$, i.e., $n - (3 + l + \tau) \geq t + log(\frac{1}{\mu})$.

  By Corollary 1 (with output length of extractor $l + \tau$), we have $t = (l + \tau + \mathcal{O}(\log(n) + \log(\frac{1}{\epsilon})))\frac{1}{1-\nu}$.

  Hence, taking ν as a very small constant close to 0, we get: for some constant ζ close to 0

$$n = (2 + \zeta)l + \mathcal{O}(\log l + \lambda) \tag{2}$$

- We now estimate the codeword length 2β, of the underlying NMC.

  The message size for this codeword is $\alpha = \tau_1 + \delta_1 + d$. By Lemma 7 and Corollary 1, we get $\alpha = \mathcal{O}(\log(l) + \lambda)$.

  By using the rate in Lemma 8, we get:

$$\beta = \mathcal{O}((\log(l))^2 + \lambda \log(\lambda) + 2\lambda \log(l)) \tag{3}$$

4.5.2 Rate

The rate of our construction of non-malleable randomness encoding is:

$$R = \frac{l + \tau}{2\beta + n}$$

By substituting n and β from Eqs. 2 and 3, respectively and τ as described above, we get:

$$R = \frac{l + \mathcal{O}(\log l + \lambda)}{\mathcal{O}((\log(l))^2 + \lambda \log(\lambda) + 2\lambda \log(l)) + (2 + \zeta)l + \mathcal{O}(\log l + \lambda)}$$

For large l, and taking $\lambda = o(\frac{l}{\log l})$, we get

$$R \geq \frac{1}{2 + \zeta}$$

Hence, the construction given achieves rate atleast $\frac{1}{2+\zeta}$, for some ζ close to 0.

4.5.3 Error

Error of the protocol, as seen in the proof, is $\varepsilon_1 + \varepsilon_2 + \varepsilon_3 = 4(2^{-\lambda})$. Since, $\lambda = o(\frac{l}{\log l})$, the error will be at least $2^{-\frac{l}{\log l}}$. For any $\rho > 0$, fixing $\lambda = \frac{l}{\log^{\rho+1} l}$, the error would be at most $4.2^{-\frac{l}{\log^{\rho+1} l}}$. Setting $\kappa = \lambda - \log 5$ the error would be $2^{-\kappa} = 2^{-\Omega(l/\log^{\rho+1} l)}$.

5 Non-malleable Codes from Non-malleable Randomness Encoders

As an application of non-malleable randomness encoding, we build a 3-state 1-augmented-non-malleable code, using non-malleable randomness encoding in black-box. For achieving an explicit constant rate and a specific error, we instantiate the construction using the construction in Sect. 4.

5.1 Construction Overview

To encode the message, we first hide the message using one part of the randomness generated in the underlying NMRE. To ensure that this ciphertext is not modified, we authenticate it using a MAC. We show that we can use NMRE's "random messages" as the keys for encryption as well authentication. The fact that the tag t does not need to be non-malleably encoded, and can instead be combined with c, is what allows us to get a 3-state NMC construction while only making a **black-box** use of the underlying NMRE. Details follow.

$\mathsf{AEnc}(m)$	$\mathsf{ADec}(\tilde{y}_1, \tilde{y}_2, \tilde{c}\|\|\tilde{t})$
– $r \in_R \{0,1\}^{r'}$	– $\tilde{k}_e\|\|\tilde{k}_a = \mathsf{NMRDec}(\tilde{y}_1, \tilde{y}_2)$
– $(k_a\|\|k_e, y_1, y_2) \leftarrow \mathsf{NMREnc}(r)$	– If $\mathsf{Vrfy}_{\tilde{k}_a}(\tilde{c}, \tilde{t}) = 1$
– $c = m \oplus k_e$	Output $\tilde{c} \oplus \tilde{k}_e$
– $t = \mathsf{Tag}_{k_a}(c)$	else Output $\perp$
– Output $(y_1, y_2, c\|\|t)$	

Theorem 2. *Let* (NMREnc, NMRDec) *be a 2-state* ϵ_1*-non-malleable randomness encoding scheme for the uniform distribution on* $\{0,1\}^{r'}$, *for messages in* $\{0,1\}^{l+\tau}$ *and* (Tag, Vrfy) *be an information theoretic* ϵ_2*-secure one-time* MAC *with key, message and tag spaces being* $\{0,1\}^{\tau}, \{0,1\}^{l}, \{0,1\}^{\delta}$. *Then* (AEnc, ADec), *as defined above, is a 3-state* $[\epsilon_1 + \epsilon_2, 1]$*-augmented non-malleable code for messages of length* l *(with the augmented state being* $c\|\|t$*).*

Further, instantiating the construction with (NMREnc, NMRDec) *achieving rate and error, as in Sect. 4.5, we can achieve a constant rate of* $\frac{1}{3+\zeta}$, *for any* $\zeta > 0$ *and an error of* $2^{-\Omega(l/\log^{\rho+1} l)}$, *for any* $\rho > 0$.

5.2 Security Proof

Let $(f_1, f_2, g) \in \mathcal{F}_3$(3-split state tampering family) where $f_1 : \{0,1\}^{\beta_1} \rightarrow \{0,1\}^{\beta_1}, f_2 : \{0,1\}^{\beta_2} \rightarrow \{0,1\}^{\beta_2}, g : \{0,1\}^{l+\delta} \rightarrow \{0,1\}^{l+\delta}$. We propose the following distribution as simulator for $(AEnc, ADec)$.

$\mathsf{ASim}_{f_1,f_2,g}$
– $k_e\|\|k_a \in_R \{0,1\}^{l+\tau}$
– $\tilde{k}_e\|\|\tilde{k}_a \leftarrow \mathsf{NMRSim}_{f_1,f_2}$
– $c = 0 \oplus k_e$
– $t = \mathsf{Tag}_{k_a}(c)$
– $\tilde{c}\|\|\tilde{t} = g(c\|\|t)$
– If $\tilde{k}_e\|\|\tilde{k}_a = same^*$
If $\tilde{c} = c$, Output $c, t, same^*$
Else output $c, t, \perp$
Else if $\mathsf{Vrfy}_{\tilde{k}_a}(\tilde{c}, \tilde{t}) = 1$
Output $c, t, \tilde{c} \oplus \tilde{k}_e$
Else Output $c, t, \perp$

We prove that $ASim_{f_1,f_2,g}$ is the simulator of $(AEnc, ADec)$ through a sequence of hybrids.

Claim 1. *If* (NMREnc, NMRDec) *is a non-malleable randomness encoding scheme, then* $\mathsf{ATamper}^m_{f_1,f_2,g} \approx_{\epsilon_1} \mathsf{Hybrid1}^m_{f_1,f_2,g}$.

Proof. By non-malleability of (NMREnc, NMRDec), we have

$$\mathsf{NMRTamper}_{f_1,f_2} \approx_{\epsilon_1} Copy(U_k, \mathsf{NMRSim}_{f_1,f_2})$$

As m is independent we have

$$m, \mathsf{NMRTamper}_{f_1,f_2} \approx_{\epsilon_1} m, Copy(U_k, \mathsf{NMRSim}_{f_1,f_2})$$

By Lemma 2 we have,

$$m, c, t, \mathsf{NMRTamper}_{f_1,f_2} \approx_{\epsilon_1} m, c, t, Copy(U_k, \mathsf{NMRSim}_{f_1,f_2})$$

The outputs of $\mathsf{ATamper}^m_{f_1,f_2,g}$, $\mathsf{Hybrid1}^m_{f_1,f_2,g}$ are determined by a deterministic function of above distributions. Therefore by Lemma 2 we have

$$m, \mathsf{ATamper}^m_{f_1,f_2,g} \approx_{\epsilon_1} m, \mathsf{Hybrid1}^m_{f_1,f_2,g}$$

$\mathsf{ATamper}^m_{f_1,f_2,g}$	$\mathsf{Hybrid1}^m_{f_1,f_2,g}$
– $k_e\|\|k_a, \tilde{k}_e\|\|\tilde{k}_a \leftarrow \mathsf{NMRTamper}_{f_1,f_2}$ – $c = m \oplus k_e$ – $t = \mathsf{Tag}_{k_a}(c)$ – $\tilde{c}\|\|\tilde{t} = g(c\|\|t)$ – If $\mathsf{Vrfy}_{\tilde{k}_a}(\tilde{c}, \tilde{t}) = 1$ Output $c, t, \tilde{c} \oplus \tilde{k}_e$ Else Output $c, t, \perp$	– $k_e\|\|k_a \in_R \{0,1\}^{l+\tau}$ – $\tilde{k}_e\|\|\tilde{k}_a \leftarrow \mathsf{NMRSim}_{f_1,f_2}$ – If $\tilde{k}_e\|\|\tilde{k}_a = same^*$ set $\tilde{k}_e\|\|\tilde{k}_a = k_e\|\|k_a$ – $c = m \oplus k_e$ – $t = \mathsf{Tag}_{k_a}(c)$ – $\tilde{c}\|\|\tilde{t} = g(c\|\|t)$ – If $\mathsf{Vrfy}_{\tilde{k}_a}(\tilde{c}, \tilde{t}) = 1$ Output $c, t, \tilde{c} \oplus \tilde{k}_e$ Else Output $c, t, \perp$

$\mathsf{Hybrid2}^m_{f_1,f_2,g}$	$\mathsf{Hybrid3}^m_{f_1,f_2,g}$
– $k_e\|\|k_a \in_R \{0,1\}^{l+\tau}$ – $\tilde{k}_e\|\|\tilde{k}_a \leftarrow \mathsf{NMRSim}_{f_1,f_2}$ – $c = m \oplus k_e$ – $t = \mathsf{Tag}_{k_a}(c)$ – $\tilde{c}\|\|\tilde{t} = g(c\|\|t)$ – If $\tilde{k}_e\|\|\tilde{k}_a = same^*$ set $\tilde{k}_e\|\|\tilde{k}_a = k_e\|\|k_a$ – If $\mathsf{Vrfy}_{\tilde{k}_a}(\tilde{c}, \tilde{t}) = 1$ Output $c, t, \tilde{c} \oplus \tilde{k}_e$ Else Output $c, t, \perp$	– $k_e\|\|k_a \in_R \{0,1\}^{l+\tau}$ – $\tilde{k}_e\|\|\tilde{k}_a \leftarrow \mathsf{NMRSim}_{f_1,f_2}$ – $c = m \oplus k_e$ – $t = \mathsf{Tag}_{k_a}(c)$ – $\tilde{c}\|\|\tilde{t} = g(c\|\|t)$ – If $\tilde{k}_e\|\|\tilde{k}_a = same^*$ If $\tilde{c} = c$ Output c, t, m Else output $c, t, \perp$ Else if $\mathsf{Vrfy}_{\tilde{k}_a}(\tilde{c}, \tilde{t}) = 1$ Output $c, t, \tilde{c} \oplus \tilde{k}_e$ Else Output $c, t, \perp$

$\mathsf{Hybrid4}^m_{f_1,f_2,g}$

- $k_e||k_a \in_R \{0,1\}^{l+\tau}$
- $\tilde{k_e}||\tilde{k_a} \leftarrow \mathsf{NMRSim}_{f_1,f_2}$
- $c = 0 \oplus k_e$
- $t = \mathsf{Tag}_{k_a}(c)$
- $\tilde{c}||\tilde{t} = g(c||t)$
- If $\tilde{k_e}||\tilde{k_a} = \mathit{same}^*$
  If $\tilde{c} = c$ Output c, t, m
  Else output $c, t, \perp$
  Else if $\mathsf{Vrfy}_{\tilde{k_a}}(\tilde{c}, \tilde{t}) = 1$
  Output $c, t, \tilde{c} \oplus \tilde{k_e}$
  Else Output $c, t, \perp$

Claim 2. $\mathsf{Hybrid1}^m_{f_1,f_2,g} \equiv \mathsf{Hybrid2}^m_{f_1,f_2,g}$.

Proof. The claim trivially follows because $\mathsf{Hybrid2}^m_{f_1,f_2,g}$ is rewriting of $\mathsf{Hybrid1}^m_{f_1,f_2,g}$.

$$k_e, k_a, \mathsf{NMRSim}_{f_1,f_2} \equiv k_e, k_a, \mathsf{NMRSim}_{f_1,f_2}$$
$$m, c, t, k_e, k_a, \mathsf{NMRSim}_{f_1,f_2} \equiv m, c, t, k_e, k_a, \mathsf{NMRSim}_{f_1,f_2}$$
$$m, \mathsf{Hybrid1}^m_{f_1,f_2,g} \equiv m, \mathsf{Hybrid2}^m_{f_1,f_2,g}$$

All equations follow by Lemma 2.

Claim 3. *If* $(\mathsf{Tag}, \mathsf{Vrfy})$ *is an* ϵ_2 *IT-secure-One-time Mac, then* $\mathsf{Hybrid2}^m_{f_1,f_2,g} \approx_{\epsilon_2} \mathsf{Hybrid3}^m_{f_1,f_2,g}$.

Proof. Let E denote the event $\tilde{k_e}, \tilde{k_a} \neq \mathit{same}^*$, and $\tilde{E}$, its compliment. Given E, both the hybrids are identical. Given $\tilde{E}$ the statistical distance of the hybrids is at most

$$\Pr_{k_a}[\mathsf{Vrfy}_{k_a}(\tilde{c}, \tilde{t}) = 1 | t = \mathsf{Tag}_{k_a}(c), \tilde{c}||\tilde{t} = f(c||t)] \leq \epsilon_2$$

Therefore claim follows.

Claim 4. *By semantic security of One Time Pad encryption*

$$\mathsf{Hybrid3}^m_{f_1,f_2,g} \equiv \mathsf{Hybrid4}^m_{f_1,f_2,g}$$

Proof. By semantic security,

$$\begin{aligned} m, m \oplus k_e &\equiv m, 0 \oplus k_e \\ m, t, m \oplus k_e, k_a &\equiv m, t, 0 \oplus k_e, k_a \end{aligned}$$

The outputs of the hybrids 3 and 4 are a randomized function of above distributions. Therefore

$$\mathsf{Hybrid3}^m_{f_1,f_2,g} \equiv \mathsf{Hybrid4}^m_{f_1,f_2,g} \equiv Copy^m_{Asim_{f_1,f_2,g}}$$

Combining the above Claims 1, 2, 3 and 4, we have

$$\mathsf{ATamper}^m_{f_1,f_2,g} \approx_{\epsilon_1+\epsilon_2} Copy^m_{Asim_{f_1,f_2,g}}$$

5.3 Rate and Error Analysis

From Sect. 4.5, we have a non-malleable randomness encoding (NMREnc, NMRDec) with a constant rate of $R \geq \frac{1}{2+\zeta}$, for any $\zeta > 0$ and an error of $\epsilon_1 = 2^{-\Omega(l/\log^{\rho+1} l)}$, for any $\rho > 0$.

5.3.1 Rate

The rate of (AEnc, ADec) is:

$$R' = \frac{l}{\frac{1}{R}.(l+\tau)+l+\delta} = \frac{l}{(2+\zeta).(l+\tau)+l+\delta}$$

where, δ is size of tag t. Hence,

$$R' = \frac{l}{(3+\zeta)l+(2+\zeta)\tau+\delta}$$

By using Lemma 7, we know that for $\lambda = o(l/\log l)$, we get $\tau + \delta \leq 3(\log l + o(l/\log l))$. Hence, we get, for large l:

$$R' \geq \frac{1}{3+\zeta}$$

5.3.2 Error

By setting $\epsilon_2 = 2^{-\lambda}$, we get that the error of (AEnc, ADec) is $\epsilon_1 + \epsilon_2 = 2^{-\Omega(l/\log^{\rho+1} l)}$, for any $\rho > 0$.

6 Conclusion

In this work, we introduced Non-malleable Randomness Encoders as a relaxation of NMCs, applicable in settings where randomness is encoded. We built a 1/2-rate, 2-state NMRE. In cases where NMREs suffice, this presents a significant advantage over using a poor-rate 2-state NMC. It would be interesting to find other applications of NMREs in addition to the ones presented in this paper i.e., to tamper-resilient security and to building 3-state (standard) with rate $\frac{1}{3}$ in a black-box. (Infact, our techniques can be generalized to show that $(t+1)$-state augmented NMCs can be constructed from t-state NMREs in black box manner.) While we know that the optimal achievable rate for 2-state NMCs is 1/2, it would be interesting to see what the optimal achievable rate for 2-state NMREs is and, more generally, for t-state NMREs. Of course, the crux of this long, compelling line of research, which is to build constant rate efficient 2-state NMCs, still remains open and would be fascinating to solve.

Acknowledgement. We thank Eshan Chattopadhyay for helpful discussions on connections between non-malleable codes and extractors. We also thank the reviewers of Eurocrypt for their useful comments.

A Proofs of Claims 2 and 3 in Sect. 4.4

A.1 Proof of Claim 2 in Sect. 4.4

We define the following events:

- Let E be the event that $same^*$ is sampled from $\mathsf{NMSim}_{f_w,g}$ and $\tilde{E}$ be its compliment.
- Let F be the event that $\tilde{w} = w$ and $\tilde{F}$ its complement.

By Proposition 1 we get:

$$\begin{aligned}\mathbf{SD}\left(\mathsf{Hybrid1}_{f,g};\mathsf{Hybrid2}_{f,g}\right) &= \Pr[E]\cdot\mathbf{SD}\left(\mathsf{Hybrid1}_{f,g}|E;\mathsf{Hybrid2}_{f,g}|E\right)\\ &+\Pr[\tilde{E}]\cdot\underbrace{\mathbf{SD}\left(\mathsf{Hybrid1}_{f,g}|\tilde{E};\mathsf{Hybrid2}_{f,g}|\tilde{E}\right)}_{=0\text{ The hybrids are identical in ``not }same^*\text{'' case}}\end{aligned}$$

So, now remains the case when $\mathsf{NMSim}_{f_w,g}$ outputs $same^*$. By using unforgeability of $(\mathsf{Tag}',\mathsf{Vrfy}')$ we show the that two hybrids are statistically close.

2. $\Pr[E].\mathbf{SD}\left(\mathsf{Hybrid1}_{f,g}|E; \mathsf{Hybrid2}_{f,g}|E\right)$

$$= \sum_{\substack{m\in\{0,1\}^{l+\tau}\\ \tilde{m}\in\{0,1\}^{l+\tau}\cup\{\perp\}}} \Pr[E] \left| \Pr[\mathsf{Hybrid1}_{f,g} = (m,\tilde{m})|E] - \Pr[\mathsf{Hybrid2}_{f,g} = (m,\tilde{m})|E] \right|$$

$$= \Pr[E] \sum_{\substack{m\in\{0,1\}^{l+\tau}\\ \tilde{m}\in\{0,1\}^{l+\tau}\cup\{\perp\}}} \Big| \Pr[F|E].$$

$$\Bigg(\underbrace{\Pr[\mathsf{Hybrid1}_{f,g} = (m,\tilde{m})|E,F] - \Pr[\mathsf{Hybrid2}_{f,g} = (m,\tilde{m})|E,F]}_{\text{=0 as given E and F both the hybrids are identical}}\Bigg) + \Pr[\tilde{F}|E].$$

$$\left(\Pr[\mathsf{Hybrid1}_{f,g} = (m,\tilde{m})|E,\tilde{F}] - \Pr[\mathsf{Hybrid2}_{f,g} = (m,\tilde{m})|E,\tilde{F}]\right)\Bigg|$$

$$= \Pr[E] \sum_{\substack{m\in\{0,1\}^{l+\tau}\\ \tilde{m}\in\{0,1\}^{l+\tau}\cup\{\perp\}}} \Pr[\tilde{F}|E]\Bigg(\Bigg| \Pr[\mathsf{Hybrid1}_{f,g} = (m,\tilde{m})|E,\tilde{F}]$$

$$- \Pr[\mathsf{Hybrid2}_{f,g} = (m,\tilde{m})|E,\tilde{F}]\Bigg|\Bigg)$$

$$= \Pr[E]\Pr[\tilde{F}|E]\Bigg(\sum_{\substack{m\in\{0,1\}^{l+\tau}\\ \tilde{m}\in\{0,1\}^{l+\tau}}} \Bigg| \Pr[\mathsf{Hybrid1}_{f,g} = (m,\tilde{m})|E,\tilde{F}]$$

$$- \underbrace{\Pr[\mathsf{Hybrid2}_{f,g} = (m,\tilde{m})|E,\tilde{F}]}_{=\text{ 0 as given E,}\tilde{F}\text{ Hybrid 2 outputs }\perp\text{ as second component}} \Bigg|$$

$$+ \sum_{m\in\{0,1\}^{l+\tau}} \Bigg| \Pr[\mathsf{Hybrid1}_{f,g} = (m,\perp)|E,\tilde{F}] - \Pr[\mathsf{Hybrid2}_{f,g} = (m,\perp)|E,\tilde{F}]\Bigg|\Bigg)$$

$$= \Pr[\tilde{F}]\Bigg(1 + \sum_{m\in\{0,1\}^{l+\tau}} \Bigg(\Bigg(\sum_{\tilde{m}\in\{0,1\}^{l+\tau}} \Pr[\mathsf{Hybrid1}_{f,g} = (m,\tilde{m})|E,\tilde{F}]\Bigg)$$

$$- \Pr[\mathsf{Hybrid1}_{f,g} = (m,\perp)|E,\tilde{F}]\Bigg)\Bigg)$$

$$= 2\Pr[\tilde{F}]\Bigg(\Pr[\text{Second component of output of } \mathsf{Hybrid1}_{f,g} \neq \perp|E,\tilde{F}]\Bigg)$$

$$= 2\Pr[\tilde{F}]\Pr[\mathsf{Vrfy}_{\tilde{k_{a_1}}}(\tilde{w},\tilde{t_1}) = 1 \wedge t_1 = \mathsf{Tag}_{k_{a_1}}(w)|E,\tilde{F}]$$

$$= 2\Pr[\tilde{F}]\Pr[\mathsf{Vrfy}_{k_{a_1}}(\tilde{w}, t_1) = 1 \wedge t_1 = \mathsf{Tag}_{k_{a_1}}(w)|\tilde{F}]$$
$$\leq 2(\varepsilon_2)$$

$$\therefore \ \mathsf{Hybrid1}_{f,g} \approx_{\varepsilon_2} \mathsf{Hybrid2}_{f,g}$$

A.2 Alternate proof of Claim 3 in Sect. 4.4

Claim 3. If Ext is an $(n, t, d, l + \tau, \varepsilon_3)$ average case extractor, then $\mathsf{Hybrid2}_{f,g} \approx_{\varepsilon_3} \mathsf{Hybrid3}_{f,g}$.

Proof. As the function modifying the state L, f_w, is dependent on W, hence $\mathsf{NMSim}_{f_w,g}$ is also dependent on W. Hence, before analyzing the auxiliary information leaked in each case, corresponding to the value of $\mathsf{NMSim}_{f_w,g}$, we define the following indicator random variables, which are also auxiliary information, w.r.t. to the source W:

$$b_{same^*} = \begin{cases} 1 & \text{if } \tilde{k_{a_1}}||\tilde{t_1}||\tilde{s} = same^* \\ 0 & \text{otherwise} \end{cases}$$

$$b_{\perp} = \begin{cases} 1 & \text{if } \tilde{k_{a_1}}||\tilde{t_1}||\tilde{s} = \perp \\ 0 & \text{otherwise} \end{cases}$$

By Proposition 1, we get:
$\mathbf{SD}\left(\mathsf{Hybrid2}_{f,g}, \mathsf{Hybrid3}_{f,g}\right)$

$$\begin{aligned} &\leq Pr[b_{same^*} = 1]\ \mathbf{SD}\left(\mathsf{Hybrid2}_{f,g}|b_{same^*} = 1, \mathsf{Hybrid3}_{f,g}|b_{same^*} = 1\right) \\ &\quad + Pr[b_{same^*} = 0 \wedge b_{\perp} = 1]\ \mathbf{SD}\left(\mathsf{Hybrid2}_{f,g}; \mathsf{Hybrid3}_{f,g}\middle| b_{same^*} = 0 \wedge b_{\perp} = 1\right) \\ &\quad + Pr[b_{same^*} = 0 \wedge b_{\perp} = 0]\ \mathbf{SD}\left(\mathsf{Hybrid2}_{f,g}; \mathsf{Hybrid3}_{f,g}\middle| b_{same^*} = 0 \wedge b_{\perp} = 0\right) \end{aligned} \tag{4}$$

Now, in order to analyze the auxiliary information leaked in each of the three cases, and use the extractor security, we first consider the conditional distribution on W, when conditioned on each of the three cases. We denote the three conditional distributions by: $W_1 := W|b_{same^*} = 1$, $W_2 := W|b_{same^*} = 0 \wedge b_{\perp} = 1$ and $W_3 := W|b_{same^*} = 0 \wedge b_{\perp} = 0$. By [Lemma 2.2a, [DORS08]], we get:

$$\Pr[\mathbf{H}_\infty(W_1) \geq \widetilde{\mathbf{H}}_\infty(W|b_{same^*}) - \lambda] \geq 1 - 2^{-\lambda}$$

which by [Lemma 2.2b, [DORS08]] further gives:

$$\Pr[\mathbf{H}_\infty(W_1) \geq n - 1 - \lambda] \geq 1 - 2^{-\lambda}$$

Similarly, we get

$$\Pr[\mathbf{H}_\infty(W_2) \geq n - 2 - \lambda] \geq 1 - 2^{-\lambda}$$
$$\Pr[\mathbf{H}_\infty(W_3) \geq n - 2 - \lambda] \geq 1 - 2^{-\lambda}$$

Now, we analyze the additional auxiliary information in each subcase:
$Case1 : b_{same^*} = 1$
In this case, the additional auxiliary information just includes a single bit, indicating whether w is modified or remains the same. So, we first define this indicator function:

$$eq(w) = \begin{cases} 0 \; if \; f_L(w) \neq w \\ 1 \; if \; f_L(w) = w \end{cases}$$

Let the auxiliary information be denoted by $E_1 \equiv eq(W)$. E_1 is independent of S because E_1 is determined given W and W is independent of S. Now, E_1 and W are correlated and E_1 can take at most two possible values.

Hence, $\widetilde{\mathbf{H}}_\infty(W_1|E_1) \geq \mathbf{H}_\infty(W_1) - 1 \geq n - 1 - \lambda - 1$ w.p. $\geq 1 - 2^{-\lambda}$. Let G_1 denote the event $\widetilde{\mathbf{H}}_\infty(W_1|E_1) \geq n - \lambda - 2$. As $n - \lambda - 2 > t$, by security of average case extractor, we get:

$$E_1, \mathsf{Ext}(W_1; S)|G_1 \approx_{\varepsilon_3} E_1, U_l|G_1$$

Now, clearly, in this case, the output of $\mathsf{Hybrid2}_{f,g}$ and $\mathsf{Hybrid3}_{f,g}$ are functions of above random variables. Hence, by Lemma 2, we get:

$$\mathsf{Hybrid2}_{f,g}|b_{same^*} = 1, G_1 \approx_{\varepsilon_3} \mathsf{Hybrid3}_{f,g}|b_{same^*} = 1, G_1$$

Hence, by further using Proposition 1, as $\Pr[G_1^c] \leq 2^{-\lambda}$, we get:

$$\mathsf{Hybrid2}_{f,g}|b_{same^*} = 1 \approx_{\varepsilon_3 + 2^{-\lambda}} \mathsf{Hybrid3}_{f,g}|b_{same^*} = 1 \tag{5}$$

$Case2 : b_{same^*} = 0$
This case is further divided into two mutually exclusive events of Case2.
$Case2a : b_\perp = 1$
Now, let G_2 denote the event $\mathbf{H}_\infty(W_2) \geq n - 2 - \lambda$. Then as $\Pr[G_2^c] \leq 2^{-\lambda}$ and using extractor security, we get:

$$\mathbf{SD}\left(\mathsf{Hybrid2}_{f,g}|b_{same^*} = 0 \wedge b_\perp = 1, \mathsf{Hybrid2}_{f,g}|b_{same^*} = 0 \wedge b_\perp = 1\right) \leq \varepsilon_3 + 2^{-\lambda} \tag{6}$$

$Case2b : b_\perp = 0$
In this case, the additional auxiliary information consists of an indicator of verification of $\tilde{w}$ and the extractor output on modified source and seed. We first define the indicator of verification bit:

$$Verify(w) = \mathsf{Vrfy'}_{\tilde{k_{a_1}}}(f_L(w), \tilde{t_1})$$

Now, let the auxiliary information be denoted by $E_2 \equiv (Verify(W), \mathsf{Ext}(\tilde{W}; \tilde{S}))$, where $\tilde{K_{a_1}}, \tilde{T_1}, \tilde{S}$ denote the distributions on the authentication key, tag spaces and the seed, when sampled from the simulator conditioned on the event Case2b. E_2 is clearly a deterministic function of $\tilde{K_{a_1}}, \tilde{W}, \tilde{T_1}, \tilde{S}$, all of which are independent of S (as we use the simulator). Hence, E_2 is independent of S. Now, E_2 and W are correlated. E_2 can take at most $2^{1+l+\tau}$ possible values.

Hence, $\widetilde{\mathbf{H}}_\infty(W_3|E_2) \geq \mathbf{H}_\infty(W_3) - (1 + l + \tau) \geq n - 2 - \lambda - (1 + l + \tau)$ w.p. $\geq 1 - 2^{-\lambda}$. Let G_3 denote the event $\widetilde{\mathbf{H}}_\infty(W_3|E_2) \geq n - (3 + \lambda + l + \tau)$. As $n - (3 + \lambda + l + \tau) > t$ (if we set parameters appropriately), by security of average case extractor and using Proposition 1, we get:

$$E_2, \mathsf{Ext}(W; S)|G_3 \approx_{\varepsilon_3} E_2, U_l|G_3$$

Now, clearly, in this case, the output of $\mathsf{Hybrid2}_{f,g}$ and $\mathsf{Hybrid3}_{f,g}$ are functions of above random variables. Hence, by Lemma 2, we get:

$$\mathsf{Hybrid2}_{f,g}|b_{same^*} = 0 \wedge b_\perp = 0, G_3 \approx_{\varepsilon_3} \mathsf{Hybrid3}_{f,g}|b_{same^*} = 0 \wedge b_\perp = 0, G_3$$

Further, since $\Pr[G_3^c] \leq 2^{-\lambda}$, using Proposition 1, we get

$$\mathsf{Hybrid2}_{f,g}|b_{same^*} = 0 \wedge b_\perp = 0 \approx_{\varepsilon_3 + 2^{-\lambda}} \mathsf{Hybrid3}_{f,g}|b_{same^*} = 0 \wedge b_\perp = 0 \quad (7)$$

Hence, by Proposition 1, Eqs. 4, 5, 6 and 7 give:

$$\mathsf{Hybrid2}_{f,g} \approx_{\varepsilon_3 + 2^{-\lambda}} \mathsf{Hybrid3}_{f,g}$$

B Appendix: From t-source Strong Non-malleable Extractors to t-state 1-augmented NMC

We generalize the connection known between seedless non-malleable extractors for t independent sources and non-malleable codes for the t-split-state family ([CG14b]), to establish a connection between strong seedless non-malleable extractors for t independent sources and augmented non-malleable codes for t-split-state family. We first define strong seedless non-malleable t-source extractor.

Definition 6 *[Li17]. A function $nmExt : (\{0,1\}^n)^t \to \{0,1\}^m$ is a (k, ϵ)-seedless strong non-malleable extractor for t independent sources w.r.t. family $\mathcal{F} = \{(f_1, \cdots f_t) : f_i : \{0,1\}^n \to \{0,1\}^n\}$, if it satisfies the following property: Let $X_1, \cdots, X_t$ be t independent (n, k)-sources and $(f_1, \cdots, f_t) \in \mathcal{F}$ be t arbitrary functions such that there exists an f_j with no fixed points, then for every i:*

$$\begin{aligned}(nmExt(X_1, \cdots, X_t), nmExt(f_1(X_1), \cdots, f_t(X_t)), X_i)\\ \approx_\epsilon (U_m, nmExt(f_1(X_1), \cdots, f_t(X_t)), X_i)\end{aligned}$$

Now, we formulate an alternate definition of a t-source relaxed strong non-malleable extractor, generalizing the definition of seedless relaxed non-malleable extractors in [CG14b]. This definition captures the property that the output of non-malleable extractor on the modified sources along with one of the source, is simulatable independent of the output of non-malleable extractor on original sources.

Definition 7. *A function* $nmExt : (\{0,1\}^n)^t \rightarrow \{0,1\}^m$ *is a* (k, ϵ)*-seedless relaxed strong non-malleable extractor for* t *independent sources w.r.t. family* $\mathcal{F} = \{(f_1, \cdots f_t) : f_i : \{0,1\}^n \rightarrow \{0,1\}^n$ *and* $\exists$ *at least one* j *s.t.* f_j *has no fixed point*$\}$*, if it satisfies the following property: Let* $X_1, \cdots, X_t$ *be* t *independent* (n, k)*-sources and* $(f_1, \cdots, f_t) \in \mathcal{F}$*, then the following hold:*

- $nmExt$ *is a* t*-source extractor for* $(X_1, \cdots, X_t)$*, i.e.,* $nmExt(X_1, \cdots, X_t) \approx_\epsilon U_m$.
- *There exists a distribution* $\mathcal{D}$ *over* $\{0,1\}^n \times (\{0,1\}^m \cup \{same^*\})$ *s.t. for an independent* $(X_1, Y) \sim \mathcal{D}$,

$$
\begin{aligned}
&(nmExt(X_1, \cdots, X_t), X_1, nmExt(f_1(X_1), \cdots, f_t(X_t))) \\
&\approx_\epsilon (nmExt(X_1, \cdots, X_t), copy((X_1, Y), (X_1, nmExt(X_1, \cdots, X_t))))
\end{aligned}
$$

Remark 1. It is clear that the non-malleability condition in Definition 6 (for $i = 1$) is sufficient for the conditions in Definition 7 to be satisfied.

But then, this relaxed notion of strong non-malleable extractor is equivalent to the following general notion of strong non-malleable extractor (where, the tampering functions can have fixed points) upto a slight loss of parameters. (This proof follows from [Lemma 5.6, [CG14b]]).

Definition 8. *A function* $nmExt : (\{0,1\}^n)^t \rightarrow \{0,1\}^m$ *is a* (k, ϵ)*-seedless strong non-malleable extractor for* t *independent sources w.r.t. family* $\mathcal{F} = \{(f_1, \cdots f_t) : f_i : \{0,1\}^n \rightarrow \{0,1\}^n\}$*, if it satisfies the following property: Let* $X_1, \cdots, X_t$ *be* t *independent* (n, k)*-sources and* $(f_1, \cdots, f_t) \in \mathcal{F}$*, then the following hold:*

- $nmExt$ *is a* t*-source extractor for* $(X_1, \cdots, X_t)$*, i.e.,* $nmExt(X_1, \cdots, X_t) \approx_\epsilon U_m$.
- *There exists a distribution* $\mathcal{D}$ *over* $\{0,1\}^n \times (\{0,1\}^m \cup \{same^*\})$ *s.t. for an independent* $(X_1, Y) \sim \mathcal{D}$,

$$
\begin{aligned}
&(nmExt(X_1, \cdots, X_t), X_1, nmExt(f_1(X_1), \cdots, f_t(X_t))) \\
&\approx_\epsilon (nmExt(X_1, \cdots, X_t), copy((X_1, Y), (X_1, nmExt(X_1, \cdots, X_t))))
\end{aligned}
$$

Hence, we take the above Definition 8 for strong non-malleable extractors and prove the following theorem.

Proposition 2. *Let* $nmExt : (\{0,1\}^n)^t \rightarrow \{0,1\}^k$ *be a* (n, ϵ)*-seedless strong non-malleable extractor for* t *independent sources (by Definition 8). Define a coding scheme* (Enc, Dec) *with message length* k *and block length* tn *as follows. The decoder* Dec *is defined by*

$$
\mathsf{Dec}(x_1, \cdots, x_t) = nmExt(x_1, \cdots, x_t)
$$

The encoder Enc *is defined as:*

$$
\mathsf{Enc}(m) := \begin{cases} x_1, \cdots, x_t \xleftarrow{\$} nmExt^{-1}(m) \\ o/p : (x_1, \cdots, x_t) \end{cases}
$$

Then, (Enc, Dec) *is a* $[\epsilon', 1]$*-augmented non-malleable code with error* $\epsilon' = \epsilon(2^k + 1)$ *for the* t*-split state family and with rate* $= \frac{k}{tn}$.

Proof. Let $m \in \{0,1\}^k$ and $f = (f_1, \cdots, f_t) \in \mathcal{F}$, the t-split-state family be arbitrary. Since Dec $= nmExt$ is a strong non-malleable extractor, by Definition 8, $\exists$ a distribution $\mathcal{D}$ s.t. for $(X_1, Y) \sim \mathcal{D}_{f_1,\cdots,f_t}$, we have:

$$\begin{aligned}&(nmExt(X_1, \cdots, X_t), X_1, nmExt(f_1(X_1), \cdots, f_t(X_t)))\\&\quad\approx_\epsilon (nmExt(X_1, \cdots, X_t), copy((X_1, Y), (X_1, nmExt(X_1, \cdots, X_t))))\end{aligned} \tag{8}$$

Claim. $\mathsf{Enc}(U_k)$ is ϵ-close to uniform.

Proof. By extractor security, we have:

$$\mathsf{Dec}(U_{tn}) \approx_\epsilon U_k$$

Further, as Enc(.) samples uniformly random element of $nmExt^{-1}(.)$, it follows that

$$\mathsf{Enc}(\mathsf{Dec}(U_{tn})) = U_{tn}$$

Hence, we get $\mathsf{Enc}(U_k) \approx_\epsilon \mathsf{Enc}(\mathsf{Dec}(U_{tn})) = U_{tn}$.

Thus, at cost of ϵ increase in error, we assume codeword is of uniform distribution.

Let $(X_1, Y) \sim \mathcal{D}_{f_1,\cdots,f_t}$. Now by Eq. 8, just by substitution, we get:

$$(M, X_1, \mathsf{Dec}(f(\mathsf{Enc}(M))) \approx_\epsilon (M, copy((X_1, Y), (X_1, M)))$$

Now, for the arbitrary m that we chose, we get:

$$(m, X_1, \mathsf{Dec}(f(\mathsf{Enc}(m))) \approx_{\epsilon 2^k} (m, copy((X_1, Y), (X_1, m)))$$

which proves the theorem.

Augmented-non-malleability of 2-state Construction in [Li17]

Corollary 2. *For any* $\beta \in \mathbb{N}$ *there exists an explicit augmented-non-malleable code with efficient encoder/decoder in 2-split state model with block length* 2β, *rate* $\Omega\left(\frac{1}{\log\ \beta}\right)$ *and error* $= 2^{-\Omega\left(\frac{\beta}{\log\ \beta}\right)}$.

Proof. As proved in [Theorem 7.9, [Li17]], the seedless 2 source non-malleable extractor constructed in [Li17] satisfies: For any (f, g) in 2-split-state family, such that atleast one of f or g has no fixed point, we have:

$$nmExt(X, Y), X, nmExt(f(X), g(Y)) \approx_\epsilon U_m, X, nmExt(f(X), g(Y))$$

which, by Remark 1, is sufficient to imply the conditions in Definition 7. Hence, by Proposition 2, it is proved that the 2-split-state construction given in [Li17] is actually a 2-split-state augmented-non-malleable code.

Further, the specific non-malleable extractor of [Li17] gives error and rate parameters for the augmented-non-malleable code, exactly as obtained in Lemma 8.

References

[AAG+16] Aggarwal, D., Agrawal, S., Gupta, D., Maji, H.K., Pandey, O., Prabhakaran, M.: Optimal computational split-state non-malleable codes. In: Kushilevitz, E., Malkin, T. (eds.) TCC 2016. LNCS, vol. 9563, pp. 393–417. Springer, Heidelberg (2016). https://doi.org/10.1007/978-3-662-49099-0_15

[ADKO15] Aggarwal, D., Dodis, Y., Kazana, T., Obremski, M.: Non-malleable reductions and applications. In: Proceedings of the Forty-Seventh Annual ACM on Symposium on Theory of Computing, STOC 2015, Portland, OR, USA, 14–17 June 2015, pp. 459–468 (2015)

[ADL14] Aggarwal, D., Dodis, Y., Lovett, S.: Non-malleable codes from additive combinatorics. In: Symposium on Theory of Computing, STOC 2014, New York, NY, USA, 31 May–03 June 2014, pp. 774–783 (2014)

[AGM+15] Agrawal, S., Gupta, D., Maji, H.K., Pandey, O., Prabhakaran, M.: A rate-optimizing compiler for non-malleable codes against bit-wise tampering and permutations. In: Dodis, Y., Nielsen, J.B. (eds.) TCC 2015. LNCS, vol. 9014, pp. 375–397. Springer, Heidelberg (2015). https://doi.org/10.1007/978-3-662-46494-6_16

[AKO15] Aggarwal, D., Kazana, T., Obremski, M.: Inception makes non-malleable codes stronger. IACR Cryptology ePrint Archive, 2015:1013 (2015)

[CG14a] Cheraghchi, M., Guruswami, V.: Capacity of non-malleable codes. In: Innovations in Theoretical Computer Science, ITCS 2014, Princeton, NJ, USA, 12–14 January 2014, pp. 155–168 (2014)

[CG14b] Cheraghchi, M., Guruswami, V.: Non-malleable coding against bit-wise and split-state tampering. In: Lindell, Y. (ed.) TCC 2014. LNCS, vol. 8349, pp. 440–464. Springer, Heidelberg (2014). https://doi.org/10.1007/978-3-642-54242-8_19

[CKR16] Chandran, N., Kanukurthi, B., Raghuraman, S.: Information-theoretic local non-malleable codes and their applications. In: Kushilevitz, E., Malkin, T. (eds.) TCC 2016. LNCS, vol. 9563, pp. 367–392. Springer, Heidelberg (2016). https://doi.org/10.1007/978-3-662-49099-0_14

[CZ14] Chattopadhyay, E., Zuckerman, D.: Non-malleable codes against constant split-state tampering. In: 55th IEEE Annual Symposium on Foundations of Computer Science, FOCS 2014, Philadelphia, PA, USA, 18–21 October 2014, pp. 306–315 (2014)

[DKO13] Dziembowski, S., Kazana, T., Obremski, M.: Non-malleable codes from two-source extractors. In: Canetti, R., Garay, J.A. (eds.) CRYPTO 2013. LNCS, vol. 8043, pp. 239–257. Springer, Heidelberg (2013). https://doi.org/10.1007/978-3-642-40084-1_14

[DKS17] Dachman-Soled, D., Kulkarni, M., Shahverdi, A.: Tight upper and lower bounds for leakage-resilient, locally decodable and updatable non-malleable codes. IACR Cryptology ePrint Archive, 2017:15 (2017)

[DLSZ14] Dachman-Soled, D., Liu, F.-H., Shi, E., Zhou, H.-S.: Locally decodable and updatable non-malleable codes and their applications. IACR Cryptology ePrint Archive, 2014:663 (2014)

[DNO17] Döttling, N., Nielsen, J.B., Obremski, M.: Information theoretic continuously non-malleable codes in the constant split-state model. Electronic Colloquium on Computational Complexity (ECCC) 24:78 (2017)

[DORS08] Dodis, Y., Ostrovsky, R., Reyzin, L., Smith, A.: Fuzzy extractors: how to generate strong keys from biometrics and other noisy data. SIAM J. Comput. **38**(1), 97–139 (2008). arXiv:cs/0602007

[DPW10] Dziembowski, S., Pietrzak, K., Wichs, D.: Non-malleable codes. In: Proceedings of Innovations in Computer Science - ICS 2010, Tsinghua University, Beijing, China, 5–7 January 2010, pp. 434–452 (2010)

[FMNV14] Faust, S., Mukherjee, P., Nielsen, J.B., Venturi, D.: Continuous non-malleable codes. In: Lindell, Y. (ed.) TCC 2014. LNCS, vol. 8349, pp. 465–488. Springer, Heidelberg (2014). https://doi.org/10.1007/978-3-642-54242-8_20

[GUV07] Guruswami, V., Umans, C., Vadhan, S.P.: Unbalanced expanders and randomness extractors from Parvaresh-Vardy codes. In: IEEE Conference on Computational Complexity, pp. 96–108 (2007)

[JKS93] Johansson, T., Kabatianskii, G., Smeets, B.: On the relation between a-codes and codes correcting independent errors. In: Helleseth, T. (ed.) EUROCRYPT 1993. LNCS, vol. 765, pp. 1–11. Springer, Heidelberg (1994). https://doi.org/10.1007/3-540-48285-7_1

[JW15] Jafargholi, Z., Wichs, D.: Tamper detection and continuous non-malleable codes. In: Dodis, Y., Nielsen, J.B. (eds.) TCC 2015. LNCS, vol. 9014, pp. 451–480. Springer, Heidelberg (2015). https://doi.org/10.1007/978-3-662-46494-6_19

[KOS17] Kanukurthi, B., Obbattu, S.L.B., Sekar, S.: Four-state non-malleable codes with explicit constant rate. In: Kalai, Y., Reyzin, L. (eds.) TCC 2017. LNCS, vol. 10678, pp. 344–375. Springer, Cham (2017). https://doi.org/10.1007/978-3-319-70503-3_11

[Li17] Li, X.: Improved non-malleable extractors, non-malleable codes and independent source extractors. In: Symposium on Theory of Computing, STOC 2017, Montreal, Canada, 19–23 June 2017 (2017)

[LL12] Liu, F.-H., Lysyanskaya, A.: Tamper and leakage resilience in the split-state model. IACR Cryptology ePrint Archive, 2012:297 (2012)

Non-malleable Codes from Average-Case Hardness: AC^0, Decision Trees, and Streaming Space-Bounded Tampering

Marshall Ball[1](✉), Dana Dachman-Soled[2], Mukul Kulkarni[2], and Tal Malkin[1]

[1] Columbia University, New York, USA
{marshall,tal}@cs.columbia.edu
[2] University of Maryland, College Park, USA
danadach@ece.umd.edu, mukul@umd.edu

Abstract. We show a general framework for constructing non-malleable codes against tampering families with average-case hardness bounds. Our framework adapts ideas from the Naor-Yung double encryption paradigm such that to protect against tampering in a class $\mathcal{F}$, it suffices to have average-case hard distributions for the class, and underlying primitives (encryption and non-interactive, simulatable proof systems) satisfying certain properties with respect to the class.

We instantiate our scheme in a variety of contexts, yielding efficient, non-malleable codes (NMC) against the following tampering classes:

- Computational NMC against AC^0 tampering, in the CRS model, assuming a PKE scheme with decryption in AC^0 and NIZK.
- Computational NMC against bounded-depth decision trees (of depth n^ϵ, where n is the number of input variables and constant $0 < \epsilon < 1$), in the CRS model and under the same computational assumptions as above.
- Information theoretic NMC (with no CRS) against a streaming, space-bounded adversary, namely an adversary modeled as a read-once branching program with bounded width.

Ours are the first constructions that achieve each of the above in an efficient way, under the standard notion of non-malleability.

1 Introduction

Non-malleable codes, introduced in the seminal work of Dziembowski et al. [31], are an extension of error-correcting codes. Whereas error-correcting codes provide the guarantee that (if not too many errors occur) the receiver can recover the original message from a corrupted codeword, non-malleable codes are essentially concerned with security. In other words, correct decoding of corrupted codewords is not guaranteed (nor required), but it is instead guaranteed that adversarial corruptions cannot influence the output of the decoding in a way that depends on the original message: the decoding is either correct or *independent* of the original message.

J. B. Nielsen and V. Rijmen (Eds.): EUROCRYPT 2018, LNCS 10822, pp. 618–650, 2018.
https://doi.org/10.1007/978-3-319-78372-7_20

The main application of non-malleable codes is in the setting of tamper-resilient computation (although non-malleable codes have also found connections in other areas of cryptography [22, 23, 36] and theoretical computer science [18]). Indeed, as suggested in the initial work of Dziembowski et al. [31], non-malleable codes can be used to encode a secret state in the memory of a device such that a tampering adversary interacting with the device does not learn anything more than the input-output behavior. Unfortunately, it is impossible to construct non-malleable codes secure against arbitrary tampering, since the adversary can always apply the tampering function that decodes the entire codeword to recover the message m and then re-encodes a related message m'. Thus, non-malleable codes are typically constructed against limited classes of tampering functions $\mathcal{F}$. Indeed, given this perspective, error correcting codes can be viewed as a special case of non-malleable codes, where the class of tampering functions, $\mathcal{F}$, consists of functions which can only modify some fraction of the input symbols. Since non-malleable codes have a weaker guarantee than error correcting codes, there is potential to achieve non-malleable codes against much broader classes of tampering functions $\mathcal{F}$ (including tampering that modifies every bit).

Exploring rich classes of tampering functions. Several works construct non-malleable codes (NMC) against general tampering classes of bounded size, but with non-explicit, existential, or inefficient constructions (cf. [19, 31, 35]). For efficient and explicit constructions, a large body of works construct NMC against bit-wise tampering (cf. [10, 21, 31]), and more generally split-state tampering (cf. [1–3, 15, 19, 20, 30, 39, 40, 44, 47]), where the adversary can tamper each part of the codeword independently of other parts, as well as NMC against permutations, flipping, and setting bits [5].

A recent line of works is shifting towards considering the construction of NMC against tampering classes $\mathcal{F}$ that correspond to well-studied complexity-theoretic classes, and may also better correspond to tampering attacks in practice. Specifically, Ball et al. [7] construct NMC against local tampering functions including NC^0, and Chattopadhyay and Li [16] construct NMC against AC^0 tampering, but inefficiently (with super-poly size codewords). Additionally, NMC with weaker notions of security are constructed by Faust et al. [32] against space-bounded tampering (in the random-oracle model), and by Chandran et al. [12] for block-wise tampering (where the adversary receives the message in a streaming fashion, block-by-block). We discuss these works in Sect. 1.3.

In this work, we continue this line of research and consider constructing non-malleable codes against various complexity classes, including: (1) AC^0 tampering, where the tampering function is represented by a polynomial size constant-depth, unbounded fan-in/fan-out circuit, (2) tampering with bounded-depth decision trees, where the tampering function is represented by a decision tree with n variables and depth n^ϵ for $\epsilon < 1$, (3) streaming tampering with quadratic space, where the tampering function is represented by a read-once, bounded-width ($2^{o(n^2)}$) branching program, (4) small threshold circuits: depth d circuits of majority gates with a quasilinear number of wires, (5) fixed polynomial time tampering: randomized turing machines running in time $O(n^k)$ for any fixed k.

Constructing non-malleable codes against a wide array of complexity classes is desirable since in practice, the capabilities of a tampering adversary are uniquely tied to the computational setting under consideration and/or the physical device being used. For example, our motivation for studying AC^0 stems from a setting wherein an attacker has limited *time* to tamper, since the tampering function must complete before race conditions take effect (e.g. before the end of a clock-cycle in a synchronous circuit). AC^0 circuits, which are constant-depth circuits, model such attackers since the propagation delay of a circuit is proportional to the length of the longest path from input to output.

1.1 Our Results

We present general frameworks for constructing non-malleable codes for encoding one and multi-bits against various tampering classes $\mathcal{F}$ for which average case hardness results are known. Our frameworks (one for single-bit and one for multi-bit) include both a generic construction, which requires that certain underlying primitives are instantiated in a suitable way, as well as a proof "template." Our frameworks are inspired by the well-known double-encryption paradigm for constructing CCA2-secure public key encryption schemes [45,48,50]. And although we rely on techniques that are typically used in the cryptographic setting, we instantiate our framework for particular tampering classes $\mathcal{F}$ in both the computational setting and in the information theoretic one. For the computational setting, our results rely on computational assumptions, and require a common-reference string (CRS), which the adversary can see before selecting the tampering function (as typical in other NMC works using CRS or random oracles). For the information theoretic setting, our results do not require CRS nor any computational assumption (as the primitives in our framework can be instantiated information theoretically). Our general theorem statements provide sufficient conditions for achieving NMC against a class $\mathcal{F}$. Somewhat informally, the main such condition, especially for the one-bit framework, is that there are sufficiently strong average-case hardness results known for the class $\mathcal{F}$. In particular, we obtain the following results, where all the constructions are efficient and, for the multi-bit NMC, the achieved rate is $1/\operatorname{poly}(m)$ where m is the length of the message being encoded.

- **Constructions for AC^0 tampering:** We obtain computational NMC in the CRS model against AC^0 tampering. Our constructions require public key encryption schemes with decryption in AC^0, which can be constructed e.g. from exponential hardness of learning parity with noise [9], as well as non-interactive zero knowledge (NIZK), which can be constructed in the CRS model from enhanced trapdoor permutations.
  Previous results by Chattopadhyay and Li [16] achieve NMC for AC^0 with information theoretic security (with no CRS), but are inefficient, with super-polynomial rate.
- **Constructions for bounded-depth decision trees:** We obtain computational NMC in the CRS model against tampering with bounded-depth decision trees. Our construction requires the same computational assumptions as

the AC^0 construction above. The depth of the decision tree we can handle is n^ϵ, where n is the number of bits being encoded, and ϵ is any constant. No results for this class were previously known.

- **Constructions for streaming, space-bounded tampering:** We obtain *unconditional* non-malleable codes against streaming, space-bounded tampering, where the tampering function is represented by a read-once, bounded-width branching program. Our construction does not require CRS or computational assumptions.
  No NMC results for this standard complexity theoretic class were previously known. However, this tampering class can be viewed as a subset (or the intersection) of the space bounded class considered by Faust et al. [32] (who don't limit the adversary to be streaming), and the block-wise tampering class considered by Chandran et al. [12] (who don't bound the adversary's space, but don't give security in the event that decoding fails). In both cases there cannot be NMC with the standard notion of security, and so those previous works must relax the security requirement (and [32] also relies on a random oracle). In contrast, we achieve standard (in fact, even stronger) notion of NMC, without random oracle (nor CRS, nor any computational assumption) for our class.
- **Additional Constructions:** We also briefly note two additional applications of our paradigm as proof of concept. Both complexity classes can be represented circuits of size $O(n^c)$ for some fixed c, a class which [35] provide non-malleable codes for in the CRS model, *without* computational assumptions. We include these results here, merely to show the applicability of our framework to general correlation bounds; for example strong correlation bounds against $\mathsf{ACC}^0[p]$ or TC^0 are likely immediately lead to non-malleable codes against the same classes using our framework.
  1. Under the same assumptions invoked in the constructions against AC^0 and bounded-depth decision trees we obtain computational NMC in the CRS model against tampering with small threshold circuits: threshold circuits with depth d and $n^{1+\epsilon}$ wires.
  2. Assuming any public key encryption scheme and zk-SNARKs, we obtain computational NMC in the CRS model against tampering by Turing Machines running in time $O(n^k)$, where k is a constant. However, we should note that these codes have weak tampering guarantees: tampering experiments with respect to different messages are only polynomially close to one another.

1.2 Technical Overview

We begin by describing our computational NMC construction (in the CRS model) for one-bit messages secure against tampering in AC^0, which will give the starting point intuition for our results. We then show how the AC^0 construction can be modified to derive a general template for constructing NMC for one-bit messages secure against a wider range of tampering classes $\mathcal{F}$, and discuss various classes $\mathcal{F}$ for which the template can be instantiated. We then discuss how the template can be extended to achieve NMC for multi-bit messages secure against a wide range of tampering classes $\mathcal{F}$. Finally, we discuss some particular instantiations of our multi-bit template, including our constructions of computational NMC (in the CRS model) against tampering in AC^0 and against bounded-depth decision trees, as well as our *unconditional* NMC (with no CRS) against streaming tampering adversaries with bounded memory.

The starting point: Computational NMC against AC^0 *for one-bit messages.* The idea is to use a very similar paradigm to the Naor and Yung paradigm for CCA1 encryption [48] (later extended to achieve CCA2 [45,50]), using double encryption with simulation-sound NIZK. The main observation is that using the tableau method, we can convert *any* NIZK proof system with polynomial verification into a NIZK proof system with a verifier in AC^0.

We also need a PKE scheme with perfect correctness and decryption in AC^0(this can be constructed using the transformation of Dwork et al. [29] on top of the scheme of Bogdanov and Lee [9]).

We now sketch (a slightly simplified version of) the NM encoding scheme:

The CRS will contain a public key PK for an encryption scheme $\mathcal{E} = (\mathsf{Gen}, \mathsf{Encrypt}, \mathsf{Decrypt})$ as above, and a CRS crs for a NIZK. For $b \in \{0,1\}$, Let $\mathcal{D}_b$ denote the distribution over $x_1, \ldots, x_n \in \{0,1\}^n$ such that $x_1, \ldots, x_n$ are uniform random, conditioned on the parity of the bits being equal to b.

To encode a bit b:

1. Randomly choose bits $x_1, \ldots, x_n$ from $\mathcal{D}_b$.
2. Compute $c_1 \leftarrow \mathsf{Encrypt}_{\mathrm{PK}}(x_1), \ldots, c_n \leftarrow \mathsf{Encrypt}_{\mathrm{PK}}(x_n)$ and $c \leftarrow \mathsf{Encrypt}_{\mathrm{PK}}(b)$.
3. Compute n NIZK proofs $\pi_1, \ldots, \pi_n$ that $c_1, \ldots, c_n$ are encryptions of bits $x_1, \ldots, x_n$.
4. Compute a NIZK proof π that there exists a bit b' such that the plaintexts underlying $c_1, \ldots, c_n$ are in the support of $\mathcal{D}_{b'}$ and b' is the plaintext underlying c.
5. Compute tableaus $T_1, \ldots, T_n$ of the computation of the NIZK verifier on $\pi_1, \ldots, \pi_n$.
6. Compute a tableau T of the computation of the NIZK verifier on proof π.
7. Output $(c_1, \ldots, c_n, c, T, (x_1, T_1), \ldots, (x_n, T_n))$.

To decode $(c_1, \ldots, c_n, c, T, (x_1, T_1), \ldots, (x_n, T_n))$:

1. Check the tableaus $T_1, \ldots, T_n, T$.
2. If they all accept, output the parity of $x_1, \ldots, x_n$.

In the proof we will switch from an honest encoding of b to a simulated encoding and from an honest decoding algorithm to a simulated decoding algorithm. At each point we will show that the decodings of tampered encodings stay the same. Moreover, if, in the final hybrid, decodings of tampered encodings depend on b, we will use this fact to build a circuit in AC^0, whose output is correlated with the parity of its input, reaching a contradiction. In more detail, in the first hybrid we switch to simulated proofs. Then we switch $c_1, \ldots, c_n, c,$

in the "challenge" encoding to encryptions of garbage $c'_1, \ldots, c'_n, c'$, and next we switch to an alternative decoding algorithm *in* AC^0, which requires the trapdoor SK (corresponding to the public key PK which is contained in the CRS).

Alternative Decoding Algorithm:

To decode $(c_1, \ldots, c_n, c, T, (x_1, T_1), \ldots, (x_n, T_n))$:

1. check the tableaus $T_1, \ldots, T_n, T$
2. If it accepts, output the decryption of c using trapdoor SK.

In the final hybrid, the simulator will not know the parity of $x_1, \ldots, x_n$ in the challenge encoding and will have received precomputed $T_1^0, T_1^1, \ldots, T_n^0, T_n^1, T$ as non-uniform advice, where T is a simulated proof of the statement "the plaintexts underlying $c'_1, \ldots, c'_n$ and the plaintext underlying c' have the same parity" and for $i \in [n], \beta \in \{0,1\}, T_i^\beta$ is a simulated proof of the statement "c'_i is an encryption of the bit β".

We will argue by contradiction that if the decoding of the tampered encoding is correlated with the parity of $x_1, \ldots, x_n$ then we can create a circuit whose output is correlated with the parity of its input in AC^0. Specifically, the AC^0 circuit will have the crs, SK, precomputed $c'_1, \ldots, c'_n, c', T, T_1^0, T_1^1, \ldots, T_n^0, T_n^1$ and adversarial tampering function f hardwired in it. It will take $x_1, \ldots, x_n$ as input. It will compute the simulated encoding in AC^0 by selecting the correct tableaus: $T_1^{x_1}, \ldots, T_n^{x_n}$ according to the corresponding input bit. It will then apply the adversarial tampering function (in AC^0), perform the simulated decoding (in AC^0) and output a guess for the parity of $x_1, ..x_n$ based on the result of the decoding. Clearly, if the decoding in the final hybrid is correlated with parity, then we have constructed a distribution over AC^0 circuits such that w.h.p. over choice of circuit from the distribution, the output of the circuit is correlated with the parity of its input. This contradicts known results on the hardness of computing parity in AC^0.

A general template for one-bit NMC. The above argument can be used to derive a template for the construction/security proof of NMC against more general classes $\mathcal{F}$. The idea is to derive a high-level sequence of hybrid distributions and corresponding *minimal* requirements for proving the indistinguishability of consecutive hybrids. We can now instantiate the tampering class $\mathcal{F}$, "hard distributions" $(\mathcal{D}_0, \mathcal{D}_1)$, encryption scheme and NIZK proof in any way that satisfies these minimal requirements. Note that each hybrid distribution is a distribution over the output of the tampering experiment. Therefore, public key encryption and NIZK against arbitrary PPT adversaries may be too strong of a requirement. Indeed, it is by analyzing the exact security requirements needed to go from one hybrid to the other that (looking ahead) we are able to remove the CRS and all computational assumptions from our construction of NMC against streaming adversaries with bounded memory. In addition, we can also use our template to obtain constructions (in the CRS model and under computational assumptions) against other tampering classes $\mathcal{F}$.

Extending the template to multi-bit NMC. The construction for AC^0 given above and the general template do not immediately extend to multi-bit messages. In particular, encoding m bits by applying the parity-based construction bit-by-bit fails, even if we use the final proof T to "wrap together" the encodings of multiple individual bits. The problem is that the proof strategy is to entirely decode the tampered codeword and decide, based on the results, whether to output 0 or 1 as the guess for the parity of some $x_1, \ldots, x_n$. But if we encode many bits, $b_1, \ldots, b_m$, then the adversary could maul in such a way that the tampered codeword decodes to $b'_1, \ldots, b'_m$ where each of b'_i is *individually* independent of the parity of the corresponding $x^i_1, \ldots, x^i_n$, but taken as a whole, the entire output may be correlated. As a simple example, the attacker might maul the codeword so that it decodes to $b'_1, \ldots, b'_m$ that are uniform subject to satisfying $b'_1 \oplus \cdots \oplus b'_m = b_1 \oplus \cdots \oplus b_m$. Clearly, there is a correlation here between the input and output, but we cannot detect this correlation in AC^0, since detecting the correlation itself seems to require computing parity!

In the case of parity (and the class AC^0), the above issue can be solved by setting m sufficiently small (but still polynomial) compared to n. We discuss more details about the special case of parity below. However, we would first like to explain how the general template must be modified for the multi-bit case, given the above counterexample. Specifically, note that the difficulty above comes into play only in the final hybrid. Thus, we only need to modify the final hybrid slightly and require that for any Boolean function F over m variables, it must be the case that the composition of F with the simulated decoding algorithm is in a computational class that still cannot distinguish between draws $x_1, \ldots, x_n$ from $\mathcal{D}_0$ or $\mathcal{D}_1$. While the above seems like a strong requirement, we show that by setting m much smaller than n, we can still obtain meaningful results for classes such as AC^0 and bounded-depth decision trees.

Multi-bit NMC against AC^0. If we want to encode m bits, for each of the underlying encodings $i \in [m]$, we will use $n :\approx m^3$ bits: $\boldsymbol{x}^i = x^i_1, \ldots, x^i_n$. To see why this works, we set up a hybrid argument, where in each step we will fix all the underlying encodings except for a single one: $\boldsymbol{x} = x_1, \ldots, x_n$, which we will switch from having parity 0 to having parity 1. Therefore, we can view C—the function computing the output of the tampering experiment in this hybrid—to be a function of variables $\boldsymbol{x} = x_1, \ldots, x_n$ only (everything else is constant and "hardwired"). For $i \in [m]$, let C_i denote the i-th output bit of C. We use $\mathsf{PAR}(\boldsymbol{x})$ to denote the parity of $\boldsymbol{x}$.

Now, for any Boolean function F over m variables, consider $F(C_1(\boldsymbol{x}), C_2(\boldsymbol{x}), \ldots, C_m(\boldsymbol{x}))$, where we are simply taking an arbitrary Boolean function F of the decodings of the individual bits. Our goal is to show that $F(C_1(\boldsymbol{x}), C_2(\boldsymbol{x}), \ldots, C_m(\boldsymbol{x}))$ is not correlated with parity of $\boldsymbol{x}$. Consider the Fourier representation of $F(y_1, \ldots, y_m)$. This is a linear combination of parities of the input variables $y_1, \ldots, y_m$, denoted $\chi_S(y_1, \ldots, y_m)$, for all subsets $S \in \{0,1\}^m$. (See here [26]).

On the other hand, $F(C_1(\boldsymbol{x}), C_2(\boldsymbol{x}), \ldots, C_m(\boldsymbol{x}))$ is a Boolean function over $n \approx m^3$ variables (i.e. a linear combination over parities of the input variables

$x_1, \ldots, x_n$, denoted $\chi_{S'}(x_1, \ldots, x_n)$, for all subsets $S' \in \{0,1\}^n$). A representation of $F(C_1(\boldsymbol{x}), C_2(\boldsymbol{x}), \ldots, C_m(\boldsymbol{x}))$ can be obtained by taking each term $\hat{F}(S)\chi_S(y_1, \ldots, y_m)$ in the Fourier representation of F and composing with $C_1, \ldots, C_m$ to obtain the term $\hat{F}(S)\chi_S(C_1(\boldsymbol{x}), C_2(\boldsymbol{x}), \ldots, C_m(\boldsymbol{x}))$. Since, by well-known properties of the Fourier transform, $|\hat{F}(S)| \leq 1$, we can get an upper bound on the correlation of $F(C_1(\boldsymbol{x}), C_2(\boldsymbol{x}), \ldots, C_m(\boldsymbol{x}))$ and $\mathsf{PAR}(\boldsymbol{x})$, by summing the correlations of each function $\chi_S(C_1(\boldsymbol{x}), C_2(\boldsymbol{x}), \ldots, C_m(\boldsymbol{x}))$ and $\mathsf{PAR}(\boldsymbol{x})$. Recall that the correlation of a Boolean function g with $\mathsf{PAR}(\boldsymbol{x})$ is by definition, exactly the Fourier coefficient of g corresponding to parity function $\chi_{[n]}$. Thus, to prove that the correlation of $\chi_S(C_1(\boldsymbol{x}), C_2(\boldsymbol{x}), \ldots, C_m(\boldsymbol{x}))$ and $\mathsf{PAR}(\boldsymbol{x})$ is low, we use the fact that $\chi_S(C_1(\boldsymbol{x}), C_2(\boldsymbol{x}), \ldots, C_m(\boldsymbol{x}))$ can be computed by a (relatively) low depth circuit. To see this, note that each C_i is in AC^0 and so has low depth, moreover, since S has size at most m, we only need to compute parity over m variables, which can be done in relatively low depth when $m \ll n$. We now combine the above with Fourier concentration bounds for low-depth circuits [51]. Ultimately, we prove that for each S, the correlation of $\chi_S(C_1(\boldsymbol{x}), C_2(\boldsymbol{x}), \ldots, C_m(\boldsymbol{x}))$ and $\mathsf{PAR}(\boldsymbol{x})$, is less than $1/2^{m(1+\delta)}$, where δ is a constant between 0 and 1. This means that we can afford to sum over all 2^m terms in the Fourier representation of F and still obtain negligible correlation.

Multi-bit NMC against bounded-depth decision trees. Our result above extends to bounded-depth decision trees by noting that (1) If we apply a random restriction (with appropriate parameters) to input $x_1, \ldots, x_n$ then, w.h.p. the AC^0 circuit used to compute the output of the tampering experiment collapses to a bounded-depth decision tree of depth $m^\varepsilon - 1$; (2) on the other hand, again choosing parameters of the random restriction appropriately, $\mathsf{PAR}(x_1, \ldots, x_n)$ collapses to parity over at least $m^{1+\varepsilon}$ variables; (3) any Boolean function over m variables can be computed by a decision tree of depth m; (4) the composition of a depth-$m^\varepsilon - 1$ decision tree and depth-m decision tree yields a decision tree of depth at most $(m^\varepsilon - 1)(m) < m^{1+\varepsilon}$. Finally, we obtain our result by noting that decision trees of depth less than $m^{1+\varepsilon}$ are uncorrelated with parity over $m^{1+\varepsilon}$ variables.

Unconditional NMC (with no CRS) against bounded, streaming tampering. Recently, Raz [49] proved that learning parity is hard for bounded, streaming adversaries. In particular, this gives rise to hard distributions $\mathcal{D}_b, b \in \{0,1\}$ such that no bounded, streaming adversary can distinguish between the two. $\mathcal{D}_b$ corresponds to choosing a random parity χ_S, outputting random examples $(\boldsymbol{x}, \chi_S(\boldsymbol{x}))$ and then outputting $\boldsymbol{x}^*$ such that $\chi_S(\boldsymbol{x}^*)$ is equal to b. The above also yields an unconditional, "parity-based" encryption scheme against bounded, streaming adversaries. Note, however, that in order to decrypt (without knowledge of the secret key), we require space beyond the allowed bound of the adversary. Given the above, we use $\mathcal{D}_b, b \in \{0,1\}$ as the hard distributions in our construction and use the parity-based encryption scheme as the "public key encryption scheme" in our construction. Thus, we get rid of the public key in the CRS (and the computational assumptions associated with the public key encryption scheme).

To see why this works, note that in the hybrid where we require semantic security of the encryption scheme, the decryption algorithm is not needed for decoding (at this point the honest decoding algorithm is still used). So essentially we can set the parameters for the encryption scheme such that the output of the Tampering experiment in that hybrid (which outputs the decoded value based on whether $x_1, .., x_n$ is in the support of $\mathcal{D}_0$ or $\mathcal{D}_1$) can be computed in a complexity class that is too weak to run the decryption algorithm. On the other hand, we must also consider the later hybrid where we show that the output of the tampering experiment can be computed in a complexity class that is too weak to distinguish $\mathcal{D}_0$ from $\mathcal{D}_1$. In this hybrid, we do use the alternate decoding procedure. But now it seems that we need decryption to be contained in a complexity class that is too weak to decide whether $x_1, \ldots, x_n$ is in the support of $\mathcal{D}_0$ or $\mathcal{D}_1$, while previously we required exactly the opposite! The key insight is that since we are in the streaming model and since (1) the simulated ciphertexts $(c'_1, \ldots, c'_n, c')$ in this hybrid contain no information about $x_1, \ldots, x_n$ and (2) the simulated ciphertexts precede $x_1, \ldots, x_n$, the output of the tampering function in blocks containing ciphertexts *does not depend on* $x_1, \ldots, x_n$ *at all.* So the decryption of the tampered ciphertexts can be given as non-uniform advice, instead of being computed on the fly, and we avoid contradiction.

In order to get rid of the CRS and computational assumption for the NIZK, we carefully leverage some additional properties of the NMC setting and the streaming model. First, we consider cut-and-choose based NIZK's (based on MPC-in-the-head), where the Verifier is randomized and randomly checks certain locations or "slots" in the proof to ensure soundness. Specifically, given a Circuit-SAT circuit C and witness w, the prover will secret share $w := w_1 \oplus \cdots \oplus w_\ell$ and run an MPC protocol among ℓ parties (for constant ℓ), where party P_i has input w_i and the parties are computing the output of $C(w_1 \oplus \cdots \oplus w_\ell)$. The prover will then "encrypt" each view of each party in the MPC protocol, using the parity-based encryption scheme described above and output this as the proof. This is then repeated λ times (where λ is security parameter). The Verifier will then randomly select two parties from each of the λ sets, decrypt the views and check that the views correspond to the output of 1 and are consistent internally and with each other.

We next note that in our setting, the NIZK simulator can actually know the randomness used by the Verifier. This is because the simulated codeword and the decoding are done by the same party in the NMC security experiment. Therefore, the level of "zero-knowledge" needed from the simulation of the NIZK is in-between honest verifier and malicious. This is because the adversary can still use the tampering function to "leak" information from the unchecked slots of the proof to the checked slots, while a completely honest verifier would learn absolutely nothing about the unchecked slots. In order to switch from a real proof to a simulated proof, we fill in unchecked slots one-by-one with parity-based encryptions of garbage. We must rely on the fact that a bounded, streaming adversary cannot distinguish real encryptions from garbage encryptions in order to argue security. Specifically, since we are in the bounded streaming model, we can argue that the adversary can only "leak" a small amount of information from the unchecked slots to the checked slots. This means that the entire output of the experiment can be simulated by a bounded, streaming adversary, which in turn means that the output of the experiment must be indistinguishable when real, unchecked encodings are replaced with encodings of garbage. Arguing simulation soundness, requires a similar argument, but more slots are added to the proof and slots in an honest proof are only filled if the corresponding position in the bit-string corresponding to the statement to be proven is set to 1. We encode the statement in such a way that if the statement changes, the adversary must switch an unfilled slot to a filled slot. Intuitively, since the bounded streaming attacker can only carry over a small amount of information from previous slots, this will be as difficult as constructing a new proof from scratch.

1.3 Related Work

The notion of NMC was formalized by Dziembowski et al. [31]. Split state classes of tampering functions introduced by Liu and Lysyanskaya [47], have subsequently received much attention with a sequence of improvements achieving reduced number of states, improved rate, or other desirable features [1–3,6, 15,17,30,39–41,44]. Recently [5,7] gave efficient constructions of non-malleable codes for "non-compartmentalized" tampering function classes.

Faust et al. [35] presented a construction of efficient NMC in CRS model, for tampering function families $\mathcal{F}$ with size $|\mathcal{F}| \leq 2^{\mathsf{poly}(n)}$, where n is the length of codeword. The construction is based on t-wise independent hashing for t proportional to $\log |\mathcal{F}|$. This gives information-theoretically secure NMC resilient to tampering classes which can be represented as poly-size circuits. While [35] construction allows adaptive selection of tampering function $f \in \mathcal{F}$ after the t-wise independent hash function h (CRS) is chosen, the bound on the size of F needs to be fixed *before* h is chosen. In particular, this means that the construction does not achieve security against the tampering functions $f \in \mathsf{AC}^0$ in general, since AC^0 contains *all* poly-size and constant depth circuit families, but rather provides tamper resilience against *specific* families in AC^0 (ACC^0, etc.) Cheraghchi and Guruswami [19] in an independent work showed the existence of information theoretically secure NMC against tampering families $\mathcal{F}$ of size $|\mathcal{F}| \leq 2^{2^{\alpha n}}$ with optimal rate $1 - \alpha$. This paper gave the first characterization of the rate of NMC, however the construction of [19] is inefficient for negligible error.

Ball et al. [7] gave a construction of efficient NMC against n^{δ}-local tampering functions, for any constant $\delta > 0$. Notably, this class includes NC^0 tampering functions, namely constant depth circuits with bounded fan-in. It should be noted however, that the results of [7] do not extend to tampering adversaries in AC^0, since even for a low depth circuit in AC^0, any single output bit can depend on *all* input bits, thus violating the n^{δ}-locality constraint.

In a recent work, Chattopadhyay and Li [16] gave constructions of NMC based on connections between NMC and seedless non-malleable extractors.

One of their results is an efficient NMC against t-local tampering functions, where the decoding algorithm for the NMC is deterministic (in contrast, the result in [7] has randomized decoding). The locality parameters of the NMC in [16] are not as good as the one in [7], but better than the deterministic-decoding construction given in the appendix of the full version of [7]. Additionally, [16] also present a NMC against AC^0 tampering functions. However, this NMC results in a codeword that is super-polynomial in the message length, namely inefficient.

A recent work by Faust et al. [32] considered larger tampering classes by considering space bounded tampering adversaries in random oracle model. The construction achieves a new notion of *leaky* continuous non-malleable codes, where the adversary is assumed to learn some bounded $\log(|m|)$ bits of information about the underlying message m. However, this result is not directly comparable to ours as the adversarial model we consider is a that of standard non-malleability (without leakage), and for a subset of this tampering class (streaming space-bounded adversary) we achieve information theoretic security without random oracles.

Chandran et al. [12] considered another variant of non-malleable codes, called *block-wise* non-malleable codes. In this model, the codeword consists of number of blocks and the adversary receives the codeword block-by-block. The tampering function also consists of various function f_is, where each f_i can depend on codeword blocks $c_1, \ldots, c_i$ and modifies c_i to ${c'}_i$. It can be observed that standard non-malleability cannot be achieved in this model since, the adversary can simply wait to receive all the blocks of the codeword and then decode the codeword as part of last tampering function. Therefore, [12] define a new notion called non-malleability with replacement which relaxes the non-malleability requirement and considers the attack to be successful only if the tampered codeword is *valid and related* to the original message.

Other works on non-malleable codes include [2,4,11,13,14,20,24,25,28,33,34,37,41]. We guide the interested reader to [38,47] for a discussion of various models for tamper and leakage resilience.

2 Definitions

Where appropriate, we interpret functions $f : S \to \{\pm 1\}$ as boolean functions (and vice-versa) via the mapping: $0 \leftrightarrow 1$ and $1 \leftrightarrow -1$. The support of vector $\boldsymbol{x}$ is the set of indices i such that $x_i \neq 0$. A *bipartite graph* is an undirected graph $G = (V, E)$ in which V can be partitioned into two sets V_1 and V_2 such that $(u, v) \in E$ implies that either $u \in V_1$ and $v \in V_2$ or $v \in V_1$ and $u \in V_2$.

Non-malleable Codes. In this section we define the notion of *non-malleable codes* and its variants. In this work, we assume that the decoding algorithm of the non-malleable code may be *randomized* and all of our generic theorems are stated for this case. Nevertheless, only our instantiation for the streaming adversary (refer Sect. 7 in full version [8]) requires a randomized decoding algorithm, while our other instantiations enjoy deterministic decoding. We note that the original definition of non-malleable codes, given in [31], required a deterministic decoding algorithm. Subsequently, in [7], an alternative definition that allows for randomized decoding was introduced. We follow here the definition of [7]. Please see [7] for a discussion on why deterministic decoding is not necessarily without loss of generality in the non-malleable codes setting and for additional motivation for allowing randomized decoding.

Definition 1 (Coding Scheme). *Let $\Sigma, \widehat{\Sigma}$ be sets of strings, and $\kappa, \widehat{\kappa} \in \mathbb{N}$ be some parameters. A coding scheme consists of two algorithms* (E, D) *with the following syntax:*

- *The encoding algorithm* (perhaps randomized) *takes input a block of message in Σ and outputs a codeword in $\hat{\Sigma}$.*
- *The decoding algorithm* (perhaps randomized) *takes input a codeword in $\hat{\Sigma}$ and outputs a block of message in Σ.*

We require that for any message $m \in \Sigma$, $\Pr[\mathsf{D}(\mathsf{E}(m)) = m] = 1$, where the probability is taken over the choice of the encoding algorithm. In binary settings, we often set $\Sigma = \{0,1\}^{\kappa}$ and $\widehat{\Sigma} = \{0,1\}^{\widehat{\kappa}}$.

We next provide definitions of non-malleable codes of varying levels of security. We present general, game-based definitions that are applicable even for NMC that are in a model with a CRS, or that require computational assumptions. The corresponding original definitions of non-malleability, appropriate for an unconditional setting without a CRS, can be obtained as a special case of our definitions when setting $\mathsf{crs} = \perp$ and taking $\mathcal{G}$ to include all computable functions. These original definitions are also presented in Appendix A.1 of the full version [8].

Definition 2 (Non-malleability). *Let $\Pi = (\mathsf{CRSGen}, \mathsf{E}, \mathsf{D})$ be a coding scheme. Let $\mathcal{F}$ be some family of functions. For each attacker A, $m \in \Sigma$, define the tampering experiment $\mathsf{Tamper}^{\Pi,\mathcal{F}}_{A,m}(n)$ (Fig. 1):*

1. Challenger samples $\mathsf{crs} \leftarrow \mathsf{CRSGen}(1^n)$ and sends crs to A.
2. Attacker A sends the tampering function $f \in \mathcal{F}$ to the challenger.
3. Challenger computes $c \leftarrow \mathsf{E}(\mathsf{crs}, m)$.
4. Challenger computes the tampered codeword $\tilde{c} = f(c)$ and computes $\tilde{m} = \mathsf{D}(\mathsf{crs}, \tilde{c})$.
5. Experiment outputs $\tilde{m}$.

Fig. 1. Non-malleability experiment $\mathsf{Tamper}^{\Pi,\mathcal{F}}_{A,m}(n)$

We say the coding scheme $\Pi = (\mathsf{CRSGen}, \mathsf{E}, \mathsf{D})$ *is non-malleable against tampering class* $\mathcal{F}$ *and attackers* $A \in \mathcal{G}$*, if for every* $A \in \mathcal{G}$ *there exists a PPT simulator* Sim *such that for any message* $m \in \Sigma$ *we have,*

$$\mathsf{Tamper}^{\Pi,\mathcal{F}}_{A,m}(n) \approx \mathbf{Ideal}_{\mathsf{Sim},m}(n)$$

where $\mathbf{Ideal}_{\mathsf{Sim},m}(n)$ *is an experiment defined as follows (Fig. 2),*

1. Simulator Sim has oracle access to adversary A and outputs $\tilde{m} \cup \{\mathsf{same}^*\} \leftarrow \mathsf{Sim}^{A(\cdot)}(n)$.
2. Experiment outputs m if Sim outputs same^* and outputs $\tilde{m}$ otherwise.

Fig. 2. Non-malleability experiment $\mathbf{Ideal}_{\mathsf{Sim},m}(n)$

Definition 3 (Strong Non-malleability). *Let* $\Pi = (\mathsf{CRSGen}, \mathsf{E}, \mathsf{D})$ *be a coding scheme. Let* $\mathcal{F}$ *be some family of functions. For each attacker* $A \in \mathcal{G}$*,* $m \in \Sigma$*, define the tampering experiment* $\mathsf{StrongTamper}^{\Pi,\mathcal{F}}_{A,m}(n)$ *(Fig. 3):*

1. Challenger samples $\mathsf{crs} \leftarrow \mathsf{CRSGen}(1^n)$ and sends crs to A.
2. Attacker A sends the tampering function $f \in \mathcal{F}$ to the challenger.
3. Challenger computes $c \leftarrow \mathsf{E}(\mathsf{crs}, m)$.
4. Challenger computes the tampered codeword $\tilde{c} = f(c)$.
5. Compute $\tilde{m} = \mathsf{D}(\mathsf{crs}, \tilde{c})$.
6. Experiment outputs same^* if $\tilde{c} = c$, and $\tilde{m}$ otherwise.

Fig. 3. Strong non-malleability experiment $\mathsf{StrongTamper}^{\Pi,\mathcal{F}}_{A,m}(n)$

We say the coding scheme $\Pi = (\mathsf{CRSGen}, \mathsf{E}, \mathsf{D})$ *is strong non-malleable against tampering class* $\mathcal{F}$ *and attackers* $A \in \mathcal{G}$ *if we have*

$$\mathsf{StrongTamper}^{\Pi,\mathcal{F}}_{A,m_0}(n) \approx \mathsf{StrongTamper}^{\Pi,\mathcal{F}}_{A,m_1}(n)$$

for any $A \in \mathcal{G}$*,* $m_0, m_1 \in \Sigma$*.*

We now introduce an intermediate variant of non-malleability, called *Medium Non-malleability*, which informally gives security guarantees "in-between" strong and regular non-malleability. Specifically, the difference is that the experiment is allowed to output same^* only when some predicate g evaluated on $(c, \tilde{c})$ is set to true. Thus, strong non-malleability can be viewed as a special case of medium non-malleability, by setting g to be the identity function. On the other hand, regular non-malleability does not impose restrictions on when the experiment is allowed to output same^*. Note that g cannot be just any predicate in order for the definition to make sense. Rather, g must be a predicate such that if g evaluated on $(c, \tilde{c})$ is set to true, then (with overwhelming probability over the random coins of D) $\mathsf{D}(\tilde{c}) = \mathsf{D}(c)$.

Definition 4 (Medium Non-malleability). *Let $\Pi = (\mathsf{CRSGen}, \mathsf{E}, \mathsf{D})$ be a coding scheme. Let $\mathcal{F}$ be some family of functions.*

Let $g(\cdot,\cdot,\cdot,\cdot)$ be a predicate such that, for each attacker $A \in \mathcal{G}$, $m \in \Sigma$, the output of the following experiment, $\mathsf{Expt}^{\Pi,\mathcal{F}}_{A,m,g}(n)$ is 1 with at most negligible probability (Fig. 4):

1. Challenger samples $\mathsf{crs} \leftarrow \mathsf{CRSGen}(1^n)$ and sends crs to A.
2. Attacker A sends the tampering function $f \in \mathcal{F}$ to the challenger.
3. Challenger computes $c \leftarrow \mathsf{E}(\mathsf{crs}, m)$.
4. Challenger computes the tampered codeword $\tilde{c} = f(c)$.
5. Challenger samples $r \leftarrow U_\ell$.
6. Experiment outputs 1 if $([g(\mathsf{crs}, c, \tilde{c}, r) = 1] \wedge [\mathsf{D}(\mathsf{crs}, \tilde{c}; r) \neq m])$.

Fig. 4. The experiment corresponding to the special predicate g

For g as above, each $m \in \Sigma$, and attacker $A \in \mathcal{G}$, define the tampering experiment
$\mathsf{MediumTamper}^{\Pi,\mathcal{F}}_{A,m,g}(n)$ *as shown in Fig. 5:*

1. Challenger samples $\mathsf{crs} \leftarrow \mathsf{CRSGen}(1^n)$ and sends crs to A.
2. Attacker A sends the tampering function $f \in \mathcal{F}$ to the challenger.
3. Challenger computes $c \leftarrow \mathsf{E}(\mathsf{crs}, m)$.
4. Challenger computes the tampered codeword $\tilde{c} = f(c)$.
5. Challenger samples $r \leftarrow U_\ell$ and computes $\tilde{m} = \mathsf{D}(\mathsf{crs}, \tilde{c}, r)$.
6. Experiment outputs same^* if $g(\mathsf{crs}, c, \tilde{c}, r) = 1$, and $\tilde{m}$ otherwise.

Fig. 5. Medium non-malleability experiment $\mathsf{MediumTamper}^{\Pi,\mathcal{F}}_{A,m,g}(n)$

We say the coding scheme $\Pi = (\mathsf{CRSGen}, \mathsf{E}, \mathsf{D})$ is medium non-malleable against tampering class $\mathcal{F}$ and attackers $A \in \mathcal{G}$ if we have

$$\mathsf{MediumTamper}^{\Pi,\mathcal{F}}_{A,m_0,g}(n) \approx \mathsf{MediumTamper}^{\Pi,\mathcal{F}}_{A,m_1,g}(n)$$

for any $A \in \mathcal{G}$, $m_0, m_1 \in \Sigma$.

We next recall some standard definitions of public-key encryption (PKE), pseudorandom generator (PRG), and non-interactive zero knowledge proof systems with simulation soundness in Sects. 2.2 and 2.3 of the full version [8].

Definition 5 (Non-interactive Simulatable Proof System). *A tuple of probabilistic polynomial time algorithms $\Pi^{\mathsf{NI}} = (\mathsf{CRSGen}^{\mathsf{NI}}, \mathsf{P}^{\mathsf{NI}}, \mathsf{V}^{\mathsf{NI}}, \mathsf{Sim}^{\mathsf{NI}})$ is a* non-interactive simulatable proof system *for language $L \in NP$ with witness relation W if $(\mathsf{CRSGen}^{\mathsf{NI}}, \mathsf{P}^{\mathsf{NI}}, \mathsf{V}^{\mathsf{NI}}, \mathsf{Sim}^{\mathsf{NI}})$ have the following syntax:*

- $\mathsf{CRSGen}^{\mathsf{NI}}$ *is a randomized algorithm that outputs* $(\mathsf{crs}^{\mathsf{NI}}, \tau_{\mathsf{sim}})$.
- *On input* crs, $x \in L$ *and witness* w *such that* $W(x, w) = 1$, $\mathsf{P}^{\mathsf{NI}}(\mathsf{crs}, x, w)$ *outputs proof* π.
- *On input* crs, x, π, $\mathsf{V}^{\mathsf{NI}}(\mathsf{crs}, x, \pi)$ *outputs either 0 or 1.*
- *On input* crs, τ_{sim} *and* $x \in L$, $\mathsf{Sim}^{\mathsf{NI}}(\mathsf{crs}, \tau_{\mathsf{sim}}, x)$ *outputs simulated proof* π'.

Completeness: We require the following completeness property: For all $x \in L$, *and all* w *such that* $W(x, w) = 1$, *for all strings* $\mathsf{crs}^{\mathsf{NI}}$ *of length* $\mathrm{poly}(|x|)$, *and for all adversaries* $\mathcal{A}$ *we have*

$$\Pr\left[\begin{array}{c}(\mathsf{crs}^{\mathsf{NI}}, \tau_{\mathsf{Sim}}) \leftarrow \mathsf{CRSGen}^{\mathsf{NI}}(1^n); (x, w) \leftarrow \mathcal{A}(\mathsf{crs}^{\mathsf{NI}});\\ \pi \leftarrow \mathsf{P}^{\mathsf{NI}}(\mathsf{crs}^{\mathsf{NI}}, x, w) : \mathsf{V}^{\mathsf{NI}}(\mathsf{crs}^{\mathsf{NI}}, x, \pi) = 1\end{array}\right] \geq 1 - \mathsf{negl}(n)$$

Soundness: We say that Π^{NI} *enjoys soundness against adversaries* $\mathcal{A} \in \mathcal{G}$ *if: For all* $x \notin L$, *and all adversaries* $\mathcal{A} \in \mathcal{G}$*:*

$$\Pr\left[\begin{array}{c}(\mathsf{crs}^{\mathsf{NI}}, \tau_{\mathsf{Sim}}) \leftarrow \mathsf{CRSGen}^{\mathsf{NI}}(1^n);\\ (x, \pi) \leftarrow \mathcal{A}(\mathsf{crs}^{\mathsf{NI}}) : \mathsf{V}^{\mathsf{NI}}(\mathsf{crs}^{\mathsf{NI}}, x, \pi) = 0\end{array}\right] \geq 1 - \mathsf{negl}(n)$$

The security properties that we require of Π^{NI} *will depend on our particular non-malleable code construction as well as the particular class,* $\mathcal{F}$, *of tampering functions that we consider. The exact properties needed are those that will arise from Theorems 2 and 4. In subsequent sections, we will show how to construct non-interactive simulatable proof systems satisfying these properties.*

Proof Systems for Circuit SAT. We now consider proof of knowledge systems for Circuit SAT, where the prover and/or verifier have limited computational resources.

Definition 6 (Proof of Knowledge Systems for Circuit SAT with Computationally Bounded Prover/Verifier). *For a circuit* C, *let* $\mathcal{L}(C)$ *denote the set of strings* x *such that there exists a witness* w *such that* $C(x, w) = 1$. *For a class* $\mathcal{C}$, *let* $\mathcal{L}(\mathcal{C})$ *denote the set* $\{\mathcal{L}(C) \mid C \in \mathcal{C}\}$. $\Pi = (\mathsf{P}, \mathsf{V})$ *is a Circuit SAT proof system for the class* $\mathcal{L}(\mathcal{C})$ *with prover complexity* $\mathcal{D}$ *and verifier complexity* $\mathcal{E}$ *if the following are true:*

- *For all* $C \in \mathcal{C}$ *and all valid inputs* (x, w) *such that* $C(x, w) = 1$, $\mathsf{P}(C, \cdot, \cdot)$ *can be computed in complexity class* $\mathcal{D}$.
- *For all* $C \in \mathcal{C}$, $\mathsf{V}(C, \cdot, \cdot)$ *can be computed in complexity class* $\mathcal{E}$.
- *Completeness: For all* $C \in \mathcal{C}$ *and all* (x, w) *such that* $C(x, w) = 1$, *we have* $\mathsf{V}(C, x, \mathsf{P}(C, x, w)) = 1$.
- *Extractability: For all* (C, x, π), *if* $\Pr_r[\mathsf{V}(C, x, \pi; r) = 1]$ *is non-negligible, then given* (C, x, π) *it is possible to efficiently extract* w *such that* $C(x, w) = 1$.

We construct Circuit SAT proof systems for the class $\mathcal{L}(\mathsf{P/poly})$ with verifier complexity AC^0 in Sect. 2.4 of full version [8]. We also construct Circuit SAT proof systems for the class. $\mathcal{L}(\mathsf{P/poly})$ with streaming verifier in Sect. 2.4 of full version [8].

Given the above, we have the following theorem:

Theorem 1. *Assuming the existence of same-string, weak one-time simulation sound NIZK with deterministic verifier, there exists same-string, weak one-time simulation sound NIZK with verifier in* AC^0.

We also recall some definitions and results related to boolean analysis and present them next. in Sect. 2.5 of full version [8].

Computational Model for Streaming Adversaries. In this section we discuss the computational model used for analysis of the streaming adversaries. This model is similar to the one used in [49].

General Streaming Adversaries. The input is represented as a stream $S_1, \ldots, S_\ell$, where for $i \in [\ell]$, each $S_i \in \{0,1\}^B$, where B is the block length. We model the adversary by a *branching program.* A branching program of length ℓ and width w, is a directed acyclic graph with the vertices arranged in $\ell+1$ layers such that no layer contains more than w vertices. Intuitively, each layer represents a time step of computation whereas, each vertex in the graph corresponds to the potential memory state learned by the adversary. The first layer (layer 0) contains a single vertex, called the *start vertex*, which represents the input. A vertex is called *leaf* if it has out-degree 0, and represents the output (the learned value of x) of the program. Every non-leaf vertex in the program has exactly 2^B outgoing edges, labeled by elements $S \in \{0,1\}^B$, with exactly one edge labeled by each such S, and all the edges from layer $j-1$ going to vertices in layer j. Intuitively, these edges represent the computation on reading S_i as streaming input. The stream $S_1, \ldots, S_\ell$, therefore, define a computation-path in the branching program.

We discuss the streaming branching program adversaries, and streaming adversaries for learning parity in Sect. 2.6 of full version [8].

3 Generic Construction for One-Bit Messages

In this section we present the generic construction for encoding a single bit.

Let $\Psi(p, c, x, y, r, z)$ be defined as a function that takes as input a predicate p, and variables c, x, y, r, z. If $p(c, x, y, r) = 1$, then Ψ outputs 0. Otherwise, Ψ outputs z.

Theorem 2. *Let* (E, D), E_1, E_2, Ext, D' *and* g *be as defined in Figs. 6, 7, 8, 9, 10 and 11. Let* $\mathcal{F}$ *be a computational class. If, for every adversary* $\mathcal{A} \in \mathcal{G}$ *outputting tampering functions* $f \in \mathcal{F}$, *all of the following hold:*

Let $\mathcal{E}$ = (Gen, Encrypt, Decrypt) be a public key encryption scheme with perfect correctness (see Definition 7 in [8]). Let Π^{NI} = ($\mathsf{CRSGen}^{\mathsf{NI}}, \mathsf{P}^{\mathsf{NI}}, \mathsf{V}^{\mathsf{NI}}, \mathsf{Sim}^{\mathsf{NI}}$) be a non-interactive simulatable proof system with soundness against adversaries $\mathcal{A} \in \mathcal{G}$ (see Definition 5). Note that in the CRS model, we implicitly assume that all algorithms take the CRS as input, and for simplicity of notation, sometimes do not list the CRS as an explicit input.

$\mathsf{CRSGen}(1^n)$:

1. Choose $(\text{PK}, \text{SK}) \leftarrow \mathsf{Gen}(1^n)$.
2. Choose $[(\mathsf{crs}_i^{\mathsf{NI}}, \tau_{\mathsf{sim}}^i)]_{i \in \{0,\ldots n\}} \leftarrow \mathsf{CRSGen}^{\mathsf{NI}}(1^n)$. Let $\overrightarrow{\mathsf{crs}}^{\mathsf{NI}} := [\mathsf{crs}_i^{\mathsf{NI}}]_{i \in \{0,\ldots n\}}$ and let $\overrightarrow{\tau}_{\mathsf{sim}} := [\tau_{\mathsf{sim}}^i]_{i \in \{0,\ldots n\}}$
3. Output $\mathsf{crs} := (\text{PK}, \overrightarrow{\mathsf{crs}}^{\mathsf{NI}})$.

Languages. We define the following languages:

- $\mathcal{L}_i^{\beta}$: For $i \in [n]$, $\beta \in \{0,1\}$, $s := (\hat{\boldsymbol{k}}, \boldsymbol{c}, c) \in \mathcal{L}_i^{\beta}$ iff the i-th ciphertext $c_i := k_i \oplus \beta$ (where $\boldsymbol{c} = c_1, \ldots, c_n$) and the i-th encryption $\hat{k}_i$ (where $\hat{\boldsymbol{k}} = \hat{k}_1, \ldots, \hat{k}_{n+1}$) is an encryption of k_i under PK (where PK is hardwired into the language).
- $\mathcal{L}$: $s := (\hat{\boldsymbol{k}}, \boldsymbol{c}, c) \in \mathcal{L}$ iff $(x_1, \ldots, x_n)$ is in the support of D_b where:
  1. For $i \in [n]$, $x_i := c_i \oplus k_i$
  2. $b := c \oplus k_{n+1}$
  3. $\hat{\boldsymbol{k}}$ is an encryption of $k_1, \ldots, k_{n+1}$ under PK (where PK is hardwired into the language).

$\mathsf{E}(\mathsf{crs}, b)$:

1. Sample $\boldsymbol{x} \leftarrow D_b$, where $\boldsymbol{x} = x_1, \ldots, x_n$.
2. Choose an $n+1$-bit key $\boldsymbol{k} = k_1, \ldots, k_n, k$ uniformly at random. For $i \in [n]$, compute $\hat{k}_i \leftarrow \mathsf{Encrypt}_{\text{PK}}(k_i)$ and compute $\hat{k}_{n+1} \leftarrow \mathsf{Encrypt}_{\text{PK}}(k)$. Let $\hat{\boldsymbol{k}} := \hat{k}_1, \ldots, \hat{k}_{n+1}$.
3. Compute $c_1 := k_1 \oplus x_1, \ldots, c_n := k_n \oplus x_n$. Let $\boldsymbol{c} := c_1, \ldots, c_n$.
4. Compute $c := b \oplus k$.
5. For $i \in [n]$, compute a non-interactive, simulatable proof T_i proving $s := (\hat{\boldsymbol{k}}, \boldsymbol{c}, c) \in \mathcal{L}_i^{x_i}$ relative to $\mathsf{crs}_i^{\mathsf{NI}}$.
6. Compute a non-interactive, simulatable proof T proving $s := (\hat{\boldsymbol{k}}, \boldsymbol{c}, c) \in \mathcal{L}$ relative to $\mathsf{crs}_0^{\mathsf{NI}}$.
7. Output $\mathsf{CW} := (\hat{\boldsymbol{k}}, c_1, \ldots, c_n, c, T, x_1, T_1, .., x_n, T_n)$.

$\mathsf{D}(\mathsf{crs}, \mathsf{CW})$:

1. Parse $\mathsf{CW} := (\hat{\boldsymbol{k}}, c_1, \ldots, c_n, c, T, x_1, T_1, .., x_n, T_n)$
2. Check that V^{NI} outputs 1 on all proofs $T_1, .., T_n, T$, relative to the corresponding CRS.
3. If yes, output b such that $x_1 \ldots x_n$ is in the support of D_b. If not, output 0.

Fig. 6. Non-malleable code $(\mathsf{CRSGen}, \mathsf{E}, \mathsf{D})$, secure against $\mathcal{F}$ tampering.

$\mathsf{E}_1(\mathsf{crs}, \overrightarrow{\tau}_{\mathsf{sim}}, r, b)$:

1. Sample $\boldsymbol{x} \leftarrow D_b$, where $\boldsymbol{x} = x_1, \ldots, x_n$.
2. Choose an $n+1$-bit key $\boldsymbol{k} = k_1, \ldots, k_n, k$ uniformly at random. For $i \in [n]$, compute $\hat{k}_i \leftarrow \mathsf{Encrypt}_{\mathrm{PK}}(k_i)$ and compute $\hat{k}_{n+1} \leftarrow \mathsf{Encrypt}_{\mathrm{PK}}(k)$. Let $\hat{\boldsymbol{k}} := \hat{k}_1, \ldots, \hat{k}_{n+1}$.
3. Compute $c_1 := k_1 \oplus x_1, \ldots, c_n := k_n \oplus x_n$. Let $\boldsymbol{c} := c_1, \ldots, c_n$.
4. Compute $c := b \oplus k$.
5. For $i \in [n]$, use τ^i_{sim} and r to simulate a non-interactive proof T'_i proving $(\hat{\boldsymbol{k}}, \boldsymbol{c}, c) \in \mathcal{L}^{x_i}_i$, relative to $\mathsf{crs}^{\mathsf{NI}}_i$.
6. Use τ^0_{sim} and r to simulate a non-interactive proof T' proving $(\hat{\boldsymbol{k}}, \boldsymbol{c}, c) \in \mathcal{L}$, relative to $\mathsf{crs}^{\mathsf{NI}}_0$.
7. Output $\mathsf{CW} := (\hat{\boldsymbol{k}}, c_1, \ldots, c_n, c, T', x_1, T'_1, .., x_n, T'_n)$.

Fig. 7. Encoding algorithm with simulated proofs.

$\mathsf{E}_2(\mathsf{crs}, \overrightarrow{\tau}_{\mathsf{sim}}, r, b)$:

1. Sample $\boldsymbol{x} \leftarrow D_b$, where $\boldsymbol{x} = x_1, \ldots, x_n$.
2. Choose $c'_1, \ldots, c'_n$ uniformly at random. Let $\boldsymbol{c}' := c'_1, \ldots, c'_n$.
3. Choose c' uniformly at random.
4. Set $\boldsymbol{k}' = c'_1, \ldots, c'_n, c'$. For $i \in [n]$, compute $\hat{k}'_i \leftarrow \mathsf{Encrypt}_{\mathrm{PK}}(k'_i)$ and compute $\hat{k}'_{n+1} \leftarrow \mathsf{Encrypt}_{\mathrm{PK}}(k')$. Let $\hat{\boldsymbol{k}}' := \hat{k}'_1, \ldots, \hat{k}'_{n+1}$.
5. For $i \in [n]$, use τ^i_{sim} and r to simulate a non-interactive proof T'_i proving $(\hat{\boldsymbol{k}}', \boldsymbol{c}', c) \in \mathcal{L}^{x_i}_i$, relative to $\mathsf{crs}^{\mathsf{NI}}_i$.
6. Use τ^0_{sim} and r to simulate a non-interactive proof T' proving $(\hat{\boldsymbol{k}}', \boldsymbol{c}', c) \in \mathcal{L}$, relative to $\mathsf{crs}^{\mathsf{NI}}_0$.
7. Output $\mathsf{CW} := (\hat{\boldsymbol{k}}', c'_1, \ldots, c'_n, c', T', x_1, T'_1, .., x_n, T'_n)$.

Fig. 8. Encoding algorithm with simulated proofs and encryptions.

$\mathsf{Ext}(\mathsf{crs}, \mathrm{SK}, \mathsf{CW})$:

1. Parse $\mathsf{CW} := (\hat{\boldsymbol{k}}, c_1, \ldots, c_n, c, T, x_1, T_1, .., x_n, T_n)$,
2. Output $\mathsf{Decrypt}_{\mathrm{SK}}(\hat{k}_{n+1})$.

Fig. 9. Extracting procedure Ext.

$\mathsf{D}'(\mathsf{crs}, k, \mathsf{CW})$:

1. Parse $\mathsf{CW} := (\hat{\boldsymbol{k}}, c_1, \ldots, c_n, c, T, x_1, T_1, .., x_n, T_n)$,
2. Check that V^{NI} outputs 1 on all proofs $T_1, .., T_n, T$, relative to the corresponding CRS,
3. If not, output 0. Otherwise, output $b := k \oplus c$.

Fig. 10. Alternate decoding procedure D', given additional extracted key k as input.

$g(\mathsf{crs}, \mathsf{CW}, \mathsf{CW}^*, r)$:

1. Parse $\mathsf{CW} = (\hat{\boldsymbol{k}}, \boldsymbol{c}, c, T, x_1, T_1, .., x_n, T_n)$, $\mathsf{CW}^* = (\hat{\boldsymbol{k}}^*, \boldsymbol{c}^*, c^*, T^*, x_1^*, T_1^*, .., x_n^*, T_n^*)$.
2. If (1) V^{NI} outputs 1 on all proofs $T^*, T_1^*, .., T_n^*$, relative to the corresponding CRS; and (2) $(\hat{\boldsymbol{k}}, \boldsymbol{c}, c) = (\hat{\boldsymbol{k}}^*, \boldsymbol{c}^*, c^*)$, then output 1. Otherwise output 0.

Fig. 11. The predicate $g(\mathsf{crs}, \mathsf{CW}, \mathsf{CW}^*, r)$.

Simulation of proofs.

1. $\Pr[g(\mathsf{crs}, \mathsf{CW}_0, f(\mathsf{CW}_0), r_0) = 1] \approx \Pr[g(\mathsf{crs}, \mathsf{CW}_1, f(\mathsf{CW}_1), r_1) = 1]$,

2. $\Psi(g, \mathsf{crs}, \mathsf{CW}_0, f(\mathsf{CW}_0), r_0, \mathsf{D}(\mathsf{crs}, f(\mathsf{CW}_0); r_0)) \approx \Psi(g, \mathsf{crs}, \mathsf{CW}_1, f(\mathsf{CW}_1), r_1, \mathsf{D}(\mathsf{crs}, f(\mathsf{CW}_1); r_1))$,

where $(\mathsf{crs}, \mathrm{SK}, \vec{\tau}_{\mathsf{sim}}) \leftarrow \mathsf{CRSGen}(1^n)$, $f \leftarrow \mathcal{A}(\mathsf{crs})$, r_0, r_1 *are sampled uniformly at random,* $\mathsf{CW}_0 \leftarrow \mathsf{E}(\mathsf{crs}, 0)$ *and* $\mathsf{CW}_1 \leftarrow \mathsf{E}_1(\mathsf{crs}, \vec{\tau}_{\mathsf{sim}}, r_1, 0)$.

Simulation of Encryptions.

1. $\Pr[g(\mathsf{crs}, \mathsf{CW}_1, f(\mathsf{CW}_1), r_1) = 1] \approx \Pr[g(\mathsf{crs}, \mathsf{CW}_2, f(\mathsf{CW}_2), r_2) = 1]$,

2. $\Psi(g, \mathsf{crs}, \mathsf{CW}_1, f(\mathsf{CW}_1), r_1, \mathsf{D}(\mathsf{crs}, f(\mathsf{CW}_1); r_1)) \approx \Psi(g, \mathsf{crs}, \mathsf{CW}_2, f(\mathsf{CW}_2), r_2, \mathsf{D}(\mathsf{crs}, f(\mathsf{CW}_2); r_2))$,

where $(\mathsf{crs}, \mathrm{SK}, \vec{\tau}_{\mathsf{sim}}) \leftarrow \mathsf{CRSGen}(1^n)$, $f \leftarrow \mathcal{A}(\mathsf{crs})$, r_1, r_2 *are sampled uniformly at random,* $\mathsf{CW}_1 \leftarrow \mathsf{E}_1(\mathsf{crs}, \vec{\tau}_{\mathsf{sim}}, r_1, 0)$ *and* $\mathsf{CW}_2 \leftarrow \mathsf{E}_2(\mathsf{crs}, \vec{\tau}_{\mathsf{sim}}, r_2, 0)$.

Simulation Soundness.

$$\Pr\left[\begin{array}{c}\mathsf{D}(\mathsf{crs}, f(\mathsf{CW}_2); r_2) \neq \mathsf{D}'(\mathsf{crs}, \mathsf{Ext}(\mathsf{crs}, \mathrm{SK}, f(\mathsf{CW}_2)), f(\mathsf{CW}_2); r_2) \\ \wedge\, g(\mathsf{crs}, \mathsf{CW}_2, f(\mathsf{CW}_2), r_2) = 0\end{array}\right] \leq \mathsf{negl}(n),$$

where $(\mathsf{crs}, \mathrm{SK}, \vec{\tau}_{\mathsf{sim}}) \leftarrow \mathsf{CRSGen}(1^n)$, $f \leftarrow \mathcal{A}(\mathsf{crs})$, r_2 *is sampled uniformly at random and* $\mathsf{CW}_2 \leftarrow \mathsf{E}_2(\mathsf{crs}, \vec{\tau}_{\mathsf{sim}}, r, 0)$.

Hardness of D_b relative to Alternate Decoding.

1. $\Pr[g(\mathsf{crs}, \mathsf{CW}_2, f(\mathsf{CW}_2), r_2) = 1] \approx \Pr[g(\mathsf{crs}, \mathsf{CW}_3, f(\mathsf{CW}_3), r_3) = 1]$,
2. $\mathsf{D}'(\mathsf{crs}, \mathsf{Ext}(\textsc{sk}, f(\mathsf{CW}_2)), f(\mathsf{CW}_2); r_2) \approx \mathsf{D}'(\mathsf{crs}, \mathsf{Ext}(\textsc{sk}, f(\mathsf{CW}_3)), f(\mathsf{CW}_3); r_3)$,

where $(\mathsf{crs}, \textsc{sk}, \overrightarrow{\tau}_{\mathsf{sim}}) \leftarrow \mathsf{CRSGen}(1^n)$, $f \leftarrow \mathcal{A}(\mathsf{crs})$, r_2, r_3 *are sampled uniformly at random,* $\mathsf{CW}_2 \leftarrow \mathsf{E}_2(\mathsf{crs}, \overrightarrow{\tau}_{\mathsf{sim}}, r_2, 0)$ *and* $\mathsf{CW}_3 \leftarrow \mathsf{E}_2(\mathsf{crs}, \overrightarrow{\tau}_{\mathsf{sim}}, r_3, 1)$.

Then the construction presented in Fig. 6 is a non-malleable code for class $\mathcal{F}$ against adversaries $\mathcal{A} \in \mathcal{G}$.

Proof (Proof of Theorem 2). We take g to be the predicate that is used in the $\mathsf{MediumTamper}^{\Pi,\mathcal{F}}_{A,m,g}(n)$ tampering experiment. We must argue that for every $m \in \{0,1\}$ and every attacker $A \in \mathcal{G}$ the output of the experiment $\mathsf{Expt}^{\Pi,\mathcal{F}}_{A,m,g}(n)$ is 1 with at most negligible probability.

Assume towards contradiction that for some $A \in \mathcal{G}$ the output of the experiment is 1 with non-negligible probability. Then this means that the probability in the last line of experiment $\mathsf{Expt}^{\Pi,\mathcal{F}}_{A,m,g}(n)$ that $g(\mathsf{crs}, \mathsf{CW}, \mathsf{CW}^*, r) = 1 \wedge \mathsf{D}(\mathsf{crs}, \mathsf{CW}^*; r) \neq m$ is non-negligible. Parse $\mathsf{CW} = (\hat{\boldsymbol{k}}, \boldsymbol{c}, c, T, x_1, T_1, .., x_n, T_n)$, $\mathsf{CW}^* = (\hat{\boldsymbol{k}}^*, \boldsymbol{c}^*, c^*, T^*, x_1^*, T_1^*, .., x_n^*, T_n^*)$.

Recall that $\mathsf{D}(\mathsf{crs}, \mathsf{CW}; r) = m$. Thus, if the above event occurs, it means that $\mathsf{D}(\mathsf{crs}, \mathsf{CW}; r) \neq \mathsf{D}(\mathsf{crs}, \mathsf{CW}^*; r)$. But since $g(\mathsf{crs}, \mathsf{CW}, \mathsf{CW}^*, r) = 1$, it means that V^{NI} outputs 1 on all proofs T^*, $[T_i^*]_{i \in [n]}$ and $(\hat{\boldsymbol{k}}, \boldsymbol{c}, c) = (\hat{\boldsymbol{k}}^*, \boldsymbol{c}^*, c^*)$.

This, in turn, means that there must be some bit x_i, x_i^* that CW and CW^* differ on. But note that by assumption $c_i = c_i^*$. Due to the fact that CW is well-formed and perfect correctness of the encryption scheme, it must mean that $c_i^* \notin \mathcal{L}_i^{x_i^*}$. But recall that by assumption, proof T_i^* verifies correctly. This means that soundness is broken by $A \in \mathcal{G}$. This contradicts the security of the proof system Π^{NI}.

Next, recall that we wish to show that for any adversary $A \in \mathcal{G}$ outputting tampering function $\{\mathsf{MediumTamper}^{\Pi,\mathcal{F}}_{A,0,g}\}_{k \in \mathbb{N}} \approx \{\mathsf{MediumTamper}^{\Pi,\mathcal{F}}_{A,1,g}\}_{k \in \mathbb{N}}$

To do so we consider the following hybrid argument:

Hybrid 0: The real game, $\mathsf{MediumTamper}^{\Pi,\mathcal{F}}_{A,0,g}$, relative to g, where the real encoding $\mathsf{CW}_0 \leftarrow \mathsf{E}(\mathsf{crs}, 0)$ and the real decoding oracle D are used.

Hybrid 1: Replace the encoding from the previous game with $\mathsf{CW}_1 \leftarrow \mathsf{E}_1(\mathsf{crs}, \overrightarrow{\tau}_{\mathsf{sim}}, r_1, 0)$ where r_1 is chosen uniformly at random and g, D use random coins r_1.

Hybrid 2: Replace the encoding from the previous game with $\mathsf{CW}_2 \leftarrow \mathsf{E}_2(\mathsf{crs}, \overrightarrow{\tau}_{\mathsf{sim}}, r_2, 0)$, where r_2 is chosen uniformly at random and g, D use random coins r_2.

Hybrid 3: Replace the decoding from the previous game, with $\mathsf{D}'(\mathsf{crs}, \mathsf{Ext}(\mathsf{crs}, \textsc{sk}, f(\mathsf{CW}_2)), f(\mathsf{CW}_2); r_2)$. where r_2 is chosen uniformly at random and g, E_2 use random coins r_2.

Hybrid 4: Same as Hybrid 3, but replace the encoding with $\mathsf{CW}_3 \leftarrow \mathsf{E}_2(\mathsf{crs}, \overrightarrow{\tau}_{\mathsf{sim}}, r_3, 1)$, where r_3 is chosen uniformly at random and g, D' use random coins r_3.

Now, we prove our hybrids are indistinguishable.

Claim. Hybrid 0 is computationally indistinguishable from Hybrid 1.

Proof. The claim follows immediately from the **Simulation of proofs** property in Theorem 2.

Claim. Hybrid 1 is computationally indistinguishable from Hybrid 2.

Proof. The claim follows immediately from the **Simulation of Encryptions** property in Theorem 2.

Claim. Hybrid 2 is computationally indistinguishable from Hybrid 3.

Proof. This claim follows from the fact that (1) if $g(\mathsf{crs}, \mathsf{CW}, \mathsf{CW}^*, r) = 1$, then the experiment outputs same^* in both Hybrid 2 and Hybrid 3; and (2) the probability that $g(\mathsf{crs}, \mathsf{CW}, \mathsf{CW}^*, r) = 0$ and the output of the experiment is different in Hybrid 2 and Hybrid 3 is at most negligible, due to the **Simulation Soundness** property in Theorem 2.

Claim. Hybrid 3 is computationally indistinguishable from Hybrid 4.

Proof. This follows from the fact that (1) for $\gamma \in \{2, 3\}$ if $g(\mathsf{crs}, \mathsf{CW}_2, f(\mathsf{CW}_2), r_2) = 1$ then $\mathsf{D}'(\mathsf{crs}, \mathsf{Ext}(\mathsf{crs}, \mathrm{SK}, f(\mathsf{CW}_\gamma)), f(\mathsf{CW}_\gamma); r_\gamma)$ always outputs 0 and so

$$\begin{aligned}&\mathsf{D}'(\mathsf{crs}, \mathsf{Ext}(\mathsf{crs}, \mathrm{SK}, f(\mathsf{CW}_\gamma)), f(\mathsf{CW}_\gamma); r_\gamma)\\&\qquad \equiv \Psi(g, \mathsf{crs}, \mathsf{CW}_\gamma, f(\mathsf{CW}_\gamma), r_\gamma, \mathsf{D}'(\mathsf{crs}, \mathsf{Ext}(\mathsf{crs}, \mathrm{SK}, f(\mathsf{CW}_\gamma)), f(\mathsf{CW}_\gamma); r_\gamma));\end{aligned}$$

and (2) the **Hardness of D_b relative to Alternate Decoding** property in Theorem 2.

4 One-Bit NMC for $\mathbf{AC^0}$

In this section, we show that our generic construction yields efficient NMC for AC^0 in the CRS model, when each of the underlying primitives is appropriately instantiated.

Theorem 3. $\Pi = (\mathsf{CRSGen}, \mathsf{E}, \mathsf{D})$ *(presented in Fig. 6) is a one-bit, computational, non-malleable code in the CRS model, secure against every* PPT *adversary $\mathcal{A}$ outputting tampering functions $f \in \mathsf{AC}^0$, if the underlying components are instantiated in the following way:*

- $\mathcal{E} := (\mathsf{Gen}, \mathsf{Encrypt}, \mathsf{Decrypt})$ *is a public key encryption scheme with perfect correctness and decryption in* AC^0.

- $\Pi^{\mathsf{NI}} := (\mathsf{CRSGen}^{\mathsf{NI}}, \mathsf{P}^{\mathsf{NI}}, \mathsf{V}^{\mathsf{NI}}, \mathsf{Sim}^{\mathsf{NI}})$ *is a same-string, weak one-time simulation-sound NIZK with verifier in* AC^0.
- *For* $b \in \{0,1\}$, D_b *is the distribution that samples bits* $x_1 \ldots x_n$ *uniformly at random, conditioned on* $x_1 \oplus \cdots \oplus x_n = b$.

Note that given Theorem 1, proof systems Π^{NI} as above exist, under the assumption that same-string, weak one-time simulation-sound NIZK with (arbitrary polynomial-time) deterministic verifier exists. Such NIZK can be constructed in the CRS model from enhanced trapdoor permutations [50]. Public key encryption with perfect correctness and decryption in AC^0 can be constructed by applying the low-decryption-error transformation of Dwork et al. [29] to the (reduced decryption error) encryption scheme of Bogdanov and Lee [9]. Refer to Sect. 4 of the full version [8] for additional details.

Proof (Proof of Theorem 3). To prove the theorem, we need to show that for every PPT adversary $\mathcal{A}$ outputting tampering functions $f \in \mathcal{F}$, the necessary properties from Theorem 2 hold. We next go through these one by one.

Simulation of proofs.

1. $\Pr[g(\mathsf{crs}, \mathsf{CW}_0, f(\mathsf{CW}_0), r_0) = 1] \approx \Pr[g(\mathsf{crs}, \mathsf{CW}_1, f(\mathsf{CW}_1), r_1) = 1]$,
2. $\Psi(g, \mathsf{crs}, \mathsf{CW}_0, f(\mathsf{CW}_0), r_0, \mathsf{D}(\mathsf{crs}, f(\mathsf{CW}_0); r_0)) \approx \Psi(g, \mathsf{crs}, \mathsf{CW}_1, f(\mathsf{CW}_1), r_1, \mathsf{D}(\mathsf{crs}, f(\mathsf{CW}_1); r_1))$,

where $(\mathsf{crs}, \mathsf{SK}, \vec{\tau}_{\mathsf{sim}}) \leftarrow \mathsf{CRSGen}(1^n)$, $f \leftarrow \mathcal{A}(\mathsf{crs})$, r_0, r_1 are sampled uniformly at random, $\mathsf{CW}_0 \leftarrow \mathsf{E}(\mathsf{crs}, 0)$ and $\mathsf{CW}_1 \leftarrow \mathsf{E}_1(\mathsf{crs}, \vec{\tau}_{\mathsf{sim}}, r_1, 0)$. This follows immediately from the zero-knowledge property of $\Pi^{\mathsf{NI}} = (\mathsf{CRSGen}^{\mathsf{NI}}, \mathsf{P}^{\mathsf{NI}}, \mathsf{V}^{\mathsf{NI}}, \mathsf{Sim}^{\mathsf{NI}})$.

Simulation of Encryptions.

1. $\Pr[g(\mathsf{crs}, \mathsf{CW}_1, f(\mathsf{CW}_1), r_1) = 1] \approx \Pr[g(\mathsf{crs}, \mathsf{CW}_2, f(\mathsf{CW}_2), r_2) = 1]$,
2. $\Psi(g, \mathsf{crs}, \mathsf{CW}_1, f(\mathsf{CW}_1), r_1, \mathsf{D}(\mathsf{crs}, f(\mathsf{CW}_1); r_1)) \approx \Psi(g, \mathsf{crs}, \mathsf{CW}_2, f(\mathsf{CW}_2), r_2, \mathsf{D}(\mathsf{crs}, f(\mathsf{CW}_2); r_2))$,

where $(\mathsf{crs}, \mathsf{SK}, \vec{\tau}_{\mathsf{sim}}) \leftarrow \mathsf{CRSGen}(1^n)$, $f \leftarrow \mathcal{A}(\mathsf{crs})$, r_1, r_2 are sampled uniformly at random, $\mathsf{CW}_1 \leftarrow \mathsf{E}_1(\mathsf{crs}, \vec{\tau}_{\mathsf{sim}}, r_1, 0)$ and $\mathsf{CW}_2 \leftarrow \mathsf{E}_2(\mathsf{crs}, \vec{\tau}_{\mathsf{sim}}, r_2, 0)$. This follows immediately from the fact that $\boldsymbol{c}, c$ and $\boldsymbol{c}', c'$ are identically distributed when generated by E_1 versus E_2 and from the semantic security of the public key encryption scheme $\mathcal{E} = (\mathsf{Gen}, \mathsf{Encrypt}, \mathsf{Decrypt})$.

Simulation Soundness.

$$\Pr\left[\begin{array}{c}\mathsf{D}(\mathsf{crs}, f(\mathsf{CW}_2); r_2) \neq \mathsf{D}'(\mathsf{crs}, \mathsf{Ext}(\mathsf{crs}, \mathsf{SK}, f(\mathsf{CW}_2)), f(\mathsf{CW}_2); r_2) \\ \wedge\, g(\mathsf{crs}, \mathsf{CW}_2, f(\mathsf{CW}_2), r_2) = 0\end{array}\right] \leq \mathsf{negl}(n),$$

where $(\mathsf{crs}, \mathsf{SK}, \vec{\tau}_{\mathsf{sim}}) \leftarrow \mathsf{CRSGen}(1^n)$, $f \leftarrow \mathcal{A}(\mathsf{crs})$, r_2 is sampled uniformly at random and $\mathsf{CW}_2 \leftarrow \mathsf{E}_2(\mathsf{crs}, \vec{\tau}_{\mathsf{sim}}, r, 0)$.

Note that $g(\mathsf{crs}, \mathsf{CW}_2, f(\mathsf{CW}_2), r_2) = 0$ only if either of the following is true: (1) V^{NI} did not output 1 on all tampered proofs $T^*, T_1^*, \ldots, T_n^*$ in

$f(\mathsf{CW}_2)$; or (2) the first 3 elements of CW_2 and $f(\mathsf{CW}_2)$ are not identical (i.e., $(\hat{\boldsymbol{k}}, \boldsymbol{c}, c) \neq (\hat{\boldsymbol{k}}^*, \boldsymbol{c}^*, c^*)$). Now in case (1), both $\mathsf{D}(\mathsf{crs}, f(\mathsf{CW}_2); r_2)$, and $\mathsf{D}'(\mathsf{crs}, \mathsf{Ext}(\mathsf{crs}, \text{SK}, f(\mathsf{CW}_2)), f(\mathsf{CW}_2); r_2)$ output 0. This is contradiction to the claim that $\mathsf{D}(\mathsf{crs}, f(\mathsf{CW}_2); r_2) \neq \mathsf{D}'(\mathsf{crs}, \mathsf{Ext}(\mathsf{crs}, \text{SK}, f(\mathsf{CW}_2)), f(\mathsf{CW}_2); r_2)$. In case (2), the extractor $\mathsf{Ext}(\mathsf{crs}, \text{SK}, f(\mathsf{CW}_2))$ outputs $k^*_{n+1} := \mathsf{Decrypt}_{\text{SK}}(\hat{k}^*{}_{n+1})$ and $\mathsf{D}'(\mathsf{crs}, \mathsf{Ext}(\mathsf{crs}, \text{SK}, f(\mathsf{CW}_2)), f(\mathsf{CW}_2); r_2)$ outputs $b^* = c^* \oplus k^*_{n+1}$. Now, if $\mathsf{D}(\mathsf{crs}, f(\mathsf{CW}_2); r_2) \neq \mathsf{D}'(\mathsf{crs}, \mathsf{Ext}(\mathsf{crs}, \text{SK}, f(\mathsf{CW}_2)), f(\mathsf{CW}_2); r_2)$ but V^{NI} outputs 1 on all tampered proofs $T^*, T^*_1, \ldots, T^*_n$ in $f(\mathsf{CW}_2)$ then one-time simulation soundness of $\Pi^{\mathsf{NI}} = (\mathsf{CRSGen}^{\mathsf{NI}}, \mathsf{P}^{\mathsf{NI}}, \mathsf{V}^{\mathsf{NI}}, \mathsf{Sim}^{\mathsf{NI}})$ does not hold.

Hardness of D_b relative to Alternate Decoding.

1. $\Pr[g(\mathsf{crs}, \mathsf{CW}_2, f(\mathsf{CW}_2), r_2) = 1] \approx \Pr[g(\mathsf{crs}, \mathsf{CW}_3, f(\mathsf{CW}_3), r_3) = 1]$,
2. $\mathsf{D}'(\mathsf{crs}, \mathsf{Ext}(\mathsf{crs}, \text{SK}, f(\mathsf{CW}_2)), f(\mathsf{CW}_2); r_2) \approx \mathsf{D}'(\mathsf{crs}, \mathsf{Ext}(\mathsf{crs}, \text{SK}, f(\mathsf{CW}_3)), f(\mathsf{CW}_3); r_3)$,

where $(\mathsf{crs}, \text{SK}, \overrightarrow{\tau}_{\mathsf{sim}}) \leftarrow \mathsf{CRSGen}(1^n)$, $f \leftarrow \mathcal{A}(\mathsf{crs})$, r_2, r_3 are sampled uniformly at random, $\mathsf{CW}_2 \leftarrow \mathsf{E}_2(\mathsf{crs}, \overrightarrow{\tau}_{\mathsf{sim}}, r_2, 0)$ and $\mathsf{CW}_3 \leftarrow \mathsf{E}_2(\mathsf{crs}, \overrightarrow{\tau}_{\mathsf{sim}}, r_3, 1)$.

Let $\boldsymbol{X}$ denote a random variable where $\boldsymbol{X}$ is sampled from D_0 with probability 1/2 and $\boldsymbol{X}$ is sampled from D_1 with probability 1/2 and let random variable CW denote the output of E_2 when $\boldsymbol{X}$ replaces $\boldsymbol{x}$.

To show (1), assume $\Pr[g(\mathsf{crs}, \mathsf{CW}_2, f(\mathsf{CW}_2), r_2) = 1]$ and $\Pr[g(\mathsf{crs}, \mathsf{CW}_3, f(\mathsf{CW}_3), r_3) = 1]$ differ by a non-negligible amount. This implies that takes as input $\boldsymbol{X}$, hardwires all other random variables, and outputs 1 in the case that $g(\mathsf{crs}, \mathsf{CW}, f(\mathsf{CW}), r) = 1$ and 0 otherwise, implying that it has non-negligible correlation to the parity of its input $\boldsymbol{X}$. We will show that the above can be computed by an AC^0 circuit with input $\boldsymbol{X}$, thus contradicting Theorem 2 from [8] which says that an AC^0 circuit has at most negligible correlation with parity of its input $\boldsymbol{X}$, denoted $\mathcal{P}(\boldsymbol{X})$.

We construct the distribution of circuits $\mathcal{C}^1_{\mathcal{F}}$, and $C \sim \mathcal{C}^1_{\mathcal{F}}$ is drawn as:

1. Sample $(\mathsf{crs}, \text{SK}, \overrightarrow{\tau}_{\mathsf{sim}}) \leftarrow \mathsf{CRSGen}(1^n)$.
2. Sample tampering function $f \leftarrow \mathcal{A}(\mathsf{crs})$.
3. Sample $\boldsymbol{c}', c'$ uniformly at random.
4. Set $\boldsymbol{k}' = c'_1, \ldots, c'_n, c$. For $i \in [n]$, compute $\hat{k}'_i \leftarrow \mathsf{Encrypt}_{\text{PK}}(k'_i)$ and compute $\hat{k}'_{n+1} \leftarrow \mathsf{Encrypt}_{\text{PK}}(k')$.
5. Sample r uniformly at random.
6. Sample simulated proofs $[T_i^{'\beta}]_{\beta \in \{0,1\}, i \in [n]}$ and T' (as described in Fig. 8).
7. Output the following circuit C that has the following structure:
   - **hardwired variables:** crs, SK, f, $\hat{\boldsymbol{k}}'$, $\boldsymbol{c}', c'$, r, $[T_i^{'\beta}]_{\beta \in \{0,1\}, i \in [n]}$.
   - **input:** $\boldsymbol{X}$.
   - **computes and outputs:** $g(\mathsf{crs}, \mathsf{CW}, f(\mathsf{CW}), r)$.

   Note that given all the hardwired variables, computing CW is in AC^0 since all it does is, for $i \in [n]$, select the correct simulated proof $T_i^{'x_i}$ based on the corresponding input bit x_i. Additionally, f in AC^0 and g in AC^0, since bit-wise comparison is in AC^0 and V^{SAT} is in AC^0. Thus, the entire circuit is in AC^0.

To show (2), assume $\mathsf{D}'(\mathsf{crs}, \mathsf{Ext}(\mathsf{crs}, \mathrm{SK}, f(\mathsf{CW}_2)), f(\mathsf{CW}_2); r_2)$ and $\mathsf{D}'(\mathsf{crs}, \mathsf{Ext}(\mathsf{crs}, \mathrm{SK}, f(\mathsf{CW}_3)), f(\mathsf{CW}_3); r_3)$ have non-negligible statistical distance. This implies that a circuit that takes as input $\boldsymbol{X}$, hardwires all other random variables, and outputs $\mathsf{D}'(\mathsf{crs}, \mathsf{Ext}(\mathsf{crs}, \mathrm{SK}, f(\mathsf{CW})), f(\mathsf{CW}); r_2)$ has non-negligible correlation to the parity of $\boldsymbol{X}$. We will show that $\mathsf{D}'(\mathsf{crs}, \mathsf{Ext}(\mathsf{crs}, \mathrm{SK}, f(\mathsf{CW})), f(\mathsf{CW}); r_2)$ can be computed by an AC^0 circuit with input $\boldsymbol{X}$, thus contradicting Theorem 2 from [8], which says that an AC^0 circuit has at most negligible correlation with the parity of its input $\boldsymbol{X}$, denoted $\mathcal{P}(\boldsymbol{X})$.

We construct the distribution of circuits $\mathcal{C}^2_{\mathcal{F}}$, and $C \sim \mathcal{C}^2_{\mathcal{F}}$ is drawn as:
1. Sample $(\mathsf{crs}, \mathrm{SK}, \overrightarrow{\tau}_{\mathsf{sim}}) \leftarrow \mathsf{CRSGen}(1^n)$.
2. Sample tampering function $f \leftarrow \mathcal{A}(\mathsf{crs})$.
3. Sample $\boldsymbol{c}', c'$ uniformly at random.
4. Set $\boldsymbol{k}' = c'_1, \ldots, c'_n, c$. For $i \in [n]$, compute $\hat{k}'_i \leftarrow \mathsf{Encrypt}_{\mathrm{PK}}(k'_i)$ and compute $\hat{k}'_{n+1} \leftarrow \mathsf{Encrypt}_{\mathrm{PK}}(k')$.
5. Sample r uniformly at random.
6. Sample simulated proofs $[T_i'^{\beta}]_{\beta \in \{0,1\}, i \in [n]}$ and T' (as described in Fig. 8).
7. Output the following circuit C that has the following structure:
   - **hardwired variables:** crs, SK, f, $\hat{\boldsymbol{k}}'$, $\boldsymbol{c}', c'$, r, $[T_i'^{\beta}]_{\beta \in \{0,1\}, i \in [n]}$.
   - **input:** $\boldsymbol{X}$.
   - **computes and outputs:** $\mathsf{D}'(\mathsf{crs}, \mathsf{Ext}(\mathsf{crs}, \mathrm{SK}, f(\mathsf{CW})), f(\mathsf{CW}); r_2)$.

   Note that $\mathsf{Ext} \in \mathsf{AC}^0$ since decryption for $\mathcal{E} := (\mathsf{Gen}, \mathsf{Encrypt}, \mathsf{Decrypt})$ in AC^0. Moreover, as above, given all the hardwired variables, computing CW is in AC^0 since all it does is, for $i \in [n]$, select the correct simulated proof $T_i'^{x_i}$ based on the corresponding input bit x_i. Additionally, f in AC^0 and D' is in AC^0, since xor of two bits is in AC^0 and V^{SAT} is in AC^0. Thus, the entire circuit is in AC^0.

Analysis for more tampering classes is presented in Sect. 4.1 of full version [8]

5 Construction for Multi-bit Messages

The construction for encoding multi-bit messages is similar to that for encoding a single bit, presented in Sect. 3. The construction repeats the procedure for encoding single bit m times, for encoding m-bit messages and binds it with a proof T.

Let $\Psi(p, c, x, y, r, z)$ be defined as a function that takes as input a predicate p, and variables c, x, y, r, z. If $p(c, x, y, r) = 1$, then Ψ outputs the m-bit string $\boldsymbol{0}$. Otherwise, Ψ outputs z.

Theorem 4. *Let* (E, D), E_1, E_2, Ext, D' *and* g *be as defined in Figs. 12, 13, 14, 15, 16 and 17. Let* $\mathcal{F}$ *be a computational class. If, for every pair of* m*-bit messages* $\boldsymbol{b}_0, \boldsymbol{b}_1$ *and if, for every adversary* $\mathcal{A} \in \mathcal{G}$ *outputting tampering functions* $f \in \mathcal{F}$*, all of the following hold:*

Let $\mathcal{E}$ = (Gen, Encrypt, Decrypt) be a public key encryption scheme with perfect correctness (see Definition 7 in [8]). Let $\Pi^{\mathsf{NI}} = (\mathsf{CRSGen}^{\mathsf{NI}}, \mathsf{P}^{\mathsf{NI}}, \mathsf{V}^{\mathsf{NI}}, \mathsf{Sim}^{\mathsf{NI}})$ be a non-interactive simulatable proof system with soundness against adversaries $\mathcal{A} \in \mathcal{G}$ (see Definition 5). Note that in the CRS model, we implicitly assume that all algorithms take the CRS as input, and for simplicity of notation, sometimes do not list the CRS as an explicit input.

$\mathsf{CRSGen}(1^n)$:

1. Choose $(\mathrm{PK}, \mathrm{SK}) \leftarrow \mathsf{Gen}(1^n)$.
2. Choose $[\mathsf{crs}^{\mathsf{NI}}_{i,j}, \tau^{i,j}_{\mathsf{sim}}]_{(i,j)=(0,0), i \in [m], j \in [n]} \leftarrow \mathsf{CRSGen}^{\mathsf{NI}}(1^n)$. Let $\overrightarrow{\mathsf{crs}}^{\mathsf{NI}} := [\mathsf{crs}^{\mathsf{NI}}_{i,j}]_{(i,j)=(0,0), i \in [m], j \in [n]}$ and let $\overrightarrow{\tau}_{\mathsf{sim}} := [\tau^{i,j}_{\mathsf{sim}}]_{(i,j)=(0,0), i \in [m], j \in [n]}$
3. Output $\mathsf{crs} := (\mathrm{PK}, \overrightarrow{\mathsf{crs}}^{\mathsf{NI}})$.

Languages. We define the following languages:

- $\mathcal{L}^{\beta}_{i,j}$: For $i \in [m], j \in [n]$, $\beta \in \{0,1\}$, $s := ([\hat{\boldsymbol{k}}^i]_{i \in [m]}, \overline{\boldsymbol{c}}, \overline{c}) \in \mathcal{L}^{\beta}_{i,j}$ iff the (i,j)-th ciphertext $c^i_j := k^i_j \oplus \beta$ (where $\overline{\boldsymbol{c}} = [c^i_j]_{i \in [m], j \in [n]}$) and the (i,j)-th encryption $\hat{k}^i_j$ (where $\hat{\boldsymbol{k}}^i = \hat{k}^i_1, \ldots, \hat{k}^i_{n+1}$) is an encryption of k^i_j under PK (where PK is hardwired into the language).
- $\mathcal{L}$: $s := ([\hat{\boldsymbol{k}}^i]_{i \in [m]}, \overline{\boldsymbol{c}}, \overline{c}) \in \mathcal{L}$ iff For each $i \in [m]$, $(x^i_1, \ldots, x^i_n)$ is in the support of D_{b^i} where:
  1. For $i \in [m], j \in [n]$, $x^i_j := c^i_j \oplus k^i_j$
  2. $b^i := c^i \oplus k^i_{n+1}$ (where $\overline{c} := c^1, \ldots, c^m$)
  3. $\hat{\boldsymbol{k}}^i$ is an encryption of $k^i_1, \ldots, k^i_{n+1}$ under PK (where PK is hardwired into the language).

$\mathsf{E}(\mathsf{crs}, \boldsymbol{b} := b^1, \ldots, b^m)$:

1. Sample $\overline{\boldsymbol{x}} := \boldsymbol{x}^1, \ldots, \boldsymbol{x}^m \leftarrow D_{\boldsymbol{b}}$, where for $i \in [m]$, $\boldsymbol{x}^i = x^i_1, \ldots, x^i_n$.
2. Choose an $m \cdot (n+1)$-bit key $\overline{\boldsymbol{k}} := [\boldsymbol{k}^i]_{i \in [m]} = [k^i_1, \ldots, k^i_n, k^i]_{i \in [m]}$ uniformly at random. For $i \in [m], j \in [n+1]$, compute $\hat{k}^i_j \leftarrow \mathsf{Encrypt}(\mathrm{PK}, k^i_j)$. For $i \in [m]$, let $\hat{\boldsymbol{k}}^i := \hat{k}^i_1, \ldots, \hat{k}^i_{n+1}$.
3. For $i \in [m], j \in [n]$, compute $c^i_j := k^i_j \oplus x^i_j$. Let $\overline{\boldsymbol{c}} := [c^i_j]_{i \in [m], j \in [n]}$.
4. For $i \in [m]$, compute $c^i := k^i \oplus b^i$. Let $\overline{c} := [c^i]_{i \in [m]}$.
5. For $i \in [m], j \in [n]$, compute a non-interactive, simulatable proof T^i_j proving $([\hat{\boldsymbol{k}}^i]_{i \in [m]}, \overline{\boldsymbol{c}}, \overline{c}) \in \mathcal{L}^{x^i_j}_{i,j}$ relative to $\mathsf{crs}^{\mathsf{NI}}_{i,j}$.
6. Compute a non-interactive, simulatable proof T proving $([\hat{\boldsymbol{k}}^i]_{i \in [m]}, \overline{\boldsymbol{c}}, \overline{c}) \in \mathcal{L}$ relative to $\mathsf{crs}^{\mathsf{NI}}_{0,0}$.
7. Output $\mathsf{CW} := ([\hat{\boldsymbol{k}}^i]_{i \in [m]}, \overline{\boldsymbol{c}}, \overline{c}, T, [(x^i_j, T^i_j)]_{i \in [m], j \in [n]})$.

$\mathsf{D}(\mathsf{crs}, \mathsf{CW})$:

1. Parse $\mathsf{CW} := ([\hat{\boldsymbol{k}}^i]_{i \in [m]}, \overline{\boldsymbol{c}}, \overline{c}, T, [(x^i_j, T^i_j)]_{i \in [m], j \in [n]})$
2. Check that V^{NI} outputs 1 on all proofs $[T^i_j]_{i \in [m], j \in [n]}, T$, relative to the corresponding CRS.
3. If yes, output $[b^i]_{i \in [m]}$ such that $x^i_1 \ldots x^i_n$ is in the support of D_{b^i}. If not, output $\mathbf{0}$.

Fig. 12. Non-malleable code (CRSGen, E, D), secure against $\mathcal{F}$ tampering.

$\mathsf{E}_1(\mathsf{crs}, \overrightarrow{\tau}_{\mathsf{sim}}, r, \boldsymbol{b} := b^1, \ldots, b^m)$:

1. Sample $\overline{\boldsymbol{x}} := \boldsymbol{x}^1, \ldots, \boldsymbol{x}^m \leftarrow D_b$, where for $i \in [m]$, $\boldsymbol{x}^i = x_1^i, \ldots, x_n^i$.
2. Choose an $m \cdot (n+1)$-bit key $\overline{\boldsymbol{k}} := [\boldsymbol{k}^i]_{i \in [m]} = [k_1^i, \ldots, k_n^i, k^i]_{i \in [m]}$ uniformly at random. For $i \in [m], j \in [n+1]$, compute $\hat{k}_j^i \leftarrow \mathsf{Encrypt}(\mathrm{PK}, k_j^i)$. For $i \in [m]$, let $\hat{\boldsymbol{k}}^i := \hat{k}_1^i, \ldots, \hat{k}_{n+1}^i$.
3. For $i \in [m], j \in [n]$, compute $c_j^i := k_j^i \oplus x_j^i$. Let $\overline{\boldsymbol{c}} := [c_j^i]_{i \in [m], j \in [n]}$.
4. For $i \in [m]$, compute $c^i := k^i \oplus b^i$. Let $\overline{c} := [c^i]_{i \in [m]}$.
5. For $i \in [m], j \in [n]$, simulate, using $\tau_{\mathsf{sim}}^{i,j}$ and r, a non-interactive proof $T_j'^i$ proving $s := ([\hat{\boldsymbol{k}}^i]_{i \in [m]}, \overline{\boldsymbol{c}}, \overline{c}) \in \mathcal{L}_{i,j}^{x_j^i}$, relative to $\mathsf{crs}_{i,j}^{\mathsf{NI}}$.
6. Simulate, using $\tau_{\mathsf{sim}}^{0,0}$ and r, a non-interactive proof T' proving $s := ([\hat{\boldsymbol{k}}^i]_{i \in [m]}, \overline{\boldsymbol{c}}, \overline{c}) \in \mathcal{L}$, relative to $\mathsf{crs}_{0,0}^{\mathsf{NI}}$.
7. Output $\mathsf{CW} := ([\hat{\boldsymbol{k}}^i]_{i \in [m]}, \overline{\boldsymbol{c}}, \overline{c}, T', [(x_j^i, T_j'^i)]_{i \in [m], j \in [n]})$.

Fig. 13. Encoding algorithm with simulated proofs.

$\mathsf{E}_2(\mathsf{crs}, \overrightarrow{\tau}_{\mathsf{sim}}, r, \boldsymbol{b} := b^1, \ldots, b^m)$:

1. Sample $\overline{\boldsymbol{x}} := \boldsymbol{x}^1, \ldots, \boldsymbol{x}^m \leftarrow D_b$, where for $i \in [m]$, $\boldsymbol{x}^i = x_1^i, \ldots, x_n^i$.
2. Choose $[c_j'^i]_{i \in [m], j \in [n]}$ uniformly at random. Let $\overline{\boldsymbol{c}}' := [c_j'^i]_{i \in [m], j \in [n]}$.
3. Choose $[c'^i]_{i \in [m]}$ uniformly at random. Let $\overline{c}' := [c'^i]_{i \in [m]}$.
4. Set the $m \cdot (n+1)$-bit key $\overline{\boldsymbol{k}}' := [\boldsymbol{k}'^i]_{i \in [m]} = [c_1'^i, \ldots, c_n'^i, c'^i]_{i \in [m]}$. For $i \in [m], j \in [n+1]$, compute $\hat{k}_j'^i \leftarrow \mathsf{Encrypt}(\mathrm{PK}, k_j'^i)$. For $i \in [m]$, let $\hat{\boldsymbol{k}}'^i := \hat{k}_1'^i, \ldots, \hat{k}_{n+1}'^i$.
5. For $i \in [m], j \in [n]$, simulate, using $\tau_{\mathsf{sim}}^{i,j}$ and r, a non-interactive proof $T_j'^i$ proving $s := ([\hat{\boldsymbol{k}}^i]_{i \in [m]}, \overline{\boldsymbol{c}}, \overline{c}) \in \mathcal{L}_{i,j}^{x_j^i}$, relative to $\mathsf{crs}_{i,j}^{\mathsf{NI}}$.
6. Simulate, using $\tau_{\mathsf{sim}}^{0,0}$ and r, a non-interactive proof T' proving $s := ([\hat{\boldsymbol{k}}^i]_{i \in [m]}, \overline{\boldsymbol{c}}, \overline{c}) \in \mathcal{L}$, relative to $\mathsf{crs}_{0,0}^{\mathsf{NI}}$.
7. Output $\mathsf{CW} := ([\hat{\boldsymbol{k}}'^i]_{i \in [m]}, \overline{\boldsymbol{c}}', \overline{c}', T', [(x_j^i, T_j'^i)]_{i \in [m], j \in [n]})$.

Fig. 14. Encoding algorithm with simulated proofs and encryptions.

Simulation of proofs.

1. $\Pr[g(\mathsf{crs}, \mathsf{CW}_0, f(\mathsf{CW}_0), r_0) = 1] \approx \Pr[g(\mathsf{crs}, \mathsf{CW}_1, f(\mathsf{CW}_1), r_1) = 1]$,

2. $\Psi(g, \mathsf{crs}, \mathsf{CW}_0, f(\mathsf{CW}_0), r_0, \mathsf{D}(\mathsf{crs}, f(\mathsf{CW}_0); r_0)) \approx \Psi(g, \mathsf{crs}, \mathsf{CW}_1, f(\mathsf{CW}_1), r_1, \mathsf{D}(\mathsf{crs}, f(\mathsf{CW}_1); r_1))$,

where $(\mathsf{crs}, \mathrm{SK}, \overrightarrow{\tau}_{\mathsf{sim}}) \leftarrow \mathsf{CRSGen}(1^n)$, $f \leftarrow \mathcal{A}(\mathsf{crs})$, r_0, r_1 *are sampled uniformly at random,* $\mathsf{CW}_0 \leftarrow \mathsf{E}(\mathsf{crs}, \boldsymbol{b}_0)$ *and* $\mathsf{CW}_1 \leftarrow \mathsf{E}_1(\mathsf{crs}, \overrightarrow{\tau}_{\mathsf{sim}}, r_1, \boldsymbol{b}_0)$.

Ext(crs, SK, CW):

1. Parse CW $:= ([\hat{\boldsymbol{k}}^i]_{i\in[m]}, \overline{\boldsymbol{c}}, \overline{c}, T, [(x^i_j, T^i_j)]_{i\in[m],j\in[n]})$,
2. Output $[\mathsf{Decrypt}(\mathrm{SK}, \hat{k}^i_{n+1})]_{i\in[m]}$.

Fig. 15. Extracting procedure Ext.

D′(crs, $[k^i]_{i\in[m]}$, CW):

1. Parse CW $:= ([\hat{\boldsymbol{k}}^i]_{i\in[m]}, , \overline{\boldsymbol{c}}, \overline{c}, T, [(x^i_j, T^i_j)]_{i\in[m],j\in[n]})$,
2. Check that V^{NI} outputs 1 on all proofs $[T^i_j]_{i\in[m],j\in[n]}, T$, relative to the corresponding CRS,
3. For $i \in [m]$, output $b^i := k^i \oplus c^i$.

Fig. 16. Alternate decoding procedure D′, given additional extracted key $[k^i]_{i\in[m]}$ as input.

g(crs, CW, CW*, r):

1. Parse CW $= ([\hat{\boldsymbol{k}}^i]_{i\in[m]}, \overline{\boldsymbol{c}}, \overline{c}, T, [(x^i_j, T^i_j)]_{i\in[m],j\in[n]})$, CW* $= ([\hat{\boldsymbol{k}}^{*i}]_{i\in[m]}, \overline{\boldsymbol{c}}^*, \overline{c}^*, T^*, [(x^{*i}_j, T^{*i}_j)]_{i\in[m],j\in[n]})$.
2. If (1) V^{NI} outputs 1 on all proofs $T^*, [T^{*i}_j)]_{i\in[m],j\in[n]}$, relative to the corresponding CRS; and (2) $([\hat{\boldsymbol{k}}^i]_{i\in[m]}, \overline{\boldsymbol{c}}, \overline{c}) = ([\hat{\boldsymbol{k}}^{*i}]_{i\in[m]}, \overline{\boldsymbol{c}}^*, \overline{c}^*)$, then output 1. Otherwise output 0.

Fig. 17. The predicate g(crs, CW, CW*, r).

Simulation of Encryptions.

1. $\Pr[g(\mathsf{crs}, \mathsf{CW}_1, f(\mathsf{CW}_1), r_1) = 1] \approx \Pr[g(\mathsf{crs}, \mathsf{CW}_2, f(\mathsf{CW}_2), r_2) = 1]$,

2. $\Psi(g, \mathsf{crs}, \mathsf{CW}_1, f(\mathsf{CW}_1), r_1, \mathsf{D}(\mathsf{crs}, f(\mathsf{CW}_1); r_1)) \approx \Psi(g, \mathsf{crs}, \mathsf{CW}_2, f(\mathsf{CW}_2), r_2, \mathsf{D}(\mathsf{crs}, f(\mathsf{CW}_2); r_2))$,

where $(\mathsf{crs}, \mathrm{SK}, \overrightarrow{\tau}_{\mathsf{sim}}) \leftarrow \mathsf{CRSGen}(1^n)$, $f \leftarrow \mathcal{A}(\mathsf{crs})$, r_1, r_2 *are sampled uniformly at random,* $\mathsf{CW}_1 \leftarrow \mathsf{E}_1(\mathsf{crs}, \overrightarrow{\tau}_{\mathsf{sim}}, r_1, \boldsymbol{b}_0)$ *and* $\mathsf{CW}_2 \leftarrow \mathsf{E}_2(\mathsf{crs}, \overrightarrow{\tau}_{\mathsf{sim}}, r_2, \boldsymbol{b}_0)$.

Simulation Soundness.

$$\Pr\left[\begin{matrix}\mathsf{D}(\mathsf{crs}, f(\mathsf{CW}_2); r_2) \neq \mathsf{D}'(\mathsf{crs}, \mathsf{Ext}(\mathsf{crs}, \mathrm{SK}, f(\mathsf{CW}_2)), f(\mathsf{CW}_2); r_2) \\ \wedge\, g(\mathsf{crs}, \mathsf{CW}_2, f(\mathsf{CW}_2), r_2) = 0\end{matrix}\right] \leq \mathsf{negl}(n),$$

where $(\mathsf{crs}, \mathrm{SK}, \overrightarrow{\tau}_{\mathsf{sim}}) \leftarrow \mathsf{CRSGen}(1^n)$, $f \leftarrow \mathcal{A}(\mathsf{crs})$, r_2 *is sampled uniformly at random and* $\mathsf{CW}_2 \leftarrow \mathsf{E}_2(\mathsf{crs}, \overrightarrow{\tau}_{\mathsf{sim}}, r, \boldsymbol{b}_0)$.

Hardness of D_b relative to Alternate Decoding.

1. $\Pr[g(\mathsf{crs}, \mathsf{CW}_2, f(\mathsf{CW}_2), r_2) = 1] \approx \Pr[g(\mathsf{crs}, \mathsf{CW}_3, f(\mathsf{CW}_3), r_3) = 1]$,
2. *For every Boolean function, represented by a circuit* F *over* m *variables,* $F \circ \mathsf{D}'(\mathsf{crs}, \mathsf{Ext}(\mathsf{crs}, \mathrm{SK}, f(\mathsf{CW}_2)), f(\mathsf{CW}_2); r_2) \approx F \circ \mathsf{D}'(\mathsf{crs}, \mathsf{Ext}(\mathsf{crs}, \mathrm{SK}, f(\mathsf{CW}_3)), f(\mathsf{CW}_3); r_3)$,

where $(\mathsf{crs}, \mathrm{SK}, \overrightarrow{\tau}_{\mathsf{sim}}) \leftarrow \mathsf{CRSGen}(1^n)$, $f \leftarrow \mathcal{A}(\mathsf{crs})$, r_2, r_3 *are sampled uniformly at random,* $\mathsf{CW}_2 \leftarrow \mathsf{E}_2(\mathsf{crs}, \overrightarrow{\tau}_{\mathsf{sim}}, r_2, \boldsymbol{b}_0)$ *and* $\mathsf{CW}_3 \leftarrow \mathsf{E}_2(\mathsf{crs}, \overrightarrow{\tau}_{\mathsf{sim}}, r_3, \boldsymbol{b}_1)$.

Then the construction presented in Fig. 12 is a non-malleable code for class $\mathcal{F}$ *against adversaries* $\mathcal{A} \in \mathcal{G}$.

We present the proof of Theorem 4 in Sect. 5.1 of the full version [8]

6 Efficient, Multi-bit NMC for $\mathbf{AC}^0$

Theorem 5. $\Pi = (\mathsf{CRSGen}, \mathsf{E}, \mathsf{D})$ *(presented in Fig. 12) is an* m*-bit, computational, non-malleable code in the CRS model against tampering by depth-*$(m^{\log^{\delta} m}/2 - c)$ *circuits with unbounded fan-in and size* $\delta \cdot \frac{\log m}{\log \log m} - p(n)$ *(where* c *is constant and* $p(\cdot)$ *is a fixed polynomial), and* m *is such that* $n = m^{3+5\delta}$*, if the underlying components are instantiated in the following way:*

- $\mathcal{E} := (\mathsf{Gen}, \mathsf{Encrypt}, \mathsf{Decrypt})$ *is a public key encryption scheme with perfect correctness and decryption in* AC^0.
- $\Pi^{\mathsf{NI}} := (\mathsf{CRSGen}^{\mathsf{NI}}, \mathsf{P}^{\mathsf{NI}}, \mathsf{V}^{\mathsf{NI}}, \mathsf{Sim}^{\mathsf{NI}})$ *is a same-string, weak one-time simulation-sound NIZK with verifier in* AC^0.
- *For* $b \in \{0,1\}$, D_b *is the distribution that samples bits* $x_1 \ldots x_n$ *uniformly at random, conditioned on* $x_1 \oplus \cdots \oplus x_n = b$.

For as in the one-bit case, given Theorem 1, proof systems Π^{NI} as above exist, under the assumption that same-string, weak one-time simulation-sound NIZK with (arbitrary polynomial-time) deterministic verifier exists. Refer to Sect. 4 of the full version [8] for a discussion of how such NIZK and public key encryption can be instantiated. The proof of the Theorem 5, is presented as proof for Theorem 11 in [8], followed by the analysis for tampering with decision trees in Sect. 6.1.

7 One-Bit NMC Against Streaming Adversaries

In this section, we show that our generic construction yields efficient *unconditional* NMC resilient against the tampering class $\mathcal{F}$ corresponding to streaming adversaries with memory $o(n'')$.

Let n be the parameter for the hard distribution described below, n' be the parameter for the semantically secure parity based encryption scheme against streaming adversaries with $o(n')$ storage (described in Sect. 7.2 of [8]), and n'' be the parameter for the non-interactive simulatable proof system with streaming verifier (described in Sect. 7.4 of [8]). Such that $n \in \omega(n'')$ and $n' \in \omega(n)$.

The Hard Distribution D_b (parameter n). Let $n = (\mu + 1)^2 - 1$. For $b \in \{0, 1\}$, a draw from the distribution D_b is defined as follows: Choose a parity χ_S uniformly at random from the set of all (non-zero) parities over μ variables ($\emptyset \neq S \subseteq [\mu]$). Choose $y_1, \ldots, y_\mu \sim \{0, 1\}^\mu$ uniformly at random. Choose y uniformly at random, conditioned on $\chi_S(y) = b$. Output the following n-bit string: $[(y_i, \chi_S(y_i)]_{i \in [\mu]} || y$.

The proof of the hardness of D_b described above, along with the details of the parity-based encryption scheme, and non-interactive simulatable proof system with streaming verifier are described in Sects. 7.1, 7.2, and 7.4 of [8] respectively.

Theorem 6. $\Pi = (\mathsf{E}, \mathsf{D})$ *(presented in Fig. 6) is a one-bit, unconditional non-malleable code against streaming adversaries with space* $o(n'')$*, if the underlying components are instantiated in the following way:*

- $\mathcal{E} := (\mathsf{Encrypt}, \mathsf{Decrypt})$ *is the parity based encryption scheme (with parameter* $n' := n'(n)$*).*
- $\Pi^{\mathsf{NI}} := (\mathsf{P}^{\mathsf{NI}}, \mathsf{V}^{\mathsf{NI}}, \mathsf{Sim}^{\mathsf{NI}})$ *the simulatable proof system with streaming verifier with parameter* $n'' := n''(n)$*.*
- *For* $b \in \{0, 1\}$*,* D_b *is the distribution described above (with parameter* n*).*

We wish to emphasize that no CRS or computational assumptions are needed for this result. Therefore, we can assume that the adversary $\mathcal{A}$ outputting tampering function f is computationally unbounded. Moreover, the result extends trivially for any number m of bits and all other parameters (n, n', n'') can remain the same and do not need to be increased. To see this, note that the only one additional property that needs to be proved in the multi-bit case (regarding hardness of D_b relative to alternate decoding in Theorem 4. But in the bounded, it can be achieved without requiring any additional memory beyond what is required in the one-bit case. We refer the interested readers to Sect. 7.5 of [8] for further details.

Acknowledgments. We are grateful to Benjamin Kuykendall for his helpful comments.

The first and fourth authors are supported in part by the Defense Advanced Research Project Agency (DARPA) and Army Research Office (ARO) under Contract #W911NF-15-C-0236, and NSF grants #CNS-1445424 and #CCF-1423306 and the Leona M. & Harry B. Helmsley Charitable Trust. The second and third authors are supported in part by an NSF CAREER Award #CNS-1453045, by a research partnership award from Cisco and by financial assistance award 70NANB15H328 from the U.S. Department of Commerce, National Institute of Standards and Technology. This work was performed, in part, while the first author was visiting IDC Herzliya's FACT center and supported in part by ISF grant no. 1790/13 and the Check Point Institute for Information Security. Any opinions, findings and conclusions or recommendations expressed are those of the authors and do not necessarily reflect the views of the the Defense Advanced Research Projects Agency, Army Research Office, the National Science Foundation, or the U.S. Government.

References

1. Aggarwal, D., Agrawal, S., Gupta, D., Maji, H.K., Pandey, O., Prabhakaran, M.: Optimal computational split-state non-malleable codes. In: [43], pp. 393–417
2. Aggarwal, D., Dodis, Y., Kazana, T., Obremski, M.: Non-malleable reductions and applications. In: Servedio, R.A., Rubinfeld, R. (eds.) 47th ACM STOC, pp. 459–468. ACM Press, June 2015
3. Aggarwal, D., Dodis, Y., Lovett, S.: Non-malleable codes from additive combinatorics. In: Shmoys, D.B. (ed.) 46th ACM STOC, pp. 774–783. ACM Press, May/June 2014
4. Aggarwal, D., Dziembowski, S., Kazana, T., Obremski, M.: Leakage-resilient non-malleable codes. In: [27], pp. 398–426
5. Agrawal, S., Gupta, D., Maji, H.K., Pandey, O., Prabhakaran, M.: Explicit non-malleable codes against bit-wise tampering and permutations. In: Gennaro, R., Robshaw, M. (eds.) CRYPTO 2015, Part I. LNCS, vol. 9215, pp. 538–557. Springer, Heidelberg (2015). https://doi.org/10.1007/978-3-662-47989-6_26
6. Agrawal, S., Gupta, D., Maji, H.K., Pandey, O., Prabhakaran, M.: A rate-optimizing compiler for non-malleable codes against bit-wise tampering and permutations. In: [27], pp. 375–397
7. Ball, M., Dachman-Soled, D., Kulkarni, M., Malkin, T.: Non-malleable codes for bounded depth, bounded fan-in circuits. In: Fischlin, M., Coron, J.-S. (eds.) EUROCRYPT 2016, Part II. LNCS, vol. 9666, pp. 881–908. Springer, Heidelberg (2016). https://doi.org/10.1007/978-3-662-49896-5_31
8. Ball, M., Dachman-Soled, D., Kulkarni, M., Malkin, T.: Non-malleable codes from average-case hardness: AC0, decision trees, and streaming space-bounded tampering. Cryptology ePrint Archive, Report 2017/1061 (2017). http://eprint.iacr.org/2017/1061
9. Bogdanov, A., Lee, C.H.: Homomorphic evaluation requires depth. In: [42], pp. 365–371
10. Chabanne, H., Cohen, G.D., Flori, J., Patey, A.: Non-malleable codes from the wire-tap channel. CoRR abs/1105.3879 (2011)
11. Chandran, N., Goyal, V., Mukherjee, P., Pandey, O., Upadhyay, J.: Block-wise non-malleable codes. Cryptology ePrint Archive, Report 2015/129 (2015). http://eprint.iacr.org/2015/129
12. Chandran, N., Goyal, V., Mukherjee, P., Pandey, O., Upadhyay, J.: Block-wise non-malleable codes. In: Chatzigiannakis, I., Mitzenmacher, M., Rabani, Y., Sangiorgi, D. (eds.) ICALP 2016. LIPIcs, vol. 55, pp. 31:1–31:14. Schloss Dagstuhl, July 2016
13. Chandran, N., Kanukurthi, B., Ostrovsky, R.: Locally updatable and locally decodable codes. In: [46], pp. 489–514
14. Chandran, N., Kanukurthi, B., Raghuraman, S.: Information-theoretic local non-malleable codes and their applications. In: [43], pp. 367–392

15. Chattopadhyay, E., Goyal, V., Li, X.: Non-malleable extractors and codes, with their many tampered extensions. In: [52], pp. 285–298
16. Chattopadhyay, E., Li, X.: Non-malleable codes and extractors for small-depth circuits, and affine functions. In: Hatami, H., McKenzie, P., King, V. (eds.) 49th ACM STOC, pp. 1171–1184. ACM Press, June 2017
17. Chattopadhyay, E., Zuckerman, D.: Non-malleable codes against constant split-state tampering. In: 55th FOCS, pp. 306–315. IEEE Computer Society Press, October 2014
18. Chattopadhyay, E., Zuckerman, D.: Explicit two-source extractors and resilient functions. In: [52], pp. 670–683
19. Cheraghchi, M., Guruswami, V.: Capacity of non-malleable codes. In: Naor, M. (ed.) ITCS 2014, pp. 155–168. ACM, January 2014
20. Cheraghchi, M., Guruswami, V.: Non-malleable coding against bit-wise and split-state tampering. In: [46], pp. 440–464
21. Choi, S.G., Kiayias, A., Malkin, T.: BiTR: built-in tamper resilience. In: Lee, D.H., Wang, X. (eds.) ASIACRYPT 2011. LNCS, vol. 7073, pp. 740–758. Springer, Heidelberg (2011). https://doi.org/10.1007/978-3-642-25385-0_40
22. Coretti, S., Dodis, Y., Tackmann, B., Venturi, D.: Non-malleable encryption: simpler, shorter, stronger. In: [42], pp. 306–335
23. Coretti, S., Maurer, U., Tackmann, B., Venturi, D.: From single-bit to multi-bit public-key encryption via non-malleable codes. In: [27], pp. 532–560
24. Dachman-Soled, D., Kulkarni, M., Shahverdi, A.: Tight upper and lower bounds for leakage-resilient, locally decodable and updatable non-malleable codes. In: Fehr, S. (ed.) PKC 2017, Part I. LNCS, vol. 10174, pp. 310–332. Springer, Heidelberg (2017). https://doi.org/10.1007/978-3-662-54365-8_13
25. Dachman-Soled, D., Liu, F.H., Shi, E., Zhou, H.S.: Locally decodable and updatable non-malleable codes and their applications. In: [27], pp. 427–450
26. De Wolf, R.: A brief introduction to fourier analysis on the boolean cube. Theory Comput. Grad. Surv. **1**, 1–20 (2008)
27. Dodis, Y., Nielsen, J.B. (eds.): TCC 2015, Part I. LNCS, vol. 9014. Springer, Heidelberg (2015). https://doi.org/10.1007/978-3-662-46494-6
28. Döttling, N., Nielsen, J.B., Obremski, M.: Information theoretic continuously non-malleable codes in the constant split-state model. Cryptology ePrint Archive, Report 2017/357 (2017). http://eprint.iacr.org/2017/357
29. Dwork, C., Naor, M., Reingold, O.: Immunizing encryption schemes from decryption errors. In: Cachin, C., Camenisch, J.L. (eds.) EUROCRYPT 2004. LNCS, vol. 3027, pp. 342–360. Springer, Heidelberg (2004). https://doi.org/10.1007/978-3-540-24676-3_21
30. Dziembowski, S., Kazana, T., Obremski, M.: Non-malleable codes from two-source extractors. In: Canetti, R., Garay, J.A. (eds.) CRYPTO 2013, Part II. LNCS, vol. 8043, pp. 239–257. Springer, Heidelberg (2013). https://doi.org/10.1007/978-3-642-40084-1_14
31. Dziembowski, S., Pietrzak, K., Wichs, D.: Non-malleable codes. In: Yao, A.C.C. (ed.) ICS 2010, pp. 434–452. Tsinghua University Press, January 2010
32. Faust, S., Hostáková, K., Mukherjee, P., Venturi, D.: Non-malleable codes for space-bounded tampering. In: Katz, J., Shacham, H. (eds.) CRYPTO 2017, Part II. LNCS, vol. 10402, pp. 95–126. Springer, Cham (2017). https://doi.org/10.1007/978-3-319-63715-0_4
33. Faust, S., Mukherjee, P., Nielsen, J.B., Venturi, D.: Continuous non-malleable codes. In: [46], pp. 465–488

34. Faust, S., Mukherjee, P., Nielsen, J.B., Venturi, D.: A tamper and leakage resilient von Neumann architecture. In: Katz, J. (ed.) PKC 2015. LNCS, vol. 9020, pp. 579–603. Springer, Heidelberg (2015). https://doi.org/10.1007/978-3-662-46447-2_26
35. Faust, S., Mukherjee, P., Venturi, D., Wichs, D.: Efficient non-malleable codes and key-derivation for poly-size tampering circuits. In: Nguyen, P.Q., Oswald, E. (eds.) EUROCRYPT 2014. LNCS, vol. 8441, pp. 111–128. Springer, Heidelberg (2014). https://doi.org/10.1007/978-3-642-55220-5_7
36. Goyal, V., Pandey, O., Richelson, S.: Textbook non-malleable commitments. In: [52], pp. 1128–1141
37. Jafargholi, Z., Wichs, D.: Tamper detection and continuous non-malleable codes. In: [27], pp. 451–480
38. Kalai, Y.T., Kanukurthi, B., Sahai, A.: Cryptography with tamperable and leaky memory. In: Rogaway, P. (ed.) CRYPTO 2011. LNCS, vol. 6841, pp. 373–390. Springer, Heidelberg (2011). https://doi.org/10.1007/978-3-642-22792-9_21
39. Kanukurthi, B., Obbattu, S.L.B., Sekar, S.: Four-state non-malleable codes with explicit constant rate. In: Kalai, Y., Reyzin, L. (eds.) TCC 2017, Part II. LNCS, vol. 10678, pp. 344–375. Springer, Cham (2017). https://doi.org/10.1007/978-3-319-70503-3_11
40. Kanukurthi, B., Obbattu, S.L.B., Sekar, S.: Non-malleable randomness encoders and their applications. Cryptology ePrint Archive, Report 2017/1097 (2017). https://eprint.iacr.org/2017/1097
41. Kiayias, A., Liu, F.H., Tselekounis, Y.: Practical non-malleable codes from l-more extractable hash functions. In: Weippl, E.R., Katzenbeisser, S., Kruegel, C., Myers, A.C., Halevi, S. (eds.) ACM CCS 2016, pp. 1317–1328. ACM Press, October 2016
42. Kushilevitz, E., Malkin, T. (eds.): TCC 2016, Part I. LNCS, vol. 9562. Springer, Heidelberg (2016). https://doi.org/10.1007/978-3-662-49096-9
43. Kushilevitz, E., Malkin, T. (eds.): TCC 2016, Part II. LNCS, vol. 9563. Springer, Heidelberg (2016). https://doi.org/10.1007/978-3-662-49099-0
44. Li, X.: Improved two-source extractors, and affine extractors for polylogarithmic entropy. In: Dinur, I. (ed.) 57th FOCS, pp. 168–177. IEEE Computer Society Press, October 2016
45. Lindell, Y.: A simpler construction of CCA2-secure public-key encryption under general assumptions. In: Biham, E. (ed.) EUROCRYPT 2003. LNCS, vol. 2656, pp. 241–254. Springer, Heidelberg (2003). https://doi.org/10.1007/3-540-39200-9_15
46. Lindell, Y. (ed.): TCC 2014. LNCS, vol. 8349. Springer, Heidelberg (2014). https://doi.org/10.1007/978-3-642-54242-8
47. Liu, F.-H., Lysyanskaya, A.: Tamper and leakage resilience in the split-state model. In: Safavi-Naini, R., Canetti, R. (eds.) CRYPTO 2012. LNCS, vol. 7417, pp. 517–532. Springer, Heidelberg (2012). https://doi.org/10.1007/978-3-642-32009-5_30
48. Naor, M., Yung, M.: Public-key cryptosystems provably secure against chosen ciphertext attacks. In: 22nd ACM STOC, pp. 427–437. ACM Press, May 1990
49. Raz, R.: Fast learning requires good memory: A time-space lower bound for parity learning. CoRR abs/1602.05161 (2016)
50. Sahai, A.: Non-malleable non-interactive zero knowledge and adaptive chosen-ciphertext security. In: 40th FOCS, pp. 543–553. IEEE Computer Society Press, October 1999

51. Tal, A.: Tight bounds on the fourier spectrum of AC0. In: O'Donnell, R. (ed.) 32nd Computational Complexity Conference, CCC 2017, Riga, Latvia, 6–9 July 2017. LIPIcs, vol. 79, pp. 15:1–15:31. Schloss Dagstuhl - Leibniz-Zentrum fuer Informatik (2017)
52. Wichs, D., Mansour, Y. (eds.): 48th ACM STOC. ACM Press, June 2016

Provable Symmetric Cryptography

Naor-Reingold Goes Public: The Complexity of Known-Key Security

Pratik Soni(✉) and Stefano Tessaro

University of California, Santa Barbara, USA
{pratik_soni,tessaro}@cs.ucsb.edu

Abstract. We study the complexity of building secure block ciphers in the setting where the key is known to the attacker. In particular, we consider two security notions with useful implications, namely public-seed pseudorandom permutations (or psPRPs, for short) (Soni and Tessaro, EUROCRYPT '17) and correlation-intractable ciphers (Knudsen and Rijmen, ASIACRYPT '07; Mandal, Seurin, and Patarin, TCC '12).

For both these notions, we exhibit constructions which make only *two* calls to an underlying non-invertible primitive, matching the complexity of building a pseudorandom permutation in the secret-key setting. Our psPRP result instantiates the round functions in the Naor-Reingold (NR) construction with a secure UCE hash function. For correlation intractability, we instead instantiate them from a (public) random function, and replace the pairwise-independent permutations in the NR construction with (almost) $O(k^2)$-wise independent permutations, where k is the arity of the relations for which we want correlation intractability.

Our constructions improve upon the current state of the art, requiring five- and six-round Feistel networks, respectively, to achieve psPRP security and correlation intractability. To do so, we rely on techniques borrowed from Impagliazzo-Rudich-style black-box impossibility proofs for our psPRP result, for which we give what we believe to be the first *constructive* application, and on techniques for studying randomness with limited independence for correlation intractability.

Keywords: Foundations · Known-key security · Pseudorandomness psPRPs · Correlation-intractability · Limited independence

1 Introduction

1.1 Overview and Motivation

Block ciphers are traditionally used within modes of operation where they are instantiated under a *secret* key. Provable security results typically assume them to be good *pseudorandom permutations* (PRPs). This has motivated a large body of theoretical works on building PRPs from weaker or less structured components, e.g., through the Feistel construction and its variants [24,28,33].

Block ciphers are however also frequently used in settings where the key is fixed, or at least *known*. We refer to this as the *known-key setting*. For instance,

J. B. Nielsen and V. Rijmen (Eds.): EUROCRYPT 2018, LNCS 10822, pp. 653–684, 2018.
https://doi.org/10.1007/978-3-319-78372-7_21

it is common to rely on *permutations*[1] or (equivalently) *fixed-key* ciphers to build hash functions [12,35], authenticated encryption [3], PRNGs [13,23], and even more involved objects, such as garbling schemes [8].

As there is no secret key to rely upon, it is less clear what kind of security properties block ciphers should satisfy in this setting. Hence, security proofs typically assume the cipher to behave like an ideal random permutation on each key. A number of design paradigms for block ciphers (cf. e.g. [1,17–19,21,25] to mention a few results) are therefore analyzed in terms of *indifferentiability* [31], an ideal-model property which implies that for single-stage security games, the cipher inherits all properties of an ideal cipher. Still, the resulting proofs are notoriously involved, and the constructions more complex than seemingly necessary for the applications in which they are used. This is in sharp contrast with hash functions, where indifferentiability has helped shaping real-world designs.

Our contributions. The only two exceptions to the above indifferentiability-based approach we are aware of are the notions of a *public-seed pseudorandom permutation* (psPRP) [37] and of *correlation-intractable block ciphers* [27,29]. Block ciphers satisfying variants of both have been shown to be sufficient to instantiate several schemes that otherwise only enjoyed security proofs assuming the cipher is ideal. Yet, the complexity of actually building these primitives from simpler objects is not understood.

In this work, we present constructions for each of the notions which only make *two* calls to an underlying non-invertible round function. All of our constructions are instantiations of the *Naor-Reingold construction* [33], which is the most efficient known approach to build a secure PRP, and we thus show that it retains meaningful properties when the seed is made *public* under appropriate assumptions on the round functions. The previously known best constructions require Feistel networks with five rounds (for psPRPs) [37] and six rounds (for correlation intractability) [29], and in both cases the security proofs relied indirectly on (weakened) forms of indifferentiability, inheriting seemingly unnecessary complexity. Here, we introduce substantially different techniques to bypass limitations of existing proofs, borrowing from areas such as black-box separations and applications of limited-independence.

We stress that our focus here is on foundations, and more specifically, breaking complexity barriers. While we follow the good practice of giving concrete bounds, we make no claims that these are suitable for practical applications. We hope however to spur quantitative research in this direction.

1.2 Public-Seed Pseudorandomness via the NR Construction

We start with our results on *public-seed pseudorandom permutations* (or psPRPs, for short), a notion recently introduced in [37], which considers a family of permutations E on n-bit strings, indexed by a seed s. (This could be obtained from a block cipher.) Ideally, we would like $\mathsf{E}_s(\cdot)$ and $\mathsf{E}_s^{-1}(\cdot)$ to be indistinguishable

[1] Permutations, as in the sponge construction, correspond to the extreme case where there is only one possible key to choose from.

from ρ and ρ^{-1} (for a random permutation ρ), *even* if the seed s is known to the distinguisher. This is obviously impossible, yet an approach to get around this borrowed from the UCE framework [6] is to *split* the distinguisher into two stages. A first stage, called the *source* S, gets access to either $(\mathsf{E}_s, \mathsf{E}_s^{-1})$ or (ρ, ρ^{-1}), but does *not* know s, and then passes on some leakage L to a second-stage PPT D, the *distinguisher*, which learns s, but has no access to the oracle any more. If E is indeed secure, D will not be able to guess which one of the two oracles S had access to. This is very similar to the security definition of a UCE H, the only difference is that there the source accesses either H_s or a random *function* f.

Clearly, nothing is gained if L is unrestricted, and thus restrictions on S are necessary. Two classes of sources were in particular considered in [37], *unpredictable* and *reset-secure* sources, inspired by analogous notions for UCEs. The former demands that when the source S accesses ρ and ρ^{-1}, an (unbounded)[2] predictor P given the leakage L cannot then guess any of S's queries (and their inverses). In contrast, the latter notion demands that a computationally unbounded distinguisher R, given L cannot tell apart whether it is given oracle access to the *same* permutation ρ, or an independent one, within a polynomial number of queries. Being a psPRP for all unpredictable sources is a potentially weaker assumption than being a psPRP for reset-secure sources, since every unpredictable source is reset-secure, but not vice versa.

Applications. PsPRPs for such restricted source classes are a versatile notion. For example, a psPRP for all reset-secure sources can be used to instantiate the permutation within permutation-based hash functions admitting indifferentiability-based security proofs, such as the sponge construction [12] (which underlies SHA-3), turning them into a UCE-secure hash function sufficient for a number of applications, studied in multiple works [5,6,11,20,30]. Also, [37] shows that psPRPs for unpredictable sources are sufficient to instantiate garbling schemes obtained from fixed-key blocks ciphers [8].

Constructing psPRPs: Previous work. But do psPRPs exist at all? Soni and Tessaro [37] show that they are implied by sufficiently strong UCEs:

Theorem (Informal) [37]. *The five-round Feistel construction, with round functions instantiated from a UCE* H *for reset-secure sources, is a psPRP for reset-secure sources.*

This left two obvious questions open, however: **(1)** Whether the number of rounds can be reduced, and **(2)** whether the same holds for unpredictable sources, too. The techniques of [37], based on proving a weaker notion of indifferentiability for five-round Feistel, fail to help answering both questions.

Our contributions. We solve both questions, and even more in fact, showing that the Naor-Reingold (NR) construction [33] solves *both* (1) and (2). In

[2] Computational versions of these notions can be defined, but the resulting notions can easily be shown impossible under the assumption that IO exists [14], and are ignored in this paper.

particular, let H be a family of functions from $n+1$ bits to n, and let P be a family of permutations on $2n$ bit strings. Then, the NR construction on seed $\boldsymbol{s} = (s, s^{\mathsf{in}}, s^{\mathsf{out}})$ and input $u \in \{0,1\}^{2n}$, outputs $v \in \{0,1\}^{2n}$, where

$$x_0 \,\|\, x_1 \leftarrow \mathsf{P}_{s^{\mathsf{in}}}(u) \;,\; x_2 \leftarrow \mathsf{H}_s(0 \,\|\, x_1) \oplus x_0,$$
$$x_3 \leftarrow \mathsf{H}_s(1 \,\|\, x_2) \oplus x_1 \;,\; v \leftarrow \mathsf{P}^{-1}_{s^{\mathsf{out}}}(x_3 \,\|\, x_2).$$

The key point here is that P only needs to satisfy a weak non-cryptographic property, namely that for a random s and for any distinct $u \neq u'$, the right halves of $\mathsf{P}_s(u)$ and $\mathsf{P}_s(u')$ only collide with negligible probability. Therefore, only two calls to a "cryptographically hard" round function H are made. Naor and Reingold [33] showed that NR is a (strong) PRP whenever H is a pseudorandom function. Here, we show the following public-seed counterparts:

Theorems 1 and 2 (Informal). *The NR construction, with round functions instantiated with a UCE* H *for* X*-sources, is a psPRP for* X*-sources, where* $X \in \{$*reset-secure, unpredictable*$\}$.

A detailed overview of our techniques is given below in Sect. 1.4. We remark here that such UCEs are of course strong, and the question of basing these on simpler assumptions is wide open. Still, we believe such results to be very important: First off, they show relations among notions, and getting a UCE (without any injectivity structure) is possibly simpler in practice than in theory (i.e., using the compression function of SHA-256). Second, even if we instantiate H from a random oracle (which gives a good UCE [6]), the result *is* useful, as this would give us a simple instantiation of a (seeded) permutation in applications which are not even known to follow from full-fledged indifferentiability, as discussed by Mittelbach [32].

1.3 Correlation Intractability

The notion of *correlation intractability* (CI) of *hash functions* was introduced by Canetti et al. [15] as a weakening of a random oracle. CI naturally extends to permutations and block ciphers [27,29]: Given the seed s, an adversary should not be able to find an input-output pair (u, v) such that $\mathsf{E}_s(u) = v$ and such that $(u, v) \in R$, where R is a hard-to-satisfy relation for a truly random permutation, a so-called *evasive* relation. This, in turn, can be generalized to k-ary relations, where k input-output pairs are to be provided. CI is well-known not to hold in the standard model for arbitrary evasive relations.[3] Therefore, here, we target constructions in ideal models.

APPLICATIONS. CI has important applications – for example, let E be a permutation family on $2n$-bit strings. Then, for $n < m < 2n$, consider the function family H from m bits to n bits such that

$$\mathsf{H}_s(x) = \mathsf{E}_s(x \,\|\, 0^{2n-m})[1 \ldots n],$$

[3] Though, of course, it could be true for specific interesting relations.

i.e., this outputs the first n bits of $\mathsf{E}_s(x \,\|\, 0^{2n-m})$. Then, it is not too hard to show that if E satisfies CI for evasive binary relations, then H is collision resistant – indeed, a collision yields two distinct pairs (u_1, v_1), (u_2, v_2) where $u_1[m+1 \ldots 2n] = u_2[m+1 \ldots 2n] = 0^{2n-m}$, whereas $v_1[1 \ldots n] = v_2[1 \ldots n]$. Along similar lines, one can prove that H can be used to instantiate the Fiat-Shamir transform [22] whenever E satisfies CI for *unary* relations. And so on.

Correlation intractability for Feistel networks. Indifferentiability is easily seen to imply CI, and therefore, by [19], the Feistel construction with 8 rounds is correlation intractable. In fact, Mandal et al. [29] observed that a weaker notion of indifferentiability, called sequential indifferentiability, is sufficient for CI, and could show that 6 rounds are enough. It is known that the 5-round Feistel construction is *not* correlation intractable for evasive 4-ary relations (an attack was given in [29]). Other weaker notions of indifferentiability are known to imply CI, but do no appear to lead to any complexity improvements for constructions from non-invertible primitives [2,16].

Our results. We study the correlation intractability of the NR construction described above where the two calls to H are replaced by two calls to (seedless) public random function f from $n+1$ to n bits, and the seed of the construction only consists of the seeds for P. (This is similar to the model of Ramzan and Reyzin [34], although they consider PRP security, and secret seeds.)

In general, this basic form of the NR construction cannot be correlation intractable – indeed, P can be instantiated by one-round Feistel with a pairwise-independent round function, and generic attacks against the correlation intractability of four-round Feistel would still apply. We show however the following result:

Theorem 3 (Informal). *For any* constant $k = O(1)$, *if* P^{-1} *is an almost* $O(k^2)$*-wise independent permutation, then the NR construction is correlation intractable for every* k*-ary evasive relation.*

For unary relations (i.e., $k = 1$), we can in fact show that instantiating P with one-round Feistel using a 10-wise independent round function suffices. As this is effectively a four-round Feistel network, this confirms that no generic attacks exist for unary relations. Our result extends to non-constant k, however under some restrictions on the class of evasive relations for which we can prove correlation intractability. We believe an important part of our result is the technique we use, which gives a surprising paradigm to amplify CI unconditionally which we discuss below in Sect. 1.5.

Limitations. In contrast to existing 6-round results, our result is weaker in that it only covers evasive relations fully if $k = O(1)$. We are not aware of counter examples showing attacks for larger k's, but our proof inherently fails. We note however that most applications of correlation intractability only require constant arity, and we leave the question of assessing whether six calls to a random function are necessary for arbitrary arity for future work.

1.4 Technical Overview – psPRPs

Let us briefly recall the setting: For some PPT source S, which queries a permutation oracle on $2n$-bit strings to produce a leakage L, we need to show that any PPT distinguisher D which learns L *and* $\boldsymbol{s} = (s, s^{\mathsf{in}}, s^{\mathsf{out}})$ cannot tell apart whether S was accessing NR using a UCE H (with seed $\boldsymbol{s}$) or a truly random permutation. We assume S is either (statistically) unpredictable (in Theorem 1) or reset-secure (in Theorem 2).

THE SOURCE $\overline{S}$. The natural approach we follow is to build another source, $\overline{S}$ from S, for which H should be a secure UCE. This source thus accesses an oracle $\mathcal{O}$ implementing a function from $n+1$ to n bits. It first samples seeds $s^{\mathsf{in}}, s^{\mathsf{out}}$ for P, and then simulates an execution of S. The oracle calls by the latter are processed by evaluating the NR construction using $s^{\mathsf{in}}, s^{\mathsf{out}}$, and the oracle $\mathcal{O}(\cdot)$ in lieu of $\mathsf{H}(s, \cdot)$. Finally, when S produces its output L, $\overline{S}$ outputs $(L, s^{\mathsf{in}}, s^{\mathsf{out}})$. We will show the following two facts:

- FACT 1. If S is unpredictable (w.r.t. the psPRP notion), then $\overline{S}$ is unpredictable (w.r.t. the UCE notion).
- FACT 2. If S is reset-secure (w.r.t. the psPRP notion), then $\overline{S}$ is reset-secure (w.r.t. the UCE notion).

Theorems 1 and 2 follow from Facts 1 and 2, respectively, by a fairly straightforward application of the (classical) indistinguishability of the NR construction with *random* round functions.[4]

THE UNPREDICTABLE CASE. Our approach to establish Fact 1 is inspired by an elegant proof of secure domain extension for UCEs via Wegman-Carter MACs [7]. (The case of reset-secure sources will be more involved and use new techniques.)

Assume, towards a contradiction, that $\overline{S}$ is *not* unpredictable; then there exists a strategy (not necessarily efficient) that given L and s^{in} and s^{out}, guesses one of the inner oracle queries of $\overline{S}$ with non-negligible probability ε, when $\overline{S}$'s oracle is a random function from $n+1$ to n bits. Imagine now that given $(L, s^{\mathsf{in}}, s^{\mathsf{out}})$ from $\overline{S}$, we *resample* an execution of $\overline{S}$ (which in particular means re-sampling the oracle used by it) consistent with outputting $(L, s^{\mathsf{in}}, s^{\mathsf{out}})$, and look at the inner oracle queries in this virtual, re-sampled execution. Then, one can show that the real and the virtual executions are likely to share an oracle query, with probability roughly at least ε^2, for our strategy to guess a query must be equally successful on the virtual execution.

We exploit this idea to build a predictor for the original source S, contradicting our hypothesis it is unpredictable. Note that S runs with a random permutation as its oracle, and produces leakage L. Imagine now we sample fresh seeds s^{in}, s^{out} for P, and for each permutation query by S defining an input-output pair (u, v), we define "fake" inputs x_0, x_1 from $x_0 \,\|\, x_1 = \mathsf{P}_{s^{\mathsf{in}}}(u)$ and

[4] A minor caveat is that we need indistinguishability even when s^{in} and s^{out} are revealed at the end of the interaction. We will show this to be true.

$x_3 \| x_2 = \mathsf{P}_{s^{\mathsf{out}}}(v)$. Then, the indistinguishability of the NR construction from a random permutation, and the construction of $\overline{S}$, implies that if we resample a virtual execution of S consistent with leakage L, and compute the resulting fake inputs using s^{in} and s^{out}, then the real and the re-sampled execution will share a fake input with probability approx. ε^2. The properties imposed on P then imply that with probability roughly ε^2 the real and the re-sampled execution must share the input (or output) of a permutation query. This leads naturally to a predictor that just re-samples an execution consistent with the leakage, and picks them as its prediction.

Reset-security. The case of reset-security is somewhat harder. Here we start from the premise that $\overline{S}$ is not reset-secure: Hence, there exists an adversary $\overline{R}$ which receives $L, s^{\mathsf{in}}, s^{\mathsf{out}}$ from $\overline{S}$, and can distinguish (with non-negligible advantage ε) being given access to the same random $f : \{0,1\}^{n+1} \to \{0,1\}^n$ used by $\overline{S}$ from being given access to an independent f'. From this, we would like to build an adversary R which receives L from S, and can distinguish the setting where R and S are given access to the same random permutation ρ, from a setting where they access independent permutations ρ, ρ'.

The challenge here is that we want to simulate $\overline{R}$ correctly, by using a permutation oracle ρ/ρ' rather than f/f'. To better see why this is tricky, say S is the source that queries its permutation oracle on a random $2n$-bit string u, obtaining output v, and leaks $L = (u, v)$. (This defines the corresponding $\overline{S}$.)[5] A clever $\overline{R}$ on input $(L = (u, v), s^{\mathsf{in}}, s^{\mathsf{out}})$ could do the following: It computes $x_0 \| x_1 \leftarrow \mathsf{P}(s^{\mathsf{in}}, u)$ and $x_3 \| x_2 \leftarrow \mathsf{P}(s^{\mathsf{out}}, v)$. Then, it queries x_1 to its oracle, and outputs 1 iff the output equals $x_0 \oplus x_2$. This should always be true when $\overline{R}$ accesses f, and almost never when it accesses f'.

The natural proof approach would now attempt to build R which runs $\overline{R}$ accessing a simulated oracle consistent with the NR construction on the permutation queries made by S. However, the problem is that generically R does not know which queries S has made. Previous work [37] handled this by requiring the construction to satisfy a weaker notion of indifferentiability, called CP-sequential indifferentiability, which essentially implies that there exists a simulator that can simulate f consistently by accessing ρ and ρ^{-1} only, and only needs to know the queries $\overline{R}$ makes to f. This would not work with NR and our $\overline{R}$, as the query x_1 is actually uniformly random, and the simulator would likely fail to set $x_0 \oplus x_2$ as the right output. This is why the approach of [37] ends up using the 5-round Feistel construction, as here $\overline{R}$'s attempt to evaluate the construction are readily detected, and answered consistently.

Our proof strategy via heavy-query sampling. Our main observation is that indifferentiability is an overkill in this setting. There is no reason $\overline{R}$ should act adversarially to the simulator. Even more so, we can use everything $\overline{R}$ knows, namely L, to our advantage! To do this, we use techniques borrowed

[5] The reader should not be confused: $\overline{S}$ is clearly not reset-secure, but remember we are in the setting of a proof by contradiction, so the reduction must work here, too.

from impossibility proofs in the random oracle model [4,26]. Namely, R, on input L from S, first performs a number of permutation queries which are meant to include all of S's likely queries to its oracle, at least when R and S are run with the same permutation oracle ρ. To do this, R samples executions of S consistent with L, and the partial knowledge of the oracle ρ acquired so far. Each time such a partial execution is sampled, all queries contained in it are made to ρ, and the process is repeated a number of times polynomial in $1/\varepsilon$. Then, R samples $s^{\mathsf{in}}, s^{\mathsf{out}}$, and internally defines an oracle $f : \{0,1\}^{n+1} \to \{0,1\}^n$ that will be used to simulate an execution of $\overline{R}^f(L, s^{\mathsf{in}}, s^{\mathsf{out}})$. To do this, R goes through all input-output pairs (u, v) for queries to ρ it has done while simulating executions of S,[6] and defines

$$f(0 \,\|\, x_1) \leftarrow x_0 \oplus x_2 \, , f(1 \,\|\, x_2) \leftarrow x_1 \oplus x_3,$$

where $x_0 \,\|\, x_1 \leftarrow \mathsf{P}_{s^{\mathsf{in}}}(u)$ and $x_3 \,\|\, x_2 \leftarrow \mathsf{P}_{s^{\mathsf{out}}}(v)$. Then, f is defined to be random on every other input (this can be simulated via lazy sampling). The final output of the simulated $\overline{R}$ is then R's final output.

The core of our proof will show that when S and R share access to ρ, then the probability that R's output is one is similar to that of $\overline{R}$ outputting one when it accesses the same oracle as $\overline{S}$. This will combine properties of the NR construction (allowing us to switch between f and ρ), and similar arguments as those used in [26] to prove that R ensures consistency on all queries that matter.[7]

1.5 Technical Overview – Correlation Intractability

Our approach towards achieving CI is based on the following blueprint. Let R be a relation which is evasive for permutations on $2n$-bit strings, and let π, σ be permutations sampled from some given distribution (this will be meant to be instantiated unconditionally below). Then, we create a new relation $R_{\pi,\sigma}$ such that $(u, v) \in R$ iff $(\pi(u), \sigma(v)) \in R_{\pi,\sigma}$. The hope is to show that *if* R is an evasive relation, then $R_{\pi,\sigma}$ is hard to satisfy for a given construction E which is only correlation-intractable for a subset of all evasive relations. Then, a new composed construction E' which outputs $\sigma^{-1}(E_s(\pi(u)))$ on input u would be correlation intractable for all evasive relations, since satisfying R for E' implies satisfying $R_{\pi,\sigma}$ for E.

Two-round Feistel. In our context, we instantiate E from a two-round Feistel network. That is, on input $x = x_0 \,\|\, x_1$, the two-round Feistel construction outputs $x_2 \,\|\, x_3$, where $x_2 \leftarrow f(0 \,\|\, x_1) \oplus x_0$ and $x_3 \leftarrow f(1 \,\|\, x_2) \oplus x_1$. In a model (as the one where we consider) where f is a random function to which the adversary

[6] The actual simulation will be slightly more involved, for the benefit of simplifying the analysis.

[7] We believe we could adapt our proof to use the better strategy of [4] to get slightly better concrete parameters, yet we found adapting it to our setting not immediate.

is given access, this construction is *not* correlation intractable. For instance, take the (unary) relation which is satisfied by all input-output pairs $(x_0 \| x_1, x_2 \| x_3)$ where $x_1 = x_2$. This is clearly evasive, but trivial to satisfy for two-round Feistel. Worse is possible with k-ary relations.

However, many relations *are* hard, even for two-round Feistel. Take for instance any relation R with the property that for all x_0, x_1, x_2, x_3, the number of x^*'s such that $(x^* \| x_1, x_2 \| x_3) \in R$ or $(x_0 \| x_1, x_2 \| x^*) \in R$ is at most $\delta \cdot 2^n$, for some negligible function δ. No adversary A making a polynomial number of queries to f will satisfy R, except with negligible probability. Indeed, when A queries (say) $y_2 \leftarrow f(1 \| x_2)$ for some x_2, the only chance to produce a pair that satisfies R is $y_1 \leftarrow f(0 \| x_1)$ was previously queried for some x_1, and additionally, $(x_2 \oplus y_1 \| x_1,\ x_2 \| x_1 \oplus y_2) \in R$. But because y_2 is being set randomly, $x_1 \oplus y_2$ is also random, this can only hold with probability at most δ by our assumption on δ. Thus, the probability that this pair satisfies R is negligible, and the union bound over all pairs of queries shows A is unlikely to *ever* satisfy R.

WHERE DOES AMPLIFICATION COME FROM? Let R be a unary evasive relation. Now, imagine, again for the sake of an oversimplified illustration, that π and σ are random permutations. Then, we want to show that with high probability over the choice of π and σ, the relation $R_{\pi,\sigma}$ is hard for two-round Feistel, even if the adversary learns the entire description of π and σ. Indeed, find any x_0, x_1, x_2, fix π, and fix $u = \pi^{-1}(x_0 \| x_1)$. Because R is evasive, there exists at most $\delta \cdot 2^{2n}$ v's (for some negligible δ) such that $(u, v) \in R$ – call the set of such v's R_u. Because σ is random, the probability that $\sigma^{-1}(x_2 \| z) \in R_u$ is at most δ, and thus the *expected* number of z such that $(x_0 \| x_1, x_2 \| z) \in R_{\pi,\sigma}$ is at most $\delta \cdot 2^n$, by linearity of expectation. A concentration bound will show that the probability that we are far from this expectation is indeed small, say at most 2^{-4n}. Taking a union bound over all x_0, x_1, x_2 shows that the probability this is true for any x_0, x_1, x_2 is at most 2^{-n}. (The symmetric argument when fixing x_1, x_2, x_3 can be handled analogously.) Thus, we have just argued that with high probability over the choice of π and σ, $R_{\pi,\sigma}$ is hard for two-round Feistel!

CHALLENGES. But obviously, this is not very useful– random permutations π, σ are inefficient to sample and describe. Also, the above result only holds for unary relations, and it is interesting to extend this to k-ary relations.

Our first insight is that the above argument only requires a bounded degree of randomness, and that (almost) t-wise independent permutations for a sufficiently small t are sufficient. We prove this using techniques for bounding sums of random variables with bounded independence [10,36], though this will require significant adaptation because almost t-wise independent permutations do not quite produce outputs which are t-wise independent, as they are required to be distinct, and also, only approximate a random permutation. We will instantiate these by using Feistel networks with sufficiently many rounds, and t-wise independent round functions, using bounds from [24]. In fact, for the case of unary relations, we will show that we can instantiate these permutations from one single Feistel round with a 10-wise independent round function.

Moving on to k-ary relations presents even more challenges. Our approach is inherently combinatorial, whereas evasiveness is defined indirectly through the inability of an adversary to win a security game. For this reason, our result will only deal with relations R that satisfy a more structured notion of evasiveness, which we refer to as *strongly* evasive. Most relations of interest that we are aware of are strongly evasive, but evasiveness does not always imply strong evasiveness. *However*, as a result of independent interest, we show that strong evasiveness and evasiveness are related, and asymptotically equivalent when k is a constant.

2 Preliminaries

NOTATIONAL PRELIMINARIES. Throughout this paper, we denote by $\mathsf{Funcs}(X, Y)$ the set of functions $X \rightarrow Y$, and in particular use the shorthand $\mathsf{Funcs}(m, n)$ whenever $X = \{0,1\}^m$ and $Y = \{0,1\}^n$. We also denote by $\mathsf{Perms}(X)$ the set of permutations on the set X, and analogously, $\mathsf{Perms}(n)$ denotes the special case where $X = \{0,1\}^n$. For $n \in \mathbb{N}$, we let $[n]$ denote the set $\{1, \ldots, n\}$.

Our security definitions and proofs will often use games, as formalized by Bellare and Rogaway [9]. Typically, our games will have boolean outputs – that is, either true or false – and we use the shorthand $\Pr[\mathsf{G}]$ to denote the probability that a certain game outputs the value true, or occasionally 1 (when the output is binary, rather than boolean). Most results in this paper will be concrete, but natural asymptotic statements can be made by allowing all parameters to be functions of the security parameter.

A *function family* with input set X and output set Y is a pair of algorithms $\mathsf{F} = (\mathsf{F.Kg}, \mathsf{F.Eval})$, where the randomized *key (or seed) generation algorithm* $\mathsf{F.Kg}$ outputs a seed s, and the deterministic *evaluation algorithm* $\mathsf{F.Eval}$ takes as inputs a valid seed s and an input $x \in X$, and returns $\mathsf{F.Eval}(s, x) \in Y$. If $X = \{0,1\}^m$ and $Y = \{0,1\}^n$, we say that F is a family of functions from m-bits to n-bits. We usually write $\mathsf{F}(s, \cdot) = \mathsf{F.Eval}(s, \cdot)$. A *permutation family* $\mathsf{P} = (\mathsf{P.Kg}, \mathsf{P.Eval})$ on n bits is the special case where $X = \{+,-\} \times \{0,1\}^n$ and $Y = \{0,1\}^n$, and for every s, there exists a permutation π_s such that $\mathsf{P.Eval}(s, (+, x)) = \pi_s(x)$ and $\mathsf{P.Eval}(s, (-, y)) = \pi_s^{-1}(y)$. We usually write $\mathsf{P}(s, \cdot) = \mathsf{P}(s, (+, \cdot))$ and $\mathsf{P}^{-1}(s, \cdot) = \mathsf{P}(s, (-, \cdot))$.

2.1 UCEs and psPRPs

We review the UCE notion introduced in [6], and the psPRP notion [37]. As explained in the latter work, they can be seen as instantiations of a general paradigm. Yet, we consider separate security games for better readability.

Concretely, let H be function family from m-bits to n-bits. Let S be an adversary called the *source* and D an adversary called the *distinguisher*. We associate with them the game $\mathsf{UCE}^{S,D}_{m,n,\mathsf{H}}$ depicted in Fig. 1. For a family E of permutations on n-bits, the psPRP-security game $\mathsf{psPRP}^{S,D}_{n,\mathsf{E}}$ differs in that $\mathcal{O}$

MAIN $\boxed{\mathsf{UCE}^{S,D}_{m,n,\mathsf{H}}}$, $\boxed{\mathsf{psPRP}^{S,D}_{n,\mathsf{E}}}$:	ORACLE $\mathcal{O}(i,x)$: // $\mathsf{UCE}^{S,D}_{m,n,\mathsf{H}}$
$(1^r, t) \xleftarrow{\$} S(\varepsilon)$, $b \xleftarrow{\$} \{0,1\}$	**if** $b = 1$ **then return** $\mathsf{H}(s_i, x)$
$s_1, \ldots, s_r \xleftarrow{\$} \boxed{\mathsf{H.Kg}}\ \boxed{\mathsf{E.Kg}}$	**else return** $f_i(x)$
$\boxed{f_1, \ldots, f_r \xleftarrow{\$} \mathsf{Funcs}(m,n)}$	ORACLE $\mathcal{O}(i,(\sigma,x))$: // $\mathsf{psPRP}^{S,D}_{n,\mathsf{E}}$
$\boxed{\rho_1, \ldots, \rho_r \xleftarrow{\$} \mathsf{Perms}(n)}$	**if** $b = 1$ **then**
$L \xleftarrow{\$} S^{\mathcal{O}}(t)$	**if** $\sigma = +$ **then return** $\mathsf{E}(s_i, x)$
$b' \xleftarrow{\$} D(s_1, \ldots, s_r, L)$	**else return** $\mathsf{E}^{-1}(s_i, x)$
return $b' = b$	**else**
	if $\sigma = +$ **then return** $\rho_i(x)$
	else return $\rho_i^{-1}(x)$

Fig. 1. Games to define UCE and psPRP security. Here, S is the source and D is the distinguisher. Boxed statements are only executed in the corresponding game.

allows for inverse queries, and the ideal object is a random permutation. The corresponding advantage metrics for an (S, D) are defined as

$$\begin{aligned} \mathsf{Adv}^{\mathsf{uce}}_{m,n,\mathsf{H}}(S,D) &= 2\Pr\left[\mathsf{UCE}^{S,D}_{m,n,\mathsf{H}}\right] - 1 \\ \mathsf{Adv}^{\mathsf{psprp}}_{n,\mathsf{E}}(S,D) &= 2\Pr\left[\mathsf{psPRP}^{S,D}_{n,\mathsf{E}}\right] - 1. \end{aligned} \tag{1}$$

Note that we adopt the multi-key versions of UCE and psPRP security, as they are the most general, and they are not known to follow from the single-key case. Our treatment scales down to the single-key version by forcing the source to always choose $r = 1$.

We say that H is UCE-secure for a class of sources $\mathcal{S}$ if $\mathsf{Adv}^{\mathsf{uce}}_{m,n,\mathsf{H}}(S,D)$ is negligible for all PPT D and all sources $S \in \mathcal{S}$. Similarly, E is psPRP secure for $\mathcal{S}$ if $\mathsf{Adv}^{\mathsf{psprp}}_{n,\mathsf{E}}(S,D)$ is negligible for all PPT D and all sources $S \in \mathcal{S}$ It is known that $\mathcal{S}$ cannot contain all PPT algorithms for security to be attainable. Next, we discuss two important classes of restrictions – unpredictable and reset-secure sources – considered in the literature [6,7,37].

Unpredictable Sources. Let S be a source and P be an adversary called the *predictor*. We associate with them games $\mathsf{f\text{-}Pred}^{P}_{m,n,S}$ and $\mathsf{p\text{-}Pred}^{P}_{n,S}$ of Fig. 2 which capture the fact that P cannot predict any of the queries of S (or their inverses), when the latter interacts with a random function from m bits to n bits, or respectively a random permutation on n-bit strings. The corresponding advantage metrics are

$$\mathsf{Adv}^{\mathsf{f\text{-}pred}}_{m,n,S}(P) = \Pr\left[\mathsf{f\text{-}Pred}^{P}_{m,n,S}\right], \ \mathsf{Adv}^{\mathsf{p\text{-}pred}}_{n,S}(P) = \Pr\left[\mathsf{p\text{-}Pred}^{P}_{n,S}\right]. \tag{2}$$

We say S is *statistically unpredictable* if $\mathsf{Adv}^{\mathsf{f\text{-}pred}}_{m,n,S}(P)$ (respectively, $\mathsf{Adv}^{\mathsf{p\text{-}pred}}_{n,S}(P)$) is negligible for all predictors P outputting a set Q' of polynomial size.

```
MAIN [f-Pred^P_{m,n,S}], (p-Pred^P_{n,S}):
Q ← ∅
(1^r, t) ←$ S(ε)
[f_1, ..., f_r ←$ Funcs(m, n)]
(ρ_1, ..., ρ_r ←$ Perms(n))
L ←$ S^O(t)
Q' ←$ P(1^r, L)
return (Q ∩ Q' ≠ ∅)

ORACLE O(i, x): // f-Pred^P_{m,n,S}
Q ← Q ∪ {(i, x)}
return f_i(x)

ORACLE O(i, (σ, x)): // p-Pred^P_{n,S}
if σ = + then y ← ρ_i(x)
else y ← ρ_i^{-1}(x)
Q ← Q ∪ {(i, x), (i, y)}
return y
```

```
MAIN [f-Reset^R_{m,n,S}], (p-Reset^R_{n,S}):
done ← false; (1^r, t) ←$ S(ε)
[f_1^0, f_1^1, ..., f_r^0, f_r^1 ←$ Funcs(m, n)]
(ρ_1^0, ρ_1^1, ..., ρ_r^0, ρ_r^1 ←$ Perms(n))
L ←$ S^O(t); done ← true
b ←$ {0,1}; b' ←$ R^O(1^r, L)
return b' = b

ORACLE O(i, x): // f-Reset^R_{m,n,S}
if ¬done then return f_i^0(x)
else return f_i^b(x)

ORACLE O(i, (σ, x)): // p-Reset^R_{n,S}
if ¬done then
   if σ = + then return ρ_i^0(x)
   else return ρ_i^{0-1}(x)
else
   if σ = + then return ρ_i^b(x)
   else return ρ_i^{b-1}(x)
```

Fig. 2. Games to define unpredictability (left) and reset-security (right) of sources. Here, S is the source, P is the predictor and R is the reset-adversary. Boxed statements are only executed in the corresponding game.

An analogous notion of computational unpredictability can be defined, but it is unachievable if IO exists [14], and is usually not needed for applications. We also note that what we formalize here is the notion of *simple* unpredictability – P is not permitted to query the underlying primitive. The notion was proved equivalent (asymptotically) for UCEs [6] to a version where we give P access to the primitive. A similar proof follows for psPRPs. (We omit it due to lack of space.)

RESET-SECURE SOURCES. Let S be a source and R be an adversary called the reset-adversary. We associate to them the games $\mathsf{f\text{-}Reset}^R_{m,n,S}$ and $\mathsf{p\text{-}Reset}^R_{n,S}$ of Fig. 2 which formalize the reset-security of S against a random function and a random permutation, respectively. The idea here is that R should not be able to tell apart whether S is accessing the same set of oracles it accesses, or not. This is captured via the advantage metrics

$$\mathsf{Adv}^{\mathsf{f\text{-}reset}}_{m,n,S}(R) = 2\Pr\left[\mathsf{f\text{-}Reset}^R_{m,n,S}\right] - 1, \quad \mathsf{Adv}^{\mathsf{p\text{-}reset}}_{n,S}(R) = 2\Pr\left[\mathsf{p\text{-}Reset}^R_{n,S}\right] - 1.$$

We say S is *statistically reset-secure* if the corresponding advantage is negligible for all reset-adversaries R making a polynomial number of *queries* to their oracle, but which are otherwise computationally unrestricted. It is known that a (statistically) unpredictable source is (statistically) reset-secure, for both UCEs [6] and psPRPs [37]. The converse is not true – S may query a fixed known input, and let L be the empty string. S is reset-secure in the strongest sense, while being easily predictable.

2.2 Evasive Relations, Correlation Intractability

In the following, a k-ary relation R over $X \times Y$ is a set of subsets $S \subseteq X \times Y$, where $1 \leq |S| \leq k$.[8] We are going to consider relations which are *evasive* with respect to a random permutation.

EVASIVE RELATIONS. Given a relation R over $\{0,1\}^m \times \{0,1\}^m$, we consider the following advantage metric, involving an adversary A:

$$\mathsf{Adv}^{\mathsf{evp}}_{R,m}(A) = \Pr_{\pi}\left[S \xleftarrow{\$} A^{\pi,\pi^{-1}} : S \in R \wedge \forall (u,v) \in S : \pi(u) = v\right],$$

where $\pi \xleftarrow{\$} \mathsf{Perms}(m)$. We say that a relation R is (q,δ)-evasive for a random permutation if $\mathsf{Adv}^{\mathsf{evp}}_{R,m}(A) \leq \delta$ for all adversaries making q queries.

CORRELATION INTRACTABILITY. Let M^f be a permutation family on m-bits which makes oracle calls to a function f from n bits to ℓ bits, to be modeled as a random function. Let R be a k-ary relation. Let A be any (possibly unbounded) adversary. We associate to A, M and R the following cri-advantage metric:

$$\mathsf{Adv}^{\mathsf{cri}}_{R,\mathsf{M}}(A) = \Pr_{s,f}\left[S \xleftarrow{\$} A^f(s) \;:\; S \in R \;\wedge\; \forall (u,v) \in S : \mathsf{M}^f(s,u) = v\right],$$

where $f \xleftarrow{\$} \mathsf{Funcs}(n,\ell)$ and $s \xleftarrow{\$} \mathsf{M.Kg}$.

3 Public-Seed Pseudorandomness of Naor-Reingold

This section revisits the Naor-Reingold construction [33] in the public-seed setting. We prove that it transforms a UCE into a psPRP, for both unpredictable (Sect. 3.2) and reset-secure sources (Sect. 3.3). Before turning to these results, however, Sect. 3.1 reviews the construction and proves a strong statement about its indistinguishability.

[8] We think of the elements as *sets*, rather than tuples – this is because looking ahead, it only makes sense in the context of correlation intractability to consider symmetric relation, as an adversary can always re-order its outputs.

3.1 The NR Construction and Its Indistinguishability

Let P be a permutation family on the $2n$-bit strings. We say that P is *α-right-universal* if $\Pr_{s \xleftarrow{\$} \mathsf{P.Kg}}[\mathsf{P}_1(s,u) = \mathsf{P}_1(s,u')] \leq \alpha$ for all distinct $u, u' \in \{0,1\}^{2n}$, where P_1 denote the second n-bit half of the output of P. Note that a pairwise-independent permutation is a good candidate of P, but a simpler approach is to employ one-round of Feistel with a pairwise independent hash function H as the round function, i.e., $\mathsf{P}(s,(u_0,u_1)) = (u_1, \mathsf{H}(s,u_1) \oplus u_0)$.

The Naor-Reingold (NR) Construction. Let H be a function family from $n+1$ bits to n bits. We define the permutation family $\mathsf{NR} = \mathsf{NR}[\mathsf{P},\mathsf{H}]$ on the $2n$-bit strings, where $\mathsf{NR.Kg}$ outputs $(s, s^{\mathsf{in}}, s^{\mathsf{out}})$ such that $s \xleftarrow{\$} \mathsf{H.Kg}$ and $s^{\mathsf{in}}, s^{\mathsf{out}} \xleftarrow{\$} \mathsf{P.Kg}$. Further, forward evaluation proceeds as follows (the inverse is obvious):

Proc. $\mathsf{NR}((s, s^{\mathsf{in}}, s^{\mathsf{out}}), U)$:
$x_0 \,\|\, x_1 \leftarrow \mathsf{P}(s^{\mathsf{in}}, U)$, $x_2 \leftarrow \mathsf{H}(s, 0 \,\|\, x_1) \oplus x_0$,
$x_3 \leftarrow \mathsf{H}(s, 1 \,\|\, x_2) \oplus x_1$, $V \leftarrow \mathsf{P}^{-1}(s^{\mathsf{out}}, x_3 \,\|\, x_2)$,
return V

Naor and Reingold [33] proved that the NR construction with random round functions is indistinguishable from a random permutation under chosen ciphertext attacks. We will need a stronger result, which we prove here, that this is true even when the seed of P is given to the adversary after it stops making queries, and when the adversary can make queries to multiple instances of the construction. It will be convenient to re-use the notation already in place for the psPRP framework, and we denote by $\mathsf{Adv}^{\mathsf{psprp}^+}_{2n,\mathsf{NR}[\mathsf{P},\mathsf{F}]}(S,D)$ the advantage obtained by (S,D) in the $\mathsf{psPRP}^{S,D}_{2n,\mathsf{NR}[\mathsf{P},\mathsf{F}]}$ game, with the modification that D is *not* given the seed for F, only the seeds used by the permutation P.

Proposition 1 (Indistinguishability of the NR construction). *Let $\mathsf{F} = \mathsf{F}[n+1,n]$ be the family of all functions from $n+1$ to n bits, equipped with the uniform distribution. Further, let P be α-right-universal. For all S, D, where S makes q queries, $\mathsf{Adv}^{\mathsf{psprp}^+}_{2n,\mathsf{NR}[\mathsf{P},\mathsf{F}]}(S,D) \leq q^2 \cdot \left(2\alpha + \frac{1}{2^{2n}}\right)$.*

The proof of Proposition 1 can be found in the full version [38, App. A.1].

3.2 The Case of Unpredictable Sources

We first prove that the NR construction transforms a UCE function family for statistically unpredictable sources into a psPRP for statistically unpredictable sources. Our proof uses a technique inspired from that of Bellare et al. [7], given originally in the setting of UCE domain extension. Concretely, we prove the following.

Theorem 1 (NR security for unpredictable sources). *Let P be a α-right universal family of permutations on $2n$-bit strings. Let H be a family of functions* *from $n+1$ bits to n bits. Then, for all distinguishers D and sources S making overall q queries to their oracle, there exists $\overline{D}$ and $\overline{S}$ such that*

$$\mathsf{Adv}^{\mathsf{psprp}}_{2n,\mathsf{NR}[\mathsf{P},\mathsf{H}]}(S,D) \leq \mathsf{Adv}^{\mathsf{uce}}_{n+1,n,\mathsf{H}}(\overline{S},\overline{D}) + q^2\left(2\alpha + \frac{1}{2^{2n}}\right). \tag{3}$$

Here, $\overline{D}$ and D are roughly as efficient, and $\overline{S}$ and S are similarly as efficient. In particular, $\overline{S}$ makes $2q$ queries. Moreover, for every predictor $\overline{P}$, there exists a predictor P such that

$$\mathsf{Adv}^{\mathsf{f\text{-}pred}}_{n+1,n,\overline{S}}(\overline{P}) \leq q^2 \cdot \left(2\alpha + \frac{1}{2^{2n}}\right) + p \cdot \sqrt{2q^2\alpha + \mathsf{Adv}^{\mathsf{p\text{-}pred}}_{2n,S}(P)}, \tag{4}$$

where p is a bound on the size of the set output by $\overline{P}$.

The asymptotic interpretation is that if $n = \omega(\log(\lambda))$ and α is negligible, if S is (statistically) unpredictable, then so is $\overline{S}$. Further, if H is a UCE for all unpredictable sources, then NR is a psPRP for all statistically unpredictable sources.

We stress that the predictor P built in the proof does not preserve the efficiency of $\overline{P}$, which is not a problem, as we only consider statistical notions. While we do not elaborate in the proof, it turns out that the running time of P is *exponential* in the length of S's leakage, thus the statement carries over to computational unpredictability if $L = O(\log \lambda)$.

Proof. We first consider three games, $\mathsf{G}_0, \mathsf{G}_1$, and G_2. Game G_0 is the game $\mathsf{psPRP}^{S,D}_{2n,\mathsf{NR}[\mathsf{P},\mathsf{F}]}$ in the case $b=1$, and modified to return true if $b'=1$. Game G_2 is the game $\mathsf{psPRP}^{S,D}_{2n,\mathsf{NR}[\mathsf{P},\mathsf{F}]}$ in the case $b=0$, and modified to return true if $b'=1$. The intermediate game G_1 is obtained by modifying G_0 as follows: Initially, r random functions $f_1, \ldots, f_r \xleftarrow{\$} \mathsf{Funcs}(n+1,n)$ are sampled, and when evaluating the NR construction within $\mathcal{O}$ queries, the evaluation of $\mathsf{H}(s_i, b \,\|\, x)$ is replaced by an evaluation of the random function $f_i(b \,\|\, x)$. Then,

$$\mathsf{Adv}^{\mathsf{psprp}}_{2n,\mathsf{NR}[\mathsf{P},\mathsf{H}]}(S,D) = (\Pr[\mathsf{G}_0] - \Pr[\mathsf{G}_1]) + (\Pr[\mathsf{G}_1] - \Pr[\mathsf{G}_2]).$$

We can directly get $\Pr[\mathsf{G}_1] - \Pr[\mathsf{G}_2] \leq q^2\left(2\alpha + \frac{1}{2^{2n}}\right)$ as a corollary of Proposition 1, since neither of G_1 and G_2 uses the seeds generated by $\mathsf{H.Kg}$.

Going on, let us consider the new source $\overline{S}$ which simulates an execution of S, and uses access to an oracle $\mathcal{O}(i, X)$, implementing for each i a function from $n+1$ bits to n bits, to internally simulate the round functions NR construction used to answer S's queries. A formal description is in Fig. 3. Also consider the distinguisher $\overline{D}$ such that

$$\overline{D}(L' = (L, \boldsymbol{s}^{\mathsf{in}}, \boldsymbol{s}^{\mathsf{out}}), \boldsymbol{s}) = D(L, (\boldsymbol{s}, \boldsymbol{s}^{\mathsf{in}}, \boldsymbol{s}^{\mathsf{out}})),$$

where $\boldsymbol{s} = (s_1, \ldots, s_r)$, $\boldsymbol{s}^{\mathsf{in}} = (s^{\mathsf{in}}_1, \ldots, s^{\mathsf{in}}_r)$, and $\boldsymbol{s}^{\mathsf{out}} = (s^{\mathsf{out}}_1, \ldots, s^{\mathsf{out}}_r)$ Therefore, G_0 and G_1 behave exactly as $\mathsf{UCE}^{\overline{S},\overline{D}}_{n+1,n,\mathsf{H}}$ with challenge bits $b=1$ and $b=0$,

Proc. $\overline{S}(\varepsilon)$:	Proc. $\overline{\mathcal{O}}(i,(\sigma,U))$:
$(1^r,t) \xleftarrow{\$} S(\varepsilon)$	**if** $\sigma = +$ **then**
return $(1^r,(1^r,t))$	$x_0 \,\|\, x_1 \leftarrow \mathsf{P}(s_i^{\mathsf{in}},U)$
	$x_2 \leftarrow \mathcal{O}(i,0 \,\|\, x_1) \oplus x_0$, $x_3 \leftarrow \mathcal{O}(i,1 \,\|\, x_2) \oplus x_1$
Proc. $\overline{S}^{\mathcal{O}}(1^r,t)$:	$V \leftarrow \mathsf{P}^{-1}(s_i^{\mathsf{out}}, x_3 \,\|\, x_2)$
$s_1^{\mathsf{in}}, s_1^{\mathsf{out}}, \ldots, s_r^{\mathsf{in}}, s_r^{\mathsf{out}} \xleftarrow{\$} \mathsf{P.Kg}$	**else**
$L \xleftarrow{\$} S^{\overline{\mathcal{O}}}(t)$	$x_3 \,\|\, x_2 \leftarrow \mathsf{P}(s_i^{\mathsf{out}},U)$
return $(L, s_1^{\mathsf{in}}, s_1^{\mathsf{out}}, \ldots, s_r^{\mathsf{in}}, s_r^{\mathsf{out}})$	$x_1 \leftarrow \mathcal{O}(i,1 \,\|\, x_2) \oplus x_3$, $x_0 \leftarrow \mathcal{O}(i,0 \,\|\, x_1) \oplus x_2$
	$V \leftarrow \mathsf{P}^{-1}(s_i^{\mathsf{in}}, x_0 \,\|\, x_1)$
	return V

Fig. 3. The source $\overline{S}$ in the proof of Theorems 1 and 2.

respectively, with the only difference of outputting true whenever the distinguisher's output is $b' = 1$. Consequently, $\mathsf{Adv}^{\mathsf{uce}}_{n+1,n,\mathsf{H}}(\overline{S},\overline{D}) = \Pr[\mathsf{G}_0] - \Pr[\mathsf{G}_1]$.

The remainder of the proof relates the unpredictability of S and that of $\overline{S}$, establishing (4) in the theorem statement. For lack of space, the argument is deferred to the full version [38, App. A.2]. □

3.3 The Case of Reset-Secure Sources

Theorem 1's importance stems mostly from the fact that it establishes the equivalence of psPRPs and UCEs for the case of (statistically) unpredictable sources. The question was left open in [37]. Many applications (e.g., instantiating the permutation within sponges, or any other indifferentiable hash construction) however require the stronger notion of reset-security. For this, [37] show that the five-round Feistel construction suffices, using a weaker variant of indifferentiability, and left open the question of whether four-rounds suffice.

We do better here: we prove that the NR construction transforms a UCE for statistically reset-secure sources into a psPRP for the same class of sources. The proof starts as the one of Theorem 1, but then shows that the source $\overline{S}$ built therein is in fact statistically reset-secure whenever S is. This step will resort to a variant of the heavy-query sampling method of Impagliazzo and Rudich [26] to simulate a random oracle from the leakage which captures "relevant correlations" with what is learnt by the source.

Theorem 2 (NR security for reset-secure sources). *Let* P *be a* α*-right universal family of permutations on* $2n$*-bit strings, and let* H *be a function family from* $n+1$ *bits to* n *bits. Then, for all distinguishers* D *and sources* S *making overall* q *queries to their oracle, there exists* $\overline{D}$ *and* $\overline{S}$ *such that*

$$\mathsf{Adv}^{\mathsf{psprp}}_{2n,\mathsf{NR}[\mathsf{P},\mathsf{H}]}(S,D) \leq \mathsf{Adv}^{\mathsf{uce}}_{n+1,n,\mathsf{H}}(\overline{S},\overline{D}) + q^2\left(2\alpha + \frac{1}{2^{2n}}\right). \quad (5)$$

Here, $\overline{D}$ and D are roughly as efficient, and $\overline{S}$ and S are similarly as efficient. In particular, $\overline{S}$ makes $2q$ queries. Moreover, for every reset-adversary $\overline{R}$ making p queries, there exists a reset-adversary R such that

$$\mathsf{Adv}^{\text{f-reset}}_{n+1,n,\overline{S}}(\overline{R}) \leq 2\mathsf{Adv}^{\text{p-reset}}_{2n,S}(R) + 4\left(q + \frac{8qp^2}{\varepsilon}\ln(4p/\varepsilon)\right)^2\left(2\alpha + \frac{1}{2^{2n}}\right), \quad (6)$$

where $\varepsilon := \mathsf{Adv}^{\text{f-reset}}_{n+1,n,\overline{S}}(\overline{R})$. In particular, R makes $4qp^2/\varepsilon \cdot \ln(4p/\varepsilon)$ queries to its oracle.

Asymptotically, (6) implies that if $\overline{R}$ exists making $p = \mathsf{poly}(\lambda)$ queries, and achieving non-negligible advantage ε, then R makes also a polynomial number of queries, and achieves non-negligible advantage, as long as α is negligible, and $n = \omega(\log \lambda)$. Thus, reset-security of R yields reset-security of $\overline{R}$.

We also believe that the technique of Barak and Mahmoody [4] can be used to reduce the $8qp^2/\varepsilon$ term to $O(qp)/\varepsilon$. We did not explore this avenue here, as the proof approach of [26] is somewhat easier to adapt to our setting.

Proof. The setup of the proof is identical to that in Theorem 1, in particular the construction of $\overline{S}$ from S (and of $\overline{D}$ from D.) The difference is in relating the reset-security of S and $\overline{S}$. In particular, let

$$\varepsilon := \mathsf{Adv}^{\text{f-reset}}_{n+1,n,\overline{S}}(\overline{R}) = \Pr\left[\text{f-Reset}^{\overline{R}}_{n+1,n,\overline{S}} \mid b = 0\right] - \Pr\left[\neg\text{f-Reset}^{\overline{R}}_{n+1,n,\overline{S}} \mid b = 1\right].$$

The RHS is the difference of the probabilities of $\overline{R}$ outputting 0 in the cases $b = 0$ and $b = 1$ respectively. We are going to build a new adversary R against S which satisfies (6). We assume without loss of generality that $\overline{R}$ is deterministic, and makes *exactly p distinct* queries to its oracle.

We start the proof with some game transitions that will lead naturally to the definition of the adversary R. Formal descriptions are found in our full version [38, Figs. 7 and 8] – our description here is self-contained.

The initial game G_1 is simply $\text{f-Reset}^{\overline{R}}_{\overline{S}}$ with the bit $b = 0$, i.e., $\overline{S}$ and $\overline{R}$ access the *same* functions $f_1, \ldots, f_r$ here. Further, G_1 returns true iff $\overline{R}$ returns 0. Thus, $\Pr[\mathsf{G}_1] = \Pr\left[\text{f-Reset}^{\overline{R}}_{\overline{S}} \mid b = 0\right]$. Game G_2 slightly changes G_1: It keeps track (in a set Q_{P}) of the triples (i, U, V) describing $\overline{\mathcal{O}}$ queries made by the simulated S within $\overline{S}$; i.e., either S queried $(i, (+, U))$, and obtained V, or queries $(i, (-, V))$, and obtained U. After $\overline{S}$ terminates with leakage $(L, \boldsymbol{s})$, where $\boldsymbol{s} = (s^{\mathsf{in}}_1, s^{\mathsf{out}}_1, \ldots, s^{\mathsf{in}}_r, s^{\mathsf{out}}_r)$, for every $(i, U, V) \in Q_{\mathsf{P}}$ we compute $x_0 \| x_1 \leftarrow \mathsf{P}(s^{\mathsf{in}}_i, U)$ and $x_3 \| x_2 \leftarrow \mathsf{P}(s^{\mathsf{out}}_i, V)$, and define table entries

$$T[i, 0 \| x_1] \leftarrow x_0 \oplus x_2, \quad T[i, 1 \| x_2] \leftarrow x_1 \oplus x_3.$$

For later reference, we denote by X the set of pairs (i, x) for which we set $T[i, x]$ using Q_{P} and $\boldsymbol{s}$. We then run $\overline{R}(L, \boldsymbol{s})$, and answer its oracle queries (i, x) using $T[i, x]$. If the entry is undefined, then we return a random value. (As we assumed all of $\overline{R}$'s queries are distinct, we do not need to remember the output.) As before, G_2 outputs true iff $\overline{R}$ outputs 0.

Note that we always have $T[i,x] = f_i(x)$ for very (i,x) such that $f_i(x)$ was queried by $\overline{S}$, and re-sampling values un-queried by $\overline{S}$ upon $\overline{R}$'s queries does not change the distribution of $\overline{R}$'s output, and hence $\Pr[\mathsf{G}_1] = \Pr[\mathsf{G}_2]$.

THE INTERSECTION SAMPLER. The game G_3 generates a surrogate for Q_{P}. This is the output of an algorithm Sam which, after $\overline{S}$ terminates with output $(L, \boldsymbol{s})$, takes as input the leakage L (crucially, not $\boldsymbol{s}$!) and an iteration parameter $\eta = 4p/\varepsilon \ln(4p/\varepsilon)$ (we let also $\tau = p \cdot \eta$). Sam queries the very same $\overline{\mathcal{O}}$ implemented by $\overline{S}$ to answer S's queries (which internally simulates the NR construction using $\overline{S}$'s own oracle), and returns a set $\widetilde{Q}_{\mathsf{P}}$ of 4-tuples (i, U, V, j) such that $j \in [p]$, and (i, U, V) is such that $\overline{\mathcal{O}}(i, (+, U))$ would return V (or equivalently $\overline{\mathcal{O}}(i, (-, V))$ would return U). Internally, Sam will make calls to another (randomized) sub-procedure $\mathcal{Q}$ which takes as input the leakage L, as well as a set Q of tuples (i, U, V, j) consistent with $\overline{\mathcal{O}}$, and returns a set Δ of at most q tuples (i, U, V), which are not necessarily consistent with $\overline{\mathcal{O}}$. We will specify in detail later below what $\mathcal{Q}$ exactly does, as some further game transitions will come handy to set up proper notation. For now, a generic understanding will suffice. In particular, given such $\mathcal{Q}$, Sam operates as in Fig. 4. As we can see, for each $(i, U, V, j) \in \widetilde{Q}_{\mathsf{P}}$, j indicates the outer iteration in which this query was added to $\widetilde{Q}_{\mathsf{P}}$. Using this information, for every $j \in [p]$, and every 4-tuple (i, U, V, j) we compute $x_0 \| x_1 \leftarrow \mathsf{P}(s_i^{\mathsf{in}}, U)$ and $x_3 \| x_2 \leftarrow \mathsf{P}(s_i^{\mathsf{out}}, V)$, define

$$\widetilde{T}[i, 0 \| x_1] \leftarrow x_0 \oplus x_2 \ , \quad \widetilde{T}[i, 1 \| x_2] \leftarrow x_1 \oplus x_3,$$

and add $(i, 0 \| x_1), (i, 1 \| x_2)$ to the set $\widetilde{X}^j$. A for now irrelevant caveat is that if one of the entries in $\widetilde{T}$ is already set, then we do not overwrite it.[9]

ALGORITHM $\mathsf{Sam}^{\overline{\mathcal{O}}}(L, \tau)$:
$\widetilde{Q}_{\mathsf{P}} \leftarrow \emptyset$
for $j = 1$ to p **do**
  for $k = 1$ to η **do**
    $\Delta_{j,k} \leftarrow \mathcal{Q}(L, \widetilde{Q}_{\mathsf{P}})$
    for all $(i, U, V) \in \Delta_{j,k}$ **do**
      $V' \leftarrow \overline{\mathcal{O}}(i, (+, U))$, $U' \leftarrow \overline{\mathcal{O}}(i, (-, V))$, $\widetilde{Q}_{\mathsf{P}} \stackrel{\cup}{\leftarrow} \{(i, U, V', j), (i, U', V, j)\}$
return $\widetilde{Q}_{\mathsf{P}}$

Fig. 4. Description of algorithm Sam.

[9] This does not matter here, as an entry can only be overwritten with the same value; below, we will change the experiment in a way that overwrites may be inconsistent, and we want to ensure we agree to keep the first value.

Then, after all of this, G_3 resumes by executing $\overline{R}(L, \boldsymbol{s})$. For $\overline{R}$'s j-th query (i, x) we do the following:

1. If $(i, x) \in \widetilde{X}^{j'}$ for some $j' \leq j$, then we respond with $\widetilde{T}[i, x]$.
2. Otherwise, if $(i, x) \in X$, but the first condition was not met, we respond with $T[i, x]$.
3. Finally, if neither of the above is true, we respond randomly.

As before, G_3 outputs true iff $\overline{R}$ outputs 0. For now, all modifications are syntactical. Indeed, up to the point we start $\overline{R}$, we satisfy the invariant that $T[i, x] = f_i(x)$ or $\widetilde{T}[i, x] = f_i(x)$ whenever these are defined, because $\overline{\mathcal{O}}$ behaves according to the NR construction using $\boldsymbol{s}$. On the other hand, if during the execution $\overline{R}(L, \boldsymbol{s})$ we respond randomly, we know for sure $f_i(x)$ was not queried by $\overline{S}$, and thus we can re-sample it. Thus, $\Pr[\mathsf{G}_3] = \Pr[\mathsf{G}_2] = \Pr[\mathsf{G}_1]$.

Moving to G_4, we now answer $\overline{\mathcal{O}}$ queries by S (within $\overline{S}$) and by Sam using random permutations $\pi_1, \ldots \pi_r$, instead of simulating the NR construction using $f_1, \ldots, f_r$, i.e., $\overline{\mathcal{O}}(i, (+, U)) = \pi_i(U)$ and $\overline{\mathcal{O}}(i, (-, V)) = \pi_i^{-1}(V)$. The seeds $\boldsymbol{s}$ are now independent of $\overline{\mathcal{O}}$. We do not change anything else. We note that the indistinguishability of G_3 and G_4 directly reduces to a suitable distinguisher for Proposition 1, as only Sam and S (within $\overline{S}$) make queries to $\overline{\mathcal{O}}$, but they do not get the keys $\boldsymbol{s}$, which are used only after all queries to $\overline{O}$ have been made to define X and $\widetilde{X}$. Therefore,

$$\Pr[\mathsf{G}_1] = \Pr[\mathsf{G}_3] \leq \Pr[\mathsf{G}_4] + (q + 2q\tau)^2 \left(2\alpha + \frac{1}{2^{2n}}\right), \tag{7}$$

where we have used the fact that Sam makes $2q\tau = 2pq\eta$ queries.

The final game is G_5 is identical to G_4, *except* that in the process of answering $\overline{R}$'s queries, if case 2 happens, we also set answer randomly. However, should such situation occur, a bad flag is set in G_5, and since up to the point this flag is set, the behavior of G_4 and G_5 is identical,

$$\Pr[\mathsf{G}_4] - \Pr[\mathsf{G}_5] \leq \Pr[\mathsf{G}_5 \text{ sets } \mathsf{bad}].$$

To analyze the probability on the RHS, we need to specify $\mathcal{Q}(L, \widetilde{Q})$ used by Sam here. (Note all statements so far were independent of it.) For a given L which appears with positive probability in G_5, consider the distribution of the input-output pairs Q_{P} defined by the interaction of S with $\overline{\mathcal{O}}$, conditioned on the leakage being L, and $\pi_1, \ldots, \pi_r$ being consistent with the triples defined by $\widetilde{Q}$. Then, $\mathcal{Q}(L, \widetilde{Q})$ outputs a sample of Q_{P} according to this distribution. Using this, we prove the following lemma in our full version [38, App. A.4], which uses ideas similar to those from [26], with some modifications due to the setting (and the fact that $\overline{R}$ makes p queries).

Lemma 1. $\Pr[\mathsf{G}_5 \textit{ sets } \mathsf{bad}] \leq \varepsilon/2$

<u>ADVERSARY $R^{\mathcal{O}}(1^r, L)$:</u>

$c \leftarrow 0$, $\widetilde{X} \leftarrow \emptyset$
$\boldsymbol{s} = (s_1^{\mathsf{in}}, s_1^{\mathsf{out}}, \ldots, s_r^{\mathsf{in}}, s_r^{\mathsf{out}}) \xleftarrow{\$} \mathsf{P.Kg}$
$\widetilde{Q}_{\mathsf{P}} \xleftarrow{\$} \mathsf{Sam}^{\mathcal{O}}(L)$
for $j = 1$ to p **do**
  for all $(i, U, V, j) \in \widetilde{Q}$ **do**
    $x_0 \| x_1 \leftarrow \mathsf{P}(s_i^{\mathsf{in}}, U)$, $x_3 \| x_2 \leftarrow \mathsf{P}(s_i^{\mathsf{out}}, V)$
    $\widetilde{X}^j \leftarrow \widetilde{X}^j \cup \{(i, 0 \| x_1), (i, 1 \| x_2)\}$
    if $\widetilde{T}[i, 0 \| x_1] = \perp$ **then** $\widetilde{T}[i, 0 \| x_1] \leftarrow x_0 \oplus x_2$
    if $\widetilde{T}[i, 1 \| x_2] = \perp$ **then** $\widetilde{T}[i, 1 \| x_2] \leftarrow x_1 \oplus x_3$
$b' \leftarrow \overline{R}^{\mathcal{O}'}(L, \boldsymbol{s})$
return b'

<u>Proc. $\mathcal{O}'(i, x)$:</u>

$c \leftarrow c + 1$, $\widetilde{X} \leftarrow \widetilde{X} \cup \widetilde{X}^c$
if $T[i, x] = \perp$ **then**
  if $(i, x) \in \widetilde{X}$ **then**
    $T[i, x] \leftarrow \widetilde{T}[i, x]$
  else $T[i, x] \xleftarrow{\$} \{0, 1\}^n$
return $T[i, x]$

Fig. 5. Adversary R in the proof of Theorem 2.

Given this, we are now ready to give our adversary R, which we build from $\overline{R}$ and Sam as described in Fig. 5. By a purely syntactical argument,

$$\Pr[\mathsf{G}_5] = \Pr\left[\mathsf{p\text{-}Reset}_S^R \mid b = 0\right], \tag{8}$$

recalling that the case $b = 0$ is the one where both S and R access the same permutations $\pi_1, \ldots, \pi_r$. Therefore, we have established, combining (8), (7), Lemma 1,

$$\Pr\left[\mathsf{p\text{-}Reset}_S^R \mid b = 0\right] \geq \Pr\left[\mathsf{f\text{-}Reset}_{\overline{S}}^{\overline{R}} \mid b = 0\right] - \frac{\varepsilon}{2} - (q + 2q\tau)^2 \left(2\alpha + \frac{1}{2^{2n}}\right). \tag{9}$$

In the full version [38, App. A.5] we also prove formally that in the case $b = 1$, R in the game $\mathsf{p\text{-}Reset}_S^R$ almost perfectly simulates an execution of $\mathsf{f\text{-}Reset}_{\overline{S}}^{\overline{R}}$, or more formally,

$$\Pr\left[\neg\mathsf{p\text{-}Reset}_S^R \mid b = 1\right] \leq \Pr\left[\neg\mathsf{f\text{-}Reset}_{\overline{S}}^{\overline{R}} \mid b = 1\right] + (q + 2q\tau)^2 \left(2\alpha + \frac{1}{2^{2n}}\right). \tag{10}$$

We can combine (10) and (9) to obtain, with $\Delta = 2(q + 2q\tau)^2 \left(2\alpha + \frac{1}{2^{2n}}\right)$,

$$\begin{aligned}
\mathsf{Adv}_{2n,S}^{\mathsf{p\text{-}reset}}(R) &= \Pr\left[\mathsf{p\text{-}Reset}_S^R \mid b = 0\right] - \Pr\left[\neg\mathsf{p\text{-}Reset}_S^R \mid b = 1\right] \\
&\geq \Pr\left[\mathsf{f\text{-}Reset}_{\overline{S}}^{\overline{R}} \mid b = 0\right] - \Pr\left[\neg\mathsf{f\text{-}Reset}_{\overline{S}}^{\overline{R}}) \mid b = 1\right] - \frac{\varepsilon}{2} - \Delta \\
&\geq \varepsilon/2 - \Delta\,.
\end{aligned}$$

This concludes the proof. □

4 Correlation Intractability of Public-Seed Permutations

In this section, we study the correlation intractability (CI) of the NR construction against k-ary evasive relations. Firstly, in Sect. 4.1, we define a stronger notion of evasiveness – *strong evasiveness* – and show that evasiveness and strong evasiveness are asymptotically equivalent when $k = O(1)$. In Sect. 4.2 we study the relations that are hard for two-round Feistel. In Sect. 4.3 we show that for $k = O(1)$ the NR construction where P^{-1} is a family of almost $O(k^2)$-wise independent permutations is correlation intractable against k-ary evasive relations. In the special case of unary evasive relations ($k = 1$), we show that (see [38, App. B.6]) P instead can be instantiated from one-round Feistel with a 10-wise independent round function.

4.1 Strong Evasiveness

Evasiveness is defined through the hardness of winning a security game. For our results, we need instead a combinatorial understanding of evasive relations. To this end, we will rely on the following notion of evasiveness, which, as we show below, is generally implied by evasiveness if $k = O(1)$.

Definition 1 (Strongly evasive relations). *Let R be a k-ary relation over $X \times X$ and $\delta \in [0, 1]$. We say that R is* δ-strongly evasive *if the following are true for all $0 \leq j < k' \leq k$:*

- *For all distinct $\mathbf{u}_1, \ldots, \mathbf{u}_{k'} \in X$, all $\mathbf{v}_1, \ldots, \mathbf{v}_j \in X$, we have*

$$|\{(\mathbf{v}_{j+1}, \ldots, \mathbf{v}_{k'}) : \{(\mathbf{u}_1, \mathbf{v}_1), \ldots, (\mathbf{u}_{k'}, \mathbf{v}_{k'})\} \in R\}| \leq \delta \cdot \prod_{i=j}^{k'-1} (|X| - i).$$

- *For all distinct $\mathbf{v}_1, \ldots, \mathbf{v}_{k'} \in X$, all $\mathbf{u}_1, \ldots, \mathbf{u}_j \in X$, we have*

$$|\{(\mathbf{u}_{j+1}, \ldots, \mathbf{u}_{k'}) : \{(\mathbf{u}_1, \mathbf{v}_1), \ldots, (\mathbf{u}_{k'}, \mathbf{v}_{k'})\} \in R\}| \leq \delta \cdot \prod_{i=j}^{k'-1} (|X| - i).$$

It is not hard to see that if a relation is δ-strongly evasive, then it is also evasive, in the sense that it is $(q, q^k\delta)$-evasive. In particular, $q^k\delta$ is negligible whenever δ is negligible, q polynomial, and $k = O(1)$.

We remark that there are relations R which are evasive, yet not strongly evasive. Consider for example the relation which contains $\{(0^{2n}, 0^{2n}), (\mathbf{u}, \mathbf{v})\}$ for *all* $\mathbf{u}, \mathbf{v} \neq 0^{2n}$. This relation is obviously evasive to start with – satisfying it requires $\pi(0^{2n}) = 0^{2n}$, which will happen with probability 2^{-2n} only, yet for $\mathbf{u}_1 = \mathbf{v}_1 = 0^{2n}$, and $\mathbf{u}_2 \neq 0^{2n}$, all strings $\mathbf{v}_2$ make $\{(\mathbf{u}_1, \mathbf{v}_1), (\mathbf{u}_2, \mathbf{v}_2)\}$ valid. Still, somehow, the intuition is that the core of R is the relation $R^* = \{\{(0^{2n}, 0^{2n})\}\}$, which *is* strongly evasive, with $\delta = 2^{-2n}$. Indeed, an attacker that satisfies the original relation R, can directly satisfy R^*, thus the fact that R^* is evasive (and in particular, strongly evasive) implies that R is evasive.

The following lemma generalizes this, and implies e.g. that for $\delta = \mathsf{negl}(\lambda)$ and $k = O(1)$, evasiveness and strong evasiveness are (qualitatively) equivalent. The proof is found in the full version [38, App. B.1].

Lemma 2 (Normalization of evasive relations). *Let $\delta > 0$, and let R be a k-ary (k^2, δ^k)-evasive relation on $X \times X$ for random permutations. Then, there exists a relation R^* which is δ-strongly evasive for random permutations, and moreover, for every $S \in R$, there exists $\emptyset \neq S^* \subseteq S$ such that $S^* \in R^*$.*

Lemma 2 now is all we need. Say E is correlation intractable for all k-ary strongly evasive relations (for some negligible δ), where $k = O(1)$. Then, E must be correlation intractable for any (k^2, δ)-evasive relation R, too. Were it not, we could take an adversary A breaking the CI for R with non-negligible advantage, and use it to break CI of R^*. To this end, we simply run A, and when it outputs $S \in R$, we outputs the corresponding $S^* \in R^*$ guaranteed by Lemma 2. (As $k = O(1)$, a random subset of S will do with constant loss in the advantage.) But since R^* is $\sqrt[k]{\delta}$-strongly evasive, this contradicts our assumption on E.

Clearly, the equivalence is merely asymptotic. If one is interested in concrete security, the best approach to use our results below is to directly assess the δ for which a specific relation R is δ-strongly evasive.

4.2 Partial Correlation Intractability of Two-Round Feistel

The two-round Feistel construction Fei_2^f, is a permutation on $2n$-bit strings that makes calls to an oracle $f : \{0,1\}^{n+1} \to \{0,1\}^n$. In particular, on input $\mathbf{x} = x_0 \,\|\, x_1$, where $x_0, x_1 \in \{0,1\}^n$, running $\mathsf{Fei}_2^f(\mathbf{x})$ outputs $\mathbf{y} = x_2 \,\|\, x_3$, where

$$x_2 \leftarrow x_0 \oplus f(0 \,\|\, x_1) \,; x_3 \leftarrow x_1 \oplus f(1 \,\|\, x_2).$$

Symmetrically, upon an inverse query, $\mathsf{Fei}_2^{f^{-1}}(\mathbf{y} = x_2 \,\|\, x_3)$ simply computes the values backwards, and outputs $\mathbf{x} = x_0 \,\|\, x_1$.

In this section, we discuss relations R on $2n$-bit strings that are hard for two-round Feistel when instantiated with a random function. In particular, we will give a combinatorial characterization which is sufficient to achieve this.

FEISTEL EVASIVENESS. We first note that in a relation R, certain sets $S \in R$ can never be satisfied by the two-round Feistel construction out of structural constraints. In particular, if we have two input-output pairs $(\mathbf{x}_1[0] \,\|\, \mathbf{x}_1[1], \mathbf{x}_1[2] \,\|\, \mathbf{x}_1[3])$ and $(\mathbf{x}_2[0] \,\|\, \mathbf{x}_2[1], \mathbf{x}_2[2] \,\|\, \mathbf{x}_2[3])$ with $\mathbf{x}_1[2] = \mathbf{x}_2[2]$ in the same set $S \in R$, then we *must* have $\mathbf{x}_1[3] \oplus \mathbf{x}_2[3] = \mathbf{x}_1[1] \oplus \mathbf{x}_2[1]$. Symmetrically, if $\mathbf{x}_1[1] = \mathbf{x}_2[1]$, then we must have $\mathbf{x}_1[0] \oplus \mathbf{x}_2[0] = \mathbf{x}_1[2] \oplus \mathbf{x}_2[2]$. It will thus be convenient to define the following.

Definition 2. *For every k-ary relation R on $2n$-bit strings, we define the relation $\overline{R} \subseteq R$ that only contains $S \in R$ if for every $(\mathbf{x}_1[0] \,\|\, \mathbf{x}_1[1], \mathbf{x}_1[2] \,\|\, \mathbf{x}_1[3])$, $(\mathbf{x}_2[0] \,\|\, \mathbf{x}_2[1], \mathbf{x}_2[2] \,\|\, \mathbf{x}_2[3]) \in S$, the following is true:*

- *If* $\mathbf{x}_1[2] = \mathbf{x}_2[2]$, *then* $\mathbf{x}_1[3] \oplus \mathbf{x}_2[3] = \mathbf{x}_1[1] \oplus \mathbf{x}_2[1]$.
- *If* $\mathbf{x}_1[1] = \mathbf{x}_2[1]$, *then* $\mathbf{x}_1[0] \oplus \mathbf{x}_2[0] = \mathbf{x}_1[2] \oplus \mathbf{x}_2[2]$.

Clearly, the significance of this is that when assessing whether R is correlation intractable for two-round Feistel, it suffices to prove that $\overline{R}$ is correlation intractable, as $S \in R \setminus \overline{R}$ can never be satisfied. We are now ready to state the following combinatorial requirement on relations, which we will prove to be evasive for two-round Feistel below.

Definition 3 (δ-2-Feistel evasive relations). *Let R be a k-ary relation over $\{0,1\}^{2n}$, and $\delta \in [0,1]$. We say that R is* δ-2-Feistel evasive *if the following are true for all $0 \le j < k' \le k$:*

- *For all distinct $\mathbf{x}_1, \ldots, \mathbf{x}_{k'} \in \{0,1\}^{2n}$, distinct $\mathbf{y}_1, \ldots, \mathbf{y}_j \in \{0,1\}^{2n}$, and $y^* \in \{0,1\}^n$ s.t. $\mathbf{x}_{j+1}[1], \ldots, \mathbf{x}_{k'}[1]$ are distinct and $y^* \notin \{\mathbf{y}_1[0], \ldots, \mathbf{y}_j[0]\}$,*

$$\Big|\{(y_{j+1}, \ldots, y_{k'})\} : \{(\mathbf{x}_1, \mathbf{y}_1), \ldots, (\mathbf{x}_j, \mathbf{y}_j), (\mathbf{x}_{j+1}, y^* \parallel y_{j+1}), \ldots, (\mathbf{x}_{k'}, y^* \parallel y_{k'})\} \in \overline{R'}\Big| \le \delta' \cdot 2^n . \quad (11)$$

- *For all distinct $\mathbf{y}_1, \ldots, \mathbf{y}_{k'} \in \{0,1\}^{2n}$, distinct $\mathbf{x}_1, \ldots, \mathbf{x}_j \in \{0,1\}^{2n}$, and $x^* \in \{0,1\}^n$ s.t. $\mathbf{y}_{j+1}[0], \ldots, \mathbf{y}_{k'}[0]$ are distinct and $x^* \notin \{\mathbf{x}_1[1], \ldots, \mathbf{x}_j[1]\}$,*

$$\Big|\{(x_{j+1}, \ldots, x_{k'})\} : \{(\mathbf{x}_1, \mathbf{y}_1), \ldots, (\mathbf{x}_j, \mathbf{y}_j), (x_{j+1} \parallel x^*, \mathbf{y}_{j+1}), \ldots, (x_{k'} \parallel x^*, \mathbf{y}_{k'})\} \in \overline{R'}\Big| \le \delta' \cdot 2^n . \quad (12)$$

Also, we let $\mathsf{FEv}(k, \delta)$ denote the set of all k-ary δ-2-Feistel evasive relations.

FEISTEL CORRELATION INTRACTABILITY. We now prove that for all relations satisfying Definition 3, two-round Feistel is indeed correlation intractable in the model where both round functions are independent random functions, to which the adversary is given oracle access.

Proposition 2 (CI of Two-round Feistel). *For $\delta \in [0,1]$ and $k \ge 1$ be an integer, let $R \in \mathsf{FEv}(k, \delta)$. For any (unbounded) adversary A making at most q queries to f, $\mathsf{Adv}^{\mathsf{cri}}_{R,\mathsf{Fei}_2^f}(A) \le 2k\delta \cdot q^{2k+1}$.*

The proof of Proposition 2 can be found in the full version [38, App. B.2].

Remark 1. For the special case of $k = 1$, that is, unary relations, one can adapt the above proof and show that $\mathsf{Adv}^{\mathsf{cri}}_{R,\mathsf{Fei}_2^f}(A) \le \delta \cdot q^2$ where A makes q queries to f and $R \in \mathsf{FEv}(1, \delta)$.

4.3 Correlation Intractability of the NR Construction

In this section we view the NR construction as a family $\mathsf{NR}^f[\mathsf{P}]$ that makes oracle calls to $f : \{0,1\}^{n+1} \rightarrow \{0,1\}^n$ and P is a family of permutations on $2n$-bits. The key generation algorithm $\mathsf{NR.Kg}$ now just outputs a tuple $(s^{\mathsf{in}}, s^{\mathsf{out}})$ where $s^{\mathsf{in}}, s^{\mathsf{out}} \xleftarrow{\$} \mathsf{P.Kg}$ and the evaluation algorithm $\mathsf{NR.Eval}$ proceeds as before but instead makes calls to f for evaluating the round function.

We show that $\mathsf{NR}^f[\mathsf{P}]$, where P^{-1} is a family of almost $O(k^2)$-wise independent permutations, is correlation intractable against strongly evasive k-ary relations when the adversary is given the seed $(s^{\mathsf{in}}, s^{\mathsf{out}})$ of the NR construction and only oracle access to f. The proof of CI proceeds by showing that P transforms a strongly evasive relation R into a 2-Feistel evasive relation $R_{\pi,\sigma}$ (see Fig. 6) and hence for the adversary to break the CI of $\mathsf{NR}^f[\mathsf{P}]$ it needs to break the CI of two-round Feistel against $R_{\pi,\sigma}$ which was studied in Sect. 4.2.

p-WISE INDEPENDENT PERMUTATIONS. For any $\varepsilon \in [0,1]$ and $p \geq 1$, we say that a family of permutations P on m-bit strings is (ε, p)-*wise independent* if for all distinct $u_1, \ldots, u_p \in \{0,1\}^m$, the distributions of $\mathsf{P}(s, u_1), \ldots, \mathsf{P}(s, u_p)$ (for $s \xleftarrow{\$} \mathsf{P.Kg}$) and of $\rho(u_1), \ldots, \rho(u_p)$ (for $\rho \xleftarrow{\$} \mathsf{Perms}(m)$) are at most ε-apart in statistical distance.

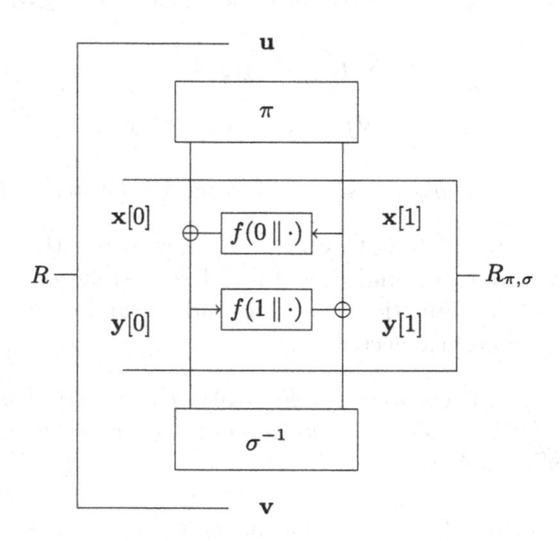

Fig. 6. The NR construction instantiated with a permutation family P on $2n$-bits such that P^{-1} is $(\varepsilon, k \cdot t)$-wise independent where $\pi = \mathsf{P}(s^{\mathsf{in}}, \cdot)$ and $\sigma = \theta(\mathsf{P}(s^{\mathsf{out}}, \cdot))$ for $s^{\mathsf{in}}, s^{\mathsf{out}} \leftarrow \mathsf{P.Kg}$. Here, θ is a permutation on $2n$-bits such that for all $\mathbf{x} \in \{0,1\}^{2n}$ we have $\theta(\mathbf{x} = x_0||x_1) = x_1||x_0$. For some k-ary strongly-evasive relation R, we construct a 2-Feistel evasive relation $R_{\pi,\sigma}$ by transforming every $(\mathbf{u}, \mathbf{v}) \in S$ where $S \in R$, by applying π to $\mathbf{u}$ and σ to $\mathbf{v}$.

FROM STRONGLY EVASIVE RELATION TO 2-FEISTEL-EVASIVE RELATION. Let R be a k-ary relation over $\{0,1\}^{2n} \times \{0,1\}^{2n}$. For $s^{\mathsf{in}}, s^{\mathsf{out}} \xleftarrow{\$} \mathsf{P.Kg}$, let $\pi = \mathsf{P}(s^{\mathsf{in}}, \cdot)$ and $\sigma = \theta(\mathsf{P}(s^{\mathsf{out}}, \cdot))$, where θ is a permutation on $2n$-bits such that for $\mathbf{x} \in \{0,1\}^{2n}$ we have $\theta(\mathbf{x} = x_0||x_1) = x_1||x_0$[10]. We define a relation $R_{\pi,\sigma}$ which is a result of transforming $\{(\mathbf{u}_1, \mathbf{v}_1), \ldots, (\mathbf{u}_{k'}, \mathbf{v}_{k'}))\} \in R$ via π and σ in the following way,

$$R_{\pi,\sigma} = \{\{(\pi(\mathbf{u}_1), \sigma(\mathbf{v}_1)), \ldots, (\pi(\mathbf{u}_{k'}), \sigma(\mathbf{v}_{k'}))\} \mid \{(\mathbf{u}_1, \mathbf{v}_1), \ldots, (\mathbf{u}_{k'}, \mathbf{v}_{k'})\} \in R\}.$$

Then, for every δ-strongly evasive k-ary relation R we show that $R_{\pi,\sigma} \in \mathsf{FEv}(k, \delta')$ for some δ' larger than δ, except with small probability, where the probability is taken over the random choice of (π, σ). This is more formally captured in the following:

Proposition 3 (CI Amplification). *For $\delta \in [0,1)$ and an integer $k \geq 1$, let R be a k-ary δ-strongly evasive relation over $\{0,1\}^{2n}$. For an even integer $t \geq 2$, let P be a family of permutations such that P^{-1} is $(\varepsilon, k \cdot t)$-wise independent. Then, for $\delta' \in [0,1]$ such that $\delta' > \delta$,*

$$\Pr_{\pi,\sigma}[R_{\pi,\sigma} \notin \mathsf{FEv}(k, \delta')] \leq 12k^2 \left(\frac{1}{\delta' - \delta}\right)^t 2^{(4k-1)n} \left[C_t \cdot \left(\frac{4\delta^* t}{2^n}\right)^{t/2} + \varepsilon \cdot (1+\delta)^t\right],$$

where $C_t = 2e^{1/6t}\sqrt{\pi t}\left(\frac{5}{2e}\right)^{t/2}$, $\delta^* = \max\left(\delta, \frac{t \cdot 2^k}{2^n}\right)$, $\pi = \mathsf{P}(s^{\mathsf{in}}, \cdot)$ and $\sigma = \theta(\mathsf{P}(s^{\mathsf{out}}, \cdot))$ for $s^{\mathsf{in}}, s^{\mathsf{out}} \xleftarrow{\$} \mathsf{P.Kg}$.

Now, Proposition 3 can be combined with Proposition 2 to establish the correlation intractability of NR against strongly evasive relations (Theorem 3).

Theorem 3 (CI of NR). *For $\delta \in [0,1)$ and an integer $k \geq 1$, let R be a k-ary δ-strongly evasive relation over $\{0,1\}^{2n}$. Further, let P be a family of permutations on $2n$-bits such that P^{-1} is $(\varepsilon, 10k^2)$-wise independent where $\varepsilon \leq 1/2^{5kn}$. Then for any (potentially unbounded) adversary A making q queries,*

$$\mathsf{Adv}^{\mathsf{cri}}_{R,\mathsf{NR}^f[\mathsf{P}]}(A) \leq \frac{24k^2 \cdot (40k\delta^*)^{5k} + 12k^2 \cdot (1+\delta)^{10k}}{2^{4kn/9}} + 2k\delta' \cdot q^{2k+1}, \tag{13}$$

where $\delta' = \delta + 2^{-n/18}$ and $\delta^* = \max\left(\delta, \frac{10k \cdot 2^k}{2^n}\right)$.

The proof of Theorem 3 can be found in the full version [38, App. B.3]. The asymptotic interpretation of Eq. (13) is that when $n = \omega(\log \lambda)$, $k = O(1)$, $\delta = \mathsf{negl}(\lambda)$ and $q = \mathsf{poly}(\lambda)$, $\mathsf{NR}^f[\mathsf{P}]$ is correlation intractable for k-ary strongly evasive relations. Combining this with Lemma 2, the CI then extends to any k-ary evasive relation. We also remark that Theorem 3 extends to the setting of

[10] It is easy to see that $\theta = \theta^{-1}$ hence $\sigma^{-1} = \mathsf{P}^{-1}(s^{\mathsf{out}}, \theta(\cdot))$. We note that θ is introduced to ensure consistency with the definition of the NR construction as $\mathsf{P}^{-1}(s^{\mathsf{out}})$ operates on $\mathbf{y}[1]||\mathbf{y}[0]$ where $\mathbf{y}$ is the output of the underlying two-round feistel.

multi-key correlation intractability introduced in [16], but to avoid notational overhead we limit ourselves to the single-key setting for this version.

On instantiating P^{-1} from Theorem 3. We detail the construction of (ε, p)-wise permutations in the full version [38, App. B.5] and show that an $O(k)$-round Feistel construction with $10k^2$-wise independent round functions can instantiate the permutation family P^{-1}. We refer the reader to the full version for more details.

4.4 Proof of Proposition 3

We will show that for every $0 \leq j < k' \leq k$ the following hold,

1. For all distinct $\mathbf{x}_1, \ldots, \mathbf{x}_{k'}$, distinct $\mathbf{y}_1, \ldots, \mathbf{y}_j$ and $y^* \in \{0,1\}^n$ such that $\mathbf{x}_{j+1}[1], \ldots, \mathbf{x}_{k'}[1]$ are distinct and $y^* \notin \{\mathbf{y}_1[0], \ldots, \mathbf{y}_j[0]\}$,
$$\mathsf{Pr}[\text{Eq. (11) does not hold for } R_{\pi,\sigma}] \leq e_1(k', j) + \varepsilon \cdot e_2(k', j). \tag{14}$$
2. For all distinct $\mathbf{y}_1, \ldots, \mathbf{y}_{k'}$, all $\mathbf{x}_1, \ldots, \mathbf{x}_j$ and all $x^* \in \{0,1\}^n$ such that $\mathbf{y}_{j+1}[0], \ldots, \mathbf{y}_{k'}[0]$ are distinct and $x^* \notin \{\mathbf{x}_1[1], \ldots, \mathbf{x}_j[1]\}$,
$$\mathsf{Pr}[\text{Eq. (12) does not hold for } R_{\pi,\sigma}] \leq e_1(k', j) + \varepsilon \cdot e_2(k', j), \tag{15}$$

where the probability is taken over the random choice of (π, σ) and

$$e_1(k', j) = 3 \cdot C_t \left(\frac{1}{\delta' - \delta}\right)^t \left(\frac{2\delta^* t}{2^n}\right)^{t/2} 2^{(k'-j)(t/2+1)},$$
$$e_2(k', j) = 2 \left(\frac{1+\delta}{\delta' - \delta}\right)^t 2^{k'-j}.$$

Given that the above hold, we then take appropriate union bounds (for Eq. (14)) over all $\mathbf{y}_1, \ldots, \mathbf{y}_{k'}, \mathbf{x}_1, \ldots, \mathbf{x}_j$ and x^* and then over all j, k'. Symmetrically, we take union bounds (for Eq. (15)). Then the following holds and this concludes the proof of Theorem 3.

$$\begin{aligned}
&\mathsf{Pr}[R_{\pi,\sigma} \notin \mathsf{FEv}(k, \delta')] \\
&\leq 2 \sum_{k'=1}^{k} \sum_{j=0}^{k'-1} 2^{n(2k'+2j+1)} \cdot (e_1(j,k') + \varepsilon \cdot e_2(j,k')) \\
&\leq 12k^2 \left(\frac{1}{\delta' - \delta}\right)^t 2^{(4k-1)n} \left[C_t \cdot \left(\frac{4\delta^* t}{2^n}\right)^{t/2} + \varepsilon \cdot (1+\delta)^t \right].
\end{aligned}$$

From now on we focus on showing Eq. (15) and the analysis for Eq. (14) is symmetrical.

Establishing Equation (15). Let us fix some arbitrary k' and j such that $0 \le j < k' \le k$. Let us also fix some distinct $\mathbf{y}_1, \ldots, \mathbf{y}_{k'} \in \{0,1\}^{2n}$, distinct $\mathbf{x}_1, \ldots, \mathbf{x}_j$ and $x^* \in \{0,1\}^n$ such that $\mathbf{y}_{j+1}[0], \ldots, \mathbf{y}_{k'}[0]$ are distinct and $x^* \notin \{\mathbf{x}_1[0], \ldots, \mathbf{x}_j[0]\}$. Then, we are interested in counting the number of tuples $((\mathbf{u}_1, \mathbf{v}_1), \ldots, (\mathbf{u}_{k'}, \mathbf{v}_{k'}))$ in R that on applying π and σ transform to $((\mathbf{x}_1, \mathbf{y}_1), \ldots, (\mathbf{x}_j, \mathbf{y}_j), (\cdot \,\|\, x^*, \mathbf{y}_{j+1}), \ldots, (\cdot \,\|\, x^*, \mathbf{y}_{k'}))$. Let us fix σ and this defines $\mathbf{v}_i = \sigma^{-1}(\mathbf{y}_i)$ for every $i \in [k']$ allowing us to focus only on the following set $\mathcal{U}$ of tuples.

$$\mathcal{U} = \{(\mathbf{u}_1, \ldots, \mathbf{u}_{k'}) \mid \{(\mathbf{u}_1, \mathbf{v}_1), \ldots, (\mathbf{u}_{k'}, \mathbf{v}_{k'})\} \in R\}.$$

Then, we are interested in counting the number of tuples $\mathbf{U} = (\mathbf{u}_1, \ldots, \mathbf{u}_{k'})$ in $\mathcal{U}$ that satisfy,

1. $\pi(\mathbf{u}_1) = \mathbf{x}_1,\ \pi(\mathbf{u}_2) = \mathbf{x}_2,\ \ldots,\ \pi(\mathbf{u}_j) = \mathbf{x}_j$.
2. $\pi_1(\mathbf{u}_{j+1}) = \pi_1(\mathbf{u}_{j+2}) \ldots = \pi_1(\mathbf{u}_{k'}) = x^*$.
3. For every $i \in \{j+1, \ldots, k'\}$, $\pi_0(\mathbf{u}_i) \oplus \pi_0(\mathbf{u}_{j+1}) = \Delta_i$, where $\Delta_i = \mathbf{y}_{j+1}[0] \oplus \mathbf{y}_i[0]$[11],

where $\pi_0(\mathbf{u})$ and $\pi_1(\mathbf{u})$ denote the first n-bits and last n-bits of $\pi(\mathbf{u})$. Or equivalently, count the number of $\mathbf{U}$'s such that $\pi(\mathbf{U})$[12] falls in $\mathcal{X}$ where,

$$\mathcal{X} = \{(\mathbf{x}_1, \ldots, \mathbf{x}_j, x \oplus \Delta_{j+1} \,\|\, x^*, \ldots, x \oplus \Delta_{k'} \,\|\, x^*) \mid x \in \{0,1\}^n\}.$$

Note that every element $\mathbf{X}$ of $\mathcal{X}$ is completely described by an n-bit string x. Now for $\mathbf{U} = (\mathbf{u}_1, \ldots, \mathbf{u}_{k'}) \in \mathcal{U}$, let $I_{\mathbf{U}}$ be an indicator random variable,

$$I_{\mathbf{U}} = \begin{cases} 1 & \text{if } (\pi(\mathbf{u}_1), \ldots, \pi(\mathbf{u}_{k'})) \in \mathcal{X}, \\ 0 & \text{otherwise.} \end{cases}$$

Then it suffices to prove that,

$$\Pr_{\pi}\left[\sum_{\mathbf{U} \in \mathcal{U}} I_{\mathbf{U}} > \delta' \cdot 2^n\right] \le e_1(k', j) + \varepsilon \cdot e_2(k', j).$$

Instead of looking at the sum $\sum I_{\mathbf{U}}$, we look at an equivalent sum $\sum I_x$ of, albeit, different indicator random variables I_x's, which will be convenient to analyse. For $x \in \{0,1\}^n$ we define an indicator random variable I_x which is 1 if π^{-1} transforms $\mathbf{X} \in \mathcal{X}$ (that corresponds to x) into some $\mathbf{U} \in \mathcal{U}$. More formally,

$$I_x = \begin{cases} 1 & \text{if } (\pi^{-1}(\mathbf{x}_1), \ldots, \pi^{-1}(x \oplus \Delta_{j+1} || x^*), \ldots, \pi^{-1}(x \oplus \Delta_{k'} || x^*)) \in \mathcal{U}, \\ 0 & \text{otherwise.} \end{cases} \tag{16}$$

[11] As the definition of δ'-2-Feistel evasiveness concerns itself with $\overline{R_{\pi,\sigma}}$.

[12] By $\pi(\mathbf{U})$ we mean the tuple $(\pi(\mathbf{u}_1), \ldots, \pi(\mathbf{u}_{k'}))$.

Then, it is easy to see that counting the number of $\mathbf{U} \in \mathcal{U}$ such that $I_{\mathbf{U}} = 1$ (or $\pi(\mathbf{U}) \in \mathcal{X}$) is the same as counting the number of $x \in \{0,1\}^n$ such that $I_x = 1$ (or $\pi^{-1}(\mathbf{X}) \in \mathcal{U}$). Therefore, $\sum_{\mathbf{U} \in \mathcal{U}} I_{\mathbf{U}} = \sum_{x \in \{0,1\}^n} I_x$ and we aim to show that,

$$\Pr_{\pi}\left[\sum_{x \in \{0,1\}^n} I_x > \delta' \cdot 2^n\right] \leq e_1(k', j) + \varepsilon \cdot e_2(k', j). \tag{17}$$

Partitioning $\{0,1\}^n$. We would like to use concentration bounds for the sum of random variables I_x's. But note that they are not independent as they may depend on the output of π^{-1} on the same input. Therefore, as a first step towards constructing independent random variables, we partition $\{0,1\}^n$ into subsets which will allow us to break the sum $\sum I_x$ into sums over these subsets.

Let us consider the following relation on $\{0,1\}^n \times \{0,1\}^n$. For any $x, x' \in \{0,1\}^n$, we say that x is related to x' (denoted as $x \sim x'$) if there exists an index set $\mathcal{B} \subseteq \{j+1, \ldots, k'\}$ where such that,

$$x = x' \oplus \bigoplus_{i \in \mathcal{B}} \Delta_i.$$

It is easy to see that the relation $\sim$ is an equivalence relation. Then, for any $x \in \{0,1\}^n$, let EQ_x denote its equivalence class, that is, $\mathsf{EQ}_x = \{x' \in \{0,1\}^n \,|x \sim x'\}$. Let $|\mathsf{EQ}_x| = l$ and it is easy to see that $l \leq 2^{k'-j}$. Let $\{\mathsf{EQ}_i\}_{i=1}^{M}$ be the M equivalence classes of $\sim$ where $|\mathsf{EQ}_i| = l$ and $M \cdot l = 2^n$. Furthermore, let $\mathsf{EQ}_i = \{x_1^i, x_2^i, \ldots, x_l^i\}$ be an enumeration of EQ_i where x_q^i is the qth member of the ith equivalence class EQ_i. Then, we can break the sum of I_x's into,

$$\sum_{x \in \{0,1\}^n} I_x = \sum_{i=1}^{M} \sum_{q=1}^{l} I_{x_q^i} = \sum_{q=1}^{l} \sum_{i=1}^{M} I_{x_q^i}.$$

For $q \in [l]$, let $X_q = \sum_{i=1}^{M} I_{x_q^i}$. In other words, X_q is the sum of qth member of each equivalence class EQ_i. We are going to show that for every $q \in [l]$,

$$\Pr[X_q > \delta' \cdot M] \leq 3C_t \cdot \frac{1}{(\delta' - \delta)^t} \left(\frac{2t\delta^*}{2^n}\right)^{t/2} l^{t/2} + 2\varepsilon \cdot \left(\frac{1+\delta}{\delta' - \delta}\right)^t \tag{18}$$

Taking union bound over all $q \in [l]$ and using $l \leq 2^{k'-j}$, we have that Eq. (17) holds.

Bounding the subsum X_q. From now on, we will focus on analysing one of the subsums X_q and the other subsums can be analogously handled. Fix some q and let $X = X_q$. Let the corresponding set of x's be $\{x^1, \ldots, x^M\}$ where each x^i comes from a different equivalence class EQ_i. For every $i_1 \neq i_2 \in [M]$,

- Firstly, $\Delta_{j+1}, \ldots, \Delta_{k'}$ are distinct as $\mathbf{y}_{j+1}[0], \ldots, \mathbf{y}_{k'}[0]$ are distinct. Therefore, $x^{i_1} \oplus \Delta_{j+1}, \ldots, x^{i_1} \oplus \Delta_{k'}$ are distinct.
- Secondly for any index set $\mathcal{B} \subseteq \{j+1, \ldots, k'\}$,

$$x^{i_1} \neq x^{i_2} \oplus \bigoplus_{i \in \mathcal{B}} \Delta_i. \tag{19}$$

This implies that for any $I_{x^{i_1}}$ and $I_{x^{i_2}}$, $\{x^{i_1}_{j+1} \oplus \Delta_{j+1}, \ldots, x^{i_1}_{k'} \oplus \Delta_{k'}\}$ and $\{x^{i_2}_{j+1} \oplus \Delta_{j+1}, \ldots, x^{i_2}_{k'} \oplus \Delta_{k'}\}$ are disjoint. Therefore, except the first j (values that correspond to $\pi^{-1}(\mathbf{x}_i)$ for $i \in [j]$), the remaining set of values in the output of π^{-1} that each $I_{x^{i_1}}$ and $I_{x^{i_2}}$ depend on are disjoint.

We will crucially exploit these two properties of I_{x^i}'s to show that the following:

Lemma 3. *For X (as defined above), there exists a random variable Z with expectation $\mu = \mathbb{E}[Z] \leq \delta \cdot M$ where Z is a sum of M independent indicator random variables, such that for any integer $a > 0$,*

$$\mathsf{Pr}[|X - \mu| > a] \leq \frac{3 \cdot \mathbb{E}[(Z-\mu)^t]}{a^t} + 2\varepsilon \cdot \frac{(M+\mu)^t}{a^t}.$$

For each indicator random variable I_{x^i}, we first define another indicator random variable $I^{\rho}_{x^i}$ where the only difference is that we replace the $k \cdot t$- wise independent permutation π^{-1} with a random permutation ρ. Note that the resulting $I^{\rho}_{x^i}$ are still not independent as they depend on the output of ρ. So, we then define a sequence of random variable $I^{*}_{x^i}$ that have the same marginal distribution as that of $I^{\rho}_{x^i}$ but are independent. Then, we show a domination argument that relates the t-th moment of $(Y - \mu)$ with the t-th moment of $(Z - \mu)$ where Y and Z are the sum of $I^{\rho}_{x^i}$ and $I^{*}_{x^i}$ respectively. The proof of Lemma 3 can be found in the full version [38, App. B.4]. Next, we apply the following concentration bound due to [10] to the random variable Z.

Lemma 4 (A.4. from [10]). *Let $t \geq 2$ be an even integer. Suppose $Z_1, \ldots, Z_n$ are independent random variables taking values in $[0,1]$. Let $Z = Z_1 + \ldots + Z_n$ and $\mu = \mathbb{E}[Z]$. Then,*

$$\mathbb{E}\left[(Z-\mu)^t\right] \leq C_t \cdot (t\mu + t^2)^{t/2}.$$

Then as $\mu \leq \delta \cdot M$ we have,

$$\mathsf{Pr}[X > \delta \cdot M + a] \leq \mathsf{Pr}[X > \mu + a] \leq 3C_t \cdot \left(\frac{t\mu + t^2}{a^2}\right)^{t/2} + 2\varepsilon \cdot \left(\frac{M+\mu}{a}\right)^t,$$

Now let $a = (\delta' - \delta) \cdot M$ and using $M \cdot l = 2^n$ and $\delta^* = \max(\delta, t \cdot 2^k / 2^n)$, we have

$$\mathsf{Pr}[X > \delta' \cdot M] \leq 3C_t \cdot \frac{1}{(\delta' - \delta)^t} \left(\frac{2t\delta^*}{2^n}\right)^{t/2} l^{t/2} + 2\varepsilon \cdot \left(\frac{1+\delta}{\delta' - \delta}\right)^t$$

which establishes that Eq. (18) holds (which establishes that Eq. (15) holds) and thereby concludes the proof of Proposition 3. □

Acknowledgments. The authors were supported by NSF grants CNS-1553758 (CAREER), CNS-1423566, CNS-1719146, CNS-1528178, and IIS-1528041, and by a Sloan Research Fellowship.

References

1. Andreeva, E., Bogdanov, A., Dodis, Y., Mennink, B., Steinberger, J.P.: On the indifferentiability of key-alternating ciphers. In: Canetti, R., Garay, J.A. (eds.) CRYPTO 2013. LNCS, vol. 8042, pp. 531–550. Springer, Heidelberg (2013). https://doi.org/10.1007/978-3-642-40041-4_29
2. Andreeva, E., Bogdanov, A., Mennink, B.: Towards understanding the known-key security of block ciphers. In: Moriai, S. (ed.) FSE 2013. LNCS, vol. 8424, pp. 348–366. Springer, Heidelberg (2014). https://doi.org/10.1007/978-3-662-43933-3_18
3. Aumasson, J.-P., Jovanovic, P., Neves, S.: NORX8 and NORX16: authenticated encryption for low-end systems. Cryptology ePrint Archive, Report 2015/1154 (2015). http://eprint.iacr.org/2015/1154
4. Barak, B., Mahmoody-Ghidary, M.: Merkle puzzles are optimal — an $O(n^2)$-query attack on any key exchange from a random oracle. In: Halevi, S. (ed.) CRYPTO 2009. LNCS, vol. 5677, pp. 374–390. Springer, Heidelberg (2009). https://doi.org/10.1007/978-3-642-03356-8_22
5. Bellare, M., Hoang, V.T.: Resisting randomness subversion: fast deterministic and hedged public-key encryption in the standard model. In: Oswald, E., Fischlin, M. (eds.) EUROCRYPT 2015. LNCS, vol. 9057, pp. 627–656. Springer, Heidelberg (2015). https://doi.org/10.1007/978-3-662-46803-6_21
6. Bellare, M., Hoang, V.T., Keelveedhi, S.: Instantiating random oracles via UCEs. In: Canetti, R., Garay, J.A. (eds.) CRYPTO 2013. LNCS, vol. 8043, pp. 398–415. Springer, Heidelberg (2013). https://doi.org/10.1007/978-3-642-40084-1_23
7. Bellare, M., Hoang, V.T., Keelveedhi, S.: Cryptography from compression functions: the UCE bridge to the ROM. In: Garay, J.A., Gennaro, R. (eds.) CRYPTO 2014. LNCS, vol. 8616, pp. 169–187. Springer, Heidelberg (2014). https://doi.org/10.1007/978-3-662-44371-2_10
8. Bellare, M., Hoang, V.T., Keelveedhi, S., Rogaway, P.: Efficient garbling from a fixed-key blockcipher. In: 2013 IEEE Symposium on Security and Privacy, pp. 478–492. IEEE Computer Society Press, May 2013
9. Bellare, M., Rogaway, P.: The security of triple encryption and a framework forcode-based game-playing proofs. In: Vaudenay, S. (ed.) EUROCRYPT 2006. LNCS, vol. 4004, pp. 409–426. Springer, Heidelberg (2006). https://doi.org/10.1007/11761679_25
10. Bellare, M., Rompel, J.: Randomness-efficient oblivious sampling. In: 35th FOCS, pp. 276–287. IEEE Computer Society Press, November 1994
11. Bellare, M., Stepanovs, I.: Point-function obfuscation: a framework and generic constructions. In: Kushilevitz, E., Malkin, T. (eds.) TCC 2016. LNCS, vol. 9563, pp. 565–594. Springer, Heidelberg (2016). https://doi.org/10.1007/978-3-662-49099-0_21
12. Bertoni, G., Daemen, J., Peeters, M., Van Assche, G.: On the indifferentiability of the sponge construction. In: Smart, N. (ed.) EUROCRYPT 2008. LNCS, vol. 4965, pp. 181–197. Springer, Heidelberg (2008). https://doi.org/10.1007/978-3-540-78967-3_11

13. Bertoni, G., Daemen, J., Peeters, M., Van Assche, G.: Sponge-based pseudo-random number generators. In: Mangard, S., Standaert, F.-X. (eds.) CHES 2010. LNCS, vol. 6225, pp. 33–47. Springer, Heidelberg (2010). https://doi.org/10.1007/978-3-642-15031-9_3
14. Brzuska, C., Farshim, P., Mittelbach, A.: Indistinguishability obfuscation and UCEs: the case of computationally unpredictable sources. In: Garay, J.A., Gennaro, R. (eds.) CRYPTO 2014. LNCS, vol. 8616, pp. 188–205. Springer, Heidelberg (2014). https://doi.org/10.1007/978-3-662-44371-2_11
15. Canetti, R., Goldreich, O., Halevi, S.: The random oracle methodology, revisited (preliminary version). In: 30th ACM STOC, pp. 209–218. ACM Press, May 1998
16. Cogliati, B., Seurin, Y.: Strengthening the known-key security notion for block ciphers. In: Peyrin, T. (ed.) FSE 2016. LNCS, vol. 9783, pp. 494–513. Springer, Heidelberg (2016). https://doi.org/10.1007/978-3-662-52993-5_25
17. Coron, J.-S., Patarin, J., Seurin, Y.: The random oracle model and the ideal cipher model are equivalent. In: Wagner, D. (ed.) CRYPTO 2008. LNCS, vol. 5157, pp. 1–20. Springer, Heidelberg (2008). https://doi.org/10.1007/978-3-540-85174-5_1
18. Dachman-Soled, D., Katz, J., Thiruvengadam, A.: 10-Round feistel is indifferentiable from an ideal cipher. In: Fischlin, M., Coron, J.-S. (eds.) EUROCRYPT 2016. LNCS, vol. 9666, pp. 649–678. Springer, Heidelberg (2016). https://doi.org/10.1007/978-3-662-49896-5_23
19. Dai, Y., Steinberger, J.: Indifferentiability of 8-round feistel networks. In: Robshaw, M., Katz, J. (eds.) CRYPTO 2016. LNCS, vol. 9814, pp. 95–120. Springer, Heidelberg (2016). https://doi.org/10.1007/978-3-662-53018-4_4
20. Dodis, Y., Ganesh, C., Golovnev, A., Juels, A., Ristenpart, T.: A formal treatment of backdoored pseudorandom generators. In: Oswald, E., Fischlin, M. (eds.) EUROCRYPT 2015. LNCS, vol. 9056, pp. 101–126. Springer, Heidelberg (2015). https://doi.org/10.1007/978-3-662-46800-5_5
21. Dodis, Y., Stam, M., Steinberger, J., Liu, T.: Indifferentiability of confusion-diffusion networks. In: Fischlin, M., Coron, J.-S. (eds.) EUROCRYPT 2016. LNCS, vol. 9666, pp. 679–704. Springer, Heidelberg (2016). https://doi.org/10.1007/978-3-662-49896-5_24
22. Fiat, A., Shamir, A.: How to prove yourself: practical solutions to identification and signature problems. In: Odlyzko, A.M. (ed.) CRYPTO 1986. LNCS, vol. 263, pp. 186–194. Springer, Heidelberg (1987). https://doi.org/10.1007/3-540-47721-7_12
23. Gaži, P., Tessaro, S.: Provably robust sponge-based PRNGs and KDFs. In: Fischlin, M., Coron, J.-S. (eds.) EUROCRYPT 2016. LNCS, vol. 9665, pp. 87–116. Springer, Heidelberg (2016). https://doi.org/10.1007/978-3-662-49890-3_4
24. Hoang, V.T., Rogaway, P.: On generalized feistel networks. In: Rabin, T. (ed.) CRYPTO 2010. LNCS, vol. 6223, pp. 613–630. Springer, Heidelberg (2010). https://doi.org/10.1007/978-3-642-14623-7_33
25. Holenstein, T., Künzler, R., Tessaro, S.: The equivalence of the random oracle model and the ideal cipher model, revisited. In: Fortnow, L., Vadhan, S.P. (eds.) 43rd ACM STOC, pp. 89–98. ACM Press, June 2011
26. Impagliazzo, R., Rudich, S.: Limits on the provable consequences of one-way permutations. In: 21st ACM STOC, pp. 44–61. ACM Press, May 1989
27. Knudsen, L.R., Rijmen, V.: Known-key distinguishers for some block ciphers. In: Kurosawa, K. (ed.) ASIACRYPT 2007. LNCS, vol. 4833, pp. 315–324. Springer, Heidelberg (2007). https://doi.org/10.1007/978-3-540-76900-2_19
28. Luby, M., Rackoff, C.: How to construct pseudorandom permutations from pseudorandom functions. SIAM J. Comput. **17**(2), 373–386 (1988)

29. Mandal, A., Patarin, J., Seurin, Y.: On the public indifferentiability and correlation intractability of the 6-round feistel construction. In: Cramer, R. (ed.) TCC 2012. LNCS, vol. 7194, pp. 285–302. Springer, Heidelberg (2012). https://doi.org/10.1007/978-3-642-28914-9_16
30. Matsuda, T., Hanaoka, G.: Chosen ciphertext security via UCE. In: Krawczyk, H. (ed.) PKC 2014. LNCS, vol. 8383, pp. 56–76. Springer, Heidelberg (2014). https://doi.org/10.1007/978-3-642-54631-0_4
31. Maurer, U., Renner, R., Holenstein, C.: Indifferentiability, impossibility results on reductions, and applications to the random oracle methodology. In: Naor, M. (ed.) TCC 2004. LNCS, vol. 2951, pp. 21–39. Springer, Heidelberg (2004). https://doi.org/10.1007/978-3-540-24638-1_2
32. Mittelbach, A.: Salvaging indifferentiability in a multi-stage setting. In: Nguyen, P.Q., Oswald, E. (eds.) EUROCRYPT 2014. LNCS, vol. 8441, pp. 603–621. Springer, Heidelberg (2014). https://doi.org/10.1007/978-3-642-55220-5_33
33. Naor, M., Reingold, O.: On the construction of pseudo-random permutations: Luby-Rackoff revisited (extended abstract). In: 29th ACM STOC, pp. 189–199. ACM Press, May 1997
34. Ramzan, Z., Reyzin, L.: On the round security of symmetric-key cryptographic primitives. In: Bellare, M. (ed.) CRYPTO 2000. LNCS, vol. 1880, pp. 376–393. Springer, Heidelberg (2000). https://doi.org/10.1007/3-540-44598-6_24
35. Rogaway, P., Steinberger, J.: Security/efficiency tradeoffs for permutation-based hashing. In: Smart, N. (ed.) EUROCRYPT 2008. LNCS, vol. 4965, pp. 220–236. Springer, Heidelberg (2008). https://doi.org/10.1007/978-3-540-78967-3_13
36. Schmidt, J.P., Siegel, A., Srinivasan, A.: Chernoff-hoeffding bounds for applications with limited independence. In: Ramachandran, V. (ed.), 4th SODA, pp. 331–340. ACM-SIAM, January 1993
37. Soni, P., Tessaro, S.: Public-seed pseudorandom permutations. In: Coron, J.-S., Nielsen, J.B. (eds.) EUROCRYPT 2017. LNCS, vol. 10211, pp. 412–441. Springer, Cham (2017). https://doi.org/10.1007/978-3-319-56614-6_14
38. Soni, P., Tessaro, S.: Naor-reingold goes public: The complexity of known-key security. Cryptology ePrint Archive, Report 2018/137 (2018). https://eprint.iacr.org/2018/137

Updatable Encryption with Post-Compromise Security

Anja Lehmann(✉) and Björn Tackmann

IBM Research – Zurich, Rüschlikon, Switzerland
{anj,bta}@zurich.ibm.com

Abstract. An updatable encryption scheme allows to periodically rotate the encryption key and move already existing ciphertexts from the old to the new key. These ciphertext updates are done with the help of a so-called update token and can be performed by an untrusted party, as the update never decrypts the data. Updatable encryption is particularly useful in settings where encrypted data is outsourced, e.g., stored on a cloud server. The data owner can produce an update token, and the cloud server can update the ciphertexts.

We provide a comprehensive treatment of *ciphertext-independent* schemes, where a single token is used to update all ciphertexts. We show that the existing ciphertext-independent schemes and models by Boneh et al. (CRYPTO'13) and Everspaugh et al. (CRYPTO'17) do not guarantee the *post-compromise* security one would intuitively expect from key rotation. In fact, the simple scheme recently proposed by Everspaugh et al. allows to recover the current key upon corruption of a single old key. Surprisingly, none of the models so far reflects the timely aspect of key rotation which makes it hard to grasp *when* an adversary is allowed to corrupt keys. We propose strong security models that clearly capture post-compromise and forward security under adaptive attacks. We then analyze various existing schemes and show that none of them is secure in this strong model, but we formulate the additional constraints that suffice to prove their security in a relaxed version of our model. Finally, we propose a new updatable encryption scheme that achieves our strong notions while being (at least) as efficient as the existing solutions.

1 Introduction

In data storage, key rotation refers to the process of (periodically) exchanging the cryptographic key material that is used to protect the data. Key rotation is considered good practice as it hedges against the impact of cryptographic keys being compromised over time. For instance, the Payment Card Industry Data Security Standard (PCI DSS) [24], which specifies how credit card data must be stored in encrypted form mandates key rotation, meaning that encrypted data must regularly be moved from an old to a fresh key. Many cloud storage providers that implement data-at-rest encryption, such as Google and Amazon, employ a similar feature [15]. The trivial approach to update an existing ciphertext

J. B. Nielsen and V. Rijmen (Eds.): EUROCRYPT 2018, LNCS 10822, pp. 685–716, 2018.
https://doi.org/10.1007/978-3-319-78372-7_22

towards a new key is to decrypt the ciphertext and re-encrypt the underlying plaintext from scratch using the fresh key. Implementing this approach for secure cloud storage applications where the data owner outsources his data in encrypted form to a potentially untrusted host is not trivial, though: Either the owner has to download, re-encrypt and upload all ciphertexts, which makes outsourcing impractical, or the encryption keys have to be sent to the host, violating security.

Updatable Encryption. A better solution for updating ciphertexts has been proposed by Boneh et al. [10]: in what they call an *updatable encryption scheme*, the data owner can produce a short update token that allows the host to re-encrypt the data himself, while preserving the security of the encryption, i.e., the token allows to migrate ciphertexts from an old to a new key, but does not give the host an advantage in breaking the confidentiality of the protected data. Boneh et al. also proposed a construction (BLMR) based on key-homomorphic PRFs, which essentially is a symmetric proxy re-encryption scheme (PRE) where one sequentially re-encrypts data from one epoch to the next.

While being somewhat similar in spirit, PRE and updatable encryption do have different security requirements: PRE schemes often keep parts of the ciphertexts static throughout re-encryption, as there is no need to make a re-encrypted ciphertexts independent from the original ciphertext it was derived from. In updatable encryption, however, the goal should be that an updated ciphertext is as secure as a fresh encryption; in particular, it should look like an independently computed ciphertext even given previous ones. Thus, any scheme that produces linkable ciphertexts, such as the original BLMR construction, cannot guarantee such a security notion capturing post-compromise security of updated ciphertexts.

Ciphertext-Independence vs. Ciphertext-Dependence. In the full version of their paper, Boneh et al. [9] provide security notions for updatable encryption, which aim to cover the desired indistinguishability of updated ciphertexts. To satisfy that notion, they have to remove the linkability from the BLMR scheme, which they achieve by moving to the setting of *ciphertext-dependent* updates. In ciphertext-dependent schemes, the owner no longer produces a single token that can update *all* ciphertexts, but produces a dedicated token for each ciphertext. Therefore, the owner has to download all outsourced ciphertexts, compute a specific token for every ciphertext, and send all tokens back to the host.

Clearly, ciphertext-dependent schemes are much less efficient and more cumbersome for the data owner than *ciphertext-independent* ones. They also increase the complexity of the update procedure for the host, who has to ensure that it applies the correct token for each ciphertext—any mistake renders the updated ciphertexts useless. Another, more subtle disadvantage of ciphertext-dependent schemes is that they require the old and new keys to be present together for a longer time, as the owner needs both keys to derive the individual tokens for all of his ciphertexts. Deleting the old key too early might risk losing the ability of decrypting ciphertexts that have not been upgraded yet, whereas keeping the old key too long makes an attack at that time more lucrative—the adversary obtains two keys at the same time. In a ciphertext-independent scheme, the old key can and should be deleted immediately after the token has been derived.

In a recent work [15], Everspaugh et al. provide a systematic treatment for such ciphertext-dependent schemes and observe that computing the token often does not require access to the full ciphertext, but only to a short ciphertext header, which allows to moderately improve the efficiency of this approach. Everspaugh et al. also show that the security notions from [9] do not cover the desired property of post-compromise security of updated ciphertexts. They provide two new security notions and propose schemes that can provably satisfy them. As a side-result, they also propose a security definition for ciphertext-independent schemes and suggest a simple xor-based scheme (XOR-KEM) for this setting.

Ambiguity of Security Models. Interestingly, both previous works phrase the algorithms and security models for updatable encryption in the flavor of normal proxy re-encryption. That leads to a mismatch of how the scheme is used and modeled—in practice, an updatable encryption scheme is used in a clear sequential setting, updating ciphertexts as the key progresses. The security model offers more and unrealistic flexibility, though: it allows to rotate keys and ciphertexts across *arbitrary* epochs, jumping back in forth in time. This flexibility gives the adversary more power than he has in reality and, most importantly, makes the security that is captured by the model hard to grasp, as it is not clear *when* the adversary is allowed to corrupt keys.

Non-intuitive security definitions increase the risk that proofs are flawed or that schemes are unintentionally used outside the security model. And in fact, the way that Everspaugh et al. [15] define security for (ciphertext-independent) schemes is ambiguous, and only the weaker interpretation of their model allows their scheme XOR-KEM to be proven secure. However, this weaker interpretation does not guarantee any confidentiality *after* a secret key got compromised, as it allows key corruption only after the challenge epoch. Thus, an updatable scheme that is secure only in such a weak model does not provide the intuitive security one would expect from key rotation: namely that after migrating to the new key, the old one becomes useless and no longer of value to the adversary. To the contrary, all previous keys still require strong protection or secure deletion.

Importance of Post-Compromise Security. Realizing secure deletion in practice is virtually impossible, as keys may be copied or moved across the RAM, swap partitions, and SSD memory blocks, and thus we consider *post-compromise security* an essential property of updatable schemes. Avoiding the assumption of securely deleted keys and re-gaining security after a temporary corruption has recently inspired numerous works on how to achieve post-compromise security in other encryption settings [6,13,14,16]. Note that an updatable encryption scheme that is not post-compromise secure can even reduce the security compared with a scheme where keys are never rotated: as one expects old keys to be useless after rotation, practitioners can be misled to reduce the safety measures for "expired" keys, which in turn makes key compromises more likely. For the example of Everspaugh et al. simple XOR-KEM scheme [15], a compromised old key allows to fully recover the fresh key.

This leaves open the important question how to design a ciphertext-independent scheme that achieves post-compromise security, capturing the full spirit of updatable encryption and key rotation.

Our Contributions. In this work we provide a comprehensive treatment for *ciphertext-independent* updatable encryption schemes that have clear advantages in efficiency and ease-of-deployment over the ciphertext-dependent solutions. We model updatable encryption and its security in the natural sequential manner that is inherent in key rotation, avoiding the ambiguity of previous works, and clearly capturing all desired security properties. We also analyze the (in)security of a number of existing schemes and finally propose a construction that provably satisfies our strong security notions.

Strong Security Models. We define updatable encryption in its natural form where keys and ciphertexts sequentially evolve over time epochs. To capture security, we allow the adversary to *adaptively* corrupt secret keys, update tokens and ciphertexts in any combination of epochs as long as this does not allow him to trivially decrypt a challenge ciphertext. In our first notion, *indistiguishability of encryptions* (IND-ENC), such a challenge ciphertext will be a fresh encryption C_d of one of two messages m_0, m_1 under the current epoch key, and the task of the adversary is to guess the bit d. This is the standard CPA game adapted to the updatable encryption and adaptive corruption setting. Our second notion, *indistiguishability of updates* (IND-UPD), returns as a challenge the re-encryption C'_d of a ciphertext either C_0 or C_1, and an adversary again has to guess the bit d.

We stress that this second property is essential for the security of updatable encryption schemes, as it captures confidentiality of *updated* encryptions, whereas IND-ENC only guarantees security for ciphertexts that originate from a fresh encryption. While IND-ENC is similar to the security of symmetric proxy re-encryption schemes, IND-UPD is a property that is special to the context of key rotation. And thus, contrary to a common belief, a symmetric PRE scheme cannot directly be used for secure updatable encryption [10,15,22]!

In the ciphertext-independent setting, capturing the information that the adversary can infer from a certain amount of corrupted tokens, keys and ciphertexts is a delicate matter, as, e.g., an update token allows the adversary to move *any* ciphertext from one epoch to the next. We observe that all existing constructions leak more information than necessary. Instead of hard-coding the behavior of the known schemes into the security model, we propose a set of leakage profiles, and define both the optimal and currently achievable leakage.

We then compare our model to the existing definition for encryption indistinguishability by Everspaugh et al. [15]. We argue that their definition can be interpreted in two ways: the weaker interpretation rules out post-compromise security, but allows the XOR-KEM construction to be secure, whereas the stronger interpretation is closer to our IND-ENC model. However, in their stronger version, as well as in our IND-ENC notion, we show that XOR-KEM cannot be secure by describing a simple attack that allows to recover the challenge secret key after compromising one old key. We further show that IND-ENC is strictly stronger than the weak interpretation of [15], but incomparable to the stronger one, due to the way both models handle adversarial ciphertexts.

Provably Secure Constructions. We further analyze several schemes according to the new definitions (Sect. 5), the results are summarized in Table 1. First, we consider a simple construction (called 2ENC) that is purely based on symmetric primitives. Unfortunately, the scheme cannot satisfy our strong security notions. Yet, instead of simply labeling this real-world solution as insecure, we formulate the additional constraints on the adversarial behavior that suffice to prove its security in relaxed versions of our IND-ENC and IND-UPD models.

Table 1. Overview of results in this work. (Corruption of secret keys in challenge epochs is forbidden by the IND-ENC and IND-UPD definitions. The symbol (✓) denotes that a schemes requires additional constraints on the tokens that can be corrupted to achieve the security notion.)

	SE-KEM	2ENC	BLMR	BLMR+	RISE
IND-ENC	(✓)	(✓)	✓	✓	✓
	No token near a challenge	Either token near challenge, or secret key			
IND-UPD	✘	(✓)	✘	(✓)	✓
		No token near a challenge		At most one token	

We then turn our attention to less efficient but more secure schemes, starting with the BLMR construction by Boneh et al. [10] that uses key-homomorphic PRFs. We show that the original BLMR scheme does satisfy our IND-ENC notion but not IND-UPD, and also propose a slight modification BLMR+ that improves the latter and achieves a weak form of update indistinguishability. While BLMR seems to be a purely symmetric solution on the first glance, any instantiation of the underlying key-homomorphic PRFs so far requires modular exponentiations or is built from lattices. The same holds for the recent ciphertext-dependent construction by Everspaugh et al. [15] that also relies on key-homomorphic PRFs and suggests a discrete-logarithm based instantiation.

Acknowledging that secure updatable encryption schemes seem to inherently require techniques from the public-key world, we then build a scheme that omits the intermediate abstraction of using key-homomorphic PRFs which allows us to take full advantage of the underlying group operations. Our construction (RISE, for Re-randomizable ciphertext-Independent Symmetric Elgamal) can be seen as the classic ElGamal-based proxy re-encryption scheme combined with a fresh re-randomization upon each re-encryption. We prove that this scheme fully achieves both of our strong security definitions.

We compare the schemes in terms of efficiency in Table 2. The costs for encryption and updates of our most secure RISE scheme are—on the owner side—even lower than the costs in the less secure BLMR scheme and the recent ciphertext-dependent scheme ReCrypt by Everspaugh et al. [15]. The solution by Everspaugh et al. shifts significantly many expensive update operations to the data owner, who has to compute two exponentiation for each ciphertext (block) that shall be updated, whereas our scheme requires the owner to compute only a single exponentiation for the update of all ciphertexts.

In Appendix A, we additionally analyze a "hybrid-encryption" scheme SE-KEM that is widely used in practical data-at-rest protection, where the encrypted plaintext is stored together with the encryption key wrapped under an epoch key. The scheme provides rather weak guarantees when viewed as an updatable encryption scheme, but may still be useful in certain scenarios due to the efficient key update.

Table 2. Comparison of computational efficiency measured by the most expensive operations for short (one-block) ciphertexts (exponentiation, symmetric cryptography). Note that the ciphertext-dependent BLMR' variant of [9] is unlikely to have a security proof [15], and BLMR and BLMR+ achieve significantly weaker security than RISE. (SE-KEM and 2ENC are omitted here as they are purely symmetric solutions.)

	Ciphertext independent	Encryption	Token derivation	Update of n Ciphertexts
BLMR' [9]		2 exp.	$2n$ sym.	$2n$ exp.
ReCrypt [15]		2 exp.	$2n$ exp.	$2n$ exp.
BLMR [10]	✔	2 exp.	2 exp.	$2n$ exp.
BLMR+ (this work)	✔	2 exp.	2 exp.	$2n$ exp.
RISE (this work)	✔	2 exp.	1 exp.	$2n$ exp.

Other Related Work. Beyond the previous work on updatable encryption [9,10,15] that we already discussed above, the most closely related line of work is on (symmetric) proxy re-encryption (PRE) [2,3,7,8,12,17,20–22]. Notably, the recent work of Berners-Lee [7] builds on the work of Everspaugh et al. [15] and views the concept of ciphertext-dependent updates as a desirable security feature of PRE in general, as it reduces the freedom of a possibly untrusted proxy. The recent work of Myers and Shull [22] studies *hybrid* PRE schemes aiming at efficient solutions for key rotation and access revocation. As stressed before, however, while being similar in the sense that PRE allows a proxy to move ciphertexts from one key to another, the desired security guarantees have subtle differences and the security property of IND-UPD that is crucial for updatable encryption is neither covered nor needed by PRE.

While this means that a secure PRE does not automatically yield a secure updatable encryption scheme, it does not prevent PREs from being secure in the updatable encryption sense as well—but this has to be proven from scratch. In fact, our schemes are strongly inspired by proxy re-encryption: For the simple double-encryption scheme discussed by Ivan and Dodis [18], we show that a weak form of security can be proven, and our most secure scheme RISE combines the ElGamal-based PRE with re-randomization of ciphertexts. We also observe similar challenges in designing schemes that limit the "power" of the token, which is related to the long-standing problem of constructing efficient PRE's that are uni-directional, multi-hop and collusion-resistant.

In the context of tokenization, which is the process of consistently replacing sensitive elements, such as credit card numbers, with non-sensitive surrogate values, the feature of key rotation has recently been studied by Cachin et al. [11]. Their schemes are inherently deterministic, and thus their results are not applicable to the problem of probabilistic encryption, but we follow their formalization of modeling key rotation in a strictly sequential manner.

Finally, a recent paper of Ananth et al. [1] provides a broader perspective on updatable cryptography, but targets generic and rather complex schemes with techniques such as randomized encodings. The definitions in their work have linkability hardcoded, as randomness has to remain the same across updates, which is in contrast to our goal of achieving efficient unlinkable schemes for the specific case of updatable encryption.

2 Preliminaries

Symmetric Encryption. A symmetric encryption scheme SE consists of a key space $\mathcal{K}$ and three polynomial-time algorithms SE.kgen, SE.enc, SE.dec satisfying the following conditions:

SE.kgen: The probabilistic key generation algorithm takes as input a security parameter and produces an encryption key $k \in \mathcal{K}$. That is, $k \xleftarrow{r} \mathsf{SE.kgen}(\lambda)$.
SE.enc: The probabilistic encryption algorithm takes a key $k \in \mathcal{K}$ and a message $m \in \mathcal{M}$ and returns a ciphertext C, written as $C \xleftarrow{r} \mathsf{SE.enc}(k, m)$.
SE.dec: The deterministic decryption algorithm SE.dec takes a key $k \in \mathcal{K}$ and a ciphertext C to return a message $(\mathcal{M} \cup \{\perp\}) \ni m \leftarrow \mathsf{SE.dec}(k, C)$

For correctness we require that for any key $k \in \mathcal{K}$, any message $m \in \mathcal{M}$ and any ciphertext $C \xleftarrow{r} \mathsf{SE.enc}(k, m)$, we have $m \leftarrow \mathsf{SE.dec}(k, C)$.

Chosen-Plaintext Security. The IND-CPA security of a symmetric encryption scheme SE is defined through the following game $\text{GAME}^{\mathsf{IND\text{-}CPA}}(\mathcal{A})$ with adversary $\mathcal{A}$. Initially, choose $b \xleftarrow{r} \{0, 1\}$ and $k \xleftarrow{r} \mathsf{SE.kgen}(\lambda)$. Run adversary $\mathcal{A}$ with oracle $\mathcal{O}_{\mathsf{enc}}(m)$, which computes $C \xleftarrow{r} \mathsf{SE.enc}(k, m)$ and returns C. When $\mathcal{A}$ outputs two messages m_0, m_1 with $|m_0| = |m_1|$ and a state *state*, compute $\tilde{C} \xleftarrow{r} \mathsf{SE.enc}(k, m_b)$ and run $\mathcal{A}(\tilde{C}, state)$, again with access to oracle $\mathcal{O}_{\mathsf{enc}}$. When $\mathcal{A}$ outputs a bit $\tilde{b}$, the game is won if $b = \tilde{b}$. The IND-CPA advantage of $\mathcal{A}$ is defined as $|2 \Pr[\text{GAME}^{\mathsf{IND\text{-}CPA}}(\mathcal{A}) \text{ won}] - 1|$, and SE is called IND-CPA-secure if for all efficient adversaries $\mathcal{A}$ the advantage is negligible in λ.

Decisional Diffie-Hellman Assumption. Our final construction requires a group $(\mathbb{G}, g, p)$ as input where $\mathbb{G}$ denotes a cyclic group $\mathbb{G} = \langle g \rangle$ of order p in which the Decisional Diffie-Hellman (DDH) problem is hard w.r.t. λ, i.e., p is a λ-bit prime. More precisely, a group $(\mathbb{G}, g, p)$ satisfies the DDH assumption if for any efficient adversary $\mathcal{A}$ the probability $\left|\Pr[\mathcal{A}(\mathbb{G}, p, g, g^a, g^b, g^{ab})] - \Pr[\mathcal{A}(\mathbb{G}, p, g, g^a, g^b, g^c)]\right|$ is negligible in λ, where the probability is over the random choice of p, g, the random choices of $a, b, c \in \mathbb{Z}_p$, and $\mathcal{A}$'s coin tosses.

3 Formalizing Updatable Encryption

We now present our formalization of updatable encryption and its desired security features, and discuss how our security model captures these properties.

An updatable encryption scheme contains algorithms for a data *owner* and a *host*. The owner encrypts data using the $\mathsf{UE.enc}$ algorithm, and then outsources the ciphertexts to the host. To this end, the data owner initially runs an algorithm $\mathsf{UE.setup}$ to create an encryption key. The encryption key evolves with *epochs*, and the data is encrypted with respect to a specific epoch e, starting with $e = 0$. When moving from epoch e to epoch $e + 1$, the owner invokes an algorithm $\mathsf{UE.next}$ to generate the key material k_{e+1} for the new epoch and an update token Δ_{e+1}. The owner then sends Δ_{e+1} to the host, deletes k_e and Δ_{e+1} immediately, and uses k_{e+1} for encryption from now on. After receiving Δ_{e+1}, the host first deletes Δ_e and then uses an algorithm $\mathsf{UE.upd}$ to update all previously received ciphertexts from epoch e to $e+1$, using Δ_{e+1}. Hence, during some epoch e, the update token from $e - 1$ to e is available at the host, but update tokens from earlier epochs have been deleted. (The host could already delete the token when all ciphertexts are updated, but as this is hard to model in the security game, we assume the token to be available throughout the full epoch.)

Definition 1 (Updatable Encryption). *An updatable encryption scheme* UE *for message space* $\mathcal{M}$ *consists of a set of polynomial-time algorithms* $\mathsf{UE.setup}$, $\mathsf{UE.next}$, $\mathsf{UE.enc}$, $\mathsf{UE.dec}$, *and* $\mathsf{UE.upd}$ *satisfying the following conditions:*

$\mathsf{UE.setup}$: The algorithm $\mathsf{UE.setup}$ is a probabilistic algorithm run by the owner. On input a security parameter λ, it returns a secret key $k_0 \xleftarrow{r} \mathsf{UE.setup}(\lambda)$.

$\mathsf{UE.next}$: This probabilistic algorithm is also run by the owner. On input a secret key k_e for epoch e, it outputs a new secret key k_{e+1} and an update token Δ_{e+1} for epoch $e + 1$. That is, $(k_{e+1}, \Delta_{e+1}) \xleftarrow{r} \mathsf{UE.next}(k_e)$.

$\mathsf{UE.enc}$: This probabilistic algorithm is run by the owner, on input a message $m \in \mathcal{M}$ and key k_e of some epoch e returns a ciphertext $C_e \xleftarrow{r} \mathsf{UE.enc}(k_e, m)$.

$\mathsf{UE.dec}$: This deterministic algorithm is run by the owner, on input a ciphertext C_e and key k_e of some epoch e returns $\{m'/\bot\} \leftarrow \mathsf{UE.dec}(k_e, C_e)$.

$\mathsf{UE.upd}$: This either probabilistic or deterministic algorithm is run by the host. On input a ciphertext C_e from epoch e and the update token Δ_{e+1}, it returns the updated ciphertext $C_{e+1} \leftarrow \mathsf{UE.upd}(\Delta_{e+1}, C_e)$.

Correctness. The correctness condition of an updatable encryption scheme ensures that an update of a valid ciphertext C_e from epoch e to $e+1$ leads again to a valid ciphertext C_{e+1} that can be decrypted under the new epoch key k_{e+1}. More precisely, we require that for any $m \in \mathcal{M}$, for any $k_0 \xleftarrow{r} \mathsf{UE.setup}(\lambda)$, for any sequence of key/update token pairs $(k_1, \Delta_1), \ldots, (k_e, \Delta_e)$ generated as $(k_{j+1}, \Delta_{j+1}) \xleftarrow{r} \mathsf{UE.next}(k_j)$ for $j = 0, \ldots, e-1$ through repeated applications of the key-evolution algorithm, and for any $C_0 \xleftarrow{r} \mathsf{UE.enc}(k_0, m)$, it holds that $m = \mathsf{UE.dec}(k_e, C_e)$ where C_e is recursively obtained through $C_{j+1} \xleftarrow{r} \mathsf{UE.upd}(\Delta_{j+1}, C_j)$.

3.1 Security Properties

The main goal of updatable encryption is twofold: First, it should enable efficient updates by a potentially corrupt host, i.e., the update procedure and compromise of the update tokens must not reduce the standard security of the encryption. Second, the core purpose of key rotation is to reduce the risk and impact of key exposures, i.e., confidentiality should be preserved or even re-gained in the presence of *temporary* key compromises, which can be split into forward and post-compromise security. Furthermore, we aim for security against adaptive and retroactive corruptions, modeling that any key or token from a current or previous epoch can become compromised.

Token Security: The feature of updating ciphertexts should not harm the standard IND-CPA security of the encryption scheme. That is, seeing updated ciphertexts or even the exposure of *all* tokens does not increase an adversary's advantage in breaking the encryption scheme.

Forward Security: An adversary compromising a secret key in some epoch e^* does not gain any advantage in decrypting ciphertexts he obtained in epochs $e < e^*$ *before* that compromise.

Post-Compromise Security: An adversary compromising a secret key in some epoch e^* does not gain any advantage in decrypting ciphertexts he obtained in epochs $e > e^*$ *after* that compromise.

Adaptive Security: An adversary can adaptively corrupt keys and tokens of the current epoch and all previous ones.

Given that updatable encryption schemes can produce ciphertexts in two ways—either via a direct encryption or an update of a previous ciphertext—we require that the above properties must hold for both settings. This inspires our split into two indistinguishability-based security notions, one capturing security of direct encryptions ($\mathsf{IND\text{-}ENC}$) and one ruling out attacks against updated ciphertexts ($\mathsf{IND\text{-}UPD}$). Both security notions are defined through experiments run between a challenger and an adversary $\mathcal{A}$. Depending on the notion, the adversary may issue queries to different oracles, defined in the next section. At a high level, $\mathcal{A}$ is allowed to adaptively corrupt arbitrary choices of secret keys and update tokens, as long as they do not allow him to trivially decrypt the challenge ciphertext.

The Importance of Post-Compromise Security. We have formalized updatable encryption in the strict sequential setting it will be used in, and in particular modeled key derivation of a new key k_{e+1} as a sequential update $(k_{e+1}, \Delta_{e+1}) \xleftarrow{r} \mathsf{UE.next}(k_e)$ of the old key k_e. Previous works [10,15] instead model key rotation by generating fresh keys via a dedicated $k_{e+1} \xleftarrow{r} \mathsf{UE.kgen}(\lambda)$ algorithm at each epoch and deriving the token as $\Delta_{e+1} \xleftarrow{r} \mathsf{UE.next}(k_e, k_{e+1})$.

One impact of our sequential model is that post-compromise security becomes much more essential, as this property intuitively ensures that new keys are independent of the old ones (which is directly ensured in the previous formalization where keys where generated independently). Without requiring post-compromise security, $\mathsf{UE.next}(k_e)$ could generate the new key by hashing the old one: $k_{e+1} \leftarrow \mathsf{H}(k_e)$. If H is modeled as a random oracle, this has no impact for standard or forward security, but any scheme with such a key update loses all security in the post-compromise setting. An adversary compromising a single secret key k_e can derive all future keys himself.

What we do not Model. The focus of this work is to obtain security against arbitrary key compromises, i.e., an adversary can steal secret keys, update tokens, and outsourced ciphertexts at any epoch. We do not consider attacks where an adversary fully takes over the owner or host and starts manipulating ciphertexts, e.g., providing adversarially generated ciphertexts to the host, or tampering with the update procedure. Thus, we model passive CPA attacks but not active CCA ones, and assume that all ciphertexts and updates are honestly generated. We believe this still captures the main threat in the context of updatable encryption, namely smash-and-grab attacks aiming at compromising the key material.

In fact, this restriction to passive attacks allows us to be more generous when it comes to legitimate queries towards corrupted epochs, as we can distinguish challenge from non-challenge ciphertexts and only prohibit the ones that allow trivial wins. Interestingly, Everspaugh et al. [15] use a similar approach in their stronger CCA-like security notion for ciphertext-dependent schemes where they are able to recognize whether a ciphertext is derived from the challenge and prevent these from being updated towards a corrupt key. They are able to recognize challenge ciphertexts as all keys are generated honestly, i.e., they are known to the challenger, and updates are required to be *deterministic*. The latter allows the challenger to trivially keep track of the challenge ciphertext, but it also makes misuse of the schemes more likely: if a scheme is implemented with probabilistic updates—which intuitively seems to only increase security—then one steps outside of the model and loses all security guarantees. In our model, we allow updates to be probabilistic, and in fact, the security of our strongest construction crucially relies on the re-randomization of updated ciphertexts.

3.2 Definition of Oracles

During the interaction with the challenger in the security definitions, the adversary may access oracles for *encryption*, for moving the key to the *next epoch*, for *corrupting the token or secret key*, and for *updating ciphertexts* into the current epoch. In the following description, the oracles may access the state of the challenger during the experiment. The challenger initializes a UE scheme with global state $(k_e, \Delta_e, \mathbf{S}, e)$ where $k_0 \leftarrow \mathsf{UE.setup}(\lambda)$, $\Delta_0 \leftarrow \perp$, and $e \leftarrow 0$, and $\mathbf{S}$ consists of initially empty sets $\mathcal{L}, \tilde{\mathcal{L}}, \mathcal{C}, \mathcal{K}$ and $\mathcal{T}$. Furthermore, let $\tilde{e}$ denote the challenge epoch, and e_{end} denote the final epoch in the game.

The sets $\mathcal{L}, \tilde{\mathcal{L}}, \mathcal{C}, \mathcal{K}$ and $\mathcal{T}$ are used to keep track of the generated and updated ciphertexts, and the epochs in which $\mathcal{A}$ corrupted a secret key or token, or learned a challenge-ciphertext (Fig. 1):

$\mathcal{L}$	List of non-challenge ciphertexts (C_e, e) produced by calls to the $\mathcal{O}_{\mathsf{enc}}$ or $\mathcal{O}_{\mathsf{upd}}$ oracle. $\mathcal{O}_{\mathsf{upd}}$ only updates ciphertexts contained in $\mathcal{L}$.
$\tilde{\mathcal{L}}$	List of updated versions of the challenge ciphertext. $\tilde{\mathcal{L}}$ gets initialized with the challenge ciphertext $(\tilde{C}, \tilde{e})$. Any call to the $\mathcal{O}_{\mathsf{next}}$ oracle automatically updates the challenge ciphertext into the new epoch, which $\mathcal{A}$ can fetch via a $\mathcal{O}_{\mathsf{upd}\tilde{\mathsf{C}}}$ call.
$\mathcal{C}$	List of all epochs e in which $\mathcal{A}$ learned an updated version of the challenge ciphertext.
$\mathcal{K}$	List of all epochs e in which $\mathcal{A}$ corrupted the secret key k_e.
$\mathcal{T}$	List of all epochs e in which $\mathcal{A}$ corrupted the update token Δ_e.

Fig. 1. Summary of lists maintained by the challenger.

$\mathcal{O}_{\mathsf{enc}}(m)$**:** On input a message $m \in \mathcal{M}$, compute $C \xleftarrow{r} \mathsf{UE.enc}(k_e, m)$ where k_e is the secret key of the current epoch e. Add C to the list of ciphertexts $\mathcal{L} \leftarrow \mathcal{L} \cup \{(C, e)\}$ and return the ciphertext to the adversary.

$\mathcal{O}_{\mathsf{next}}$**:** When triggered, this oracle updates the secret key, produces a new update value as $(k_{e+1}, \Delta_{e+1}) \xleftarrow{r} \mathsf{UE.next}(k_e)$, and updates the global state to $(k_{e+1}, \Delta_{e+1}, \mathbf{S}, e+1)$. If the challenge query was already made, this call will also update the challenge ciphertext into the new epoch, i.e., it runs $\tilde{C}_{e+1} \xleftarrow{r} \mathsf{UE.upd}(\Delta_{e+1}, \tilde{C}_e)$ for $(\tilde{C}_e, e) \in \tilde{\mathcal{L}}$ and sets $\tilde{\mathcal{L}} \cup \{(\tilde{C}_{e+1}, e+1)\}$.

$\mathcal{O}_{\mathsf{upd}}(C_{e-1})$**:** On input a ciphertext C_{e-1}, check that $(C_{e-1}, e-1) \in \mathcal{L}$ (i.e., it is an honestly generated ciphertext of the previous epoch $e-1$), compute $C_e \xleftarrow{r} \mathsf{UE.upd}(\Delta_e, C_{e-1})$, add (C_e, e) to the list $\mathcal{L}$ and output C_e to $\mathcal{A}$.

$\mathcal{O}_{\mathsf{corrupt}}(\{\mathsf{token}, \mathsf{key}\}, e^*)$**:** This oracle models adaptive corruption of the host and owner keys, respectively. The adversary can request a key or update token from the current or any of the previous epochs.
- Upon input token, $e^* \leq e$, the oracle returns Δ_{e^*}, i.e., the update token is leaked. Calling the oracle in this mode sets $\mathcal{T} \leftarrow \mathcal{T} \cup \{e^*\}$.
- Upon input key, $e^* \leq e$, the oracle returns k_{e^*}, i.e., the secret key is leaked. Calling the oracle in this mode sets $\mathcal{K} \leftarrow \mathcal{K} \cup \{e^*\}$.

$\mathcal{O}_{\mathsf{upd}\tilde{\mathsf{C}}}$**:** Returns the current challenge ciphertext $\tilde{C}_e$ from $\tilde{\mathcal{L}}$. Note that the challenge ciphertext gets updated to the new epoch by the $\mathcal{O}_{\mathsf{next}}$ oracle, whenever a new key gets generated. Calling this oracle sets $\mathcal{C} \leftarrow \mathcal{C} \cup \{e\}$.

Fine-grained corruption modeling. Note that in the case of key-corruption in an epoch e^*, the oracle $\mathcal{O}_{\mathsf{corrupt}}(\mathsf{key}, e^*)$ only reveals the secret key k_{e^*}, but not the update token of the epoch. This assumes erasure of the token as an ephemeral value on the owner side. If the adversary also wants to learn the token, he can make a dedicated query for token-corruption in the same epoch. This allows to capture more fine-grained corruption settings.

Moreover, we have chosen to give the adversary a dedicated challenge-update oracle $\mathcal{O}_{\mathsf{upd}\tilde{\mathsf{C}}}$ that simply returns the updated challenge ciphertext of the current epoch, i.e., it does not require knowledge of the challenge ciphertext from the previous epoch. This gives the adversary more power compared with the definition in earlier models [10,15]: Therein, an adversary wanting to know an updated version of the challenge ciphertext for some epoch $e' > \tilde{e}$ (with $\tilde{e}$ denoting the challenge epoch) had to make update queries in *all* epochs from $\tilde{e}$ to e', which in turn is only allowed if $\mathcal{A}$ has not corrupted any secret key between $\tilde{e}$ and e'. Consequently, $\mathcal{A}$ could not receive an updated challenge ciphertext after a single key corruption, which we consider too restrictive. Therefore, we internally update the challenge ciphertext with every key rotation and allow the adversary to selectively receive an updated version at every epoch of his choice. Thus, in every epoch after $\tilde{e}$, the adversary $\mathcal{A}$ can choose whether he wants to learn the secret key or an updated version of the challenge ciphertext.

3.3 "Leakage" Profiles

The main benefit of *ciphertext-independent* updatable encryption schemes is that a single token can be used to update all ciphertexts from one epoch to the next. However, the generality of the token also imposes a number of challenges when modeling the knowledge of the adversary after he has corrupted a number of keys, tokens and updated challenge ciphertexts. For instance, if the adversary knows a challenge ciphertext $\tilde{C}$ from epoch $\tilde{e}$ and an update token for epoch $\tilde{e}+1$, he can derive an updated version of $\tilde{C}$ himself, which is not captured in the set $\mathcal{C}$ that only reflects the challenge ciphertexts that $\mathcal{A}$ has directly received from the challenger. This inference of updated ciphertexts via an update token is clearly inherent in ciphertext-independent schemes.

Practical schemes often enable the adversary to derive even more information, e.g., a token might allow not only to update but also to "downgrade" a ciphertext into the previous epoch, i.e., the updates are bi-directional, or even allow to update and downgrade a secret key via a token. While these features are present in all current solutions, we do not see a reason why they *should* be inherent in updatable encryption in general. Thus, we model different inference options outside of the game by defining extended sets $\mathcal{T}^*$, $\mathcal{C}^*$ and $\mathcal{K}^*$ that capture the information an adversary can infer from the directly learned tokens, ciphertexts or keys. In the security games defined in the next section, we will require the intersection of the extended sets of known challenge ciphertexts $\mathcal{C}^*$ and known secret keys $\mathcal{K}^*$ to be empty, i.e., there must not exist a single epoch where the adversary knows both the secret key and the (updated) challenge. We give an example of such direct and inferable information in Fig. 2.

Note that such inference is less an issue for *ciphertext-dependent* schemes where the owner has to a derive a dedicated token for each ciphertext. This naturally limits the power of the token to the ciphertext it was derived for, and prevents the adversary from using the token outside of its original purpose.

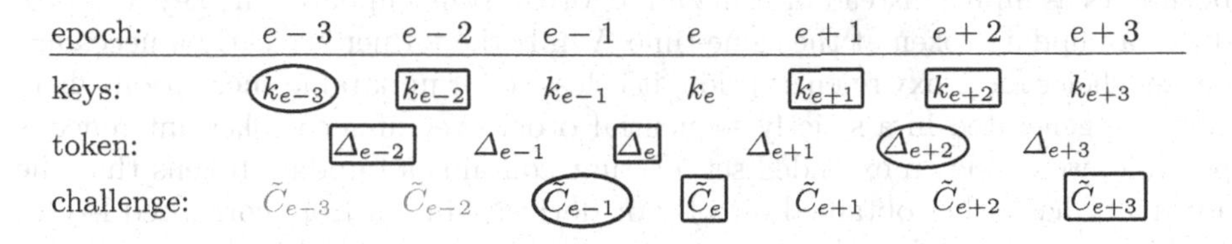

Fig. 2. Example of direct and indirect knowledge of an adversary. The boxed values denote $\mathcal{A}$'s directly received information as captured in $\mathcal{K}, \mathcal{T}$ and $\mathcal{C}$, whereas the circled ones denote the inferable values for a scheme with token-inference and bi-directional updates of ciphertexts and keys.

Capturing Key Updates. In many schemes (in fact all the ones we will consider), an update token does not only allow to update ciphertexts, but also the secret key itself. That is, if an adversary has learned a key k_e of epoch e and the update token Δ_{e+1} of the following epoch, then he can also derive the new key k_{e+1}. If that is the only possible derivation, we call this an *uni-directional key update.* If in addition also key downgrades are possible, i.e., a key k_e can be derived from k_{e+1} and Δ_{e+1}, we call this *bi-directional key updates.*

In the context of proxy re-encryption, a similar property is known as "collusion-resistance". So far only uni-directional and single-hop schemes satisfy this property, though [2,3,12,17,20,21], indicating that preventing keys to be updatable in a more flexible setting is a challenging property.

For defining uni- and bi-directional key updates we use the information contained in $\mathcal{K}$ and $\mathcal{T}$ to derive the inferable information. Recall that $\mathcal{K}$ denotes the set of epochs in which the adversary has obtained the secret key. The sets $\mathcal{K}^*_{\mathsf{uni}}$ and $\mathcal{K}^*_{\mathsf{bi}}$ are then defined via the recursive predicate corrupt-key as follows:

Uni-directional key updates:

$$\mathcal{K}^*_{\mathsf{uni}} \leftarrow \{e \in \{0,\ldots,e_{\mathsf{end}}\} \mid \mathsf{corrupt\text{-}key}(e) = \mathsf{true}\}$$
and $\mathsf{true} \leftarrow \mathsf{corrupt\text{-}key}(e)$ iff:
$$(e \in \mathcal{K}) \lor (\mathsf{corrupt\text{-}key}(e-1) \land e \in \mathcal{T})$$

Bi-directional key updates:

$$\mathcal{K}^*_{\mathsf{bi}} \leftarrow \{e \in \{0,\ldots,e_{\mathsf{end}}\} \mid \mathsf{corrupt\text{-}key}(e) = \mathsf{true}\}$$
and $\mathsf{true} \leftarrow \mathsf{corrupt\text{-}key}(e)$ iff:
$$(e \in \mathcal{K}) \lor (\mathsf{corrupt\text{-}key}(e-1) \land e \in \mathcal{T})$$
$$\lor (\mathsf{corrupt\text{-}key}(e+1) \land e+1 \in \mathcal{T})$$

Capturing Token Inference from Subsequent Secret Keys. The second indirect knowledge we model is the derivation of an update token from two subsequent secret keys. This is possible in all existing schemes where a token Δ_{e+1} is deterministically derived from the keys k_e and k_{e+1}. In fact, all previous definitions explicitly model the token computation as an algorithm that receives both keys as input, instead of using an algorithm that updates the key and produces an update token at the same time. While the former is clearly a necessary design choice for proxy re-encryption, it is less so for updatable encryption where keys are generated in a strictly sequential order. Yet, if such token inference is possible, we define an extended set $\mathcal{T}^*$ that contains all update tokens that the adversary has either obtained directly or derived himself from corrupted keys.

More, precisely, for schemes with *token-inference*, the adversary can derive from any two subsequent keys k_e and k_{e+1} the update token Δ_{e+1} from epoch e to $e+1$. We capture this by defining $\mathcal{T}^*$ via the sets $\mathcal{T}$ of corrupted token epochs and $\mathcal{K}^*$ denoting the extended set of corrupted key epochs as defined above.

$$\mathcal{T}^* \leftarrow \{e \in \{0, \ldots, e_{\mathsf{end}}\} \mid (e \in \mathcal{T}) \lor (e \in \mathcal{K}^* \land e-1 \in \mathcal{K}^*)\}$$

On a first glance it might look like we could run into a definitional loop between inferred tokens and keys, as the extended set $\mathcal{T}^*$ based on $\mathcal{K}^*$ could now also impact the definition of $\mathcal{K}^*$ (which we build from $\mathcal{T}$). This is not the case though: the additional epochs e that will be contained in $\mathcal{T}^*$ are epochs where the adversary already knew k_e and k_{e-1}. Thus the additional tokens Δ_e where $e \in \mathcal{T}^* \backslash \mathcal{T}$ would have no impact on a (re-definition) of $\mathcal{K}^*$ as all inferable keys from Δ_e are already in $\mathcal{K}^*$.

Capturing Challenge Ciphertext Updates. For capturing all the epochs in which the adversary knows a version of the challenge ciphertext, we define the set $\mathcal{C}^*$ containing all *challenge-equal* epochs. Informally, a challenge-equal epoch is every epoch in which the adversary knows a current version of the challenge ciphertext. This can be either obtained via a direct call to the challenge-ciphertext oracle $\mathcal{O}_{\mathsf{upd}\tilde{\mathsf{C}}}$, or by the adversary computing it himself via a (sequence of) updates. We have to distinguish between two cases, depending on whether the updates are uni- or bi-directional. In schemes with *uni-directional* updates, an update token Δ_e can only move ciphertexts from epoch $e-1$ into epoch e, but not vice versa. Note that uni-directional updates are by definition possible in all ciphertext-independent schemes. A scheme where a token Δ_e also allows to downgrade ciphertexts from epoch e to $e-1$, is called *bi-directional.*

Clearly, for security, uni-directional schemes are desirable, as the bi-directional property does not provide additional useful features but only allows the adversary to trivially derive more information. However, bi-directional schemes are easier to build, as this is related to the problem of designing uni-directional and multi-hop proxy re-encryption schemes, for which a first (compact) lattice-based solution was proposed only recently [25].

In both cases, we define $\mathcal{C}^*_{\mathsf{uni}}$ and $\mathcal{C}^*_{\mathsf{bi}}$ by using the information contained in $\mathcal{C}$, $\mathcal{T}^*$ and $\tilde{e}$ to derive the inferable information. Recall that $\tilde{e}$ denotes the challenge epoch, $\mathcal{C}$ denotes the set of epochs in which the adversary has obtained an updated version of the ciphertext (via $\mathcal{O}_{\mathsf{upd}\tilde{\mathsf{C}}}$), and $\mathcal{T}^*$ is the augmented set of tokens known to the adversary. The sets $\mathcal{C}^*_{\mathsf{uni}}$ and $\mathcal{C}^*_{\mathsf{bi}}$ of all challenge-equal ciphertexts are then defined via the recursive predicate challenge-equal as follows:

Uni-directional ciphertext updates:

$$\mathcal{C}^*_{\mathsf{uni}} \leftarrow \{e \in \{0, \ldots, e_{\mathsf{end}}\} \mid \mathsf{challenge\text{-}equal}(e) = \mathsf{true}\}$$
$$\text{and } \mathsf{true} \leftarrow \mathsf{challenge\text{-}equal}(e) \text{ iff:}$$
$$(e = \tilde{e}) \lor (e \in \mathcal{C}) \lor (\mathsf{challenge\text{-}equal}(e-1) \land e \in \mathcal{T}^*)$$

Bi-directional ciphertext updates:

$$\mathcal{C}^*_{\mathsf{bi}} \leftarrow \{e \in \{0, \ldots, e_{\mathsf{end}}\} \mid \mathsf{challenge\text{-}equal}(e) = \mathsf{true}\}$$
$$\text{and } \mathsf{true} \leftarrow \mathsf{challenge\text{-}equal}(e) \text{ iff:}$$
$$(e = \tilde{e}) \lor (e \in \mathcal{C})$$
$$\lor (\mathsf{challenge\text{-}equal}(e-1) \land e \in \mathcal{T}^*)$$
$$\lor (\mathsf{challenge\text{-}equal}(e+1) \land e+1 \in \mathcal{T}^*)$$

Optimal Leakage. The optimal leakage, capturing only the inference minimally necessary to perform ciphertext-independent updates would be $\mathcal{T}^* = \mathcal{T}$, $\mathcal{K}^* = \mathcal{K}$ and $\mathcal{C}^* = \mathcal{C}^*_{\mathsf{uni}}$. That is, there is no token inference, keys cannot be updated via a token and ciphertext updates are only uni-directional. All our schemes have leakage $(\mathcal{T}^*, \mathcal{C}^*_{\mathsf{bi}}, \mathcal{K}^*_{\mathsf{bi}})$, and we leave it as an interesting open problem whether efficient schemes with less leakage exist. Interestingly, the extended set of corrupted tokens $\mathcal{T}^*$ does not give the adversary more power in our IND-ENC and IND-UPD definitions, compared with definitions that are based only on $\mathcal{T}$.

3.4 Security Notions for Updatable Encryption

We are now ready to formally define the security notions for updatable encryption schemes in the remainder of this section. We propose two indistinguishability-based notions—the first capturing the security of fresh encryptions in the presence of key evolutions and adaptive corruptions, and the second defining the same security for updated ciphertexts.

Adaptive Encryption Indistinguishability (IND-ENC). Our IND-ENC notion ensures that ciphertexts obtained from the UE.enc algorithm do not reveal any information about the underlying plaintexts even when $\mathcal{A}$ adaptively compromises a number of keys and tokens before and after the challenge epoch. Thus this definition captures forward and post-compromise security.

Definition 2 (IND-ENC). *An updatable encryption scheme* UE *is said to be* IND-ENC-secure *if for all probabilistic polynomial-time adversaries* $\mathcal{A}$ *it holds that* $|\Pr[\mathsf{Exp}^{\mathsf{IND\text{-}ENC}}_{\mathcal{A},\mathsf{UE}}(\lambda) = 1] - 1/2| \le \epsilon(\lambda)$ *for some negligible function* ϵ.

Experiment $\mathsf{Exp}^{\mathsf{IND\text{-}ENC}}_{\mathcal{A},\mathsf{UE}}(\lambda)$:
$k_0 \xleftarrow{r} \mathsf{UE.setup}(\lambda)$
$e \leftarrow 0;\ \ \tilde{e} \leftarrow \bot;\ \ \mathcal{L} \leftarrow \emptyset$ // *these variables are updated by the oracles*
$(m_0, m_1, state) \xleftarrow{r} \mathcal{A}^{\mathcal{O}_{\mathsf{enc}}, \mathcal{O}_{\mathsf{next}}, \mathcal{O}_{\mathsf{upd}}, \mathcal{O}_{\mathsf{corrupt}}}(\lambda)$
proceed only if $|m_0| = |m_1|$
$\tilde{e} \leftarrow e;\ \ d \xleftarrow{r} \{0,1\}$
$\tilde{C} \xleftarrow{r} \mathsf{UE.enc}(k_{\tilde{e}}, m_d),\ \ \ \tilde{\mathcal{L}} \leftarrow \{(\tilde{C}, \tilde{e})\}$
$d' \xleftarrow{r} \mathcal{A}^{\mathcal{O}_{\mathsf{enc}}, \mathcal{O}_{\mathsf{next}}, \mathcal{O}_{\mathsf{upd}}, \mathcal{O}_{\mathsf{corrupt}}, \mathcal{O}_{\mathsf{upd}\tilde{\mathsf{C}}}}(\tilde{C}, state)$
return 1 if $d' = d$ and the following condition holds:
$\mathcal{A}$ has not learned k_{e^*} in any *challenge-equal* epoch e^*, i.e., let $\mathcal{C}^*$ denote the set of all *challenge-equal epochs* and $\mathcal{K}^*$ the set of epochs in which $\mathcal{A}$ learned the secret key, then it must hold that $\mathcal{C}^* \cap \mathcal{K}^* = \emptyset$

This experiment follows the typical IND-CPA definition, but additionally grants the adversary access to the $\mathcal{O}_{\mathsf{next}}$, $\mathcal{O}_{\mathsf{upd}}$, $\mathcal{O}_{\mathsf{corrupt}}$ and $\mathcal{O}_{\mathsf{upd}\tilde{\mathsf{C}}}$ oracles defined in Sect. 3.2. To exclude trivial wins, we require that $\mathcal{A}$ has not learned the secret key in any challenge-equal epoch. Recall that a "challenge-equal" epoch is every epoch in which the adversary knows a current version of the challenge ciphertext. This can be either obtained via a direct call to the challenge-ciphertext oracle or by the adversary computing it himself via a (sequence of) updates. The exact set of challenge-equal epochs ($\mathcal{C}^*$) and secret keys that are known to the adversary ($\mathcal{K}^*$) depends on the leakage profile, which has to specified when proving IND-ENC security. For all schemes proven secure in this work, the leakage profile is the one defined in Sect. 3.3.

Insufficiency of IND-ENC *for Full Post-Compromise Security.* It is often claimed that symmetric proxy re-encryption (PRE) can be used for updatable encryption, indicating that security of symmetric PRE is sufficient for the security of key-evolving schemes [10, 15, 22]. In fact, the security definition for ciphertext-independent schemes given by Boneh et al. [10] and Everspaugh et al. [15] coincides with the security of symmetric PRE. Our IND-ENC definition can be seen as a strengthened version (as it allows adaptive corruptions) of such PRE security adapted to the sequential setting of an updatable encryption scheme. However, an updatable scheme only satisfying IND-ENC would not necessarily provide the security properties one expects. Note that in the IND-ENC definition above, the challenge is a *fresh* encryption of one of the two challenge messages m_0, m_1, but not an updated ciphertext. Thus, IND-ENC security cannot guarantee anything about the security of *updates.* In fact, a scheme where the update algorithm UE.upd includes all the old ciphertexts $C_0, \dots C_e$ in the updated ciphertext C_{e+1} could be considered IND-ENC secure, but clearly lose all security if a single old key gets compromised.

We therefore also propose a second definition that requires indistinguishability of updates, and in combination with IND-ENC guarantees the security properties one expects from updatable encryption.

Adaptive Update Indistinguishability (IND-UPD). The IND-UPD notion ensures that an updated ciphertext obtained from the UE.upd algorithm does not reveal any information about the previous ciphertext, even when $\mathcal{A}$ adaptively compromises a number of keys and tokens before and after the challenge epoch. Thus this definition again captures forward and post-compromise security in an adaptive manner. We will informally refer to this notion also as *unlinkability.*

Definition 3 (IND-UPD). *An updatable encryption scheme* UE *is said to be* IND-UPD-secure *if for all probabilistic polynomial-time adversaries* $\mathcal{A}$ *it holds that* $|\Pr[\mathsf{Exp}^{\mathsf{IND\text{-}UPD}}_{\mathcal{A},\mathsf{UE}}(\lambda) = 1] - 1/2| \le \epsilon(\lambda)$ *for some negligible function* ϵ.

Experiment $\mathsf{Exp}^{\mathsf{IND\text{-}UPD}}_{\mathcal{A},\mathsf{UE}}(\lambda)$:
$k_0 \xleftarrow{r} \mathsf{UE.setup}(\lambda)$
$e \leftarrow 0;\ \ \tilde{e} \leftarrow \perp;\ \ \mathcal{L} \leftarrow \emptyset$ // *these variables are updated by the oracles*
$(C_0, C_1, state) \xleftarrow{r} \mathcal{A}^{\mathcal{O}_{\mathsf{enc}}, \mathcal{O}_{\mathsf{next}}, \mathcal{O}_{\mathsf{upd}}, \mathcal{O}_{\mathsf{corrupt}}}(\lambda)$
proceed only if $(C_0, \tilde{e}-1) \in \mathcal{L}$ and $(C_1, \tilde{e}-1) \in \mathcal{L}$ and $|C_0| = |C_1|$
$\tilde{e} \leftarrow e;\ \ d \xleftarrow{r} \{0,1\}$
$\tilde{C} \xleftarrow{r} \mathsf{UE.upd}(\Delta_{\tilde{e}}, C_d),\ \ \tilde{\mathcal{L}} \leftarrow \{(\tilde{C}, \tilde{e})\}$
$d' \xleftarrow{r} \mathcal{A}^{\mathcal{O}_{\mathsf{enc}}, \mathcal{O}_{\mathsf{next}}, \mathcal{O}_{\mathsf{upd}}, \mathcal{O}_{\mathsf{corrupt}}, \mathcal{O}_{\mathsf{upd}\tilde{\mathsf{C}}}}(\tilde{C}, state)$
return 1 if $d' = d$ and all of the following conditions hold
1) $\mathcal{A}$ has not learned $\Delta_{\tilde{e}}$, i.e., $\tilde{e} \notin \mathcal{T}^*$
2) $\mathcal{A}$ has not learned k_{e^*} in any *challenge-equal* epoch e^*, i.e., let $\mathcal{C}^*$ denote the set of all *challenge-equal epochs* and $\mathcal{K}^*$ the set of epochs in which $\mathcal{A}$ learned the secret key, then it must hold that $\mathcal{C}^* \cap \mathcal{K}^* = \emptyset$
3) if UE.upd is deterministic, then $\mathcal{A}$ has neither queried $\mathcal{O}_{\mathsf{upd}}(C_0)$ nor $\mathcal{O}_{\mathsf{upd}}(C_1)$ in epoch $\tilde{e}$

This experiment is similar to IND-ENC, but instead of requiring a fresh encryption to be indistinguishable, we let the adversary provide two ciphertexts C_0 and C_1 and return the update $\tilde{C}$ of one of them. The task of the adversary is to guess which ciphertext got updated. Note that the adversary is allowed to corrupt the secret key $k_{\tilde{e}-1}$, i.e., from right before the challenge epoch. Similar as in IND-ENC we exclude trivial wins where the adversary learned the secret key of a challenge-equal epoch. Moreover, if the update algorithm is deterministic, $\mathcal{A}$ is also not allowed to update any of the two challenge ciphertexts into the challenge epoch himself.

4 Comparison with Existing Models

We now compare our security notion with the definition proposed by Everspaugh et al. [15], which in turn builds upon the work by Boneh et al. [9]. We also discuss the XOR-KEM scheme that was claimed to be a secure ciphertext-independent scheme [15]. Note that, for ciphertext-independent schemes, only the property of encryption indistinguishability (UP-IND-BI in [15]) was previously defined but not the additional update indistinguishability, and thus our comparison focuses on IND-ENC.

The UP-IND-BI definition by Everspaugh et al. [15] is ambiguous, and we show that one can either interpret the model such that it excludes any key compromises before the challenge (i.e., it does not cover post-compromise security), or it is closer to our model and allows a restricted form of key corruptions before the challenge. We refer to the former as weakUP-IND-BI and to the latter as strongUP-IND-BI model. We stress that neither weakUP-IND-BI nor strongUP-IND-BI used in our comparison is the verbatim definition presented in [15]. Both are adaptions of the UP-IND-BI model to the sequential setting that we use in our work and in which updatable schemes are naturally used. This adaptation revealed an ambiguity in the UP-IND-BI model w.r.t. whether it allows key corruptions before the challenge.

One reason for the ambiguity is that the XOR-KEM scheme, which is claimed secure, is secure only in the weakUP-IND-BI model, but not in strongUP-IND-BI: we show that it loses all security if the adversary can corrupt an old key, which is allowed in the stronger model, as well as in our IND-ENC game.

Overall we show the following:

Theorem 1. IND-ENC $\Longrightarrow$ weakUP-IND-BI, IND-ENC $\not\Longleftrightarrow$ strongUP-IND-BI.

4.1 weakUP-IND-BI vs. strongUP-IND-BI

The key reason for the ambiguity of the security definition by Everspaugh et al. [15] is that the security game does not convey the notion of epochs and thus it is not clear *when* the adversary is allowed to corrupt secret keys. The definition considers static corruptions, and assumes a known threshold t that separates honest from corrupted keys. That is, all keys $k_1, \ldots, k_t$ are assumed to be uncorrupted, whereas the keys $k_{t+1}, \ldots, k_\kappa$ are considered corrupted and are given to the adversary. Jumping ahead, the security notion then allows challenge queries for all keys k_i where $i \leq t$ and disallows any update or token corruption queries towards a corrupt key k_j, i.e., where $j > t$.

One interpretation is that the threshold t strictly separates honest from corrupt epochs, i.e., the uncorrupted keys $k_1, \ldots, k_t$ belong to the first t epochs in which the adversary can request the challenge. We call this the weakUP-IND-BI model, as all corrupted keys $k_{t+1}, \ldots, k_\kappa$ must occur after the challenge epoch(s).

The second interpretation is that $k_1, \ldots, k_t$ merely refer to *some* t honest keys, but not necessarily to the first t epochs. That is, the corrupt keys could belong to arbitrary epochs, and key compromises before the challenge epoch(s) would be allowed. We call this the strongUP-IND-BI model.

Honest vs. Adversarial Ciphertexts. The weakUP-IND-BI and strongUP-IND-BI definitions do not distinguish between challenge and non-challenge ciphertexts in the responses to the update oracle, and allow $\mathcal{O}_{\mathsf{upd}}$ to be called with arbitrary ciphertexts. Thus, in contrast to our definition that only allows updates of honestly generated ciphertexts, the oracle $\mathcal{O}_{\mathsf{upd}}(C_e)$ omits the check whether

$(C_e, e) \in \mathcal{L}$ and simply returns the updated ciphertext for any input. Consequently, the adversary is not allowed to make *any* update query towards a corrupted epoch, as the query could be the challenge ciphertext. We show that for strongUP-IND-BI security, this difference of updating also adversarially crafted ciphertexts prevents our IND-ENC notion to be strictly stronger than strongUP-IND-BI. For weakUP-IND-BI this does not give the adversary any additional advantage though.

The weakUP-IND-BI Model. We follow the original definition by Everspaugh et al. [15] (in its weaker sense) and adopt it to our notation. As our scheme is strictly sequential, we cannot give the adversary all corrupted keys $k_{t+1}, \ldots, k_\kappa$ already at the beginning of the game, but rather let $\mathcal{A}$ corrupt them via the $\mathcal{O}_{\mathsf{corrupt}}(\mathsf{key}, \cdot)$ oracle. Further, we consider a single challenge query in some epoch $\tilde{e} \leq t$, whereas [15] granted the adversary a dedicated left-or-right oracle for all keys before t.

Experiment $\mathsf{Exp}_{\mathcal{A},\mathsf{UE}}^{\mathsf{weakUP\text{-}IND\text{-}BI}}(\lambda)$:
$k_0 \xleftarrow{r} \mathsf{UE.setup}(\lambda)$
$e \leftarrow 0;\ \ \tilde{e} \leftarrow \perp$ // *these variables are updated by the oracles*
$(m_0, m_1, state) \xleftarrow{r} \mathcal{A}^{\mathcal{O}_{\mathsf{enc}}, \mathcal{O}_{\mathsf{next}}, \mathcal{O}_{\mathsf{upd}}, \mathcal{O}_{\mathsf{corrupt}}}(\lambda)$
proceed only if $\tilde{e} \leq t$ and $|m_0| = |m_1|$
$\tilde{e} \leftarrow e;\ \ d \xleftarrow{r} \{0,1\}$
$\tilde{C} \xleftarrow{r} \mathsf{UE.enc}(k_{\tilde{e}}, m_d)$
$d' \xleftarrow{r} \mathcal{A}^{\mathcal{O}_{\mathsf{enc}}, \mathcal{O}_{\mathsf{next}}, \mathcal{O}_{\mathsf{upd}}, \mathcal{O}_{\mathsf{corrupt}}}(\tilde{C}, state)$
return 1 if $d' = d$ and the following condition holds:
1) no query $\mathcal{O}_{\mathsf{corrupt}}(\mathsf{key}, e')$ was made where $e' < t+1$
2) no query $\mathcal{O}_{\mathsf{corrupt}}(\mathsf{token}, t+1)$ was made in epoch $t+1$
3) no query $\mathcal{O}_{\mathsf{upd}}(\cdot)$ was made in epoch $t+1$

The winning condition requires that $\mathcal{A}$ does not learn the update token towards the first corrupted epoch e_{t+1}, nor makes any update query in e_{t+1}, as both would enable the adversary to update the challenge ciphertext into a corrupted epoch.

This weaker interpretation does not guarantee any confidentiality *after* a secret key got compromised, as it allows key corruption only after the challenge epoch. Thus, an updatable scheme that is secure only in the weakUP-IND-BI model does not provide the intuitive security one would expect from key rotation: namely that after migrating to the new key, the old one becomes useless and no longer of value to the adversary. To the contrary, all previous keys still require strong protection or secure deletion.

The strongUP-IND-BI Model. In the stronger interpretation, $\mathcal{A}$ can corrupt a set of arbitrary epochs, i.e., also before he makes the challenge query, but has to commit to them upfront. Whereas Everspaugh et al. [15] hand the adversary all keys already at the beginning, we let $\mathcal{A}$ retrieve them sequentially via the $\mathcal{O}_{\mathsf{corrupt}}(\mathsf{key}, \cdot)$ oracle in all epochs that he announced as corrupted in the beginning of the game.

Experiment $\mathsf{Exp}^{\mathsf{strongUP\text{-}IND\text{-}BI}}_{\mathcal{A},\mathsf{UE}}(\lambda)$:
$k_0 \xleftarrow{r} \mathsf{UE.setup}(\lambda)$
$e \leftarrow 0;\ \ \tilde{e} \leftarrow \perp$ // *these variables are updated by the oracles*
$(\mathcal{K}^*, state) \xleftarrow{r} \mathcal{A}(\lambda)$
$(m_0, m_1, state) \xleftarrow{r} \mathcal{A}^{\mathcal{O}_{\mathsf{enc}},\mathcal{O}_{\mathsf{next}},\mathcal{O}_{\mathsf{upd}},\mathcal{O}_{\mathsf{corrupt}}}(state)$
proceed only if $\tilde{e} \notin \mathcal{K}^*$ and $|m_0| = |m_1|$
$\tilde{e} \leftarrow e;\ \ d \xleftarrow{r} \{0,1\}$
$\tilde{C} \xleftarrow{r} \mathsf{UE.enc}(k_{\tilde{e}}, m_d)$
$d' \xleftarrow{r} \mathcal{A}^{\mathcal{O}_{\mathsf{enc}},\mathcal{O}_{\mathsf{next}},\mathcal{O}_{\mathsf{upd}},\mathcal{O}_{\mathsf{corrupt}}}(\tilde{C}, state)$
return 1 if $d' = d$ and the following condition holds:
1) no query $\mathcal{O}_{\mathsf{corrupt}}(\mathsf{key}, e)$ was made where $e \notin \mathcal{K}^*$
2) no query $\mathcal{O}_{\mathsf{corrupt}}(\mathsf{token}, e')$ was made where $e' \in \mathcal{K}^*$ or $e' - 1 \in \mathcal{K}^*$
3) no query $\mathcal{O}_{\mathsf{upd}}(\cdot)$ was made in an epoch e'' where $e'' \in \mathcal{K}^*$

The second winning condition forbids the adversary to receive any token that is connected to an epoch where $\mathcal{A}$ knows the secret key. This can be seen as the bi-directionality of key updates hard-coded in the experiment, which is captured in our IND-ENC definition via the definition of $\mathcal{K}^*_{\mathsf{bi}}$. The third condition forbids any ciphertext updates towards a corrupted epoch.

4.2 Insecurity of XOR-KEM in the strongUP-IND-BI and IND-ENC Model

Everspaugh et al. [15] proposed a simple construction, termed XOR-KEM, as a secure ciphertext-independent updatable encryption scheme. We now show that this scheme is neither secure in the stronger interpretation of their model nor in our IND-ENC definition.

The XOR-KEM scheme relies on a standard symmetric encryption scheme SE which it uses in a simple hybrid construction. Therein, every message gets encrypted under a fresh key x and x gets xor'd under the epoch key k_e. For updating a ciphertext, only the part depending on k_e gets updated via the token $\Delta_{e+1} \leftarrow (k_e \oplus k_{e+1})$.

$\mathsf{XOR\text{-}KEM.setup}(\lambda)$: return $k_0 \xleftarrow{r} \mathsf{SE.kgen}(\lambda)$
$\mathsf{XOR\text{-}KEM.next}(k_e)$: $k_{e+1} \xleftarrow{r} \mathsf{SE.kgen}(\lambda)$, $\Delta_{e+1} \leftarrow (k_e \oplus k_{e+1})$, return (k_{e+1}, Δ_{e+1})
$\mathsf{XOR\text{-}KEM.enc}(k_e, m)$: $x \xleftarrow{r} \mathsf{SE.kgen}(\lambda)$, return $C_e \leftarrow ((k_e \oplus x), \mathsf{SE.enc}(x, m))$
$\mathsf{XOR\text{-}KEM.upd}(\Delta_{e+1}, C_e)$: parse $C_e = (C^1, C^2)$, return $C_{e+1} \leftarrow ((C^1 \oplus \Delta_{e+1}), C^2)$
$\mathsf{XOR\text{-}KEM.dec}(k_e, C_e)$: parse $C_e = (C^1, C^2)$, return $\mathsf{SE.dec}(k_e \oplus C^1, C^2)$

Attack against XOR-KEM. We now present a simple attack against the XOR-KEM scheme, for which we only require the adversary to learn one key in some epoch before the challenge epoch. Let this epoch be $e < \tilde{e}$, to which $\mathcal{A}$ commits before the game starts. In epoch e, $\mathcal{A}$ requests the secret key k_e via $\mathcal{O}_{\mathsf{corrupt}}(\mathsf{key}, e)$. and also makes a standard encryption query $\mathcal{O}_{\mathsf{enc}}(m)$ receiving a ciphertext $C_e = ((k_e \oplus x), \mathsf{SE.enc}(x, m))$. The adversary then computes $x \leftarrow C_e^1 \oplus k_e$, where C_e^1 denotes the first part $(k_e \oplus x)$ of the ciphertext. Then, in all epochs from e to $\tilde{e}$, the adversary requests an updated version of C_e via $\mathcal{O}_{\mathsf{upd}}(\cdot)$. Note that strongUP-IND-BI forbids updates only *towards* but not *from* a corrupt key, and thus these queries are legitimate. Finally, in the challenge epoch $\tilde{e}$, $\mathcal{A}$ uses the updated (non-challenge) ciphertext $C_{\tilde{e}} = ((k_{\tilde{e}} \oplus x), \mathsf{SE.enc}(x, m))$ and its previously computed x to derive the secret key $k_{\tilde{e}}$ of the challenge epoch. Clearly, he can now trivially win the strongUP-IND-BI game, and did not violate any of the winning restrictions. The same attack applies in our IND-ENC game.

In the weakUP-IND-BI game, however, this attack is not possible, as $\mathcal{A}$ does not see a secret key before the challenge epoch, and is also not allowed to update any ciphertext into a corrupt epoch (i.e., he cannot perform the same attack by updating a non-challenge ciphertext into a corrupt epoch after $\tilde{e}$).

Weakening the strongUP-IND-BI *Model.* A tempting easy "fix" would be to forbid *any* updates from a corrupted epoch into an honest epoch in the strongUP-IND-BI model. This would allow the XOR-KEM scheme to be proven secure, and at the same time preserve $\mathcal{A}$'s capability of corrupting keys before the challenge epoch.

However, this "fix" would significantly weaken the guaranteed security, as it essentially disallows the adversary to see any updated ciphertexts after an attack. For instance, the following attack would be excluded by the model: Assume the adversary at some epoch e corrupts the secret key k_e and *one* ciphertext C_e from a large set of outsourced ciphertexts. Then, the key gets rotated into k_{e+1} and all ciphertexts get re-encrypted to the new key. In that new epoch $e+1$, the adversary learns neither k_{e+1} nor the update token, but steals *all* ciphertexts from the database. Intuitively, confidentiality of these updated ciphertexts should be guaranteed, as the adversary never compromised the key and all ciphertexts in the same epoch. This attack would not be covered by the model though, and the XOR-KEM scheme becomes entirely insecure if such an attack happens, as it allows the adversary to decrypt all re-encrypted ciphertexts even though he never corrupted k_{e+1}.

4.3 IND-ENC vs. strongUP-IND-BI (and weakUP-IND-BI)

XOR-KEM serves as a separating example between the weakUP-IND-BI and the two stronger strongUP-IND-BI, IND-ENC models, and both models are in fact strictly stronger than weakUP-IND-BI. Such a strict relation does not exist between strongUP-IND-BI and IND-ENC though: we show that both models are incomparable. Below we give the high-level ideas for the two separating examples, and refer to the full version [19] for their detailed descriptions as well as for the argumentation why IND-ENC implies weakUP-IND-BI.

Separating Example I (strongUP-IND-BI $\not\Rightarrow$ IND-ENC). The first separating example exploits the fact that in strongUP-IND-BI, the adversary is not allowed to update any ciphertext into a corrupt epoch, whereas IND-ENC allows such updates for non-challenge ciphertexts. We derive a scheme UE' from a secure UE where we let the token Δ_{e+1} also contain an encryption C_{key} of the old key k_e under the new key k_{e+1}. Further, when updating, UE' appends C_{key} to the updated ciphertext.

In the strongUP-IND-BI game, this change cannot increase $\mathcal{A}$'s advantage as he is not allowed to see any token towards a corrupt epoch e^*, nor make any updates towards e^*. In all other epochs, C_{key} is an encryption under a key unknown to the adversary. However, in the IND-ENC game, $\mathcal{A}$ can corrupt the secret key $k_{\tilde{e}+1}$ in the epoch after he makes the challenge query, and update an arbitrary non-challenge ciphertext from $\tilde{e}$ to $\tilde{e}+1$ using the $\mathcal{O}_{\mathsf{upd}}$ oracle. From there he extracts C_{key}, decrypts $k_{\tilde{e}}$ and can now trivially win the IND-ENC game as he knows the secret key of the challenge epoch.

Separating Example II (IND-ENC $\not\Rightarrow$ strongUP-IND-BI). Our model is not strictly stronger than strongUP-IND-BI, due to fact that we are more restrictive for ciphertexts that can be updated. Whereas we only allow honestly generated ciphertexts C_e to be updated (which is enforced by $\mathcal{O}_{\mathsf{upd}}$ checking whether $C_e \in \mathcal{L}$), strongUP-IND-BI is more generous and returns the update of *any* ciphertext (as they aim for *authenticated* encryption). This can be exploited to turn a secure scheme UE into UE'' that is secure in our IND-ENC model, but insecure according to strongUP-IND-BI. The idea is to modify the update algorithm, such that it returns the update token when it gets invoked with a special ciphertext, that would never occur for an honest encryption.

In the strongUP-IND-BI game, the update oracle then enables the adversary to obtain tokens in epochs where he would not be allowed to learn a token directly, leading to trivial wins without violating the winning condition. An adversary in our IND-ENC game cannot benefit from this modification, as it cannot poke the update oracle on the adversarially crafted ciphertext. We explain in the full version of this paper why the same idea does not apply for the weakUP-IND-BI model, which is in fact strictly weaker than IND-ENC.

4.4 IND-UPD vs. UP-REENC

Our IND-UPD definition is similar in spirit to the re-encryption indistinguishability notion UP-REENC by Everspaugh et al. [15], which captures post-compromise security of updates as well. However, the UP-REENC notion was only proposed for ciphertext-*dependent* schemes. Note that the difference between ciphertext-dependent and independent schemes has a significant impact on the achievable security: a single update token in the ciphertext-independent setting has much more functionality than in ciphertext-dependent schemes, which in turn gives the adversary more power when he compromises such tokens. Thus, no ciphertext-independent scheme can satisfy the UP-REENC definition. Our IND-UPD definition formalizes this extra power in ciphertext-independent schemes in a way that carefully excludes trivial wins but still captures strong post-compromise guarantees. The aspect that IND-UPD allows adaptive corruptions, whereas UP-REENC only considers static ones, makes both definitions incomparable.

Interestingly, in the ciphertext-dependent setting, this property has a somewhat "esoteric" flavor as it got motivated by an exfiltration attack where the adversary fully breaks into both the host and owner, compromising all ciphertexts and keys but is only able to extract a small amount of information at that time. The re-encryption indistinguishability should then guarantee that when the key gets rotated and the adversary compromises all the updated ciphertexts again (but not the new key), the previously extracted information becomes useless. This seems to be a somewhat contrived attack scenario, and might lead to the impression that such update indistinguishability is rather an optional feature. This is not the case for ciphertext-independent schemes: without the dedicated IND-UPD property an updatable encryption scheme does not guarantee *any* security of the updated ciphertexts when an old key gets compromised!

5 Constructions

We analyze several constructions of updatable encryption with respect to our security notions of indistinguishability of encryptions (IND-ENC) and updates (IND-UPD). First, we analyze the simple double-encryption construction that is purely based on symmetric primitives (Sect. 5.1). Unfortunately, the scheme cannot satisfy our strong security notions. We formulate the additional constraints on the adversarial behavior that suffices to prove its security in relaxed versions of our IND-ENC and IND-UPD models.

We then proceed to less efficient but more secure schemes, starting with the BLMR construction by Boneh et al. [10] based on key-homomorphic PRFs (Sect. 5.2). We show that the original BLMR scheme satisfies IND-ENC but not IND-UPD, and also propose a slight modification BLMR+ that improves the latter and achieves a weak form of update indistinguishability.

In Sect. 5.3, we introduce a new ElGamal-based scheme RISE and show that it fully achieves both of our strong security definitions. While proposing a "public-key solution" for a symmetric key primitive might appear counter-intuitive at first, we stress that the efficiency is roughly comparable to that of BLMR under known instantiations for the key-homomorphic PRF (same number of exponentiations). Also, taking advantage of the underlying group operations allows us to get full IND-UPD security.

All of our schemes allow to infer token from two subsequent keys and bi-directional updates of the ciphertexts and keys. Thus, all theorems are with respect to the leakage profile $(\mathcal{T}^*, \mathcal{K}^*_{\mathsf{bi}}, \mathcal{C}^*_{\mathsf{bi}})$ as defined in Sect. 3.3.

In the Appendix A, we additionally describe and analyze a symmetric KEM construction SE-KEM, which is widely used in practice since it does not require an (expensive) re-encryption of the payload data upon key rotation. This scheme is, however, better suited for deployment within the cloud infrastructure, because it requires the encryption keys to be sent to the host performing the re-encryption.

Furthermore, the fact that the data is not re-encrypted makes ciphertexts fully linkable. We therefore show only basic encryption security and under a weak adversary model.

5.1 Double Encryption (2ENC)

An approach that is based only on symmetric encryption is to first encrypt the plaintext under an "inner key," and subsequently encrypt the resulting ciphertext under a second, "outer key." In each epoch, the outer key is changed, and the ciphertext is updated by decrypting the outer encryption and re-encrypting under the new key. This scheme has been proposed by Ivan and Dodis [18] as symmetric uni-directional proxy re-encryption.[1] It has also appeared in other contexts, such as so-called "over-encryption" for access revocation in cloud storage systems [4]. More formally, this scheme can be phrased as an updatable encryption scheme 2ENC as follows.

$\mathsf{2ENC.setup}(\lambda)$: $k_0^o \stackrel{r}{\leftarrow} \mathsf{SE.kgen}(\lambda)$, $k^i \stackrel{r}{\leftarrow} \mathsf{SE.kgen}(\lambda)$, return $k_0 \leftarrow (k_0^o, k^i)$
$\mathsf{2ENC.next}(k_e)$: parse $k_e = (k_e^o, k^i)$, create $k_{e+1}^o \stackrel{r}{\leftarrow} \mathsf{SE.kgen}(\lambda)$,
  $\Delta_{e+1} \leftarrow (k_e^o, k_{e+1}^o)$, $k_{e+1} \leftarrow (k_{e+1}^o, k^i)$,
  return (k_{e+1}, Δ_{e+1})
$\mathsf{2ENC.enc}(k_e, m)$: parse $k_e = (k_e^o, k^i) \leftarrow k_e$,
  return $C_e \leftarrow \mathsf{SE.enc}(k_e^o, \mathsf{SE.enc}(k^i, m))$
$\mathsf{2ENC.upd}(\Delta_{e+1}, C_e)$: parse $\Delta_{e+1} = (k_e^o, k_{e+1}^o)$,
  return $C_{e+1} \leftarrow \mathsf{SE.enc}(k_{e+1}^o, \mathsf{SE.dec}(k_e^o, C_e))$
$\mathsf{2ENC.dec}(k_e, C_e)$: parse $k_e = (k_e^o, k^i)$, return $\mathsf{SE.dec}(k_e^o, \mathsf{SE.dec}(k^i, C))$

Clearly this scheme does not achieve our desired IND-ENC security: A ciphertext can be decrypted if an adversary sees the secret key of *some* epoch and one of the tokens relating to the epoch where he learned the ciphertext. However, we show that this is the only additional attack, i.e., if the adversary never sees such a combination of tokens and keys, then the scheme is secure, which is formalized by the following theorem.

Theorem 2 **(2ENC is weakly IND-ENC secure).** *Let* SE *be an* IND-CPA*-secure encryption scheme, then* 2ENC *is (weakly)* IND-ENC*-secure if the following additional condition holds: If $\mathcal{A}$ makes* any *query to $\mathcal{O}_{\mathsf{corrupt}}(\mathsf{key}, \cdot)$, then, for any challenge-equal epoch $e \in \mathcal{C}^*$, $\mathcal{A}$ must not call $\mathcal{O}_{\mathsf{corrupt}}(\mathsf{token}, \cdot)$ for epochs e or $e+1$.*

The proof of this theorem turns out to be surprisingly subtle and is provided in the full version [19]. As intuitively expected, it consists of two reductions to the IND-CPA security of SE, but the reduction for the outer encryption part is complicated by the fact that $\mathcal{A}$ may call either $\mathcal{O}_{\mathsf{corrupt}}$ or $\mathcal{O}_{\mathsf{upd}\tilde{\mathsf{C}}}$ adaptively and in multiple epochs. Instead of guessing all epochs, which would lead to a large

[1] It is uni-directional in a proxy re-encryption scheme; the proxy removes the outer layer. As an updatable scheme, which replaces the outer layer, it is bi-directional.

loss in tightness, we devise a specific hybrid argument and formalize the intuition that only epochs with a query to $\mathcal{O}_{\mathsf{upd}\tilde{\mathsf{C}}}$ can help $\mathcal{A}$ in gaining advantage.

It is also easy to see that the double encryption scheme is not IND-UPD secure: The inner ciphertext remains static and an adversary seeing tokens that allow him to unwrap the outer encryption can trivially link ciphertexts across epochs. But we again show that this is the only attack, i.e., 2ENC achieves a weak form of IND-UPD security if the adversary is restricted to learn at most one update token Δ_e for an epoch e for which he also obtained the challenge ciphertext in epochs e or $e-1$.

Theorem 3 (**2ENC is weakly IND-UPD secure**). *Let* SE *be an* IND-CPA-*secure encryption scheme, then* 2ENC *is (weakly)* IND-UPD-*secure if the following additional condition holds: For any challenge-equal epoch* $\mathsf{e} \in \mathcal{C}^*$, $\mathcal{A}$ *must not call* $\mathcal{O}_{\mathsf{corrupt}}(\mathsf{token}, \cdot)$ *for epochs* e *or* $\mathsf{e}+1$.

The proof follows along the lines of that for Theorem 2, with the main difference that we have to distinguish between the cases where the single special query $\mathcal{O}_{\mathsf{corrupt}}(\mathsf{token}, e)$ occurs before or after the challenge epoch $\tilde{e}$. The proof is given in the full version of this paper [19].

5.2 Schemes from Key-Homomorphic PRFs (BLMR and BLMR+)

Boneh et al. [10] proposed an updatable encryption scheme based on key-homomorphic pseudorandom functions, to which we will refer to as BLMR-scheme. We first recall the notion of key-homomorphic PRFs and then present the BLMR and our improved BLMR+ scheme.

Definition 4 (Key-homomorphic PRF [9]**).** *Consider an efficiently computable function* $\mathsf{F} : \mathcal{K} \times \mathcal{X} \rightarrow \mathcal{Y}$ *such that* $(\mathcal{K}, \oplus)$ *and* $(\mathcal{Y}, \otimes)$ *are both groups. We say that* F *is a key-homomorphic PRF if the following properties hold:*

1. F *is a secure pseudorandom function.*
2. *For every* $k_1, k_2 \in \mathcal{K}$*, and every* $x \in \mathcal{X}$*:* $\mathsf{F}(k_1, x) \otimes \mathsf{F}(k_2, x) = \mathsf{F}((k_1 \oplus k_2), x)$

A simple example of a secure key-homomorphic PRF is the function $\mathsf{F}(k, x) = \mathsf{H}(x)^k$ where $\mathcal{Y} = \mathbb{G}$ is an additive group in which the DDH assumption holds, and H is a random oracle [23].

Based on such a key-homomorphic PRF F, the BLMR construction is described as the following scheme:

BLMR.setup(λ): compute $k_0 \xleftarrow{r} \mathsf{F.kgen}(\lambda)$, return k_0
BLMR.next(k_e): $k_{e+1} \xleftarrow{r} \mathsf{F.kgen}(\lambda)$, return $(k_{e+1}, (k_e \oplus k_{e+1}))$
BLMR.enc(k_e, m): $N \xleftarrow{r} \mathcal{X}$, return $((\mathsf{F}(k_e, N) \otimes m), N)$
BLMR.dec(k_e, C_e): parse $C_e = (C_1, N)$, return $m \leftarrow C_1 \otimes \mathsf{F}(k_e, N)$.
BLMR.upd(Δ_{e+1}, C_e): parse $C_e = (C_1, N)$, return $((C_1 \otimes \mathsf{F}(\Delta_{e+1}, N)), N)$

Indeed, the subsequent theorem shows that BLMR is IND-ENC-secure.

Theorem 4 (BLMR **is** IND-ENC**-secure**). *Let* F *be a key-homomorphic PRF where* F.kgen(λ) *returns uniformly random elements from* $\mathcal{K}$, *then* BLMR *is* IND-ENC*-secure.*

The proof uses an alternative characterization of PRF (as in the original proof in [10]) together with the techniques already used in the proofs of the 2ENC scheme. The proof is given in the full paper [19]. The BLMR scheme does not achieve the notion IND-UPD of update-indistinguishability though, as the second part of the ciphertext remains static throughout the updates. This might have inspired the change to the ciphertext-dependent setting in the full version of Boneh et al.'s paper [9]. Ciphertext-dependent updates, however, have the disadvantage that the key owner must produce one update token for each ciphertext to be updated. We show that a mild form of IND-UPD security can be achieved in the ciphertext-independent setting via a simple modification to the BLMR scheme.

The BLMR+ *scheme.* The BLMR+ scheme follows the basic structure of BLMR, but additionally encrypts the nonce. In more detail, in every epoch the owner also generates a second key $k_e' \xleftarrow{r} \mathsf{SE.kgen}(\lambda)$ of a symmetric encryption scheme and encrypts the nonce-part N of each ciphertext under that key. In BLMR+, we simply include the old and new symmetric key into the update token and let the host re-encrypt the nonce.

The choice to simply reveal both keys might seem odd, but (in certain attack scenarios) it does not reveal more information to a corrupt host than what every updatable encryption scheme leaks anyway. Looking at two consecutive epochs, a corrupt host knows which updated and old ciphertext belong together – as he generated them – and thus letting him re-encrypt a static nonce does not reveal any additional information. The main advantage of BLMR+ over BLMR is that an adversary seeing only (updated) ciphertexts of different epochs cannot tell anymore which of them belong together. Clearly, this unlinkability is limited, though, as an adversary can still link ciphertexts whenever he also learned a related token which allows him to decrypt the static nonce.

In more detail, this modification results in the following scheme BLMR+:

BLMR+.setup(λ): $k_0^1 \xleftarrow{r} \mathsf{F.kgen}(\lambda)$, $k_0^2 \xleftarrow{r} \mathsf{SE.kgen}(\lambda)$, return $k_0 \leftarrow (k_0^1, k_0^2)$
BLMR+.next(k_e): parse $k_e = (k_e^1, k_e^2)$,
  create $k_{e+1}^1 \xleftarrow{r} \mathsf{F.kgen}(\lambda)$, $k_{e+1}^2 \xleftarrow{r} \mathsf{SE.kgen}(\lambda)$,
  $k_{e+1} \leftarrow (k_{e+1}^1, k_{e+1}^2)$, $\Delta_{e+1} \leftarrow (k_e^1 \oplus k_{e+1}^1, (k_e^2, k_{e+1}^2))$,
  return (k_{e+1}, Δ_{e+1})
BLMR+.enc(k_e, m): parse $k_e = (k_e^1, k_e^2)$, draw $N \xleftarrow{r} \mathcal{X}$,
  $C^1 \leftarrow \mathsf{F}(k_e^1, N) \otimes m$, $C^2 \xleftarrow{r} \mathsf{SE.enc}(k_e^2, N)$, return $C_e \leftarrow (C^1, C^2)$
BLMR+.dec(k_e, C_e): parse $k_e = (k_e^1, k_e^2)$ and $C_e = (C^1, C^2)$,
  $N \leftarrow \mathsf{SE.dec}(k_e^2, C^2)$, return $m \leftarrow C^1 \otimes \mathsf{F}(k_e^1, N)$
BLMR+.upd(Δ_{e+1}, C_e): parse $\Delta_{e+1} = (\Delta_{e+1}', (k_e^2, k_{e+1}^2))$ and $C_e = (C_e^1, C_e^2)$,
  $N \leftarrow \mathsf{SE.dec}(k_e^2, C^2)$, $C_{e+1}^1 \leftarrow C_e^1 \otimes \mathsf{F}(\Delta_{e+1}', N)$, $C_{e+1}^2 \xleftarrow{r} \mathsf{SE.enc}(k_{e+1}^2, N)$,
  return $C_{e+1} \leftarrow (C_{e+1}^1, C_{e+1}^2)$

We first state the following corollary as an easy extension of Theorem 4 on BLMR. The encryption of the nonce can be easily simulated in the reduction.

Corollary 1. *The* BLMR+ *scheme is* IND-ENC *secure.*

We then prove that the modified BLMR+ scheme described above indeed achieves a weak form of IND-UPD security. The intuition behind the level of security specified in the following theorem is that knowing either a token or the key of the ciphertexts later used in the challenge in a round *before* the challenge allows the adversary to decrypt the nonce. Also, obtaining the challenge ciphertext and a related token *after* the challenge query allows the adversary to decrypt the nonce. To obtain unlinkability, we cannot allow the adversary to access the nonce both before and after the challenge query in epoch $\tilde{e}$. The theorem formalizes that we have security unless the adversary gains this access.

Theorem 5 (BLMR+ **is weakly** IND-UPD **secure**). *Let* F *be a key-homomorphic PRF, and assume that all elements of* $\mathcal{X}$ *are encoded as strings of the same length. Let* SE *be a* IND-CPA*-secure symmetric encryption scheme. Then, the scheme* BLMR+ *is (weakly)* IND-UPD*-secure if the following additional condition holds: Let* e_{first} *denote the epoch in which the first ciphertext that is later used as challenge* C_0 *or* C_1 *was encrypted. If there exist some* $e^* \in \{e_{\mathsf{first}}, \ldots, \tilde{e} - 1\}$ *where* $e^* \in \mathcal{K}^* \cup \mathcal{T}^*$*, i.e.,* $\mathcal{A}$ *knows the secret key* k_{e^*} *or token* Δ_{e^*}*, then for any challenge-equal epoch* $e \in \mathcal{C}^*$*,* $\mathcal{A}$ *must not call* $\mathcal{O}_{\mathsf{corrupt}}(\mathsf{token}, \cdot)$ *for epochs* e *or* $e + 1$*.*

The proof of this theorem is essentially a combination of the techniques used in the proofs of Theorems 3 and 4. It is provided in the full version [19].

5.3 Updatable Encryption Based on ElGamal (RISE)

We finally present a scheme that achieves both strong notions of indistinguishability of encryptions (IND-ENC) and updates (IND-UPD). This scheme uses the classical proxy re-encryption idea based on ElGamal that was originally proposed by Blaze et al. [8], but uses it in the secret-key setting. This alone would not be secure though, as parts of the ciphertext would remain static. What we additionally exploit is that ElGamal ciphertexts can be *re-randomized* by knowing only the public key. Thus, we add the "public-key" element of the epoch to the token and perform a re-randomization whenever a ciphertext gets updated. This makes it the first of the considered schemes where the update algorithm is probabilistic. Interestingly, probabilistic updates are not allowed in the work by Everspaugh et al. [15] which require updates to be deterministic such that the challenger in the security game can keep track of the challenge ciphertexts. Further, in the security proof we also rely on the *key anonymity* property [5] of ElGamal, which guarantees that ciphertexts do not leak information about the public key under which they are encrypted.

The use of public-key techniques for secret-key updatable encryption may appear unnecessary. We emphasize, however, that previous constructions are based on key-homomorphic PRFs, all instantiations of which are based on such techniques as well. By contrast, the direct use of the group structure without the intermediate abstraction allows us to implement the re-randomization and thereby achieve full IND-UPD security.

In fact, in terms of exponentiations, an encryption in our RISE scheme is as efficient as in BLMR and in Everspaugh et al.'s. ReCrypt scheme [15], whereas the computations of update tokens and ciphertext updates are even more efficient than in [15] due to the ciphertext-independent setting of our work.

Let $(\mathbb{G}, g, q)$ be system parameters available as CRS such that the DDH problem is hard w.r.t. λ, i.e., q is a λ-bit prime. The scheme RISE is described as follows.

RISE.setup(λ): $x \xleftarrow{r} \mathbb{Z}_q^*$, set $k_0 \leftarrow (x, g^x)$, return k_0
RISE.next(k_e): parse $k_e = (x, y)$, draw $x' \xleftarrow{r} \mathbb{Z}_q^*$,
 $k_{e+1} \leftarrow (x', g^{x'})$, $\Delta_{e+1} \leftarrow (x'/x, g^{x'})$ return (k_{e+1}, Δ_{e+1})
RISE.enc(k_e, m): parse $k_e = (x, y)$, $r \xleftarrow{r} \mathbb{Z}_q$, return $C_e \leftarrow (y^r, g^r m)$
RISE.dec(k_e, C_e): parse $k_e = (x, y)$ and $C_e = (C_1, C_2)$, return $m' \leftarrow C_2 \cdot C_1^{-1/x}$
RISE.upd(Δ_{e+1}, C_e): parse $\Delta_{e+1} = (\Delta, y')$ and $C_e = (C_1, C_2)$,
 $r' \xleftarrow{r} \mathbb{Z}_q$, $C_1' \leftarrow C_1^{\Delta} \cdot y'^{r'}$, $C_2' \leftarrow C_2 \cdot g^{r'}$, return $C_{e+1} \leftarrow (C_1', C_2')$

The keys x for the encryption scheme are chosen from $\mathbb{Z}_q^*$ instead of $\mathbb{Z}_q$ as usual. The reason is that the update is multiplicative, and this restriction makes sure that each key is uniformly random in $\mathbb{Z}_q^*$. As this changes the distribution only negligibly, the standard Diffie-Hellman argument still applies. (However, the adaptation simplifies the security proof.)

The detailed proofs of the following theorems are provided in the full version of this paper [19].

Theorem 6 (RISE is IND-ENC secure). *The updatable encryption scheme RISE is IND-ENC secure under the DDH assumption.*

On a high-level, the proof exploits two properties of ElGamal encryption. First, a re-randomized ciphertext has the same distribution as a fresh encryption of the same plaintext. Second, as ElGamal encryption is key-anonymous [5], i.e., encryptions under two different public keys are indistinguishable, the adversary cannot distinguish between encryptions under the actual round key and encryptions under an independent, random key. These observations are used in game hops to make the challenge ciphertext independent from the information that the adversary learns by querying the other oracles. The remainder is a reduction to the DDH assumption, which underlies the security of ElGamal.

We also show that the scheme RISE is unlinkable. This property is mainly achieved by the re-randomization of the updates, but also leverages the key anonymity of ElGamal ciphertexts.

Theorem 7 (RISE is IND-UPD secure). *The updatable encryption scheme RISE is IND-UPD secure under the DDH assumption.*

The proof follows roughly along the same lines as that of Theorem 6. It is complicated a bit by the fact that, in contrast to IND-ENC, non-updated versions of the challenge-ciphertexts exist in the game even prior to the actual challenge epoch, which means that in the reduction we have to guess certain parameters, such as the epochs directly preceding the challenge epoch in which the adversary obtains update tokens, to keep the simulation consistent. Nevertheless, we show that, with a proper construction of the hybrid argument, the loss remains polynomial.

One might wonder whether one could more generally build a secure updatable encryption scheme from any secure symmetric proxy re-encryption with key-anonymity that additionally allows public re-randomization of ciphertexts. For that analysis one would need a security notion for such a primitive schemes that also allows *adaptive* corruptions as in our models. However, so far, even for plain (symmetric) proxy re-encryption adaptive corruptions have only been considered for schemes that are uni-directional and single-hop, i.e., where the re-encryption capabilities would not be sufficient for updatable encryption.

6 Conclusion and Open Problems

We have provided a comprehensive model for ciphertext-independent updatable encryption schemes, complementing the recent work of Everspaugh et al. [15] that focuses on ciphertext-dependent schemes. Ciphertext-independent schemes are clearly superior in terms of efficiency and ease-of-use when key rotation is required for large volumes of ciphertexts, whereas ciphertext-dependent solutions give a more fine-grained control over the updatable information.

We formalized updatable encryption and its desired properties in the strict sequential manner it will be used, avoiding the ambiguity of previous security models. Our two notions IND-ENC and IND-UPD guarantee that fresh encryptions and updated ciphertext are secure even if an adversary can adaptively corrupt several keys and tokens before and after learning the ciphertexts.

Somewhat surprisingly, and contradictory to the claim in [15], we have shown that the XOR-KEM scheme is not a secure ciphertext-independent schemes in such a strong sense. For the (existing) schemes 2ENC, BLMR, BLMR+, and SE-KEM, we formalized the security of the schemes by specifying precisely the conditions on the adversary under which a weak form of IND-ENC and IND-UPD security is achieved. We also specified a scheme that builds on ElGamal encryption. By additionally exploiting the algebraic structure of the underlying groups, instead of using the key-homomorphic PRF abstraction as in previous works, we were able to build a scheme that fully achieves our strong security notions while being at least as efficient as existing schemes that are either weaker or require ciphertext-dependent tokens.

All schemes we analyze allow to infer tokens from keys, and enable bi-directional updates of ciphertexts and keys, whereas an ideal updatable encryption scheme should only allow uni-directional updates of ciphertexts. Building such an ideal scheme is related to the open challenge of building proxy re-encryption schemes that are uni-directional, multi-hop and collusion-resistant.

Yet, while most proxy re-encryption work is in the public-key setting, updatable encryption has secret keys, so the construction of schemes with similar properties may be easier and is an interesting and challenging open problem.

Acknowledgments. This work has been supported in part by the European Commission through the Horizon 2020 Framework Programme (H2020-ICT-2014-1) under grant agreements number 644371 WITDOM and 644579 ESCUDO-CLOUD, and through the Seventh Framework Programme under grant agreement number 321310 PERCY, and in part by the Swiss State Secretariat for Education, Research and Innovation (SERI) under contract numbers 15.0098 and 15.0087.

A Symmetric Key-Encapsulation (SE-KEM)

We additionally analyze a scheme that can be considered as a symmetric key-encapsulation mechanism (KEM) together with a standard symmetric encryption scheme. The KEM has one key k_e per epoch e, and for each ciphertext it wraps an "inner" key x under which the actual message is encrypted. During an update, where the token is given by two keys (k_e, k_{e+1}) of subsequent epochs, all inner keys are simply un-wrapped using k_e and re-wrapped under the new key k_{e+1}.

This scheme is used in practical data-at-rest protection at cloud storage providers. The keys are, however, managed within the cloud storage systems. Not all nodes are equal; there are nodes that have access to the keys, and nodes that store the encrypted data.[2] In this scenario, it is acceptable to have the proxy nodes perform the updates. We stress that the scheme is not applicable for outsourcing encrypted data, as it fully reveals the secret keys in the update procedure!

We describe the algorithms in a slightly different way to consider SE-KEM as a ciphertext-independent updatable encryption scheme. The algorithms are described in more detail as the scheme SE-KEM as follows.

$\mathsf{SE\text{-}KEM.setup}(\lambda)$: return $k_0 \stackrel{r}{\leftarrow} \mathsf{SE.kgen}(\lambda)$
$\mathsf{SE\text{-}KEM.next}(k_e)$: $k_{e+1} \stackrel{r}{\leftarrow} \mathsf{SE.kgen}(\lambda)$, $\Delta_{e+1} \leftarrow (k_e, k_{e+1})$, return (k_{e+1}, Δ_{e+1})
$\mathsf{SE\text{-}KEM.enc}(k_e, m)$: $x \stackrel{r}{\leftarrow} \mathsf{SE.kgen}(\lambda)$, return $C_e \leftarrow (\mathsf{SE.enc}(k_e, x), \mathsf{SE.enc}(x, m))$
$\mathsf{SE\text{-}KEM.upd}(\Delta_{e+1}, C_e)$: parse $C_e = (C^1, C^2)$, and $\Delta_{e+1} = (k_e, k_{e+1})$, return $C_{e+1} \leftarrow (\mathsf{SE.enc}(k_{e+1}, \mathsf{SE.dec}(k_e, C^1)), C^2)$
$\mathsf{SE\text{-}KEM.dec}(k_e, C_e)$: parse $C_e = (C^1, C^2)$, return $\mathsf{SE.dec}(\mathsf{SE.dec}(k, C^1), C^2)$

While this scheme is very similar to the hybrid AE as described by Everspaugh el al. [15], our description differs in that the token is independent of the ciphertext, and consists of the keys (k_e, k_{e+1}) used for encryption in epochs e and $e+1$. In cloud storage systems where the keys for data-at-rest encryption are managed within the cloud, this is a faithful description of the real behavior.

[2] In OpenStack Swift, for instance, the "proxy server" nodes have access to the keys, whereas the role of the "object server" nodes is to store the ciphertext.

The security that can be offered by such a solution is necessarily limited. First, if the adversary obtains a challenge in epoch e and also sees one of the tokens in epochs e or $e+1$, the $\mathsf{IND\text{-}ENC}$ security is immediately broken. Furthermore, as the ciphertext update does not re-encrypt the second component, the ciphertexts are linkable through the epochs, i.e., $\mathsf{SE\text{-}KEM}$ cannot achieve any form of $\mathsf{IND\text{-}UPD}$ security. Still, we show that under the described (strict) constraints, the scheme guarantees a mild form of $\mathsf{IND\text{-}ENC}$ security.

Theorem 8 **($\mathsf{SE\text{-}KEM}$ is weakly $\mathsf{IND\text{-}ENC}$ secure).** *Let* SE *be an* $\mathsf{IND\text{-}CPA}$*-secure encryption scheme, then* $\mathsf{SE\text{-}KEM}$ *is (weakly)* $\mathsf{IND\text{-}ENC}$*-secure if the following additional condition holds: For any challenge-equal epoch* $e \in \mathcal{C}^*$*,* $\mathcal{A}$ *must not call* $\mathcal{O}_{\mathsf{corrupt}}(\mathsf{token}, \cdot)$ *for epochs* e *or* $e+1$.

The proof is very similar to the one of Theorem 2 and is provided in the full paper [19].

References

1. Ananth, P., Cohen, A., Jain, A.: Cryptography with updates. In: Coron, J.-S., Nielsen, J.B. (eds.) EUROCRYPT 2017, Part II. LNCS, vol. 10211, pp. 445–472. Springer, Cham (2017). https://doi.org/10.1007/978-3-319-56614-6_15
2. Ateniese, G., Benson, K., Hohenberger, S.: Key-private proxy re-encryption. In: Fischlin, M. (ed.) CT-RSA 2009. LNCS, vol. 5473, pp. 279–294. Springer, Heidelberg (2009). https://doi.org/10.1007/978-3-642-00862-7_19
3. Ateniese, G., Fu, K., Green, M., Hohenberger, S.: Improved proxy re-encryption schemes with applications to secure distributed storage. ACM Trans. Inf. Syst. Secur. **9**(1), 1–30 (2006)
4. Bacis, E., De Capitani di Vimercati, S., Foresti, S., Paraboschi, S., Rosa, M., Samarati, P.: Access control management for secure cloud storage. In: Deng, R., Weng, J., Ren, K., Yegneswaran, V. (eds.) SecureComm 2016. LNICST, vol. 198, pp. 353–372. Springer, Cham (2017). https://doi.org/10.1007/978-3-319-59608-2_21
5. Bellare, M., Boldyreva, A., Desai, A., Pointcheval, D.: Key-privacy in public-key encryption. In: Boyd, C. (ed.) ASIACRYPT 2001. LNCS, vol. 2248, pp. 566–582. Springer, Heidelberg (2001). https://doi.org/10.1007/3-540-45682-1_33
6. Bellare, M., Singh, A.C., Jaeger, J., Nyayapati, M., Stepanovs, I.: Ratcheted encryption and key exchange: the security of messaging. In: Katz, J., Shacham, H. (eds.) CRYPTO 2017, Part III. LNCS, vol. 10403, pp. 619–650. Springer, Cham (2017). https://doi.org/10.1007/978-3-319-63697-9_21
7. Berners-Lee, E.: Improved security notions for proxy re-encryption to enforce access control. Cryptology ePrint Archive, Report 2017/824 (2017). http://eprint.iacr.org/2017/824
8. Blaze, M., Bleumer, G., Strauss, M.: Divertible protocols and atomic proxy cryptography. In: Nyberg, K. (ed.) EUROCRYPT 1998. LNCS, vol. 1403, pp. 127–144. Springer, Heidelberg (1998). https://doi.org/10.1007/BFb0054122
9. Boneh, D., Lewi, K., Montgomery, H., Raghunathan, A.: Key homomorphic PRFs and their applications. Cryptology ePrint Archive, Report 2015/220 (2015). http://eprint.iacr.org/2015/220

10. Boneh, D., Lewi, K., Montgomery, H., Raghunathan, A.: Key homomorphic PRFs and their applications. In: Canetti, R., Garay, J.A. (eds.) CRYPTO 2013, Part I. LNCS, vol. 8042, pp. 410–428. Springer, Heidelberg (2013). https://doi.org/10.1007/978-3-642-40041-4_23
11. Cachin, C., Camenisch, J., Freire-Stoegbuchner, E., Lehmann, A.: Updatable tokenization: Formal definitions and provably secure constructions. Cryptology ePrint Archive, Report 2017/695 (2017). http://eprint.iacr.org/2017/695
12. Chow, S.S.M., Weng, J., Yang, Y., Deng, R.H.: Efficient unidirectional proxy re-encryption. In: Bernstein, D.J., Lange, T. (eds.) AFRICACRYPT 2010. LNCS, vol. 6055, pp. 316–332. Springer, Heidelberg (2010). https://doi.org/10.1007/978-3-642-12678-9_19
13. Cohn-Gordon, K., Cremers, C., Dowling, B., Garratt, L., Stebila, D.: A formal security analysis of the signal messaging protocol. In: EuroS&P (2017)
14. Cohn-Gordon, K., Cremers, C., Garratt, L.: On post-compromise security. Cryptology ePrint Archive, Report 2016/221 (2016). http://eprint.iacr.org/2016/221
15. Everspaugh, A., Paterson, K., Ristenpart, T., Scott, S.: Key rotation for authenticated encryption. In: Katz, J., Shacham, H. (eds.) CRYPTO 2017, Part III. LNCS, vol. 10403, pp. 98–129. Springer, Cham (2017). https://doi.org/10.1007/978-3-319-63697-9_4
16. Günther, F., Mazaheri, S.: A formal treatment of multi-key channels. In: Katz, J., Shacham, H. (eds.) CRYPTO 2017. LNCS, vol. 10403, pp. 587–618. Springer, Cham (2017). https://doi.org/10.1007/978-3-319-63697-9_20
17. Hohenberger, S., Rothblum, G.N., shelat, A., Vaikuntanathan, V.: Securely obfuscating re-encryption. In: Vadhan, S.P. (ed.) TCC 2007. LNCS, vol. 4392, pp. 233–252. Springer, Heidelberg (2007). https://doi.org/10.1007/978-3-540-70936-7_13
18. Ivan, A., Dodis, Y.: Proxy cryptography revisited. In: NDSS 2003. The Internet Society, February 2003
19. Lehmann, A., Tackmann, B.: Updatable encryption with post-compromise security. Cryptology ePrint Archive, Report 2018/118 (2018). http://eprint.iacr.org/2018/118
20. Libert, B., Vergnaud, D.: Multi-use unidirectional proxy re-signatures. In: Ning, P., Syverson, P.F., Jha, S. (eds.) ACM CCS 2008, pp. 511–520. ACM Press, October 2008
21. Libert, B., Vergnaud, D.: Tracing malicious proxies in proxy re-encryption. In: Galbraith, S.D., Paterson, K.G. (eds.) Pairing 2008. LNCS, vol. 5209, pp. 332–353. Springer, Heidelberg (2008). https://doi.org/10.1007/978-3-540-85538-5_22
22. Myers, S., Shull, A.: Efficient hybrid proxy re-encryption for practical revocation and key rotation. Cryptology ePrint Archive, Report 2017/833 (2017). http://eprint.iacr.org/2017/833
23. Naor, M., Pinkas, B., Reingold, O.: Distributed pseudo-random functions and KDCs. In: Stern, J. (ed.) EUROCRYPT 1999. LNCS, vol. 1592, pp. 327–346. Springer, Heidelberg (1999). https://doi.org/10.1007/3-540-48910-X_23
24. PCI Security Standards Council: Requirements and security assessment procedures. PCI DSS v3.2 (2016)
25. Polyakov, Y., Rohloff, K., Sahu, G., Vaikuntanthan, V.: Fast proxy re-encryption for publish/subscribe systems. Cryptology ePrint Archive, Report 2017/410 (2017). http://eprint.iacr.org/2017/410

Author Index

Printed in the United States
By Bookmasters